Springer Geochemistry

The Springer Geochemistry series seeks to publish a broad portfolio of scientific books, aiming at researchers, students, and everyone interested in geochemistry. The series includes peer-reviewed monographs, edited volumes, textbooks, and conference proceedings. It covers the entire research area including, but not limited to, Isotope Geochemistry, Biogeochemistry, and Environmental Geochemistry.

Giulio Armando Ottonello

Quantum Geochemistry

Volume 1

Springer

Giulio Armando Ottonello
Fellow of the Mineralogical Society
of America
Corresponding Member of the Accademia
Nazionale dei Lincei
Rome, Italy

ISSN 2366-6285 ISSN 2366-6293 (electronic)
Springer Geochemistry
ISBN 978-3-031-21836-1 ISBN 978-3-031-21837-8 (eBook)
https://doi.org/10.1007/978-3-031-21837-8

This Springer imprint is published by the registered company Springer Nature Switzerland AG
The registered company address is: Gewerbestrasse 11, 6330 Cham, Switzerland

Paper in this product is recyclable.

In Memory of Elisabeth

Preface

Warum ich zuletzt am liebsten mit der Natur verkehre, ist,weil sie immer recht hat und der Irrtum bloß auf meiner Seite sein kann

This beautiful aphorism is due to Goethe (Natur Wissenschaftliche Schriften, Band 1, Allgemeines Aphorismen: Warheit und Irrtum; Leipzig Im Insel Verlag, 1946) "*Why, at the very end, I like to traffic with the things of Nature? Because I know She is always right and the Error can be only by my side*". This is the leading concept that a Geochemist should always keep in his mind when trying to translate in formal reactions the "*facts*" that he "*sees*" happening on Earth, and, more generally in the Cosmos.

But "*seeing*" is not synonymous of "*looking at*" because epistemology requires the existence of "*Observables*" and the (intellectual) capability to interpret them. The Observables we encounter in this book are those masterfully described by Dirac in "*The Principles of Quantum Mechanics*" (Chap. 1, Pars. 10 and 11). In Par. 10, Dirac notes "*We have made a number of assumptions about the way in which states and dynamical variables are to be represented mathematically in the theory. These assumptions are not, by themselves, laws of nature, but become laws of nature when we make some further assumptions that provide a physical interpretation of the theory. Such further assumptions must take the form of establishing connexions between the results of observation, on one hand, and the equations of the mathematical formalism on the other*" (Dirac, 1957).

The critical point is thus the capability of "*establishing connections*": we must first "*see*" (experimentally, conceptually, not just physically) the facts of Nature, and then translate them into a mathematical formalism (which requires not only some smartness but also a sound knowledge of the most advanced theories).

If this book helps some young Geochemists to develop the idea that Quantum Mechanics is essential to reach this task, then I did partly fulfill my objective.

The work is divided into three main parts: "*Principles*", "*Methods*" and "*Applications*". The Principles may be identified with the "*Assumptions*" in Dirac's definition, the Methods are the "*Connexions*", and the "*Applications*" must be taken as a promotional exercise.

Renato Colle, Carlo Gatti and Benedetta Mennucci reviewed selected sections of the book. I gratefully acknowledge their help, and also, indirectly, the help of all co-authors with whom I cooperated in the last two decades. I take full responsibility for all errors and misinterpretations that might still be present in the text. Humble apologies for the unavoidable self-referencing, which sound disturbing not only for the reader but for the writer as well.

Calci (PI), Italy Giulio Armando Ottonello

Contents

Part II Methods

Part III Applications

Part I
Principles

Chapter 1
Particles and Waves

This short chapter is intended to give an introductory appraisal of some fundamental concepts of Wave Mechanics which are at the base of the Quantum procedures. As Geochemists we are mainly interested to the Quantum Mechanical implications in terms of thermophysical and thermodynamic "*observables*" (in the sense assigned to this term by Dirac 1957). We are thus not much concerned with the dual nature of a radiation, but rather with the fact that a wavelike representation of it ensures a satisfactory capability of quantifying the energy emitted by a material body, in whatever aggregation state, while the "*quasi-particle*" acceptation permits to scale the energy in molar terms (hence something implicitly associated to mass). Seeing the "*wave*" as a sort of "*disturbance*" of the medium associated to the particle traveling through it (McWeeny 1972) is, in this dual sense, particularly helpful and reassuring.

1.1 Premises

At the heart of the Quantum methods, there are the strenuous attempts of Planck to reconcile the Wien–Kirchhoff "*magic*" function describing the intensity of the blackbody radiation, i.e. $f(\lambda, T) = a \Big/ \left\{\lambda^5 \left[\exp\left(\frac{b}{\lambda T}\right)\right]\right\}$, valid at short wavelength (λ) and temperature T, with the Rayleigh–Jeans function, $f(\lambda, T) = 8\pi k_B / \lambda^4$ (with k_B as Boltzmann constant; $k_B = 1.380658 \times 10^{-23}\ J{\cdot}K^{-1}$) for the low-intensity region, valid at long wavelengths, but predicting an "*ultraviolet catastrophe*" (i.e. an exponential increase of the radiation intensity) at short wavelengths.

Planck's "*inspired guess*" (Montwill and Breslin 2012) was initially empirical, with form $f(\lambda, T) = a \Big/ \left\{\lambda^5 \left[\exp\left(\frac{b}{\lambda T}\right) - 1\right]\right\}$, (Planck 1900) and only later it became apparent to him that the empirical constants (a, b) in the above expression were related to a physical constant (h) establishing that the energy E is transferred in

G. A. Ottonello, *Quantum Geochemistry*, Springer Geochemistry,
https://doi.org/10.1007/978-3-031-21837-8_1

packets (Quanta) related to the frequency $v(\Delta E = hv; E = nhv)$ and that only energy states obeying this principle are allowed (Planck 1901). The constant h was actually related by Planck to an "*energy element*" ε associated with a resonator of frequency ν and the assigned magnitude was $h = 6.55 \times 10^{-27}$ *erg·s* (the accepted value of Planck's constant in SI units is $6.6260755 \times 10^{-34}$ *J·s*). Also, for the first time, in this contribution, Planck sketched the calculation of the entropy, as related to the probability of resonator complexions via the Boltzmann principle (Carson 2011).

When expressed in terms of spectral radiance B, Planck's law has form (Carson 2011)

$$B_\lambda(\lambda, T) = \left(8\pi hc/\lambda^5\right) \Big/ \left[\exp\left(\frac{hc}{\lambda k_B T}\right) - 1\right] \tag{1.1.1}$$

when the radiance is related to the wavelength of emission or

$$B_v(v, T) = \left(8\pi hv^3/c^3\right) \Big/ \left[\exp\left(\frac{hv}{k_B T}\right) - 1\right] \tag{1.1.2}$$

when the radiance is related to the frequency of emission v and c, in both equations, is the speed of light in vacuum ($c = 299{,}792{,}458\ m{\cdot}s^{-1}$). It may be easily seen that, at short wavelengths, the exponential term at denominator in Eq. (1.1.1) becomes dominant, so that the -1 term may be omitted and Planck's equation reduces to the Wien–Kirchoff formulation. Also, if we expand the exponential term in a Taylor's polynomial:

$$\exp(x) = 1 + x + \frac{x^2}{2!} + \frac{x^3}{3!} + \ldots \frac{x^n}{n!} \tag{1.1.3}$$

and consider that at long wavelengths (or high-T) the exponential term becomes small, then

$$\exp\left(\frac{hc}{\lambda k_B T}\right) \cong 1 + \frac{hc}{\lambda k_B T} \tag{1.1.4}$$

hence

$$\left[\exp\left(\frac{hc}{\lambda k_B T}\right)\right]^{-1} \cong \left[\frac{hc}{\lambda k_B T}\right]^{-1} = \frac{\lambda k_B T}{hc} \tag{1.1.5}$$

and Eq. (1.1.1) reduces to

$$B_\lambda(\lambda, T) = \frac{2ck_B T}{\lambda^4} \tag{1.1.6}$$

which is formally identical to the Rayleigh–Jeans equation.

Indeed, the basic assumptions leading to the Rayleigh–Jeans and Planck equations are the same: the radiation emitted by a blackbody (i.e. a body adsorbing all energies coming from the exterior and releasing all energies generated inside) is caused by the electrons of the material body atoms. However, while in the Rayleigh–Jeans derivation there is the (implicit) assumption that the electronic oscillators responsible of the internally generated radiation may assume whatsoever value, Planck revolutionary assumed that the energies of the oscillators were discrete and had to be proportional to an integral multiple of the frequency.

In differential terms Eq. (1.1.2) is re-expressed as

$$d\rho(\nu, \mathrm{T}) = \rho_\nu(\mathrm{t})d\nu = \frac{8\pi h}{c^3} \times \frac{\nu^3 d\nu}{\exp(h\nu/\mathrm{k_B T}) - 1} \tag{1.1.7}$$

where $\rho_\nu(\mathrm{t})d\nu$ is the radiant energy density observed in the frequency interval between ν and $\nu + d\nu$. This differential form better helps to visualize the actual nature of the experimental observations, such as the original plot shown in Fig. 1.1.

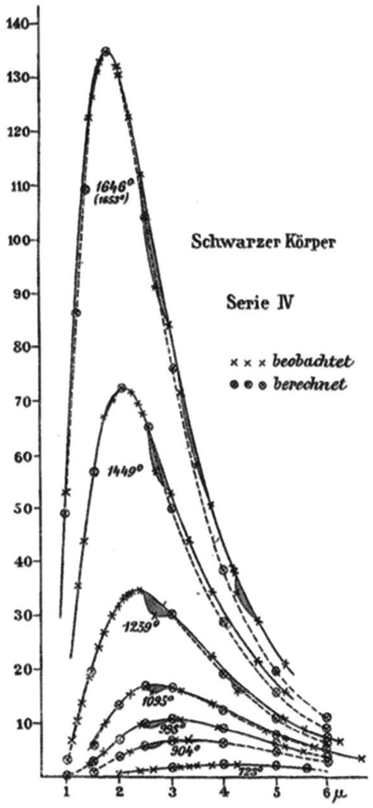

Fig. 1.1 Spectrum of the thermal radiation emitted by a black body. The measurements of Lummer and Pringsheim are fitted by continuous functions and compared with predictions based on Wien's radiation law. Ordinate axis: arbitrary units. From Lummer and Pringsheim (1901)

Incidentally, from the above development, we learnt that also to electrons (which have a definite mass, depending on their linear or rotational speed) it is associated a wavelike "*disturbance*" (see later). While Planck believed that, once emitted, the light behaves like a classical wave, Einstein proposed that indeed it is constituted in terms of small packets of energy (i.e. "*photons*" with $E = h\nu$) and explained the photoelectric effect by applying simple energy conservation arguments. This led him to propose that the minimum energy required to remove an electron from the surface of a metal is related to a characteristic parameter (ϕ), now known as "*work function*" (i.e. $h\nu \geq \phi$). For our purposes, it is, however, also important to recall that Einstein (1907) related again the Planck theory to Thermodynamics. Stemming from the consideration that, according to Planck's theory, the mean energy of a great number of oscillators is given by

$$\langle\langle E \rangle\rangle = \frac{\hbar\omega_t}{\exp(\hbar\omega_t/k_B T) - 1} \tag{1.1.8}$$

where ω_t is the characteristic angular frequency and $\hbar = \frac{h}{2\pi}$, Einstein (1907) proposed that the internal energy of a solid composed of an Avogadro number of atoms N_0 is given by

$$U = \frac{3N_0}{\exp(\hbar\omega_E/k_B T) - 1} \tag{1.1.9}$$

C_V was then obtained from Eq. (1.1.9) by partial derivation in dT:

$$C_V = \left(\frac{\partial U}{\partial T}\right) = \frac{3N_0 k_B X^2 e^x}{(e^x - 1)^2} = 3RE_{(X)} \tag{1.1.10}$$

with $X = \frac{h\nu_E}{k_B T} = \frac{\hbar\omega_E}{k_B T}$, and

$$E_{(X)} = \frac{X^2 e^x}{(e^x - 1)^2} = \frac{X^2}{a \sin^2\left(\frac{X}{2}\right)} \tag{1.1.11}$$

Equation (1.1.11), known as the "*Einstein function*", is tabulated for various X-values. In the Einstein function, the characteristic frequency ω_E (and the corresponding characteristic temperature $\theta_E = \hbar\omega_E/k_B$) has an arbitrary value that optimizes function (1.1.10) on the basis of high-T experimental data. In Fig. 1.2, we see the original drawing of Einstein (1907) showing the performance of function (1.1.10) with respect to experiments. The extrapolation of Eq. (1.1.10) at low temperature results in notable discrepancies from experimental values. These discrepancies found few years later a reasonable explanation after the studies of Debye (1912) and Born and Von Karman (1913).

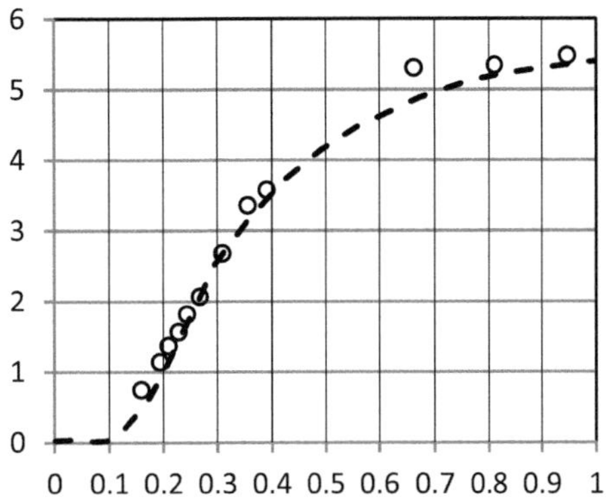

Fig. 1.2 Isochoric heat capacity according to the Einstein (1907) model. The abscissa axis is X and the ordinate axis is the isochoric heat capacity in *cal/(mol* $\times$ *K)*. Modified from Einstein (1907)

As a final historical premise, I want to stress here the role that de Broglie (1924) played in clarifying the wave–particle duality of light. Stemming from Einstein's relativistic observation that the wavelength λ and the momentum $\mathbf{p}$ of a photon are related by $\lambda = h/\mathbf{p}$, de Broglie argued that both light and matter obey this equation and predicted that a particle of mass m moving at velocity v has a (de Broglie) wavelength $\lambda = h/\mathrm{mv}$. This deduction has humongous consequences: for example, electrons traveling at 1% of the speed of light have a wavelength of 2.43 *pm*; those ones accelerated through a potential field of 10,000 *V* attain a wavelength of 12 *pm*; if the potential field drops to 100 *V* then the wavelength increases to 120 *pm*, and so on.

1.2 Photons and Electrons

In the first introductory chapter to *Wave–Mechanics*, Sommerfeld (1930) wrote "*From the point of view of the wave-theory, at least for isotropic media, light rays may be defined as the orthogonal trajectories of wave-surfaces*". Several decades later, McWeeny (1972) defined the wave surfaces as "*disturbances*" associated to particles moving at high speed along a linear trajectory. Both definitions have the virtue of clearing the mind from apparent contradictions concerning the dual nature of light.

1.2.1 Harmonic Waves

The light is a radiation transmitted in localized packets of photons exhibiting "*particle properties*". To these particles, traveling at speed (*c*) it is associated a wavelike "*disturbance*" (McWeeny 1972) whose frequency (ν) is related to energy E and momentum **p** by

$$\mathrm{E} = h\nu \tag{1.2.1}$$

$$\mathbf{p} = \frac{h\nu}{c} \tag{1.2.2}$$

It follows from Eq. (1.2.1) that the wavelength of the *disturbance* is

$$\lambda = \frac{1}{\nu} \equiv \frac{h}{\mathrm{E}} \tag{1.2.3}$$

Let me now assign to the *disturbance* the form of a harmonic wave traveling along the x-axis in the positive direction:

$$\Psi(\mathrm{x},\mathrm{t}) = \mathrm{A}\exp\left\{\pm\frac{2\pi \mathrm{i}}{\lambda}(\mathrm{x}-\mathrm{ut})\right\} \tag{1.2.4}$$

The wave travels with a constant velocity u at all times t. If λ is the *wavelength* and A is the *amplitude*, the *wavenumber* (k) is $k = 1/\lambda$, the *frequency* (ν) is $\nu = \mathrm{ku}$, and the *period* ($\mathcal{T}$) is $\mathcal{T} = 1/\nu$.

If x increases by 1/k the values of Ψ and of its derivative are unchanged: hence the disturbance is periodic in space (at any given time t). If t increases by, $\mathcal{T}, \Psi$ and its derivative are again unchanged: the disturbance is periodic in time (at any point in space). Moreover, the function oscillates between +A and –A. Because $\exp(\mathrm{i}\theta) = \cos\theta + \mathrm{i}\sin\theta$, a disturbance which is everywhere real can be regarded as the real part of either functions (1.2.4) (i.e. positive and negative).

1.2.2 Wave Packets

Any *disturbance* traveling in the positive direction with velocity u can be represented, at the light of the Fourier's theorem, in terms of combined waves of type (1.2.4) with variable $k = 1/\lambda$ and suitably chosen amplitudes A:

$$\Psi(\mathrm{x},\mathrm{t}) = \int_{-\infty}^{+\infty} \mathrm{A}(k)\exp\{\pm 2\pi i k(\mathrm{x}-\mathrm{ut})\}\mathrm{d}k \tag{1.2.5}$$

At the limit, the integral may be represented as a summation of individual terms $\mathrm{A}_1\exp\{\pm 2\pi \mathrm{i}k_1(\mathrm{x}-\mathrm{ut})\} + \mathrm{A}_2\exp\{\pm 2\pi \mathrm{i}\mathrm{k}_2(\mathrm{x}-\mathrm{ut})\} + \ldots$.... Taking two terms with $k_1 = -k_2 = k$ and $\mathrm{A}_1 = \mathrm{A}_2$ we obtain a "*traveling wave*" with $\psi \propto \cos 2\pi k(\mathrm{x}-\mathrm{ut})$ and corresponding to the real part of (1.2.4).

To depict a "*localized disturbance*" (McWeeny 1972) which travels with the moving particle, we must choose amplitude factors A(k) able to generate a constructive interference at a specified point (and destructive elsewhere). This may be obtained adopting a very narrow range of k's. Consider in fact a Gaussian wave packet with

$$\mathrm{A}(k) = \mathrm{a}\ \exp\left\{-\sigma(k-k_0)^2\right\} \tag{1.2.6}$$

When σ is large we have a superposition of wave trains with wavenumbers differing inappreciably from k_0 and amplitudes falling off rapidly, according to the Gaussian law. In this case (McWeeny 1972),

$$\Psi(\mathrm{x},\mathrm{t}) = \int_{-\infty}^{+\infty} \mathrm{a}\ \exp\left\{-\sigma(k-k_0)^2 + 2\pi i k(\mathrm{x}-\mathrm{ut})\right\} dk \tag{1.2.7}$$

Because, for any p = positive and q = complex, we have the identity

$$\int_{-\infty}^{+\infty} \exp\left(-\mathbf{p}x^2 - qx\right)dx = \sqrt{\frac{\pi}{\mathbf{p}}}\exp\left[\frac{q^2}{4\mathbf{p}}\right], \tag{1.2.8}$$

Equation (1.2.5) may be rewritten in the form:

$$\Psi(\mathrm{x},\mathrm{t}) = \mathrm{a}\sqrt{\frac{\pi}{\sigma}}\exp\left[-\frac{\pi^2(\mathrm{x}-\mathrm{ut})^2}{\sigma}\right]\exp\{2\pi i k_0(\mathrm{x}-\mathrm{ut})\} \tag{1.2.9}$$

As shown schematically in Fig. 1.3, Eq. (1.2.9) describes an oscillation of wavelength $\lambda_0 = 1/k_0$ whose amplitude rapidly decreases in both propagation directions about the point x = 0 (part a of the figure). Because the speed of photons is constant in vacuum (u = c) the packed propagates without dispersion and Ψ(x, t) at distance ut ≡ ct is identical to Ψ(x, 0).

If we apply the above reasoning to a particle, whose disturbance has a wavelength related to the particle momentum **p** = mv by de Broglie's law, then $\lambda = h/\mathbf{p}$, and remembering that the energy E is related to frequency by E = $h\nu$, we must deduce that the velocity of propagation is affected by the field into which it propagates. In fact, if the total energy E is $\mathrm{E} = \frac{1}{2}\mathrm{mv}^2 + \mathrm{V}$ with V = potential energy, we get

$$\mathrm{u} = \lambda\nu = \frac{hE}{\mathbf{p}h} = \frac{E}{\sqrt{2\mathrm{m}(\mathrm{E}\text{-}\mathrm{V})}} \tag{1.2.10}$$

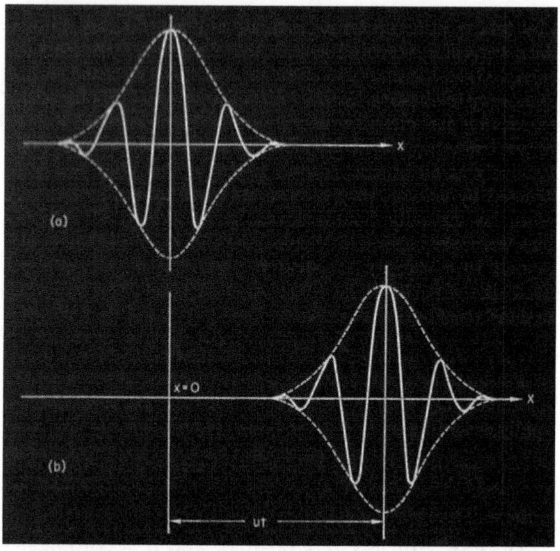

Fig. 1.3 Motion of a wave packet. (**a**) At time t = 0 the maximum amplitude is observed at x = 0. (**b**) at time *t* and distance ut the packet is reproduced without dispersion. Modified from McWeeny (1972)

If the potential V is non-uniform, the velocity of propagation depends then on frequency (through E) and on the particle's position (through V). Thus a dispersion will be observed, with the *phase velocity* (u) no longer coincident with the velocity of the packet itself (or "*group velocity*" v_g).

We demonstrate now that, by virtue of Eq. (1.2.10) the group velocity v_g coincides with the particle's velocity v. We assume to this purpose that the relationship between v and *k* is non-linear (as implicit in Eq. 1.2.10) and that it may be represented by a polynomial expansion of the type (McWeeny, 1972):

$$\mathrm{v} = \mathrm{v}_0 + \alpha(k - k_0) + \frac{1}{2}\beta(k - k_0)^2 + \ldots \tag{1.2.11}$$

with $\alpha = \left[\frac{\mathrm{dv}}{\mathrm{d}k}\right]_0$ and $\beta = \left[\frac{\mathrm{d}^2\mathrm{v}}{\mathrm{d}k^2}\right]_0$ valid at $k = k_0$. Consider now the real part of the integral (1.2.7). The main contribution comes from values around k_0, hence $\delta = k - k_0$ is small in this region. If we put $k = k_0 + \delta$, expand the imaginary term, and truncate to the first order we obtain

$$2\pi \mathrm{i}k\{\mathrm{x} - \mathrm{u}(k)\mathrm{t}\} \cong 2\pi \mathrm{i}k_0(\mathrm{x} - \mathrm{u}_0\mathrm{t}) + 2\pi \mathrm{i}\delta(\mathrm{x} - \alpha\mathrm{t}) \tag{1.2.12}$$

with u_0 = phase velocity at $k = k_0$.

We may now integrate Eq. (1.2.7) and the result is (Eq. 1.12 in McWeeny 1972):

Table 1.1 Photons and particles. Main features and duality of wave and particle nature (from McWeeny 1972; with modifications)

Subject	Photon	Particle of mass m
Wave train	$\exp\{2\pi i(kx - \nu t)\}$	$\exp\{2\pi i(kx - \nu t)\}$
Frequency	$\nu = E/h$ (frequency of radiation)	$\nu = E/h$ (frequency in wave function)
Energy	E (energy of photon)	$E = \frac{1}{2}mv^2 + V$ (classical)
Phase velocity	$u = c$	$u = \frac{E}{\sqrt{2m(E-V)}} = \frac{E}{mv}$ (unobservable)
de Broglie wavelength	$\lambda = c/\nu = h/p$	$\lambda = u/\nu = h/mv$
Group velocity	$v_g = c$ (photon velocity)	$v_g = v$ (particle velocity)
Dispersion	None	Spreading of packet

$$\Psi(x,t) \cong a\sqrt{\frac{\pi}{\sigma}}\exp\left[\frac{-\pi^2(x-\alpha t)^2}{\sigma}\right]\exp\{2\pi i k_0(x - u_0 t)\} \qquad (1.2.13)$$

From Eq. (1.2.13), we may deduce that the wave train travels with the mean phase velocity u_0 (2nd exponential) but the wave profile is centered at $x = \alpha t$ (1st exponential), hence it moves with a group velocity $v_g = \alpha = (d\nu/dk)_0 (\text{at}\, k = k_0)$. Finally, by relating the frequency to the particle's velocity through de Broglie's relation we may see that $v_g = v$, hence the packet "*stays with the particle whose behaviour it is supposed to describe*" (McWeeny 1972):

$$v_g = \frac{d\nu}{dk} = \frac{1}{h}\frac{dE}{dk} = \frac{1}{h}\frac{d}{dk}\left(\frac{p^2}{2m} + V\right) = \frac{1}{h}\frac{d}{dk}\left(\frac{h^2k^2}{2m} + V\right) = \frac{hk}{m} = v \qquad (1.2.14)$$

Although there are similarities between the equations describing the behavior of photons and electrons, we have seen that real particles imply disturbances which spread as they travel along a particular positive trajectory, while photons do not. Table 1.1 resumes the resemblances between photons and particles and the main features of the duality between waves and particles.

1.2.3 de Broglie Waves

The existence of wavelike disturbances of appropriate wavelength (i.e. atomic scale) associated to particles of limited mass and traveling at high speed furnishes an independent explanation of the quantized Bohr's orbits. In fact, for a circular

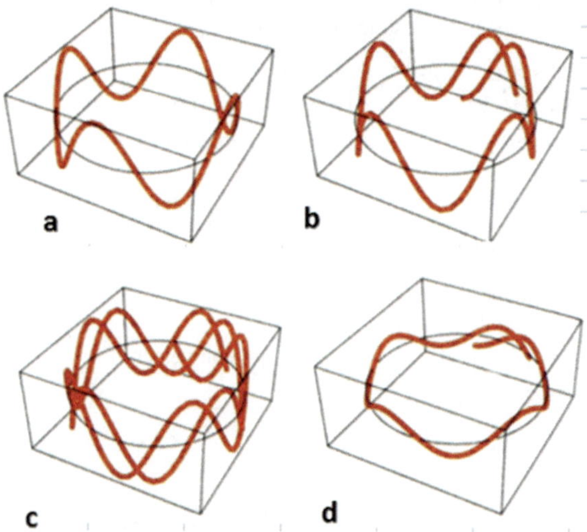

Fig. 1.4 **a** The wave is in phase and matches after a complete revolution; **b** wave mismatch; **c** two non-in-phase revolutions; **d** total amplitude after two revolutions. Modified from McQuarrie (2008)

orbit such as that one covered by a 1 s electron around a proton, the traveled distance must be a multiple integer of the wavelength. That is, we have

$$2\pi r = n\lambda \quad n = 1, 2, 3\ldots \tag{1.2.15}$$

Because a particle of mass m moving at velocity v has a (de Broglie) wavelength $\lambda = h/mv$, Eq. (1.2.15) brings immediately to Bohr's quantization condition

$$mvr = n\hbar \tag{1.2.16}$$

Bohr's radius (a_0) for $n = 1$ is 0.529 $\mathring{A}$ (5.29×10^{-11} *m*), and with $\hbar = 1.055 \times 10^{-34}$ $J{\cdot}s$ and $m = 9.11 \times 10^{-31}$ *kg* we get the speed of the traveling electron $v = 2.19 \times 10^6$ $m{\cdot}s^{-1}$. This speed corresponds roughly to 1% of the speed of light.

Obviously, Eq. (1.2.15) implies that the disturbance must be in phase with the circular orbit after a complete revolution (Fig. 1.4a). If this does not occur (Fig. 1.4b), the wave progressively disappears after few orbits (Fig. 1.4c) and the sum of the amplitudes decreases (Fig. 1.4d).

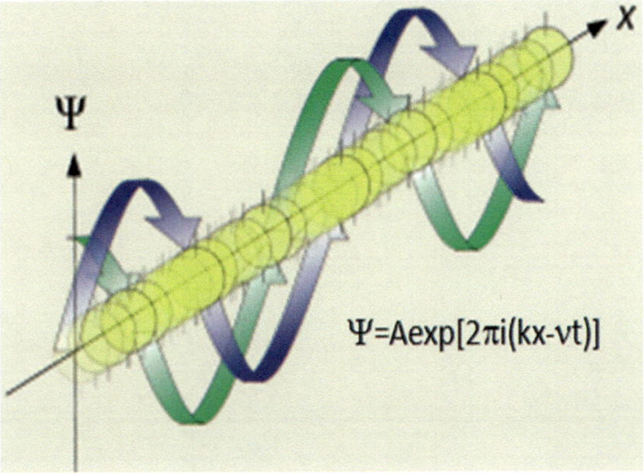

Fig. 1.5 Propagation of a de Broglie wave (from Maschen Wikipedia Commons 2012; modified)

In Fig. 1.5, we see schematically the propagation of a de Broglie wave along x with the real (blue) and imaginary (green) components split up. The color opacity of the yellow dots is intended to return the probability concept associated to the squared function (see later). The maximum probability is associated to minimum amplitude.

Chapter 2
The Schrödinger Equation

In this chapter we enter in more detail in the significance of a wave function, when this is associated to a "*free particle*" or to a particle subjected to external constraints. We analyze it in terms of space and time components and give a first appreciation of the concept of "*probability density function*". We finally see that not all functional forms of a wave equation are acceptable, but only those which do not lead to an "*infinity catastrophe*".

2.1 Free Particle

Consider the positive term of a *disturbance* as depicted in terms of an harmonic wave traveling along the x-axis:

$$\Psi(\mathrm{x},\mathrm{t}) = \exp\left\{\frac{2\pi \mathrm{i}}{\lambda}(\mathrm{x}-\mathrm{ut})\right\} = \phi_{\frac{\lambda}{2\pi}}(\mathrm{x})\cdot\phi^{*}_{\frac{\lambda}{2\pi}}(\mathrm{ut}) \qquad (2.1.1)$$

Taking into account the fact that the energy E is related to velocity u through the momentum $\mathbf{p}$ by $\mathrm{u} = \lambda v = \frac{\mathrm{E}}{\mathbf{p}}$ we may split the above equation into two exponential contributions, one related to momentum and one to energy

$$\Psi(\mathrm{x},\mathrm{t}) = \exp\left(\mathrm{i}\frac{\mathbf{p}}{\hbar}\mathrm{x}\right)\exp\left(-\mathrm{i}\frac{\{\mathrm{E}\}}{\hbar}\mathrm{t}\right) = \phi_{\frac{\mathbf{p}}{2\pi}}(\mathrm{x})\cdot\phi^{*}_{\frac{\mathbf{p}}{2\pi}}(\mathrm{t}) \qquad (2.1.2)$$

where $\hbar = \frac{h}{2\pi}$ is the *rationalized* Planck's constant.

Making the partial derivatives on distance x and time t

$$-\hbar^{2}\frac{\partial^{2}}{\partial \mathrm{x}^{2}}\Psi = \mathbf{p}^{2}\Psi(\mathrm{x},\mathrm{t}) \qquad (2.1.3)$$

G. A. Ottonello, *Quantum Geochemistry*, Springer Geochemistry,
https://doi.org/10.1007/978-3-031-21837-8_2

$$i\hbar\frac{\partial}{\partial t}\Psi = E\Psi(x,t) \tag{2.1.4}$$

and combining with

$$E = \mathbf{p}^2/2m \tag{2.1.5}$$

we get the one-dimensional form of the Schrödinger equation for a free electron

$$i\hbar\frac{\partial\Psi}{\partial t} = -\frac{\hbar^2}{2m}\frac{\partial^2\Psi}{\partial x^2} = \mathsf{T}_x\,\Psi \tag{2.1.6.1}$$

$$\mathsf{T}_x = -\frac{\hbar^2}{2m}\frac{\partial^2}{\partial x^2} \tag{2.1.6.2}$$

The three-dimensional form is obtained expanding the momentum **p** in its axial components p_x, p_y, p_z:

$$E = \frac{1}{2m}\left(p_x^2 + p_y^2 + p_z^2\right) \tag{2.1.7}$$

$$i\hbar\frac{\partial\Psi}{\partial t} = -\frac{\hbar^2}{2m}\left(\frac{\partial^2\Psi}{\partial x^2} + \frac{\partial^2\Psi}{\partial y^2} + \frac{\partial^2\Psi}{\partial z^2}\right) \equiv -\frac{\hbar^2}{2m}\nabla^2\Psi = \mathsf{T}_r\,\Psi(r) \tag{2.1.8}$$

2.2 Particle in a Conservative Potential Field

Consider now the particle as moving in an external field of classical potential energy V(x,y,z).

The bulk energy of the traveling particle is, in this case,

$$E = \frac{1}{2m}\left(p_x^2 + p_y^2 + p_z^2\right) + V(x,y,z) \tag{2.2.1}$$

and Eq. (2.1.8) is replaced by

$$i\hbar\frac{\partial\Psi}{\partial t} = -\frac{\hbar^2}{2m}\nabla^2\Psi + \mathsf{V}\Psi = \mathsf{H}\Psi \tag{2.2.2}$$

The right-hand side of Eq. (2.2.2) is the Hamiltonian operator H

$$\mathsf{H} = -\frac{\hbar^2}{2m}\nabla^2 + \mathsf{V} \tag{2.2.3}$$

and Eq. (2.2.3) is usually written in the form

$$\mathsf{H}\Psi = i\hbar\frac{\partial\Psi}{\partial t} \tag{2.2.4}$$

It should be stressed that the Hamiltonian operator H may be applied to any function ψ and must not be confused with the Hamiltonian function H ($\mathbf{p}$,x) relating the energy to the momentum components and to position in Newtonian systems.

2.3 Time-Independent Equation

The time-independent Schrödinger equation is crucial in understanding the deep significance of "*stationary states*" and, more generally, in the quantum theory of valence and molecular structure.

Let me denote the time-dependent wave function in the form

$$\Psi(x, y, z, t) = \psi(x, y, z)\varphi(t) \tag{2.3.1}$$

In vectorial notation the above equation may be expressed as

$$\Psi(\mathbf{r}, t) = \psi(\mathbf{r})\varphi(t) \tag{2.3.2}$$

Rewriting Eq. (2.2.2) in the form

$$\frac{i\hbar}{\Psi(\mathbf{r}, t)}\frac{\partial\Psi(\mathbf{r}, t)}{\partial t} = -\frac{\hbar^2}{2m}\nabla^2 + V(\mathbf{r}) \tag{2.3.3}$$

and combining with Eq. (2.3.2) we obtain

$$\frac{i\hbar}{\varphi(t)}\frac{d\varphi(t)}{dt} = \frac{1}{\psi(\mathbf{r})}\left(-\frac{\hbar^2}{2m}\nabla^2\psi(\mathbf{r}) + V(\mathbf{r})\psi(\mathbf{r})\right) \tag{2.3.4}$$

While the left-hand side of Eq. (2.3.4) is time-dependent, the right-hand side is space-dependent. The x, y, z values (i.e. $\mathbf{r}$ in vectorial notation) must be assigned (and constant) in order to satisfy the left-hand side as well. Assume now the constant to be the energy E of the particle. In this case we may write

$$\frac{d\varphi(t)}{dt} = -i\frac{E}{\hbar}\varphi \tag{2.3.5}$$

hence

$$\varphi(t) = \exp\left(-i\frac{E}{\hbar}t\right) \tag{2.3.6}$$

which is the second exponential term on the right in Eq. (2.1.2) previously written for the free particle. E is thus effectively the energy of the system in the state with wave function (2.3.2). Moreover, because Eq. (2.2.2) contains only the first derivative with respect to time, the ambiguity in sign of the traveling wave solution of the classical wave Eq. (1.2.4) is removed, and this justifies the choice of selecting only the positive value of the harmonic wave Eq. (2.1.1). Introducing the Hamiltonian operator H we may finally write the *time-independent Schrödinger equation* as

$$\mathsf{H}\psi = E\psi \tag{2.3.7}$$

From any solution of Eq. (2.3.7), we may build a particular solution of the time-dependent equation $\psi(\mathbf{r}, t)$ as well, such as (Eq. 1.25 in McWeeny, 1972):

$$\Psi(\mathbf{r}, t) = \psi(\mathbf{r}) \exp\left(\frac{-iEt}{\hbar}\right) \tag{2.3.8}$$

2.4 Probability Density Function

Equations (2.2.4) and (2.3.7) may be defined as "*linear*" (McWeeny, 1972) in the sense that the solution is arbitrary within a constant factor, i.e. if ψ is a solution, so is $c\psi$ with c as arbitrary constant $\neq 0$. To remove this ambiguity it is introduced the concept of "*probability density function*" $|\psi|^2$ which is the product of the function itself by its complex conjugate (see par A.4). Because the probability of finding the particle somewhere in space is unity, we may pose, in vectorial notation:

$$\int |\Psi|^2 d\mathbf{r} = 1 \tag{2.4.1}$$

where the integration is carried out over the entire space ($d\mathbf{r}$ is a differential volume in vector notation when $\mathbf{r}$ defines the space in terms of x, y, z variables).

Alternative solutions of the type $c\psi$ are no longer valid and the function is said to be "*normalized*".

At the light of Eq. (2.4.1), the significance of "*stationary state*" is clear: the probability density function, describing where the particle is likely to be found, persists indefinitely without change:

$$|\Psi(\mathbf{r},t)|^2 = |\psi(\mathbf{r})|^2 \tag{2.4.2}$$

and the normalization condition reduces to

$$\int |\psi(\mathbf{r})|^2 d\mathbf{r} = 1 \tag{2.4.3}$$

The fact that the function must be "*normalizable*" implies other important consequences:

- Even before normalization the function must obey the condition

$$\int |\Psi|^2 d\mathbf{r} = \text{finite} \tag{2.4.4}$$

i.e. the function must be "*quadratically integrable*".

- The square of the wave must be single valued, because two distinct values of the probability of finding the particle in a given place are a nonsense and cannot thus be admitted.
- Ψ may become infinite at a finite number of points (i.e. "*singularities*"), but this must happen in such a way that the integral converges.
- Since the probability of finding the particle somewhere when searching over the entire space must be unity at all times, it follows that the normalization integral (2.4.4) must be time-independent.

2.5 The Amplitude Equation

Let me rewrite Eq. (2.3.4) in the form of a system

$$\left.\begin{array}{r} \frac{d\varphi(t)}{dt} = -\frac{i}{\hbar}E\varphi(t) \rightarrow \varphi_E(t) = \exp(-iEt/\hbar) \\ -\frac{\hbar^2}{2m}\frac{d\psi(x)}{dx^2} + V(x)\psi(x) = E\psi(x) \rightarrow \mathsf{H}\left(|\mathbf{p}|^2 + \mathsf{V}\right)\psi(x) = E\psi(x) \end{array}\right\} \tag{2.5.1}$$

Multiplying the second equation by $-2m/\hbar^2$ and transposing the term in E, one gets the *amplitude equation* written in terms of rationalized Planck's constant

$$\frac{d^2\psi}{dx^2} + \frac{2m}{\hbar^2}\{E - V(x)\}\psi = 0 \tag{2.5.2}$$

i.e. $\psi(x)$ determines the amplitude of the function $\psi(x, t)$.

2.6 Solutions of the Wave Equation

To be a satisfactory wave function, a solution of the Schrödinger wave equation must be "*continuous, single-valued and finite throughout the configuration space of the system*" (Pauling and Bright Wilson,1935). With "*configuration space of the system*" it is meant "*for all values of the coordinate x which the system may assume*" (ibid.).

Equation (2.5.2) has various solutions corresponding to distinct energy terms E_n, so it is better written with the amplitude function $\psi_n(x)$ in the form

$$\frac{d^2\psi_n}{dx^2} + \frac{8\pi^2 m}{h^2}\{E_n - V(x)\}\psi_n = 0 \qquad (2.6.1)$$

The corresponding integrated time equation is

$$\varphi_n(t) = \exp\left(-\frac{2\pi i E_n}{h}t\right) = \exp\left(-\frac{iE_n}{\hbar}t\right) \qquad (2.6.2)$$

The general solutions of Eqs. (2.6.1) and (2.6.2) are the sum of all the particular solutions corresponding to the various (stationary) energy states. The mono-dimensional general expression thus takes the form

$$\Psi(x,t) = \sum_n a_n\Psi_n(x,t) = \sum_n a_n\psi_n(x)\exp\left(-\frac{iE_n}{\hbar}t\right) \qquad (2.6.3)$$

where the a_n terms are constants. The values of the functions $\psi_n(x)$ that satisfy the general Eq. (2.6.3) are the "*eigenfunctions*" (*Eigenfunktionen*; Sommerfeld, 1930) of the wave equation and the E_n values are the characteristic energy values, or "*eigenvalues*" (*Eigenwerte*; Sommerfeld, 1930) of the system.

The analysis of the deep intrinsic significance of the eigenfunctions (and eigenvalues) of Eq. (2.6.3) in quantum chemistry made by Pauling and Bright Wilson (1935) is particularly stringent and must be stressed again here. Consider then a system with one degree of freedom, i.e. a potential energy function increasing without limit for large positive and negative values of x (Fig. 2.1). For a given value of E, the wave equation is

$$\frac{d^2\psi}{dx^2} = \frac{2m}{\hbar^2}\{V(x) - E\}\psi \qquad (2.6.4)$$

For $x > a$ $V(x) > E$, hence $V(x)–E$ positive and, if ψ is positive, the curvature $\frac{d^2\psi}{dx^2}$ is positive as well (and obviously negative if ψ is negative). Select now an arbitrary point $x = c$ where the function has a certain value and a certain slope as illustrated by Curve 1 in Fig. 2.2.

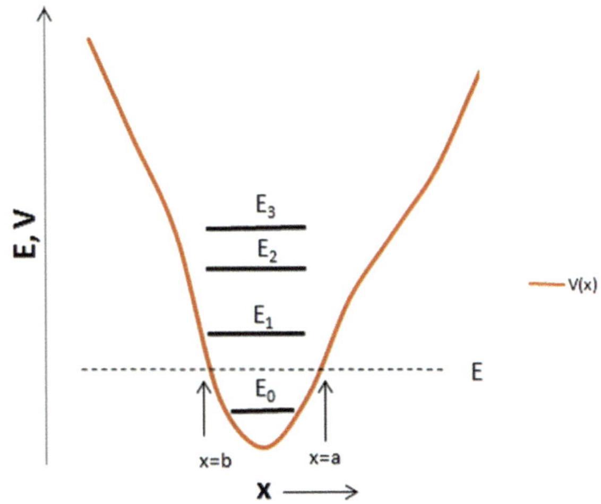

Fig. 2.1 Potential energy function for a general system with one degree of freedom. Modified from Pauling and Bright Wilson (1935)

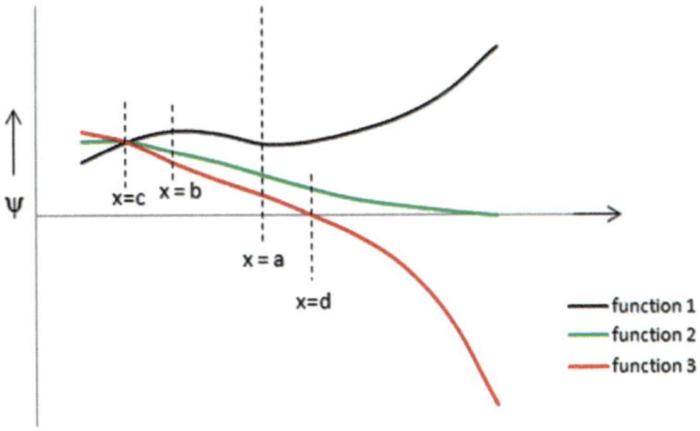

Fig. 2.2 The behavior of ψ for x > a. Modified from Pauling and Bright Wilson, 1935)

With the increase of x the difference between V(x) and E decreases and hence the slope in the function $\psi(x)$ decreases. At the point x = b where V(x) < E the slope becomes negative and then again positive at x = a when V(x) > E. For x > a the function will increase however without limit as schematically indicated in Fig. 2.2 and the result will be an "*infinity catastrophe*". The function cannot be thus accepted as a wave function. We make thus a second attempt (curve 3 in Fig. 2.2): the slope remain negative, and, as drawn, the curve intersects at x = d. For x > d the function decreases more and more rapidly, suffering thus again of the infinity catastrophe. We see then that only a very accurate selection of the slope of the function at point x = c will bring the function to approach asymptotically the value zero with increasing x (function 2 in Fig. 2.2).

I conclude stating that for a satisfactory value of E there is only a satisfactory value of the slope at x = c by the use of which a satisfactory wave function can be built up. Customarily, the values of E_n which define the characteristic energy of the system are indicated as depicted in Fig. 2.1, with n = principal quantum number (and indeed we will define n in this way at Sect. 3.4).

Writing the one-dimensional space- and time-dependent wave function $\Psi(x, t)$ in the form

$$-\frac{\hbar^2}{2m}\frac{\partial^2\Psi(x,t)}{\partial x^2} + V(x)\Psi(x,t) = \frac{\hbar}{i}\frac{\partial\Psi(x,t)}{\partial t} \rightarrow [\mathsf{T}_x + \mathsf{V}(x)]\Psi(x,t)$$
$$= -i\hbar\frac{\partial}{\partial t}\Psi(x,t) \quad (2.6.5)$$

where T_x and $\mathsf{V}(x)$ are, respectively, the kinetic and potential energy operators:

$$\mathsf{T}_x + \mathsf{V}(x) = \mathsf{H} \quad (2.6.6)$$

and $\Psi(x,t)$ is normalized to 1:

$$\int_{-\infty}^{+\infty} |\Psi(x,t)|^2 dx = 1 \quad (2.6.7)$$

also the amplitude equations must be normalized, so that each individual amplitude function $\psi_n(x)$ satisfies

$$\int_{-\infty}^{+\infty} \psi_n^*(x)\psi_n(x) = 1 \quad (2.6.8)$$

The amplitude functions are frequently real, in which case $\psi_n^*(x) = \psi_n(x)$.

2.7 The Particle in a One-Dimensional Box

A "*box*" is a useful model to reproduce in an easy fashion several kinds of constraints such as those occurring to a molecule in an empty container or to an electron into a metal. Consider first the case of a one-dimensional cage where the potential is zero inside and rises steeply at the edges to a constant value (Fig. 2.3).

The Schrödinger equation inside the box is

$$-\frac{\hbar^2}{2m}\frac{\partial^2\psi}{\partial x^2} = E\psi \quad (2.7.1)$$

and, outside the box

$$-\frac{\hbar^2}{2m}\frac{\partial^2\psi}{\partial x^2} = (E - V_0)\psi \quad (2.7.2)$$

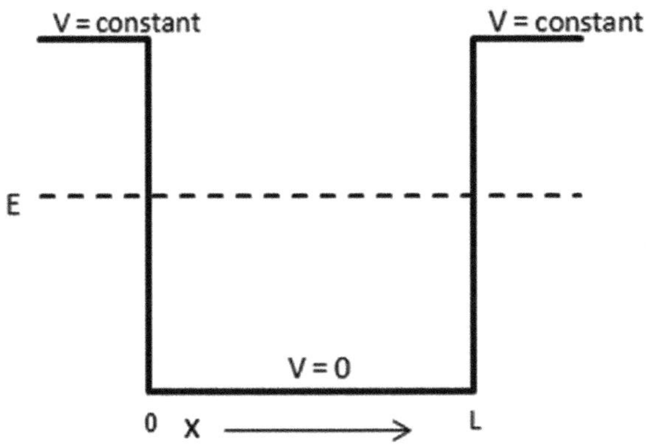

Fig. 2.3 One-dimensional potential box with identical limits. The potential is arbitrarily set to zero inside the box

The solution of (2.7.1) is sinusoidal with arbitrary constants A and B

$$\psi = A\ \sin\left(x\sqrt{\frac{2mE}{\hbar^2}}\right) + B\cos\left(x\sqrt{\frac{2mE}{\hbar^2}}\right) \tag{2.7.3}$$

Outside the box it assumes the exponential form

$$\psi = C\exp\left(x\sqrt{\frac{2m(V_0 - E)}{\hbar^2}}\right) + D\exp\left(-x\sqrt{\frac{2m(V_0 - E)}{\hbar^2}}\right) \tag{2.7.4}$$

The values of the constants A, B, C, D inside and outside the box depend upon the value assigned to V_0 outside the box. As shown in Fig. (2.4), the perturbation may penetrate into regions where it would not be expected, i.e. beyond the limits of the one-dimensional box at x = 0 and x = L. The A, B, C and D constant must then be resolved numerically or graphically in such a way to match at the limits of the boundary.

Let me now consider the case of the "*infinitely deep*" potential box ($V_0 \rightarrow -\infty$). In this case, ψ attains zero exactly at the boundaries of the box. This implies that B = 0 because cos(0) = 1. The second boundary implies that

$$\psi(L) = A\sin\left(L\sqrt{\frac{2mE}{\hbar^2}}\right) \tag{2.7.5}$$

Denoting $k = \sqrt{\frac{2mE}{\hbar^2}}$ I may anticipate that besides the trivial solution A = 0, all the other choices must respect

$$k(L) = n\pi \quad n = 1, 2, \ldots \tag{2.7.6}$$

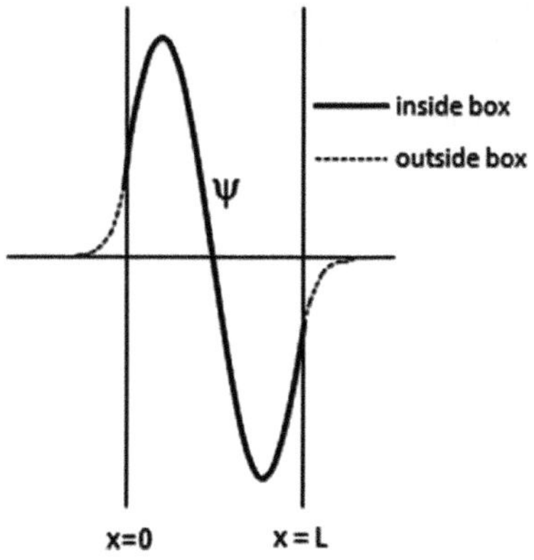

Fig. 2.4 Typical eigenfunction for the one-dimensional box with finite non-zero potential at the limits. At the box boundaries x = 0 and x = L he functions (2.7.3) and (2.7.4) join smoothly

hence

$$E_n = \frac{h^2 n^2}{8mL^2} \quad n = 1, 2, \ldots \tag{2.7.7}$$

We may thus state that the energy of the particle is "*quantized*" and the quantization arises naturally from the boundary condition. In other terms, with the words of Schrodinger (1926) "*The introduction of integers arises in the same natural way as, for example, in a vibrating string, for which the number of nodes is integral. The new conception can be generalized, and I believe that it penetrates deeply into the true nature of the quantum rules*".

In Fig. 2.5, we see the wave functions and the probability densities for the particle in a box of length L for the first four quantum numbers n.

Simple applications of a one-dimensional box (i.e. the "*free electron model*") have been provided for the π electrons in linear conjugated hydrocarbons. A nice example is given by McQuarrie (2008) for butadiene $H_2C = CHCH = CH_2$. The first absorption band corresponding to the $1^1B_u^+ \leftarrow 1^1A_g^-$ transition experimentally observed by Direct Absorption Spectroscopy (DAS) to occur at $4.626 \times 10^4\ cm^{-1}$ (Leopold et al., 1984) has been shown to correspond approximately to the expected energy involved in the transition of one of the four π electrons from the ground (n = 2) state to the n = 3 excited state once a reasonable box length L is introduced in the calculation, i.e.:

$$\Delta E_{3-2} = \frac{h^2(3^2 - 2^2)}{8mL^2} \tag{2.7.8}$$

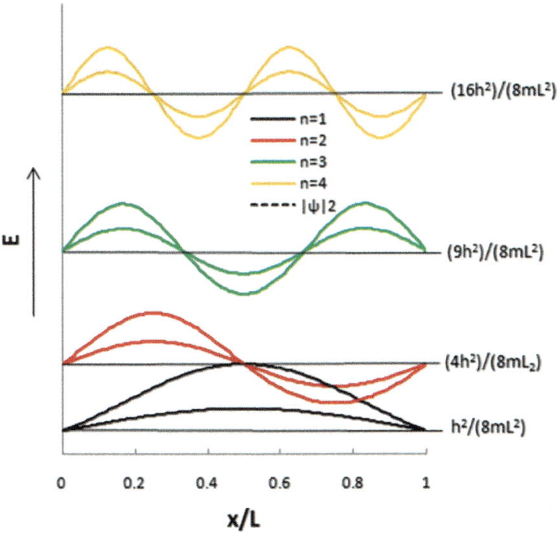

Fig. 2.5 Energy levels, wave functions (solid lines), and probability densities (dotted lines) for the first four quantum numbers of a particle in a box of length L. The quantum number corresponds to the number of loops in the region $0 \leq x \leq L$

Adopting L = 5.78 Å = 5.78×10^{-12} *m*), replacing to m the electron mass at rest ($m_e = 9.109 \times 10^{-31}$ *kg*) and remembering that Planck's constant is 6.626×10^{-34} $J \cdot s$ one gets $\Delta E_{3-2} = 9.02 \times 10^{-19} J$, which corresponds to 5.62 *eV* (1 *eV* = 1.60218×10^{-19} *J*) and to 4.54×10^4 cm^{-1}. (1 *eV* = 8065.54 cm^{-1}), not far from the observed absorption band, which is by the way affected by a large vibronic bandwidth (see Fig. 2.3 and Table IV in Leopold et al.,1984).

2.8 The Particle in a Three-Dimensional Box

It is a simple extension of the one-dimensional case. Consider a box with edges a, b, and c oriented along the x, y, and z axes and origin at (0, 0, 0). The Schrödinger equation may be in this case written as follows:

$$\frac{-\hbar^2}{2m}\left(\frac{\partial^2\psi}{\partial x^2} + \frac{\partial^2\psi}{\partial y^2} + \frac{\partial^2\psi}{\partial z^2}\right) = E\psi(x, y, z) \quad \begin{cases} 0 \leq x \leq a \\ 0 \leq y \leq b \\ 0 \leq z \leq c \end{cases} \tag{2.8.1}$$

The wave function must vanish at the walls (boundaries) of the box (i.e. at 0, x = a, y = b, and z = c). To find a satisfactory solution, we may expand the wave function separating the variables, i.e.:

$$\psi(x, y, z) = X(x)Y(y)Z(z) \tag{2.8.2}$$

Replacing Eq. (2.8.2) in Eq. (2.8.1) and dividing through by $\psi(x, y, z) = X(x)Y(y)Z(z)$, we obtain

$$\frac{-\hbar^2}{2m}\frac{1}{X(x)}\frac{d^2X}{dx^2} + \frac{-\hbar^2}{2m}\frac{1}{Y(y)}\frac{d^2Y}{dy^2} + \frac{-\hbar^2}{2m}\frac{1}{Z(z)}\frac{d^2Z}{dz^2} = E_X + E_Y + E_Z = E \tag{2.8.3}$$

Introducing the quantum number n_x as the number of loops along x in the region between 0 and a (cf. Fig. 2.5) the wavelength becomes $2a/n_x$ and the (normalized) X(x) function reduces to

$$X(x) = \sqrt{\frac{2}{a}}\sin\left(\frac{n_x \pi x}{a}\right), \quad n_x = 1, 2, 3, \ldots, \quad 0 \le x \le a \tag{2.8.4}$$

with

$$E_{n_x} = \frac{n_x^2 h^2}{8ma^2} \tag{2.8.5}$$

The complete wave function $\psi_{n_x n_y n_z}(x, y, z)$ for values of x, y, and z inside the box has the form

$$\psi_{n_x n_y n_z}(x, y, z) = \sqrt{\frac{8}{abc}}\sin\left(\frac{n_x \pi x}{a}\right)\sin\left(\frac{n_y \pi y}{b}\right)\sin\left(\frac{n_z \pi z}{c}\right) \tag{2.8.6}$$

with

$$E_{n_x n_y n_z} = \frac{h^2}{8m}\left(\frac{n_x^2}{a^2} + \frac{n_y^2}{b^2} + \frac{n_z^2}{c^2}\right) \tag{2.8.7}$$

Figure 2.6 represents the geometrical analog of the various stationary states in Cartesian coordinates where the lattice is defined in one octant about the origin (0, 0, 0) by translations 1/a, 1/b, and 1/c. The lattice points at the various nodes have the coordinates n_x/a, n_y/b, n_z/c with n_x = 1,2,…; n_y = 1,2,… and n_z = 1,2,… Each node defines a wave function with energy $E_{n_x n_y n_z} = \frac{h^2}{8m} L^2_{n_x n_y n_z}$ with the distance from the origin $L_{n_x n_y n_z}$ given by $L_{n_x n_y n_z} = \sqrt{\frac{n_x^2}{a^2} + \frac{n_y^2}{b^2} + \frac{n_z^2}{c^2}}$.

The wave function $\Psi_{n_x n_y n_z}$ can be considered as consisting of standing waves along the x, y, and z-directions, with $n_x + 1$, $n_y + 1$, and $n_z + 1$ nodal planes perpendicular, respectively, to the x–, y–, and z-axes from x = y = z = 0 to x = a, y = b, and z = c.

In the special case of a "*cubic box*", i.e. a box with edges a = b = c, while the lowest energy level 111 with quantum numbers $n_x = n_y = n_z = 1$ is "*non-degenerate*", all the successive levels are "*degenerate*" with the "*quantum*

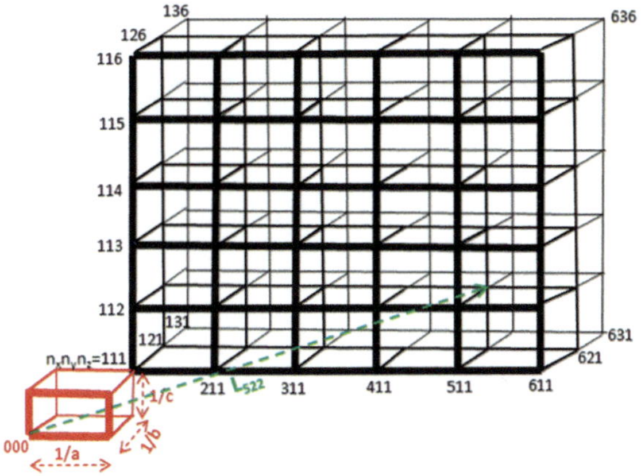

Fig. 2.6 Geometrical representation of the energy levels for a particle in a rectangular box. Modified from Pauling and Bright Wilson (1935)

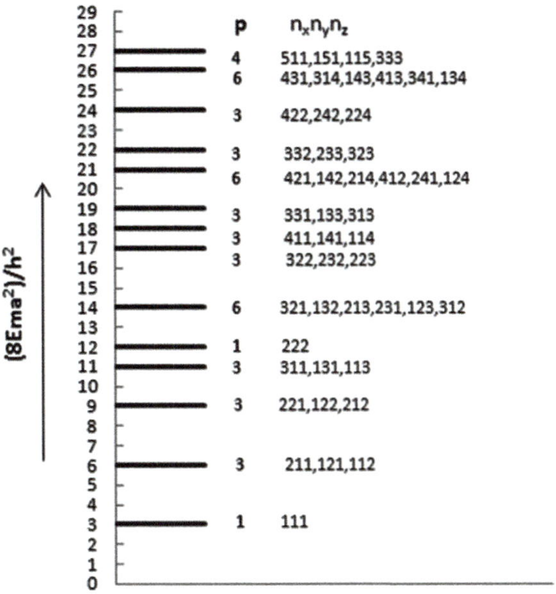

Fig. 2.7 Non-degenerate and degenerate energy levels for a particle in a cubic box

weight" of the level increasing with n. The degree of degeneracy is schematically shown in Fig. 2.7 for the various energy levels, and details on the corresponding quantum states.

Chapter 3
The Harmonic Oscillator

After a short premise on the mathematical implications arising from the adoption of the harmonic oscillator in the wave-mechanical interpretation of the vibrational energy (i.e. the significance of "*singularities*" and the need for a "*normalization*"), we look, in some detail, at the "*form*" of the potential well ("*sampled*", as we will see later, but not determined by the vibrational motion). This well, which is intrinsically anharmonic (due, as we will see later, to the various functional forms of the static energy acting on particles at distance), becomes approximately harmonic at the bottom of the well. We finally see, through a functional analysis, how the "*amplitude equation*" may be rewritten, when Hooke's law is applied, analyzing its expected *eigenfunctions* for a linear and a 3D oscillator.

3.1 Premises

The Harmonic Oscillator is a fundamental tool in understanding how thermal energy is transferred to a system in the form of vibration of nuclear centers about the center of mass of individual moieties such as molecules, clusters or crystalline aggregates. The quantum mechanical procedure is essentially a reasoned application of wave mechanics, based on the pioneering work done by Sommerfeld (1930).

The analysis begins from the vibration equation in a cartesian coordinate system, which may be represented in the form

$$\nabla^2 u + k^2 u = 0 \tag{3.1.1}$$

G. A. Ottonello, *Quantum Geochemistry*, Springer Geochemistry,
https://doi.org/10.1007/978-3-031-21837-8_3

with u as displaement factor and k as force constant. Introducing the Laplacian operator in spherical coordinates:

$$\nabla^2 = \frac{1}{r^2}\left[\frac{\partial}{\partial r}\left(r^2\frac{\partial}{\partial r}\right)\right] + \frac{1}{r^2}\left[\frac{1}{\sin\theta}\frac{\partial}{\partial\theta}\left(\sin\theta\frac{\partial}{\partial\theta}\right) + \frac{1}{\sin^2\theta}\frac{\partial^2}{\partial\phi^2}\right]$$
$$= \frac{1}{r^2}\left[\mathsf{T}(r) + \mathsf{L}^2(\theta,\phi)\right] \tag{3.1.2}$$

one has for the free particle equation:

$$\frac{1}{r^2}\left[\mathsf{T}(r) + \mathsf{L}^2(\theta,\phi)\right] + k^2u(\mathbf{r}) = 0 \tag{3.1.3}$$

To integrate the equation with the method of separation of variables, Sommerfeld (1930) set

$$u(\mathbf{r}) = R(r)\Theta(\theta)\Phi(\phi) \tag{3.1.4}$$

The solution of (3.1.4) must be "*uniform*" (i.e. single-valued and continuous) in the coordinate region $0 \le \theta \le \pi$, $-\pi \le \phi \le +\pi$. Because ϕ does not occur explicitly in Eq. (3.1.4) it is called "*cyclic coordinate*". Its function $\Phi(\phi)$ may be set to $\Phi(\phi) = \exp(\pm im\phi)$, and the uniformity postulate imposes integer values of m. Inserting Φ in the differential equation, dividing by $R\Theta\Phi$ and multiplying by r^2, Sommerfeld obtained

$$\frac{r^2}{R}\left(\frac{d^2R}{dr^2} + \frac{2}{r}\frac{dR}{dr} + k^2R\right) = -\frac{1}{\Theta}\left[\frac{1}{\sin\theta}\frac{d}{d\theta}\left(\sin\theta\frac{d\Theta}{d\theta}\right) - \frac{m^2\Theta}{\sin^2\theta}\right] = \lambda \tag{3.1.5}$$

Since the common value of the two sides of Eq. (3.1.5) must be equal to a constant (that Sommerfeld denoted "*separation constant*" with symbol λ) the differential equation for Θ reduces to:

$$\left[\frac{1}{\sin\theta}\frac{d}{d\theta}\left(\sin\theta\frac{d\Theta}{d\theta}\right) - \frac{m^2\Theta}{\sin^2\theta}\right] + \left(\lambda - \frac{m^2}{\sin^2\theta}\right)\Theta = 0 \tag{3.1.6}$$

Introducing x = $\cos\theta$ as an independent variable, posing $\Theta(\theta) = y(x)$ and accounting for the fact that $\sin\theta d\theta = -dx$, $\sin\theta\frac{d\Theta}{d\theta} = -(1-x^2)y$, Eq. (3.1.6) is transformed into the differential equation of the general (or "*associated*") Legendre function (see Appendix A.15)

$$(1-x^2)\ddot{y} - 2x\dot{y} + \left(\lambda - \frac{m^2}{1-x^2}\right)y = 0 \tag{3.1.7}$$

The name given to values of the independent variable x for which one of the coefficients becomes infinite is "*singular point*". The standard form in which the coefficient of the highest derivative of y is reduced to unity is said to be "*normalized*".

To determine the exponent α of the initial term at the point $z = x - x_0 = 0$, the term *y* is then expanded in a power series of type:

$$y = z^\alpha (a_0 + a_1 z + a_2 z^2 + \cdots) \tag{3.1.8}$$

The series is inserted in the left-hand side of the differential equation and the factor of the lowest power z^α is set to zero. Repeating the operation for the following higher powers gives a recurrence formula for the a_k's. This procedure, which is based on the termination of a power series, has been denoted by Sommerfeld (1930) "*the polynomial method*". The proper value of the spherical functions is determined by the choice of λ. Denoting $n = (m + k)$ the proper value is simply

$$\lambda = n(n + 1) \tag{3.1.9}$$

The above method of determining a proper value and proper function is applicable in all cases in which the differential equation leads to a recurrence formula with two terms. This is so with all quantum problems that are exactly resolvable (i.e. that do not require methods of approximation or perturbations).

The polynomial method (or "*power series method*") has been severely criticized by McQuarrie (2008) as "*tedious at best*" and better replaced by the "*operator method*" extensively discussed in his Appendix to Chapter 5. For the purposes of this book the polynomial method is however quite appropriate because it stresses the significance of singularities and the relation of the force constant with the quantum mechanical description of observables.

By introducing the concept of "*reduced mass*", we will see that potential wells may be obtained also for diatomic molecules, and, more generally, for polyatomic molecules and crystalline substances applying "*Symmetry Preserving Relaxation Procedures*" (see Sect. 8.9.1).

The potential well of a polyatomic molecule or a crystal is not "*harmonic*", and *anharmonicity* is more or less correctly depicted even at the simples theory level, such as in the case of the diatomic molecule H_2 in Fig. 3.1 where the results of single-point RHF/STO-3G calculations are compared with the results of a Morse potential conformed in such a way to reproduce the experiments. Also in this simple case it is however evident that the harmonic condition applies rather well at the athermal limit and, more generally, in the region of the minimum that may encompass room conditions.

The fact that the Harmonic Oscillator is a good approximation for small vibrations has been elegantly stressed by McQuarrie (2008) by expressing the potential well through a Taylor expansion of the potential energy V(l) around the equilibrium bond length $l = l_0$:

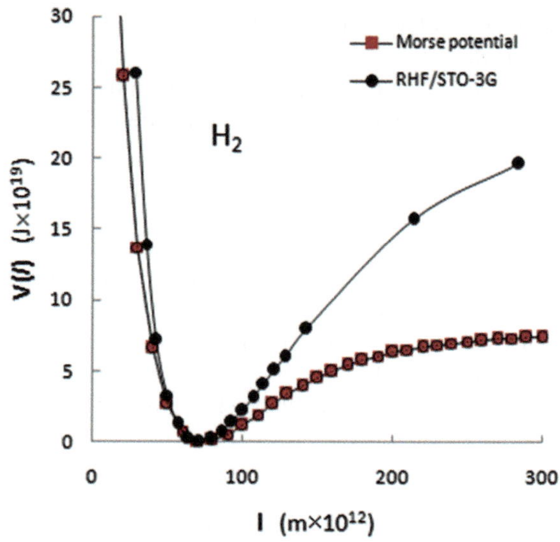

Fig. 3.1 Morse-potential energy curve for the hydrogen diatomic molecule $V(l) = E_D\{1 - \exp[-\beta(l - l_0)]\}^2$ Compared with ab-initio RHF/STO-3G calculations. The adopted values of the dissociation energy (E_D), the curvature at minimum (β), and the equilibrium distance l_0 are taken from McQuarrie (2008) ($E_D = 7.61 \times 10^{-19}\ J$, $\beta = 0.0193\ pm^{-1}$ and $l_0 = 74.1\ pm$

$$V(l) = V(l_0) + \frac{(l - l_0)}{1}\left(\frac{dV}{dl}\right)_{l=l_0} + \frac{(l - l_0)^2}{2!}\left(\frac{d^2V}{dl^2}\right)_{l=l_0} + \frac{(l - l_0)^3}{3!}\left(\frac{d^3V}{dl^3}\right)_{l=l_0} + \cdots \tag{3.1.10}$$

Denoting $(l - l_0) = x$, $(d^2V/dl^2)_{l=l_0} = k$, $(d^3V/dl^3)_{l=l_0} = \gamma_3$, Eq. (3.1.10) becomes

$$V(x) = \frac{k}{2}x^2 + \frac{\gamma_3}{6}x^3 + \cdots \tag{3.1.10}$$

The force constant k is equal to the curvature of V(l) at the minimum and, if we limit ourselves to small values of x, then the first term in the series is dominant and the influence of the higher order terms is negligible, hence, the potential is *harmonic* in the neighborhood of the bottom.

Obviously the perception of the eventual *asymmetricity* depends upon the capability of expanding the analysis far away from the minimum. For example, the range of inter-atomic distances encompassed in Fig. 3.1 extends (on the expansion side) up to more than four times the equilibrium distance. If we limit ourselves to more restricted ranges, such as in the case of the pyrope garnet in Fig. 3.2 (where the most expanded state is 1.03 times the cell edge at minimum), the perception of anharmonicity is lost, while it becomes evident the effect of elastic stiffness. In any case the asymmetricity may be easily reproduced by changing opportunely the value of the force constant or applying the "*Quasi-Harmonic*" procedure (see Sect. 8.9.1).

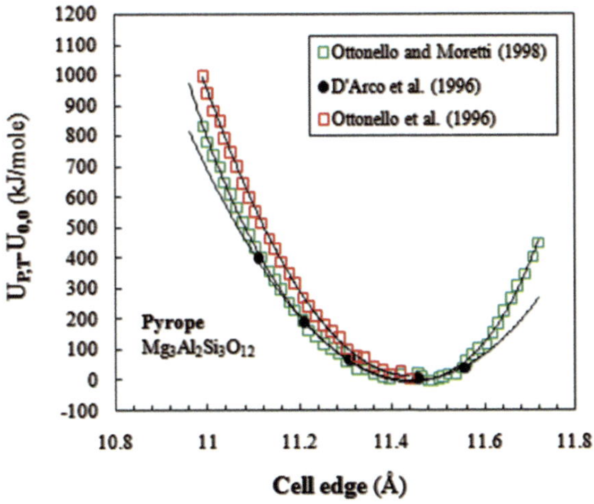

Fig. 3.2 Potential well for the pyrope garnet. The results of ab-initio calculations by D'Arco et al. (1996) are compared with static Born–Mayer parameterizations (Ottonello et al. 1996; Ottonello and Moretti 1998)

3.2 Application of Hooke's Law

Consider for simplicity the particle to be attracted toward the origin by a force F(x) obeying Hooke's law (i.e. simply proportional to distance). With $k > 0$ as proportionality constant, the force and the potential are given by $F(x) = -kx$ and $V(x) = -\frac{1}{2}kx^2$, respectively. The amplitude equation (2.5.2) may be then rewritten in the form

$$\frac{d^2\psi}{dx^2} + \frac{2m}{\hbar^2}\left(E - \frac{1}{2}kx^2\right)\psi = 0 \qquad (3.2.1)$$

To simplify the equation, we may introduce two energy-related terms: $\lambda = \frac{2m}{\hbar^2}E$ and $\alpha = \sqrt{\frac{mk}{\hbar^2}}$.

The amplitude equation then becomes

$$\frac{d^2\psi}{dx^2} + \left(\lambda - \alpha^2x^2\right)\psi = 0 \qquad (3.2.2)$$

The force constant is directly related to the frequency of the motion v_0 through:

$$k = 4\pi^2 m v_0^2 \qquad (3.2.3)$$

and the potential V(x) may be then also written as

$$V(x) = 2\pi^2 m v_0^2 x^2 = \frac{1}{2} k x^2 \tag{3.2.4}$$

In this case, the magnitude α assumes the form $\alpha = \frac{2\pi m v_0}{\hbar} = \frac{\sqrt{mk}}{\hbar}$. The amplitude equation for the particle is then

$$\frac{d^2\psi}{dx^2} + \frac{2m}{\hbar^2}\left(E - 2\pi^2 m v_0^2 x^2\right)\psi = 0 \tag{3.2.5}$$

which, after substitution, reproduces identically Eq. (3.2.2).

3.3 Functional Analysis

The functional analysis begins considering regions where $|x|$ is very large. In these regions λ is negligibly small compared to $-\alpha^2 x^2$ (see Eq. 3.2.2), hence this last term is dominant and, consequently, we may pose

$$\frac{d^2\psi}{dx^2} \cong \alpha^2 x^2 \psi \tag{3.3.1}$$

Consider now an exponential function of type

$$\psi(x) = \exp\left(\pm\frac{1}{2}\alpha x^2\right) \tag{3.3.2}$$

Differentiating the above equation we obtain

$$\frac{d\psi}{dx} = \pm\alpha x \exp\left(\pm\frac{1}{2}\alpha x^2\right) \tag{3.3.3}$$

$$\frac{d^2\psi}{dx^2} = \pm\alpha \exp\left(\pm\frac{1}{2}\alpha x^2\right) + \alpha^2 x^2 \exp\left(\pm\frac{1}{2}\alpha x^2\right) \tag{3.3.4}$$

When $|x|$ is very large the first term may be neglected. Concerning the second term, only the negative one is satisfactory because the positive one tends rapidly to infinity when $|x|$ increases. We accept thus the asymptotic solution $\psi = \alpha^2 x^2 \exp\left(\pm\frac{1}{2}\alpha x^2\right)$ and depict the difference between the asymptotic solution and the actual one over the entire configuration space in terms of a delta function $\delta(x)$ (see Appendix A.5) expanded in a power series of x, such that $\psi(x) = \delta(x)\exp\left(-\frac{1}{2}\alpha x^2\right)$ for $-\infty \leq x \leq \infty$.

$$\frac{d\psi}{dx^2} = \left(\alpha^2 x^2 \delta - \alpha\delta - 2\alpha x\dot{\delta} + \ddot{\delta}\right) \exp\left(\pm\frac{1}{2}\alpha x^2\right) \tag{3.3.5}$$

Dividing by $\exp\left(-\frac{1}{2}\alpha x^2\right)$ Eq. (3.3.2), the terms in $\alpha^2 x^2 \delta$ cancel out and the wave function reduces to

$$\delta - 2\alpha x\dot{\delta} + (\lambda - \alpha)\delta = 0 \tag{3.3.6}$$

Changing the independent variable from x to $\xi = \sqrt{\alpha}x$ and putting $\delta(x) = y(\xi)$ one has

$$\frac{d^2y}{d\xi^2} - 2\xi\frac{dy}{d\xi} + (\lambda - \alpha)y = 0 \tag{3.3.7}$$

which may be solved by expanding $y(\xi)$ in power series and determining the coefficients.

The series however terminates becoming a finite polynomial of degree n for particular values of λ given by

$$\lambda = (2n + 1)\alpha, \quad n = 0, 1, 2, \ldots \tag{3.3.8}$$

The finite polynomial that satisfies Eq. (3.3.7) is the Hermite polynomial $H_n(\xi)$ (see Appendix A.14). Remembering the significance of λ and α, the admitted stationary state energies of the harmonic oscillator are

$$E_n = \left(n + \frac{1}{2}\right)\hbar\sqrt{\frac{k}{m}} \tag{3.3.9}$$

According to classical mechanics the vibration frequency is

$$\nu = \frac{1}{2\pi}\sqrt{\frac{k}{m}} \tag{3.3.10}$$

hence

$$E_n = \left(n + \frac{1}{2}\right)h\nu \tag{3.3.11}$$

The energies for the ground state (n = 0) and the two excited states (n = 1, 2) are then $E_0 = \frac{1}{2}h\nu$, $E_1 = \frac{3}{2}h\nu$, $E_2 = \frac{5}{2}h\nu$, respectively.

The Hermite orthogonal functions $H_n(\xi)$ are (see Appendix A.14.3)

$$\psi_n(x) = N_n \cdot H_n(\xi) \cdot \exp\left(-\frac{\xi^2}{2}\right)\sqrt{2} \tag{3.3.12}$$

where N_n is the normalization factor that reduces to one the probability of finding the particle within the entire configuration space:

$$N_n = \left(\frac{\alpha}{\pi}\right)^{1/2} \frac{1}{\sqrt{2^n n!}} \tag{3.3.13}$$

$$\int_{-\infty}^{+\infty} \psi_n(x)\psi_m(x)dx = \delta_{nm} \tag{3.3.14}$$

with Eq. (3.3.14) satisfying the orthogonality condition over the configuration space between any two n, m functions.

3.4 Eigenfunctions

The first three Hermite polynomials (n = 0, 1, 2) are $H_0(\xi) = 1$, $H_1(\xi) = 2\xi$, $H_2(\xi) = 4\xi^2 - 2$. Replacing them into Eq. (3.3.12) we get the eigenfunctions of the first three states:

$$\psi_0(x) = \left(\frac{\alpha}{\pi}\right)^{1/4} \exp\left(-\frac{\alpha x^2}{2}\right) \tag{3.4.1}$$

$$\psi_1(x) = \sqrt{2}\left(\frac{\alpha}{\pi}\right)^{1/4} x\sqrt{\alpha} \exp\left(-\frac{\alpha x^2}{2}\right) \tag{3.4.2}$$

$$\psi_2(x) = \frac{\sqrt{2}}{2}\left(\frac{\alpha}{\pi}\right)^{1/4} \left(2\alpha x^2 - 1\right) \exp\left(-\frac{\alpha x^2}{2}\right) \tag{3.4.3}$$

The three eigenfunctions $\psi_n(x)$ are plotted in Fig. 3.3.

The number of nodes corresponds to n, and the energy increases with the increase of n. The corresponding probability density functions are shown in Fig. 3.4. The probability distribution is markedly different from the classical one. While according to classical mechanics the maximum probability is found at the boundary where the motion is inverted and the boundaries are sharp, in quantum mechanics, the probability density penetrates into classically forbidden regions.

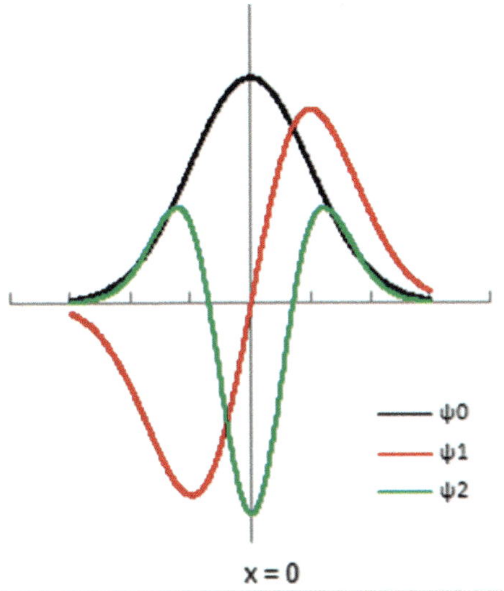

Fig. 3.3 Eigenfunctions of the linear harmonic oscillator for n = 0, 1 and 2. $\alpha = 1$; $-4 \leq x \leq 4$

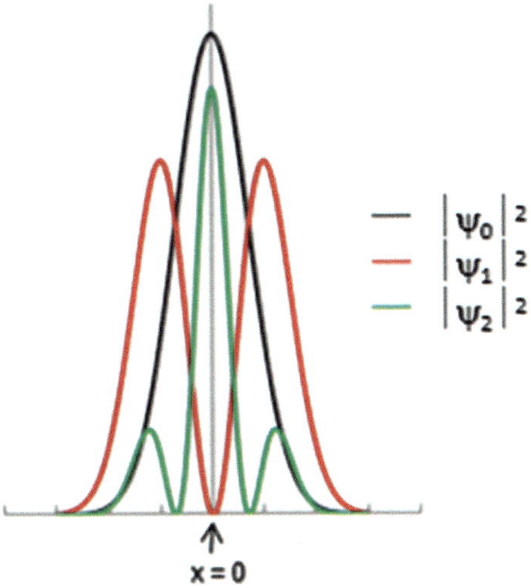

Fig. 3.4 Probability distribution functions of the linear harmonic oscillator for n = 0, 1 and 2. $\alpha = 1$; $-4 \leq x \leq 4$

With the increase of n, however, this sharp distinction progressively vanishes and, as we see for instance for n = 10 in Fig. 3.5 the quantum mechanical function approximates rather closely (in its averaged value) the classical probability distribution of the harmonic oscillator.

As anticipated, the above treatment heavily depends on the pioneering developments of Sommerfeld (1930). To add an historical flavor I add thus here the original drawing of Sommerfeld (his Fig. 3) of "*The first five Proper Functions of the oscillator*" (Fig. 3.6). These are based on the equation $\psi_n = \frac{1}{2^{n/2}\sqrt{n!}} H_n(\xi) \cdot \exp\left(-\frac{\xi^2}{2}\right)$ with the first term on the right before the Hermite polynomial as normalization factor.

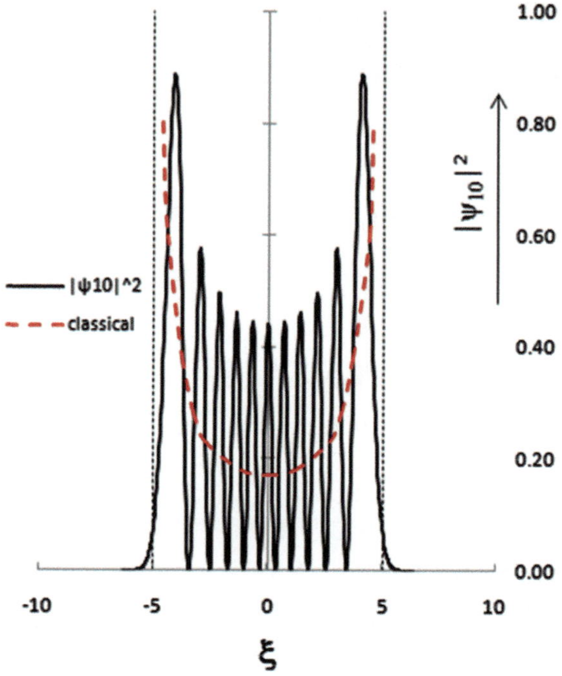

Fig. 3.5 Probability distribution functions of the linear harmonic oscillator for n = 10. $\alpha = 10$; $H(\xi) = 1024\xi^{10} - 23040\xi^8 + 161280\xi^6 - 403299\xi^4 + 302400\xi^2 - 30240$. The classical probability distribution is superimposed (red dashed curve)

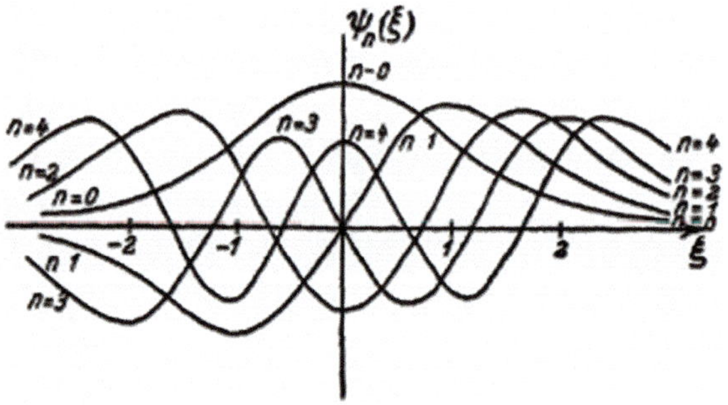

Fig. 3.6 The first five Proper Functions of the oscillator, after Arnold Sommerfeld (from Sommerfeld 1930)

3.5 Tridimensional Harmonic Oscillator

3.5.1 Cartesian Coordinates

I follow hereafter the guidelines exposed by Pauling and Bright Wilson (1935) to whom the reader is referred for a more detailed outline. The particle is bound to the origin by a Hookian force whose constant along the x–, y– and z-axes are k_x, k_y and k_z, respectively. The potential energy V is thus

$$V = \frac{1}{2}\left(k_x x^2 + k_y y^2 + k_z z^2\right) \tag{3.5.1}$$

Replacing to constants the classical frequencies (i.e. $\nu_x = \sqrt{\frac{k_x}{4\pi^2 m}}$; $\nu_y = \sqrt{\frac{k_y}{4\pi^2 m}}$; $\nu_z = \sqrt{\frac{k_z}{4\pi^2 m}}$).

Equation (3.5.1) becomes

$$V = 2\pi^2 m\left(\nu_x^2 x^2 + \nu_y^2 y^2 + \nu_z^2 z^2\right) \tag{3.5.2}$$

and the Schrödinger wave Eq. (2.6.1) takes now the form

$$\frac{\partial^2\psi}{\partial x^2} + \frac{\partial^2\psi}{\partial y^2} + \frac{\partial^2\psi}{\partial z^2} + \frac{8\pi^2 m}{h^2}\left\{E - 2\pi^2 m\left(\nu_x^2 x^2 + \nu_y^2 y^2 + \nu_z^2 z^2\right)\right\}\psi = 0 \tag{3.5.3}$$

To simplify Eq. (3.5.3), it is necessary to make the following replacements:

$$\lambda = \frac{2m}{\hbar}E \tag{3.5.4.1}$$

$$\alpha_x = \frac{2\pi m}{\hbar}\nu_x; \quad \alpha_y = \frac{2\pi m}{\hbar}\nu_y; \quad \alpha_z = \frac{2\pi m}{\hbar}\nu_z \tag{3.5.4.2}$$

obtaining

$$\frac{\partial^2\psi}{\partial x^2} + \frac{\partial^2\psi}{\partial y^2} + \frac{\partial^2\psi}{\partial z^2} + \left(\lambda - \alpha_x^2 x^2 - \alpha_y^2 y^2 - \alpha_z^2 z^2\right)\psi = 0 \tag{3.5.5}$$

This is a partial differentiation equation in three independent variables. To solve it one must proceed as done for the *Amplitude Equation* (2.5.2) and separate the variables. Assuming that the solution may have form

$$\psi(x, y, z) = X(x) \cdot Y(y) \cdot Z(z) \tag{3.5.6}$$

where X(x) is function of x alone, etc., and applying this provisional procedure to Eq. (3.5.5) we obtain, after division by ψ (Eq. 15–8 in Pauling and Bright Wilson 1935):

$$\left(\frac{1}{X}\frac{d^2X}{dx^2} - \alpha_x^2 x^2\right) + \left(\frac{1}{Y}\frac{d^2Y}{dy^2} - \alpha_y^2 y^2\right) + \left(\frac{1}{Z}\frac{d^2Z}{dz^2} - \alpha_z^2 z^2\right) + \psi = 0 \tag{3.5.7}$$

Each one of the three terms in brackets on the left of Eq. (3.5.7) depends upon one single variable. If we vary x and held constant y and z only the first term can vary. However, since the sum of all the terms is also a constant we must conclude that also $\frac{1}{X}\frac{d^2X}{dx^2}$ is independent of x and is itself also a constant (see to this purpose Par. 13 in Pauling and Bright Wilson 1935). Equation (3.5.7) may then be split into three total differential equations of the type

$$\frac{d^2X(x)}{dx^2} + (\lambda_x - \alpha_x^2 x^2)X(x) = 0 \tag{3.5.8.1}$$

$$\frac{d^2Y(y)}{dy^2} + \left(\lambda_y - \alpha_y^2 y^2\right)Y(y) = 0 \tag{3.5.8.2}$$

$$\frac{d^2Z(z)}{dz^2} + (\lambda_z - \alpha_z^2 z^2)Z(z) = 0 \tag{3.5.8.3}$$

where λ_x, λ_y and λ_z are three separation constants, such that

$$\lambda_x + \lambda_y + \lambda_z = \lambda \tag{3.5.9}$$

The analogy of each one of the three Eqs. (3.5.8.1–3.5.8.3) with the amplitude equation for the one-dimensional harmonic oscillator (3.2.2) is obvious.
X(x) is given by (see Eq. 3.3.12):

$$X(x) = N_{nx} \cdot H_n(\xi) \cdot \exp\left(-\frac{\xi^2}{2}\right) \tag{3.5.10}$$

and λ_x obeys the constraint

$$\lambda_x = (2n_x + 1)\alpha_x \tag{3.5.11}$$

where n_x may assume the values 0, 1, 2, …
Because similar relations hold for Y(y) and Z(z), the total energy is (see Eq. 3.3.11)

$$E_{n_x,n_y,n_z} = h\left[\left(n + \frac{1}{2}\right)\nu_x + \left(n + \frac{1}{2}\right)\nu_y + \left(n + \frac{1}{2}\right)\nu_z\right] \tag{3.5.12}$$

and the complete wave function is

$$\psi_{n_x,n_y,n_z}(x, y, z) = N_{n_x,n_y,n_z} H_{n_x}(\xi_x) H_{n_y}(\xi_x) H_{n_x}(\xi_z) \exp\left(\frac{-\xi_x^2}{2}\right) \exp\left(\frac{-\xi_y^2}{2}\right) \exp\left(\frac{-\xi_z^2}{2}\right) \tag{3.5.13}$$

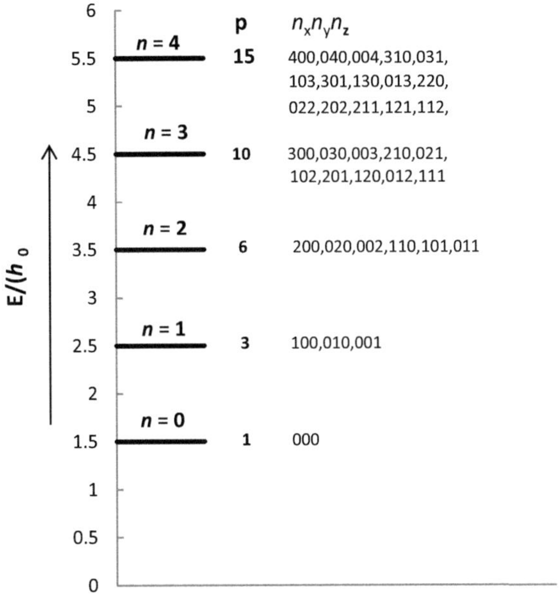

Fig. 3.7 Quantum weights, quantum numbers, and energy levels for the three-dimensional isotropic harmonic oscillator. Modified from Pauling and Bright Wilson (1935)

where the normalizing factor has form

$$N_{n_x,n_y,n_z} = \left[\frac{(\alpha_x\alpha_y\alpha_z)^{1/2}}{\pi^{3/2}2^{n_x+n_y+n_z}n_x!n_y!n_z!}\right]^{1/2} \tag{3.5.14}$$

The adoption of Planck's constant in the discretization of energy (Eqs. 3.3.9 and 3.3.11) inplies that what we treated until now as simple integers (n) assume indeed a precise quantum mechanical significance. I represent thus them from here on out with a different font. Denoting as n the "*total quantum number*" $n = n_x + n_y + n_z$ (Pauling and Bright Wilson, 1935) and assuming identical oscillation frequencies along the three axes, i.e. $\lambda_x = \lambda_y = \lambda_z = \lambda_0$ all the energy levels (except the lowest one) are degenerate with quantum weight $\frac{(n+1)(n+2)}{2}$ The quantum weights and quantum numbers of the first four energy levels are shown in Fig. 3.7 which is a formal representation originally proposed by Pauling and Bright Wilson (1935).

3.5.2 *Cylindrical Coordinates*

The equations of transformation are as follows (see also Appendix A.2.1):

$$\left.\begin{aligned} x &= \rho\cos\varphi \\ y &= \rho\sin\varphi \\ z &= z \end{aligned}\right\} \tag{3.5.15}$$

where φ is the angle between the x-axis and the projection of the point with x-, y-, z-cartesian coordinates on the *xy*-plane.

The Laplacian operator ∇^2 valid for Cartesian coordinates:

$$\nabla^2 = \frac{\partial}{\partial x^2} + \frac{\partial}{\partial y^2} + \frac{\partial}{\partial z^2} \tag{3.5.16}$$

is here replaced by

$$\nabla^2 = \frac{1}{\rho}\frac{\partial}{\partial \rho}\left(\rho\frac{\partial}{\partial \rho}\right) + \frac{1}{\rho^2}\frac{\partial^2}{\partial \varphi^2} + \frac{\partial^2}{\partial z^2} \tag{3.5.17}$$

Assuming that $\lambda_x = \lambda_y = \lambda_0$ (which is the only case where the separation into the three coordinates is possible), the wave Eq. (3.5.3) is replaced by

$$\left[\frac{1}{\rho}\frac{\partial}{\partial \rho}\left(\rho\frac{\partial \psi}{\partial \rho}\right) + \frac{1}{\rho^2}\frac{\partial^2 \psi}{\partial \varphi^2} + \frac{\partial^2 \psi}{\partial z^2}\right] + \frac{8\pi^2 m}{h^2}\left\{E - 2\pi^2 m\left(v_0^2 x^2 + v_z^2 z^2\right)\right\}\psi = 0 \tag{3.5.18}$$

Denoting $\alpha = \frac{2\pi m}{\hbar} v_0$ and keeping unvaried the other replacements (see Eqs. 3.5.3 and 3.5.4.1 and 3.5.4.2) we get

$$\frac{1}{\rho}\frac{\partial}{\partial \rho}\left(\rho\frac{\partial \psi}{\partial \rho}\right) + \frac{1}{\rho^2}\frac{\partial^2 \psi}{\partial \varphi^2} + \frac{\partial^2 \psi}{\partial z^2} + (\lambda - \alpha^2\rho^2 - \alpha_z^2 z^2)\psi = 0 \tag{3.5.19}$$

We proceed then as done with the Cartesian coordinates (cf. Eq. 3.5.6) and split the wave function into three parts dependent each on one single coordinate, i.e. $\psi = P(\rho) \cdot \Phi(\phi) \cdot Z(z)$:

$$\frac{1}{P\rho}\frac{d}{d\rho}\left(\rho\frac{dP}{d\rho}\right) + \frac{1}{\Phi}\frac{1}{\rho^2}\frac{d^2\Phi}{\partial \varphi^2} + \frac{1}{Z}\frac{d^2 Z}{\partial z^2} + (\lambda - \alpha^2\rho^2 - \alpha_z^2 z^2) = 0 \tag{3.5.20}$$

Isolating the Z-dependent part from the rest of the equation, because the sum is constant and the separated components are constant, we may write (Pauling and Bright Wilson 1935)

$$\frac{d^2 Z}{dz^2} + (\lambda_z - \alpha_z^2 z^2)Z = 0 \tag{3.5.21}$$

which is analogous to Eq. (3.5.8.3) and

$$\frac{1}{P\rho}\frac{d}{d\rho}\left(\rho\frac{dP}{d\rho}\right) + \frac{1}{\Phi}\frac{1}{\rho^2}\frac{d^2\Phi}{d\varphi^2} - \alpha^2\rho^2 + \lambda' = 0 \tag{3.5.22}$$

with $\lambda' = \lambda - \lambda_z$ and the λ_z parameter restricted to $\lambda_z = (2n_z + 1)\alpha_z \quad n_z = 0, 1, 2, \ldots$.

Multiplying through the terms in Eq. (3.5.22) by ρ^2 it turns out that the second term is independent of ρ

Hence, denoting $m = \sqrt{-\frac{1}{\Phi}\frac{d^2\Phi}{d\varphi^2}}$ we have

$$\frac{d^2\Phi}{d\varphi^2} + m^2\Phi = 0 \tag{3.5.23}$$

$$\frac{1}{P\rho}\frac{d}{d\rho}\left(\rho\frac{dP}{d\rho}\right) + \left(\lambda' - \alpha^2\rho^2 - \frac{m^2}{\rho^2}\right)P = 0 \tag{3.5.24}$$

The normalized solution of Eq. (3.5.23) is

$$\Phi(\varphi) = \frac{1}{2\pi}\exp(im\varphi) \tag{3.5.25}$$

To be single-valued, m, in Eq. (3.5.25), must be a positive or negative integer or zero.

For large values of ρ, Eq. (3.5.24) becomes approximately (Pauling and Bright Wilson 1935)

$$\frac{d^2P}{d\rho^2} - \alpha^2\rho^2P = 0 \tag{3.5.26}$$

whose asymptotic solution is $\exp\left(\pm\frac{\alpha}{2}\rho^2\right)$.

We proceed then as done for the mono-dimensional harmonic oscillator in Cartesian coordinates, and make the following substitutions:

$$P(\rho) = \exp\left(-\frac{\alpha}{2}\rho^2\right)f(\rho) \tag{3.5.27}$$

$$\xi = \sqrt{\alpha}\rho \tag{3.5.28}$$

and replace $f(\rho)$ by $F(\xi)$, obtaining:

$$\frac{d^2F}{d\xi^2} - 2\xi\frac{dF}{d\xi} + \frac{1}{\xi}\frac{dF}{d\xi} + \left(\frac{\lambda'}{\alpha} - 2 - \frac{m^2}{\xi^2}\right)F = 0 \tag{3.5.29}$$

To avoid a rapid increase of the wave function at large ξ, λ' must be restricted to

$$\lambda' = 2(|m| + n' + 1)\alpha \tag{3.5.30}$$

with n' even integer.

The admitted energy eigenvalues are then (see Eqs. 17–25 in Pauling and Bright Wilson 1935):

$$\mathrm{E}_{m,n',\mathrm{n}_z} = (|m| + n' + 1)h\nu_0 + \left(n_z + \frac{1}{2}\right)h\nu_0 \tag{3.5.31}$$

which, in the case of the isotropic harmonic oscillator with $\nu_z = \nu_0$, reduces to

$$\mathrm{E}_n = \left(n + \frac{3}{2}\right)h\nu_0 \quad n = |m| + n' + n_z \tag{3.5.32}$$

with $m = 0, \pm 1, \pm 2, \ldots;\ n' = 0, 2, 4, 6, \ldots;\ n_z = 0, 1, 2, \ldots$.

The wave function has the following form (see Pauling and Bright Wilson 1935; Eqs. 17–27):

$$\psi_{n'_x,m,n_z}(\rho, \varphi, \mathrm{z}) = \mathrm{N}_{n',m,n_z} \exp(im\varphi) \exp\left(-\frac{\alpha}{2}\rho^2\right)\mathrm{F}_{|m|,n'}(\xi) \exp\left(-\frac{\alpha_z}{2}\mathrm{z}^2\right)H_{n_z}(\xi) \tag{3.5.33}$$

With N_{n',m,n_z} as normalization factor.

Chapter 4
The Uncertainty Principle

There is nowadays a somewhat disturbing attitude to emphasize the apparently magic aspects of Quantum Mechanics. Heisenberg's uncertainty principle is among the most quoted ones. It is however opportune to stress that this principle concerns essentially the impossibility of quantifying position and momentum of an "*observable*" (in Dirac's sense of the term) without introducing a perturbation during the observation act.[1] Moreover, the validity of this principle is not restricted to the realm of particles, but has a universal validity.[2] It is also quite obvious that, because energy goes with quanta (and no quanta-fractions are supposed to exist), it is hopeless to search something within the quantum limit. What would happen if we could do it? At a much higher conceptual level, this has been elegantly expressed by Feynman "*The uncertainty principle protects quantum mechanics*".

4.1 Premises

Heisenberg's uncertainty principle (Heisenberg, 1927b) asserts, in its elementary form, that the probable errors in the measurements of the position of a free particle (Δx) and its corresponding momentum component (Δp_x) are nearly equivalent to the reduced Planck's constant, i.e.:

$$\Delta x \Delta p_x \cong \hbar \tag{4.1.1}$$

---

[1] At his Sect. 2 Heisenberg (1927b) introduces an elegant treatment of the problems in terms of the Dirac—Jordan theory.

[2] In a nice exemplificative calculation McQuarrie (2008) shows that a baseball thrown at 90 *mph* has a momentum of $5.6\,kg.m.s^{-1}$ and even this involves an uncertainty in the position of its center of mass (1.2×10^{-16} *Å*). Conversely, to know the position of an electron with respect to the nucleus with an uncertainty of 0.01 *Å* we would know its momentum at the precision level of $2.6 \times 10^{-25}\ kg.m.s^{-1}$.

G. A. Ottonello, *Quantum Geochemistry*, Springer Geochemistry,
https://doi.org/10.1007/978-3-031-21837-8_4

Based on Eq. (4.1.1), the more the precision in specifying the position of a particle increases, the less precise is the specification of the associated momentum. Consider, for instance, the wave packets discussed at Sect. 1.2.2 and, particularly, Fig. 1.3: if Δx corresponds to the width of the packet (proportional to $\sqrt{\sigma}$) and the spread of momentum values around the mean is of the order $\hbar\Delta k$ (or, $\hbar\sqrt{1/\sigma}$, by Eq. 1.2.6), the product of the two is of the order $\hbar$.

4.2 General Form

In a most general form, the uncertainty principle relates any two observables A and B through a corollary of the *Expectation Value Postulate:*

"*The root mean square deviations ΔA and ΔB, obtained by repeated measurements of observables A and B for a system in state Ψ, are related by*[3]:

$$\Delta A \Delta B \geq \frac{1}{2}\langle \Psi | \mathsf{C} | \Psi \rangle \tag{4.2.1}$$

where C is the "*commutator of the operators associated with A and B*" (McWeeny, 1972; see also Sect. A.10)

$$i\mathsf{C} = AB - BA = [A, B] \tag{4.2.2}$$

Applying the above equation to position x and momentum p_x in Schrödinger notation, we have

$$i\mathsf{C}\Psi = -i\hbar\left(x\frac{\partial \Psi}{\partial x} - \frac{\partial}{\partial x}(x\Psi)\right) \tag{4.2.3}$$

Differentiating the second term in Eq. (4.2.3) as a product and discarding the arbitrary operand, one has

$$i\mathsf{C} = xp_x - p_x x = -i\hbar\mathsf{I} \tag{4.2.4}$$

where I is the identity operator which leaves any Ψ unchanged. From Eqs. (4.2.1) and (4.2.2) one gets a lower bound to (4.1.1), i.e.:

$$\Delta x \Delta p_x \geq \frac{1}{2}\hbar \tag{4.2.5}$$

[3] The symbols $\langle|$ and $|\rangle$ are referred to as "bra" and "ket", respectively. When bra and ket come together in the order $\langle\,|\,\rangle$ they form a bra(c)ket which implies a scalar product (see Sect. A.8. for a more extended outline).

4.3 Application

I report hereafter three exemplificative applications of the uncertainty principle: the first one, due to Feynman et al. (1965) and the other two to Pauli (1973).

4.3.1 A Modified Double-Slit Experiment

Feynman's treatment begins discussing the superposition of probability amplitudes. There is a source of particles (*s* in Fig. 4.1a; actually an *electron gun* simply consisting of a tungsten wire, heated by an electric current and surrounded by a metal box with a hole on it; the wire is at negative voltage with respect to the box so that electrons are ejected through the hole).

There is then a wall (a simple metal plate) with two slits in it. After the wall there is another wall which serves as "*backstop*" and in front of it a movable detector located in the example at position x (a Geiger counter or an electron multiplier). The

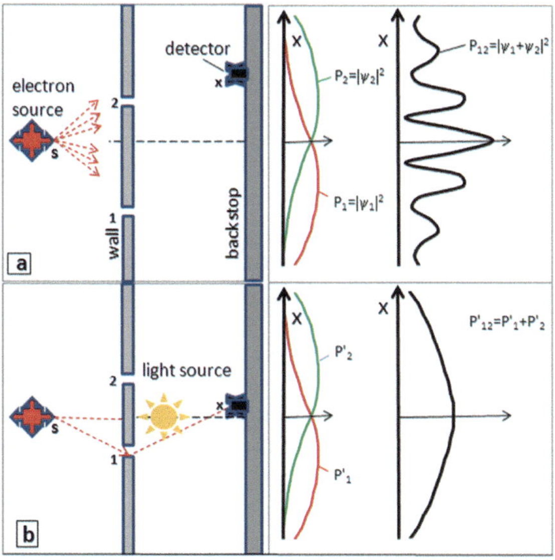

Fig. 4.1 Modified double-slit experiment after Feynman et al. (1965). In part **a** of the figure, we see the results of the double-slit experiment showing that an unperturbed registration detects the interference effects leading to a probability function P_{12} which corresponds to the squared modulus of the summation of the single waves. In part **b**, it is shown that a light source placed on the path of the electrons toward the detector detects a number of flashes P'_1 corresponding nearly (i.e. within the efficiency limits of the detectors) to the partial probability P_1, independently of the fact that the other hole is open or closed. Summing however the number of flashes one has an events curve P'_{12} substantially different from the probability curve P_{12}

quantum mechanical probability that a particle will arrive at x is the absolute square of the probability amplitude. In Dirac's notation, the probability amplitude may be written as follows:

$$\langle \text{particles arrive at x} \mid \text{particle leaves s} \rangle \equiv \langle x|s \rangle \tag{4.3.1}$$

(note that the starting condition is placed at the right of the vertical bar and the final condition at the left). When both slits are open the probability that an electron arrives at the detector is

$$P_{12} = |\psi_1 + \psi_2|^2 \tag{4.3.2}$$

The total amplitude of the process is the sum of the amplitudes for the two routes considered separately, i.e.

$$\langle x|s \rangle_{\text{both holes open}} = \langle x|s \rangle_{\text{through 1}} + \langle x|s \rangle_{\text{through 2}} \tag{4.3.3}$$

We may now write the probability that the particle goes from s to x through hole 1 as

$$\langle x|s \rangle_{\text{via1}} = \langle x|s \rangle \langle 1 \mid s \rangle \tag{4.3.4}$$

This notation tells us that events occur in succession, the resultant amplitude for the particular route is obtained by multiplying in succession the amplitude of the successive events. Applying the notation to both routes, we have

$$\langle x|s \rangle_{\text{both}} = \langle x|1 \rangle \langle 1|s \rangle + \langle x|2 \rangle \langle 2|s \rangle = \psi_1 + \psi_2 \tag{4.3.5}$$

This can be generalized for a free particle with a definite energy going in an empty space from $\mathbf{r}_1$ to $\mathbf{r}_2$ as

$$\langle r_2|r_1 \rangle_{\text{both}} = r_{12} \exp(i\mathbf{p}r_{12}/\hbar) \tag{4.3.6}$$

where $r_{12} = r_2 - r_1$ and $\mathbf{p}$ is the momentum.

Let us imagine now a modification of the experimental set of Fig. 4.1, with the first wall mounted on rollers, so that it can move freely upward or downward and the detector on the backstop placed at x = 0 (Fig. 4.2).

We would expect that an electron traveling through hole 1 kicks against the border of the wall and is deflected toward the detector. Since the vertical component of the electron momentum is changed, the wall must recoil in the opposite direction with the same momentum, i.e. a downward kick. If the electron goes through hole 2 an upward kick will be experienced by the wall. For every position of the detector, the momentum received by the plate will have a different value for a travel via $s \rightarrow 1 \rightarrow x$ or $s \rightarrow 2 \rightarrow x$. Measuring the plate momentum after the electron goes by we can estimate how much the momentum of the plate has changed, but, if we

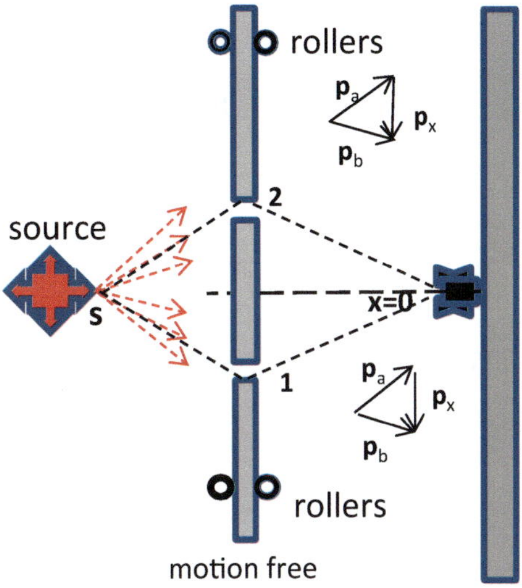

Fig. 4.2 An experiment to measure the recoil of the wall induced by hitting electrons. After Feynman et al. (1973), redrawn

do not know exactly where the plate is, we cannot identify precisely the position of the holes. They will be in a different place for every electron that goes through. The center of the interference pattern will have a different location for each electron that goes through and the wiggles of the interference pattern will be smeared out. In Feynman's words "*Quantum mechanics maintains its perilous but still correct existence*".

4.3.2 Doppler Effect and Momentum Measurements

Consider a particle whose velocity v_x is to be measured by the effect of an impinging wave train of light of finite length L (Fig. 4.3).

The wave train contains so many light quanta that one of them is certainly scattered from the particle. Assume this light quantum to approach the particle along the negative x-direction and be scattered along the positive one. The energy and momentum balances before and after the collision are given, respectively, by

$$h\nu + \mathrm{E} = h\nu' + \mathrm{E}' \Rightarrow h\nu' = h\nu + \mathrm{E} - \mathrm{E}' \tag{4.3.7}$$

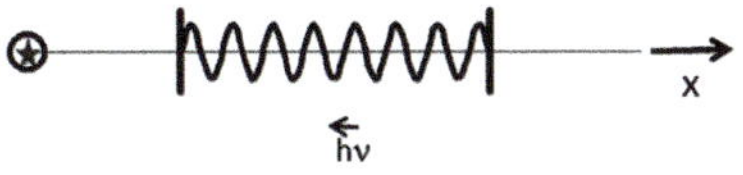

Fig. 4.3 Wave train of light impinging on a particle. From Pauli (1973), redrawn

and

$$\frac{h\nu}{c} + \mathrm{p_x} = \frac{h\nu'}{c} + \mathrm{p'_x} \quad \Rightarrow \mathrm{p'_x} = \mathrm{p_x} - \frac{h}{c}(\nu + \nu') \tag{4.3.8}$$

Because

$$v_\mathrm{x} = \frac{\partial \mathrm{E}}{\partial \mathrm{p_x}}; \qquad v'_\mathrm{x} = \frac{\partial \mathrm{E'}}{\partial \mathrm{p'_x}} \tag{4.3.9}$$

differentiating Eq. (4.3.8) by p_x, one gets

$$h\frac{\partial \nu'}{\partial p_x} = v_x - v'_x \frac{\partial p'_x}{\partial p_x} \tag{4.3.10}$$

$$\frac{\partial \mathrm{p'_x}}{\partial \mathrm{p_x}} = 1 - \frac{h}{c}\frac{\partial \nu'}{\partial \mathrm{p_x}} \tag{4.3.11}$$

$$h\frac{\partial \nu'}{\partial p_x} = v_x - v'_x\left(1 - \frac{h}{c}\frac{\partial \nu'}{\partial p_x}\right) \tag{4.3.12}$$

Rearranging Eq. (4.3.12), one gets

$$h\frac{\partial \nu'}{\partial p_x}\left(1 - \frac{v'_x}{c}\right) = v_x - v'_x \tag{4.3.13}$$

Relating the deltas (i.e. uncertainties)—in momenta to deltas in frequency through the $\frac{\partial \nu'}{\partial p_x}$ differential:

$$\Delta \nu = \frac{\partial \nu'}{\partial \mathrm{p_x}} \cdot \Delta \mathrm{p_x} \tag{4.3.14}$$

we obtain[4]

$$h\Delta\nu = \frac{v_x - v'_x}{1 - v'_x/c} \cdot \Delta p_x \approx \left(v_x - v'_x\right)\Delta p_x \tag{4.3.15}$$

Because the wave train passes the particle in a finite time t, the temporal limitation generates an uncertainty in frequency proportional to $1/t$, i.e. $\Delta\nu \approx {}^{1}/_{\mathrm{t}}$. Hence, from Eq. (4.3.15) we get an uncertainty in the momentum of type

[4] Note that I omitted the prime from the original formulation.

$$\Delta p_x \approx \frac{h}{\left(v_x - v'_x\right)t} \tag{4.3.16}$$

Moreover, because we cannot determine when the particle changed its velocity during the time interval t, without ruining the arrangement for measuring momentum, the position uncertainty is of the type:

$$\Delta x \approx \left(v_x - v'_x\right)t \tag{4.3.17}$$

Combining the two uncertainties, we get finally

$$\Delta p_x \cdot \Delta x \approx h. \tag{4.3.18}$$

4.3.3 Coherence Property of Light

The light emitted by a single atom at position A and collected on a screen when the light passes through two holes (Fig. 4.4) is a system of bright and dark bands which depend on the path difference:

$$D = \frac{2\pi}{\lambda}(AL_2 - AL_1) \tag{4.3.19}$$

To determine through which whole the quantum of light passed then we must know the momentum of the atom in the x-direction to a precision

$$\Delta p_x \leq \frac{2h}{\lambda}\sin\left(\frac{\theta}{2}\right) \tag{4.3.20}$$

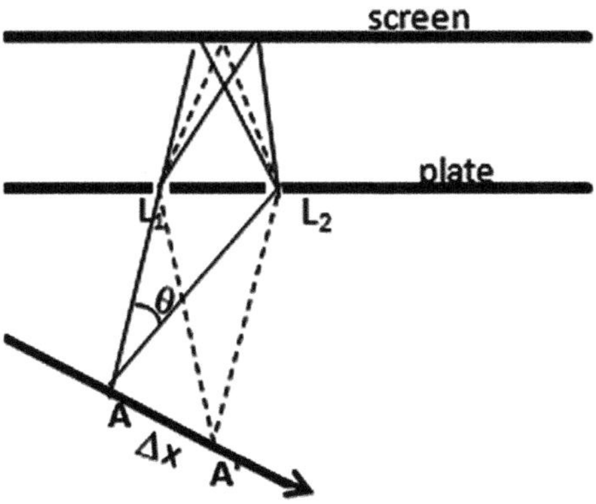

Fig. 4.4 Diffraction of the light emitted by a single atom. From Pauli (1973), redrawn

This implies an uncertainty in position

$$\Delta x \geq \frac{h}{2} \cdot \frac{1}{\Delta p_x} = \frac{h}{2} \Big/ \left[\frac{2h}{\lambda} \sin\left(\frac{\theta}{2}\right)\right] = \frac{\lambda}{4\sin(\theta/2)} \tag{4.3.21}$$

If the atom instead of being at position A is supposed to be at position A′ the new path difference would be

$$D' = \frac{2\pi}{\lambda}(A'L_2 - A'L_1) \tag{4.3.22}$$

and

$$D - D' = \frac{2\pi}{\lambda} \cdot 2\sin\left(\frac{\theta}{2}\right) \cdot \Delta x > \pi \tag{4.3.23}$$

The interference pattern is thus "*washed out*" (Pauli, 1973), as soon as we have determined through which hole the light quantum passed.

Chapter 5
The Hydrogen Atom

The quantum mechanical treatment of the hydrogen atom is central in the development of Quantum Chemistry, from the comprehension of the spectral series (the Balmer series of hydrogen was the first one expressed through a simple formula) to that of the atomic structure under the hydrogen-like approximation (i.e. one electron occupying progressively orbits of higher energy in all possible configurations). Sommerfeld (1930) introduces the quantum mechanical treatment of the hydrogen atom as "*The Kepler Problem*" i.e. the spatial quantisation of the orbital eccentricity. The intrinsic beauty of the treatment rests on the mathematical deconvolution of the wave equation in terms of discrete components, whose nature is intimately associated to the magnetic–azimuthal and radial components of the total quantum number. It is basically a powerful development operated by Pauling and Bright Wilson (1935) on what imagined by Sommerfeld (1930) and briefly anticipated at Chaps. 2 and 3.

5.1 Premises

Sommerfeld's treatment begins considering the potential energy exerted by a nucleus of charge Ze and an electron of charge $-e$ placed at radial distance r.

$$V = -\frac{Ze^2}{r} \tag{5.1.1}$$

The normalization of this potential implies that at infinite distance the potential is zero (i.e. V = 0 for r = ∞). Assuming a similar normalization for the bulk energy E, the amplitude equation (2.6.4) may be written in the form

G. A. Ottonello, *Quantum Geochemistry*, Springer Geochemistry,
https://doi.org/10.1007/978-3-031-21837-8_5

$$\nabla\psi + \frac{8r^2m}{h^2}\left(E + \frac{Ze^2}{r}\right)\psi = 0 \tag{5.1.2}$$

Treating the equation in polar coordinates r, θ, ϕ it is assumed that its solution is in the form

$$\psi = RP_l^m(\cos\theta)\exp(im\phi) \tag{5.1.3}$$

with R function of r alone and l, m integers with $l \geq 0$.

The problem reduces thus to the treatment already seen at Sect. 3.1, about the use of spherical harmonics and separation of variables. Adopting Eq. (3.1.6) with a proper value of λ (see Eq. 3.1.9), one derives the expression of $\nabla\psi$ from Eq. (3.1.2). One gets then for R the differential equation

$$\frac{d^2R}{dr^2} + \frac{2}{r}\frac{dR}{dr} + \left(A + 2\frac{B}{r} + \frac{C}{r}\right)R = 0 \tag{5.1.4}$$

with

$$A = \frac{8\pi^2 m}{h^2}E, \quad B = \frac{8\pi^2 m}{h^2}Ze^2, \quad C = -l(l-1) \tag{5.1.5}$$

Hereafter I follow the treatment provided by Pauling and Bright Wilson (1935), which is essentially based on Sommerfeld's pioneering work (Sommerfeld 1930).

5.2 Separating the Wave Equation

Let me denote with subscripts 1 and 2 the Cartesian coordinates and masses of nucleus and electron, i.e. x_1, y_1, z_1, m_1 and x_2, y_2, z_2, m_2, respectively. The wave equation, with the above provisos, has the form

$$\frac{1}{m_1}\left(\frac{\partial^2\psi_T}{\partial x_1^2} + \frac{\partial^2\psi_T}{\partial y_1^2} + \frac{\partial^2\psi_T}{\partial z_1^2}\right) + \frac{1}{m_2}\left(\frac{\partial^2\psi_T}{\partial x_2^2} + \frac{\partial^2\psi_T}{\partial y_2^2} + \frac{\partial^2\psi_T}{\partial z_2^2}\right) + \frac{8\pi^2}{h^2}(E_T - V)\psi_T \tag{5.2.1}$$

where

$$V = -\frac{Ze^2}{\sqrt{(x_2-x_1)^2 + (y_2-y_1)^2 + (z_2-z_1)^2}} \tag{5.2.2}$$

and the subscript T is meant to indicate that the E_T and ψ_T are those of the entire system (i.e. electron + nucleus).

Introducing the coordinates of the center of mass of the system as

$$x = \frac{m_1 x_1 + m_2 x_2}{m_1 + m_2}, \quad x = y\frac{m_1 y_1 + m_2 y_2}{m_1 + m_2}, \quad z = \frac{m_1 z_1 + m_2 z_2}{m_1 + m_2} \tag{5.2.3}$$

and describing the position of the electron with respect to the nucleus in terms of polar coordinates r, θ, ϕ:

$$r \sin\theta \cos\phi = x_2 - x_1, \quad r \sin\theta \sin\phi = y_2 - y_1, \quad r\cos\theta = z_2 - z_1 \tag{5.2.4}$$

the wave equation (5.2.1) transforms into

$$\begin{aligned} &\frac{1}{m_1 + m_2}\left(\frac{\partial^2 \psi_T}{\partial x^2} + \frac{\partial^2 \psi_T}{\partial y^2} + \frac{\partial^2 \psi_T}{\partial z^2}\right) \\ &+ \frac{1}{\mu}\left[\frac{1}{r^2}\frac{\partial}{\partial r}\left(r^2 \frac{\partial \psi_T}{\partial r}\right) + \frac{1}{r^2 \sin^2\theta}\frac{\partial^2 \psi_T}{\partial \phi^2} + \frac{1}{r^2 \sin\theta}\frac{\partial}{\partial \theta}\left(\sin\theta \frac{\partial \psi_T}{\partial \theta}\right)\right] \\ &+ \frac{8\pi^2}{h^2}[E_T - V(r, \theta, \phi)]\psi_T = 0 \end{aligned} \tag{5.2.5}$$

where the terms in parentheses and square brackets are, respectively, the Laplacians in Cartesian and polar coordinates (see Sect. A.17) and μ is the reduced mass of the system, i.e. $\mu = \frac{m_1 m_2}{m_1 + m_2}$.

The wave equation of the total system ψ_T is now separated into two parts, one as a function of the Cartesian coordinates x, y, z of the center of mass of the system only, and the other one as a function of the polar coordinates r, θ, ϕ of the second particle relative to the first one only:

$$\psi_T(x, y, z, r, \theta, \phi) = F(x, y, z)\psi(r, \theta, \phi) \tag{5.2.6}$$

5.2.1 Translational Motion

Substituting Eq. (5.2.6) into (5.2.5), and dividing through by $\psi_T F_\psi$ we make explicit the contribution of translational energy E_{tr} to the bulk energy of the system E_T, i.e.

$$E_T = E + E_{tr} \tag{5.2.7}$$

$$\nabla^2 F + \frac{8\pi^2(m_1 + m_2)}{h^2} E_{tr} F = 0 \tag{5.2.8}$$

$$\nabla^2\psi + \frac{8\pi^2\mu}{h^2}[E - V(r, \theta, \phi)]\psi = 0 \tag{5.2.9}$$

(with the two ∇^2 operators in Eqs. (5.2.8) and (5.2.9) referring, respectively, to Cartesian and polar coordinates). Equation (5.2.8) basically states that the energy contribution due to the translational motion of the system is the same of that of a particle with mass $m_1 + m_2$. In most applications, this energy contribution may be ignored and the energy of the system may be considered to correspond simply to the energy term E (i.e. ignoring E_{tr}). Moreover, the wave equation (5.2.9) corresponds to the wave equation of a single particle with a mass μ equal to the reduced mass of the system under the influence of a potential field $V = V(r, \theta, \phi)$. Assuming now that V is indeed function of r alone (i.e. $V = V(r)$) we can split the function into three parts, i.e.

$$\psi(r, \theta, \phi) = R(r) \cdot \Theta(\theta) \cdot \Phi(\phi) \tag{5.2.10}$$

Replacing Eq. (5.2.10) into (5.2.9) and dividing through by $R\Theta\Phi$ we get

$$\frac{1}{r^2 R}\frac{d}{dr}\left(r^2\frac{dR}{dr}\right) + \frac{1}{r^2\sin^2\theta\Phi}\frac{d^2\Phi}{d\phi^2} + \frac{1}{r^2\sin\theta\,\Theta}\frac{d}{d\theta}\left(\sin\theta\frac{d\Theta}{d\theta}\right) + \frac{8\pi^2\mu}{h^2}[E - V(r)] = 0 \tag{5.2.11}$$

Multiplying all terms in Eq. (5.2.11) by $r^2\sin^2\theta$, the second term reduces to $\frac{1}{\Phi}\frac{d^2\Phi}{d\phi^2}$ which is function only of the independent variable ϕ. Because the term must balance all terms independent of ϕ, it must be equal to a constant that we may denote as $-m^2$ (Pauling and Bright Wilson 1935):

$$\frac{d^2\Phi}{d\phi^2} = -m^2\Phi \tag{5.2.12}$$

Equation (5.2.11) may be then rewritten as follows:

$$\frac{1}{R}\frac{d}{dr}\left(r^2\frac{dR}{dr}\right) - \frac{m^2}{\sin^2\theta} + \frac{1}{\sin\theta\,\Theta}\frac{d}{d\theta}\left(\sin\theta\frac{d\Theta}{d\theta}\right) + \frac{8\pi^2\mu r^2}{h^2}[E - V(r)] = 0 \tag{5.2.13}$$

The first and the last terms on the left of Eq. (5.2.13) are independent of θ, and the remaining terms do not depend on r, so that both groups may be equated to a constant. Equating the θ dependent terms to $-\beta$ and the r-dependent terms to $+\beta$ and multiplying them by Θ and R^2/r^2, respectively, we get (Sommerfeld 1930; Pauling and Bright Wilson 1935):

$$-\frac{m^2}{\sin^2\theta}\Theta + \frac{1}{\sin\theta}\frac{d}{d\theta}\left(\sin\theta\frac{d\Theta}{d\theta}\right) + \beta\Theta = 0 \quad (5.2.14)$$

$$\frac{1}{r^2}\frac{d}{dr}\left(r^2\frac{dR}{dr}\right) - \frac{\beta}{r^2}R + \frac{8\pi^2\mu}{h^2}[E - V(r)]R = 0 \quad (5.2.15)$$

In order to find the allowed values of energy, we must now solve Eqs. (5.2.12), (5.2.14) and (5.2.15). Equation (5.2.12) has acceptable solutions only for certain values of m. Introducing the m eigenvalues in Eq. (5.2.14) we may see that only certain values of β are allowed. Finally, introducing the eigenvalues of β in Eq. (5.2.15) we find the eigenvalues of E, which are then the allowed energy values for the stationary states of the system.

5.2.2 *Solving the ϕ Equation: Significance of the Magnetic Quantum Number*

Equation (5.2.12) involving the polar coordinate ϕ has solutions for

$$\Phi_m(\phi) = \frac{1}{\sqrt{2\pi}}\exp(im\phi) \quad (5.2.16)$$

And must be single-valued at the cyclic points $\phi = 0, 2\pi, \ldots$. It follows that m (the "*magnetic quantum number*") must be an integer (elsewhere we will denote the magnetic quantum number as m_l to distinguish it from the spin quantum number m_s; here I omit the subscript for graphical simplicity). The solutions are $m = 0, \pm 1, \pm 2, \ldots$(note that positive and negative terms must be seen as distinct solutions). The pre-exponential term has the duty of normalizing the function $\Phi_m(\phi)$, satisfying thus the probability condition

$$\int_0^{2\pi} \Phi_m^*(\phi)\Phi_m(\phi)d\phi = 1 \quad (5.2.17)$$

For an identical modulus of m, the two functions $\Phi_{|m|}(\phi)$ and $\Phi_{-|m|}(\phi)$ satisfy the same differential equation. Hence, any linear combination of the two has the same property. Moreover, their sum and their difference correspond, respectively, to cosine and sine functions. Replacing the trigonometric equivalents, the allocated solutions (normalized and mutually orthogonal) are

$$\Phi_0(\phi) = \frac{1}{\sqrt{2\pi}}$$

$$\Phi_{|m|}(\phi) = \begin{cases} \frac{\cos|m|\phi}{\sqrt{\pi}} & |m| = 1, 2, 3, \ldots \\ \\ \frac{\sin|m|\phi}{\sqrt{\pi}} & m = 1, 2, 3, \ldots \end{cases} \tag{5.2.18}$$

5.2.3 *Solving the θ Equation: Significance of the Azimuthal Quantum Number*

One proceeds by introducing a new variable z = cosθ, which is implicitly constrained into the limits −1 < = z < = 1. The function $\Theta(\theta)$ is then replaced by the function P (z) = $\Theta(\theta)$. Because

$$\sin^2\theta = 1 - z^2 \tag{5.2.19}$$

and

$$\frac{d\Theta}{d\theta} = \frac{dP}{dz} \cdot \frac{dz}{d\theta} = -\frac{dP}{dz} \sin\theta \tag{5.2.20}$$

we have

$$\frac{d}{dz}\left[\left(1 - z^2\right)\frac{dP(z)}{dz}\right] + \left[\beta - \frac{m^2}{1 - z^2}\right]P(z) = 0 \tag{5.2.21}$$

The polynomial method previously adopted to solve the problem of the harmonic oscillator (Sect. 3.1) cannot be applied here but requires a further substitution (Sommerfeld 1930; Pauling and Bright Wilson 1935):

$$P(z) = x^{|m|/2} y^{|m|/2} G(z) = \left(1 - z^2\right)^{|m|/2} G(z) \tag{5.2.22}$$

The differential equation to be satisfied by the polynomial method is

$$(1 - z^2)\ddot{G} - 2(|m| + 1)z\dot{G} + [\beta - |m|(|m| + 1)]G = 0 \tag{5.2.23}$$

G is then expanded as a polynomial function of z, of the type

$$G = a_0z^0 + a_1z^1 + a_2z^2 + a_3z^3 + \cdots \tag{5.2.24}$$

Introducing the above expansion in Eq. (5.2.23), one gets

$$\begin{aligned}
&1 \cdot a_2z^2 + 2 \cdot 3 \cdot a_3z + 3 \cdot 4 \cdot a_4z^2 + 4 \cdot 5 \cdot a_5z^3 + \cdots \\
&\quad - 1 \cdot 2 \cdot a_2z^2 - 2 \cdot 3 \cdot a_3z^3 - \cdots \\
&\quad - 2 \cdot 1(|m| + 1) \cdot a_1z - 2 \cdot 2(|m| + 1) \cdot a_2z^2 - 2 \cdot 3(|m| + 1) \cdot a_3z^3 - \cdots \\
&\quad + b \cdot a_0 + b \cdot a_1z + b \cdot a_2z^2 + b \cdot a_3z^3 + \cdots = 0
\end{aligned} \tag{5.2.25}$$

where $b = [\beta - |m|(|m| + 1)]$.

To satisfy the equality to zero each term in z must vanish, hence:

$$\begin{aligned}
1 \cdot 2 \cdot a_2 + b \cdot a_0 &= 0 \\
2 \cdot 3 \cdot a_3 + [b - 2 \cdot 1 \cdot (|m| + 1)] \cdot a_1 &= 0 \\
3 \cdot 4 \cdot a_4 + [b - 2 \cdot 2 \cdot (|m| + 1) - 1 \cdot 2] \cdot a_2 &= 0 \\
4 \cdot 5 \cdot a_5 + [b - 2 \cdot 3 \cdot (|m| + 1) - 2 \cdot 3] \cdot a_3 &= 0
\end{aligned} \tag{5.2.26}$$

Expressing as v the power of z, the above system may be generalized as follows:

$$(v + 1) \cdot (v + 2) \cdot a_{v+2} + [b - 2v(|m| + 1) - v(v - 1)] \cdot a_v = 0 \tag{5.2.27}$$

Rearranging the terms in Eq. (5.2.27), we arrive to a two-term recursion formula of the type

$$a_{v+2} = a_v \frac{(v + |m|)(v + |m| + 1) - \beta}{(v + 1)(v + 2)} \tag{5.2.28}$$

It may be shown by functional analysis (Courant and Hilbert 1931) that an equation of this type converges for $-1 < z < 1$ but it diverges at the limits $z = -1$ and $z = +1$ and it is not an acceptable wave function. However, posing

$$\beta = (v' + |m|)(v' + |m| + 1) \qquad v' = 0, 1, 2, \ldots \tag{5.2.29}$$

it is quite obvious that the even and odd series brake off at the terms in z^{v+}. We link thus the $|m|$ and v' terms by a new parameter l, i.e. the "*azimuthal quantum number*":

$$l = |m| + v' \tag{5.2.30}$$

The allowed values of l are $|m|$, $|m| + 1$, $|m| + 2$, … and the characteristic values of β for which G(z) may be defined through the recursion formula (5.2.28) are

$$\beta = l(l+1) \tag{5.2.31}$$

Resuming briefly: the allowed solutions of the θ equation are of the type

$$\Theta(\theta) = \left(1 - z^2\right)^{|m|/2} \mathrm{G}(z) \tag{5.2.32}$$

with G(z) defined in terms of a recursion formula (Eq. 5.2.28) and β obeying (5.2.31). Functions of the type (5.2.32) are known as "*Associated Legendre's functions*". Their nature is discussed in some detail at Sect. A.15.

5.2.4 Solving the r Equation: Significance of the Radial (n′) and Total (n) Quantum Numbers

By taking into account Eq. (5.2.31) we may now rewrite the r-dependent term in Eq. (5.2.13) as follows:

$$\frac{1}{\mathrm{r}^2}\frac{\mathrm{d}}{\mathrm{dr}}\left(\mathrm{r}^2\frac{\mathrm{dR}}{\mathrm{dr}}\right) + \left\{-\frac{l(l+1)}{\mathrm{r}^2} + \frac{8\pi^2\mu}{\mathrm{h}^2}[\mathrm{E} - \mathrm{V(r)}]\right\}\mathrm{R} = 0 \tag{5.2.33}$$

Replacing the r-dependent potential in the form $\mathrm{V(r)} = -\frac{Ze^2}{\mathrm{r}}$ where Z is the number of protons (i.e. the "*atomic number*") we may build the hydrogen-like approximation and explore the admitted energy levels for an atom composed of a nucleus of variable charge surrounded by a single electron. We introduce to this purpose the symbols

$$\alpha = \sqrt{-\frac{8\pi^2\mu\mathrm{E}}{\mathrm{h}^2}} \tag{5.2.34}$$

$$\lambda = \frac{4\pi^2\mu Z e^2}{\mathrm{h}\alpha} \tag{5.2.35}$$

Denoting as ρ the new independent variable $\rho = 2\alpha\mathrm{r}$ we transpose R(r) = S(ρ) and the wave equation becomes

$$\frac{1}{\rho^2}\frac{\mathrm{d}}{\mathrm{d}\rho}\left(\rho^2\frac{\mathrm{dS}}{\mathrm{d}\rho}\right) + \left[-\frac{1}{4} - \frac{l(l+1)}{\rho^2} + \frac{\lambda}{\rho}\right]\mathrm{S} = 0 \quad 0 \le \rho \le \infty \tag{5.2.36}$$

By applying the procedure adopted to describe the harmonic oscillator, we may see that when ρ is large a good approximation of the above equation is $\frac{\mathrm{d}^2\mathrm{S}}{\mathrm{d}\rho^2} = \frac{1}{4}\mathrm{S}$, which has solutions for $\mathrm{S} = \exp\left(+\frac{\rho}{2}\right)$ and $\mathrm{S} = \exp\left(-\frac{\rho}{2}\right)$. Because only the second one is acceptable as a wave function we place

$$F(\rho) = \frac{S(\rho)}{\exp\left(-\frac{\rho}{2}\right)} \tag{5.2.37}$$

$$\ddot{F} + \left(\frac{2}{\rho} - 1\right)\dot{F} + \left[\frac{\lambda}{\rho} - \frac{l(l+1)}{\rho^2} - \frac{1}{\rho}\right]F = 0 \qquad 0 \geq \rho \leq \infty \tag{5.2.38}$$

By posing then $F(\rho) = \rho^l L(\rho)$, we obtain

$$\rho\ddot{L} + [2(l+1) - \rho]\dot{L} + (\lambda - l - 1)L = 0 \tag{5.2.39}$$

which can be translated into the recursion formula

$$(\lambda - l - 1 - v) \cdot \cdot a_v + [2(v+1)(l+1) + v(v+1)] \cdot a_{v+1} = 0 \tag{5.2.40}$$

(see Sommerfeld 1930; Pauling and Bright Wilson 1935; for a more extended treatment). With the same arguments adopted to describe the properties of the harmonic oscillator, it may be shown that for any value of λ and l, the $S(\rho)$ function is unacceptable as a wave function unless it breaks off. Moreover, the series must break off at a finite number of terms. The two conditions are satisfied only for $\lambda = n$, with $n = n' + l - 1$. n is then the "*total quantum number*" and n' its "*radial component*".

Resuming briefly: the allowed solutions of the r-dependent wave equation have form

$$R(r) = \exp\left(-\frac{\rho}{2}\right)\rho^{l}L(\rho) \tag{5.2.41}$$

with L(ρ) (the "*Associated Laguerre function*") defined in terms of the recursion formula (5.2.40) with $\lambda = n$ (see Sect. A.16.2).

5.3 Hydrogen-Like Wave Functions

The discrete stationary states for a one-electron atom result, as we have seen, from a combination of three components, each one dependent on one or more quantum numbers. We may then generalize them in the form

$$\psi_{n,l,m}(r, \theta, \phi) = R_{n,l}(r)\Theta_{l,m}(\theta)\Phi_m(\phi) \tag{5.3.1}$$

where

$$\Phi_m(\phi) = \frac{1}{\sqrt{2\pi}}\exp(im\phi) \tag{5.3.2}$$

$$\Theta_{l,m}(\theta) = \left[\frac{(2l+1)(l+|m|)!}{2(l+|m|)!}\right]^{1/2} \mathrm{P}_l^{|m|} \cos\theta \tag{5.3.3}$$

$$R_{n,l}(r) = -\left\{\left(\frac{2Z}{na_0}\right)^3 \frac{(n-l-1)!}{2n[(n+l)!]^3}\right\}^{1/2} \exp\left(-\frac{\rho}{2}\right)\rho^l L_{n+l}^{2l+1}(\rho) \tag{5.3.4}$$

with $\rho = \left(\frac{2Z}{na_0}\right)\mathrm{r}$, and $\mathrm{a}_0 = \frac{\hbar^2}{\mu e^2}$.

a_0 is the radius of the smallest orbit of the hydrogen atom of the old quantum theory (i.e. the "*Bohr radius*") and corresponds to 0.529 *Å*. The minus sign in Eq. (5.3.4) is conventionally adopted to return positive values of the function for small values of r. The nature of the *Associated Legendre Functions* $\mathrm{P}_l^{|m|} \cos\theta$ and of the *Associated Laguerre Polynomials* $\mathrm{L}_{n+1}^{2l+1}(\rho)$ are discussed at Sects. A.15.3 and A.16.2, respectively. The radial wave functions for all quantum combinations up to n = 6 are listed in Table 5.1, the functions $\Phi_m(\phi)$ are listed in Table 5.2, and the associated Legendre functions normalized to unity are listed in Table 5.3. The terms in parentheses appearing in the radial wave functions are related to the *Associated Laguerre Polynomials* $\mathrm{L}_{n+l}^{2l+1}(\rho)$ by the factor $-\frac{(n+l)!}{(n-l-1)!}$ which has been combined with the normalizing factor and reduced to the simplest form (Pauling and Bright Wilson 1935). All functions are separately normalized to unity, obeying

$$\left.\begin{array}{l} \int_0^\infty [R_{n,l}(\mathrm{r})]^2 \mathrm{r}^2 \mathrm{dr} = 1 \\ \int_0^\pi [\Theta_{l,m}(\theta)]^2 \sin\theta \mathrm{d}\theta = 1 \\ \int_0^{2\pi} \Phi *_m (\phi)\Phi_m(\phi)\mathrm{d}\phi = 1 \end{array}\right\} \tag{5.3.5}$$

and collectively normalized, obeying also

$$\int_0^\infty \int_0^\pi \int_0^{2\pi} \psi *_{n,l,m} (r,\theta,\phi)\psi_{n,l,m}(r,\theta,\phi) r^2 \sin\theta \mathrm{d}\phi \mathrm{d}\theta \mathrm{d}\phi r = 1 \tag{5.3.6}$$

They are finally also mutually orthogonal, with the integral

$$\int_0^\infty \int_0^\pi \int_0^{2\pi} \psi_{n,l,m}(r,\theta,\phi)\psi_{n',l',m'}(r,\theta,\phi) r^2 \sin\theta \mathrm{d}\phi \mathrm{d}\theta \mathrm{d}\phi r$$

vanishing whenever $n \neq n', l \neq l', m \neq m'$

Table 5.1 Hydrogen-like radial wave functions

n	l	Orbital	$R_{n,l}(r)$
1	0	1s	$R_{1,0}(r) = (Z/a_0)^{3/2} 2\exp\left(-\frac{\rho}{2}\right)$
2	0	2s	$R_{2,0}(r) = \frac{(Z/a_0)^{3/2}}{2\sqrt{2}}(2-\rho)\exp\left(-\frac{\rho}{2}\right)$
2	1	2p	$R_{2,1}(r) = \frac{(Z/a_0)^{3/2}}{2\sqrt{6}}\rho\exp\left(-\frac{\rho}{2}\right)$
3	0	3s	$R_{3,0}(r) = \frac{(Z/a_0)^{3/2}}{9\sqrt{3}}(6-6\rho+\rho^2)\exp\left(-\frac{\rho}{2}\right)$
3	1	3p	$R_{3,1}(r) = \frac{(Z/a_0)^{3/2}}{9\sqrt{6}}(4-\rho)\rho\exp\left(-\frac{\rho}{2}\right)$
3	2	3d	$R_{3,2}(r) = \frac{(Z/a_0)^{3/2}}{9\sqrt{30}}\rho^2\exp\left(-\frac{\rho}{2}\right)$
4	0	4s	$R_{4,0}(r) = \frac{(Z/a_0)^{3/2}}{96}(24-36\rho+12\rho^2-\rho^3)\exp\left(-\frac{\rho}{2}\right)$
4	1	4p	$R_{4,1}(r) = \frac{(Z/a_0)^{3/2}}{32\sqrt{15}}(20-10\rho+\rho^2)\rho\exp\left(-\frac{\rho}{2}\right)$
4	2	4d	$R_{4,2}(r) = \frac{(Z/a_0)^{3/2}}{96\sqrt{5}}(6-\rho)\rho^2\exp\left(-\frac{\rho}{2}\right)$
4	3	4f	$R_{4,3}(r) = \frac{(Z/a_0)^{3/2}}{96\sqrt{35}}\rho^2\exp\left(-\frac{\rho}{2}\right)$
5	0	5s	$R_{5,0}(r) = \frac{(Z/a_0)^{3/2}}{300\sqrt{5}}(120-240\rho+120\rho^2-20\rho^3+\rho^4)\exp\left(-\frac{\rho}{2}\right)$
5	1	5p	$R_{5,1}(r) = \frac{(Z/a_0)^{3/2}}{150\sqrt{30}}(120-90\rho+18\rho^2-\rho^3)\rho\exp\left(-\frac{\rho}{2}\right)$
5	2	5d	$R_{5,2}(r) = \frac{(Z/a_0)^{3/2}}{150\sqrt{70}}(42-14\rho+\rho^2)\rho^2\exp\left(-\frac{\rho}{2}\right)$
5	3	5f	$R_{5,3}(r) = \frac{(Z/a_0)^{3/2}}{300\sqrt{70}}(8-\rho)\rho^3\exp\left(-\frac{\rho}{2}\right)$
5	4	5g	$R_{5,4}(r) = \frac{(Z/a_0)^{3/2}}{900\sqrt{70}}\rho^4\exp\left(-\frac{\rho}{2}\right)$
6	0	6s	$R_{6,0}(r) = \frac{(Z/a_0)^{3/2}}{2160\sqrt{6}}(720-1800\rho+1200\rho^2-300\rho^3+30\rho^4-\rho^5)\exp\left(-\frac{\rho}{2}\right)$
6	1	6p	$R_{6,1}(r) = \frac{(Z/a_0)^{3/2}}{432\sqrt{210}}(840-840\rho+252\rho^2-28\rho^3+\rho^4)\rho\exp\left(-\frac{\rho}{2}\right)$
6	2	6d	$R_{6,2}(r) = \frac{(Z/a_0)^{3/2}}{864\sqrt{105}}(336-168\rho+242\rho^2-\rho^3)\rho^2\exp\left(-\frac{\rho}{2}\right)$
6	3	6f	$R_{6,3}(r) = \frac{(Z/a_0)^{3/2}}{2592\sqrt{35}}(72-18\rho+\rho^2)\rho^3\exp\left(-\frac{\rho}{2}\right)$
6	4	6g	$R_{6,4}(r) = \frac{(Z/a_0)^{3/2}}{12960\sqrt{7}}(10-\rho)\rho^4\exp\left(-\frac{\rho}{2}\right)$
6	5	6h	$R_{6,5}(r) = \frac{(Z/a_0)^{3/2}}{12960\sqrt{77}}\rho^5\exp\left(-\frac{\rho}{2}\right)$

Table 5.2 The $\Phi_m(\phi)$ functions

m	Complex form	Real form
0	$\Phi_0(\phi) = \frac{1}{\sqrt{2\pi}}$	$\Phi_0(\phi) = \frac{1}{\sqrt{2\pi}}$
−1	$\Phi_{-1}(\phi) = \frac{1}{\sqrt{2\pi}}\exp(-i\phi)$	$\Phi_{-1,\cos\phi} = \frac{1}{\sqrt{2\pi}}\cos\phi$
1	$\Phi_1(\phi) = \frac{1}{\sqrt{2\pi}}\exp(i\phi)$	$\Phi_{1,\sin\phi} = \frac{1}{\sqrt{2\pi}}\sin\phi$
−2	$\Phi_{-2}(\phi) = \frac{1}{\sqrt{2\pi}}\exp(-i2\phi)$	$\Phi_{-2,\cos\phi} = \frac{1}{\sqrt{2\pi}}\cos(2\phi)$
2	$\Phi_2(\phi) = \frac{1}{\sqrt{2\pi}}\exp(i2\phi)$	$\Phi_{2,\sin\phi} = \frac{1}{\sqrt{2\pi}}\sin(2\phi)$
−5	$\Phi_{-5}(\phi) = \frac{1}{\sqrt{2\pi}}\exp(-i5\phi)$	$\Phi_{-5,\cos\phi} = \frac{1}{\sqrt{2\pi}}\cos(5\phi)$
5	$\Phi_5(\phi) = \frac{1}{\sqrt{2\pi}}\exp(i5\phi)$	$\Phi_{5,\sin\phi} = \frac{1}{\sqrt{2\pi}}\sin(5\phi)$

Table 5.3 Associated Legendre functions normalized to unity $\Theta_{l,m}(\theta)$ up to $n = 3$

l	Orbital	$\Theta_{l,,m}(\theta)$
0	s	$\Theta_{0,0}(\theta) = \frac{\sqrt{2}}{2}$
1	p	$\Theta_{1,0}(\theta) = \frac{\sqrt{6}}{2}\cos\theta$
1	p	$\Theta_{1,\pm1}(\theta) = \frac{\sqrt{3}}{2}\sin\theta$
2	d	$\Theta_{2,0}(\theta) = \frac{\sqrt{10}}{4}(3\cos^2\theta - 1)$
2	d	$\Theta_{2,\pm1}(\theta) = \frac{\sqrt{15}}{2}\sin\theta\cos\theta$
2	d	$\Theta_{2,\pm2}(\theta) = \frac{\sqrt{15}}{4}\sin^2\theta$
3	f	$\Theta_{3,0}(\theta) = \frac{3\sqrt{14}}{4}\left(\frac{5}{3}\cos^2\theta - \cos\theta\right)$
3	f	$\Theta_{3,\pm1}(\theta) = \frac{\sqrt{42}}{8}\sin\theta(5\cos^2\theta - 1)$
3	f	$\Theta_{3,\pm2}(\theta) = \frac{\sqrt{105}}{4}\sin^2\theta\cos\theta$
3	f	$\Theta_{3,\pm3}(\theta) = \frac{\sqrt{70}}{8}\sin^2\theta$
4	g	$\Theta_{4,0}(\theta) = \frac{9\sqrt{2}}{16}\left(\frac{35}{3}\cos^4\theta - 10\cos^2\theta + 1\right)$
4	g	$\Theta_{4,\pm1}(\theta) = \frac{9\sqrt{10}}{8}\sin\theta\left(\frac{7}{3}\cos^2\theta - \cos\theta\right)$
4	g	$\Theta_{4,\pm2}(\theta) = \frac{3\sqrt{5}}{8}\sin^2\theta(7\cos^2\theta - 1)$
4	g	$\Theta_{4,\pm3}(\theta) = \frac{3\sqrt{70}}{8}\sin^2\theta\cos\theta$
4	g	$\Theta_{4,\pm4}(\theta) = \frac{3\sqrt{35}}{16}\sin^4\theta$
5	h	$\Theta_{5,0}(\theta) = \frac{15\sqrt{22}}{16}\left(\frac{21}{5}\cos^5\theta - \frac{14}{3}\cos^3\theta + \cos\theta\right)$
5	h	$\Theta_{5,\pm1}(\theta) = \frac{\sqrt{165}}{16}\sin\theta(21\cos^4\theta - 14\cos^2\theta + 1)$
5	h	$\Theta_{5,\pm2}(\theta) = \frac{\sqrt{1155}}{8}\sin^2\theta(3\cos^3\theta - \cos\theta)$
5	h	$\Theta_{5,\pm3}(\theta) = \frac{\sqrt{770}}{32}\sin^2\theta(9\cos^2\theta - 1)$
5	h	$\Theta_{5,\pm4}(\theta) = \frac{3\sqrt{385}}{16}\sin^4\theta\cos\theta$
5	h	$\Theta_{4,\pm5}(\theta) = \frac{3\sqrt{154}}{32}\sin^5\theta$

Table 5.4 Hydrogen-like wave functions $\psi_{n,l,m}(\mathrm{r},\theta,\phi)$ (real form) up to n = 3

n	l	m_l	Orbital	$\psi_{n,l,m}(\mathrm{r},\theta,\phi)$
1	0	0	1s	$\psi_{1s}=\frac{1}{\sqrt{\pi}}(\mathrm{Z/a_0})^{3/2}\exp(-\sigma)$
2	0	0	2s	$\psi_{2s}=\frac{1}{4\sqrt{2\pi}}(\mathrm{Z/a_0})^{3/2}(2-\sigma)\exp\left(-\frac{\sigma}{2}\right)$
2	1	0	$2p_x$	$\psi_{2p_x}=\frac{1}{4\sqrt{2\pi}}(\mathrm{Z/a_0})^{3/2}\sigma\exp\left(-\frac{\sigma}{2}\right)\cos\theta$
2	1	±1	$2p_y$	$\psi_{2p_y}=\frac{1}{4\sqrt{2\pi}}(\mathrm{Z/a_0})^{3/2}\sigma\exp\left(-\frac{\sigma}{2}\right)\sin\theta\sin\phi$
2	1	±1	$2p_z$	$\psi_{2p_z}=\frac{1}{4\sqrt{2\pi}}(\mathrm{Z/a_0})^{3/2}\sigma\exp\left(-\frac{\sigma}{2}\right)\sin\theta\cos\phi$
3	0	0	3s	$\psi_{3s}=\frac{1}{81\sqrt{3\pi}}(\mathrm{Z/a_0})^{3/2}(27-18\sigma+2\sigma^2)\exp\left(-\frac{\sigma}{3}\right)$
3	1	0	$3p_x$	$\psi_{3p_x}=\frac{\sqrt{2}}{81\sqrt{\pi}}(\mathrm{Z/a_0})^{3/2}(6-\sigma)\sigma\exp\left(-\frac{\sigma}{3}\right)\cos\theta$
3	1	±1	$3p_y$	$\psi_{3p_y}=\frac{\sqrt{2}}{81\sqrt{\pi}}(\mathrm{Z/a_0})^{3/2}(6-\sigma)\sigma\exp\left(-\frac{\sigma}{3}\right)\sin\theta\sin\phi$
3	1	±1	$3p_z$	$\psi_{3p_y}=\frac{\sqrt{2}}{81\sqrt{\pi}}(\mathrm{Z/a_0})^{3/2}(6-\sigma)\sigma\exp\left(-\frac{\sigma}{3}\right)\sin\theta\cos\phi$
3	2	0	$3d_x$	$\psi_{3d_x}=\frac{1}{81\sqrt{\pi}}(\mathrm{Z/a_0})^{3/2}\sigma^2\exp\left(-\frac{\sigma}{3}\right)(3\cos^2\theta-1)$
3	2	±1	$3d_{x+z}$	$\psi_{3d_{x+z}}=\frac{\sqrt{2}}{81\sqrt{\pi}}(\mathrm{Z/a_0})^{3/2}\sigma^2\exp\left(-\frac{\sigma}{3}\right)\sin\theta\cos\theta\cos\phi$
3	2	±1	$3d_{y+z}$	$\psi_{3d_{y+z}}=\frac{\sqrt{2}}{81\sqrt{\pi}}(\mathrm{Z/a_0})^{3/2}\sigma^2\exp\left(-\frac{\sigma}{3}\right)\sin\theta\cos\theta\sin\phi$
3	2	±2	$3d_{xy}$	$\psi_{3d_{xy}}=\frac{1}{81\sqrt{2\pi}}(\mathrm{Z/a_0})^{3/2}\sigma^2\exp\left(-\frac{\sigma}{3}\right)\sin^2\theta\cos2\phi$
3	2	±2	$3d_{x+y}$	$\psi_{3d_{x+y}}=\frac{1}{81\sqrt{2\pi}}(\mathrm{Z/a_0})^{3/2}\sigma^2\exp\left(-\frac{\sigma}{3}\right)\sin^2\theta\sin2\phi$

The complete wave functions $\psi_{n,l,m}(\mathrm{r},\theta,\phi)$ functions (real form) up to n = 3 are listed in Table 5.4. The parameter σ replacing ρ is related to this last by $\sigma=\rho\frac{n}{2}=\frac{\mathrm{Zr}}{a_0}$.

Table 5.1 Hydrogen-like wave functions ψ_{nlm} [illegible]

Chapter 6
Many-Electron Atoms

The quantum mechanical treatment of atomic energies received a strong impulse by the powerful analysis of the multiplets due to Slater (1929). Indeed, the introduction of the so-called "*Slater determinant*" is preliminary to any method of resolution of the quantum mechanical Hamiltonian.

It must be stressed, however, with Scholtz (2006), that already before Pauli's mathematization of spin was known, Heisenberg started to consider the consequences of the new phenomenon for many-electron systems (Heisenberg 1926). Heisenberg understood that the interaction of the orbital magnetic momentum of electrons with the (still hypothetical) spin might play a crucial role in this phenomenon.

Before treating to some extent Slater's method and the effects of spin–orbit interactions, I recall here briefly the nature of the quantum numbers l, m_l, and m_s and their relations with the total angular momentum **L,** the magnetic angular momentum **J,** and the spin angular momentum **S**.

We see then in some detail the atomic spectra and the electronic configurations of the many-electron atoms, with details on the various coupling schemes at the ground state.

6.1 Quantum Numbers

6.1.1 Azimuthal Quantum Number

Slater started where Bohr finished, with an atom composed of a number of electrons, each characterized by a total quantum number n, and an azimuthal quantum number l, the latter determining the angular momentum, and regarded as a vector (normal to the plane of the orbit, in Bohr's theory). The relationship between l and the total angular momentum operator L is:

G. A. Ottonello, *Quantum Geochemistry*, Springer Geochemistry,
https://doi.org/10.1007/978-3-031-21837-8_6

$$L^2\Psi = \hbar^2 l(l+1)\Psi \tag{6.1.1}$$

where Ψ is the electron wave function. If no torque acts on the angular momentum vector l, its direction is arbitrary and one may choose any axis in space. The component of the vector in the arbitrary direction can take any integer value between the parallel and anti-parallel orientations (i.e. for any integer l, the component of l along the arbitrary axis (m_l) can be l, $l-1$, $l-2,\ldots-l$). There are $2\,l + 1$ separate stationary states with the same energy (degenerate). The azimuthal quantum number 0 is historically denoted by the letter s (for "*sharp*") and depicts a spherical shape (no planar nodes passing through the nucleus). For $l = 1$, the identification is p (for "*principal*"): three dumbbell-shaped polar aligned orbitals (one node passing through the nucleus). For $l = 2$, the historical identification is d (for "*diffuse*") and for $l = 3$ is f (for "*fundamental*"). The remaining letters follow alphabetically. The spectral series are related to n and L in the following way: $n = 1$, $L = 0$ (Lyman series (ultraviolet); $n = 2$, $L = \hbar\sqrt{2}$ Balmer series (visible); $n = 3$ $L = \hbar\sqrt{6}$ Ritz–Paschen series (near infrared); $n = 4$ $L = 2\hbar\sqrt{3}$ Ritz–Brackett series (short wavelength infrared); $n = 5$ $L = 2\hbar\sqrt{5}$ Pfund series (mid wavelength infrared).

A schematic illustration of the quantum mechanical orbital angular momentum may be found in Fig. 6.1. For each value of l and m_l, the vector spans a conical shape around the quantization axis, depicting a circle centered on the z-axis.

6.1.2 Electron Spin

In discussing the energy levels of the hydrogen atom, I neglected the effect of electron spin because this would have required some anticipations on the perturbation methods. The fine structure of the hydrogen spectrum lines (Pauling and Goudsmith 1930) imply however the existence of a new property of the electron (i.e. an "*angular momentum*" due to spinning on its own axis).[1] Electrons are classified as *Fermions*, because they possess half-integer spin [i.e. $m_s = \pm (1/2)$]. As such, they obey the Pauli exclusion principle (there cannot be two identical fermions with the same quantum numbers or in a more familiar way, for electrons, "*no more than two electrons may occupy the same orbital*"). Two opposite oriented spins on the same orbital are graphically represented by arrows in opposite directions (↑↓) and usually denoted as "*spin up*" and "spin down", respectively. The first experimental confirmation of the existence of the electron spin is due to Gerlach and Stern (1922a, b). The notion of electron spin was introduced by Uhlembeck and

[1] Actually, neither the spin nor the magnetic moment are associated to the internal structure of the electron, but, rather, with the structure of its wave field. The view of an electron as a spinning ball of charge is thus rather simplistic (Ohanian, 1986).

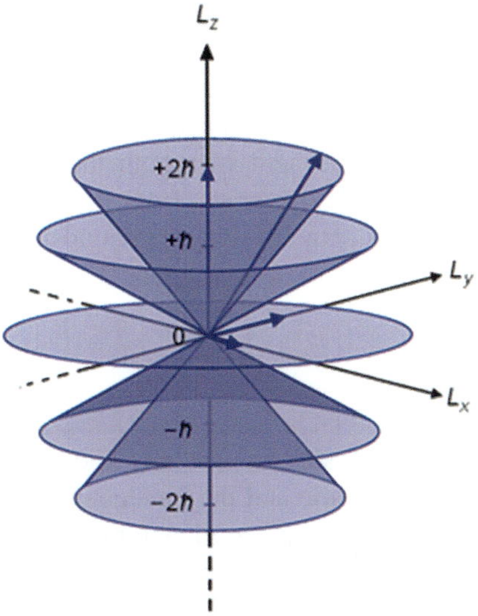

Fig. 6.1 Projection along the z-axis of the orbital angular momentum. The cones and plane represent possible orientations of the angular momentum vector for l = 2 and m_l = 2, 1, 0, −1, −2. The axis used for the polar coordinates in this analysis is chosen arbitrarily. The quantum number m_l refers to the projection of the angular momentum in this arbitrarily chosen direction, conventionally called the z direction or "*quantization axis*" L_Z. The magnitude of the angular momentum in the z-direction is given by the formula $L_Z = m_l h/2\pi$. Modified from Maschen—own work, Public Domain, https://commons.wikimedia.org/w/index.php?curid=17763200

Goudsmit (1925, 1926) and Goudsmit (1928) to explain the energy level diagram of alkali-like metals (where most levels appear in closely spaced groups of two) and alkaline earth metals. An independent proposal (affected by a numerical error) was carried out almost contemporaneously by Bichowscky and Urey (1926). The existence of electron spin was postulated by many others, though with unconvincing arguments. We may consider, for simplicity, a neutral alkali atom in its normal state as the combination of an alkali ion with one valence electron. The electronic structure of the ion is particularly simple: a completed outer group of two or eight electrons. The interaction of the ion and the valence electron depends (neglecting small relativistic effects) essentially on the total and the azimuthal quantum numbers of the valence electron (n and l, respectively). The levels corresponding to the given values of the two quantum numbers were observed to split into two levels (or "*doublets*"). According to the Uhlembeck–Goudsmit hypothesis, the electron has angular momentum and magnetic moment associated with the spinning motion of a charged particle about an axis through it. The magnitude of the total angular momentum of the electron is $\frac{h}{2\pi}\sqrt{m_s(m_s+1)}$ and the component of the angular momentum along any prescribed axis is either $+h/4\pi$ or $-h/4\pi$ which indicates that m_s can assume only the values +1/2 and −1/2. The total magnetic moment is $\sqrt{3}$ Bohr magnetons and the component along a prescribed axis is either +1 or −1 Bohr magnetons$\left(\frac{eh}{4\pi mc}\right)$. The magnetic moment associated to spin interacts with the magnetic field generated by the rotation of the electron about the nucleus along the orbit and this interaction gives rise to the observed splitting. The Zeeman effect (splitting of spectral lines under the influence of a magnetic field) results from the interaction of the field with both the orbital and spin magnetic moments m_l and m_s. The complexity arises from the anomalous value 2 of the Landé g factor for the electron spin (Pauling and Goudsmit 1930).

6.1.3 Hund's Rules

Hund (1927) built what are nowadays universally known as Hund's Rules, stemming from Sommerfeld's systematic and the Landé analysis of the Zeeman effect. His interpretation was framed within the Quantum Mechanics formulation with explicit reference to the work of Heisenberg, Schrödinger, and the Uhlembeck–Goudsmith electron spin theory. Hund's rules apply essentially to the ground state of an atom. The first rule[2] states that, before two opposite spin electrons can occupy the same orbital of a given sublevel, all the other orbitals of the same sublevel must have been occupied by single electrons of equal spin. This rule derives from the consideration that the lowest energy atomic state is the one that maximizes the total spin quantum number S in the open sublevel. The reason for this rule stands on the fact that the electrons in singly occupied orbitals are less effectively screened from the nucleus, so that such orbitals contract and electron–nucleus attraction energy becomes greater in magnitude (i.e. the configuration is stabilized to some extent). As an example, consider the ground state of silicon. The electronic configuration of Si is $1s^2\ 2s^2\ 2p^6\ 3s^2\ 3p^2$ and in the outer shell there are $3p^2$ electrons. The possible terms allowed by the Pauli exclusion principles are ^1D, ^3P, and ^1S (Russell–Saunders notation, see later). Hund's first rule now states that the ground state term is the triplet ^3P (multiplicity = $2S + 1 = 3$).

Hund's second rule states that, for a given multiplicity, the term with the largest value of the total orbital angular momentum L has the lowest energy. In fact, L is the higher when all electrons orbit in the same direction. They have less opportunity to meet each other and the Coulomb repulsion is in this way minimized. Consider as an example titanium which has configuration $1s^2\ 2s^2\ 2p^6\ 3s^2\ 3p^6\ 3d^2\ 4s^2$. In this case, the open shell is $3d^2$ and the allowed terms include three singlets (^1S, ^1D, and ^1G with $L = 0$, 2, and 4, respectively) and two triplets (^3P and ^3F with $L = 1$ and 3, respectively). The configuration ^1G has the lowest energy based on rule #2.

The third rule states that in an atom with the outermost sub-shell half-filled or less, the level with the lowest value of the total angular momentum quantum

[2] Also popularized as "*bus seat rule*": bus passengers preferentially occupy all double seats singly before double occupation is compulsory dictated by the crowd.

number J has also the lowest energy. If, on the contrary, the outermost shell is more than half-filled, the lowest energy is that one of the highest J. As an example consider again silicon. Based on the first rule we deduced that the ground state term is the triplet ^{3}P (multiplicity = $2S + 1 = 3$). With $L = 1, 0, -1$ we have $J = L + S. L + S-1, \ldots |L-S| = 2, 1, 0$. Because there are only 2 electrons out of the 6 necessary to complete the shell, this is less than half-filled, hence the ground state is that one with $J = 0$ (i.e. $^3P_0$).

6.1.4 Atomic Units

Bohr (1913) assumed that only orbits with an angular momentum which is an integral multiple of $h/2\pi$ are allowed; hence, the allowed values of r are those ones obeying

$$\mu r^2 \frac{d\theta}{dt} = \frac{nh}{2\pi} \tag{6.1.2}$$

with n integer. It follows that the allowed values of energy are

$$E = -T = -\frac{1}{2}\frac{1}{\mu r^2}\frac{n^2h^2}{4\pi^2} = -\frac{e^2}{2r} \tag{6.1.3}$$

and eliminating r from the terms:

$$E_n = -\frac{2\pi^2 \mu e^4}{n^2 h^2} \tag{6.1.4}$$

In Bohr's own words: "*In any molecular system consisting of positive nuclei and electrons in which the nuclei are at rest relative to each other and the electrons move in circular orbits, the angular momentum of every electron round the centre of its orbit will in the permanent state of the system be equal to h/2π, where h is Planck's constant*" (Bohr 1913).

We have seen in Chap. 5 that, we arrive at the same results by treating the solution of the r equation. We introduced to this purpose the symbols $\alpha = \sqrt{-\frac{8\pi^2 \mu E}{h^2}}$ and $\lambda = \frac{4\pi^2 \mu Z e^2}{h\alpha}$ (Eqs. 5.2.34, 5.2.35) and deduced that the condition for the recursion formula (5.2.40) to break off after a finite number of terms implied that $\lambda = n$. Solving for E, we get

$$E_n = -\frac{1}{2}\frac{\mu Z^2 e^4}{n^2 \hbar^2} \tag{6.1.5}$$

Table 6.1 Atomic units (a.u.) and their SI equivalents. m_0 is the rest mass of electron, e is the electron charge, and $\hbar$ is also known as *rationalized* Planck's constant. a_0 is the Bohr radius. Velocity v is $\sim c/137$, with c = speed of light. The a.u. energy unit is known as *Hartree*. Modified from Pisani (1996)

Magnitude	a.u.	Corresponding SI
Mass	m_0	$9.1094 \times 10^{-31} kg$
Charge	e	$1.6022 \times 10^{-19} C$
Angular momentum	$\hbar = h/2\pi$	$1.0546 \times 10^{-34} J \cdot s$
Length	$a_0 = \hbar^2/(m_0 e^2)$	$5.2918 \times 10^{-11} m$
Energy	$E_h = e^2/a_0$	$4.3598 \times 10^{-18} J$
Time	$t_0 = \hbar/E_h$	$2.4189 \times 10^{-17} s$
Velocity	$v = E_h/(m_0 e^2)$	$2.2579 \times 10^6 m \cdot s^{-1}$

which is identical to Eq. (6.1.4). Denoting the Rydberg constant $\mathcal{R}$ as $\mathcal{R} = \frac{1}{2}\frac{\mu e^4}{\hbar^2 c}$ we have $E_n = -\frac{Z^2}{n^2} E_H$ where $E_H = \mathcal{R}hc$. The energy of the hydrogen-like atom depends thus only on the value of the total quantum number $n = n' + l + 1$. Table 6.1 resumes the significance of the various a.u. and their relationships with the SI system.

Let me now take the proton at the origin of the coordinates of the hydrogen atom and apply the one-particle Schrödinger equation for the electron. Replacing V with the classical potential

$$V = -\frac{e^2}{4\pi\varepsilon_0 r} \tag{6.1.6}$$

with ε_0 = vacuum permittivity and r = electron to proton distance, we have

$$H\psi = -\frac{\hbar^2}{2m_0}\nabla^2\psi - \frac{e^2}{4\pi\varepsilon_0 r}\psi = E_0\psi \tag{6.1.7}$$

where m_0 is the rest mass of electron.

Consider now a "*trial*" function apt to reproduce function (6.1.7) for large values of r. The simplest form is an exponential function of the type

$$\psi(r) = A\exp(-br) \tag{6.1.8}$$

Differentiating twice Eq. (6.1.8) in dr we get

$$\nabla^2\psi = Ab\left(b - \frac{2}{r}\right)\exp(-br) \tag{6.1.9}$$

which, replaced in (6.1.7), gives

$$\left[\frac{\hbar^2}{2\mathrm{m}_0}\mathrm{b}\left(\mathrm{b}-\frac{2}{\mathrm{r}}\right)+\frac{e^2}{4\pi\varepsilon_0\mathrm{r}}+\mathrm{E}\right]\mathrm{A}\exp(-\mathrm{br})=0 \qquad (6.1.10)$$

Let me separate now the terms in r within the square bracket from those which do not contain explicitly r:

$$\left[\left(\frac{\hbar^2\mathrm{b}^2}{2\mathrm{m}_0}+\mathrm{E}\right)-\frac{1}{\mathrm{r}}\left(\frac{\hbar^2\mathrm{b}}{\mathrm{m}_0}-\frac{e^2}{4\pi\varepsilon_0}\right)\right]\mathrm{A}\exp(-\mathrm{br})=0 \qquad (6.1.11)$$

Both terms in parentheses must go to zero in order that the expression in square brackets to vanish. This implies in turn that

$$\mathrm{b}=\frac{\mathrm{m}_0e^2}{\hbar^2 4\pi\varepsilon_0},\quad \mathrm{E}=-\frac{\mathrm{m}_0e^4}{32\hbar^2\pi^2\varepsilon_0^2},\quad \mathrm{A}=\mathrm{b}^{3/2}\sqrt{\pi} \qquad (6.1.12)$$

the last equality arising from the normalization condition.

Replacing the various magnitudes with their SI numerical values (i.e. $\mathrm{m}_0 = 9.10939 \times 10^{-31}$ kg; $e = 1.60218 \times 10^{-19}$ *C*; $\hbar = 1.05457 \times 10^{-34}$ *J·s*; $\varepsilon_0 = 8.85419 \times 10^{-12}$ $C^2 \cdot J^{-1} \cdot m^{-1}$), we get an energy eigenvalue of -2.17987×10^{-18} *J*, corresponding to -13.605 *eV* nearly similar to the observed ionization potential of H changed of sign (13.595 *eV*, i.e. the energy required to remove the electron from its localized orbit about the proton and bring it to infinite distance). In atomic units, this amount of energy corresponds to minus half *Hartree* ($\mathrm{E_h} = \frac{\hbar^2}{a_0\mathrm{me}^2} = 4.35975 \times 10^{-18} J \equiv 27.21\,14 \quad eV$) The reciprocal of b is the "*Bohr radius*" $\mathrm{b}^{-1} = \frac{\hbar^2 4\pi\varepsilon_0}{\mathrm{me}^2} = a_0 = 0.52917 \times 10^{-10} m \approx 0.529\ \AA$ corresponding to the radius of the first orbit in Bohr's theory. Considering that the probability of finding the particle over the entire space at disposal is 1, the probability of finding it within a restricted part of volume corresponding to the radial distance from r to r + dr is $\mathrm{P(r)dr} = 4\pi\mathrm{r}^2\mathrm{dr} \times |\psi|^2$. The radial probability density P(r) has its maximum at dP/dr = 0 and at this condition $\mathrm{r} = \mathrm{a}_0$. The Bohr radius is thus interpreted as the most probable distance of the electron from the nucleus at the ground state. We may then select $\mathrm{r} = \mathrm{a}_0$ as a peculiar distance to depict the "*form*" (i.e. the angular factors) of the various atomic orbitals. "*Contour maps*" represent more generally a region of the space around the nucleus where the electron density is higher than a selected threshold (disregarding the immediate proximity of the nucleus where the electron density goes to zero).

In Figs. 6.2 to 6.5, we see contour maps generated for the Highest Occupied Molecular Orbital (HOMO; actually in this context MO≡AO) with basis sets internally consistent and sufficiently approximate to reproduce the shielding effects of inner electrons. Their "*form*" and angular orientation are greatly dependent on the angular factors, while the radial extensions of the lobes depend upon the selected iso-value of the probability density (i.e. the square of wave function).

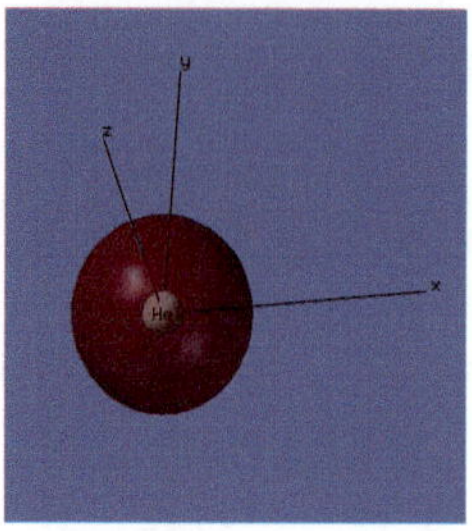

Fig. 6.2 Isodensity contour map of the 1s AO of He (0.02 $e/Å^3$). The orbital energy resolved at HF/CEP-121G level is −0.917 *Hartree*; the HF electronic energy of the atom is −2.859895 *Hartree*. The contour maps are generated at the HF/SDD level

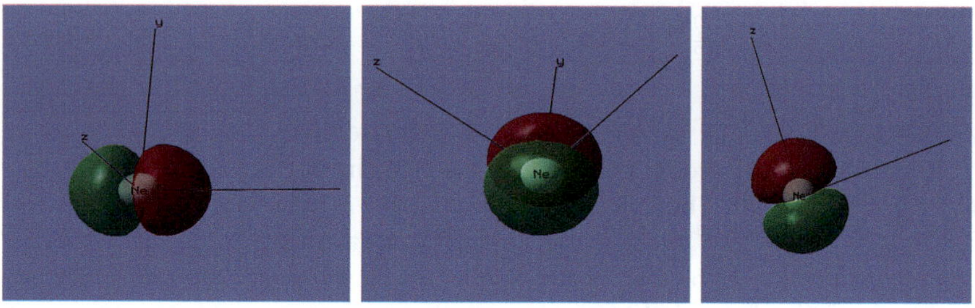

Fig. 6.3 Isodensity contour maps of the $2p_x$, $2p_y$ and $2p_z$ AOs of Ne (0.02 $e/Å^3$). The orbital energy resolved at HF/CEP-121G level is −0.846 *Hartree*; the HF electronic energy of the atom is −34.495373 *Hartree*. The contour maps are generated at the HF/SDD level

We see thus that, while for the two $1s^2$ electrons of He ($n = 1$; $l = 0$; $m_l = 0$; $m_s = +1/2, -1/2$) the HOMO has a isodensity surface spherically symmetrical with respect to the nuclear center (Fig. 6.12), the three (triply degenerate) $2p^6$ HOMOs of Ne ($n = 2$, $l = 1$, $m_l = +1, 0, -1$; $m_s = +1/2, -1/2, +1/2, -1/2, +1/2, -1/2$) have lobes of different signs symmetrically oriented with respect to the orientation axis (x, y, and z, respectively) (Fig. 6.3).

In Fig. 6.4 we see the HOMOs of Cu^+. These are the 3d contour maps generated at the same theory level of He and Ne but with a meshed surface instead of a transparent one, for a better perception of the angular symmetry.

Figure 6.5 shows finally the HOMOs of Yb. The 4f contour maps were generated at the same theory level of the previous one but in this case the isodensity surface is located at one order of magnitude lower (i.e. 0.002 $e/Å^3$). It should be noted for this purpose that the 5 s, 5p, and 6 s electrons, when resolved at HF/SDD level, have a higher energy with respect to 4f electrons and occupy an inner portion of the electron cloud, concurring significantly to the overall shielding effect.

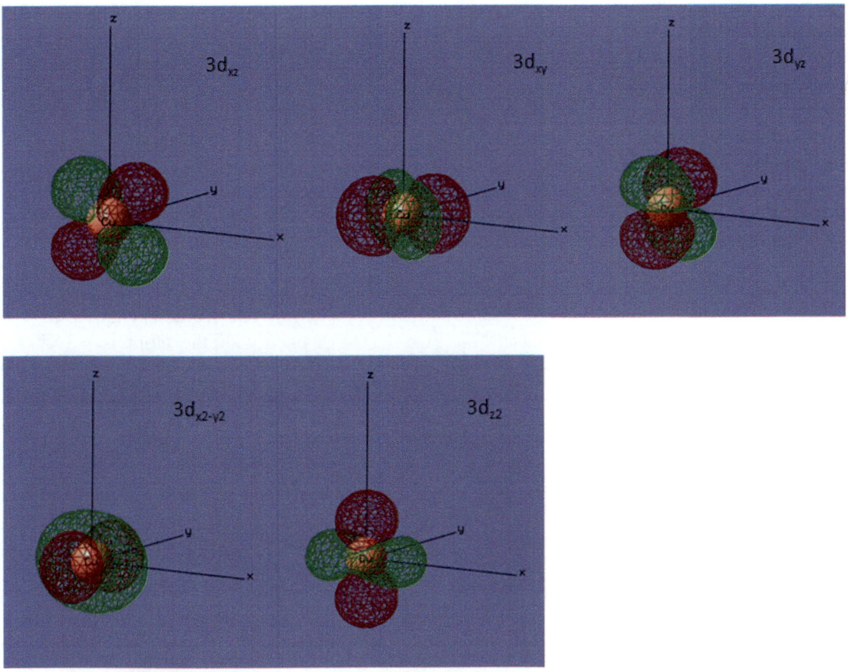

Fig. 6.4 Isodensity contour maps (meshes) of the $3d_{xz}$, $3d_{xy}$, $3d_{y2}$, $3d_{x2\text{-}y2}$, $3d_{z2}$, AOs of Cu^+ (0.02 *e/Å*^{3}). The orbital energy resolved at HF/CEP-121G level is −0.807 *Hartree*; the HF electronic energy of the atom is −195.0493665 *Hartree*

6.1.5 *Atomic Spectra*

The optical spectrum of a many-electron atom is made up of hundreds of lines. For a given atom, the optical emission lines are related to transitions between *m, n* couples of discrete energy levels (E_1, E_2,….E_m,….E_n,… "*stationary states*"), the frequency of the emitted radiation being related to them through Planck's constant, i.e. $\nu = \frac{E_m - E_n}{h}$. Only certain transitions are allowed. The energy gaps are generally interpreted as due to different interactions of the various angular momenta of the parts of an atom and, more precisely, to different relative orientations of these parts (see later on).

Empirically, it was found that all the emission lines in the optical spectrum of hydrogen, represented in terms of wavenumbers $\overline{\nu} = \frac{1}{\lambda}(cm^{-1})$, are reproduced by

$$\overline{\nu} = \mathcal{R}_{\mathrm{H}}\left(\frac{1}{n^2} - \frac{1}{m^2}\right) \tag{6.1.13}$$

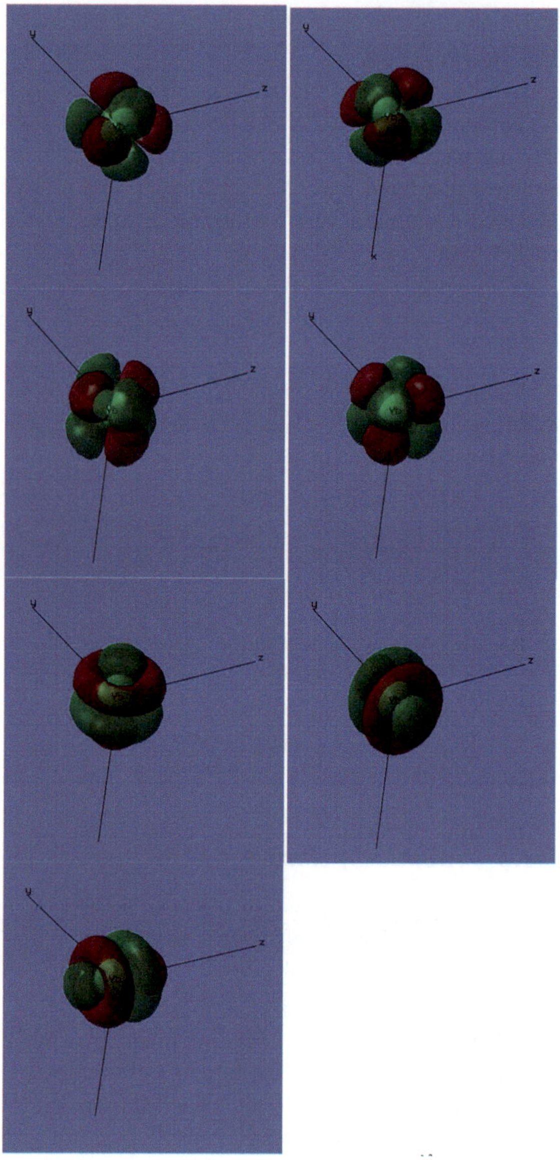

Fig. 6.5 Isodensity contour maps of the 4f AOs of Yb (0.002 $e/\AA^3$). The orbital energy resolved at HF/SDD level with 28 frozen core electrons attains −1.805 *Hartree*: The HF electronic energy of the atom at HF/CEP-121G level is −242.534172217 *Hartree*

where R_H is the Rydberg constant for hydrogen (R_H = 109,677.581) and *n,m* are integers with $n < m$ and n = 1,2,3,4,5 for the -Lyman, -Balmer, -Paschen, -Brackett, and -Pfund series of lines, respectively.

Denoting m the mass of an electron and M that of a proton, the movement of the electron about the proton is equivalent to that of a particle of mass $\mu = \mathrm{m} \times \mathrm{M}/(\mathrm{m} + \mathrm{M})$. If the movement of the electron about the nucleus is along a spherical orbit of radius r, the Coulomb attractive force between the two (M and m) charges must be counterbalanced by a centrifugal force such that

$$\frac{e^2}{\mathrm{r}^2} = \mu\mathrm{r}\left(\frac{\mathrm{d}\theta}{\mathrm{dt}}\right)^2 \tag{6.1.14}$$

where θ is the azimuthal angle. Denoting T the kinetic energy, V the potential energy, and E the total energy, we have

$$\mathrm{T} = \frac{1}{2}\mu\mathrm{v}^2 = \frac{1}{2}\mu\mathrm{r}^2\left(\frac{\mathrm{d}\theta}{\mathrm{dt}}\right)^2 = \frac{e^2}{2\mathrm{r}} \tag{6.1.15}$$

$$\mathrm{V} = -\frac{e^2}{\mathrm{r}} \tag{6.1.16}$$

$$\mathrm{E} = \mathrm{T} + \mathrm{V} = -\frac{e^2}{2\mathrm{r}} \equiv -\mathrm{T} \tag{6.1.17}$$

6.1.6 Quantum Numbers and Electron Configuration of All Elements at Ground State

I conclude this preliminary treatment by recalling the quantum numbers for the first four levels (Table 6.2) and the electron configurations of all elements in their ground state as arising from Bohr's model (Table 6.3).

The values of n and l are denoted as we have seen by symbols, as 1s, 2s, 2p, 3p, 3d, 4d, 4f,…5f referring, respectively, to the cases where n and l are 1—0, 2—0, . ; 2—1, 3—1, 3—2, 4—2, etc. Then the structure of an entire atom in a given state can be given, as far as Bohr's theory is concerned, by giving the n and l of each electron, denoted as in the (Mg) example: $1s^1 2s^2 2p^6 3s^2$, meaning that there are two (1s) electrons, two (2s), six (2p), and two (3s). An excited state of the same atom could be $1s^2 2s^2 2p^6 3s^1 3p^1$; meaning that one electron moved from the s to the p-sublevel.

With the increase of Z, the electronic configuration at the ground state is governed by the following principles:

(1) the *Aufbau or "build-up" principle:* Electrons occupy orbitals of progressively increasing energy;
(2) the *Pauli exclusion principle:* No more than two electrons may occupy the same orbital and they must have opposite spin. Based on this principle, 2n is the maximum number of electrons compatible with a given level.

Table 6.2 Quantum numbers (first four levels)

n	l	m_l	m_s	
1 (level K)	0 (sublevel s)	0	+1/2	−1/2
2 (level L)	0 (sublevel s)	0	+1/2	−1/2
	1 (sublevel p)	+1	+1/2	−1/2
		0	+1/2	−1/2
		−1	+1/2	−1/2
3 (level M)	0 (sublevel s)	0	+1/2	−1/2
	1 (sublevel p)	+1	+1/2	−1/2
		0	+1/2	−1/2
		−1	+1/2	−1/2
	2 (sublevel d)	+2	+1/2	−1/2
		+1	+1/2	−1/2
		0	+1/2	−1/2
		−1	+1/2	−1/2
		−2	+1/2	−1/2
4 (level N)	0 (sublevel s)	0	+1/2	−1/2
	1 (sublevel p)	+1	+1/2	−1/2
		0	+1/2	−1/2
		−1	+1/2	−1/2
	2 (sublevel d)	+2	+1/2	−1/2
		+1	+1/2	−1/2
		0	+1/2	−1/2
		−1	+1/2	−1/2
		−2	+1/2	−1/2
	3 (sub-level f)	+3	+1/2	−1/2
		+2	+1/2	−1/2
		+1	+1/2	−1/2
		0	+1/2	−1/2
		−1	+1/2	−1/2
		−2	+1/2	−1/2
		−3	+1/2	−1/2

(3) *Hund's rules:*

- Before two opposite spin electrons can occupy the same orbital of a given sublevel, all the other orbitals of the same sublevel must have been occupied by single electrons of equal spin.
- For a given multiplicity, the term with the largest value of the total orbital angular momentum L has the lowest energy.
- In an atom with the outermost sub-shell half-filled or less, the level with the lowest value of the total angular momentum quantum number J has also the lowest energy. If, on the contrary, the outermost shell is more than half-filled, the lowest energy is that one of the highest J.

Table 6.3 Electron configuration of elements

		1	2		3			4				5				6				7
Z	Elements		s	p	s	p	d	s	p	d	f	s	p	d	f	s	p	d	f	s
1	H	1																		
2	He	2																		
3	Li	2	1																	
4	Be	2	2																	
5	B	2	2	1																
6	C	2	2	2																
7	N	2	2	3																
8	O	2	2	4																
9	F	2	2	5																
10	Ne	2	2	6																
11	Na	2	2	6	1															
12	Mg	2	2	6	2															
13	Al	2	2	6	2	1														
14	Si	2	2	6	2	2														
15	P	2	2	6	2	3														
16	S	2	2	6	2	4														
17	Cl	2	2	6	2	5														
18	Ar	2	2	6	2	6														
19	K	2	2	6	2	6		1												
20	Ca	2	2	6	2	6		2												
21	Sc	2	2	6	2	6	1	2												
22	Ti	2	2	6	2	6	2	2												
23	V	2	2	6	2	6	3	2												

(continued)

Table 6.3 (continued)

		1	2		3			4				5				6				7
Z	Elements		s	p	s	p	d	s	p	d	f	s	p	d	f	s	p	d	f	s
24	Cr	2	2	6	2	6	5	1												
25	Mn	2	2	6	2	6	5	2												
26	Fe	2	2	6	2	6	6	2												
27	Co	2	2	6	2	6	7	2												
28	Ni	2	2	6	2	6	8	2												
29	Cu	2	2	6	2	6	10	1												
30	Zn	2	2	6	2	6	10	2												
31	Ga	2	2	6	2	6	10	2	1											
32	Ge	2	2	6	2	6	10	2	2											
33	As	2	2	6	2	6	10	2	3											
34	Se	2	2	6	2	6	10	2	4											
35	Br	2	2	6	2	6	10	2	5											
36	Kr	2	2	6	2	6	10	2	6											
37	Rb	2	2	6	2	6	10	2	6			1								
38	Sr	2	2	6	2	6	10	2	6			2								
39	Y	2	2	6	2	6	10	2	6	1		2								
40	Zr	2	2	6	2	6	10	2	6	2		2								
41	Nb	2	2	6	2	6	10	2	6	4		1								
42	Mo	2	2	6	2	6	10	2	6	5		1								
43	Tc	2	2	6	2	6	10	2	6	6		1								
44	Ru	2	2	6	2	6	10	2	6	7		1								
45	Rh	2	2	6	2	6	10	2	6	8		1								
46	Pd	2	2	6	2	6	10	2	6	10										

(continued)

Table 6.3 (continued)

		1	2		3			4				5				6				7
Z	Elements		s	p	s	p	d	s	p	d	f	s	p	d	f	s	p	d	f	s
47	Ag	2	2	6	2	6	10	2	6	10		1								
48	Cd	2	2	6	2	6	10	2	6	10		2								
49	In	2	2	6	2	6	10	2	6	10		2	1							
50	Sn	2	2	6	2	6	10	2	6	10		2	2							
51	Sb	2	2	6	2	6	10	2	6	10		2	3							
52	Te	2	2	6	2	6	10	2	6	10		2	4							
53	I	2	2	6	2	6	10	2	6	10		2	5							
54	Xe	2	2	6	2	6	10	2	6	10		2	6							
55	Cs	2	2	6	2	6	10	2	6	10		2	6			1				
56	Ba	2	2	6	2	6	10	2	6	10		2	6			2				
57	La	2	2	6	2	6	10	2	6	10		2	6	1		2				
58	Ce	2	2	6	2	6	10	2	6	10	2	2	6			2				
59	Pr	2	2	6	2	6	10	2	6	10	3	2	6			2				
60	Nd	2	2	6	2	6	10	2	6	10	4	2	6			2				
61	Pm	2	2	6	2	6	10	2	6	10	5	2	6			2				
62	Sm	2	2	6	2	6	10	2	6	10	6	2	6			2				
63	Eu	2	2	6	2	6	10	2	6	10	7	2	6			2				
64	Gd	2	2	6	2	6	10	2	6	10	7	2	6	1		2				
65	Tb	2	2	6	2	6	10	2	6	10	9	2	6			2				
66	Dy	2	2	6	2	6	10	2	6	10	10	2	6			2				
67	Ho	2	2	6	2	6	10	2	6	10	11	2	6			2				
68	Er	2	2	6	2	6	10	2	6	10	12	2	6			2				
69	Tm	2	2	6	2	6	10	2	6	10	13	2	6			2				

(continued)

Table 6.3 (continued)

		1	2		3			4				5				6				7
Z	Elements		s	p	s	p	d	s	p	d	f	s	p	d	f	s	p	d	f	s
70	Yb	2	2	6	2	6	10	2	6	10	14	2	6			2				
71	Lu	2	2	6	2	6	10	2	6	10	14	2	6	1		2				
72	Hf	2	2	6	2	6	10	2	6	10	14	2	6	2		2				
73	Ta	2	2	6	2	6	10	2	6	10	14	2	6	3		2				
74	W	2	2	6	2	6	10	2	6	10	14	2	6	4		2				
75	Re	2	2	6	2	6	10	2	6	10	14	2	6	5		2				
76	Os	2	2	6	2	6	10	2	6	10	14	2	6	6		2				
77	Ir	2	2	6	2	6	10	2	6	10	14	2	6	7		2				
78	Pt	2	2	6	2	6	10	2	6	10	14	2	6	9		1				
79	Au	2	2	6	2	6	10	2	6	10	14	2	6	10		1				
80	Hg	2	2	6	2	6	10	2	6	10	14	2	6	10		2				
81	Tl	2	2	6	2	6	10	2	6	10	14	2	6	10		2	1			
82	Pb	2	2	6	2	6	10	2	6	10	14	2	6	10		2	2			
83	Bi	2	2	6	2	6	10	2	6	10	14	2	6	10		2	3			
84	Po	2	2	6	2	6	10	2	6	10	14	2	6	10		2	4			
85	At	2	2	6	2	6	10	2	6	10	14	2	6	10		2	5			
86	Rn	2	2	6	2	6	10	2	6	10	14	2	6	10		2	6			
87	Fr	2	2	6	2	6	10	2	6	10	14	2	6	10		2	6		1	
88	Ra	2	2	6	2	6	10	2	6	10	14	2	6	10		2	6		2	
89	Ac	2	2	6	2	6	10	2	6	10	14	2	6	10		2	6	1	2	
90	Th	2	2	6	2	6	10	2	6	10	14	2	6	10		2	6	2	2	
91	Pa	2	2	6	2	6	10	2	6	10	14	2	6	10	2	2	6	1	2	
92	U	2	2	6	2	6	10	2	6	10	14	2	6	10	3	2	6	1	2	

(continued)

Table 6.3 (continued)

		1	2		3			4				5				6				7
Z	Elements		s	p	s	p	d	s	p	d	f	s	p	d	f	s	p	d	f	s
93	Np	2	2	6	2	6	10	2	6	10	14	2	6	10	4	2	6	1	2	
94	Pu	2	2	6	2	6	10	2	6	10	14	2	6	10	6	2	6		2	
95	Am	2	2	6	2	6	10	2	6	10	14	2	6	10	7	2	6		2	
96	Cm	2	2	6	2	6	10	2	6	10	14	2	6	10	7	2	6	1	2	
97	Bk	2	2	6	2	6	10	2	6	10	14	2	6	10	8	2	6	1	2	
98	Cf	2	2	6	2	6	10	2	6	10	14	2	6	10	10	2	6		2	
99	Es	2	2	6	2	6	10	2	6	10	14	2	6	10	11	2	6		2	
100	Fm	2	2	6	2	6	10	2	6	10	14	2	6	10	12	2	6		2	
101	Md	2	2	6	2	6	10	2	6	10	14	2	6	10	13	2	6		2	
102	No	2	2	6	2	6	10	2	6	10	14	2	6	10	14	2	6		2	
103	Lw	2	2	6	2	6	10	2	6	10	14	2	6	10	14	2	6	1	2	

6.2 Slater's Coupling Scheme

We may define quantum-mechanically a "*multiplet*" as an ensemble of states differing only for a single quantum number. Their energy levels are quite near each other and this, in the atomic spectroscopy, is revealed by nearly grouped emission lines detected by spectrometric devices with high resolution.

Slater (1929) in his analysis of the multiplets considered the interactions of the various angular momenta of the parts of an atom and treated the various energy levels resulting from different relative orientations of these parts, as done by others before him. However, he considered the whole degeneracy, i.e. that coming from orbital as well as spin angular momentum. In doing this, he introduced in his analysis the electronic spin of Uhlenbeck and Goudsmit (1925): each electron has an angular momentum (with consequent magnetic moment) entirely apart from the momentum l which it acquires by its motion. This additional angular momentum, denoted by m_s, (which like l is a vector), can likewise be oriented arbitrarily in space, without reference to the direction of l. Slater (1929) then noted that the angular momentum vectors l and m_s actually are coupled together; there are differences in energy depending on orientation and a consequent splitting up of each energy level into a number of different ones. In the commonest case, the coupling can be considered in several stages. First, the l's of the various electrons group themselves into a vector sum **L**, which takes on only integer values, 0, 1, 2, as the separate the l's do. Each such arrangement gives a different energy and a different term in the spectrum.[3] Second, the spins of the electrons group themselves into a vector sum **S**, taking on the integral or half-integral values. The different values of **S** lead to terms of different multiplicities and different energies. Finally, **L** and **S** are coupled together (their sum being called **J** total angular momentum) and are integral or half-integral to correspond with **S**.

In modern terms, we say now that the total angular momentum commutes with the Hamiltonian and its action on the wave function may be defined in terms of the j and m_j quantum numbers related, respectively, to the norm of the total angular momentum and to its projection along a specified axis:

$$\mathrm{J}^2\Psi = \hbar^2 j(j+1)\Psi \tag{6.2.1}$$

$$\mathrm{J_z}\Psi = \hbar m_j\Psi \tag{6.2.2}$$

The vectorial coupling between **L** and **S** is schematically depicted in Fig. 6.6.

Let me now return to the previously discussed quantum number notation which characterizes the various electrons and let me consider in how many ways two p electrons may couple. It may be rapidly seen that for each p electron there are 2 $(2\,l + 1) = 6$ different m_l, m_s sets, hence 36 combinations of one state of the first

[3] The terms with **L** equal, respectively, to 0, 1, 2, 3,… are called S, P, D, F,… terms.

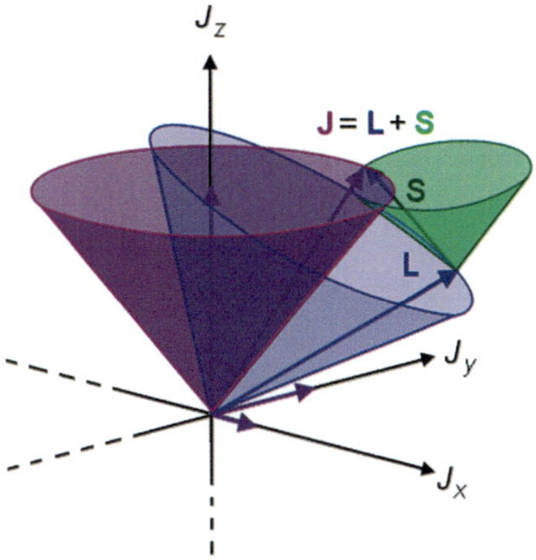

Fig. 6.6 Vector cones of total angular momentum **J** (purple), orbital **L** (blue), and spin **S** (green). From Maschen—Own work, Public Domain, https://commons.wikimedia.org/w/index.php?curid=17763201

Table 6.4 Slater notation, preliminary to the coupling scheme

Usual notation	Slater notation		$\sum m_l$	$\sum m_s$	Notes
(n 1 1 ½) (n' 1 1 ½)	{(n 1 1) (n' 1 1)}	{ }	2	1	Exc
(n 1 0 ½) (n' 1 1 ½)	{(n 1 0) (n' 1 1)}	{ }	1	1	A
(n 1 −1 ½) (n' 1 1 ½)	{(n 1 −1) (n' 1 1)}	{ }	0	1	
(n 1 1 −½) (n' 1 1 ½)	{(n' 1 1)	{(n 1 1)}	2	0	
(n 1 0 −½) (n' 1 1 ½)	{(n' 1 1)	{(n 1 0)}	1	0	
(n 1 −1 −½) (n' 1 1 ½)	{(n' 1 1)	{(n 1 −1)}	0	0	
(n 1 1 ½) (n' 1 0 ½)	{(n 1 1) (n' 1 0)}	{ }	1	1	A
(n 1 0 ½) (n' 1 0 ½)	{(n 1 1) (n' 1 0)}	{ }	0	1	Exc
(n 1 −1 ½) (n' 1 0 ½)	{(n 1 1) (n' 1 0)}	{ }	−1	1	
(n 1 1 −½) (n' 1 0 ½)	{(n 1 1)	{(n' 1 0)}	1	0	
(n 1 0 −½) (n' 1 0 ½)	{(n' 1 0)	{(n 1 0)}	0	0	
(n 1 −1 −½) (n' 1 0 ½)	{(n' 1 0)	{(n 1 −1)}	−1	0	
n 1 1 ½) (n' 1 −1 ½)	{(n 1 1) (n' 1 −1)}	{ }	0	1	
....			....		
(n 1 −1 −½) (n' 1 −1 −½)	{(n 1 −1) (n' 1 −1)}	{ }	−2	1	Exc

electron with one of the second. These states are listed in Table 6.4 which is an extended form of Slater's (1929) Table 6.1.

The first bracket in Table 6.4 groups the terms n, l, and m_l with $m_s = 1/2$ and the second those ones with $m_s = -1/2$. The third and fourth columns give the summation of m_l and m_s, respectively. The note Exc means that the combination is excluded and A that it corresponds to an identity.[4]

In graphical form, the couplings are shown in Fig. 6.7. The abscissa axis in Fig. 6.7a is $\sum m_s$ and the ordinate axis is $\sum m_l$. Each dotted circle depicts one of the six degenerate states. In Fig. 6.7b, we see the points corresponding to the sum

[4] Because $l = 1$ for both p electrons it means that $n = n' = 2$.

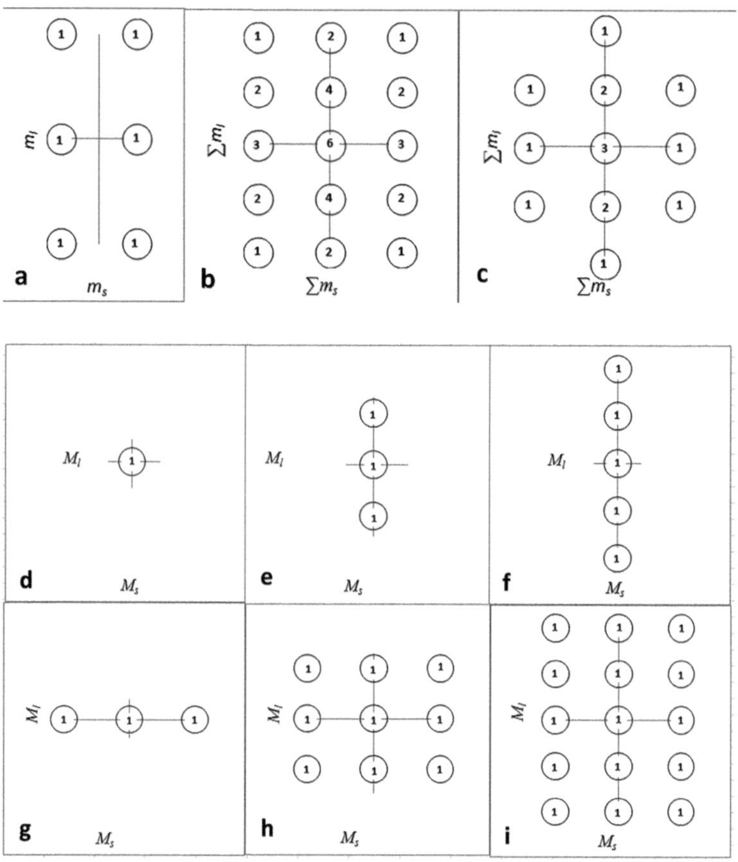

Fig. 6.7 Slater coupling scheme: **a** single *p* electron; **b** two *p* electrons; **c** two equivalent *p* electrons; **d** ^{1}S; **e** ^{1}P; **f** ^{1}D; **g** ^{3}S; **h** ^{3}P; **i** ^{3}D. Modified from Slater (1929)

of two p electrons (i.e. 6 × 6 = 36 microstates). The number inside the circles is the number of states with the same $\sum m_s$ and $\sum m_l$. Moreover, each point in Fig. 6.7b is intended to also represent the M_L and M_S of one of the multiplets existing in the coupled atom. Hence, the number of states in Fig. 6.7b may be seen as a summation of Hund's combinations corresponding, respectively, to ^{1}S; ^{1}P; ^{1}D; ^{3}S; ^{3}P; and ^{3}D (lower part of the figure). If their patterns are superimposed, they just return the 4b case. Figure 6.7c takes into account the fact that some combinations are excluded (hence the actual occurrences are much less). When the terms not allowed are removed, then the new pattern in place of Fig. 6.7b proves to be Fig. 6.7c, and this is the superposition of 'P 'S 'D, which then are the actual multiplets allowed with two equivalent p electrons (that is, two electrons, each with $l = 1$, and with the same n). In Slater's words "*Thus it must be that two p electrons produce these multiplets. But that is just what our other method gives; for two l' s, each equal to unity, can add to give L = 0, 1, 2,* (S, P, D *terms); while two s's, each*

1/2, can give S = 0, 1 (singlet, triplet). In every case, the two methods lead to the same result, just as here".

In an actual atom thus the realized couplings are much less than those predicted by a simple combinatory scheme based on the quantum numbers of the single electrons. This is due to the Pauli exclusion principle: no two electrons can have identical sets of numbers (n, l. m_l, m_s). Also, two combinations of electrons that differ only in the interchange of the quantum numbers of two electrons are to be treated as identical. As stated in terms of the bracket method of writing the electronic combinations, this means that no two sets of quantum numbers (n, l, m_l) within a bracket can be identical and that the order of arrangement within a bracket is immaterial. By these restrictions, the number of terms is often enormously decreased.

6.3 Spin-Multiplicity and Russell–Saunders Coupling Scheme

Consider now the many-electron atom. The orbital angular momentum **L** and the spin angular momentum **S** are given by the sum of the momenta of all i electrons of an atom, i.e.

$$\mathbf{L} = \sum_i l_i \quad (6.3.1)$$

$$\mathbf{S} = \sum_i s_i \quad (6.3.2)$$

In the following, I will refer to the scalar components of the vector L along the z-axis, which add scalarly:

$$L_z = \sum_i l_{z_i} \quad (6.3.3)$$

$$S_z = \sum_i s_{z_i} \quad (6.3.4)$$

and to the M_{L} and M_{S} summations:

$$M_L = \sum_i m_{l_i} \quad (6.3.5)$$

$$M_s = \sum_i m_{s_i} \quad (6.3.6)$$

M_{L} and M_{S} are representatives of an electron population, but, just as for one electron, the allowed values of M_{L} are restricted to $L, L-1,\ldots,-L+1, -L$ and M_{S} is

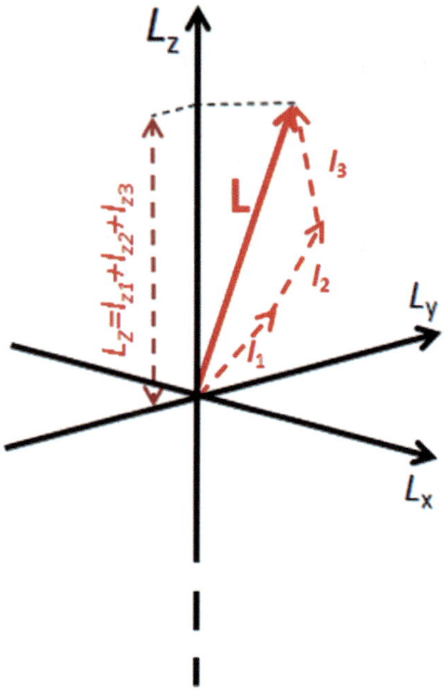

Fig. 6.8 L_z scalar component and its relations with the $\mathbf{L} = l_1 + l_2 + l_3$ vector

similarly restricted to S, $S-1, \ldots, -S+1, -S$. Moreover, while L_z and S_z determine the energy, all attainable values of M_L and M_S are degenerate.

The "*spin multiplicity*" $2S_z + 1$ is simply the projection along the *z*-axis of the **S** vector augmented of +1. Figure 6.8 is a schematic illustration returning graphically the relationships between **L**, ***l***, L_z, and l_z. Note however that the true vectorial nature of **L** implies that the vector may actually span a conical region around the *quantization axis* z (see Fig. 6.1).

The total angular momentum **J** is given by the vectorial summation of **L** and **S**:

$$\mathrm{J} = \mathrm{L} + \mathrm{S} \tag{6.3.7}$$

The summation holds also for the scalar components, i.e.

$$J_z = L_z + S_z \tag{6.3.8}$$

In the Russell–Saunders notation, the multiplicity is placed at the apex preceding the symbol that corresponds to the value assumed by *L*. The symbol corresponds to the upper case of the symbol returning the electronic configuration, i.e.

$$l = \begin{matrix} 0 & 1 & 2 & 3 & 4 & 5 \\ s & p & d & f & g & h \end{matrix} \tag{6.3.9}$$

Table 6.5 Set of quantum numbers for the np^2 configuration

μστατε			M_L	M_L
1	(n 1 1 ½)	(n 1 1 −½)	+2	0
2	(n 1 1 ½)	(n 1 0 ½)	+1	+1
3	(n 1 1 ½)	(n 1 0 −½)	+1	0
4	(n 1 1 −½)	(n 1 0 ½)	+1	0
5	(n 1 1 −½)	(n 1 0 −½)	+1	−1
6	(n 1 1 ½)	(n 1 −1 ½)	0	+1
7	(n 1 1 ½)	(n 1 −1 −½)	0	0
8	(n 1 1 −½)	(n 1 −1 ½)	0	0
9	(n 1 0 ½)	(n 1 0 −½)	0	0
10	(n 1 1 −½)	(n 1 −1 −½)	0	−1
11	(n 1 −1 ½)	(n 1 0 ½)	−1	+1
12	(n 1 −1 −½)	(n 1 0 ½)	−1	0
13	(n 1 −1 ½)	(n 1 0 −½)	−1	0
14	(n 1 −1 −½)	(n 1 0 −½)	−1	−1
15	(n 1 −1 ½)	(n 1 −1 −½)	−2	0

$$L = \begin{matrix} 0 & 1 & 2 & 3 & 4 & 5 \\ S & P & D & F & G & H \end{matrix} \tag{6.3.10}$$

After the symbol, at subscript, one finds the value attained by the total quantum number **J** (actually its scalar projection J_z), i.e. $^{2S+1}L_J$.

Consider now a many-electron atom such as carbon. The ground state electronic configuration of C is $1s^2 2s^2 2p^2$. The completely filled sublevels $1s^2$ and $2s^2$ have $L = 0$ and $S = 0$; hence, they are in a closed shell condition. They are thus non-influent in determining the L,S,J values of the bulk population. The number of microstates in the p-sublevel is easily obtained by the formula

$$\mu_{\text{states}} = \frac{\nu_{\max}!}{\nu!(\nu_{\max} - \nu)!} \tag{6.3.11}$$

with $\nu_{\max} = 6$ and $\nu = 2$ we get $\mu_{\text{states}} = 15$.[5]

What is stated above is generally valid for whatever np^2 configuration (Table 6.5).

The highest value of M_L in the Table is +2 and the corresponding value of M_S is only 0. Because the highest value of M_L is +2, this means implicitly that $L = 2$, and, because $M_S = 0$, then also $S = 0$. There are thus 5 degenerate states corresponding

[5] It should be noted that actually the possible M_L-M_S combinations are 21 but 6 of them conflict with the Pauli exclusion principle namely $M_L = 2$, $M_S = 1$ ($m_{l1} = 1, m_{s1} = \uparrow, m_{l2} = 1, m_{s2} = \uparrow$); $M_L = 2$, $M_S = -1$ ($m_{l1} = 1, m_{s1} = \downarrow$ $m_{l2} = 1, m_{s2} = \downarrow$); $M_L = 0$, $M_S = 1$ ($m_{l1} = 0, m_{s1} = \uparrow, m_{l2} = 0, m_{s2} = \uparrow$); $M_L = 0$, $M_S = -1$ ($m_{l1} = 0, m_{s1} = \downarrow, m_{l2} = 0, m_{s2} = \downarrow$); $M_L = -2$, $M_S = 1$ ($m_{l1} = -1, m_{s1} = \uparrow, m_{l2} = -1, m_{s2} = \uparrow$); $M_L = -2$, $M_S = -1$ ($m_{l1} = -1, m_{s1} = \downarrow, m_{l2} = -1, m_{s2} = \downarrow$).

Table 6.6 Reduced set of quantum numbers for the np^2 configuration ($L = 2$ excluded)

$\mu_{\sigma\tau\alpha\tau\epsilon}$			M_L	M_S
2	$(n\ 1\ 1\ ½)$	$(n\ 1\ 0\ ½)$	+1	+1
4	$(n\ 1\ 1\ -½)$	$(n\ 1\ 0\ ½)$	+1	0
5	$(n\ 1\ 1\ -½)$	$(n\ 1\ 0\ -½)$	+1	−1
6	$(n\ 1\ 1\ ½)$	$(n\ 1\ -1\ ½)$	0	+1
8	$(n\ 1\ 1\ -½)$	$(n\ 1\ -1\ ½)$	0	0
9	$(n\ 1\ 0\ ½)$	$(n\ 1\ 0\ -½)$	0	0
10	$(n\ 1\ 1\ -½)$	$(n\ 1\ -1\ -½)$	0	−1
11	$(n\ 1\ -1\ ½)$	$(n\ 1\ 0\ ½)$	−1	+1
13	$(n\ 1\ -1\ ½)$	$(n\ 1\ 0\ -½)$	−1	0
14	$(n\ 1\ -1\ -½)$	$(n\ 1\ 0\ -½)$	−1	−1

to $L = 2$; $M_L = +2, +1, 0, -1, -2$; $S = 0$; $M_S = 0$. We may thus select from the above table five degenerate microstates corresponding to this condition. We will choose $\mu_{state} =$ 1, 3, 7, 12, 15. With $L = 2$ and $S = 0$, because $J = L + S, L + S-1, L + S-2, |L-S|$ there is only a condition for J, i.e. $J = 2$ and the corresponding Russell–Saunders notation is $^1D_2$. We are now left with the following population (Table 6.6):

The highest value of M_L is now +1, which implies $L = 1$ and occurs with $M_S = 1$, 0, and −1. There are 9 microstates obeying this configuration $L = 1$; $S = 1$; $M_L = 1, 0, -1$; $M_S = 1, 0, -1$ ($\mu_{state} = 2, 4, 5, 6, 8, 10, 11, 13, 14$). With $L = 1$ and $S = 1$, J may attain 2, 1, 0. The Russell–Saunders corresponding terms will be $^3P_2$, $^3P_1$, $^3P_0$. We are left now with a single microstate ($\mu_{state} = 9$) corresponding to the condition $L = 0$, $S = 0$, $M_L = 0$, $M_S = 0$ and to the Russell–Saunders term $^1S_0$. Because the degeneracy of the various states is simply given by $2J + 1$, we will have 5, 5, 3, 1, 1 degenerate states which adds up to 15.

Table 6.7 resumes the various electron configurations (with the exclusion of the J subscript) for all s, p, and d sublevels.

Table 6.7 Electron configurations for equivalent electrons in the s, p, and d sublevels

Electron configuration	Russell–Saunders term symbol (excluding the J subscript)
s^1	^2S
p^1	^2P
p^2, p^4	^1S, ^1D, ^3P
p^3	^2P, ^2D, ^4S
p^1, p^5	^2P
d^1, d^9	^2D
d^2, d^8	^1S, ^1D, ^1G, ^3P, ^3F
d^3, d^7	^2P, ^2D, ^2D, ^2F, ^2G, ^2H, ^4P, ^4F
d^4, d^6	^1S, ^1S, ^1D, ^1D, ^1F, ^1G, ^1G, ^1I, ^3P, ^3P, ^3D, ^3F, ^3F, ^3G, ^3H, ^5D
d^5	^2S, ^2P, ^2D, ^2D, ^2D, ^2G, ^2G, ^2H, ^2I, ^4P, ^4D, ^4F, ^4G, ^6S

Table 6.8 Ground states of the p and d sublevels

p^v	1	0	−1	L_z	$2S_z + 1$	Ground state
p^1	↑			1	2	^2P
p^2	↑	↑		1	3	^3P
p^3	↑	↑	↑	0	4	^4S
p^4	↑↓	↑	↑	1	3	^3P
p^5	↓	↑↓	↑	1	2	^2P
p^6	↑↓	↑↓	↑↓	0	1	^1S

d^v	2	1	0	−1	−2	L_z	$2S_z + 1$	Ground state
d^1	↑					2	2	^2D
d^2	↑	↑				3	3	^3F
d^3	↑	↑	↑			3	4	^4F
d^4	↑	↑	↑	↑		2	5	^5D
d^5	↑	↑	↑	↑	↑	0	6	^6S
d^6	↑↓	↑	↑	↑	↑	2	5	^5D
d^7	↑↓	↑↓	↑	↑	↑	3	4	^4P
d^8	↑↓	↑↓	↑↓	↑	↑	3	3	^3F
d^9	↑↓	↑↓	↑↓	↑↓	↑	2	2	^2D
d^{10}	↑↓	↑↓	↑↓	↑↓	↑↓	0	1	^1S

If we limit ourselves to depict the ground state corresponding to a given number of electron within a sublevel, we may apply the "*fill-in the box*" procedure (implicitly based on Hund's rules). For a p-sublevel, $l = 1$ and m_l may attain three $(2\,l + 1)$ states (+1, 0, −1). Moreover, the maximum number (ν_{max}) of p electrons is 6. We may thus build up a table with 6 rows and 3 columns representing the value of m_l. The procedure for the d sublevel is analogous (Table 6.8).

I will return later to the significance of the term configuration when dealing with the factorization of the secular equation (Sect. 6.4.3).

6.4 The He Atom

6.4.1 Ground State

The electron configuration of He in the ground state is $1s^2$: there are two electrons of charge $-e$ orbiting around a nucleus of charge $+Ze$ at distances r_1 and r_2, respectively. The potential energy is

$$V = \frac{-Ze^2}{r_1} - \frac{-Ze^2}{r_2} + \frac{e^2}{r_{12}} \qquad (6.4.1)$$

with the third term on the right expressing the mutual interaction of the two electrons at distance r_{12}.

Considering the nucleus at rest, the wave equation for the two electrons is

$$\mathsf{H}\psi = \frac{-h^2}{8\pi m_0}\left(\frac{\partial^2\psi}{\partial x_1^2} + \frac{\partial^2\psi}{\partial y_1^2} + \frac{\partial^2\psi}{\partial z_1^2} + \frac{\partial^2\psi}{\partial x_2^2} + \frac{\partial^2\psi}{\partial y_2^2} + \frac{\partial^2\psi}{\partial z_2^2}\right) + \left(\frac{-Ze^2}{r_1} + \frac{-Ze^2}{r_1} + \frac{e^2}{r_{12}}\right)\psi = W\psi \tag{6.4.2}$$

Omitting the mutual interaction term e^2/r_{12}, the wave equation may be solved exactly. This term may thus be seen as a "*perturbation function*" $H' = e^2/r_{12}$. The unperturbed equation may be separated into two terms

$$\psi^0(x_1, y_1, z_1, x_2, y_2, z_2) = u_1^0(x_1, y_1, z_1)u_2^0(x_2, y_2, z_2) \tag{6.4.3}$$

where the symbol u refers to a single electron of a many-electron atom in its own configuration (e.g. 1 s,2 s,2p…).

The energy of the system will be

$$W^0 = W_1^0 + W_2^0 \tag{6.4.4}$$

and the wave equation for u_1^0, u_2^0 will be

$$\frac{\partial^2 u_1^0}{\partial x_1} + \frac{\partial^2 u_1^0}{\partial y_1} + \frac{\partial^2 u_1^0}{\partial z_1} + \frac{8\pi^2 m_0}{h^2}\left(W_1^0 + \frac{Ze^2}{r_1}\right)u_1^0 = 0 \tag{6.4.5.1}$$

$$\frac{\partial^2 u_2^0}{\partial x_2} + \frac{\partial^2 u_2^0}{\partial y_2} + \frac{\partial^2 u_2^0}{\partial z_2} + \frac{8\pi^2 m_0}{h^2}\left(W_1^0 + \frac{Ze^2}{r_2}\right)u_2^0 = 0 \tag{6.4.5.2}$$

These are nothing else than the *hydrogen-like wave equations* (Sect. 5.3) whose solutions in terms of polar coordinates are $\psi_{n,l,m}(r_1, \theta_1, \varphi_1)$ and $\psi_{n,l,m}(r_2, \theta_2, \varphi_2)$, respectively.

We may now write the unperturbed wave function for the lowest *n*, *l*, *m* level (100, 100) as

$$\psi_{100,100}^0 = \psi_{100}(r_1, \theta_1, \varphi_1)\psi_{100}(r_2, \theta_2, \varphi_2) = u_{1s}(r_1, \theta_1, \varphi_1)u_{1s}(r_2, \theta_2, \varphi_2) \tag{6.4.6}$$

and the bulk unperturbed energy as

$$W_{100,100}^0 = W_1^0 + W_2^0 = -2Z^2W_H \tag{6.4.7}$$

Introducing the appropriate hydrogen-like wave function (Table 5.1):

$$\psi_{100} = u_{1s} = \sqrt{\frac{Z^3}{\pi a_0}}\exp\left(-\frac{\rho}{2}\right) \tag{6.4.8}$$

With a_0 = Bohr's radius and $\rho = {}^{2Zr}/_{a_0}$ we find

$$\psi^0_{100,100} = \frac{Z^3}{\pi a_0^3} \exp\left(-\frac{\rho_1}{2}\right) \exp\left(-\frac{\rho_2}{2}\right) \tag{6.4.9}$$

The perturbation term W′ is the average value of the perturbation function and is given by the integral

$$W' = \int \psi_0^* H' \psi_0 dr = \int \frac{e^2}{r_{12}} \psi^2_{100,100} dr \tag{6.4.10}$$

where the volume element dr is

$$dr = r_1^2 dr_1 \sin\theta_1 d\theta_1 d\varphi_1 \cdot r_2^2 dr_2 \sin\theta_2 d\theta_2 d\varphi_2 \tag{6.4.11}$$

For two spherically symmetrical distributions of electricity with density functions $e^{-\rho 1}$, $e^{-\rho 2}$, respectively, this term reduces to [see Appendix V in Pauling and Wilson (1935)]

$$W' = \frac{5}{4} Z W_H \tag{6.4.12}$$

The perturbed energy is thus

$$W' = W_H\left(-2Z^2 + \frac{5}{4}Z\right) \tag{6.4.13}$$

In Table 6.9 are listed the electron energies obtained with Eq. (6.4.13) for He and the isoelectronic 1s ions Li^+, Be^{2+}, B^{3+}, and C^{4+}. When compared with the sum of ionization energies (1st + 2nd), the difference between experimental and calculated values remains roughly constant (an overestimation of roughly 4 *eV*) but the percentual error decreases with the increase of the nuclear charge from roughly 5% for He to 0.4% for C^{4+}. This is due to the fact that the correlation effect

Table 6.9 Calculated vs observed (exp; *eV*) values of the energy of He and isoelectronic ions in the $1s^2$ configuration (From Pauling and Bright Wilson 1935)

	$-W_{exp}$	$-W^0$	$-W^0 - W'$	Δ^0	Δ'	$-\Delta^0/\Delta'$
He	78.62	108.24	74.42	29.62	−4.2	0.142
Li^+	197.14	243.54	192.8	46.4	−4.34	0.094
Be^{2+}	369.96	432.96	365.31	63	−4.65	0.056
B^{3+}	596.4	676.5	591.94	80.1	−4.5	0.056
C^{4+}	876.2	974.16	872.69	98	−3.5	0.036

decreases progressively in importance with the progressive increase of the nucleus–electron interaction. Nevertheless, the perturbation is never negligible and attains for C^{4+} $\sim 10^2$ *eV.*

6.4.2 Excited States $1s^1 2s^1$, $1s^1 2p^1$, Spin Neglected

One electron has the total quantum number $n = 1$ and the other $n = 2$. In the preceding section, the unperturbed hydrogen-like wave function was depicted as the product of two functions $\psi_{n_1 l_1, m_1}(1)\psi_{n_2 l_2, m_2}(2)$ for electrons with polar coordinates $(1) = (r_1, \theta_1, \phi_1)$; $(2) = (r_2, \theta_2, \phi_2)$.

In the first excited state $1s^1 2s^1$ we have $n_1 = 1$, $n_2 = 2$ and $n_1 = 2$, $n_2 = 1$. The corresponding zeroth-order energy is

$$W^0_{n_1, n_2} = -4\mathcal{R}hc\left(\frac{1}{n_1^2} + \frac{1}{n_2^2}\right) = -5\mathcal{R}hc \tag{6.4.14}$$

Neglecting the electron spin the excited level appears as eightfold degenerate

$$\left.\begin{array}{ll} 1s(1) & 2s(2) \\ 2s(1) & 1s(2) \\ 1s(1) & 2p_x(2) \\ 2p_x(1) & 1s(2) \\ 1s(1) & 2p_y(2) \\ 2p_y(1) & 1s(2) \\ 1s(1) & 2p_z(2) \\ 2p_z(1) & 1s(2) \end{array}\right\} \tag{6.4.15}$$

The perturbation integrals have the following form (From Pauling and Bright Wilson, 1935):

$$J_s = \int\int 1s(1)2s(2)\frac{e^2}{r_{12}}1s(1)2s(2)dr_1 dr_2 \tag{6.4.16.1}$$

$$K_s = \int\int 1s(1)2s(2)\frac{e^2}{r_{12}}2s(1)1s(2)dr_1 dr_2 \tag{6.4.16.2}$$

$$J_{p_x} = \iint 1s(1)2p_x(2)\frac{e^2}{r_{12}}1s(1)2p_x(2)dr_1dr_2 \quad (6.4.16.3)$$

$$J_{p_y} = \iint 1s(1)2p_y(2)\frac{e^2}{r_{12}}1s(1)2p_y(2)dr_1dr_2 \quad (6.4.16.4)$$

$$J_{p_z} = \iint 1s(1)2p_z(2)\frac{e^2}{r_{12}}1s(1)2p_z(2)dr_1dr_2 \quad (6.4.16.5)$$

$$K_{p_x} = \iint 1s(1)2p_x(2)\frac{e^2}{r_{12}}2p_x(1)1s(2)dr_1dr_2 \quad (6.4.16.6)$$

$$K_{p_y} = \iint 1s(1)2p_y(2)\frac{e^2}{r_{12}}2p_y(1)1s(2)dr_1dr_2 \quad (6.4.16.7)$$

$$K_{p_z} = \iint 1s(1)2p_z(2)\frac{e^2}{r_{12}}2p_z(1)1s(2)dr_1dr_2 \quad (6.4.16.8)$$

Inasmuch as the functions $J_{p_x}, J_{p_y}, J_{p_z}$ and $K_{p_x}, K_{p_y}, K_{p_z}$ differ from one another in terms of orientation but not in energy, we may represent them simply as J_p and K_p, respectively. The "*Coulomb Integrals*" J_s and J_p represent the average Coulomb interaction with probability distribution $\{1\ s(1)\}^2$ and $\{2\ s(2)\}^2$. The "*Resonance Integrals*" K_s and K_p have wave functions differing from one another by the interchange of the electrons. Hence, they are also called "*Exchange Integrals*" or "*Interchange Integrals*" (Pauling and Bright Wilson 1935).

All the remaining perturbation integrals vanish by symmetry-related properties and the secular determinant may be written as (Pauling and Bright Wilson, 1935)

$$\begin{vmatrix} J_s - W' & K_s & 0 & 0 & 0 & 0 & 0 & 0 \\ K_s & J_s - W' & 0 & 0 & 0 & 0 & 0 & 0 \\ 0 & 0 & J_p - W' & K_p & 0 & 0 & 0 & 0 \\ 0 & 0 & K_p & J_p - W' & 0 & 0 & 0 & 0 \\ 0 & 0 & 0 & 0 & J_p - W' & K_p & 0 & 0 \\ 0 & 0 & 0 & 0 & K_p & J_p - W' & 0 & 0 \\ 0 & 0 & 0 & 0 & 0 & 0 & J_p - W' & K_p \\ 0 & 0 & 0 & 0 & 0 & 0 & K_p & J_p - W' \end{vmatrix} = 0 \quad (6.4.17)$$

The solution of the secular determinant allows us to determine the perturbation energies as $W' = J_s + K_s$; $J_s - K_s$; $J_p + K_p$ (3 roots); $J_p - K_p$ (3 roots).

6.4.3 Spin–orbit Interaction

In 1927, pursuing his investigation of the nature of spectral complexities in many-electron atoms, Heisenberg proposed the existence of a "*resonance phenomenon*" between the spin states of the different electrons and, perhaps, their orbital momenta.[6] In his investigation, Heisenberg (1927a) studied states of n-electron systems in an atom or molecule. Abstracting at first from spin, he described a composite system as a kind of non-commutative product of the eigenfunctions. He considered the result as a state of the "*unperturbed*" composite system, while the spin coupling ("*resonance*") had to be taken into account as a perturbation due to the "*more subtle interplay of the electrons*". In Heisenberg's (1927a) words: "*In the unperturbed case, the eigenfunction of the total system can be written as product of all functions of the single electrons, e.g.* $l_1 m_2 \cdots p_n$. *The unperturbed problem is n!-fold degenerate, because a permutation of the electrons leads to equal energy values of the total system*" (see Scholtz 2006; for a more extensive account of Heisenberg's approach).

In Fig. 6.9 (*Grotrian diagram*), we see the (experimental) energy level diagram of the He atom. There are 191 accessible energy levels and 2291 lines (Morton et al. 2006). The diagram is a rearranged and simplified version of the graphical output available at the NIST site (https://physics.nist.gov/cgi-bin/ASD/lines). There are two separate sets of transitions: one among singlet states (S = 0; multiplicity = 2S + 1 = 1) and one among triplet states (S = 1; 2S + 1 = 3). Three "*selection rules*"[7] establish that the allowed transitions are those ones involving $\Delta L = 0, \pm 1$; $\Delta S = 0$; $\Delta J = 0, \pm 1$. Hence, no transitions between singlet and triplet states are possible because the second rule would be in this case violated.

The splitting arises partly from the differences in the Coulomb integrals J_s and J_p and partly from differences in the integrals K_s and K_p. For the state with $W' = J_s + K_s$, the zeroth-order wave function is

$$\psi = \frac{1}{\sqrt{2}}[1s(1)2s(2) + 2s(1)1s(2)] \qquad (6.4.18)$$

and generates a "*resonance*" between a structure where the first electron (1) is in the 1s orbit and the second (2) in the 2s orbit and a structure where the occupancies are interchanged [i.e. (2) in 1s and (1) in 2s]. A wave function of type (6.4.18) is "*symmetric in the positional coordinates of the two electrons*" (the interchange of

[6] Heisenberg made clear that here the word "*resonance*" was not to be understood in the sense of classical physics, but as an expression of a physical intuition of the "*more subtle interplay of the electrons in an atom*" (Heisenberg, 1927a).

[7] The selection rules have been deduced experimentally and are valid only for light atoms. For example, the spectrum of Hg shows many singlet–triplet transitions, one of these being the familiar intense purple light emission (Mc Quarrie, 2008).

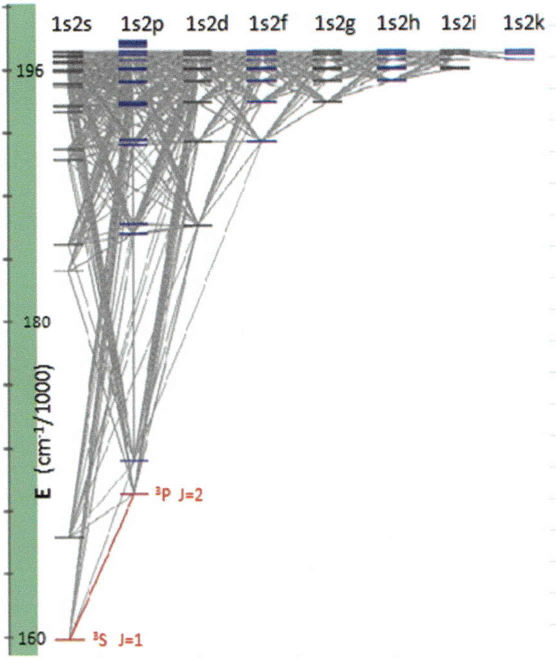

Fig. 6.9 Grotrian diagram of the spectrum of atomic Helium. It is a simplified version of the output created at the NIST site (https://physics.nist.gov/cgi-bin/ASD/lines). The data are from Morton et al. (2006)

coordinates leaves the function unchanged; Pauling and Bright Wilson 1935). On the contrary, the function

$$\psi = \frac{1}{\sqrt{2}}[1s(1)2s(2) - 2s(1)1s(2)] \qquad (6.4.19)$$

is said to be "*antisymmetric in the positional coordinates of the two electrons*" because interchange causes the function to change sign. The functions (6.4.18) and (6.4.19) lead, respectively, to singlet and triplet states.

To take into account the effect of spin–orbit interaction, we must multiply each of the eight orbital functions (6.4.15) by each one of the four spin functions

$$\left.\begin{array}{ll} \alpha(1) & \alpha(2) \\ \alpha(1) & \beta(2) \\ \beta(1) & \alpha(2) \\ \beta(1) & \beta(2) \end{array}\right\} \qquad (6.4.20)$$

We obtain in this way 32 functions. To solve the system, however, the spin functions (6.4.20) are better replaced by a linear combination of normalized and mutually orthogonal forms:

$$\left.\begin{array}{l}\alpha(1)\quad\alpha(2)\\ \frac{1}{\sqrt{2}}[\alpha(1)\beta(2)+\beta(1)\alpha(2)]\\ \beta(1)\quad\beta(2)\\ \frac{1}{\sqrt{2}}[\alpha(1)\beta(2)-\beta(1)\alpha(2)]\end{array}\right\} \tag{6.4.21}$$

Multiplying the zeroth-order wave functions (6.4.18, 6.4.19) by the four spin functions for the $1s^12s^1$ configuration, one obtains (Pauling and Bright Wilson 1935)

$$\left.\begin{array}{ll}\frac{1}{\sqrt{2}}[1s(1)2s(2)+2s(1)1s(2)]\cdot\alpha(1)\alpha(2) & (1)\\ \frac{1}{\sqrt{2}}[1s(1)2s(2)+2s(1)1s(2)]\cdot\frac{1}{\sqrt{2}}[\alpha(1)\beta(2)+\beta(1)\alpha(2)] & (2)\\ \frac{1}{\sqrt{2}}[1s(1)2s(2)+2s(1)1s(2)]\cdot\beta(1)\beta(2) & (3)\\ \frac{1}{\sqrt{2}}[1s(1)2s(2)-2s(1)1s(2)]\cdot\frac{1}{\sqrt{2}}[\alpha(1)\beta(2)-\beta(1)\alpha(2)] & (4)\\ \frac{1}{\sqrt{2}}[1s(1)2s(2)-2s(1)1s(2)]\cdot\alpha(1)\alpha(2) & (5)\\ \frac{1}{\sqrt{2}}[1s(1)2s(2)-2s(1)1s(2)]\cdot\frac{1}{\sqrt{2}}[\alpha(1)\beta(2)+\beta(1)\alpha(2)] & (6)\\ \frac{1}{\sqrt{2}}[1s(1)2s(2)-2s(1)1s(2)]\cdot\beta(1)\beta(2) & (7)\\ \frac{1}{\sqrt{2}}[1s(1)2s(2)+2s(1)1s(2)]\cdot\frac{1}{\sqrt{2}}[\alpha(1)\beta(2)-\beta(1)\alpha(2)] & (8)\end{array}\right\} \tag{6.4.22}$$

Functions (6.4.22,1,4) are symmetric in the coordinates of the two electrons (symmetric in the spin and orbit the first 3; antisymmetric in both the spin part and orbital part the fourth). Functions (6.4.22,5,8) are antisymmetric in the two electrons [antisymmetric in the orbital part and symmetric in the spin part (5–7); antisymmetric in the orbital or in the spin parts (6–8)]. The wave properties described for the $1s^12s^1$ configuration apply to any other configuration as well. Figure 6.10 resumes the levels resulting from spin–orbit interactions for the $1s^2$, $1s^12s^1$, and $1s^12p^1$ of the He atom: each filled circle represents a spin-symmetric wave function, and each open circle a spin-antisymmetric one. Three circles on a row indicate a triply degenerate state.

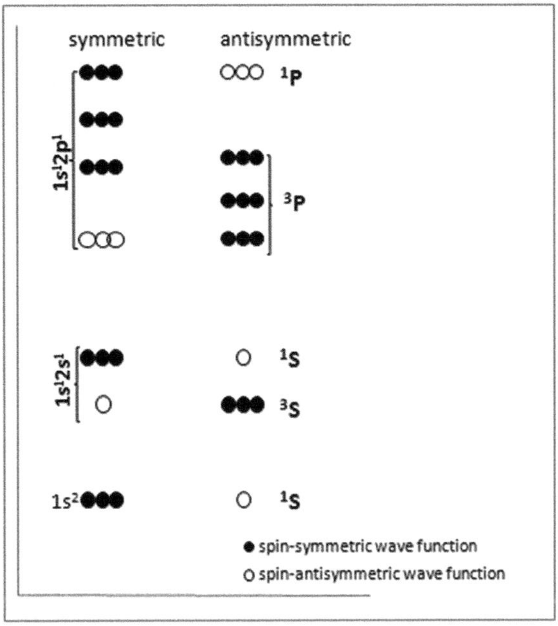

Fig. 6.10 Levels for the $1s^2$, $1s^12s^1$ $1s^12p^1$ configurations of the He atom. Modified from Pauling and Bright Wilson (1935)

6.4.4 *Factorization of the Secular Equation*

Let α, β....v denote the particular l, m_l, m_s set of quantum numbers of the separate single electron (1) with $u_\alpha(1)$. Omitting the correlation term $\sum e^2/r_{ij}$ the solution of the equation for the atom will be of the type $\psi_1^0 = u_\alpha(1)u_\beta(2)...u_v(N)$. Written as a determinant, and accounting for the Pauli exclusion principle, it takes the form

$$\psi^0 = \begin{vmatrix} u_\alpha(1) & u_\beta(1) & \dots & u_v(1) \\ u_\alpha(2) & u_\beta(2) & \dots & u_v(2) \\ \dots & \dots & \dots & \dots \\ u_\alpha(N) & u_\beta(N) & \dots & u_v(N) \end{vmatrix} \tag{6.4.23}$$

Consider now the complete set of quantum numbers for the configuration $1s^22p^1$ (Table 6.10).

Each row on the table corresponds to a set of functions $u_\alpha(1)u_\beta(2)...u_v(N)$ in the determinant (6.4.23). Being $1s^2$ a completed (*closed*) shell, the microstate is entirely determined by the 2p-sublevel. The ψ^0 antisymmetric wave function is composed just of the set of six functions of the accessible microstates because no other set satisfies the Pauli exclusion principle. The order followed in denoting the six 2p functions is non-influent.

Table 6.10 Set of quantum numbers for the $1s^22p^1$ configuration

μ_{state}				M_L	M_S
1	(1 0 0 ½)	(1 0 0 −½)	(2 1 1 ½)	+1	+½
2	(1 0 0 ½)	(1 0 0 −½)	(2 1 1 −½)	+1	−½
3	(1 0 0 ½)	(1 0 0 −½)	(2 1 0 ½)	0	+½
4	(1 0 0 ½)	(1 0 0 −½)	(2 1 0 −½)	0	−½
5	(1 0 0 ½)	(1 0 0 −½)	(2 1 −1 ½)	−1	+½
6	(1 0 0 ½)	(1 0 0 −½)	(2 1 −1 −½)	−1	−½

Generally speaking, denoting H_{11} the true Hamiltonian for the α set in position (1), the corresponding secular equation will be of order k, with k corresponding to the number of allowed sets of functions

$$\begin{vmatrix} \mathsf{H}_{11}-\mathrm{W} & \mathsf{H}_{12} & \cdots & \mathsf{H}_{1k} \\ \mathsf{H}_{21} & \mathsf{H}_{22}-\mathrm{W} & \cdots & \mathsf{H}_{2k} \\ \cdots & \cdots & \cdots & \cdots \\ \mathsf{H}_{k1} & \mathsf{H}_{k2} & \cdots & \mathsf{H}_{kk}-\mathrm{W} \end{vmatrix} = 0 \tag{6.4.24}$$

$$\mathsf{H}_{nm} = \int \psi_n^* \mathsf{H} \psi_m dr \tag{6.4.25}$$

ψ_n in Eq. (6.4.25) is an antisymmetric wave function composed of the functions u corresponding to the nth row of a table such as Table 6.10. We have already seen that for the configuration $1s^22p^1$ k is equal to 6. The secular equation factors into six linear factors because no other functions have the same values of M_L and M_S. Consider now again the 15 microstates accessible for the configuration np^2 (Table 6.5). The solution of the secular equation is greatly simplified by the theorem stating that "*The integral H_{nm} is zero unless ψ_m and ψ_n have the same value of $\sum m_s$ and the same value of $\sum m_l$, these quantities*[8] *being the sums of quantum numbers m_s and m_l of the functions u making up ψ_m and ψ_n*" (Pauling and Bright Wilson 1935). In the light of this theorem, the 15th degree equation is reduced to a cubic factor ($M_L = 0$, $M_S = 0$, μ_{states}7,8,9), two quadratic factors ($M_L = 1$, $M_S = 0$, μ_{states} 3,4), and 8 linear factors (all the remaining μ_{states} in Table 6.5). The diagrammatic representation of the configuration np^2 is shown in Fig. 6.11.

Because ψ_1 alone has $M_L = 2$ and $M_S = 0$ then $\mathsf{H}_{11}-W$ is a linear factor of the secular equation. This root must appear five times in the secular determinant corresponding to the degenerate states $L = 2$, $S = 0$, $M_S = 0$, $M_L = 2, 1, 0, -1, -2$ (i.e. in two of the linear factors $M_L = 2, -2$, $M_S = 0$; in two of the quadratic factors $M_L = 1, -1$, $M_S = 0$; and in the cubic factor $M_L = 0$, $M_S = 0$). Likewise, the $\mathsf{H}_{22}-W$ linear factor must correspond to nine states (i.e. six roots of the linear

[8] M_S and M_L in our notation.

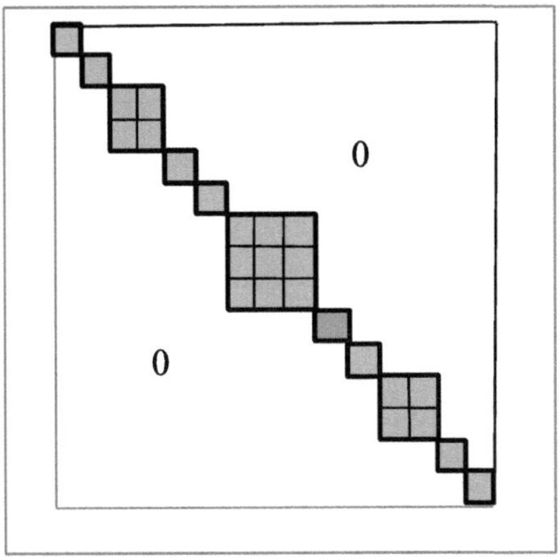

Fig. 6.11 Diagrammatic representation of the secular determinant for the configuration np^2. Modified from Pauling and Bright Wilson (1935)

factors $M_L = \pm 1, M_S = \pm 1$; $M_L = 0, M_S = \pm 1$; two roots of the quadratic factors $M_L = \pm 1$, $M_S = 0$; and one root of the cubic factor $M_L = 0$, $M_S = 0$). Since we have two roots for the cubic factor, the third one may be obtained by applying the theorem stating that "*the sum of the roots of a secular equation, or of one of its factors, is equal to the sum of the diagonal elements of the equation*" (Pauling and Bright Wilson 1935). The three energy levels found in this way for the np^2 configuration are as follows:

$$\left.\begin{array}{ll} W = H_{11} & {}^1D \\ W = H_{22} & {}^3P \\ W = H_{77} + H_{88} + H_{99} - H_{11} - H_{22} & {}^1S \end{array}\right\} \tag{6.4.26}$$

6.4.5 *Evaluation of Integrals*

We have seen that only certain diagonal non-vanishing H_{nm} terms need to be evaluated in order to obtain the energy levels. Unless a common constant term, these energies may be expressed in terms of integrals (I_i, F^k, G^k) involving the radial factors of the wave functions (Slater 1929).

The energy operator may be represented as

$$\frac{\int u^*_{(n/x)} H u_{(n/x)} dv}{\int u^*_{(n/x)} u_{(n/x)} dv} \tag{6.4.27}$$

where $\int dv \equiv \int dxdydz$ is meant to represent an integration over the volume.

The functions u are determinants; that is, sums of the N! terms formed from the product $u_{(n1/x1)}\ldots u_{(nN/xN)}$ carrying out all permutations of the coordinates $x_1\ldots x_N$, even permutations having the coefficient +1, and odd ones the coefficient −1. The product u*u is then also a double sum, while each of the integrals (in consequence of orthogonality relations) reduces to single sums, of which all the terms are identical (Slater 1929).

The energy operator H consists of three parts:

(i) a constant term;
(ii) a sum of terms each depending on the coordinates of a single electron which may be written as $\mathsf{H}_{(x1)} + \mathsf{H}_{(x2)} + \ldots \mathsf{H}_{(xN)}$;
(iii) a sum of Coulomb repulsion terms each depending on the coordinates of two (k,l) electrons $\sum_{k\neq 1} e^2/r_{kl}$.

The whole integral then consists of a constant, and two triple sums, two of the summations being over the permutations found in u and u*, and the third over the terms in the energy operator.

The term (ii) (a sum of terms $\mathsf{H}(x_i)$, each depending on the coordinates of one electron) may be translated into a summation of single I_n terms depending on a single set of quantum numbers, i.e. (Slater 1929)

$$\begin{aligned} I_{n_1} &= \int u^*_{(n1/x1)} \quad \mathsf{H}_{(x1)} u_{(n1/x1)} dx \\ &\ldots \\ I_{n_N} &= \int u^*_{(nN/xN)} \quad \mathsf{H}_{(xN)} u_{(nN/xN)} dx \end{aligned} \tag{6.4.28}$$

The term (iii) is composed of two parts: if the same permutation occurs both in n and u*, then the integral reduces to

$$J_{n,n'} = \int u^*_{(n/x_k)} u^*_{(n'/x_l)} \frac{e^2}{r_{kl}} u_{(n/x_k)} u_{(n'/x_l)} dx_k dx_l \tag{6.4.29}$$

If the permutation in u* differs from that in u by the interchange of two electron coordinates and the term e^2/r_{kl} refers to the particular coordinates x_k, x_l which have been interchanged, the integral reduces to

$$K_{n,n'} = \int u^*_{(n/x_l)} u^*_{(n'/x_k)} \frac{e^2}{r_{kl}} u_{(n/x_k)} u_{(n'/x_l)} dx_k dx_l \tag{6.4.30}$$

The $J_{n,n'}$, $K_{n,n'}$ contributions give rise to N! pairs of n's each, with all $K_{n,n'}$ terms having a negative sign. Finally, the normalization factor $\int u^*_{(n/x)} u_{(n/x)} dx)$ (Eq. 6.4.27) corresponds just to N! (Slater 1929). Hence, the energy summation reduces to constant + $(n's)I_n$ + P(pairs of $n's$) $J_{n,n}$—P(pairs of $n's$)$K_{n,n'}$.

To evaluate the integrals *I*, *J*, and *K*, Slater then considers the exact form of the wave function. All the individual functions $u_{(n/x)}$ are solutions of Schrodinger's equation for an electron in a central field, denoted by Slater (1929) U(xyz):

$$H_1u = \left(-\frac{h^2}{8\pi^2 m}\nabla^2 + U\right)u = \varepsilon u \qquad (6.4.31)$$

u in Eq. (6.4.31) is a function of x, y, z, and of certain coordinates representing the spin, depending parametrically on n, l, m_l, and m_s while ε is a function of the quantum numbers alone.

Following Pauli (1927), Slater then splits u as a product of a function of position (x,y,z) and spin (m_s)[9]:

$$u(n, l, m_l, m_s/x, y, z, m_s) \equiv u(n, l, m_l/x, y, z)u(m_s/m_s) \qquad (6.4.32)$$

Expressing the positional parameters in spherical coordinates r, θ, ϕ, through several passages, he arrives at

$$u(n, l, m_l/x, y, z) \equiv R(n, l/r)\Theta(l, m_l/\theta)\Phi(l, m_l/\varphi) \qquad (6.4.33)$$

$$u(n, l, m_l, m_s/x, y, z, m_s) = R(n, l/r)\Theta(l, m_l/\theta)\Phi(l, m_l/\varphi)\delta(m_s/m_s) \qquad (6.4.34)$$

Replacing then U in Eq. (6.4.31) with a Coulomb potential accounting for the central charge and the mutual electronic interactions in the form

$$V = \sum_{i=1}^{N}\left(-\frac{Ze^2}{r_i}\right) + \sum_{1}^{N}\frac{e^2}{r_{ij}} \quad [i > j] \qquad (6.4.35)$$

and integrating over the space and spin coordinates, i.e. $\int dv = \sum_{-1/2}^{1/2}(m_s)\int dxdydz$, it is then shown that the *I* integrals reduce to

$$I(n, l, m_l, m_s) = I(n, l) = -\int R^2(n, l/r)\left(V(r) + \frac{Ze^2}{r}\right)4\pi r^2 dr \qquad (6.4.36)$$

The *J* and *K* integrals are then shown to be independent of the spin coordinate, i.e.

$$J(n, l, m_l, m_s/n', l', m_1', m_s') \equiv J(n, l, m_l/n', l', m_l') \qquad (6.4.37.1)$$

---

[9] The notation $u(m_s/m_s)$ is meant to indicate that m_s appear both as spin and coordinate. Symbolically, this is represented by the δ function $\delta(m_s/m_s) = u(m_s/m_s)$ (see Slater, 1929; for a more extended explanation).

$$K(n,l,m_l,m_s/n',l',m_l',m_s') \equiv K(n,l,m_l/n',l',m_l') \qquad (6.4.37.2)$$

Using the expansion

$$\frac{1}{\mathrm{r(xx')}} = \sum(k,m)\frac{(k-|m|)!}{(k+|m|)!}\frac{\mathrm{r(a)}^k}{\mathrm{r(b)}^{k+1}}P_k^{|m|}(\cos\theta)P_k^{|m|}(\cos\theta')\exp[im(\varphi-\varphi')] \qquad (6.4.38)$$

where $P_k^{|m|}$ is an Associated Legendre's Function of degree k and order $|m|$ (see Par. A.15); r(xx') is the distance between (xyz) and (x'y'z'); and r(a), r(b) are, respectively, the smaller and the greater of r and r'. Slater (1929) shows then that the J and K integrals depend parametrically upon constant terms (a^k, b^k) depending each on the radial part of the wave function:

$$\mathrm{a}^k(l,m_l/l',m_l') = \frac{(2l+1)(l-|m_l|)!}{(l+|m_l|)!}\frac{(2l'+1)(l'-|m_l'|)!}{(l'+|m_l'|)!}\times$$
$$\int_0^\pi \left[P_l^{|m_l|}(\cos\theta)\right]^2[P_k^0(\cos\theta)]\frac{\sin\theta}{2}d\theta \times \int_0^\pi \left[P_{l'}^{|m_l'|}(\cos\theta')\right]^2[P_k^0(\cos\theta')]\frac{\sin\theta'}{2}d\theta' \qquad (6.4.39)$$

$$F^k(n,l/n',l') = (4\pi)^2e^2\int_0^\infty\int_0^\infty R^2(n,l/\mathrm{r})R^2(n',l'/\mathrm{r}')\frac{\mathrm{r(a)}^k}{\mathrm{r(b)}^{k+1}}\mathrm{r}^2\mathrm{r}'^2\mathrm{drdr}' \qquad (6.4.40)$$

$$J(n,l,m_l/n',l',m_l') = \sum(k)\mathrm{a}^k(l,m_l/l',m_l')F^k(n,l/n',l') \qquad (6.4.41)$$

$$G^k(n,l/n',l') = (4\pi)^2e^2\int_0^\infty\int_0^\infty R(n,l/\mathrm{r})R(n',l'/\mathrm{r})\mathrm{R(n,1/r')R(n',1'/r')}\frac{\mathrm{r(a)}^k}{\mathrm{r(b)}^{k+1}}\mathrm{r}^2\mathrm{r}'^2\mathrm{drdr}' \qquad (6.4.42)$$

$$K(n,l,m_l/n',l',m_l') = \sum(k)\mathrm{b}^k(l,m_l/l',m_l')G^k(n,l/n',l') \qquad (6.4.43)$$

While the G^k, F^k factors are characteristic of each external electron interacting with electrons in the internal (closed) shells (see later), a^k and b^k may be computed once forever. The integral I (Eq. 6.4.36) is needed only in finding the center of gravity of a multiplet system because it depends only on n and l, and so is the same for all the various degenerate states. In finding the energy differences between multiplets, one needs thus only the integrals J and K, which do not depend explicitly on the central field U.

In Tables 6.11 and 6.12 are listed the values computed by Slater (1929) for $l \leq 2$ (i.e. s, p, d electrons). All pairs, both of which are in closed shells, will give

Table 6.11 Values of $a^k(l,m_l/l',m'_l)$. Positive and negative values of m_l, m'_l are grouped under the symbol ± . In cases with two ± signs (pp, pd, and dd interactions) the two can be combined in any of the four possible ways (from Slater 1929)

Electrons	L	m_l	l'	m'_l	$k = 0$	$k = 2$	$k = 4$
Ss	0	0	0	0	1	0	0
Sp	0	0	1	±1	1	0	0
	0	0	1	0	1	0	0
Pp	1	±1	1	±1	1	1/25	0
	1	±1	1	0	1	−2/25	0
	1	0	1	0	1	4/25	0
Sd	0	0	2	±2	1	0	0
	0	0	2	±1	1	0	0
	0	0	2	0	1	0	0
Pd	1	±1	2	±2	1	2/35	0
	1	±1	2	±1	1	−1/35	0
	1	±1	2	0	1	−2/35	0
	1	0	2	±2	1	−4/35	0
	1	0	2	±1	1	2/35	0
	1	0	2	0	1	4/35	0
Dd	2	±2	2	±2	1	4/49	1/441
	2	±2	2	±1	1	−2/49	−4/441
	2	±2	2	0	1	−4/49	6/441
	2	±1	2	±1	1	1/49	16/441
	2	±1	2	0	1	2/49	−24/441
	2	0	2	0	1	4/49	36/441

identical contributions to each of the unperturbed terms. All sets of pairs, in which one is an outer electron and the other one of the electrons of a closed shell, will, when summed over the electrons of the closed shell, give the same result for each unperturbed state. That is, "*as far as multiplet separations are concerned, closed shells exert no influence at all; they affect only the position of the whole set of multiplet terms*" (Slater 1929).

6.5 Application to Correlation and Exchange Energy

Before showing some applications, it is important to stress again that all the terms of any multiplet have the same energy, moreover, the direct computation for a p electron interacting with the P shell shows that the interaction energy of a single electron with a closed shell is independent of the m_l of the outer electron

Table 6.12 Values of $b^k(l,m_l/l',m'_l)$. Symbols as in Table 6.10 (from Slater 1929)

Electrons	l	m_l	l'	m'_l	$k = 0$	$k = 1$	$k = 2$	$k = 3$	$k = 4$
Ss	0	0	0	0	1	0	0	0	0
Sp	0	0	1	±1	0	1/3	0	0	0
	0	0	1	0	0	1/3	0	0	0
Pp	1	±1	1	±1	1	0	1/25	0	0
	1	±1	1	0	0	0	3/25	0	0
	1	±1	1	±1	0	0	6/25	0	0
	1	0	1	0	1	0	4/25	0	0
Sd	0	0	2	±2	0	0	1/5	0	0
	0	0	2	±1	0	0	1/5	0	0
	0	0	2	0	0	0	1/5	0	0
Pd	1	±1	2	±2	0	2/5	0	3/245	0
	1	±1	2	±1	0	1/5	0	9/245	0
	1	±1	2	0	0	1/15	0	18/245	0
	1	±1	2	±1	0	0	0	30/245	0
	1	±1	2	±2	0	0	0	9/245	0
	1	0	2	±2	0	0	0	15/245	0
	1	0	2	±1	0	1/5	0	24/245	0
	1	0	2	0	0	4/15	0	27/245	0
Dd	2	±2	2	±2	1	0	4/49	0	1/441
	2	±2	2	±1	0	0	6/49	0	5/441
	2	±2	2	0	0	0	4/49	0	15/441
	2	±2	2	±1	0	0	0	0	35/441
	2	±2	2	±2	0	0	0	0	70/441
	2	±1	2	±1	1	0	1/49	0	16/441
	2	±1	2	0	0	0	1/49	0	30/441
	2	±1	2	±1	0	0	6/49	0	40/441
	2	0	2	0	1	0	4/49	0	36/441

(Slater 1929).[10] Consider in fact the case of a single 3p electron interacting with electrons in the completed (closed) K and L shells (n = 1 and 2, respectively, i.e. the $1s^22s^22p^63p^1$ configuration). The 3p electron may attain the degenerate states (3 1 1 ½), (3 1 0 ½), (3 1 −1 ½), (3 1 1 −½), (3 1 0 −½), (3 1 −1 −½). Grouping electrons with parallel and anti-parallel spins as in Slater's notation and detailing the electron stacking of the completed shells we have the set of (Table 6.13).

There are six wave functions corresponding to the six microstates. The diagonal terms of the energy in a determinant built on these functions must give the energy of the six degenerate microstates. Because the I integrals have no effect on the

[10] This is not restricted to p electrons but may be extended to the general case of a certain number of electrons outside closed shells, and the central closed shells.

Table 6.13 Set of quantum numbers for the $1s^2 2s^2 2p^6 3p^1$ configuration (Slater's 1929 notation)

μ_{state}		Σm_l	Σm_s
1	{ (1 0 0) (2 0 0) (2 1 1) (2 1 0) (2 1 − 1) (3 1 1)} { (1 0 0) (2 0 0) (2 1 1) (2 1 0) (2 1 − 1)}	+1	½
2	{ (1 0 0) (2 0 0) (2 1 1) (2 1 0) (2 1 − 1) (3 1 0)} { (1 0 0) (2 0 0) (2 1 1) (2 1 0) (2 1 − 1)}	0	½
3	{ (1 0 0) (2 0 0) (2 1 1) (2 1 0) (2 1 − 1) (3 1 − 1)} { (1 0 0) (2 0 0) (2 1 1) (2 1 0) (2 1 − 1)}	−1	½
4	{ (1 0 0) (2 0 0) (2 1 1) (2 1 0) (2 1 − 1)} { (1 0 0) (2 0 0) (2 1 1) (2 1 0) (2 1 − 1) (3 1 1)}	+1	−½
5	{ (1 0 0) (2 0 0) (2 1 1) (2 1 0) (2 1 − 1)} { (1 0 0) (2 0 0) (2 1 1) (2 1 0) (2 1 − 1) (3 1 0)}	0	−½
6	{ (1 0 0) (2 0 0) (2 1 1) (2 1 0) (2 1 − 1))} { (1 0 0) (2 0 0) (2 1 1) (2 1 0) (2 1 − 1) (3 1 − 1)}	−1	−½

multiplet separations and all J and K integrals arising from interactions of inner electrons in the K and L shells are ineffective as well, we may limit ourselves to evaluate interactions where the 3p electron is one of the members. Consider now a $n = 3$, $l = 1$, $m_l = 1$ electron (hereafter 311): the interaction with 1 s electrons will result in a 2 J(311/100) − K(311/100) correlation energy; that one with 2 s electrons will be 2 J(311/200) − K(311/200) correlation energy; finally, that one with 2p electrons will be 2 J(311/211) − K(311/211) + 2 J(311/210) − K(311/210) + 2 J(311/21–1) − K(311/21–1).

To render more explicit the calculus let me follow all steps.

6.5.1 3p → 1s Interaction

3p: $n = 3$, $l = 1$, $m_l = 1$.

1 s: $n = 1$, $l = 0$, $m_l = 0$.

$J(n,l,m_l/n',l',m'_l) \equiv J(311/100)$

$$J(n,l,m_l/n',l',m_l') = \sum (k)\mathrm{a}^k(l,m_l/l',m_l')F^k(n,l/n',l')$$

$$\mathrm{a}^k(l,m_l/l',m_l') \equiv \mathrm{a}^k(11/00)$$

$$F^k(n,l/n',l') \equiv F^k(31/10)$$

$$\sum_k \mathrm{a}^k(\mathrm{l},\mathrm{m_l}/\mathrm{l}',\mathrm{m_l'}) = 1(11/00) + 0(11/00) + 0(11/00)$$

$$\sum_k \mathrm{a}^k F^k = 1F^0(31/10)$$

$$K(n, l, m_l/n', l', m_l') = K(311/100) = \sum (k)\mathrm{b}^k(11/00)G^k(31/10)$$
$$= 0G^0 + \frac{1}{3}G^1 + 0G^2 + 0G^3 + 0G^4$$

$$2J(311/100) - K(311/000) = 2F^0(31/10) - \frac{1}{3}G^1(31/10)$$

It may be easily verified that the interaction integrals of an outer p electron with an s shell are the same whether the p electron has m ∼= 1 or 0 (or—1), as one immediately verifies, i.e.

$$2J(310/100) - K(310/000) = 2F^0(31/10) - \frac{1}{3}G^1(31/10)$$

$$2J(31-1/100) - K(31-1/000) = 2F^0(31/10) - \frac{1}{3}G^1(31/10)$$

The same result holds for the interaction with the 2p shell, which is a further proof of the degeneracy of the multiplets.

6.5.2 *Two Electrons Outside Closed Shells, One in an s State (Helium, Alkaline Earths)*

Their electronic stacking is resumed in Table 6.14.

In symbolizing the unperturbed wave function I do not adopt Slater's notation but I prefer to explicit the spins of the interacting electrons. The scheme is thus of the type $(ns)(n'p)$ (Table 6.15).

The microstates 1,2,3,10,11,12 with Σm_l, Σm_s = (11),(01),(−11),(1−1),(0. −1), (−1−1) are triply degenerate and belong to a ^{3}P state. All remaining terms are singly degenerate and belong to a ^{1}P state. For the energies we have

Table 6.14 Electronic stacking of He and alkaline earths

		1	2		3			4				5				6			
Z	Elements		s	p	s	P	d	s	p	d	f	s	p	d	f	s	p	d	f
2	He	2																	
12	Mg	2	2	6	2														
20	Ca	2	2	6	2	6		2											
28	Sr	2	2	6	2	6	10	2	6			2							
56	Ba	2	2	6	2	6	10	2	6	10		2	6			2			

Table 6.15 Set of quantum numbers for the (ns) (n'p) configuration

μ_{state}			Σm_l	Σm_s
1	$(n\ 0\ 0\ ½)$	$(n'\ 0\ 0\ ½)$	+1	+1
2	$(n\ 0\ 0\ ½)$	$(n'\ 1\ 0\ ½)$	0	+1
3	$(n\ 0\ 0\ ½)$	$(n'\ 1\ -1\ ½)$	−1	+1
4	$(n\ 0\ 0\ ½)$	$(n'\ 1\ 1\ -½)$	+1	0
5	$(n\ 0\ 0\ ½)$	$(n'\ 1\ 0\ -½)$	0	0
6	$(n\ 0\ 0\ ½)$	$(n'\ 1\ -1\ -½)$	−1	0
7	$(n\ 0\ 0\ -½)$	$(n'\ 1\ 1\ ½)$	+1	0
8	$(n\ 0\ 0\ -½)$	$(n'\ 1\ 0\ ½)$	0	0
9	$(n\ 0\ 0\ -½)$	$(n'\ 1\ -1\ ½)$	−1	0
10	$(n\ 0\ 0\ -½)$	$(n'\ 1\ 0\ -½)$	+1	−1
11	$(n\ 0\ 0\ -½)$	$(n'\ 1\ 0\ -½)$	0	−1
12	$(n\ 0\ 0\ -½)$	$(n'\ 1\ -1\ -½)$	−1	−1

^{1}P: $J(n\ 0\ 0/n'\ 1\ 1) + K(n\ 0\ 0/n'\ 1\ 1)$
^{3}P: $J(n\ 0\ 0/n'\ 1\ 1) - K(n\ 0\ 0/n'\ 1\ 1)$
The two states are thus separated by the $\pm K(n\ 0\ 0/n'\ 1\ 1)$ exchange term.

6.5.3 Shell of Equivalent P Electrons (Elements C, N, O, F, *Etc.*)

Their electronic stacking is resumed in Table 6.16.

p^2—This is the case discussed at Par. 6.2. There are 15 wave functions (Table 6.5). The states belong to ^{1}S, ^{1}D, ^{3}P (Table 6.6). The energy terms are as follows:

^{1}S: $F^0(n\ 1/n\ 1) + (10/25)F^2(n\ 1\ /n\ 1)$
^{1}D: $F^0(n\ 1/n\ 1) + (1/25)F^2(n\ 1\ /n\ 1)$

Table 6.16 Elements with equivalent outer shell p stacking

		1	2		3	
Z	Elements	s	s	p	s	p
6	C	2	2	2		
7	N	2	2	3		
8	O	2	2	4		
9	F	2	2	5		
14	Si	2	2	6	2	2
15	P	2	2	6	2	3
16	S	2	2	6	2	4
17	Cl	2	2	6	2	5

^{3}P: $F^0(n\ 1/n\ 1) - (5/25)F^2(n\ 1\ /n\ 1)$

The term $F^0(n\ 1/\ n\ 1)$ is positive, and it occurs in all the levels and does not affect thus the separation which reduces to $(6/25)F^2$ (^{1}D-^{3}P) and $(9/25)F^2$ (^{1}S-^{1}D). In accordance with Hund's rule, the terms of largest L and S lie lowest, i.e. ^{3}P is the lowest, then we have ^{1}D, and finally ^{1}S.

According to McLennan and Shaver (1924), for Si ($1s^22s^22p^63s^2$) the term values are ^{3}P = 65,615 cm^{-1}; ^{1}D = 59,466 cm^{-1}; ^{1}S = 50,370 cm^{-1}. This gives ^{3}P-^{1}D = 6149 cm^{-1} and ^{1}D-^{1}S = 9096 cm^{-1}. These values are consistent, respectively, with $F^2(3\ 1/\ 3\ 1)$ = 25, 602 cm^{-1} and $F^2(3\ 1/\ 3\ 1)$ = 25, 267 cm^{-1} in good agreement between themselves.

p^3—The multiplets are ^{4}S, ^{2}D, ^{3}P (Table 6.6). The separations are ^{4}S – ^{2}D = $(9/25)F^2(n\ 1/n\ 1)$ and ^{2}D – ^{2}P – = $(6/25)F^2(n\ 1/n\ 1)$. The comparison with the normal spectrum of N ($1s^22s^22p^3$) based on the observations of Compton and Boyce (1929) (^{4}S = 117,435; ^{2}D = 98,143; ^{2}P = 88,537) is not encouraging, though Hund's rule is apparently obeyed (^{4}S is the lowest and ^{2}P the highest energy level).

p^4—The multiplets are ^{3}P, ^{1}D, and ^{1}S. (Table 6.6). The separations (concerning energy) are as n p^3 (i.e. are ^{1}D – ^{1}S = $(9/25)F^2(n\ 1/n\ 1)$ and ^{3}P – ^{1}D = $(6/25)$ $F^2(n\ 1/n\ 1)$. An example of application is the normal spectrum of O ($1s^22s^22p^4$). Mc Lennan et al. (1928) found ^{1}D = 25,500 + ^{3}P and ^{1}S = 39,500 + 25,500 + ^{1}P. From the 3P – 1S split we should have 106,000 and from the 1D – 1S split 109,000 in very good agreement with Slater's method.

p^5 and p^6 yield only one multiplet (2P) and need no consideration.

6.5.4 Shell of Equivalent D Electrons (Iron Group)

Their electronic stacking is resumed in Table 6.17

d^2—The normal spectrum of Titanium shows that it does not obey Hund's rule, the electronic stacking being better explained as $1s^22s^22p^63s^23p^64s^23d^2$. The multiplets are thus ^{3}F, ^{3}P, ^{1}G, ^{1}D, ^{1}S (Table 6.6).[11] The separations depend on the two parameters $F^2(n\ 2/\ n\ 2)$ and $F^4(n\ 2/n\ 2)$, which bear no fixed relation to each other:

^{3}F – ^{3}P = $(135/441)F^2(n\ 2/n\ 2) - (75/441)F^4(n\ 2\ /n\ 2)$

^{3}F – ^{1}G = $(108/441)F^2(n\ 2/n\ 2) + (10/441)F^4(n\ 2\ /n\ 2)$

^{3}F – ^{1}D = $(45/441)F^2(n\ 2/n\ 2) + (45/441)F^4(n\ 2\ /n\ 2)$

^{3}F – ^{1}S = $(198/441)F^2(n\ 2/n\ 2) + (135/441)F^4(n\ 2\ /n\ 2)$

[11] In most atoms, the energies of the various orbitals increase in the order 1, 2s, 2p, 3s, 3p, 4s, 3d, 4p, 5s, 4d, 5p, 6s, 4f, 5d, 6p, 7s, 6d (Eyring et al., 1944).

Table 6.17 Electronic stacking of iron group elements

		1	2		3			4
Z	Elements		s	p	s	p	d	s
21	Sc	2	2	6	2	6	1	2
22	Ti	2	2	6	2	6	2	2
23	V	2	2	6	2	6	3	2
25	Mn	2	2	6	2	6	5	2
26	Fe	2	2	6	2	6	6	2
27	Co	2	2	6	2	6	7	2
28	Ni	2	2	6	2	6	8	2
30	Zn	2	2	6	2	6	10	2

Assuming provisionally F^2 to be half of F^4 (Slater 1929), the separations are in the ratio ^3F-^3P = 98; ^3F-^1D = 67; ^3F-^1G = 113; and ^3F-^1S = 165. The expected order of terms is ^3F, ^1D, ^3P, ^1G, ^1S as experimentally observed (Russell 1927).

d^3—d^7—d^8: These, together with d^2, are the only cases where there are significant separations in the multiplets of the highest multiplicity (^3F-^3P for d^2 and d^8 and ^4F-^4P for d^3 and d^7). In practice, these separations are given by the same formula, $(135/441)F^2(n\,2/n\,2) - (75/441)\,F^4(n\,2/n\,2)$. Experimentally, the observed separation is about 8000 for Ti (d^2), 9500 for V(d^3), 14,000 for Co (d^7), and 15,000 for Ni (d^8), i.e. a fairly uniform increase of about 1000 to 1500 for the addition of one d electron (Slater 1929).

Chapter 7
Molecules

In these short notes we encounter first the conceptually simplest molecule, the H_2^+ ion, i.e. a single electron rotating about two nuclear centers, and we move then toward a more complex case: the H_2 molecule. After the perturbation treatments of Heisenberg (1926, 1927a) and Dirac (1927) on the degeneracy associated to the permutation of two electrons, the problem of treating two electrons moving in an electrical field generated by two nuclear centers (and by themselves) was successfully afforded by Heitler and London (1927). The Heitler–London treatment of the H_2 molecule is of paramount importance in the history of Quantum Chemistry. This was realized from the very beginning. Already in 1929 Van Vleck wrote "*Is it too optimistic to hazard the opinion that this is perhaps the beginnings of a science of mathematical chemistry in which chemical heats of reaction are calculated by quantum mechanics just as are the spectroscopic frequencies of the physicist?*" (Van Vleck 1929). In the treatment we encounter some essential tools and concepts, such as the Born–Hoppenheimer approximation, the nature of the vibrational and rotational motion of diatomic molecules and the Morse function. Moving toward more complex cases we afford then some anticipations on another homo-nuclear diatomic molecule (O_2), then a linear polyatomic molecule (CO_2), and a non-planar molecule such as di-hydrogen monoxide (H_2O). These last three molecules have a great importance in Geochemistry and are treated more extensively in the application sections. Finally, the complexities are discussed arising when molecules are immerged in a reaction field or are depicted as structural parts of crystalline aggregates, able to rotate and vibrate in the medium. Electromagnetic and relativistic effects are disregarded in the whole chapter.

G. A. Ottonello, *Quantum Geochemistry*, Springer Geochemistry,
https://doi.org/10.1007/978-3-031-21837-8_7

7.1 Premises

The nomenclature principle that I will follow assigns to atomic properties Roman letters and to molecular properties their Greek counterpart. Thus, while for an atom the quantized angular momentum l (0, 1, 2,…,$n-1$) is represented by the symbols s, p, d,…; the component along the internuclear axis of a diatomic molecule has symbols σ, π, δ,…

When the two separated atoms that are going to compose the diatomic molecule approach each other, the orbits of their electrons are affected by the fact that an axial direction is imposed. Consider now the three p orbitals p_x, p_y, p_z. While they are identical for an isolated atom (spherical symmetry), the orbit projecting along the internuclear axial direction has zero angular momentum and is labeled as σp to distinguish it from those ones elongated on other two directions. These are energetically non-equivalent in the cylindrical field and are denoted πp orbitals ($l = 1$ along the internuclear axis). The electronic state of a diatomic molecule is characterized by the net orbital angular momentum Λ along the internuclear axis, and this is easily computed by summing up the angular components of the contributing AOs, to σ, π, δ, … ($l = 0$, 1, 2,…). It is thus denoted by the corresponding upper case Greek letters (Σ, Π, Δ,…). The net spin angular momentum (along the same axis) is indicated by a superscript preceding the Greek letter in terms of multiplicity (i.e. ^{1} for a singlet, ^{3} for a triplet, and so on). If the diatomic molecule is in a Σ state, the molecular orbitals of this particular configuration are either symmetric (Σ^+) or antisymmetric (Σ^-) with respect to a plane orthogonal to the internuclear axis. This may happen when the angular momentum vectors of π or δ molecular orbitals are opposed in the →← or the ←→ sense (Barrow 1962). The net symmetry of the molecule (even = gerade = g) or odd (ungerade = u) is indicated as a subscript. I will return again on the significance of this nomenclature at Sect. (7.2.1).

7.2 The Hydrogen Molecular Ion

The simplest molecule one may conceive is the molecular ion H_2^+: a single electron moving in the field of two homo-nuclear centers of charge 1. The motion of a particle in the field of two centers, each attracting it with an inverse-square law may be afforded by separating the variables. The problem was first solved by Burrau (1927) by numerical integration of the wave equation recurring to prolate spherical coordinates (see Sect. A.2.2). Denoting respectively r_A and r_B the distance of the electron from the A and B nuclear centers and by R_{AB} the distance between the centers themselves, the variables λ, μ relating the position of the electron with respect to the distance between the foci, i.e.

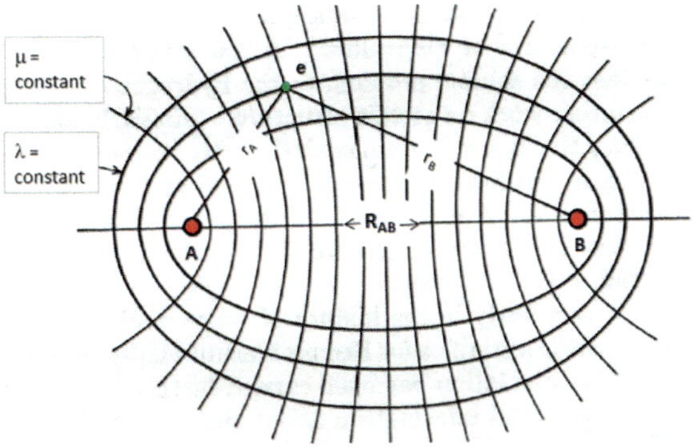

Fig. 7.1 The two-center problem in prolate-spheroidal coordinates

$$\lambda = \frac{r_A + r_B}{R_{AB}} \tag{7.2.1}$$

$$\mu = \frac{r_A - r_B}{R_{AB}} \tag{7.2.2}$$

result to be respectively constant on the ellipses and on the hyperbolas (Fig. 7.1). λ may attain values comprised between 1 and +∞ while μ attains values comprised between −1 and +1. Denoting as ϕ the rotational angle about the axis joining the foci, the system is completely determined.

Burrau's treatment begins with the Schrödinger equation written in the usual form for Cartesian coordinates

$$\frac{\partial^2\psi}{\partial x^2} + \frac{\partial^2\psi}{\partial x^2} + \frac{\partial^2\psi}{\partial x^2} + \frac{8\pi^2 m}{h^2}\left(E + \frac{e^2}{r_1} + \frac{e^2}{r_2}\right)\psi = 0 \tag{7.2.3}$$

With the new coordinates the equation is transformed in such a way to depend upon the two-electron-nuclei distances and the azimuthal angle:

$$\begin{aligned} &\frac{\partial^2\psi}{\partial\phi^2} + \frac{\partial^2\psi}{\partial r_1^2} + \frac{2}{r_1}\frac{\partial\psi}{\partial r_1} + \frac{r_1^2 + r_2^2 - 4c^2}{r_1 r_2}\frac{\partial^2\psi}{\partial r_1 \partial r_2} \\ &+ \frac{\partial^2\psi}{\partial r_2^2} + \frac{2}{r_2}\frac{\partial\psi}{\partial r_2} + \left(\frac{\beta}{r_1} + \frac{\beta}{r_2} - \alpha\right)\psi = 0 \end{aligned} \tag{7.2.4}$$

with $r_1 = r_A$; $r_2 = r_B$; $c = R_{AB}/2$; $\beta = \frac{8\pi^2 m_0 e^2}{h^2} \equiv \frac{2}{a_0}$; $\alpha = \frac{\beta^2}{4}\frac{E}{E_H} \equiv \frac{1}{a_0^2}\frac{E}{E_H}$;

$E_H = -\frac{2\pi^2 m_0 e^4}{h^2}$

Through variable separation in λ and μ, Burrau then obtains (his Eq. 7.2.4):

$$
\begin{aligned}
&\left(\lambda^2-1\right)\frac{\partial^2\psi}{\partial\lambda^2}-2\lambda\frac{\partial\psi}{\partial\lambda}+\left(2\varepsilon\lambda-\gamma\lambda^2\right)\psi\\
&\quad+\left(1-\mu^2\right)\frac{\partial^2\psi}{\partial\mu^2}-2\mu\frac{\partial\psi}{\partial\mu}+\gamma\mu^2\psi+c^2\left(\lambda^2-\mu^2\right)\frac{\partial^2\psi}{\partial\phi^2}=0
\end{aligned}
\tag{7.2.5}
$$

$\varepsilon = \beta c = \frac{2c}{a_0}$ corresponds to the distance between the nuclei when expressed in terms of units of a_0 (Bohr's radius $a_0 = \dfrac{h^2}{4\pi^2 m_0 e^2} = 0.52917725\text{Å}$); $\gamma = \alpha c^2$; $E = \dfrac{4\gamma}{\varepsilon^2} E_H$.

Through few passages Burrau (1927) arrives then to determine a "*separation constant*" (A) through the ratios of exact differentials:

$$
\frac{d\sigma_\lambda}{d\lambda} = \sigma_\lambda^2 - \frac{\gamma\lambda^2 - 2(\varepsilon - \sigma_\lambda)\lambda - A}{\lambda^2 - 1} \tag{7.2.6.1}
$$

$$
\frac{d\sigma_\mu}{d\mu} = \sigma_\mu^2 + \frac{\gamma\mu^2 + 2\sigma_\mu\mu - A}{1 - \mu^2} \tag{7.2.6.2}
$$

with $\sigma_\lambda = \frac{1}{\psi_\lambda}\frac{\partial\psi_\lambda}{\partial\lambda}$; and $\sigma_\mu = \frac{1}{\psi_\mu}\frac{\partial\psi_\mu}{\partial\mu}$.

Through numerical treatment of the two Eqs. (7.2.6.1, 7.2.6.2), γ and E are then solved as a function of internuclear distance. Table 7.1 lists the estimates of Burrau (1927) for different values of internuclear distance ε. From γ the energy ratio E/E_H is then obtained by applying $\frac{E}{E_H} = \frac{4\gamma}{\varepsilon^2}$ (3rd column). By adding the potential energy of the two interacting nuclei $E_{nuclei} = \frac{2}{\varepsilon} E_H$ Burrau obtains finally the total Energy $\frac{E_{total}}{E_H} = \left(\frac{4\gamma}{\varepsilon^2} - \frac{2}{\varepsilon}\right)$ (4th column[1]).

The E_{total}/E_H energy ratio depicts a potential well with the minimum localized at $\varepsilon = 2.0 \pm 0.1$ and $E_{total}/E_H = 1.204 \pm 0.002$ (Fig. 7.2). Posing $E_H = 13.53$ *eV* the estimated total energy at minimum is 16.29 *eV*. By adding the Zero-Point Correction (E_{ZPC}) on the basis of the single frequency:

$$
\nu = \frac{1}{2\pi}\sqrt{\frac{1}{2m}\frac{d^2E}{d\varepsilon^2}} \tag{7.2.7}
$$

$$
E_{ZPC} = \frac{1}{2} h\nu = 0.07 eV \tag{7.2.8}
$$

Burrau (1927) gets $E_{minimum} = 16.22 \pm 0.03$ *eV*.

[1] Note that in the following I will change sign to the original data.

Table 7.1 Energy estimates for H_2^+ at different values of internuclear distance ε according to Burrau (1927). Energies of interacting nuclei are in Rydberg's units (*Ry*)

ε	γ	E/E_H	E_{nuclei}	E_{total}/E_H
0.00	0.000	4.000	–	–
1.00	0.724	2.896	2.000	0.896
1.30	1.119	2.648	1.538	1.110
1.60	1.559	2.436	1.250	1.186
1.80	1.870	2.309	1.111	1.198
2.00	2.204	2.204	1.000	1.204
2.20	2.552	2.109	0.909	1.200
2.40	2.917	2.025	0.833	1.192
2.95	3.995	1.836	0.678	1.158
∞	–	1.000	0.000	1.000

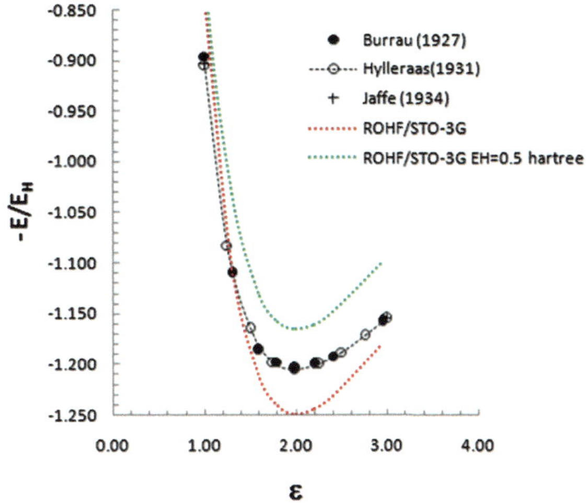

Fig. 7.2 Potential well of H_2^+ according to Burrau (1927). Estimates obtained with similar procedures by Hylleraas (1931) and Jaffe (1934) are superimposed for comparative purposes. The two dotted lines refer to calculations obtained with a variational procedure adopting LCAO and Slater-Type Orbitals (STO) reproduced by 3 primitive Gaussians

The procedure followed by Hylleraas (1931) and Jaffé (1934) is based on the solution of a Hamiltonian function $\mathsf{H}(\mu)$ in terms of associated Legendre's functions $P_l^{|m|}$ (see Sect. A.15):

$$\mathsf{H}(\mu) = \sum_{l=|m|}^{\infty} c_l P_l^{|m|}(\mu) \tag{7.2.9}$$

where c_l are constant coefficients of the azimuthal quantum number. Because the variable μ spans the range $-1 \leq \mu \leq 1$ which is the range covered by the argument $z = r \times \sin\phi$ and we are interested in the normal states of the system with $m = 0$ and l even, the determinantal equation reduces to

$$\begin{vmatrix} \frac{1}{3}C-\mu & \frac{2}{15}C & 0 & 0 & \dots \\ \frac{2}{3}C & \frac{11}{21}C-6-\mu & \frac{4}{21}C & 0 & \dots \\ 0 & \frac{12}{35}C & \frac{39}{77}C-20-\mu & \dots & \dots \\ 0 & 0 & 0 & \dots & \dots \\ \dots & \dots & \dots & \dots & \dots \end{vmatrix} = 0 \quad (7.2.10)$$

where C = −0.25*E* × R_{AB} and the only non-vanishing terms are the principal diagonal and the immediately adjacent diagonals. Neglecting the adjacent diagonals, the roots of the equation are obviously $\mu=\frac{1}{3}C$; $\mu=\frac{11}{21}C-6$; $\mu=\frac{39}{70}C-20$; and so on. Focusing the attention on the first root, including the first two non-diagonal terms, and replacing μ in the second diagonal term by $\mu=\frac{1}{3}C$, the determinant (7.2.10) transforms into

$$\begin{vmatrix} \frac{1}{3}C-\mu & \frac{2}{15}C \\ \frac{2}{3}C & \frac{11}{21}C-6-\frac{1}{3}C \end{vmatrix} = 0 \quad (7.2.11)$$

with a 3rd degree solution $\mu={}^{1}\!/_{3}C+{}^{2}\!/_{135}C^{2}+{}^{4}\!/_{8505}C^{3}$(Pauling and Bright Wilson 1935) or a 5th degree solution $\mu={}^{1}\!/_{3}C+{}^{2}\!/_{135}C^{2}+{}^{4}\!/_{8505}C^{3}-0.000013\,C^{4}-0.0000028C^{5}$ (Hylleraas 1931). Denoting the energy of the electron in the field of the two nuclei as

$$E'=E-\frac{e^{2}}{R_{AB}} \quad (7.2.11)$$

and considering that the ellipsoidal coordinates are related by

$$\frac{e^{2}}{r_{A}}+\frac{e^{2}}{r_{B}}=\frac{4e^{2}\lambda}{(\lambda^{2}-\mu^{2})R_{AB}} \quad (7.2.12)$$

One proceeds exploiting the functional dependency of μ on C for the normal state, eliminate μ and obtains the characteristic value of C and hence the energy as a function of R_{AB} (see Hylleraas 1931; Jaffé 1934).

7.2.1 Energy Levels and Wave Functions

As already seen when discussing the effect of rotational symmetry on the wave function of the hydrogen atom (Sect. 5.2.2), also in the case of a diatomic molecule the dependence of the potential energy on the rotational angle may be expressed in terms of the magnetic quantum number m_l through the function $\exp(im_l\phi)$ (see also Eq. 5.2.16). The system of two nuclei exerts torques which modify the total angular momentum of the electron, however, the component along the axis joining the nuclear centers is constant in classical mechanics (hence quantized in quantum applications). We have already seen that being single-valued at the cyclic points $\phi = 0$, 2π, m_l must be an integer. The solutions are $m_l = 0, \pm 1, \pm 2, \ldots$.(doubly degenerate besides $m_l = 0$). Band spectroscopists denote the states with $m_l = 0$ as σ (rotational symmetry about the axis), those with $m_l = \pm 1$ as π states, and those with $m_l = \pm 2$ as δ states. Performing an inversion about the midpoint of the axis connecting the foci, the solution of the wave equation is unchanged or merely changes sign. The functions which do not change on inversion are said to be "*even*" are denoted by the symbol *g* (for the german equivalent "*gerade*") while those ones changing sign are said to be "*odd*" and denoted by the symbol *u* (for "*ungerade*"). We can thus have σ_g, σ_u, π_g, π_u, δ_g, δ_u functions, and so on.

Separating the variables with the aid of prolate-spheroidal coordinate, the wave function can be seen as a product of two functions depending respectively on λ and μ:

$$u = L(\lambda)M(\mu)\exp(im_l\phi) \tag{7.2.13}$$

By changing the sign of μ and rotating ϕ by 180° (π radiants) the exponential factor $\exp(im_l\phi)$may remain unchanged or change sign. This means that $M(\mu)$ is an even or an odd function of μ (see note[2]). This property holds for any homo-nuclear molecule and not just for H_2^+. Consider now in detail Fig. 7.1. When $\lambda = 1$ we are on the axis joining the two foci and $r_A + r_B = R_{AB}$. Moreover, μ varies linearly with r_A or r_B. More specifically, when $r_A = R_{AB}$ (and $r_B = 0$) $\mu = 1$ and, when $r_A = 0$ (and $r_B = R_{AB}$) $\mu = 0$. If we are dealing with σ states, the exponential factor σ $\exp(im_l\phi)$equals 1 and the wave function, which can be denoted as $L(1)M(\mu)$, shows the behavior of the M function Let me now denote as z the distance along the axis connecting the foci when expressed in atomic units of a_0 (the Bohr's hydrogen radius). For $R_{AB} = 2a_0$ the point at $z = -1$ corresponds to $\mu = -1$ and the point at $z = 1$ to $\mu = 1$. The function $M(\mu)$ between the two limiting points attains the values shown in Fig. 7.3.

On the right of the right-hand nucleus we have $r_A - r_B = R_{AB} \Rightarrow \mu = 1$; $r_A + r_B = R_{AB} + 2r_B \Rightarrow \lambda = 1 + 2r_B/R_{AB}$. λ is thus a linear function of z, and varies from 1 on the B focus to infinity at $z = \infty$. Hence, along this part of the path the wave function may be written as $L(\lambda)M(1)$ and gives specifically the picture of L. The two peaks located at the foci represent the 1s wave function [exp(-r)] times a normalizing factor. Because

$$r_A = \frac{R_{AB}}{2}(\lambda + \mu) \tag{7.2.14.1}$$

$$r_B = \frac{R_{AB}}{2}(\lambda - \mu) \tag{7.2.14.2}$$

---

[2] It is even for the states σ_g, π_u, δ_g and odd for the states σ_u, π_g, δ_u.

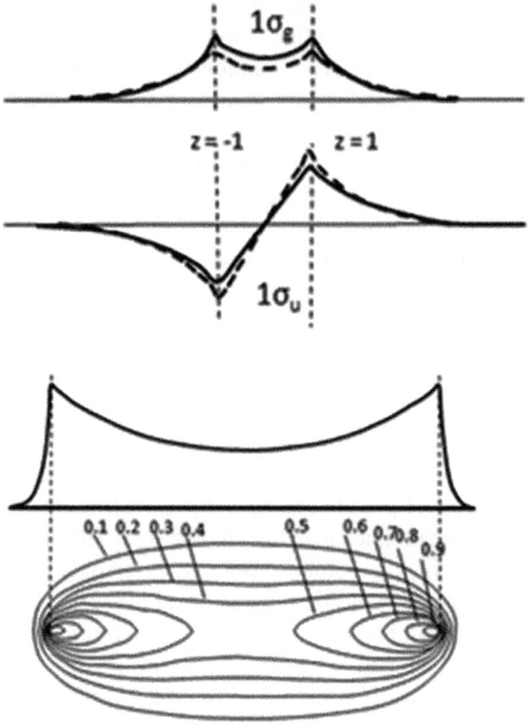

Fig. 7.3 Upper part: Wave functions $1\sigma_g$ and $1\sigma_u$ of H_2^+ along the internuclear axis in units of a_0. The results of the LCAO approximation (dashed lines) are compared with the exact values (solid lines; from Slater 1963; with modifications). Lower part: The electron distribution function for the hydrogen molecular ion according to Burrau (1927). The isodensity contour lines in the lower part are drawn multiplying by 0.9, 0.8,.... 0.1 the maximum value (set to one). The projection of the values at the various distances along the axis which connects the foci is not much dissimilar in the two cases (the distance between the foci in the upper and lower figure do not correspond and the comparison is made only for illustrative purposes)

for large values of R_{AB} the wave function approaches the values

$$1\sigma_g = \exp(-r_A) + \exp(-r_B) = 2\cosh\left(\frac{R_{AB}\mu}{2}\right)\exp\left(\frac{-R_{AB}\lambda}{2}\right) \quad (7.2.15.1)$$

$$1\sigma_u = -\exp(-r_A) - \exp(-r_B) = 2\sinh\left(\frac{R_{AB}\mu}{2}\right)\exp\left(\frac{-R_{AB}\lambda}{2}\right) \quad (7.2.15.2)$$

The function $L(\lambda)$ is in this case $\exp(-\lambda R_{AB}/2)$ and the function $M(\mu)$ is either $2\cosh(-\mu R_{AB}/2)$ or $2\sinh(-\mu R_{AB}/2)$, for gerade (*g*) or ungerade (*u*) states, respectively.

Functions (7.2.15.1, 7.2.15.2) are plotted in Fig. 7.4 for various values of the R_{AB} internuclear distance. As we may see at large distance there are two peaks

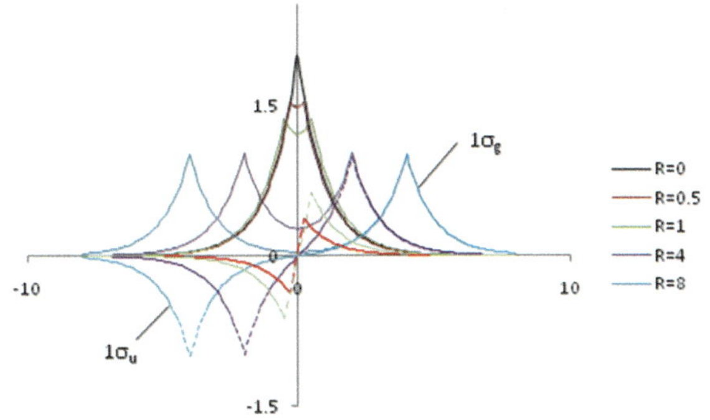

Fig. 7.4 Un-normalized wave functions as simple summation of AO for different values of the R_{AB} internuclear distance

centered on the nuclear foci, however for the $1\sigma_g$ molecular orbital a single peak is observed when the internuclear distance falls to zero. Disregarding the normalization factor[3], this is the condition attained by the He^+ ion (a nucleus of charge 2 with a single electron completely localized on it).

To better understand the significance of what we have seen for the $1\sigma_g$ and $1\sigma_u$ wave functions we may plot a diagram of the potential energy without including the internuclear repulsion and expanding the analysis to other excited states (Fig. 7.5). We may thus see that, when expressed in *Ry*, the energy converges at R = 0 toward $-4/1^2 = -4$; $-4/2^2 = -1$; $-4/3^2 = -0.444$; $-4/4^2 = -0.25$

These are obviously the values corresponding to a nucleus of charge 2 (i.e. He^+), because an electron moving around a nucleus of charge Z has energy levels corresponding to $-Z^2/n^2$ *Ry*. As R approaches infinity the energy levels attain again –Z but here the charge is evidently 1 and the limits correspond to one or another of the levels of the hydrogen atom (i.e. -1; $-1/4$; $-1/9$; …).

Prolate ellipsoidal coordinates are useful also in estimating the exact potential energy of the hydrogen molecular ion in excited states. Stemming from Eq. (7.2.13) it has been shown by Baber and Hassé (1935) that Eq. (7.2.6.1) may be expanded in the form

$$\frac{d}{d\lambda}\left[\left(\lambda^2-1\right)\frac{dL}{d\lambda}\right]+\left(A+2R\lambda-p^2\lambda^2-\frac{m^2}{\lambda^2-1}\right)L=0 \tag{7.2.16}$$

where

$$p^2=-\frac{1}{4}R^2E \tag{7.2.17}$$

With the distance R expressed in units of a_0 and E in *Ry*. $L(\lambda)$ may be then expanded as shown by Jaffé (1934):

[3] To get the normalization constant we must make the square of the wave function and integrate over all space. A quick account of the procedure is given elsewhere.

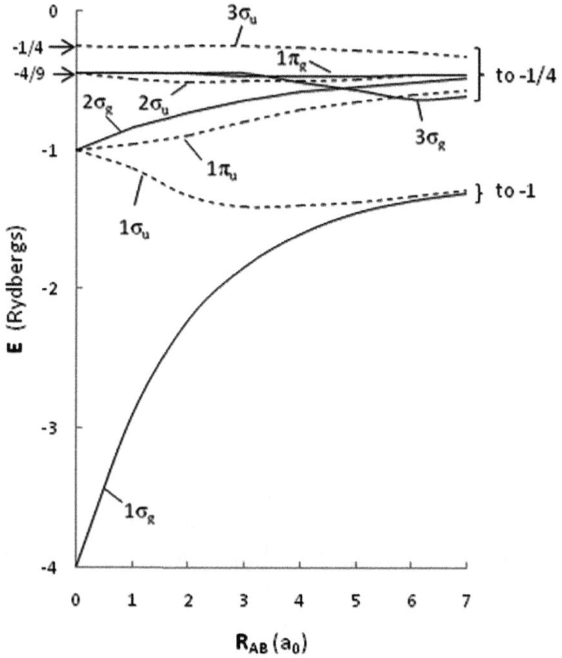

Fig. 7.5 Energy levels of H_2^+ as a function of internuclear distance when internuclear repulsion is omitted. Modified from Slater (1963)

$$L(\lambda) = \left(\lambda^2 - 1\right)^{m/2} (\lambda + 1)^{\sigma} \exp(-p\lambda) y(\zeta) \qquad (7.2.18)$$

with

$$\sigma = \frac{R}{p} - m - 1 \qquad (7.2.19)$$

and

$$\zeta = \frac{\lambda - 1}{\lambda + 1} \qquad (7.2.20)$$

On the basis of the above equations Bates et al. (1953) computed the exact potential energies for H_2^+ in various excited states corresponding to the electronic states

Table 7.2 exact potential energies (E + 2/R) for some excited states of the H_2^+ molecule (Bates et al. 1953). Distances are in atomic units (a_0) units and energies are in *Ry*

R	$2\sigma_g$	$2\sigma_u$	$3\sigma_g$	$4\sigma_u$
1	+1.154	+1.521	+1.550	+1.749
2	+0.278	+0.480	+0.528	+0.747
3	+0.029	+0.160	+0.152	+0.409
4	−0.077	+0.010	−0.071	+0.235
6	−0.162	−0.122	−0.292	+0.041
8	−0.194	−0.175	−0.347	−0.085
10	−0.209	−0.201	−0.346	−0.170
∞	−0.250	−0.250	−0.250	−0.250

$$m = 0 \quad \begin{cases} l = 0; & n = 1, 2, 3 \\ l = 1; & n = 2, 3, 4 \\ l = 2; & n = 3 \\ l = 3; & n = 4 \end{cases}$$

$$m = 1 \quad \begin{cases} l = 1; & n = 2 \\ l = 2; & n = 3 \end{cases}$$

Exact potential energies for the $2\sigma_g$, (unite atom notation corresponding to $m = 0; l = 0; n = 2$), $3\sigma_u$, ($m = 0; l = 1; n = 3$), $3\sigma_g$ ($m = 0; l = 2; n = 3$), and $4\sigma_u$($m = 0; l = 3; n = 4$) according to Bates et al. (1953) are listed in Table 7.2.

The contour diagrams for some excited states of H_2^+ according to Bates et al. (1953) are drawn in Fig. 7.6. The analogy with 1s, 2p, and 3d degenerate atomic orbitals is obvious when comparing Figs. 7.6 and 6.2.

The procedure of writing "*Molecular Orbitals*" (MO) as a simple summation of functions of the type $\exp(-r_A)$, $\exp(-r_B)$ deemed to represent atomic wave

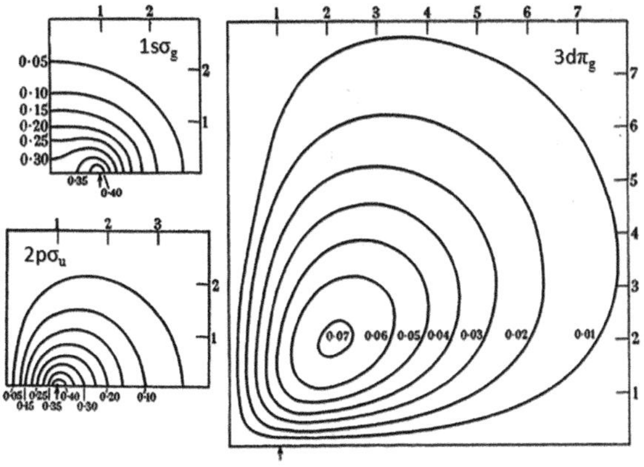

Fig. 7.6 Contour diagrams of normalized wave functions for some excited states of the H_2^+ molecule (Modified from Bates et al. 1953). Arrows mark the position of one of the nuclear centers. The z-axis is along the internuclear line. The original unite atom notation of Bates et al. (1953) is adopted in identifying the electronic states

functions and denoted as "*Atomic Orbitals*" (AO) (known as LCAO or "*Linear Combination of Atomic Orbitals*") is only approximate at best, especially at a relatively short distance between the nuclear centers.

Bates et al. (1953) in comparing the results of exact calculations [obtained developing Eqs. (7.2.6.1, 7.2.6.2)] with the LCAO procedure, adopted, for the wave functions of the H_2^+ molecule in the ground and excited states, the approximations:

$$\Psi_g = \frac{\exp(-r_A) + \exp(-r_B)}{[2\pi(1+S)]^{1/2}} \tag{7.2.21.1}$$

$$\Psi_u = \frac{\exp(-r_A) - \exp(-r_B)}{[2\pi(1-S)]^{1/2}} \tag{7.2.21.2}$$

Differently from (7.2.15.1, 7.2.15.2), in Eqs. (7.2.21.1, 7.2.21.2) appears at the denominator of the superposition the integral S, which considers the effect of the space volume common to both atomic orbitals:

$$S = \int d^3r\psi_a(r_a)\psi_b(r_b) = \left(1 + R_{AB} + \frac{1}{3}R_{AB}^2\right)\exp(-R_{AB}) \tag{7.2.22}$$

As we may see in Table 7.3, the agreement is moderately good only at a high distance.

The lowest energy levels of H_2^+ as a function of internuclear distance (comprehensive of internuclear repulsion) are drawn in Fig. 7.7. The analogy with the potential well of Fig. 7.2 is evident. The nuclear repulsion cause all energy levels to go to infinity as R_{AB} approaches zero. Moreover, as we may see, only the $1\sigma_g$ state and, though less evidently, the $3\sigma_g$ state conform to a potential well. We may conceive these wells as due to the presence of an electronic cloud attracting the two nuclei and preventing nuclear repulsion through shielding the positive charges. This is obviously prodromal to the concept itself of chemical bond. Also, in the

Table 7.3 exact and LCAO wave functions $1\sigma_g$, $2\sigma_u$ for the hydrogen molecular ion according to Bates et al. (1953). *R* is the internuclear separation. Distances are in atomic units (a_0)

R	R_{AB}	Ψ	exact	LCAO	R_{AB}	Ψ	Exact	LCAO
0	2	$1\sigma_g$	0.315	0.233	4	$1\sigma_g$	0.127	0.099
1			0.458	0.360			0.175	0.153
2			0.120	0.132			0.378	0.373
3			0.030	0.049			0.123	0.137
4			0.007	0.018			0.039	0.050
0	2	$2\sigma_u$	0.000	0.000	4	$2\sigma_u$	0.000	0.000
1			0.453	0.536			0.146	0.141
2			0.199	0.197			0.436	0.435
3			0.078	0.073			0.158	0.160
4			0.029	0.027			0.055	0.059

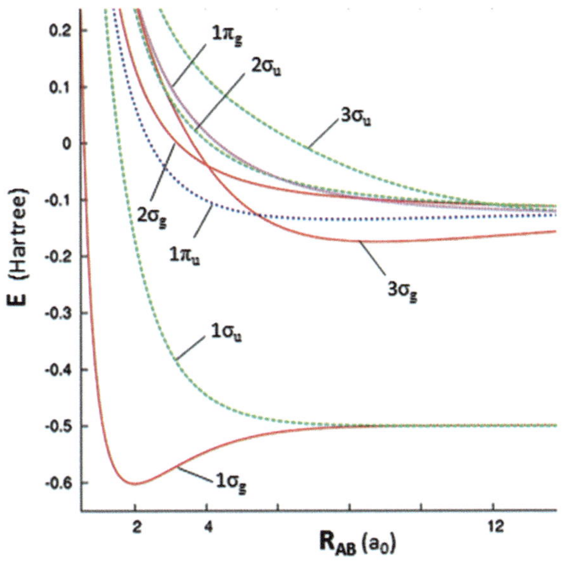

Fig. 7.7 Energy levels of H_2^+ as a function of internuclear distance when internuclear repulsion is taken into account. Modified from Slater (1963)

conceptual limit that assigns to the nuclear centers a fixed position corresponding to a peculiar R_{AB} distance (leading to a certain energy state), we may consider the energy levels leading to an attractive state as "*potential energy curves which limit and control the motion of the nuclei within the well*". This rather rough statement is a prerequisite for the introduction of the Born–Hoppenheimer approximation (Born and Hoppenheimer 1927), necessary to unravel the (separate) contributes of the electronic and nuclear motion to the stability of the molecule.

7.2.2 The Separation of Nuclear and Electronic Motion (Born–Hoppenheimer Approximation)

Imagine now to separate the two nuclei from the r_0 condition corresponding to the minimum of the potential well, moving them progressively at higher R_{AB} distance. If the rate of increase of R_{AB} is slow enough with respect to electronic motion, the electrons will rearrange themselves at each distance and will conform their movement as if the two nuclear centers were immobile at that particular interionic distance. To pull apart the two nuclei we must make a work against the force that would put them precisely at the bottom of the well. The force we must apply is precisely dE/dR and the force times the displacement [(dE/dR) $\times$ dR] is the increase of energy experienced by the system. By releasing the two nuclei they will be attracted by the electronic cloud between them toward the minimum and will be respectively repelled when nuclear repulsion is predominant. By stating that the electrons "*rearrange themselves*" at all distances with respect to the nuclear position it is meant that we consider their movements extremely fast with respect to the movements of the nuclei along the axis chosen to represent the interionic distance. The energy E(R) is then the potential energy of nuclear motion and contains (implicitly) also the kinetic energy of the electrons (itself a function of R). Denoting X_i the (X,Y,Z) coordinates of the nuclei and as x_j those of the electrons, the coulomb interaction between all pairs of charges (nuclei and electrons) may be denoted as V (X_i,x_j) and the Schrödinger equation for the system is then

$$\left[-\sum_i \frac{\hbar}{2M_i}\nabla_i^2 - \sum_j \frac{\hbar}{2m_0}\nabla_j^2 + V(X_i, x_j)\right]\psi(X_i, x_j) = \varepsilon\psi(X_i, x_j) \qquad (7.2.23)$$

where ε is the energy, M_i are the nuclear masses and m_0 is the mass of the electron at rest (relativistic effects disregarded). The Laplacians ∇_i^2, and ∇_j^2 are the kinetic energy operators of nuclei and electrons. If we held constant the position of the nuclei, the Laplacian operator ∇_i^2 reduces to zero and the Schrödinger equation becomes

$$\left[\sum_j \frac{\hbar}{2m_0}\nabla_j^2 + V(X_i, x_j)\right]u(X_i, x_j) = E(X_i)u(X_i, x_j) \qquad (7.2.24)$$

The wave function $u(X_i, x_j)$is different from $\psi(X_i, x_j)$ because it does not satisfy the same Schrödinger equation. Examples of $u(X_i, x_j)$and $E(X_i)$ for the H_2^+ molecule are those seen in Figs. 7.4 and 7.7, respectively. For this simple diatomic molecule, the orientation in space is immaterial, hence we may state that $E(X_i) \equiv E(R_{AB})$. If we use $E(X_i)$as a potential energy function that governs the nuclear motion, the Schrödinger equation becomes finally:

$$\left[\sum_j \frac{\hbar}{2m_0}\nabla_j^2 + E(X_i)\right]v(X_i) = \varepsilon v(X_i) \qquad (7.2.25)$$

The wave function $v(X_i)$is only a function of the nuclear positions and the energy ε does not depend on any parameter. Moreover ε, arising from the equality (7.2.25) is a good approximation of the exact Schrödinger equation (7.2.23). It may be also shown that the function of the exact problem is represented by the product[4]):

$$u(X_i, x_j)v(X_i) = \psi(X_i, x_j) \qquad (7.2.26)$$

[4] See Appendix 2 in Slater, (1963); for a proof of this theorem.

Electrostatics tells us that the force acting on each nucleus is a net effect of the attraction between the nuclear charge and the electron cloud and of the mutual repulsion between neighboring nuclei. This force may be obtained (and actually it is, as we will see later on) by a partial derivative of the potential energy for nuclear motion $E(X_i)$. However, $E(X_i)$ is itself obtained by solving Eq. 7.2.25, which includes the potential plus the kinetic energy of the electrons. Moreover, the electrons are assumed to be distributed according to the wave function $u(X_i, x_j)$ which describes a dependency on both nuclear and electronic coordinates. The correctness of this dual approach is testified by the Feynman–Hellmann theorem (Hellmann 1937; Feynman 1939).

7.2.3 The Feynman Approach

Feynman abstract Feynmann (1939) wrote "*The force on a nucleus in an atomic system is shown to be just the classical electrostatic force that would be exerted on this nucleus by other nuclei and by the electrons*". and also "*This permits an independent calculation of the slope of the curves of energy versus position of the nuclei, and may thus increase the accuracy, or decrease the labor involved in the calculation of these curves*".

When the molecule is in a steady state of energy $E(X_i)$ then a force f_k is associated to any displacement with respect to one of the X_k's, such that

$$f_k = -\frac{dE}{dX_k} \tag{7.2.27}$$

Being this the case, it may be proved that

$$f_k = -\int u^*(X_i x_j)\frac{\partial V}{\partial X_k}u(X_i x_j)dv \tag{7.2.28}$$

as well.

To do this we multiply both sides of Eq. 7.2.24 by $u^*(X_i x_j)$and integrate over the electronic coordinates x_j:

$$E(X_i) = \int u^*(X_i x_j)\left[-\sum_j \frac{\hbar^2}{2m_0}\nabla_j^2 + V(X_i, x_j)\right]u(X_i x_j)dv \tag{7.2.29}$$

Then we must differentiate each term which is function of X_k in (7.2.29) with respect to X_k (this means, u^*, u, and V) proof that the first two (u^* and u) must cancel, i.e. that

$$\int \frac{\partial u^*}{\partial X_k}\left[-\sum_j \frac{\hbar^2}{2m_0}\nabla_j^2 + V\right] u\,dv = \int u^*\left[-\sum_j \frac{\hbar^2}{2m_0}\nabla_j^2 + V\right]\frac{\partial u}{\partial X_k}dv \quad (7.2.30)$$

Since $-\sum_j \frac{\hbar^2}{2m_0}\nabla_j^2 + V$ is a self-adjoint operator, we may rewrite the term on the right in Eq. 7.2.29 in the form $\int \frac{\partial u}{\partial X_k}\left[-\sum_j \frac{\hbar^2}{2m_0}\nabla_j^2 + V\right] u^* dv$. The summation of the two terms in (7.2.29) is then $E\int\left(u\frac{\partial u^*}{\partial X_k} + u^*\frac{\partial u^*}{\partial X_k}\right)dv$. The integral is merely zero because by normalization $\int uu^* dv = 1$ and its derivative with respect to any parameter vanishes; whence (7.2.28) is proven.

7.2.4 Vibrational and Rotational Motion of Diatomic Molecules

Consider now a simple diatomic molecule composed of two nuclear centers of mass M_A and M_B, respectively (Fig. 7.8). The center of mass (x) is related to the positions of M_A and M_B through the lever rule (i.e. the segments r_A and r_B have relative lengths opposite to the ratio of relative masses). Moreover the two nuclei rotate in the plane of the figure with an angular velocity ω such that the linear velocity of the two atoms is $\omega \times r_A$ and $\omega \times r_B$, respectively.

The relative motion of the two atoms with respect to the center of mass does not depend on the effective motion of the whole molecule in the space. Moreover, the relative positions with respect to the center of mass obey

$$M_A r_A = M_B r_B \quad (7.2.31.1)$$

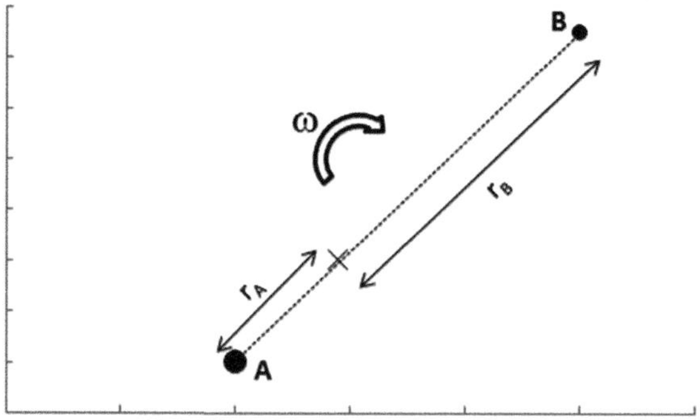

Fig. 7.8 A diatomic molecule with nuclei of mass A and B placed at distance r_A and r_B from the center of mass x and rotating on the plane of the figure with angular velocity ω

$$r_A + r_B = R \tag{7.2.31.2}$$

The energy is E(R) and the force pulling away either of the nuclei from the center of mass is −dE/dR. From Newton's second law we have

$$M_A \frac{d^2 r_A}{dt^2} = M_B \frac{d^2 r_B}{dt^2} = -\frac{dE}{dR} \tag{7.2.32}$$

and from (7.2.31.1, 7.2.21.2)

$$M_A r_A = M_B r_B = \frac{M_A M_B}{M_A + M_B} R \tag{7.2.33}$$

The ratio $\frac{M_A M_B}{M_A + M_B}$ is called "*reduced mass*" of the molecule. Denoting this parameter as $\mathcal{M}$ we easily obtain

$$\mathcal{M} \frac{d^2 R}{dt^2} = -\frac{dE}{dR} \tag{7.2.34}$$

We transformed in this way Eq. (7.2.32) into the equation of motion of a single particle of mass $\mathcal{M}$ which moves at distance R from the origin under the influence of a conservative potential energy field E.

Denoting as R_0 the distance at the A and B nuclei at the minimum of the potential well, any displacement from R_0 will generate a restoring force which may be approximately treated as simply Hookian, provided that the displacement is not severe:

$$\frac{dE}{dR} = k_H (R - R_0) \tag{7.2.35}$$

The oscillation of the nuclear centers about the minimum will be in this case simply harmonic, with a frequency directly dependent on the square root of the constant k_H and inversely dependent on the reduced mass of the molecule $\mathcal{M}$:

$$\nu = \frac{1}{2\pi} \sqrt{\frac{k_H}{\mathcal{M}}} \tag{7.2.36}$$

Denoting as $-E_D$ the dissociation energy of the molecule at the minimum (R_0), the vibrational energy E_{vib} will be given by

$$E_{vib} = -E_D + \left(n_{vib} + \frac{1}{2} \right) h\nu \tag{7.2.37}$$

where n_{vib} (the vibrational number of the nuclei) is equal to zero at the lowest vibrational state. We will accordingly denote as "*zero-point energy*" E_{ZP} the vibrational energy of the system observed when all vibrational modes are grouped on the lowest vibrational number accessible to the system itself:

$$E_{ZP} = \frac{1}{2}hv \quad (7.2.38)$$

If the molecule rotates with an angular velocity ω (Fig. 7.5), the linear velocities of atoms A and B with respect to the center of mass x will be $\omega \times r_A$ and $\omega \times r_B$, respectively. The rotational (kinetic) enegy E_{rot} associated to the movement will be

$$E_{rot} = \frac{1}{2}\left[M_A(\omega r_A)^2 + M_B(\omega r_B)^2\right] = \frac{1}{2}I\omega^2 \quad (7.2.39)$$

where $I = M_A r_A^2 + M_B r_B^2$ is the ordinary moment of inertia. Moreover, the angular momentum $\mathbf{p}$ will be

$$\mathbf{p} = \hbar\sqrt{l(l+1)} \quad (7.2.40)$$

and the component of the angular momentum along the quantization axis z will be

$$p_z = \hbar m_l \quad (7.2.41)$$

Replacing Eqs. (7.2.40, 7.2.41) in Eq. (7.2.39) one finally obtains

$$E_{rot} = \frac{\mathbf{p}^2}{2I} = \frac{l(l+1)\hbar^2}{2I} \quad (7.2.42)$$

(see Sect. 5.2 for the quantum-mechanical significance of the azimuthal l and the magnetic m_l quantum numbers).

For each electronic energy level accessible to a given molecule such as those ones displayed in Fig. 7.4 for H_2^+ there will then be a set of E_{vib} + E_{rot} levels whose quantization depends on the vibrational (v), the azimuthal (l) and the magnetic (m_l) quantum numbers:

$$E_{vib} + E_{rot} = -D + \left(v + \frac{1}{2}\right)\hbar v + \frac{l(l+1)\hbar^2}{2I} \quad (7.2.43)$$

For each set of n and l there will be $m_l = 2l + 1$ rotational + vibrational degenerate states (same energy though different wave functions). Accounting for the fact that rotation induces stretching in an actual (non-rigid) molecule, Eq. (7.2.43) may be modified as follows (Pauling and Bright Wilson 1935):

$$E_{vib} + E_{rot} = -D + \left(v + \frac{1}{2}\right)\hbar v + \frac{l(l+1)\hbar^2}{2I} - \frac{l^2(l+1)^2\hbar^4}{8\pi^2 v^2 I^3} \quad (7.2.44)$$

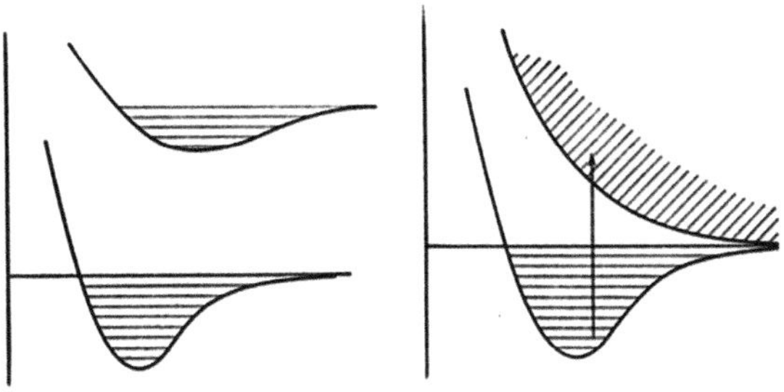

Fig. 7.9 Left side: vibrational energy levels associated to two different stable electronic states in a diatomic molecule (schematic). Right side: Transition from a quantized vibrational level to a continuum level of a repulsive electronic state. Modified from Slater (1963)

The spacing of the non-degenerate vibrational states is small with respect to the dissociation energy D, and the rotational levels are closely spaced with respect to the vibrational levels, so that superimposing the rotational, vibrational, and electronic energy levels on a same energy diagram is unpractical. We may however provide a schematic representation of the transition from one electronic energy state to another. We may group, to this purpose, the transitions into three main types:

(1) Transitions involving a change of the electronic quantum number n. This involves also changes in v and l. The resulting spectrum is a band spectrum[5]): (infrared, visible, near-ultraviolet);
(2) Transition in which n does not change, but changes involve v and l (infrared);
(3) Transitions in which only l changes (far infrared).

In Fig. 7.9 (left side) we see schematically a transition between two stable states [such as the $1\sigma_g$ and $3\sigma_g$ of H_2^+ in Fig. (7.7)]. The energy levels are different and different are also the curvatures of the bounding limits and the depths of the potential wells. Also, because the minima are observed at different values of R_0, the spacing between the vibrational states will be different (though they are schematically all drawn as identical). On the right side of the same figure, we see a transition from a stable-state of the type $1\sigma_g$ to an unstable state of the type $1\sigma_u$. In this state, there is no net attraction of the two nuclei because the electron probability density along the internuclear axis is virtually zero. The nuclei repel thus each other. Hence, there cannot be any sort of quantized vibration between them but only a continuous set of energies. One can have nevertheless a transition, with absorption of energy (represented by the upward directed arrow) and a consequent continuous absorption over a considerable frequency range in the spectrum, up to dissociation of the molecule.

[5] l changes by ± 1 units or 0 and this limits greatly the number of observed lines.

7.2.5 The Morse Function

The simple treatment we have seen insofar fails in generating equally spaced vibrational levels, while experimental observations show that the observed vibrational levels show a decrease in spacing with the increase of energy, for a given value of *n*. To overcome this ambiguity Morse (1929) proposed a potential of form:

$$E(r) = A + D\exp[-2a(r - r_0)] - 2D\exp[-a(r - r_0)] \tag{7.2.44}$$

A in (7.2.44) is meant to recast all potentials generated by (7.2.44) for neutral molecules on the lowest vibrational level of the lowest electronic state. r_0 is related to the lowest vibrational state by

$$r_0 = \sqrt[3]{\frac{3000A}{\omega_0}} \quad (cm^{-1}) \tag{7.2.45}$$

D is found in wave numbers by

$$D = \frac{\omega_0^2}{4\omega_0 x} \tag{7.2.46}$$

where x is an empirically determined constant and a is given by

$$a = 0.2454\sqrt{\mathcal{M}\omega_0 x} \tag{7.2.47}$$

To solve the problem Morse begun with the generalized form of the wave equation for the hydrogen atom

$$\psi_{n,l,m}(r, \theta, \phi) = R_{n,l}(r)\Theta_{l,m}(\theta)\Phi_m(\phi) \tag{7.5.2.42}$$

and analyzed in detail the r-dependent term. We may recall then, in solving the r equation for the hydrogen atom (Sect. 5.2.4) we wrote the r-dependent term in Eq. (5.2.13) as follows:

$$\frac{1}{r^2}\frac{d}{dr}\left(r^2\frac{dR}{dr}\right) + \left\{-\frac{l(l+1)}{r^2} + \frac{8\pi^2\mu}{h^2}[E - V(r)]\right\}R = 0 \tag{7.5.2.13}$$

I replace now here the symbols to render them consistent with the notation adopted in discussing the position of nuclei and electrons in a diatomic molecule:

$$\frac{1}{R^2}\frac{d}{dR}\left(R^2\frac{d\mathcal{R}}{dR}\right) + \left\{-\frac{l(l+1)}{r^2} + \frac{8\pi^2\mu}{h^2}[E - V(R)]\right\}\mathcal{R} = 0 \tag{7.2.48}$$

The above equation implies that V is indeed function of R alone [i.e. $V = V(R)$] and that the wave function can be split the function into three parts, i.e.

$$\psi(r, \theta, \phi) = \mathcal{R}(R) \cdot \Theta(\theta) \cdot \Phi(\phi) \tag{7.2.49}$$

Through the substitution $\mathcal{R}(R) = (1/R)S(R)$ Eq. (7.2.48) transforms into

$$\frac{d^2S}{dR^2} + \left\{ -\frac{l(l+1)}{R^2} + \frac{8\pi^2\mu}{h^2}[E - V(R)] \right\} S = 0 \tag{7.2.50}$$

Insertion of the Morse potential into the radial equation yields

$$\frac{d^2S}{dR^2} + \left\{ -\frac{l(l+1)}{R^2} + \frac{8\pi^2\mu}{h^2}\{E - D - D\exp[-2a(R - R_0)] + 2D\exp[-a(R - R_0)]\} \right\} S = 0 \tag{7.2.51}$$

Replacing $\exp[-a(R - R_0)]$ with y and $l(l+1)\frac{h^2}{8\pi^2\mu R_0^2}$ with a, the radial equation becomes

$$\frac{d^2S}{dy^2} + \frac{1}{y}\frac{dS}{dy} + \frac{8\pi^2\mu}{a^2h^2}\left(\frac{E - D}{y^2} + \frac{2D}{y} - D - \mathcal{A}\frac{R_0^2}{y^2R^2} \right) S = 0 \tag{7.2.52}$$

With a Taylor expansion of the R_0^2/R^2 ratio in powers of $(y-1)$ (Pekeris 1924):

$$\frac{R_0^2}{R^2} = \frac{1}{\left(1 - \frac{\ln y}{aR_0}\right)} = 1 + \frac{2}{aR_0}(y - 1) + \left(-\frac{1}{aR_0} + \frac{3}{a^2R_0^2} \right)(y - 1)^2 + \ldots \tag{7.2.53}$$

we get

$$\frac{d^2S}{dy^2} + \frac{1}{y}\frac{dS}{dy} + \frac{8\pi^2\mu}{a^2h^2}\left(\frac{E - D - c_0}{y^2} + \frac{2D - c_1}{y} - D - c_2 \right) S = 0 \tag{7.2.54}$$

where

$$c_0 = \mathcal{A}\left(1 - \frac{3}{aR_0} + \frac{3}{a^2R_0^2} \right) \tag{7.2.55.1}$$

$$c_1 = \mathcal{A}\left(\frac{4}{aR_0} - \frac{6}{a^2R_0^2}\right) \tag{7.2.55.2}$$

$$c_2 = \mathcal{A}\left(-\frac{1}{aR_0} + \frac{3}{a^2R_0^2}\right) \tag{7.2.55.3}$$

Through the substitutions (Pekeris 1924; Pauling and Bright Wilson 1935)[6]:

$$z = 2dy \tag{7.2.56.1}$$

$$d^2 = \frac{8\pi^2\mathcal{M}}{a^2h^2}(D + c_2) \tag{7.2.56.2}$$

$$b^2 = -\frac{32\pi^2\mathcal{M}}{a^2h^2}(E - D + c_0) \tag{7.2.56.3}$$

$$S(y) = \exp(-z/2)z^{b/2}F(z) \tag{7.2.56.4}$$

Eq. (7.2.51) simplifies into:

$$\frac{d^2F}{dz^2} + \left(\frac{b+1}{z} - 1\right)\frac{dF}{dz} + \frac{v}{z}F = 0 \tag{7.2.57}$$

where

$$v = \frac{4\pi^2\mathcal{M}}{a^2h^2d}(2D - c_1) - \frac{1}{2}(b + 1) \tag{7.2.58}$$

Equation (7.2.57) is closely related to the radial equation of the hydrogen atom and may be solved in the same manner [see Eqs. (5.2.38) and (5.2.39)]. In order to obtain a polynomial solution the values attained by v must be integer (v = 0,1,2...) and the boundary conditions must satisfy F→0 as r→ −∞.

The vibrational + rotational energies obtained from the above equations accounting for the limits imposed by Eq. (7.2.58) are

$$E_{vib} + E_{rot} = D + c_0 - \frac{(D - c_1/2)^2}{(D + c_2)} + \frac{ah(D - c_1/2)}{\pi\sqrt{2\mathcal{M}}\sqrt{D + c_2}}\left(v + \frac{1}{2}\right) - \frac{a^2h^2}{8\pi^2\mathcal{M}}\left(v + \frac{1}{2}\right)^2 \tag{7.2.59}$$

[6] Note that in Pekeris(1934) the development is directly on $\mathcal{R}(R)$; moreover Eq. (7.2.56.4) is replaced by $\mathcal{R}(\mathcal{Y}) = N\exp(-z/2)z^{b/2}F(\overline{z})$.

The form usually employed in the study of observed spectra (with very accurate results for nearly all molecules) is obtained by expanding Eq. (7.2.59) in terms of powers of c_1/D and c_2/D (Pauling and Bright Wilson 1935):

$$\frac{E_{vib}+E_{rot}}{hc}=\overline{\nu}_0\left(v+\frac{1}{2}\right)-x_0\,\overline{\nu}_0\left(v+\frac{1}{2}\right)^2 + l(l+1)B_0+D_0l^2(l+1)^2-\alpha_0\left(v+\frac{1}{2}\right)l(l+1) \tag{7.2.60}$$

where c is the speed of light and

$$\overline{\nu}_0=\frac{a}{2\pi c}\sqrt{\frac{2D}{\mathcal{M}}} \tag{7.2.61.1}$$

$$x_0=\frac{h\,\overline{\nu}_0\,c}{4D} \tag{7.2.61.2}$$

$$B_0=\frac{h}{8\pi^2 I_0 c} \tag{7.2.61.3}$$

$$D_0=-\frac{h^3}{128\pi^6\mathcal{M}^3\,\overline{\nu}_0^2\,c^3R_0^6} \tag{7.2.61.4}$$

$$\alpha_0=-\frac{3h^2\,\overline{\nu}_0}{16\pi^2\mathcal{M}R_0^2D}\left(\frac{1}{aR_0}-\frac{1}{a^2R_0^2}\right) \tag{7.2.61.5}$$

The Morse potential which is, by nature "*adaptive*", conforms rather well to the exact energies (Fig. 7.10), while simple procedures based on a linear combination of atomic orbitals do not (Table 7.4).

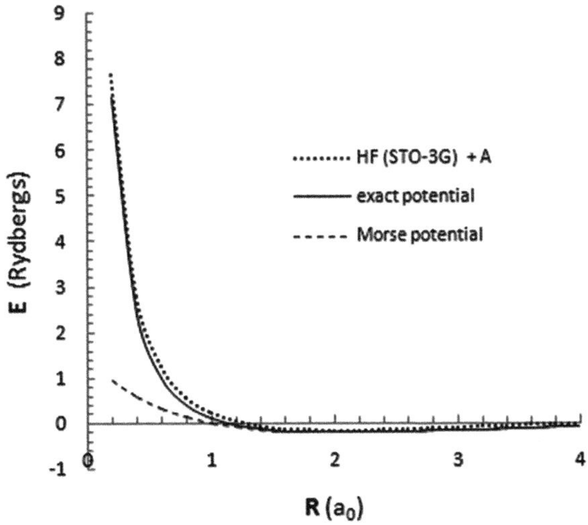

Fig. 7.10 $1\sigma_g$energies for H_2^+ arising from ROHF STO −3G calculations compared to exact values and results of the Morse potential. Distances are in atomic units (a_0). Energies are in *Ry*

Table 7.4 $1\sigma_g$ energies for H_2^+ arising from ROHF/STO−3G calculations (HF + A) compared to exact values and results of the Morse potential. Distances are in atomic units (a_0). Energies are in *Ry*. The A term to be added to the HF energy is barely the dissociation energy of the molecule, which depends upon the choice of the value that we assign to the isolated atoms

R	HF + A	Exact(*)	Morse(*)
0.2	7.6567	7.1416	0.9654
0.4	2.7689	2.3984	0.5821
0.6	1.2489	0.9903	0.3093
0.8	0.5697	0.3910	0.1184
1	0.2196	0.0964	−0.0122
1.2	0.0283	−0.0579	−0.0989
1.4	−0.0770	−0.1399	−0.1536
1.6	−0.1326	−0.1819	−0.1854
1.8	−0.1583	−0.2005	−0.2010
2	−0.1654	−0.2053	−0.2053
2.2	−0.1606	−0.2017	−0.2020
2.4	−0.1486	−0.1931	−0.1937
2.6	−0.1324	−0.1817	−0.1824
2.8	−0.1141	−0.1687	−0.1693
3	−0.0951	−0.1551	−0.1556
3.2	−0.0764	−0.1415	−0.1417
3.4	−0.0587	−0.1282	−0.1282
3.6	−0.0422	−0.1154	−0.1153
3.8	−0.0272	−0.1034	−0.1033
4	−0.0137	−0.0922	−0.0922

(*) exact and Morse function values according to Slater (1963). Note that the Morse function adopted by Slater (1963) does not include the A term in Eq. (7.2.43) and the other sensitive parameters are expressed in atomic units.

In this kind of calculations, Slater-Type-Orbitals (STO) are replaced by a linear combination of 3 primitive Gaussians (3G) on each nuclear center, contracted to generate 2 basis sets. For the simple calculation of H_2^+ only the ϕ_{1s} is needed, with orbital exponent $\zeta = 1$, i.e.:

$$\phi_{1s}(\zeta_1, r) = \left(\frac{\zeta_1^3}{\pi}\right)^{1/2} \exp(-\zeta_1 r) = \left(\frac{1}{\pi}\right)^{1/2} \exp(-r) \qquad (7.2.62)$$

The STO is then replaced by an orbital $\phi_{1s'}$ generated by a set of (K = 3) Gaussians (each with optimized coefficient ($d_{1s,k}$) and exponent (α_{1k}) (Hehre et al. 1969; see also Sect. 12.4):

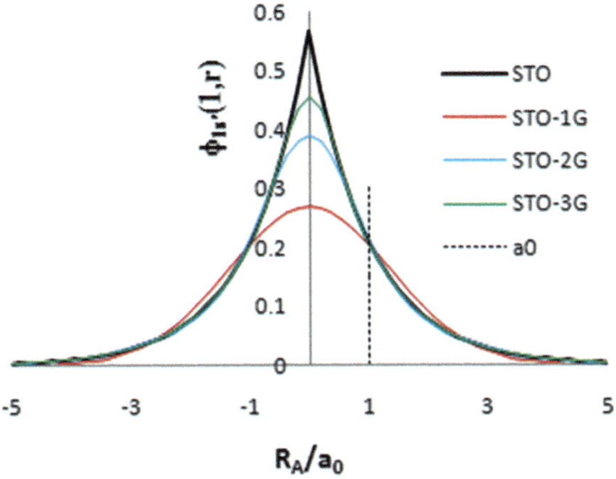

Fig. 7.11 Slater-Type-Orbitals for the 1s state and its reproduction trough summations of K Gaussians with appropriate coefficients and exponents

$$\phi_{1s'}(1, r) = \sum_{k}^{K} d_{1s,k} g_{1s}(\alpha_{1k}, r) \tag{7.2.63}$$

$$g_{1s}(\alpha, r) = \left(\frac{2\alpha}{\pi}\right)^{3/4} \exp\left(-\alpha r^2\right) \tag{7.2.64}$$

The need to replace Slater orbitals (STO) with Gaussian functions arises from the fact that multicentered integrals are more easy to evaluate. However, Slater orbitals and Gaussian functions behave very differently for small values of R. Combining several primitive Gaussian functions to curve fit the Slater orbital reduces the problem but does not eliminate it completely. As we may see in Fig. (7.11) a 3G expansion is apparently quite appropriate to reproduce the Slater orbital at $R = a_0$, but it cannot reproduce the cusp at $R = 0$, and the probability density is too low in proximity of the nucleus.

When we superimpose the results of our computational exercise to the "*exact*" calculations of Burrau (1927), Hylleraas (1931), and Jaffe (1934) in terms of E/E_H energy ratios, we see that, while the procedure is sufficiently accurate in reproducing the correct internuclear distance (i.e. the location of the minimum), the dissociation energy (hence the potential well) depends largely on the value assigned to the energy of the Hydrogen atom E_H (hence to *A*). Adopting the 3G coefficients and exponents returns an energy of −0.4665 *Hartree* (and generates the red-dotted line in Fig. 7.2), while the exact energy of the 1s electron of the hydrogen atom is −0.5 *Hartree* (i.e. a single Slater orbital with exponent equal to 1 (green dotted line in Fig. 7.2).

The calculated dissociation energy in the first case is 2.9896 *eV*, and in the second case is 2.0802 *eV* (see note[7]). The dissociation energy estimate of Burrau (1927) is 2.69±0.03 *eV*, almost intermediate between the two energies obtained at simple ROHF/STO−3G level. All the above listed values are higher than the nowadays accepted value (1.8 *eV*). Computing the electronic energy and the ZP energy of the H_2 molecule at the same theory level and subtracting the obtained value from the corresponding energy of H_2^+ we get an ionization potential of the H_2 molecule of 14.383 *eV*, decidedly lower than the experimental value (15.603 *eV*).

7.3 The H_2 Molecule and the Many Electrons Problem

The H_2^+ molecular ion introduced us to the treatment of molecular orbitals in the absence of inter-electronic interaction. The next step is the analysis of electron correlation effects in many electrons molecules. The simplest case is obviously represented by the molecular hydrogen H_2 where two electrons interact between themselves and with two nuclei of homologous charge.

The problem was afforded already by Heisemberg (1926a, b; 1927) in a series of seminal contributions that appeared shortly after Schrödinger's wave mechanics treatment (Schrödinger 1926). Considering initially the two electrons as non-interacting, one may describe the wave function of the system as the product of the individual wave functions u_1 and u_2 pertaining to the two electrons at their respective positions x_1 and x_2. We may represent this wave function by adopting the usual notation $u_1(\mathrm{x}_1)$, $u_2(\mathrm{x}_2)$. The interaction of the two electrons may be then treated as a "*perturbation*". Denoting the wave unperturbed functions of the system as $u_1(\mathrm{x}_1)u_2(\mathrm{x}_2)$ and $u_2(\mathrm{x}_1)u_1(\mathrm{x}_2)$, the correct wave functions which take into account of the perturbation are the symmetric and antisymmetric combinations:

$$u_1(\mathrm{x}_1)u_2(\mathrm{x}_2) + u_2(\mathrm{x}_1)u_1(\mathrm{x}_2) \qquad (7.3.1.1)$$

$$u_1(\mathrm{x}_1)u_2(\mathrm{x}_2) - u_2(\mathrm{x}_1)u_1(\mathrm{x}_2) \qquad (7.3.1.2)$$

The energy of the two functions differs by an amount defined in terms of the "*exchange*" integral

$$\int u_1^*(\mathrm{x}_1)u_2^*(\mathrm{x}_2)\mathsf{H}u_2^*(\mathrm{x}_1)u_1^*(x_2)\mathrm{d}\mathrm{x}_1\mathrm{d}\mathrm{x}_2 \qquad (7.3.2)$$

[7] H_2^+ dissociates into H and H^+; the theoretical electronic energy of H is −0.5 *Hartree* while H^+ has no electrons, hence no electron energy. The dissociation energy of H_2^+ is therefore the minimum energy attained by the molecule with the nuclear centers placed at equilibrium distance (i.e. bottom of potential well), plus the Zero Point Correction, less the theoretical (or calculated) energy of H.

where H is the Hamiltonian operator. The symmetric function (7.3.1.1) was shown by Heisenberg to be associated with a singlet term in the spectrum of a 2-electrons atom, while the antisymmetric function (7.3.1.2) involved a triplet (Heisenberg 1926, 1927a). As noted by Slater (1963), from these apparently simple considerations arose both the treatment of complex spectra and the theory of covalent bonding. Moreover, to generate a triplet, the spin must be parallel, however, based on Pauli exclusion principle, no two electrons with parallel spin can have the same orbital quantum number. Hence, Heisenberg anticipated that the antisymmetric function must vanish if u_2 becomes identical to u_1. The generalized relation between antisymmetric function and exclusion principle was formalized by Dirac (1926) and, in a later paper devoted to the general n-electrons problem, by Heisemberg (1928). Slater (1929) proposed the utilization of a determinantal form combining coordinates and spin (i.e. the "*spin orbitals*"). The Slater determinant was written in the form:

$$\begin{vmatrix} u_1(1) & u_1(2) & \dots & u_1(3) \\ u_2(1) & u_2(2) & \dots & u_2(3) \\ \dots & \dots & \dots & \dots \\ u_n(1) & u_n(2) & \dots & u_n(n) \end{vmatrix} \quad (7.3.3)$$

In this form the u_i's denote the spin orbitals and their coordinates and spin are indicated as arguments (1,…n). Each member identifies a function corresponding to the specific assignment of the quantum numbers to them pertaining. The solution of the above determinant or "*secular equation of the energy*" defines the energy levels of the multiplets terms. Some hints on this determinant are given in the following gray box. We will see a practical application of the determinant later on.

Background information: The Slater Determinant In 1929 Slater wrote a fundamental contribution on the "*Theory of Complex Spectra*" where, starting from Hund's scheme for multiplet classification (Hund 1927) he computed the energy distances between multiplets by wave mechanics without the need of group theory. The article has the clarity of the pioneering developments and is highly formative. The process of building up wave functions of the proper symmetry by using determinants was not new (see Wailer and *Hartree*, 1929) but Slater (1929) introduces the spin function into the determinant in a more satisfactory way (at the very beginning of the calculation) considering that the other symmetry characters can be avoided and one needs only consider antisymmetric wave functions.

Slater (1929) begins stressing that the principle of antisymmetry shows itself in the exclusion principle and the Fermi-Dirac statistics. The first states that only those energy levels are possible which are characteristic numbers of Schrodinger's equation for the system; the second further restricts the possible energies to those connected with characteristic functions antisymmetric in the electrons. He then anticipates "*Our whole object is to find such characteristic functions and numbers*".

Slater considers then an electron moving in a central field of force, according to wave mechanics, characterized by the same two quantum numbers n and l. The arbitrary direction of the angular momentum is described by quantizing the component m_l in a fixed direction. Similarly the orientation of the spin is given by specifying its component m_s in the same fixed direction. Thus each electron in an atom, neglecting the interactions of their rotations, is specified by the four quantum numbers n, l, m_l, m_s, but now each electron has a wave function which is "*a function of its spatial coordinates and of a coordinate representing its spin, depending on the numbers n, l, m_l, m_s, which is a solution of Schrodinger's equation for a particle in a central field*" (Slater 1929).

The next step in Slater's approach is to assume that the separate electrons move in fields of force with spherical symmetry, so that there is no tendency for orientation of the various angular momenta. Differently from *Hartree* (1928) which assumed each electron to move in a field of force slightly different from the others, Slater (1929) neglects the difference, assuming that all the electrons move in precisely the same field "*and this field is to be so chosen as to give the best agreement with the correct values even without further corrections*".

Slater denotes then the function for the ith electron by $u(n_i/x_i)$, where n_i stands for the four numbers n, l, m_l, m_s and x symbolizes the four coordinates (three Cartesian, one of spin) of the ith electron. He notes then that the product of these functions, for all the electrons (1…..N) of the atom, gives a function which approximately satisfies Schrodinger's equation. That is, $u(n_1/x_1)u(n_2/x_2)\ldots.\ u(n_N/x_N)$ is an approximate solution. But it is not antisymmetric in the electrons, so that it does not satisfy the exclusion principle. "*To build up an antisymmetric solution, we first note that we still have an approximate solution, connected with the same energy value, if we interchange any two x's, obtaining for example* $u(n_1/x_2)\ u(n_2/x_1)\ldots.u(n_N/x_N)$*. We still have an approximation with the same energy if we make a linear combination of any such solutions. Then we can make the one possible combination which is antisymmetric, and it will both satisfy the exclusion principle, and will be an approximate solution of Schrodinger's equation. This combination is conveniently written as a determinant* " (Slater 1929):

$$u = \begin{vmatrix} u(n_1/x_1) & u(n_1/x_2) & \ldots & u(n_1/x_N) \\ \ldots & \ldots & \ldots & \ldots \\ \ldots & \ldots. & \ldots & \ldots \\ u(n_N/x_1) & u(n_N/x_2) & & u(n_N/x_N) \end{vmatrix} \tag{7.3.4.1}$$

If H represents the energy operator, the diagonal term of the energy with respect to one of the approximate wave functions, written as a determinant is

$$\frac{\int (u^* \mathrm{H} u)\mathrm{dv}}{\int (u^* u)\mathrm{dv}} \tag{7.3.4.2}$$

The functions u are determinants; that is, sums of the N! terms formed from the product $u(n_1/x_1)\ldots.\ u(n_N/x_N)$ by carrying out all permutations of the coordinates $x_1 \ldots x_N$, (even permutations having the coefficient +1, odd ones the coefficient −1). The product u^*u is then also a double sum, and each of the integrals would be likewise; except that, in consequence of orthogonality relations, they really reduce to single sums, of which all the terms are identical. The normalization integral $\int (u^* \mathrm{H} u)\mathrm{dv}$ is just N!

Each individual function $u(n_i/x_i)$ is a solution of Schrodinger's equation for an electron in a central field. Denoting $V(x,y,z)$ the potential of such a field, the Schrodinger *s* equation for the individual function is

$$\mathrm{H}_1 u = \left[\nabla \frac{\mathrm{h}^2}{8\pi^2 \mathrm{m}} + \mathrm{V}(\mathrm{x}_1, \mathrm{x}_2, \ldots, \mathrm{x}_N)\right] u = \varepsilon u \tag{7.3.4.3}$$

where u is a function of x, y, z, and certain coordinates representing the spin, depending parametrically on n, l, m_l, and m_s and ε is a function of the quantum numbers alone.

In short, if we assume that the potential V does not depend upon the particles' coordinates, we could expand the Schrödinger equation into single particles contributions and the wavefunction could be written as an *Hartree Product*, i.e. a product of single particles solutions:

$$\boldsymbol{\Psi}(\mathrm{x}_1, \mathrm{x}_2, \mathrm{x}_N) = \psi_1(\mathrm{x}_1)\psi_2(\mathrm{x}_2)\ldots\psi_N(\mathrm{x}_N) \tag{7.3.4.4}$$

This product, however, when applied to fermions does not satisfy the anti-symmetric requirement (hence nor the Pauli Exclusion Principle) because it does not change sign on an interchange of coordinates. I.e. for a single orbital Eq. (7.3.4.4) would reduce to an identity

$$\boldsymbol{\Psi}(\mathrm{x}_1, \mathrm{x}_2) = \psi(\mathrm{x}_1)\psi(\mathrm{x}_2) = \psi(\mathrm{x}_2)\psi(\mathrm{x}_1) \tag{7.3.4.5}$$

To make the function antisymmetric for two electrons we must write:

$$\boldsymbol{\Psi}(\mathrm{x}_1, \mathrm{x}_2) = \frac{1}{\sqrt{2}}[\psi_1(\mathrm{x}_1)\psi_2(\mathrm{x}_2) - \psi_1(\mathrm{x}_2)\psi_2(\mathrm{x}_1)] \tag{7.3.4.6}$$

or, as a (Slater) determinant:

$$\Psi(\mathrm{x}_1, \mathrm{x}_2) = \frac{1}{\sqrt{2}}\begin{vmatrix} \psi_1(\mathrm{x}_1) & \psi_2(\mathrm{x}_1) \\ \psi_1(\mathrm{x}_2) & \psi_2(\mathrm{x}_2) \end{vmatrix} \tag{7.3.4.7}$$

For N fermions the Slater determinant has form

$$\Psi = \frac{1}{\sqrt{N!}} \begin{vmatrix} \psi_1(x_1) & \psi_2(x_1) & \ldots & \psi_N(x_1) \\ \psi_1(x_2) & \psi_2(x_2) & \ldots & \psi_N(x_2) \\ \ldots & \ldots & \ldots & \ldots \\ \psi_1(x_N) & \psi_2(x_N) & & \psi_N(x_N) \end{vmatrix} \tag{7.3.4.8}$$

For a closed shell of electrons only one eigenfunction may be written, hence, the eigenfunction is completely determined by the configuration. For example, for a shell of six 2p electrons, denoting in brackets the values assumed by the n,l,m_l and m_s quantum numbers we will have

$$\frac{1}{\sqrt{6!}} \begin{vmatrix} \psi(211\tfrac{1}{2}|1) & \ldots\ldots & \psi(211\tfrac{1}{2}|6) \\ \psi(211\tfrac{\bar{1}}{2}|1) & \ldots\ldots & \psi(211\tfrac{\bar{1}}{2}|6) \\ \psi(210\tfrac{1}{2}|1) & \ldots\ldots & \psi(210\tfrac{1}{2}|6) \\ \psi(210\tfrac{\bar{1}}{2}|1) & \ldots\ldots & \psi(210\tfrac{\bar{1}}{2}|6) \\ \psi(21\bar{1}\tfrac{1}{2}|1) & \ldots\ldots & \psi(21\bar{1}\tfrac{1}{2}|6) \\ \psi(21\bar{1}\tfrac{\bar{1}}{2}|1) & \ldots\ldots & \psi(21\bar{1}\tfrac{\bar{1}}{2}|6) \end{vmatrix} \tag{7.3.4.9}$$

where the first part in the parentheses are listed the four quantum numbers and in the other part the number of the electron occupying the orbital and $\bar{1} \equiv -1$; $\frac{\bar{1}}{2} \equiv -\frac{1}{2}$ (Eyring's notation).

For an incomplete (open) shell the number of eigenfunctions depends upon the configuration. As an example, the $1s^2 2p^1$ configuration has twelve eigenfunctions whose determinants may be written as

$$\begin{aligned}
D_1 &= \tfrac{1}{\sqrt{2!}}\left|\psi(100\tfrac{1}{2}\,|1)\psi(211\tfrac{1}{2}|2)\right| & D_7 &= \tfrac{1}{\sqrt{2!}}\left|\psi(100\tfrac{\bar{1}}{2}\,|1)\psi(211\tfrac{1}{2}|2)\right| \\
D_2 &= \tfrac{1}{\sqrt{2!}}\left|\psi(100\tfrac{1}{2}\,|1)\psi(211\tfrac{\bar{1}}{2}|2)\right| & D_8 &= \tfrac{1}{\sqrt{2!}}\left|\psi(100\tfrac{\bar{1}}{2}\,|1)\psi(211\tfrac{\bar{1}}{2}|2)\right| \\
D_3 &= \tfrac{1}{\sqrt{2!}}\left|\psi(100\tfrac{1}{2}\,|1)\psi(210\tfrac{1}{2}|2)\right| & D_9 &= \tfrac{1}{\sqrt{2!}}\left|\psi(100\tfrac{\bar{1}}{2}\,|1)\psi(210\tfrac{1}{2}|2)\right| \\
D_4 &= \tfrac{1}{\sqrt{2!}}\left|\psi(100\tfrac{1}{2}\,|1)\psi(210\tfrac{\bar{1}}{2}|2)\right| & D_{10} &= \tfrac{1}{\sqrt{2!}}\left|\psi(100\tfrac{\bar{1}}{2}\,|1)\psi(210\tfrac{\bar{1}}{2}|2)\right| \\
D_5 &= \tfrac{1}{\sqrt{2!}}\left|\psi(100\tfrac{1}{2}\,|1)\psi(21\bar{1}\tfrac{1}{2}|2)\right| & D_{11} &= \tfrac{1}{\sqrt{2!}}\left|\psi(100\tfrac{\bar{1}}{2}\,|1)\psi(21\bar{1}\tfrac{1}{2}|2)\right| \\
D_6 &= \tfrac{1}{\sqrt{2!}}\left|\psi(100\tfrac{1}{2}\,|1)\psi(21\bar{1}\tfrac{\bar{1}}{2}|2)\right| & D_{12} &= \tfrac{1}{\sqrt{2!}}\left|\psi(100\tfrac{\bar{1}}{2}\,|1)\psi(21\bar{1}\tfrac{\bar{1}}{2}|2)\right|
\end{aligned} \tag{7.3.4.10}$$

where just the principal diagonal is listed between bars (Eyring et al. 1944).

Posing $H_{ij} = \int (u^* H u) dv$ and $S_{ij} = \int (u^* u) dv$ according to the first order perturbation theory for degenerate systems, the energy levels for a system of N fermions are given by the roots of the equation

$$\begin{vmatrix} H_{11}-S_{11}E & H_{12}-S_{12}E & \ldots\ldots & H_{1N}-S_{1N}E \\ H_{21}-S_{21}E & H_{22}-S_{22}E & \ldots\ldots & H_{2N}-S_{2N}E \\ \ldots\ldots & \ldots\ldots & \ldots\ldots & \ldots\ldots \\ H_{N1}-S_{N1}E & H_{N2}-S_{N2}E & \ldots\ldots & H_{NN}-S_{NN}E \end{vmatrix} = 0 \qquad (7.3.4.11)$$

The H_{ij} and S_{ij} vanish except for those combinations which have the same eigenvalue for all operators of the set (including angular momentum).

7.3.1 *The Heitler and London Treatment*

Already in 1916 Lewis suggested that the chemical bond between two atoms in a diatomic molecule arises from pairs of electrons held jointly by the two nuclei. The treatment of Heitler and London (1927) provides a rationale to this anticipation. Their approach was later reprised, extended, and generalized by Slater (1931) and Pauling (1931; 1939), generating the ideas of directed valence, tetrahedral carbon, treatments of multiple bonds, hybridization, partial ionic character, and many other concepts of paramount importance in Geochemistry (see Bright Wilson 1977; for a more extensive account of the importance of the Heitler–London approach).

Here I give some outline of the procedure, after the slight modifications introduced by Pauling and Bright Wilson (1935).

Consider a system composed of two nuclei (A and B in Fig. 7.12) and two electrons (1 and 2 in the same figure).

Denoting the inter-particle distances as outlined in Fig. (7.12), the wave equation obeys

$$\nabla_1^2\psi + \nabla_2^2\psi + \frac{8\pi^2 m_0}{h^2}\left[W + e^2\left(\frac{1}{r_{A1}} + \frac{1}{r_{B1}} + \frac{1}{r_{A2}} + \frac{1}{r_{B2}} - \frac{1}{r_{12}} - \frac{1}{r_{AB}}\right)\right]\psi = 0 \qquad (7.3.5)$$

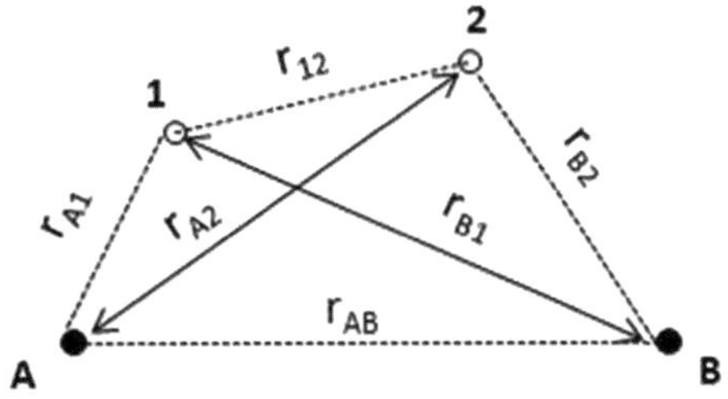

Fig. 7.12 Diagrammatical representation of inter-particle distances in a diatomic molecule with two electrons at a fixed position. From Pauling and Bright Wilson (1935), redrawn

At large values of r_{AB} the function becomes that of a system consisting of two (non-interacting) hydrogen atoms. The system is two-fold degenerate and represented by any linear independent combination of $u_{1s_A}(1)u_{1s_B}(2)$ and $u_{1s_A}(2)u_{1s_B}(1)$with u_{1s_A} indicating the hydrogen-like 1s wave function for electron 1 about nucleus A. For smaller values of r_{AB} one may adopt a linear combination of the two product functions as a variational contribution. It is then posed:

$$\psi_I = u_{1s_A}(1)u_{1s_B}(2) \tag{7.3.6.1}$$

$$\psi_{II} = u_{1s_B}(1)u_{1s_A}(2) \tag{7.3.6.2}$$

and

$$H_{I\ I} = \int\int \psi_I H \psi_I dr_1 dr_2 \tag{7.3.7.1}$$

$$H_{I\ II} = \int\int \psi_I H \psi_{II} dr_1 dr_2 \tag{7.3.7.2}$$

$$\Delta^2 = \int\int \psi_I \psi_{II} dr_1 dr_2 \tag{7.3.7.3}$$

Δ^2 in (7.3.7.3) is the square of the "*orthogonality integral*" according to Pauling and Bright Wilson (1935).

In the original work of Heitler and London (1927) this integral, represented as S, arises from the consideration that a linear combination of two purely orthogonal wave functions $\psi_1\varphi_2$ (with electron 1 localized on nucleus a and electron 2 localized on nucleus (b) and $\psi_2\varphi_1$ (with electron 1 localized on nucleus a and electron 2 localized on nucleus (b) demands

$$\alpha = a\psi_1\varphi_2 + b\psi_2\varphi_1 \tag{7.3.8.1}$$

$$\beta = c\psi_1\varphi_2 + d\psi_2\varphi_1 \tag{7.3.8.2}$$

$$a^2 + b^2 + 2abS = 1 \tag{7.3.8.3}$$

$$c^2 + d^2 + 2cdS = 1 \tag{7.3.8.4}$$

$$ac + bd + (ad + bc)S = 0 \tag{7.3.8.5}$$

hence

$$\alpha = \frac{1}{\sqrt{2+2S}}(a\psi_1\varphi_2 + b\psi_2\varphi_1) \tag{7.3.9.1}$$

$$\beta = \frac{1}{\sqrt{2 - 2S}}(a\psi_1\varphi_2 - b\psi_2\varphi_1) \tag{7.3.9.2}$$

Adopting the Pauling–Bright Wilson notation, the corresponding secular equation is

$$\begin{vmatrix} H_{I\ I} - W & H_{I\ II} - \Delta^2 W \\ H_{II\ I} - \Delta^2 W & H_{II\ II} - W \end{vmatrix} = 0 \tag{7.3.10}$$

The determinant is solved for the condition $H_{I\ I} = H_{II\ II}$, giving the symmetric and antisymmetric variational energy terms

$$W_S = \frac{H_{I\ I} + H_{I\ II}}{1 + \Delta^2} \tag{7.3.11.1}$$

$$W_A = \frac{H_{I\ I} - H_{I\ II}}{1 + \Delta^2} \tag{7.3.11.2}$$

The corresponding wave functions are, respectively

$$\psi_S = \frac{1}{\sqrt{2(1 + \Delta^2)}}[u_{1s_A}(1)u_{1s_B}(2) + u_{1s_B}(1)u_{1s_A}(2)] \tag{7.3.12.1}$$

$$\psi_A = \frac{1}{\sqrt{2(1 + \Delta^2)}}[u_{1s_A}(1)u_{1s_B}(2) - u_{1s_B}(1)u_{1s_A}(2)] \tag{7.3.12.2}$$

ψ_S is symmetric in both the positional coordinates of the electrons and of the nuclei, while ψ_A is antisymmetric in both sets of coordinates.

For the "*Coulomb integral*" $H_{I\ I}$ it is found on evaluation the following form (Pauling and Bright Wilson 1935):

$$\begin{aligned} H_{I\ I} = & \int\int u_{1s_A}(1)u_{1s_B}(2)(2W_H \\ & - \frac{e^2}{r_{B1}} - \frac{e^2}{r_{A2}} + \frac{e^2}{r_{12}} + \frac{e^2}{r_{AB}})u_{1s_A}(1)u_{1s_B}(2)dr_1dr_2 \\ = & 2W_H + 2J + J' + \frac{e^2}{r_{AB}} \end{aligned} \tag{7.3.13}$$

where W_H is the term appearing in the $1s_A$ wave equation of the hydrogen atom $-\frac{h^2}{8\pi^2 m_0}\nabla^2 u_{1s_A} - \frac{e^2}{r_A}u_{1s_A} = W_H u_{1s_A}$, and *J* and *J'* are, respectively

$$J = \frac{e^2}{a_0}\left[-\frac{1}{D} + \left(1 + \frac{1}{D}\right)\exp(-2D)\right] \quad (7.3.14)$$

$$J' = \frac{e^2}{a_0}\left[\frac{1}{D} - \left(\frac{1}{D} + \frac{11}{8} + \frac{3}{4}D + \frac{1}{6}D^2\right)\exp(-2D)\right] \quad (7.3.15)$$

For the "*exchange integral*" $H_{I\ II}$ the same treatment yields

$$\begin{aligned} H_{I\ II} = & \int\int u_{1s_A}(1)u_{1s_B}(2)(2W_H \\ & -\frac{e^2}{r_{A1}} - \frac{e^2}{r_{B2}} + \frac{e^2}{r_{12}} + \frac{e^2}{r_{AB}})u_{1s_B}(1)u_{1s_A}(2)dr_1dr_2 \\ = & 2\Delta^2 W_H + 2\Delta K + K' + \Delta^2\frac{e^2}{r_{AB}} \end{aligned} \quad (7.3.16)$$

with

$$K = -\frac{e^2}{a_0}\left(1 + \frac{1}{D}\right)\exp(-D) \quad (7.3.17)$$

and

$$\begin{aligned} K' = & \frac{e^2}{5a_0}\{\left(\frac{25}{8} - \frac{23}{4}D - 3D^2 - \frac{1}{3}\right)\exp(-2D) \\ & + \frac{6}{D}[\Delta^2(\gamma + \log D) + \Delta'^2 \mathcal{E}i(-4D) - 2\Delta\Delta'\mathcal{E}i(-2D)]\} \end{aligned} \quad (7.3.18)$$

The K' integral was first evaluated by Sugiura (1927), after a first approximate evaluation operated by Heitler and London (1927).

$\mathcal{E}i$ in Eq. (7.3.18) is a function known as "*integral logarithm*" (with t as argument; see Jahnke and Emde 1945):

$$\mathcal{E}i(-x) = -\int_x^\infty \frac{\exp(-t)}{t}dt \quad (7.3.19)$$

γ is the Euler constant ($\gamma = 0.5772...$) and $\Delta' = \left(1 - D + \frac{1}{3}D^2\right)\exp(D)$.

The values attained by the coulomb and exchange integrals of the Heitler–London treatment are listed in Table 7.5 for various values of the internuclear distance r_{AB}, together with the energies of the attractive and the repulsive states W_A and W_S, respectively.

When plotting the various energies as a function of internuclear distance (Fig. 7.13) we may see that the antisymmetric state results always in a net repulsion,

Table 7.5 Energy values of the Heitler–London treatment of the H_2 molecule. Energies are in Ry

r_{AB}/a_0	W_A	W_S	$H_{I\ I}$	$H_{I\ II}$
0	∞	∞	∞	∞
0.5	1.7979	−0.4629	−0.3751	−0.6702
1	−0.5904	−1.993	−1.8084	−2.1794
1.5	−1.3619	−2.2255	−2.0208	−2.3233
2	−1.6921	−2.2073	−2.0383	−2.2403
2.5	−1.848	−2.1412	−2.0254	−2.1455
3	−1.9295	−2.0834	−2.0139	−2.0726
4	−1.9824	−2.0228	−2.0033	−2.0203

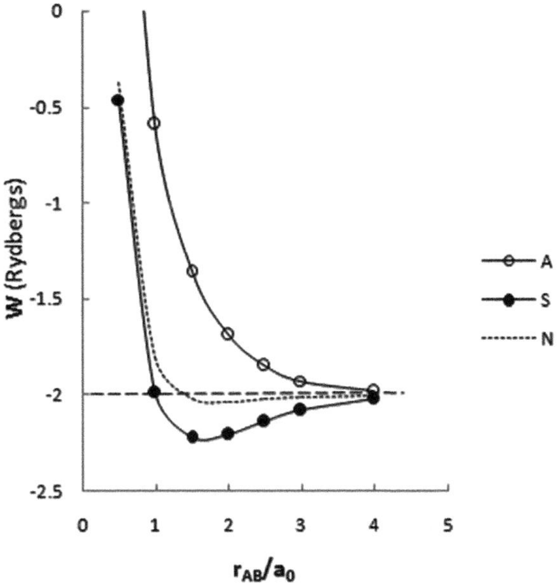

Fig. 7.13 Repulsive (A), attractive (S) and non-interacting (N) states in the Heitler–London treatment of the H_2 molecule

while the symmetric state defines a potential well, identifying an equilibrium distance at ~0.8 *Å* (not far from the observed one which is 0.74 *Å*).

Proceeding as done for the H_2^+ molecular ion we may calculate the dissociation energy, which result to be slightly underestimated with respect to the experimental observation (0.115 against 0.173 *Hartree*). The curvature of the least-squares fitting potential function near the minimum implies a stretching frequency of the nuclei of 4800 cm^{-1}, while the band spectral value is 4317.9 cm^{-1} (Pauling and Bright Wilson 1935). More interestingly, the small attractive state depicted by the $H_{I\ I}$ integral [and corresponding to the hypothetical case of non-interacting atoms, where the wave function is represented barely by $\psi_I = u_{1s_A}(1)u_{1s_B}(2)$] tells us that the bonding energy of the H_2 molecule is in the main "*resonance*" or "*interchange*" energy.

Introducing the spin (so far neglected), and obeying the Pauli exclusion principle, the orbital wave functions must be multiplied by spin functions in order to get normalized spin-orbit functions. The terms on the left of equations (7.3.12.1, 7.3.12.2) are then replaced by

$$\psi_S \cdot \frac{1}{\sqrt{2}}[\alpha(1)\beta(2) - \beta(1)\alpha(2)] \tag{7.3.20.1}$$

$$\psi_A \cdot \alpha(1)\alpha(2) \tag{7.3.20.2}$$

$$\psi_A \cdot \frac{1}{\sqrt{2}}[\alpha(1)\beta(2) + \beta(1)\alpha(2)] \tag{7.3.20.3}$$

$$\psi_A \cdot \beta(1)\beta(2) \tag{7.3.20.4}$$

We see thus that the normal (attractive) state of the H_2 molecule is a singlet, with the spins of the electrons opposed, while the antisymmetric (repulsive) condition leads to a triplet. We may also infer that, being actually 4 the energy conditions attainable when the two nuclei approach each other, the probability of forming an H_2 molecule through an attractive state is 1/4.

7.3.2 *The First Variational Approaches*

As shown by Ekhart (1930), the variational approach is conveniently represented in terms of an integral of form

$$E = \frac{\int \varphi^* H \varphi d\tau}{\int \varphi^* \varphi d\tau} \tag{7.3.36}$$

where H is the Hamiltonian operator and φ is a function which satisfies certain boundary conditions but is otherwise arbitrary. Because it contains, so to say, "*some variable parameters*" (Weinbaum 1933), it has the property that the lowest value of energy W obtained from varying the numerical parameters is the best approximation to the value of E, and that E-W is always positive or zero.

Adopting the scheme devised by Slater:

	I	II	III	IV
ψ	+	−	+−	
φ	−	+		+−

Weinbaum (1933) developed a trial function of the type

$$\psi = c(\psi_I - \psi_{II}) + (\psi_{III} + \psi_{IV}) \tag{7.3.37}$$

The first term $(\psi_I - \psi_{II})$is the atomic Heitler–London contribution and the second term $(\psi_{III} + \psi_{IV})$is the ionic Heitler–London contribution. The results obtained by varying c would then be comparable with the Heitler–London–Sugiura results. Weinbaum (1933) however treated the function as an implicit 3 parameters function (i.e. $1/r_{AB}$, Z = shielding constant and $\rho = Z \times r_{AB}$) and then reduced it to a two-parameter function letting the shielding constant to be Z = 1.

Weinbaum (1933) developed then, at first, expressions of the ψ-ψ and ψ-φ interactions as independent from the internuclear distance r_{AB}:

$$\begin{aligned}\int \psi_I H_I \psi_I d\tau_I &= \int \psi_I\left(-\frac{Z^2}{2} + \frac{Z-1}{r_{A1}} - \frac{1}{r_{B1}}\right)\psi_I d\tau_I \\ &= -\left[\frac{Z^2}{2} + (1-Z)Z + I_1\right]\end{aligned} \tag{7.3.38.1}$$

$$\int \psi_I H_I \varphi_I d\tau_I = \left[-\frac{Z^2}{2}s + (2-Z)I_2\right] \tag{7.3.38.2}$$

where “s” represent an integrals of type $\int \psi\varphi d\tau$. Then, through various represent an integral involving developments in series of I = ZF with F = F(ρ), he arrived to

$$W = \frac{(\lambda Z^2 + \mu Z)}{\nu} + \frac{Z}{\rho} \tag{7.3.39}$$

where

$$\lambda = (c^2+1)(1+2sF_2 - s^2) + 4cF_2 \tag{7.3.40.1}$$

$$\mu = (c^2+1)(F_6 - 2 - 2F_1 - 4sF_2) + F_3 + c^2F_4 - 4c(s + sF_1 + 2F_2 - s) \tag{7.3.40.2}$$

$$\nu = \left[(c^2+1)(s^2+1) + 4cs\right] \tag{7.3.40.3}$$

The lowest value of W was found in this way by Weinbaum (1933) with p = 1.67 and c = 6.322 (W = −1.1187 *Ry*, corresponding to a dissociation energy ΔE of 4.00 *eV*.

Minimizing further W with respect to the shielding constant Z (Eq. 7.3.39) Weinbaum (1933) obtained

$$\frac{dW}{dZ} = \left(\frac{2\lambda}{\nu}\right)Z + \frac{\mu}{\nu} + \frac{1}{\rho} \tag{7.3.41.1}$$

$$Z_{min} = -\left(\frac{\nu}{2\lambda}\right)\left(\frac{\mu}{\nu} + \frac{1}{\rho}\right) \tag{7.3.41.2}$$

$$W_{Z-min} = -\left(\frac{\nu}{4\lambda}\right)\left(\frac{\mu}{\nu} + \frac{1}{\rho}\right)^2 \tag{7.3.41.3}$$

Equation (7.3.41.3) is again a function of c and ρ, but the lowest value of W (−1.148 *Ry*) is attained with p = 1.69, c = 3.9, and Z = 1.193. As both the Wang-ionic type (Wang 1928) and the Rosen-ionic type (Rosen 1931) treatments gave the same internuclear distance, Weinbaum (1933) performed a last calculation posing $\rho = 1.7$ and varying c and s until the lowest value of W was attained (W = −1.1515 *Ry*, corresponding to a dissociation energy of 4.10 *eV*; with c = 5.7, s = 0.07 and Z = 1.190). This last energy is only 0.58 *eV* lower than what was experimentally observed.

7.3.3 The James–Coolidge and Kolos–Wolniewicz Procedures

Following the procedure adopted for helium by Hylleraas (1931), James and Coolidge (1933) introduced the inter-electronic distance r_{12} as an additional variational parameter. As done by Burrau (1927) for the molecular ion H_2^+, they adopted the prolate ellipsoidal coordinates. They then argued that the ground state of H_2 being a Σ state, the true function representing it:

- must have rotational symmetry about the nuclear axis, hence, only the difference between the azimuthal angles of the two electrons is necessary (and not their individual values);
- must be symmetrical in the two electrons and also in the nuclei;
- when one electron is at a large distance from the nuclei the wave function must approximate the product of an H-like function for the distant electron and an $H_2{}^+$-like function for the close electron.

The last consideration suggested them to insert in the trial function an exponential factor reducing to exp(-r) for either electron at a large distance.

Expressing the distances in terms of the Borh's radius ($R = R_{AB}/a_0$) the prolate ellipsoidal coordinates take the form

$$\lambda_1 = \frac{r_{1A} + r_{1B}}{R}; \quad \lambda_2 = \frac{r_{2A} + r_{2B}}{R} \tag{7.3.42.1}$$

$$\mu_1 = \frac{r_{1A} - r_{1B}}{R}; \quad \mu_2 = \frac{r_{2A} - r_{2B}}{R} \tag{7.3.42.2}$$

$$\rho = \frac{2r_{12}}{R} \tag{7.3.43}$$

The trial function is a power series in the five variables, with coefficients to be determined by variation in order to return the minimum energy (by definition always above the true value, experimentally observed):

$$\psi = \sum \mathrm{mnjkp} C_{\mathrm{mnjkp}} [\mathrm{mnjkp}] \tag{7.3.44}$$

where [mnjkp] stands for

$$\frac{1}{2\pi} \exp[-\delta(\lambda_1 + \lambda_2)] \left[\lambda_1^m \lambda_2^n \mu_1^j \mu_2^k \rho^v + \lambda_1^n \lambda_2^m \mu_1^k \mu_2^j \rho^v\right] \tag{7.3.45}$$

The summation is extended over positive or zero values of the indices, with the unique restriction that j + k must be even.

The variational limit is found by computing the minimum value of the integral $\int\int \psi^* H \psi dV_1 dV_2$ extended over the space occupied by the two electrons (V_1, V_2). This is achieved by changing opportunely the C coefficients, which are subjected to the normalizing condition:

$$\int\int \psi^* \psi dV_1 dV_2 = 1 \tag{7.3.46}$$

For a given choice of ρ and R, the values of the C's minimizing the energy are those satisfying the system:

$$\begin{aligned}(H_{11} - LS_{11})C_1 + (H_{12} - LS_{12})C_2 \\ + \ldots + (H_{1s} - LS_{1s})C_s = 0\end{aligned} \tag{7.3.47.1}$$

$$(H_{12} - LS_{12})C_1 + (H_{22} - LS_{22})C_2 + \ldots + (H_{2s} - LS_{2s})C_s = 0 \tag{7.3.47.2}$$

$$\ldots\ldots\ldots\ldots\ldots\ldots\ldots\ldots\ldots\ldots \tag{7.3.47.3}$$

$$(H_{1s} - LS_{1s})C_1 + (H_{2s} - LS_{2s})C_2 + \ldots + (H_{ss} - LS_{ss})C_s = 0 \tag{7.3.47.4}$$

where H and S are the matrix components of the Hamiltonian energy operator and of unity, respectively and L is a Lagrangian multiplier. Adopting the same symbols, the secular equation is

Table 7.6 Variational total energy (a.u.) and binding energy (*eV*) of H_2 compared to experimental observation [4.46 ± 0.04 *eV*, according to Richardson and Davidson (1929) plus a zero-point contribution of 0.27 *eV*]. The internuclear distance R is expressed in Bohr's radii. From James and Coolidge (1933)

Function	Total energy	Binding energy	R
One term	−2.189	−2.56	1.40
5 terms	−2.33290	−4.507	1.40
11 terms	−2.34580	−4.682	1.40
13 terms	−2.34693	−4.697	1.40
Without r_{12}	−2.3154	−4.27	1.40
Heitler-London	−2.21	−2.9	1.40
Heitler-London	−2.24	−3.2	1.51
Wang	−2.278	−3.76	1.42
Rosen	−2.297	−4.02	1.416
Observed		−4.73 ± 0.04	1.40

$$\begin{vmatrix} H_{11} - LS_{11} & H_{11} - LS_{11} & \cdots & H_{11} - LS_{11} \\ H_{12} - LS_{12} & H_{22} - LS_{22} & \cdots & H_{2s} - LS_{2s} \\ \cdots\cdots & \cdots\cdots & \cdots & \cdots\cdots \\ H_{1s} - LS_{1s} & H_{2s} - LS_{2s} & \cdots\cdots & H_{ss} - LS_{ss} \end{vmatrix} = 0 \tag{7.3.48}$$

It is important here to stress that, although the exponent δ should attain the limiting value R/2 (at least at a large distance), it is treated by James and Coolidge (1933) as an arbitrary parameter to be varied as a power series function in λ_1 and λ_2. It is thus seen that the series rapidly converges after a few terms (see Table 7.6).

Kolos and Wolniewicz (1965) developed further the James–Coolidge approach, starting from the consideration that the Heitler–London wave function

$$\Psi = \psi_A(1)\psi_B(2) + \psi_B(1)\psi_A(2) \tag{7.3.49}$$

represents the exact solution of the electronic Schrödinger equation for two H atoms at an infinite distance while it fails at a finite distance because it does not take into account the correlation effects. The James and Coolidge wave function is adequate for small and intermediate distance but it must be modified to attain the asymptotic form (7.3.49) for R →∞.

In prolate elliptical coordinates, adopting the Kolos–Wolniewicz symbolism, the asymptotic wave function (7.3.49) has the form

$$\Psi = \exp\left[-\frac{1}{2}R(\xi_1 + \xi_2)\right] \cosh\left[-\frac{1}{2}R(\eta_1 - \eta_2)\right] \tag{7.3.50}$$

while, for the ground state of H_2, the James–Coolidge wave function is

$$\Psi = \sum_i c_i[\Phi_i(1,2) + \Phi_i(2,1)] \tag{7.3.51}$$

with the basis sets are represented by

$$\Phi_i(1,2) = \exp[-\alpha(\xi_1 + \xi_2)]\xi_1^{r_i}\eta_1^{s_i}\xi_2^{\overline{r}_i}\eta_2^{\overline{s}_i}\rho^{\mu_i} \tag{7.3.52}$$

and with c_i and α as variational parameters.

To achieve generalization, Kolos and Wolniewicz (1965) replaced Eq. (7.3.52) with

$$\Phi_i = \exp[-\alpha\xi_1 - \overline{\alpha}\xi_2]\cosh(\beta\eta_1 + \overline{\beta}\eta_2)\xi_1^{r_i}\eta_1^{s_i}\xi_2^{\overline{r}_i}\eta_2^{\overline{s}_i}\rho^{\mu_i} \tag{7.3.53}$$

where $\alpha, \overline{\alpha}, \beta, \overline{\beta}$ are variational parameters.

This new form is flexible enough to reproduce at best the dissociation energy of the molecule and also the energy of the excited states. In fact, in Cartesian coordinates the total wave function (7.3.51) takes the form

$$\Psi = \sum_i c_i\left[\Phi_i(1,2)x_1^l \pm \Phi_i(2,1)x_2^l\right] \tag{7.3.54}$$

where x_i denotes the Cartesian coordinate of the ith electron perpendicular to the molecular axis. For l = 0 and l = 1, Eq. (7.3.54) represents wavefunctions for Σ; and Π states, respectively. The + or - sign denotes a singlet or a triplet state. The gerade (*g*) or ungerade (*u*) character of the wavefunction is determined by the parity of $s_i + \overline{s}_i + l$ in (7.3.53 and 7.3.54). It is of interest to recall that the total energy estimated by Kolos and Wolniewicz (1965) was 13 cm^{-1} below the experimental value (a bias corresponding to $\sim 1 \times 10^{-4}$ *Ry*), violating apparently the basic variational principle. This discrepancy was finally proven to be artificial, by refinement of the experimental measurements (Bright Wilson 1977). In Fig. 7.14 we see a comparative plot of the Kolos–Wolniewicz and the modified Heitler–London approaches (after Wang 1928; and Rosen 1931). The lowering of the potential energy when approaching the minimum from large interionic distance arises from the overlapping fields of electrons and nuclei. At short distance the nuclear effects are more sensible and E_P (PE in Fig. 7.14) rises sharply. The kinetic energy E_K (KE in Fig. 7.14) first falls and then rises again steeply with the decrease of internuclear distance. Its most stabilizing effect (in terms of binding energy E) is seen at internuclear distances sensibly higher with respect to the well bottom. Heitler and London showed that the major part of the binding energy in their treatment came from "*electron exchange*." The wave function contains two terms differing by the exchange of the two electrons (or equivalently, by the interchange of the two nuclear identifications). The expectation value of the energy thus has a cross-term proportional to

Fig. 7.14 The Kolos–Wolniewickz total (E), Kinetic (KE) and Potential (PE) scaled electronic energies (black lines), compared with the results of the Heitler–London approach (red lines)

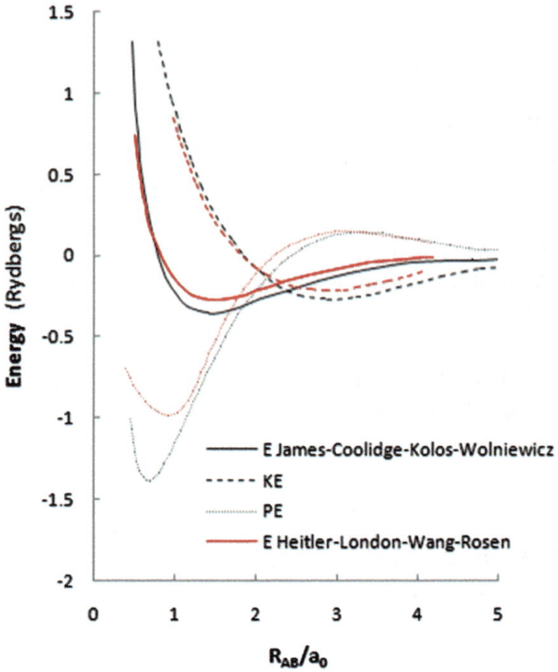

$$\int a(1)b(2)Ha(2)b(1)d\tau \quad (7.3.55)$$

called the "*resonance energy*" by Heisenberg (1926, 1927a). This is the major stabilizing term in the Heitler–London treatment (Bright Wilson 1977).

7.3.4 Molecular Orbitals and the Concept of "Self-Consistent Field"

A potential field is "*self-consistent*" whenever "*the wave function of the electron, determined by solving Schrödinger's equation for the field of nuclei and electron itself leads to a charge distribution which, together with that of nuclei leads, by the principles of electrostatics, to the potential function already assumed*" (Slater, 1963). Except for the special case of H_2^+, where each center attracts the single electron according to the inverse-square law, the two-center problem cannot be solved precisely (even in the simplest case represented by H_2), due to the presence of a smeared-out charge arising from the other electron. We are thus forced to use approximations, such as those adopted in the LCAO procedure, refining them at the light of the variational principle. We may apply the variational principle by setting up a determinantal function formed by spin orbitals and varying arbitrarily them in order to minimize the energy. I resume hereafter the treatment of Slater (1963)

Let us denote u_g the wave function of the electron in the two-center field, symmetric, in the same way of the $1\sigma_g$ is symmetric in H_2^+, and satisfying the condition of the self-consistent field. If we have two electrons in this orbital, with opposed spins, the two occupied spin orbitals can be denoted $u_{g\alpha}$ and $u_{g\beta}$. The determinant will then be

$$\frac{1}{\sqrt{2}}\begin{vmatrix} u_g(1)\alpha(1) & u_g(2)\alpha(2) \\ u_g(1)\beta(1) & u_g(2)\beta(2) \end{vmatrix} = u_g(1)u_g(2)\frac{\alpha(1)\beta(2)-\beta(1)\alpha(2)}{\sqrt{2}} \tag{7.3.56}$$

In Eq. (7.3.56) (Slater, 1963) depicted the determinant as a product of coordinates $[u_g(1)u_g(2)]$ and spins $[\frac{\alpha(1)\beta(2)-\beta(1)\alpha(2)}{\sqrt{2}}]$. The function of coordinates is symmetric on interchange of coordinates, hence we may vary the molecular orbital u_g in order to minimize the energy. I recall now that the normalized atomic orbital for the electron about nucleus (a) has the form $\sqrt{\frac{\alpha^3}{\pi}}\exp(-\alpha r_a)$ and that the "*overlap integral*" (usually denoted S) has the form

$$S = \frac{\alpha^3}{\pi}\int \exp[-\alpha(r_a + r_b)dv] \tag{7.3.57}$$

A good trial function for the two-center problem is then (Slater, 1963):

$$u_u = \left[\frac{2\pi}{\alpha^3}(1+S)\right]^{-1/2}[\exp(-\alpha r_a) + \exp(-\alpha r_b)] \tag{7.3.58}$$

The value of α appropriate for the hydrogen atom is $\alpha = 1$, but we may eventually change it, in order to minimize the energy. For the moment we will keep the variational term α constant at 1 and will proceed to evaluate all the states (besides u_g, already treated).

The antisymmetric function corresponding to u_g and arising from the LCAO procedure is

$$u_u = \left[\frac{2\pi}{\alpha^3}(1-S)\right]^{-1/2}[\exp(-\alpha r_a) - \exp(-\alpha r_b)] \tag{7.3.59}$$

Combining the functions u_g and u_u with the spin functions α or β one gets four spin orbitals. As shown by Slater (1932), there are 6 ways of assigning the two electrons to the 4 spin orbitals, in order to be consistent with the exclusion principle. We may denote them g^+g^-, u^+u^-, g^+u^+, g^+u^-, g^-u^+, g^-u^-. The first two represent a

Table 7.7 Wave functions of Configurations. From Slater (1963) $a = \sqrt{\frac{\pi}{\alpha^3}}\exp(-\alpha r_{1a})$; $b = \sqrt{\frac{\pi}{\alpha^3}}\exp(-\alpha r_{1b})$

Configuration	State	Function
g^+g^-	$^1\Sigma_{g+}$	$\frac{a(1)a(2)+b(1)b(2)+a(1)b(2)+b(1)a(2)}{2(1+S)} \times \frac{\alpha(1)\beta(2)-\beta(1)\alpha(2)}{\sqrt{2}}$
u^+u^-	$^1\Sigma_{g-}$	$\frac{a(1)a(2)+b(1)b(2)-a(1)b(2)-b(1)a(2)}{2(1-S)} \times \frac{\alpha(1)\beta(2)-\beta(1)\alpha(2)}{\sqrt{2}}$
g^+u^+	$^3\Sigma_u$	$\frac{b(1)a(2)-a(1)b(2)}{\sqrt{2(1-S^2)}} \times \alpha(1)\alpha(2)$
$(g^+u^- + g^-u^+)$	$^3\Sigma_u$	$\frac{b(1)a(2)-a(1)b(2)}{\sqrt{2(1-S^2)}} \times \frac{\alpha(1)\beta(2)+\beta(1)\alpha(2)}{\sqrt{2}}$
g^-u^-	$^3\Sigma_u$	$\frac{b(1)a(2)-a(1)b(2)}{\sqrt{2(1-S^2)}} \times \beta(1)\beta(2)$
$(g^+u^- - g^-u^+)$	$^1\Sigma_u$	$\frac{a(1)a(2)-b(1)b(2)}{\sqrt{2(1-S^2)}} \times \frac{\alpha(1)\beta(2)-\beta(1)\alpha(2)}{\sqrt{2}}$

singlet, and the third one and the last one are two components of a triplet with M_S = 1 and $M_S = -1$, respectively. The fourth and the fifth can be composed to give the third component of the triplet for $M_S = 0$ and another singlet:

$$g^+u^- + g^+u^+ = [g(1)u(2) \mp u(1)g(2)][\alpha(1)\beta(2) \pm \beta(1)\alpha(2)] \tag{7.3.60}$$

The "*Wave functions of Configurations*" (Slater 1963) formed from 1s hydrogen orbitals are listed in Table (7.7). The functions are written in terms of the atomic orbitals a and b.

To set up the matrix elements of the energy the following integrals are needed (see Table 3.1 in Slater 1963) and his Eqs. (4−5, 6):

$$\int a(1)(-\nabla_i^2)a(1)dv_1 = \alpha^2 \tag{7.3.61.1}$$

$$\int a^2(1)\left(-{}^2\!/_{r_{1a}}\right)dv_1 = -2\alpha \tag{7.3.61.2}$$

$$\int a(1)b(1)dv_1 = S = \left(1 + w + \frac{w^2}{3}\right)\exp(-w) = 1 - \frac{1}{6}w^2 + \frac{1}{24}w^4 - \ldots \tag{7.3.61.3}$$

$$\int a^2(1)\left(-2/r_{1b}\right)dv_1 = \alpha J = \alpha\left[-\frac{2}{w} + \left(2 + \frac{2}{w}\right)\exp(-2w)\right]$$
$$= \alpha\left(-2 + \frac{4}{3}w^2 - \frac{4}{3}w^3 + \frac{4}{5}w^4 - \ldots\right) \tag{7.3.61.4}$$

$$\int a(1)b(1)\left(-2/r_{1b}\right)dv_1 = \alpha K = -\alpha(2 + 2w)\exp(-w)$$
$$= \alpha\left(-2 + w^2 - \frac{2}{3}w^3 + \frac{1}{4}w^4 - \ldots\right) \tag{7.3.61.5}$$

$$\int a(1)\left(-\nabla_1^2/r_{1b}\right)b(1)dv_1 = -\alpha^2(K + S) = \alpha^2\left(1 + w - \frac{1}{3}w^2\right)$$
$$= \alpha^2\left(1 - \frac{5}{6}w^2 + \frac{2}{3}w^3 - \frac{3}{24}w^4 + \ldots\right) \tag{7.3.61.6}$$

$$\int a^2(1)b^2(2)\left(2/r_{12}\right)dv_1dv_2 = [aa|bb] = \alpha J' = \alpha\left[2w^{-1} - \left(2w^{-1} + \frac{11}{4} + \frac{3}{2}w + \frac{1}{3}w^2\right)\exp(-2w)\right]$$
$$= \alpha\left(\frac{5}{4} - \frac{1}{6}w^2 + \frac{1}{30}w^4 - \ldots\right) \tag{7.3.61.7}$$

$$\int a(1)b(1)a(2)b(2)\left(2/r_{12}\right)dv_1dv_2 = [ab|ab] = \alpha K'$$
$$= \frac{2}{5}\alpha\left\{\begin{array}{c} -\left(-\frac{25}{8} + \frac{23}{4}w + 3w^2 + \frac{1}{3}w^3\right)\exp(-2w) \\ + 6w^{-1}\left[S^2(C + \ln w) + S'^2\mathrm{Ei}(-4w) - 2SS'\mathrm{Ei}(-2w)\right]\end{array}\right\}$$
$$= \alpha\left[\frac{5}{4} - \frac{1}{6}w^2 + \left(\frac{3}{50} + \frac{8}{75}\ln 4\right)w^4 - \ldots\right] \tag{7.3.61.8}$$

$$\int a^2(1)a(2)b(2)\left(2/r_{1b}\right)dv_1dv_2 = [aa|ab]$$
$$= \alpha\left[\left(\frac{5}{8}w^{-1} + \frac{1}{4} + 2w\right)\exp(-w) + \left(-\frac{5}{8}w^{-1} - \frac{1}{4}\right)\exp(-3w)\right] \tag{7.3.61.9}$$
$$= \alpha\left(\frac{5}{4} - \frac{7}{24}w^2 + \frac{3}{32}w^4 - \ldots\right) = \alpha L$$

$$\int a^2(1)a^2(2)\left(2/r_{1b}\right)dv_1dv_2 = \frac{5}{4}\alpha \tag{7.3.61.10}$$

Besides the variational parameter α, also the internuclear distance r_{AB} plays a role in terms of $w = \alpha\frac{r_{AB}}{a_0}$. Integrals operate over the volume space occupied by one or both electrons. Their numerical values are listed in Table 7.8 for various values of the internuclear distance.

Table 7.8 Numerical values of the (Heitler–London) integrals for various values of internuclear distance (from Slater 1963)

r_{AB}/a_0	S	J	K	J'	K'	L
0	1.0000	−2.0000	−2.0000	1.2500	1.2500	1.2500
0.5	0.9603	−1.7927	−1.8196	1.2103	1.1353	1.1816
1	0.8584	−1.4587	−1.4715	1.1090	0.8733	1.0141
1.5	0.7252	−1.1674	−1.1157	0.9807	0.5937	0.8107
2	0.5865	−0.9451	−0.8120	0.8519	0.3683	0.6162
2.5	0.4583	−0.7811	−0.5746	0.7368	0.2132	0.4512
3	0.3485	−0.6601	−0.3983	0.6396	0.1170	0.3215
4	0.1893	−0.4992	−0.1832	0.4951	0.0312	0.1540

To appreciate the significance of the various terms, we must remind that the energy E of the system may be written in terms of α and w as follows:

$$E = E_T + E_V = \alpha^2 F_1(w) + \alpha F_2(w) \tag{7.3.62}$$

where the first term on the right represents the kinetic energy (E_T) and the second one is the potential energy (E_V). E depends thus on α both directly and indirectly through w. If we want to apply the variational principle and find the value of α that minimize E for a given value of $w = \alpha \frac{r_{AB}}{a_0} = \alpha R$, we must perform a partial derivative as follows (Slater, 1963):

$$2\alpha F_1 + \alpha^2 R \frac{dF_1}{dw} + F_2 + \alpha R \frac{dF_2}{dw} = 0 \tag{7.3.63}$$

obtaining:

$$\alpha = -\left(F_2 + w\frac{dF_2}{dw}\right) \Big/ \left(2F_1 + w\frac{dF_1}{dw}\right) \tag{7.3.64}$$

Through various passages Slater (1963) found then that the relation of the two forms of energy with the various integrals are as follows:

$$E_T = \frac{2\alpha^2}{1 \pm S^2}\left(1 \mp KS \mp S^2\right) \tag{7.3.65}$$

$$E_V = \frac{\alpha}{1 \pm S^2}\left(-4 + 2J + J' \pm 4KS \pm K'\right) \tag{7.3.66}$$

Denoting H_{11}, H_{22} the diagonal terms of the energy matrix, and H_{12} the non-diagonal term, we have (Slater, 1963):

$$H_{11} = \alpha^2\left[\frac{2(1-S-K)}{1+S}\right] + \alpha\left[\frac{2(-2+J+2K)}{1+S} + \frac{5/8+J'/2+K'+2L}{(1+S)^2} + \frac{2}{w}\right] \tag{7.3.67.1}$$

$$H_{22} = \alpha^2\left[\frac{2(1+S+K)}{1-S}\right] + \alpha\left[\frac{2(-2+J-2K)}{1-S} + \frac{5/8+J'/2+K'-2L}{(1-S)^2} + \frac{2}{w}\right] \tag{7.3.67.2}$$

$$H_{12} = \alpha^2\left(\frac{5/8-J'/2}{1-S^2}\right) \tag{7.3.67.3}$$

where, again, the terms proportional to α^2 represent the kinetic energy and those proportional to α the potential energy. Moreover, within these last, the first term in square bracket is the attraction between electrons and nuclei, the second one is the repulsion between electrons, and the third one is the repulsion between nuclei.

7.3.5 The Secular Matrix with Molecular Orbitals

The diagonal matrix elements are plotted in Fig. 7.15 against the internuclear distance for $\alpha = 1$.

Their numerical values are listed in Table 7.9.

The first diagonal tern of the energy matrix (H_{11}) represents the Hartree–Fock ground-state energy. We may see in Fig. (7.17) that it depicts a potential well (with the minimum at about the experimentally observed equilibrium distance), but it rises too rapidly with the increase of R. The second diagonal term (H_{22}) does not depict

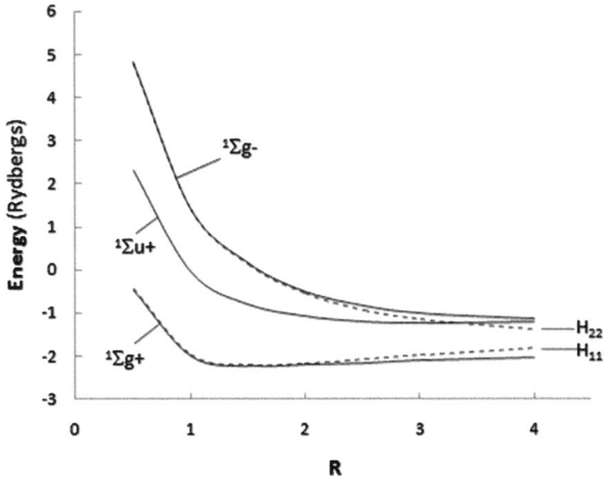

Fig. 7.15 Energy of the H_2 molecule at various internuclear distances for the three states $^1\Sigma_{g+}$, $^1\Sigma_{g-}$, $^1\Sigma_{u+}$ (black solid lines) and for the diagonal elements H_{11}, H_{22} (red-dashed lines). Modified from Slater (1963)

Table 7.9 Energies (Ry) of various states of H_2 (Molecular orbitals) at $\alpha = 1$ and different internuclear distances (Slater 1963)

R	H_{11}	H_{22}	H_{12}	$1\Sigma_{g+}$	$1\Sigma_{g-}$	$1\Sigma_{u+}$
0.5	−0.455	4.8109	0.2553	−0.4674	4.8232	2.3089
1	−1.9719	1.4077	0.2679	−1.993	1.4288	−0.0547
1.5	−2.1945	0.1173	0.284	−2.2289	0.1517	−0.7937
2	−2.1616	−0.5538	0.3034	−2.217	−0.4984	−1.0854
2.5	−2.0676	−0.9256	0.3248	−2.1535	−0.8397	−1.1984
3	−1.9656	−1.1505	0.3474	−2.0935	−1.0225	−1.2384
4	−1.7908	−1.373	0.3915	−2.0256	−1.1382	−1.1994

any minimum and decreases rapidly with the increase of R. Both terms approach asymptotically −2 + 5/8 *Ry* at an infinite distance while the correct energy would be just −2 *Ry* (i.e. 1 *Hartree*; the energy of two isolated H atoms). This is basically due to the fact that all integrals (S, J, K, J', K', L) attain zero at the limit while w becomes infinite (see Eqs. 7.3.67.1, 7.3.67.2). Having the diagonal (H_{11}, H_{22}) and non-diagonal (H_{12}) elements of the energy matrix, to get the values of $^1\Sigma g+$ and $^1\Sigma g-$ is "*simply a matter of solving a quadratic equation in E*" (Slater 1963).

It is of interest now to compare this behavior with that of the corresponding Heitler–London components of the $^1\Sigma g+$, $^1\Sigma g-$ energies. In the limit where the overlap integral S goes to zero (infinite distance) the two states are composed in equal part of the Heitler–London non-ionic wave function a(1)b(2) + b(1)a(2) and the Heitler–London ionic function a(1)a(2) + b(1)b(2). The energy of the Heitler–London non-ionic state is −2 *Ry* (i.e; that of two non-interacting H atoms) while that of the ionic state attains −2 + 5/4 *Ry*. In fact, in the ionic state two electrons are localized on the same nucleus [i.e. a(1)a(2) or b(1)b(2)], and the molecule is constituted by a proton (H^+) and an ion (H^-). H^+ has obviously no electronic energy, while for H^- we obtain from the computed integrals:

$$E_H = 2\alpha^2 - 4\alpha + \frac{5}{4}\alpha = 2\alpha^2 - \frac{11}{4}\alpha \tag{7.3.68}$$

and, with $\alpha = 1$

$$E_H = 2 - \frac{11}{4} = -2 + \frac{5}{4} \tag{7.3.69}$$

Removing the electron from the neutral atom H requires 1 *Ry* of ionization energy and, combining the electron with the neutral atom to form H^- releases 0.0555 *Ry* of electron affinity (Pekeris 1955). Hence the energy of the ionic state would be −2 + 1−0.0555 = −1.0555 *Ry*. This is much lower than what predicted by the combination of Heitler–London functions above described.

7.3.6 Some Anticipations on the Harthree–Fock–Roothan Theory

The determinantal approach has a close relationship with the Self-Consistent-Field (SCF) procedure introduced by Hartree (1928). Hartree's idea was that the motion of an electron in the electrical field of the nuclei and of the other electrons resents of an "*averaged*" charge distribution of the electrons (averaged over the wave function). The further development of Hartree's approach made by Fock (1930) led to the SCF-with-exchange procedure (generally known as Hartree–Fock method). Imagining to have N-electrons in the systems, we will have N spin orbitals. Setting up the determinantal function and computing the average value of the Hamiltonian we may demand that the average energy be set at minimum by an opportune choice of the (normalized and orthogonal) spin orbitals. This condition may be expressed in terms of differential equations (the Hartree–Fock or HF equations). For one of the spin orbitals, such, for instance $\Phi_i(x_1)$(with x_1 symbolizing the coordinates and spin of the first electron, the HF equation is (Slater, 1963):

$$\begin{aligned}&\left[-\nabla_1^2-\sum_a\frac{2Z_a}{r_{1\alpha}}\right]\Phi_i(x_1)+\left[\sum_j\int\Phi_j^*(x_2)\frac{2}{r_{12}}\Phi_j(x_2)dv_2\right]\Phi_i(x_1)\\&-\sum_j\left[\int\Phi_j^*(x_2)\frac{2}{r_{12}}\Phi_i(x_2)dv_2\right]\Phi_j(x_1)=E_i\Phi_i(x_1)\end{aligned}\quad(7.3.70)$$

The above equation is, in a general way, the Schrödinger equation for an electron which is not "*acting on itself*" but, rather, is constrained in the field of all (a) nuclei and (j) electrons (Slater 1963). The quantity is the kinetic energy operator. The summation in "a" is a sum over all "a" nuclei of the Coulomb attractions between the first electron and the nucleus (with Z_a as nuclear charge and r_{1a} as electron-nucleus distance). The summation $\sum_j \int \Phi_j^*(x_2)\frac{2}{r_{12}}\Phi_j(x_2)dv_2$ returns the potential energy (experienced at the position of the first electron) arising from the complete electron distribution, with a charge density $\sum_j \Phi_j^*(x_2)\Phi_j(x_2)$. The "*exchange term*" is represented by the summation over j: $-\sum_j\left[\int\Phi_j^*(x_2)\frac{2}{r_{12}}\Phi_i(x_2)dv_2\right]\Phi_j(x_1)$. The effect of the exchange correction is "*to decrease the charge density acting on the wave function*" (Slater 1963) so that, at the point where the electron is located (i.e. in the spin-orbital Φ_i), "*all charge of the same spin is removed, leaving only charge of the opposite spin*" (Slater 1963).

We may now generally state that, denoting χ_μ an atomic orbital and as $c_{i,k}$ the coefficients necessary to build up the ith molecular orbital ϕ_i, the linear combination of atomic orbitals (LCAO) may be represented as

$$\phi_i = \sum_{\mu} c_{i,\mu}\chi_\mu \tag{7.3.71}$$

If there are N orbitals, then the determinant generating the molecular wavefunction $\Psi(\mathbf{r})$ is of Nth degree and has N roots (eigenvalues). For each of these eigenvalues, there will be a unique set of solutions for the $c_{i,\mu}$'s determining the appropriate linear combination of the eigenfunctions of Eq. 7.3.70. The eigenfunctions may thus be defined as the best approximation to the solutions of Schrödinger's equation at the adopted theory level (Slater, 1963).

This is expressed by the Fock equation:

$$\mathsf{F}\Phi_i = \varepsilon_i\Phi_i \tag{7.3.72}$$

where Φ_i is the ith spin-orbital, the Fock operator F, is an effective one-electron Hamiltonian, and the eigenvalues, ε_i are the orbital energies of the various (i) spin orbitals.

The Fock operator is itself the summation of four different operators (Slater, 1963; cf. Eq. 7.3.70):

$$\mathsf{F}\Phi_i = \mathsf{T}\Phi_i + \mathsf{V}_{N-e} + \mathsf{J}\Phi_i + \mathsf{K}\Phi_i = \varepsilon_i\Phi_i \tag{7.3.73}$$

which, in atomic units, may be represented as follows:

$$\mathsf{T} = -\frac{1}{2}\nabla^2 \tag{7.3.74.1}$$

$$\mathsf{V}_{a-e} = \sum_{a}^{\text{nuclei}} -\frac{Z_a}{r_{ia}} \tag{7.3.74.2}$$

$$\mathsf{J}\Phi_j(1) = \left\{ \sum_{j}^{\text{occupied}} \int \Phi_j(2)\frac{1}{r_{12}}\Phi_j(2)dv_2 \right\}\Phi_j(1) \tag{7.3.74.3}$$

$$\mathsf{K}\Phi_j(1) = \left\{ \sum_{i}^{\text{occupied}} \int \Phi_i(2)\frac{1}{r_{12}}\Phi_j(2)dv_2 \right\}\Phi_j(1) \tag{7.3.74.4}$$

with T, V, J and K as kinetic, electrons-nuclei Coulomb attraction, electron-electron Coulomb repulsion, and "*exchange*" operators. Self-interaction and self-exchange are included in the summations, but cancel out one another, since the two integrals are equivalent when $i = j$.

Expanding the Fock operator in terms of N basis functions for n electrons a set of n × N matrix equations is formed (Roothan 1951; Hall 1951). Using a notation similar to that one adopted in GAUSSIAN (Foresman and Frisch 2015) for a closed shell (each MO doubly occupied) the Fock matrix may be written in the form:

$$F_{\mu\nu} = TH_{\mu\nu}^{core} + \sum_{\lambda}^{N}\sum_{\sigma}^{N} P_{\lambda\sigma}\left[(\mu\nu|\lambda\sigma) - \frac{1}{2}[(\mu\lambda|\sigma\nu)]\right] \tag{7.3.75}$$

with the core Hamiltonian accounting for the movement of individual electrons in the field of the bare nuclei (kinetic + nucleus-electron attraction operators):

$$H_{\mu\nu}^{core} = \int \chi_\mu(T + V_{a-e})\chi_\nu \tag{7.3.76}$$

and $P_{\lambda\sigma}$ as density matrix, whose elements are a sum over the occupied MOs:

$$P_{\lambda\sigma} = 2\sum_{i}^{n/2} c_{\lambda,i}c_{\sigma,i} \tag{7.3.77}$$

It must be stressed that the bi-electronic integrals in Eq. (7.3.75) follows the notation:

$$\begin{aligned}(\mu\nu|\lambda\sigma) &= \langle\mu\lambda \mid \nu\sigma\rangle \\ &= \int \chi_\mu^*(\mathbf{r}_1)\chi_\nu(\mathbf{r}_1)\frac{1}{r_{12}}\chi_\lambda^*(\mathbf{r}_2)\chi_\sigma(\mathbf{r}_2)d\mathbf{r}_1 d\mathbf{r}_2\end{aligned} \tag{7.3.78}$$

where the parenthesized term, replacing the ket-bra notation (see Sect. A.8), is intended to indicate both functions depending on electron 1 on the left and those depending on electron 2 on the right, with functions indicated solely by their indices (Foresman and Frisch 2015).

The Roothan–Hall matrix equation, to be solved self-consistently, is

$$FC = \varepsilon SC \tag{7.3.79}$$

with C as a matrix of MO coefficients, $\boldsymbol{\varepsilon}$ as a diagonal matrix of orbital energies and S as an overlap matrix of the terms:

$$S_{\mu\nu} = \int \chi_\mu\chi_\nu d\tau \tag{7.3.80}$$

The Hartree–Fock energy is

$$E_{total}^{HF} = E_{NR} + \frac{1}{2}\sum_{\mu\nu} P_{\mu\nu}\left(H_{\mu\nu}^{core} + F_{\mu\nu}\right) \tag{7.3.81}$$

with E_{NR} as nuclear repulsion energy.

7.3.7 A Simple Application of the Harthree–Fock–Roothan Procedure

In Fig. 7.16 we see the isodensity surface (0.08 e/$\mathring{A}^3$) obtained when resolving the isolated H_2 molecule at the Harthree-Fock theory level, with Slater-Type Orbitals (STO) replaced by a linear combination of 3 primitive Gaussian functions (i.e. STO −3G; see Sect. 12.4).

The adopted basis set for each H atom is composed of three Gaussians (3G) combined to mimic the radial part of the (exponential) Slater Type Orbital (STO) 1s of H. This replacement allows integrals and integral derivatives to be calculated efficiently. The atomic orbitals so obtained (AO's) are then combined linearly to build up the molecular orbital (MO). Denoting as $g_p(\zeta, r)$ each "*primitive*" Gaussian function, the construction may be represented as (see for instance Foresman and Frish 2015):

$$g_P(\varsigma, r) = Nx^l y^m z^n \exp(-\alpha r^2) \quad (7.3.82.1)$$

$$\chi_\mu = \sum_p d_{\mu,p} g_p(\alpha, r) \quad (7.3.82.2)$$

$$\phi_i = \sum_\mu c_{i,\mu} \chi_\mu \quad (7.3.82.3)$$

where ϕ_i is the ith molecular orbital, $c_{i,\mu}$ are the molecular orbital coefficients adopted to combine linearly the atomic orbitals χ_μ; $d_{\mu,p}$ is the fixed coefficient for the Gaussian primitive $g_p(\zeta, r)$; ζ is the orbital exponent controlling the tightness or

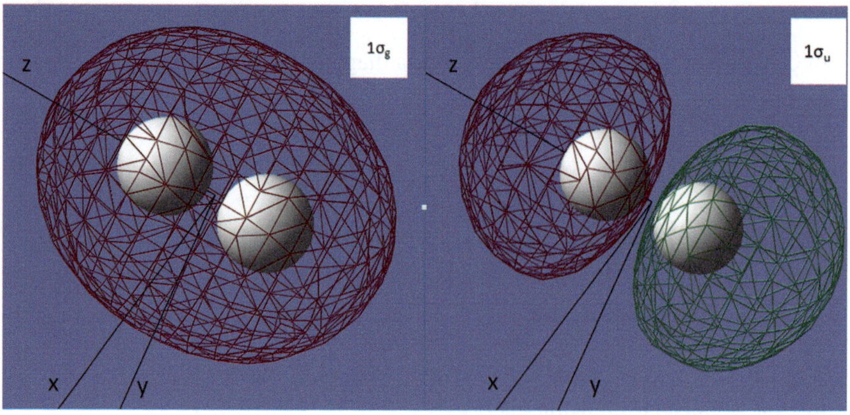

Fig. 7.16 The HF/STO−3G 0.08 $e/\mathring{A}^3$ isodensity surface (meshes; fine grid) of H_2 . The HOMO ($1\sigma_g$) and LUMO ($1\sigma_u$) are represented respectively at the left and right part of the figure. Calculations carried out with the GAUSSIAN03 code (Frisch et al. 2003)

diffuseness of each primitive function; the exponents on x, y, and z dictate its angular form (s. p. d, etc.) and the coefficient N is a normalization factors, such that the integral over all space of its square is 1.

The computation begins with the construction of the Harthree–Fock wavefunction $\Psi(\mathbf{r})$ combining the molecular orbitals and fulfilling the Pauli exclusion principle (i.e. an antisymmetric combination). In our case this is quite simple: we have two molecular orbitals and, being in a closed-shell, two spin conditions; i.e. a 2×2 matrix:

$$\Psi(\mathbf{r}) = \frac{1}{\sqrt{n!}} \begin{vmatrix} \phi_1(r_1)\alpha(1) & \phi_1(r_1)\beta(1) \\ \phi_1(r_2)\alpha(2) & \phi_1(r_2)\beta(2) \end{vmatrix} \tag{7.3.83}$$

The first term in Eq. (7.3.83) is the normalization factor, depending on the number n of electron in the molecule (in our case n = 2). The subscripts on the MO's run from 1 to n/2 while those on the position vectors from 1 to n. $\alpha(m)$ and $\beta(m)$ are the alpha and beta spin functions for the mth electron. Since each orbital is doubly occupied (closed-shell) there are as many rows and columns in the determinant as there are electrons in the system.

The wavefunction still is not precisely known because we do not know the molecular orbital coefficients $c_{i,\mu}$ (Eq. 7.3.82.3), thus a trial wavefunction is created[8] and the determinant is solved for the energy, by applying the variational theorem in an iterative way, until a predefined precision limit is attained:

$$\int \frac{\Psi^* H \Psi d\tau}{\Psi^* \Psi d\tau} = E_{var} \geq E_{true} \tag{7.3.84}$$

In practice, this corresponds to finding a minimum in the partial derivative of the energy, with respect to the selected set of orthonormal spin orbitals:

$$\frac{\partial E_{var}}{\partial \Phi_i} = 0 \tag{7.3.85}$$

At the variational limit, the overlap, kinetic, potential energy, and core Hamiltonian matrixes are

$$\underset{\text{Overlap}}{\begin{vmatrix} 1.000000 & 0.678069 \\ 0.678069 & 1.000000 \end{vmatrix}} \quad \underset{\text{Kinetic energy}}{\begin{vmatrix} 0.760032 & 0.253868 \\ 0.253868 & 0.760032 \end{vmatrix}} \quad \underset{\text{Potential energy}}{\begin{vmatrix} 1.899370 & 1.246140 \\ 1.246140 & 1.899370 \end{vmatrix}} \quad \underset{\text{Core Hamiltonian}}{\begin{vmatrix} -1.113934 & -0.992271 \\ -0.992271 & -1.113934 \end{vmatrix}}$$

[8] There are several methods of estimating the initial $c_{i,\mu}$ coefficients. The most common is using the eigenvectors from a truncated form of the actual Hamiltonian.

The eigenvalues are obtained by solving for ε Eq. (7.3.79), i.e.:

$$\varepsilon = \mathsf{FC}(\mathsf{SC})^{-1} \tag{7.3.86}$$

and result to be −0.59022 and 0.70065 *Hartree*. The molecule is a Symmetrical Top (i.e. a rigid body in which two of the moments of inertia along the three principal axes are equal (y, z rotational constants = 1977 × 10^9 *Hertz*; rotational T = 94.9 *K*). Its reduced mass is 1.0078 amu. The distance between the nuclear foci along the *z*-axis at the stationary point is, at the adopted theory level, 0.712 *Å* (i.e. ~1.36 a_0), the Hartree–Fock energy of the molecule is −1.1175059 *Hartree* (1 *Hartree* = e^2/a_0; i.e. the Coulomb repulsion between two electrons separated by 1 *bohr*; see Table 6.1 and note[9]). The nuclear repulsion energy is 0.4976 *Hartree*. The Highest Occupied Molecular Orbital HOMO ($1\sigma_g$) has an orbital energy eigenvalue (arising from the SCF density) of −0.590 *Hartree* and a kinetic energy of 0.604 *Hartree*. The Lowest Unoccupied Molecular Orbital (i.e. the first excited state, LUMO; $1\sigma_u$) has an orbital energy eigenvalue of 0.700 *Hartree* and a kinetic energy of 1.572 *Hartree*. The nuclei vibrate about the reduced mass center with a single stretching mode of frequency of 5481 cm^{-1} due to a force constant of 17.8401 *mDyne/Å*. The Zero-Point Correction is 0.012487 *Hartree* (0.34 *eV*). The vibrational contribution to the thermal energy of the molecule is 32.786 *kJ/mol* (much higher than the translational and rotational contributions which are, respectively 3.720 and 2.477 *kJ/mol*). Adopting for the dissociated atoms an electron energy of 0.5 *Hartree*, the binding energy is E = (−1.1175059 + 0.012487 + 2 × 0.5) = −0.1026359 *Hartree* (−2.793 *eV*). Adopting the 3G HF energy of the isolated H atoms (−0.4665) we get E = −0.172 *Hartree*) (−4.680 *eV*; quite close to the results of the 11 terms function of James and Coolidge 1933; cf. Table 7.6). We face thus also here the problem already encountered for the H_2^+ molecular ion. It should be noted for this purpose that the –V/T ratio between the potential and kinetic energies attains 1.9248, well below the theoretical value imposed by the Virial theorem, i.e. $\langle E_V\rangle = -2\langle E_T\rangle$.

7.4 The O_2 Molecule as an Example of Open-Shell Configuration

7.4.1 Premises

O_2 has symmetry $D_{\infty h}$ (linear with inversion center) but unusual properties arising from the fact that its outer electrons are not in a closed-shell. Its ground state is $^3\Sigma^-_g$ (i.e. triply degenerate, even, with sigma symmetry) as theoretically and experimentally observed. Due to this fact, the oxygen molecule in its ground state has a

[9] The *Hartree* energy is often denoted as Eh and corresponds to 4.359748×10^{-18} *J*; it corresponds also to 27.2114 *eV* and to 2 *Ry*.

magnetic moment arising from the spin. These magnetic moments are observed not only in the spectrum of the molecule but also in large-scale properties in the O_2 gas ($O_{2,g}$) and in the O_2 liquid ($O_{2,l}$). If $O_{2,l}$ is poured between the poles of a large magnet, it is oriented in the magnetic field "*in the same way that iron filings would be*" (Slater 1963). The theoretical study of O_2 was first afforded by Meckler (1953) and Kotani et al. (1957) in the framework of the Extended Configuration Interaction procedure. The two studies differ mainly in the functional form assigned to the atomic orbitals: a product of powers of r and Gaussian functions [i.e. $\exp(-\alpha r^2)$] for Meckler (1953) and the ordinary [i.e. $\exp(-\alpha r)$] expansion for Kotani et al. (1957).

The procedure in both cases begins with the construction of symmetry orbitals stemming from the 1s, 2s, and 2p AOs of the two atoms in the molecule.

7.4.2 The Extended Configuration Interaction Approach

We have already seen at Sect. (7.3.4) how the "*Wave functions of Configurations*" may be formed from 1s hydrogen orbitals by suitably combining the functions u_g and u_u with the spin functions α or β (see Table 7.7).

A diatomic molecule is symmetric upon reflection whenever a plane passing through the internuclear axis (and causing ϕ to pass to $-\phi$) cause the wave function to change sign or to remain unchanged.

We have also seen that in the case of a single electron, separating the variables with the aid of prolate-spheroidal coordinate, the wave function can be depicted as a product of two functions depending respectively on λ and μ:

$$\mathrm{u} = \mathrm{L}(\lambda)\mathrm{M}(\mu)\exp(\mathrm{i}m_l\phi) \tag{7.2.12}$$

By changing the sign of μ and rotating ϕ by 180° (i.e. by π radiants) the exponential factor $\exp(\mathrm{i}m_l\phi)$may remain unchanged or change sign. This means that $\mathrm{M}(\mu)$ is an even or an odd function of μ.

We need now to build up two-electron (orbital) functions of particular symmetry properties. This could be done at first sight by setting up a combined function in the form (Slater, 1963):

$$\begin{aligned} &\mathrm{F}_1(\lambda_1\mu_1)(\mathrm{i}m_l\phi_1)\mathrm{F}_2(\lambda_2\mu_2)(\mathrm{i}m_l\phi_2) \\ &\quad + \mathrm{F}_2(\lambda_1\mu_1)(\mathrm{i}m_l\phi_1)\mathrm{F}_1(\lambda_2\mu_2)(\mathrm{i}m_l\phi_2) \end{aligned} \tag{7.4.1.1}$$

This function is symmetric upon interchange of the coordinates of the two electrons (1 and 2) but does not have the proper symmetry upon change of signs of ϕ_1 and ϕ_2 (it does not become negative nor it transforms into itself). The correct expansion (Slater, 1963) requires accounting for the sign of the magnetic quantum number m_l:

Table 7.10 Ground configuration of O_2. Assignment of electrons to Spin-orbitals (from Slater 1963)

Assignment	π^{+}_{g+}	π^{-}_{g+}	π^{+}_{g-}	π^{-}_{g-}	M_L	M_S
1	1	1	0	0	2	0
2	1	0	1	0	0	1
3	1	0	0	1	0	0
4	0	1	1	0	0	0
5	0	1	0	1	0	−1
6	0	0	1	1	−2	0

$$\begin{aligned}&[\mathrm{F}_1(\lambda_1\mu_1)\mathrm{F}_2(\lambda_2\mu_2) \pm \mathrm{F}_2(\lambda_1\mu_1)\mathrm{F}_1(\lambda_2\mu_2)\\&\quad \times \{\exp[im_l(\phi_1 - \phi_2)] \pm \exp[-im_l(\phi_1 - \phi_2)]\}]\end{aligned} \tag{7.4.1.2}$$

The function with the plus signs depends on the ϕ through the factor $[\cos m_l(\phi_1\text{-}\phi_2)]$ and when the ϕ's change sign the function is unchanged. The function with the minus signs depends on the ϕ through the factor $[\sin m_l(\phi_1\text{-}\phi_2)]$ and, when the ϕ's change sign, the function changes of sign.

Applying the above equation to any pair of spin orbitals in the molecule (assuming them to be both occupied by electrons), and making their determinantal functions results in complete set of antisymmetric two-electron functions. A suitable linear combination of them can represent the (exact) solution of the Schrödinger equation. With the term "*configuration*" it is meant the particular linear combination of Slater's determinants adopted to conform the wave function. With "*interaction*" it is meant that the occupation arises actually from the selected configurational state functions (CSFs) built from spin orbitals. If the expansion accounts of all the possible CSF of appropriate symmetry, then it is an "*extended (or full) configuration interaction*".

As an example I may anticipate here that, in the particular case of O_2, at ground state the $1\sigma_g$, $1\sigma_u$, $2\sigma g$, $2\sigma_u$ and $3\sigma_g$ spin orbitals are fully occupied with 2 electrons. Moreover, there are four electrons in $1\pi_u$ but only 2 in the $1\pi_g$, which is capable of hosting 4 electrons. There are thus 4!/(2!2!) possible ways of electrons to occupy the spin orbitals (Table 7.10), resulting in 6 determinantal functions ψ to be suitably combined to depict the contribution of the $1\pi_g$ to the ground state energy arising from:

$$\Psi = \sum_{i=1}^{6} c_i \psi_i \tag{7.4.2}$$

Functions 2 and 5 in Table (7.10) are triplets ($M_S = 1,-1$) and the remaining ones are singlets ($M_S = 0$). Functions with $M_L = 0$ are Σ states and those with $M_L = 2$ are Δ states. Since all spin orbitals are of g type then also the corresponding function is even. As anticipated the ground state of O_2 is a triplet ($^3\Sigma_g^-$) and corresponds to the assignment 2 in the above table:

$$\psi_2 = \frac{1}{\sqrt{2}}\left[\pi_{g+}(1)\pi_{g-}(2) - \pi_{g-}(1)\pi_{g+}(2)\right]\alpha(1)\alpha(2) \tag{7.4.3}$$

The $2^{-1/2}$ term in (7.4.3) is the normalization factor. $\pi_{g+}(1)$is the corresponding function of electron (1) and $\alpha(1)$ its Pauli spin function (corresponding to + spin).

Besides the different functional form adopted to describe AOs, the main differences between the treatments of Meckler (1953) and Kotani et al. (1957) resides in the fact that, while Meckler explored the effect of internuclear distance on the molecular energy, Kotani and coworkers fixed arbitrarily the distance to a value near to that one experimentally observed. In both treatments the oxygen 1s, 2s, and 2p atomic orbitals were used to build three σ_g symmetry orbitals ($1\sigma_g$, $2\sigma_g$, $3\sigma_g$), their ungerade counterparts ($1\sigma_u$, $2\sigma_u$, $3\sigma_u$) one π_g, and one π_u. However, while Meckler (1953) disregarded the hybridization of 2s and 2p σ atomic orbitals, Kotani et al. (1957) took into account this combination, as well as a larger amount of other excited states.

7.4.3 The Meckler Procedure

Meckler's treatment (Meckler 1953) begins considering a set of N single-particle functions of spatial and spin coordinates [10] $[u_i(\overline{x})]$

$$u_i(\overline{x}) = \varphi_i(x)\begin{cases}\alpha\\ \beta\end{cases} \tag{7.4.4}$$

where α and β denote, as usual, respectively $+\frac{1}{2}; -\frac{1}{2}$ spin components. The next step involves selecting n functions of the type (7.4.4) and the assignment of electrons to the subset. Denoting with N the electrons in the molecule there are $\frac{N!}{n!(N-n)!}$ selections. The bi-electronic product (antisymmetric, in accordance with Pauli's exclusion principle) is depicted as

$$\psi = (n!)^{-1/2}\det\{u_1(1)u_2(2)\ldots u_n(n)\} \tag{7.4.5}$$

Denoting now $v_j(\overline{x})$ $(1 \le i \le N)$ a set of N independent functions whose members are linear combinations of the set $u_i(\overline{x})$ $(1 \le i \le N)$, then the entire set of $\frac{N!}{n!(N-n)!}$ configurations formed with the u_i's elements is equivalent to the entire set of $\frac{N!}{n!(N-n)!}$ configurations formed with the v_i's elements and neither representation can describe the system more completely than the other ("*equivalence*

[10] More precisely, a product of a spatial function and a spin function with a spin-free Hamiltonian (cf. Meckler, 1953).

theorem", see Appendix I in Meckler 1953). The uncombined AO of the two atoms is then chosen to form a primitive set and the set is simplified by orthogonalization and symmetrization according to the geometry of the investigated molecule. In the case of O_2 the primitive set is composed of the $1s^2 2s^2 2p^4$ normal atomic configuration of the two atoms. This leads to 10 functions for each atom represented by Meckler as

$$1s\begin{cases}\alpha\\ \beta\end{cases} \quad 2s\begin{cases}\alpha\\ \beta\end{cases} \quad 2p_0\begin{cases}\alpha\\ \beta\end{cases} \quad 2p_+\begin{cases}\alpha\\ \beta\end{cases} \quad 2p_-\begin{cases}\alpha\\ \beta\end{cases} \tag{7.4.6}$$

where the terms at subscript in the p orbitals denote the component of the orbital angular momentum m_l about the internuclear axis (either 0, + 1, or –1). Considering only the configurations with completely filled 1s and 2s shells the permutations are operated only by interchanging 8 electrons on the 12 2p orbitals, leading to 495 possible configurational states. The atomic functions adopted by Meckler are

$$s = \left(\frac{\pi}{2a}\right)^{-3/4} \exp\left[-a\left(r + \frac{1}{2}R\right)^2\right] \tag{7.4.7.1}$$

$$s' = \left(\frac{\pi}{2a}\right)^{-3/4} \exp\left[-a\left(r - \frac{1}{2}R\right)^2\right] \tag{7.4.7.2}$$

$$\sigma = \left(\frac{\pi}{2b}\right)^{-3/4} \exp\left[-b\left(r + \frac{1}{2}R\right)^2\right] \tag{7.4.7.3}$$

$$\sigma' = \left(\frac{\pi}{2b}\right)^{-3/4} \exp\left[-b\left(r - \frac{1}{2}R\right)^2\right] \tag{7.4.7.4}$$

$$p_0 = 2b^{1/2}\left(\frac{\pi}{2b}\right)^{-3/4}\left(z + \frac{1}{2}R\right) \exp\left[-b\left(r + \frac{1}{2}R\right)^2\right] \tag{7.4.7.5}$$

$$p_0' = 2b^{1/2}\left(\frac{\pi}{2b}\right)^{-3/4}\left(z - \frac{1}{2}R\right) \exp\left[-b\left(r - \frac{1}{2}R\right)^2\right] \tag{7.4.7.6}$$

$$p_\pm = (2b)^{1/2}\left(\frac{\pi}{2b}\right)^{-3/4}(x \pm iy) \exp\left[-b\left(r + \frac{1}{2}R\right)^2\right] \tag{7.4.7.7}$$

$$p_\pm' = (2b)^{1/2}\left(\frac{\pi}{2b}\right)^{-3/4}(x \pm iy) \exp\left[-b\left(r - \frac{1}{2}R\right)^2\right] \tag{7.4.7.8}$$

where the a and b factors for the s and p functions (a = 20; b = 0.8) were optimized to represent at best the atomic functions 1s, 2s, and 2p; i.e

$$\exp\left(-\mathrm{ar}^2\right) \quad (1\mathrm{s}) \tag{7.4.8.1}$$

$$\exp\left(-\mathrm{br}^2\right) - \gamma \exp\left(-\mathrm{ar}^2\right) \quad (2\mathrm{s}) \tag{7.4.8.2}$$

$$\left\{ \begin{array}{c} \mathrm{x} + \mathrm{iy} \\ \mathrm{z} \\ \mathrm{x} - \mathrm{iy} \end{array} \right\} \exp\left(-\mathrm{br}^2\right) \quad (2\mathrm{p}) \tag{7.4.8.3}$$

r is the position of the electron with respect to the center of the coordinate system and R is the internuclear distance (Fig. 7.17).

The AOs were then combined by Meckler into MOs of definite axial angular momentum and definite parity:

$$\varphi_\mathrm{s} = \frac{1}{\sqrt{2}}(\mathrm{s} + \mathrm{s}') \tag{7.4.9.1}$$

$$\chi_\mathrm{s} = \frac{1}{\sqrt{2}}(\mathrm{s} - \mathrm{s}') \tag{7.4.9.2}$$

$$\varphi_\sigma = \frac{\mathrm{N}}{\sqrt{2}}[(\sigma + \sigma') + \mathrm{A}(\mathrm{s} + \mathrm{s}')] \tag{7.4.9.3}$$

$$\chi_\sigma = \frac{\mathrm{P}}{\sqrt{2}}[(\sigma - \sigma') + \mathrm{D}(\mathrm{s} - \mathrm{s}')] \tag{7.4.9.4}$$

$$\varphi_0 = \frac{\mathrm{M}}{\sqrt{2}}\left[(\mathrm{p}_0 - \mathrm{p}_0') + \mathrm{B}(\sigma + \sigma') + \mathrm{C}(\mathrm{s} + \mathrm{s}')\right] \tag{7.4.9.5}$$

$$\chi_0 = \frac{\mathrm{J}}{\sqrt{2}}\left[(\mathrm{p}_0 + \mathrm{p}_0') + \mathrm{E}(\sigma - \sigma') + \mathrm{F}(\mathrm{s} - \mathrm{s}')\right] \tag{7.4.9.6}$$

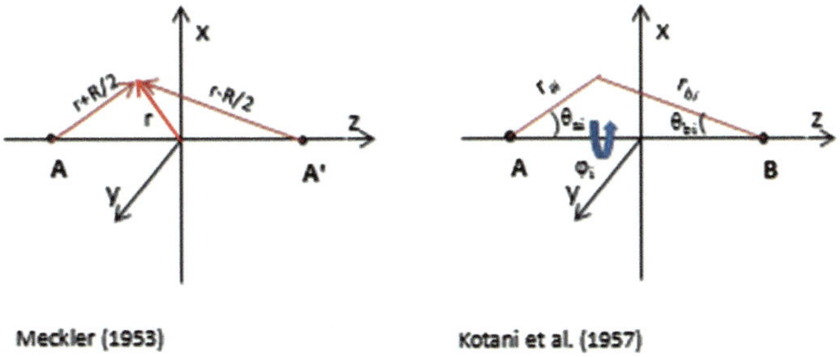

Fig. 7.17 Coordinate systems in the Meckler (1953) and Kotani et al. (1957) approaches

Table 7.11 Assignment of the electron population to the 2p shell in O_2 (from Meckler 1953; with modifications)

m_l	Number of electrons of differing m_l														
+1	0	1	1	2	2	**2**	3	**3**	3	3	**4**	4	4	4	4
0	4	3	4	2	3	**4**	1	**2**	3	4	**0**	1	2	3	4
−1	4	4	3	4	3	**2**	4	**3**	2	1	**4**	3	2	1	0
M_L	−4	−3	−2	−2	−1	**0**	−1	**0**	1	2	**0**	1	2	3	4

$$\varphi_{\pm} = \frac{L}{\sqrt{2}}\left(p_{\pm} + p'_{\pm}\right) \tag{7.4.9.7}$$

$$\chi_{\pm} = \frac{K}{\sqrt{2}}\left(p_{\pm} - p'_{\pm}\right) \tag{7.4.9.8}$$

(see Table I in Meckler 1953; for the values attained by the N,A,P,D,M,J.L.K molecular coefficients with the increase at different internuclear distances R).

The next step was the assignment of 8 electrons to the 2p orbitals (the 1s and 2s shells were assumed as completely filled thus having even parity, no spin angular momentum, and zero axial angular momentum). As shown in Table 7.11 each 2p level is filled with 4 electrons but the one-electron magnetic angular momentum m_l may attain different values (i.e. + 1,0,−1) and, as a net result, the total axial angular momentum of the molecule M_L spans from + 4 to −4.

Take now as an example the first conditions of zero total angular momentum (6th column in the electronic assignment; occupancies in bold) and expand the configuration taking into account the spin. We would get the first 10 configurations listed in Table (7.12).

The original drawing of Meckler (1953) used to depict the 3−2−3 assignment is reproduced in Fig. 7.18.

The ground state of the oxygen molecule $^3\Sigma g^-$ arises from the interaction of 9 states corresponding to the space orbitals listed in Table 7.13 and the ground state of the singlet $^1\Sigma g^+$ is the combination of the 12 different orbital populations listed in the same table. As usual, in this table the lower case Greek letter symbolizes the absolute value of the axial angular momentum (i.e. $\sigma = 0$; $\pi = 1$) and the subscript indicates the parity (g = even; u = odd).

The corresponding wave functions arising from the spin algebra connecting the many-electron spin system to the one-electron spins (cf. Appendix II in Meckler 1953) are[11]

[11] I recall here again that Σ is the MO counterpart symbolic representation of the σ AO symbolic representation; the-at apex denotes change of sign under reflection through an axial plane. The 1,3 at apex preceding the symbols denote the multiplicity, i.e. "*singlet*" or "*triplet*"). A Σ state arises whenever $M_L = 0$, a π state whenever $M_L = +1$ or -1 and a Δ state whenever $M_L = +2$ or -2.

Table 7.12 Configuration tabulation for the $M_L = 0$ conditions. Only states with even total parity are listed (transforming into themselves upon inversion) (from Meckler 1953; with modifications)

ψ	φ_0	χ_0	φ_+	χ_+	φ_-	χ_-	$m_l(+1-0--1)$
I*	+−	+−	+	+	−	−	2–4–2
I_s*	+−	+−	−	−	+	+	
II*	+−	+−	+	−	+	−	
II_s*	+−	+−	−	+	−	+	
III*	+−	+−	+	−	−	+	
III_s*	+−	+−	−	+	+	−	
IV*	+−	+−	+−		+−		
V*	+−	+−	+−			+−	
VI*	+−	+−		+−	+−		
VII*	+−	+−		+−		+−	
I	+−		+−	+	+−	−	3–2–3
I_s	+−		+−	−	+−	+	
II	+−		+	+−	−	+−	
II_s	+−		−	+−	+	+−	
III	+−		+	+−	−	+−	
III_s	+−		−	+−	+	+−	
IV	+	−	+−	+	−	+−	
IV_s	−	+	+−	+	−	+−	
V	+	+	+−	−	−	+−	
V_s	−	−	+−	+	+	+−	
VI	+	−	+	+−	+−	+	
VI_s	−	+	−	+−	+−	+	
VII	+	−	−	+−	+−	+	
VII_s	−	+	+	+−	+−	−	
VIII	+	+	−	+−	+−	−	
VIII_s	−	−	+	+−	+−	+	
IX		+−	+−	+	+−	−	
IX_s		+−	+−	−	+−	+	
X		+−	−	+−	+	+−	
X_s		+−	+	+−	−	+−	
XI			+−	+−	+−	+−	4–0–4

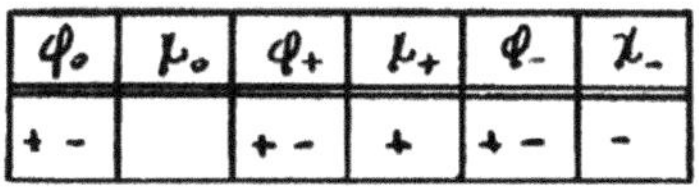

Fig. 7.18 The original drawing of Meckler (1953) (readapted) corresponding to the m_l (+1 – 0 – −1) condition (3−2−3) leading to a total axial angular momentum $M_L = 0$

Table 7.13 Tabulation by space orbital (Meckler 1953)

$^3\Sigma g^-$						$^1\Sigma g^+$		
a	$\sigma_g{}^2$	$\pi_u{}^2$	$\pi_g{}^2$	$\sigma_u{}^2$	$\sigma_g{}^2$	$\pi_u{}^2$	$\pi_g{}^2$	$\sigma_u{}^2$
b	$\sigma_g{}^2$	$\pi_u{}^2$	$\pi_g{}^2$	$\sigma_u{}^2$	$\sigma_g{}^2$	$\pi_u{}^2$	$\pi_g{}^2$	$\sigma_u{}^2$
c	$\sigma_g{}^2$	$\pi_u{}^4$	$\pi_g{}^2$	…	$\sigma_g{}^2$	$\pi_u{}^4$	…	$\sigma_u{}^2$
d	$\sigma_g{}^2$	$\pi_u{}^2$	$\pi_g{}^4$	…	$\sigma_g{}^2$	…	$\pi_g{}^4$	$\sigma_u{}^2$
e	…	$\pi_u{}^4$	$\pi_g{}^2$	$\sigma_u{}^2$	$\sigma_g{}^2$	$\pi_u{}^2$	$\pi_g{}^2$	$\sigma_u{}^2$
f	…	$\pi_u{}^2$	$\pi_g{}^4$	$\sigma_u{}^2$	$\sigma_g{}^2$	$\pi_u{}^4$	$\pi_g{}^2$	…
g	σ_g	$\pi_u{}^3$	$\pi_g{}^3$	σ_u	$\sigma_g{}^2$	$\pi_u{}^2$	$\pi_g{}^4$	…
h	σ_g	$\pi_u{}^3$	$\pi_g{}^3$	σ_u	σ_g	$\pi_u{}^3$	$\pi_g{}^3$	σ_u
i	σ_g	$\pi_u{}^3$	$\pi_g{}^3$	σ_u	σ_g	$\pi_u{}^3$	$\pi_g{}^3$	σ_u
j	…	…	…	…	…	$\pi_u{}^4$	$\pi_u{}^2$	$\sigma_u{}^2$
k	…	…	…	…	…	$\pi_u{}^2$	$\pi_u{}^4$	$\sigma_u{}^2$
l	…	…	…	…	…	$\pi_u{}^4$	$\pi_u{}^4$	…

$$
{}^3\Sigma g^- \quad \begin{cases}
\psi_a = \frac{1}{\sqrt{2}}\left(\mathrm{I}^* - \mathrm{I}_s^*\right) \\
\psi_b = \frac{1}{\sqrt{2}}\left(\mathrm{III}^* - \mathrm{III}_s^*\right) \\
\psi_c = \frac{1}{\sqrt{2}}\left(\mathrm{I} + \mathrm{I}_s\right) \\
\psi_d = \frac{1}{\sqrt{2}}\left(\mathrm{II} + \mathrm{II}_s\right) \\
\psi_e = \frac{1}{\sqrt{2}}\left(\mathrm{IX} + \mathrm{IX}_s\right) \\
\psi_f = \frac{1}{\sqrt{2}}\left(\mathrm{X} + \mathrm{X}_s\right) \\
\psi_g = \frac{1}{2}\left(\mathrm{III} - \mathrm{III}_s + \mathrm{VII} - \mathrm{VII}_s\right) \\
\psi_h = \frac{1}{2}\left(\mathrm{IV} - \mathrm{IV}_s + \mathrm{VI} - \mathrm{VI}_s\right) \\
\psi_i = \frac{1}{2}\left(\mathrm{V} - \mathrm{V}_s + \mathrm{VIII} - \mathrm{VIII}_s\right)
\end{cases} \tag{7.4.10.1}
$$

$$
{}^1\Sigma g^+ \quad \begin{cases}
\psi_a = \frac{1}{\sqrt{12}}\left(2\mathrm{I}^* + 2\mathrm{I}_s^* - \mathrm{II}^* - \mathrm{II}_s^* - \mathrm{III}^* - \mathrm{III}_s^*\right) \\
\psi_b = \frac{1}{2}\left(\mathrm{III}^* + \mathrm{III}_s^* - \mathrm{II}^* - \mathrm{II}_s^*\right) \\
\psi_c = \mathrm{IV}^* \\
\psi_d = \mathrm{VII}^* \\
\psi_e = \frac{1}{\sqrt{2}}\left(\mathrm{V}^* + \mathrm{VI}^*\right) \\
\psi_f = \frac{1}{\sqrt{2}}\left(\mathrm{I} - \mathrm{I}_s\right) \\
\psi_g = \frac{1}{\sqrt{2}}\left(\mathrm{II} - \mathrm{II}_s\right) \\
\psi_h = \frac{1}{\sqrt{24}}\left(2\mathrm{III} + 2\mathrm{III}_s - 2\mathrm{VII} - 2\mathrm{VII}_s - \mathrm{IV} - \mathrm{IV}_s + \mathrm{VI} + \mathrm{VI}_s - \mathrm{V} - \mathrm{V}_s + \mathrm{VIII} + \mathrm{VIII}_s\right) \\
\psi_i = \frac{1}{\sqrt{8}}\left(\mathrm{V} + \mathrm{V}_s - \mathrm{VIII} - \mathrm{VIII}_s - \mathrm{IV} - \mathrm{IV}_s + \mathrm{VI} + \mathrm{VI}_s\right) \\
\psi_j = \frac{1}{\sqrt{2}}\left(\mathrm{IX} - \mathrm{IX}_s\right) \\
\psi_k = \frac{1}{\sqrt{2}}\left(\mathrm{X} - \mathrm{X}_s\right) \\
\psi_l = \mathrm{XI}
\end{cases} \tag{7.4.10.2}
$$

The molecular eigenfunction for the selected (μ) configuration has the form:

$$\Psi_\mu = \sum_j a_j^\mu \psi_j \qquad (7.4.11)$$

where j are the orthonormalized states belonging to μ and ψ_j are the wavefunctions belonging to it (i.e. for the $^3\Sigma g^-$ configuration of the 7 wavefunctions of Eq. 7.4.9.1).

The time-independent Schrödinger Eq. (2.3.7) is then rewritten in the form

$$H\Psi_\mu = E_\mu \Psi_\mu \qquad (7.4.12)$$

and expanded as

$$H\sum_j a_j^\mu \psi_j = E_\mu \sum_j a_j^\mu \psi_j \qquad (7.4.13)$$

Both sides of (7.4.13) are then multiplied by ψ_i* and integrated over all the electronic coordinates, including spin.

Table 7.14 lists the energy levels which are the eigenvalues of the matrices that describe the configuration interaction (visualized in Fig. 7.19).

I anticipated earlier (Sect. 7.2) the significance of the Morse potential (Morse, 1929) and its "*adaptive nature*". It is now of interest to recall that Meckler (1959) applied Morse curves of type

Table 7.14 Energy values in Ry (from Meckler 1953; modified). R is expressed in a.u

	R							
	Atomic multiplet	1.5	2	2.5	4	6	9	∞
$^3\Sigma g^-$	O O $^3P^3P$	2.72829	−0.23468	−0.31694	0.08042	0.00149	−0.00007	0
	O O $^3P^1D$	6.40877	1.54484	0.8305	0.2794	0.20262	0.20169	0.20185
	O O $^3P^1D$	7.76336	2.38823	1.22172	0.30399	0.20344	0.20179	0.20185
	O O $^3P^1S$	8.29122	3.01394	1.87352	0.65839	0.50607	0.50456	0.50463
	O^+ O^- $4S^2P$	8.78366	3.46764	1.0488	0.9706	1.0196	1.1251	1.34567
	O^+ O^- $2D^2P$	10.4106	4.00478	2.27794	1.263	1.3133	1.42527	1.64844
	O^+ O^- $2D^2P$	10.532	4.46958	2.5358	1.2724	1.3224	1.42787	1.64844
	O^+ O^- $2P^2P$	10.93	4.70011	3.11485	1.5104	1.516	1.62718	1.8503
	O^{++} O^{--} $2P^2P$	14.4078	6.54955	4.62401	3.951	4.5439	4.99493	5.88731
	O O $^3P^3P$	2.79791	−0.05417	−0.18362	0.078	0.00147	−0.00004	0
	O O $^3P^3P$	2.9923	1.62544	0.96468	0.18686	0.00234	0	0
	O O $^1D^1D$	5.73553	1.92591	1.04601	0.45875	0.405	0.4037	0.4037
	O O $^1D^1D$	6.73148	2.46215	1.39478	0.48964	0.40564	0.4037	0.4037
	O O $^1D^1D$	6.84833	2.67301	1.52243	0.59051	0.40635	0.40371	0.4037

(continued)

Table 7.14 (continued)

	R							
	Atomic multiplet	1.5	2	2.5	4	6	9	∞
$^1\Sigma g^+$	O O $^1S^1D$	8.07864	3.7913	1.73912	0.79866	0.70916	0.70657	0.70648
	O O $^1S^1S$	9.17578	3.85523	2.21663	1.1319	1.0107	1.00919	1.0093
	$O^+ O^-$ $^2D^2P$	10.5099	4.37816	2.62522	1.2745	1.3133	1.42526	1.64844
	$O^+ O^-$ $^2D^2P$	10.691	4.82006	3.03125	1.4799	1.5159	1.62718	1.8503
	$O^+ O^-$ $^2D^2P$	10.7902	5.36583	3.28579	1.5122	1.5243	1.62972	1.8503
	$O^{++} O^{--}$ $^1D^1S$	11.8867	6.75097	4.82246	4.2524	4.7679	5.20356	6.08917
	$O^{++} O^{--}$ $^1S^1S$	14.7282	7.0137	5.09405	4.5536	5.061	5.50306	6.39193

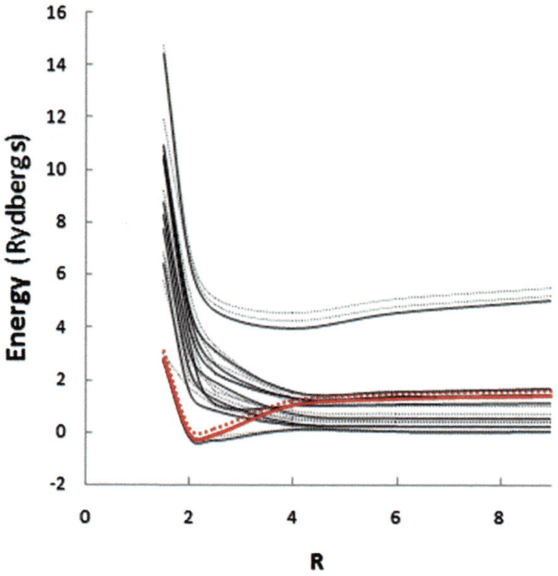

Fig. 7.19 Energy levels of the $^3\Sigma g^-$ (black solid lines) and $^1\Sigma g^+$ (black dotted lines) states of O_2 as a function of internuclear distance according to Meckler (1953). The red lines are the single configuration approximation energies for the states c and f in Table 7.12 (single configuration approximation; see Meckler, 1953 for an extended discussion)

$$E(r) = A + D\{1 - \exp[-a(r - r_0)]\}^2 \tag{7.4.14}$$

to the states $^3\Sigma g^-$and$^1\Sigma g^+$ obtaining A = −0.3728; D = 0.5079; a = 1.649 and r_0 = 2.255 for the triplet and A = −0.2109; D = 0.3205; a = 1.760 and r_0 = 2.30 for the singlet state. Stemming from the Morse potential (7.4.14) the corresponding dissociation constant D_0 is expressed in atomic units by

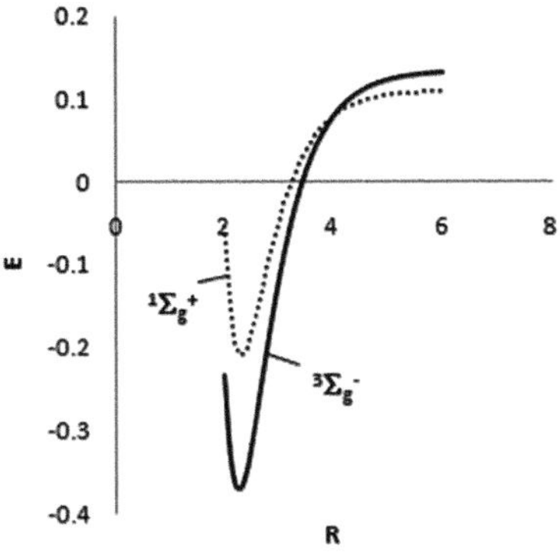

Fig. 7.20 Morse curve fit to electronic energy. The distance R is in atomic units (a.u.) and the energy R in *Ry*

$$D_0 = A + a\left(\frac{D}{m}\right)^{1/2} - \frac{a^2}{4m} \tag{7.4.15}$$

where m is the mass of an oxygen atom. The stretching frequency is given by

$$\nu_1 = 2a\left(\frac{D}{m}\right)^{1/2} \tag{7.4.16}$$

The Morse curve fit (Fig. 7.20) actually does not represent the true electronic energy but the difference in electronic energy with respect to the two isolated atoms and is strictly valid only below R = 4 (Morse 1953).

The minimum well for the triplet $^3\Sigma g^-$ is located at $r_0 = 2.255$, $D_0 = -0.3600$ *Ry*. The stretching frequency ν_1 is 1503 cm^{-1}. These values are not far from what experimentally observed.[12] The ($r_0 = 2.28$; a.u.; $D_0 = -0.3735$ *Ry*; $\nu_1 = 150.4$ cm^{-1}; Herzberg 1950). The agreement is poorer for the $^1\Sigma g^+$ state (cf. Table VII in Morse).

7.4.4 The Kotani-Mizuno-Kayama-Ishiguro Procedure

The coordinate system adopted by the Authors is shown in Fig. (7.17). The two oxygen nuclei are denoted A and B. The distances of the ith electron from the nuclear centers are r_{ai} and r_{bi}, respectively. θ_{ai} , θ_{bi} are the positional angles on the x-z plane and φ_i is the rotational angle about the internuclear axis. The AO's

[12] D_0 is, except for the zero-point correction, the energy difference of the stable molecule and the dissociated atoms. It is this quantity which is experimentally determined by analysis of band spectra.

belonging to A and B are denoted χa and χb, respectively. For the 1s, 2s, and 2p AO's the Authors adopt normalized Slater functionals of type:

$$1s_a(1) = \sqrt{\frac{\delta_1^3}{\pi}} \exp(-\delta_1 r_{a1}) \tag{7.4.17.1}$$

$$2s_a(1) = \sqrt{\frac{\delta_2^5}{3\pi}} r_{a1} \exp(-\delta_2 r_{a1}) \tag{7.4.17.2}$$

$$2p\sigma_a(1) = \sqrt{\frac{\delta_2^5}{\pi}} r_{a1} \exp(-\delta_2 r_{a1}) \cos\theta_{a1} \tag{7.4.17.3}$$

$$2p\pi_a^+(1) = \sqrt{\frac{\delta_2^5}{\pi}} r_{a1} \exp(-\delta_2 r_{a1}) \sin\theta_{a1} \exp(+i\varphi_1) \tag{7.4.17.4}$$

$$2p\pi_a^-(1) = \sqrt{\frac{\delta_2^5}{\pi}} r_{a1} \exp(-\delta_2 r_{a1}) \sin\theta_{a1} \exp(-i\varphi_1) \tag{7.4.17.5}$$

where δ_1 and δ_2 are the orbital exponents.

Because the molecular integrals involving the above Slater's type AOs (STO) are expressed in terms of R, $\alpha = \delta_1 R$, $\beta = \delta_2 R$ the values of R, α,β selected for numerical convenience are, respectively 2.30 a.u. (near to the observed value for $^3\Sigma g^-$ which is 2.282 a.u.), $\alpha = 17.75$, $\beta = 5.25$. These values bring to $\delta_1 = 7.717$ and $\delta_2 = 2.283$, not far from the values expected from a strict application of Slater's rule (i.e. $\delta_1 = 7.70$ and $\delta_2 = 2.275$).

The symmetry orbitals were then built up from the STO as follows:

$$(\sigma 1s)_{g,u} = 1s_a \pm 1s_b \tag{7.4.18.1}$$

$$(\sigma 2s)_{g,u} = 2s_a \pm 2s_b \tag{7.4.18.2}$$

$$(\sigma 2p)_{g,u} = 2p\sigma_a \pm 2p\sigma_b \tag{7.4.18.3}$$

$$\left(\pi^{\pm} 2p\right)_{g,u} = 2p\pi_a^{\pm} \pm 2p\pi_b^{\pm} \tag{7.4.18.4}$$

From the symmetry orbitals (7.4.18.1−7.4.18.4) Orthogonalized Molecular Orbitals were built adopting the mathematical procedure of Schmidt (Courant and Hilbert 1953) and are generally known as Schmidt-OMOs:

$$\sigma_{1g} = \alpha_{11} (\sigma 1s)_g \tag{7.4.19.1}$$

$$\sigma_{2g} = \alpha_{21}(\sigma 1s)_g + \alpha_{22}(\sigma 2s)_g \tag{7.4.19.2}$$

$$\sigma_{3g} = \alpha_{31}(\sigma 1s)_g + \alpha_{32}(\sigma 2s)_g + \alpha_{33}(\sigma 2p)_g \tag{7.4.19.3}$$

$$\pi_g^{\pm} = \gamma_g(\pi^{\pm} 2p)_g \tag{7.4.19.4}$$

$$\sigma_{1u} = \beta_{11}(\sigma 1s)_u \tag{7.4.19.5}$$

$$\sigma_{2u} = \beta_{21}(\sigma 1s)_u + \beta_{22}(\sigma 2s)_u \tag{7.4.19.6}$$

$$\sigma_{3u} = \beta_{31}(\sigma 1s)_u + \beta_{32}(\sigma 2s)_u + \beta_{33}(\sigma 2p)_u \tag{7.4.19.7}$$

$$\pi_u^{\pm} = \gamma_u(\pi^{\pm} 2p)_u \tag{7.4.19.8}$$

The values of α_{ij}, β_{ij},γ_g and γ_u in these orbitals are determined in order to fulfill the orthonormality condition and are listed in Table (7.15).

From the point of view of calculus, the most laborious procedure was the evaluation of molecular integrals involving STOs. The unnormalized symmetry orbital integrals and the Schmidt-OMOs integrals were then assembled (this work was actually carried out in previous contributions of Kotani and coworkers; see Kotani et al. 1957 and references therein).

Because there are 16 electrons in the O_2 molecule and the net charge of the nuclear centers is 8, the (spin-free) Hamiltonian operator was depicted as

$$\mathsf{H} = -\frac{1}{2}\sum_{i=1}^{16}\Delta_i - \sum_{i=1}^{16}\left(\frac{8}{r_{ai}} + \frac{8}{r_{bi}}\right) + \sum_{i>j}^{16}\frac{1}{r_{ij}} + \frac{64}{R} \tag{7.4.20}$$

The Hamiltonian (7.4.20) has symmetry $D_{\infty h}$ (linear with inversion center) and its eigenfunctions may be classified specifying:

- the component of the angular momentum of the electrons along the internuclear axis (Λ);
- the character of the inversion at the center (I);
- the character of the reflection at any plane passing through both nuclei (v)

Table 7.15 Coefficients of the Schmidt-type Ortogonalized Molecular Orbitals of O_2 (Kotani et al. 1957)

α_{11}	0.70710 59	β_{11}	0.70710 76
α_{21}	−0.16451 50	β_{21}	−0.18141 32
α_{22}	0.64420 46	β_{22}	0.85443 12
α_{31}	0.01656 28	β_{31}	−0.05226 28
α_{32}	−0.15659 10	β_{32}	0.40917 14
α_{33}	0.63726 88	β_{33}	0.94154 01
γ_g	0.76305 78	γ_u	0.66189 67

Kotani et al. (1957) denoted each Λ, I, v possible combination as a Ω-set, which, when coupled with M_L and M_S completely defines both symmetry and configuration.

Because, generally, many primitive functions (A, B, C) exist for a given Ω, they were denoted as $\psi_\Omega^A, \psi_\Omega^B, \psi_\Omega^C, \ldots$ etc.

They then assumed that the orbital energies of the Schmidt-OMOs increase in the order σ_{1g}, σ_{1u}, σ_{2g}, σ_{2u}, σ_{3g}, $\pi_u^{\pm}$, $\pi_g^{\pm}$, σ_{3u}. They assumed also that the ground state and the normal excited states of O_2 are composed mainly of configurations with σ_{1g}, σ_{1u}, σ_{2g}, σ_{2u}, σ_{3g}, filled orbitals and no electrons in σ_{3u}. With these provisos, the possible assignments are quite limited because they involve only 6 electrons in the $\pi_u^{\pm}$, $\pi_g^{\pm}$, orbitals (we have already seen six of them in Table 7.10).

For each Ω-M_L, M_S set, the Schmidt-OMOs whose product has appropriate Λ, I symmetry were selected, considering that each orbital can accommodate two electrons at most (primitive functions). The products were then multiplied by various spin functions with appropriate M_L, M_S symmetry (Table 7.16). The results were finally antisymmetrized.

The number of spin functions corresponding to given values of M_S and M_L is independent of M_L and may be thus indicized as $f_{M_S}^N$, denoting with N the total number of electrons:

$$f_{M_S}^N = \frac{(2M_S + 1) \times N!}{\left(\frac{N}{2} + M_S + 1\right)! \times \left(\frac{N}{2} - M_S\right)!} \tag{7.4.21}$$

They were thus denoted by $\Theta_{M_S,M_L,k}$ where $k = 1, 2, \ldots, f_{M_S}^N$.

Because the assignment involves 16 electrons and 10 Schmidt-OMOs (each of which capable of hosting 2 electrons) the number of orbitals which contain only one electron ("*unpaired orbitals*") is at most 4. All in all, there are r "*paired orbitals*" and N−2r unpaired ones. In writing a generic primitive function Ψ_Ω^F, the paired orbitals precede the unpaired ones:

Table 7.16 Spin Functions Θ_h(1,2,3,4) adopted by Kotani et al. (1957) for the O_2 orbitals

M_{SL}	M_L	h	θ
1	1	1	$\frac{1}{\sqrt{2}}\alpha(1)\alpha(2)[\alpha(3)\beta(4) - \beta(3)\alpha(4)]$
1	1	2	$\frac{1}{\sqrt{2}}\alpha(3)\alpha(4)[\alpha(1)\beta(2) - \beta(1)\alpha(2)]$
1	1	3	$\frac{1}{2}[\alpha(1)\beta(2)\alpha(3)\alpha(4) + \beta(1)\alpha(2)\alpha(3)\alpha(4)$ $- \alpha(1)\alpha(2)\alpha(3)\beta(4) - \alpha(1)2(2)\beta(3)\alpha(4)]$
0	0	1	$\frac{1}{2}[\alpha(1)\beta(2) - \beta(1)\alpha(2)][\alpha(3)\beta(4) - \beta(3)\alpha(4)]$
0	0	2	$\frac{1}{\sqrt{3}}\{\alpha(1)\alpha(2)\beta(3)\beta(4) - \beta(1)\beta(2)\alpha(3)\alpha(4)$ $-\frac{1}{2}[\alpha(1)\beta(2) + \beta(1)\alpha(2)][\alpha(3)\beta(4) + \beta(3)\alpha(4)]\}$

$$\Psi^{F}_{\Sigma_g^-} = \frac{1}{\sqrt{2}} \begin{bmatrix} \pi_g^+(1)\pi_u^-(2)\sigma_{3g}(3)\sigma_{3u}(4)\pi_g^-(5)\pi_g^-(6)\pi_u^+(7)\pi_u^+(8) \\ -\pi_g^-(1)\pi_u^+(2)\sigma_{3g}(3)\sigma_{3u}(4)\pi_g^+(5)\pi_g^+(6)\pi_u^-(7)\pi_u^-(8) \end{bmatrix} \times \sigma_{2u}(9)\sigma_{2u}(10)\sigma_{2g}(11)\sigma_{2g}(12)\sigma_{1u}(13)\sigma_{1u}(14)\sigma_{1g}(15)\sigma_{1g}(16) \quad (7.4.22)$$

The explicit forms of the spin functions adopted by Kotani et al. (1957) for the N−2r electrons [$\Theta_h(1,2,3,4)$] are listed in Table 7.16.

From the primitive function Ψ^A_Ω, g antisymmetrized wave functions Φ^A_{Ω,M_S,M_L} (with quantum numbers Ω, M_S, M_L) were built:

$$\Phi^{A,k}_{\Omega,M_S,M_L}(1,2,\ldots N) = \frac{1}{\sqrt{N!2^r}}\sum_P\sum_h \overline{U}^{M_S}_{hk}(P)P\Psi^A_\Omega\Theta_{M_S,M_L,h} \quad (1 \le k \le g) \quad (7.4.23)$$

where P is any permutation among the N! permutations of N-electrons and $U^{M_S}(P)$ is

$$U^{M_S}(P) = (-1)^P V^{M_S}(P^{-1})^T \quad (7.4.24)$$

with $V^{M_S}(P)$ as a representation matrix (T meaning transposition).

A suitable linear combination of the various functions of type (7.4.24) was then considered to represent the best wave function for the Ω, M_S, M_L quantum state:

$$\Phi_{\Omega,M_S,M_L}(1,2,\ldots N) = \sum_{A,k} c_{A,k}\Phi^{A,k}_{\Omega,M_S,M_L}(1,2,\ldots N) \quad (7.4.25)$$

Combining the Schmidt-OMOs (7.4.19.1−7.4.19.8) with the spin functions of Table (7.16), thirty wave functions of the type (7.4.23) may be built up. Each one of these functions is a "*configuration.*" Seventeen of them and their coefficients leading to the appropriate *extended configuration interaction* are listed in Table 7.17.

Configuration A corresponds to the usual antisymmetrized OMO built from the LCAO procedure. Its total energy denoted by the authors "*single configuration*" arises to −149.00734 a.u. (see Appendix 1. In Kotani et al. 1957; for details on the calculation). We may now note that in the first 9 configurations the molecular orbitals $1\sigma_g$, $1\sigma_u$, $2\sigma_g$ and $2\sigma_u$ are fully occupied by 8 electrons) but this is not the case for the remaining 21 configurations (only 8 of them shown in Table 7.17). Thus, if we consider the 1s and 2s AOs in O_2 to be fully occupied, the CI would reduce to an appropriate combination of the first nine configurations (i.e. "*p electrons configurational interaction*", which is the case investigated by Meckler as well). The fact that Meckler's and Kotani's coefficients do not differ appreciably for this CI is fortuitous, as the functional forms of AOs are inherently different in the

Table 7.17 Configurations used in CI calculations. The coefficients c_i are those adopted in normalized wave function for the ground state $^3\Sigma g^-$. (a) Kotani et al. (1957); (b) Meckler (1957). When k is omitted it means that suffices other than 1 in (7.4.23) vanish. The coefficients attributed to Meckler (1953) were actually obtained by Kotani et al. (1957) by interpolation, fixing the internuclear distance at R = 2.3 a.u. (see Table I in Meckler (1953). From Kotani et al. (1957) with modifications

Configurations											Coefficients c_i		
i,k	π_u^+	π_u^-	π_g^+	π_g^-	σ_{3g}	σ_{3u}	σ_{2g}	σ_{2u}	σ_{1g}	σ_{1u}	(a)	(b)	(a)
A	2	2	1	1	2		2	2	2	2	0.9629	0.9749	0.949
B	2	2	1	1		2	2	2	2	2	−0.0959	−0.0826	−0.0735
C	1	1	2	2	2		2	2	2	2	−0.1719	−0.1352	−0.1521
D	1	1	2	2		2	2	2	2	2	0.0515	0.0329	
E,1	↑	↑	↓	↓	2	2	2	2	2	2	0.0027	0.0026	
E,2	↓	↓	↑	↑	2	2	2	2	2	2	0.0259	0.0233	0.0272
F,1	↑	2	2	↑	↓	↓	2	2	2	2	0.1716	0.1473	0.1465
F,2	↓	2	2	↓	↑	↑	2	2	2	2	−0.0309	−0.0263	−0.0226
F,3	↑	2	2	↑	↑	↑	2	2	2	2	0.0179	0.0178	0.0184
H	2	2	1	1	2	2	2		2	2			−0.0005
J,1	2	2	↑	↑	2	↓	2	↓	2	2			−0.2118
J,3	2	2	↑	↑	2	↑	2	↑	2	2			−0.0438
L	2	2	1	1	2	2		2	2	2			−0.0162
M,1	↑	2	2	↑	2	↓	↓	2	2	2			0.0202
M,2	↓	2	2	↓	2	↑	↑	2	2	2			0.0132
M,3	↑	2	2	↑	2	↑	↑	2	2	2			−0.0073
O,1	2	2	↑	↑	↓	2	↓	2	2	2			0.0156

Table 7.18 Ground and lower excited molecular states of O_2 arising from permutations of four electrons over $\pi_u^{\pm}$ and $\pi_g^{\pm}$orbitals (and no electrons in σ_{3u}). For the sake of conciseness the states are specified by assigning the number of electron holes rather than occupancies. States in parenthesis not observed experimentally. From Kotani et al. (1957) slightly modified

π_g^+	π_g^-	π_u^+	π_u^-	States
1	1			${}^3\Sigma_g^-$; ${}^1\Sigma_g^+$
2				${}^1\Delta_g$
	2			${}^1\Delta_g$
1			1	${}^1\Sigma_u^-$; ${}^3\Sigma_u^-$; ${}^3\Sigma_u^+$; $({}^1\Sigma_u^+)$
	1	1		${}^1\Sigma_u^-$; ${}^3\Sigma_u^-$; ${}^3\Sigma_u^+$; $({}^1\Sigma_u^+)$
1		1		$({}^1\Delta_u)$
	1		1	$({}^3\Delta_u)$

two procedures. With this CI, the total energy decreases of 3.424 *eV* with respect to the single configuration A.

Theoretically, the best estimate of the actual CI would consider all 30 configurations. Because the computational work behind such a procedure was, at that time, extremely demanding, Kotani et al. (1957) limited themselves to restricted selections (15 configurations out of 30) corresponding to those ones having non-vanishing matrix elements between themselves and configuration A. The coefficients of the selected configurations are reported in the last column of Table 7.17. With this extended configuration the total energy is 5.000 *eV* below the energy of configuration A.

Besides the ground state, Kotani et al. (1957) investigated also the energy of some of the lower excited states of O_2. The investigated molecular states are represented in Table 7.18.

To appreciate the amount (and the significance) of excitation energy, the diagonal elements of the Hamiltonian (7.4.20) are expanded with reference to the ground state:

$$
\begin{aligned}
\mathsf{H}_{AA}({}^3\Sigma_{g^-}) &= \sum_i (\varepsilon_i + h_i) + \frac{64}{R} \\
&= \sum_i (2h_i) + \{\sum_k \sum_j [2J(k,j) - K(k,j)] \\
&+ \sum_k \sum_a [2J(k,a) - K(k,a)]\} \\
&+ \left[J\left(\pi_g^+ \pi_g^-\right) - K\left(\pi_g^+ \pi_g^-\right)\right] + \frac{64}{R}
\end{aligned}
\tag{7.4.26}
$$

J and K in the above equation are, respectively, the Coulomb integral and the Exchange integral, i.e.:

$$J(k,l) = (\varphi_k\varphi_l|\varphi_k\varphi_l) \quad (7.4.27.1)$$

$$K(k,l) = (\varphi_k\varphi_l|\varphi_l\varphi_k) \quad (7.4.27.2)$$

with φ representing Schmidt-OMOs, and

$$l = \sigma_{1g}, \sigma_{1u}, \sigma_{2g}, \sigma_{2u}, \sigma_{3g}, \pi_u^{\pm}, \text{and } \pi_g^{+} \text{ or } \pi_g^{-} \quad (7.4.28.1)$$

$$\varepsilon_j = h_j + \sum_k [2J(k,j) - K(k,j)] + \frac{1}{2}\left[\sum_a [2J(a,j) - K(a,j)]\right] \quad \left(j,k \neq \pi_g^{\pm}\right) \quad (7.4.28.2)$$

$$\varepsilon_a = h_a + \sum_k [2J(k,ja) - K(k,a)] + \frac{1}{2}\left[\sum_a [2J(b,a) - K(b,a)]\right] \quad (7.4.28.3)$$

$$\left(j,k = \pi_g^{\pm}, k \neq \pi_g^{\pm}\right)$$

$$h = -\frac{1}{2}\Delta - \left(\frac{8}{r_a} + \frac{8}{r_b}\right) \quad (7.4.28.4)$$

$$h_l = (\varphi_l|h|\varphi_l) \quad (7.4.28.5)$$

$$(\varphi_l|h|\varphi_l) = \int \varphi_i^*(1)\varphi_k^*(2)\frac{1}{r_{12}}\varphi_j(1)\varphi_l(2)dv_1dv_2 \quad (7.4.28.6)$$

The state $^1\Delta_g$ differs from the ground state $^3\Sigma_g^-$ essentially in terms of the exchange integral $K(\pi_g^+\pi_g^-)$ and the state $^1\Sigma_g^+$ differs from the ground state in terms of $2K(\pi^+\pi^-)$. The energy relations between the other investigated excited states and $^3\Sigma_g^-$ are more complex (see Appendix 1 in Kotani et al. 1957; for details).

The vertical excitation energies computed by Kotani et al. (1957) with the CI procedure are compared in Table (7.19) and Figure (7.21) with the results of spectroscopic observations and with the SCF procedure. The agreement between computation and experimental observation is moderately good only for the two lowest excited states $^1\Delta_g$ and $^1\Sigma_g^+$.

7.4.5 Hybridization in O_2

Hybridization will be treated more amply later on, when dealing with the CH_4 molecule and the $[SiO_4]^{4-}$ cluster. It is however of interest to recall here that one of

Table 7.19 Vertical excitation energies relative to ground state ${}^3\Sigma_g^-$ (Kotani et al. 1957)

State	SCF	CI (π)	Observed
${}^3\Sigma_g^-$	0.000	0.000	0.000
${}^1\Delta_g$	1.484	0.764	1.000
${}^1\Sigma_g^+$	2.969	1.423	1.600
${}^1\Sigma_u^+$	3.683	4.570	7.900
${}^3\Sigma_u^+$	3.936	4.824	7.500
${}^3\Sigma_u^-$	13.699	14.587	9.500

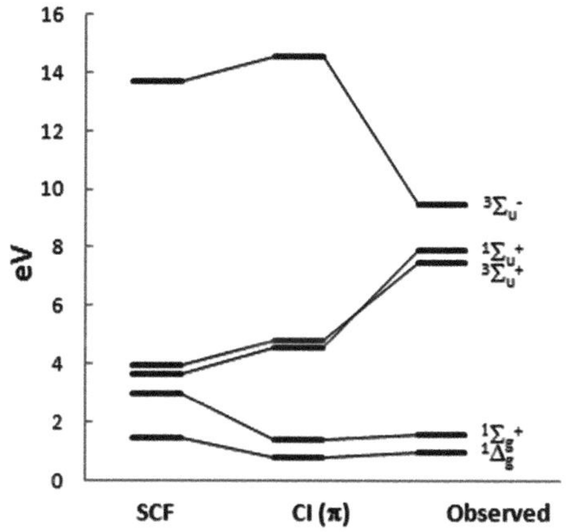

Fig. 7.21 The lowest excited states of O_2 after Kotani et al. (1957) compared with spectroscopic observation. The zero on the ordinate axis corresponds to the ground state ${}^3\Sigma_g^-$. The nominal values of the other states correspond thus to the vertical excitation energy calculated for an internuclear distance of 2.3 a.u. The agreement between computation and experiment is good only for the two lowest states

the first attempts to compute the energy of hybrid orbitals concerns the O_2 molecule. Among the lower energy states, the molecular orbital σ_{3g} arising from the combination of (σ2s)g and (σ2p)g (plus a limited amount of (σ1s)g (cf. Eq. 7.4.19.3 and Table 7.15)) has the highest overlap charge density in the region between the two nuclear centers (Fig. 7.22).

Assuming that charge overlap results in binding, then, the σ_{3g} orbital contributes to the stability of the molecule and must be taken into account. Kotani et al. (1957) considered explicitly this possibility taking into account the configuration J_1 in Table (7.17). In this configuration σ_{2u} and σ_{3u} are coupled in a singlet and (in this sense) the mixed configuration "*may be taken in some sense as the s-p hybridization in the molecular orbital scheme*" (Kotani et al. 1957; Case 3). The fact that only the J_1 state was selected, stems from the consideration that all the other configurations differing by A in the σ orbital part (B, H, etc.) involve σ_{3u}, whose energy is far above those of σ_{3g}, σ_{2u}, and σ_{2g} (see also Fig. 7.23). The CI taking into account A

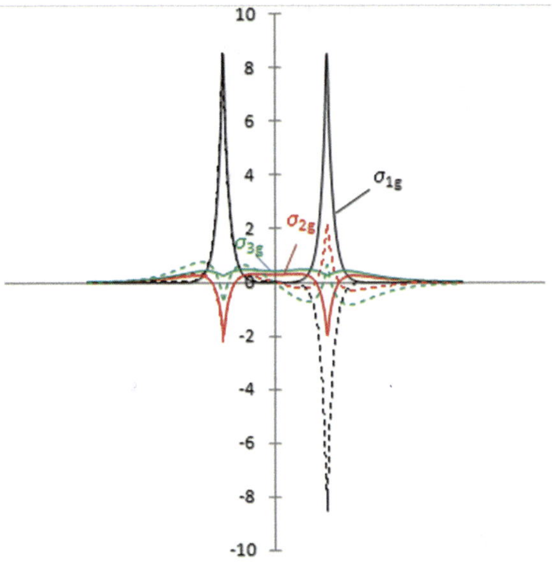

Fig. 7.22 Schmidt-OMOs of O_2 along the internuclear axis. The dashed lines are the odd (u) counterparts of the even (g) configurations

and J_1 is 0.9704A−0.2414J_1. The total energy is lowered with respect to A by 2.1687 *eV*.

7.5 The CO_2 Molecule as an Example of Linear Heteronuclear Closed-Shell Configuration

7.5.1 Premises

Also CO_2, like O_2, has symmetry $D_{\infty h}$ (linear with inversion center). The first quantum-mechanical characterization is due to Mulliken (1935) and Moffitt (1949). Both approaches were semi-empirical but gave a correct appraisal of the ground state, the bond order, and the ionization potential. The first non-empirical approach, based on LCAO and Slater-type AOs is due to Mulligan (1951).

7.5.2 The Mulligan Treatment

There are in CO_2 22 electrons in a closed-shell configuration at the ground state. Mulligan (1951) adopted for all 1s, 2s, 2p electrons STOs of the usual form:

$$1s = \sqrt{\frac{\delta_1^3}{\pi}} \exp(-\delta_1 r) \tag{7.5.1.1}$$

$$2s = \sqrt{\frac{\delta_2^5}{3\pi}} r \exp(-\delta_2 r) \tag{7.5.1.2}$$

$$2p_x = \sqrt{\frac{\delta_2^5}{\pi}} r \exp(-\delta_2 r) \sin\theta \cos\varphi \tag{7.5.1.3}$$

$$2p_y = \sqrt{\frac{\delta_2^5}{\pi}} r \exp(-\delta_2 r) \sin\theta \sin\varphi \tag{7.5.1.4}$$

$$2p_z = \sqrt{\frac{\delta_2^5}{\pi}} r \exp(-\delta_2 r) \cos\theta \tag{7.5.1.5}$$

where angles are identified as already seen for O_2 in Fig. 7.17, and the orbital exponents δ_1 and δ_2 assume the values 7.70 – 2.275 for oxygen and 5.70 – 1.625 for carbon. The $2p_z$ orbitals are directed along the internuclear axis. They are all positive in the positive z-direction.

Since the original nodeless STO 2s is not orthogonal to 1s, to achieve orthogonality it was modified by Mulligan as follows:

$$2s_O^* = \frac{2s_O - 0.2334 \times 1s_O}{0.9724} \tag{7.5.2.1}$$

$$2s_C^* = \frac{2s_C - 0.2225 \times 1s_C}{0.9754} \tag{7.5.2.2}$$

where $2s_O$, $1s_O$, $2s_C$, and $1s_C$ are the original STOs for oxygen and carbon, respectively.[13] All in all, there are in the molecule 15 AOs. From these, 15 MOs were built applying:

$$\phi_i = \sum_p a_{i,p} \chi_p \tag{7.5.3}$$

with $a_{i,p}$ as undetermined coefficients of the pth AO in the ith MO.

[13] The notation adopted in the following treatment by Mulligan assigns *o*, *s*, *z*, *x*, *y*, in the place of 1s, 2s, $2p_z$, $2p_x$,$2p_y$ to the first oxygen atom; then *o*', *s*', *z*', *x*', *y*', for the corresponding orbitals of the second oxygen atom. The carbon orbitals are designated by c, s_c, z_c, x_c, y_c.

Table 7.20 Symmetry orbitals for CO_2 according to Mulligan (1950)

σ_g	σ_u	π_u	π_g
$\sigma_1 = \frac{1}{\sqrt{2}}(o + o')$	$\sigma_6 = \frac{1}{\sqrt{2}}(o - o')$	$\sigma_{10} = \frac{1}{\sqrt{2}}(x + x')$	$\sigma_{14} = \frac{1}{\sqrt{2}}(x - x')$
$\sigma_2 = c$	$\sigma_7 = \frac{1}{\sqrt{2}}(s - s')$	$\sigma_{11} = x_c$	$\sigma_{15} = \frac{1}{\sqrt{2}}(y - y')$
$\sigma_3 = \frac{1}{\sqrt{2}}(s + s')$	$\sigma_8 = \frac{1}{\sqrt{2}}(0 + 0')$	$\sigma_{12} = \frac{1}{\sqrt{2}}(y + y')$	
$\sigma_4 = s_c$	$\sigma_9 = s_c$	$\sigma_{13} = y_c$	
$\sigma_5 = \frac{1}{\sqrt{2}}(s - s')$			

The MOs belonging to the set of irreducible representation of the symmetry group $D_{\infty h}$ (Sponer and Teller 1941) were then transformed to unnormalized symmetry orbitals by applying:

$$\sigma_p = \sum_g U_{p,g}\chi_g \tag{7.5.3}$$

where $U_{p,g}$ is a real, orthogonal (15×15) matrix. There are in this representation five σ_g, four σ_u, two π_g, and four π_u symmetry orbitals, related to the MOs by

$$\phi_i = \sum_n b_{i,n}\sigma_n \tag{7.5.4}$$

In Table 7.20 we see the LCAO leading to the four groups of symmetry orbitals.

To obtain the ground state energy with the variational procedure, Mulligan (1950) developed the matrix equation

$$(\mathsf{H} + \mathsf{G})a_i = \varepsilon_i \mathsf{S} a_i \tag{7.5.5}$$

S in Eq. (7.5.5) is a Hermitian matrix whose elements are overlap integrals, defined by

$$S_{pq} = \overline{S}_{qp} = \int \overline{\chi}_p \chi_q dv \tag{7.5.6}$$

H is the bare nuclear-field orbital Energy matrix, with elements

$$H_{pq} = \overline{H}_{qp} = \int \overline{\chi}_p \mathsf{H} \chi_q dv \tag{7.5.7}$$

and G is the electronic interaction matrix with elements

$$G_{pq} = \overline{G}_{qp} = \sum_{rt} \overline{a}_{rt} G_{pqgt} a_{rt} \tag{7.5.8}$$

with

$$G_{prqt} = 2J_{prqt} - J_{prtq} \quad (7.5.9)$$

and

$$J_{prqt} = e^2 \int \frac{\overline{\chi}_p^{\mu} \overline{\chi}_r^{\nu} \chi_q^{\mu} \chi_t^{\nu}}{r^{\mu\nu}} dv^{\mu\nu} \quad (7.5.10)$$

written for the μth and νth electron. Denoting as L the Hamiltonian operator $\mathsf{L} = (\mathsf{H} + \mathsf{G})$ Eq. (7.5.5) reduces to

$$\mathsf{L}a_i = \varepsilon_i \mathsf{S}a_i \quad (7.5.11)$$

"*The problem of finding the best LCAO MO's for the ground state thus reduces to the solution of the secular Eq.* (7.5.11) *for the eigenvalues, or LCAO orbital energies,* ε_{ij} *and the eigenvectors, or undetermined LCAO coefficients,* a_i*. Since L depends on the* a_i*'s inasmuch as G depends on the ai's, the need of a self-consistent procedure is obvious, hence the name LCAO SCF theory*" (Mulligan 1951).

Actually, what was solved by Mulligan, was not Eq. (7.5.11) but rather its counterpart

$$\mathsf{H}'b_i = \varepsilon_i \mathsf{S}'b_i \quad (7.5.12)$$

Expressed in terms of symmetry orbital and related to Eq. (7.5.11) by the transformation

$$b_i = Ua_i \quad (7.5.13)$$

where a_i and b_i are column vectors whose components are the coefficients of the AO's and of the symmetry orbitals in the MO ϕ_i, respectively; U are the terms of the matrix (7.5.3) and L', S' are given by

$$\mathsf{L}' = U\mathsf{L}U^{\mathsf{T}} \quad (7.5.14.1)$$

$$\mathsf{S}' = U\mathsf{S}U^{\mathsf{T}} \quad (7.5.14.2)$$

with U^{T} meaning transposed U.

In Table (7.21) the eleven LCAO MOs of lowest energy obtained by the self-consistent-field procedure are listed. Each orbital is filled by two electrons with paired spins to yield the known $^1\Sigma_g^+$ ground state of CO_2. The remaining four orbitals are unfilled and correspond to the excited states of the molecule.

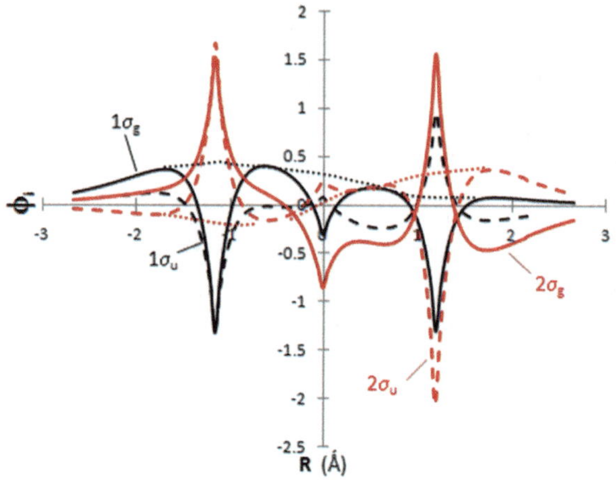

Fig. 7.23 LCAO-SCF $1\sigma_g$, $1\sigma_u$, $2\sigma g$, and $2\sigma_u$ MOs of CO_2 along internuclear axis according to Mulligan (1951). The dotted lines are only an eye-guide meant to mild the spikes located on the nuclear centers. The C nucleus is located at R = 0 and the two O nuclei at −0.162 and 0.162 *Å*, respectively

Table 7.21 LCAO SCF molecular orbitals for the ground state of CO_2 (Mulligan 1951)

$K(\sigma_g) = \frac{1}{\sqrt{2}}(o + o')$	$K(\sigma_u) = \frac{1}{\sqrt{2}}(o - o')$
$K*(\sigma_g) = c$	
$1\sigma_g = 0.48(s + s') + 0.32s_c + 0.21(z - z')$	$1\sigma_u = 0.38(s - s') + 0.11(z + z') - 0.51z_c$
$2\sigma_g = -0.51(s + s') + 0.40s_c + 0.46(z - z')$	$2\sigma_u = -0.64(s - s') + 0.37(z + z') - 0.65z_c$
$1\pi_g = \frac{1}{\sqrt{2}}(x - x')$	$1\pi_u = 0.49(x - x') + 0.53x_c$
$1\pi_g* = \frac{1}{\sqrt{2}}(y - y')$	$1\pi_u* = 0.49(y + y') + 0.53y_c$

The LCAO SCF results for the $1\sigma_g$ and $1\sigma_u$ orbitals indicate rather large contributions of the carbon orbitals to the 2s orbitals of oxygen, and, on the basis of the criterion of maximum overlap they suggest a bonding character for $1\sigma_g$ (Fig. 7.23).

Since the removal of an electron from the $2\sigma_u$ MO changes the internuclear distance from 1.1615 *Å* to 1.18 *Å*, (Bueso-Sankkehi 1941) although there is no sensible charge overlap in the internuclear regions between the carbon center and the two oxygen nuclei, this MO may be considered as weakly bonding.

The dotted lines on the $1\sigma_g$ and $2\sigma_u$ wave functions are meant to help the visualization of the general trend along the internuclear axis of this heteronuclear chain. These general trends are quite similar (in form and number of nodes) to the wave functions depicted by a particle in the box (or a "*free-electron model*"; cf Sect. 2.7 and Figs. 2.4 and 2.5). They bear analogies also with wave functions for a hypothetical one-dimension model crystal, as we will see later on.

7.6 The H_2O Molecule as an Example of Non-linear Heteronuclear Closed-Shell Configuration

7.6.1 *Premises*

The di-hydrogen monoxide molecule has a triangular shape. Placing the oxygen atom at the center of the Cartesian system and taking the plane of the molecule to be the y-z plane, then the z-axis is the axis of rotation and the internuclear lines make an angle approximately intermediate between the right angle and the tetrahedral angle. The molecule thus has C_{2v} symmetry (i.e. a 2-fold axis, associated with the symmetry operation C_2 and two σ_v vertical planes associated with two mirror reflections).

As historical premise it is worthy to note that the Quantum Mechanical treatments of H_2O include Valence-Bond (VB) approaches (Coolidge 1932; Van Vleck and Cross 1934), the Slater-Pauling Directed Valency theory (Slater 1931) and LCAO-SCF methods (Mulliken 1934; Duncan and Pople 1953; Ellison and Shull 1954). Other theoretical approaches to the electronic structure of H_2O were carried out by Linnett and Poe (1951). Pople (1950) adopted as a basis the Lennard-Jones perturbation procedure (Lennard-Jones 1949).

7.6.2 *The Ellison-Shull Model*

It is quite similar in many respects to Mulligan's treatment for CO_2. Ellison and Shull (1955) begun with establishing STOs of the type (7.5.1.1, 7.5.1.5) for oxygen and hydrogens with $\delta = 1$ for hydrogen and $\delta_1 = 7.7$, $\delta_2 = 2.275$ for oxygen. The AOs of oxygen were then denoted o (1s), s' (2s'), z ($2p_z$), x (2px), and y ($2p_y$), and those of the two hydrogens h' ($1s_{h'}$) and h'' ($1s_{h''}$). They also noted that the original nodeless STO 2s is not orthogonal to 1s. To achieve orthogonality it was modified as follows:

$$2s = s = \frac{(2s' - Q1s)}{\sqrt{1 - Q^2}} \tag{7.6.1}$$

where Q is the overlap integral S(1s,2s'), i.e. $[Q = S(\chi_p^\delta \chi_q^\delta) = \int \chi_p^\delta \chi_q^\delta dv^\delta]$.

The SCF calculations were carried out for 6 different values of the H-O-H bond angle: 90, 100, 105, 110, 120, and 180 degrees but keeping the O-H internuclear distance as constant and corresponding to what experimentally observed (i.e. R = 0.9581 $\AA \equiv 1.8103$ a.u.).

From the selected AOs those ones belonging to irreducible representations of the symmetry group of the molecule were selected and MO's (possessing the same symmetry properties) were built. The normalized symmetry orbitals were designated as follows:

$$\sigma_1 = \frac{1}{\sqrt{2}}(\mathrm{h}' + \mathrm{h}''); \quad \sigma_2 = 1\mathrm{s}_\mathrm{O}; \quad \sigma_3 = 2\mathrm{s}_\mathrm{O}; \quad \sigma_4 = 2\mathrm{p}_{\mathrm{z}_\mathrm{O}};$$

$$\sigma_5 = \frac{1}{\sqrt{2}}(\mathrm{h}' - \mathrm{h}''); \quad \sigma_6 = 2\mathrm{p}_{\mathrm{y}_\mathrm{O}}; \quad \sigma_7 = 2\mathrm{p}_{\mathrm{x}_\mathrm{O}}$$

By taking linear combinations of symmetry orbitals of the same symmetry, a like number of MO's having that symmetry were formed:

$$\phi(1\mathrm{a}_1) = \mathrm{c}_{11}\sigma_1 + \mathrm{c}_{12}1\mathrm{s}_\mathrm{O} + \mathrm{c}_{13}2\mathrm{s}_\mathrm{O} + \mathrm{c}_{14}2\mathrm{p}_{\mathrm{z}_\mathrm{O}} \tag{7.6.2.1}$$

$$\phi(2\mathrm{a}_1) = \mathrm{c}_{21}\sigma_1 + \mathrm{c}_{22}1\mathrm{s}_\mathrm{O} + \mathrm{c}_{23}2\mathrm{s}_\mathrm{O} + \mathrm{c}_{24}2\mathrm{p}_{\mathrm{z}_\mathrm{O}} \tag{7.6.2.2}$$

$$\phi(3\mathrm{a}_1) = \mathrm{c}_{31}\sigma_1 + \mathrm{c}_{32}1\mathrm{s}_\mathrm{O} + \mathrm{c}_{33}2\mathrm{s}_\mathrm{O} + \mathrm{c}_{34}2\mathrm{p}_{\mathrm{z}_\mathrm{O}} \tag{7.6.2.3}$$

$$\phi(4\mathrm{a}_1) = \mathrm{c}_{41}\sigma_1 + \mathrm{c}_{42}1\mathrm{s}_\mathrm{O} + \mathrm{c}_{43}2\mathrm{s}_\mathrm{O} + \mathrm{c}_{44}2\mathrm{p}_{\mathrm{z}_\mathrm{O}} \tag{7.6.2.4}$$

$$\phi(1\mathrm{b}_2) = \mathrm{c}_{55}\sigma_5 + \mathrm{c}_{56}2\mathrm{p}_{\mathrm{y}_\mathrm{O}} \qquad (7.6.2.1)$$

$$\phi(2\mathrm{b}_2) = \mathrm{c}_{65}\sigma_5 + \mathrm{c}_{66}2\mathrm{p}_{\mathrm{y}_\mathrm{O}} \qquad (7.6.2.1)$$

$$\phi(1\mathrm{b}_1) = 2\mathrm{p}_{\mathrm{x}_\mathrm{O}} \qquad (7.6.2.1)$$

the a_1 orbitals having A_1 symmetry, the a_2 orbitals B_2 symmetry and the b_1 orbital B_1 symmetry.

The procedure for setting up the secular equations for H_2O is analogous to that one adopted by Mulligan (1950) for CO_2.(i.e. the James–Coolidge procedure; cf. Sect. 7.3.3).

Taking advantage of the symmetry constraints the 7×7 secular matrix reduces to one 4×4, one 2×2 and one 1×1 equation. Table 7.22 lists the Ellison–Shulls coefficients obtained for an H-O-H bond angle of 105°, corresponding to the experimental observation.

Table 7.22 LCAO-SCF MOs for the ground state of H_2O after Ellison and Shull (1955) for an H-O-H bond angle of 105° (only the coefficients of the "*Complete treatment*" are listed

Occupied orbitals						Unoccupied orbitals	
c_{11}	−0.0033	c_{21}	0.1781	c_{31}	0.3341	c_{41}	1.061
c_{12}	1.0002	c_{22}	−0.0286	c_{32}	−0.0258	c_{42}	−0.086
c_{13}	0.0163	c_{23}	0.845	c_{33}	−0.4601	c_{43}	−0.833
c_{14}	0.0024	c_{24}	0.1328	c_{34}	0.8277	c_{44}	−0.642
c_{55}	0.7759					c_{45}	1.230
c_{56}	0.5428					c_{46}	−1.013

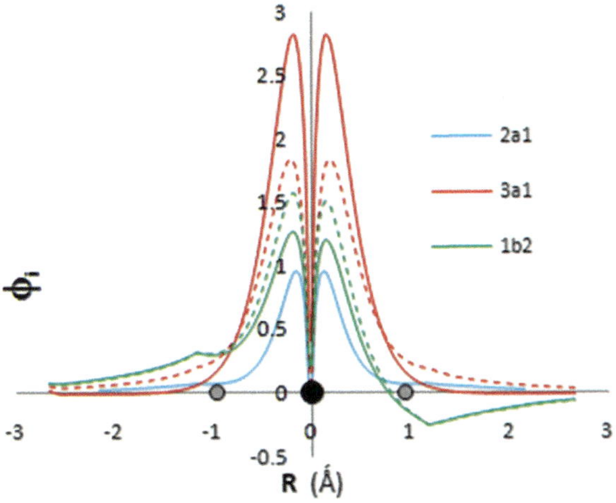

Fig. 7.24 LCAO-SCF bonding MOs of H_2O along internuclear axis according to Ellison and Shull (1955) drawn for a H-O-H bond angle of 180°. The dashed lines represent the most intensive bonds for a bond angle corresponding to the experimental observation (i.e. 105°)

The stable structure is a $^1\Delta_1$ singlet and the bonding is attributed to the $\phi(1b_2)$ and $\phi(3a_1)$ orbitals. This latter is partly constituted by a "*negative oxygen hybrid*" (i.e. plus $2p_z$ minus 2s character; cf. Mulliken 1951). As we may see in Fig. 7.24, at 180° there is indeed a conspicuous charge overlap in the internuclear regions. Diminishing the bond angle the amount of overlap on the two bonding orbitals varies in opposite fashion.

The calculated dipole moment (m = 1.52 10^{-18} *esu·cm*) is slightly lower than what was experimentally observed and slightly lower is also the computed electronic energy (i.e. −2314.4 vs −2030.5 *eV*), corresponding to a weaker dissociation energy as well (i.e. 7.7 vs 10.06 *eV*).

7.6.3 The Duncan-Pople Model

Actually, Duncan and Pople (1953) were not directly concerned to establishing the ground state properties of the H_2O molecule, but, rather, to the evaluation of the role of *lone-pair* electrons in determining substantially the dipole and quadrupole moments of the molecule electrons.[14] In fact, lone-pair electrons, just as bonding electrons, could be described either in terms of molecular orbitals "*forming irreducible representations of the symmetry group and distributed throughout the*

[14] A *lone pair* is an *electron* pair in the outermost shell of an atom that is not shared or bonded to another atom. It is also called "*unshared pair*". Lone pairs affect substantially the polarity of charge of a molecule, reinforcing or reducing its dipole and quadrupole moment.

molecule" (Duncan and Pople 1953) or in terms of equivalent orbitals "*transforming into one another under operations of the symmetry group and generally of a more localized character*" (ibid.). The theory of inner shell and lone-pair electrons was anticipated by Lennard-Jones and Pople (1950) and more specifically applied to H_2O by Pople (1950). It was shown that a transformation can be applied to the molecular orbitals to give sets of equivalent bond orbitals and sets of lone-pair orbitals. In H_2O there are two such sets, each with two members pertaining to MOs of the same (A_1) symmetry. The two sets can be obtained applying

$$[\Psi_1(\cos\lambda) + \Psi_2(\sin\lambda)] \tag{7.6.3.1}$$

$$[\Psi_2(\cos\lambda) - \Psi_1(\sin\lambda)] \tag{7.6.3.2}$$

where λ is an arbitrary parameter obeying a suitable, convenient criterion of localization.

The geometrical representation of H_2O adopted by Duncan and Pople (1953) differs from what anticipated at Sect. (7.6.1). The x-direction is taken along the molecular axis and the z-direction is perpendicular to the molecular plane, so that $\Psi_{(b')}$, $\Psi_{(b'')}$ are mirror images in the plane y = 0. Also, the atomic functions adopted for the 2s and 2p orbitals of oxygen do not correspond to the form already seen in the previous treatments but are analytical approximations of the form:

$$2s = \sqrt{\frac{Z^5}{96\pi}} r \exp\left(-\frac{Z}{2r}\right) \tag{7.6.4.1}$$

$$2p = \sqrt{\frac{Z^5}{32\pi}} r \exp\left(-\frac{Z}{2r}\right) \cos\theta \tag{7.6.4.2}$$

where Z is Slater's value of the screening constant (Z = 4.55 for oxygen).

Based on the model of Duncan and Pople (1953), two molecular orbitals (b', b") may form as a linear combination of hybrid orbitals 2p, 2s of oxygen, and orbital 1s of hydrogen:

$$\Psi_{(b')} = \lambda\left[\cos\varepsilon_b\psi_{O(2s)} + \sin\varepsilon_b\psi_{O(2p,b')}\right] + \xi\psi_{H'(1s)} \tag{7.6.4.1}$$

$$\Psi_{(b'')} = \lambda\left[\cos\varepsilon_b\psi_{O(2s)} + \sin\varepsilon_b\psi_{O(2p,b'')}\right] + \xi\psi_{H''(1s)} \tag{7.6.4.2}$$

and two "*lone-pair*" orbitals (l', l") may form as a combination of hybrid orbitals 2s, 2p of oxygen:

$$\Psi_{(l')} = \cos\varepsilon_l\psi_{O(2s)} + \sin\varepsilon_l\psi_{O(2p,l')} \tag{7.6.4.3}$$

Table 7.23 Orbital constants and the dipole moments for H_2O according to Duncan and Pople (1953)

Magnitude	Value	Magnitude	Value
λ/ε	1.06	β	120.2°
λ	0.617	γ	107.8°
ε	0.582	M_b	-6.82×10^{-18} e.s.u
$\cos\varepsilon_b$	0.093	M_l	3.03×10^{-18} e.s.u
$\cos\varepsilon_l$	0.578	M_p	5.63×10^{-18} e.s.u
		M	1.84×10^{-18} e.s.u

$$\Psi_{(l'')} = \cos \varepsilon_l \psi_{O(2s)} + \sin \varepsilon_l \psi_{O(2p,l'')} \tag{7.6.4.4}$$

The ratio λ/ξ defines the polarity of the bond and ε_b describes the hybridization state of the orbitals. Since $\cos\varepsilon_b = 0.093$ (Duncan and Pople 1953), molecular hybrid orbitals $\Psi_{(b')}, \Psi_{(b'')}$ are composed essentially of the wave functions of atomic orbitals 2p of oxygen and 1s of hydrogen. The value of 0.578 obtained for $\cos\varepsilon_l$ also indicates that hybrid orbitals $\Psi_{(l')}, \Psi_{(l'')}$ are essentially of type sp^3

The electron density corresponding to a bond orbital is given by

$$\begin{aligned}[\chi(b')]^2 &= \lambda^2\left[\cos \varepsilon_b \psi_{O(2s)} + \sin \varepsilon_b \psi_{O(2p,b')}\right]^2 \\ &\quad + 2\lambda\xi\left[\cos \varepsilon_b \psi_{O(2s)} + \sin \varepsilon_b \psi_{O(2p,b')}\right]\psi_{H(1s,l)} + \xi^2\left[\psi_{H(1s,l)}\right]^2\end{aligned} \tag{7.6.5}$$

And the dipole moment of a molecule with N doubly occupied equivalent orbitals and lone pairs is given by

$$M = M_b + M_l + M_p = -2e\sum_{j=1}^{N} \overline{\chi}_j \chi_j \mathbf{r} d\tau + e\sum_{\alpha} Z_\alpha \mathbf{r} \tag{7.6.6}$$

where $\mathbf{r}$ is the position vector and eZ_α the charge of the nucleus.

The orbital constants and the dipole moments for H_2O obtained by Buncan and Pople are listed in Table 7.23.

7.7 Molecules in a Reaction Field

7.7.1 Premises

When a molecule, or a charged cluster of atoms, are immersed in a dielectric medium, their properties (structure and energy at ground state and in excited states, vibrational modes) are considerably affected. Electrostatic effects, which are often important in the gas phase, are appreciably reduced when the moiety is placed in a solvent with a high dielectric constant. For instance, while in the gas phase, conformations with no dipole moment are usually preferred over those leading to dipoles, other factors may become more important once in solution. This "*mediation*" operated by the solvent has been the object of accurate investigation in determining the observed modifications in conformational equilibria and rotational barriers of simple organic compounds (cf. Wong et al. 1991; and references therein). As long as specific solvent effects (such as hydrogen bonding and donor-acceptor interactions) are not present, the effects of solvents on the energies of organic compounds are eventually related to the dielectric constant of the solvent in a simple manner. In Geochemistry, solvent effects are particularly important. As an example, I may anticipate here that even the building block of the so-called "*silicates*", i.e. the orthosilicate group $[SiO_4]^{4-}$, while intrinsically unstable as an isolated cluster, becomes stable at all effects when immersed in a medium with the dielectric properties of molten silica (Richet and Ottonello 2014). More generally, whatever "*solvent*" dictates a peculiar effect on the "*solute*". Evaluating precisely these effects (in terms of energy and structure) is prodromal to the quantitative appraisal of solute-solvent interactions. It is moreover essential for a correct interpretation of complexation-hydrolysis reactions and in the quantitative assessment of the activity coefficient, in solutions of whatever salinity and ionic strength. We will see in some details some of the most important *Self-Consistent Reaction Field Models* (SCRF). The "*solvent*" is seen, in these models, as a continuum with a discrete (and geometrically invariant) dielectric constant, and the solute moiety is placed into a cavity of the solvent of appropriate form. Figure 7.25 shows in a schematic but efficient way the differences in the three main classes of approach.

In the Onsager Model (Fig. 7.25a) the solute moiety occupies a spherical cavity of radius a_0. The dipole moment of the solute molecule (represented as an arrow) induces a dipole moment in the solvent medium (versus-arrow). The electric field applied to solvent dipole interacts with the solute dipole, leading eventually to stabilization (i.e. gain in negative energy). In the Polarized Continuum Model (PCM) of Tomasi (Fig. 7.25b) the cavity in the solvent is built up by the solute atoms and it is seen as a series of interlocking spheres of appropriate dimension. An "*apparent charge distribution*" (σ, Cancès et al. 1997) is spread on the cavity surface and discretized by point charges each associated with a small portion "*tessera*" of the cavity surface (i.e. the tessellated surface representing the limiting region pervaded by the solvent itself (i.e. the "*Solvent Accessible Surface*" or SAS,

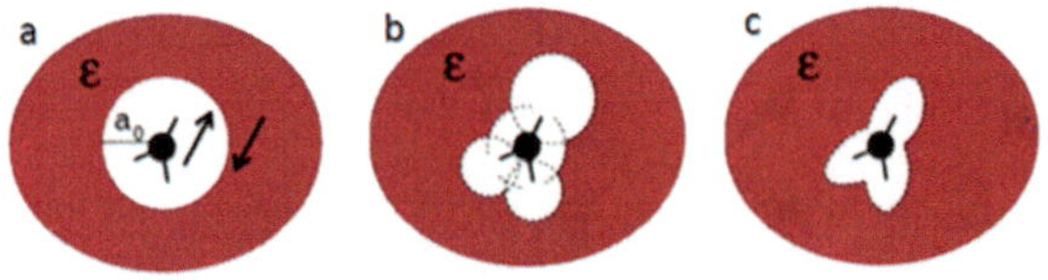

Fig. 7.25 Schematic representation of SCRF models. **a** Onsager; **b** Tomasi's Polarized Continuum; **c** Isodensity Polarized Continuum. Modified from Foresman and Frisch (1993)

see later on). In the isodensity-PCM (Fig. 7.25c) the interlocked spheres are replaced by an isodensity ipersurface obtained by resolving iteratively various SCF cycles, updating progressively its form, up to convergency (i.e. a variational state) and adding finally (in a later cycle) the coupling effect.

7.7.2 *The Onsager Model*

In his development Onsager (1936) begins considering the implications of the Debye theory of the "*internal field*". According to this theory, the "*internal field*" which polarizes a molecule in the dielectric equals the *external field*, augmented by $(4/3)\pi$ times the electric moment induced in a unit volume of the dielectric. Debye's implicit assumption is that the force-couple which tends to orient an electrically asymmetric molecule in a polarized dielectric is proportional to the same "*internal field*", i.e.:

$$\frac{\varepsilon - 1}{\varepsilon + 2} = \frac{4}{3}\pi \sum \mathrm{N}\left(\alpha + \frac{\mu^2}{3k_B T}\right) \tag{7.7.1}$$

with ε = dielectric constant; α = polarizability; μ = permanent electric moment; N = numeral density (molecules/cc in the original formulation).

Debye's Equation (7.7.1) derives from the more general Clausius–Mossotti relation (Rysselberghe 1932).

$$\frac{\varepsilon_r - 1}{\varepsilon_r + 2} = \frac{\mathrm{N}\alpha}{3\,\epsilon_0} \tag{7.7.2}$$

where ϵ_0 is the vacuum permittivity (ϵ_0 = 8.5411878176 $\times$ 10^{-12} $C^2 \times N^{-1} \times m^{-2}$) and $\varepsilon_r = \varepsilon/\epsilon_0$

It should be noted in passing that, in the common practice, the dielectric constant of the medium is replaced by the relative permittivity ε_r, which is dimensionless. Also, the subscript "r" is dropped and, somehow erroneously, the term "*dielectric constant*" is assigned to what actually is a "*relative permittivity*". Hence, ε = 1 in vacuum, ~78.6 in water, and so on.

Because all real molecules have a positive polarizability α, the dipole moment in a medium where several molecules are present does not remain constant.

The total field **F** which acts upon a spherical molecule in a polarized dielectric medium may be written as a summation of internal contribution (i.e. a "*cavity field*" **G,** proportional to the total electric field intensity **E**) and a "*reaction field*" **R**, proportional to the total electric moment **m** (and, as such, depending on the instantaneous orientation of the molecule; Eq. 12 in Onsager 1936):

$$\mathbf{F} = \mathbf{G} + \mathbf{R} = \frac{3\varepsilon}{2\varepsilon + 1}\mathbf{E} + \frac{2(\varepsilon - 1)}{(2\varepsilon + 1)\alpha^3}\mathbf{m} \tag{7.7.3}$$

F, the total electric moment is thus the vector sum of the permanent and the induced dipole moments "*in the environment of the molecule*" (cf. Onsager 1936)

$$\mathbf{m} = \mu_0\mathbf{u} + \alpha\mathbf{F} \tag{7.7.4}$$

where **u** denotes a unit vector in the direction of the dipole axis.

The mean orientation of a molecule is determined by the orienting force-couple exerted by the cavity field **G** upon the electric moment of the molecule.

The explicit formula for the internal field **I** is

$$\mathbf{I} = \mathbf{F} - \mathbf{E} = \frac{2(\varepsilon - 1)}{(2\varepsilon + 1)}\frac{\mu}{\alpha^2}\mathbf{u} + \left[1 + \frac{n^2(\varepsilon - 1)}{(2\varepsilon + n^2)}\right] \tag{7.7.5}$$

where n^2 represents the "*internal dielectric constant*" (Onsager 1936).

Since all real molecules are electrically deformable, the reaction field **R** causes an enhancement of the electric moment of any molecule immersed in a dielectric. The induced moment due to the cavity field (**G**) is in a similar way enhanced by the corresponding component of **R**.

The electric moment μ of a spherical molecule is thus computed as a function of the dielectric constant **E** of the environment. For computational purposes, it is convenient to consider an "*external*" electric moment μ^*, which determines the interaction of the molecule with all distant charges and long-range fields in the dielectric.[15]

The work done by the solvent on the solute molecule is seen quantum-mechanically as a perturbation Hamiltonian $\mathsf{H}_1 = -\hat{\mu}\mathbf{R}$ which describes the coupling between the molecular dipole operator $\hat{\mu}$ and the reaction field **R**, itself proportional to the dipole moment μ^*.

Posing:

$$\mu^* = \frac{3\varepsilon}{2\varepsilon + 1}\mu \tag{7.7.6}$$

$$R = \frac{2(\varepsilon - 1)}{2\varepsilon + 1}\frac{m}{a_0^3} \tag{7.7.7}$$

where a_0 is the radius of the spherical cavity hosting the solute molecule (see. Eq. 7 in Onsager 1936), the proportionality is obtained through the coefficient g (Wong et al. 1991):

[15] For spherical molecules, the ratio μ^*/μ increases from unity to 3/2 with the increase of the dielectric constant from 1 to ∞.

$$g = \frac{2(\varepsilon - 1)}{2\varepsilon + 1} \frac{1}{a_0^3} \tag{7.7.8}$$

In SCF calculations the effect of the reaction field is incorporated as an additional term of the Fock matrix with ϕ_λ, ϕ_σ as basis functions (Wong et al. 1991):

$$F_{\lambda\sigma} = F_{\lambda\sigma}^0 - g\mu^* \langle \phi_\lambda | \hat{\mu} | \phi_\sigma \rangle \tag{7.7.9}$$

The energy of the system, inclusive of the solvent polarization, is given by (Wong et al. 1991):

$$E = \langle \Psi | H_0 | \Psi \rangle - \frac{1}{2} \mu^* R \tag{7.7.10}$$

where H_0 is the Hamiltonian of the unperturbed system and the 2nd term on the right is the electrostatic contribution to the free energy of solvation.

For an ionic solute of charge Q one must account for ion-dipole interaction through the Born term (Wong et al. 1991):

$$E_{ion} = E - \frac{1}{2}\left(1 - \frac{1}{\varepsilon}\right)\frac{Q^2}{a_0} \tag{7.7.11}$$

Because in the case of a charged molecule the dipole moment is not translationally invariant, one must also account for the actual displacement of the electronic and nuclear components of the dipole moment (μ_e, μ_n) with respect to the center of charge through (Wong et al. 1991):

$$\mu' = \mu_e \frac{Q + n_e}{n_e} + \mu_n \tag{7.7.12}$$

As a first application of the Onsager model I report the results of Wong et al. (1991) on the 1,2 Dichloroethane conformers (gauche and trans; Fig. 7.26) in various media.

The cavity radius was calculated by Wong et al. (1991) on the basis of the greatest internuclear distance with van der Waals (vdW) radii. A more direct approach implies the experimental knowledge of the molar volume (V_m) of the solute, by applying:

$$a_0 = \sqrt[3]{\frac{3V_m}{4\pi N_0}} \tag{7.7.12}$$

with the volume expressed in $\mathring{A}^3$ and N_0 = Avogadro number.

In Table 7.24 we see the marked "*mediating effect*" exerted by solvents of increasing permittivity, which reduces progressively the energy difference between the two conformers. This effect is well accounted for by the Onsager procedure,

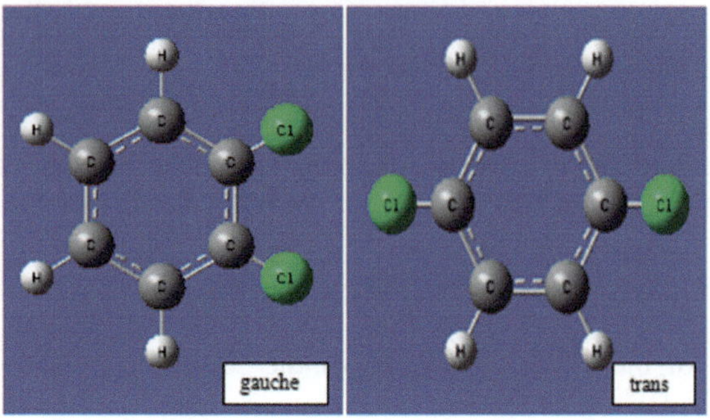

Fig. 7.26 Gauche and trans conformers of Dichloroethane

Table 7.24 Calculated (Wong et al. 1991) versus observed (Abraham and Bretschneider 1974) gauche-trans conformer energy difference (*ΔE*; *kJ/mol*) for 1,2 Dichloromethane in solvents of increasing permittivity (adapted, from Wong et al. 1991). The adopted basis set in HF calculations is 6−31G(d). The cavity radius is a_0 = 3.65 *Å*

		Calculated ΔE		
Medium	ε	HF	MP2	Observed ΔE
Vapor	1.00	8.20	6.86	5.02
Cyclohexane	2.00	5.52	3.68	3.81
Carbon disulfide	2.60	4.73	2.89	3.47
Diethyl ether	4.30	3.43	2.26	2.89
Ethyl acetate	6.00	2.80	1.72	1.76
Pure liquid	10.10	2.05	1.09	1.30
Acetone	20.70	1.51	0.59	0.75
Acetonitrile	35.90	1.26	0.38	0.63

especially when the adopted theory level is appropriate to the electronic complexity of the investigate moiety (i.e. MP2 against HF) and the dielectric constant of the medium is high.

7.7.3 The Polarized Continuum Model (PCM)

This extremely successful and important model has been implemented through the years by the research group lead by Jacopo Tomasi (Tomasi and Persico 1994; Cancès et al. 1997; Mennucci, and Tomasi 1997; Mennucci et al. 1997; Cossi et al. 1998; Cancès and Mennucci 1998; Tomasi et al. 1999;).The basic formulation of the method is due to Miertuš et al. (1981), Miertuš and Tomasi (1982). The polarization of the infinite dielectric is recast into a system of virtual charges placed on the cavity surface with density $\sigma(\mathbf{s})$.[16]

The total electrostatic potential acting on S [$V(\mathbf{r})$] is seen as the summation of two distinct contributions, $V_\rho(\mathbf{r})$, $V_\sigma(\mathbf{r})$ arising from the solute charge distribution and from the virtual charge distribution, respectively:

$$V(\mathbf{r}) = V_\rho(\mathbf{r}) + V_\sigma(\mathbf{r}) \tag{7.7.13}$$

In classical electrostatics the potential (7.7.13) in the charge free part of the system obeys the Laplace equation

$$\nabla^2 V(\mathbf{r}) = 0 \tag{7.7.14}$$

On the other hand, the boundary conditions imply on the cavity surface

$$V(s)_- = V(s)_+ \tag{7.7.15.1}$$

$$\left[\frac{\partial V(s)}{\partial n}\right]_{s-} = \left[\frac{\partial V(s)}{\partial n}\right]_{s+} \tag{7.7.15.2}$$

with the – and + indexes indicative of points lying at infinitely small distance from the boundary (inside and outside the cavity), being n the normal at the surface cavity at the points.

Moreover, at infinite distance from the cavity one may place (Miertuš et al.1981):

$$\lim_{r \to \infty} rV(r) = 0 \tag{7.7.16.1}$$

$$\lim_{r \to \infty} r^2 \left[\frac{\partial V(r)}{\partial r}\right] = 0 \tag{7.7.16.2}$$

Because $\sigma(s)$ is related to the normal components of the electric displacement vector $\mathbf{p} = \mathbf{E}[(\varepsilon - 1)/4\pi]$ by

$$\sigma(s) = \frac{\varepsilon - 1}{4\pi}\left[\frac{\partial V(s)}{\partial n}\right]_{\delta+} \tag{7.7.17}$$

at the light of Eq. (7.7.15.2), one may use the electric field calculated in vacuum ($E(S)_{n-}$) at the place of the electric displacement calculated outside the cavity (Miertuš et al. 1981):

[16] The bold character of **s** implies a vectorial notation with a charge distribution which attains zero only at the cavity surface.

$$\sigma(s) = -\frac{\varepsilon - 1}{4\pi\varepsilon}E(s)_{n-} \tag{7.7.18}$$

The surface charge distribution produces an electrostatic potential, $V_\sigma(r)$, which can be added to the solute Hamiltonian H_0 to take into account the polarization produced by the solvent in the continuum model, as we have already seen in the Onsager procedure.

The surface S of the cavity is subdivided into portions ΔS small enough to permit σ(s) to be considered as constant at their interior. The cavity is represented as a system of interlocking spheres, centered on appropriate nuclei of the solute with radii R_1, R_2,...R_k. Each (spherical) surface K defining a portion is further divided into elements ΔS_{ki} of areal extension determined by fixed increments $\Delta\varphi$ and $\Delta\theta$ in polar coordinates having k as center:

$$\Delta S_{ki} = 2R_k^2 \sin\theta_{ki} \sin\left(\frac{\Delta\theta}{2}\right)\Delta\varphi \tag{7.7.19}$$

Because the electric field **E** depends on σ(s) (Eqs. 7.7.13 and 7.7.18) an iterative procedure is compulsory.

Consider now the Onsager treatment of a non-polarizable dipole μ in a spherical cavity of radius a_0. The solution of the Laplace equation for this simple case in terms of surface charge distribution leads to:

$$\sigma(s) = \frac{\varepsilon - 1}{4\pi}E_{n+} = -\frac{3\varepsilon}{2\varepsilon + 1}\left(\frac{1}{4\pi}\frac{\varepsilon - 1}{\varepsilon}\frac{2\mu\cos\theta}{a_0^3}\right) \tag{7.7.17}$$

If we use $V_\rho^{\ 0}$ to calculate the normal component of the electric field on the cavity surface, we have

$$E_{\rho,n+} = \frac{2\mu\cos\theta}{\varepsilon a_0^2} \tag{7.7.18}$$

$$\sigma^0(s) = -\frac{1}{4\pi}\frac{\varepsilon - 1}{\varepsilon}\frac{2\mu\cos\theta}{a_0^3} \tag{7.7.19}$$

Hence, the factor in parentheses on the right-hand side of Eq. (7.7.17) is that one we would have obtained if the mutual polarization of the surface charges was not considered, and the term before parentheses may be seen as a corrective factor f = 3ε/(2ε + 1) ranging between 1 and 1.5 according to the value assigned to ε.

As we may see in Table (7.25) the iterative process converges in a few steps.

Table 7.25 Iterative calculation of the corrective factor which introduces the reaction field effect on the surface charge distribution of a dipole in a spherical cavity (Miertuš et al. 1981)

ε	Cycle				Exact
	1st	2nd	3rd	4th	
1	1.0	1.000	1.000	1.000	1.000
2	1.0	1.167	1.194	1.199	1.250
5	1.0	1.267	1.337	1.357	1.400
10	1.0	1.300	1.390	1.417	1.450
20	1.0	1.316	1.417	1.449	1.475
80	1.0	1.329	1.437	1.473	1.494
∞	1.0	1.333	1.444	1.482	1.500

Actually, each iterative process involves itself several steps:

(1) a set of values for the surface charge density $\sigma^{00}(S_{Ki})$, is first obtained by replacing $E(s)_n$ in Eq. (7.7.18) with the electric field produced by the unperturbed solute charge distribution

$$\rho^0(r),\ \text{i.e.}\ \sigma^{00}(s_{ki}) = -\left[\frac{(\varepsilon - 1)}{4\pi\varepsilon}\right] E^0_\rho(s_{ki})_{n-} \qquad (7.7.20.1)$$

(2) $\rho^0(r)$ is still kept constant and the mutual polarization of the surface charges is introduced through *ad-hoc* point charges $q^{00}_{ki} = \sigma^{00}(s_{ki})\Delta S_{ki}$. These charges generate at the center of the surface elements an additional contribution to the electric field, $E^0_\sigma(s_{ki})_{n-}$. The new definition of charge densities

$$\sigma^{01}(s_{ki}) = -\left[\frac{(\varepsilon - 1)}{4\pi\varepsilon}\right]\left(E^0_\rho + E^{00}_\sigma\right) \qquad (7.7.20.2)$$

is iteratively improved by substituting in it the electric field deriving from the q^{01}_{ki} values, etc.

(3) once the polarization of $\rho^0(r)$ is appropriately evaluated imposing a final step q^{0f}_{ki} of point charges, a new solute charge distribution $\rho^1(r)$ is obtained by solving the Hamiltonian $\mathsf{H} = \mathsf{H}_0 + V_\sigma$, where the perturbation V_σ is due to the charges q^{0f}_{ki}.
(4) self-consistency is attained by repeated readjustments of $\rho(r)$ and $\sigma(s)$ through Steps 2 and 3.

The accuracy of the procedure depends upon various parameters:

- the shape of the cavity, i.e. the number of spheres and the value of the corresponding radii,
- the actual size of the surface elements,
- the level of approximation (I,II,III),

(I) interaction energy arising from the unperturbed solute charge distribution ρ^0 and surface charge distribution without self-polarization, σ^{00},
(II) interaction energy arising from ρ^0 and self-polarized distribution σ^{0f},
(III) interaction energy after mutual polarization of ρ and σ.

One of the test cases investigated by Miertuš et al. (1981) is particularly important in Geochemistry: the H_2O molecule, and it is useful to report it in some detail.

The investigation was carried out with a fixed geometry, i.e. an O-H internuclear distance of 0.99 Å and a bond angle of 108°. The cavity was conformed by three spheres centered on the solute nuclei with radii $R_O = 1.4$ Å; and $R_{H1} = R_{H2} = 1.2$ Å. The dielectric permittivity of the medium was set at $\varepsilon = 80$. All calculations were performed with the 4−31G basis set.

Changing progressively the angular increments $\Delta\theta$, $\Delta\varphi$, the set of surface point charges q_{ki}^{0f} and the corresponding V_σ were included in the solute Hamiltonian. An energy W_1 was thus obtained for each couple and the interaction energy $W_{int} = W_1 - W_0$ was evaluated as difference with respect to the solute energy in W_0. At the same time Miertus et al. (1981) obtained a new wave function from which the expectation value of one-electron observables could have been evaluated.

Two of these observables (i.e. the interaction work W_{int} and the bulk moment μ are listed in Table 7.26. Convergence on the shape of ΔS was achieved with angular increments larger than 22.5°.

As we may see in this table, the self-polarization of surface charges (Approximation II) has a remarkable influence on W_{int}. The elimination of excess surface charges (still present at the level of approximation II) through a compensation factor which distributes a charge $-e$ over the surface charges (according to the sign and magnitude they have; see Miertus et al. 1981; for the computational details) improves further the method, changing appreciably the results in terms of interaction energy (Approximation III).

Also the selected a_0 radii influence greatly the results, as we may see in Table 7.27. Radial increments of 0.25 and 0.50 Å with respect to the vdW values produce a sharp reduction of Q, which for vdW radii augmented of 0.5 Å is less

Table 7.26 Effect of the adopted number of point charges and Approximation level on the estimated W_{int} (*kJ/mol*), μ (*debyes*, D), and Q (*e* units). The value of μ in vacuo is 2.69 D. From Miertus et al. (1981) with modifications

Number of points	$\Delta\Theta$	$\Delta\varphi$	Approximation I			Approximation II			Approximation III		
			W_{int}	μ	Q	W_{int}	μ	Q	W_{int}	μ	Q
14	90	90	−90.4	3.02	−0.197	–	–	–	–	–	–
32	60	60	−65.7	2.92	−0.229	−37.4	2.83	−0.111	−43.6	2.86	0.000
196	22.5	22.5	−67.8	2.94	−0.160	−39.6	2.84	−0.083	−42.8	2.86	0.000
338	18	18	−67.8	2.93	−0.220	−39.8	2.83	−0.109	−45.6	2.86	0.000
652	12.87	12.87	−68.2	2.93	−0.207	−40.0	2.84	−0.104	−45.7	2.86	0.000

Table 7.27 Dependence of W_{int}, μ and Q on the radii of the spheres defining the cavity. $\Delta\theta = \Delta\varphi = 18°$ Units as in Table 7.26. From Miertus et al. (1981) with modifications

Radii	R_O	R_H	Approximation I			Approximation II			Approximation III		
			W_{int}	μ	Q	W_{int}	μ	Q	W_{int}	μ	Q
vdW	1.40	1.20	−67.8	2.93	−0.220	−39.8	2.84	−0.109	−45.6	2.86	0.000
vdW + 0.25 Å	1.65	1.45	−50.0	2.87	−0.056	−31.2	2.82	−0.033	−32.0	2.82	0.000
vdW + 0.50 Å	1.90	1.70	−35.1	2.84	−0.011	−21.2	2.78	−0.007	−21.3	2.78	0.000

than 0.01 *e* (cf. Approximation II in Table 7.27). Augmenting the vdW radii of this constant amount reduces however considerably also W_{int} and μ.

Assuming each atom to have its own dimension (proportional to the vdW radius; as in the cases resumed in Table 7.27) the surface is seen as an interlocking of tessellated spheres (Fig. 7.21a) while with the Unified Atom Topological Model (UATM; Barone et al. 1997; Cossi and Barone 2000; Cossi et al. 2002) hydrogen atoms are enclosed in the sphere of the atom to which they are bonded, returning, in the case of H_2O, a single pentakis dodecahedral tessellation (b).[17]

In both cases the charge distribution respects the dipolar nature of the moiety. The colors in Fig. 7.27 return in a chromatic scale the value of the fictitious charges describing (with appropriate positions and values) the reaction potential generated by a dielectric on a given charge distribution (i.e. "*image charge method*"; Tomasi and Persico 1994; Cancès et al. 1997).

The "*tesserae*" (i.e. small tiles) are triangular and correspond to the projections on the surface of the faces of a pentakis dodecahedron inscribed in this sphere. Tesserae completely buried by some other spheres are discarded, while those cut by intersections with one or more other spheres are replaced by new polygonal tesserae whose areas are calculated using the Gauss–Bonnet theorem (Cossi et al. 1996). The size of tesserae can be chosen according to the solute dimensions, and to the desired accuracy; a good compromise is usually an average area of 0.4 $Å^2$ per tessera (Barone et al. 1997).

The discretization of the cavity surface by polygonal tesserae does not introduce any sort of ambiguity in the Quantum mechanical treatment of the solute-solvent interaction because these geometrical elements are characterized by their area, their central point, and the unit vector normal to the surface. Moreover, all these geometrical parameters are analytically differentiable with respect to nuclear displacements (Barone et al. 1997). The tessellation procedure has been constantly improved through the years and much effort has been devoted to find the best "*dimension*" to assign to atomic constituents. In standard PCM, the tessellation of the cavity is built with the GEPOL or GEPOL−93 procedures (Pascual-Ahuir et al.

[17] Calculations can be performed also with Pauling radii, Bondi radii (Bondi, 1964), Universal Force Field radii (UFF; Rappè et al., 1992).

Fig. 7.27 The Van der Waals surface of the IEFPCM model for H_2O, with atom radii effectively proportional to Van der Waals radii (**a**) and atom radii of the Unified Atom Topological Model (UATM) (**b**). The radii adopted in this case are the UAHF (United Atom Topological Model applied on radii optimized for the HF/6−31G(d) theory level; Cossi et al. 2002). The color returns in a chromatic scale the value of the single fictitious charge q' = −q, placed at a distance −d (i.e. inside the conductor). The conductor surface acts as a mirror (i.e. "blue" negative fictitious charge in the neighborhood of a solute cation and "red" positive ones in the neighborhood of an anion). Inside the cavity the electrostatic potential obeys the Poisson equation $-\nabla^2 V = 4\pi\rho_{n+e}$ where ρ_{n+e} is the solute nuclear + electronic charge. (From Ottonello 2018)

1987; Tomasi and Persico 1994; Cossi et al. 1996). The sphere radii are set according to the atomic number, the charge, and the hybridization of the atom, possibly corrected for first neighbors' effects.

The molecular cavity is formed by interlocking spheres centered on the nuclei of the solute atoms (or atomic groups, as we have seen for H_2O in case a). The cavity is then smoothed by adding some more spheres (not centered on solute atoms), to approximate the reentrant parts of the cavity surface.

Also the nature of the region limiting the cavity has been the object of continuous investigation. As an example, Fig. 7.28 shows schematically the cavity depicted for a spherical solute molecule i in a liquid solvent s with a solvation diameter σ_s. The dotted line is the "*van der Waals*" surface depicted by a probe sphere rolling on the surface generated by the interlocking spheres, which represent the atoms with unscaled vdW atomic radii (here coinciding for simplicity with a single sphere) and the dashed line is the envelope of the volume formed from scaled vdW atomic radii. Scaling, through the electrostatic factor γ_{el}, displaces to a more external region the solute-solvent limit and defines the actual "*Solvent Accessible Surface*" (SAS). The "*Solvent Excluding Surface*" (SES) is that one generated by the atomic or group spheres and by the spheres created automatically to smooth the surface.

The precision attained by PCM in reproducing the actual "*shape*" of complex molecules in solution is astonishing, and, as already mentioned, it is one of the reasons for its widespread utilization in Material Science. As an example we see in Fig. 7.29 the UATM reproduction of taxol, a large organic molecule of formula $C_{47}H_{51}NO_{14}$, and the GEPOL93 mapping of the surface with tesserae of area 0.4 $Å^2$ (Cossi et al. 1997).

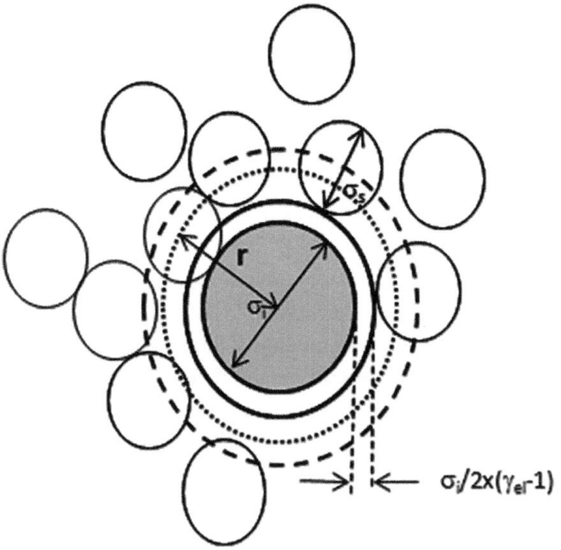

Fig. 7.28 Spherical cavity of radius r defining the Solvent Accessible Surface (SAS; dashed line) originated by a solute of diameter σ_i in a hard spheres solvent of molecules of radius σ_s. The electrostatic scaling parameters γ_{el} act on the nominal vdW radius of the solute, expanding (or contracting) it. With $\gamma_{el} = 1$ and absence of added spheres one has an unscaled vDW surface (dotted line). The Solvent Excluding Surface (SES, solid lines) is generated by the atomic or group spheres and by the spheres created automatically to smooth the surface. Electrostatic scaling operates only on the inductive energy terms. From Ottonello et al. (2015). With the permission of AIP Publishing LLC

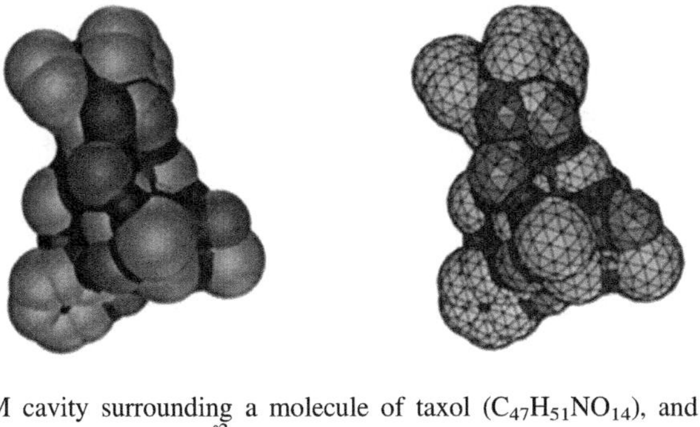

Fig. 7.29 UATM cavity surrounding a molecule of taxol ($C_{47}H_{51}NO_{14}$), and mapping of the surface with *tesserae* of area 0.4 $Å^2$. From Cossi et al. (2002). With the permission of AIP publishing LLC)

7.7.4 The Integral Equation Formalism of the Polarized Continuum Model (IEFPCM)

It is an improvement of PCM aimed at treating with equal efficiency intrinsically anisotropic media with a tensorial permittivity, and ionic solutions, besides the standard isotropic liquids treated by PCM. This new formulation, of paramount importance in material science, is quite innovative also in Quantum Chemistry because the standard formalism (Poisson operator) is here replaced by the use of other operators derived from the theory of integral equations (Hackbusch 1995).

The name "*Integral Equation Formalism*" (IEF) besides reflecting the adoption of this mathematical formalism, was meant by the Authors to stress the fact that the same theoretical approach "*allows to reach a unified analytical solution of the corresponding electrostatic problem for all the quoted systems (namely, one or more solutes either in a common isotropic solvent, in an anisotropic dielectric, like liquid crystals and solid matrices, or in an ionic solution)*" (Mennucci et al. 1997).

The system of interest here is a solute M in a cavity Ω surrounded by a dielectric with permittivity ε. The regions of the system inside and outside the cavity Ω (C in Mennucci et al. 1997) are identified by the subscripts i and e, respectively.

The unified formalism implies (Mennucci et al. 1997):

$$L_i V = 4\pi\rho_M \qquad \text{in}\, \Omega \tag{7.7.21.1}$$

$$L_e V = 0 \qquad \text{outside}\, \Omega \tag{7.7.21.2}$$

$$[V] = 0 \qquad \text{on}\, \Gamma \tag{7.7.21.3}$$

$$\partial_L [V] = 0 \qquad \text{on}\, \Gamma \tag{7.7.21.4}$$

with Γ as cavity boundary (cavity surface Σ in Mennucci et al.,1997).

L in the above equations is an operator that returns the appropriate form of potential. The presence of the charge distribution F_M located inside a cavity of surface Γ generates a potential V. The operator L_i is simply the negative of the nabla operator (i.e. $L_i = -\nabla$). The operator L_e changes of significance with the imposed conditions, more precisely:

$$L_e V = -\varepsilon\Delta V \qquad \text{isotrop.} - \text{diel.} \tag{7.7.22.1}$$

$$L_e V = -\text{div}(\varepsilon \cdot \nabla\Delta) \quad \text{anisotrop.} - \text{diel.} \tag{7.7.22.2}$$

$$L_e V = -\varepsilon\Delta V + \varepsilon\|^2 V \quad \text{ionic} - \text{sol.} \tag{7.7.22.3}$$

Equations (7.7.21.3, 7.7.21.4) are "*jump conditions*" (Mennucci et al. 1997) indicating that external and internal potentials across the surface Γ are identical (i.e.

$V_e = V_i$) as well as identical are their derivatives in an homogeneous isotropic medium (see also Eqs. 7.7.15.1, 7.7.15.2):

$$\left(\frac{\partial V}{\partial n}\right)_i - \varepsilon\left(\frac{\partial V}{\partial n}\right)_e \tag{7.7.23}$$

Equation (7.7.22.3) is the "*Linearized Poisson Boltzmann*" (LPB) equation. $\mathcal{K}$ in Eq. 7.7.22.3 is the inverse of the Debye length and accounts for the effect of ionic screening:

$$\mathcal{K} = \frac{F}{\varepsilon_2}\sqrt{\frac{8\pi I}{RT}} \tag{7.7.24}$$

$$I = \frac{1}{2}\sum_i m_i Z_i^2 \tag{7.7.25}$$

where I is the ionic strength of the solvent with dielectric constant ε_2, and F is the Faraday constant.

The form of ionic strength here represented is that one usually adopted in Geochemistry, where the summation is extended to all solute charges of individual molality m_i.[18]

The electrostatic interaction between two charge distributions ρ and ρ' of the solute molecule (both located inside the cavity Ω_i) is given by (Cancès et al. 1997):

$$E_I(\rho, \rho') = \int_{R^3} \rho'(x)V(x)dx \tag{7.7.26}$$

V(x) in Eq. (7.7.26) is the electrostatic potential created by the charge distribution ρ at x. This potential obeys, as we have already seen, the Poisson equation:

$$\rho(x) = -\mathrm{div}[\varepsilon(x) \cdot \nabla V(x)] \tag{7.7.27}$$

where $\varepsilon(x) = \varepsilon_i$ inside the cavity and $\varepsilon(x) = \varepsilon_e$ outside the cavity. The internal (ε_i) and external (ε_e) relative permittivities are depicted respectively as a 3 × 3 unit tensor and a 3 × 3 positive-symmetric tensor (Cancès et al. 1997):

$$\varepsilon_i = \begin{pmatrix} 1 & 0 & 0 \\ 0 & 1 & 0 \\ 0 & 0 & 1 \end{pmatrix} \qquad \varepsilon_e = \begin{pmatrix} \varepsilon_1 & 0 & 0 \\ 0 & \varepsilon_2 & 0 \\ 0 & 0 & \varepsilon_3 \end{pmatrix} \tag{7.7.28.1,7.2}$$

[18] In Mennucci et al. (1997) m_i is replaced by c_i the concentration of the *i*th ionic species in moles per cubic meter.

Consider now the operators $-\Delta$ and $-\mathrm{div}(\varepsilon\nabla V)$ of the Poisson equation. After denoting as G_i and G_e their Green's functions[19]

$$G_i = \frac{1}{4\pi|x - y|} \tag{7.7.29.1}$$

$$G_e = \frac{1}{4\pi\sqrt{\det \varepsilon_e \left[\varepsilon_e^{-1} \cdot (x - y)\right] \cdot (x - y)}} \tag{7.7.29.2}$$

Cancès et al. (1997) define the following components of the Calderon projector[20]

$$(\mathsf{S}_i \cdot u)(x) = \int_\Gamma G_i(x, y)u(y)dy \tag{7.7.30.1}$$

$$(\mathsf{S}_e \cdot u)(x) = \int_\Gamma G_e(x, y)u(y)dy \tag{7.7.30.2}$$

$$(\mathsf{D}_i \cdot u)(x) = \int_\Gamma \partial_y G_i(x, y)u(y)dy \tag{7.7.30.3}$$

$$(\mathsf{D}_e \cdot u)(x) = \int_\Gamma \partial_y G_e(x, y)u(y)dy \tag{7.7.30.4}$$

$$(\mathsf{D}_i^* \cdot u)(x) = \int_\Gamma \partial_x G_i(x, y)u(y)dy \tag{7.7.30.5}$$

Γ in the above equations denotes again the cavity boundary and x belong to it (i.e. $x \in \Gamma$) while $u(y) \in L^2(\Gamma)$. Denoting $n(y)$ the outward pointing normal at y one has (Cancès et al. 1997):

$$\partial_y G_i(x, y) = \nabla_y G_i(x, y) \cdot n(y) \tag{7.7.31.1}$$

[19] Green's function are used as a convenient method for solving more complicated differential equations. They are numerical solutions to an inhomogenous differential equation with a "*driving term*" given by a delta function. In the case of an electromagnetic field they describe how the field behaves according to the laws of quantum mechanics. In this case the Green's functions are quantum-mechanical operators having no longer simple numerical values. See Qin (2014) for an extended treatment. Few notes are also available in the Mathematics section of this book (Par. A.20).

[20] The Calderón projector is a pseudo-differential operator used widely in applied mathematics to solve linear partial differential equations which have been formulated as integral equations (i.e. Boundary Element Method, BEM). See Steinbach (2008) for an extended treatment.

$$\partial_x G_i(x,y) = \nabla_x G_i(x,y) \cdot n(y) \qquad (7.7.31.2)$$

$$\partial_y G_e(x,y) = \left[\varepsilon_e \nabla_y G_e(x,y)\right] \cdot n(y) \qquad (7.7.31.3)$$

With the operator S_i as self-adjoint, D_i^* as the adjoint of D_i and $\mathsf{S}_i\mathsf{D}_i^* = \mathsf{D}_i\mathsf{S}_i$ the interaction energy (7.7.26) is now rewritten as

$$E_I(\rho,\rho') = E_1 + E_2 = \int_{R^3} \rho'(x)f(x)dx + \int_{R^3} \rho'(x)W(x)dx \qquad (7.7.32)$$

where $f(x) = \int_{R^3} G(x,y)\rho(y)dy$ and W is the apparent potential (W = V–*f*). Introducing the "*equivalent surface charge*" $\sigma = W_i/S_i$ one has

$$E_2 = \int_{\Gamma} \sigma(x)\phi_i'(x)dx \qquad (7.7.33)$$

where ϕ'_i is the electrostatic potential generated by the distribution ρ' in the vacuum:

$$\phi_i' = \int_{R^3} G_i(x,y)\rho'(y)dy \qquad (7.7.34)$$

The equivalent surface charge σ is the unique solution to the equation

$$A \cdot \sigma = g \qquad (7.7.35)$$

with

$$A = \left(\frac{\mathsf{I}}{2} - \mathsf{D}_e\right)\mathsf{S}_i + \left(\frac{\mathsf{I}}{2} + \mathsf{D}_i^*\right)\mathsf{S}_e \qquad (7.7.36.1)$$

$$g = -\left(\frac{\mathsf{I}}{2} - \mathsf{D}_e\right)f_i + \mathsf{S}_e \cdot \partial f_i \qquad (7.7.36.2)$$

$\partial f_i = -E_n$ is the normal component of the electric field generated by ρ in the vacuum and I is the identity operator (Mennucci, personal communication; see also

Cancès and Mennucci 1998b). Because in an isotropic dielectric ε_e is a scalar, one has $S_e = S_i/\varepsilon_e$; $f_e = f_i$; $D_e = D_i$. Equation (7.7.36.1) may be then transformed into

$$A = \mathsf{S}_i\left[\left(\frac{\mathsf{I}}{2} - \mathsf{D}_i^*\right)\cdot f_i + \frac{1}{\varepsilon_e}\left(\frac{\mathsf{I}}{2} + \mathsf{D}_i^*\right)\right] \tag{7.7.37}$$

and Eq. (7.7.36.2) into

$$g = -\frac{\varepsilon_e - 1}{\varepsilon_e}\mathsf{S}_i \cdot E_n \tag{7.7.38}$$

Multiplying both terms by $\frac{\varepsilon_e}{\varepsilon_e - 1}\mathsf{S}_i^{-1}$ one finally gets

$$\left[\frac{\varepsilon_e + 1}{\varepsilon_e - 1}\frac{\mathsf{I}}{2} - \mathsf{D}_i^*\right]\cdot\sigma = -E_n \tag{7.7.39}$$

The computational practice of replacing a continuous distribution by a set of discrete point charges q_i (each associated with a small portion of the cavity surface) is mathematically representable as a set of linear equations written in a matrix formulation (Cancès et al. 1997):

$$\mathbf{q} = \Sigma\boldsymbol{\sigma} = -\Sigma\mathbf{D}^{-1}\mathbf{E}_n \tag{7.7.40}$$

where the bold characters are intended to represent the vectorial nature of the represented magnitudes, $\Sigma\sigma$ is the matrix of surface charges, $\mathbf{D}$ is a non-symmetric square matrix with dimension equal to the number of tesserae and $\mathbf{E}_n$ is the column vector containing the normal components of the electric field due to the solute.

Comparing Eqs. (7.7.39) and (7.7.40) we may easily see that (7.7.39) is exactly the operator-like form of the matrix Eq. (7.7.40) adopted for the standard PCM procedure. Hence IEFPCM reduces to PCM in an isotropic medium.

7.7.5 Isodensity Polarized Continuum Model (IPCM)

In this model the cavity, instead of being described by vdW radii, is based upon a total electron density isosurface calculated at the level of theory being applied. The surface is smooth and easily integrable, greatly improving the computational cost. The integration is carried out on basins defined by the gradient vector field, or simply by using the center of nuclear charge. In this way one avoids "*concern over the arbitrary choice of van der Waals radii for the spheres and the surface grid*" (Foresman et al. 1996). Nevertheless, instead of a set of radii for the spheres, one needs to specify the isosurface level (typically in the range 0.0004 – 0.001 e/$\AA^3$). The cavity derived in this way is based uniquely on the electronic environment.

Defining $\mathbf{r}_s$ the position vector of the integration point on the isodensity surface relative to the origin, the outward normal, **n**, at any point on the surface of the (convex) cavity is

$$\mathbf{n}(\mathbf{r}_s) = \frac{\nabla\rho(\mathbf{r}_s)}{|\nabla\rho(\mathbf{r}_s)|} \tag{7.7.41}$$

and the infinitesimal isodensity surface elements ds($\mathbf{r}_s$), defined in spherical polar coordinates, is

$$ds(\mathbf{r}_s) = \frac{(\mathbf{r}_s)^2 \sin[\theta(\mathbf{r}_s)]}{\Lambda_s^i \cdot \mathbf{n}(\mathbf{r}_s)} d\theta(\mathbf{r}_s) d\varphi(\mathbf{r}_s) \tag{7.7.42}$$

where Λ_s^i is the unit vector corresponding to $\mathbf{r}_s$.

The radial coordinate for the integration point at $\mathbf{r}_s$ is found by tracing the ray defined by the angular coordinates. The weights needed for the integration of the surface in the new coordinate system are

$$w_i = \frac{(\mathbf{r}_s^i)^2 w_\theta^i x_\varphi^i}{\Lambda_s^i \cdot \mathbf{n}(\mathbf{r}_s^i)} \tag{7.7.43}$$

After this, the usual PCM scheme is followed. Because the solute-solvent interaction alters the solute density, the procedure is iteratively repeated to convergence. The process is however not fully self-consistent because terms that couple the isodensity to the solute Hamiltonian are missing.

7.7.6 The Conductor-Like Screening Model (COSMO)

This model has been promoted by Klamt and Schüürmann (1993), Klamt (1995), Eckert and Klamt (2002). Apparently, in origin, it was primarily intended for an aqueous solution; in fact Klamt (1995) wrote: "*It still appears strange that water is able to screen molecular polarities nearly as well as electric conductors, i.e. to screen them almost perfectly*". To explain this behavior he imagined a virtual experiment: a set of hypothetical cubic solvent molecules S and a cubic solute X placed in a conducting medium (Fig. 7.30a). Both X and S are perfectly screened by surface charges and gained their ideal screening energies Δ^S and Δ^X. The screening prevents molecular interaction so that they can be arranged in any way without any change in energy. Denoting σ_i the surface charge density of the solute X and -σ_i that of the solvent molecules S, one may arrange the solvent molecules around the solute X in such a way that the net charge on adjacent surfaces is zero (Fig. 7.30b). Then "*it does not make any difference whether there is a conducting surface between the molecules or not, and we can remove these without any change in*

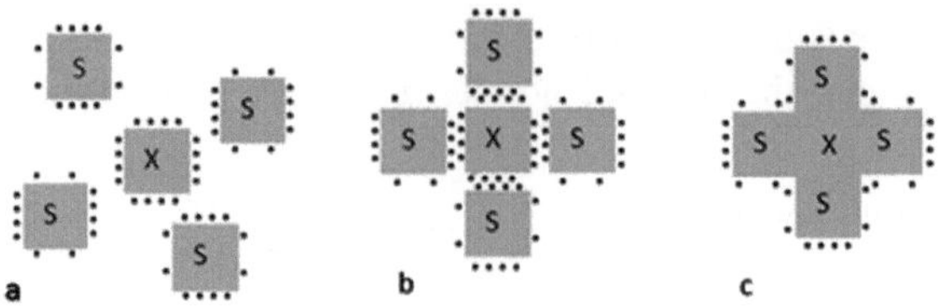

Fig. 7.30 COSMO conceptual scheme. **a** ensemble of ideally screened molecules; **b** ideal pairing of surface charges; **c** solute X screened by four solvent S molecules. Modified from Klamt (1995)

energy" (Fig. 7.30c). In this condition, the solute is screened by the solvent molecules but it still has the same energy as if it would be screened by a conductor. One may thus conclude: "*If a solvent can offer the opposite ideal surface charge densities for all faces of a solute molecule, then it is able to screen the solute as good as a conductor"(*Klamt 1995*).*

As we have already seen, the response of a homogeneous dielectric continuum to any charge distribution of the solute consists of a surface charge distribution $\sigma(\mathbf{r})$ on the SAS arising from the polarization of the dielectric medium and is related to the total electric field at the inner side of the SAS ($\mathbf{E}^{-}$) through

$$\sigma(\mathbf{r}) = \frac{\varepsilon - 1}{4\pi\varepsilon} \mathrm{n}(\mathbf{r})\mathbf{E}^{-}(\mathbf{r}) \tag{7.7.44}$$

where $\mathrm{n}(\mathbf{r})$ is the surface normal vector at a point $\mathbf{r}$ and the bold character denotes the vectorial nature of the involved magnitudes. COSMO provides an approximate, but very accurate, non-iterative approach for the solution of Eq. (7.7.44) for a perfect conductor.

The authors first considered the problem of the total screening energy $\Delta \mathrm{E}$ generated by N point charges $\mathrm{Q_i}$ at positions $\mathbf{r}_\mathrm{i}$ within a sphere of radius R:

$$\Delta x = -\frac{1}{2} \mathbf{QDQ} \tag{7.7.45}$$

In Eq. (7.7.45) D is the Green function operator of the sphere

$$\mathrm{D_{ij}} = \frac{\mathrm{R}}{\sqrt{\mathrm{R}^4 - 2\mathrm{R}^2\mathrm{r_i r_j} + \mathrm{r_i^2 r_j^2}}} \tag{7.7.46}$$

They noticed then that Eq. (7.7.45) bears a close analogy to the Coulomb interaction of the charges itself:

$$\mathrm{E_{coulomb}} = -\frac{1}{2} \mathbf{QCQ} \tag{7.7.47}$$

where $C_{ij} = \frac{1}{\sqrt{\|r_i - r_j\|}}$.

Because Eq. (7.7.45) is valid only for perfect conductors, Klamt and Schüürmann (1993) introduced first a weak ε-dependent factor of form $f(\varepsilon) = (\varepsilon - 1)/\left(\varepsilon + {}^1\!/_2\right)$. Then they developed a sort of Green's function for "*realistic cavity shapes*" in perfect conductors. For an arbitrary closed surface S subdivided into an arbitrary number of small segments s_μ the electrostatic interaction of a unit charge at $\mathbf{r}_j$ with a unit charge on s_μ centered at t_μ is given by

$$b_{i,\mu} = \frac{1}{|s_\mu|} \int_{s_\mu} \|\mathbf{r} - \mathbf{r}_i\|^{-1} d^2r \approx \|\mathbf{t}_\mu - \mathbf{r}_i\|^{-1} \tag{7.7.48}$$

The charge of the segment s_μ for a constant surface charge density σ_μ on an area $|s_\mu|$ is

$$q_\mu = |s_\mu| \sigma_\mu \tag{7.7.49}$$

and the electrostatic interaction of unit charges on s_μ and s_v is for $\mu \neq v$

$$a_{\mu,v} = \frac{1}{|s_\mu||s_v|} \int_{s_\mu} \int_{s_v} \|\mathbf{r} - \mathbf{r}'\|^{-1} d^2r' d^2r \approx \|\mathbf{t}_\mu - \mathbf{t}_v\|^{-1} \tag{7.7.50}$$

For a diagonal element $a_{\mu,\mu}$, one has moreover:

$$a_{\mu,\mu} \approx \frac{3.8}{\sqrt{|s_\mu|}} \tag{7.7.51}$$

(see Appendix B in Klamt and Schüürmann 1993). The total energy of the system may be then written in vectorial notation as

$$E(\mathbf{q}) = \frac{1}{2}\mathbf{QCQ} + \mathbf{QBq} + \frac{1}{2}\mathbf{qAq} \tag{7.7.52}$$

where $\mathbf{q}$ is the vectorial notation for the M surface charges q_j and $\mathbf{A}$ and $\mathbf{B}$ are respectively the matrices of the $a_{\mu v}$ and $b_{\iota\mu}$.

Because the actual screening charge of the system at equilibrium is such to minimize the total energy, one may pose (Eq. 9 in Klamt and Schüürmann 1993):

$$\nabla_q E(\mathbf{q})|_{\mathbf{q}^*} = \mathbf{BQ} + \mathbf{Aq}^* = 0 \tag{7.7.53}$$

hence

$$\mathbf{q}^* = -\mathbf{A}^{-1}\mathbf{BQ} \tag{7.7.54}$$

The total energy of the screened system E(σ*) is thus

$$E(\sigma^*) = \frac{1}{2}\mathbf{Q}(\mathrm{C}-\mathbf{BA}^{-1}\mathbf{B})\mathbf{Q} \tag{7.7.55}$$

and

$$\Delta E = -\frac{1}{2}\mathbf{Q}(\mathbf{BA}^{-1}\mathbf{B})\mathbf{Q} = -\frac{1}{2}\mathrm{QDQ} \tag{7.7.56}$$

The Green function of the arbitrarily shaped cavity is thus $\mathbf{BA}^{-1}\mathbf{B}$ and ΔE has a close analogy to the solution for the simply spherical shape.

Denoting now **Φ** the vector of potentials Φ_μ on the segments, i.e.:

$$\mathbf{\Phi} = \mathbf{BQ} + \mathbf{Aq} \tag{7.7.57}$$

where **BQ** is the potential arising on the segments from the *source charges* Q, and **Aq** is that one arising from the *surface charges* q, the condition of vanishing potential on the surface of a conductor may be written as **Φ**= 0, and this condition is just equivalent to that one already obeyed in Eq. (7.7.53).

In the practical MO calculations, the solute/solvent interaction potential arising from screening and to be introduced in the Hamiltonian may be described in terms of φ_κ(r), φ_λ(r) basis functions replacing the positions r_i. The overlap of the basis functions generates an additional charge density distribution of type $\rho_{\kappa\lambda}(\mathbf{r}) = P_{\kappa\lambda}\varphi_\kappa(\mathbf{r})\varphi_\lambda(\mathbf{r})$, where $P_{\kappa\lambda}$ are the density matrix elements replacing Q_i (in practice the index i is replaced by a double indexation κλ denoting the combinations of basis functions; see Appendix C in Klamt and and Schüürmann 1993).

Exploiting further this analogy, Eq. (7.7.48) is conveniently replaced by

$$b_{i,\mu} \approx \int \frac{\varphi_\kappa(\mathbf{r})\varphi_\lambda(\mathbf{r})}{\|\mathbf{t}_\mu - \mathbf{r}\|} d^3r \tag{7.7.58}$$

The integral in Eq. (7.7.58) is of the core-electron interaction type and thus readily available in any MO code.

7.7.7 Other Procedures

The surface charge density $\sigma^{01}(r_{Ki})$, obtained iteratively by the IEFPCM procedure (cf. Eq. 7.7.20.2) is commonly defined Apparent Surface Charge (ASC). In a more recent formulation (Cossi et al. 2002; Scalmani and Frisch 2010) a Continuous Surface Charge (CSC) formalism replaces ASC and proves to be more smooth, and free of the singularities and discontinuities.

Adopting the CSC formulation Lipparini et al. (2020) proposed a variational formulation of IEFPCM where the polarization charges q_i are treated, simultaneously with the atomic coordinates of the solute and its electronic density, as independent variables, through a proper free energy functional of the solute and continuum dielectric system. In the various numerical examples proposed by the authors one may appreciate a notable reduction of computational time, while the energy and geometry of the solute are virtually unaffected (cf. Tables I, II, III, IV in Lipparini et al. 2010).

Marenich et al. (2009) presented a continuum solvation model (SMD) which couples the IEFPCM treatment of the electrostatic contribution (ΔG_{ENP}) with a cavity-dispersion-solvent-structure term arising from short-range interactions between the solute and solvent molecules in the first solvation shell (ΔG_{CDS}) and adding a final term ($\Delta G^0{}_{conc}$) arising from concentration change between the *gas* and the *liquid* standard states.

$$\Delta G_{solv} = \Delta G_{ENP} + G_{CDS} + \Delta G^0_{conc} \tag{7.7.59}$$

The ENP subscript in Eq. (7.7.59) denotes the electronic (E), nuclear (N), and polarization (P) components of the free energy. The nuclear relaxation component of the ENP term is equal to the difference between the gas-phase total energy (calculated at the gas-phase equilibrium structure) and the total energy calculated when the moiety is in a liquid solvent. G_{CDS} is nominally associated with the free energy change associated with solvent cavitation (C), changes in dispersion (D) energy, and possible changes in local solvent structure (S).

The SMD model employs a single set of parameters (intrinsic atomic Coulomb radii and atomic surface tension coefficients) optimized over six electronic structure methods. The parameterization was achieved by Marenich et al. (2009) with a training set of 2821 solvation data including 112 aqueous ionic solvation free energies, 220 solvation free energies for 166 ions in acetonitrile, methanol, and dimethyl sulfoxide, 2346 solvation free energies for 318 neutral solutes in 91 solvents (90 nonaqueous organic solvents and water), and 143 transfer free energies for 93 neutral solutes between water and 15 organic solvents. The elements present in the solutes are H, C, N, O, F, Si, P, S, Cl, and Br. The adopted electronic structure methods were M05−2X/MIDI!6D, M05−2X/6−31G*, M05−2X/6−31+G**, M05−2X/cc-pVTZ, B3LYP/6−31G*, and HF/6−31G*. With the 6−31G* basis set, the SMD model achieves mean unsigned errors of 2.5 – 4.2 *kJ/mol* in the solvation free energies of tested neutrals and mean unsigned errors of 16.7 *kJ/mol* on average.

7.8 Molecular Vibrations

7.8.1 Premises

As we have already seen in Chap. 3, the Harmonic Oscillator is a fundamental tool in understanding how thermal energy is transferred to a system in the form of vibration of nuclear centers about the center of mass of individual moieties such as molecules, clusters, or crystalline aggregates. As already stressed, the Harmonic Oscillator is a good approximation for small vibrations. In fact, by expressing the potential well through a Taylor expansion of the potential energy $V(l)$ about the equilibrium bond length $l = l_0$ and recalling that the force constant k is equal to the curvature of the potential $V(l)$ at the minimum, it may be easily seen that, for small values of x, then the first term in the series is dominant and the influence of the higher order terms is negligible, hence, the potential is *harmonic* in the neighborhood of the well bottom.

By application of Hooke's Law we saw then that the amplitude Eq. (2.5.3) may be rewritten in the form:

$$\frac{d^2\psi}{dx^2} + \frac{2m}{\hbar^2}\left(E - \frac{1}{2}kx^2\right)\psi = 0 \tag{7.8.1}$$

and, by introducing the energy-related terms: $\lambda = \frac{2m}{\hbar^2}E$ and $\alpha = \sqrt{\frac{mk}{\hbar^2}}$ we modified it to get

$$\frac{d^2\psi}{dx^2} + (\lambda - \alpha^2x^2)\psi = 0 \tag{7.8.2}$$

We anticipated that the force constant is directly related to the frequency of the motion ν_0 through

$$k = 4\pi^2 m\nu_0^2 \tag{7.8.3}$$

We saw then that the potential V(x) may be then also written as

$$V(x) = 2\pi^2 m\nu_0^2 x^2 \tag{7.8.4}$$

so that the magnitude α assumes the form $\alpha = \frac{2\pi m\nu_0}{\hbar}$ and the amplitude equation for the particle becomes

$$\frac{d^2\psi}{dx^2} + \frac{2m}{\hbar^2}(E - 2\pi^2 m\nu_0^2 x^2)\psi = 0 \tag{7.8.5}$$

The functional analysis showed us that the finite expansions that satisfy the above equation are Hermite polynomials and that the admitted stationary state energies of the harmonic oscillator are

$$E_n = \left(n + \frac{1}{2}\right)\hbar\sqrt{\frac{k}{m}} \quad (7.3.3.9)$$

obeying a classical vibrational frequency $\nu = \frac{1}{2\pi}\sqrt{\frac{k}{m}}$, whence

$$E_n = \left(n + \frac{1}{2}\right)h\nu \quad (7.3.3.11)$$

We introduced then (Sect. 7.2.4) the concept of reduced mass $\mathcal{M}$ and showed that the 2nd equation of motion of Newton brings to

$$\mathcal{M}\frac{d^2R}{dt^2} = -\frac{dE}{dR} \quad (7.2.33)$$

and

$$\nu = \frac{1}{2\pi}\sqrt{\frac{k_H}{\mathcal{M}}} \quad (7.2.35)$$

7.8.2 Normal Modes

Consider now a molecule with N atoms. In terms of normal coordinates we need 3N Cartesian components to specify it completely (i.e. 3 for each nucleus). Adopting three coordinates to specify the center of mass, any motion along these three coordinates will be a pure translation. There are thus 3 translational degrees of freedom for any molecule. If the molecule rotates about its center of mass, then also 3 rotational degrees of freedom exist about the axial components. But these are reduced to 2 if the molecule is linear. Subtracting from the bulk freedom (i.e. 3N) the translational and rotational terms, the remaining terms are purely vibrational (i.e. N_{vib} = 3N−6 for a non-linear molecule and N_{vib} = 3N−5 for a linear one). Because, in the absence of an external field, the energy of the molecule does not depend upon its orientation or upon the position of its center of mass, then it is obvious that its vibrational energy will depend only upon the 3N−6 or 3N−5 coordinates.

Let me consider now again the Taylor expansion of the potential energy V (l) about the equilibrium bond length $l = l_0$:

$$V(l) = V(l_0) + \frac{(l-l_0)}{1}\left(\frac{dV}{dl}\right)_{l=l_0} + \frac{(l-l_0)^2}{2!}\left(\frac{d^2V}{dl^2}\right)_{l=l_0} + \frac{(l-l_0)^3}{3!}\left(\frac{d^3V}{dl^3}\right)_{l=l_0} + \dots \tag{7.3.1.10}$$

Its multidimensional generalization may be written as follows:

$$V\left(q_1, q_2, \dots, q_{N_{vib}}\right) = V_0 + \sum_i \left(\frac{\partial V}{\partial q_i}\right)_0 + \frac{1}{2}\sum_{ij} b_{ij} q_i q_j + \dots \tag{7.8.6}$$

where

$$b_{ij} = \left(\frac{\partial^2 V}{\partial q_i \partial q_j}\right)_0 \tag{7.8.7}$$

and the coordinates q_i, q_j are related to the actual coordinates of the i,j nuclei through the scaling factors (Bright Wilson et al. 1955):

$$q_i = x_i \sqrt{M_i} \tag{7.8.8}$$

with M_i as mass of the ith nucleus and x_i as its displacement about the equilibrium value.

If we pose V = 0 when $q_1, q_2, \dots q_{Nvib} = 0$, then $V_0 = 0$. Also, because (by choice of coordinate axes) the equilibrium is attained at $q_1 = 0$, $q_2 = 0$, $q_{Nvib} = 0$, then also the 2nd term in Eq. (7.8.6) is zero and, neglecting higher terms, Eq. 7.8.6 may be rewritten as follows:

$$V\left(q_1, q_2, \dots, q_{N_{vib}}\right) = \frac{1}{2}\sum_{ij} b_{ij} q_i q_j \tag{7.8.9}$$

At the light of transformation (7.8.8), the kinetic energy T is simply given by

$$T = \frac{1}{2}\sum_i \dot{q}_i^2 \tag{7.8.10}$$

Because the potential energy V depends on the mutual position of the nuclei (and not on masses), these are eliminated from the bulk energy expression. The classical Lagrangian equations of motion have this form:

$$\frac{d}{dt}\left(\frac{\partial T}{\partial \dot{q}_k}\right)+\frac{\partial V}{\partial q_k}=0 \quad k=1,2,\ldots N_{vib} \tag{7.8.11}$$

and, because $\left(\frac{\partial V}{\partial q_i}\right)=0$ at equilibrium:

$$\ddot{q}_k+\sum_i b_{ik}=0 \quad k=1,2,\ldots N_{vib} \tag{7.8.12}$$

If b_{ij} vanishes for $i\neq j$ and we limit ourselves to squared terms q_i^2, then:

$$\ddot{q}_k+b_{kk}q_k=0 \quad k=1,2,\ldots N_{vib} \tag{7.8.13}$$

and the solutions of the equation of motion are (Pauling and Bright Wilson 1935):

$$q_k=q_k^0\sin\left(t\sqrt{b_{kk}}+\delta_k\right) \quad k=1,2,\ldots N_{vib} \tag{7.8.14}$$

where the q_k^0,s are amplitude constants and the δ_k,s are integration constants.

To eliminate formally the cross products (instead of simply neglecting them) we must benefit of matrix algebra and find a new set of coordinates $Q_n(n = 1,2,\ldots,N_{vib})$, related to the q's by

$$q_k=\sum_{n=1}^{N_{vib}}B_{kn}Q_n \tag{7.8.15}$$

The kinetic energy now becomes

$$T=\frac{1}{2}\sum_n\dot{Q}_n^2 \tag{7.8.16}$$

and the potential energy is

$$V=\frac{1}{2}\sum_n\lambda_nQ_n^2 \tag{7.8.17}$$

The equation of motion has now solutions

$$Q_n=Q_n^0\sin\left(t\sqrt{\lambda_l}+\delta_l\right) \quad n=1,2,\ldots N_{vib} \tag{7.8.18}$$

The Q_n coordinates are the *normal coordinates* or "*normal modes*" of the molecule.

7.8.3 Vibrational Amplitude and Frequency

If all amplitude constants Q_n^0 are zero except one (Q_1^0), introducing the product $A_k = B_{k1}Q_1^0$ and posing $\lambda = \lambda_1$ we obtain (Pauling and Bright Wilson, 1953):

$$-\lambda A_k + \sum_{i=1}^{N_{vib}} b_{ik}A_i = 0 \quad k = 1, 2, \ldots N_{vib} \tag{7.8.19}$$

This set of simultaneous equations has a solution only when (besides the trivial case $A_1 = A_2 = \ldots = 0$) it is satisfied the corresponding determinant:

$$\begin{vmatrix} b_{11} - \lambda & b_{12} & \ldots & b_{1N_{vib}} \\ b_{21} & b_{112} - \lambda & \ldots & b_{2N_{vib}} \\ \ldots & \ldots & \ldots & \ldots \\ b_{N_{vib}1} & b_{N_{vib}2} & \ldots & b_{N_{vib}N_{vib}} - \lambda \end{vmatrix} = 0 \tag{7.8.20}$$

The values of the B_{kn} can be determined posing $A_{kn} = B_{kn}Q_n^0$ and $\sum_k B_{kn}^2 = 1$, obtaining[21] N_{vib} particular solutions of the equation of motion. By adding together all these particular solutions one gets

$$q_k = \sum_{n=1}^{N_{vib}} Q_n^0 B_{kn} \sin\left(t\sqrt{\lambda_n} + \delta_n\right) \tag{7.8.21}$$

where Q_n^0 are the *amplitudes* and δ_n the *phases*. If the molecule vibrates in such a way that all Q_n^0 except one are zero, the summation reduces to the particular selected case. In the case that the non-zero value is Q_1^0 Eq. (7.8.16) reduces to

$$q_k = Q_1^0 B_{k1} \sin\left(t\sqrt{\lambda_1} + \delta_1\right) \quad k = 1, 2, \ldots, N_{vib} \tag{7.8.22}$$

and each of the nuclei undergoes a simple harmonic oscillation about its equilibrium position with phase δ_1 and frequency

$$\nu_1 = \frac{1}{2\pi}\sqrt{\lambda_1} \tag{7.8.23}$$

i.e. a "*normal mode of vibration*".

The wave equation for the nuclear motion of the molecule is (Pauling and Bright Wilson, 1953):

[21] The subscript n specifies which root λ_n of the secular equation has been used.

$$\sum_{n=1}^{N_{vib}} \frac{\partial^2 \psi}{\partial Q_n^2} + \frac{8\pi^2}{h^2}\left(W - \frac{1}{2}\sum_{n=1}^{N_{vib}} \lambda_n Q_n^2\right)\psi = 0 \tag{7.8.24}$$

where ψ is the nuclear wave function and W is the total energy (nuclear + electronic). Putting:

$$\psi = \psi_1(Q_1)\psi_{a2}(Q_2)\ldots\ldots\psi_{N_{vib}}(Q_{N_{vib}}) \tag{7.8.25}$$

we may see that the nuclear wave function is separable into N_{vib} one-dimensional equations:

$$\frac{\partial^2 \psi_k}{\partial Q_k^2} + \frac{8\pi^2}{h^2}\left(W_k - \frac{1}{2}\lambda_k Q_k^2\right)\psi_k = 0 \tag{7.8.26}$$

and the total energy may be expressed as a summation of the energies W_k associated to each normal coordinate:

$$W = \sum_{k=1}^{N_{vib}} W_k \tag{7.8.27}$$

The vibrational Hamiltonian operator H_{vib} is given by

$$\mathsf{H}_{vib} = -\sum_{n=1}^{N_{vib}} \left(\frac{\hbar^2}{2\mu_n}\frac{d}{dQ_n^2} + \frac{1}{2}\lambda Q_n^2\right) \tag{7.8.28}$$

and the vibrational energy is

$$E_{vib} = \sum_{n=1}^{N_{vib}} h\nu_n\left(n_{vib} + \frac{1}{2}\right) \qquad n_{vib} = 0, 1, 2, \ldots \tag{7.8.29}$$

with n_{vib} as a vibrational quantum number.

7.8.4 IR and Raman Spectra

Each one of the nuclear wave functions (7.8.25) may be written in the form:

$$\psi_n = \mathcal{N}_n \exp\left(-\frac{1}{2}\alpha Q_n^2\right) H_n(Q_n\sqrt{\alpha}) \tag{7.8.30}$$

where $\mathcal{N}_n$ is the normalization factor $\mathcal{N}_n = \left(\frac{\sqrt{\alpha}}{2^n n! \sqrt{\pi}}\right)^{1/2}$, H_n is the Hermite polynomial of degree n and $\alpha = \sqrt{\frac{\mathcal{M}k}{\hbar^2}}$.

The first four solutions of the nuclear wave function are

$$\psi_0 = \left(\frac{\alpha}{\pi}\right)^{1/2} \exp\left(-\frac{1}{2}\alpha Q_0^2\right) \tag{7.8.31.1}$$

$$\psi_1 = \sqrt{2\alpha}\left(\frac{\alpha}{\pi}\right)^{1/2} Q_1 \exp\left(-\frac{1}{2}\alpha Q_1^2\right) \tag{7.8.31.2}$$

$$\psi_2 = \frac{1}{\sqrt{2}}\left(\frac{\alpha}{\pi}\right)^{1/2} \left(2\alpha Q_2^2 - 1\right) \exp\left(-\frac{1}{2}\alpha Q_2^2\right) \tag{7.8.31.3}$$

$$\psi_3 = \sqrt{\frac{\alpha}{3}}\left(\frac{\alpha}{\pi}\right)^{1/2} \left(2\alpha^{3/2} Q_3^3 - 3\alpha^{1/2} Q_3\right) \exp\left(-\frac{1}{2}\alpha Q_3^2\right) \tag{7.8.31.4}$$

Transitions between vibrational energy states occur by the absorption of photons of the appropriate energy in the infrared portion of the spectrum (IR) and involve only adjacent vibrational states (i.e. $\Delta n = \pm 1$). In fact, transitions occur only whenever the dipole transition moment is non-null. Posing $\xi = Q\sqrt{\alpha}$ the dipole transition moment is given by the integral

$$I_{0\to n} \propto \int_{-\infty}^{\infty} \psi_n(\xi)\xi\psi_0(\xi)d\xi \tag{7.8.32}$$

For a 0→1 transition we have

$$I_{0\to 1} \propto \left(\frac{2\alpha}{\pi}\right)^{1/2} \int_{-\infty}^{\infty} \xi^2 \exp(-\xi^2)d\xi \tag{7.8.33}$$

where the integrand is always positive, while, for a 0→2 transition we have

$$I_{0\to 2} \propto \left(\frac{\alpha}{2\pi}\right)^{1/2} \int_{-\infty}^{\infty} \left(2\xi^3 - \xi\right) \exp(-\xi^2)d\xi = 0 \tag{7.8.34}$$

Transitions between vibrational energy states may occur also whenever an incident photon, instead of being adsorbed, is simply scattered. The inelastic scattering of light within or near the visible range cause the energies of the photons to be shifted up or down in characteristic ways (i.e. *Stokes* or *anti-Stokes*). The amount of scattering is known as "*Raman activity*". An induced dipole moment within the molecule is created, based on its polarizability. For a molecule to exhibit a Raman effect, there must be thus a change in its dipole polarizability with respect to the vibrational coordinate corresponding to the roto-vibronic state. The intensity of the Raman scattering is proportional to this polarizability change. Vibrational modes are said to be *Raman active* or *Raman inactive*, depending on whether their Raman activity is non-zero. The term "*forbidden mode*" is reserved to a transition which is both IR inactive and Raman inactive.

To understand the selection rules for the Raman effect let me assume that in the absence of incident photons the harmonic oscillator is initially in a state a with quantum number n. Let me also denote the dipole moment associated with the displacement r of an electron e from the initial position as $\mathbf{R} = e\mathrm{r}$. The matrix elements $(a|\mathbf{R}|j)$ will be different from zero only if state j has the quantum number $n \pm 1$. Consider now the state b with quantum number m. Similarly to a, $(j|\mathbf{R}|b)$will be different from zero only if state j has the quantum number $m\pm1$. Both matrix elements will be simultaneously different from zero only if $m = n$ or $m = n\pm2$. The frequency of the first excitation ν_{aj} is connected to the excitation energy E_{aj} by $\nu_{aj} = E_{aj}/h$. if the incident frequency is much higher than the scattering frequency ($\nu >> \nu_{aj}$) then we may write (Eyring et al. 1944):

$$
\begin{aligned}
&\sum \frac{\nu_{aj} - \nu_{jb}}{\nu}(a|\mathbf{R}|j)(j|\mathbf{R}|b) \\
&= \frac{\nu_{ab}}{\nu}\sum_j (a|\mathbf{R}|j)(j|\mathbf{R}|b) = \frac{\nu_{ab}}{\nu}\sum_j (a|\mathbf{RR}|b)
\end{aligned}
\quad (7.8.35)
$$

For any matrix element of the type $(a|x_1x_2|b)$ with x_1,x_2 equal x,y, or z and different from zero, a corresponding frequency $\nu + \nu_{ab}$ will appear in the Raman spectrum.

In practice, vibrational frequencies are computed at the stationary point by determining the second derivatives of the energy with respect to the Cartesian nuclear coordinates and then transforming them to mass-weighted coordinates. Raman intensities are obtained by numerical differentiation of dipole derivatives with respect to the electric field. The polarizability derivatives are obtained numerically by differentiating the analytic polarizability with respect to nuclear coordinates.

Consider for instance a molecule with N nuclei. The energy E will be a function of the 3N nuclear coordinates $(x_1,y_1,y_1,x_2,y_2,z_2,\ldots,x_N,y_N,z_N)$. Let me denote for simplicity these coordinates as R_i (i = 1,2,…,3N). The energy derivatives with respect to nuclear coordinates will be

$$
f_i = \frac{\partial E}{\partial R_i} \quad (7.8.36)
$$

$$f_{ij} = \frac{\partial^2 E}{\partial R_{ij}} \tag{7.8.37}$$

Following Pople et al. (1981) I will denote m_i as the mass associated with the coordinate R_i (so that $m_1 = m_2 = m_3$ is the mass of the first nucleus, $m_4 = m_5 = m_6$ the mass of the second nucleus, etc.). For a small displacement r_i, the classical equations of motion are

$$m_i \ddot{r}_i = -\sum_j \mathit{f}_{ij} r_j \tag{7.8.38}$$

The mass-weighted force constants are

$$f_{ij} = \mathit{f}_{ij} m_i^{-1/2} m_i^{-1/2} \tag{7.8.39}$$

and the normal mode displacements are

$$r_i = a_i \exp(2\pi \nu_i it) \tag{7.8.40}$$

with a_i as amplitude. The eigenvalues of the f_{ij} matrix are related to the harmonic frequencies ν via:

$$\sum_j f_{ij} a_j = 4\pi^2 \nu_i^2 a_i \tag{7.8.41}$$

and the eigenvectors give the amplitude of the same coordinates. Six of the 3N eigenvalues (5 for a linear molecule) will be zero, corresponding to the rotational and translational degrees of freedom. The remaining 3N−6 eigenvalues would then correspond to the internal vibrations. The procedure is thus based on the location of the energy hypersurface minimum where $f_i = 0$ (i = 1,2,...,3N) and the second derivatives (f_{ij}) represent the Cartesian force constants of the normal modes.

In practice, by diagonalizing the 3N × 3N matrix, it is found that the 6 modes corresponding to translation and rotation are never zero although close to it (especially the translational ones). To avoid confusion, the coordinates r_i are then orthogonalized to the known normal coordinates for translation and rotation and the f_{ij} matrix is transformed into a (3N−6) × (3N−6) matrix for the remaining 3N−6 coordinates. The eigenvalues of this matrix are the best approximation to the true harmonic frequencies.

The comparative appraisal made by Pople et al. (1981) at HF theory level with the 3−21G basis set and the 6−31G(d) basis set on 37 organic molecules containing H, C, N, O, and F, with no more than ten atoms and no more than two (6−31G*) or four (3−21G) non-hydrogen atoms revealed that when compared to experiments (Shimanouchi 1967) the calculated harmonic 3−21G frequencies are consistently too large (by up to 400 cm^{-1}), with a mean difference, ν(3−21G) - ν(expt.) of

Table 7.28 Calculated versus observed harmonic frequencies (cm^{-1}) of the gaseous molecules H_2O, CO_2, and CH_4. The experimental observations are after Shimanouchi (1967).

H_2O (C_{2V})					
Mode	HF/3−21G	HF/6−31G(d)	B3LYP/6−31G(d)	G2	Experimental
a_1	1799	1827	1713	1825	1591
a_1	3943	4069	3728	4070	3657
a_1	3809	4187	3850	4189	3756
CO_2 ($D_{\infty h}$)					
Mode	HF/3−21G	HF/6−31G(d)	B3LYP/6−31G(d)	G2	Exper
π_u	659	745	640	745	667
σ_g	1428	1519	1372	1519	1333
σ_u	2464	2586	2436	2586	2349
CH_4 (T_d)					
Mode	HF/3−21G	HF/6−31G(d)	B3LYP/6−31G(d)	G2	Exper
t_2	1520	1488	1356	1487	1306
e	1740	1703	1578	1703	1534
a_1	3186	3197	3047	3197	2917
t_2	3280	3202	3163	3302	3019

179 cm^{-1}. The ratio ν(3−21G)/ν(expt.) on 486 frequencies is systematically higher than identity (1.123) implying that the HF/3−21G frequencies are about 12% too large. When calculations are carried out at HF/6−31G(d) theory level the error is somewhat larger, but, according to Pople et al. (1981) this is basically due to the fact that the 6−31G* sample set (83 frequencies) contains smaller molecules which have fewer low-frequency torsion-type modes. This systematic overestimation of the force constants (and of the ensuing harmonic frequencies) is basically due to the fact that the HF theory does not give a proper account of the electronic correlation.

I carried out a few calculations on three simple gaseous molecules of paramount importance in Geochemistry, selected from the above set (H_2O, CO_2, and CH_4), to give some further hint of the effects of theory and basis set on predictions (Table 7.28).

The computations confirm that the shift is systematic, though less severe than expected, and does not involve intercept terms (Fig. 7.31). G2 computations are internally consistent ($R^2 = 0.999$) but quite overestimated ($\nu_{calc} = 1.109 \times \nu_{obs}$). Computations carried out at HF level give the worst results ($\nu_{calc} = 1.069 \times \nu_{obs}$, $R^2 = 0.991$ with the 3−21G basis set; $\nu_{calc} = 1.059 \times \nu_{obs}$, $R^2 = 0.991$ with the polarized functions 6−31G*). The adoption of the hybrid DFT functional B3LYP improves consistently the predictions both in terms of absolute values and internal consistency ($\nu_{calc} = 1.033 \times \nu_{obs}$, $R^2 = 0.998$).

In Fig. 7.32 we see the IR and Raman spectra of CO_2 arising from calculation. The molecule is linear, there are thus 3N−5 = 4 vibrational harmonic modes. The most IR intense mode is the σ_u asymmetric stretching, while the intensity of the (doubly degenerate) π_u bending mode is quite limited. It must be noted to this

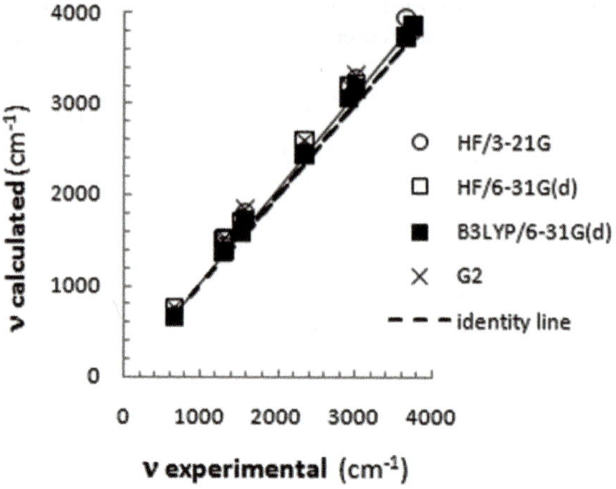

Fig. 7.31 Calculated versus observed harmonic frequencies of the H_2O, CO_2, and CH_4. Gaseous molecules. The solid line is the best interpolant of the B3LYP/6−31G(d) procedure (Table 7.28). Experimental observations after Shimanouchi 1967)

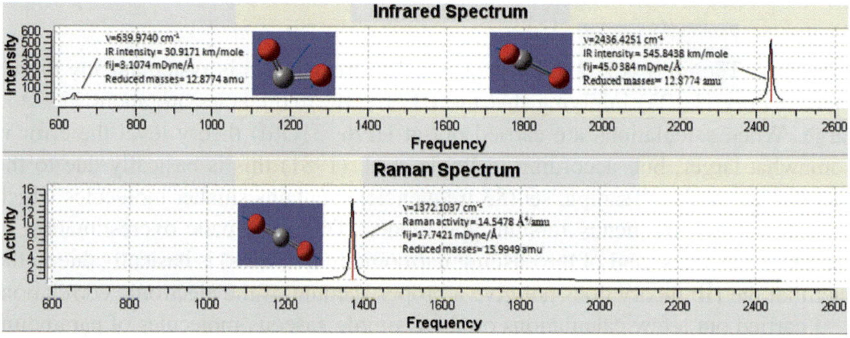

Fig. 7.32 IR and Raman spectra of the CO_2 gaseous molecule

purpose that, while according to IUPAC the adopted units for the integrated absorption coefficient are 10 × *m/mole*, in IR spectroscopy the most adopted unit is a multiple of it, i.e. 1000 × *m/mole* = 1 *km/mole*. Both modes are Raman inactive. The σ_g Raman active mode is IR-inactive.

In Table 7.29 are resumed the results obtained on a more computationally demanding molecule, the Fe-diacetate metalorganic complex (Ottonello and Vetuschi Zuccolini 2008), the first ab initio investigation of this geochemically important molecule.

In $[Fe(CH_3COO)_2]^0$ the metal is placed on a plane that also contains a hydrogen of the CH_3 group. The in-plane hydrogen has a shorter bond length than the other two out-of-plane hydrogens. The variational structure, although computed without

Table 7.29 Unscaled DFT/B3LYP 6−31G(d,p) and DFT/B3LYP 6−31G(d) + VTZ gas-phase wavenumbers (cm^{-1}) of the bidentate Fe-diacetate complex (Otttonello and Vetuschi Zuccolini 2008) compared with the experimental observations of Edwards and Lewis (1993). Approximate assignments after Edwards and Lewis (1993) are also listed. Symbols: ν = stretching; δ = deformation; ρ = rocking; ip = in-plane. Values in bold are the most distinctive features of the bidentate complex according to Edwards and Lewis (1993)

IR-active			Raman-active			
6−31G (d,p)	6−31G(d)-VTZ	Observed	6−31G (d,p)	6−31G(d)-VTZ	Observed	Assignment
58	77	94				
			119	104	116	
128	132	141				
134	134	164				
			146	138	157	
202	188	189				
210	220	208				
			302	306		
			321	319		
392	401	320				
498	494	456				
			507	515		
561	550	510				ρ(COO)ip
			616	597	617	ρ(COO)
597	623	531				
717	712	614			669	δ(COO)
723	724	671				
964	972	**955**				ν(C–C)
			973	964	**957**	ν(C–C)
			1035	1031		
1033	1039	1050				ρ(CH_3)
			1081	1076	1060	ρ(CH_3)
1076	1083					
1395	1416	1337				
			1418	1397	1353	δ(CH_3)
			1485	1460	**1461**	δ(CH_3)
1462	1486					
			1510	1481		
1482	1511					
1520	1520	**1519**				ν(C-O)
			1528	1529		
			1585	1563	1597	
1576	1587	1621				
3044	3062	2947				ν(C-H)
			3066	3048	2945	ν(C-H)
			3135	3129	3019	ν(C- H)

(continued)

Table 7.29 (continued)

IR-active			Raman-active			
3130	3136	3024				
			3180	3167		
3167	3182	3125				

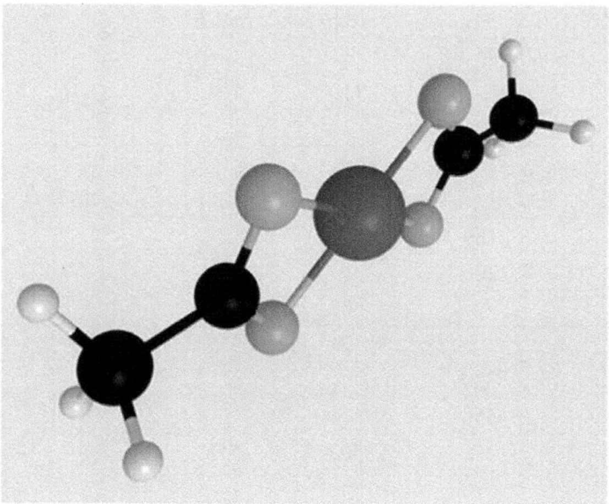

Fig. 7.33 Ball-and-stick representation of the B3LYP/6−31G(d,p) ground-state structure of $[Fe(CH_3COO)_2]^0$. $R_{C\text{-}O}$ = 1.278–1.279 Å; $R_{C\text{-}C}$ = 1.501 Å; $R_{O\text{-}Fe}$ = 2.033–2.036 Å; $R_{C\text{-}H}$ = 1.091 −1.096 Å; $\alpha_{(O\text{-}C\text{-}O)}$ = 118.5°; $\alpha_{(C\text{-}O\text{-}Fe)}$ = 88.1°. From Ottonello and Vetuschi Zuccolini (2008). With the permission of Geochimica et Cosmochimica Acta

symmetry constraints, is very close to the C_s symmetry. Iron is placed between the two deprotonated acetate ions as a highly charged atom with a central potential. With this center-symmetrical position the cation induces some instability from the computational point of view. With the adopted basis sets, very flat (symmetry-constrained) potential energy surfaces are generated, leading to the appearance of an imaginary frequency in the numerical computation of the second derivative of the energy. To bypass the lack of convergence we initially eliminated the symmetry constraints, allowing a larger degree of freedom. The resulting variational dimer acetate joined to an iron atom is an elongated molecule corresponding to type C of Edwards and Lewis (1993) (Fig. 7.33). An interesting feature of the methyl group is that the Fe–acetate appear to be rotated with respect to the symmetry plane. In particular, the two CH_3 groups of the dimeric species are rotated in opposite directions, thus showing the existence of a principal axis of rotation orthogonal to the plane of alignment of the heavy atoms and corresponding to the C_2 point group symmetry.

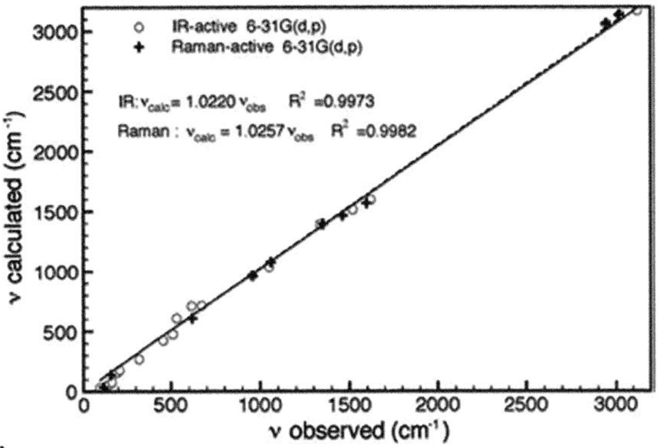

Fig. 7.34 Comparison of the computed B3LYP/6−31G (d,p) gas-phase wavenumbers (cm^{-1}) of the bidentate Fe–diacetate complex and the experimental observations of Edwards and Lewis (1993). From Ottonello and Vetuschi Zuccolini (2008). With the permission of Geochimica et Cosmochimica Acta

Experimentally observed frequencies and the provisional assignments of Edwards and Lewis (1993) are compared with our results of *ab-initio* computations for the bidentate geometry in Table 7.29 and Fig. 7.34.

The scaling factor necessary to reproduce measured wave numbers from calculated ones is somewhat higher than the canonical one for the adopted theory level and wave expansions, but the internal correlation leaves little doubt about the consistency of the variational structure with the existing vibrational investigations [observed = 0.9763 × calculated; R^2 = 0.9981 for IR-active frequencies and observed = 0.9765 × calculated; R^2 = 0.9988 for Raman-active frequencies with a 6 −31G(d,p) basis set]. The most distinctive features of the bidentate complex, which, according to Edwards and Lewis (1993) are the two bands at 1519 (IR) cm^{-1} assigned to asymmetric and 1461 (Raman) cm^{-1} symmetric stretching of the C–O bond and the associated $\nu_{(C-C)}$ stretching features at 942 (IR) and 945 (Raman) cm^{-1}, are satisfactorily reproduced within the limits of the method. Combining the 6 −31G(d,p) set of C, O, and H with the VTZ triple split set of Ahlrics for iron has little effect on the vibrational results.

7.9 Nuclear Mass and Isotopological Exchanges

Isotope geochemistry is probably the branch of Earth Sciences where the impact of Quantum Chemistry has been most beneficial. To better appreciate the significance of the above statement I must briefly recall some fundaments of the theory.

7.9.1 Separative Effect and Equilibrium Exchanges: The Bigeleisen-Mayer Approximation

Let me consider the vibrational partition function Q_{vib} arising from i Einstein oscillators:

$$Q_{vib} = \prod_{i} \frac{\exp(-X_i/2)}{1 - \exp(-X_i)} \tag{7.9.1}$$

Because $X_i = \frac{h\nu_i}{k_B T} = \frac{\hbar\omega_i}{k_B T}$ at high T (low X_i), Q_{vib} reduces to $\prod_i (1/X_i)$. Since the potential energies for molecules differing only in isotopic constituents are alike, one may define a "*separative effect*" based on the partition function ratio f of isotopically heavy and light molecules ($Q^{\bullet}, Q^{\circ}$, respectively) in such a way that the rotational and translational contributions to the partition function cancel out:

$$\begin{aligned}\left(\frac{s^{\bullet}}{s^{\circ}}\right) f &= \left(\frac{s^{\bullet}}{s^{\circ}}\right)\left(\frac{Q^{\bullet}}{Q^{\circ}}\right)\left(\frac{m^{\circ}}{m^{\bullet}}\right)^{3/2} \\ &= \prod_{i} \frac{X_i^{\bullet} \exp(-X_i^{\bullet}/2)/\left[1 - \exp(-X_i^{\bullet})\right]}{X_i^{\circ} \exp(-X_i^{\circ}/2)/\left[1 - \exp(-X_i^{\circ})\right]}\end{aligned} \tag{7.9.2}$$

The ratio of the symmetry numbers $\frac{s^{\bullet}}{s^{\circ}}$ in Eq. (7.9.2) merely represents the relative probabilities of forming symmetrical and unsymmetrical molecules and $m^{\bullet}, m^{\circ}$ are the masses of exchanging molecules (the translational contribution to the partition function ratio is at all T equal to the 3/2 power ratio of the inverse molecular weight). Denoting ΔX_i the vibrational frequency shift from isotopically heavy to light molecules (i.e. $\Delta X_i = X_i^{\circ} - X_i^{\bullet}$) and being ΔX_i intrinsically positive, Eq. (7.9.2) may be translated into:

$$\left(\frac{s^{\bullet}}{s^{\circ}}\right) f = \prod_{i} \frac{X_i^{\bullet}}{X_i^{\bullet} + \Delta X_i} \exp(\Delta X_i/2) \frac{\left\{1 - \exp\left[-(X_i^{\bullet} + \Delta X_i)\right]\right\}}{\left[1 - \exp(-X_i^{\bullet})\right]} \tag{7.9.3}$$

The Helmholtz free energy of each component in the reaction is related to the partition function Q:

$$F = k_B T \ln Q \tag{7.9.4}$$

Based on Eq. (7.9.3) the difference in the Helmholtz free energy of formation of two isotopic molecules with respect to their gaseous atoms depends on the shift of vibrational frequencies between heavy and light isotope-bearing compounds. That is, according to Bigeleisen and Mayer (1947):

$$\frac{\Delta F^{\bullet} - \Delta F^{\circ}}{RT} = \sum_i [-\frac{1}{2}\Delta X_i + \ln\left(1 + \frac{\Delta X_i}{X_i^{\bullet}}\right) + \ln\frac{1 - \exp(-X_i^{\bullet})}{1 - \exp(-X_i^{\circ})}] + \ln\left(\frac{s^{\bullet}}{s^{\circ}}\right) \tag{7.9.5}$$

where $\Delta F^{\bullet}, \Delta F^{\circ}$ the Helmholtz free energies of formation of isotopically heavy and light molecules. If ΔX_i is small, which is the case for all isotopes except hydrogen, Eq. (7.9.5) may be reduced to:

$$\frac{\Delta F^{\bullet} - \Delta F^{\circ}}{RT} = -\sum_i \left[\frac{1}{2} - \frac{1}{X_i^{\bullet}} + \frac{1}{\exp(X_i^{\bullet}) - 1}\right]\Delta X_i + \ln\left(\frac{s^{\bullet}}{s^{\circ}}\right) \tag{7.9.6}$$

and the separative effect $\left(\frac{s^{\bullet}}{s^{\circ}}\right)f$ becomes

$$\left(\frac{s^{\bullet}}{s^{\circ}}\right)f = 1 + \sum_i \left[\frac{1}{2} - \frac{1}{X_i^{\bullet}} + \frac{1}{\exp(X_i^{\bullet}) - 1}\right]\Delta X_i \tag{7.9.7}$$

If $X_i^{\bullet}$ is also small, then:

$$\left(\frac{s^{\bullet}}{s^{\circ}}\right)f = 1 + \sum_i \frac{X_i^{\bullet}\Delta X_i}{12} = 1 + \sum_i \frac{\Delta\left(X_i^{\bullet 2}\right)}{24} \tag{7.9.8}$$

Since ΔX_i is positive, $\left(\frac{s^{\bullet}}{s^{\circ}}\right)f$is always greater than one and the heavy isotope preferentially stabilizes in the condensed phase while the light isotope favors the gaseous phase.

At low temperature and light frequencies, the separative effect per unit shift (the terms in brackets in Eq. 7.9.7) approaches 1/2 (Fig. 7.35) and the Helmholtz free energy difference approaches the differences in zero-point energies. At high T (low frequencies), the separative effect per unit shift approaches zero and the total separative effect $(s^{\bullet}/s^{\circ})f$approaches one, so that no isotopic fractionation is observed.

A useful approximation for the isotopes of heavy elements is given by (Bigeleisen and Mayer 1947):

$$\left(\frac{s^{\bullet}}{s^{\circ}}\right)f = 1 + \frac{m\Delta M}{24M^2}X_s^2 n \tag{7.9.9}$$

where M is the mass of the central atom (isotope) surrounded by n ligands of mass m, ΔM is the difference in isotopic masses, and a-dimensional frequency X_s is related to the totally symmetric frequency arising from the stretching of bands with central atom n_s by

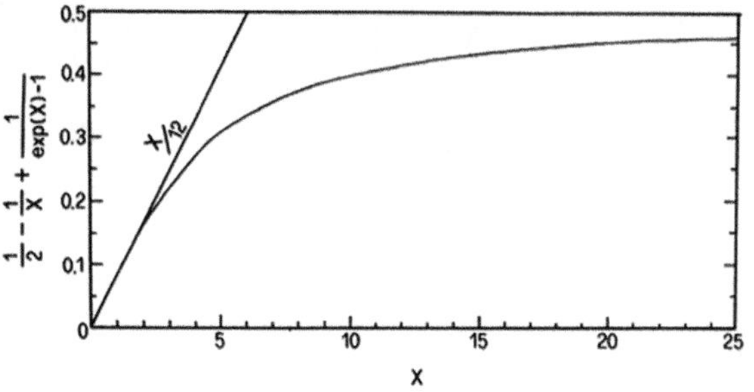

Fig. 7.35 Separative effect per unit shift, plotted against non-dimensionalized frequencies $X = h\nu/k_B T = \hbar\omega/k_B T$ From Bigeleisen and Mayer (1947). With the permission of AIP publishing

$$x_s = \frac{h\nu_s}{k_B T} = \frac{h}{k_B T}\sqrt{\frac{2\pi C}{m}} \tag{7.9.10}$$

with C as force constant.

Table 7.30 lists the result of calculations carried out by Bigeleisen and Mayer (1947) and based on Eq. (7.9.9) for isotopic exchange reactions at T = 300 K for tetrahedrally and octahedrally coordinated "*heavy*" isotopes:

$$^{28}SiF_4 + {}^{30}SiF_6^{2-} \Leftrightarrow {}^{30}SiF_4 + {}^{28}SiF_6^{2-} \tag{7.9.11}$$

$$^{120}SnCl_4 + {}^{118}SnCl_6^{2-} \Leftrightarrow {}^{118}SnCl_4 + {}^{120}SnCl_6^{2-} \tag{7.9.12}$$

Note that the separative effect is quite significant, even for the relatively "*heavy*" isotopes of tin. Note also the large cancellation effect on the isotopic fractionation constant, when isotopic exchanges of the following type are written.

$$^{120}SnCl_4 + {}^{118}Sn \rightarrow {}^{118}SnCl_4 + {}^{120}Sn \tag{7.9.13}$$

$$^{118}SnCl_6^{2-} + {}^{120}Sn \rightarrow {}^{120}SnCl_6^{2-} + {}^{118}Sn \tag{7.9.14}$$

Table 7.30 Separative effect f and isotopic fractionation constant K for heavy isotopes. ω_s is the angular totally symmetric stretching frequency derived from Raman spectra (from Bigeleisen and Mayer 1947)

Molecule	ω_s	f	K
SiF_4	800	1.111	
			1.002
$SiF_6{}^{2-}$	600	1.109	
$SnCl_4$	367	1.00256	
			1.00025
$SnCl_6{}^{2-}$	314	1.00281	

This cancellation (always marked in exchanges of acid-base type) may be ascribed to the decrease in force constant C of the central atom with the increase in coordination number *n* and is sometimes not so significant in other types of isotopic exchange reactions.

7.9.2 *Isotopic Fractionation: A Simple Application to Sr Isotopes*

Now we can ask ourselves: "*how marked is the cancellation effect ?*" and, more important: "*at what decimal place in a fractionation scale is operative ?*". The question is not peregrine, because in Geochemistry one often assign precise significances to the fractionation factors observed in the so-called heavy isotopes. I will take this opportunity also to show how powerful is the Quantum Mechanical procedure in establishing the correct magnitudes. Let me consider the isotopic fractionation of Sr between two gaseous molecules such as $SrCl_2$ and SrO. $SrCl_2$ is an asymmetric top with rotational symmetry number 2. When solved at simple HF/STO−3G theory level it returns three vibrational frequencies: a librational mode (ν_1 = 48.4783 cm^{-1}) a symmetric stretching (ν_2 = 314.245 cm^{-1}) and an asymmetric stretching (ν_3 = 405.399 cm^{-1}). The three operating force constants are, respectively, f_1 = 0.0630, f_2 = 2.0739 and f_3 = 4.4854 *mDyne/Å*. SrO is a prolate symmetric top with rotational symmetry number 1. When resolved at the same theory level it returns a single vibrational (stretching) mode with ν_1 = 284.8279 cm^{-1} and f_1 = 0.8784 *mDyne/Å*. The complete vibrational analysis carried out with the GAUSSIAN98 computational package is resumed in Table 7.31. It required 2' of cpu for $SrCl_2$ and 1'40" for SrO on a common laptop.

Consider now the isotopic exchange

$$^{86}SrCl_2 + {}^{87}SrO \Leftrightarrow {}^{87}SrCl_2 + {}^{86}SrO \tag{7.9.15}$$

Denoting the activities in terms of the fractional amount of substituted isotopes we have

$$K_{7.9.15} = \frac{\left({}^{87}SrCl_2/{}^{86}SrCl_2\right)}{\left({}^{87}SrO/{}^{86}SrO\right)} \tag{7.9.16}$$

The separative effects may be easily interpolated as a reciprocal function of T:

$$1000\ln\left[\left(\frac{s^\bullet}{s^\circ}\right)f\right] = \left[(a_{\bullet\circ}) + (b_{\bullet\circ}) \times 10^6/T^2 + (c_{\bullet\circ}) \times 10^{12}/T^4\right] \tag{7.9.17}$$

Table 7.31 Ab initio vibrational properties of $SrCl_2$ and SrO resolved at simple HF/STO−3G level. (Ottonello 2018)

	$[SrCl_2]$ gas RHF/STO−3G (cartesian)				Units
	^{88}Sr	^{87}Sr	^{86}Sr	^{84}Sr	
v_1	48.0143	48.1267	48.2417	48.4783	cm^{-1}
v_2	313.9903	314.0518	314.1149	314.2454	cm^{-1}
v_3	401.2908	402.285	403.3026	405.3991	cm^{-1}
E_{HF}	−4009.8349955				*Hartree*
E_{ZPC}	0.001739	0.001742	0.001744	0.001750	*Hartree*
Therm corr U	0.006122	0.006123	0.006124	0.006127	*Hartree*
Therm corr H	0.007066	0.007067	0.007069	0.007072	*Hartree*
Therm corr G	−0.027756	−0.027739	−0.027722	−0.027686	*Hartree*
U	16.071	16.075	16.079	16.087	*kJ/mol*
C_V	46.250	46.242	46.233	46.212	*J/(mol·K)*
S	306.637	306.503	306.365	306.085	*J/(mol·K)*
	[SrO] gas RHF/STO−3G (cartesian)				Units
	^{88}Sr	^{87}Sr	^{86}Sr	^{84}Sr	
v_1	283.7906	284.0410	284.2977	284.8279	cm^{-1}
E_{HF}	−3174.1535732				*Hartree*
E_{ZPC}	0.0006470	0.0006470	0.0006480	0.0006490	*Hartree*
Therm corr U	0.0034480	0.0034480	0.0034480	0.0034490	*Hartree*
Therm corr H	0.0043920	0.0043920	0.0043930	0.0043930	*Hartree*
Therm corr G	−0.0228150	−0.0227990	−0.0227830	−0.0227490	*Hartree*
U	9.054	9.054	9.054	9.054	*kJ/mol*
C_V	27.916	27.911	27.911	27.907	*J/(mol·K)*
S	239.588	239.446	239.304	239.011	*J/(mol·K)*

Table 7.32 $^{87}Sr/^{86}Sr$ separative effects (1000lnβ) and $[SrCl_2]/[SrO]$ fractionation factor (Ottonello 2018)

T (K)	$^{87}Sr/^{86}Sr$ $[SrCl_2]$	$^{87}Sr/^{86}Sr$ [SrO]	1000 ln α	$\alpha \equiv K_{7.9.15}$
273.15	0.874179	0.150945	0.724497	1.0007248
373.15	0.481924	0.082128	0.400366	1.0004004
473.15	0.303637	0.05143	0.252362	1.0002524
573.15	0.208359	0.035176	0.173213	1.0001732
673.15	0.151672	0.025555	0.126123	1.0001261
773.15	0.115278	0.019399	0.095894	1.0000959
873.15	0.090547	0.015224	0.075356	1.0000754
973.15	0.072986	0.012264	0.060775	1.0000608
1073.15	0.060074	0.010090	0.050054	1.0000501
1173.15	0.050304	0.008446	0.041943	1.0000419
1273.15	0.042736	0.007173	0.035660	1.0000357

and the separative coefficients of the two exchanging molecules may be subtracted, generating the expected fractionation factor. Because the fractionation factor is $\alpha = K^{1/r}$ with r = number of substituted isotopes per formula unit, having in our case single isotopic substitution, we may pose α = K and, because the separative effects are not too much dissimilar the isotopic delta of the reaction obeys $\Delta \cong 1000 \ln \alpha$.We may thus see that the fractionation factor operates at the 4th decimal scale only below T ≅ 500 °C (Table 7.32).

7.10 From Absolute to Conventional Magnitudes

All the magnitudes we have treated until now for ions, atoms, and molecules were expressed in terms of atomic units (a.u.) or their energy counterparts. In Geochemistry, however, which is mainly devoted to the quantitative description of the reactive properties of Earth's materials, there is an extensive need for internally consistent energy datasets, which, by their intrinsic nature, adopt "*conventional*" transpositions of the absolute magnitudes. Conventional thermodynamic (and thermophysical) magnitudes apply to any aggregation state of the matter. The transposition often involves the concept of "*formation*", i.e. the amount of energy involved in a process is referenced to a particular original state, which, in most cases, is the "*stable-state*" of the matter, or its constituent elements, at P, T of interest, or at room conditions.

7.10.1 Electron-Convention and Ion-Convention

In the ion-convention (IC) all the thermodynamic properties of the electron are set to zero (Table 7.33). In the Electron Convention (EC) the electron is considered to be equivalent to an element to all extents and therefore its enthalpy of formation (H_f) and entropy of formation (S_f) are defined as equal to zero to all temperatures. In the EC, which is adopted in most part of thermochemical tabulations (Wagman et al. 1982; Robie and Hemingway 1995; Chase 1998) the enthalpy of the electron is assumed to be non-zero at 298.15 K and to correspond to the integrated heat capacity of an ideal gas following Boltzmann statistics ($-5/2 \times RT$). Integration between the limiting 0 *K* and 298.15 *K* temperatures gives an additional energy term of 6.197 *kJ/mol* which must be multiplied by the net charge of the molecule and added to the thermochemical cycle (see later), in order to compare the IC computations with thermochemical tables. It must be noted that the integrated heat capacity of the electron obeying to the Fermi-Dirac (FD) statistics, is 0.033 *eV* (3.146 *kJ/mol*), (Bartmess 1994). The entropy at T in EC is simply translational (plus a degeneracy effect with G = 2) and is evaluated through the Sackur–Tetrode equation:

Table 7.33 Thermochemistry of the electron in the EC, IC, and FD conventions (Bartmess 1994). Data are in *kJ/mol* (H,G,E) and *J/(mol × K)* (S, C_P). Parentheses denote assumed values (Bartmess 1994)

	EC	IC	FD
H_f	0	0	0
G_f	0	(0)	0
S_{298}	20.979	(0)	22.734
H_{298}-H_0	6.197	0	3.146
C_P	20.786	17.129	
E_{298}	3.730	3.720	

$$S_{e,EC} = \frac{5}{2}R + R\ln\left[\frac{(2\pi m_e k_B T)^{3/2} k_B G}{P_0 h^3}\right] \tag{7.10.1}$$

Table (7.33) lists the various magnitudes in the three conventions.

7.10.2 *Dissociation Energy*

The zero-point dissociation energy of the gaseous molecule into gaseous atoms is obtained by subtracting from the sum of the electronic energies of the individual gaseous atoms at the athermal limit (0 *K*) the total energy and zero-point correction (ZPC) of the gaseous molecule:

$$D_0 = \sum_{i=1}^{n} \nu_i E_{e,A_i} - \left(E_{e,M} + \Delta E' + E_{ZPC,M}\right) \tag{7.10.2}$$

It is however sometimes necessary to introduce also a correction to this algebraic summation [ΔE' in Eq. (7.10.2)], whose nature will be detailed to some extent later on. The remaining energy terms ($E_{e,Ai}$, $E_{e,M}$ and $E_{ZPC,M}$ in Eq. 7.10.2) are all obtained by experiment, and, as such, available in literature (*e.g.* the NIST-JANAF or the NBS thermochemical tables).

7.10.3 *Internal Energy and Enthalpy*

Calculations necessary to retrieve absolute properties from *ab-initio* magnitudes are resumed in the thermochemical cycle of Fig. 7.36.

$H_{corr,T}$ in this cycle is the thermal correction to the enthalpy of the isolated gaseous (non-linear) molecule arising from internal energy:

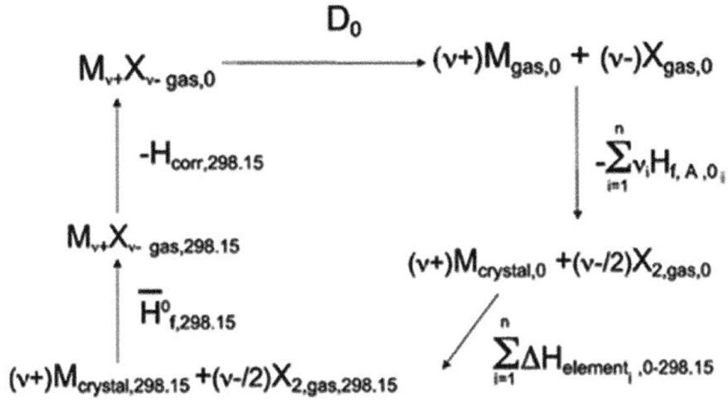

Fig. 7.36 Thermochemical cycle adopted for the determination of the enthalpy of formation from the elements at T = 298.15 *K*, P = 1 *bar*. D_0 is the zero-point dissociation energy of the gaseous molecule. $H_{f,Ai,0}$ is the zero-point enthalpy of the formation of gaseous atoms from stable elements. $\Delta H_{element,0-298.15}$ is the 0 *K* to 298.15 *K* enthalpy difference. From Ottonello (2018)

$$\begin{aligned} H_{corr,T} &= Nk_BT^2\left[\left(\frac{\partial \ln q_{trans}}{\partial T}\right)_V + \left(\frac{\partial \ln q_{rot}}{\partial T}\right)_V + \left(\frac{\partial \ln q_{vib}}{\partial T}\right)_V\right] + PV \\ &= \frac{3}{2}Nk_BT + \frac{3}{2}Nk_BT + Nk_B\sum_i X_iT\left(\frac{1}{2} + \frac{1}{e^{-X_i}-1}\right) + Nk_BT \end{aligned} \tag{7.10.3}$$

where q_{trans} , q_{rot}, q_{vib} are the translational, rotational, and vibrational contributions to the microscopic partition function,[22,23] *h* is Planck's constant, k_B is Boltzmann's constant, N is the number of gaseous molecules (i.e. N = 1 for an isolated gaseous molecule and N = N_0 Avogadro's number for one mole of substance) , X terms are vibrational temperatures ($X_i = h\varpi_i/k_BT$, with ϖ_i = ith vibrational normal mode), and the last term (PV product) arises from a simple application of the perfect gas law. Adopting the integrated heat capacity of gas following Boltzmann statistics, the enthalpy of formation from the elements at standard state (stable element, T = 298.15 *K*, P = 1 *bar*) of the gaseous species is

$$\begin{aligned} \overline{H}^0_{f,298.15} &= H_{corr,298.15} - D_0 \\ &+ \sum_{i=1}^{n} \nu_i H_{f,A_i,0} - \sum_{i=1}^{n} \nu_i \Delta H_{element_i,0\rightarrow 298.15} + \frac{5}{2}RT\xi_M \end{aligned} \tag{7.10.4}$$

where ξ_M is the net charge of the gaseous molecule.

---

[22] Assuming the first and higher excited states to be inaccessible at any T, the electronic contribution to the partition function (q_e) corresponds to the electronic spin multiplicity of the molecule, which gives no contribution to the internal energy of the molecule.

[23] For a single atom the rotational partition function is identically 1 at all T, so that there are no contributions or rotational terms to internal energy, heat capacity and entropy; for a linear molecule the rotational contribution varies linearly with T so that the partial derivative of the natural logarithm of the rotational term reduces to T^{-1}.

7.10.4 Entropy and Gibbs Free Energy of Formation from the Elements

The entropy of any isolated molecule can be calculated from first principles by applying (McQuarrie and Simon 1997):

$$S = Nk_B + Nk_B \ln\left(\frac{q(V,T)}{N}\right) + Nk_B T\left(\frac{\partial \ln q(V,T)}{\partial T}\right)_V \tag{7.10.5}$$

Splitting the microscopic partition function q(V,T) in its translational, rotational, vibrational, and electronic components and scaling in molar notation we may write

$$\overline{S}^0 = S_{trans} + S_{rot} + S_{vib} + S_e = R\left(\ln q_{trans} + \frac{3}{2}\right)$$
$$+ R\left(\ln q_{rot} + \frac{3}{2}\right) + R\sum_i \left(\frac{X_i}{e^{X_i} - 1} - \ln\left(1 - e^{-X_i}\right)\right) + R\ln q_e + R \tag{7.10.6}$$

In estimating the translational partition function of isolated gaseous molecules, the V dependency is converted into a P dependency by application of perfect gas law (McQuarrie and Simon,1997; Ochterski 2000):

$$q_{trans} = \left(\frac{2\pi m k_B T}{h^2}\right)^{3/2} V = \left(\frac{2\pi m k_B T}{h^2}\right)^{3/2} \frac{k_B T}{P} \tag{7.10.7}$$

The rotational partition function is

$$q_{rot} = \frac{\pi^{1/2}}{\sigma_{rot}}\left(\frac{T^{3/2}}{\left(\Theta_{rot,x} \times \Theta_{rot,y} \times \Theta_{rot,z}\right)^{1/2}}\right) \tag{7.10.8}$$

where $\Theta_{rot} = \frac{h^2}{8\pi^2 I k_B}$, with I = momentum of inertia and σ_{rot} = rotational symmetry factor.

The Gibbs free energy of formation of the molecule from the stable elements at standard state is now readily obtained by applying:

$$\overline{G}^0_{f,298.15} = \overline{H}^0_{f,298.15} - 298.15 \times \left(\overline{S}^0_{298.15} - \sum_{i=1}^{n} \nu_i S^0_{A,i}\right) \tag{7.10.9}$$

where $S^0{}_{A,i}$ is the standard state entropy of the monatomic stable elements (i.e. $S^0_H = \frac{1}{2} S^0_{H_{2,gas}}$).

7.10.5 Dissociation Energy of Some Geologically Relevant Gaseous Species

Assume to know the total energy and the Zero-Point Correction (ZPC) of a diatomic gas such as O_2. Subtracting from the sum of these two terms the electron energy of the two O atoms that compose the molecule and changing sign to the result we get the "*dissociation energy*" of the gaseous molecule at the athermal limit (*i.e.* 0 *K*). It should be noted that the dissociation energy differs from the dissociation enthalpy which is usually referred to 298.15 *K*. The value one gets at B3LYP/6−31G(d) theory level for O_2 is 512 *kJ/mol* (Vetuschi Zuccolini et al. 2011); 18.3 *kJ* higher than the experimental value (493.7 *kJ/mol*). Adopting the G2 procedure (Curtiss et al. 1991) we would have obtained a dissociation energy of 483.7 *kJ/mol*, and with the computationally more demanding CSB-Q procedure we would have got 495.0 *kJ/mol* (cf. Table 7.34). The origin of the discrepancy in our case stems on the fact that we adopted the same computational parameters in computing the total energy of the diatomic molecule and of the isolated gaseous atoms. Indeed the valence electrons in the isolated (gaseous) atom should be allowed to occupy wider portions of the space around the nucleus and the adoption of diffuse functions for the isolated atom could be more appropriate. Indeed if we were to adopt the 6−31+G(d) diffuse functions for O we would have got a dissociation energy of 475.3 *kJ/mol*, i.e. 18.4 *kJ/mol* lower than what experimentally observed. The appropriate electron energy of the isolated gaseous atom O is thus almost exactly intermediate between the value obtained with the 6−31G(d) and the 6−31+G(d) basis set. A more consistent practice using localized basis set is to assign to the isolated atom the total energy that reproduces exactly a null (conventional) enthalpy of formation from the elements, for the various elements in their stable aggregation state at 298.15 *K*, 1 bar.[24] In the case of elements that form stable diatomic molecules at 298.15 *K*, 1 *bar*, such as O_2, F_2, Cl_2, N_2, H_2, S_2, and P_2, the calculation is straightforward (subtract the ZPC correction and divide by two). In the case of elements that form condensed phases (alkalies, alkaline earths, transition metals, etc.) the computation is more demanding and requires the utilization of codes able to account for the appropriate symmetry constraints in periodic systems (see later).

7.10.6 Thermodynamic Properties of Gaseous Species

Application of Eqs. (7.10.4, 7.10.6, 7.10.9) to gaseous species allow to compute the enthalpy of formation from the elements at standard state (H°_f ; ideal gas, at T = 298.15 *K*, P = 1 *bar*), the Gibbs free energy of formation from the elements (G°_f), and the entropy (S°). In Table 7.35 are listed thermodynamic values obtained for

---

[24] Examples of practical application of the procedure to crystalline compounds are given later on.

Table 7.34 Calculated and experimental dissociation energies (*kJ/mol*)

Molecule	B3LYP	G2	CBS-Q	Exp
Reference	[*]	[**]	[**]	[***]
CH	335.8	333.9	336.8	334.3
CH_2 ($^3B_1$)	759.5	747.3	747.3	751.4
CH_3	1223.5	1209.2	1209.6	1210.0
CH_4	1658.3	1641.4	1645.1	1642.2
CO	1043.7	1074.9	1079.5	1070.3
CO_2	1580.1	1607.1	1609.2	1597.9
NH	341.7	325.1	325.9	330.5
NH_2	721.9	709.6	711.7	711.3
NH_3	1150	1153.5	1156.9	1157.7
NO	626.2	629.3	630.1	628.0
H_2O	885.3	918.4	918.8	917.6
HF	531.2	569.9	570.3	566.1
HCl	413.6	431.8	429.3	427.6
Cl_2	200.1	246.4	233.5	239.3
H_2	440.6	–	–	448.3
N_2	917.7	933.5	936.4	941.8
F_2	170	157.3	153.1	154.4
O_2	512	495.0	483.7	493.7
P_2	448.8	484.1	479.9	485.8
S_2	397.2	427.6	407.5	421.3
SO_2	930.6	1061.1	1041.4	1062.7
SO	483.9	519.7	505.4	516.7
H_2S	711.5	727.2	723.8	724.7
MAD Error	23.5	3.5	5.5	
AVE Error	15.2	−0.8	2.3	
MAX Error	132.1	−9.2	21.3	

[*] Vetuschi Zuccolini et al. (2011)
[**] Ochterski et al. (1996)
[***] Chase et al. (1985); Huber and Herzberg (1979) (LiF, CO)

some geochemically relevant gaseous species with the G2 Theoretical Model Chemistry Algorithm (TMCA) (Vetuschi Zuccolini et al. 2011). For all atoms H are adopted the energies of the isolated atoms computed at the same theory level while for H is adopted the theoretical Hartree–Fock (HF) limit (0.5 *Hartree*). We may see that proceeding in this way the (conventional) $H°_f$ and $G°_f$ for elements forming stable-state gaseous diatomic molecules (H_2, O_2, N_2) is not exactly zero as it should be. I stress also that, being all the investigated molecules neutral, the distinction between EC and IC is ineffective.

By knowing the partition function, we may depict how a T-dependent magnitude evolves with the intensive variable.

Table 7.35 Thermodynamic parameters for some gaseous species of geochemical interest obtained with the G2 TMCA. $\overline{H}_f^0$, $\overline{G}_f^0$ in *kJ/mole*; $\overline{S}^0$ in *J/(mole × K)*

	*Ab-initio**			Experimental**		
Species	$\overline{H}_f^0$	$\overline{S}^0$	$\overline{G}_f^0$	$\overline{H}_f^0$	$\overline{S}^0$	$\overline{G}_f^0$
H_2O	−237.6	188.2	−224.2	−241.8	188.8	−228.6
CO_2	−392.5	213.4	−393.2	−393.5	213.8	−394.4
CO	−112.7	197.3	−139.2	−110.5	197.3	−137.1
O_2	4.9	204.4	5.1	0	205.15	0
H_2	−4.5	130.0	−4.3	0	130.68	0
CH_4	−74.5	185.9	−50.2	−74.8	186.26	−50.7
H_2S	−23.6	205.3	−36.2	−20.6	205.8	−33.4
N_2	4.7	191.2	4.8	0	191.61	0
NH_3	−41.2	201.1	−14.2	−45.9	192.77	−16.4

(*) Vetuschi Zuccolini et al. (2011); (**) Robie and Hemingway (1995)

Let me consider the entropy of the substance. In Table (7.36) we see the isochoric heat capacity and the entropy of H_2O in the gaseous state computed at the B3LYP/6−31G(d,p) theory level at various T conditions (Vetuschi Zuccolini et al. 2011). Rotational and translational contributions to the isochoric heat capacity are identical and correspond to (3/2)RT. The rotational and translational contributions to the entropy of the substance above T of reference (T_{ref}) both simply correspond

Table 7.36 Ab-initio isochoric heat capacity and entropy of gaseous H_2O (Vetuschi Zuccolini et al. 2011)

T (*K*)	C_V (*J/mol × K*)		S (*J/(mol × K)*				
	Total	Vibrational	Total	Traslational	Rotational	Vibrational	Woolley (1980)
298.15	25.117	0.176	188.765	144.800	43.940	0.025	188.834
400	25.698	0.753	198.661	150.909	47.606	0.146	198.788
600	27.619	2.678	212.794	159.339	52.660	0.795	213.052
800	29.861	4.916	223.434	165.318	56.250	1.866	223.825
1000	32.217	7.272	232.204	169.954	59.032	3.217	232.738
1200	34.531	9.586	239.802	173.745	61.304	4.749	240.485
1400	36.652	11.707	246.567	176.950	63.229	6.389	247.407
1600	38.505	13.565	252.697	179.724	64.894	8.075	253.69
1800	40.087	15.142	258.303	182.176	66.362	9.770	259.451
2000	41.413	16.468	263.475	184.364	67.676	11.435	264.769

to (3/2) $R\ln(T/T_{ref})$. The slight differences between Woolley's values (Woolley 1980) and those of Vetuschi Zuccolini et al. (2011) depend upon the different harmonic frequencies. Moreover, additional corrections (Stripp-Kirkwood and centrifugal effects) are provided by (Woolley 1980) and the NIST-JANAF tabulation (Chase 1998).

Chapter 8
3D and 2D Periodic Structures

Again, in this chapter, we encounter many fundamental concepts such as the Metric Tensor, the Reciprocal Lattice, the Voronoi Cells and Brillouin zones, the Bloch Theorem, the Thomas–Fermi model, and the uniform electron gas (UEG) treatment. We see also the ab initio treatment of phonon frequencies (with details on the Einstein, Debye, and Kieffer models) and the ab initio implications on the nature of Band structures and projected bands. Only ideal crystals are considered initially. Some defect chemistry applications are developed only later on, together with some concepts concerning the "*surface energy*" (with application of Green's functions). In treating the Lattice Geometry, it is adoped the construction and the nomenclature proposed by Viterbo (1996), with minimal modifications.

8.1 Premises

When investigating the thermochemical properties of crystalline substances, we, more or less consciously, adopt an idealization of the phase of interest: we imagine a crystal as being composed of a regular network of atoms, with inter-ionic distances dictated by the symmetry operations of the spatial group. In this idealized picture, a crystal is an unperturbed tridimensional replication of the asymmetric unit. Actually, this hypothetical homogeneity-structural continuity is, in real phases, interrupted by local irregularities called *lattice defects*. Lattice defects are furthermore subdivided into two main groups: *extended defects* and *point defects*. Extended defects are primarily composed of linear dislocations, shear planes, and intergrowth phenomena. Extended defects interrupt the continuity of the crystal, generating crystal sub-grains whose dimensions depend, in a complex fashion, on the density of extended defects per unit area. The presence of extended defects does not significantly affect the energy properties of crystalline compounds, because the concentration of extended defects necessary to create significant energy

G. A. Ottonello, *Quantum Geochemistry*, Springer Geochemistry,
https://doi.org/10.1007/978-3-031-21837-8_8

modifications is too high. Extended defects such as dislocations *cannot exist in the crystal at thermodynamic equilibrium* (Friedel 1964), and their energy contribution must be evaluated through the elastic theory. Calculations in this sense (Holder and Granato 1969) lead to energies between 1 and 22 *eV* per atomic unit length. At a defect energy of about 20 *eV* and a dislocation density of $10^9\ cm^{-2}$, the resulting energy is about 50 *J/mol*. This is lower than experimental error in most experimental measurements of lattice energy (see Sect. 8.5) and may virtually be neglected in calculations. Nevertheless, point defects exist in crystals *at equilibrium conditions*. They primarily affect some chemical-physical features of crystalline substances, such as electronic and ionic conductivity, diffusivity, light absorption and light emission, etc., and, in a subordinate fashion, their thermodynamic properties. As already noted, our idealized picture of a crystal has little to do with reality. As extended defects interrupt the symmetry of crystals, point defects modify their chemistry significantly, inducing defect equilibria with coexisting phases and modifying stoichiometry. The disorder induced by the presence of point defects is grouped into two main categories: *intrinsic disorder* and *extrinsic disorder*. Intrinsic disorder is observed in conditions of perfect stoichiometry of the crystal, while the extrinsic disorder is associated to external parameters (such as oxygen fugacity fO_2, in the case of O-based minerals, Lux–Flood acidity of the intergrowth medium, and so on).

8.2 Lattice Geometry

8.2.1 Translational Symmetry

In an ideal crystal, generated by an ordered repetition of groups of atoms along the three coordinates of the Euclidean space, all repeated units are related by translational symmetry operations represented by a vectorial set **T**:

$$\mathbf{T} = u\mathbf{a} + v\mathbf{b} + w\mathbf{c} \tag{8.2.1}$$

with u, v, w real integer.

The set of points generated by the vector **T** are the "*lattice nodes*" of the three-dimensional structure with coordinates u, v, and w. By setting to 1 these coordinates one has the "*unit cell*". The moduli a, b, and c of the vectors and the angles $\alpha = \mathbf{b/c}$, $\beta = \mathbf{a/c}$, $\gamma = \mathbf{a/b}$ are the "*lattice constants*" which define the geometry of the structure with respect to the crystallographic axes X, Y, and Z.

By choosing (arbitrarily) a particular lattice point as "*origin*" any other point is uniquely defined by the vector (8.2.1).

The choice of an origin implies the selection of a "*primitive cell*", whose volume is defined by the cross product of the vectors (see Sect. A.5.3.5):

$$V = |\mathbf{a} \cdot (\mathbf{b} \times \mathbf{c})| \tag{8.2.2}$$

The crystal structure is completely determined by the lattice constants and by the coordinates of all atoms in the unit cell:

$$\mathbf{r}_j = x_j\mathbf{a} + y_j\mathbf{b} + z_j\mathbf{c} \quad (j = 1, 2, \ldots, N) \tag{8.2.3}$$

The vector $\mathbf{r}_j$ links the origin of the cell to the nuclear position of atom j in Cartesian coordinates.

The Cartesian coordinates of each atom (nucleus) in the unit cell (*fractional coordinates*) are represented by real-positive numbers with values comprised between 0 and 1:

$$x = \frac{X}{a}; \quad y = \frac{Y}{b}; \quad z = \frac{Z}{c} \tag{8.2.4}$$

Having chosen a particular origin (hence a *primitice cell*), application of the **T** vector (8.2.1) to constant u, v, w generates a row of atoms with indices [uvw]. The three rows corresponding to the base vectors **a**, **b**, and **c** have symbols [100], [010], and [001], respectively.

A plane intersecting the crystallographic axes at points (p, 0, 0), (0, q, 0), (0, 0, r) with integer p, q, r obeys the equation

$$\frac{x}{p} + \frac{y}{q} + \frac{z}{r} = 1 \tag{8.2.5}$$

Posing h = qr, k = pr, l = pq and m = pqr, the plane equation may be rewritten as:

$$hx + ky + lz = m \tag{8.2.6}$$

where h, k, and l are the *Miller indices*. The plane passing through the origin will have $m = 0$ and the plane passing closest to it will have $m = 1$. Crystallographic planes parallel to one of the axes X, Y, or Z will have indices (*0kl*), (*h0l*), or (*hk0*), respectively.

8.2.2 The Metric Tensor

Adopting the ket notation (Sect. A.8), the matrix of the basic vectors may be represented as:

$$|e\rangle = \begin{pmatrix} \mathbf{a} \\ \mathbf{b} \\ \mathbf{c} \end{pmatrix} \tag{8.2.7}$$

The scalar product of the vector **T** by its transpose gives its modulus T as[1]:

$$\mathrm{T} = \langle \mathbf{T} | \mathbf{T}^{-1} \rangle = \langle \mathrm{u} | \mathrm{e} \rangle \langle \mathrm{e} | \mathrm{u} \rangle = \langle \mathrm{u} | \mathrm{G} | \mathrm{u} \rangle \tag{8.2.8}$$

where **G** is the *metric tensor*:

$$\mathbf{G} = \begin{pmatrix} \mathbf{a}\cdot\mathbf{a} & \mathbf{a}\cdot\mathbf{b} & \mathbf{a}\cdot\mathbf{c} \\ \mathbf{b}\cdot\mathbf{a} & \mathbf{b}\cdot\mathbf{b} & \mathbf{b}\cdot\mathbf{c} \\ \mathbf{c}\cdot\mathbf{a} & \mathbf{c}\cdot\mathbf{b} & \mathbf{c}\cdot\mathbf{c} \end{pmatrix} \tag{8.2.9}$$

8.2.3 The Reciprocal Lattice

Given the basis of a *direct lattice* through the basic operator $|\mathrm{e}\rangle$ (Eq. 8.2.7) the reciprocal lattice is that one defined by the operator $|\mathrm{e}^*\rangle$ which satisfies the identity condition $\mathrm{I} = |\mathrm{e}^*\rangle\langle \mathrm{e}|$:

$$\begin{pmatrix} 1 & 0 & 0 \\ 0 & 1 & 0 \\ 0 & 0 & 1 \end{pmatrix} = \begin{pmatrix} \mathbf{a}*\cdot\mathbf{a} & \mathbf{a}*\cdot\mathbf{b} & \mathbf{a}*\cdot\mathbf{c} \\ \mathbf{b}*\cdot\mathbf{a} & \mathbf{b}*\cdot\mathbf{b} & \mathbf{b}*\cdot\mathbf{c} \\ \mathbf{c}*\cdot\mathbf{a} & \mathbf{c}*\cdot\mathbf{b} & \mathbf{c}*\cdot\mathbf{c} \end{pmatrix} \tag{8.2.10}$$

with **a*** normal to the plane (**b**, **c**), **b*** normal to the plane (**a**, **c**), and **c*** normal to the plane (**a**, **b**).

Posing $\mathrm{p} = \mathrm{V}^{-1}$, with V = unit cell volume, the null dot products imply that[2]:

$$\mathbf{a}* = \mathrm{p}(\mathbf{b} \wedge \mathbf{c}) \tag{8.2.11.1}$$

$$\mathbf{b}* = \mathrm{p}(\mathbf{c} \wedge \mathbf{a}) \tag{8.2.11.2}$$

$$\mathbf{c}* = \mathrm{p}(\mathbf{a} \wedge \mathbf{b}) \tag{8.2.11.3}$$

Multiplying both sides of Eq. (8.2.11.1) by **a**, one obtains in fact an identity:

$$\mathbf{a}*\cdot\mathbf{a} = \mathrm{p}(\mathbf{b} \wedge \mathbf{c})\cdot\mathbf{a} = 1 = \mathrm{pV} \tag{8.2.12}$$

The family of planes generated by reciprocal lattice vector **T***:

$$\mathbf{T}* = h\mathbf{a}* + k\mathbf{b}* + l\mathbf{c}* \tag{8.2.13}$$

---

[1] Note that here I conform strictly to Dirac's notation (cf. Sect. A.8). Because bra and kets concern vectors, the vector symbol inside bra and kets is not in bold. Moreover I identify a vectorial transpose as the symbol elevated to −1.

[2] ∧ in Eq. (8.2.11.1) means the wedge product of **b** and **c** vectors.

is orthogonal to the family of planes generated by the direct lattice vector **T**, and the reciprocal metric matrix is the transpose of the metric tensor, i.e. $\mathbf{G}^* = \mathbf{G}^{-1}$:

$$\mathbf{G}^* = \begin{pmatrix} \mathbf{a}*\cdot\mathbf{a}* & \mathbf{a}*\cdot\mathbf{b}* & \mathbf{a}*\cdot\mathbf{c}* \\ \mathbf{b}*\cdot\mathbf{a}* & \mathbf{b}*\cdot\mathbf{b}* & \mathbf{b}*\cdot\mathbf{c}* \\ \mathbf{c}*\cdot\mathbf{a}* & \mathbf{c}*\cdot\mathbf{b}* & \mathbf{c}*\cdot\mathbf{c}* \end{pmatrix} \tag{8.2.14}$$

For a Bravais lattice defined in terms of primitive vectors **a**, **b**, **c** the matrix inversion would actually imply[3]:

$$[\mathbf{a}*\mathbf{b}*\mathbf{c}*]]^{\mathrm{T}} = 2\pi[\mathbf{abc}]]^{-1} \tag{8.2.15}$$

the factor 2π coming out naturally from the study of periodic structures through Fourier transforms (see Sect. A.12). The crystallographic definition implies the reciprocal lattice vectors to be

$$\mathbf{a}* = \frac{\mathbf{b}\times\mathbf{c}}{\mathbf{a}\cdot(\mathbf{b}\times\mathbf{c})} \tag{8.2.16}$$

The crystallographic definition has the advantage that **a*** is just the reciprocal magnitude of **a** in the direction of $\mathbf{b} \times \mathbf{c}$. This simplifies greatly the mathematical manipulation and allows expressing the reciprocal lattice dimensions in units of spatial frequency.

8.3 Crystal Symmetry

8.3.1 Laue Classes and Point Groups

The combination of proper and improper symmetry axes intersecting in a point generates a *point group*. The terminology arises from the consideration that the symmetry operators form a mathematical *group* leaving one *point* fixed. Crystals having the same point group belong to the same crystal class. Out of the 32 possible point groups (Table 8.1), 13 have a single axis. The coexistence of proper and improper axes simultaneously present in the same direction generates new groups formally indicated with symbols of the type $n/\overline{n}$.

The *Neumann principle* establishes that the symmetry elements of any physical property must include the symmetry elements of the crystal point group. Based on this principle "*...the physical properties of a crystal may present a higher, but not a lower, symmetry of the point group. As a consequence, physical experiments do not normally reveal the true symmetry of the crystal*" (Viterbo 1996). The 11 Laue

[3] Dropping the factor 2π implies assuming $\exp(2\pi i\mathrm{K}\cdot\Re) = 1$

Table 8.1 Crystal systems, Laue symmetry, and point groups (from Viterbo 1996; with modifications)

Crystal system	Laue class	Point groups
Triclinic	$\overline{1}$	$1, \overline{1}$
Monoclinic	$2/m$	$2, m, 2/m$
Orthorombic	mmm	$222, mm2, mmm$
Tetragonal	$4/m$	$4, \overline{4}, 4/m$
	$4/mmm$	$422, 4mm, \overline{4}2m, 4/mmm$
Trigonal/ Rhombohedral	$\overline{3}$	$3, \overline{3}$
	$\overline{3}m$	$32, 3m, \overline{3}m$
Hexagonal	$6/m$	$6, \overline{6}, 6/m$
	$6/mmm$	$622, 6mm, \overline{6}2m, 6/mmm$
Cubic	$m\overline{3}$	$23, m\overline{3}$
	$m\overline{3}m$	$432, \overline{4}3m, m\overline{3}m$

Classes (Table 8.1) collect together Point Groups not experimentally differentiable through physical experiments such as diffraction.

Because the crystal periodicity is compatible only with axes of order 1, 2, 3, 4 and 6 this implies restrictions on the possible geometry. There are 7 crystal systems representing the groups that may be actually formed. These collect the symmetry classes with common lattice features, which may be described by cells of the same type (Table 8.1). Other types of lattice may be however formed by combining *non-primitive* cells. Altogether, there are 14 possible Bravais Lattices. I remand to Aroyo et al. (2006) for a proper outline of the development of theory of the classical crystallographic groups, the relations between crystal structures in crystal chemistry, and the correct determination of the symmetry of a crystal structure.

8.3.2 Voronoi Cells and Brillouin Zones

Tracing an orthogonal plane through the midpoint of a segment connecting two lattice points and collecting all possible planes generated by intersections (i.e. a "*Dirichlet Construction*") one generates a *Voronoi Cell* (or "*Wigner–Seitz Cell*" in the common crystallographic terminology). The equivalent of the *Wigner–Seitz Cell* in the reciprocal space is the *Brillouin Zone*. As we will see later on in the Applications, the Brillouin Zone is extremely important in the investigation of lattice dynamics. There are 14 of them corresponding to the 14 reciprocal *Bravais Lattices*.

In Fig. 8.1 we see, as an example, the first Brillouin zone of the Face Centered Cubic System (FCC), with the center of the Brillouin zone (Γ), the high-symmetry points (*L-U-X-W-K*) and the symmetry lines (Λ, Δ, Σ).

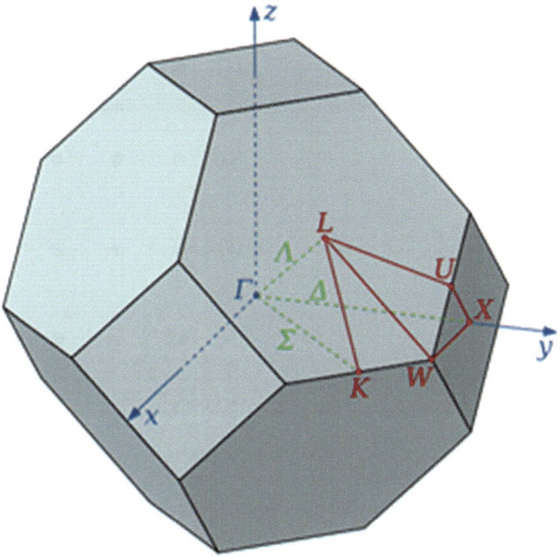

Fig. 8.1 First Brillouin zone of the Face Centered Cubic System (FCC). *Γ* is the center, *K* is the middle of an edge joining two hexagonal faces, *L* is the center of a hexagonal face, *U* is the middle of an edge joining a hexagonal and a square face, *W* is a corner pointy, *X* is the center of a square face. From Wikimedia Commons, the free media repository

8.4 Bloch Theorem

Felix Bloch gave, almost a century ago (Block 1928), fundamental theoretical contributions to the understanding of electron behavior in crystal lattices, ferromagnetism, and nuclear magnetic resonance. A Bloch wave (also called Bloch function) is a wave composed of a plane wave modulated by a periodic function. Hence, The Bloch's Theorem may be stated as follows: "*The energy eigenstates of Block waves determine the basis of the electron wave function in crystals*".

To understand why this is so, we must consider the states of definite energy and the nature of time-dependent states for an electron in a crystal. Following Feynman et al. (1965), we begin with a one-dimensional lattice.

Let us consider a linear array of atoms, such as that one depicted in Fig. 8.2, and let us suppose that an electron can jump from one atom to the neighbor with a certain energy (i.e. a certain amplitude). Suppose moreover that the inter-atomic spacings are constant.

The base states of the electron in three neighboring atomic positions will be depicted as $|n-1\rangle$, $|n\rangle$ and $|n+1\rangle$. Using these base states, any state $|\phi\rangle$ of the electron in the one-dimensional crystal may be written as a superposition of the base states:

$$|\phi\rangle = \sum_{n} |n\rangle\langle n \mid \phi\rangle \tag{8.4.1}$$

Let us call A_n the amplitude for an electron to be at x_n. Equation (8.4.1) will be in this case replaced by

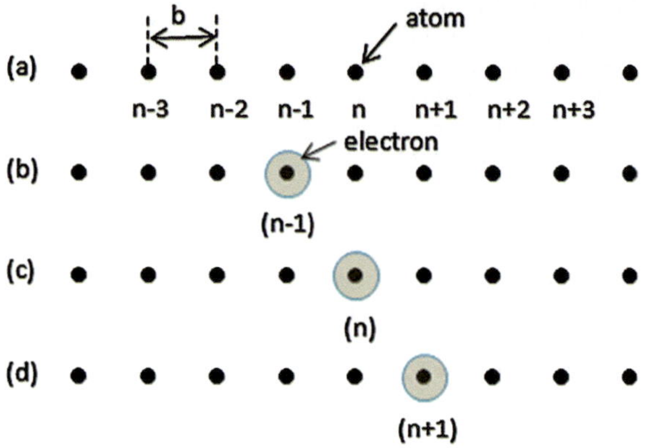

Fig. 8.2 The base states of an electron in a one-dimensional crystal. Modified from Feynman et al. (1965)

$$|\phi\rangle = \sum_{n} |n\rangle A_n \tag{8.4.2}$$

The amplitude is given by

$$A_n = \exp(ikx_n)\exp[-(i/\hbar)E_t] \tag{8.4.3}$$

where the energy at time t (E_t) varies with respect to the zero energy (E_0; i.e. the energy that the electron would have if "*it couldn't leak away from one of the atoms*"; Feynman et al. 1965):

$$E = E_0 - 2A\cos kb \tag{8.4.4}$$

with b = atom spacing, and A is a constant independent of t. The space dependence of the amplitudes is represented by $a(x_n) = \exp(ikx_n)$. The amplitudes oscillate as we go from one atom to the next, as shown in Fig. 8.3. This is a complex oscillation. The *magnitude* is the same at every atom but the *phase* at a given time advances by the amount (ikb) from one atom to the next.

We may see also in this figure that the change of amplitude depends upon the value assigned to k, i.e. it becomes smaller with the decrease of k. For small enough k we have $\cos kb \approx 1 - k^2b^2/2$ and, posing $E_0 = 2A$, Eq. (8.4.4) becomes

$$E = Ak^2b^2 \tag{8.4.5}$$

Or, in other terms, "*the energy of the state is proportional to the square of the wave number which describes the spatial variations of the amplitudes*" (Feynman et al. 1965).

With an amplitude A_n at x_n, the probability of finding an electron at x_n is $|A_n|^2$. Equation (8.4.3) describes thus "*stationary states*" i.e. a probability not changing with time. To understand why electrons actually proceed along the atom's alignment with time we must admit that electrons, like all particles, move as *wave*

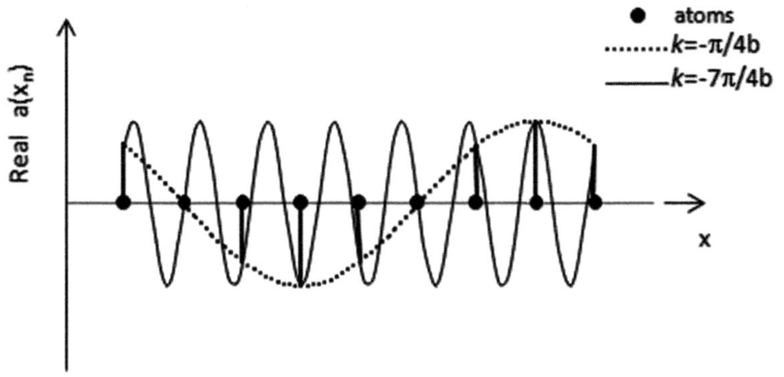

Fig. 8.3 Space dependent amplitudes (Real part) for different values of k. The physical situation at the atomic positions is the same, regardless of the k periodicity. Modified from Feynman et al. (1965)

packets (or "*combined waves*", see Fig. 1.3 and Sect. 1.2.2), with a "*group velocity*" v_g related to the variation of k with frequency. This is a "*localized disturbance*" (McWeeny 1972) which travels with the moving particle. To describe it we must choose k-dependent amplitude factors able to generate a constructive interference at a specified point (and destructive elsewhere). This may be obtained by adopting a very narrow range of k's (see Sect. 1.2.2).

Considering in fact a Gaussian wave packet with

$$\mathrm{A}(k) = \mathrm{a}\exp\left\{-\sigma(k-k_0)^2\right\} \tag{1.2.6}$$

when σ is large we have a superposition of wave trains with wavenumbers differing inappreciably from k_0 and amplitudes falling off rapidly, according to the Gaussian law. In this case, through various passages (McWeeny 1972) we may arrive to a complete (i.e. x and t dependent) wave function, in the form:

$$\Psi(\mathrm{x},\mathrm{t}) = \mathrm{a}\sqrt{\frac{\pi}{\sigma}}\exp\left[-\frac{\pi^2(\mathrm{x}-\mathrm{ut})^2}{\sigma}\right]\exp\{2\pi i k_0(\mathrm{x}-\mathrm{ut})\} \tag{1.2.9}$$

Replacing the wavenumber k with the wave vector $\mathbf{k}$ and the position x with the space vector $\mathbf{r}$ and modulating the resulting plane wave [exp(i$\mathbf{kr}$)] through a periodic function u($\mathbf{r}$) which has the same periodicity as the crystal one has a *Block function*:

$$\Psi(\mathbf{r}) = \exp(i\mathbf{kr})\mathrm{u}(\mathbf{r}) \tag{8.4.6}$$

The real part of a typical Bloch function in one dimension is depicted in Fig. 8.4.

The energy eigenstates of Eq. (8.4.6) are usually depicted as $\psi_{n\mathbf{k}}$ where n (the band index) indicates the type of Block wave [same $\mathbf{k}$ but different periodicity u(r)].

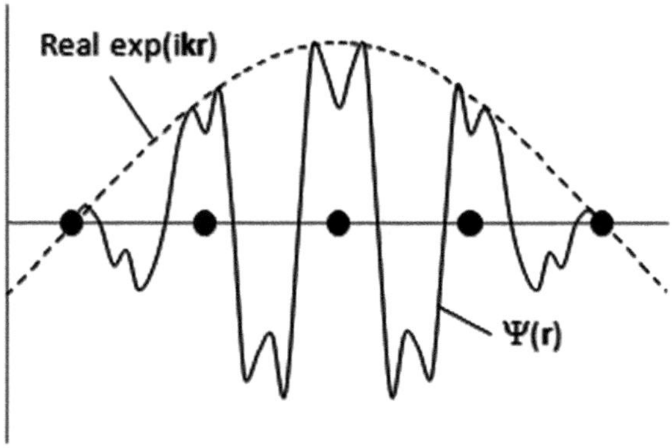

Fig. 8.4 Schematic drawing of a typical Bloch wave in one dimension. The dotted line is the real part of the plane wave

All (n) Block waves occur for values of **k** falling in the Brillouin zone of the reciprocal lattice. Hence they are translationally invariant in the reciprocal lattice, i.e.

$$\Psi_{n\mathbf{k}} = \Psi_{n(\mathbf{k}+\mathbf{T})}, \tag{8.4.7}$$

To proof that a Block function represents actually the energy eigenstates for an electron in a crystal we make first a little change of notation and denote the translational operator working in the primitive lattice as:

$$\mathsf{T}_{\mathrm{n}_1,\mathrm{n}_2,\mathrm{n}_3} = \mathrm{n}_1\mathbf{a}_1 + \mathrm{n}_2\mathbf{a}_2 + \mathrm{n}_3\mathbf{a}_3 \tag{8.4.8}$$

Analogously, the vector operating in the reciprocal space will be depicted as

$$\mathsf{k}_{\theta_1,\theta_2,\theta_3} = \theta_1\mathbf{b}_1 + \theta_2\mathbf{b}_2 + \theta_3\mathbf{b}_3 \tag{8.4.9}$$

with $\mathbf{b}_1$, $\mathbf{b}_2$ and $\mathbf{b}_3$ as reciprocal lattice vectors.

We reverse then Eq. (8.2.1) and write:

$$u(\mathbf{r}) = \exp(-i\mathbf{k}\mathbf{r})\Psi(\mathbf{r}), \tag{8.4.10}$$

If our wave function is an eigenstate of all the translational operators we may write:

$$\Psi(\mathbf{r} + \mathbf{a}_\mathrm{j}) = \mathrm{C}_\mathrm{j}\Psi(\mathbf{r}), \tag{8.4.11}$$

where C_j are the (three) eigenvalues not depending on **r**.

Posing

$$\mathrm{C}_\mathrm{j} = \exp(2\pi \mathrm{i}\theta_\mathrm{j}), \tag{8.4.12}$$

we may write:

$$\begin{aligned} u(\mathbf{r}+\mathbf{a}_j) &= \exp\left[-i\mathbf{k}\cdot(\mathbf{r}+\mathbf{a}_j)\right]\psi(\mathbf{r}+\mathbf{a}_j) \\ &= \left[\exp(-i\mathbf{k}\cdot\mathbf{r})\exp(-i\mathbf{k}\cdot\mathbf{a}_j)\right]\left[\exp(2\pi i\theta_j)\psi(\mathbf{r})\right] \\ &= \exp(-i\mathbf{k}\cdot\mathbf{r})\exp(-2\pi i\theta_j)\exp(2\pi i\theta_j)\psi(\mathbf{r}) = u(\mathbf{r}), \end{aligned} \tag{8.4.13}$$

Equation (8.4.13) proves that u(**r**) is the periodicity of the lattice, and, because of Eq. (8.4.11), it proves also that the Bloch wave is an eigenstate of all the translational operators.

8.5 Lattice Energy

The *lattice energy* is basically the static part of the energy of a crystalline solids (disregarding the zero point energy). It is by far the most important energy of the solids, exceeding by order of magnitudes the vibrational one (see later). If we consider for simplicity an ionic crystal, then we may define the lattice energy as *the amount of energy required at absolute zero temperature to convert one mole of crystalline component into constituent ions in a gaseous state at an infinite distance*. We may consider it as composed of coulombic + repulsive + dispersive terms (+ the zero point energy).

8.5.1 Coulomb Attraction

The attractive force exerted by a cation of charge Z_+ on an anion of charge Z_- at distance r is given by:

$$F_c = Z_+Z_-e^2/r^2 \tag{8.5.1}$$

The coulombic potential at distance r is obtained by integration of (8.5.1) in dr from infinite distance to r:

$$\Phi_c = \int -F_c dr = \frac{Z_+Z_-e^2}{r} \tag{8.5.2}$$

8.5.2 Short-Range Repulsion

First of all, let us note that short-range coulombic repulsion exists, due to the homologous nature of the electronic charges of the two atoms at a short distance from each other. However, the main repulsive force is a manifestation of the exclusion principle. The increase of energy with decreasing distance is exponential. Born first suggested a potential in the form (Born 1920):

$$\Phi_R = \frac{b}{r^n} \tag{8.5.3}$$

where b is a constant and n (Born exponent) is an integer number (between 5 and 12) which depends on the nature of interacting ions. Equation (8.5.3) was then replaced by the form:

$$\Phi_R = b \times \exp\left(\frac{-r}{\rho}\right) \tag{8.5.4}$$

which expresses in a better way the exponential dependency of the radial functions. b and ρ are two constants, usually defined respectively as "*repulsion factor*" and "*hardness factor*".

8.5.3 *Dispersive Forces*

If electron clouds of cations and anions are deformable in an electric field and thus have polarizabilities α_+, and α_-, then the value of the dipole moment induced in the cations by the presence of anions (μ_{-+}) and that of anions, induced by cations (μ_{+-}) are respectively proportional to $\frac{\alpha_+}{r^3}$ and $\frac{\alpha_-}{r^3}$. The electric field acting on the cations is proportional to $-\alpha_+\overline{\mu}^2_{-+}/_{r^6}$ where $\overline{\mu}^2_{-+}$ is the mean square dipole moment of the cations.

Polarization may be been formally represented as the movement of a charge Z_e through a displacement distance d. The displacement is due to field F, and the drop in potential energy caused by movement in the field is F $\times$ d. If the force acting on charge Z_e by virtue of the parent nucleus is

$$F = -K_F d \tag{8.5.5}$$

then the resultant fall in potential energy will be:

$$-K_F d^2 + \frac{1}{2}K_F d^2 = -\frac{1}{2}K_F d^2 = -\frac{1}{2}\alpha F^2 \tag{8.5.6}$$

The interaction of the two dipoles is thus associable to a potential energy proportional to the square of the electric field F. F itself is moreover proportional to r^{-3}, so that the functional dependency of the dispersive energy over the distance will be proportional to r^{-6}.

Accounting for the instantaneous higher moments of the charge distributions of the atoms, leads to an inverse eight power functional form (dipole-quadrupole interactions). The bulk dispersive potential is represented as shown by Mayer (1933):

$$\Phi_D = \frac{dd_{+-}}{r^6} + \frac{dq_{+-}}{r^8} \tag{8.5.7}$$

The dipole–dipole (dd_{+-}) and dipole–quadrupole coefficients (dq_{+-}) can be derived from free ion polarizability α and the *mean excitation* $\overline{E}$ by applying:

$$dd_{+-} = \frac{3}{2}\alpha_+\alpha_-\left(\frac{\overline{E}_+\overline{E}_-}{\overline{E}_+ + \overline{E}_-}\right) \tag{8.5.8}$$

$$dq_{+-} = \frac{27}{8}\left(\frac{\alpha_+\alpha_-}{e^2}\right)\left(\frac{\overline{E}_+\overline{E}_-}{\overline{E}_+ + \overline{E}_-}\right)\left[\frac{\alpha_+\overline{E}_+}{n_+} + \frac{\alpha_-\overline{E}_-}{n_-}\right] \tag{8.5.9}$$

(Mayer 1933; Boswara and Franklyn 1968) where e is the electronic charge and n_+, n_- are *effective electrons* (see Tosi 1964, for an extended discussion on the significance of the various terms).

8.5.4 Madelung Constant

The Madelung constant ($\mathcal{M}$) represents the distribution of virtual point charges around a central charge selected as origin and obeying the crystal symmetry. For a simple isometric crystal such as Halite (NaCl, cubic, Laue class $m\overline{3}m$, space group $Fm\overline{3}m$), the Madelung constant quite easy to compute. Taking the central charge on a Na^+ ion, we find 6 Cl^- ions as first neighbors at distance r_0, corresponding to 1/2 the cell edge a_0. The second neighbors are 12 Na^+ ions at distance $r_0\sqrt{2}$, the third neighbours are 8 Cl^- ions at distance $r_0\sqrt{3}$, the fourth neighbors are 6 Na^+ ions at distance $2r_0$, the fifth neighbours are 24 Cl^- ions at distance $r_0\sqrt{5}$, and so on. Collecting the common terms, $\mathcal{M}$ is simply the summation of terms in brackets (extended to infinity), which, in this particular case, reduces to ln2. The Coulomb potential acting on the cation is:

$$\Phi_{C,Na^+} = -\frac{Z_{+-}^2 e^2}{r_0}\left(6 - \frac{12}{\sqrt{2}} + \frac{8}{\sqrt{3}} - \frac{6}{\sqrt{4}} + \frac{24}{\sqrt{5}} - \ldots\ldots\right) \tag{8.5.10}$$

where Z_{+-}^2 is the largest common factor of the point-like charges.

If we now repeat the calculation adopting Cl^- as the central charge, we obtain a Coulomb potential which (for this particular structure) is identical to that acting at the cationic site. The bulk Coulomb potential is given by the summation of the cationic and anionic parts divided by two:

$$\Phi_C = \frac{\Phi_{C,Na+} + \Phi_{C,Cl-}}{2} = -\frac{Z_{+-}^2 e^2 \mathcal{M}}{r_0} \tag{8.5.11}$$

In the case of a general crystal with no particular structural constraints, the Coulomb potential of a unit cell obeys (Catti 1978):

$$\Phi_C = \frac{1}{2} e^2 \sum_{i=1}^{N} \sum_{j=1}^{N} \sum_{-\infty \mathbf{l}}^{+\infty} \frac{Z_i Z_j}{|\mathbf{x}_{ij} + \mathbf{l}|} \quad (\mathbf{l} \neq 0 \quad \text{for} \quad i = j\,) \tag{8.5.12}$$

where N is the number of ions in the cell, $\mathbf{x}_{ij} = \mathbf{x}_i - \mathbf{x}_j$ is an inter-atomic vector between ions contained in the cell, and $\mathbf{l}$ is a vector of the direct lattice. Because the Coulomb potential decreases slowly for large distances the direct calculation through (8.5.12) is not convenient. A much faster converging series is obtained by replacing the point charges with charge distributions of spherical symmetry (Bertaut 1952). If the charge densities do not overlap and fall to zero at some distances from the nuclear centers one has only a reciprocal lattice sum (Bertaut series). In the charge densities never go to zero but have a Gaussian shape (and hence they overlap) a direct lattice sum is also necessary (Ewald double sum; Ewald 1921). Catti (1978) investigated the optimal convergence for the Ewald's double series a proposed a very efficient computational tool (MADEWA, Fortran IV; Catti 1978). Writing the Ewald formula in the form:

$$\Phi_C = \frac{1}{2} e^2 \sum_{i=1}^{N} \sum_{j-1}^{N} Z_i Z_j F(\mathbf{x}_{ij}) \tag{8.5.13}$$

Catti (1978) expanded $F(\mathbf{x}_{ij})$ as:

$$\begin{aligned} F(\mathbf{x}_{ij}) = & \frac{2}{\pi V} \sum_{-\infty \mathbf{h}}^{+\infty} \frac{1}{h^2} \exp\left(-\frac{\pi^2 h^2}{A^2}\right) \cos(2\pi \mathbf{h} \cdot x_{ij}) \\ & + \sum_{-\infty \mathbf{l}}^{+\infty} \frac{\operatorname{erfc}(A|\mathbf{x}_{ij} + \mathbf{l}|)}{|\mathbf{x}_{ij} + \mathbf{l}|} - \frac{2A}{\sqrt{\pi}} \delta_{ij} \quad (\mathbf{l} \neq 0 \quad \text{for} \quad i = j\,) \end{aligned} \tag{8.5.14}$$

where A is a convergence parameter inversely proportional to the width of the Gaussian representing the ionic charge distribution; erfc is the complementary error function, V is the volume of the unit cell, $\mathbf{h}$ is a vector of the reciprocal lattice, δ_{ij} is the Kronecker delta and Σ is a sum over non-centrosymmetric lattice vectors. The electrostatic part of the lattice energy E_C is then expressed as:

$$E_C = \sum_{r=1}^{N}\sum_{s=r}^{N} Z_r Z_s C_{rs} \tag{8.5.15}$$

$$C_{rr} = \frac{N_0 e^2}{2Z} p_r \sum_{m=1}^{p_r} F(\mathbf{x}_{r,m} - \mathbf{x}_{r,1}) \tag{8.5.16}$$

$$C_{rs} = \frac{N_0 e^2}{2Z} p_s \sum_{m=1}^{p_s} F(\mathbf{x}_{s,m} - \mathbf{x}_{s,1}) \tag{8.5.17}$$

where p_r and p_s are the multiplicities of the crystallographic sites occupied by the rth and the sth ions, respectively.

Going back so the simple Halite crystal, the repulsive potential at the central charge is:

$$\Phi_R = 2b\left[\exp\left(\frac{-r}{\rho}\right) + \exp\left(\frac{-2r}{\rho}\right) + \exp\left(\frac{-3r}{\rho}\right) + \ldots\right] \approx 2b\exp\left(\frac{-r}{\rho}\right) \tag{8.5.18}$$

(the effect of repulsive forces vanishes rapidly with increasing distance and, for practical purposes, extended effects beyond the first neighbors at first sight can be neglected). We then have:

$$\Phi_{total} = \frac{-1.38Z_{+-}^2 e^2}{r} + 2b\exp\left(\frac{-r}{\rho}\right) \tag{8.5.19}$$

The ions in the array arrange themselves in order to minimize the bulk energy of the substance. The equilibrium condition is analytically expressed by equating to zero the first derivative of the potential energy over distance:

$$\left(\frac{\partial \Phi_{total}}{\partial r}\right)_{r=r_0} = 0 \tag{8.5.20}$$

Applying the above condition and solving over b we obtain:

$$\Phi_{total(r=r_0)} = \frac{-1.38Z_{+-}^2 e^2}{r_0}\left(1 - \frac{\rho}{r_0}\right) \tag{8.5.21}$$

Equation (8.5.21) is known as *Born–Landé equation.* If we now consider the effect of the crystal lattice, by introducing Madelung constant $\mathcal{M}$ and multiplying by Avogadro's number N_0 (i.e. number of atoms per weight formula unit), we get:

$$\Phi_{\mathrm{total}(r=r_0)} = \frac{-\mathcal{M}N_0Z_{+-}^2e^2}{r_0}\left(1-\frac{\rho}{r_0}\right) \tag{8.5.22}$$

Equation (8.5.22) is the *Born–Mayer equation*.

8.5.5 Lattice Energy and Enthalpy

The enthalpy of crystalline substances is a magnitude of paramount importance in Geochemistry, whose main duty is the computation of chemical reactions among Earth's materials. Understanding the close physical–chemical relationships between enthalpy, and, more specifically the "*enthalpy of formation from the elements at their stable standard state*" and the lattice energy is prodromal to a better comprehension of the eventual ambiguities that may arise from a Quantum Mechanical treatment of the argument.

In calculating the lattice energy of a crystal, we adopt an arbitrary reference condition (two isolated ions in the gaseous state and at an infinite distance), assigning a zero potential to this condition. It is worth stressing that this condition is not equivalent to the standard state commonly adopted in thermochemical calculations, which is normally that of an *element at the stable state at reference P, T.*

The relationships between the two different states and between the "*enthalpy of formation from the elements at standard state*" (H_f^0) and the "*lattice energy*" (U) are easily understood by referring to the Born–Haber–Fayans thermochemical cycle. In this cycle, the formation of a crystalline compound MX from isolated atoms at gaseous state (M and X) is visualized as a stepwise process connecting the various transformations:

$$\begin{array}{ccc} M_{(\mathrm{metal})} + \frac{1}{2}X_{2(\mathrm{gas})} & \xrightarrow{E_s + \frac{1}{2}E_D} & M_{(\mathrm{gas})} + X_{(\mathrm{gas})} \\ \uparrow -H_f^0 & & \downarrow I_M - E_X \\ MX_{(\mathrm{crystal})} & \xleftarrow{-U} & M^+_{(\mathrm{gas})} + X^-_{(\mathrm{gas})} \end{array} \tag{8.5.23}$$

where E_S is the sublimation energy of the metal, E_D is the dissociation energy of the gaseous diatomic molecule X_2, I_M is the ionization potential of the metal M, E_X is the electron affinity of the gaseous atom and U is the Lattice Energy of the MX crystal. Applying the Hess additivity rule to the involved energy terms. we have:

$$H_f^0 = E_S + \frac{1}{2}E_D + I_M - E_X - U \tag{8.5.24}$$

Equation (8.5.24) can be generalized for polyatomic molecules and crystals as follows:

$$H^0_{f,MX_n} = E_S + \frac{n}{2}E_D + \sum_1^n I_M - nE_X - U + (1+n)RT \quad (8.5.25)$$

where $\sum_1^n I_M$ is the summation of the first n ionization potentials of M.

Lastly note that Eq. (8.5.25) includes the expansion work [(1 + n)R)], operated on the gaseous ions, when moving from absolute zero to standard state temperature.

Adopting the equation of Catti (1981), the lattice energy of a ionic crystal may be expressed through the double summations:

$$U = \sum_{i=1}^n \sum_{j=i}^n Z_i Z_j C^{EL}_{ij} + \sum_{i=1}^n \sum_{j=i}^n b_{ij} C^R_{ij} + \sum_{i=1}^n \sum_{j=i}^n dd_{ij} C^{DD}_{ij} + \sum_{i=1}^n \sum_{j=i}^n dq_{ij} C^{DQ}_{ij} \quad (8.5.26)$$

extended to all ions in the lattice. Coefficients b_{ij}, dd_{ij} and dq_{ij} have the same significance shown in Eqs. (8.5.4) and (8.5.7). Coefficients $C^{EL}_{ij}, C^R_{ij}, C^{DD}_{ij}, C^{DQ}_{ij}$ depend on crystal symmetry and on inter-ionic distances and are obtained through standard mathematical routines which cannot be detailed here (see Appendix A in Tosi 1964 for exhaustive treatment). Table 8.2 lists as examples the lattice energy values obtained through application of Eq. (8.5.26) to orthorhombic silicates (olivine, spatial group *Pbnm*). Values are expressed in *kJ/mol*. We note that lattice energy is mainly (about 90%) composed of the attractive coulombic term. Repulsive energy is about 15% of bulk energy and is of opposite sign. Dispersive energy is greatly subordinate (1–3%).

Table 8.3 lists the various energy terms of the Born–Haber–Fayans thermochemical cycle in alkaline-halides (Grenwood 1970). We may note that in this case lattice energy of enthalpy of formation from the elements are not as dissimilar as in silicates due to the lower bonding energy of alkaline-halides.

Table 8.2 Lattice energy of olivine end-members; values in *kJ/mol* (Ottonello 1987). E_{BHF} = Energy of the thermochemical cycle; E_C = coulomb energy; E_R = repulsive energy; E_{DD}, E_{DQ} = dispersive energies

Compound	E_{BHF}	$\overline{H}_{f,T_r,P_r}$	U	E_C	E_R	E_{DD}	E_{DQ}
Mg_2SiO_4	18,456.4	−2175.7	−20,632.1	−24,131.3	3835.4	−259.4	−76.0
Fe_2SiO_4	19,268.4	−1481.6	−20,750.3	−23,934.6	3779.3	−442.1	−152.9
Ca_2SiO_4	17,608.5	−2316.2	−19,925.1	−23,105.4	3546.9	−282.4	−84.3
Mn_2SiO_4	18,825.4	−1728.1	−20,553.5	−23,734.8	3641.3	−348.5	−111.6
Co_2SiO_4	19,443.9	−1406.7	−20,850.6	−24,040.8	3855.4	−490.5	−174.6
Ni_2SiO_4	19,609.9	−1405.2	−21,015.1	−24,219.8	4027.4	−596.4	−226.2
Zn_2SiO_4	19,326.0	−1390.4	−20,716.4	−24,068.0	3875.2	−393.9	−129.6

Table 8.3 Born–Haber–Fayans energy terms for alkali halides. Values in kJ/mol (from Grenwood 1970)

Compound	$-H_f^0$	E_S	$\frac{1}{2}E_D$	$\sum_1^n I_M$	E_X	U
LiCl	408.8	138.9	121.3	520.1	351.5	837.6
NaCl	410.9	100.4	121.3	495.8	360.7	767.8
KCl	436.0	81.2	121.3	418.8	354.8	702.4
RbCl	430.5	77.8	121.3	402.9	354.8	677.8
CsCl	433.0	70.3	121.3	375.7	359.8	640.6

In passing, I emphasize that the electron affinity of chlorine should be constant in all compounds and that observed differences must be imputed to experimental approximation.

8.5.6 Force Field Expressions

The kinds *of* interactions we have seen in the previous paragraphs are essentially appropriate for ionic solids with negligible covalent contributions. In a more accurate model potential reference framework the total potential Φ acting on the crystal lattice may be seen as a function of an inter-atomic spacing vector **R**, accounting for three-body and four-body energy terms as well, besides the simple two-body contributions previously treated. Adopting the formalism of Catti (1996), these contributions may be depicted as:

$$\Phi(\mathbf{R}) = \sum_{ij} E_{ij}(r_{ij}) + \sum_{ijk} E_{ijk}(\varphi_{ijk}) + \sum_{ijkl} E_{ijkl}(\theta_{ijkl}) \tag{8.5.27}$$

with r_{ji} as inter-atomic distance, φ_{ijk} as bond-angle, and θ_{ijkl} as torsional-angle.

In covalent solids the Born–Mayer or Huggins–Mayer two-body potential $E_{ij}(r_{ij})$ is often replaced by a simple quadratic term:

$$E_{ij} = \frac{1}{2} k_r \left(r_{ij} - r_{ij}^0 \right)^2 \tag{8.5.28}$$

where r_{ij}^0 is the (equilibrium) bond distance at the athermal limit and k_r is the (harmonic) force constant for the bond length deformation.

A more precise treatment involves a Morse potential (see Sect. 7.2.5) of the type:

$$E_{ij} = D_{ij}\left\{1 - \exp\left[1 - \alpha\left(r_{ij} - r_{ij}^0\right)\right]\right\}^2 - D_{ij} \tag{8.5.29}$$

where D_{ij} is the two-body bond energy and α is a parameter related again to the force constant k_r (Catti 1996).

The three-body (angular) potential is again in most cases harmonic, with the quadratic parameter φ_{ijk} related to the equilibrium bond-angle φ^0_{ijk} through the bond-bending force constant k_φ:

$$E_{ijk} = \frac{1}{2} k_\varphi \left(\varphi_{ijk} - \varphi^0_{ijk} \right)^2 \tag{8.5.30}$$

8.6 The Quantum Mechanical Treatment

8.6.1 *From the Classical-Heuristic Toward the Quantum–Mechanical Approach*

We have already seen that the Madelung energy is the most relevant form of energy contribution to the stability of a ionic crystal (see Sects. 8.5.4, 8.5.5 and Table 8.2). The Madelung energy decreases algebraically with decreasing lattice constant and "*tends to contract the lattice*" (Born and Huang 1954), or, in other terms, can be "*described as attractive*" (ibid). At short inter-ionic distances however, overlap effects become relevant. These can be depicted as *perturbative* terms. The first attempts to quantify quantum mechanically the effects of overlap in crystals may be ascribed to Lenz (1932), Jensen (1932, 1936a,b), and Gombàs (1943). Jensen, particularly, developed in this sense the Thomas–Fermi–Dirac procedure, introducing in the computation of lattice energy (besides the classical coulomb energy and the zero point kinetic energy of the electron $-\frac{3^{5/3}h^2}{40m_0}\left(\frac{1}{\pi}\right)^{2/3}\rho^{5/3}$ with m_0 as rest mass of the electron) the exchange energy, through the Dirac's exchange term per unit volume $-\frac{3^{4/3}e^2}{4}\left(\frac{1}{\pi}\right)^{1/2}\rho^{4/3}$. This is perhaps the first attempt to bring together the quantum–mechanical approach with the classical views of the two-body interaction in crystals. According to Jensen, two distinct effects combine in determining the overlap energy per unit volume (E_{ov}, hereafter):

(1) The zero-point kinetic energy is given by the superposed densities of ions 1 and 2 (ρ_1, ρ_2), based on

$$E_0 = \frac{3^{5/3}h^2}{40m_0}\left(\frac{1}{\pi}\right)^{2/3}(\rho_1 + \rho_2)^{5/3} \tag{8.6.1}$$

instead of the simple summation

$$E_0' = \frac{3^{5/3}h^2}{40m_0}\left(\frac{1}{\pi}\right)^{2/3}\left(\rho_1^{5/3} + \rho_2^{5/3}\right) \tag{8.6.2}$$

hence:

$$E_{0V} = E_0 - E_0' = \frac{3^{5/3}h^2}{40m_0}\left(\frac{1}{\pi}\right)^{2/3}\left[(\rho_1 + \rho_2)^{5/3} - \left(\rho_1^{5/3} + \rho_2^{5/3}\right)\right] \tag{8.6.3}$$

E_{0V} is positive, leading to an interionic repulsion the more important the wider is the overlap region.

(2) In a similar way, the difference in the exchange energy per unit volume (E_{xV}) is given by:

$$E_{xV} = E_x - E_x' = -\frac{3^{4/3}e^2}{4}\left(\frac{1}{\pi}\right)^{1/2}\left[(\rho_1 + \rho_2)^{4/3} - \left(\rho_1^{4/3} + \rho_2^{4/3}\right)\right] \tag{8.6.4}$$

This form of energy is negative and contributes to the stability of the crystal.

It should be noted that distinct overlap regions may occur, i.e. between positive and negative 1st neighbors an between negative 2nd neighbors as shown schematically for RbI in Fig. 8.5.

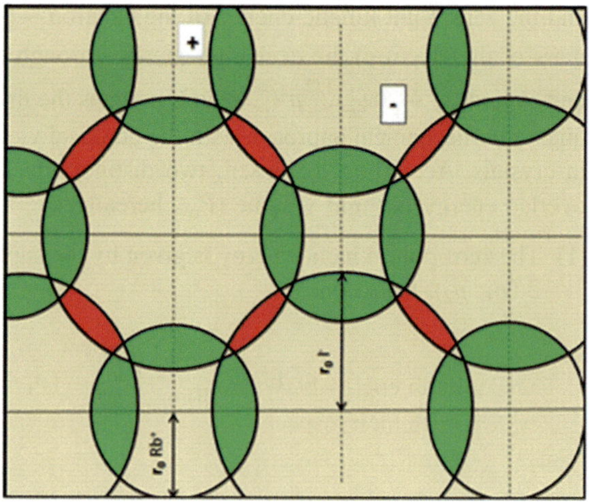

Fig. 8.5 Ions in cubic planes of RbI. The green zones are 1st neighbors (cation–anion) overlap zones and the red ones are 2nd neighbors (anion–anion) overlap zones. Modified from Jensen (1932)

Table 8.4 Radii of free ions (r_0) and "*additive radii*" (r_j) after Jensen (1932). Values are in a.u. (i.e. multiples of Bohr's radius a_0)

	Na^+	K^+	Rb^+	Cs^+
r_0 (A^+)	2.85	3.25	3.65	3.85
r_j (A^+)	2.08	2.3	2.53	2.65
	F^-	Cl^-	Br^-	I^-
r_0 (B^-)	5.1	5.35	5.5	5.65
r_j (B^-)	4.06	4.41	4.64	4.85

The "*penetration distance*" between ions A^+ and B^- (s_{A+B-}) is simply given by (Jensen 1932):

$$s_{A^+B^-} = r_j(A^+) + r_j(B^-) - r_0(A^+B^-) \tag{8.6.5}$$

with r_0 as nearest neighbor distance. (s_{A+B-}) may be easily computed for alkali-halides adopting the values of free ions and additive-radii of Jensen (Table 8.4).

All energies, besides the Madelung contribution, have a very short range and their summation at a short distance is always repulsive. They are basically an expression of Pauli's exclusion principle. But the application of Pauli's principle requires a rearrangement in the free ion wave functions, so that the separation of the various physical factors in terms of discrete contributions, as above outlined, has only a symbolic value.

In the case of *van der Waals* crystals (i.e. crystals conceived as consisting of electrically neutral molecules), the Madelung term disappears but, if the molecules are not spherically symmetrical, the electrostatic interactions are not absent and the cohesion is assured by multipole interactions of 2nd (dipole–dipole), 3rd (dipole–quadrupole) and 4th order (quadrupole–quadrupole).

Denoting e_i the element of charge at the position x^i (taken with respect to a suitable center as origin) the multipole moments can be defined as a vector and tensors of higher orders:

$$m_\alpha = \sum_i e_i x^i_\alpha \tag{8.6.6}$$

$$q_{\alpha\beta} = \sum_i e_i x^i_\alpha x^i_\beta \tag{8.6.7}$$

The Coulomb energy between two neutral molecules 1 and 2 with moments m^1_α, $q^1_{\alpha\beta}$, m^2_α, $q^2_{\alpha\beta}$ with their respective centers connected by vector $\mathbf{R}$ is (Born and Huang 1954):

$$
\begin{aligned}
E_C = & \sum_{\alpha\beta} m_\alpha^1 m_\beta^2 \left[\frac{\partial^2}{\partial y_\alpha \partial y_\beta}\frac{1}{|y|}\right]_{y=\mathbf{R}} - \frac{1}{2} \\
& \sum_{\alpha\beta\gamma}\left(m_\alpha^1 q_{\beta\gamma}^2 - m_\alpha^2 q_{\beta\gamma}^1\right)\left[\frac{\partial^3}{\partial y_\alpha \partial y_\beta \partial y_\gamma}\frac{1}{|y|}\right]_{y=\mathbf{R}} \\
& + \frac{1}{4}\sum_{\alpha\beta\gamma\lambda} q_{\alpha\beta}^1 q_{\gamma\lambda}^2 \left[\frac{\partial^4}{\partial y_\alpha \partial y_\beta \partial y_\gamma \partial y\lambda}\frac{1}{|y|}\right]_{y=\mathbf{R}} - \ldots
\end{aligned}
\tag{8.6.8}
$$

Considering e_i and x^i as referred to the charge and position vector of the electrons, E_C can be eventually regarded as a quantum mechanical operator and the mean value of Eq. (8.6.8) may be obtained through perturbation methods with respect to the unperturbed wave function of the system composed of the molecules 1 and 2. The 1st order perturbed wave function of the molecule 1 placed in a constant electric field $\mathcal{E}$ is (Born and Huang 1954):

$$
\psi_i + \sum\nolimits'_{j'} \frac{-\left(\sum_\alpha |\langle i \mid m_\alpha^1 | i'\rangle| \mathcal{E}_\alpha\right)}{E_i - E_{i'}}
\tag{8.6.9}
$$

with

$$
\sum_\alpha m_\alpha^1 \mathcal{E}_\alpha = \sum_{\alpha,i} e_i x_\alpha^i \mathcal{E}_\alpha
\tag{8.6.10}
$$

and corresponds to the static Coulomb interaction. The 2nd order energy $E^{(2)}$ attains the value (Born and Huang 1954):

$$
E^{(2)} = -\frac{2B}{3R^6} \quad (R = |\mathbf{R}|)
\tag{8.6.11}
$$

with

$$
B = -\sum\nolimits'_{i'} \sum\nolimits'_{j'} \frac{\left(\sum_\alpha |\langle i \mid m_\alpha^1 | i'\rangle|^2\right)\left(\sum_\beta \left|\langle j \mid m_\beta^2 | j'\rangle\right|^2\right)}{E_i + E_j - E_{i'} - E_{j'}}
\tag{8.6.12}
$$

i and j in Eq. (8.6.12) designate the electronic states of the unperturbed molecules and i′, j′ those of the perturbed ones. Moreover, the primes exclude i′ = i and j′ = j. E_i, E_j, $E_{i'}$, $E_{j'}$ are eigenvalues of the unperturbed molecules and $\langle i \mid m_\alpha^1 | i'\rangle$ and $\langle j \mid m_\beta^2 | j'\rangle$ are the matrix elements of the dipoles between the states i and i′ in molecule 1 and between the states j and j′ in molecule 2.

Considering only the component of the induced moment in the direction of the field and averaging its value over various orientations of the molecule one obtains (Born and Huang 1954):

$$\frac{2}{3}+\frac{\sum_{j'}'\sum_{\alpha}\left|\langle i \mid m_{\alpha}^{1}|i'\rangle\right|^{2}}{\Delta_1}|\mathcal{E}|=\alpha_1|\mathcal{E}| \tag{8.6.13}$$

where Δ_1 is the excitation energy needed to pass from state i to state i′ and α_1 is the polarizability of molecule 1.

Simplifying Eq. (8.6.12) as

$$B\cong\sum_{i'}'\sum_{j'}'\frac{\left(\sum_{\alpha}\left|\langle i \mid m_{\alpha}^{1}|i'\rangle\right|^{2}\right)\left(\sum_{\beta}\left|\langle j \mid m_{\beta}^{2}|j'\rangle\right|^{2}\right)}{\Delta_1+\Delta_2} \tag{8.6.14}$$

one obtains the London approximation for the van der Waals potential E_{vdW} (London 1930; Born and Huang 1954):

$$E_{vdW}=\frac{C}{R^6}\quad C=-\tfrac{2}{3}B=-\frac{3}{2}\left(\frac{\Delta_1\Delta_2}{\Delta_1+\Delta_2}\right)\alpha_1\alpha_2 \tag{8.6.15}$$

All the above resumed treatments have their own intrinsic validity and the quantum–mechanical approach must lead to mutually consistent computational results. To achieve this goal, however, it is necessary to change partially the reference framework in terms of nuclear and electronic charges.

At the light of the Born–Oppenheimer approximation, the electrons conform instantaneously to changes in the relative positions of the nuclei. It follows that the motions of electrons and nuclei are uncoupled and the total energy of the system may be depicted in a parametric way as a summation of mutual electron–electron (Φ_{ee}), electron-nucleus (Φ_{en}) and nucleus-nucleus (Φ_{nn}) terms dependent only on the nuclear position vector **R**:

$$\Phi(\mathbf{R})=\Phi_{ee}+\Phi_{en}+\Phi_{nn} \tag{8.6.16}$$

As we have already seen for many-electron atoms and molecules, also in the Quantum Mechanical investigation of periodic structures, the first concern is the choice of the electronic Hamiltonian. In most circumstances this Hamiltonian is non-relativistic, though, in some cases, this is a poor approximation, especially for substances containing heavy elements. In Geochemistry, for example, native Hg and Hg-compounds are important. Hg, with Z = 80, has 1s electrons with a mean square velocity which is $v \approx 0.58 \times c$. The corresponding relativistic mass of the 1*s* electrons is $m = 1.23 \times m_0$ and hence the 1s orbital is contracted of the same amount (see also Sect. 6.1.4).

The non-relativistic Hamiltonian may be represented as:

$$\mathsf{H}^{\mathsf{nr}} = -\sum_{i}\frac{\nabla_i^2}{2} - \sum_{i,A}\frac{Z_A}{r_{iA}} + \sum_{i>j}\frac{1}{r_{ij}} + \sum_{A>B}\frac{Z_A Z_B}{r_{AB}} \tag{8.6.17}$$

where the first term is the kinetic energy, the second one is the electrons-nuclei interaction, the third one is the inter-electronic interaction and the fourth one is the inter-nuclear interaction. The knowledge of the Hamiltonian and of the total number of electrons is sufficient to depict both static and time-dependent properties of the crystal. In many circumstances, moreover, it may be assumed the crystal to be neutral; i.e. the number of electronic charges compensates the nuclear charges of translationally inequivalent nuclear centers.

If one could solve the Schrödinger equation

$$\mathsf{H}^{\mathsf{nr}}\Psi_0 = E_0(\mathbf{R})\Psi_0 \tag{8.6.18}$$

and obtain the lowest eigenvalue of a given **R** configuration then, he would know in a straightforward way both the ground state energy and the structure at the same time. However, being Ψ_0 an anti-symmetric function of 3nN′ spatial coordinates and nN′ spin coordinates (with N′ = number of crystalline cells), and being the inter-electronic interaction non-factorizable into nN′ problems for individual electrons, this is unfeasible (see Pisani 1996a; for a more extended discussion).

8.6.2 The Thomas–Fermi Model and the Uniform Electron Gas (UEG) Treatment

The simplest form one may conceive to reproduce the effects of a periodic spacing of nuclear charges on the electronic charges is the uniform electron gas treatment, which, in some characteristics, is a reasonable approximation for light metals. In this model, not only the nuclear charges are regularly spaced, but also the electron density is uniform in space. This simplification allows to put in evidence the effects arising from the mutual repulsive interaction of electrons among themselves disregarding the actual atomic lattice. It is thus often adopted as a simple reference model for metals. The UEG is also known as Jellium (i.e. "*a positively charged jelly-like background*"; according to Conyers Herring). An extended discussion on the UEG may be found in Hughes (2007). The formalism itself provides the basis for the Local Density Approximation of the exchange–correlation energy in the Density Functional Theory that we will briefly visit later on.

In the Thomas–Fermi model (Thomas 1927; Fermi 1927, 1928), the electrons are assumed to be distributed uniformly in each small volume element ΔV of the jellium (i.e. locally) but can vary from one small volume to another. According to the Thomas–Fermi model, the bulk kinetic energy of the electron gas (E_K) is given by

$$E_K = C_K \int [\rho(\mathbf{r})]^{5/3} d^3\mathbf{r} \tag{8.6.19}$$

$$C_K = \frac{3\hbar^2}{40m_e}\left(\frac{3}{\pi}\right)^{2/3} \tag{8.6.20}$$

The potential energy of an electron arising from the nuclear attraction (E_{eN}) is:

$$E_{eN} = \int \rho(\mathbf{r}) V_N(\mathbf{r}) d^3\mathbf{r} \tag{8.6.21}$$

where V_N is the potential energy exerted on the electron at position r by a nucleus of charge *Ze*, i.e.:

$$V_N(\mathbf{r}) = -\frac{Ze^2}{\mathbf{r}} \tag{8.6.22}$$

Finally, the potential energy of the electron due to their mutual electric repulsion (E_{ee}) is:

$$E_{ee} = \frac{e^2}{2}\int \frac{\rho(\mathbf{r})\rho(\mathbf{r}')}{|\mathbf{r}-\mathbf{r}'|} d^3\mathbf{r} d^3\mathbf{r}' \tag{8.6.23}$$

and the bulk energy is $E = E_K + E_{eN} + E_{ee}$.

For this highly idealized scheme, the unique parameter characterizing the electrons is the (uniform) density ρ expressed in terms of Wigner–Seitz radius (r_s):

$$\rho = \frac{4}{3}\pi r_s^3 \Rightarrow r_s = \frac{1}{\sqrt{\frac{4}{3}\pi\rho}} \tag{8.6.24}$$

Classically, for a gas composed of two electrons moving in a uniform positive charge field, the electrostatic self-interaction and the interaction with the positive charges cancel out. Taking into account that the wave functions of the two electrons must be anti-symmetric, the energy of the individual electrons reduce to (Pisani 1996a):

$$\left[-\frac{1}{2}\nabla^2 + (v_x)_i\right]\psi_i(\mathbf{r}) = \varepsilon_i \psi_i(\mathbf{r}) \tag{8.6.25}$$

where the exchange term $(v_x)_i$ is zero for the free electron approximation or, else, a constant (independent of **r** for symmetry reasons) in a HF approach (see later).

The eigenvectors of the above equation are plane waves (Sect. 8.4) and the eigenvalues obey:

$$\varepsilon_i \equiv \varepsilon(\mathbf{k}) = \frac{1}{2}k^2 + v_x(k) \tag{8.6.26}$$

Orbitals are progressively occupied following the Aufbau principle, filling a spherical zone centered in $\mathbf{k} = 0$ with radius $k_F = 1/(0.521 \times r_s)$. The HF value of $v_x(\mathbf{k})$ is (Pisani 1996a):

$$v_x(\mathbf{k}) = \frac{k_F}{2\pi}F\left(\frac{k}{k_F}\right) \tag{8.6.27}$$

The ground state energy in the free electron gas approximation is purely kinetic:

$$\varepsilon_{kin}(\rho) = \frac{3}{10}\left(3\pi^2\right)^{2/3} \tag{8.6.28}$$

while in the HF approach, there is in addition an exchange term $\varepsilon_x(r)$, such that:

$$\varepsilon_{HF}(\rho) = \varepsilon_{kin}(\rho) + \varepsilon_x(\rho) = \frac{3}{10}\left(3\pi^2\right)^{2/3} - \frac{3}{4}\left(\frac{3\rho}{\pi}\right)^{1/3} \tag{8.6.29}$$

The application to alkali metals, using the free electron levels [$\varepsilon(\mathbf{k}) = k^2/2$], gives a satisfactory reproduction of many observables, such as the low-T heat capacity, the electrical conductivity and the spin susceptibility but, in general, the fact that correlation terms are neglected leads to serious discrepancies in the evaluation of the lattice energy (Fig. 8.6). Estimates of the correlation energy for the electron gas have been provided for both high densities ($r_s < 1$) and low density ($r_s > 1$) metals (Gell-Mann and Bruckner 1957; Ceperley 1978, 1980).

8.6.3 Hartree–Fock Approach

In the Hartree–Fock approach (HF) Ψ_0 of Eq. (8.6.18) is approximated through a Slater determinant corresponding to the anti-symmetrized product of nN′ one-electron functions. The one-electron Fock Hamiltonian f is (Pisani 1996a):

$$f\psi_i(\mathbf{r}) = \left[-\frac{\nabla^2}{2} + v(\mathbf{r}) + \int \frac{d\mathbf{r}'\rho(\mathbf{r}')}{|\mathbf{r}-\mathbf{r}'|}\right] - \sum_j \int \frac{d\mathbf{r}'\left[\psi_j(\mathbf{r}')^*\psi_i(\mathbf{r}')\right]}{|\mathbf{r}-\mathbf{r}'|}\psi_j(\mathbf{r}) = \varepsilon_i\psi_i(\mathbf{r}) \tag{8.6.30}$$

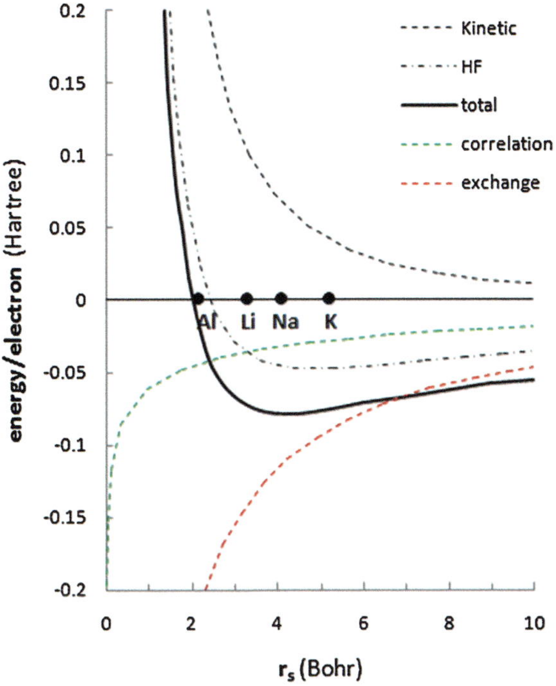

Fig. 8.6 The various forms of energy contributing to the bulk energy per electron in the electron gas approach, plotted as a function of the Wigner–Seitz radius r_s. The r_s value of the valence electrons in a few simple metals is also reported. Modified from Pisani (1996a)

with the first three terms replacing the two terms of the electron gas treatment, and the fourth one representing the exact non-local exchange operator.

The HF energy (E_{HF}) may be defined as "*the expectation value of the non-relativistic Hamiltonian* H^{nr}" (Pisani 1996a):

$$\begin{aligned} E_{HF} = \langle \Psi_{HF} | \mathsf{H}^{nr} | \Psi_{HF} \rangle = 2\sum_{i} \int d\mathbf{r} \{ \psi_i(\mathbf{r})^* + \left[-\nabla^2/2 + V(\mathbf{r}) \right] \psi_i(\mathbf{r}) \} + \\ + \sum_{ij} \int d\mathbf{r} d\mathbf{r}' [\psi_i(\mathbf{r})^* \psi_i(\mathbf{r})] \left[\psi_j(\mathbf{r}')^* \psi_j(\mathbf{r}') \right] / |\mathbf{r} - \mathbf{r}'| + \\ - \frac{1}{2} \sum_{ij} \int d\mathbf{r} d\mathbf{r}' \left[\psi_i(\mathbf{r})^* \psi_j(\mathbf{r}) \right] \left[\psi_j(\mathbf{r}')^* \psi_i(\mathbf{r}')^* \right] / |\mathbf{r} - \mathbf{r}'| \end{aligned} \tag{8.6.31}$$

In ferromagnetic and anti-ferromagnetic solids one cannot obviously adopt the closed-shell condition and it is necessary to choose a linear combination of Slater determinants in order to get the correct spin eigenfunctions. In the Restricted Open Hartree–Fock approach (ROHF) the wave function is a sum of Slater determinants containing a *closed-shell* subset (double electronic occupancy) and an *open-shell* subset formed by orbitals containing a single electron. Whenever it is possible to define two sets of orbitals: the first occupied by paired electrons and the second by

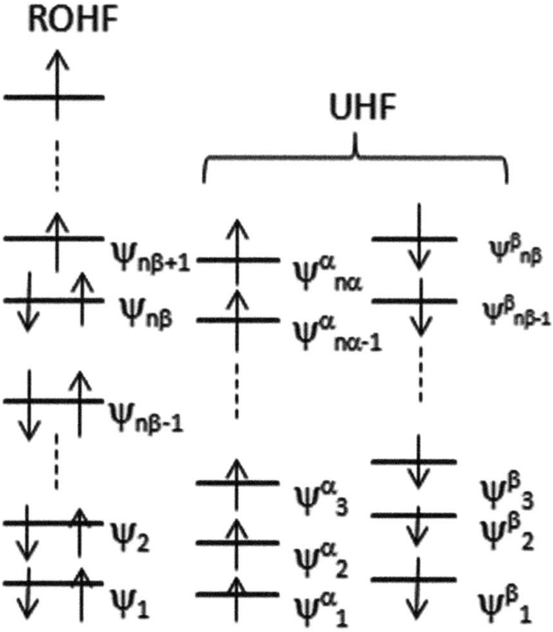

Fig. 8.7 Molecular orbital diagram for the Restricted Open-shell Hartree–Fock method (ROHF, left) and Unrestricted Hartree–Fock method (UHF, right). Modified from Aprà (1996)

electrons with parallel spin (*half-closed-shell* cases), then the ROHF wave function reduces to a single determinant (Aprà 1996). In the Unrestricted Hartree–Fock approach (UHF) the constraint of double occupancy is absent and the α electrons are allowed to occupy orbitals non-occupied by β electrons. This greater variational freedom produces wave functions that are energetically more stable than the corresponding ROHF. Moreover, the UHF procedures allow the description of anti-ferromagnetic materials possessing a locally negative spin density (Aprà 1996). A schematic description of the energy levels in a ROHF and UHF approach is given in Fig. 8.7.

In the UHF the (mono-determinantal) wave function involves n_α MOs and n_β MOs, corresponding to the electrons with α and β spin, respectively (with wave functions ψ_α and ψ_β). The wave function Ψ describing a system with $N = n_\alpha + n_\beta$ electrons has the form (Aprà 1996):

$$\Psi = (N!)^{-1/2}\det\left\{\begin{array}{c} \psi_1^\alpha(1)\alpha(1),\ldots,\psi_{n\alpha}^\alpha(n_\alpha)\alpha(n_\alpha)\cdot \\ \psi_1^\beta(n_\alpha+1)\beta(n_\alpha+1),\ldots,\psi_{n\beta}^\beta(N)\beta(N) \end{array}\right\} \tag{8.6.32}$$

The UHF energy is given by:

$$\begin{aligned} E_{UHF} &= \langle \Psi_{UHF}|H^{nr}|\Psi_{UHF}\rangle \\ &= \sum_k \psi_k^\alpha \Big| h + \frac{1}{2}(J - K^\alpha)|\psi_k^\beta + \sum_m \psi_m^\beta \Big| h + \frac{1}{2}(J - K^b)\Big)|\psi_m^\beta \end{aligned} \tag{8.6.33}$$

where the J and K^{α}, K^{β} functionals may be defined as Coulomb and resonance integrals, respectively (cf. Sects. 6.4.2 and 7.3.6):

$$J(\mathbf{r}_1)\psi(\mathbf{r}_1) = \sum_i \left[\int d\mathbf{r}_2 \psi_i^*(\mathbf{r}_2)\frac{1}{r_{12}}\psi_i(\mathbf{r}_2)\right]\psi(\mathbf{r}_1) \tag{8.6.34}$$

$$K^{\alpha}(\mathbf{r}_1)\psi(\mathbf{r}_1) = \sum_i \left[\int d\mathbf{r}_2 \psi_i^{\alpha *}(\mathbf{r}_2)\frac{1}{r_{12}}\psi(\mathbf{r}_2)\right]\psi_i^{\alpha}(\mathbf{r}_1) \tag{8.6.35}$$

$$K^{\beta}(\mathbf{r}_1)\psi(\mathbf{r}_1) = \sum_i \left[\int d\mathbf{r}_2 \psi_i^{\beta *}(\mathbf{r}_2)\frac{1}{r_{12}}\psi(\mathbf{r}_2)\right]\psi_i^{\beta}(\mathbf{r}_1) \tag{8.6.36}$$

In the investigation of crystalline materials the Molecular Orbitals (MO) are conveniently replaced by Crystalline Orbitals $\psi_i(\mathbf{r}, \mathbf{k})$ (CO; Roetti 1996). A CO is a linear combination of Block Functions $\phi_{\mu}(\mathrm{r}, \mathrm{k})$, defined in terms of local functions (which may be referred to as "*atomic orbitals*" AO; Roetti 1996) which, in their turn are generated by a certain number n_G of individually-normalized Gaussian functions G:

$$\psi_i(\mathbf{r}, \mathbf{k}) = \sum_{\mu} a_{\mu,i}(\mathbf{k})\phi_{\mu}(\mathbf{r}, \mathbf{k}) \tag{8.6.37}$$

$$\phi_{\mu}(\mathbf{r}, \mathbf{k}) = \sum_{\mathbf{T}} \varphi_{\mu}(\mathbf{r} - \mathbf{A}_{\mu} - \mathbf{T}) \times \exp(i\mathbf{k} \cdot \mathbf{T}) \tag{8.6.38}$$

$$\varphi_{\mu}(\mathbf{r} - \mathbf{A}_{\mu} - \mathbf{T}) = \sum_{j=1}^{n_G} d_j G(\alpha_j; \mathbf{r} - \mathbf{A}_{\mu} - \mathbf{T}) \tag{8.6.39}$$

k in Eqs. (8.6.37) and (8.6.38) is a reciprocal lattice vector; $a_{\mu,i}$ in Eq. (8.6.37) is the weighting coefficient for the μth function in the ith orbital; A_{μ} is the coordinate of the nucleus in the zero reference cell (on which φ_{μ} is centered). d_j and α_j are, respectively, coefficient and exponent of the jth Gaussian set. The summation in **T** in Eq. (8.6.38) is extended to the set of lattice vectors.

The bulk Coulomb bi-electronic contribution to the total energy (E_{ee}^{coul}) of the crystal may be seen an extended summation of bi-electronic integral terms (Roetti 1996):

$$E_{ee}^{coul} = \frac{1}{2}\sum_{1,2}\sum_{\mathbf{T}} P_{12}^{\mathbf{T}} \sum_{3,4}\sum_{\mathbf{Q}} P_{34}^{\mathbf{Q}} \sum_{\mathbf{S}} \left[\varphi_1^{\mathbf{0}}\varphi_2^{\mathbf{T}} | \varphi_3^{\mathbf{S}}\varphi_4^{\mathbf{S}+\mathbf{Q}}\right] \tag{8.6.40}$$

The first four summation terms (1, 2, 3 and 4) involve the AOs of the unit cell and the other three (**T**, **S**, and **Q**) span the infinite set of translation vectors). P is the density matrix. In the adopted notation, moreover, $\varphi_2^{\mathbf{T}}(\mathrm{r})$ is the AO2 in cell **T**. The **S** summation is related to the distance between two interacting distributions and is treated through a multipole expansion (Roetti 1996). Denoting as $C_{12}^{\mathbf{T}}$ and $Z_{12}^{\mathbf{T}}$, respectively the electron–electron and electron–nuclei series, and rearranging, one has:

$$C_{12}^{\mathbf{T}} + Z_{12}^{\mathbf{T}} = \sum_{\lambda}\sum_{\mathbf{S}} \int d\mathbf{r} \int d\mathbf{r}' \varphi_1^0(\mathbf{r})\varphi_2^{\mathbf{T}}(\mathbf{r})|\mathbf{r} - \mathbf{r}' - \mathbf{S}|^{-1} \rho_\lambda(\mathbf{r}' - \mathbf{S}) \quad (8.6.41)$$

where $\rho_\lambda(\mathbf{r} - \mathbf{S})$ is the Mulliken shell net charge distribution:

$$\rho_\lambda(\mathbf{r} - \mathbf{S}) \equiv \{\lambda\}' \equiv \{\lambda\} - Z_\lambda = \sum_{3\in\lambda}\sum_{4\mathbf{Q}} P_{34}^{\mathbf{Q}} \varphi_3(\mathbf{r} - \mathbf{S})\varphi_4(\mathbf{r} - \mathbf{S} - \mathbf{Q}) - Z_\lambda \quad (8.6.42)$$

In (8.6.42), $\{\lambda\}$ is the electronic charge distribution of shell λ and Z_λ is the fraction of nuclear charge formally attributed to it.[4]

The exchange–correlation contribution to the total energy is (Roetti 1996):

$$E_{xc} = \frac{1}{2}\sum_{12}\sum_{\mathbf{T}} P_{12}^{\mathbf{T}} \left[-\frac{1}{2}\sum_{34}\sum_{\mathbf{Q}} P_{34}^{\mathbf{Q}} \sum_{\mathbf{S}} [\varphi_1^{\mathbf{0}}\varphi_2^{\mathbf{T}}|\varphi_3^{\mathbf{S}}\varphi_4^{\mathbf{S}+\mathbf{Q}}] \right] \quad (8.6.43)$$

The term in square brackets is the exchange contribution to the 12 **T** element of the direct space Fock matrix. This series is rapidly converging and does not need particular rearrangements to be brought to completion (Roetti 1996).

8.6.4 Density Functional Approach

Interestingly, the Hohenberg–Kohn theorem (Hohenberg and Kohn 1964) which is at the core of the Density Functional Theory (DFT) was presented by the authors as the correct approach to an "*inhomogeneous electron gas*". It follows basically the pioneering Thomas–Fermi method, in which the electronic density $\rho(\mathbf{r})$ plays a central role and the system of eletrons is pictured more like a classical liquid. The theorem concerns the ground state of an interacting electron gas in an external potential $v(r)$ and proves that "*there exists a universal functional of the density, F* $[\rho(\mathbf{r})]$, *independent of* $v(\mathbf{r})$, *such that the expression* $E = \int v(\mathbf{r})\rho(\mathbf{r})d\mathbf{r} + F[\rho(\mathbf{r})]$ *has as its minimum value the correct ground state energy associated with* $\rho(\mathbf{r})$" (Hohenberg and Kohn 1964).

[4] The AOs belonging to a given atom are grouped into shells λ containing all AOs with the same quantum numbers n and l or all the AOs with the same quantum number n in the case of sp shells (sp constraint).

To compute the ground state energy (E_0), which is uniquely determined by the electron density at ground state [$\rho_0(\mathbf{r})$], one must follow the variational Kohn–Sham procedure (Kohn and Sham 1965). Following Pisani (1996a) we may define the procedure in few steps:

(1) Find the *n*/2 lowest eigenvalues (and the corresponding eigenfunctions of a one-electron effective Hamiltonian h_{eff}

$$\mathsf{h}_{eff}\psi_i(\mathbf{r}) \equiv \left[-\frac{\nabla^2}{2} + V_{eff}(\mathbf{r})\right]\psi_i(\mathbf{r}) = \varepsilon_i\psi_i(\mathbf{r}) \tag{8.6.44}$$

(2) Calculate the density:

$$\rho(\mathbf{r}) = 2\sum_i |\psi_i(\mathbf{r})|^2 \tag{8.6.45}$$

(3) Re-compute $V_{eff}(\mathbf{r})$ as a functional of the density $\rho(\mathbf{r})$:

$$V_{eff}(\mathbf{r}) = V(\mathbf{r}) + \int \frac{\rho(\mathbf{r}')}{|\mathbf{r}-\mathbf{r}'|}d\mathbf{r}' + \mu_{exc}(\mathbf{r};[\rho]) \tag{8.6.46}$$

(4) Replace $V_{eff}(\mathbf{r})$ in (8.6.44) and repeat the calculus, until convergence is attained.

The procedure correspond to minimizing an energy expression with respect to an arbitrary function $\rho(\mathbf{r})$ which represent a n-electrons density:

$$\begin{aligned} E_{[\rho(\mathbf{r})]} &= \int \rho(\mathbf{r})V(\mathbf{r})d\mathbf{r} + \frac{1}{2}\int\frac{[\rho(\mathbf{r})\rho(\mathbf{r}')]}{|\mathbf{r}-\mathbf{r}'|}d\mathbf{r}d\mathbf{r}' + \int \rho(\mathbf{r})g(\mathbf{r};[\rho])d\mathbf{r} + \sum_{A>B}\frac{Z_A Z_B}{r_{AB}} \\ &= E_{Ext} + E_J + E_{k-e-c} + E_{NR} \end{aligned} \tag{8.6.47}$$

The first term (E_{Ext}) in Eq. (8.6.47) is the interaction energy with the external potential. The second term in the same equation (E_J) is the self-interaction, and the third term (E_{k-e-c}), which contains the g($\mathbf{r}$) function "*may be interpreted as a density of kinetic, exchange and correlation energy per unit electron density at* **r**" (Pisani 1996a). The g($\mathbf{r}$) function is also known as "*universal functional*" of the density ρ, because it uniquely depends upon the whole charge distribution (Pisani 1996a). The fourth term in (8.6.47) (E_{NR}) is the nuclear repulsion. There are in general two kinds of approximations to g($\mathbf{r}$), whose analytical form is not known. The Local Density approximation (LDA) and the General Gradient Approximation (GGA). In LDA, g($\mathbf{r}$) is considered to depend only on the value of ρ at **r**.

The minimum energy E_0 is obtained in correspondence of $\rho_0(\mathbf{r})$:

$$E_0 = \int \rho_0(\mathbf{r})V(\mathbf{r})d\mathbf{r} + \frac{1}{2}\int \frac{[\rho_0(\mathbf{r})\rho_0(\mathbf{r}')]}{|\mathbf{r}-\mathbf{r}'|}d\mathbf{r}d\mathbf{r}' + E_{kps} + E_{xc}^{LDA} + E_{Rep\,Nuc} \tag{8.6.48}$$

$$E_{kps} = \sum_i \int \psi_i(\mathbf{r})^* \left(-\frac{\nabla^2}{2}\right)\psi_i(\mathbf{r})d\mathbf{r} \tag{8.6.49}$$

$$E_{xc}^{LDA}(\rho) = \int \rho_0(\mathbf{r})\varepsilon_{xc}(\mathbf{r}[\boldsymbol{\rho_0}])d\mathbf{r} \tag{8.6.50}$$

In magnetic system E_{xc}^{LDA} is replaced by the Local Spin Density Approximation E_{xc}^{LSDA} which is a function of the local electron spin densities $\rho\uparrow$, $\rho\downarrow$.

$$E_{xc}^{LSDA}(\rho_\uparrow, \rho_\downarrow) = \int \rho_0(\mathbf{r})\varepsilon_{xc}(\mathbf{r};[\rho_\uparrow, \rho_\downarrow])d\mathbf{r} \tag{8.6.51}$$

In many circumstances the LDA and LSDA approximations do not work properly and are conveniently replaced by GGA, which include gradient corrections to the density:

$$E_{xc}^{GGA}(\rho) = \int f[\rho(\mathbf{r}), \nabla\rho(\mathbf{r};)]d\mathbf{r} \tag{8.6.52}$$

In comparison with LDA and LSDA, GGA's tend to improve total energies, atomization energies, energy barriers, and structural energy differences (Perdew et al. 1996; and references therein), while the relative effect on bond strength is somewhat contradictory (Ozoliņš and Körling 1993; Filippi et al. 1994). Various GGA functional of large interest in Geochemistry are introduced at Sect. 10.5.

8.7 Lattice Dynamics

8.7.1 Vibrations of a Diatomic Chain

I begin obviously with the Born–von Karman treatment (Born and von Karman 1912), due to its seminal importance in solid state physics, and, particularly, with the pictorial representation of a linear diatomic chain subjected to linear displacements provided by Born and Huang (1954) (Fig. 8.8).

The chain is composed of particles of the same kind placed at distance *s* and evenly altered by particles of a different kind. The first neighbors are thus placed at a distance *s*/2.

Considering the chain as an elastic string, the attractive force between first neighbors may be equated to the tension $\dot{\Phi}(s/2)$. Extending uniformly the chain by

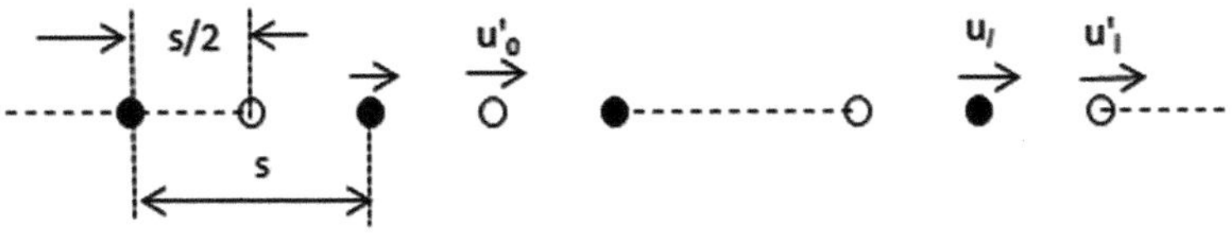

Fig. 8.8 Linear diatomic chain. Modified from Born and Huang (1954)

δs, the tension is increased by $\frac{1}{2}\delta s\ddot{\Phi}(s/2)$ The Young's modulus E_Y corresponds to the increase of tension divided by the extension per unit length (δ/s):

$$E_Y = \frac{1}{2}s\ddot{\Phi}(s/2) \tag{8.7.1}$$

The chain may be seen as a linear lattice with two particles per unit cell. Labeling the different cells by an integer index l and denoting the displacements of the two particles from the evenly spaced configuration in cell l as u_l and u'_l, their equations of motion take the form (Born and Huang 1954):

$$m\ddot{u}_l = \dot{\Phi}\left(u'_l - u_l + \frac{1}{2}s\right) - \dot{\Phi}\left(u_l - u'_{l-1} + \frac{1}{2}s\right) \tag{8.7.2.1}$$

$$m'\ddot{u}'_l = \dot{\Phi}\left(u_{l+1} - u'_l + \frac{1}{2}s\right) - \dot{\Phi}\left(u'_l - u_l + \frac{1}{2}s\right) \tag{8.7.2.2}$$

with m and m′ as masses of the two particles and $\ddot{u} \equiv \partial^2 u/\partial t^2$.

Developing the potential function in Taylor's series and neglecting the 2nd and higher order terms, which are appropriate for small vibrations, Born and von Karman (1912) obtained:

$$m\ddot{u}_l = \ddot{\Phi}\left(\frac{1}{2}s\right)\left[(u'_l - u_l) - (u_l - u'_{l-1})\right] \tag{8.7.3.1}$$

$$m'\ddot{u}'_l = \ddot{\Phi}\left(\frac{1}{2}s\right)\left[(u_{l+1} - u'_l) - (u'_l - u_l)\right] \tag{8.7.3.2}$$

In order to represent the displacements as progressive wave solutions, they introduced the substitutions:

$$u_l = u \times \exp(2\pi i\eta l - i\omega t) \tag{8.7.4.1}$$

$$u'_l = u' \times \exp(2\pi i\eta l - i\omega t) \tag{8.7.4.2}$$

where ω is the angular frequency and $|s/\eta|$ the wavelength. Substituting the u_l and u'_l terms in Eqs. (8.7.3.1) and (8.7.3.2), the infinite set of Eqs. (8.7.2.1) and (8.7.2.2) is replaced by two linear homogeneous equations:

$$\left[m\omega^2 - 2\ddot{\Phi}\left(\frac{1}{2}s\right)\right]u + \ddot{\Phi}\left(\frac{1}{2}s\right)[1 + \exp(-2\pi i\eta)u'] = 0 \tag{8.7.5.1}$$

$$\ddot{\Phi}\left(\frac{1}{2}s\right)[1 + \exp(-2\pi i\eta)u] + \left[m'\omega^2 - 2\ddot{\Phi}\left(\frac{1}{2}s\right)\right]u' = 0 \tag{8.7.5.2}$$

which have solution only for.[5]

$$\begin{vmatrix} m\omega^2 - 2\ddot{\Phi}\left(\frac{1}{2}s\right) & \ddot{\Phi}\left(\frac{1}{2}s\right)[1 + \exp(-2\pi i\eta)] \\ \ddot{\Phi}\left(\frac{1}{2}s\right)[1 + \exp(-2\pi i\eta)] & m'\omega^2 - 2\ddot{\Phi}\left(\frac{1}{2}s\right) \end{vmatrix} = 0 \tag{8.7.6}$$

The two admissible solutions of ω^2 are the acoustic and the optical frequencies (Born and Huang 1954):

$$\omega^2 = \left\{ \begin{array}{ll} \dfrac{\ddot{\Phi}\left(\frac{1}{2}s\right)\left\{(m+m') - \left[(m+m')^2 - 4,,' \sin^2 \pi\eta\right]^{1/2}\right\}}{mm'} & \text{(acoustic)} \\ \dfrac{\ddot{\Phi}\left(\frac{1}{2}s\right)\left\{(m+m') + \left[(m+m')^2 - 4,,' \sin^2 \pi\eta\right]^{1/2}\right\}}{mm'} & \text{(optical)} \end{array} \right\} \tag{8.7.7}$$

The two amplitude ratios *u/u′* are obtained replacing the obtained ω in Eqs. (8.7.5.1) and (8.7.5.2):

$$\frac{u}{u'} = \left\{ \begin{array}{ll} \dfrac{-m'[1 + \exp(-2\pi i\eta)]}{(m - m') - \left[(m+m')^2 - 4mm' \sin^2 \pi\eta\right]^{1/2}} & \text{(acoustic)} \\ \dfrac{-m'[1 + \exp(-2\pi i\eta)]}{(m - m') + \left[(m+m')^2 - 4mm' \sin^2 \pi\eta\right]^{1/2}} & \text{(optical)} \end{array} \right\} \tag{8.7.8}$$

Taking the real part of the complex solutions as representative of the actual displacements one gets a sinusoidal wave with the same wavelength and frequency for both particles. Both ω^2 and u/u′ are periodic in η with period 1. The period is conveniently restricted to a unit range comprised between −1/2 and 1/2, i.e.:

[5] Often the second derivative of the potential over the distance $\ddot{\Phi}$ is replaced by the force constant C, which implies the adoption of Hooke's law. See for instance Eq. 21 of Chap. 4 in Kittel (1976).

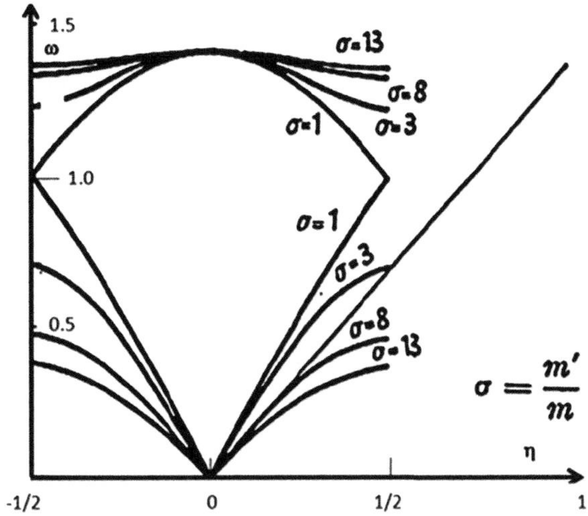

Fig. 8.9 Vibrational frequencies of a diatomic chain (arbitrary units). Modified from Born and Huang (1954)

$$-\frac{1}{2} \leq \eta < \frac{1}{2} \tag{8.7.9}$$

This corresponds to a cutoff of the wave number at (1/2 s) and waves with a longer wavenumber do not correspond to any further distinct vibrational mode. Plotting the angular frequency against η for different m/m' mass ratios, the frequencies fall into two different groups, the lower corresponding to the *acoustic* modes and the higher to the *optical* modes as shown in Fig. 8.9, which is the original drawing of Born and Huang (1954) (their Fig. 7).

The term "*acoustic*" derives from the fact that the long waves of the acoustic branch are identical with the longitudinal elastic vibration and the two particles in the same cell move at unison as a rigid unit without causing dispersion (see later).

Through a Taylor expansion of the acoustic solution in Eq. (8.7.7), we may see that for small values of η, ω approaches

$$\omega \cong \sqrt{\frac{2\pi^2\eta^2\ddot{\Phi}(s/2)}{(m+m')}} \tag{8.7.10}$$

and the frequency is linearly proportional to the wavenumber. The phase velocity v of the acoustic vibrations approaches

$$v \cong \sqrt{\frac{s^2\ddot{\Phi}(s/2)}{2(m+m')}} \tag{8.7.11}$$

which is the same velocity of the elastic vibration of a string having a Young modulus E_Y given by Eq. (8.7.1).

As $\eta \rightarrow 0$ the optical vibrations approach

$$\omega \cong \sqrt{\frac{2(m+m')\ddot{\Phi}(s/2)}{mm'}} \tag{8.7.12}$$

and the amplitude ratio for the two particles is inversely proportional to their masses (cf. Eq. 8.7.8):

$$\frac{u}{u'} = -\frac{m'}{m} \quad \rightarrow \quad mu + m'u' = 0 \tag{8.7.13}$$

It follows that the two masses move exactly in opposition and the center of mass of the cell remains stationary.

As far as I know, Fig. 8.9 is the first representation of the so-called "*dispersion relation*". In subsequent years the abscissa axis has been replaced by the wave vector $\mathbf{k} = 2\pi/\lambda$ (see for instance Fig. 4 in Chap. 4 of Kittel 1976).

8.7.2 Density of Modes in a Tridimensional Lattice: The Continuum Limit

Consider for simplicity a cubic lattice. Applying periodic boundary conditions over N^3 primitive cells and assuming the volume V to be L^3, the application of periodic boundary conditions demands that the wave vector $\mathbf{k}$ obeys (Kittel 1976):

$$\exp\left[i\left(k_x x + k_y y + k_z z\right)\right] \equiv \exp\left\{i\left[k_x(x+L) + k_y(y+L) + k_z(z+L)\right]\right\} \tag{8.7.14}$$

The running wave solution demands moreover, for each direction:

$$u_s = u_0 \times \exp(iska - i\omega_k t) \tag{8.7.15}$$

with a as inter-particle spacing. The possible solutions are:

$$k_x = k_y = k_z = 0, \quad \pm\frac{2\pi}{L}, \quad \pm\frac{4\pi}{L}, \ldots, \quad \pm\frac{N\pi}{L} \tag{8.7.16}$$

Being the volume $V = L^3$, the allowed values of k per unit volume of $\mathbf{k}$ space are

$$\left(\frac{L}{2\pi}\right)^3 = \frac{V}{8\pi^3} \tag{8.7.17}$$

There is thus only one allowed value of $\mathbf{k}$ per unit volume of $\mathbf{k}$ space, for each polarization and for each branch. The total number of modes with wave vector less than $\mathbf{k}$ is, for each polarization type:

$$N = \left(\frac{L}{2\pi}\right)^3 \frac{4}{3}\pi k^3 = \left(\frac{L}{2\pi}\right)^3 \frac{4}{3}\pi \left(\frac{\omega}{v}\right)^3 = \frac{V}{6\pi^2}\left(\frac{\omega}{v}\right) \tag{8.7.18}$$

where v is the sound velocity. The substitution $k = \omega/v$ implies moreover the "*continuum approximation*" (implicit also in the Debye model, see later).

For each polarization type the density of modes $\mathcal{D}(\omega)$ is:

$$\mathcal{D}(\omega) = \frac{dN}{d\omega} = \frac{V\omega^2}{2\pi^2 v^3} \tag{8.7.19}$$

8.7.3 Tridimensional Lattice in the Harmonic Approximation

Let me consider again the bulk potential acting on the crystal lattice as a function of the inter-atomic spacing vector $\Phi(\mathbf{R})$ (Eq. 8.2.11.1). Truncating the Taylor expansion of $\Phi(\mathbf{R})$ around the equilibrium configuration $\mathbf{R}_0$ corresponds to assume harmonic motions of all particles about their equilibrium position. Consider now a lattice with n particles, each labeled by the s index (s = 1, 2, …, n). Assuming that the translational vector $\mathbf{T}$ operates with translational symmetry on the unit cell, the equilibrium atomic positions are expressed by the relation:

$$\mathbf{R}^0_{s,\mathbf{T}} = \mathbf{R}^0_s + \mathbf{T} \tag{8.7.20}$$

with the vector $\mathbf{u}_{s,\mathbf{T}}(\mathbf{T})$ as a dynamical variable representing the shifts of the particles from their equilibrium position as a function of time. We may pose:

$$\mathbf{R}_{s,\mathbf{T}}(t) = \mathbf{R}^0_{s,\mathbf{T}} + \mathbf{u}_{s,\mathbf{T}}(t) \tag{8.7.21}$$

Indexing with j the Cartesian components (j = 1, 2, 3) and indicating as $\Phi(\mathbf{R}_0)$ the static component of the bulk potential ($\mathbf{u} = 0$), the potential energy referred to a unit cell of the crystal takes the form (Catti 1996):

$$\Phi(\mathbf{R}) = \Phi(\mathbf{R}_0) + \frac{1}{2}\sum_{j,j',s,s',} V_{j,j'}(s,s',\mathbf{T}) u_{j,s} u_{j',s',\mathbf{T}} \tag{8.7.22}$$

where $V_{j,j'}(s, s', \mathbf{T})$ are the force constants which, under the harmonic approximation, are those ones coupling the shifts for pairs of different atoms in the crystal (see also Footnote 5):

$$V_{j,j'}(s,s',\mathbf{T}) \equiv \left(\frac{\partial^2\Phi}{\partial u_{j,s}\partial u_{j',s',\mathbf{T}}}\right)_{\mathbf{u}=0} \tag{8.7.23}$$

The equations of motion (8.7.2.1, 8.7.2.2) have now their correspondent in:

$$m_s\ddot{u}_{j,s} = -\frac{\partial\Phi}{\partial u_{j,s}} = -\sum_{j',s',\mathbf{T}} V_{j,j'}(s,s',\mathbf{T})u_{j',s',\mathbf{T}} \tag{8.7.24}$$

with solutions:

$$u_{j,s,T}(t) = A_{j,s}\exp[i(\mathbf{k}\cdot\mathbf{T} - \omega t)] \tag{8.7.25}$$

The amplitudes $A_{j,s}$ are different for different particles, and, from cell to cell, the particles undergo a phase shift equal to the scalar product **k**·**T**. The phase factor exp (i**k**·**T**), appearing in the equation of motion of the particle, is identical to the phase factor induced in the electronic wave function for an electron subjected to the same translation (cf. Sect. 8.4). The independent wave vectors **k** are only those contained in the Brillouin zone.

To better understand what is above stated, we may consider a monatomic cubic lattice (i.e. one atom in the primitive cell). When a wave propagates along the [100], [110], or [111] direction, the planes move in phase with a directional motion either perpendicular or parallel to the direction of the wave vector **k**. Assuming the elastic forces to obey Hooke's law with force constant C_p, the inter-planar spacing depends on the direction of **k**. In fact, the equation of motion of the plane is:

$$m\ddot{u} = \sum_p C_p(u_{s+p} - u_s) \tag{8.7.26}$$

where p is the displacement. For displacements having the time dependence $\exp(-i\omega t)$ we have $\ddot{u} = -\omega^2 u$ and Eq. (8.7.26) reduces to:

$$-m\omega u_s = \sum_p C_p(u_{s+p} - u_s) \tag{8.7.27}$$

The traveling wave has solutions

$$u_{s+p} = u\exp[i(s+p)\mathbf{k}a] \tag{8.7.28}$$

where *a* is the spacing between planes. Introducing in Eq. (8.7.27) the time dependence of Eq. (8.7.28) we have:

$$\omega^2 m = -\sum_p C_p[\exp(ip\mathbf{k}a) - 1] \tag{8.7.29}$$

Because, for translational symmetry $C_p = C_{-p}$, Eq. (8.7.29) leads to (Kittel 1976):

$$\omega^2 m = -\sum_p C_p[\exp(ip\mathbf{k}a) + \exp(-ip\mathbf{k}a) - 2] \tag{8.7.30}$$

and, because $[\exp(ip\mathbf{k}a) + \exp(-ip\mathbf{k}a)] = 2\cos p\mathbf{k}a$, we obtain finally (Kittel 1976):

$$\omega^2 = \frac{2}{m}\sum_{p>0} C_p(1 - \cos p\mathbf{k}a) \tag{8.7.31}$$

This form of the dispersion relation is particularly informative. It shows in fact that the slope of the wave vector is zero at the zone boundaries $\mathbf{k} = \pm \pi/a$.

If the interaction involves only nearest neighbor planes, then Eq. (8.7.31). further reduces to:

$$\omega^2 = \frac{2C_1}{m}(1 - \cos \mathbf{k}a) = \frac{4C_1}{m}\sin^2\left(\frac{\mathbf{k}a}{2}\right) \tag{8.7.32}$$

Thus

$$\omega = \left(\frac{4C_1}{m}\right)^{1/2} \sin\left(\frac{\mathbf{k}a}{2}\right) \tag{8.7.33}$$

The ratio is $\omega/\sqrt{\frac{4C_1}{m}}$ is simply sinusoidal (Fig. 8.11) and, for $\mathbf{k} \ll 1/a$ (or $\mathbf{k} \gg a$) ω is directly proportional to k. For two adjacent planes the ratio of displacements is:

$$\frac{u_{s+1}}{u_s} = \frac{u\exp[i(s+1)\mathbf{k}a]}{u\exp(is\,\mathbf{k}a)} = \exp(i\mathbf{k}a) \tag{8.7.34}$$

As we may see in Fig. 8.10, there is no need to push the analysis beyond the first Brillouin zone, because the range $-\pi$ to $+\pi$ for the phase $\mathbf{k}a$ covers all independent values of the exponential, i.e.:

$$-\pi \le \mathbf{k}a \le \pi \quad \rightarrow \quad -\tfrac{\pi}{a} \le \mathbf{k} \le \tfrac{\pi}{a} \tag{8.7.35}$$

8.7.4 Born Matrix

The acceptable solutions of the traveling wave Eq. (8.7.29) are those obeying a set of homogeneous linear equations in the complex amplitudes $A_{j,s}$:

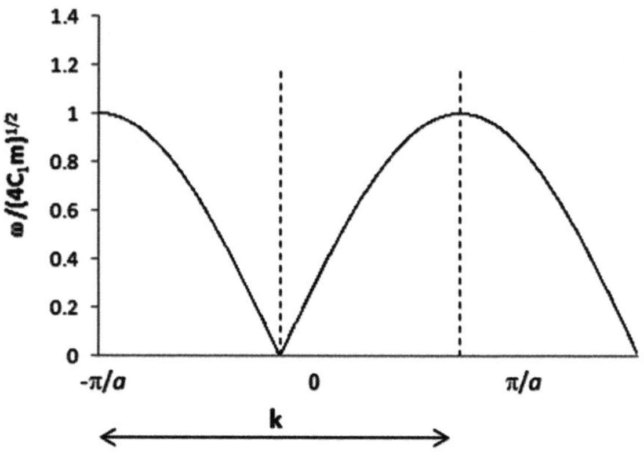

Fig. 8.10 Dispersion relation in a cubic monoatomic lattice of atoms with mass m and spacing a between planes. Transversal acoustic mode

$$\sum_{j'=1}^{3}\sum_{s'=1}^{p}\left[B_{j,j'}(s,s',k) - m_s\omega^2\delta_{j,j'}\delta_{s,s'}\right]A_{j',s'} \tag{8.7.36}$$

with $B_{j,j'}$ as Fourier transforms of the force constant (see Sect. A.12.2):

$$B_{j,j'}(s,s',\mathbf{k}) = \sum_{\mathbf{T}} V_{j,j'}(s,s',\mathbf{T})\exp(i\mathbf{k}\cdot\mathbf{T}) \tag{8.7.37}$$

Contracting the pair of indexes j, s into a single one, the amplitudes reduce to a 3p × 1 column matrix **A** and the $B_{j,j'}(s,s',\mathbf{k})$ to a 3p × 3p square matrix **B(k)**, known as "*Born matrix*" or also "*dynamical matrix*". In matrix form Eq. (8.7.36) reduces to $(\mathbf{B} - \mathbf{m}\omega^2)\mathbf{A} = 0$ and the *generalized secular equation* requires:

$$\det\left[\mathbf{B}(\mathbf{k}) - \mathbf{m}\omega^2\right] = 0 \tag{8.7.38}$$

There are 3p solutions of the secular equations in the squared angular frequency ω^2, constituting the branches of the dispersion relation $\omega(\mathbf{k})$.

The absolute complex amplitudes $A_{j,s,n}(\mathbf{k})$ are related to the energy eigenvalues involved in each displacement $e_{j,s,n}(\mathbf{k})$ by:

$$A_{j,s,n}(\mathbf{k}) = \zeta_n(\mathbf{k}) \times e_{j,s,n}(\mathbf{k}) \tag{8.7.39}$$

where $\zeta_n(\mathbf{k})$ are coefficients obtained from the solution of Eq. (8.7.25) in terms of a linear combination of all normal modes (Catti 1996):

$$u_{j,s,T}(t) = \sum_{n}\sum_{\mathbf{k}} \zeta_n(\mathbf{k})e_{j,s,n}(\mathbf{k})\exp[i(\mathbf{k}\cdot\mathbf{T} - \omega_n(\mathbf{k})t)] \tag{8.7.40}$$

8.8 Ab Initio Phonon Frequencies

The ab initio calculations of phonon frequencies fall into several classes. In the *linear response method,* the dynamical matrix is expressed in terms of the inverse dielectric matrix describing the response of the valence electron density to a periodic lattice perturbation (see Resta 1985, for a review). In the *frozen-phonon* method (Kunc and Martin 1983), the phonon energy is calculated as a function of the displacement amplitude in terms of the difference in the energies of the distorted and ideal lattices. The *direct method* uses the forces calculated via the Hellmann–Feynman theorem assuming a finite range of interaction in the total energy calculations. The values of the force constant matrices (and hence the dynamical matrix and phonon dispersion curves) are then readily derived (Frank et al. 1995; Parlinski et al. 1997). In the *small displacements method,* (Kresse et al. 1993a, b; Alfé 2009) it is exploited the fact that the proportionality between displacements and forces (expressed by the force constant matrix) may be considered to be linear, when the displacements are small enough. Other methods adopt numerical procedures to generate the Hessian matrix, stemming from the analytical gradient of the energy with respect to the atomic positions (Pascale et al. 2003).

8.8.1 The Density Functional Linear Response Approach

The linear response method has been promoted by Baroni et al. (1987) in the framework of density functional perturbation theory (DFPT). Within the adiabatic approximation, the lattice distortion associated with a phonon can be seen as a static perturbation acting on the electrons. The harmonic force constants of crystals are thus determined by their static linear electronic response. The linear variation of the electron density upon application of an external, static, perturbation determines the energy variation up to the third order in the perturbation (cf. Giannozzi et al. 1991; and references therein).

The force constant associated to each atomic displacement $u_{j,s,T}$ is obtained quantum mechanically by application of the Hellmann–Feynman theorem (Feynman 1939) stating that the force associated with the variation of external parameters, is given by the ground state expectation value of the derivative of V. In terms of atomic displacements, this may be written as (Giannozzi et al. 1991):

$$\frac{\partial \Phi}{\partial u_{j,s,T}} = \int \rho(\mathbf{r}) \frac{\partial V(\mathbf{r})}{\partial u_{j,s,T}} d\mathbf{r} \tag{8.8.1}$$

with $\rho(\mathbf{r})$ and $V(\mathbf{r})$ representing respectively the electron density and the external potential acting on the electrons at position $\mathbf{r}$. The 2nd order derivative of the potential with respect to the atomic displacements gives the force constants over the distance in terms of bi-linear coefficients:

$$\begin{aligned} C_{j,j'}(s,s',\mathbf{T}) &= \left(\frac{\partial^2\Phi}{\partial u_{j,s}\partial u_{j',s',\mathbf{T}}}\right)_{\mathbf{u}=0} \\ &= C^{ion}_{j,j'}(s,s',\mathbf{T}) + \int\left[\frac{\partial V(\mathbf{r})}{\partial u_{j,s,\mathbf{T}}}\frac{\partial \rho(\mathbf{r})}{\partial u_{j,s}} + \rho_0(r)\frac{\partial^2 V(r)}{\partial u_{j,s}\partial u_{j',s',\mathbf{T}}}\right]d\mathbf{r} \end{aligned} \quad (8.8.2)$$

where C^{ion} is the ionic contribution to the force constants, which is essentially the second derivative of an Ewald sum (see Appendix A in Giannozzi et al. 1991) and the second term on the right is the electronic contribution.

The phonon frequencies are then obtained by diagonalization of the dynamical matrix, **B** (Giannozzi et al. 1991):

$$\mathbf{B}_{j,j'}(s,s',\mathbf{T}) = \frac{C_{j,j'}(s,s',\mathbf{T})}{\left(m_j m_{j'}\right)^{1/2}} \quad (8.8.3)$$

The atomic displacements induce changes in the electron potential energy and this causes modifications in the electron density. The self-consistent-field (SCF) DFT potential V_{SCF} for a system of electrons moving in the external potential of the ions is given by (Giannozzi et al. 1991):

$$V_{SCF}(\mathbf{r}) = V_{ion}(\mathbf{r}) + e^2\int\frac{\rho(\mathbf{r}')}{|\mathbf{r}-\mathbf{r}'|}dr' + v_{xc}(\rho(\mathbf{r})) \quad (8.8.4)$$

where $v_{xc}(\rho(\mathbf{r}))$ is the exchange contribution. This potential can be determined by standard techniques of self-consistent band structure calculations, allowing the computation of the electronic linear response of periodic systems in the framework of DFT.

Superimposing to the ionic potential a perturbation ΔV_{bare} of periodicity **T**, the self-consistent potential changes accordingly by the amount $\Delta V_{SCF,}$ and this induces an electron density variation $\Delta\rho$. If $\Delta\rho$ is known ΔV_{SCF} can be obtained by linearization of Eq. (8.8.4):

$$\Delta V_{SCF}(\mathbf{r}) = \Delta V_{bare}(\mathbf{r}) + e^2\int\frac{\Delta\rho(\mathbf{r}')}{|\mathbf{r}-\mathbf{r}'|}d\mathbf{r}' + \Delta\rho(\mathbf{r})\left(\frac{dv_{xc}}{d\rho}\right)_{\rho=\rho_0(\mathbf{r})} \quad (8.8.5)$$

with $\rho_0(\mathbf{r})$ as unperturbed density. If, on the other hand, it is known ΔV_{SCF}, a first-order perturbation method will give the Fourier transform of the electron density delta $[\Delta\tilde{\rho}(\mathbf{T}+\mathbf{G})]$:

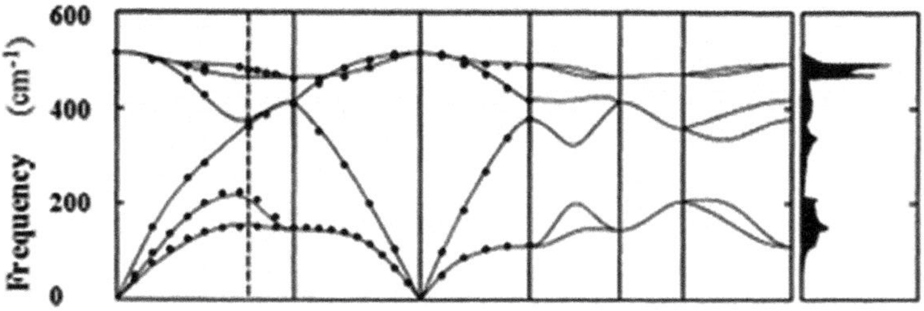

Fig. 8.11 Calculated phonon dispersions (solid lines) and densities of state (DOS) of the elemental semiconductors Si. The solid diamonds are the experimental data of Nilsson and Nelin (1972). Modified from Giannozzi et al. (1991)

$$\Delta\tilde{\rho}(\mathbf{T}+\mathbf{G}) = \frac{4}{N\Omega}\sum_{\mathbf{k}}\sum_{c,v}\frac{\langle\psi_{v,k} \mid \exp[-i(\mathbf{T}+\mathbf{G})]\cdot\mathbf{r}|\psi_{c,k+\mathcal{T}}\rangle\langle\psi_{c,k+\mathcal{T}} \mid \Delta V_{SCF}|\psi_{v,k}\rangle}{\varepsilon_{v,k} - \varepsilon_{k+\mathcal{T}}} \tag{8.8.6}$$

In Eq. (8.8.6), Ω is the volume of the unit cell, v and c indicate valence and conduction bands, respectively, and the sum over **k** covers the first Brillouin zone (Giannozzi et al. 1991).[6] Equations (8.8.5 and 8.8.6) are resolved iteratively until convergence is attained.

The linear response theory has been applied by Giannozzi et al. (1991), in a very successful way, to the semiconductors Si, Ge, GaAs, AlAs, GaSb, and AlSb. We see in Fig. 8.11, as an example, the comparison of their calculations (solid curves) with the neutron inelastic scattering experiments of Nilsson and Nelin (1972) (solid diamonds). The density of states (DOS), returning the frequency histograms of the various vibrational modes is shown on the right part of the figure.

8.8.2 The Direct Approach: Ab Initio Force Constant Method

The ab initio force constant approach is based on the supercell with periodic boundary conditions (Frank et al. 1995). A displacement $\mathbf{u}(0, \nu)$ in such a supercell, induces forces $\mathbf{F}(\mathbf{n}, \mu)$ acting on all other atoms (Parlinski et al. 1997):

$$\mathbf{F}(\mathbf{n}, \mu) = -\sum_{\mathbf{L}}\mathbf{B}(\mathbf{n}, \mu; \mathbf{L}.\nu)\cdot\mathbf{u}(0, \nu) \tag{8.8.7}$$

[6] The linear response to a perturbation of given **T** only contains Fourier components of wave vector **T** + **G**, different **T**'s do not mix at this order of perturbation theory (Giannozzi et al. 1991).

where **n** is the index of the primitive unit cell; **L** is the index of the lattice constants of the supercell; μ and v are indices of atoms in the primitive unit cell and $\mathbf{B}(\mathbf{n},\mu;\mathbf{L}.v)$ are the 3 × 3 harmonic force constant matrices relating atoms (**n**, μ) and (**L**, v).

On each atom of the supercell acts a force coming from the atom displaced inside the supercell (and from the equivalent atoms displaced in all images of the supercell; which is implicit in the periodic boundary conditions.

Parlinski et al. (1997) introduce a "*cumulant force constant matrix*" (Maradudin and Vosko 1968), with the same symmetry of $\mathbf{B}(\mathbf{n},\mu;\mathbf{L}.v)$:

$$\mathbf{B}_{\Sigma}(\mathbf{n},\mu;0.v) = -\sum_{\mathbf{L}} \mathbf{B}(\mathbf{n},\mu;\mathbf{L}.v) \qquad (8.8.8)$$

The 3 × 3 $\mathbf{B}_{\Sigma}$ matrix is then recast into a 9 × 1 column matrix $\mathcal{B}_{\Sigma}$ such that:

$$(\mathcal{B}_{\Sigma})_{\mathrm{k}} = \left(\mathbf{B}_{\Sigma})_{\mathrm{i,j}}\right) \quad \mathrm{k} = (\mathrm{i}-1)+\mathrm{j} \qquad (8.8.9)$$

$\mathcal{B}_{\Sigma}$ is then decoupled into a part $\mathcal{A}$ of dimension 9 × p and a part $\mathcal{P}_{\Sigma}$ of dimension p × 1:

$$\mathcal{B}_{\Sigma}(\mathbf{n},\mu;0,v) = \mathcal{A}(\mathbf{n},\mu;0,v) + \mathcal{P}_{\Sigma}(\mathbf{n},\mu;0,v) \qquad (8.8.10)$$

While the matrix $\mathcal{A}$ is independent of the potential strength and is determined by symmetry considerations only, the matrix $\mathcal{P}_{\Sigma}$ is "*characterized by the potential*" (Parlinski et al. 1997).

Combining Eqs. (8.8.9) and (8.8.7) it is found:

$$\mathbf{F}(\mathbf{n},\mu) = -\mathcal{U}(0,v)\cdot\mathcal{A}(\mathbf{n},\mu;0,v)\cdot\mathcal{P}_{\Sigma}(\mathbf{n},\mu;0,v) \qquad (8.8.11)$$

$\mathcal{U}(0, v)$ in Eq. (8.8.11) is a 3 × 9 matrix with elements of the displacement vector **u** (0, v) introduced to fulfill the combinatory rules **B·u** in Eq. (8.8.7).

Equation (8.8.11) is further simplified by introducing:

$$\mathcal{C}(\mathbf{n},\mu;0,v) = -\mathcal{U}(0,v)\cdot\mathcal{A}(\mathbf{n},\mu;0,v) \qquad (8.8.12)$$

and reduces to:

$$\mathbf{F}(\mathbf{n},\mu) = \mathcal{C}(\mathbf{n},\mu;0,v)\cdot\mathcal{P}_{\Sigma}(\mathbf{n},\mu;0,v) \qquad (8.8.13)$$

which can be finally written in global matrix form:

$$\underline{\mathcal{F}} = \underline{\mathcal{C}}\cdot\underline{\mathcal{P}_{\Sigma}} \qquad (8.8.14)$$

$\underline{\mathcal{F}}$, $\underline{\mathcal{C}}$ and $\underline{\mathcal{P}_\Sigma}$ in Eq. (8.8.14) are, respectively (3ns × 1), (3ns × p) and (p × 1) dimensional matrices, where n is the number of atoms in the supercell and s is the number of force fields, i.e. "*a set of 3n Hellmann–Feynman forces obtained for a run with a single displaced atom*" (Parlinski et al. 1997).

Because the number of force fields (3ns) exceeds by far the number of parameters (p), Eq. (8.8.14) is reversed (i.e. $\underline{\mathcal{P}_\Sigma} = \underline{\mathcal{C}}^{-1} \cdot \underline{\mathcal{F}s}$) and solved by a least square fitting treatment of the singular matrix $\underline{\mathcal{C}}$.

Using the cumulant force constants (Eq. 8.8.8), the conventional dynamical matrix of the crystal **D(k**;$\mu\nu$):

$$\mathbf{D}(\mathbf{k}, \mu\nu) = \frac{1}{\sqrt{\mathrm{M}_\mu \mathrm{M}_\nu}} \sum_{\mathrm{m}} \mathbf{B}(0, \mu; \mathbf{m}, \nu) \times \exp\{-2\pi i\mathrm{k} \cdot [\mathbf{R}(0, \mu) - \mathbf{R}(\mathbf{m}, \nu)]\} \tag{8.8.15}$$

where M are the atomic masses and $\mathbf{R}(0, \mu), \mathbf{R}(\mathbf{m}, \nu)$ their positions, is replaced by the "*approximate dynamical matrix*" $\mathcal{D}$(**k**;μν):

$$\mathcal{D}(\mathbf{k}, \mu\nu) = \frac{1}{\sqrt{\mathrm{m}_\mu \mathrm{m}_\nu}} \sum_{\mathrm{m}'} \mathrm{w}_{\mathbf{m}'} \mathbf{B}(0, \mu; \mathbf{m}', \nu) \times \exp\{-2\pi i\mathrm{k} \cdot [\mathbf{R}(0, \mu) - \mathbf{R}(\mathbf{m}', \nu)]\} \tag{8.8.16}$$

The summation in m running over all atoms in the supercell in Eq. (8.8.15) is here replaced by a summation in an "*extended supercell*" which includes atoms on all its surfaces and runs over all neighbors inside the extended supercell to ensure Hermitian properties (i.e. real eigenvalues; see Sect. A.14). The weighting factors $\mathrm{w}_{\mathbf{m}'}$ have the duty of assuring that at the wave vector $\mathbf{k}_\mathrm{L}$ fulfilling the condition $\exp(2\pi i\mathbf{k}_\mathrm{L} \cdot \mathbf{L}) = 1$ and the approximate and conventional dynamical matrices are equal. I.e. $\mathbf{D}(\mathbf{k},\mu\nu) = \mathcal{D}(\mathbf{k}, \mu\nu)$ (Parlinski et al. 1997).

The direct approach has been applied by Parlinski et al. (1997) and Sternik and Parlinski (2005) to zirconia (ZrO_2). At ambient pressure, cubic zirconia (space group *Fm*3*md*) transforms to the tetragonal form (space group $P4_2/nmc$) at about 2570 *K*. A further phase transition to a monoclinic structure occurs when lowering the temperature below 1400 (space group $P2_1/c$). In the cubic to tetragonal transition, the lattice periodicity along the c direction changes from *a*/2 in the cubic cell to *a* in the tetragonal cell. According to Group theory (Negita 1989), such a cell doubling can be induced by the condensation of a phonon mode at the cubic Brillouin zone boundary X point, $\mathbf{k}_\mathrm{X} = (0, 0, 2\pi/a)$, since a function characterized by this wave vector $\mathbf{k}_\mathrm{X}$ has a translational periodicity of *a* in the *c* direction with same periodicity as the cubic lattice in *a* and *b* directions. I reproduce here the calculated phonon dispersion diagram of Sternik and Parlinsky (2005), slightly

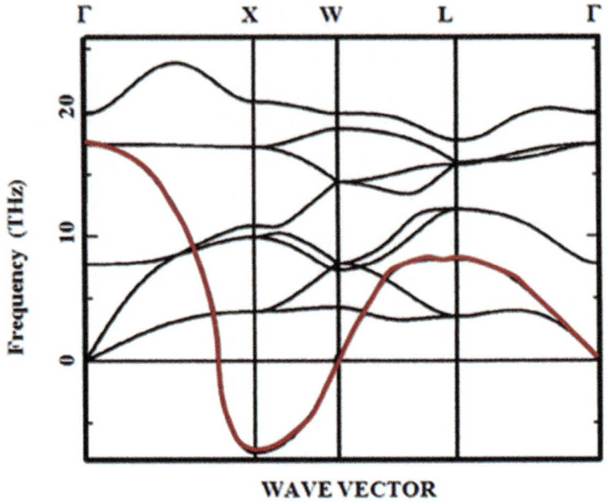

Fig. 8.12 Calculated phonon dispersion in cubic zirconia calculated with the direct method. The soft mode is drawn in red. Γ(0,0,0), X(½,0,½), W(½,¼,¾), L(½,½,½). Modified from Sternik and Parlinski (2005)

modified to put in evidence the soft mode[7] occurring at the X point of the Brillouin zone, with an imaginary frequency of ~i5.9 *THz* (Fig. 8.12).

Undoubtedly, the capability of testifying the existence of this soft mode is remarkable. Releases of the Phonon and PhononA software, based on the direct method, are available on the net (http://computingformaterials.com/). PhononA, particularly, is able to account for anharmonicity effects. As an input, these softwares require a file specifying the crystal structure, in terms of positional parameters of non-equivalent atoms in the unit cell and the cell edges, and a corresponding file of the corresponding Hellnann–Feynman forces.

8.8.3 The "Small Displacements" Method

The "*small displacements method*" (Kresse et al. 1995; Alfé et al. 2001; Alfé 2009) exploits the fact that when the thermal displacement of atoms from their ground state position is small enough, then the proportionality between displacements and forces may be seen as simply linear (see Eq. 8.8.1).

Following the notation of Kresse et al. (1995), the displacement of a single atom μ in the cell l by a vector **u** is labeled $\mathbf{u}\begin{pmatrix} l \\ \mu \end{pmatrix}$. The induced force field, calculated via the Hellmann–Feynman theorem is $\mathbf{F}\begin{pmatrix} ml \\ \mu\nu \end{pmatrix}$. The force constant matrix is depicted as

[7] Soft modes are phonons of vanishing energy at the long wavelength limit. They are typical of excited states such as those attained at solid state transitions.

$$\Phi_{\alpha\beta}\begin{pmatrix} ml \\ \nu\mu \end{pmatrix} \approx -\frac{\tilde{\mathbf{F}}_{\beta}\begin{pmatrix} ml \\ \nu\mu \end{pmatrix}}{\tilde{\mathbf{u}}_{\alpha}\begin{pmatrix} l \\ \nu \end{pmatrix}} \qquad (8.8.17)$$

where $\tilde{\mathbf{u}}_{\alpha}\begin{pmatrix} l \\ \nu \end{pmatrix}$ is a set of orthogonal displacement vectors oriented along the Cartesian axes:

$$\tilde{\mathbf{u}}_{\alpha}\begin{pmatrix} l \\ \nu \end{pmatrix} = \sum_i w_{\alpha i}\overline{\mathbf{u}}_i\begin{pmatrix} l \\ \nu \end{pmatrix} \quad i = 1,3 \qquad (8.8.18)$$

with $\overline{\mathbf{u}}_i\begin{pmatrix} l \\ \nu \end{pmatrix}$ it as linear independent displacement vectors; and

$$\tilde{\mathbf{F}}_{\alpha}\begin{pmatrix} ml \\ \mu\nu \end{pmatrix} = \sum_i w_{\alpha i}\overline{\mathbf{F}}_i\begin{pmatrix} ml \\ \mu\nu \end{pmatrix} \qquad (8.8.19)$$

with $\overline{\mathbf{F}}_i\begin{pmatrix} ml \\ \mu\nu \end{pmatrix}$ as force fields.

The dynamical matrix is given by

$$\mathbf{D}_{\alpha\beta}\begin{pmatrix} \mathbf{k} \\ \mu\nu \end{pmatrix} = \frac{1}{\sqrt{\mathrm{M}_{\mu}\mathrm{M}_{\nu}}}\sum_l \Phi_{\alpha\beta}\begin{pmatrix} ml \\ \mu\nu \end{pmatrix} \times \exp\left\{i\mathbf{k}\cdot\left[\mathbf{R}\begin{pmatrix} m \\ \mu \end{pmatrix} - \mathbf{R}\begin{pmatrix} l \\ \mathbf{k} \end{pmatrix}\right]\right\} \qquad (8.8.20)$$

In principle, the approach gives always a linear combination of the exact inter-atomic force constants: (Kresse et al. 1995):

$$\Phi_{\alpha\beta}\begin{pmatrix} ml \\ \mu\nu \end{pmatrix} = \sum_L \Phi_{\alpha\beta}^{\mathrm{exact}}\begin{pmatrix} ml + L \\ \mu\nu \end{pmatrix} \qquad (8.8.21)$$

where L labels the periodically repeated supercells.

Only a very small set of supercell calculations (a single calculation for a primitive cell with full cubic symmetry) is sufficient to determine the dynamical matrix for arbitrary **k**, whereas the response calculation has to be repeated many times for crystals of lower symmetry (at most, three times the number of atoms in the primitive cell, due to translational invariance). This number, however, can be reduced substantially in most materials using symmetry relations. In ionic materials, convergence can be slow. This is because in the limit of zero wave vector the displacement of charges creates dipoles which interact with long range forces, producing a macroscopic electric field in the limit of zero wave vector, which is responsible for the so-called "*LO-TO splitting*". I.e. a splitting of the frequencies of

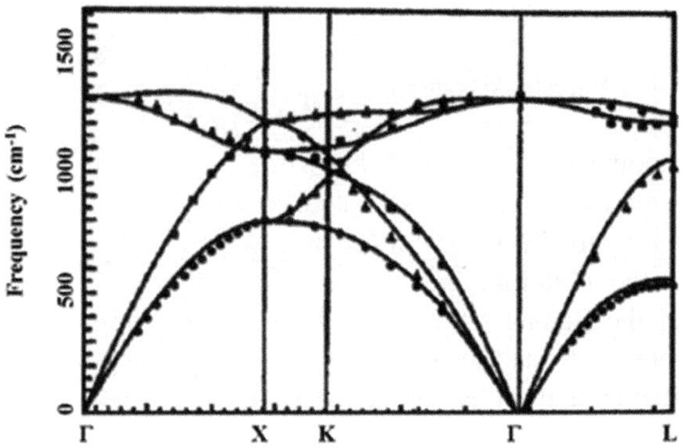

Fig. 8.13 Phonon dispersion relations for diamond. The full lines show the results of the ab initio small displacement approach of Kresse et al. (1995). The solid symbols are the inelastic neutron scattering data of Warren et al. (1967). Modified from Kresse et al. (1995)

the optical vibrational modes parallel and perpendicular to the electric field (Alfé 2009).

I report hereafter a few examples of phonon dispersion relations obtained on materials of particular interest in Geochemistry. The first one is due to Kresse et al. (1995) and concerns diamonds (Fig. 8.13). The Helmann–Feynman force field was calculated using the Vienna ab-initio molecular-dynamics program (VAMP; Kresse and Hafner 1993a, b) on a 64-atom (2 × 2 × 2) simple cubic supercell. The amplitude of the atomic displacements was limited to 0.5% of the lattice vectors of the primitive cell. The phonon frequencies were calculated at the LDA theoretical equilibrium lattice constants (a = 3.530 *Å*; against the experimental a = 3.567 *Å*). The calculated bulk modulus was K = 4.60 *Mbar*, against the experimental K = 4.52 *Mbar*. These over-binding effects are typical of LDA. The calculated phonon dispersion relations when compared with the inelastic neutron scattering data (Warren et al. 1967) show a very good agreement, except for small differences of the order of 2% in the optical phonon at the Γ-point (i.e. the usual LDA error; Kresse et al. 1995).

In Fig. 8.14 we see the phonon dispersion relations computed by Alfé et al. (2001) for hexagonal-close-packed (hcp) iron,[8] which constitutes the major end-member of the solid part of the Earth's core, in solid mixture with Ni. Calculations were performed at the DFT-GGA theory level with the Projector Augmented Wave (PAW) approach (VASP code), accounting for the electron's thermal excitation using the standard methods of finite temperature DFT developed by Mermin (1965). Harmonic phonon frequencies were computed over the entire Brillouin zone. All the calculations employed a 15 × 15 × 9 electronic Monkhorst–Pack set,[9] (Monkhorst and Pack 1976), which gives 135 k-points in the

[8] According to Vočadlo et al. (2000) hcp is thermodynamically slightly more stable than double-hcp at Earth's core pressures and temperatures. [i.e. 100 $\leq$ P (*GPa*) $\leq$ 400: 3000 T $\leq$ (*K*) $\leq$ 7000 ± 1000].

[9] A rectangular grid of points evenly spaced throughout the Brillouin zone.

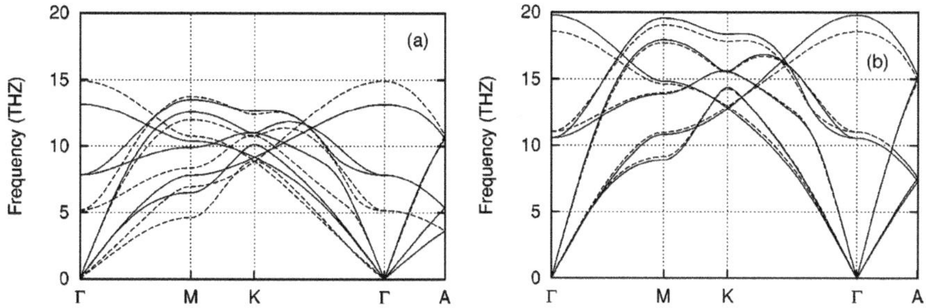

Fig. 8.14 Phonon dispersion relations of hcp Fe calculated at V = 58.67 $Å^3$ (**a**) and V = 6.97 $Å^3$ (**b**). DFT-GGA frequencies calculated directly from DFT at the two volumes are shown as solid curves. In **a** the dashed lines give frequencies obtained from an empirical inverse-power model (see Alfé et al. 2001, for details). In **b**, the dashed lines are the DFT-GGA frequencies for V = 8.67 $Å^3$ scaled by a factor of 1.409. From Alfé et al. (2001). With permission from the American Physical Society (APS)

irreducible wedge of the Brillouin zone. The adopted electronic temperature was T_{el} = 4000 *K*.

In the method of finite temperature DFT developed by Mermin, the fundamental quantity is the electronic free energy F_{el} calculated at electronic temperature T_{el}

$$F_{el}(\mathbf{R}_1,\ldots\mathbf{R}_N, T_{el}) = E_{DFT} - TS_{el} \tag{8.8.22}$$

where the electronic entropy S_{el} is given by

$$S_{el} = -2k_B T_{el} \sum_i [f_i \ln f_i + (1 - f_i)\ln(1 - f_i)] \tag{8.8.23}$$

The DFT energy (comprehensive of all kinetic, electron-nucleus, Hartree, and exchange–correlation terms) is calculated with the N nuclei fixed at positions $\mathbf{R}_1$,... $\mathbf{R}_N$. f_i in the expression of the electronic entropy is the thermal Fermi–Dirac occupation number of orbital i.

Treating the statistical mechanics of the nuclei in the classical limit, corresponds to assuming that the electronic and nuclear degrees of freedom are in thermal equilibrium with each other, so that T_{el} = T (Alfé et al. 2001) The Helmholtz free energy of the whole system is then:

$$F = -k_B T \ln\left\{\frac{1}{N!\Lambda^{3N}} \int d\mathbf{R}_1,\ldots d\mathbf{R}_N \times \exp\left[-\frac{F_{el}(\mathbf{R}_1,\ldots\mathbf{R}_N, T_{el})}{k_B T}\right]\right\} \tag{8.8.24}$$

where Λ is the thermal wavelength:

$$\Lambda = -\frac{h}{\sqrt{2k_B MT}} \tag{8.8.25}$$

with M as nuclear mass.

A FORTRAN90 code to calculate phonons using the small displacement method has been provided by Alfé (2009). The program summary is available also on the web at: http://cpc.cs.qub.ac.uk/summaries/AEDP_v1_0.html.

8.9 Thermophysical Properties

8.9.1 The "Symmetry Preserving, Variable Cell-Shape Structure Relaxation Procedure"

The first step in the assessment of the various thermophysical parameters of interest is the conformation of the potential well at the athemal limit. The number of points necessary to conform the static potential depends upon the structural class of the investigated substance and is determined by the attainment of a reasonable internal consistency with Equation of State (EoS) predictions based on a Birch-Murnaghan 3rd order finite-strain equation (or other EoS as well). As an example, we see in Fig. 8.15 the results of this procedure, applied by Ottonello et al. (2009a) to Forsterite, one of the most important rock-forming mineral components in the Earth's upper mantle. Forsterite crystallizes in the orthorhombic bipyramidal class (space group *Pbnm*; non-conventional setting of *Pnma* n. 62). There are 6 irreducible atoms in the conventional cell and 28 atoms in the unit cell (four formula units per unit cell). All-electron calculations were performed within the linear combination of atomic orbitals (LCAO) approach with the hybrid B3LYP functional. The equilibrium structure was determined by using a quasi-Newton algorithm with a Broyden–Fletcher–Goldfarb–Shanno (BFGS) Hessian updating scheme (see Fletcher 1987; and references therein) and following geometry optimization strategy as proposed by Schlegel (1982). Convergence in the geometry optimization process was tested on the root-mean-square (RMS) and the absolute value of the largest component of both the gradients and nuclear displacements. The variational structure at the athermal limit in the absence of external pressure returned a molar (zero point) volume $V^{\circ} = 44.682$ (*cc/mol*) ($a_0 = 4.7935$; $b_0 = 10.2941$; $c_0 = 6.0126$) and an electronic energy $E_0 = -991.055720$ *Hartree*. Provisional cell edges were then assigned to each compressional state and the minimum energy c/b and a/b ratios over a wide P range at the athermal limit were refined. This exercise is for Forsterite computationally very demanding because 152 runs (86 for the c/b ratios and 66 for the a/b ratios) are necessary to conform satisfactorily the potential well of the substance on 14 different compressional

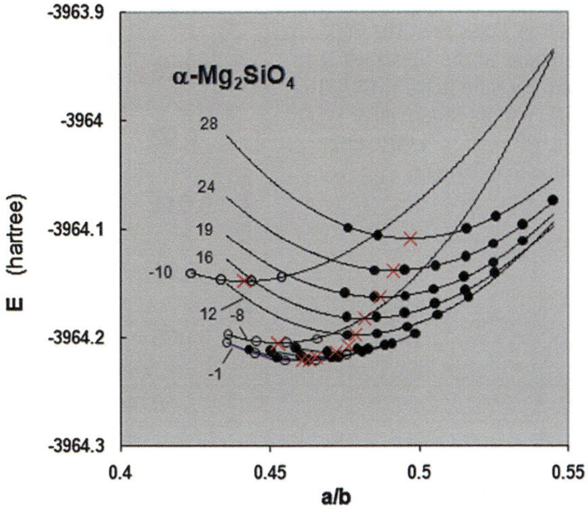

Fig. 8.15 Energy versus a/b ratio at given molar volumes corresponding to compressional (filled circles) and expanded states (open circles) of Forsterite. The values of static potential pressure listed on the left of the various curves are determined by the partial derivative of static potentials with respect to volume, i.e. P = −(dE/dV). From Ottonello (2018)

states.[10] In Fig. 8.15 we see, as an example, in the Forsterite case study, the identification of the local a/b minima on the various a/b curves corresponding to different nominal pressures.

8.9.2 *Compressibility at the Athermal Limit and Seismic Velocities*

The partial derivative of a third-order polynomial fitted over the molar electronic energy (E, *J/mol*)–molar volume (V, *J/bar*) couples allowed to establish the ab initio potential value of static pressure P_{st} corresponding to a given compressional state. The bulk modulus of the substance at zero pressure and its P-derivative ($\partial K/\partial P = K'$) were then obtained by Gaussian integration of the Birch–Murnaghan third-order finite elasticity isothermal EoS:

$$P_{st} = -\left(\frac{\partial E_{ai}}{\partial V}\right)_{T=0} = \frac{3}{2}K_0\left[\eta^{-7/3} - \eta^{-5/3}\right]\left\{1 - \frac{3}{4}(4 - K')\left[\eta^{-2/3} - 1\right]\right\} + P_0 \tag{8.9.1}$$

where η is the dimensionless variable "*dilaton*" (η = V/V_0) and P_0 is the reference pressure (usually $P_0 = 0$) We see an example, for the Forsterite case, in Fig. 8.16.

As we have already seen, the propagation of a wave in a continuum may be conveniently described in terms of elongation **u** and wave vector **q**:

[10] In the newest releases of CRYSTAL the procedure is fully automated.

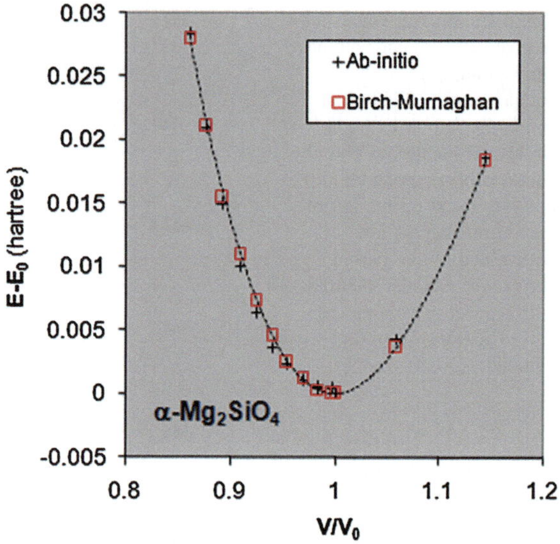

Fig. 8.16 Ab initio potential well of Forsterite. The bulk modulus of the substance at zero pressure is $K_0 = 130.810$ *GPa* and its P-derivative is is $\partial K/\partial P = K' = 4.001$. From Ottonello (2018)

$$\mathbf{u} = \mathbf{u}_0 \exp(2\pi i \nu t) \exp(-2\pi i \mathbf{q} \cdot r) \tag{8.9.2}$$

with ν = frequency. The velocity of propagation of the wave is

$$v = {}^{\nu}/_{\mathbf{q}} \tag{8.9.3}$$

Relating the position to the (time-dependent) elongation in vectorial terms:

$$\mathbf{r} = r_0 + \mathbf{u} \tag{8.9.4}$$

with r_0 = mean position, the equilibrium position at each time t is defined in terms of elastic constants:

$$c_{ijkl} \frac{\partial \mathbf{u}_k}{\partial x_l \partial x_j} = \rho \frac{\partial^2 \mathbf{u}_i}{\partial t^2} \tag{8.9.5}$$

where ρ is density and c_{ijkl} are the coefficients of the dynamical matrix. Accounting for the elongation expression the above equation is simplified to:

$$c_{ijkl} \mathbf{u}_k q_i q_l = \rho v^2 \mathbf{u}_i \tag{8.9.6}$$

The wave vector ρv^2 collects the eigenvalues of the dynamical matrix, solved by the Christoffel determinant. The Christoffel determinant has three roots which depend upon the relative orientation of the wave vector **q** with respect to the axes of symmetry of the crystal. By use of the Voigt's notation we replace, in the case of Forsterite, the indexes of the fourth rank elastic tensor c_{ijkl} with elastic stiffnesses

$c_{\alpha\beta}$: $c_1 = c_{11}$; $c_2 = c_{22}$; $c_3 = c_{33}$; $c_4 = c_{23} + c_{32}$; $c_5 = c_{13} + c_{31}$; $c_6 = c_{12} + c_{21}$ (Nye 1969).

The Christoffel determinant is then resolved by introducing Cauchy relations and accounting for the internal dependencies imposed by the symmetry of the crystal (Autier and Zarembovich 2006). Because the elastic constants are the 2nd derivative of the energy density with respect to the strain components:

$$c_{ij} = \frac{1}{V}\frac{\partial^2 E}{\partial \varepsilon_i \partial \varepsilon_j}\bigg|_0 \tag{8.9.7}$$

we may evaluate their discrete values by imposing a certain amount of strain ε along the crystallographic direction corresponding to the component of the dynamical matrix. Denoting as c′ the arithmetic mean of the diagonal terms c_{11}, c_{22}, c_{33}; c″ the arithmetic mean of the terms c_{12}, c_{23}, c_{33} and c‴ the arithmetic mean of the diagonal terms c_{44}, c_{55}, c_{66}, the Voight (upper, μ_V) and Reuss (lower, μ_R) bounds are given by:

$$\mu_V = \frac{1}{5} \times (c' - c'' + 3c''') \tag{8.9.8.1}$$

$$\mu_R = \left[\frac{1}{15} \times \left(\frac{12}{c' - c''} + \frac{9}{c'''}\right)\right]^{-1} \tag{8.9.8.2}$$

The average shear modulus is found from the Hill average (Hill 1952):

$$\overline{\mu}_{VRH} = \frac{1}{2}(G_V + G_R) \tag{8.9.9}$$

Young's modulus E and Poisson's ratio ν_P are respectively:

$$E = \frac{9K_0\overline{\mu}_{VRH}}{3K_0 + \overline{\mu}_{VRH}} \tag{8.9.10}$$

$$\nu_P = \frac{3K_0 - 2\overline{\mu}_{VRH}}{2 \times (3K_0 + \overline{\mu}_{VRH})} \tag{8.9.11}$$

The averaged shear and longitudinal wave velocities are given by:

$$\overline{v}_s = \sqrt{\overline{\mu}_{VRH}/\rho} \tag{8.9.12}$$

$$\overline{v}_L = \sqrt{\frac{K_0 + \frac{4}{3} \times \overline{\mu}_{VRH}}{\rho}} \tag{8.9.13}$$

The analysis may be carried out in a similar manner for all crystals taking properly into account their symmetry and the obtained seismic velocities may be adopted to constrain adequately the acoustic cutoff limits of vibrational frequencies in the absence of experimental data (see Sect. 8.10.4). We will see in the application section further examples of the procedure for a tetragonal crystal (Stishovite).

8.9.3 Mode-Gamma Analysis and Thermal Expansion

The "*quasi-harmonic*" mode-gamma analysis is based on the assumption that, although the nuclear centers behave, in the quantum–mechanical Hamiltonian, as harmonic oscillators, "external" zones may be explored as well by expanding the substance through symmetry preserving operations. This in practice corresponds to changing progressively the force constants of the various bonds. Repeating the Hessian analysis on the various expanded states one obtains for each normal mode linearized frequencies v_i that vary progressively with the imposed volume V. The so-called γ is given by the ratio:

$$\gamma_{\rm i} = -\frac{\partial \ln v_{\rm i}}{\partial \ln {\rm V}} \tag{8.9.14}$$

and the thermal expansion of the substance at T ($\alpha_{\rm T}$), multiplied by the bulk modulus at the same temperature (K_T) is given by:

$$\alpha_{\rm T}{\rm K}_{\rm T} = -\left(\frac{\partial {\rm V}}{\partial {\rm T}}\right)_{\rm P} \times \left(\frac{\partial {\rm V}}{\partial {\rm P}}\right)_{\rm T}^{-1} = \frac{\rm R}{\rm ZV}\sum_{\rm i=4}^{3n} \gamma_{\rm i} e^{\rm X_i}\left(\frac{\rm X_i}{e^{\rm X_i}-1}\right)^2 \tag{8.9.15}$$

where Z are the formula units in the primitive unit cell. The thermal dependency of the bulk modulus is obtained by applying (Anderson 1995):

$$\left(\frac{\partial {\rm K}}{\partial {\rm T}}\right)_{\rm P} = -\overline{\alpha {\rm K}} \times \left[{\rm K}' - {\rm q}^{\rm ht} + 1\right] \tag{8.9.16}$$

where $\overline{\alpha K}$ is the limiting αK product at $T \rightarrow \infty$ and

$$\overline{\gamma}_{\rm ht} = \left(\frac{1}{3n-3}\right)\sum_{\rm i=4}^{3n} \gamma_{\rm i} \tag{8.9.17}$$

$${\rm q}^{\rm ht} = \frac{\ln \overline{\gamma}_{\rm ht} - \ln \gamma_0}{\ln {\rm V}_{\rm ht} - \ln {\rm V}_0} \tag{8.9.18}$$

Values of γ_0 and q may be obtained only by adiabatic decompression methods (Boehler and Ramakrishnan 1980; Boehler 1982). There is however a strict

Table 8.5 Quasi-harmonic analysis of vibrational modes, and the ensuing thermodynamic parameters. $\overline{\alpha K}$ and $(dK/dT)_P$ are in *bar/K*. $\alpha_T = \alpha_0 T + \alpha_1 + \alpha_2 T^{-1} + \alpha_3 T^{-2} + \alpha_4 T^{-3}$. Thermal expansion expression valid in the T range $200 \leq T\ (K) \leq 3000$. Mg_2SiO_4 polymorphs after Ottonello et al. (2009a); SiO_2-stishovite after Ottonello et al. (2009b); Phase Anhydrous B ($Mg_{14}Si_5O_{24}$) after Ottonello et al. (2010). Sapphirine-351 ($Mg_3Al_{10}SiO_{20}$) and Sapphirine-442 ($Mg_4Al_8Si_2O_{20}$) after Belmonte et al. (2014)

Phase	q^{ht}	$\overline{\alpha K}$	K′	$(dK/dT)_P$	$\alpha_0 \times 10^7$	$\alpha_1 \times 10^7$	$\alpha_2 \times 10^3$	α_3	α_4
γ-Mg_2SiO_4	3.283	51	4.322	−104	0.0000	323.000	−7.308	1.374	−150.406
β-Mg_2SiO_4	3.262	55	4.415	−118	0.0000	464.000	−13.800	3.620	-395.094
α-Mg_2SiO_4	1.257	47	4.001	−175	0.1003	269.309	5.608	−2.616	243.828
SiO_2-stishovite	1.32	55	6.588	−345	0.0590	167.770	2.287	−2.152	237.025
$Mg_{14}Si_5O_{24}$	1.434	51	4.140	−189	0.0863	230.860	8.313	−4.345	528.645
$Mg_3Al_{10}SiO_{20}$	1.007	44	4.280	−188	0.0608	157.090	8.284	−4.355	547.232
$Mg_4Al_8Si_2O_{20}$	1.320	42	4.680	−210	0.0903	104.110	14.034	−6.447	814.288

interdependence between γ_0 and q,[11] and, in the absence of data, it is reasonable to assume $\gamma_0 = 1$. The power law relation (8.9.18) is in this way recast into

$$q^{ht} \cong \frac{\ln \overline{\gamma}_{ht}}{\ln\left(V_{ht}/V_0\right)} \tag{8.9.19}$$

By knowing the thermal effect on K we may easily retrieve the thermal expansion at all T conditions and recast it eventually as a polynomial over reciprocal temperature with negligible error in reproducibility. Examples of mode-gamma analysis carried out by our laboratory on some important rock-forming minerals are resumed in Table 8.5.

If a full phonon dispersion of the substance is available, then the $\alpha_T \times K_T$ product (Eq. 8.9.15) is better represented as (Belmonte 2017):

$$\alpha_T K_T(V,T) == \frac{R}{ZV} \sum_{\mathbf{k},i=1}^{3n} \gamma_i(\mathbf{k},V) e^{x_i} \left(\frac{x_i(\mathbf{k},V)}{e^{x_i(\mathbf{k},V)} - 1} \right)^2 \tag{8.9.20}$$

Here the summation is extended to the acoustic branches and the notation emphasizes that the adimensional frequencies are subjected to phonon dispersion (besides being V-dependent).

Belmonte (2017) carried out a very detailed ab initio investigation on Periclase (MgO with *fcc* or B1-type cubic structure) which has always been considered to be a milestone for computational investigation, due to the possibility of comparing

[11] "*we apply the frequently used power law relationship $\gamma = \gamma_0 (V/V_0)^q$ (6), where γ_0 is γ at zero pressure. Using (6), various sets of γ_0 and q are chosen, and γ values are calculated at pressures where dT/dP measurements were carried out*" (Boehler and Ramakrishnan 1980).

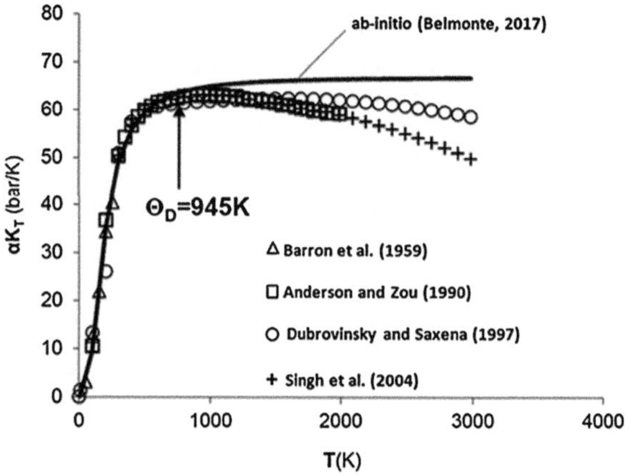

Fig. 8.17 The $\alpha_T K_T$ product of MgO. The ab initio DFT/B3LYP results of Belmonte (2017) (Eq. 8.9.21) are compared to experiments (Barron et al. 1959; Dubrovinsky and Saxena 1997) and assessments (Anderson and Zou 1990; Singh et al. 2004)

calculations with a great variety of experimental data in a broad range of P–T conditions.

Based on the investigation of Belmonte (2017), the $\alpha_T K_T$ product of MgO attains at high T a constant value of roughly 65 *bar/K* (Fig. 8.17) while both experiments (Barron et al. 1959; Dubrovinsky and Saxena 1997) and assessments (Anderson and Zou 1990; Singh et al. 2004) seem to indicate that $\alpha_T K_T$ should decrease above the Debye temperature (θ_D).

Instead of using the mode-gamma procedure, and a 3rd order Birch–Murnaghan EoS (Eq. 8.9.1), Belmonte (2017) adopted a Mie-Grüneisen Equation of State (EoS) (see later) to depict the P–V–T relations and to obtain the appropriate thermal derivative of the bulk modulus at zero pressure (and at higher pressures as well, see later).

As we may see in Fig. 8.18 the computed values of thermal expansion reproduce the experimental observations (Barron et al. 1959; Suzuki 1975; Dubrovinsky and Saxena 1997). reasonably well up to $\sim 2000\ K$, then are overestimated. Overestimation is more severe, and begins at lower T, when the ab initio calculations are carried out at DFT-LDA theory level (Karki et al. 2000).

8.9.4 Ab Initio Equation of State (EoS)

P–V–T relations in crystalline solids of geochemical interest are of paramount importance in establishing their extrinsic stability limits in the wide thermobaric regimes involved in planetary investigations (pressure plays almost always a minor role in material science). In discussing some aspects of EoS in rock-forming minerals I will thus conform to the notation of Anderson (1995) (a classical text of reference in Earth Sciences), with limited modifications.

Ignoring magnetization and optical effects, the Helmholtz free energy may be expressed in terms of three contributions E_{st}, F_{vib}, and $E_{f\text{-}el}$. where E_{st} is the

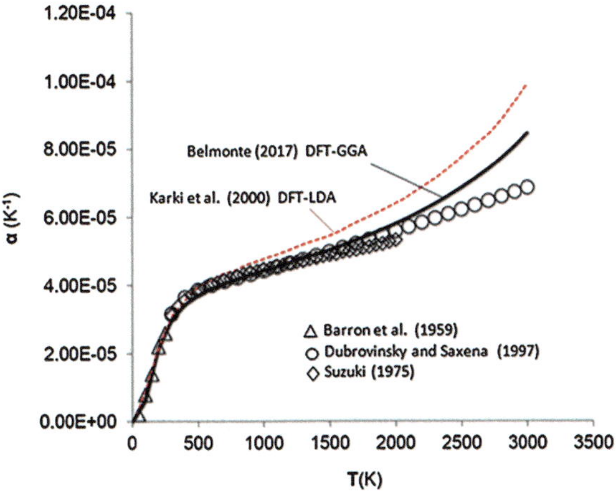

Fig. 8.18 Thermal expansion of MgO. Ab initio DFT-GGA computations of Belmonte (2017) (Eq. 8.9.20) are compared with DFT-LDA results of Karki et al. (2000) and with the experiments of Barron et al. (1959), Suzuki (1975), Dubrovinsky and Saxena (1997)

potential of a static lattice at absolute zero, F_{vib} is the vibrational part of the Helmholtz free energy and $E_{f\text{-}el}$ is the potential arising from delocalized (free) electrons:

$$F = E_{st} + F_{vib} + E_{f-el} \tag{8.9.21}$$

The two terms E_{st} and $E_{f\text{-}el}$ may be grouped into a single ab initio (ai) calculated contribution arising from the motion of localized + delocalized electrons about nuclei:

$$F = E_{ai} + F_{vib} \tag{8.9.22}$$

Because pressure P is the partial derivative of Helmholtz free energy with respect to volume V:

$$P = -\left(\frac{\partial F}{\partial V}\right)_T \tag{8.9.23}$$

we have, from Eq. (8.9.22):

$$P = -\left(\frac{\partial E_{ai}}{\partial V}\right)_{T=0} - \left(\frac{\partial F_{vib}}{\partial V}\right)_T \tag{8.9.24}$$

Besides the V-dependent zero point static contributions, P_{st}, pressure contains thus a thermal contribution (i.e. the "*thermal pressure*" P_{th}) which is dependent upon both V and T (cf. Eq. 1.17 in Anderson 1995):

$$P(V,T) = P_{st}(V) + P_{th}(V,T) \tag{8.9.25}$$

Replacing V with the dimensionless variable "*dilaton*" $\eta = V/V_0$ with the subscript 0 indicating the reference pressure (usually $P_0 = 0$), we get (see Eq. 1.17 in Anderson 1995):

$$P(\eta,T) = P_{st}(\eta) + P_{th}(\eta,T) \tag{8.9.26}$$

We can now write

$$K_T(\eta,T) = K_0(\eta) + K_{th}(\eta,T) \tag{8.9.27}$$

where K_T is the isothermal bulk modulus that is defined as

$$K_T = -V\left(\frac{\partial P}{\partial V}\right)_T \tag{8.9.28}$$

Equation (8.9.27) stresses that only at the athermal limit the bulk modulus is a simple function of the dilaton.

According to Brillouin (1946), P_{th} may be considered as the pressure required to keep the volume constant as T rises. Integrating the identity:

$$\alpha_T K_T = \left(\frac{\partial P}{\partial T}\right)_V \tag{8.9.29}$$

and recalling (8.9.25) we get:

$$P_{th} = \int_0^T \alpha_T K_T dT + c = P_{th,harm} + P_{th,anharm} \tag{8.9.30}$$

where c is an integration constant representing the virtual thermal pressure eventually present at the athermal limit and V_0.

In Eq. (8.9.30) one identifies a harmonic and an anharmonic contribution to the bulk thermal pressure. The first term coincides with the static pressure generated by the EoS for dilaton values >1 while the second one is obtained by the difference between bulk thermal pressure and its harmonic component.

A correct evaluation of the above integral requires knowledge of both the isobaric thermal expansion α_T and the intrinsic temperature derivative $(\partial K_T/\partial T)_V$. The most correct way to proceed is to explore the P–V–T space with an ab initio full phonon dispersion calculation, through the Mie-Gruneisen EoS (Landau and Liftshitz 1980):

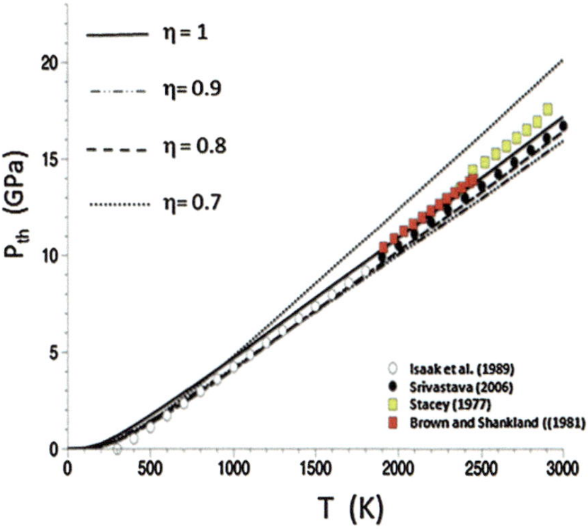

Fig. 8.19 Ab initio B3LYP thermal pressure (P_{th}) calculated for MgO along different isochors (η = V/V$_0$), compared with experimental observations at ambient pressure (η = 1; Isaak et al. 1989), thermodynamic assessments (Srivastava 2006) and geophysical models (Stacey 1977; Brown and Shankland). Modified from Belmonte (2017)

$$P(V,T) = -\left(\frac{\partial U_{st}}{\partial V}\right)_T + \frac{k_B T}{2V}\sum_{i=1}^{3n}\gamma_i(\mathbf{k},V)\cdot x_i(\mathbf{k},V) + \frac{k_B T}{V}\sum_{i=1}^{3n}\gamma_i(\mathbf{k},V)\cdot\frac{x_i(\mathbf{k},V)}{\exp[x_i(\mathbf{k},V)-1]} \tag{8.9.31}$$

In Fig. 8.19 we see the results of the full phonon dispersion calculations carried out by Belmonte (2017) for Periclase.

Splitting the vibrational contribution to P(V,T) into Zero Point [P_{ZP}(V)] and thermal [P_{th}(V,T)] components, the lattice dynamics analog of Eq. (8.9.28) is (Born and Huang 1954):

$$P(V,T) = -\left(\frac{\partial F}{\partial V}\right)_T = P_{st}(V,T) + P_{ZP}(V) + P_{th}(V,T) \tag{8.9.32}$$

Equating Eqs. (8.9.31) and (8.9.32), Belmonte (2017) calculated the thermal pressure up to very high compressional states of Periclase (η = 0.70) (Fig. 8.19).

We may immediately appreciate that, contrarily to thermal expansion, thermal pressure doesn't display at room pressure (η = 1) any deviation from linearity at high T (i.e. T > $\Theta_D \cong 945$ *K*). Apparently, linearity at high T is maintained also at high compression. According to Belmonte (2017), the slight differences with respect to the low-T data of Isaak et al. (1989) and Srivastava (2006) are due to the poorly adopted constraint [P_{th} (300 *K*, 1 *bar*) = 0], which lacks a theoretical justification, because thermal vibrations obviously are present already at the athermal limit.

8.10 Thermodynamics

8.10.1 Simple Approximate Treatment

Following the treatment of Born and Huang (1954), it is worth to anticipate that the normal coordinates q_i describing the atomic displacements in a crystal follow a sinusoidal function of time:

$$q_i = A_i \sin(2\pi \nu_i t + \delta_i) \tag{8.10.1}$$

with amplitude A_i, (linear) frequency ν_i and δ_i as arbitrary constant. The number of *normal vibrations* associated to Eq. (8.10.1) (or "*normal modes*" cf. Sect. 8.7.1 and following) is obviously equal to the degrees of freedom of the system, corresponding to 3N, with N as the number of particles. In many ways "*the vibratory system is completely equivalent to a collection of independent simple harmonic oscillators with the respective frequencies* ν_i*. In particular, the possible quantum–mechanical energy levels are identical in the two cases*" (Born and Huang 1954). These "*energy levels*" determine the thermodynamic functions of the crystal. The vibrational part of the Helmholtz free energy F_{vib} is determined by the "*partition function*"; i.e. the sum of the Boltzmann factors corresponding to all possible energy levels:

$$Q_{vib} = \sum_i \exp(-\varepsilon_i/k_B T) \tag{8.10.2}$$

with ε_i as eigenvalue of the energy operator, and

$$F_{vib} = -k_B T \ln Q_{vib} \tag{8.10.3}$$

The eigenvalues of an oscillator with frequency ν_i are $\frac{1}{2}h\nu_i$, $\frac{3}{2}h\nu_i$, $\frac{5}{2}h\nu_i$, ...; and, collecting the energies of the independent oscillators separately, and summing up, it is obtained:

$$Q_i = \exp\left(-\frac{1}{2}h\nu_i \Big/ k_B T\right) \sum_{s=0}^{\infty} \exp(-sh\nu_i/k_B T) = \frac{\exp(-\frac{1}{2}h\nu_i/k_B T)}{1 - \exp(-h\nu_i/k_B T)} \tag{8.10.4}$$

$$F_{vib,i} = \frac{1}{2}h\nu_i + k_B T \ln[1 - \exp(-h\nu_i/k_B T)] \tag{8.10.5}$$

Adding the static lattice free energy U_0 (deprived of the zero point vibrational contribution) to the vibrational part of the Helmholtz free energy it is obtained the internal energy of the system (U):

$$U = U_0 + \frac{1}{2}\sum_i h\nu_i + k_B T \ln \sum_i [1 - \exp(-h\nu_i/k_B T)] \tag{8.10.6}$$

The static energy U_0 introduced in Eq. (8.10.6) is at the minimum of the potential well, when all atoms are at equilibrium positions. Any shift from these positions, induced by thermal stress, alters both frequencies and static energy. Because the vibrational entropy is related to the vibrational Helmholtz free energy by

$$S_{vib} = -\left(\frac{\partial F_{vib}}{\partial T}\right)_V \tag{8.10.7}$$

the energy of the system of oscillators is given by:

$$U = F + TS_{vib} = F - T\left(\frac{\partial F_{vib}}{\partial T}\right)_V = U_0 + \frac{1}{2}\sum_i h\nu_i + k_B T \sum_i \frac{(h\nu_i/k_B T)}{[\exp(h\nu_i/k_B T) - 1]} \tag{8.10.8}$$

At the athermal limit, the last term on the right in Eq. (8.10.8) vanishes and the energy reduces to the static term U_0 plus the zero point energy $\frac{1}{2}\sum_i h\nu_i$.

Differentiating the internal energy with respect to T, it is obtained the isochoric heat capacity:

$$C_V = \left(\frac{\partial U}{\partial T}\right)_V = k_B \sum_i \frac{(h\nu_i/k_B T)^2 \times \exp(h\nu_i/k_B T)}{[\exp(h\nu_i/k_B T) - 1]^2} \tag{8.10.9}$$

At high T, where $k_B T \gg h\nu_i$, the summations in Eqs. (8.10.8) and (8.10.9) can be expanded in terms of $(h\nu_i/k_B T)$ obtaining:

$$U \cong U_0 + \frac{1}{2}\sum_i h\nu_i + 3k_B TN \tag{8.10.10}$$

$$C_V \cong 3k_B TN = 3\left(\frac{N}{N_0}\right)k_B N_0 = 3nR \tag{8.10.11}$$

where n is the number of atoms per formula unit.

8.10.2 Planck Distribution

For a set of identical harmonic oscillators in thermal equilibrium, the relative occupancies of the various quantum states are given by the Boltzmann factor:

$$\frac{N_{n+1}}{N_n} = \exp(-h\nu/k_B T) \equiv \exp(-\hbar\omega/k_B T) \tag{8.10.12}$$

The fraction of the number of oscillators in the *n*th quantum state with respect to their total number is:

$$\frac{N_n}{\sum_{s=0}^{\infty} N_s} = \frac{\exp(-n\hbar\omega/k_B T)}{\sum_{s=0}^{\infty} \exp(-s\hbar\omega/k_B T)} \tag{8.10.13}$$

The average excitation quantum number of an oscillator is (Kittel 1976):

$$\langle n \rangle = \frac{\sum_s s \exp(-s\hbar\omega/k_B T)}{\sum_s \exp(-s\hbar\omega/k_B T)} \tag{8.10.14}$$

Posing $x = \exp(-\hbar\omega/k_B T)$, the terms on the right in Eq. (8.10.14) become:

$$\sum_s x^s = \frac{1}{1-x} \tag{8.10.15.1}$$

$$\sum_s s x^s = x \frac{d}{dx} \sum_s x^s = \frac{x}{(1-x)^2} \tag{8.10.15.2}$$

Replacing the two summations in Eq. (8.10.14) the Planck distribution reduces to (Kittel 1976):

$$\langle n \rangle = \frac{x}{1-x} = \frac{1}{\exp(\hbar\omega/k_B T) - 1} \tag{8.10.16}$$

8.10.3 The Einstein and Debye Models

I anticipated the Einstein model (Einstein 1907) in the first pages of this book, due to its paramount importance in Thermodynamics. Actually, this model is based on the simplest approximation, namely that all the vibrational frequencies are equal. As noted by Born and Huang (1954) "*One could imagine this to be the case only if the lattice particles in a simple lattice were to vibrate independently on one another, each isotropically about its mean position*".

Stemming from the consideration that, according to Planck's theory, the mean energy of a great number of oscillators is given by

$$\langle\langle E \rangle\rangle = \frac{\hbar\omega_t}{\exp(\hbar\omega_t/k_B T) - 1}, \tag{1.1.8}$$

where ω_t is the characteristic angular frequency, Einstein (1907) proposed that the internal energy of a solid composed of an Avogadro number of atoms N_0 is given by:

$$U = \frac{3N_0}{\exp(\hbar\omega_E/k_BT) - 1}, \qquad (1.1.9)$$

C_V was then obtained from Eq. (1.1.9) by partial derivation in dT:

$$C_V = \left(\frac{\partial U}{\partial T}\right) = \frac{3N_0k_BX^2e^X}{(e^X - 1)^2} = 3RE_{(X)}, \qquad (1.1.10)$$

with $X = \frac{h\nu_E}{k_BT} = \frac{\hbar\omega_E}{k_BT}$, and

$$E_{(X)} = \frac{X^2e^X}{(e^X - 1)^2} = \frac{X^2}{a\ \sin^2\left(\frac{X}{2}\right)}, \qquad (1.1.11)$$

In the Einstein function ($E_{(x)}$), the characteristic frequency ω_E (and the corresponding characteristic temperature $\theta_E = \hbar\omega_E/k_B$) has an arbitrary value that optimizes Eq. (1.1.11) on the basis of high T experimental data.

We have also seen at Sect. 8.7.2 that, for each polarization type, the density of modes $\mathcal{D}(\omega)$ is:

$$\mathcal{D}(\omega) = \frac{dN}{d\omega} = \frac{V\omega^2}{2\pi^2v^3} \qquad (8.7.19)$$

In the continuum (Debye) approximation (Debye 1912), the sound velocity is constant and there is one allowed value of **k** per unit volume of **k** space, for each polarization and for each branch. For N primitive cells there are N acoustic phonon modes and, because the total number of modes with wave vector less than **k** is:

$$N = \left(\frac{L}{2\pi}\right)^3\frac{4}{3}\pi k^3 = \left(\frac{L}{2\pi}\right)^3\frac{4}{3}\pi\left(\frac{\omega}{v}\right)^3 = \frac{V}{6\pi^2}\left(\frac{\omega}{v}\right)^3 \qquad (8.7.18)$$

this implies the existence of a cutoff frequency:

$$\omega_D = \left(\frac{6\pi^2v^3N}{V}\right)^{1/3} \qquad (8.10.17)$$

and a maximum wave vector: $\mathbf{k}_D = \omega_D/V$; i.e. "*Die Linien kleinster Schwingungszahl sind die gewohnlichen akustischen Schwingungen*" (Debye 1912).

While the internal energy of a discrete collection **k** of oscillators in thermal equilibrium with angular frequencies $\omega_{\mathbf{k}}$ is, for each polarization type:

$$U = \sum_{\mathbf{k}} n_{\mathbf{k}} \hbar \omega_{\mathbf{k}} \tag{8.10.18}$$

where each n is related to ω in the terms seen at Sect. 8.10.2. In the presence of dispersion, the above equation is better replaced by:

$$U = \int \mathcal{D}(\omega) \hbar \omega n(\omega, T) d\omega = \int_0^{\omega_D} \left(\frac{V\omega^2}{2\pi^2 v^3} \right) \left[\frac{\hbar\omega}{\exp(\hbar\omega/k_B T) - 1} \right] d\omega \tag{8.10.19}$$

Assuming that the phonon velocity is the same in the 3 crystallographic directions and extracting from the integral the constant terms, it is obtained (Kittel 1976):

$$U = \left(\frac{3V\hbar}{2\pi^2 v^3} \right) \int_0^{\omega_D} \frac{\omega^3}{\exp(\hbar\omega/k_B T) - 1} d\omega = \left(\frac{3V k_B^4 T^4}{2\pi^2 v^3 \hbar^3} \right) \int_0^{x_D} \frac{x^3}{\exp(x) - 1} d\omega \tag{8.10.20}$$

where $x = \hbar\omega/k_B T$ and $x_D = \hbar\omega_D/k_B T = \theta_D/T$.

Differentiating Eq. (8.10.20) in dT, we obtain the isochoric heat capacity (Kittel 1976):

$$\begin{aligned} C_V &= \left(\frac{3V\hbar^2}{2\pi^2 v^3 k_B T^2} \right) \int_0^{\omega_D} \frac{\omega^4 \exp(\hbar\omega/k_B T)}{[\exp(\hbar\omega/k_B T) - 1]^2} d\omega \\ &= 9Nk_B \left(\frac{T}{\theta} \right)^3 \int_0^{x_D} \frac{x^4}{[\exp(x) - 1]^2} d\omega \end{aligned} \tag{8.10.21}$$

Moreover, taking into account Eq. (8.10.17) we may simplify Eq. (8.10.20) into:

$$U = 9Nk_B T \left(\frac{T}{\theta_D} \right)^3 \int_0^{x_D} \frac{x^3}{[\exp(x) - 1]} d\omega \tag{8.10.22}$$

It may be easily seen that Eqs. (8.10.20) and (8.10.22) are mutually consistent.

In the case of a cubic crystal, the Debye approximation is quite reasonable. We see for example in Fig. 8.20 the comparison between the Debye model and the calculations of Iona (1941) for KCl.

To calculate the distribution function for the simplest cubic crystal (KCl), the masses of the ions were assumed by Iona (1941) to be equal, and, under the assumption of equal forces between the ions (except for the difference in sign of the

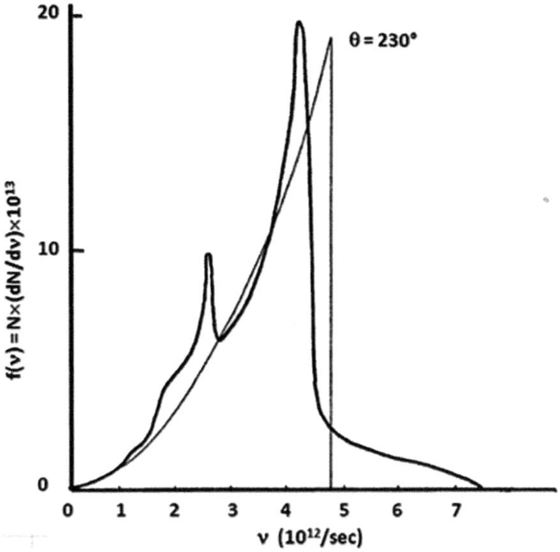

Fig. 8.20 Density distribution of the lattice vibrations of the KC1 crystal. The distribution of characteristic vibrations of the KCl crystal (heavy solid line) was calculated by Iona (1941) on the basis of the Born–von Karman theory (Born and von Karman 1912, 1913) with repulsive forces based on the calculated elastic constants (Ljungquist, unpublished) and the observed infrared dispersion frequencies. The thin line is the Debye model. Modified from Iona (1941)

charge) the crystal was considered as a simple cubic lattice. When the masses of the ions are equal the lattice "*behaves somewhat like a monatomic solid with one atom to each cell; the optical branches join smoothly with the acoustic branches so that we have apparently only three acoustic branches, the upper halves of which correspond to the optical vibrations*" (Born and Huang 1954).

8.10.4 The Kieffer Model

Figure 8.21 shows a schematic representation of the density of states in a crystal. Acoustic modes generally prevail at low frequencies, the optical continuum occurs at intermediate frequencies, and Einstein's oscillators are found at high frequencies.

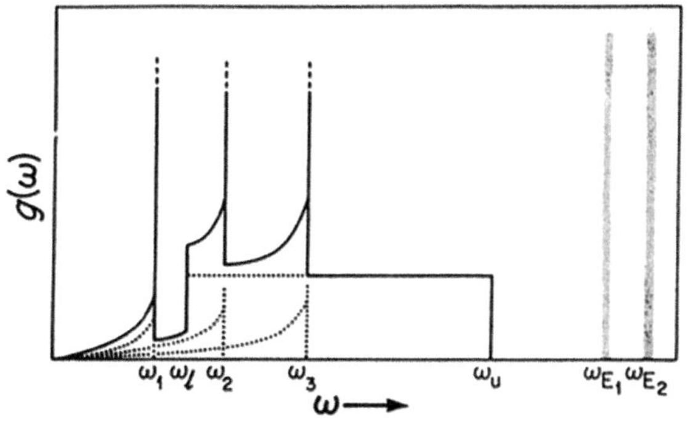

Fig. 8.21 Schematic representation of a vibrational spectrum of a crystalline phase. Dotted curves: acoustic branches and optical continuum. Solid line: total spectrum. E_1 and E_2: Einstein oscillators. From Kieffer (1979c). With permission from John Wiley and Sons

According to Kieffer (1979a, b, c, 1980, 1982, 1985), the density of states for acoustic branches (*dispersed acoustic function*) may be represented by three sinusoidal functions:

$$\mathcal{S}(\omega) = \left(\frac{2}{\pi}\right)^3 \int_0^{x_i} \frac{[\arcsin(x/x_i)]^2 x^2 e^2}{\left(x_i^2 - x^2\right)^{1/2} (e^x - 1)^2} dx \tag{8.10.23}$$

where i is the maximum frequency of each acoustic branch (see Kieffer 1979c).

In the optical continuum, the frequency distribution is approximately constant (see Fig. 8.17). The corresponding vibrational function has the form (Kieffer 1979c, 1985):

$$\mathcal{K}(x_l, x_u) = \int_{x_l}^{x_u} \frac{x^2 e^x}{(x_u - x_l)^x (e^x - 1)^2} dx \tag{8.10.24}$$

where x_l and x_u correspond, respectively, to the lower and upper bounds. For the Einstein oscillators the vibrational function is (cf. Eq. 1.1.11):

$$\mathcal{E}(x_E) = \frac{x_E^2 e^{x_E}}{(e^{x_E} - 1)^2} \tag{8.10.25}$$

The isochoric molar heat capacity, based on vibrational functions (8.10.23–8.10.25), is

$$C_V = \frac{3R}{\mathcal{Z}} \sum_{i=1}^{3} \mathcal{S}(x_i) + 3nR\left(1 - \frac{1}{n\mathcal{Z}} - q\right) \mathcal{K}(x_l, x_u) + 3nRq\mathcal{E}(x_E) \tag{8.10.26}$$

where q is the proportion of vibrational modes due to Einstein oscillators and $\mathcal{Z}$ is the number of formula units per unit cell.

Because the dispersed acoustic function (8.10.23), the optic continuum function (8.10.24), and the Einstein function (8.10.25) may be tabulated for the limiting values of adimensional frequencies (see Tables 1, 2, and 3 in Kieffer 1979c), the evaluation of C_V reduces to the appropriate choice of lower and upper cutoff frequencies for the optic continuum (i.e. x_l and x_u limits of integration in Eq. 8.10.24), of the three zone edges for the dispersed acoustic function (i.e., three values of the x_i upper limit of integration in Eq. 8.10.23), and of the frequency bands representing Einstein oscillators (i.e., x_E in Eq. 8.10.25). Acoustic branches for the various phases of interest may be derived from acoustic velocities through the guidelines outlined by Kieffer (1980). Vibrational modes at higher frequencies may be derived by infrared and Raman spectra. Note incidentally that the tabulated values of the dispersed sine function in Kieffer (1979c) are 3 times the real ones (i.e.

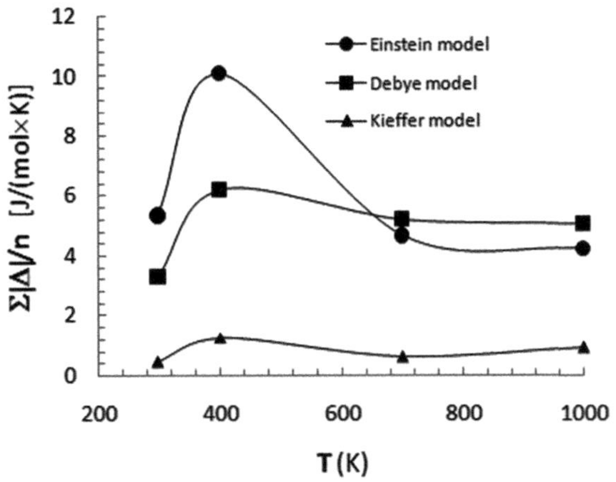

Fig. 8.22 Reduced residuals in the application of the Einstein, Debye, and Kieffer models to rock-forming minerals at various T (T = 298.15 *K*: 28 values; T = 400 *K*: 2 values; T = 700 *K*: 25 values; T = 1000 *K*: 23 values). See Kieffer (1985) for the list of investigated phases and computational details

the listed values must be divided by 3 to obtain the appropriate $S(x_i)$ value for each acoustic branch; see also Kieffer 1985).

To appreciate the predictive properties of Kieffer's model, it is sufficient to compare the reduced residual in the entropy estimates at various T with the results of the Einstein and Debye models for a set of rock-forming minerals (Fig. 8.22).

Besides the evident superiority of the procedure, one advantage of Kieffer's model with respect to the two preceding formulations is its wider T range of applicability (the Debye's model is appropriate to low frequencies and hence to low T, whereas the Einstein model is appropriate to high frequencies and hence to high T).

The biases still observed between experimental measurements and Kieffer's model predictions are due to the relative paucity of experimental data concerning cutoff frequencies of acoustic branches, and also to the assumption that the frequencies of the lower optical branches are constant with **k** and equivalent to those detected by Raman and IR spectra (corresponding only to vibrational modes at **k** = 0). Indeed, several of these vibrational modes, and often the most important ones, are Raman and/or IR inactive.

Notwithstanding the remarkable model achievements, the approach has lost most of its appeal with respect to ab initio procedures, with the giant progresses in parallel computing in the last two decades. It keeps however utility in avoiding the ab initio problem of undetectable acoustic frequencies at the Γ point (resolved only through time-expensive computations on other special points in extended supercells) or in systematic investigations of structural classes.

8.10.5 *Magnetic Coupling Effects*

Magnetic couplings add important contributions to the bulk heat capacity of crystalline substances at low T (Schottky anomalies) and to the bulk entropy of the substance at all T (as *quenched* and *unquenched* terms, see later). As far as I know, these contributions are not accounted for by any exchange–correlation term in pre-arranged density functionals, and deserve a separate treatment. As an example, I report hereafter the case of REE-Fe garnets, of general formula $X_3Y_5O_{12}$. Like other garnets, their structure is best described by the *Ia3d* space group. Their body-centered cell contains eight formula units. In the Wyckoff notation, Rare Earth cations X occupy special position *c* (24 dodecahedrally coordinated sites), whereas the Y cation (Fe^{3+}) share special positions *a* and *d* (16 tetrahedrally and 24 octahedrally coordinated sites, respectively). All cations are thus in fixed position, with no degrees of freedom, while oxygen anions are in general position *h*, with three degrees of positional freedom.

REE-Fe garnets display strong magnetic contributions to the heat capacity function, due to couplings between X-sublattice and Y_a, Y_d-sublattice magnetic dipoles (as indicated by Néel 1948), Fe^{3+} ions in Y_a sites are strongly coupled antiferromagnetically to Fe^{3+} ions in Y_d sites. The resulting magnetization of the Y-sublattices is in turn antiferromagnetically coupled to the spins of the trivalent rare earths on the X site). Coupling between the Y_a and Y_d sublattices is responsible for the Curie point, which is about 550 *K* for all Fe garnets. Harris and Meyer (1962) quantified energy splitting in the framework of the Weiss Molecular Field model (WMF) and the spin-wave approximation (SW). To address the problem, the above authors subtracted a "*lattice contribution*", evaluated through the Debye T^3 law, from the bulk heat capacity by the selection of appropriate θ_{Dcal} values:

$$C_V = \frac{12}{5}\pi^4 n N_0 k_B \left(\frac{T}{\theta_{Dcal}}\right)^3 \qquad (8.10.27)$$

For the sake of clarity, it must be emphasized that S-state rare earth ions (J = 0) deserve separate treatment with respect to non-S-state (J $\neq$ 0) ions. The spin-wave approximation (SW) adopted by Harris and Meyer (1962) for $Y_3Fe_5O_{12}$ and $Lu_3Fe_5O_{12}$ involves the following assumptions:

(*i*) the exchange magnetic field is isotropic; (*ii*) the external field is zero; (*iii*) only nearest neighbor interactions are important, and (*iv*) magnetization therefore does not decrease substantially with T from the value attained at absolute zero. With these provisos, magnetic heat capacity is represented by a Bloch-type $T^{3/2}$ law (Kittel 1976):

$$C_{magn} = 0.1175\left(k_B T/D\right)^{3/2} \qquad (8.10.28)$$

where:

$$D = \frac{1}{S}\left[50J_{aa} - \frac{125}{4}J_{ad} + \frac{75}{4}J_{dd}\right] \qquad (8.10.29)$$

and J_{aa}, J_{dd}, and J_{ad} are exchange integrals for Y_a-Y_d, Y_d-Y_d, and Y_a-Y_d sublattice interactions and S is total spin. Term D should attain almost the same value for $Y_3Fe_5O_{12}$ and $Lu_3Fe_5O_{12}$, but Eq. (8.10.28) does not reproduce the expected Schottky anomaly (i.e. the magnetic heat capacity, being linearized, does not decrease progressively at high T; cf. Tinkham 1961). Moretti and Ottonello (1998) thus preferred to simulate the magnetic effect of the Y_a, Y_d sublattices through the WMF approach, even for S-state X ions, introducing appropriate average values for splitting terms E1 (Table 8.6). For non-S-state X ions, the magnetic interactions

Table 8.6 Magnetic splitting terms for REE-Fe garnets (From Moretti and Ottonello 1998)

Compound	$\bar{E}1$	$\bar{E}2$	References
$Y_3Fe_5O_{12}$	99.022	–	1
$La_3Fe_5O_{12}$(*)	(99.022)	–	1
$Ce_3Fe_5O_{12}$(#)	(38.873)	–	1
$Pr_3Fe_5O_{12}$($)	(25–30)	–	1
$Nd_3Fe_5O_{12}$($)	(20–25)	–	1
$Sm_3Fe_5O_{12}$	53.05	–	2
	38.873	–	1
$Eu_3Fe_5O_{12}$(*)	(99.022)	–	1
$Gd_3Fe_5O_{12}$	27.800	–	2
	26.826	–	1
	24.9	–	3
	19.2	–	4
$Tb_3Fe_5O_{12}$	36.000	50.000	2
	37.040	109.340	1
$Dy_3Fe_5O_{12}$	32.400	62.500	2
	28.764	72.396	1
$Ho_3Fe_5O_{12}$	32.400	71.000	2
	30.312	53.999	1
$Er_3Fe_5O_{12}$	20.700	50.000	2
	20.011	37.576	1
	18.2	21.6	5
$Tm_3Fe_5O_{12}$(&)	(37.040)	(109.340)	1
$Yb_3Fe_5O_{12}$	25.000	–	1
	24.478	–	2
	25.5	–	3
	108	–	4
	24.9	–	5
	23.7	–	6
$Lu_3Fe_5O_{12}$	67.753	–	1

(*) assumed identical to $Y_3Fe_5O_{12}$; (#) assumed identical to $Sm_3Fe_5O_{12}$; (&) assumed identical to $Tb_3Fe_5O_{12}$; ($) adopted range. References: (1) Moretti and Ottonello (1998); (2) Harris and Meyer (1962); (3) Pauthenet (1959); (4) Aleonard (1960); (5) Sievers and Tinkam (1961); (6) Wickersheim (1961)

among the various sublattices were represented by a summation of two terms (nuclear spin effects neglected):

$$C_{magn} = \frac{3}{2}C_1 + \frac{3}{2}C_2 \tag{8.10.30}$$

where:

$$C_i = RT\frac{\partial^2}{\partial T^2}(T \ln Z_i) \tag{8.10.31}$$

and:

$$Z_i = \sum_m \exp\left(-E^i_m/k_B T\right) \tag{8.10.32}$$

E^i_m in Eq. (8.10.32) is the mth energy level in the ith site (see Table 8.6). The resulting magnetic heat capacity is given by the Schottky equation:

$$C_{magn} = 3R\left\{\left(\frac{\bar{E}1}{k_B T}\right)^2\left[\frac{\exp(-\bar{E}1/k_B T)}{[1+\exp(-\bar{E}1/k_B T)]^2}\right] + \left(\frac{\bar{E}2}{k_B T}\right)^2\left[\frac{\exp(-\bar{E}2/k_B T)}{[1+\exp(-\bar{E}2/k_B T)]^2}\right]\right\} \tag{8.10.33}$$

The magnetic contributions to the heat capacity (Schottky anomalies), computed by Moretti and Ottonello (1998) for REE-Fe garnets via Eq. (8.10.32) are shown in Fig. 8.23. These contributions must be added to the vibrational terms and anharmonic corrections to obtain the bulk heat capacity of the substance at constant pressure.

Concerning the magnetic contributions to entropy, it is worthy to recall that all possible quantum states (2S + 1) and (2J + 1) are energetically degenerate and, at first approximation, equally populated, as long as the energy differences between the several states are small compared with $k_B T$. At sufficiently low temperatures, $k_B T$ attains the magnitude of the energy gaps, so that those states with lower energies are selectively populated, giving rise to what is usually called the "*ordering discontinuity*". Besides this, magnetic couplings give rise to '*unquenched*' contributions, appearing as a sort of continuum over the entire heat capacity curve.

Concerning more properly REE-ferrites, since we have the complete expression of the magnetic heat capacity function, entropy at the T of reference (298.15 *K*) may readily be obtained by integration of the heat capacity expression (8.10.33). Nevertheless, we know that each non-S-state ion (at sufficient high T) contributes to the bulk entropy of the substance with spin entropy of the form (Kittel 1986):

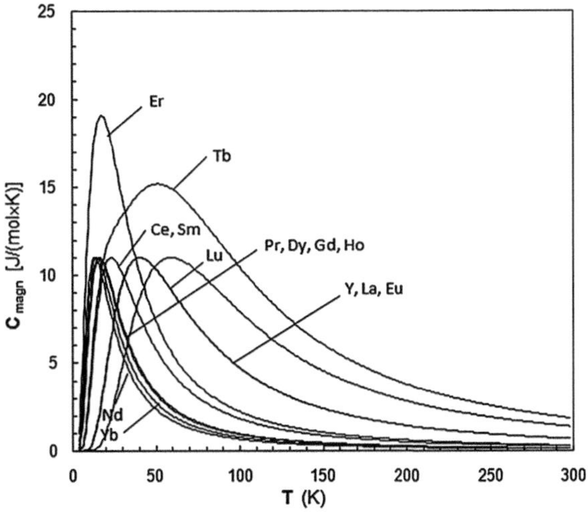

Fig. 8.23 Schottky anomalies in the magnetic contributions to the heat capacity of REE ferrites

$$S_m = N_0 k_B \ln(2J + 1) \tag{8.10.34}$$

where J is the total angular momentum of the ion.

The values of orbital angular momentum **L**, total spin **S**, total angular momentum **J**, ground state and magnetic spin entropy for non S-state ions of interest are listed in Table 8.7.

Table 8.8 lists calculated entropy contributions and the total entropy for the 15 REE ferrite garnets. We may note that the magnetic contributions attain in some cases values which are roughly one order of magnitude higher than the anharmonic

Table 8.7 Magnetis spin entropies for trivalent rare earth ions and ferric iron (J/mol × K). (From Moretti and Ottonello 1998)

Ion	L	S	J	Ground state	S_m
Ce^{3+}	3	1/2	5/2	$^2F_{5/2}$	3.561
Pr^{3+}	5	1	4	$^3H_4$	4.366
Nd^{3+}	6	3/2	9/2	$^4I_{9/2}$	4.576
Sm^{3+}	5	5/2	5/2	$^6H_{5/2}$	3.561
Eu^{3+}	3	3	0	$^7F_0$	0
Gd^{3+}	0	7/2	7/2	$^8S_{7/2}$	4.132
Tb^{3+}	3	3	6	$^7F_6$	5.097
Dy^{3+}	5	5/2	15/2	$^6H_{15/2}$	5.510
Ho^{3+}	6	2	8	$^5I_8$	5.630
Er^{3+}	6	3/2	15/2	$^4H_{15/2}$	5.510
Tm^{3+}	5	1	6	$^3H_6$	5.097
Yb^{3+}	3	1/2	7/2	$^2F_{7/2}$	4.132
Fe^{3+}	0	5/2	5/2	$^6S_{5/2}$	3.561

J = |L − S| when the shell is less than half-full
J = L + S when the shell is more then half-full

Table 8.8 Entropy terms for the various REE garnets (*J/mol* × *K*). S_v: vibrational contribution; S_{an}: anharmonic contribution; S_m: magnetic contribution; $S^0_{298.15}$: third law entropy at reference conditions (T_r = 298.15 *K*, P_r = 1 *bar*). From Moretti and Ottonello (1998)

Compound	S_v		S_{an}	S_m	$S^0_{298.15}$			
	(1)	(2)			(1)	(2)	(3)	(4)
$Y_3Fe_5O_{12}$	425.069	373.775	3.602	16.601	445.272	393.978	–	–
$La_3Fe_5O_{12}$	427.536	404.104	3.725	16.601	447.862	424.430	–	–
$Ce_3Fe_5O_{12}$	425.975	400.844	3.621	17.179	446.774	421.643	–	–
$Pr_3Fe_5O_{12}$	424.573	397.759	3.285	17.221	445.079	418.265	–	–
$Nd_3Fe_5O_{12}$	422.732	393.912	3.736	17.184	443.652	414.833	–	–
$Sm_3Fe_5O_{12}$	419.742	388.074	3.530	17.179	440.451	408.783	420.974 ± 4.184	435.1
$Eu_3Fe_5O_{12}$	420.135	387.841	3.568	16.601	440.304	408.011	409.894 ± 4.184	432.2
$Gd_3Fe_5O_{12}$	417.993	384.976	3.732	17.221	438.946	405.929	418.054 ± 4.184	450.8
$Tb_3Fe_5O_{12}$	416.991	382.146	3.513	33.640	454.144	419.300	438.134 ± 4.184	446.9
$Dy_3Fe_5O_{12}$	415.792	379.006	3.562	34.173	453.526	416.740	440.019 ± 4.184	427.3
$Ho_3Fe_5O_{12}$	413.771	376.617	3.721	34.349	451.841	414.687	481.864 ± 4.184	473.1
$Er_3Fe_5O_{12}$	413.263	374.990	3.512	34.297	451.073	412.800	507.374 ± 4.184	483.9
$Tm_3Fe_5O_{12}$	413.309	374.085	3.532	33.640	450.481	411.257	487.719 ± 4.184	–
$Yb_3Fe_5O_{12}$	412.208	372.240	3.728	17.210	433.145	393.178	465.119 ± 4.184	–
$Lu_3Fe_5O_{12}$	411.164	370.449	3.698	16.953	431.816	391.101	463.954 ± 4.184	–

(1) calculated with band G in Table 8.9; (2) calculated with band E in Table 8.9; (3) from Kanke and Navrotsky (1998) based on calorimetric experiments by Kimizuka et al. (1983) and references therein; (4) from Parida et al. (2008)

terms. Vibrational entropy (Kieffer model) calculations were conducted by Moretti and Ottonello (1998) adopting two different values for the upper cutoff limit of the optical continuum (ω_{u1} coincident with band G and ω_{u2} coincident with band E in Table 8.9). The best agreement with Kanke and Navrotsky's (1998) estimates of standard state entropy were obtained with band G. Band E, identified by Hofmeister and Campbell (1992) on powder spectra, does not look like a true upper cutoff frequency, due to its position (almost at the top of a big shoulder) and the very low transmittance value. The calorimetric data of Parida et al. (2008) are also listed in the same table. These data do not exhibit the volume-based regularities envisaged by Glasser (2014) in a recent comparative appraisal, and are thus of difficult interpretation (see Glasser 2014; for an extended discussion). The vibrational parameters adopted for Kieffer model calculations are shown in Table 8.9.

8.10.6 Crystal Field Effects

The crystal field quantum mechanical theory was developed by Bethe (1929) to account for the magnetic properties of transition metal complexes. It is a pioneer paper in Quantum Chemistry because it utilizes the perturbation theory, coupled to a point group analysis of the symmetry in terms of associated Legendre functions.[12] In Bethe's words "*We shall accordingly unite in a 'society' all these spherical harmonics that transform under symmetry operations of the crystal, and then, by a suitable choice of linear combinations instead of the original spherical harmonics, we would try to make these 'societies' as small as possible. This corresponds to finding the irreducible representations*".

Bethe (1929) considered in detail the restrictions arising from the various symmetry groups.

Inserting an atom into a crystal, the symmetry of the potential energy is reduced from spherical symmetry to the symmetry of the position that the atom occupies in the crystal. For example, in NaCl, it will be reduced to the cubic holohedral symmetry. To accomplish the reduction of the $(2l + 1)$ dimensional representations of the rotation group, Bethe (1929) applied the fundamental theorem of the group theory, stating that "*every irreducible representation of a group can be decomposed into its irreducible constituents in one and only one way, and, in the irreducible representation, the character of each group element is equal to the sum of the characters that belong to the element in the irreducible representation*" (Bethe 1929). Consider again the cubic holohedral symmetry. The 24 pure rotations (which are identical with the rotations of the octahedral group Oh) fall into five classes: E (identity), C_2, C_3, C_4, and C_5, which, by inversion, transform into J (inversion),

[12] The associated Legendre functions or "*spherical harmonics*" are basis functions for irreducible representations of SO(3), the group of rotations in three dimensions, and thus play a central role in the point group theoretic discussion of rotational momenta (cf. Sect. A.15).

Table 8.9 IR absorption bands of REE-Fe garnets (cm^{-1}). References: (1) McDevitt (1969); (2) Hofmeister and Campbell (1992) "classical dispersion analysis" values; asterisks indicate "powder data position" values. From Moretti and Ottonello (1998). Values im brackets obtained by linear regression techniques (see Table 8 in Moretti and Ottonello 1998)

Band	Type	Y (1)	Y (2)	La	Ce	Pr	Nd	Sm	Eu	Gd	Tb	Dy	Ho	Er	Tm	Yb	Lu
A		–	–	–	–	–	–	–	–	–	–	–	–	–	–	–	–
B	E1	655	655.3	(606.4)	(611.7)	(616.7)	(623.0)	630	635	638	(643.2)	647	655	656	658	658	665
C	E2	605	594.7	(556.5)	(561.7)	(566.7)	(572.9)	580	588	590	(592.8)	592	600	606	607	612	616
D	E3	566	565.4	(535.5)	(538.7)	(541.8)	(545.7)	550	553	555	(558.1)	562	565	567	568	568	570
E	ω_{u2}	–	470*	(398.9)	(404.5)	(409.8)	(416.5)	426	427	435	(437.7)	(443.3)	(446.6)	(450.7)	(453.9)	(456.9)	(460.1)
F		–	403.2	–	–	–	–	–	–	–	–	–	–	–	–	–	–
G	ω_{u1}	381	379.9	(361.5)	(364.0)	(366.2)	(369.1)	373	373	379	(378.3)	380	(382.2)	(384.0)	(385.3)	(386.6)	(388.0)
H		369	353.5	–	–	–	–	352	358	361	–	365	368	370	371	374	383
I		338	327.1	–	–	–	–	326	326	326	–	330	330	328	329	331	333
		318	308.5	–	–	–	–	311	309	307	–	309	308	305	303	305	–
J		–	296.8	–	–	–	–	–	–	–	–	–	–	–	–	–	–
K		259	256.7	–	–	–	–	252	252	254	–	255	255	254	251	253	–
L		227	223.5	–	–	–	–	–	–	–	–	–	–	–	–	–	–
M		110	108.2	–	–	–	–	115	115	112	–	111	107	110	108	108	106
N		–	208.8	–	–	–	–	214	214	213	–	215	213	212	208	211	211
		–	156.7*	–	–	–	–	–	–	–	–	–	–	–	–	–	–
O		144	144	–	–	–	–	–	–	–	–	–	–	–	–	–	–
P	ω_{l}	–	100.3*	(112.1)	(111.5)	(111)	(110.3)	110	109	107	(108.1)	108	108	107	106	106	(105.8)
Q		–	–	–	–	–	–	–	–	–	–	–	–	–	–	–	–

Table 8.10 Character table of the octahedral group O_h. A_1, A_2, E, T_1, T_2 are the conventional names for the irreducible representations. In brackets are listed the conventional names assigned by Bethe (1929). Note also that in Bethe (1929) $C_3 \rightarrow C_5$; $C_2 \rightarrow C_4$; $C_4 \rightarrow C_3$

O_h	{e}	$\{3C_2 = 3C_4^2\}$	$\{8C_3\}$	$\{6C_2\}$	$\{6C_4\}$
A_1 (Γ_1)	1	1	1	1	1
A_2 (Γ_1)	1	1	1	−1	−1
E (Γ_1)	2	2	−1	0	0
T_1 (Γ_1)	3	−1	0	−1	1
T_2 (Γ_1)	3	−1	0	1	−1

JC_2, JC_3, JC_4, JC_5. The octahedral group, consisting of 24 elements, possesses accordingly five irreducible representations. The only possible decomposition is as follows (Bethe 1929):

$$24 = 3^2 + 3^2 + 2^2 + 1^2 + 1^2 \tag{8.10.35}$$

i.e. the octahedral group possesses two three-dimensional, one two-dimensional, and two one-dimensional representations. The character table of the group is resumed in (Table 8.10).

In terms of Associated Legendre Functions (Sect. A.15), the potential V inside a sphere is defined as:

$$V(r, \theta, \phi) = \sum_{l=0}^{\infty} \sum_{m=-l}^{l} P_{l,m} R_l^m(r, \theta, \phi) \tag{8.10.36}$$

where each R_l^m is a solution of the Laplace equation demanding

$$V(r, \theta, \phi) = 0 \tag{8.10.37}$$

The Schrödinger equation of the free ion is invariant with respect to arbitrary rotations of the coordinate system. The rotation group of the sphere possesses one irreducible representation of each dimensionality $2l+1 \quad (l = 0, 1, 2, \ldots)$ (Wigner 1927). Inserting the ion into a crystal, the symmetry of the potential energy is reduced from the spherical symmetry to the symmetry of the position that the ion occupies in the crystal. The new group is a sub-group of the former.

To every k-dimensional irreducible component of the $(2l + 1)$ dimensional representation of the rotation group must correspond to one irreducible "*society*" of spherical harmonics of lth order. The eigenfunctions of the irreducible representation of the crystal group must obey the requirement of the general perturbation theory

$$\int V\psi_i\psi_k dx = 0 \quad i \neq k(i = -l, \ldots, +l) \tag{8.10.38}$$

with ψ_i and ψ_k as wave functions to transform under operations of the group of the crystal and x as all coordinates.

From the nk associated Legendre functions of the n "*societies*", Bethe formed new eigenfunctions ψ_i by taking linear combinations

$$\Psi_i = \sum_{i=1}^{nk} b_{ij}\psi_i \quad (i = 1, \ldots, nk) \tag{8.10.39}$$

and determined the coefficients by solving the usual determinantal equation of degree nk. This gave for the term value n k-fold roots and, at the same time, he obtained the k zeroth order eigenfunctions belonging to each. These eigenfunctions are however not the final ones, because one has to form from them linear combinations that satisfy the condition of the perturbation theory

$$\int V\psi_i\psi_k d\tau = \delta_{ik}\varepsilon_i \tag{8.10.40}$$

For tetragonal symmetry ($1A_4$), for example, there are four linear combinations ψ_i (i = 1, 2, 3, 4) satisfying the above condition:

$$\psi_i = \psi_{nl}(r)\left[\alpha_{i1}\sqrt{2}P_3^1\cos\theta + \alpha_{i1}\sqrt{2}P_3^1\sin\theta + \alpha_{i3}\sqrt{2}P_3^3\cos 3\theta + \alpha_{i4}\sqrt{2}P_3^3\sin 3\theta\right] \tag{8.10.41}$$

$$\sum_{j=1}^{4}\alpha_{ij}\left(\varepsilon_{jk} - \delta_{jk}\varepsilon_i\right) = 0 \quad (k = 1, 2, 3, 4) \tag{8.10.42}$$

The determinantal equation assigning the displacements of the terms is:

$$\begin{vmatrix} \varepsilon_{11}-\varepsilon & 0 & \varepsilon_{13} & 0 \\ 0 & \varepsilon_{11}-\varepsilon & & -\varepsilon_{13} \\ \varepsilon_{13} & 0 & \varepsilon_{33}-\varepsilon & 0 \\ 0 & -\varepsilon_{13} & 0 & \varepsilon_{33}-\varepsilon \end{vmatrix} = \left[(\varepsilon_{11}-\varepsilon)(\varepsilon_{33}-\varepsilon) - \varepsilon_{13}^2\right]^2 = 0 \tag{8.10.43}$$

The two eigenvalues are:

$$\varepsilon', \varepsilon'' = \frac{\varepsilon_{11}+\varepsilon_{33}}{2} \pm \sqrt{\left(\frac{\varepsilon_{11}-\varepsilon_{33}}{2}\right)^2 + \varepsilon_{13}^2} \tag{8.10.44}$$

The eigenfunctions for ε' are:

$$\psi'_1 = \sqrt{1-\delta}P_3^1 \cos\phi + \sqrt{1+\delta}P_3^3 \cos 3\phi \quad (8.10.45.1)$$

$$\psi'_2 = \sqrt{1-\delta}P_3^1 \sin\phi + \sqrt{1+\delta}P_3^3 \sin 3\phi \quad (8.10.45.2)$$

with

$$\delta = \frac{\varepsilon_{13} - \varepsilon_{11}}{\sqrt{(\varepsilon_{33} - \varepsilon_{11})^2 + 4\varepsilon_{13}^2}} \quad (8.10.46)$$

Similarly, the eigenfunctions for ε'' are:

$$\psi''_1 = \sqrt{1+\delta}P_3^1 \cos\phi - \sqrt{1-\delta}P_3^3 \cos 3\phi \quad (8.10.47.1)$$

$$\psi''_2 = \sqrt{1+\delta}P_3^1 \sin\phi - \sqrt{1-\delta}P_3^3 \sin 3\phi \quad (8.10.47.2)$$

In 1932, Van Vleck extended the Bethe's theory to ligand orbitals with the aim of predicting the contributions of the crystal field to the bulk lattice energy of the crystal. These magnitudes are also obtainable experimentally by optical spectroscopy.

Let me go now to the practical applications of the crystal field theory and consider a co-ordination polyhedron, in which the cation (with its nucleus at the center of the spherical polar coordinates) is surrounded by 6 anionic ligands, located 2 by 2 on the polar axes ("*octahedral field*"; Fig. 8.26). If we move the six ligands from infinite distance toward the nucleus of the central cation, short-range repulsion between the electron clouds of the anionic ligands and the electrons of the d orbitals of the central atom gives rise to a generalized increase in orbital energies. The repulsion will be more intense for orbitals oriented along the Cartesian axes (d_{z^2} and $d_{x^2-y^2}$; see Fig. 8.24) and less marked for orbitals transversally oriented with respect to the cartesian axes (d_{xy}, d_{xz}, d_{yz}).

The degeneracy of the five d orbitals is thus partly eliminated. The energy gap between the first and second groups of orbitals (defined by convention d_γ, d_ε, or e_g, t_{2g} respectively) is known as "*crystal field splitting*" and is usually denoted by the symbol Δ (or 10 Dq).

According to electrostatic theory, the magnitude of crystal field splitting is given by:

$$\Delta = 5e\mu a^4/d^6 \quad (8.10.48)$$

where e is the electronic charge, μ the dipole moment of the ligand, a the mean radial distance of d orbitals, and d the cation-to-anion distance. With respect to a spherical field, where the degeneracy of d orbitals is maintained, the increase in

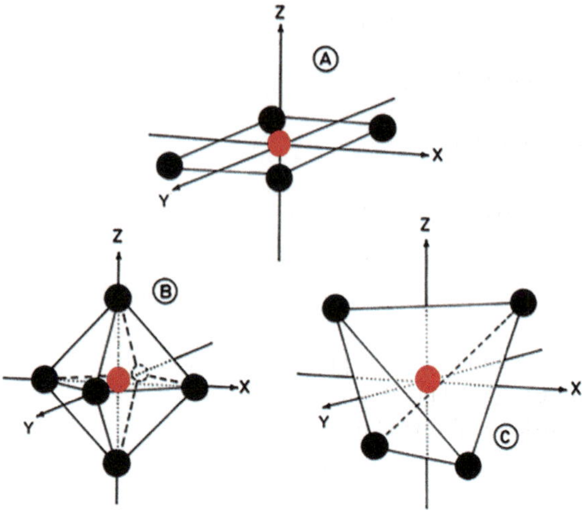

Fig. 8.24 Positions of anionic ligands (black spheres) around a central cation (red sphere) for various configurational states: A = square-planar co-ordination; B = octahedral co-ordination C = tetrahedral co-ordination. Cation nuclei located in center of Cartesian coordinates

energy of orbitals e_g is $\frac{3}{5}\Delta$, and the decrease in energy of orbitals t_{2g} is $\frac{2}{5}\Delta$. Orbitals e_g are therefore called "*destabilized*" and orbitals t_{2g} "*stabilized*".[13]

Cation and co-ordination number being equal, the magnitude of Δ depends on the type of ligand and increases according to the spectrochemical series:

$$Br^- < Cl^- < F^- < OH^- < O^{2-} < H_2O < EDTA^{4-} < NH^{2-} < NO^- < CN^- \quad (8.10.49)$$

A "*weak field*" is usually defined as one generated by ligands at the beginning of the spectrochemical series; a "*strong field*" is caused by ligands at the end of the series.

The value of Δ may be deduced by light absorbance measurements on the various complexes. The frequency of the maximum of the first absorption band corresponds to the transference of an electron from a t_{2g} to an e_g orbital and directly gives the value of Δ expressed in cm^{-1}:

$$\nu_a = {}^{\Delta}\!/_{h} \quad (8.10.50)$$

Figure 8.25 shows energy levels for *d* orbitals in crystal fields of differing symmetry. The splitting operated by the octahedral field is much higher than that of the tetrahedral field ($\Delta_t = \frac{4}{9}\Delta_o$), and lower than the effect imposed by the square-planar field. If we define as "*crystal field stabilization energy*" the decrease in energy with respect to a spherical field and compare the values for the tetrahedral and octahedral fields, we see that, for all configurations (besides $3d^0$, $3d^5$, and $3d^{10}$

[13] It is customary to label the orbitals with the lower case letters corresponding to the labels of the irreducible representations (See Table 8.10). The additional subscript g declares the "*even*" parity.

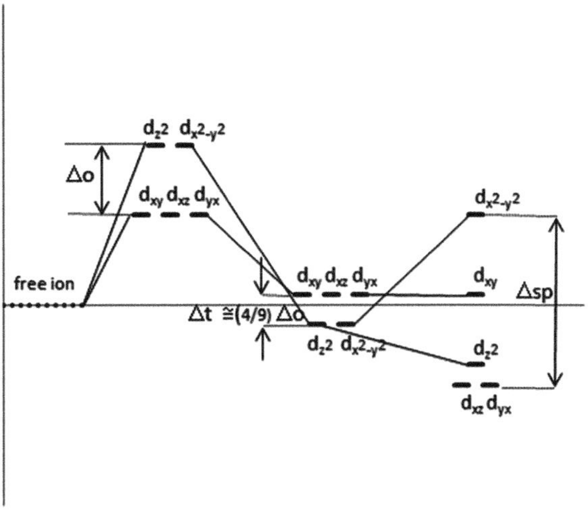

Fig. 8.25 Crystal field splitting for d orbitals: Δo = octahedral field; Δ*t* = tetrahedral field; Δsp = square-planar field

for which crystal field stabilization is obviously zero, as it coincides with the spherical field), the octahedral field has a higher stabilization effect, called "*Octahedral Site Preference Energy*" (OSPE) (cf. Table 8.11).

In 1957, almost contemporaneously, McClure and Dunitz and Orgel proposed precise calculations of crystal field stabilization energies in tetrahedral and octahedral fields, and the corresponding OSPE values. Their data are summarized in Table 8.12. The Δ_o values reported by McClure (1957) were actually derived for hydrated ions (although the OH^- groups are near O^{2-} in the spectrochemical series 8.10.49), while those listed by Dunitz and Orgel (1957) were obtained for crystalline oxides and glasses from spectroscopic data of light absorption. As noted by Dunitz and Orgel (1957), the Δ_o values for O^{2-} and OH^- ligands are identical for bivalent ions but very different for trivalent ions (8–12% lower for O^{2-} with respect to OH^- ligands). In both works the Δ_t value was not experimentally obtained but was assumed to correspond to $\frac{4}{9}\Delta_o$, in agreement with theory. The estimated OSPE values thus have a high level of incertitude. In some instances, for trivalent ions Dunitz and Orgel (1957) imposed the "*strong field*" condition. Both McClure (1957) and Dunitz and Orgel (1957) derived their OSPE values with the aim of explaining the intracrystalline distribution of transition elements observed in multiple oxides with cubic structures (spinels). According to the above authors, a more or less high OSPE value is crucial in intracrystalline partitioning. Their hypothesis was later taken up again by several authors. In particular, Burns (1970) furnished several examples of natural compounds, calculating the crystal field stabilization energy terms from adsorption spectra under polarized light.

Spinels are particularly apt to the investigation of CFSE. This class of substances will be treated in some detail at Sect. 20.3. Here I limit myself to present lattice energy terms ($U = E_c + E_r$), the modulus of the bulk energy gap between normal and inverse conditions ($|\Delta U|$), and the modulus of the crystal field stabilization energy gap (|ΔCFSE|) for some spinel compounds (Table 8.13).

Table 8.11 Electronic configuration and crystal field stabilization energies for 3d electrons in transition elements. (Ar) = argon core: $1s^2 2s^2 2p^6 3s^2 3p^6$

Ion	Electronic configuration	Spin					Octahedral field	Tetrahedral field	OSPE
Sc^{3+}; Ti^{4+}	$(Ar)3d^0$						0	0	0
Ti^{3+}	$(Ar)3d^1$	↑					2/5	12/45	6/45
V^{3+}	$(Ar)3d^2$	↑	↑				4/5	24/45	12/45
Cr^{3+}	$(Ar)3d^3$	↑	↑	↑			6/5	16/45	38/45
Mn^{2+}; Fe^{3+}	$(Ar)3d^5$	↑	↑	↑	↑	↑	0	0	0
Fe^{2+}	$(Ar)3d^6$	↑↓	↑	↑	↑	↑	2/5	12/45	6/45
Co^{2+}	$(Ar)3d^7$	↑↓	↑↓	↑	↑	↑	4/5	24/45	12/45
Ni^{2+}	$(Ar)3d^8$	↑↓	↑↓	↑↓	↑	↑	6/5	16/45	38/45
Cu^{2+}	$(Ar)3d^9$	↑↓	↑↓	↑↓	↑↓	↑	3/5	8/45	19/45
Zn^{2+}	$(Ar)3d^{10}$	↑↓	↑↓	↑↓	↑↓	↑↓	0	0	0

Table 8.12 Crystal field stabilization energies for 3d transition elements according to McClure (1957) (1) and Dunitz and Orgel (1957) (2); (*) valid for strong field conditions

Ion	Reference	Δ_o (cm^{-1})	Δ_t (cm^{-1})	Oct.Stab *(kJ/mol)*	Tet.stab *(kJ/mol)*	OSPE
Ti^{3+}	(1)	20,300	9000	96.7	64.4	32.3
Ti^{3+}	(2)	18,300	8130	87.4	58.6	28.8
V^{3+}	(1)	18,000	8400	128.4	120.0	8.4
V^{3+}	(2)	16,700	7400	160.2(*)	106.7	53.5(*)
V^{2+}	(1)	11,800	5200	168.2	36.4	131.8
Cr^{3+}	(1)	17,600	7800	251.0	55.6	195.4
Cr^{3+}	(2)	15,700	6980	224.7	66.9(*)	157.8(*)
Cr^{2+}	(1)	14,000	6200	100.4	29.3	71.1
Mn^{3+}	(1)	21,000	9300	150.2	44.4	105.8
Mn^{3+}	2)	18,900	8400	135.6	40.2	95.4
Mn^{2+}	(1)	7500	3300	0	0	0
Fe^{3+}	(1)	14,000	6200	0	0	0
Fe^{2+}	(1)	10,000	4400	47.7	31.4	16.3
Fe^{2+}	(2)	10,400	4620	49.8	33.1	16.7
Co^{3+}	(1)	–	7800	188.3	108.8	79.5
Co^{2+}	(1)	10,000	4400	71.5	62.7	8.8
Co^{2+}	(2)	9700	4310	92.9(*)	61.9	31.0(*)
Ni^{2+}	(1)	8600	3800	122.6	27.2	95.4
Ni^{2+}	(2)	8500	3780	122.2	36.0(*)	86.2(*)
Cu^{2+}	(1)	13,000	5800	92.9	27.6	65.3
Cu^{2+}	(2)	12,600	5600	90.4	26.8	63.6

Table 8.13 Lattice energy and CFSE for some spinel compounds. Values of U, E_c, E_r from Ottonello (1986). CFSE values calculated with values proposed by McClure (1957). Data in *kJ/mol*. x: degree of inversion

Compound	x	U	E_c	E_r	$\lvert\Delta U\rvert$	$\lvert\Delta CFSE\rvert$
$CoFe_2O_4$	1	−18,712	−21,730	3018	270	8.8
$CoCr_2O_4$	0	−18,918	−22,090	3172	300	186.6
$CuFe_2O_4$	1	−18,805	−21,697	2892	150	65.3
$FeAl_2O_4$	0	−19,029	−22,844	3815	400	16.3
$FeFe_2O_4$	1	−18,655	−21,664	3009	220	16.3
$FeCr_2O_4$	0	−18,810	−22,029	3219	260	179.1
$NiCr_2O_4$	0	−18,948	−22,116	3168	200	100.0
$NiFe_2O_4$	0	−18,793	−21,829	3036	220	95.4

Consider now a metallic ion with d^5 configuration. Based on Hund's rule, the first three electrons occupy the t_{2g} degenerate orbitals with parallel spin. The remaining two electrons then have two possibilities:

(1) to occupy the two e_g degenerate orbitals with parallel spin, or
(2) to fill two of the three t_{2g} degenerate orbitals with opposed spin.

The two configurations are energetically determined by the influence of the crystal field and are denoted respectively by conditions of "*high*" or "*low*" spin. In the presence of a weak octahedral field (generated by ligands at the beginning of the spectrochemical series 8.10.49), the value of Δ_o is low.

In the presence of a strong field (ligands at the end of the spectrochemical series), crystal field splitting is high and electrons tend to occupy the t_{2g} stabilized orbitals, opposing the spin and leaving the destabilized e_g orbitals free. The magnetic properties of the compound (which depend on the electron spin condition) are thus influenced by the crystal field.

8.10.7 *Ab Initio Computations: Phonon Set at Γ Point*

As we have already seen, at the Γ point the acoustic frequencies attain zero but we know that a sinusoidal phonon dispersion takes place throughout the crystal. The remaining (optical modes) may be treated for simplicity as Einstein oscillators (i.e. phonon dispersion neglected). Adopting the Kieffer model to depict the acoustic contribution to the bulk isochoric heat capacity of the substance takes the form:

$$C_V = \frac{3R}{Z}\left(\frac{2}{\pi}\right)^3 \sum_{i=1}^{3} \int_0^{x_i} \frac{\left[\arcsin(x/x_i)^2\right] x^2 e^x dx}{\left(x_i^2 - x^2\right)^{1/2}\left(e^x - 1\right)^2} + \frac{R}{Z}\sum_{4}^{3n} e^{x_i}\left(\frac{x_i}{e^{-x_i} - 1}\right)^2 \quad (8.10.51)$$

where n is the number of atoms in the unit cell and Z is the number of formula units per unit cell. The first term on the right of Eq. (8.10.51) is the acoustic contribution at $\mathbf{k}_{max}$ represented in terms of a sine-wave dispersion relation. The second term is the contribution of all remaining (optic) modes. The ab initio isochoric heat capacity attains always the Dulong–Petit limit at high T.

The isobaric heat capacity C_P at room pressure (1 bar) is readily derived from C_V by applying:

$$C_P = C_V + T\alpha_T^2 K_T V_T \tag{8.10.52}$$

The computation of the isothermal bulk modulus K_T and the (isobaric) thermal expansion α_T will be detailed later on.

In the absence of configurational disorder, the entropy of one mole of a crystalline substance is reduced to the following summation:

$$S = \frac{3R}{Z}\left(\frac{2}{\pi}\right)^3 \sum_{i=1}^{3} \int_0^{x_i} \frac{\left[\arcsin(x/x_i)^2\right] x dx}{\left(x_i^2 - x^2\right)^{1/2}(e^x - 1)} - \frac{3R}{Z}\left(\frac{2}{\pi}\right)^3 \sum_{i=1}^{3} \int_0^{x_i} \frac{\left[\arcsin(x/x_i)^2\right]}{\left(x_i^2 - x^2\right)^{1/2}} \ln(1 - e^{-x}) dx$$
$$+ \frac{R}{Z}\sum_{i=4}^{3n}\left(\frac{x_i}{e^{x_i} - 1} - \ln(1 - e^{-x_i})\right) + \frac{R}{Z}\ln(Q_e) + \int_0^T V_T K_T \alpha_T^2 dT \tag{8.10.53}$$

The first two terms on the right in Eq. (8.10.53) are the Kieffer model sine-wave dispersion contributions (acoustic modes). The third term is the optical contribution of the remaining 3*n*−3 (optic) modes. The fourth term is the electronic contribution arising from spin multiplicity and the last term is the anharmonic contribution to the entropy of the substance.

In the harmonic approximation, the internal energy of the substance is given by:

$$U = E_{ai,crystal} + \frac{3RT}{Z}\left(\frac{2}{\pi}\right)^3 \sum_{i=1}^{3} \int_0^{x_i} \frac{\left[\arcsin(x/x_i)^2\right] x dx}{\left(x_i^2 - x^2\right)^{1/2}(e^x - 1)}$$
$$+ \frac{RT}{Z}\sum_{i=4}^{3n} x_i\left(\frac{1}{2} + \frac{1}{e^{-x_i} - 1}\right) \tag{8.10.54}$$

where the 2nd and 3rd terms on the right constitute the thermal correction to U (Therm-U), comprehensive of zero point energy). Adding the PV product to Therm-U, one obtains the thermal correction to enthalpy (Therm-H), and subtracting the TS product from Therm-H at the same P–T, one has the thermal correction to the Gibbs free energy (Therm-G).

In many circumstances, this approximated treatment may be quite successful. We see as an example in Fig. 8.26 for α-Al_2O_3. Pure α-alumina (corundum) has been promoted since 1948 as a heat capacity standard because it does not

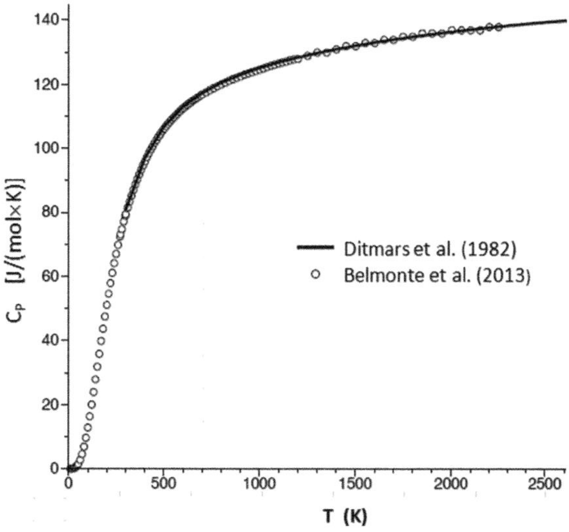

Fig. 8.26 Ab initio bulk isobaric heat capacity of -Al_2O_3 (Belmonte et al. 2013; solid line) compared with smoothed values arising from calorimetric measurements on the Standard Reference Material 720 (Ditmars et al. 1982). The bias in the T range from room conditions to the melting temperature does not exceed one *J/(mol × K)* in most cases. Modified from Belmonte et al. (2013)

experience any solid state transformation from the athermal limit to the melting point which, at room pressure, takes place at $T_f = 2327 \pm 6$ *K*.

The $[(Al_2O_3)_6]°$ *D3d6 (R-3c)* unit was resolved by Belmonte et al. (2013), within the LCAO approximation and in the framework of Bloch's theorem. Calculations were carried out with the Density Functional Theory and the hybrid B3LYP functional. The adopted basis set corresponds to the 85-11G(d) and 8-411G(d) contractions for Al and O adopted by Montanari et al. (2006). The vibrational properties of α-$Al_2O_3\alpha$ corundum were calculated with the CRYSTAL09 program (Dovesi et al. 2009), adopting more stringent computational parameters than the previous investigation (Montanari et al. 2006). The threshold of the SCF energy was set to 10^{-8} *Hartree* for the geometry optimization and 10^{-10} *Hartree* for the vibrational frequency calculations. The reciprocal space was sampled according to a regular sublattice with a shrinking factor equal to 6 and corresponding to 32 independent **k**-points in the irreducible part of the Brillouin zone. The discrete values of C_P calculated at various T were then translated into a Shomate function:

$$C_P = a + b \times T + c \times T^{-2} + d \times T^{-1/2} + e \times T^{-3} + f \times T^2 + g \times T^3 \quad (8.10.55)$$

with a, b, c, d, e, f, g as interpolation parameters. The ab initio C_P is so accurate to be hardly distinguished from the best interpolant of calorimetric data (Ditmars et al. 1982). This means that not only the vibrational calculations are accurate, but also the geophysical parameters concurring to determine the anharmonicity of the compound are accurate as well.

Knowing U and S, the enthalpy of the substance (H) and its Helmholtz and Gibbs free energies (F and G, respectively) are readily obtained by the usual closed-system thermodynamic relations (H = U + PV; F = U − TS; G = H − TS). These absolute magnitudes are not however those available in thermodynamic tabulations, which rather refer to a delta with respect to a stable state of reference (i.e. molar enthalpy of formation from the elements and molar Gibbs free energy of formation from the elements; $\overline{H}^0_{f,298.15}$, $\overline{G}^0_{f,298.15}$, respectively). As already seen for neutral molecules, comparison of the ab initio enthalpy of formation from the elements of the substance with experimental observations requires the assessment of a thermochemical cycle of the type (here the thermal correction is made more explicit; see the analogy with Eq. 7.10.4):

$$\overline{H}^0_{f,298.15} = \left(\Delta U_{298.15-0} + P\Delta V_{298.15-0}\right) - D_0 + \sum_{i=1}^{n/Z} n_i H_{f,A_i,0} - \sum_{i=1}^{n/Z} n_i \Delta H_{element_i,0\rightarrow 298.15} \quad (8.10.56)$$

where the term in brackets is the thermal correction to the enthalpy at 298.15 *K*, $H_{f,A_i,0}$ is the enthalpy of formation of the ith gaseous atom from the stable element at T = 0 *K*, P = 1 *bar* (with the summation extended to the n/Z atoms in the molecule), $\Delta H_{element_i,0\rightarrow 298.15}$ is the enthalpy difference between T = 298.15 *K* and T = 0 *K* for the monatomic elements and D_0 is the zero point dissociation energy of the gaseous molecule into gaseous atoms at 0 K:

$$D_0 = \sum_{i=1}^{n/Z} n_i E_{A_i} - \left(E_{HF,crystal} + E_{ZPE,crystal}\right) \quad (8.10.57)$$

This energy term corresponds to the sum of the electronic energies of the gaseous atoms at 0 K less the electronic + zero point energy of the crystal.

The estimate of appropriate electronic energies for an isolated gaseous atom is still a matter of debate. One may adopt either the theoretical complete basis set limit or at least add diffuse functions with respect to the basis set adopted in the bulk. A more practical way is to assign to the gaseous atoms electronic energies returning a correct dissociation energy (D_0) of the element at its stable state.

It is worth to stress here that, in a chemically balanced reaction, adding thermal corrections to ab initio electron energies and summing up algebraically the various terms one obtains directly enthalpy and Gibbs free energy changes of the reaction at 1 *bar*, 298.15 *K*, without the need of assessing enthalpy and Gibbs free energy of formation from the elements, which are subject to considerable uncertainties, due to the difficulties in estimating the appropriate electronic energy for isolated gaseous atoms (see Ottonello et al. 2007; for a more detailed account of the problem).

8.10.8 *Ab Initio Computations: Full Phonon Dispersion*

If a full phonon dispersion calculation is available, the isochoric heat capacity, the entropy, and the internal energy of the crystal are readily evaluated by applying:

$$C_V = \frac{R}{Z}\sum_{i=1}^{3n} e^{x_i(\mathbf{k},V)}\left(\frac{x_i(\mathbf{k},V)}{e^{x_i(\mathbf{k},V)}-1}\right)^2 \tag{8.10.58}$$

$$S = \frac{R}{Z}\sum_{i=1}^{3n}\left(\frac{x_i(k,V)}{e^{x_i(\mathbf{k},V)}-1} - \ln\left(1-e^{-x_i(\mathbf{k},V)}\right)\right) + \frac{R}{Z}\ln(Q_e) + \int_0^T V_T K_T \alpha_T^2 dT \tag{8.10.59}$$

$$U = U_0 + \frac{RT}{Z}\sum_{i=1}^{3n} x_i(\mathbf{k},V)\left(\frac{1}{2} + \frac{1}{e^{x_i(\mathbf{k},V)}-1}\right) \tag{8.10.60}$$

where the summations are taken over all the vibrational modes sampled at discrete **k**-points in the first Brillouin Zone in order to include phonon dispersion effects in the calculation.

As an example of full phonon dispersion calculation on Earth's materials, I report hereafter (Table 8.14) the results of Belmonte (2017) on periclase, the simple *fcc* form of MgO (B1-type cubic). The investigation was carried out at DFT/B3LYP theory level under LCAO assumption. The adopted basis set expansions were quite extended [i.e. (8s)-(511sp)-(1d) for Mg and (8s)-(411sp)-(1d) contractions for O] and proved quite appropriate in previous investigations (Ottonello et al. 2010a; De La Pierre et al. 2011; Belmonte et al. 2014; Erba et al. 2014).

We may see that the Ab initio full phonon computations improve substantially with the size of the investigated supercell. The 5 × 5 × 5 supercell (125 k-points sampling of the first BZ) gives an accuracy of ~1% or less in the simulation of the heat capacity (and other thermodynamic properties) in the low-T range. The slight difference at high T is most likely due to an overestimation of thermal expansivity (Belmonte 2017).

Table 8.14 Isobaric heat capacity [*J/(mol × K)*] of MgO, obtained at DFT/B3LYP theory level (LCAO approximation). **(1)** = 2 × 2 × 2; **(2)** = 4 × 4 × 4; **(3)** = 5 × 5 × 5 supercells; **(4)** = NIST-JANAF (Belmonte 2017)

T (K)	(1)	(2)	(3)	(4)
298.15	34.4	36.5	36.6	37.1
700	45.9	48.5	48.6	48.7
1000	48.5	51.1	51.3	51.2
1800	53.8	56.5	56.7	54.9

8.11 Isotopological Exchanges

Isotopological exchanges in crystals follow the same rules anticipated at Sect. 7.9 for molecules. In the case of crystals, however, the application of Ab initio procedures is even more beneficial, due to the high number of frequencies concurring to define the separative effect.

Consider the $MgCr_2O_4$ cubic spinel phase (picrocromite; space group $Fd\overline{3}mZ$) as an example. $MgCr_2O_4$ was investigated at DFT/B3LYP theory level in the LCAO approximation by Ottonello et al. (2007) together with the tetragonal polymorph $I4_1/amd$. The adopted basis set was the same of a previous investigation by Catti et al. (1999) for the same substance, i.e. [8-411G*] for O; [8-511G*] for Mg; and [8–6-411(41)G] for Cr. Two more diffuse shells were added to those of the crystal set in order to get the atomic energies for single atoms. Cutoff limits were set to 10^{-7} (coulomb overlap tolerance), 10^{-7} (coulomb penetration tolerance), 10^{-7} (exchange overlap tolerance), 10^{-7} (exchange pseudo-overlap in the direct space), and 10^{-14} (exchange pseudo-overlap in the reciprocal space). The DFT exchange–correlation contribution was evaluated by numerical integration over the cell volume (Pascale 2004). Radial and angular points of the atomic grid were generated through Gauss–Legendre and Lebedev quadrature schemes. A (75,594)*p* grid pruning was applied (75 radial points and a variable number of angular points, with a maximum of 594 points on the Lebedev surface in the most accurate integration region; Pascale et al. 2004). The reciprocal space was sampled according to a regular sublattice with a shrinking factor IS equal to 8 (Saunders et al. 2003), corresponding to 29 independent **k** vectors in the irreducible Brillouin zone. The gradient with respect to the atomic coordinates was evaluated analytically (Doll 2001a, b; Orlando et al. 2002). The equilibrium atomic positions were determined by using a modified conjugate gradient algorithm as proposed by Schlegel (1982) (see Civalleri et al. 2001). Convergence in the geometry optimization process was tested on the root-mean-square (RMS) and the absolute value of the largest component of both the gradients and nuclear displacements. The threshold for the maximum and the RMS forces and the maximum and the RMS atomic displacements on all the atoms were set to 0.00045 and 0.00030 au and 0.00180 and 0.00120 au, respectively.

As we may see in Table 8.15, due to the presence of a heavy atom in the structure, frequencies are sensibly affected by the sampling grid, while the effect is less evident on the bulk electronic energy and optimized geometry. Increasing the number of maximum angular points from 434 to 594 while keeping constant the number of radial points (75), one observes a generalized increase in linearized frequencies which may attain + 66 cm^{-1} in the worst case. The zero point energy increases of roughly 2×10^{-3} *Hartree* but this increase is counterbalanced by a similar decrease in the total electronic energy of the substance. The cell edge at the variational limit does not change appreciably (i.e. from 8.4192 to 8.4180 $\mathring{A}$). Changing the sampling grid to the computationally more demanding XXLGRID extension (99,1280p) brings no practical benefits with respect to the XLGRID computation.

Table 8.15 Effect of the sampling grid on energies, structure and frequencies of the *Fd*$\bar{3}$*mZ* cubic $Mg^{52}Cr_2O_4$ isotopologues calculated at the DFT/B3LYP level of theory. E_{el} = total elecronic energy; ZPE = zero point energy (optical modes); E_0 = total electronic + zero point energy. (From Ottonello et al. 2007)

Name	LGRID	XLGRID	XXLGRID
n-rad, n-ang	75,434	75,594	99,1280
Cell edge	8.4192	8.4180	8.4179
E_{el}	−2590.2523433	−2590.2537550	−2590.2537510
ZPE	0.0184067	0.0201014	0.0201184
E_0	−2590.2339366	−2590.2336536	−2590.2336326
Mode			
F_{2U}	150.78	216.58	217.11
F_{1U}	247.95	275.62	275.22
F_{2G}	286.44	309.63	310.01
E_U	302.04	331.07	332.09
F_{1U}	378.15	430.67	430.45
F_{1G}	406.62	436.12	436.39
E_G	422.83	436.12	436.39
F_{2U}	423.89	459.10	459.65
F_{1U}	438.12	468.56	468.93
A_{2U}	470.63	510.28	510.06
F_{2G}	481.33	531.94	531.22
E_U	516.46	549.74	550.88
F_{1U}	574.11	614.40	616.18
F_{2G}	586.95	628.57	629.12
A_{2U}	590.85	628.57	629.12
A_G	643.62	696.79	697.51

In Table 8.16 are listed the vibrational frequencies and the mode-gamma Grüneisen parameters obtained by Ottonello et al. (2007) for the various isotopologues of the cubic phase *Fd*$\bar{3}$*mZ* at both UHF and B3LYP level of theory and XLGRID integration. In the same Table, computed frequencies are compared with the assignment of Sinha (1999) based on the de Launay model (de Launay 1956) and with experimental observations.

Apparently, vibrational frequencies computed at the UHF level are slightly overestimated with respect to model frequencies and experimental observations, while correlation effects seem to produce a weakening of force constants that results in vibrations of lower frequency in better agreement with experimental evidences.

Although translational and rotational terms are not normal vibrational modes, they nevertheless appear in the product rule equation when considering the result of

Table 8.16 Calculated (UHF and *B3LYP*) versus observed and model wave numbers for the $Fd\bar{3}mZ$ cubic polymorph. Species in bold are IR active. $\gamma_i = -(\partial \ln v_i/\partial \ln V)_T$. From Ottonello et al. (2007)

^{50}Cr	^{52}Cr	^{53}Cr	^{54}Cr	^{52}Cr exper. (1)	^{50}Cr–^{54}Cr exper. (2)	DAF model (3)	Species	γ_i
241.7324	237.8444	235.9651	234.1264			280	F_{2U}	
220.0081	***216.4787***	***214.7717***	***213.1015***					*−0.0703*
303.6328	301.4671	300.3869	299.3087	245	254–251.5	266	$\mathbf{F_{1U}}$	
277.6204	***275.5586***	***274.5312***	***273.5063***					*1.4168*
334.3501	334.3501	334.3501	334.3501			253	F_{2G}	
309.6326	*309.6326*	*309.6326*	*309.6326*					*0.9853*
375.7449	368.4565	364.9676	361.5760			339	E_U	
337.4248	***330.8777***	***327.7437***	***324.6971***					*1.3670*
464.3522	464.3522	464.3522	462.4183	426	435–427.5	354	$\mathbf{F_{1U}}$	
440.8840	*435.9740*	*433.6241*	*431.3413*					*1.4168*
472.2143	467.1925	464.7748	464.3522			457	F_{1G}	
430.6678	*430.6678*	*430.6678*	*430.6678*					*1.2405*
487.2336	487.2336	487.2336	487.2336	457		457	E_G	
451.4583	***451.4583***	***451.4583***	***451.4583***					*1.1105*
523.3338	521.5589	520.7304	519.9376			525	F_{2U}	
460.6323	***459.0513***	***458.3140***	***457.6091***					*1.3733*
533.1287	529.9935	528.5038	527.0622	512	527–526	520	$\mathbf{F_{1U}}$	
471.3426	*468.4740*	*467.1072*	*465.7830*					*1.3506*
566.3114	555.3143	550.0506	544.9338			394	A_{2U}	
519.9852	*509.9871*	*505.1952*	*500.5335*					*1.0366*
572.3777	572.3777	572.3777	572.3777			559	F_{2G}	
549.7388	*549.7388*	*549.7388*	*549.7388*					*1.2803*

(continued)

Table 8.16 (continued)

^{50}Cr	^{52}Cr	^{53}Cr	^{54}Cr	^{52}Cr exper. (1)	^{50}Cr–^{54}Cr exper. (2)	DAF model (3)	Species	γ_i
603.4363	603.4227	603.4165	603.4170			582	E_U	
531.9501	*531.9414*	*531.9375*	*531.9338*					*1.0827*
655.8143	652.5827	651.1023	649.7037	638	650–649	666	$\mathbf{F_{1U}}$	
616.7788	***614.3355***	***613.2216***	***612.1699***					*1.4565*
692.5948	692.5988	692.5948	692.5948			659	F_{2G}	
628.5710	*628.5710*	*628.5710*	*628.5710*					*0.9472*
696.1788	696.1786	696.1784	696.1783			692	A_{2U}	
644.1968	*644.0708*	*644.0171*	*643.9683*					*1.1822*
754.1905	754.1905	754.1905	754.1905	692		692	A_G	
696.7874	*696.7874*	*696.7874*	*696.7874*					*1.1084*

(1) Lutz et al. (1991); (2) Preudhomme and Tarte (1971c); (3) Sinha (1999)

Fig. 8.27 Isotopic ergodicity diagram for the cubic polymorph of $MgCr_2O_4$. Modified from Ottonello and Vetuschi Zuccolini (2005)

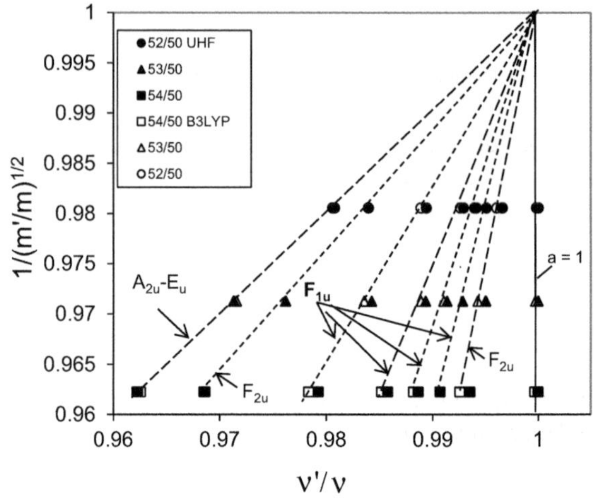

weak forces which convert the motions of translation and rotation into oscillatory motions of low frequency (see Bright Wilson et al. 1955; for a detailed treatment). We cannot thus expect a simple inverse proportionality between frequencies of isotopic molecules and masses of substituted isotopes. Nevertheless, since the momenta of inertia are not affected by significant changes of symmetry and potential energy function, their bulk effect on isotopic frequency shifts may be expected to be represented by a simple functional form. It is this sort of reasoning that lead to promoting the utilization of "*isotopic ergodicity diagrams*" aimed to discriminate the intrinsic separative effect dictated by each normal vibrational mode (Ottonello and Vetuschi Zuccolini 2005). Such a diagram, built for the cubic polymorph of $MgCr_2O_4$ is depicted in Fig. 8.27.

In this plot, the frequency ratios computed for the solid phase are plotted against the reciprocal of the square root of the masses of the substituted isotopes. The frequency ratios obtained at DFT/B3LYP level are, in this diagram, grouped into eight classes, of increasing ergodicity. The two classes with the lowest ergodicity factor group are the IR inactive A_{2u} and E_u asymmetric modes, then there are four intermediate classes corresponding to IR active asymmetric modes F_{1u} and the remaining class contributing to the overall separative effect is again a F_{2u} IR inactive mode. This fact explains why Preudhomme and Tarte (1971c) observed only four IR active modes affected by isotopic substitution of the Cr atom (cf. Table 8.16). The last class groups all ineffective (from an isotopic fractionation point of view) normal modes. The picture arising from the UHF calculations is nearly identical.

Separative effects (Eq. 7.9.8) may be computed at various T and represented in the form of interpolated functions over reciprocal temperature:

Table 8.17 Coefficients of the T-dependent separative effects for the $MgCr_2O_4$ polymorphs (single substitution). From Ottonello et al. (2007)

Species	Couples	a	b × 10^{-6}	c × 10^{-12}
(1)	$^{52}Cr/^{50}Cr$	0.0100	1.4877	−0.0095
	$^{53}Cr/^{50}Cr$	0.0145	2.1901	−0.0138
	$^{54}Cr/^{50}Cr$	0.0188	2.8660	−0.0180
	$^{53}Cr/^{52}Cr$	0.0045	0.7024	−0.0044
	$^{54}Cr/^{52}Cr$	0.0088	1.3783	−0.0086
	$^{54}Cr/^{53}Cr$	0.0043	0.6759	−0.0042
(2)	$^{52}Cr/^{50}Cr$	0.0059	1.2231	−0.0067
	$^{53}Cr/^{50}Cr$	0.0086	1.8002	−0.0097
	$^{54}Cr/^{50}Cr$	0.0111	2.3560	−0.0127
	$^{53}Cr/^{52}Cr$	0.0027	0.5771	−0.0031
	$^{54}Cr/^{52}Cr$	0.0052	1.1329	−0.0060
	$^{54}Cr/^{53}Cr$	0.0025	0.5558	−0.0029
(3)	$^{52}Cr/^{50}Cr$	0.0060	1.2305	−0.0067
	$^{53}Cr/^{50}Cr$	0.0087	1.8111	−0.0098
	$^{54}Cr/^{50}Cr$	0.0113	2.3702	−0.0128
	$^{53}Cr/^{52}Cr$	0.0027	0.5806	−0.0031
	$^{54}Cr/^{52}Cr$	0.0053	1.1397	−0.0061
	$^{54}Cr/^{53}Cr$	0.0026	0.5591	−0.0030

(1) = $Fd\bar{3}mZ$ – UHF (2) = $Fd\bar{3}mZ$ – B3LYP;
(3) = $I4_1/amd$ – B3LYP

$$1000\ln\left[\left(\frac{s^{\bullet}}{s^{\circ}}\right)f\right] = \left[(a_{\bullet\circ}) + (b_{\bullet\circ}) \times 10^6/T^2 + (c_{\bullet\circ}) \times 10^{12}/T^4\right] \qquad (8.11.1)$$

where $a_{\bullet\circ}, b_{\bullet\circ}, c_{\bullet\circ}$ are the interpolation coefficient for a single heavy-light isotopological couple (Table 8.17)

In Fig. 8.28 we may see that the separative effect is, in $MgCr_2O_4$, quite sensible even at room temperature and nearly identical to what has been observed in gaseous Cr(III) compounds (i.e. about 6 per mil per *a.m.u.* at room T; Ottonello and Vetuschi Zuccolini 2005a). We may also appreciate that the structural distortion (tetragonal vs cubic) has no sensible counterparts on the computed separative factors, while a sensible effect is envisaged to electron correlation phenomena (UHF vs DFT). The effect of the crystal field is virtually negligible when the valence of the central metal is held constant (Ottonello et al. 2007).

In Table 8.18 we see finally how the different isotopic masses of ^{50}Cr, ^{52}Cr, ^{53}Cr, and ^{54}Cr affect the thermodynamic properties of picrocromite at T = 298.15 *K* and P = 1 *bar*, through the different thermal corrections. ^{50}Cr is produced artificially in nuclear reactors but it is quite stable, with a half-life of 1.8×10^{17} *anna.* Because ^{52}Cr is the most abundant stable isotope of Cr (83.789%) the values tabulated for $Mg^{52}Cr_2O_4$ may be taken as representative of the natural picrocromite compound.

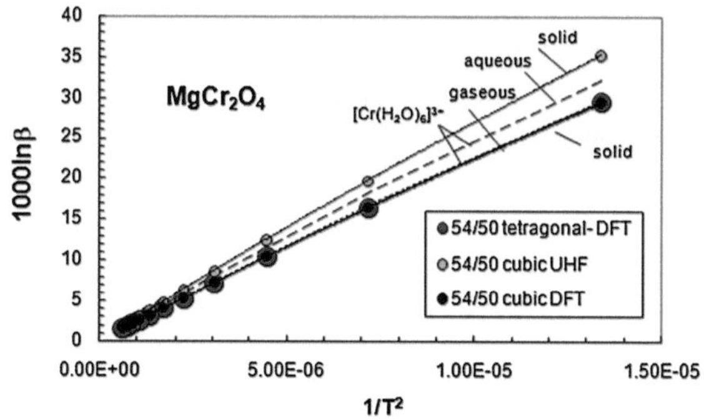

Fig. 8.28 Separative effects of chromium [$\beta \equiv (^{\bullet}s/^{\circ}s) \times f$] in $MgCr_2O_4$. The couple $^{54}Cr/^{50}Cr$ in tetragonal and cubic forms of $MgCr_2O_4$ (i.e. *$Fd\bar{3}mZ$* and *$I4_1/amd$*) Ottonello et al. 2007) is compared with computed fractionations in an aqueous and a gaseous Cr(III) species (hexa-aqua-chromium) (Ottonello and Vetuschi Zuccolini 2005a)

Table 8.18 Ab-initio DFT/B3LYP standard state (T = 298.15 *K*; P = 1 *bar*; pure component) thermodynamic functions (Ottonello et al. 2007)

Magnitude	$Mg^{50}Cr_2O_4$	$Mg^{52}Cr_2O_4$	$Mg^{53}Cr_2O_4$	$Mg^{54}Cr_2O_4$	$Mg^{52}Cr_2O_4$	Units
C_V	121.639	121.959	122.111	122.259	121.839	*J/(mol × K)*
C_P	122.337	122.657	122.809	122.957	122.537	*J/(mol × K)*
S_{vib}	90.8815	91.529	91.845	92.156	91.3095	*J/(mol × K)*
$S°$	114.167	114.814	115.130	115.441	114.595	*J/(mol × K)*
U_{vib}	69.936	69.808	69.747	69.689	69.860	*kJ/mol*
U	−2590.240	−2590.240	−2590.240	−2590.240	−2590.240	*Hartree*
ΔU_{corr}	0.026637	0.026588	0.026565	0.026543	0.026608	*Hartree*
H	−2590.240	−2590.240	−2590.240	−2590.240	−5180.481	*Hartree*
ΔH_{corr}	0.026639	0.026590	0.026567	0.026545	0.026610	*Hartree*
$H°_f$	−1650.439	−1650.567	−1650.628	−1650.686	−1650.553	*kJ/mol*
F_{vib}	42.840	42.519	42.364	42.212	33.412	*kJ/mol*
F	−2590.247	−2590.247	−2590.247	−2590.247	−2590.246	*Hartree*
G	−2590.234	−2590.234	−2590.234	−2590.234	−2590.234	*Hartree*
ΔG_{corr}	0.013674	0.013552	0.013493	0.013435	0.013597	*Hartree*
$G°_f$	−1538.322	−1538.643	−1538.798	−1538.950	−1538.563	*kJ/mol*

8.12 Loss of Symmetry

8.12.1 Premises

I borrowed the title of this section from Pisani (1996b) because quite appropriate and exhaustive. With "*loss of symmetry*" it is intended here that, we are concerned with quantum mechanical calculations implying the formation of a two-dimensional (2D) slab at the expense of a 3D structure; or with a defect process that interrupts locally the 3D continuity of a crystal. The comprehension of the subtilities involved in a 3D $\rightarrow$ 2D process is prodromal to understanding the properties of the surface of a 3D crystal, invariant upon translation. In Geochemistry, chemisorption processes acquire more and more interest, and it is desirable the development of appropriate tools for the first principles investigation of surface energies in crystals. To some extent also extended defects such as staking faults and planes of intergrowth may be seen as a loss of symmetry (3D $\rightarrow$ 2D), as well as dislocations (3D $\rightarrow$ 1D). The case of point defects (0D) is quite different. The quantum chemical effort is still at the very beginning, and often plagued by a poor knowledge of the appropriate Kröger–Vink notation (Kröger 1964) to be adopted to describe efficiently the nature of the defect itself. This is basically the reason why I anticipate in three short paragraphs the nature of point defects in natural crystals.

8.12.2 Intrinsic Defects

The disorder induced by the presence of point defects is grouped into two main categories: "*intrinsic disorder*" and "*extrinsic disorder*".

The Intrinsic disorder is observed in conditions of perfect stoichiometry of the crystal. It is related to two main defect equilibria: "*Schottky defects*" and "*Frenkel defects*".

Let us consider a generic oxide MO; in conditions of perfect stoichiometry, one mole of MO is composed of one mole of M and half a mole of O_2. If a cation is subtracted from a cationic position, creating a doubly ionized cationic vacancy (V_M''), then, in order to maintain crystal electroneutrality, one oxygen ion must be subtracted from a lattice site pertaining to the anion, thus creating an anionic vacancy ($V_O^{\cdot\cdot}$). Using the Kröger–Vink notation, the process, defined as "*Schottky defectuality*", is represented as follows:

$$MO \rightarrow V_M'' + V_O^{\cdot\cdot} \tag{8.12.1}$$

or:

$$M_M^{\times} + O_O^{\times} \rightarrow V_M'' + V_O^{\cdot\cdot} \tag{8.12.2}$$

or also:

$$0 \rightarrow V_M'' + V_O^{\cdot\cdot} \tag{8.12.3}$$

where the superscript $^{\times}$ indicates neutrality relative to the ideal crystal, $'$ stands for a negative excess charge taken by the cation vacancy, $^{\cdot}$ stands for a positive excess charge taken by the oxygen vacancy, and the subscripts indicate the lattice position of the defect.

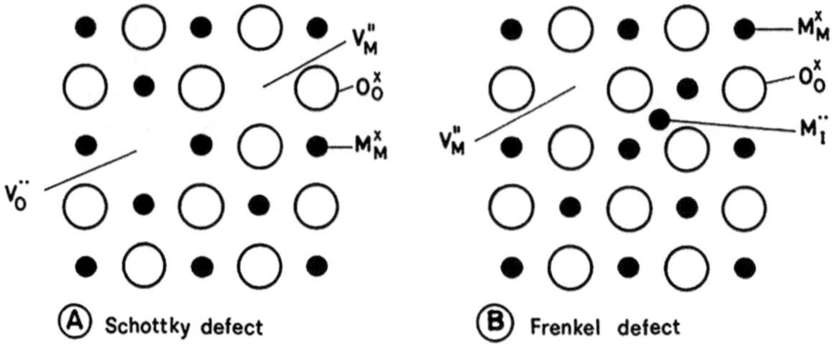

Fig. 8.29 Intrinsic Schottky (A) and Frenkel defectualities (B) in a MO crystal. From Ottonello (1997). With the permission of Columbia University Press

Let us now imagine moving a generic cation M from its normal position M ($M_M^{\times}$) to an interstitial position i, leaving the cationic site previously occupied empty:

$$M_M^{\times} \rightarrow M_i^{\cdot\cdot} + V_M^{\prime\prime} \tag{8.12.4}$$

The equilibrium thus established is a "*Frenkel defect*". In both the Schottky and Frenkel equilibria the stoichiometry of the crystal is unaltered (Fig. 8.29).

Assuming that the thermodynamic activity of the various species obeys Raoult's law, thus corresponding to their molar concentration (denoted hereafter by square brackets), the constant of the Schottky process reduces to:

$$K_S = \left[V_M^{\prime\prime}\right]\left[V_O^{\cdot\cdot}\right] \tag{8.12.5}$$

and that of the Frenkel process is:

$$K_F = \left[M_i^{\cdot\cdot}\right]\left[V_M^{\prime\prime}\right] \tag{8.12.6}$$

(as the defect concentrations are extremely low, i.e. generally lower than 10^{-4} in terms of molar fractions, the activity of ionic species at their normal position is assumed to be one).

8.12.3 Extrinsic Defects

To understand the significance of the "*extrinsic disorder*" we must imagine to immerse the crystal MO in an atmosphere in which the partial pressure of oxygen (P_{O_2}) is lower than the partial pressure of intrinsic stability (i.e. the partial pressure of oxygen at which crystal MO is perfectly stoichiometric; $P_{O_2}^{*}$). We observe two distinct defect processes:

$$O_O^{\times} \rightarrow \frac{1}{2} O_{2(gas)} + V_O^{\cdot\cdot} + 2e' \qquad (8.12.7)$$

$$M_M^{\times} + O_O^{\times} \rightarrow \frac{1}{2} O_{2(gas)} + M_i^{\cdot\cdot} + 2e' \qquad (8.12.8)$$

The first process produces doubly ionized positive oxygen vacancies and electrons (e′) and the second produces doubly ionized positive cation interstitials and electrons. The equilibrium constants of the two processes are given by:

$$K_7 = \left[V_O^{\cdot\cdot}\right][e']^2 P_{O_2}^{1/2} \qquad (8.12.9)$$

$$K_8 = \left[M_i^{\cdot\cdot}\right][e']^2 P_{O_2}^{1/2} \qquad (8.12.10)$$

As the bulk crystal is electroneutral, we may apply the "*electroneutrality condition*". For equilibrium (8.12.7) we write:

$$2\left[V_O^{\cdot\cdot}\right] = [e'] \qquad (8.12.11)$$

(i.e. the molar concentration of electrons must double the molar concentration of doubly ionized positive vacancies). From Eqs. (8.12.9) and (8.12.11) we obtain

$$[e'] = 2^{1/3} K_7^{1/3} P_{O_2}^{-1/6} \qquad (8.12.12)$$

$$\left[V_O^{\cdot\cdot}\right] = \frac{K_7^{1/3}}{4^{1/3}} P_{O_2}^{-1/6} \qquad (8.12.13)$$

Applying the electroneutrality condition to the equilibrium of Eq. (8.12.8) we have:

$$2\left[M_i^{\cdot\cdot}\right] = [e'] \qquad (8.12.14)$$

which, combined with Eq. (8.12.10)) gives:

$$[e'] = 2^{1/3} K_8^{1/3} P_{O_2}^{-1/6} \qquad (8.12.15)$$

$$\left[M_i^{\cdot\cdot}\right] = \frac{K_8^{1/3}}{4^{1/3}} P_{O_2}^{-1/6} \qquad (8.12.16)$$

At P_{O_2} higher than $P_{O_2}^*$, we observe the formation of doubly ionized cationic vacancies and positive electronic vacancies (i.e. electron holes with symbol $h^{\cdot}$):

$$\frac{1}{2}O_{2(gas)} \rightarrow O_O^{\times} + V_M'' + 2h^{\cdot} \quad (8.12.17)$$

The constant of equilibrium (8.12.17) is:

$$K_{17} = \left[V_M''\right]\left[h^{\cdot}\right]^2 P_{O_2}^{-1/2} \quad (8.12.18)$$

and defect concentrations are given by:

$$\left[h^{\cdot}\right] = 2^{1/3} K_{17}^{1/3} P_{O_2}^{1/6} \quad (8.12.19)$$

$$\left[V_M''\right] = \frac{K_{17}^{1/3}}{4^{1/3}} P_{O_2}^{1/6} \quad (8.12.20)$$

Figure 8.30 plots defect concentrations as a function of the partial pressure of oxygen, ranging from low P_{O_2}, to intrinsic pressure $P_{O_2}^*$ and to high P_{O_2}.

Diagrams such as those presented in Fig. 8.30 are commonly used in the interpretation of the defectuality of crystalline substances. Based on the above equations, the defect concentration depends on P_{O_2} in an exponential fashion (proportional to $P_{O_2}^{-1/6}$ and $P_{O_2}^{1/6}$ in low P_{O_2} and high P_{O_2} ranges respectively; as the plot is logarithmic, the exponent becomes the slope of the function). Outside the intrinsic stability range the stoichiometry of the crystal, as we have seen, is altered by exchanges with the surrounding atmosphere.

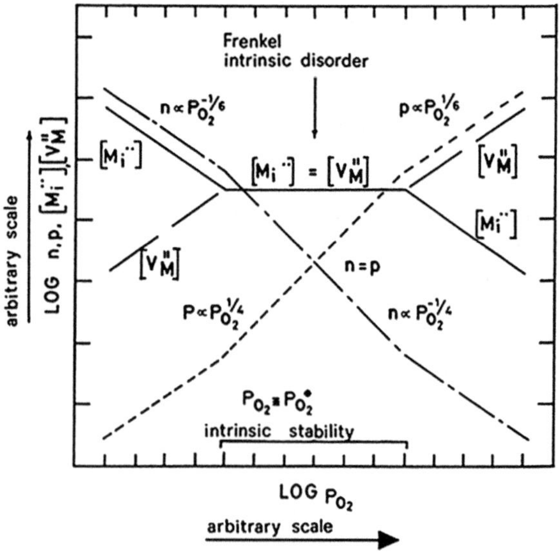

Fig. 8.30 Defect concentration as a function of partial pressure of oxygen. Magnesium Frenkel defects are assumed to predominate at stoichiometric composition. At low P_{O_2}, electronic conductivity is of type n (negative charge transport) and proportional to $P_{O_2}^{-1/6}$. At high P_{O_2}, electronic conductivity is of type p (positive charge transport) and proportional to $P_{O_2}^{1/6}$ From Morlotti and Ottonello (1982)

Background information: Point Impurities Defect processes are not simply ruled by the partial pressure of oxygen, but by the chemistry of all phases present in the system. For example, in the presence of a in silica-rich phase, Forsterite (Mg_2SiO_4) can incorporate SiO_2 in excess with respect to its normal stoichiometry through the defect equilibrium:

$$2SiO_2 \Leftrightarrow 2V''_{Mg} + Si_i^{\cdot\cdot\cdot\cdot} + Si_{Si}^{\times} + 4O_O^{\times} \tag{8.12.21}$$

or through the equilibrium (cf. Smith and Stocker 1975):

$$SiO_2 \Leftrightarrow 2V''_{Mg} + Si_{Si}^{\times} + 2V_O^{\cdot\cdot} + 2O_O^{\times} \tag{8.12.22}$$

Based on the experimental data of Pluschkell and Engell (1968), the silica excess in this way reaches 2.4% at 1400 °*C*.

To understand the significance of the term "*point impurity*" let us imagine equilibrating a fayalite crystal (Fe_2SiO_4) in a sufficiently oxidizing atmosphere to allow a defect equilibrium of the following type to proceed toward the right:

$$2Fe_{Fe}^{\times} \rightarrow V''_{Fe} + 2Fe_{Fe}^{\cdot} \tag{8.12.23}$$

or consider that a limited amount of trace component $Sm_4(SiO_4)_3$ may dissolve into a forsteritic olivine according to the process:

$$Sm_4(SiO_4)_3 + 4Mg_{Mg}^{\times} + 2Si_{Si}^{\times} + 8O_O^{x} \Leftrightarrow 4Sm_{Mg}^{\cdot} + 2V''_{Mg} + SiO_2 \tag{8.12.24}$$

In both cases, the crystal maintains its electroneutrality, but locally we observe the presence of cations whose charge (3+ in this case) is not that of the normal constituent (2+) and which, in the case of equilibrium (8.12.24), do not correspond to a major component. These "*point impurities*", normally present in the crystal at equilibrium, sensibly affect the chemical-physical properties of the phase, acting for instance as activators in light emission processes or as centers of heterogeneous nucleation. Processes like (8.12.24) also regulate the solubility of trace components in rock-forming minerals (Morlotti and Ottonello 1984).

Consider now a Schottky equilibrium taking place in a forsterite crystal and involving Mg^{2+} cations and O^{2-} anions, moving them from their lattice positions to the surface of the crystal:

$$Mg_{Mg}^{\times} + O_O^{\times} \rightarrow V''_{Mg} + V_O^{\cdot\cdot} + MgO_{(surface)} \tag{8.12.25}$$

To subtract the cation Mg^{2+} from its lattice position in the crystal and to bring it to the surface, we must work against the static potentials (coulombic + repulsive + dispersive) at the Mg site. In terms of energy, this work corresponds to half the lattice contribution of Mg^{2+} (in the Mg site of interest; i.e. M1 or M2) to the bulk static energy of the phase:

$$\Delta E_{U,i} = \frac{1}{2}\left(E_{C,i} + E_{R,i} + E_{D,i}\right) \tag{8.12.26}$$

Subtracting a charge in a given lattice position also contributes to the defect energy with an induced polarization term (which, in some circumstances, is the main form of energy in the bulk process):

$$\Delta E_P = -\frac{1}{2} e\Phi_e \tag{8.12.27}$$

Table 8.19 lists defect energies calculated with the above method in fayalite and forsterite crystals. Note that the obtained energies are of the order of some *eV*, i.e. substantially lower than the energies connected with extended defects. Note also that, ionic species being equal, the defect energies depend on the lattice position of the defects themselves. For instance, to create a cationic vacancy on site M (V''_{M1}), only requires energy lower than that necessary to create V''_{M2}. Also, to create an oxygen vacancy on site O3 ($V^{\cdot\cdot}_{O3}$), the energy is lower than that necessary to create $V^{\cdot\cdot}_{O1}$ or $V^{\cdot\cdot}_{O2}$.

Background information: Polarization energy We may note in Table 8.19 that, besides the Coulomb interaction, the polarization energy also plays a dominant role in the defect creation process.

Let us imagine now subtracting a −e charge at a given point in the crystal lattice: at that point effective charge +e exerts a polarization energy of the type depicted in Eq. (8.12.27). In this equation, Φ_e is the electric potential induced by the polarized medium at charge +e taken at origin. If the medium is composed of i ions, each at distance r_i and with polarizability α_i, Eq. (8.12.27) will be expressed as:

$$E_P = -\frac{1}{2} e \sum_i \frac{\mu_i}{r_i^2} = -\frac{1}{2} e \sum_i \alpha_i \frac{F_i}{r_i^2} \tag{8.12.28}$$

where F_i is the component directed along r_i (towards effective charge +e) of total electric field Φ_e on the *i*th ion and μ_i is the corresponding dipole moment. Applying differential operators we can write:

$$F_i = \left(\frac{e}{r_i^2} + F_j \frac{\gamma_i}{r_i}\right) \tag{8.12.29}$$

Table 8.19 Defect energies in forsterite and fayalite based on lattice energy calculations. ΔE_C = coulomb term; ΔE_R = repulsive term; ΔE_D = dispersive term; Δ_{CFSE} = crystal field term; ΔE_P = polarization energy; I = ionization potential; E = electron affinity; E_D = dissociation energy of O_2; ΔH = enthalpy of defect process. From Ottonello et al. (1990)

Process	ΔE_C	ΔE_R	ΔE_D	Δ_{CFSE}	ΔE_P	I	E	E_D	ΔH
FAYALITE									
(1)	22.50	−4.89	0.89	0.31	−15.31	–	–	–	3.50
(2)	24.60	−4.81	0.87	0.28	−14.84	–	–	–	6.10
(3)	−27.34	6.14	−1.07	–	12.44	–	7.46	2.79	0.42
(4)	−27.32	7.11	−1.15	–	12.65	–	7.46	2.79	1.54
(5)	−52.20	12.8	−2.06	–	24.68	–	14.92	5.58	3.72
(6)	11.60	5.59	−0.87	0.28	−42.94	30.64	–	–	4.30
(7)	15.80	5.75	−0.72	0.22	−42.00	30.64	–	–	9.77
(8)	−18.67	−6.70	1.01	0.62	−7.66	30.64	–	–	−0.76
(9)	−20.17	−6.59	0.99	0.56	−7.42	30.64	–	–	−1.99
(10)	−1.07	–	–	–	–	–	–	–	−1.07
(11)	−1.03	–	–	–	–	–	–	–	−1.03
FORSTERITE									
(12)	23.44	−4.59	0.11	–	−17.14	–	–	–	1.82
(13)	23.44	−6.29	2.08	–	−17.14	–	–	–	2.09
(14)	25.16	−4.46	0.13	–	−16.46	–	–	–	4.37
(15)	55.39	−12.71	1.48	–	−26.22	–	−7.46	−2.79	7.69
(16)	54.73	−14.51	1.54	–	−26.22	–	−7.46	−2.79	4.89
(17)	52.73	−13.10	1.50	–	−26.00	–	−7.46	−2.79	4.89

(1) $Fe_{M1}^{\times} \rightarrow V_{M1}'' + Fe^{2+}_{(surface)}$	(10) $Fe_{M1}^{\cdot} + Fe_{Si}' \rightarrow \{Fe_{M1}^{\cdot}Fe_{Si}'\}^{\times}$
(2) $Fe_{M2}^{\times} \rightarrow V_{M2}'' + Fe^{2+}_{(surface)}$	(11) $Fe_{M2}^{\cdot} + Fe_{Si}' \rightarrow \{Fe_{M2}^{\cdot}Fe_{Si}'\}^{\times}$
(3) $\frac{1}{2}O_{2(gas)} \rightarrow O^{2-}_{(O1,surface)}$	(12) $Mg_{M1}^{\times} \rightarrow V_{M1}'' + Mg^{2+}_{(surface)}$
(4) $\frac{1}{2}O_{2(gas)} \rightarrow O^{2-}_{(O2,surface)}$	(13) $Fe_{M1}^{\times} \rightarrow V_{M1}'' + Fe^{2+}_{(surface)}$
(5) $\frac{1}{2}O_{2(gas)} \rightarrow O^{2-}_{(O3,surface)}$	(14) $Mg_{M2}^{\times} \rightarrow V_{M2}'' + Mg^{2+}_{(surface)}$
(6) $Fe_{M1}^{\times} + h^{\cdot} \rightarrow V_{M1}'' + Fe_{Si}' + Si^{4+}_{(surface)}$	(15) $O_{O1}^{\times} \rightarrow V_{O1}'' + \frac{1}{2}O_{2(gas)}$
(7) $Fe_{M2}^{\times} + h^{\cdot} \rightarrow V_{M2}'' + Fe_{Si}' + Si^{4+}_{(surface)}$	(16) $O_{O2}^{\times} + 2e' \rightarrow V_{O2}'' + \frac{1}{2}O_{2(gas)}$
(8) $Fe_{M1}^{\times} + h^{\cdot} \rightarrow Fe_{M1}^{\cdot}$	(17) $O_{O3}^{\times} + 2e' \rightarrow V_{O3}'' + \frac{1}{2}O_{2(gas)}$
(9) $Fe_{M2}^{\times} + h^{\cdot} \rightarrow Fe_{M2}^{\cdot}$	

where γ_i is the modulus of dipole ij at distance r_i from charge +e and F_j is the summation of the gradients of the electric potential from the various j ions toward i; i.e.

$$F_j = \sum_j \nabla_i \Phi_{e,j} \tag{8.12.30}$$

and

$$\nabla_i \Phi_{e,j} = \left(\frac{\partial \Phi_{e,j}}{\partial x}, \frac{\partial \Phi_{e,j}}{\partial y}, \frac{\partial \Phi_{e,j}}{\partial z} \right) \tag{8.12.31}$$

The solution of the system (8.12.28–8.12.31) leads to a high number of equations (cf. Rittner et al. 1949) and simplifying assumptions must be introduced to solve the calculations properly. The generally adopted method is that of Mott and Littleton (1938), which solves the polarization energy between two ions (i and j) through the system:

$$E_P = -\frac{1}{2} e \Phi_e \tag{8.12.32}$$

$$\Phi_e = -ea^3 \left[M_i \sum_i \frac{1}{r^4} + M_j \sum_j \frac{1}{r^4} \right] \tag{8.12.33}$$

$$M_i = \frac{2\alpha_i}{\alpha_i + \alpha_j} \frac{1}{4\pi} \left(1 - \frac{1}{\varepsilon_0} \right) \tag{8.12.34}$$

$$M_j = \frac{2\alpha_j}{\alpha_i + \alpha_j} \frac{1}{4\pi} \left(1 - \frac{1}{\varepsilon_0} \right) \tag{8.12.35}$$

where α^3 is the volume occupied by the pair of ions and ε_0 is the dielectric constant of the medium.

Among the various energy terms, Table 8.19 lists energy values assigned to point impurities such as $Fe^{\cdot}_{M1}$, $Fe^{\cdot}_{M2}$. In the calculation of their energies, a ionization term necessary to subtract one electron from Fe^{2+} has also been added (ionization potential). With defect notation, this process can be expressed as:

$$Fe^{\times}_{M1} + h^{\cdot} \rightarrow Fe^{\cdot}_{M1} \tag{8.12.36}$$

An alternative way of calculating defect energies on the basis of static potentials is that outlined by Fumi and Tosi (1957) for alkali-halides; the energy of the Schottky process is seen as an algebraic summation of three terms:

$$\mathrm{Na}_{\mathrm{Na}}^{\times} \rightarrow V'_{\mathrm{Na}} + \mathrm{Na}^{+}_{(\mathrm{gas})} \qquad (+4.80\ eV) \tag{8.12.37}$$

$$\mathrm{Cl}_{\mathrm{Cl}}^{\times} \rightarrow V^{\cdot}_{\mathrm{Cl}} + \mathrm{Cl}^{-}_{(\mathrm{gas})} \qquad (+5.14\ eV)) \tag{8.12.38}$$

$$\mathrm{Na}^{+}_{(\mathrm{gas})} + \mathrm{Cl}^{-}_{(\mathrm{gas})} \rightarrow \mathrm{Na}_{\mathrm{Na}}^{\times} + \mathrm{Cl}_{\mathrm{Cl}}^{\times} \qquad (-7.82\ eV) \tag{8.12.39}$$

$$0 \rightarrow V'_{\mathrm{Na}} + V^{\cdot}_{\mathrm{Cl}} \qquad (+2.12\ eV) \tag{8.12.40}$$

Note that, in this case, the energies of processes (8.12.37) and (8.12.38) are doubled with respect to the energy necessary to bring the ion on the surface of the crystal; moreover, the energy of process (8.12.39) is equal to the lattice energy of the crystal.

Background Information: Associate Defects Point defects of opposite charge tend to be reciprocally attracted, forming the so-called "*associate defects*".

Let me consider the general case of two point defects A and B, forming an associate defect (AB) according to the process:

$$\mathrm{A} + \mathrm{B} \rightarrow (\mathrm{AB}) \tag{8.12.41}$$

If we imagine that both defects A and B and associate (AB) are distributed in conditions of complete configurational disorder and substitute thermodynamic activity for the molar concentration, we have:

$$\mathrm{K}_{\mathrm{AB}} = \frac{[(\mathrm{AB})]}{[\mathrm{A}][\mathrm{B}]} \tag{8.12.42}$$

The Gibbs free energy modification due to associative process (disregarding volumetric effects) is:

$$\Delta \mathrm{G}_{\mathrm{AB}} = -\mathrm{k}_{\mathrm{B}}\mathrm{T} \ln \mathrm{K}_{\mathrm{AB}} = \Delta \mathrm{H}_{\mathrm{AB}} - \mathrm{T}\Delta \mathrm{S}_{\mathrm{AB}} \tag{8.12.43}$$

The enthalpy change related to associative process ($\Delta \mathrm{H}_{\mathrm{AB}}$) is due essentially to Coulomb interactions and, subordinately, to polarization, repulsion, covalent bonding, elastic interactions, and vibrational effects. The latter two causes are generally negligible and may have some effect only at low T.

The Coulombic energy of associative process is

$$\Delta H_{AB,coulombic} = -\frac{q^2}{\varepsilon_s r} \qquad (8.12.44)$$

where q is the charge of interacting defects, r is the distance of shortest approach, and ε_s the static dielectric constant of the crystal.[14] Evaluation of the coulombic interaction through Eq. (8.12.44) usually leads to an overestimation of the energy of the associative process.

The entropy variation connected with associative processes (ΔS_{AB}) may be ascribed to two different contributions: configurational and vibrational. The configurational term is given by:

$$\Delta S_{AB,conf} = k_B \ln\left(\frac{Z}{\sigma}\right) \qquad (8.12.45)$$

where Z is the number of ways in which the associate may be formed and σ is a symmetry factor. For an associate (AB), Z corresponds to the number of equivalent sites that can be occupied by B at the shortest distance from A (or by A at the shortest distance from B). For the same associate (AB), σ is one, but becomes two for an associate dimer (AA), three for a trimer (AAA), and so on.

The vibrational amount contributing to the entropy of defect processes ($\Delta S_{AB,vibr}$) arises from the variation of the vibrational spectrum of the crystal in the neighborhood of the associate:

$$\Delta S_{AB,vibr} = k_B \ln f \qquad (8.12.46)$$

$$f = \prod_i \frac{\nu_{0,i}}{\nu'_i} \qquad (8.12.47)$$

where $\nu_{0,i}$, ν'_i are vibrational frequencies. Due respectively to free and associated defects (see Kröger 1964, for more detailed treatment).

For a pair of Schottky defects with respectively x and y first neighbors and ν', ν'' vibrational frequencies, vibrational term f has the form:

$$f = \left(\frac{\nu_0}{\nu'}\right)^x \left(\frac{\nu_0}{\nu''}\right)^y \qquad (8.12.48)$$

[14] Actually, as pointed out by Kröger (1964), at small distances the use of the static dielectric constant based on a continuum approach to interaction energy is not justified, as one would rather adopt the permittivity of a vacuum. However, using the static dielectric constant leads to underestimation of the binding energy by a factor of 10–15%, counterbalancing the fact that short-range repulsive forces are neglected.

For instance, for a Schottky defect pair in NaCl, we have:

$$\frac{\nu_0}{\nu'} = \frac{\nu_0}{\nu''} = 2 \tag{8.12.49}$$

$$x = y = 6 \tag{8.12.50}$$

from which we obtain f = 4096. If the vibrational frequency of the defects increases due to association, then f > 1, and the vibrational entropy also increases. The opposite happens with f < 1.

In the case of a Frenkel defect pair, if ν_i is the frequency of interstitial atom i and ν'_i is the vibrational frequency of the Z first neighbors surrounding it, we have:

$$f = \left(\frac{\nu_0}{\nu_i}\right)\left(\frac{\nu_0}{\nu'_i}\right)^z\left(\frac{\nu_0}{\nu'}\right)^x \tag{8.12.51}$$

Combining the various equations we obtain:

$$K_{AB} = \frac{Zf}{\sigma}\exp\left(-\frac{\Delta H_{AB}}{k_B T}\right) \tag{8.12.52}$$

8.12.4 Surface Energy

The energy parameters investigated quantum mechanically in surface studies are of several kinds. Besides the energy bands of electrons at the surface, they comprise five distinct magnitudes (Pisani 1996b):

1. surface formation energy per unit area (A) = $(\gamma - n\alpha)/(2nA)$;
2. surface relaxation energy per unit area = $(\beta - \gamma)/(2A)$;
3. surface reconstruction energy per unit area = $(\delta - t\gamma)/(2tA)$;
4. adsorption energy with respect to isolated atoms or molecules = $\sum \mu \in_\mu - \eta$; $\left(\sum \mu \in_\mu - \eta\right)/2$
5. adsorption energy with respect to an isolated monolayer = $\zeta - \eta$; $(2\zeta - \theta)/2$

   denoting with:

- α the energy of the bulk holo-crystal (HC);
- β the energy of the un-relaxed (*hkl*)n slab (S);
- γ the energy of the relaxed slab;
- δ the energy of the relaxed-reconstructed slab;
- $\in_\mu$ the energy of the isolated ad-species;
- ζ the energy of the isolated ad-layer (M);

- η the energy of an optimized M–S system of reference;
- θ the energy of an optimized M–S–M system of reference.

Besides the above listed processes, it is also important to stress that the "*surface energy*" affects the reactivity and extrinsic stability of Earth's materials, through the relative Surface/Volume (S/V) ratio.

From the definition of Jura and Garland (1952), the thermal heat content of the surface of the crystal may be expressed as:

$$h_{298} = \Delta H_{298}/A \tag{8.12.53}$$

with ΔH_{298} = difference in the heat of solution of a solid of surface area A with respect to a solid with A = 0 (Jura and Garland 1952; Lipsett et al. 1927, 1928). If we denote ΔC_p the difference in the heat content of an infinite crystal (i.e. area = 0) with respect to a microscopic crystal and we assume that the entropy goes to zero at 0 K for both crystals, then we have (Giauque 1949; Jura and Garland 1952):

$$s = \frac{1}{A}\int_0^T \frac{\Delta C_P}{T} dT \tag{8.12.54}$$

$$h = h_{298} + \frac{1}{A}\int_{298}^T \Delta C_P dT \tag{8.12.55}$$

From Eqs. (8.12.53)–(8.12.55) it follows:

$$\begin{aligned} g &= \frac{\Delta H_{298}}{A} + \frac{1}{A}\int_{298}^T \Delta C_P dT + \frac{T}{A}\int_0^T \frac{\Delta C_P}{T} dT \\ &= \frac{1}{A}\left(\Delta H_{298} + \int_{298}^T \Delta C_P dT + E\int_0^T \frac{\Delta C_P}{T} dT\right) \end{aligned} \tag{8.12.56}$$

We may pose thus:

$$\Delta G^* = g\left(A^* - A^0\right) \tag{8.12.57}$$

$$\gamma^* = \exp\left(\frac{\Delta G^*}{RT}\right) \tag{8.12.58}$$

$$\mu^* = \mu^0 + \mathrm{RT}\ln\gamma^* \tag{8.12.59}$$

$$\mu = \mu^* + \mathrm{RT}\ln \mathrm{x} \tag{8.12.60}$$

with $\Delta G*$ = Gibbs free energy difference between the (macroscopic) condensed phase and the condensed phase at the dimensional state of interest (but identical structure); and with

A^0 = molar surface at standard state;

A^* = molar surface at the dimensional state of interest;

μ^0 = standard state chemical potential;

μ^* = potential referred to the surface tension contribution;

μ = potential under the dimensional conditions of interest;

x = molar fraction of the pure component in mixture (assumed ideal for simplicity).

I may offer an emblematic application of the S/V effect by referring to periclase MgO, since we have in this phase both theoretical and experimental appraisals of his calorimetric properties under different dimensional states (Lennard–Jones and Taylor 1925; Lennard–Jones and Dent 1929; Jura and Pitzer 1952). The macroscopic condensed form of MgO is the stable cubic face-centered periclase with:

$$G^0_{f,298.15} = -569300\ (J/mol);\ V^0 = 11.25\ (cm^3/mol);\ A^0 \approx 30\ (cm^2/mol)$$

Burning metallic Mg ribbon in an oxygen flow one obtains the so-called "*Virgin smoke*" (small cubes of MgO, 45–85 *nm*; same structural state) with a surface area = 10.5 ± 1 m^2/g. Since the molar weight of MgO is 40.31 *g/mol*, we have $A^* = 4.23 \times 10^6\ cm^2/mol$. From the existing appraisals of g, we take g = 1000 erg/cm^2 (Jura and Garland 1952; Lennard–Jones and Taylor 1925; Lennard–Jones and Dent 1929). Then we may apply our method, to obtain:

$$\Delta G* = g\left(A^* - A^0\right) = 0.0001\frac{J}{cm^2}\left(4.23 \times 10^6 - 30.12\right)\frac{cm^2}{mol} = 423\frac{J}{mol} \tag{8.12.61}$$

$$\gamma^* = \exp\left(\frac{\Delta G^*}{RT}\right) = 1.186 \tag{8.12.62}$$

We see thus that the virgin smoke is more reactive than the macroscopic periclase of about half *kJ* per mole and this is conveniently translated into an adimensional activity coefficient. Obviously, with grain dimensions progressively vanishing, the surface tension becomes dominant and the activity coefficient may be expected to attain higher values, up to a limit that may be thought to correspond to the dimension of the asymmetric unit (4 formula units in the case of MgO) and eventually calculated by first principles. What above exposed is of general validity for any condensed phase.

8.12.5 Energy Band Structures and Projected Bands

To understand the significance of energy band structures in crystals we must reconsider briefly what we have seen at Sect. 8.4 concerning the Block theorem. Namely, a Bloch wave (also called Bloch function) is a wave composed of a plane wave modulated by a periodic function. The energy eigenstates of Block waves determine the basis of the electron wave functions in crystals. Replacing the wavenumber k with the wave vector **k** and the position x with the space vector **r** and modulating the resulting plane wave [exp(i**kr**)] through a periodic function u (**r**) which has the same periodicity of the crystal one has a *Block function*:

$$\Psi(\mathbf{r}) = \exp(\mathrm{i}\mathbf{k}\mathbf{r})\mathrm{u}(\mathbf{r}) \tag{8.12.63}$$

The energy eigenstates of Eq. (8.12.63)) are usually depicted as $\psi_{\mathrm{n}\mathbf{k}}$ where n (the band index) indicates the type of Block wave [same **k** but different periodicity u(r)]. All (n) Block waves occur for values of **k** falling in the Brillouin zone of the reciprocal lattice. Hence they are translationally invariant in the reciprocal lattice, i.e.

$$\Psi_{\mathrm{n}\mathbf{k}} = \Psi_{\mathrm{n}(\mathbf{k}+\mathbf{T})}, \tag{8.12.64}$$

Because, for the same **k**, different (n) eigenfunctions can satisfy the Block theorem, the acceptable basis functions are labeled $\Psi_{\mathrm{n}}^{\mathbf{k}}(\mathbf{r})$, each of them belonging to a different irreducible representation of the group of one-electron Hamiltonians h.

To find the energy levels $\mathrm{E_n}(\mathbf{k})$ associated to $\Psi_{\mathrm{n}}^{\mathbf{k}}(\mathbf{r})$, each Block functions is conveniently expanded in a finite set:

$$\Psi_{\mathrm{n}}^{\mathbf{k}}(\mathbf{r}) = \sum_{\mu} c_{\mu \mathrm{n}}^{\mathbf{k}} \phi_{\mu}^{\mathbf{k}}(\mathbf{r}), \tag{8.12.65}$$

where the various coefficients $c_{\mu\mathrm{n}}^{\mathbf{k}}$ are determined by factorizing the problem into separate parts and solving the matrix equation (Pisani 1996a):

$$\mathsf{H}^{\mathbf{k}}\mathrm{C}^{\mathbf{k}} = \mathrm{S}^{\mathbf{k}}\mathrm{C}^{\mathbf{k}}\mathrm{E}^{\mathbf{k}} \tag{8.12.66}$$

where

$$\mathsf{H}_{\mu\nu}^{\mathbf{k}} = \left\langle \phi_{\mu}^{\mathbf{k}} \middle| \mathsf{h} \middle| \phi_{\nu}^{\mathbf{k}} \right\rangle \tag{8.12.67}$$

$$\mathrm{S}_{\mu\nu}^{\mathbf{k}} = \left\langle \phi_{\mu}^{\mathbf{k}} \middle| \phi_{\nu}^{\mathbf{k}} \right\rangle \tag{8.12.68}$$

$S^{\mathbf{k}}$ in the above equations is the overlap matrix (i.e. the identity matrix I, if the basis functions are orthogonal) and $C^{\mathbf{k}}$ is a column matrix with the coefficients of the crystalline orbitals.

The Block functions entering Eq. (8.12.64) may be either based on AO's:

$$\phi_{\mu}^{\mathbf{k}}(\mathbf{r}) = \sum_{\mathbf{T}} \exp(\mathbf{ik} \cdot \mathbf{T})\varphi_{\mu}\left(\mathbf{r} - \mathbf{A}_{\varphi} - \mathbf{T}\right) \tag{8.12.69}$$

where A_{φ} if the position of the atom in the reference cell on which the atomic orbital is centered (Dovesi 1996), or be based on plane waves:

$$\phi_{\parallel}^{\mathbf{k}}(\mathbf{r}) = \exp[\mathbf{ir} \cdot (\mathbf{k} + \mathcal{K})] \tag{8.12.70}$$

where k is the "translationally equivalent vector in the first Brillouin Zone" (Pisani, 1996a). Denoting $\varepsilon_{\mathrm{i}}^{\mathbf{k}}$ the energy eigenvalues associated to each **k** point and introducing the step function $\theta\left(\mathrm{e} - \varepsilon_{\mathrm{i}}^{\mathbf{k}}\right)$ (a function which returns 1 if $\varepsilon_{\mathrm{i}}^{\mathbf{k}}$ is less than e and 0 otherwise), the number of states below a certain energy level i(e) is given by (Pisani 1996a):

$$\mathrm{i(e)} = \frac{2}{\mathrm{V_{BZ}}} \sum_{\mathrm{i}} \int_{\mathrm{BZ}} \mathrm{dk}\theta\left(\mathrm{e} - \varepsilon_{\mathrm{i}}^{\mathbf{k}}\right) \tag{8.12.71}$$

and the matrix I(e) of the integrated density of states (IDOS) is given by:

$$\mathrm{I(e)} = \frac{2}{\mathrm{V_{BZ}}} \sum_{\mathrm{i}} \int_{\mathrm{BZ}} \mathrm{dk} c_{\mu\mathrm{i}}^{\mathbf{k}*} c_{\nu\mathrm{i}}^{\mathbf{k}} \exp(\mathrm{ik} \cdot \mathrm{T})\theta\left(\mathrm{e} - \varepsilon_{\mathrm{i}}^{\mathbf{k}}\right) \tag{8.12.72}$$

The Fermi energy E_F is found by requiring that there are exactly 2n spin orbitals per cell:

$$\mathrm{i(E_F)} = 2\mathrm{n} \tag{8.12.73}$$

The total density of states DOS is given by:

$$\mathrm{n(e)} = {}^{\mathrm{di(e)}}\!/_{\mathrm{de}} \tag{8.12.74}$$

and the projected density of states (PDOS) is given by the IDOS derivative:

$$\mathrm{N(e)} = {}^{\mathrm{dI(e)}}\!/_{\mathrm{de}} \tag{8.12.75}$$

Each energy level $E_n(\mathbf{k})$ is associated to $\Psi_{\mathrm{n}}^{\mathbf{k}}(\mathbf{r})$ evolves smoothly with changes in **k**, forming a band of states more or less dispersed moving along the various special points in the Brillouin zone. The number of non-equivalent **k** point increases with the actual number of cells N in the crystal and $E_n(\mathbf{k})$ becomes progressively more and more continuous. Plotting the values of $E_n(\mathbf{k})$ along straight lines connecting the special points **k**, "*band structure plots*" are created. In these plots it is possible to observe, along special crystallographic directions, white lenses (i.e. lack of bands), corresponding to the energy gap between the *conduction band* and the *valence band*. The band gaps are generally classified into two types, i.e. "*direct*" or "*indirect*" depending on whether the lowest-energy state above the band gap has or has not the same **k** as the highest-energy state beneath the band gap. The iso-surface of states with energy equal to the Fermi level E_F is known as "*Fermi surface*". Conductors, semiconductors, and ionic crystals differ appreciably in their band structures. Hereafter I resume briefly what was found by Dal Corso (1996) investigating the band structure of Al, Si, and NaCl as prototypical examples of the three categories of crystalline solids. The investigation of Dal Corso was carried out with the Pseudopotential Plane Waves program PWSCF. The (freeware) code works in the framework of DFT using the LDA approximation. The orbitals are expanded in Plane Waves (PW) and the cores are described by norm-conserving Pseudo-Potentials (PP). Details on the structure of the code and its operationality may be found in Dal Corso (1996) together with a lucid, concise, and quite comprehensible outline of the limits and potentiality of the various computational procedures. As stressed by Dal Corso (1996) "*the eigenvalues of the Kohn–Sham equations are not to be considered as the energy of the electrons*" although it is common practice to compare the computational energy results with the results of photoemission experiments (which are generally precise within ~0.1 *eV*).

The energy bands of Al, Si, and NaCl are shown in Fig. 8.31.

The 8 lowest energy bands of silicon were reproduced by Dal Corso (1996) for 6 k points with an energy cutoff of 10 *Hartree*. Crystalline silicon is a group 4 covalent semiconductor. It has a *fcc* Bravais lattice with two atoms per unit cell (hence 8 electrons, filling the four valence bands). The experimental energy gap between the conduction band and valence band is 1.17 *eV*. This gap is quite evident in the computed energy bands (Fig. 8.31) which have also the appropriate form and bandwidth. The electronic configuration is [Ne] $3s^23p^2$. The LDA approximation is particularly good in this case because the atoms are held together by a covalent bond between sp^3 hybrid orbitals. The PP frozen-core approximation is also quite appropriate because the Ne-like core electrons of the single Si atoms are tightly bound to the nuclei and do not relax appreciably in the solid (Del Corso 1996). Crystalline aluminum is a group IIIA metal conductor whose atoms have configuration [Ne] $3s^23p^1$. Again the frozen-core approximation for Ne-like core electrons is quite appropriate and also the LDA approximation for outer electrons works well. In the case of metals, the convergence of the total energy as a function of the cutoff energy and of the special point sampling requires special procedures (i.e. "*smearing parameters*"; cf. Methfessel and Paxton 1989) to avoid the effects of the Fermi surface. The conductor has an *fcc* Bravais lattice with one atom per cell. The energy bands were obtained by Dal Corso (1996) with a SCF calculation involving 60 special points. The energy cutoff was 15 *Hartree*. In Fig. 8.31 are shown the lowest 10 bands. The zero energy is set on the Fermi surface. NaCl is an ionic compound essentially held by coulomb forces between Na^+ and Cl^- ions. Again the crystal may be described by a *fcc* Bravais lattice, with two atoms per unit cell. The adopted

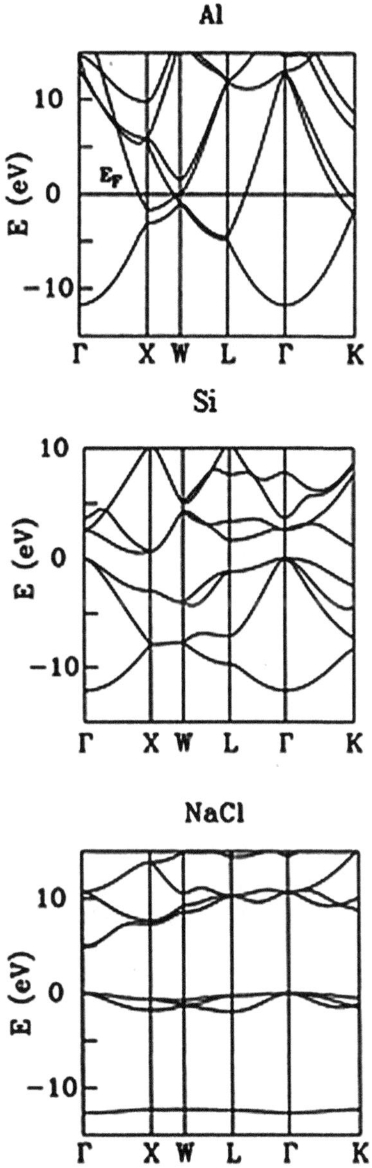

Fig. 8.31 Band structures of aluminum (lowest 10 bands), silicon (lowest 8 bands), and sodium chloride (lowest 8 bands) along several symmetry lines. Modified from Dal Corso (1996). The energy zero of Al is set at the Fermi surface energy level

Pseudo-Potentials (PP) assumes that only $3s^1$ is a valence electron. The 8 lowest energy bands of the solid shown in Fig. 8.31 were obtained with a cutoff energy of 30 *Hartree* and 2k points. They exhibit a quite limited dispersion with a "*direct*" gap at the **Γ** point consistent with the experimental observation (4.7 *eV*).

In the next paragraphs, we will see how the surface energy may be approximated quantum mechanically creating slabs of appropriate thickness from the 3D crystal. The electronic structure of a thick slab is similar to that of the holo-crystal but some peculiar surface characteristics become evident in its band structure. To make a comparative appraisal of the 3D band structure with the band structure of a slab, a general **k** vector of the Brillouin zone is decomposed into its parallel and perpendicular components:

$$\mathbf{k} = \mathbf{k}_{\|} + \mathbf{k}_{\perp} \tag{8.12.76}$$

A "*projected band structure*" is one where all the energy eigenvalues $\varepsilon_i(\mathbf{k}_{\|} + \mathbf{k}_{\perp})$ are attributed to $\mathbf{k}_{\|}$ independently of $\mathbf{k}_{\perp}$ I remand the interested reader to Pisani (1996b) for an appropriate treatment of "*projected bands*".

8.12.6 A Classical Quantum–mechanical Approach to Surface Energy

Figure 8.32 represents the formation of a new surface in a crystal of the NaCl-type (LiCl), after Van der Hoff and Benson (1954). The lattice energy of the original crystal at the athermal limit is denoted U_{holo}. Dividing the crystal into two hemi-crystals by a plane parallel to a {100} plane of the lattice, the energy of each part becomes U_{hemi}. The surface energy is the difference between U_{holo} and the two U_{hemi}:

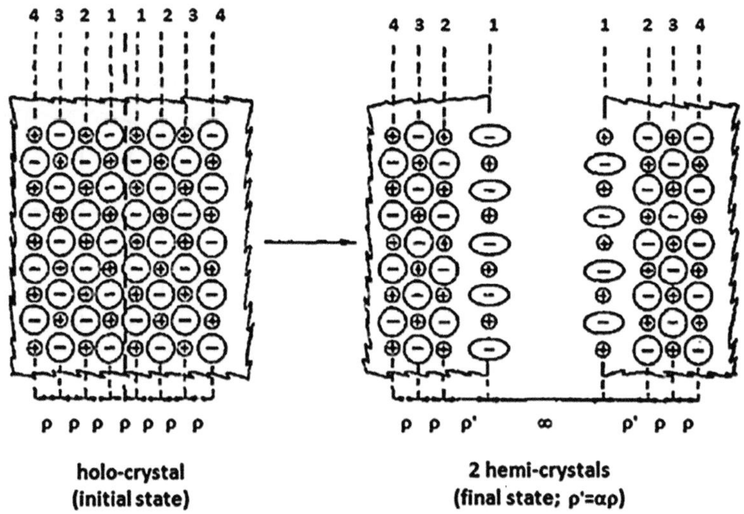

Fig. 8.32 Formation of two new surfaces by division of a holo-crystal into two hemi-crystals. Modified from Van der Hoff and Benson (1954). The distortion of the hemi-crystal due to polarization of surface F^- ions and change in lattice spacing is indicated diagrammatically on the right part of the figure. The polarization of Li^+ is neglected

$$U_{surf} = -\frac{1}{2}(U_{holo} - 2U_{hemi}) \tag{8.12.77}$$

Near the two surfaces, the lattice of the hemi-crystals is distorted due to relaxation phenomena associated to dangling bonds at the surfaces. The change in lattice spacing decreases exponentially on proceeding inwards from the surface. The main contribution to the surface energy arising from this distortion is due to the change in spacing between the surface and the second layer (Lennard–Jones and Dent 1928; Dent 1929). Also, the change in polarization of the ions upon formation of the surface affects considerably the surface energy. The calculations of Dent (1929) for eight monovalent alkaline-halides show that the distortion decreases the surface energy by about 20%, and it is due almost entirely to the polarization of the surface ions.

In the Van der Hoff–Benson model, developed for the LiF crystal, the Li^+ ion is treated as a point charge while the F^- ion is represented as a nucleus of charge +5 surrounded by a shell of 6 electrons. The spacing $\alpha\rho$ between the surface and the second layer is allowed to vary. The deviation from spherical symmetry is accounted for by the introduction of an excited state and covalent bonding with the nearest neighbor Li^+ ions of the second layer. The electrostatic field is the classical one arising from point charges located at nuclear centers plus a set of dipoles with moment μ situated at the F^- sites of the first layer and directed perpendicular to the surface, the inward normal being considered positive. The crystal lattice spacing in the holo-crystal is assumed to be $\rho = 2.01$ Å (Benson and Wyllie 1951). The surface area occupied by an ion pair is:

$$\sigma = (2\rho_2)^{-1}\left(U_{hemi} - \frac{1}{2}U_{holo}\right) \tag{8.12.78}$$

In Fig. 8.33 it is shown a F^- ion of the nth layer, surrounded by its nearest neighbors. For any value of n, $\rho_1 = \rho_2 = \rho_3 = \rho_4 = \rho$. If n = 1, $\rho_1 = \rho' = \alpha\rho$ and $\rho_6 = \infty$. If n = 2, $\rho_5 = \rho$ and $\rho_6 = \rho' = \alpha\rho$. For all other layers $\rho_5 = \rho_6 = \rho$.

Each F^- ion is considered to be situated in a potential field $\left[-V_-^{(n)}(x, y, z)\right]$ due to the rest of the crystal arising from the contributions of the point charges at all the other lattice sites $\left[P_-^{(n)}(x, y, z)\right]$ and the dipole of the 1st layer at F sites $\left[Q_-^{(n)}(x, y, z)\right]$:

$$-V_-^{(n)}(x, y, z) = P_-^{(n)}(x, y, z) + Q_-^{(n)}(x, y, z) \tag{8.12.79}$$

The potential at the origin arising from non nearest neighbor charges $\left[V_-^{(n)}(0)\right]$ is:

$$V_-^{(n)}(0) = -\left|P_-^{(n)}(0) + Q_-^{(n)}(0)\right| \tag{8.12.80}$$

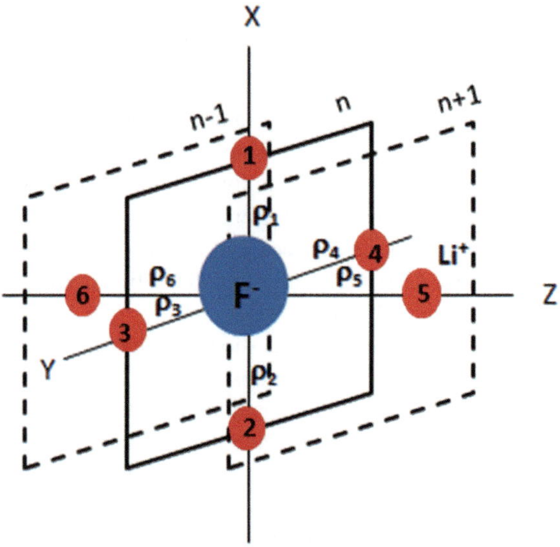

Fig. 8.33 Coordinate system of a F^{-} ion in the nth layer

and, "*by analogy*" (Van der Hoff and Benson 1954), the potential field at a positive site is:

$$V_{+}^{(n)}(0) = -\left|P_{+}^{(n)}(0) + Q_{+}^{(n)}(0)\right| \tag{8.12.81}$$

It follows, for symmetry reasons, that

$$P_{-}^{(n)}(0) = -P_{+}^{(n)}(0) \tag{8.12.82}$$

The cohesive energy of a holo- or hemi-crystal is regarded as made up from a summation of classical electrostatic terms $\left[\varepsilon_{F^{-(n)}(\text{classical})}\right]$ plus a contribution associated with each F^{-} ion and representing the energy of the F^{-} ion in the crystal, in excess of that calculated from the classical treatment $\left[\varepsilon_{F^{-(n)}(\text{quantal})}\right]$:

$$U_{\text{hemi}} = \frac{1}{2}\sum_{n}^{\infty}{}_{1}\left[V_{-}^{(n)}(0) - V_{+}^{(n)}(0)\right] + \sum_{n}^{\infty}{}_{1}\left[\varepsilon_{F^{-(n)}(\text{quantal})} - \varepsilon_{F^{-(n)}(\text{classical})}\right]$$

(8.12.83.1)

$$U_{\text{holo}} = \sum_{n}^{\infty}{}_{1}\left[V_{-}^{(\infty)}(0) - V_{+}^{(\infty)}(0)\right] + \sum_{n}^{\infty}{}_{1}\left[\varepsilon_{F^{-(\infty)}(\text{quantal})} - \varepsilon_{F^{-(\infty)}(\text{classical})}\right]$$

(8.12.83.2)

Combining Eqs. (8.12.80)–(8.12.82) and (8.12.83.1)–(8.12.83.2), after few passages Van der Hoff and Benson (1954) obtained:

$$\sigma = \frac{1}{2\rho^2}\left[\frac{1}{\rho}\{B + 2[S(1) - S(\alpha)]\} + \frac{\mu}{2\rho^2}D(\alpha) + \sum_n^\infty {}_1\left\{\left(\varepsilon_{F^-(n)} - \varepsilon_{F^-(n)}\right)_{quantal} - \left(\varepsilon_{F^-(n)} - \varepsilon_{F^-(n)}\right)_{classical}\right\}\right] \tag{8.12.84}$$

where S(1) and S(α) are the sums of the dipole networks over plane squares and cubic lattices. These vary in a complex manner with the indices of the substance. For example, the sum adopted by MacKenzie (1950) for (100) planes of a cubic lattice is

$$S = \sum_{k,l,m}{}' k\left(k^2 + l^2 + m^2\right)^{-n} \tag{8.12.85}$$

where k, l, m are the Miller indices of the substance, n is an integer (or a half-integer) and the dash on the summation sign indicates that the case $k = l = m = 0$ is excluded. The procedure transforms the lattice sums into sums of Bessel functions or e-powers that converge' rather rapidly. $k + l + m$ may be restricted to even or odd values, Replacing $(k^2 + l^2 + m^2)^{-n}$ with $\left[\left(k + a\right)^2 + (l + b)^2 + \left(m + c\right)^2\right]^{-n}$, the method is applicable to off-lattice points.[15]

Moreover,

$$B = \sum_n^\infty {}_1(-1)^{n+1}S(n) = 0.06524647 \tag{8.12.86.1}$$

$$D(\alpha) = \frac{\mu}{\rho^2}\sum_n^\infty {}_1\left[Q_+^{(n)}(0) - Q_-^{(n)}(0)\right] \tag{8.12.86.2}$$

To calculate $\left[\varepsilon_{F^-(n)(quantal)}\right]$ the six electrons of a F^- ion in the nth layer are assigned in pairs to the orbitals:

$$\Psi_x^{(n)} = \psi_x^{(\infty)} = C^{(\infty)}[\psi_x - S\phi_1 + S\phi_2] \quad (\text{all} \quad n) \tag{8.12.87.1}$$

$$\Psi_y^{(n)} = \psi_y^{(\infty)} = C^{(\infty)}\left[\psi_y - S\phi_3 + S\phi_4\right] \quad (\text{all} \quad n) \tag{8.12.87.2}$$

$$\Psi_z^{(n)} = \psi_z^{(\infty)} = C^{(\infty)}[\psi_z - S\phi_5 + S\phi_6] \quad (\text{for} \quad n \geq 3) \tag{8.12.87.3}$$

[15] In this case a, b and c are the coordinates of the point under consideration with respect to a lattice point defined by k, l, m (Van der Hoff and Benson 1953).

and this last is replaced by

$$\Psi_z^{(2)} = \psi_z^{(2)} = C^{(2)}\left[\psi_z - S\phi_5 + S'\phi_6'\right] \tag{8.12.88}$$

This orbital is analogous to $\psi_z^{(\infty)}$ but modified to maintain orthogonality with the orbital ϕ_6' which now represents a 1 s-wave function for Li^+ in position "6" at a distance $\rho_6 = \rho' = \alpha\rho$. S′ is an overlap integral defined by

$$S' = \int \phi_5' \psi_z d\tau \tag{8.12.89}$$

For n = 1:

$$\Psi_z^{(1)} = N_{\lambda k}\left[\psi_z^{(1)} + \lambda\chi^{(1)} + k\omega_5'\right] \tag{8.12.90}$$

$$\psi_z^{(1)} = C^{(1)}\left[\psi_z - S'\phi_5'\right] \tag{8.12.91}$$

$C^{(2)}$ in Eq. (8.12.88) and $N_{\lambda k}$ in Eq. (8.12.90) are normalizing factors. The wave functions ψ and ϕ represent 2p orbitals for F^- and 1s orbitals for Li^+ ions, respectively (Benson and Wyllie 1951).

The Hamiltonian for an electron of a F^- ion in the nth layer is:

$$\mathsf{H}^{(n)} = \mathsf{H} + V_-^{(n)}(x, y, z) \tag{8.12.92}$$

with H as a single electron Hamiltonian for the free ion F^-. The dipole moment of a surface F^- ion is:

$$\mu = -2\left(\Psi_z^{(1)}|z|\Psi_z^{(1)}\right) \tag{8.12.93}$$

The quantum energy in excess of the classical one is:

$$\begin{aligned}\varepsilon_{F^{-(n)}(\text{quantal})} = {} & V_-^{(n)}(0) + 6\sum_i{}^{(n)}\frac{1}{\rho_i} + 4\left[\left(\psi_z^{(\infty)}\middle|\mathsf{H} + \sum_i{}^{(n)} v_i\middle|\psi_z^{(\infty)}\right) - (\psi_z|\mathsf{H}|\psi_z)\right] \\ & + 2\left[\left(\psi_z^{(n)}\middle|\mathsf{H} + \sum_i{}^{(n)} v_i\middle|\psi_z^{(n)}\right) - (\psi_z|\mathsf{H}|\psi_z)\right] + \mu\left(p_-^{(1)} + q_-^{(1)}\delta(n-1)\right)\end{aligned} \tag{8.12.94}$$

where v_i is the potential energy of the single metal ion i (seen as a perturbation) and δ is the Dirac function defined by $\delta(x) = 0$ for $x \neq 0$; $\delta(0) = 1$

The surface energy is:

$$\sigma = \frac{1}{2\rho^2}\left[\begin{array}{l} \frac{1}{\rho}\left\{B + 2[S(1) - S(\alpha)] + 6\left(\frac{2}{\alpha} - 3\right)\right\} \\ +4\left(\psi_z^{(\infty)}\middle|\sum_i{}^{(1)}v_i + \sum_i{}^{(2)}v_i - 2\sum_i{}^{(\infty)}v_i\middle|\psi_z^{(\infty)}\right) \\ +2\left\{\left(\psi_z^{(1)}\middle|\mathsf{H} + \sum_i{}^{(1)}v_i\middle|\psi_z^{(1)}\right) + \left(\psi_z^{(2)}\middle|\mathsf{H} + \sum_i{}^{(2)}v_i\middle|\psi_z^{(2)}\right) - 2\left(\psi_z^{(\infty)}\middle|\mathsf{H} + \sum_i{}^{(\infty)}v_i\middle|\psi_z^{(\infty)}\right)\right\} \\ -2\left(p_-^{(1)}(0) + q_-^{(1)}(0)\left(\psi_z^{(1)}|z|\psi_z^{(1)}\right) - \frac{1}{2}q_-^{(1)}(0)\mu + 2\Delta\right) \end{array}\right] \tag{8.12.95}$$

For any value of α the parameters λ, k and Δ are those ones which minimize σ.

B/ρ in the above expression is the classical Coulomb energy necessary to separate the two parts of the crystal when the spacing remains unchanged. The term $(2/\rho)[S(1) - S(\alpha)]$ is a classical correction for the change in spacing at the surface and the remaining terms arise from the quantum–mechanical approach. The computed surface energy at the athermal limit is $\sigma = 557$ *erg/cm*^{2}; $\mu = 0.079$ *Debye* and $\rho' = 2.02$ *Å*.

The surface energy of LiF received recently a renewed interest in the investigation of electrode/electrolyte interface in Li batteries. I remand to Ozhabes et al. (2015) for a modern application of the surface energy of this substance.

8.12.7 Surface Energy Through Green's Functions

I resume hereafter the treatment of Kalkstein and Soven (1971) for a hemi-crystal generated by a cleavage plain. The "*surface*" here is either of the planes of atoms immediately adjacent to the cleavage plane. Localized orbitals (assumed as mutually orthogonal) are generated through Wannier functions (see Sect. A.21) and it is also assumed that only a single s-like orbital is associated with each atom.

The DOS is obtained by calculating the Green's function for the cleaved crystal. G is given in terms of G^0 the Green function of the holo-crystal (see Sect. A.20)

$$\mathsf{G} = \mathsf{G}^0\mathsf{V}\mathsf{G}\delta(x - s) \tag{8.12.96}$$

with V as a perturbation term:

$$\mathsf{V} = \mathsf{H} - \mathsf{H}^0 \tag{8.12.97}$$

The operators satisfy:

$$(E - i\delta - \mathsf{H})\mathsf{G} = 1 \tag{8.12.98}$$

where E is the energy and δ a positive infinitesimal.

Although the Hamiltonian of the hemi-crystal H does not have matrix elements between localized functions centered on different sides of the cleavage plane, denoting a localized function centered on lattice site **R**, as a ket vector $|\mathrm{R}\rangle$[16] one may pose (Kalkstein and Soven 1971):

$$\langle \mathrm{R} \mid \mathsf{V}|\mathrm{R}'\rangle\rangle = -\langle \mathrm{R} \mid \mathsf{H}^0|\mathrm{R}'\rangle\rangle \tag{8.12.99}$$

where R belongs to the hemi-crystal at right hand side of the cleavage and R′ to the hemi-crystal at left. Denoting U as the difference in potentials of the holo- and hemi-crystals one may write:

$$\langle \mathrm{R} \mid \mathsf{V}|\mathrm{R}'\rangle\rangle = \langle \mathrm{R} \mid \mathsf{U}|\mathrm{R}'\rangle\rangle \tag{8.12.100}$$

In this way, it may be easily seen that the perturbation U arises from a change in the electronic distribution near the surfaces, and "*cannot be written down in any general way*" (Kalkstein and Soven 1971).

The hemi-crystal preserves however the lattice translational symmetry of the holo-crystal, parallel to the cleavage plane. It may be thus introduced a mixed Bloch-Wannier representation of the coordinate vector **R** of an arbitrary atom, in the form:

$$\mathbf{R} = \mathrm{n}\mathbf{d} + \mathbf{R}_{||} + \boldsymbol{\tau}_{\mathrm{n}} \tag{8.12.101}$$

where **d** is the vector normal to the cleavage plane whose magnitude is the spacing between adjacent crystallographic planes parallel to the cleavage plane; $\mathbf{R}_{||}$ is the coordinate vector parallel to the cleavage plane; $\boldsymbol{\tau}_{\mathrm{n}}$ is a translation vector, parallel to the cleavage plane; and, n = 0,1,... (or n = -1,-2,...) are atoms on the right (or left) hemi-crystal.The translation vector $\boldsymbol{\tau}_{\mathrm{n}}$ transposes the atoms of the nth plane on the surface. It may thus be defined as a Bloch-like function for translations parallel to the cleavage plane in the form:

$$|\mathrm{k}_{||}, \mathrm{n}\rangle = \frac{1}{\sqrt{\mathrm{N}_{||}}} \sum_{\mathbf{R}_{||}} |\mathrm{n}\mathbf{d} + \mathbf{R}_{||} + \boldsymbol{\tau}_{\mathrm{n}}\rangle \exp\left[\mathrm{i}\mathbf{k}_{||} \cdot \left(\mathbf{R}_{||} + \boldsymbol{\tau}_{\mathrm{n}}\right)\right] \tag{8.12.102}$$

where $\mathbf{k}_{||}$ is a wavevector parallel to the cleavage plane and $\mathrm{N}_{||}$ is the number of atoms in the surface. The functions of type (8.12.102) are localized near the nth crystallographic plane. It is shown by Kalkstein and Soven (1971) that all distinct functions of this type are obtained when $\mathbf{k}_{||}$ assumes all allowed values in the two-dimensional Brillouin zone corresponding to the arrangement of atoms in the surface planes or "*Surface Brillouin Zone*" (SBZ) (see Appendix in Kalkstein and Soven 1971).

[16] I do not use here a bold character because the vectorial nature of R is implicit in the *bra* and *ket* notation (see Sect. A.8).

For two m and n planes parallel to the surface, the existence of translational symmetry in terms of basis sets of the type (8.12.102) is implicit in the relations:

$$\left\langle \mathrm{m}, \mathrm{k}_{||} \,\middle|\, \mathsf{G}^0 \middle| \mathrm{n}, \mathrm{k}'_{||} \right\rangle\rangle = \delta\left(\mathbf{k}_{||} - \mathbf{k}'_{||}\right) \mathsf{G}^0\left(\mathrm{m} - \mathrm{n}, \mathbf{k}_{||}\right) \tag{8.12.103.1}$$

$$\left\langle \mathrm{m}, \mathrm{k}_{||} \,\middle|\, \mathsf{G} \middle| \mathrm{n}, \mathrm{k}'_{||} \right\rangle\rangle = \delta\left(\mathbf{k}_{||} - \mathbf{k}'_{||}\right) \mathsf{G}\left(\mathrm{m}, \mathrm{n}, \mathbf{k}_{||}\right) \tag{8.12.103.2}$$

$$\left\langle \mathrm{m}, \mathrm{k}_{||} \,\middle|\, \mathsf{V} \middle| \mathrm{n}, \mathrm{k}'_{||} \right\rangle\rangle = \delta\left(\mathbf{k}_{||} - \mathbf{k}'_{||}\right) \mathsf{V}\left(\mathrm{m}, \mathrm{n}, \mathbf{k}_{||}\right) \tag{8.12.103.3}$$

Because G^0 is the Green's function of the holo-crystal, it depends upon the difference m–n, while G and V depend upon m and n separately. Suppressing the argument $\mathbf{k}_{||}$, $\mathsf{G}(\mathrm{m}, \mathrm{n}, \mathbf{k}_{||})$ may be written as $\mathsf{G}(\mathrm{m}, \mathrm{n})$ and $\mathsf{V}(\mathrm{m}, \mathrm{n}, \mathbf{k}_{||})$ as V (m, n). Based on Eq. (8.12.102) the perturbation term reduces to a sum of matrix elements in $\mathbf{R}'_{||}$:

$$\mathsf{V}(\mathrm{m}, \mathrm{n}) = \sum_{\mathbf{R}'_{||}} \left| \mathbf{md} \middle| \mathsf{V} \middle| \mathbf{nd} + \mathbf{R}'_{||} + \tau_{\mathrm{n}} \right\rangle \exp\left[\mathrm{i}\mathbf{k}_{||} \cdot \left(\mathbf{R}'_{||} + \tau_{\mathrm{n}} - \tau_{\mathrm{m}} \right) \right] \tag{8.12.104}$$

and the operator Eq. (8.12.96) reduces to a set of algebraic difference equations of the type:

$$\mathsf{G}(\mathrm{m}, \mathrm{n}) = \mathsf{G}^0(\mathrm{m}, \mathrm{n}) + \sum_{\mathrm{p,q}} \mathsf{G}^0(\mathrm{m}, \mathrm{n}) \mathsf{V}(\mathrm{p}, \mathrm{q}) \mathsf{G}(\mathrm{q}, \mathrm{n}) \tag{8.12.105}$$

(one for each $\mathbf{k}_{||}$).

Under the simplifying assumption that the Hamiltonian of the holo-crystal has only one-center and nearest neighbor two-center matrix elements, and that the perturbation U is completely confined to the surface planes, any Bloch function for the perfect crystal may be written in the form[17]:

$$|\mathbf{k}\rangle = \left|\mathbf{k}_{||} + \mathbf{k}_{\perp}\right\rangle = \frac{1}{\sqrt{\mathrm{N}_{\perp}}} \sum_{\mathrm{n}} \left| \mathbf{nd}, \mathbf{k}_{||} \right\rangle \exp[\mathrm{in}\mathbf{k}_{\perp}\mathbf{d}] \tag{8.12.106}$$

with $\mathrm{N}_{\perp}$ as number of planes parallel to the cleavage plane in the holo-crystal (i.e. parallel to the surface).

The Green's function of the unperturbed crystal (i.e. the holo-crystal) is:

$$\mathsf{G}^0(\mathrm{n}) = \frac{1}{\mathrm{N}_{\perp}} \sum_{\mathbf{k}_{\perp}} \frac{\exp[\mathrm{in}\mathbf{k}_{\perp}\mathbf{d}]}{\left[\mathrm{E} - \mathrm{i}\delta - \mathrm{E}^0(\mathbf{k})\right]} \tag{8.12.107}$$

[17] The full set of wave vectors for the perfect crystal is represented by the expression $\mathbf{k} = \mathbf{k}_{||} + \mathbf{k}_{\perp}$ with $\mathbf{k}_{\perp}$ perpendicular to the cleavage plane and satisfying the condition $\mathbf{k}_{\perp} \leq \pi/\mathbf{d}$.

with:

$$E^0(\mathbf{k}) = W(\mathbf{k}_{||}) + 2T(\mathbf{k}_{||})\cos\left[\mathbf{k}_{\perp}d + \theta(\mathbf{k}_{||})\right] \tag{8.12.108}$$

$$W(\mathbf{k}_{||}) = H^0(n, n, \mathbf{k}_{||}) \tag{8.12.109}$$

$$T(\mathbf{k}_{||}) = H^0(n, n \pm 1, \mathbf{k}_{||})/\exp\left[i\theta(\mathbf{k}_{||})\right] \tag{8.12.110}$$

Because, for fixed $\mathbf{k}_{||}$, the unperturbed energy varies from W − 2T to W + 2T, the replacement of Eq. (8.12.108) into (8.12.107) yields:

$$G^0(n) = \frac{1}{\sqrt{4T^2 - \omega^2}}\left[\frac{\omega + i\mu}{2T}\right] \times \exp(-in\theta) \tag{8.12.111}$$

with $\mu = \sqrt{4T^2 - \omega^2}$

The Green's function for the cleaved crystal [G (m, m)] is:

$$G(m, m) = \frac{1}{\mu}\left[1 + \left(\frac{\omega + \mu}{2T}\right)^{2|m|} \times \frac{\mu - i(\omega - 2U^0)}{\mu + i(\omega - 2U^0)}\right] \tag{8.12.112}$$

and the electronic Local Density of States (LDS) on the surface [ρ_m(E)] is:

$$\rho_m(E) = \frac{1}{\pi N_{||}} \cdot \mathrm{Im}\sum_{\mathbf{k}_{||}} G(m, m, \mathbf{k}_{||}) \tag{8.12.113}$$

Kalkstein and Soven (1971) applied their procedure to a cubic lattice with (100) and (111) surfaces. For a cleavage on the (100) plane, the SBZ is a square of edge length 2x/a, where a is the lattice constant. Applying Eq. (8.12.109) one has:

$$W(\mathbf{k}_{||}) = 2E_1\left[\cos(ak_y) + \cos(ak_Z)\right] \tag{8.12.114.1}$$

with k_y and k_z as axial components of $\mathbf{k}_{||}$ and

$$T(\mathbf{k}_{||}) = E_1 \tag{8.12.114.2}$$

With a cleavage on the (111) plane, the atoms on the surface form a hexagonal structure. Applying Eq. (8.12.109) one has:

$$W(\mathbf{k}_{||}) = 0 \tag{8.12.115.1}$$

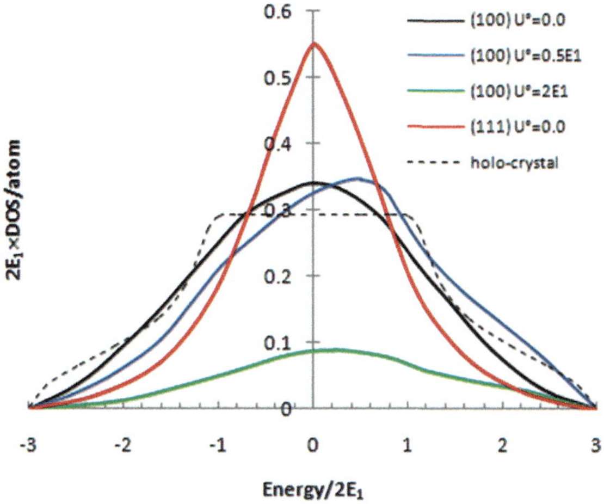

Fig. 8.34 Bulk-state LDS for the (100) and (111) surfaces of a cubic lattice and different U^0 after Kalkstein and Sover (1971). The dashed line is the infinite crystal density of states per atom

$$T^2\left(\mathbf{k}_{||}\right) = E_1^2\left[1 + 4\cos^2\left(\frac{ak_y}{\sqrt{2}}\right) + 4\cos\left(\frac{ak_y}{\sqrt{2}}\right)\cos\left(\frac{3ak_y}{\sqrt{6}}\right)\right] \quad (8.12.115.2)$$

Figure 8.34 resumes the results of Kalkstein and Soven (1971) in terms of bulk-state electronic LDS for different values of perturbation.

The various electronic LDS (solid lines in Fig. 8.34) bear little resemblance to the infinite crystal density of states (dashed line). The LDS for the (100) surface in the absence of surface perturbation (i.e. identical potential in the cleaved and uncleaved crystals) is "*a somewhat more sharply peaked function*" (Kalkstein and Soven 1971) with respect to the holo-crystal density. The qualitative effects of a nonzero U, are different depending upon whether or not it is strong enough to produce bound states. Bound states can occur only for $U^0 > E_1$. For values of U^0 smaller than E_1 (i.e. $U^0 = 0.5E_1$ in Fig. 8.34), the bulk-state LDS is essentially a distorted version of that occurring for $U^0 = 0$. For $U^0 = 2E_1$ the area under the bulk LDS is less than unity because of the existence of bound states. For the (111) surface in the absence of surface perturbation, the peak is more sharp and the root mean square width at half height is only $1/\sqrt{2}$ that of the holo-crystal.

A similar approach to electronic LDS may be found in Allan and Lenglart (1968) who investigated the (001) surface of a cubic crystal and obtained (besides an LDS essentially similar to that of Kalkstein and Soven 1971; for $U^0 = 0$) the electronic contribution to the heat capacity and the surface electronic entropy of the crystal. In doing this they assumed that the crystal has a half-filled band, so that the Fermi level does not change with the temperature in both cases of an infinite crystal and a crystal with a (001) surface. The variation of the isochoric electronic heat capacity between these two cases is:

$$\Delta C_V = \int_{-\infty}^{+\infty} E\Delta n(E) \frac{\partial f(E,T)}{\partial T} dE \qquad (8.12.116)$$

where $\Delta n(E)$ is the variation of the density of states with energy and f(E, T) is the Fermi function for $E_F = 0$. The electronic contribution to the surface entropy (S_{el}) is given by integration between the athermal limit and T of interest:

$$S_{el} = \frac{N^3}{2N^2} \int_0^T \frac{\Delta C_V}{\partial T} dT \equiv \frac{N^2}{2} \int_0^T \frac{\Delta C_V}{\partial T} dT \qquad (8.12.117)$$

Application to Ag and Mo (which are not however simple cubic lattices) shows that the variation of the surface entropy with T is weak and most part of it is due to phonons (S_{ph}). The electronic contribution for silver (which has a broad s band with only one s electron) is weak even at 1000 *K* ($S_{el}/(S_{el} + S_{ph}) \approx 1.85\%$). For molybdenum (which has a narrower d band with 5d electrons) the electronic contribution is small at 350 *K* but becomes not negligible at 1000 *K* [$S_{el}/(S_{el} + S_{ph}) \approx 12.5\%$; see Table 1 in Allan and Lenglart 1968].

8.12.8 Supercell and Embedded Cluster Quantum Mechanical Procedures for Point Defects

In the Supercell method a unit cell, containing the defect and its first neighbors, is periodically repeated to form an infinite lattice (Car et al. 1984, 1985; Nichols et al. 1989). The actual number of cells is however limited to the minimum space required to take into account relaxation phenomena. From the generated supercell, slabs are eventually cut out as already seen at Sect. 8.12.6 for the investigation of surface effects. The number of atoms in the 2D slab does not increase with cluster size as dramatically as in 3D structures and many phenomena that occur on surfaces do not involve long-range polarization effects (Nada et al. 1993).

The first applications of supercell procedures were devoted to the investigation of impurity diffusion pathways and mechanisms. For example, Car et al. (1984) investigated the (interstitial) migration of aluminum in silicon through the utilization of Green's functions. In this study, vacancy processes were not taken into consideration. However Car et al. (1985) considered in detail defect-mediated diffusion mechanisms. Their work established that native point defects mediate impurity diffusion with activation energies comparable to experimental values.

It is worth to recall here that diffusion processes in crystals are markedly conditioned by point defects. The presence of vacancies (or interstitials) allows ions to jump from one site to a vacant neighboring position, leaving behind a vacancy: in this way, an *elemental diffusion process* is created (together with complementary *vacancy migration*) as a summation of single displacement steps. During its movement from the occupied to the vacant site, the ion performs a saddle-shaped

Table 8.20 Migration enthalpy in alkali halides and simple oxides. Values in *eV*. Data from Barr and Liliard (1971) (1) and Greenwood (1970) (2)

Compound	H_{V_M}	H_{V_X}	H_{M_i}	H_{X_i}
LiF	0.70(1)	0.65(1)	–	–
CaF_2	–	–	–	1.65(1)
SrF_2	–	–	–	1.00(1)
BaF_2	–	–	–	0.78(1)
LiCl	0.39(1)	–	–	–
NaCl	0.71(1)	1.00(1)	–	–
AgBr	–	–	0.17(1)	–
KCl	0.71(1)	1.00(1)	–	–
KBr	0.65(1)	0.91(1)	–	–
NaBr	0.78(1)	1.17(1)	–	–
KI	1.21(1)	–	–	–
AgCl	–	–	0.13(1)	–
$PbCl_2$	–	0.35(1)	1.56(2)	–
$PbBr_2$	–	0.30(1)	–	–
MgO	3.38(1)	2.69(2)	–	–
CaO	3.51(1)	–	–	–
PbO	2.91(1)	–	–	–
BeO	2.43(1)	–	–	–
UO_2	1.21(2)	–	–	–
ZrO_2	1.08(2)	–	–	–

H_{V_M} = migration enthalpy of cationic vacancy
H_{V_X} = migration enthalpy of anionic vacancy
H_{M_i} = migration enthalpy of cationic interstitial
H_{X_i} = migration enthalpy of anionic interstitial

trajectory to overcome the repulsive barriers of neighboring ions. Work against the repulsive potential requires a certain amount of energy (mainly accessible in thermal form). This energy, which is the only energy amount required during displacement) is defined as "*migration energy*" (actually, not only the repulsive potential but all the static energy forms present in the crystal field play a role in determining the amount of migration energy, although the repulsive term is the dominant one).

Table 8.20 lists the *migration enthalpies* of some halides and simple oxides for *vacancy* and *interstitial migration*.

Let me now consider for instance a crystal MO; if the diffusion takes place by migration of cationic vacancies, the number of atoms which undergo the process depends on the vacancy concentration $[V''_M]$ and the thermal state of single atoms M (the jump takes place only whenever atom M in the neighborhood of the vacancy has sufficient energy to perform it). The diffusion coefficient associated with the vacancy migration process is given by (Jost 1960; Lasaga 1981):

$$D = \frac{d^2 \nu_0}{3}\left[V_M''\right] \exp\left(-\frac{H_{V_M''}}{kT}\right) \quad (8.12.118)$$

where d is the inter-ionic distance between vacant and occupied sites, and ν_0 is a specific vibrational frequency of the crystal (Einstein frequency, or a fraction of the Debye frequency; see Lasaga 1981).

A similar relation holds for anionic vacancies and for diffusion through interstitial migration (substitute $\left[V_M''\right], H_{V_M''}$ by $\left[V_O^{\cdot\cdot}\right], H_{V_O^{\cdot\cdot}}$ and so on).

As at high T, the vacancy concentration depends exponentially on T, i.e.:

$$\left[V_M''\right] = K_S^{1/2} = \exp\left(-\frac{\Delta H_S}{2k_B T}\right) \exp\left(\frac{\Delta S_S}{2k_B}\right) \quad (8.12.119)$$

Equation (8.12.119) can be re-expressed as:

$$D = D_0 \exp\left(-\frac{Q_D}{k_B T}\right) \quad (8.12.120)$$

where Q_D represents the "*bulk activation energy of the diffusion process*", composed of the enthalpy of vacancy formation plus the enthalpy of vacancy migration:

$$Q_D = \frac{\Delta H_S}{2} + H_{V_M''} \quad (8.12.121)$$

and D_0 groups the pre-exponential terms:

$$D_0 = \frac{d^2 \nu_0}{3} \exp\left(\frac{\Delta S_S}{2k_B}\right) \quad (8.12.122)$$

Equation (8.12.122) finds its application in the region of intrinsic disorder (a similar equation can be developed for Frenkel defects), where Schottky and Frenkel defects are dominant with respect to point impurities and non-stoichiometry.

With decreasing temperature, the intrinsic defect population decreases exponentially, and, at low T, the extrinsic disorder becomes dominant. Moreover, extrinsic disorder for oxygen-based minerals (such as silicates and oxides) is sensibly affected by the partial pressure of oxygen in the system and, in the region of intrinsic pressure, by the concentration of point impurities. In this new region, term Q_D does not embody the enthalpy of defect formation, but simply the enthalpy of migration of the defect, i.e.:

$$Q_D = H_{V_M''} \quad (8.12.123)$$

One of the defective materials most extensively investigated through quantum–mechanical procedures is, for obvious commercial reasons, the silicon semiconductor. Bernholc et al. (1978) adopted the Koster-Slater procedure[18] (Koster and Slater 1954) for the utilization of Wannier's functions, and combined them with a Green's functions treatment.

Denoting $G^0(E)$ the Green's function operator of the perfect crystal, the bound states in the gap between valence and conduction band, associated to a given perturbation U, are resolved by the determinant (Bernholc et al. 1978):

$$D(E) = \det\left|1 - G^0(E)U\right| \tag{8.12.124}$$

The change in the charge density which is needed for self-consistency is given by (Bernholc et al. 1978):

$$\Delta\rho = \frac{2}{\pi}\int_{occ} dE\left\{1 - \left[1 - G^0(E)U\right]^{-1}\right\}G^0(E) \tag{8.12.125}$$

In Fig. 8.35 we see the computed charge densities for a localized vacancy in the silicon lattice. The total change in the charge density integrates to its full value of −4.0, as an atom with four valence electrons has been removed (Bernholc et al. (1978). The contour plot in part (a) of the figure shows that the charge is localized almost entirely within the cavity defined by the nearest neighbors. Bernholc et al. (1978) found, within the fundamental gap, only one bound state, with T_2 symmetry (threefold degenerate), lying at 0.8 eV above valence bands. Since the crystal is neutral, this state contains only two electrons. Part (b) of the figure shows the contour plot of the contributions of the first three shells to the T_2 bound state. The change in the density of states in the valence bands integrates to −5.2 electrons instead of −6 (and to −0.1 electrons, instead of 0, for A_1 symmetry). Part (c) of the figure shows the total charge density around the vacancy. All the bonds in the crystal remain virtually unchanged, and the dangling bonds point toward the missing atom. The contour plots reveal moreover that the dangling bonds have a rather constant charge density near the atoms and extend appreciably into the vacant site (Bernholc et al. 1978).

With an essentially similar procedure, Car et al. (1984) computed the formation enthalpy of the silicon vacancy as a function of the Fermi level energy E_F in the gap between conduction band and valence band. The symbolism V^0 adopted by Car et al. (1984) could actually refer to $V_{Si}^{''''}$ taking into account that, as we have seen above, the total change in the charge density integrates to its full value −4.0. It follows that a "++ *charged vacancy*" is most likely $V_{Si}^{''}$; a+ charged vacancy is $V_{Si}^{'''}$, and a– charged vacancy is $V_{Si}^{'''''}$. The same is true for silicon interstitials and presumably a 0-charged interstitial is $Si_i^{\cdot\cdot\cdot\cdot}$ and so on. The fact that the formation energy of vacancies rises with E_F is probably due to the fact that the main (stabilizing) form

[18] If the matrix of the total perturbed Hamiltonian is set up using the Wannier functions as a basis, the coefficients of the Wannier functions in the expansion of the perturbed wave function satisfy a set of difference equations which can be converted to an approximate differential equation.

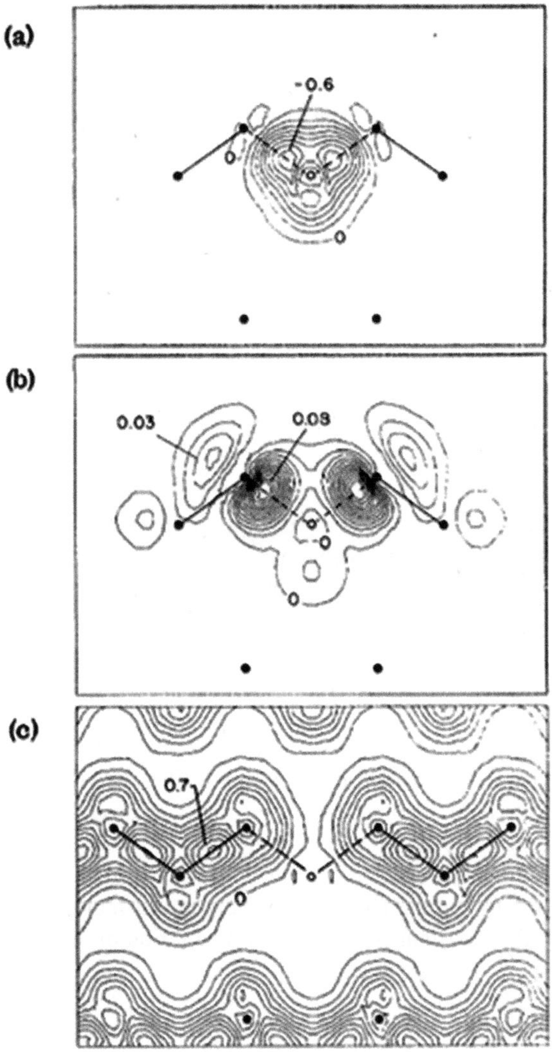

Fig. 8.35 Contours of constant electron density (a.u.) near a vacancy in the silicon semiconductor. In part **a** it is shown the total change in the charge density (negative contours). In part **b** it is shown the T_2 bound state (see text) and in part **c** the total charge density. From Bernholc et al. (1978). With permission from the American Physical Society

of energy of the defect is polarization, which is severely affected by the net charge of the defect itself with respect to its next neighbors. Concerning the silicon interstitials Car et al. (1984) observed a quite high energy (i.e. from + 6 to + 8 *eV*) for tetrahedral positions while energies were shown to decrease slightly but progressively for hexagonal and bond-centered positions (see their Fig. 1).

The initial computations of Car et al. (1984) were later revised by Car et al. (1985) which repeated them with more flexible basis sets, and reduced the overall error bar. The resulting formation energies are ~1 *eV* lower than the previous ones. Car et al. (1985) investigated quantum mechanically the diffusion processes of various impurities in silicon. Their data, which I resume here in Table 8.21

Table 8.21 Theoretical and experimental (in bold) activation energies of diffusion (*eV*) for Si, P and Al in silicon. Q_i = intrinsic activation energy; Q_n and Q_p = n-type and p-type transport, respectively

Species (original notation)	Species (interpreted)	Q_i	Q_n	Q_p	References
V^{++}	V_{Si}''	5.1	–	–	(1)
V^{+}	V_{Si}'''	4.7	–	–	(1)
V^{0}	V_{Si}''''	4.2	–	–	(1)
V^{-}	V_{Si}'''''	4.5	–	–	(1)
I^{++}	$Si_i^{\cdot\cdot\cdot\cdot\cdot\cdot}$	5.6	–	–	(1)
I^{+}	$Si_i^{\cdot\cdot\cdot\cdot\cdot}$	5.5	–	–	(1)
I^{0}	$Si_i^{\cdot\cdot\cdot\cdot}$	5.7	–	–	(1)
$(P_i)^0$	$P_i^{\cdot\cdot\cdot\cdot\cdot}$	5.1	5.1	–	(1)
$(P_i)^-$	$P_i^{\cdot\cdot\cdot\cdot}$	5.2	4.6	–	(1)
$(P_i)^-$	$P_i^{\cdot\cdot\cdot}$	5.6	4.5	–	(1)
$(P_SV)^+$	$\left(P_{Si}^{\cdot}V_{Si}''''\right)'''$	3.2	3.8	–	(1)
$(P_SV)^0$	$\left(P_{Si}^{\cdot}V_{Si}''''\right)''''$	3.0	3.0	–	(1)
$(P_SV)^-$	$\left(P_{Si}^{\cdot}V_{Si}''''\right)'''''$	3.2	3.6	–	(1)
P	–	**3.7**	**3.1**	–	(2)
$(Al_SV)^+$	$\left(Al_{Si}'V_{Si}''''\right)''''$	3.9	–	3.2	(1)
$(Al_SV)^0$	$\left(Al_{Si}'V_{Si}''''\right)'''''$	3.5	–	3.5	(1)
$(Al_SV)^-$	$\left(Al_{Si}'V_{Si}''''\right)'''''$	3.5	–	4.1	(1)
$(Al_SI)^0$	$\left(Al_{Si}'Al_i^{\cdot\cdot\cdot}\right)^{\cdot\cdot}$	4.0	–	4.0	(1)
$(Al_i)^{++}$	$Al_i^{\cdot\cdot\cdot\cdot\cdot}$	4.3	–	3.0	(1)
$(Al_i)^{+}$	$Al_i^{\cdot\cdot\cdot\cdot}$	3.9	–	3.1	(1)
$(Al_i)^{0}$	$Al_i^{\cdot\cdot\cdot}$	3.9	–	3.9	(1)
Al	–	**3.5**	–	–	(3)

(1) Car et al. (1985); (2) Makris and Masters (1973); (3) Fuller and Ditzemberger (1956)

concern also associated defects. The adopted notation is valid as stated above, so I list in the table both the original and the "*interpreted*" notation. For the meaning of "*intrinsic*" (Q_i), n-type (Q_n), and p-type (Q_p) activation energies I remand the reader to the original paper and to the few notes on defects in this book.

Along the same pathway, Nichols et al. (1989) investigated the diffusion mechanisms of B, P, As, and Sb in silicon and found that vacancies and interstitials mediate the diffusion of B, P, and As with comparable activation energies, while, for Sb, the interstitial component has a high activation energy. Besides giving the activation energy of the diffusion process, Nichols et al. (1989) provided also the formation enthalpy and migration enthalpy components of Q_D (cf. Eq. 8.12.121) and the binding energy of the associated defects (cf. Eq. 8.12.44) which are quite in line with the few experimental observations available (cf. Table 1 in Nichols et al.

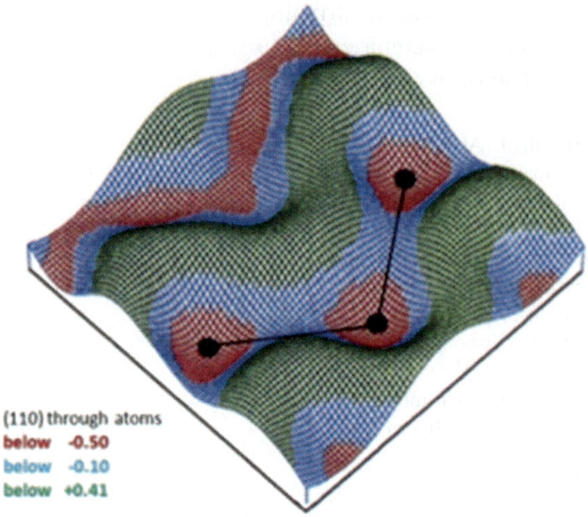

Fig. 8.36 Perspective total-energy (110) surface plot depicting the migration of a neutral B interstitial through the Si crystal (black dots and solid back lines. The areas colored red are lowest in energy. Modified from Nichols et al. (1989)

1989). An important novelty introduced by Nicholls et al. (1989) is the construction of a "*global total energy surface*" depicting the interactions between N host atoms and a single impurity in an X-atom supercell. This construction is essential to ascertain the lowest-energy diffusion pathways (i.e. "*interstitial mediated*"; cf. Nichols et al. 1989) and the migrating species. Such a surface results from the collection of the total energies of N + 1 atoms competing for N substitutional sites, with the relaxation of the surrounding host crystal included for each configuration. The energy surface can be displayed either as a contour plot (the contours depicting constant energies) or as a perspective plot. We see for example in Fig. 8.36 the perspective plot for the diffusion of B in silicon (actually a slab of the supercell along the selected crystallographic directions). The in-plane coordinates are the chosen (diamond-structure) coordinates and the third axis is the energy. The zero energy was chosen at the saddle point of the "*B kick-out process*" (cf. Nichols et al. 1989). Regions of different energy are color-coded. The lowest-energy migration pathway for neutral B is along the low-electron-density channels.

Concerning the "*embedded cluster*" techniques, Pisani (1996b) stressed that the treatment of the holo-crystal is always prodromal to the application of Green's functions to point defects. This is the reason that basically led the Quantum Chemistry group of Torino to develop an ab initio "*embedding code*", for the construction of supercells (SC) where the neutral point defect is embedded (*EMBED*; Pisani et al. 1994a, 1996). The size of the SC must be large enough to contain the "*perturbed zone*" (i.e. the one-electron eigenfunctions associated with the defect are localized within the SC). If the defect is charged (such as in the case of vacancies and/or interstitials) the problem is not simply the local perturbation created by the defect itself, because the infinite repetition of charged units leads to "*unphysical divergent contributions to the Hamiltonian*" (Pisani 1996b). In this case, the charged defect is compensated by a uniform background of opposite charge, and the contribution of the background to the bulk coulomb energy is treated semi-empirically through an a-posteriori correction. Leslie and Gillian (1985) for example proposed for the cubic lattice a correction in the form:

$$E'' = -{}^{\alpha Q^2}/_{\varepsilon_0 L} \tag{8.12.126}$$

where α is the appropriate Madelung constant, Q is the charge of the point defects, ε_0 is the static dielectric constant of the holo-crystal and L is the lattice parameter of the SC. The method of Leslie and Gillian (1985) has been applied to Frenkel defects in Li_2O by De Vita et al. (1992) at LCAO-DFT theory level and by Pisani et al. (1994b) to MgO at LCAO-HF theory level with reasonable results.

EMBED works with the perturbed cluster (PC) equations (Pisani et al. 1990). In the PC approach, the perfect host crystal (S^f)[19] and the imperfect system containing the defect (S) are considered in parallel. S^f and S differ only for the location of a few atoms in a strict neighborhood of an origin o (and hence they differ also slightly in composition). The symmetry group of S, with symmetry operations $G\{R_1,\ldots,R_h\}$ is selected as a finite sub-group of the point group of S^f (G^f). The atoms of S^f and S are organized into *stars* of atoms which are equivalent with respect to the h operations of G. The standard ordering of the stars is determined according to increasing distance from o. Between the stars of S^f and S, there is a close one-to-one correspondence, with the exception of the first m stars of S, which constitute the *proper defect zone* A, and the corresponding first m′ ones of S^f, which occupy the regular lattice positions in the absence of defects.

The system is divided into two parts: an "*inner zone*" C with the first n stars of atoms of S($n \geq m$) and an "*outer zone*" D grouping all the remaining atoms in the system. Interposed between the two sets there may exist a *boundary zone,* B, comprising n-m stars. For each calculation on the C cluster (m, m′, n), a "*reference calculation*" is performed on a C' cluster (0, 0, n + m′−m), with the same origin and the same tolerances. This reference calculation is conventionally designated as "*auto-embedding*", because "*it concerns an embedded cluster cut out from the perfect crystal which defines the same outer zone D as the cluster C*" (Pisani et al. 1994a). The computational procedure is quite complex, and I try here to resume it briefly in its main aspects.

The basis set of localized n-GTO defining the AO's of the atoms in the imperfect system coincides with that of the HC, except for the atoms in the proper defect region. This set is partitioned into a finite subset (C) that describes the molecular cluster, and the complementary, infinite set (D). This is reflected in the partitioning of the general representative matrix X into blocks, with a further subdivision into a "local" (X_1), and an "outer" (X_2) matrix.

[19] The f superscript designates mathematical or physical objects of the defect-free crystal.

$$X = \begin{pmatrix} X_{CC} & X_{CD} \\ X_{DC} & X_{DD} \end{pmatrix} = \begin{pmatrix} X_{CC} & X_{CD} \\ X_{DC} & 0_{DD} \end{pmatrix} + \begin{pmatrix} 0_{CC} & 0_{CD} \\ 0_{DC} & X_{DD} \end{pmatrix} = X_1 + X_2 \quad (8.12.127)$$

Indicating with S the overlap matrix and with F the Fock matrix it is also posed:

$$Q(e) = eS - F \quad (8.12.128)$$

$$\left(S_{CC}^{-1/2} F_{CC} S_{CC}^{-1/2}\right) |j\rangle = e_j |j\rangle \quad (8.12.129)$$

with $|j\rangle$ as eigenvector and e_j as eigenvalues of the molecular cluster problem.

To each set of localized GTO's defining the AO, it is associated with an *adjoined Gaussian* (i.e. an s-type normalized GTO whose exponent equals that of the most diffuse Gaussian of the contracted set). Two AO's are said to be m-neighbors if the overlap between their adjoined Gaussians is larger than 10^{-m}. It is then constructed as a basis set of orthogonalized symmetry orbitals (OSO) in the cluster subspace. By representing the F_{CC} matrix in this basis set, the cluster eigenvectors are generated and classified according to the different irreducible representations. Only the irreducible part of the X_{CD} matrices in the AO basis set is calculated and stored, corresponding to the set of AO couples ($CD \equiv \gamma\delta$) which are unique by symmetry.

Denoting $\rho_{\mu\nu}(\lambda)$ the $\mu\nu$ element of the PDOS of the imperfect system:

$$\rho_{\mu\nu}(\lambda) = \sum_i a^*_{\nu i} a_{\mu i} \delta(e_i - \lambda) \quad (8.12.130)$$

with e_i as eigenvalue and $a_{\nu i}$ as eigenvector coefficients, and defining $g(\lambda, e)$ the 2D thermalized step-function:

$$g(\lambda, e) = f(\lambda) - f(e) \quad (8.12.131)$$

the elements of the energy-dependent coupling matrices M_{DD} (e;1) are:

$$M_{\mu\nu}(e; 1) = -fg(\lambda, e)\left[\frac{\rho_{\mu\nu}(\lambda)}{\lambda - e}\right] d\lambda \quad (8.12.132)$$

At the athermal limit where w = 0, Eq. (8.12.131) reduces to:

$$g(\lambda, e) = \theta(\lambda - E_F) - \theta(e - E_F) \quad (8.12.133)$$

where $\theta(x)$ is the Heaviside step-function and E_F is the Fermi energy of the system.

Assuming that the PDOS of the imperfect system in the D region is not much dissimilar with respect to the defect free PDOS ("*fundamental approximation*" of the PC method; cf. Pisani et al. 1994a) one may replace in Eq. (8.12.132) $\rho_{\mu\nu}(\lambda)$, with $\rho^f_{\mu\nu}(\lambda)$.

The density matrix is subdivided into two blocks P_{CC} and P_{CD} and each block is the sum of "*cluster*" (P^{clu}), "*coupling*" (P^{cou}), and "*constant*" (P^{cst}) contributions. A general element jk of the P_{CD} block is thus:

$$P_{jk} = P^{clu}_{jk} + P^{cou}_{jk} + P^{cst}_{jk} \quad (8.12.134)$$

where

$$P^{clu}_{jk} = 2\delta_{jk} f(e_j) \quad (8.12.135.1)$$

$$P^{cst}_{jk} = -\sum_{\mu} P^{cst}_{j\mu} S_{\mu k} \quad (8.12.135.2)$$

$$P^{cou}_{jk} = \mathcal{N}_{jk} + \mathcal{N}_{kj} \quad (8.12.135.3)$$

In Eq. (8.12.135.1) e_j is the cluster eigenvalue (see Eq. 8.12.130) and $\mathcal{N}_{jk}$, $\mathcal{N}_{kj}$ in Eq. (8.12.135.3) are members of the $\mathcal{N}_{CD}$ matrix and its transpose $\mathcal{N}_{DC}$. (see Pisani et al. 1994a; for details concerning these matrix).For the P_{CC} the cluster contributions are:

$$P^{clu}_{j\mu} = 0 \quad (8.12.136.1)$$

$$P^{cst}_{j\mu} = -\sum_{\nu} \left\{ S_{j\nu} P^{f}_{\nu\mu} \right\} \quad (8.12.136.2)$$

$$P^{cou}_{jk} = -\sum_{\nu} Q_{j\nu}(e_j) M_{\nu\mu}(e_j; 1) \quad (8.12.136.3)$$

The general Fock matrix elements $F_{\gamma\mu}$ with γ belonging to the cluster C is written in the standard form:

$$F_{\gamma\mu} = T_{\gamma\mu} + \sum_{\lambda} Z^{\lambda}_{\gamma\mu} + \sum_{\lambda\nu} P_{\lambda\nu}(\gamma\mu|\lambda\nu) - \frac{1}{2}\sum_{\lambda\nu} P_{\lambda\mu}(\gamma\lambda|\mu\nu) \equiv M_{\gamma\mu} + B_{\gamma\mu} \quad (8.12.137)$$

with the same label λ denoting nuclei and AO's. $T_{\gamma\mu}$ is the kinetic term (neglected unless γ and μ are m_3 neighbors); the second and third are the Coulomb terms (see Pisani et al. 1994a; for details about the way of treating them); the last is the exchange term (neglected unless γ and μ are m_5 neighbors, or γ and λ are m_4 neighbors, or μ and ν are m_4 neighbors, or λ and ν are m_5 neighbors). The one-electron (kinetic plus nuclear, M), and the two-electron (B) part of the Hamiltonian are distinguished in the last passage.

The HF energy of the system S is formally expressed as follows:

$$E(S) = Tr\left[\left(M_1 + \frac{1}{2}B_1\right)P_1\right] + Tr\left[\left(M_2 + \frac{1}{2}B_2\right)P_2\right] + E_{nrep} \tag{8.12.138}$$

where subscripts denote local (1) and outer (2) matrices (see Eq. 8.12.127), Tr denotes truncation, and E_{nrep} is the nuclear repulsion term. The above equation is replaced by a suitable "*local energy expression*" (Pisani et al. 1994a):

$$E'(S) = Tr\left[\left(M_1 + \frac{1}{2}B_1\right)P_1\right] + Tr\left[\left(Z_2 + \frac{1}{2}B'_2\right)P^f_2\right] + E_{nrep} \tag{8.12.139}$$

where Z_2 and B'_2 are the "*local*" contributions to the "*outer*" Hamiltonian. With reference to the notation of Eqs. (8.12.138) and (8.12.139), this means that "*the indices* ($\gamma\mu$) *of Z and B are both in D, while the sum over* λ *runs only over nuclei in C, and in the sum over* ($\lambda\nu$) *at least one of the two indices must belong to C*" (Pisani et al. 1994a). The truncation criteria in Eqs. (8.12.138) and (8.12.139) are the same adopted for the Fock matrix (8.12.137) as well as the approximate treatment of Coulomb terms.

An interesting investigation has been carried out by Nada et al. (1993) in order to assess the relative performances of supercell and perturbed-cluster (PC) techniques in determining the formation energy of vacancies on the surface of crystals. The investigation involved the (100) surface of LiF, simulated by a single layer of LiF. The lithium vacancy, that I denote V'_{Li}, was created by displacing vertically a Li ion from its site to a distance of 5 Å above the surface. The charge neutrality (surface + displaced Li) was in this way preserved. The Authors tested supercells of different sizes (Fig. 8.37).

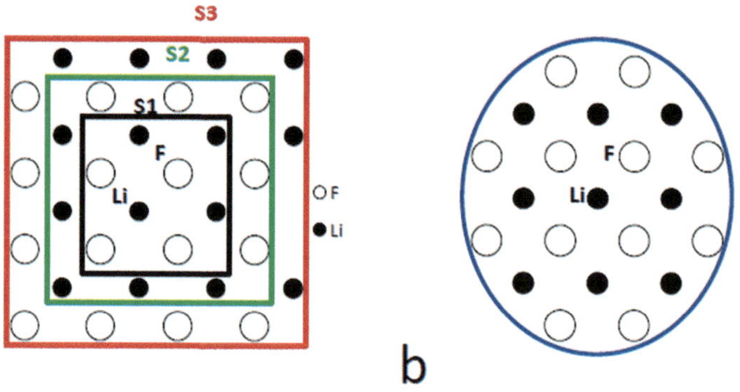

Fig. 8.37 Comparison between supercells (S1, S2, S3) (**a**) and embedded cluster calculations (first four cells of neighbors). **b** In the vacancy calculation, the labeled Li atom is displaced perpendicularly from the surface. Modified from Nada et al. (1993)

Table 8.22 Uncorrected and corrected vacancy formation energy in a.u., for the three supercells and the embedded cluster. A_x is the variation (in Å) of the distance between the defect and its first neighbors. A_E is the contribution to the energy (in a.u.) due to the relaxation. From Nada et al. (1993)

	E^f_{uncorr}	E^f_{corr}	A_x	A_E
S1	0.576	0.374	0.071	0.008
S2	0.367	0.371	0.156	0.019
S3	0.309	0.369	0.156	0.022
Cluster	0.387	0.349	0.170	0.014

The smallest (S1) includes the displaced Li, 4 first neighbor F, and 3 second neighbor Li (8 atoms). An expanded sampling of 18 atoms (S2) involved five additional F and five additional Li (18 atoms). The largest cell (S3) involved 7 additional F and 7 additional Li (32 atoms). For the embedded cluster calculations Nada et al. (1993) adopted a cluster with a size comparable to S3 (i.e. 21 atoms; the displaced Li ions and the first four shells of neighbors). The primary aim of the Authors was to investigate "*how the short-range polarization effects are handled by the two approaches*" (Nada et al. 1993) (see Sect. 8.12.3 for a short account of the nature of polarization energy). Also in the PC approach, the same displacement of the Li atom from the surface was applied, in order to make comparisons easier.

The results of the comparative calculations are resumed in Table 8.22.

In the supercell approach the uncorrected energy (E^f_{uncorr}) is simply the difference between the total energy per unit supercell of two calculations performed with the displaced and the non-displaced ion respectively. As we may see, the obtained energies depend significantly from the size of the supercell. The corrected formation energy of the vacancy (E^f_{corr}) was obtained subtracting from E^f_{uncorr} the attraction between vacancy and displaced ion (−0.106 a.u.) within each cell. The corrected values are almost identical among themselves. Concerning the embedded cluster calculations, the Authors make reference to "*local energy*" values E′ (see Eq. 8.12.139). The total charge (electrons + nuclei) in the *local space* is not the same in the perfect and defective crystal cases. In the defective case the cluster is more positive by a quantity $q = 0.1636|e|$. Two corrections must thus be added: the first one [$e_1 = -(q - 1)/d$] correcting for the attraction between the Li^+ ion at a distance d from the surface, and the effective charge on the vacancy site ($q' = q - 1$); the second one ($e_2 = \mu q$), with μ as chemical potential of the monolayer, accounting for the contribution from the missing charge:

$$E^f_{corr} = E^f_{uncorr} + \frac{1-q}{d} + \mu q \qquad (8.12.140)$$

The corrected formation energy is within 10% of the supercell results. The slight decrease of E^{f}_{corr} in the supercell calculations with increasing cluster size can be explained by considering that the vacancy charge ($-e$) is better described by the largest supercell and, in the PC calculation, this charge is allowed to flow into the outer region (Nada et al. 1993). Thus the two procedures converge with the increase in the size of the supercell.

Part II
Methods

Chapter 9
Molecular Orbitals in the HF Theory

The passage from atomic orbitals (AOs) to Molecular Orbitals (MOs) is not so simple as it could appear at first glance. The best sentence apt to resume the task is, in my opinion, that one expressed long ago by Roothan: "*The basic concept of the Molecular Orbital method is to find approximate electronic wave functions for a molecule by assigning to each electron a one-electron wave function which in general extends over the whole molecule*" (Roothan 1951).

The *Variational Principle* establishes that, for any antisymmetric normalized function of the electronic coordinates, the expectation value of the energy will be always greater than the energy of the exact wave function. In the context of the Hartree–Fock (HF) method, the best possible solution is at the *Hartree–Fock limit*; i.e. the limit of the HF energy as the basis set approaches completeness. The HF energy is the minimal energy for a single Slater determinant. For a molecular orbital or crystalline calculation, the initial approximate one-electron wave functions are typically a linear combination of atomic orbitals (LCAO). Based on the above guidelines, one may handle the molecular orbital expansion coefficients in such a way to minimize the energy.

9.1 Premises

Let me assign to each electron a one-electron wave function depending on the space coordinates of that electron only and assume that the function extends over the whole molecule (i.e. *Molecular Orbital* MO):

$$\varphi_i^n = \varphi_i(x^n, y^n, z^n) \quad (9.1.1)$$

where x^n, y^n, z^n are the coordinates of the nth electron and i the MO. Based on Pauli exclusion principle, no more than two electrons may occupy each MO, hence φ_i may occur no more than twice in the N-electrons wave function obtained as a

G. A. Ottonello, *Quantum Geochemistry*, Springer Geochemistry,
https://doi.org/10.1007/978-3-031-21837-8_9

product of MO's. To distinguish the two electrons occupying a single MO, we must assign to each electron a wave function which contains, in addition to space coordinates, also the spin coordinates of that electron. The orbital is now identified as a "*Molecular Spin Orbital*" (MSO). If magnetic contributions (i.e. m_l) are neglected, each MSO factors into an MO function and a spin function:

$$\psi_k^n = \psi_k(x^n, y^n, z^n, s^n) = \varphi_{i(k)}(x^n, y^n, z^n)\eta_k(s^n) = \varphi_{i(k)}\eta_k \tag{9.1.2}$$

In (9.1.2) the superscript n stands for both space and spin coordinates on the nth electron, while the subscripts k and i identify MSO and MO, respectively. The spin factors are taken to be

$$\eta_k^n = \begin{cases} \alpha^n \\ \beta^n \end{cases}$$

The total N-electron wave function is the *Antisymmetrized Product* (AP) of MSO's[1]

$$\Phi = \frac{1}{\sqrt{N!}} \begin{vmatrix} \psi_1^1 & \psi_2^1 & \cdots & \psi_N^1 \\ \psi_1^2 & \psi_2^2 & \cdots & \psi_N^2 \\ \cdots & \cdots & \cdots & \cdots \\ \psi_1^N & \psi_2^N & & \psi_N^N \end{vmatrix} \tag{9.1.3}$$

All MSOs in (9.1.3) must be linearly independent (otherwise the determinant vanishes). It follows that no two MSOs can be the same, or, that only two MOs may be the same, provided that they have opposite spin.

The wave functions are geometrical objects, which, when subjected to transformation, have the properties of base vectors. Consider now as an example the $2p_x$, $2p_y$, and $2p_z$ functions of an atom. Because these behave as unit vectors **i**, **j**, **k** and because it is customary to write the components v_1, v_2, v_3 of an arbitrary vector **v** as a column vector, it follows that **i**, **j**, **k** must be written as row vector, so that:

$$\mathbf{v} = \mathbf{i}v_1 + \mathbf{j}v_2 + \mathbf{k}v_3 = (\mathbf{ijk}) \begin{pmatrix} v_1 \\ v_2 \\ v_3 \end{pmatrix} \tag{9.1.4}$$

We collect thus now the set of MSO's in a row vector $\boldsymbol{\Psi}$:

$$\boldsymbol{\Psi} = (\psi_1 \psi_2 \ldots \psi_N) \tag{9.1.5}$$

---

[1] The antisymmetric product was first applied by Goepper–Mayer and Skar (1938) to the six carbon π electron of benzene and Roothan (1951) extended the procedure to include all electrons in the molecule.

and subject each MSO to a linear transformation:

$$\psi'_k = \sum_{\lambda} \psi_\lambda \mathrm{A}_{\lambda\kappa} \tag{9.1.6}$$

or

$$\boldsymbol{\Psi}' = \boldsymbol{\Psi}\mathrm{A} \tag{9.1.7}$$

where **A** is a non-singular N × N matrix. If Φ and Φ' are the antisymmetrized products (AP's) built from $\boldsymbol{\Psi}$ and $\boldsymbol{\Psi}'$, then:

$$\Phi' = \Phi \mathrm{Det}(\mathrm{A}) \tag{9.1.8}$$

Since the ψ_k are linearly independent and the transformation matrix A may be chosen in such a way that the transformed MSOs (ψ'_k) form an orthonormal set, then "*it may be assumed that also the initial MSOs are orthonormal*" (Roothan 1951), i.e.:

$$\int \overline{\psi}_k \psi_\lambda \mathrm{d}\tau = \delta_{k\lambda} \tag{9.1.9}$$

where $\mathrm{d}\tau$ is the one-electron volume element (including spin). The above equation implies that the AP (9.1.3) is normalized, i.e.:

$$\int \overline{\Phi}\Phi \mathrm{d}\tau = 1 \tag{9.1.10}$$

9.2 Roothan–Hall Equations

The Roothaan–Hall equations (Roothaan 1951; Hall 1951) are a representation of the HF equation in a non-orthonormal basis set which can be of Gaussian-type or Slater-type. The molecular orbital expansion coefficients ($c_{\nu i}$) are related to the elements of the Fock matrix ($F_{\mu\nu}$) and to the elements of the overlap matrix ($S_{\mu\nu}$) by

$$\sum_{\nu=1}^{\mathrm{N}} \left(\mathrm{F}_{\mu\nu} - \varepsilon_i \mathrm{S}_{\mu\nu}\right) \times \mathrm{c}_{\nu i} = 0 \tag{9.2.1}$$

Written in matrix form, Eq. (9.2.1) is

$$\mathrm{FC} = \mathrm{SC}\varepsilon \tag{9.2.2}$$

Each element in Eq. (9.2.2) is itself a matrix: C is the matrix of the coefficients; ε is a diagonal matrix of the ε_i elements corresponding each to the one-electron molecular energy of the ith molecular orbital χ_i; S is the overlap matrix of the basis functions; F is the Fock matrix (electron–electron interactions), which, for a closed-shell, has elements obeying:

$$F_{\mu\nu} = \sum_{\lambda=1}^{N}\sum_{\sigma=1}^{N} P_{\lambda\sigma}\left[(\mu\nu|\lambda\sigma) - \frac{1}{2}(\mu\lambda|\nu\sigma)\right] + H_{\mu\nu}^{core} \tag{9.2.3}$$

$H_{\mu\nu}^{core}$ is the energy matrix of single electrons in the field of the nuclei and $P_{\mu\nu}$ is the density matrix:

$$P_{\lambda\sigma} = 2 \times \sum_{i=1} c_{\lambda i}^{*} c_{\sigma i} \tag{9.2.4}$$

with the summation extended only over the occupied orbitals.

The Roothan–Hall equation is solved iteratively (*Self-Consistent Field*, SCF method) until the minimum is attained (*Convergence*).

Chapter 10
DFT Functionals

The Density Functional Theory is at the core of all modern ways of computing accurate electron energies under all aggregation states of the matter. To a superficial observer, DFT could appear as a sort of cook-book with ingredients appropriate to the various *typical dishes*. It is thus worthy to describe to some extent the progressive developments obtained through the years by theorists and to present the most important recipes. Again, fundamental arguments are brought to the surface, such as the deep significance of the "*correlation*" and "*exchange*" energies, the pioneering investigations of Slater and Wigner–Seitz and the prodromal importance of the Colle-Salvetti development. Of course, the interested reader is warmly recommended (as for all other parts of the book) to visit the original sources, for a better perception of the subtleties involved in the theory.

10.1 Premises

As we have seen at Sect. 8.6.4, the DFT energy is composed of several terms, some of which are arbitrary functions of the n-electron density $\rho(\mathbf{r})$:

$$E_{[\rho(\mathbf{r})]} = \int \rho(\mathbf{r})V(\mathbf{r})d\mathbf{r}') + \frac{1}{2}\int \frac{[\rho(\mathbf{r})\rho(\mathbf{r}')]}{|\mathbf{r}-\mathbf{r}'|}d\mathbf{r}d\mathbf{r}' + \int \rho(\mathbf{r})g(\mathbf{r};[\rho])d\mathbf{r} + \sum_{A>B}\frac{Z_A Z_B}{r_{AB}} = \\ = E_{Ext} + E_J + E_{kxc} + E_{NR} \tag{8.6.24}$$

The first term (E_{Ext}) in Eq. 8.6.24 is the interaction energy with the external potential. The second term in the same equation (E_J) is the self-interaction, the third term ($E_{k\text{-}e\text{-}c}$), which contains the g($\mathbf{r}$) function "*may be interpreted as a density of*

G. A. Ottonello, *Quantum Geochemistry*, Springer Geochemistry,
https://doi.org/10.1007/978-3-031-21837-8_10

kinetic, exchange and correlation energy per unit electron density at **r**" (Pisani 1996). The g(**r**) function is also known as "*universal functional*" of the density ρ, because it uniquely depends upon the whole charge distribution (Pisani 1996a). The fourth term in (8.6.24) (E_{NR}) is the nuclear repulsion.
The minimum energy E_0 is obtained in correspondence of $\rho_0(\mathbf{r})$:

$$E_0 = \int \rho_0(\mathbf{r})V(\mathbf{r})d\mathbf{r} + \frac{1}{2}\int \frac{[\rho_0(\mathbf{r})\rho_0(\mathbf{r}')]}{|\mathbf{r}-\mathbf{r}'|}d\mathbf{r}d\mathbf{r}' + E_{kps} + E_{xc}^{LDA} + E_{NR} \quad (8.6.25)$$

$$E_{kps} = \sum_i \int \psi_i(\mathbf{r})^* \left(-\frac{\nabla^2}{2}\right)\psi_i(\mathbf{r})d\mathbf{r} \quad (8.6.26)$$

$$E_{xc}^{LDA}(\rho) = \int \rho_0(\mathbf{r})\varepsilon_{xc}(\mathbf{r};[\rho_0])d\mathbf{r} \quad (8.6.27)$$

with E_{kps} representing the kinetic energy of n pseudo-independent electrons (Pisani 1996a) and E_{xc}^{LDA} the exchange–correlation term under Local Density Approximation.

In magnetic system E_{xc}^{LDA} is replaced by the Local Spin Density Approximation E_{xc}^{LSDA} which is function of the local electron spin densities $\rho\uparrow$, $\rho\downarrow$.

$$E_{xc}^{LSDA}(\rho_\uparrow,\rho_\downarrow) = \int \rho_0(\mathbf{r})\varepsilon_{xc}(\mathbf{r};[\rho_\uparrow,\rho_\downarrow])d\mathbf{r} \quad (8.6.28)$$

In many circumstances the LDA approximation does not work properly and is conveniently replaced by the General Gradient Approximation (GGA) which include gradient corrections to the density:

$$E_{xc}^{GGA}(\rho) = \int f[\rho(\mathbf{r}),\nabla\rho(\mathbf{r})]d\mathbf{r} \quad (8.6.29)$$

Basically, thus the functionals normally used in DFT are integrals of some function of the density and (possibly) of the density gradient.

10.2 Slater's Local Spin Density Exchange Functional

Already in 1951 Slater wrote "*It is shown that the Hartree–Fock equations can be regarded as ordinary Schrodinger equations for the motion of electrons, each electron moving in a slightly different potential field, which is computed by electrostatics from all the charges of the system, positive and negative, corrected by the removal of an exchange charge, equal in magnitude to one electron, surrounding the electron whose motion is being investigated*". The corrected charge distribution seen by an electron at a certain position equals, according to Slater (1951), the charge of e-1 electrons. The exchange charge density equals ρ^{+} at the position of the electron in question, gradually falling off as one moves away from that point. One gets a rough idea of the distance in which it has fallen to a small value by replacing it with a constant density ρ^{+} inside a sphere of radius r_0, zero outside the sphere. One has thus:

$$e = \frac{4}{3}\pi r_0^3 |\rho^{+}| \quad \Rightarrow r_0 = \left(\frac{3e}{4\pi}|\rho^{+}|\right)^{1/3} \tag{10.2.1}$$

Assuming the wave functions to be approximated by plane-waves, the exchange potential results to be

$$v_{x,K,i}(\mathbf{r}) = \frac{-k_F(\mathbf{r})}{\pi}\left[1 + \frac{k_F^2(\mathbf{r}) - k^2}{2k_F(\mathbf{r})}\ln\left|\frac{k + k_F(\mathbf{r})}{k - k_F(\mathbf{r})}\right|\right] \tag{10.2.2}$$

where k_F is the local value of the Fermi momentum, which, in the electron-gas treatment, is linked to the local density $\rho(\mathbf{r})$ through $k_F(\mathbf{r}) = [3\pi^2\rho(\mathbf{r})]^{1/3}$. Averaging over the occupied state k one gets

$$v_{x,i}(\mathbf{r}) = -\frac{3}{2\pi}\left[3\pi^2\rho(\mathbf{r})\right]^{1/3} \tag{10.2.3}$$

while in the Kohn–Sham procedure the functional turns out to be

$$v_{x,i}(\mathbf{r}) = -\frac{1}{\pi}\left[3\pi^2\rho(\mathbf{r})\right]^{1/3} \tag{10.2.4}$$

The two equations differ thus for the factor 3/2.

The Coulomb correlation (see later) was accounted for by Slater through a variational parameter α. The final Slater LDA exchange–correlation potential is thus

$$v_{xc}(\mathbf{r}) = -\alpha\frac{3}{2\pi}\left[3\pi^2\rho(\mathbf{r})\right]^{1/3} \equiv \alpha\frac{3}{2\pi}k_F(\mathbf{r}) \tag{10.2.5}$$

We must remind now that the (Dirac) exchange energy is given by the integral:

$$E_x(\rho) = C_X\int v_X[\rho(\mathbf{r})]\rho(\mathbf{r})d\mathbf{r} \tag{10.2.6}$$

where

$$C_X = \frac{3}{2}\left(\frac{3}{4\pi}\right)^{1/3} = 0.7386 \tag{10.2.7}$$

Replacing the potential v_x in Eq. 10.2.6 with the potential v_{xc} (Eq. 10.2.5) and solving self-consistently by minimizing the energy E_{xc} with respect to the α parameter, the effective exchange–correlation is readily obtained. This procedure is also known as Xα method (see later).

10.3 Correlation Energy

10.3.1 Wigner–Seitz Procedure

Wigner and Seitz begun to investigate the properties of metallic sodium (lattice constant, heat of vaporization, compressibility, etc.), under the free electron approximation in 1933. In 1934 they published other two seminal papers (Wigner 1934; Wigner and Seitz 1934). In that period the advancements in theory were quite fast, and it is actually difficult to unravel the merits of the various contributors, which are often intimately interconnected. As an example, to calculate the ground state energy of free electrons through application of the Schrodinger equation, Wigner and Seitz (1933) anticipated that "*It will not be necessary to solve it for the entire lattice, because it will have the same symmetry as the crystal and hence will merely repeat itself a great number of times. Because of this symmetry, the derivative of the wave function at every crystallographic symmetry plane will be zero perpendicular to this plane…*" (Wigner and Seitz 1933), and "*If we draw lines connecting the nearest atoms and consider the planes bisecting these perpendicularly, we have every atom surrounded by a truncated octahedron*" (ibid.) The *Voronoi Cell* in this way generated (a truncated octahedron, reproduced here in Fig. 10.1 to add some historical flavor to the treatment) is generally depicted in quantum mechanical applications as a *Wigner–Seitz Cell* (see also Sect. 8.3.2).

The derivative of the wave function must vanish (by symmetry reasons) at the middle points of the planes of the polyhedron. Hence, it is possible to replace the polyhedron by a sphere of equal volume and "*to take as boundary conditions that the derivative of the wave function vanishes at the boundary of this sphere*" (Wigner and Seitz 1933); from where the concept of "*Wigner–Seitz Radius*" derives.

Assuming that in a metal two electrons are never around the same ion, is equivalent to assuming that every nucleus is surrounded by a spherical electron cloud which will "*exactly*" cancel its potential outside of the sphere.

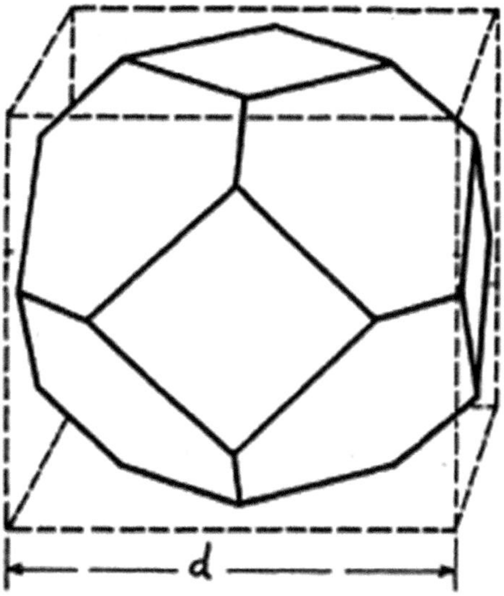

Fig. 10.1 The original drawing of Wigner and Seitz (1933) representing the lattice of metallic Na

The statistical connections between the positions of two electrons in an ideal Fermi gas are depicted in terms of determinant of a complete wave function of the form:

$$\Psi(1,2,\ldots,n)=\frac{1}{\sqrt{n!}}\begin{vmatrix}\psi_1(1) & \psi_1(2) & \psi_1(n)\\ \psi_2(1) & \psi_2(2) & \psi_2(n)\\ & & \\ \psi_n(1) & \psi_n(2) & \psi_2(n)\end{vmatrix} \tag{10.3.1}$$

where ψ are the wave functions of the different states and the numbers in parentheses represent both Cartesian and spin coordinates.

Because of the orthogonality relations of the ψ's, squaring (10.3.1) and integrating over the coordinates of all electrons except the considered ones (1 and 2) leads (apart from a constant) to

$$\sum_{k=1}^{n}\sum_{\lambda=1}^{n}\left[|\psi_k(1)|^2|\psi_\lambda(2)|^2-\psi_k(1)\psi_\lambda(1)^*\psi_k(2)^*\psi_\lambda(2)\right] \tag{10.3.2}$$

If the edge of the cubic-shaped crystal is L, assuming that the line joining two electrons at distance r lies in the x direction, leads to[1]:

$$\sum_{\nu_1\nu_2\nu_3}\sum_{\mu_1\mu_2\mu_3}[1-\exp\{2\pi i(\nu_1-\mu_1)r/L\}] \tag{10.3.3}$$

[1] The probability does not depend on the direction.

with

$$\nu = \left(\frac{3n}{8\pi}\right)^{1/3} \tag{10.3.4.1}$$

$$\nu_1^2 + \nu_2^2 + \nu_3^2 \leq \nu^2 \tag{10.3.4.2}$$

$$\mu_1^2 + \mu_2^2 + \mu_3^2 \leq \nu^2 \tag{10.3.4.3}$$

and the probability of the electrons with parallel spin being at distance r is (Wigner and Seitz 1933):

$$P(r) = \frac{9}{\pi^2 \nu^2 (4\nu^2 - 1)^2} \sum_{\nu_1=-\nu}^{\nu} \sum_{\mu_1=-\nu}^{\nu} \pi^2 \left(\nu^2 - \nu_1^2\right)\left(\nu^2 - \mu_1^2\right)\left[1 - \exp\{2\pi i(\nu_1 - \mu_1)r/L\}\right] \tag{10.3.5}$$

From these short notes we see that Wigner and Seitz anticipated the utilization of plane waves in depicting the wave functions behind the motion of electron in crystals. A notion that will be later extended and formalized by Felix Block (see Sect. 8.4).

The interaction of the electrons with anti-parallel spin, contains, in addition to the interaction of uniformly distributed space charges, another term, due to the fact that the electrons repel each other and try to keep as far apart as possible. This "*correlation energy*" decreases the total energy of the system by an amount which may be approximately calculated through a Rayleigh-Schrodinger perturbation theory in a power series of e^2 (Wigner 1934; Wigner and Seitz 1934).

The correlation effect modifies the form of the wave function, which, for a system of 2n electrons, occupying n doubly-degenerate states is initially (Wigner 1934):

$$\frac{1}{n!}\begin{vmatrix} \psi_1(x_1) & \cdots & \psi_1(x_n) \\ \cdot & \cdot & \cdot \\ \cdot & \cdot & \cdot \\ \cdot & \cdot & \cdot \\ \psi_n(x_1) & \cdots & \psi_n(x_n) \end{vmatrix} \cdot \begin{vmatrix} \psi_1(y_1) & \cdots & \psi_1(y_n) \\ \cdot & \cdot & \cdot \\ \cdot & \cdot & \cdot \\ \cdot & \cdot & \cdot \\ \psi_n(y_1) & \cdots & \psi_n(y_n) \end{vmatrix} \tag{10.3.6}$$

x in the above Fock equation represents the three Cartesian coordinates of electrons with upward spin, and y those of electrons with downward spin.

The modified wave function has the form (Wigner and Seitz 1934):

$$\frac{1}{n!}\begin{vmatrix} \psi_1(y_1...y_n;x_1) & ... & \psi_1(y_1...y_n;x_n) \\ \cdot & \cdot & \cdot \\ \cdot & \cdot & \cdot \\ \cdot & \cdot & \cdot \\ \psi_n(y_1...y_n;x_1) & ... & \psi_n(y_1...y_n;x_n) \end{vmatrix} \cdot \begin{vmatrix} \psi_1(y_1) & ... & \psi_1(y_n) \\ \cdot & \cdot & \cdot \\ \cdot & \cdot & \cdot \\ \cdot & \cdot & \cdot \\ \psi_n(y_1) & ... & \psi_n(y_n) \end{vmatrix} \tag{10.3.7}$$

The functions $\psi_1(y_1...y_n;x_1)$ are solutions of the Schrödinger equation for x with different configurations of the electrons $y_1...y_n$ with opposite spin [instead of $\psi_1(x_1)$ in Eq. 10.3.6].

Under the assumption that free electrons behave as plane waves, i.e. $\psi_\nu(x) = \psi_0(x)\exp\{2\pi i\nu \cdot x/L\}$ where ν is a positive or negative integer and L is the length of the crystal-edge, corresponding to free electrons with non-vanishing eigenvalues of momentum (Wigner and Seitz 1934; cf. also Sect. 8.4)], the change of mutual potential energy of the electrons with upward spin arising from the transition from (10.3.6) to (10.3.7) is

$$E_{2-1} = \frac{1}{n!}\int dy|y|^2 \sum_{\nu'} \varepsilon'_\nu(y_1...y_n) \tag{10.3.8}$$

with

$$\varepsilon' v(y_1...y_2) = \int dx_1 dx_2 \left(\frac{ne^2}{2r}\right)\left[|\psi_\nu(y;x_1)\psi_\nu(y;x_2)|^2 - |\psi_\nu(x_1)\psi_\nu(x_2)|\right] \cdot g(r) \tag{10.3.9}$$

where

$$g(r) = 1 - 9\left[\frac{\sin(r/d) - (r/d)\cos(r/d)}{(r/d)^3}\right]^2 \tag{10.3.10}$$

and d is the probability of the distance for free electrons with parallel spins:

$$d = \left(\frac{v_0}{r\pi^2}\right)^{1/3} = \left(\frac{4}{9\pi}\right)^{1/3} r_s = 0.521 r_s \tag{10.3.11}$$

with v_0 = atomic volume and r_s as the radius of the sphere with this volume.

10.3.2 The Colle-Salvetti Developement

Colle and Salvetti (1975) extended and generalized the Wigner–Seitz procedure for closed shells, expressing the correlation energy density as a function of the electron density and the Laplacian of the second-order Hartree–Fock density matrix.

I summarize hereafter in a very concise way the Colle-Salvetti procedure because it is of paramount importance in Quantum Chemistry, being at the core of the Lee–Yang–Parr (LYP) functional, which, in conjunction with the Becke—3-parameters exchange functional, constitutes the most widely adopted DFT method in practical calculations (i.e. B3LYP; see later).

The correlated function of the system was expressed by Colle and Salvetti (1975) in the form

$$\Psi(\mathrm{x}_1, \mathrm{x}_2, \ldots \mathrm{x}_\mathrm{N}) = \Psi_{\mathrm{HF}}(\mathrm{x}_1, \mathrm{x}_2, \ldots \mathrm{x}_N) \prod_{\mathrm{i}>\mathrm{j}} \left[1 - \phi\left(\mathrm{r}_\mathrm{i}, \mathrm{r}_\mathrm{j}\right)\right] \quad (10.3.12)$$

where $\Psi_{\mathrm{HF}}(\mathrm{x}_1, \mathrm{x}_2, \ldots \mathrm{x}_\mathrm{N})$ corresponds to (10.3.1) and the function $\phi(\mathrm{r}_\mathrm{i}, \mathrm{r}_\mathrm{j})$ is

$$\phi(\mathrm{r}_\mathrm{i}, \mathrm{r}_\mathrm{j}) = \exp(-\beta^2 r^2)\left[1 - \Phi(\mathbf{R})\left(1 + \frac{r}{2}\right)\right] \quad (10.3.13)$$

with $\mathrm{r} = \left|\mathrm{r}_\mathrm{i} - \mathrm{r}_\mathrm{j}\right|; \mathbf{R} = \frac{1}{2}\left(\mathrm{r}_\mathrm{i} + \mathrm{r}_\mathrm{j}\right)$ and β is a trial function (see later).

When electrons are far apart, the product in Eq. 10.3.12 reduces to 1 and $\Psi = \Psi_{\mathrm{HF}}$, while, when two electrons collide, the correlated wave function obeys:

$$\Psi = \Psi_{\mathrm{HF}} \Phi(\mathbf{R}) \quad (10.3.14)$$

The region in which the function $\phi\left(\mathrm{r}_\mathrm{i}, \mathrm{r}_\mathrm{j}\right)$ is appreciably different from zero depends on the value of the trial function β. The volume of this region may be depicted as inversely proportional to the electron density $\rho(\mathbf{r})$:

$$\mathrm{V} = 4\pi \int_0^\infty \exp(-\beta^2 \mathbf{r}^2)\mathbf{r}^2 \mathrm{dr} = \frac{\mathcal{K}}{\rho(\mathbf{r})} \quad (10.3.15)$$

with $\mathcal{K}$ as averaged number of electrons (i.e. $\mathcal{K} = \frac{\pi^{3/2}}{\beta^3 \rho(\mathbf{r})}$).

Moreover, the value of Φ at each point $\mathbf{R}$ depends only on value of β and thus on the electronic density at the point $\mathbf{R}$. Expanding $\mathbf{R}$ as a function of β one obtains (Eq. 12 in Colle and Salvetti 1975):

$$\Phi^2(\mathbf{R}) = \left(\frac{3}{64}\sqrt{\frac{\pi}{8}}\frac{1}{\beta^2} + \frac{1}{8\beta^2} + \frac{1}{8}\sqrt{\frac{\pi}{2}}\right) + 2\Phi(\mathbf{R})\left(\frac{3}{16\beta} - \frac{1}{8}\sqrt{\frac{\pi}{2}} + \frac{\sqrt{\pi}}{4}\right)$$
$$- \frac{\sqrt{\pi}}{2} + \frac{1}{8}\sqrt{\frac{\pi}{2}} = 0 \tag{10.3.16}$$

From the above equation it is rapidly seen that, as β varies between 0 and ∞, Φ varies between 0 and 1, i.e.:

$$\lim_{\beta \to 0} \Phi = -1.715\beta \tag{10.3.17}$$

$$\lim_{\beta \to \infty} \Phi = 1 - \frac{1}{\sqrt{\pi}}\frac{1}{\beta} \tag{10.3.18}$$

An excellent approximation of β is given by Colle and Salvetti (1975):

$$\Phi \cong \frac{\sqrt{\pi}\beta}{1+\sqrt{\pi}\beta} \tag{10.3.19}$$

Stemming from the exact two-electron density matrix without spin (McWeeny 1960):

$$P_2(r_1, r_2; r_1', r_2') = N(N-1)\int \Psi_{HF}(x_1, x_2, x_3, \ldots, x_N) \prod_{i>j}\left[1 - \phi\left(r_i, r_j\right)\right] \cdot$$
$$\left[1 - \phi\left(r'_i, r'_j\right)\right]\Psi_{HF}(x_1', x_2', x_3, \ldots, x_N) dr_3 \ldots drX_N ds_1 \ldots ds_N \tag{10.3.20}$$

Colle and Salvetti (1975) arrived to the approximation:

$$P_2(r_1, r_2; r_1', r_2') = P_{2HF}(r_1, r_2; r_1', r_2') \cdot \left[1 - \phi(r_1, r_2) - \phi(r_1', r_2') + \phi(r_1, r_2)\phi(r_1', r_2')\right] \tag{10.3.21}$$

and, under the reasonable assumption that

$$P_1(r_1, r_1') = P_{1HF}(r_1, r_1') \tag{10.3.22}$$

they expressed the correlation energy as the difference between the total electron energy and the HF exchange term:

$$E_C = E - E_{HF} = \frac{1}{2}\int P_{2HF}(r_1, r_2)\cdot\left[\phi^2(r_1, r_2) - 2\phi(r_1, r_2)\right]\frac{1}{r}dr_1 dr_2 \quad (10.3.23)$$

From Eqs. 10.3.13 and 10.3.22 and imposing the values of $P_{2HF}(r_1{}'r_2)$ calculated at the points $r_1 = r'_1$ and $r_2 = r'_2$, Colle and Salvetti arrived finally to:

$$E_C = -0.04918\int \rho(\mathbf{R})\frac{1 + 0.173W\exp(-0.158/\beta)}{1 + \frac{0.8}{\beta}}d\mathbf{R} \quad (10.3.24)$$

with

$$\beta = 2.29\times\rho(\mathbf{R})^{1/3} \quad (10.3.25)$$

$$W = 0.3814\rho(\mathbf{R})^{-8/3}\left[\nabla^2 P_{2HF}(\mathbf{R} - r/2; \mathbf{R} + r/2)\right]_{r=0} \quad (10.3.26)$$

Correlation energies (E_c) and Hartree–Fock energies (E_{HF}) calculated for closed shell systems by Colle and Salvetti (1975) are listed in Table 10.1. The mean percent error declared by the Authors is at the 2.5% level, but this is a rather pessimistic view since the highest bias (8% for H_2O) reduces to 1% when the datum of Dunning et al. (1973) (−0.364 *a.u.*) is replaced with the value indicated by Hollister and Sinanoglu (1966) (−0.340 *a.u.*).

The coefficient (q = 2.29) which multiplies the electron density raised to 1/3 in Eq. 10.3.25 was obtained by Colle and Salvetti (1975) by assuming that the volume of electron collision is proportional to the volume of the exclusion in Wigner's formula. Colle and Salvetti (1979) improved the goodness of the correlation–corrected wave function ψ_0 assuming that the volume in which the correlation function (10.3.12) works must decrease monotonically and be zero when ψ_0 is the exact wave function. The new value of β is then

$$\beta = q\left\{1 + a\left[\frac{(-h(1) - h(2))P_2^0(r_1, r_2; r'_1, r'_2)}{P_2^0(r_1, r_2; r'_1, r'_2)} - \frac{(-h(1) - h(2))P_{2HF}(r_1, r_2; r'_1, r'_2)}{P_{2HF}(r_1, r_2; r'_1, r'_2)}\right]^2_{\substack{r'_1 = r_1\\ r'_2 = r_2\\ r = 0}}\right\}\rho(\mathbf{R})^{1/3} \quad (10.3.27)$$

Table 10.1 HF and correlation energy for closed shell systems after Colle and Salvetti (1975). Data are in *Hartree*

System	E_{HF}	E_c
He	−2.861678	−0.0416
Li^+	−7.236413	−0.0438
Be^{2+}	−13.611256	−0.0441
Be	−14.573014	−0.0926
B^+	−24.237555	−0.106
CH_4	−40.20452	−0.289
H_2O	−76.0384	−0.336

where:

$$h(i) = -\frac{1}{2}\nabla_i^2 - \sum_{\alpha} \frac{Z_\alpha}{|\mathbf{R}_\alpha - r_i|} \tag{10.3.28}$$

$$\rho(\mathbf{R}) = P_1^0(\mathbf{R}, \mathbf{R}) \tag{10.3.29}$$

and $P_2^0(r_1, r_2; r_1', r_2')$ and $P_{2HF}^0(r_1, r_2; r_1', r_2')$ are, respectively, the two-electron density matrices, without spin, derived respectively from ψ_0 and ψ_{HF}.

Restricting the calculation of the correlation energy only to those parts of the space in which the electronic collisions really take place allows to improve progressively the wave function ψ_0 and eventually to centre this improvement in those parts of the space more interesting for particular purposes (for instance, a dissociation involving a special bond) (Colle and Salvetti 1979).

In Table 10.2 and Fig. 10.2 we see as an example the excellent fit of the Colle-Salvetti procedure in the case of two closed shell diatomic molecules (H_2 and Li_2) at various inter-atomic distances.

Early application of the Colle–Salvetti functional to excited states may be found in Colle et al. (1977, 1978).

Table 10.2 HF, E_0, correlation, total and exact non-relativistic energies for H_2 and Li_2 at various bond lengths. The experimental data are from Kolos and Wolniewicz (1965) for H_2 (upper part of the table) and Clementi (1974) for Li_2 (lower part). Energies and distances in a.u. (from Colle and Salvetti 1979)

R (a.u.)	E_{HF}	E_0	E_c	E_0+E_c	Experim
1	−1.0849	−1.0984	−0.0243	−1.1227	−1.1245
1.4	−1.1334	−1.1519	−0.0229	−1.1748	−1.1745
2	−1.0914	−1.1208	−0.0194	−1.1402	−1.1381
2.5	−1.01986	−1.06998	−0.01415	−1.08413	−1.08154
3	−0.9892	−1.0482	−0.0119	−1.0601	−1.0573
8	−0.7860	−0.99991	−0.00131	−1.0012	−1.0001
4	−14.8523	−14.8598	−0.1208	−14.9806	−14.9830
5.4	−14.8708	−14.8812	−0.1139	−14.9951	−14.9958
6	−14.8674	−14.8805	−0.1106	−14.9911	−14.9915
8.5	−14.8395	−14.8698	−0.1016	−14.9714	−14.9689

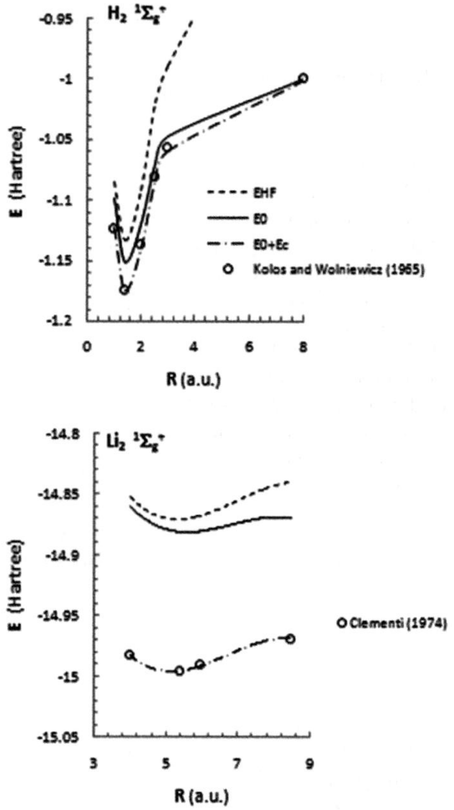

Fig. 10.2 Behaviors of E_{HF}, E_0, $E + E_c$ as functions of the interatomic distance R in H_2 and Li_2. Modified from Colle and Salvetti (1979)

10.3.3 The Lee–Yang–Parr Functional

The Colle-Salvetti correlation Energy may be also expressed in the form:

$$E_C = -4a \int \frac{\rho_2^{HF}(\mathbf{R},\mathbf{R})}{\rho(\mathbf{R})} \left\{ R \frac{1 + b\rho(\mathbf{R})^{-8/3}\left[\nabla_s^2 \rho_2^{HF}(\mathbf{R},r)\right]_{r=0} \exp\left[-c\rho^{-1/3}(\mathbf{R})\right]}{1 + d\rho(\mathbf{R})^{-1/3}} \right\} d\mathbf{R} \tag{10.3.12}$$

$$\rho_2^{HF}(\mathbf{R},r) = \frac{1}{2}\rho(\mathbf{R}+r/2)\rho(\mathbf{R}-r/2) - \frac{1}{4}\left[\rho_1^{HF}\left(\mathbf{R}+\frac{r}{2};\mathbf{R}-\frac{r}{2}\right)\right]^2 \tag{10.3.13}$$

were $\rho(\mathbf{R})$ is the Hartree–Fock electron density, $\rho_2^{HF}(\mathbf{R}, r)$ are the diagonal elements of the second-order HF reduced density matrix, a, b, c, d are the constants obtained by the fitting procedure on the HF orbital for He and ρ_1^{HF} is the Hartree–Fock 6st-order reduced density matrix.

Lee et al. (1988) expanded the Laplacian of the second-order HF density matrix in terms of a local "*Weizsacher*" kinetic energy density [$t_W(\mathbf{R})$] plus the local HF kinetic energy density [$t_{HF}(\mathbf{R})$]:

$$\nabla_r^2 \rho_2^{HF}(\mathbf{R}, r)_{r=0} = \rho(\mathbf{R})\left[t_{HF}(\mathbf{R}) - 2t_W(\mathbf{R})\right] \quad (10.3.14)$$

$$t_W(\mathbf{R}) = \frac{1}{8}\frac{|\nabla\rho(\mathbf{R})|^2}{\rho(\mathbf{R})} - \frac{1}{8}\nabla^2\rho \quad (10.3.15.1)$$

$$t_{HF}(\mathbf{R}) = \frac{1}{2}\left[\nabla_{r_1}\nabla_{r_2}\rho(r_1, r_2)\right]_{r_1=r_2=\mathbf{R}} - \frac{1}{8}\sum_i \frac{|\nabla\rho_i(\mathbf{R})|^2}{\rho_i(\mathbf{R})} - \frac{1}{8}\nabla^2\rho \quad (10.3.15.2)$$

The correlation energy in terms of electron density, [$t_W(\mathbf{R})$] and [$t_{HF}(\mathbf{R})$] is thus (Lee et al. 1988):

$$E_C = -a\int \frac{\rho(\mathbf{R}) + b\rho(\mathbf{R})^{-2/3}\left[t_{HF}(\mathbf{R}) - 2t_W(\mathbf{R})\right]\exp\left[-c\rho(\mathbf{R})^{-1/3}\right]}{1 + d\rho(\mathbf{R})^{-1/3}} d\mathbf{R} \quad (10.3.16)$$

valid for closed-shell conditions.
For open-shell conditions an excess of α- or β-spin orbitals exists in the open shell part. $\rho_2^{HF}(\mathbf{R}, \mathbf{R})$ is then replaced as follows:

$$\rho_2^{HF}(\mathbf{R},\mathbf{R}) = \frac{1}{2}\rho^2(\mathbf{R}) - \frac{1}{2}\left[\rho_\alpha^2(\mathbf{R}) + \rho_\beta^2(\mathbf{R})\right]^2 \quad (10.3.17)$$

where $\rho_\alpha(\mathbf{R})$ and $\rho_\beta(\mathbf{R})$ are the α-spin and β-spin electron densities.
Posing then:

$$\gamma(\mathbf{R}) = 2\left[1 - \frac{\rho_\alpha^2(\mathbf{R}) + \rho_\beta^2(\mathbf{R})}{\rho^2(\mathbf{R})}\right]^2 \quad (10.3.18)$$

Equation 10.3.14 is replaced by

$$\nabla_r^2 \rho_2^{HF}(\mathbf{R}, r)_{r=0} = 2\left[\rho_\alpha(\mathbf{R})t_{HF}^\alpha(\mathbf{R}) + \rho_\beta(\mathbf{R})t_{HF}^\beta(\mathbf{R}) - \rho(\mathbf{R})t_W(\mathbf{R})\right] \quad (10.3.19)$$

and Eq. 10.3.16 by:

$$E_C = -a\int \frac{\rho(\mathbf{R}) + 2b\rho(\mathbf{R})^{-5/3}\left[\rho_\alpha(\mathbf{R})t_{HF}^\alpha(\mathbf{R}) + \rho_\beta(\mathbf{R})t_{HF}^\beta(\mathbf{R}) - \rho(\mathbf{R})t_W(\mathbf{R})\right]\exp\left[-c\rho(\mathbf{R})^{-1/3}\right]}{1 + d\rho(\mathbf{R})^{-1/3}} \gamma(\mathbf{R})d\mathbf{R}$$

(10.3.20)

The presence of local kinetic energy densities ($t_{HF}, t^{\alpha}_{HF}, t^{\beta}_{HF}$) in Eqs. 10.3.16 and 10.3.20 implies that these are not pure density functional expressions, unless Kohn–Sham orbitals are utilized. To make these expressions explicit density functionals, Lee et al. (1988) adopted gradient expansions over the Thomas–Fermi kinetic energy density (t_{TF}):

$$t_{TF}(r) = C_F \rho(\mathbf{R})^{5/3} \tag{10.3.21}$$

with

$$C_F = \frac{3}{10}\left(3\pi^2\right)^{2/3} \tag{10.3.22}$$

The conventional closed-shell gradient expansion of t_{TF} gives:

$$t_{HF}(\mathbf{R}) = t_{TF}(\mathbf{R}) = C_F \rho(\mathbf{R})^{5/3} \tag{10.3.23}$$

to zero-order, and

$$t_{HF}(\mathbf{R}) = t_{TF}(\mathbf{R}) + \left[\frac{1}{9} t_W(R) + \frac{1}{18}\nabla^2 \rho\right] \tag{10.3.24}$$

to second-order

The second-order correlation energy for closed shell conditions then becomes (Lee et al. 1988):

$$E_C = -a \int \frac{1}{1 + d\rho(\mathbf{R})^{-1/3}} \left\{\rho + b\rho(\mathbf{R})^{-2/3}\left[C_F \rho(\mathbf{R})^{5/3} - 2t_W + \left(\frac{1}{9} t_W + \frac{1}{18}\nabla^2 \rho(\mathbf{R})\right)\right] \exp\left[-c\rho(\mathbf{R})^{-1/3}\right]\right\} d\mathbf{R} \tag{10.3.25}$$

The corresponding functional for open-shell conditions is readily derived through the substitutions:

$$t^{\alpha}_{HF} = \frac{1}{2} t_{TF}(2\rho_\alpha(\mathbf{R}), \mathbf{R}) \tag{10.3.26.1}$$

$$t^{\beta}_{HF} = \frac{1}{2} t_{TF}(2\rho_\beta(\mathbf{R}), \mathbf{R}) \tag{10.3.26.2}$$

Omitting the functional dependence on **R**, the open-shell second-order expansion is:

Table 10.3 Correlation energies of atoms and molecules (a.u.) with the LYP functionals. [1] = experimental; [2] = Colle-Salvetti formulations; [3] Zeroth-order LYP functional; [4] second-order LYP functional. From Lee et al. (1988) with modifications

Species	[1]	[2]	[3]	[4]
Closed shell				
He	−0.042	−0.0416	−0.039	−0.0437
Li^+	−0.0435	−0.0438	−0.0388	−0.0475
Be^{2+}	−0.0443	−0.0442	−0.0368	−0.049
Be	−0.094	−0.0926	−0.083	−0.095
B^+	−0.111	−0.106	−0.092	−0.107
Ne	−0.387	−0.375	−0.313	−0.383
Ar	−0.790	−0.743	−0.710	−0.751
Kr	–	−1.735	−1.683	−1.748
Xe	–	−2.733	−2.659	−2.743
H_2O	−0.372	−0.336	−0.317	−0.338
CH_4	−0.293	−0.290	−0.275	−0.294
Open shell				
Li (^{2}S)	−0.045	−0.050	−0.045	−0.053
B (^{2}P)	−0.125	−0.128	−0.114	−0.128
C (^{3}P)	−0.156	−0.161	−0.146	−0.161
N (^{4}S)	−0.189	−0.188	−0.176	−0.193
Cl (^{2}P)	−0.735	−0.688	−0.656	−0.695

$$E_C = -a\int \frac{\gamma}{1+d\rho^{-1/3}}\left[\rho + 2b\rho^{-5/3}f(\rho)\exp\left(-c\rho^{-1/3}\right)\right]d\mathbf{R} \tag{10.3.27}$$

with:

$$\begin{aligned} f(\rho) &= 2^{2/3}C_F\rho_\alpha^{8/3} + 2^{2/3}C_F\rho_\beta^{8/3} \\ &\quad - \rho t_W + \frac{1}{9}\left(\rho_\alpha t_W^\alpha + \rho_\beta t_W^\beta\right) + \frac{1}{18}\left(\rho_\alpha\nabla^2\rho_\alpha + \rho_\beta\nabla^2\rho_\beta\right) \end{aligned} \tag{10.3.28}$$

Table 10.3 resumes some computational results of the LYP functional, compared to the original Colle-Salvetti formulation.

10.4 Correlation Energy in LSDA: The Vosko–Wilk–Nusair Functional

In the Vosko–Wilk–Nusair functional (Vosko et al. 1980) a Padé approximant technique is used to accurately interpolate the Monte Carlo results of Ceperley and Alder (1980) for paramagnetic and ferromagnetic substances within the range of densities for atoms, molecules, and metals. These results are combined with the Random Phase Approximation (RPA)[2] spin-dependence so as to produce a correlation energy for a spin-polarized homogeneous electron gas with an estimated maximum error of 0.5×10^{-3} *Hartree*.

The assessment of Vosko et al. (1980) of the correlation energy in the LSDA begins with a factorization of the correlation energy in terms of a Δ deviation of the correlation energy per particle observed in magnetic systems $\varepsilon_C(\rho_s,\zeta)$ with respect to paramagnetic ones $\varepsilon_C^P(\rho_s)$:

$$\varepsilon_C(\rho_s,\zeta) = \varepsilon_C^P(\rho_s) + \Delta\varepsilon_c(\rho_s,\zeta) \tag{10.4.1}$$

Equation 10.4.1 is meant to emphasize the independence of the spin polarization part from the "*paramagnetic background*" and accentuates that $\varepsilon_C^P(\rho_s)$ and $\Delta\varepsilon_C(\rho_S,\zeta)$ can be taken from different sources, with the condition that $\Delta\varepsilon_C(\rho_S,0) = 0$ (Vosko et al. 1980).

The difference of $\varepsilon_C(\rho_S,\zeta)$ from the dominant RPA term $\varepsilon_{RPA}(\rho_S,\zeta)$ is expressed as:

$$\varepsilon_C(\rho_s,\zeta) = \varepsilon_{RPA}(\rho_s,\zeta) + \varepsilon_b^{(2)} + \delta\varepsilon_c(\rho_s,\zeta) \tag{10.4.2}$$

where $\varepsilon_b^{(2)}$, the second-order exchange, is the leading correction [at the same magnitude order of $\varepsilon_{RPA}(\rho_S,\zeta)$, and attaining 2.418×10^{-2} *Hartree* at the high density limit]. $\delta\varepsilon_c(\rho_s,\zeta)$ is a smaller term approaching, at the high density limit:

$$\delta\varepsilon_c(\rho_s,0) = \rho_s(A_{1x}\ln\rho_s + C_{1x}) + D\rho_s + 0(\rho_s^2) \tag{10.4.3}$$

with $0(\rho_s^2)$ as ground state energy; $A_{1x} = 6.5\times10^{-3}$ *Hartree* and $C_{1x} = -1.05\times10^{-2}$ *Hartree* (Carr and Maradudin 1964).

[2] In the RPA, electrons are assumed to respond only to the total electric potential V(**r**) which is the sum of an external perturbing potential $V_{ext}(\mathbf{r})$ and a screening potential $V_{sc}(\mathbf{r})$. $V_{ext}(\mathbf{r})$ is assumed to oscillate, so that the model yields a dynamic dielectric function. However, the contribution to the dielectric function is assumed to average out, so that only the potential at wave vector **k** is effective.

The constant D is unknown, thus Vosko et al. (1988) posed:

$$\delta_1\varepsilon_C(\rho_S,\zeta) = \varepsilon_C(\rho_S,\zeta) - \varepsilon_{RPA}(\rho_S,\zeta) \tag{10.4.4}$$

$$\delta_2\varepsilon_C(\rho_S,0) = \delta_1\varepsilon_C(\rho_S,0) - \varepsilon_b^{(2)} - \rho_S(A_{1x}\ln\rho_S + C_{1x}) \tag{10.4.5}$$

with $\delta_1\varepsilon_C(\rho_S,\zeta)$ and $\delta_2\varepsilon_C(\rho_S,0)$ approaching respectively $\varepsilon_b^{(2)}$ and 0 as $\rho_s \to 0$.

Because $\varepsilon_b^{(2)}$ does not depend on ρ_s and ζ Eq. 10.4.4 brings to

$$\delta_1\Delta\varepsilon_c(\rho_s,\zeta) = \Delta\varepsilon_c(\rho_s,1) - \Delta\varepsilon_{RPA}(\rho_s,1) \tag{10.4.6}$$

By using this scaling relation the analogs of A_{1x} and C_{1x} for ζ=1 were obtained posing:

$$\delta_2\Delta\varepsilon_C(\rho_S,1) = \delta_1\Delta\varepsilon_C(\rho_S,1) - \rho_S(\Delta A_{1X}\ln\rho_S + \Delta C_{1x}) \tag{10.4.7}$$

$$\Delta A_{1X} = \left(\frac{1}{2^{4/3}} - 1\right)A_{1x} \tag{10.4.8}$$

$$\Delta C_{1X} = \left(\frac{1}{2^{4/3}} - 1\right)C_{1x} - \frac{\frac{4}{3}A_{1x}\ln 2}{2^{4/3}} \tag{10.4.9}$$

Consider now an intermediate condition i = P/F between a pure Paramagnetic (P) and a pure Ferromagnetic (F) state.

For $\rho_s << 1$:

$$\varepsilon_C^i(\rho_s) = A_0^i\ln\rho_s + C_0^i + \rho_s\left[A_1^i\ln\rho_s + C_1^i\right] + \cdots \tag{10.4.10}$$

while for $\rho_s >> 1$ $\varepsilon_C^i(\rho_S)$ is approached by a series in the form

$$\varepsilon_C^i(\rho_S) = \frac{g_0^i}{\rho_S} + \frac{g_1^i}{\rho_S^{3/2}} + \frac{g_2^i}{\rho_S^2} + \cdots \tag{10.4.11}$$

A function combining satisfactorily the two limiting conditions is obtained through a two-point Padé approximant for $\rho_S\frac{d\varepsilon_C}{d\rho_S}$:

$$\rho_S\frac{d\varepsilon_C}{d\rho_S} = A\frac{1 + b_1x}{1 + b_1x + b_2x^2 + b_3x^3} \tag{10.4.12}$$

with $x = \rho_S^{1/2}$.

Equation 10.4.12 was integrated with the limiting condition that $\varepsilon_C(\rho_S) = 0 \quad as\, \rho_S \to \infty$ and gave (Eq. 4.4 in Vosko et al. 1980):

$$\varepsilon_C(\rho_S) = A\left\{\ln\frac{x^2}{X(x)} + \frac{2b}{Q}\tan^{-1}\left(\frac{Q}{2x+b}\right) - \frac{bx_0}{X(x_0)}\left[\ln\frac{(x-x_0)^2}{X(x)} + \frac{2(b+2x_0)}{Q}\tan^{-1}\left(\frac{Q}{2x+b}\right)\right]\right\} \tag{10.4.13}$$

The functional (10.4.13) satisfies Eq. 10.4.4 by condition.

x_0, b and c in Eq. 10.4.13 replace b1, b2 and b3 in Eq. 10.4.12 when posing:

$$X(x) = x^2 + bx + c \tag{10.4.14.1}$$

$$Q = \left(4c - b^2\right)^{1/2} \tag{10.4.14.2}$$

Table 10.4 compares the results of Ceperley and Alder (1980) for ferromagnetic and paramagnetic substances at various values of ρ_s with the interpolation results based on Eq. 10.4.13.

The last colum lists the differences from the RPA calculations of von Barth and Hedin (1972) based on the equation:

$$\Delta\varepsilon_{RPA}(\rho_S, \zeta) = \Delta\varepsilon_{RPA}(\rho_S, 1)\cdot f(\zeta) \tag{10.4.15}$$

where the f(ζ) is given by:

Table 10.4 Results of the Vosko-Wilk-Nusair functional (VWN) compared with Ceperley and Alder's (CA) correlation energies and with the RPA evaluations of von Barth and Hedin (1972) $\delta = \Delta\varepsilon_c(\rho_s, 1) - \Delta\varepsilon_{RPA}(\rho_s, 1)$. Data are in *Hartree*$\times 10^3$. The adopted interpolation parameters of the VWN functional are $x_0 = -0.409286$, b = 13.0720, and c = 42.7198 for the paramagnetic case, and $x_0 = -0.743294$, b = 20.1231, c = 101.578 for the ferromagnetic case, respectively

ρ_s	ε_c^P[CA]	ε_c^F[CA]	ε_c^P[VWN]	ε_c^F[VWN]	δ
0.5	–	–	−77.065	−40.12	1.46
1	–	–	−60.02	−31.53	1.57
2	−45.1	−24	−44.785	−23.855	1.525
3	–	–	−36.885	−19.87	1.425
4	–	–	−31.785	−17.29	1.32
5	−28.15	−15.6	−28.135	−15.435	1.225
6	–	–	−25.355	−14.015	1.135
7.5	–	–	−22.2	−12.395	1.02
10	−18.61	−10.5	−18.545	−10.5	0.87
15	–	–	−14.15	−8.19	0.665
20	−11.5	−6.775	−11.55	−6.795	0.53
50	−5.7	−3.545	−5.705	−3.55	0.21
100	−3.1895	−2.073	−3.185	−2.065	0.085

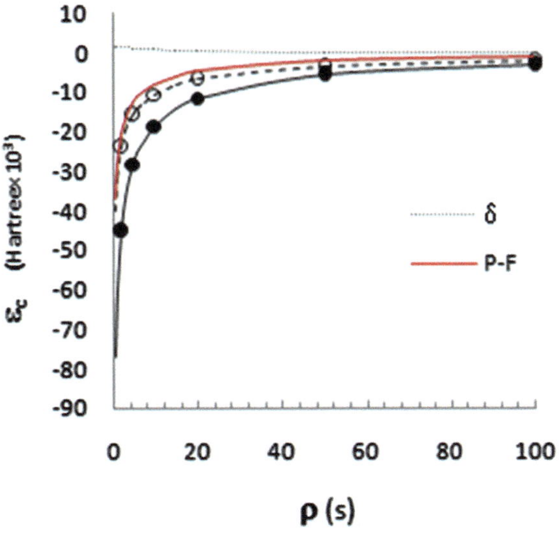

Fig. 10.3 correlation energy in LSDA. The paramagnetic state is represented by the solid line interpolant [VWN] and the solid circles [CA]. The ferromagnetic state is represented by the dashed line interpolant [VWN] and the open circles [CA]. The differences between the VWN functional and the RPA calculations of von Barth and Hedin (1972) are represented by the dotted line. The red line collects the differences between the correlation energies of the Paramagnetic and Ferromagnetic states for constant values of $\rho_{(s)}$

$$f(\zeta) = \frac{\left[(1+\zeta)^{4/3} + (1-\zeta)^{4/3} - 2\right]}{2(2^{1/3}-1)} \tag{10.4.16}$$

In Fig. 10.3 we may appreciate the goodness of the procedure, the non-relevant differences from the RPA and, more important, the fact that for all density conditions the correlation effects favor the paramagnetic state.

10.5 Correlation-Exchange Functionals

10.5.1 Becke's Functional

The Becke's gradient-corrected density-functional approximation (Becke 1986, 1988) was aimed at reproducing the correct 1/r asymptotic behavior of the exchange energy density with a single parameter. He begun with the Hartree–Fock LDA approximation, written, in a.u. as follows (Becke 1988):

$$\mathrm{E}_{\mathrm{X}}^{\mathrm{LDA}} = -\mathrm{C}_{\mathrm{X}} \sum_{\sigma} \int \rho_{\sigma}^{4/3} \mathrm{d}^3\mathbf{r} \tag{10.5.1}$$

In the above equation σ is equivalent to $\rho\uparrow, \rho\downarrow$., so that practically it is a LSDA approximation.

The functional was then corrected on semi-empirical ground to give the complete (exchange+correlation) energy (Becke 1986) as follows:

$$E_{xc}^{SE} = E_X^{LDA} - \beta \sum_\sigma \int \rho_\sigma^{4/3} \frac{x_\sigma^2}{\left(1 + \gamma x_\sigma^2\right)} d^3\mathbf{r} \tag{10.5.2}$$

where SE at superscript stands for "*semi-empirical*" x_σ is the dimensionless ratio

$$x_\sigma = \frac{|\nabla \rho_\sigma|}{\rho_\sigma^{4/3}} \tag{10.5.3}$$

and β and γ are adjustable parameters. The functional was named $X_{\alpha\beta\gamma}$ by Becke (1986) due to its analogy with the Lowest-order Gradient Correction (LGC) functional $X_{\alpha\beta}$ of Herman et al. (1969):

$$E_{xc}^{LGC} = E_X^{LDA} - \beta \sum_\sigma \int \frac{\left(\nabla \rho_\sigma\right)^2}{\rho_\sigma^{4/3}} d^3\mathbf{r} \tag{10.5.4}$$

to which it reduces in the case of small density gradients (Becke 1988).

The LGC approximation may be also expressed in terms of an "*enhancement factor*" over the LSDA exchange energy.

For the Becke functional, the enhancement factor has form (Gill 1996):

$$F(x) = \frac{bx^2}{1 + 6bx \sinh^{-1}(x)} \tag{10.5.5}$$

with b = β.

The accuracy of the Becke functional in remarkable in reproducing the exact exchange energies in various atomic series with a single constant value of the β parameter (β = 0.0042 a.u.). This explains his success in returning correct observables, when coupled with the Lee-Yan-Parr functional. The β parameter was determined by a least-squares fit to the exchange energies of the six noble gas atoms helium through radon. The HF densities were obtained by Becke (1988) from the wave functions of Clementi and Roetti (1974) and McLean and McLean (1981). The best-fit value, to two significant figures, is β = 0.0042 a.u., and yields a rms relative deviation from the exact exchange energies of only 0.11% (Table 10.5).

In Table 10.6 we see the $X_{\alpha\beta\gamma}$ functional performance for atoms hydrogen through argon, and in Table 10.7, for the first transition series. In both cases the value β = 0.0042 a.u. was used throughout, and in open-shell cases the input Hartree–Fock densities were spherically averaged. As stressed by Becke (1988), the agreement between the $X_{\alpha\beta\chi}$ functional results and the exact exchange energies is excellent.

The exchange functional of Becke (1986, 1988) was later revisited by Handy and Cohen (2001) by allowing the coefficient of the Dirac exchange term (C_X) to be

Table 10.5 The exchange energy fit of Becke (1988). Energies are in a.u. LDA values obtained by applying Eq. (10.5.1)

	Exact	LDA	$X_{\alpha\beta\chi}$
He	−1.026	−0.884	−1.025
Ne	−12.11	−11.03	−12.14
Ar	−30.19	−27.86	−30.15
Kr	−93.89	−88.62	−93.87
Xe	−179.2	−170.6	−179.0
Rn	−387.5	−373.0	−387.5

Table 10.6 Atomic exchange energies for atoms hydrogen through argon, reproducd by the $X_{\alpha\beta\chi}$ functional. From Becke (1988)

	Exact	LDA	$X_{\alpha\beta\gamma}$
H	−0.313	−0.268	−0.310
He	−1.026	−0.884	−1.025
Li	−1.781	−1.538	−1.775
Be	−2.667	−2.312	−2.658
B	−3.744	−3.272	−3.728
C	−5.045	−4.459	−5.032
N	−6.596	−5.893	−6.589
O	−8.174	−7.342	−8.169
F	−10.00	−9.052	−10.02
Ne	−12.11	−11.03	−12.14
Na	−14.02	−12.79	−14.03
Mg	−15.99	−14.61	−16.00
Al	−18.07	−16.53	−18.06
Si	−20.28	−18.59	−20.27
P	−22.64	−20.79	−22.62
S	−25.00	−23.00	−24.89
Cl	−27.51	−25.35	−27.49
Ar	−30.19	−27.86	−30.15

Table 10.7 Atomic exchange energies for the 3d transition series. From Becke (1988)

	Exact	LDA	$X_{\alpha\beta\chi}$
Sc	−38.03	−35.28	−38.03
Ti	−41.04	−38.14	−41.05
V	−44.20	−41.15	−44.22
Cr	−47.76	−44.64	−47.87
Mn	−50.98	−47.66	−51.06
Fe	−54.38	−50.93	−54.48
Co	−57.97	−54.36	−58.08
Ni	−61.68	−57.95	−61.84
Cu	−65.79	−62.00	−66.06
Zn	−69.64	−65.64	−69.86

optimized. The new functional, proposed for atoms in the first two rows of the periodic table has the form:

$$E_X(\rho) = \sum_\sigma \int -\left(1.05151 \times C_X + 1.43169 \times u_\sigma^2\right)\rho_\sigma^{4/3} d^3\mathbf{r} \tag{10.5.6}$$

where

$$u_\sigma = \frac{0.006x^2}{1 + 0.006x^2} \tag{10.5.7}$$

Although the utilization of this new functional reduces considerably (i.e. roughly to ½) the discrepancies with respect to exact (HF) calculations, the performance is much worst for heavier atoms.

10.5.2 Perdew–Wang Functionals

The electronic exchange energy as a functional of the density $E_X^{GGA}(\rho)$ is approximated by Perdew (1986), Perdew and Wang (1986, 1992) as follows:

$$E_X^{GGA}(\rho) = A_X \int d^3\mathbf{r}, n^{4/3} F(s) \tag{10.5.8}$$

$$A_X = -\frac{3}{4}(3\pi)^{1/3} \tag{10.5.9}$$

$$k_F = \left(3\pi^2\rho\right)^{1/3} \tag{10.5.10}$$

$$s = \frac{|\nabla\rho|}{2k_F\rho} \tag{10.5.11}$$

$$F(s) = \left(1 + 296s^2 + 14s^4 + 0.2s^6\right)^{1/15} \tag{10.5.12}$$

where k_F is the local Fermi vector and s is the density gradient.

This functional, denoted hereafter as PW GGA-I is simple enough to be applied routinely in self-consistent calculations for atoms, molecules, and solids. Calculated exchange energies for atoms fall within 1% of Hartree–Fock values. Some exemplificative applications to spherical atoms are shown in Table 10.8.

The PW GGA-I exchange-energy functional, which has a semi-empirical parameter [F(s)], has been later refined by Perdew et al. (1992). The new functional (hereafter PW GGA-II) adds theoretical refinements in the small- and large-gradient limits, it includes the correlation energy, and ensures a greater accuracy with respect to PW GGA-I.

Table 10.8 Exchange energies Ex (in a.u.) of neutral atoms with nonrelativistic Hartree–Fock densities (from Perdew and Wang 1986, modified)

Atom	LDA	GGA-I	Exact
H	−0.268	−0.311	−0.3125
He	−0.884	−1.033	−1.026
Li	−1.538	−1.789	−1.781
Be	−2.31	−2.68	−2.67
Ne	−11.03	−12.22	−12.11
Ar	−27.86	−30.29	−30.18
Zn	−65.63	−69.93	−69.7
Kr	−88.6	−93.8	−93.9
Xe	−170.6	−178.6	−179.1

The PW GGA-II exchange energy is:

$$E_X^{GGA-II}(\rho_\uparrow \rho_\downarrow) = \frac{1}{2} E_{XX}[2\rho_\uparrow] + \frac{1}{2} E_{XX}[2\rho_\downarrow] \tag{10.5.13}$$

with

$$E_X^{GGA-II}(\rho) = \int d^3r(\rho)\varepsilon_{XX}(r_S, 0)F_S \tag{10.5.14}$$

$$\varepsilon_X(r_S, 0) = -3k_F/4\pi \tag{10.5.15}$$

where k_F is the local Fermi wave vector, (expressed in a.u. $k_F = 1.91916/r_s$) and the function F_s is

$$F(s) = \frac{1 + 0.19645s \cdot \sinh^{-1}(7.7956s) + [0.2743 - 0.1508\exp(-100s^2)]s^2}{1 + 0.19645s \cdot \sinh^{-1}(7.7956s) + 0.004s^4} \tag{10.5.16}$$

The PW GGA-II correlation energy is

$$E_C^{GGA-II}(\rho_\uparrow \rho_\downarrow) = \int d^3r\rho[\varepsilon_C(r_S, \zeta) + H(t, r_S, \zeta)] \tag{10.5.17}$$

where t is another scalar gradient:

$$t = \frac{|\nabla\rho|}{2gk_S\rho} \tag{10.5.18}$$

r_s is the Wigner–Seitz radius (cf. Eq. 8.6.8), ζ is the local polarization:

$$\zeta = \frac{(\rho_\uparrow - \rho_\downarrow)}{\rho} \tag{10.5.19}$$

ε_{xc} is the exchange-polarization energy for a uniform electron gas and

$$g = g(\zeta) = \frac{1}{2}\left[(1+\zeta)^{2/3} + (1-\zeta)^{2/3}\right] \tag{10.5.20}$$

is the Wang–Perdew spin-scaling factor (Wang and Perdew 1991).

The function H equals $H_0 + H_1$, with

$$H_0 = g^3 \frac{\beta^2}{2\alpha} \ln\left[1 + \frac{2\alpha}{\beta} \frac{t^2 + At^4}{1 + At^2 + A^2t^4}\right] \tag{10.5.21}$$

and $\alpha = 0.09$; $\beta = vC_c(0) = 0.066725$; $\nu = (16/\pi)(3\pi^2)^{1/3}$; $C_c = 0.004235$; $C_x = -0.001667$.

The function A is:

$$A = \frac{2\alpha}{\beta} \frac{1}{\exp\left[\frac{-2\alpha r_C(r_S,\zeta)}{g^3\beta^2}\right] - 1} \tag{10.5.22}$$

and H_1 is

$$H_1 = \nu[C_C(r_S) - C_C(0) - 3C_X/7]g^3t^2 \times \exp\left[-100g^4\left(k_S^2/k_F^2\right)t^2\right] \tag{10.5.23}$$

In Table 10.9 we see the differences in the DFT exchange–correlation energy of the LSD, the PW-GGA-I and the PW-GGA-II functionals with respect to the "*exact*" HF exchange (only) energy. The last column report the experimental value after Hotop and Lineberger (1985) less the correlation term. The LSD approximation overestimates the magnitude of the correlation energy (by about a factor of 2; Perdew et al. 1992), which leads to over-binding in a molecular or crystalline aggregate (ibid). This tendency is corrected by both GGA functionals.

Also the GGA approximation may be expressed in terms of an enhancement factor over the LSDA exchange energy:

$$E_C^{GGA-I}(\rho_\uparrow\rho_\downarrow) = E_X^{LSDA} - \sum_\sigma \int F[x_\sigma]\rho_\sigma(\mathbf{r})^{4/3} d\mathbf{r} \tag{10.5.24}$$

In the PW-GGA-I functional, the enhancement factor has the form (Gill 1996):

$$F[x_\sigma] = \frac{bx^2 - (b - \beta')x^2 \exp(-cx^2) - 10^{-6}x^d}{1 + 6bx \sinh^{-1}(x) - 10^{-6}x^d A_{XX}^{-1}} \tag{10.5.25}$$

Table 10.9 Differences in the DFT exchange–correlation energy of the LSD, the PW-GGA-I and the PW-GGA-II functionals with respect to the "*exact*" HF exchange (only) energy $E_X^{HF} - E_{xc}^{DFT}$ for neutral atoms H through Ar. The last column report the experimental value after Hotop and Lineberger (1985) less the correlation term. The LSD exchange correlation energy for a uniform gas is given by $E_{CC}^{LSD}(\rho_\uparrow\rho_\downarrow) = \int d^3r\rho\varepsilon_{xc}(r_S, \zeta)$. All energy data are in *eV*. From Perdew et al. (1992) with modifications

Atom	$-E_x^{HF}$	LSD	PW GGA-I	PW GGA-II	$E_{expt} - E_c$
H	8.50	−0.61	0.004	0.003	0.00
He	27.91	−0.80	1.39	1.00	1.14
Li	48.47	−2.52	1.64	1.08	1.24
Be	72.57	−3.55	2.80	1.97	2.57
B	101.98	−4.93	4.06	2.71	2.67
C	137.40	−6.20	5.30	3.65	4.26
N	179.52	−7.53	6.51	4.66	5.13
O	222.64	−8.07	8.93	6.53	7.02
F	272.42	−8.49	11.36	8.55	8.76
Ne	329.49	−9.04	13.63	10.59	10.61
Na	381.44	−11.72	14.03	10.78	10.82
Mg	435.23	−13.48	15.64	11.85	12.07
Al	491.77	−15.67	16.76	12.64	13.07
Si	551.95	−17.79	17.96	13.57	14.17
P	616.10	−20.03	19.04	14.46	15.06
S	680.46	−21.38	21.04	16.10	17.26
Cl	748.77	−22.27	22.96	17.77	19.44
Ar	821.38	−24.43	24.65	19.29	21.42

where $\beta' = 5(36\pi)^{-5/3}$; b = 0.0042; c = 1.6455; and[3]:

$$A_{xx} = -\frac{3}{2}\left(\frac{3}{4\pi}\right)^{1/3} \tag{10.5.26}$$

Adamo and Barone (1998) modified the enhancement factor of the PW-GGA-I functional to reproduce better the long-range dispersive interactions. They begun from the consideration that total atomic exchange energies computed by the PW functional are relatively independent to changes in the exponent of the x^d term, while the interaction energies for rare-gas atom dimers (He_2, Ne_2) are extremely sensitive to such changes. The d exponent was in this way reduced to d = 3.73. They passed then to the optimization of the b term by computing the total exchange energies for the atoms of the first and second row of the periodic table. Their results,

[3] Note that here I changed slightly the symbols to distinguish them from analogous terms in the Becke and Perdew-Wang functionals, i.e. $\beta' \neq \beta$ of Eq. 10.5.4 and $A_{xx} \neq A_x$ of Eq. 10.5.7.

Table 10.10 Magnitude of the exact (HF) exchange energy and difference between exact exchange energy and various density functionals for the exchange energy, for neutral atoms. All calculations employ Hartree–Fock densities for the computed ground-state configuration and term and the ccpVQZ basis set or the 6–311++G(3df, 3pd) expansion (Li, Be, Na, Mg). Data are in a.u. (*Hartree*). From Adamo and Barone, 1998; slightly modified. B = Becke (1988) functional; PW = Perdew and Wang (1992) functional. mPW is the modified PW after Adamo and Barone (1998). G96 is the functional of Gill (1996)

Atom	HF	Δ-LSD	Δ-B	Δ-PW	Δ-mPW	Δ-G96
H	−0.313	+0.045	+0.003	+0.006	+0.004	+0.001
He	−1.026	+0.142	+0.001	+0.009	+0.004	−0.003
Li	−1.781	+0.243	+0.006	+0.018	+0.010	+0.003
Be	−2.666	+0.354	+0.009	+0.001	+0.013	+0.009
B	−3.770	+0.474	+0.010	+0.015	+0.013	+0.008
C	−5.077	+0.590	+0.009	+0.027	+0.013	+0.007
N	−6.607	+0.706	+0.011	+0.030	+0.018	+0.008
O	−8.219	+0.835	−0.005	+0.016	−0.001	−0.009
F	−10.045	+0.356	−0.019	+0.002	−0.017	−0.025
Ne	−12.110	+1.076	−0.023	+0.006	−0.027	−0.037
Na	−14.018	+1.234	−0.010	+0.013	+0.010	−0.017
Mg	−15.992	+1.383	−0.006	+0.016	−0.007	−0.010
Al	−18.051	+1.541	+0.003	+0.026	0.000	−0.005
Si	−20.304	+1.696	+0.012	+0.036	+0.008	−0.002
P	−22.642	+1.849	+0.020	+0.046	+0.016	+0.004
S	−25.035	+2.012	+0.029	+0.053	+0.020	+0.001
Cl	−27.545	+2.168	+0.029	+0.057	+0.022	−0.003
Ar	−30.185	+2.322	+0.031	+0.061	+0.024	+0.001

with an optimized value b = 0.0046 (close to the value b = 0.0042 adopted in the Becke and the PW-GGA-I functional) are listed in Table 10.10. The average absolute error is 12×10^{-3} *Hartree* and the maximum error is only 27×10^{-3} *Hartree*.

10.5.3 The Gill Functional

This functional was inspired by a "*minimalist search for simplicity in DFT*" (Gill 1996), and its development is quite instructive.

Gill began considering in detail the importance of the Coulomb potential in the classically forbidden outer regions of a molecule where the spin density decays exponentially. For an hydrogen atom this density is effectively exponential

$$\rho_{\mathrm{H}}(\mathbf{r}) = \frac{1}{\pi}\exp(-2\mathbf{r}) \tag{10.5.27}$$

An *exact* exchange potential which cancels its Coulomb potential in the forbidden zone is Gill (1996):

$$\mathrm{V}_{\mathrm{H}}^{\mathrm{exact}}(\mathbf{r}) = \exp(-2\mathbf{r})\left(1+\frac{1}{\mathbf{r}}\right) - \frac{1}{3} \tag{10.5.28}$$

while a general improvement over a density-based exchange functional may be represented in terms of a gradient-dependent function [g(x)] present in the integrand:

$$\mathrm{E_X}(\rho) = \int \rho^{4/3}(\mathbf{r})\mathrm{g}(\mathrm{x})\mathrm{d}\mathbf{r} \tag{10.5.29}$$

The comparison between an exact (10.5.28) and an approximate functional "*may be seen as a gauge of the performance of the approximate functional itself*" (Gill 1996).

The spin-polarized Dirac functional is a special form of (10.5.29) where g(x) is constant:

$$\mathrm{g}(\mathrm{x}) = -\frac{3}{2}\left(\frac{3}{4\pi}\right)^{1/3} = \mathrm{C_X} \equiv \alpha \tag{10.5.30}$$

(see also Eq. 10.2.7), and

$$\mathrm{V_H}(\mathbf{r}) = -\left(\frac{6}{\pi^2}\right)^{1/3}\exp\left(-\frac{2\mathrm{r}}{3}\right) \tag{10.5.31}$$

The gradient-dependent function g(x) proposed by Gill (1996) in the new functional has the form:

$$\mathrm{g}(\mathrm{x}) = \alpha - \mathrm{bx}^{3/2} \tag{10.5.32}$$

$$\mathrm{V}_{\mathrm{H}}^{\mathrm{G96}}(\mathbf{r}) = \mathrm{V_H}(\mathbf{r}) = 3\sqrt{2}\pi^{1/6}\mathrm{b}\times\exp\left(\frac{\mathrm{r}}{3}\right)\left(\frac{1}{\mathrm{r}}-\frac{5}{18}\right). \tag{10.5.33}$$

The value of b was set to 1/137 in order to reproduce the exact (HF) exchange energy for the Ar atom with the basis set expansion 6-31G++. Gill (1996) noted that the function (10.5.32) is non analytic at the origin and its second and higher derivatives at x = 0 are infinite. The non-analyticity at x = 0 is however necessary for a high accuracy. In fact, as shown by Gill and Pople (1993) the unique functional of the form (10.5.29) that returns the correct HF energy (i.e. 0.5 *Hartree*) and the correct density has a g(x) with a singularity at x = 0.

The exchange energy computed through this functional has the form:

$$E_X^{G96} = \int \rho^{4/3}\left(\alpha - \frac{|\nabla\rho|^{3/2}}{137\rho^2}\right)\mathbf{dr} \tag{10.5.34}$$

and provides effectively quite accurate results (Table 10.10).

10.5.4 The Perdew–Burke–Ernzerhof Functional

This functional follows the trend of search for simplicity, with the added value that all parameters (other than those in LSDA) are fundamental constants. It improves the PW-GGA-I functional through an accurate description of the linear response of the uniform electron gas and corrects the behavior under uniform scaling, and a smoother potential (Perdew et al. 1996).

The construction begins with the GGA correlation, in the form:

$$E_C^{GGA-I}(\rho_\uparrow\rho_\downarrow) = \int d^3\mathbf{r}\rho\left[\varepsilon_C^{unif}(r_S,\zeta) + H(r_S,\zeta,t)\right]\mathbf{dr} \tag{10.5.35}$$

where ε_C^{unif} is the correlation energy per particle of a uniform electron gas, and r_s, ζ, t maintain the same meaning of the PW-GGA-II functional (i.e. r_s is the local Seitz radius; ζ is the relative spin polarization and t is a dimensionless density gradient; cf. Eqs. 10.5.18 and 10.5.19).

The function H, which has the same significance attributed to it in the PW-GGA-II functional and corresponds to H_0 when $\gamma = 0.025$ (Perdew et al. 1992) is here replaced by

$$H = \left(\frac{e^2}{a_0}\right)\gamma g^3 \times \ln\left[1 + \frac{\beta}{\gamma}t^2\left(\frac{1+At}{1+At^2+A^2t^4}\right)\right] \tag{10.5.36}$$

where

$$A = \frac{\beta}{\gamma}\left[\exp\left(-\frac{a_0\varepsilon_C^{unif}}{\gamma g^3 e^2}\right) - 1\right]^{-1} \tag{10.5.37}$$

$g = g(\zeta)$ is the Wang-Perdew spin-scaling factor of Eq. 10.5.20 and

$$\gamma = \frac{1 - \ln 2}{\pi^2} = 0.031091 \tag{10.5.38}$$

This new function grows monotonically from t = 0 where it attains the value established from the second-order gradient expansion $H \to \left(\frac{e^2}{a_0}\right)\beta g^3 t^2$ (Wang and

Perdew 1991) to $H \rightarrow -\varepsilon_c^{unif}$ in the rapidly varying $t \rightarrow \infty$ limit, causing thus the correlation to vanish.

Under uniform scaling to the high density limit ($r_s \rightarrow 0$; $\rho \rightarrow \infty$) the correlation energy tends to

$$E_c^{GGA-III}(\rho_\uparrow \rho_\downarrow) \rightarrow -\frac{e^2}{a_0}\int d^3\mathbf{r}\rho\gamma g \times \ln\left[1 + \frac{1}{\left(\chi s^2/g^2\right) + \left(\chi s^2/g^2\right)^2}\right] \quad (10.5.39)$$

where s is the density gradient of the PW-GGA-I functional (cf. Eq. 10.5.4); $\gamma = (2\pi^2/16)^{1}/3$; and $\chi = 0.72161$.

Based on the above equation, the correlation energy of a 2-electrons ion of charge $Z = \infty$, attains −0.0482 *Hartree*, not far from the "*exact*" value estimated by Ivanov and coworkers ($E_c = -0.0467$ *Hartree*; Ivanov et al. unpublished). For an infinite *jellium*, the density gradient s goes to zero, hence $E_C^{GGA-III}(\rho_\uparrow \rho_\downarrow) = -\infty$.

Levy and Perdew (1993) have shown that, for an electronic charge $\sqrt{\alpha e}$ the exchange–correlation energy $E_{xc}^{\alpha}[\rho]$ must obey the coordinate scaling identities

$$E_{xc}^{\alpha}(\rho) = \alpha^2 E_{xc}(\rho_\lambda) \quad (10.5.40)$$

$$E_{xc}^{\lambda}(\rho) = \lambda^2 E_{xc}(\rho_\lambda) \quad (10.5.41)$$

where $E_{xc}[\rho]$ is the exact exchange–correlation energy and $\lambda = \alpha^{-1}$.

Moreover, Lieb and Oxford (1981) have shown that the exchange–correlation energy for a Coulomb system of particles of charge *e* has a lower bound in the form

$$E_{xc}(\rho) \geq -Ce^{2/3}\int \rho(x)^{4/3}dx \quad (10.5.42)$$

with C = 1.68.

Based on the above observations, Perdew et al. (1996) deduced that also the exchange part of the exchange–correlation energy must obey a uniform density scaling and should scale like λ.

For zero relative spin polarization ($\zeta = 0$) the exchange energy must everywhere obey:

$$E_X^{GGA} = \int d^3\mathbf{r}\rho\varepsilon_X^{unif}(\rho)d\mathbf{r} \quad (10.5.43)$$

with $\varepsilon_C^{unif} = -3e^2 k_F/4\pi$.

and, in order to recover the correct uniform gas limit $F_x(0) = 1$.

To mimic the LSDA approximation, which is excellent for small density variations around the uniform density, while the gradient expansion is not, Perdew et al. (1996) posed:

$$F_X(s) \rightarrow 1 + \mu s^2 \qquad (s \rightarrow 0) \tag{10.5.44}$$

where the effective gradient coefficient for exchange $\mu = \beta(\pi^2/3) = 0.21951$ cancels out that for correlation.

They observed then, that to obey the Lieb–Oxford bound, the spin-polarized enhancement factor $F_X(\zeta)$ must also obey

$$F_X(\zeta = 1, s) = 2^{1/3} F_X\left(s/2^{1/3}\right) \tag{10.5.45}$$

and, in order to satisfy both Eqs. (10.5.44 and 10.5.45) they proposed that $F_x(s)$ obeys:

$$F_x(s) = 1 + k - \frac{k}{(1 + \mu s^2/k)} \tag{10.5.46}$$

where k = 0.804.

They finally represented the complete exchange–correlation energy (E_{xc}^{PBE}) in terms of the enhancement factor F_{xc} over local (i.e. non s-dependent) exchange:

$$E_{xc}^{PBE}(\rho) = \int d^3\mathbf{r} \rho \varepsilon_X^{unif} F_{xc}(r_S, \zeta, s) d\mathbf{r} \tag{10.5.47}$$

The non-locality of the above function is clearly depicted by the enhancement factor, plotted in Fig. 10.4 as function of the density gradient for $\zeta = 0$ and $\zeta = 1$.

10.5.5 Hybrid Functionals

The utilization of "*hybrid functionals*" has been promoted by Becke (1993a, b). A hybrid functional is a functional that mix a DFT exchange–correlation with an Hartree–Fock exchange term in fixed proportions. Moreover *local* terms (i.e. directly dependent on density) and *non-local* terms (i.e. dependence on mathematical derivatives of the density) are distinguished.

Becke's analysis begins by considering attentively the simplicity and the conceptual significance of the Kohn–Sham DFT: a system of independent non-interacting electrons in a common, one-body potential V_{KS}, yielding the same density as the "*real*" fully interacting system.

In practice this corresponds to stating that the non-interacting density:

Fig. 10.4 Enhancement factors of the Perdew-Becke-Ernzerhof (PBE) exchange–correlation functional (solid lines) compared with the values obtained with the PW-GGA-II functional (open circles). The abscissa axis covers the range of interest for real systems. Modified from Perdew et al. (1996)

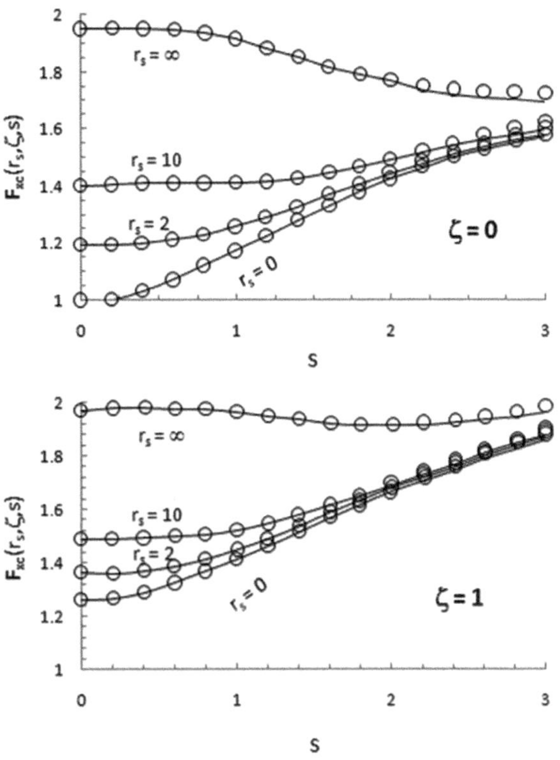

$$\rho = \sum_{i=1}^{N} |\psi_i|^2 \tag{10.5.48}$$

arises from independent reference orbitals ψ_i which satisfy the independent-particle Schrödinger equation:

$$-\frac{1}{2}\nabla_2\psi_i + V_{KS}\psi_i = \varepsilon_i\psi_i \tag{10.5.49}$$

The Kohn–Sham density-functional *exchange–correlation* energy E_{xc} represents moreover the difference between the total electronic energy of the real, fully-interacting system (E_{total}) and the summation of the kinetic energy of the non-interacting reference system (T_0) plus the nuclear interaction energy and the classical Coulomb self-energy:

$$E_{xc} = E_{total} - \left(T_0 + \int \rho V_{nuclear} d^3\mathbf{r} + \frac{1}{2}\int\int \frac{\rho(\mathbf{r}_1)\rho(\mathbf{r}_2)}{r_{12}} d^3\mathbf{r}_1 d^3\mathbf{r}_2\right) \tag{10.5.50}$$

From the above equation it is evident that "*although innocuous in appearance, E_{xc} contains, buried within it, all the details of two-body exchange and dynamical* *correlation and a kinetic-energy component as well*" (Becke 1993a). Notwithstanding this fact, E_{xc} depends uniquely on the total electronic density distribution (Hohenberg and Kohn 1964; Kohn and Sham 1965).

The Kohn–Sham local potential in Eq. 10.5.49 is given by:

$$V_{KS} = V_{nuclear} + V_{el} + V_{xc} \tag{10.5.51}$$

where V_{el} (the usual electronic Coulomb potential) is:

$$V_{el}(\mathbf{r}_1) = \int \frac{\rho(\mathbf{r}_2)}{r_{12}} d^3\mathbf{r}_2 \tag{10.5.52}$$

and the exchange–correlation potential V_{xc} is nothing else than the partial derivative of the exchange–correlation energy with respect to the local density[4]:

$$V_{xc} = \frac{\partial E_{xc}}{\partial \rho} \tag{10.5.53}$$

The exchange–correlation energy may be then written in the form:

$$E_{xc} = E_x + E_c^{LSDA} \tag{10.5.54}$$

Since exchange is by far the dominant component of the exchange–correlation energy, it is commonly argued that exchange should be treated *exactly*. However, the conventional HF E_x, is not exactly equivalent to the Kohn–Sham E_x, since Kohn–Sham and HF orbitals are not the same. Moreover, the splitting of E_{xc} into separate exchange and correlation parts (Eq. 10.5.54) is artificial and potentially misleading. Only exchange and correlation *together* has ultimate physical meaning. Finally, HF exchange, although "*exact*" in a mathematical sense, is a *very poor* representation of *exchange–correlation* in molecular systems, and LSDA for E_c "*cannot repair the fundamental defects of the Hartree–Fock approximation in molecular bonds*" (Becke 1993a).

Becke's analysis continues then considering attentively the significance of the "*adiabatic connection*", a rigorous ab initio formula, which is at the basis of the Kohn–Sham DFT exchange–correlation energy. This function is expressed by Becke (1993a, b) in the form:

$$E_{xc}(\rho) = \int_{\lambda=0}^{1} U_{xc}^{\lambda} d\lambda \tag{10.5.55}$$

[4] The entire formalism is easily generalized to spin unrestricted problems also, in terms of local spin densities ρ_α, ρ_β.

U_{xc}^{λ} in the above equation is the *potential energy* of exchange–correlation at the intermediate coupling-strength λ. The importance of this formula resides on the fact that it "*connects*" the non-interacting Kohn–Sham reference system ($\lambda = 0$) to the fully interacting real system ($\lambda = 1$) through a continuum of *partially* interacting systems where λ is an inter-electronic coupling-strength parameter.

It must be noted that, only the *kinetic* part of the exchange–correlation energy is generated, in effect, by the λ integration, although the integrand of Eq. 10.5.55 refers explicitly to *potential* energy.

The conventional exchange–correlation LSDA substitutes for each U_{xc}^{λ} in Eq. 10.5.55 a model value from local uniform-electron–gas theory. Unfortunately, the local electron–gas model is physically inappropriate near the $\lambda = 0$ exchange-only limit in molecular bonds, and it is actually the principal source of the over-binding error of the LSDA at the $\lambda = 0$ end of the coupling-strength integration.

A first approximation for the λ dependence of U_{xc}^{λ} is a *linear interpolation:*

$$E_{xc} \cong \frac{1}{2}U_{xc}^{0} + \frac{1}{2}U_{xc}^{1} \tag{10.5.56}$$

While, as already noted, U_{xc}^{0} is just the Kohn–Sham exchange energy E_x, at the other extreme (i.e. $\lambda = 1$) U_{xc}^{1} may be approximated as:

$$U_{xc}^{1} \cong U_{xc}^{LSDA} = u_{xc}\left[\rho_{\alpha}(\mathbf{r}), \rho_{\beta}(\mathbf{r})\right]d^3\mathbf{r} \tag{10.5.57}$$

where u_{xc}, the exchange–correlation potential-energy density of a spin-polarized electron gas (Perdew and Wang 1992) replaces the bulk potential of the LSDA equivalent formulation (cf. Eq. 8.6.29).

The exchange–correlation energy of Eq. (10.5.56) is thus replaced by:

$$E_{xc} \cong \frac{1}{2}E_X + \frac{1}{2}U_{xc}^{LSDA} \tag{10.5.58}$$

This "*half-and-half*" theory (Becke 1993b) employs the electron–gas model at full interaction strength only, and discard its problematic at the $\lambda = 0$ limit, corresponding to the non-interacting Kohn–Sham reference system, where the exchange-correlation potential energy U_{xc}^{0} is the "*pure exchange energy of the Slater determinant of the Kohn–Sham orbitals*" (E_x), essentially, though not exactly, equal in value to the conventional Hartree–Fock exchange energy.

Based on the above arguments, Becke (1993b) proposes the following expansion:

$$E_{xc} = E_{xc}^{LSDA} + a_0\left(E_X^{exact} - E_X^{LSDA}\right) + a_x\Delta E_X^{B88} + a_C\Delta E_C^{PW91} \tag{10.5.59}$$

which is "*the simplest possible mixture of exact exchange, the LSDA for exchange–correlation, and gradient corrections of exchange and correlation type, that exactly recovers the uniform electron–gas limit*" (Becke 1993b).

E_{xc}^{LSDA} in the above equation is the leading term, whose correlation component adopts the electron gas parametrization of Perdew and Wang (1992). a_0, a_x and a_c in the same equation are semi-empirical coefficients obtained by an appropriate fit to experimental data. a_0 (= 0.20) reflects the relative importance of a system's independent-particle character, while a_x (= 0.72) and a_c (= 0.81) allow optimum admixtures of exchange and correlation-type gradient corrections.

It is important to stress that E_X^{exact} is the exact exchange energy, ΔE_X^{B88} is Becke's 1988 gradient correction (to the LSDA) for exchange, and ΔE_C^{PW91} is the gradient correction for correlation of Perdew and Wang.

The functional is commonly identified in the computational routines as B3LYP. It should be noted to this purpose that, although in the B3LYP hybrid functional the parameters values are those obtained by Becke (1993b) through linear fitting of the G1 data set of Pople and coworkers (Pople et al. 1989; Curtiss et al. 1990, 1991), they are apparently valid also when different functional are replaced in Eq. (10.5.59). In the GAUSSIAN package for instance, the adopted expansion is (Foresman and Frisch 1993):

$$E_{xc} = E_{xc}^{LSDA} + a_0\left(E_X^{exact} - E_X^{LSDA}\right) + a_X \Delta E_X^{B88} + a_C E_C^{LYP} + (1 - a_C)E_C^{VWN} \tag{10.5.60}$$

The magnitude of non-local corrections to the local spin density approximation in real systems is in this way reproduced. In fact, because the Lee–Yang–Parr (LYP) functional contains a local term E_C^{LYP} equivalent to VWN, this last may be used to provide the excess local correlation (cf. Sects. 10.3.3 and 10.4).

The B3LYP hybrid functional is by far the most adopted one in geochemical investigations. Thus it is important to give an idea about its performance in reproducing observables of high geochemical interest such as dissociation energies, ionization potentials, proton affinities, and first-row atomic energies, here reproduced in Tables 10.11 and 10.15. Exchange–correlation energies of Eq. (10.5.59) were computed adopting the Hartree-Fock orbitals of Clementi and Roetti (1974) and spherically symmetrized densities in open-shell cases (Becke 1993a).

Other hybrid functional adopt the 3-parameters expansion above described. For example, B3P86 specifies the same functional with the non-local correlation provided by Perdew and Wang (1986), while B3PW91 adopts the functional with the non-local correlation of Perdew and Wang (1992).

Becke (1997) and Schmider and Becke (1998) revisited the original B3LYP procedure on the basis of a more extended dataset (G2) (Curtiss et al. 1991) and the analysis of the dimensionless, or "*reduced*" spin density gradient

$$s_\sigma = \nabla\rho_\sigma / \rho_\sigma^{4/3} \tag{10.5.61}$$

Table 10.11 Dissociation energies computed via the B3LYP functional (Becke 1993b), compared to experimental observations (Curtiss et al. 1990; Gill et al. 1992). Data are in *kJ/mol*

Species	Observed	Calculated	Δ
H_2	433.0	425.1	−7.9
LiH	234.3	221.3	−13.0
BeH	196.2	228.0	31.8
CH	334.3	334.3	0.0
CH_2 ($^3B_1$)	751.4	770.3	18.8
CH_2 ($^1A_1$)	713.8	703.7	−10.0
CH_3	1210.0	1224.2	14.2
CH_4	1642.2	1646.4	4.2
NH	330.5	340.2	9.6
NH_2	711.3	724.3	13.0
NH_3	1157.7	1158.1	0.4
OH	423.8	426.3	2.5
H_2O	917.6	907.9	−9.6
HF	565.7	557.7	−7.9
Li_2	100.4	74.9	−25.5
LiF	575.7	551.0	−24.7
C_2H_2	1627.2	1627.6	0.4
C_2H_4	2225.5	2235.5	10.0
C_2H_6	2787.8	2797.8	10.0
CN	738.9	739.3	0.4
HCN	1262.7	1265.2	2.5
CO	1071.9	1060.2	−11.7
HCO	1130.9	1145.2	14.2
H_2CO	1494.5	1497.5	2.9
CH_3OH	2011.7	2011.7	0.0
N_2	941.8	933.0	−8.8
N_2H_4	1696.2	1703.7	7.5
NO	628.0	633.9	5.9
O_2	493.7	515.1	21.3
H_20_2	1055.6	1045.2	−10.5
F_2	154.4	149.0	−5.4
CO_2	1597.9	1611.3	13.4
SiH_2 ($^1A_1$)	604.2	597.5	−6.7
SiH_2 ($^3B_1$)	516.3	528.9	12.6
SiH_3	895.4	892.4	−2.9
SiH_4	1266.9	1255.2	−11.7
PH_2	605.4	614.2	8.8
PH_3	951.4	943.9	−7.5
H_2S	724.7	722.6	−2.1
HCI	427.6	426.8	−0.8
Na_2	69.5	55.2	−14.2

(continued)

Table 10.11 (continued)

Species	Observed	Calculated	Δ
Si_2	309.6	319.2	9.6
P_2	485.8	469.4	−16.3
S_2	421.3	442.7	21.3
Cl_2	239.3	245.2	5.9
NaCI	407.9	387.4	−20.5
SiO	797.1	770.7	−26.4
CS	709.2	698.3	−10.9
SO	516.7	529.3	12.6
ClO	264.8	278.7	13.8
ClF	252.3	254.0	1.7
Si_2H_6	2092.4	2078.2	−14.2
CH_3Cl	1552.3	1561.5	9.2
CH_3SH	1862.3	1866.9	4.6
HOCI	654.0	653.5	−0.4
SO_2	1062.7	1051.9	−10.9

They re-expressed the exchange part of the GGA (E_X^{GGA}) in the form

$$E_X^{GGA} = \sum_\sigma \int e_{x\sigma}^{LSDA}(\rho_\sigma) g_{x\sigma}\left(s_\sigma^2\right) d^3\mathbf{r} \tag{10.5.62}$$

where $e_{x\sigma}^{LSDA}$ is the σ-spin exchange-energy density per unit volume of a *uniform electron gas* with spin density equal to its local value r_σ and $g_{x\sigma}$ is the gradient "*enhancement*" factor which corrects for the LSDA underestimation of the exchange energies of inhomogeneous systems.

The "*dynamical*" correlation for opposite spins was expressed as:

$$E_{c\alpha\beta}^{GGA} = \int e_{c\alpha\beta}^{LSDA}(\rho_\alpha, \rho_\beta) g_{c\alpha\beta}\left(s_{avg}^2\right) d^3\mathbf{r} \tag{10.5.63}$$

$$s_{avg}^2 = \frac{1}{2}\left(s_\alpha^2 + s_\beta^2\right) \tag{10.5.64}$$

and, for parallel spins as:

$$E_{c\sigma\sigma}^{GGA} = \int e_{c\sigma\sigma}^{LSDA}(\rho_\sigma) g_{c\sigma\sigma}\left(s_\sigma^2\right) d^3\mathbf{r} \tag{10.5.65}$$

The total dynamical correlation was then:

$$E_C^{GGA} = E_{c\alpha\beta}^{GGA} + E_{c\sigma\sigma}^{GGA} \tag{10.5.66}$$

The optimum correction factors g in Eqs. 10.5.62, 10.5.63 and 10.5.65 where obtained by expanding them in a power series of the "*remapped*" variable $u(s^2)$:

$$g(s^2) = g(u) = \sum_{i=0}^{m} c_i u^i(s^2) \tag{10.5.67}$$

Each correction factor g(u) was then treated separately through the adoption of non-linear coefficients γ:

$$u_{x\sigma}(s_\sigma^2) = \frac{\gamma_{x\sigma} s_\sigma^2}{1 + \gamma_{x\sigma} s_\sigma^2} \tag{10.5.68.1}$$

$$u_{c\alpha\beta}(s_{avg}^2) = \frac{\gamma_{c\alpha\beta} s_{avg}^2}{1 + \gamma_{c\alpha\beta} s_{avg}^2} \tag{10.5.68.2}$$

$$u_{c\sigma\sigma}(s_\sigma^2) = \frac{\gamma_{c\sigma\sigma} s_\sigma^2}{1 + \gamma_{c\sigma\sigma} s_\sigma^2} \tag{10.5.68.3}$$

Figure 10.5 illustrates the mapping from the reduced gradient s to the finite variable u, in the case of the correction factor $u_{x\sigma}$, with $\gamma_{x\sigma} = 0.004$ (Eq. 10.5.68.1). The comparison with a typical exchange correction factor $[g_{x\sigma}(s_\sigma)]$ justifies the adoption of the power series (10.5.67).

The values of γ were atom-optimized and kept constant (Schmider and Becke 1998), i.e. $\gamma_{x\sigma} = 0.004$; $\gamma_{c\alpha\beta} = 0.006$; $\gamma_{\sigma\sigma} = 0.2$. The expansion constants c_i of Eq. (10.5.67) were instead determined from experimental thermo-chemical data directly by means of straightforward linear least-squares fitting. Expansion to the second-order on the 2c sub-set[5] of G2 (Curtiss et al. 1997) gave rise to the following values: $c_{x0} = 0.790194$; $c_{x1} = 0.400221$; $c_{x2} = 0.832857$; $c_{c\alpha\beta 0} = 0.934715$; $c_{c\alpha\beta 1} = 1.14105$; $c_{c\alpha\beta 2} = -5.33398$; $c_{c\sigma\sigma 0} = -0.120163$; $c_{c\sigma\sigma 1} = 2.82332$; $c_{c\sigma\sigma 2} = -2.59142$. To the 9 coefficients a further linear exact-exchange parameter $c_{exact} = 0.219847$ was added, and the final expression for the exchange–correlation energy was:

$$E_{xc} = E_X + E_C + c_{exact} E_X^{exact} \tag{10.5.69}$$

The keyword normally adopted for this hybrid functional is B98.

[5] The 2c sub-set is composed of 56 heats of formation at 298 *K* + 42 ionization potentials + 25 electron affinities + 10 total energies.

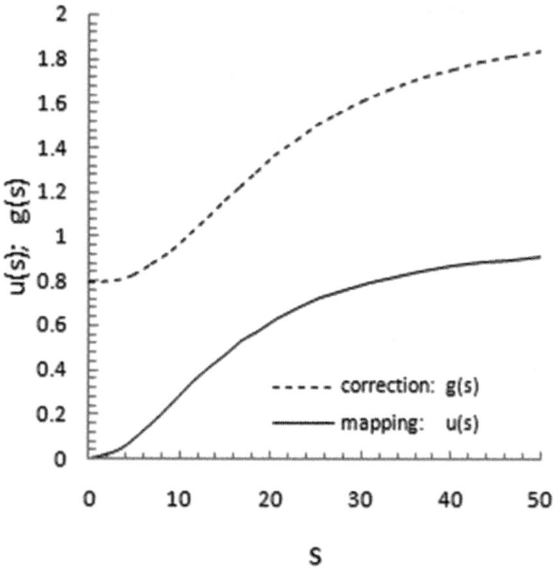

Fig. 10.5 The finite variable u, in the case of the correction factor $u_{x\sigma}$, with $\gamma_{x\sigma} = 0.004$ (Eq. 10.5.68.1) compared with the bulk exchange correction factor [g(s)] for the fit of the 2c subset of the G2 database. Modified from Schmider and Becke (1998)

As we may see in Tables 10.12, 10.13 and 10.14, the performance of the two procedures (B3LYP and B98) is quite similar, with some improvements for B98. The Mean Absolute Deviation (MAD) on the ionization potential, which is ~ 0.14 *eV* with the B3LYP functional lowers to ~ 0.12 *eV* with B98. As noted by Schmider and Becke (1998), the performance of B98 is biased by the strongly overestimated IP for the oxygen molecule. This was ascribed by the Authors to the poor geometries for O_2 and O_2^+ optimized at MP2 level.

Concerning the proton affinities, the MAD, which is ~ 4.97 *kJ/mol* with the B3LYP functional lowers to ~ 4.60 *kJ/mol* with B98. Concerning the estimates of bulk atomic energies, the MAD, which is 6×10^{-3} *Hartree* with the B3LYP functional lowers to 4×10^{-3} *Hartree* with B98.

In Table 10.15 we see the enthalpies of formation from the elements at standard state calculated via the B98 functional for the 148 substances of the G2 data set (Curtiss et al. 1997).

The bar-graph of Fig. 10.6 shows that the errors are well behaved, i.e. the distribution is almost Gaussian, centered on zero, notwithstanding a slight skewing toward positive values.

10.5.6 Double-Hybrid Functionals

These functional were introduced by Grimme (2006a, b) to take better into account the effects of the dynamic electron correlation. They are based on Becke's power-series and are explicitly parameterized by including damped atom-pairwise

Table 10.12 Ionization potential computed via the B3LYP functional (Becke 1993b), and the B98 functional (Schmider and Becke, 1989) compared to experimental observations (Curtiss et al. 1991; Gill et al. 1992). Data are in *eV*

Species	Observed	B3LYP	Δ-B3LYP	B98	Δ-B98
H	13.6	13.71	0.11	13.67	0.07
Li	5.39	5.56	0.17	5.47	0.08
Be	9.32	9.02	−0.3	9.11	−0.21
B	8.3	8.71	0.41	8.56	0.26
C	11.26	11.58	0.32	11.44	0.18
N	14.54	14.78	0.24	14.64	0.1
O	13.61	13.95	0.34	13.87	0.26
F	17.42	17.58	0.16	17.54	0.12
Ne	21.56	21.6	0.04	21.6	0.04
Na	5.14	5.27	0.13	5.11	−0.03
Mg	7.65	7.57	−0.08	7.79	0.14
Al	5.98	6.12	0.14	5.97	−0.01
Si	8.15	8.25	0.10	8.11	−0.04
P	10.49	10.57	0.08	10.42	−0.07
S	10.36	10.48	0.12	10.45	0.09
Cl	12.97	13.04	0.07	13.03	0.06
Ar	15.76	15.8	0.04	15.79	0.03
CH_4	12.62	12.47	−0.15	12.52	−0.10
NH_3	10.18	10.12	−0.06	10.11	−0.07
OH	13.01	13.09	0.08	13.07	0.06
H_2O	12.62	12.54	−0.08	12.54	−0.08
HF	16.04	15.99	−0.05	15.99	−0.05
SiH_4	11	10.85	−0.15	10.91	−0.09
PH	10.15	10.31	0.16	10.17	0.02
PH_2	9.82	10.03	0.21	9.9	0.08
PH_3	9.87	9.81	−0.06	9.81	−0.06
SH	10.37	10.43	0.06	10.43	0.06
SiH_2 ($^3B_1$)	10.47	10.42	−0.05	10.42	−0.05
SiH_2 ($^1A_1$)	12.78	12.64	−0.14	12.63	−0.15
HCl	12.75	12.74	−0.01	12.73	−0.02
C_2H_2	11.4	11.23	−0.17	11.18	−0.22
C_2H_4	10.51	10.36	−0.15	10.3	−0.21
CO	14.01	14.05	0.04	10.14	−3.87
N_2 ($^2\Sigma_g$)	15.58	15.77	0.19	15.83	0.25
N_2 ($^2\Sigma_u$)	16.7	16.65	−0.05	16.53	−0.17
O_2	12.07	12.46	0.39	12.73	0.66
P_2	10.53	10.41	−0.12	10.34	−0.19
S_2	9.36	9.58	0.22	9.58	0.22
Cl_2	11.5	11.35	−0.15	11.36	−0.14
ClF	12.66	12.55	−0.11	12.55	−0.11
CS	11.33	11.34	0.01	11.42	0.09

Table 10.13 Proton affinities computed via the B3LYP functional (Becke 1993b) and the B98 functional (Schider and Becke 1998), compared to experimental observations (Curtiss et al. 1991; Gill et al. 1992). Data are in *kJ/mol*

Species	Observed	B3LYP	Δ-B3LYP	B98	Δ-B98
H_2	421.7	422.2	0.4	417.1	−4.6
C_2H_2	637.2	656.9	19.7	655.2	18.0
NH_3	847.3	855.2	7.9	856.9	9.6
H_2O	690.8	693.3	2.5	692.5	1.7
SiH_4	644.3	643.9	−0.4	643.9	−0.4
PH_3	782.8	778.6	−4.2	782.8	0.0
H_2S	706.3	706.7	0.4	705.4	−0.8
HCI	559.0	563.2	4.2	560.7	1.7

Table 10.14 Total atomic energies computed via the B3LYP functional (Becke 1993b), and the B98 functional (Schmider and Becke 1998) compared to exact values (Gill et al. 1992). Data are in *Hartree*

Species	Exact	B3LYP	Δ-B3LYP	B98	Δ-B98
H	−0.500	−0.504	−0.004	−0.502	−0.002
He	−2.904	−2.908	−0.004	−2.909	−0.005
Li	−7.478	−7.484	−0.006	−7.485	−0.007
Be	−14.667	−14.657	0.010	−14.664	0.003
B	−24.654	−24.644	0.010	−24.649	0.005
C	−37.845	−37.837	0.008	−37.840	0.005
N	−54.590	−54.582	0.008	−54.583	0.007
O	−75.067	−75.066	0.001	−75.068	−0.001
F	−99.731	−99.735	−0.004	-99.738	−0.007
Ne	−128.937	−128.935	0.002	−128.940	−0.003

dispersion corrections of the form $c_6 \times r^{-6}$ (see later). They adopt for correlation the LYP functional with HF exchange and a perturbative second-order correlation part (MP2) obtained from the Kohn–Sham orbitals and eigenvalues.

Grimme's motivation arose from the following considerations (Grimme 2006b):

- Common exchange-only GGAs introduce a localized exchange hole around each electron that mimics effects of static electron correlation. These effects are absent in HF and are at least in part necessary for an accurate description of molecules.
- Exact (nonlocal) exchange is necessary to describe situations in which the exchange–correlation hole has a long-range component.
- The long-range ECs that are responsible for dispersive van der Waals interactions require consideration of virtual KS orbitals and are thus essentially nonexistent in all current GGAs. More accurate DFT functionals must include virtual orbital-dependent terms.

Table 10.15 Heats of formation from the elements at standard state (P = 10^5 *Bar*; T = 298.15 *K*) computed via the B98 functional (Schmider and Becke 1998) compared to experimental data. The data set is G2 (Curtiss et al. 1997). Data are in *kJ/mol*

N	System	Exper	B98	Δ	N	System	Exper	B98	Δ
1	H_2	0.00	7.15	7.15	75	C_2F_4	−658.56	−688.39	−29.83
2	LiH	139.33	145.27	5.94	76	C_2Cl_4	−12.43	−14.39	−1.97
3	BeH	341.83	325.81	−16.02	77	CF_3CN	−495.39	−488.69	6.69
4	CH	596.22	595.13	−1.09	78	CH_3CCH (propyne)	184.93	190.96	6.02
5	CH_2 ($^3B_1$)	392.04	390.87	−1.17	79	CH_2CCH_2 (allene)	190.37	179.54	−10.84
6	CH_2 ($^1A_1$)	429.91	434.22	4.31	80	C_3H_4 (cyclopropene)	276.98	288.49	11.51
7	CH_3	146.44	143.18	−3.26	81	CH_3CHCH_2 (propylene)	20.00	22.84	2.85
8	CH_4	−74.89	−74.14	0.75	82	$(CH_2)_3$ (cyclopropane)	53.14	58.62	5.48
9	^3NH	356.48	349.07	−7.41	83	C_3H_8 (propane)	−104.60	−101.67	2.93
10	NH_2	188.70	176.48	−12.22	84	$CH_2CHCHCH_2$ (butadiene)	110.04	107.95	−2.09
11	NH_3	−45.90	−48.49	−2.59	85	CH_3CCCH_3 (2-butyne)	145.60	151.42	5.82
12	OH	39.33	35.82	−3.51	86	C_4H_6 (methylene-cyclopropane)	200.41	193.30	−7.11
13	H_2O	−241.84	−234.89	6.95	87	C_4H_6 (bicyclo-butane)	217.15	233.26	16.11
14	HF	−272.55	−266.56	5.98	88	C_4H_6 (cyclo−butene)	156.48	166.19	9.71
15	SiH_2 ($^1A_1$)	272.80	267.23	−5.56	89	$(CH_2)_4$ (cyclo−butane)	28.45	30.33	1.88
16	SiH_2 ($^3B_1$)	360.66	353.38	−7.28	90	$(CH_3)_2CCH_2$ (iso-butene)	−16.74	−7.49	9.25
17	SiH_3	200.41	192.17	−8.24	91	C_4H_{10} (trans-butane)	−125.52	−122.01	3.51
18	SiH_4	34.31	30.63	−3.68	92	$(CH_3)_3CH$ (iso-butane)	−134.18	−117.65	16.53
19	PH_2	138.49	124.26	−14.23	93	$(C_2H_4)_2C$ (spiro-pentane)	185.35	193.18	7.82
20	PH_3	5.44	2.09	−3.35	94	C_6H_6 (benzene)	82.59	69.75	−12.84
21	H_2S	−20.50	−20.08	0.42	95	CH_2F_2	−450.66	−453.25	−2.59
22	HCl	−92.30	−90.79	1.51	96	CHF_3	−697.05	−696.68	0.38

(continued)

Table 10.15 (continued)

N	System	Exper	B98	Δ	N	System	Exper	B98	Δ
23	Li_2	215.89	224.85	8.95	97	CH_2Cl_2	−95.52	−95.94	−0.42
24	LiF	−335.14	−327.65	7.49	98	$CHCl_3$	−103.18	−97.57	5.61
25	C_2H_2	226.73	237.82	11.09	99	CH_3NH_2	−23.01	−27.28	−4.27
26	C_2H_4	52.47	51.88	−0.59	100	CH_3CN	75.31	77.82	2.51
27	C_2H_6	−84.01	−82.89	1.13	101	CH_3NO_2	−74.48	−79.16	−4.69
28	CN	438.90	454.59	15.69	102	CH_3ONO	−66.53	−70.71	−4.18
29	HCN	131.80	138.83	7.03	103	CH_3SiH_3	−29.29	−26.86	2.43
30	CO	−110.54	−101.46	9.08	104	HCOOH	−378.65	−380.49	−1.84
31	HCO	41.84	33.39	−8.45	105	$HCOOCH_3$	−355.64	−355.64	0.00
32	H_2CO	−108.62	−109.54	−0.92	106	CH_3CONH_2	−238.49	−248.03	−9.54
33	CH_3OH	−200.83	−199.16	1.67	107	$C2H_4NH$ (aziridine)	126.36	123.43	−2.93
34	N_2	0.00	16.86	16.86	108	NCCN	306.69	315.98	9.29
35	N_2H_4	95.35	82.59	−12.76	109	$(CH_3)2NH$	−18.41	−18.16	0.25
36	NO	90.29	83.30	−6.99	110	$C_2H_5NH_2$	−47.28	−53.97	−6.69
37	$^3O_2$	0.00	−14.35	−14.35	111	H_2CCO	−47.49	−62.09	−14.60
38	H_2O_2	−136.11	−130.42	5.69	112	C_2H_4O (oxirane)	−52.59	−50.58	2.01
39	F_2	0.00	4.69	4.69	113	CH_3CHO	−166.10	−166.94	−0.84
40	CO_2	−393.51	−400.12	−6.61	114	$(CHO)_2$ (glyoxal)	−212.13	−213.26	−1.13
41	Na_2	142.09	117.44	−24.64	115	C_2H_5OH	−235.18	−227.74	7.45
42	Si_2	585.22	573.88	−11.34	116	CH_3OCH_3	−184.10	−177.57	6.53
43	P_2	143.55	148.16	4.60	117	C_2H_4S (thioxirane)	82.01	79.20	−2.80
44	S_2	128.62	108.87	−19.75	118	$(CH_3)_2SO$	−151.46	−142.76	8.70
45	Cl_2	0.00	−7.07	−7.07	119	C_2H_5SH	−46.44	−46.07	0.38

(continued)

Table 10.15 (continued)

N	System	Exper	B98	Δ	N	System	Exper	B98	Δ
46	NaCl	−182.26	−181.46	0.79	120	$(CH_3)_2S$	−37.24	−40.21	−2.97
47	SiO	−103.09	−89.37	13.72	121	CH_2CHF	−138.91	−147.49	−8.58
48	CS	279.91	289.03	9.12	122	C_2H_5Cl	−112.13	−114.22	−2.09
49	SO	5.02	−6.23	−11.25	123	CH_2CHCl	37.24	18.66	−18.58
50	ClO	101.21	91.55	−9.67	124	CH_2CHCN	180.75	188.24	7.49
51	ClF	−55.40	−58.62	−3.22	125	$(CH_3)_2CO$	−217.28	−218.45	−1.17
52	Si_2H_6	79.91	70.71	−9.20	126	CH_3COOH	−432.63	−428.36	4.27
53	CH_3Cl	−81.84	−83.97	−2.13	127	CH_3COF	−442.25	−441.08	1.17
54	CH_3SH	−23.01	−23.93	−0.92	128	CH_3COCl	−246.73	−242.67	−4.06
55	HOCl	−74.48	−76.36	−1.88	129	C_3H_7Cl	−127.07	−131.88	4.81
56	SO_2	−296.85	−279.20	17.66	130	$(CH_3)_2CHOH$	−256.27	−272.80	16.53
57	BF_3	−1135.58	−1128.05	7.53	131	$C_2H_5OCH_3$	−219.16	−216.31	-2.85
58	BCl_3	−402.92	−408.90	−5.98	132	$(CH_3)_3N$	−20.38	−23.85	3.47
59	AlF_3	−1209.30	−1180.01	29.29	133	C_4H_4O (furane)	−33.72	−34.73	1.00
60	$AlCl_3$	−584.59	−585.89	−1.30	134	C_4H_4S (thiophene)	108.87	115.06	−6.19
61	CF_4	−933.20	−936.42	−3.22	135	C_4H_4NH (pyrole)	98.07	108.37	−10.29
62	CCl_4	−95.98	−87.24	8.74	136	C_5H_5N (pyridine)	115.14	140.58	−25.44
63	OCS	−138.41	−154.60	−16.19	137	HS	138.53	143.01	−4.48
64	CS_2	116.94	97.28	−19.66	138	CCH	582.54	565.26	17.28
65	COF_2	−638.90	−613.83	25.06	139	$CHCH_2$	289.95	299.57	−9.62
66	SiF_4	−1614.94	−1576.82	38.12	140	CH_3CO	−20.38	−10.04	−10.33
67	$SiCl_4$	−662.75	−657.64	5.10	141	H_2COH	−23.72	−17.07	−6.65
68	N_2O	82.05	80.12	−1.92	142	CH_3O	11.76	17.15	−5.40

(continued)

Table 10.15 (continued)

N	System	Exper	B98	Δ	N	System	Exper	B98	Δ
69	ClNO	51.71	41.17	−10.54	143	C_2H_5O	−0.42	−15.48	15.06
70	NF_3	−132.09	−150.16	−18.07	144	CH_3S	114.14	124.68	−10.54
71	PF_3	−958.43	−947.30	11.13	145	C_2H_5	116.44	120.92	−4.48
72	O_3	142.67	173.13	30.46	146	$(CH_3)_2CH$	87.99	89.96	−1.97
73	F_2O	24.52	20.84	−3.68	147	$(CH_3)_3C$	46.15	51.46	−5.31
74	ClF_3	−158.87	−180.67	−21.80	148	NO_2	14.85	33.10	−18.24

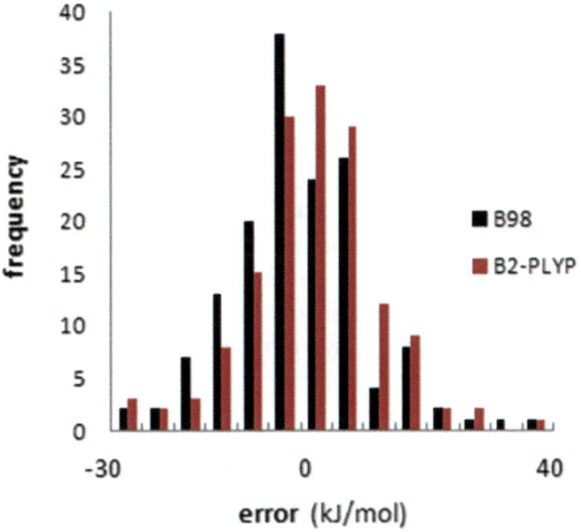

Fig. 10.6 Bar-graph of the B98 and B2-PLYP errors. For graphical purposes errors have been subdivided into 14 classes and the bars centered (i.e. class one covers errors in the range −30 *kJ/mol* to −25 *kJ/mol* and is centered on −27.5 *kJ/mol*)

Grimme (2006a) adopted Becke's Equation (10.5.67) with m = 3 and re-determined the linear parameters c_i by a least-squares fit procedure including vdW corrections (see later) while keeping unvaried the nonlinear parameters γ of Becke's work (Eqs. 10.5.68.1, 10.5.68.2 and 10.5.68.3).

The dispersion corrected total energy is:

$$E_{DFT-D} = E_{KS-DFT} + E_{disp} \tag{10.5.70}$$

where $E_{KS\text{-}DFT}$ is the usual self-consistent mean-field Kohn–Sham energy obtained from the chosen density functional and E_{disp} is an empirical dispersion correction given by:

$$E_{disp} = -s_6 \sum_{i=1}^{N_{at}-1} \sum_{j=i+1}^{N_{at}} \frac{c_6^{ij}}{R_{ij}^6} f_{dmp}(R_{ij}) \tag{10.5.71}$$

N_{at} in the above equation is s the number of atoms in the system, $c_6{}^{ij}$ denotes the dispersion coefficient for atom pair ij, s_6 is a global scaling factor (see below) and R_{ij} is the interatomic distance. f_{dmp} in the same equation is a dumping function necessary to avoid near-singularities for small R (Grimme 2004):

$$f_{dmp}(R) = \frac{1}{1 + \exp\left[-\alpha\left({}^{R}/_{R_0} - 1\right)\right]} \tag{10.5.72}$$

R_0 in Eq. (10.5.72) is the sum of a atomic van der Waals radii. This and other functions of similar type have been discussed in detail in Grimme (2004). Because

$f_{dmp}(R)$ decays at small R fast enough to zero, the dispersion corrections between atoms positioned well below typical van der Waals distances are negligible and thus, "*normal*" bonds are not significantly affected by the correction.

Consistent van der Waals radii R_0 for arbitrary elements were derived from theory, i.e. the radius of the 0.01 a.u. electron density contour from ROHF/TZV computations of the atoms in the ground state. The radii were initially scaled by a factor of s_e = 1.22 (Grimme 2004; appropriate for first and second-row atoms), then by a factor s_e = 1.10 (Grimme 2006a, b; more appropriate for heavier atoms). The scaling problem[6] was circumvented by adoption of simple computational scheme derived from the London formula for dispersion. The inter-atomic dispersion coefficients were shown to be consistent with a simple geometric mean:

$$c_6^{ij} = \sqrt{c_6^i c_6^j} \tag{10.5.72}$$

with individual coefficients based on DFT/PBE0 calculations (Adamo and Barone 1999) of atomic ionization potentials I_P and static dipole polarizabilities α:

$$c_6^i = 0.05 N I_p^i \alpha^i \tag{10.5.72}$$

where N = 2, 10, 18, 36 and 54 for atom of the 1st, 2nd, 3rd, 4th and 5th row, respectively.

The computed c_6^i parameters are listed in Table 10.16, together with the corresponding van der Waals radii.

Having determined the c_6 and R_0 dispersive terms, the expansion parameters of Eq. (10.5.67) with the included dispersion energy had to be re-determined by least-squares optimization with respect to energetic deviations of a training set of systems (Grimme 2006a; Table 10.17).

The new double-hybrid functional (B2-PLYP) was represented by Becke (2006b) as

$$E_{xc} = (1 - a_X)E_X^{GGA} + a_X E_X^{HF} + bE_C^{GGA} + cE_C^{PT2} \tag{10.5.73}$$

where E_c^{PT2} is the usual second-order Møller–Plesset-type expression for the correlation energy but evaluated with the Kohn–Sham orbitals (and the corresponding eigenvalues ε):

[6] Part of the problems of the original approach (Grimme 2004), resides in the combination rule employed for the composed coefficients that gave too much weight to the smaller coefficient (lighter atom).

Table 10.16 Dispersion coefficients (c_6) (derived from UDFT-PBE0/QZVP computations) (and van der Waals Radii (R_0) (derived from atomic ROHF/TZV computations) for Elements H–Xe. The dispersion coefficients are in $J \times nm^6 \times mol^{-1}$ and the vdW Radii in Å (From Grimme (2006a) slightly modified

Element	c_6	R_0	Element	c_6	R_0
H	0.14	1.001	K	10.80[a]	1.485
He	0.08	1.012	Ca	10.80[a]	1.474
Li	1.61	0.825	Sc-Zn	10.80[a]	1.562[b]
Be	1.61	1.408	Ga	16.99	1.650
B	3.13	1.485	Ge	17.1	1.727
C	1.75	1.452	As	16.37	1.760
N	1.23	1.397	Se	12.64	1.771
O	0.70	1.342	Br	12.47	1.749
F	0.75	1.287	Kr	12.01	1.727
Ne	0.63	1.243	Rb	24.67[a]	1.628
Na	5.71[a]	1.144	Sr	24.67[a]	1.606
Mg	5.71[a]	1.364	Y-Cd	24.67[a]	1.639[b]
Al	10.79	1.639	In	37.32	1.672
Si	9.23	1.716	Sn	38.71	1.804
P	7.84	1.705	Sb	38.44	1.881
S	5.57	1.683	Te	31.74	1.892
Cl	5.07	1.639	I	31.5	1.892
Ar	4.61	1.595	Xe	29.99	1.881

[a]Average of preceding group VIII and following group III element
[b]Average of preceding group II and following group III element

Table 10.17 Expansion Parameters (Grimme 2006a)

i	c_x	$c_{\alpha\beta}$	$c_{\sigma\sigma}$
0	1.08662	0.69041	0.22340
1	−0.52127	6.30270	−1.56208
2	3.25429	−14.97120	1.94293

$$E_C^{PT2} = \frac{1}{4} \sum_{ia} \sum_{jb} \frac{\left[(ia|jb) - (ib|ja)\right]^2}{\varepsilon_i + \varepsilon_j - \varepsilon_a - e_b} \tag{10.5.74}$$

The indices ia and jb represent single occupied-virtual replacements and the terms in brackets denote regular two-electron integrals over the KS orbitals (Grimme 2006b).

The virtual orbital-dependent functional parameters that describe the mixture of HF and GGA exchange (a_x) and of the PT2 and GGA correlation (c), respectively were obtained in a least-squares-fit procedure to the G2 set of heat of formations ($a_x = 0.53$; $c = 0.27$). The parameter b is 0.73. The MAD of the new functional on the 148 substances of the G2 set is 7.53 *kJ/mol*. The bar-graph of Fig. 10.6 shows that the errors are very well distributed, i.e. the distribution is Gaussian, centered on zero, without skewing.

10.6 Comparative Appraisal

We have seen that, in general, the GGA apparently give better results in terms of molecular properties with respect to LSDA and that hybrid functional performs even better, perhaps because they introduce some mathematical flexibility when calibrated against an observational set of data. Through the years several systematic comparisons of DFT theories with experiments appeared, but it is often difficult to ascertain the role of the functional and the role of the adopted basis set expansion in determining eventual errors. A quite interesting comparative appraisal of the performance of the various DFT functional was proposed by Johnson et al. (1993), adopting for orbital expansion the standard 6-31G(d) basis set. The functionals were separated into exchange and correlation parts. The exchange part was either Slater (S; Slater 1974), corresponding to the free-electron gas, or Becke (B; Becke 1988), which includes a gradient correction. The correlation part was either ignored (N = null), treated by LSDA theory using the VWN parameterization (Vosko et al. 1980) or treated using the LYP gradient-corrected functional (Lee et al. 1988). The correlation and exchange parts were then combined to give the six possible cases: S-null (corresponding to Hartree–Fock-Slater or Xα theory, α = 2/3), B-null, S-VWN (corresponding to regular LSDA theory), B-VWN, S-LYP, and B-LYP (with gradient corrections for both exchange and correlation). The investigate functional are schematically represented in Fig. 10.7.

Moving upward along the vertical axis replaces a zeroth-order exchange functional with a gradient-corrected one. Moving along the horizontal axis corresponds to increasing the sophistication of the correlation treatment, from none (N), to zeroth-order (0), to first-order (1).

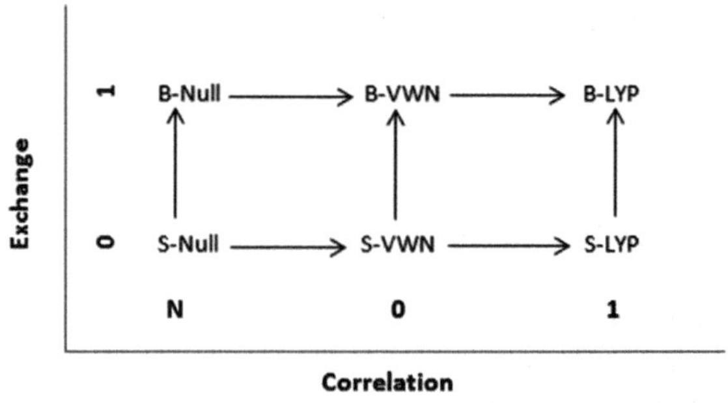

Fig. 10.7 DFT functionals classified by treatment of exchange and correlation. Modified from Johnson et al. (1993)

The investigated compounds belong to a subset of the G2 data set including all neutral molecules with only first-row atoms, plus H_2. The comparative assessment of Johnson et al. (1993) includes the atomic energies, the equilibrium geometries, the dipole moments, the harmonic frequencies and the dissociation energies. Here, for conciseness, only the atomic energies and the dissociation energies are reproduced (i.e. Tables I and V in Johnson et al. 1993).

Absolute atomic energies of the six functionals investigated by Johnson et al. (1993) are listed in Table 10.18.

While the B-LYP total energies do approach the estimated exact values quite well, the other tested functionals do not give satisfactory total energies. S-null fails to reproduce atomic energies of sufficient magnitude. B-null, which adds a gradient correction to S-null exchange, gives energies which are generally within a few *mHartree* of HF. B-VWN, while giving reasonable exchange energies, overestimates the correlation energy by using the free-electron gas approximation and hence gives total energies which are far too negative for every atom listed. This is due to the fact that KS energies are not variational with respect to the Schrodinger Hamiltonian, and hence are not guaranteed to lie above the exact values (Johnson et al. 1993).

The computation of accurate dissociation energies (D_0) is particularly demanding and is a good test of the robustness of the procedure. It involves in fact an accurate optimization of the structure at the athermal limit and a frequency calculation in order to establish the contribution of the Zero Point Energy (ZPE) to the bulk D_0. I reproduce hereafter (Table 10.19) the results obtained for CO_2 by Foresman and Frisch (1993) with the 6-31G(d) basis set and five different DFT functional. Results are compared with two ab initio procedures (HF and MP2) and with experiments. I recall here that the dissociation energy is determined by subtracting the bulk energy of the substance (E_{el}+ZPE) to the electron energy of the constituent atoms. That is in the case of CO_2:

$$D_0^{CO_2} = E_{el}^{C} + 2E_{el}^{O} - \left(E_{el}^{CO_2} + ZPE\right) \tag{10.6.1}$$

(see Sect. 8.5.5 for a comparison of D_0 with the lattice energy U and the enthalpy of formation from the elements at standard state H_0^{f}).

We may see that the structure is approximately good with all procedures (but HF) while the dissociation energies are predicted sufficiently well only by Becke's 3-parameters hybrid functionals. The HF procedure under-bounds the complex severely.

The G2 set of dissociation energies after Johnson et al. (1993) is shown in Table 10.20. The functionals S-null and B-null give dissociation energies which are generally too low. The dissociation energies of the S-VWN and S-LYP functionals are almost always too high. The two functionals B-VWN and B-LYP perform very well and give dissociation energies with a mean error (relative to experiment) close to zero. The molecules which are under-bound at this theory level are the simple hydrides with lone-pair electrons (H_2O and NH_3, for example). H_2O, F_2, and

Table 10.18 Absolute total energies of neutral atoms computed with the six DFT functional depicted in Fig. 10.7 and compared with HF, MP2, QCISD and with the exact values after Davidson et al. (1991). The adopted basis set is 6-31G(d). Data are expressed in *Hartree* (From Johnson et al. 1993)

Atom	S-Null	S-VWN	S-LYP	B-Null	B-VWN	B-LYP	HF	MP2	QCISD	Exact
H	0.4540	0.4760	0.4540	0.4954	0.5178	0.4954	0.4982	0.4982	0.4982	0.5000
He	2.7146	2.8267	2.7582	2.8540	2.9671	2.8978	2.8552	2.8664	2.8702	2.9037
Li	7.1896	7.3410	7.2433	7.4264	7.5781	7.4801	7.4314	7.4319	7.4319	7.4781
Be	14.2164	14.4420	14.3119	14.5606	14.7869	14.6563	14.5669	14.5964	14.6165	14.6674
B	24.0532	24.3441	24.1798	24.5145	24.8061	24.6413	24.5220	24.5625	24.5826	24.6539
C	37.0950	37.4537	37.2}27	37.6723	38.0318	37.8320	37.6809	37.7365	37.7552	37.8450
N	53.6840	54.1125	53.8760	54.3761	54.8053	54.5684	54.3854	54.4595	54.4754	54.5893
O	73.9544	74.4884	74.2105	74.7890	75.3238	75.0470	74.7839	74.8820	74.8977	75.0670
F	98.4097	99.0490	98.7290	99.3803	100.0220	99.7021	99.3650	99.4890	99.4995	99.7340
Ne	127.3950	128.1419	127.7777	128.4964	129.2442	128.8796	128.4744	128.6262	128.6285	128.9390

Table 10.19 Dissociation energy of CO_2 (*kJ/mol*) computed with different DFT functional and compared with HF and MP2 procedures (Foresman and Frisch 1993). The experimental interionic distance R(C-O) is 1.162 Å and the dissociation energy is 1597.9 *kJ/mol*

	CO_2		C	O		D_0	Δ (*kJ*)
Method	R(C-O)	E	ZPE	E	E		
S-VWN	1.171	−187.61677	0.0116	−37.56616	−74.64334	1975.3	−377.4
S-VWN5	1.171	−187.18193	0.0116	−37.45370	−74.48842	1942.2	−344.3
B-LYP	1.183	−188.56306	0.0112	−37.83202	−75.04696	1643.5	−45.6
B3-LYP	1.169	−188.58094	0.0114	−37.84628	−75.06062	1580.7	17.2
B3-PW91	1.168	−188.50695	0.0115	−37.82569	−75.03113	1594.1	3.8
HF	1.143	−187.63418	0.0114	−37.68086	−74.38393	982.0	615.9
MP2	1.180	−188.10775	0.0111	−37.73297	−74.88004	1584.9	13.0

molecules with multiple bonds, tend to be over-bound. This is partly due to the adopted basis set expansion, which is quite limited. The ab initio methods (HF, MP2, and QCISD) give binding energies which are too low, due mainly to inadequate treatment of electron correlation.

10.7 DFT + *U*

I introduce this class of functionals at the end of the chapter because they find extensive application only for a limited (although important) group of substances known as "*Mott Insulators*". In non-interacting electron systems metals and insulators differ mainly in the filling of their electronic bands and in the relative position of the Fermi level: while for insulators the upper band is completely filled and the Fermi level lies in the band gap, in metals the upper band is filled only partially and the Fermi level lies inside the band. Insulators with a small energy gap behave as semiconductors due to thermal excitation of the electrons (transistors). As early noted by de Boer and Verwey (1937), many transition-metal oxides with a partially filled d-electron band (hence expected to be conductors) are indeed insulators due to strong Coulomb repulsion between correlated electrons. These prototypical "*false conductors*" which, according to the conventional band theories, are predicted to be conductive, while exhibiting insulating properties, are ubiquitously known as "*Mott Insulators*", because Sir Nevil Mott was the first to propose that their bandgap (responsible of the insulating behavior) is due to inter-electronic forces (Mott and Peierls, 1937; Mott, 1949). For the NiO insulator Mott (1949) proposed that the conduction process

$$\left(Ni^{2+}O^{2-}\right)_2 \rightarrow Ni^{3+}O^{2-} + Ni^{1+}O^{2-} \tag{10.7.1}$$

is prevented by the energy gap arising from the competition between the Coulomb potential *U* between 3d electrons and the transfer integral t of 3d electrons between

Table 10.20 Theoretical dissociation energies obtained by Johnson et al. (1993) with the six functionals of Fig. 10.7 (*kJ/mol*) with the 6-31G* basis set. Experimental dissociation energies are from Curtiss et al. (1991)

Molecule	S-Null	S-VWN	S-LYP	B-Null	B-VWN	B-LYP	HF	MP2	QCISD	Exper
H_2	321.3	449.8	419.2	332.6	463.6	431.8	317.6	362.3	381.6	432.2
LiH	131.4	240.6	219.7	140.6	252.3	229.7	127.2	166.5	184.5	234.3
BeH	194.1	241.4	227.6	192.9	241.0	225.5	204.2	189.5	160.2	196.2
CH	255.2	362.8	357.3	236.8	344.8	334.3	209.2	266.1	278.7	334.3
CH_2 ($^3B_1$)	685.3	841.0	828.9	605.4	760.7	743.5	597.9	676.6	672.4	751.4
CH_2 ($^1A_1$)	567.4	774.5	760.2	506.3	714.6	695.0	469.9	590.8	606.7	713.8
CH_3	1067.8	1341.0	1325.1	950.6	1223.8	1202.1	934.3	1085.3	1085.3	1210.0
CH_4	1453.9	1827.6	0.1	1282.8	1656.9	1631.3	1256.9	1482.0	1480.7	1642.2
NH	246.0	366.9	360.7	227.6	348.5	342.3	179.9	253.6	261.5	330.5
NH_2	557.3	791.2	780.7	492.0	725.5	714.6	413.4	576.6	581.6	711.3
NH_3	941.0	1280.3	1266.1	806.3	1144.3	1130.1	712.1	972.4	965.2	1157.7
OH	357.7	472.4	472.0	302.9	417.1	412.5	241.4	354.4	351.0	423.8
OH_2	787.0	1007.5	1003.7	655.2	874.9	867.3	551.0	789.9	768.6	917.6
FH	499.2	611.7	609.2	416.3	524.3	520.5	343.5	494.5	477.0	565.7
Li_2	23.4	94.1	91.2	14.2	85.8	82.8	9.2	59.0	87.4	100.4
LiF	526.8	633.0	638.5	462.3	564.4	567.4	356.5	541.8	516.7	575.7
HCCH	1522.6	1835.1	1857.7	1278.2	1590.3	1604.1	1137.6	1529.7	1469.4	1627.2
H_2CCH_2	2049.3	2514.2	2523.0	1746.8	2211.2	2209.6	1649.3	2047.6	2015.4	2225.5
H_3CCH_3	2530.1	3146.8	3143.0	2166.1	2781.9	2765.2	2117.1	2546.0	2523.4	2787.8
CN	723.4	893.3	926.8	580.7	749.4	778.2	353.1	630.1	650.2	738.9
HCN	1178.2	1449.8	1475.3	989.5	1260.2	1281.1	773.6	1202.1	1128.4	1262.7
CO	1080.3	1227.6	1263.1	903.3	1049.8	1077.0	704.2	1064.0	993.3	1071.9
HCO	1142.7	1353.1	1386.6	930.5	1139.7	1164.4	713.8	1079.5	1019.2	1130.9

(continued)

Table 10.20 (continued)

Molecule	S-Null	S-VWN	S-LYP	B-Null	B-VWN	B-LYP	HF	MP2	QCISD	Exper
H_2CO	1443.9	1747.2	1771.1	1196.6	1499.5	1513.8	995.0	1403.7	1346.8	1494.5
H_3COH	1846.8	2306.2	2318.8	1530.1	1987.4	1988.7	1387.0	1819.2	1779.5	2011.7
N_2	847.7	1076.5	1105.8	710.9	938.5	967.8	439.7	887.4	804.6	941.8
H_2NNH_2	1425.9	1969.0	1977.8	1139.7	1677.4	1683.6	928.8	1422.1	1391.2	1696.2
NO	642.2	810.9	848.5	480.7	647.7	681.2	194.1	564.0	522.2	628.0
O_2	623.4	730.5	777.4	429.7	533.9	572.4	120.9	492.0	414.2	493.7
HOOH	1006.7	1298.7	1327.6	751.4	1038.5	1057.7	457.7	918.8	865.3	1055.6
F_2	287.9	349.8	384.1	147.7	198.3	227.6	−143.5	154.0	116.7	154.4
CO_2	1700.0	1942.6	2012.5	1346.4	1586.6	1643.9	982.0	1594.1	1455.2	1597.9
Mean deviation	−79.9	149.0	157.7	−226.8	0.4	4.2	−359.0	−93.7	−120.5	–
MAD	104.6	149.4	76.6	226.8	18.4	23.4	359.4	93.7	120.5	–

neighboring atoms (in the tight-binding approximation; see Sect. 20.1.1.1). Whenever the Coulomb potential exceeds the transfer integral of the z nearest neighbors an effective gap is created (Cococcioni et al. 2012):

$$E_{gap} = U - 2zt \tag{10.7.2}$$

It is almost superfluous to recall that HF cannot take into account short-range correlation between electrons with different spins, because up-spin and down-spin electrons are treated independently. It is not so in DFT, where the problem stems rather from the fact that exchange–correlation functionals (xc) tendentially over-delocalize valence electrons, leading to apparent metallic ground states (Cococcioni 2012; Himmetoglu et al. 2014). The over-delocalization is due to the inability of xc to effectively cancel out the *Hartree* self-interaction because "*a remaining fragment of the same electron is still there that can induce added self-interaction, consequently inducing an excessive delocalization of the wave functions*" (Tolba et al. 2018).

In the framework of DFT, band structures generated by LDA suffer from the drastic approximation in the xc energy, which is determined only from the local electron density $n(\mathbf{r})$ (Imada et al. 1998).[7]

Even the Local Spin Density Approximation (LSDA; cf. Sects. 10.1 and 10.2) fails to reproduce the conductive properties (and, in some cases, the ground state) in materials possessing a strong electronic correlation. 3d transition metal oxides such as CoO and FeO are predicted by LSDA to be metals, while they are insulators (with developed correlation gaps) and the LSDA-predicted gap for NiO and MnO is by far too small with respect to what experimentally observed (see Anisimov et al. 1991; and references therein).

In the sixties of the twentieth-century Hubbard developed a seminal treatment of electron correlation in narrow s- (Hubbard 1963) and d- and f-bands of transition and rare earth metals and their alloys and salts. The Hubbard model is now at the base of the DFT+*U* treatment of Mott–Hubbard insulators (Fig. 10.8).

In a Mott–Hubbard insulator the d band is split into lower and upper Hubbard bands, separated by the on-site d–d Coulomb interaction *U*. Charge excitations across that band gap involves a charge transfer of the type of Eq. 10.7.1, at the Energy cost *U*, which may be generalized as (Imada et al. 1998):

$$(d^n)_1 + (d^n)_1 \rightarrow (d^{n-1})_1 + (d^{n+1})_1 \tag{10.7.3}$$

The two-components Hubbard Hamiltonian for an m-fold degenerate atomic electron shell, such as a 3d shell is (Hubbard 1964):

[7] When the electron density n(**r**) has a marked spatial dependence, such as in the case of strongly fluctuating spin and charge degrees of freedom in correlated metals, this approximation breaks down (Imada et al. 1998).

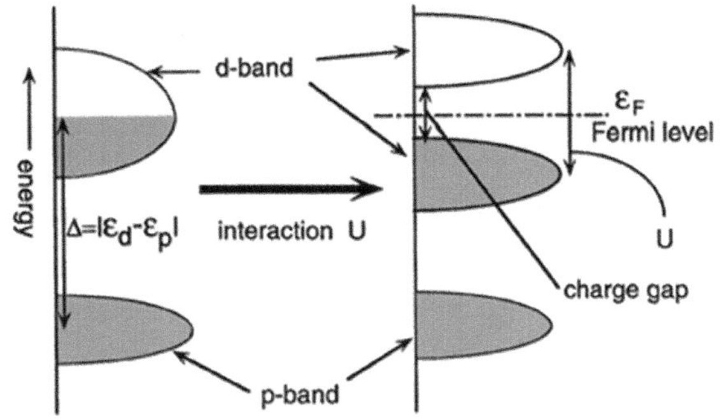

Fig. 10.8 Schematic illustration of energy levels for a Mott–Hubbard insulator. Modified from Imada et al. (1998)

$$H_{Hub} = \sum_{i,j}\sum_{\mu,\nu} T_{i,j}^{\mu,\nu} c_{i\mu}^{+} c_{j\nu} + \frac{1}{2}\sum_{i,j,k,l}\sum_{\mu,\nu,\sigma,\tau} (i\mu j\nu|1/r|k\sigma l\tau) c_{i\mu}^{+} c_{j\nu}^{+} c_{k\sigma} c_{l\tau} \tag{10.7.4}$$

μ in Eq. 10.7.4 is a label distinguishing degenerate wave functions of the shell and including the spin label; $c_{i\mu}$ is the destruction operator for an electron in the state μ on the atom labeled i, and $c_{i\mu}^{+}$ is the corresponding creation operator. The first term in Eq. 10.7.4 is just the Hamiltonian for the electrons moving in the appropriate potential field with

$$T_{1,j}^{\mu\nu} = \int \phi_{\mu}^{*}(\mathbf{x} - \mathbf{R}_1)\left[-\frac{\hbar^2}{2m}\nabla^2 + V\right]\varphi_{\nu}(\mathbf{x} - \mathbf{R}_i)d\mathbf{x} \tag{10.7.5}$$

$\phi_\nu(\mathbf{x}\text{-}\mathbf{R}_i)$ represents here a degenerate 3d function in the state ν for atom i at position $\mathbf{R}_i$ and V is the nuclear potential acting on the electrons together with the self-consistent potential due to electrons in other bands.

The second term represents the interactions of the electrons of the band, where $(i\mu j\nu|1/r|k\sigma l\tau)'$ are the matrix elements of the electrostatic interaction, which would be used in a calculation of the energy levels of the atomic spectrum (Hubbard 1964):

$$(i\mu j\nu|1/r|k\sigma l\tau) = e^2 \int \frac{\varphi_{\mu}^{*}(\mathbf{x} - \mathbf{R}_i)\varphi_{\sigma}(\mathbf{x} - \mathbf{R}_k)\varphi_{\nu}^{*}(\mathbf{x}' - \mathbf{R}_j)\varphi_{\tau}(\mathbf{x}' - \mathbf{R}_l)}{\mathbf{x} - \mathbf{x}'} d\mathbf{x}d\mathbf{x}' \tag{10.7.6}$$

Neglecting all the interaction terms in Eq. 10.7.4, apart from the terms for which i = j = k = l, Hubbard obtained the approximate Hamiltonian:

$$H_{Hub} = \sum_{i,j}\sum_{\mu,\nu} T_{i,j}^{\mu,\nu} c_{i\mu}^{+} c_{j\nu} + \frac{1}{2}\sum_{i}\sum_{\mu,\nu,\sigma,\tau} (i\mu j\nu|1/r|k\sigma l\tau) c_{i\mu}^{+} c_{j\nu}^{+} c_{k\sigma} c_{l\tau} \tag{10.7.7}$$

In its simplest "*one-band incarnation*" (Cococcioni 2012), the Hubbard Hamiltonian reduces to:

$$H_{Hub} = t \sum_{\langle i,j \rangle, \sigma} \left(c^{+}_{i,\sigma} c_{j,\sigma} + H.c. \right) + U \sum_{i} n_{i,\uparrow} n_{i,\downarrow} \tag{10.7.8}$$

where $\langle i,j \rangle$ denotes nearest-neighbors and $n_{i,\uparrow}, n_{i,\downarrow}$ are number operators for electrons of spin $\uparrow, \downarrow$ on site i, and H.c. is the Hermitian conjugate.[8] The first term in Eq. 10.7.8 accounts for the hopping energy between neighboring sites. The amplitude *t* is proportional to the bandwidth (dispersion) of the valence electronic states, and, in practice, it represents the "*single-particle term of the total energy*" (Cococcioni 2012). The second term, proportional to the product of the occupation numbers of atomic states on the same site, represents the Coulomb repulsion between electrons of the same atom, with strength *U* and corresponds in practice to the energy cost of double occupancy of the same site (i.e. the Hubbard *U*). When single-particle terms of the energy are overcome by short-range Coulomb interactions ($t << U$) the system becomes insulator (even at half-filling conditions, when band theory would predict a metal): "*the electrons cannot hop around because they don't have sufficient energy to overcome the repulsion from other electrons on neighbor sites*" (Cococcioni 2012). Whenever single-particle terms of the energy dominate with respect to Coulomb repulsion between electrons of the same atom ($t >> U$) even simple DFT without Hubbard terms are appropriate.

With the acronymic DFT+*U* one defines a *U* correction applied to a generic approximate DFT functional operating in Local or General Gradient approximations (i.e. LDA, GGA). The computational cost of DFT+*U* is not much higher than that of normal DFT computations. The Hubbard Hamiltonian operates only for strongly correlated electronic states (such as localized 3d or 4f orbitals), while the rest of the valence electrons are treated at the "*standard*" level of approximation.

When operating under LDA the total energy of a system can be written as follows (Cococcioni 2012; Himmetoglu et al. 2013):

$$E_{LDA+U}[\rho(\mathbf{r})] = E_{LDA}[\rho(\mathbf{r})] + E_{Hub}\left[\left\{n^{1\sigma}_{mm'}\right\}\right] - E_{dc}\left[\left\{n^{1\sigma}\right\}\right] \tag{10.7.9}$$

where $n^{i\sigma}_{mm'}$ are the occupation numbers of localized orbitals identified by the atomic site index i, state indexes m, m′ and spin σ; E_{Hub} is the term that contains electron–electron interactions as modeled in the Hubbard Hamiltonian and E_{dc} is a "*double-counting*" (dc) term that models the contribution to the DFT energy from correlated electrons as a mean-field approximation to E_{Hub}. This term must be subtracted from E_{Hub} because already present in E_{LDA}. Due to the lack of a precise diagrammatic expansion of the DFT total energy, the E_{dc} term is not uniquely

[8] The Hamiltonian must be Hermitian: if there is an amplitude for hopping between i and j nearest neighbors, then there must be also an amplitude for the reverse process (i.e. hopping between j and i)., and the H.c. term is compulsory.

defined, and different possible formulations and implementations exist (see Cococcioni 2012; Himmetoglu et al. 2013; for an extensive treatment).

As anticipated, the Hubbard correction is only applied to the localized states of the system, most affected by correlation effects. Their occupation numbers are defined as projections of occupied Kohn–Sham orbitals $\left(\psi_{\mathbf{k}v}^{\sigma}\right)$ on the states of a localized basis set $\left(\phi_{m}^{i}\right)$:

$$n_{mmm'}^{i\sigma} = \sum_{\mathbf{k},v} f_{\mathbf{k}v}^{\sigma} \langle \psi_{\mathbf{k}v}^{\sigma} | \phi_{m'}^{i} \rangle / \phi_{m}^{i} | \psi_{\mathbf{k}v}^{\sigma} \rangle \tag{10.7.10}$$

$f_{\mathbf{k}v}^{\sigma}$ in Eq. 10.7.10 are the Fermi–Dirac occupations of the Kohn–Sham (KS) states (with $\mathbf{k}$ as k-point and v as band index).

An extensive review of the performance of the DFT+U functionals has been recently proposed by Tolba et al. (2018) to whom I remand the interested readers. In this book we will see some application of DFT+U to the orthosilicates Fayalite (Sect. 20.4.2.1) and Ahrensite (Sect. 20.4.2.2).

GPSR Compliance

The European Union's (EU) General Product Safety Regulation (GPSR) is a set of rules that requires consumer products to be safe and our obligations to ensure this.

If you have any concerns about our products, you can contact us on ProductSafety@springernature.com

In case Publisher is established outside the EU, the EU authorized representative is:

Springer Nature Customer Service Center GmbH
Europaplatz 3
69115 Heidelberg, Germany

Batch number: 10732594

Printed by Printforce, the Netherlands

Springer Geochemistry

The Springer Geochemistry series seeks to publish a broad portfolio of scientific books, aiming at researchers, students, and everyone interested in geochemistry. The series includes peer-reviewed monographs, edited volumes, textbooks, and conference proceedings. It covers the entire research area including, but not limited to, Isotope Geochemistry, Biogeochemistry, and Environmental Geochemistry.

Giulio Armando Ottonello

Quantum Geochemistry

Volume 2

Springer

Giulio Armando Ottonello
Fellow of the Mineralogical Society
of America
Corresponding Member of the Accademia
Nazionale dei Lincei
Rome, Italy

ISSN 2366-6285 ISSN 2366-6293 (electronic)
Springer Geochemistry
ISBN 978-3-031-21836-1 ISBN 978-3-031-21837-8 (eBook)
https://doi.org/10.1007/978-3-031-21837-8

This Springer imprint is published by the registered company Springer Nature Switzerland AG
The registered company address is: Gewerbestrasse 11, 6330 Cham, Switzerland

Paper in this product is recyclable.

In Memory of Elisabeth

Preface

Warum ich zuletzt am liebsten mit der Natur verkehre, ist,weil sie immer recht hat und der Irrtum bloß auf meiner Seite sein kann

This beautiful aphorism is due to Goethe (Natur Wissenschaftliche Schriften, Band 1, Allgemeines Aphorismen: Warheit und Irrtum; Leipzig Im Insel Verlag, 1946) "*Why, at the very end, I like to traffic with the things of Nature? Because I know She is always right and the Error can be only by my side*". This is the leading concept that a Geochemist should always keep in his mind when trying to translate in formal reactions the "*facts*" that he "*sees*" happening on Earth, and, more generally in the Cosmos.

But "*seeing*" is not synonymous of "*looking at*" because epistemology requires the existence of "*Observables*" and the (intellectual) capability to interpret them. The Observables we encounter in this book are those masterfully described by Dirac in "*The Principles of Quantum Mechanics*" (Chap. 1, Pars. 10 and 11). In Par. 10, Dirac notes "*We have made a number of assumptions about the way in which states and dynamical variables are to be represented mathematically in the theory. These assumptions are not, by themselves, laws of nature, but become laws of nature when we make some further assumptions that provide a physical interpretation of the theory. Such further assumptions must take the form of establishing connexions between the results of observation, on one hand, and the equations of the mathematical formalism on the other*" (Dirac, 1957).

The critical point is thus the capability of "*establishing connections*": we must first "*see*" (experimentally, conceptually, not just physically) the facts of Nature, and then translate them into a mathematical formalism (which requires not only some smartness but also a sound knowledge of the most advanced theories).

If this book helps some young Geochemists to develop the idea that Quantum Mechanics is essential to reach this task, then I did partly fulfill my objective.

The work is divided into three main parts: "*Principles*", "*Methods*" and "*Applications*". The Principles may be identified with the "*Assumptions*" in Dirac's definition, the Methods are the "*Connexions*", and the "*Applications*" must be taken as a promotional exercise.

Renato Colle, Carlo Gatti and Benedetta Mennucci reviewed selected sections of the book. I gratefully acknowledge their help, and also, indirectly, the help of all co-authors with whom I cooperated in the last two decades. I take full responsibility for all errors and misinterpretations that might still be present in the text. Humble apologies for the unavoidable self-referencing, which sound disturbing not only for the reader but for the writer as well.

Calci (PI), Italy Giulio Armando Ottonello

Contents

Part II Methods

Part III Applications

Chapter 11
Electron Correlation Methods

This chapter is essentially an introductory treatment of the advanced procedures apt to solve molecular equilibrium structures and to characterize at best their ground state energy and the energy of excited states as well: the Møllller Plesset (MP) Perturbation theory and the Configuration Interaction (CI) procedures. We see then some aspects of the Complete Basis Set (CBS) Extrapolation, necessary to avoid truncation effects in the choice of the one-electron basis set.

11.1 Møllller Plesset Perturbation

Møllller and Plesset (1934) developed a perturbation theory for a many electrons system in which the unperturbed HF solution appears as the zero-order approximation. In this development the first-order correction for the energy and the charge density of the system turns out to be zero, hence, the second-order perturbation (MP2) is the first effective correction to HF results.

Assigning to a system of n electron the HF determinantal form

$$\Phi^0 = \frac{1}{\sqrt{n!}}\begin{vmatrix} \phi_1(q_1) & \phi_1(q_2) & \cdots & \phi_1(q_n) \\ \phi_2(q_1) & \cdots & \cdots & \cdots \\ \cdots & \cdots & \cdots & \cdots \\ \phi_n(q_1) & \cdots & \cdots & \phi_n(q_n) \end{vmatrix} \tag{11.1.1}$$

where the q's accounts for space and spin coordinates, the $\Phi_r(n)$ functions are orthogonal-normalized solutions of the equation

$$i\hbar\frac{\partial}{\partial t}\phi_r(q) = (H_0 + B - A)\phi_r(q) \tag{11.1.2}$$

H_0 is the Hamiltonian in an external potential field V.

G. A. Ottonello, *Quantum Geochemistry*, Springer Geochemistry,
https://doi.org/10.1007/978-3-031-21837-8_11

The equation of motion for a density matrix ρ obeying

$$\langle q|\rho|q'\rangle = \sum_{r=1}^{n} \phi_r(q)\phi_r^*(q') \tag{11.1.3}$$

is

$$i\hbar\rho = (H_0 + B - A)\rho - \rho(H_0 + B - A) \tag{11.1.4}$$

Assuming that H_0 does not contain the time explicitly it is equivalent to assume that we are in a stationary state μ, such that (Mølller and Plesset 1934):

$$F_\mu \phi_r^{(\mu)}(q) \equiv (H_0 + B_\mu - A_\mu)\phi_r^{(\mu)}(q) \tag{11.1.5}$$

$$F_\mu \rho_\mu - \rho_\mu F_\mu = 0 \tag{11.1.6}$$

where F_μ is the one-electron operator and ρ_μ is the charge density. The energy of the system is, in this approximation:

$$E_\mu^{(0)} = D\left\{\rho\left(H_0 + \frac{1}{2}B_\mu - \frac{1}{2}A_\mu\right)\right\} \tag{11.1.7}$$

where D denotes the diagonal sum.

The corresponding wave function of the stationary state is given by the summation $\sum_{i=1}^{n} F_\mu^{(i)}$ of the operators:

$$F_\mu = H_0 + B_\mu - A_\mu \tag{11.1.8}$$

operating only on functions of the coordinates of the ith electron.

Denoting:

$$\lambda_\mu = \sum_{r=1}^{n} \lambda_r^{(\mu)} \tag{11.1.9}$$

through various passages it is obtained, from Eq. (11.1.7):

$$\lambda_\mu = E_\mu^{(0)} + \frac{1}{2}D[\rho_\mu(B_\mu - A_\mu)] \tag{11.1.10}$$

$$H_\mu \Phi_\mu^{(0)} = E_\mu^{(0)} \tag{11.1.11}$$

Considering the HF approximation as the zero-order approximation to the exact theory, and the deviation Γ_μ of the Hartree-Fock operator H_μ from the exact operator K as a small perturbation, the exact equation is:

$$(\mathsf{H}_\mu + \Gamma_\mu)\Phi_\mu = E_\mu \Phi_\mu \quad (11.1.12)$$

with

$$\Gamma_\mu = \mathcal{K} - \mathsf{H}_\mu \quad (11.1.13)$$

The exact eigenfunction Φ_μ, supposed to lie near the HF solution $\Phi_\mu^{(0)}$, and the exact eigenvalue E_μ, supposed to lie near the corresponding $E_\mu^{(0)}$, are expanded in the following series:

$$\Phi_\mu = \Phi_\mu^{(0)} + \Phi_\mu^{(1)} + \Phi_\mu^{(2)} + \ldots \quad (11.1.14.1)$$

$$E_\mu = E_\mu^{(0)} + E_\mu^{(1)} + E_\mu^{(2)} + \ldots \quad (11.1.14.2)$$

Introducing the above expansions into the exact equation (11.1.8) and collecting the terms of the same magnitude it is found, at the light of the Rayleigh–Schrödinger perturbation theory:

$$\mathsf{H}_\mu \Phi_\mu^{(0)} = E_\mu^{(0)} \Phi_\mu^{(0)} \quad (11.1.15)$$

$$\left[\mathsf{H}_\mu - E_\mu^{(0)}\right]\Phi_\mu^{(1)} = \left[E_\mu^{(1)} - \Gamma_\mu\right]\Phi_\mu^{(0)} \quad (11.1.16)$$

$$\left[\mathsf{H}_\mu - E_\mu^{(0)}\right]_\mu^{(2)} = E_\mu^{(2)}\Phi_\mu^{(0)}\left[E_\mu^{(1)} - \Gamma_\mu\right]\Phi_\mu^{(1)} \quad (11.1.17)$$

Considering the complete set of eigenfunctions $\Phi_{\mu\nu}^{(0)}$ of the HF operator H_μ associated with a definite zero-order state μ:

$$\mathsf{H}_\mu \Phi_{\mu\nu}^{(0)} = E_{\mu\nu}^{(0)} \Phi_{\mu\nu}^{(0)} \quad (11.1.18)$$

and choosing ν in such a way that $\Phi_{\mu 0}^{(0)} = \Phi_\mu^{(0)}$, the functions $\Phi_\mu^{(1)}, \Phi_\mu^{(2)}, \ldots$ may be expressed in terms of the complete set of functions $\Phi_{\mu\nu}^{(0)}$, i.e.:

$$\Phi_\mu^{(1)} = \sum_\nu a_{\mu\nu}^{(1)} \Phi_{\mu\nu}^{(0)} \quad (11.1.19.1)$$

$$\Phi_{\mu}^{(2)} = \sum_{\nu} a_{\mu\nu}^{(2)} \Phi_{\mu\nu}^{(0)} \tag{11.1.19.2}$$

$$.... = ... \tag{11.1.19.3}$$

substituting Eq. (11.1.19.1) in (11.1.16), one has, at the light of (11.1.18), (Mølller and Plesset 1934):

$$\sum_{\nu} A_{\mu\nu}^{(1)} \left(E_{\mu\nu}^{(0)} - E_{\mu}^{(0)} \right) \Phi_{\mu\nu}^{(0)} = \left(E_{\mu}^{(1)} - \Gamma_{\mu} \right) \Phi_{\mu}^{(0)} \tag{11.1.20}$$

Expanding the right side of Eq. (11.1.20) in terms of the functions $\Phi_{\mu\nu}^{(0)}$ and equating the corresponding coefficients, for $\nu = 0$, it is obtained the first-order correction:

$$E_{\mu}^{(1)} = \int \ldots \int \Phi_{\mu}^{(0)*} \Gamma_{\mu} \Phi_{\mu}^{(0)} dq \tag{11.1.21}$$

and, for $\nu \neq 0$:

$$a_{\mu\nu}^{(1)} = \frac{\int \ldots \int \Phi_{\mu}^{(0)*} \Gamma_{\mu} \Phi_{\mu}^{(0)} dq}{E_{\mu}^{(0)} - E_{\mu\nu}^{(0)}} \tag{11.1.22}$$

The normalization condition for Φ_{μ} gives moreover:

$$a_{\mu\nu}^{(1)} = 0 \tag{11.1.23}$$

Through various passages it is then shown that the first-order correction $E^{(1)} = 0$ and this means that "*the perturbation method shows that the theory of the self-consistent field including exchange is accurate in the determination of energy to the second approximation*" (Mølller and Plesset 1934).

To understand better the significance of what is stated we must consider again the Hellmann-Feynmann theorem, which equates the derivative of the total energy to the unperturbed Hamiltonian (Sects. 7.2.2,3; see also Raghavachari and Pople 1981); i.e:

$$\left(\frac{dE_{\mu}}{d_{\mu}} \right)_{\mu=0} = \left\langle \Phi^{(0)} \mid \Gamma_0 \middle| \Phi^{(0)} \right\rangle \tag{11.1.24}$$

with the term on the right representing the unperturbed wave function.

The second-order correction in the energy was obtained by Mølller and Plesset (1934) in the same manner of the first-order correction, by substituting Eq. (11.1.19.2) in (11.1.17):

$$E_{\mu}^{(2)} = \sum_{v \neq 0} \frac{\int \ldots \int \Phi_{\mu}^{(0)*} \Gamma_{\mu} \Phi_{\mu v}^{(0)} \, dq \cdot \int \ldots \int \Phi_{\mu v}^{(0)*} \Gamma_{\mu} \Phi_{\mu}^{(0)} dq \cdot}{E_{\mu}^{0} - E_{\mu v}^{0}} \tag{11.1.25}$$

The functions $\Phi_{\mu v}^{(0)}$ were obtained from the determinant

$$\Phi_{\mu}^{(0)} = \frac{1}{\sqrt{n!}} \begin{vmatrix} \phi_1^{(\mu)}(q_1) & \phi_1^{(\mu)}(q_2) & \ldots & \phi_1^{(\mu)}(q_n) \\ \phi_2^{(\mu)}(q_1) & \ldots & \ldots & \ldots \\ \ldots & \ldots & \ldots & \ldots \\ \phi_n^{(\mu)}(q_1) & \ldots & \ldots & \phi_n^{(\mu)}(q_n) \end{vmatrix} \tag{11.1.26}$$

by replacing one or more of the functions $\Phi^{(\mu)}$ by functions $\psi^{(\mu)}$.

Equation (11.1.24) reduces to:

$$E^{(2)} = \sum_{rs:p\sigma} \frac{\left| \iiint \int \psi_{\rho}^{*}(q) \psi_{\sigma}^{*}(q')(qq'|V|q''q''')\{\varphi_r(q'')\varphi_s(q''') - \varphi_r(q''')\varphi_s(q'')\} dq dq' dq'' dq''' \right|^2}{\lambda_r + \lambda_s - \lambda_{\rho} - \lambda_{\sigma}} \tag{11.1.27}$$

which can be eventually rewritten in the usual form:

$$E^{(2)} = -\frac{1}{4} \sum_{ij}^{virtual} \sum_{ab}^{occ} \frac{|\langle rs|V|\rho\sigma\rangle|}{\epsilon_i + \epsilon_j - \epsilon_a - \epsilon_b} \tag{11.1.28}$$

where i and j replace r and s; a and b replace ρ and σ; and ε stands for λ.

A systematic investigation on the accuracy of molecular equilibrium structures of 19 small closed-shell molecules containing first-row atoms (HF, H_2O, NH_3, CH_4, N_2, CH_2, CO, HCN, CO_2, HNC, C_2H_2, CH_2O, HNO, N_2H_2, O_3, C_2H_4, F_2, HOF, H_2O_2) carried out by Helgaker et al. (1997) has shown that using the correlation-consistent polarized valence triple-zeta (cc-pVTZ) basis, set the MP2 geometries are remarkably accurate, with a mean deviation Δ relative to experiment in the calculated bond distances of −0.0015 *Å* (calculated—observed). Somewhat surprisingly, using the more extended cc-PVQZ basis set, the predictions do not improve (Δ = −0.0026 *Å*) and with the cc-PVQZ basis set the bond distances are decidedly overestimated (Δ = 0.0129 *Å*). In any case, the MP2 procedure greatly ameliorates the HF results, which are severely underestimated (Δ = −0.0201, −0.0280, −0.029 for HF/cc-pVDZ, HF/cc-pVTZ, HF/cc-pVQZ, respectively).

In Fig. 11.1 are plotted the differences between the calculated and experimental bond lengths for the different wave functions calculated at HF and MP2 theory level

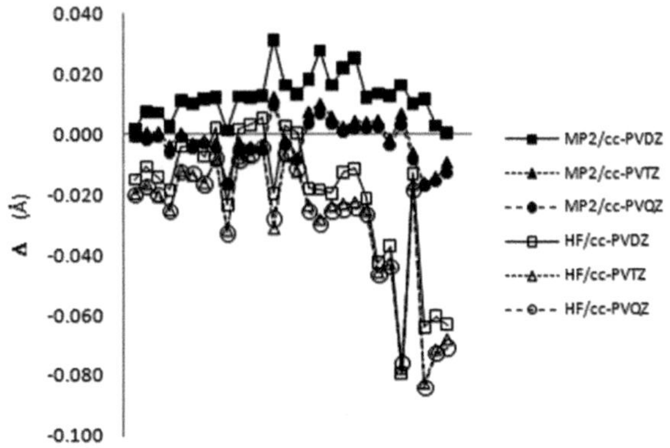

Fig. 11.1 Errors relative to experiment in the calculated bond lengths at the MP2 and HF theory level with cc-pVDZ, cc-pVTZ and cc-pVQZ basis functions. Modified from Helgaker et al. (1997)

Table. 11.1 Errors in the ground state electronic energy for the investigated molecules (calculated—observed). Data in *Hartree.*

Molecule	cc-PVDZ	cc-PVTZ	cc-PVQZ
HF	0.22914	0.12040	0.08499
H_2O	0.18833	0.09834	0.06936
HOF	0.41134	0.23206	–
H_2O_2	0.37295	0.20533	0.15372
HNC	0.25428	0.16891	–
NH_3	0.14791	0.07743	0.05576
N_2H_2	0.28094	0.17158	0.13532
C_2H_2	0.22701	0.15049	0.12622
HNO	0.32239	0.19736	–
HCN	0.27779	0.19242	–
C_2H_4	0.22273	0.13872	0.11274
CH_4	0.11061	0.05912	0.04448
N_2	0.27470	0.17989	0.14743
CH_2O	0.28854	0.17458	0.13707
CH_2	0.09862	0.06242	0.06242
CO	0.28397	0.18512	0.15087
CO_2	0.45450	0.27968	0.22002
O_3	0.52791	0.30928	0.23717
F_2	0.44760	0.25262	0.18930
Mean Δ	0.28533	0.17136	0.12846

by Helgaker et al. (1997). In this plot, the 28 distinct bonds found in the 19 sample molecules are arranged in order of increasing experimental bond length.

In Table 11.1 are listed the errors in the electronic energy at the ground state for the same investigated molecules. In estimating the errors, I adopted the NIST data for the basis set cc-pVDZ, cc-pVTZ, and cc-pVQZ at the MP2 theory level and the zero-point energies of the 19 molecules, then I subtracted the experimental atomization energy and the exact atomic energies of Davidson et al. (1991). As expected, the errors progressively reduce with the increase in the number of adopted Gaussians.

A third-order Møller and Plesset correlation enegy (MP3) may be obtained by applying a Rayleigh-Schrödinger third-order perturbation (Bartlett and Silver 1975; Pople et al. 1986):

$$E_{MP3} = E_{MP2} + E^{(3)} \tag{11.1.28}$$

$$\begin{aligned} E^{(3)} = & \frac{1}{8}\sum_{ij}^{virtual}\sum_{abcd}^{occ} a^*_{ijab} a_{ijcd}(ab||cd) + \\ & \frac{1}{8}\sum_{ijkl}^{virtual}\sum_{ab}^{occ} a^*_{ijab} a_{klabd}(kl||ij) + \\ & \frac{1}{8}\sum_{ijk}^{virtual}\sum_{abc}^{occ} a^*_{ijab} a_{ikjcb}(ak||ic) \end{aligned} \tag{11.1.29}$$

MP3 correlation energies compared with various Configuration Interaction theories are shown in (Table 11.2).

11.2 Configuration Interaction

With the term "*configuration*", it is meant the particular linear combination of Slater's determinants adopted to conform a wave function. With "*interaction*" it is meant that the occupation arises actually from the selected configurational state functions (CSFs) built from spin orbitals. If the expansion accounts of all the possible CSF of appropriate symmetry, then it is an "*extended (or full) configuration interaction*". We encountered a first time the concept of "*configuration interaction*" when discussing the possible ways of combining the functions u_g and u_u with the spin functions α or β in forming the "*wave functions of Configurations*" for the 1s hydrogen orbitals (see Par 7.3 and Table 7.6). We have also seen that, in the particular case of O_2 at ground state, the $1\sigma_g$, $1\sigma_u$, $2\sigma_g$, $2\sigma_u$ and 3σ spin orbitals are fully occupied with two electrons. Moreover, there are four electrons in $1\sigma_u$ but only two in the π_g, which is capable of hosting four electrons. There are thus 4!/(2! 2!) possible ways of electrons to occupy the spin orbitals (Table 7.9), resulting in six determinantal functions ψ to be suitably combined to depict the contribution of the $1\pi_g$ to the ground state energy arising from:

$$\Psi = \sum_{i=1}^{6} c_i \psi_i \tag{7.4.2}$$

Table. 11.2 Correlation energies obtained with various CI procedures for some selected molecules. The adopted basis set is 8-41 for the elements Li to He and 4-1 fo H. Data in *Hartree*. MP3 results after Pople et al. (1976). All other procedures from Pople et al. (1977).

Molecule	MP3	CIS (Cycle 1)	CID	CISD	Size correlation
Li	−0.03221	−0.03213	−0.03276	−0.03281	−0.03284
Be (^1S)	−0.07064	−0.06866	−0.07527	−0.07562	−0.07870
Be (^3P)	−0.03738	−0.03726	−0.03820	−0.03824	−0.03837
B (^2P)	−0.08862	−0.08646	−0.09363	−0.09430	−0.09794
B (^4P)	−0.04724	−0.04699	−0.04835	−0.04841	−0.04870
C (^3P)	−0.10925	−0.10688	−0.11231	−0.11283	−0.11618
C (^5S)	−0.06410	−0.06361	−0.06486	−0.06494	−0.06545
N	−0.13102	−0.12843	−0.13094	−0.13118	−0.13387
O	−0.16794	−0.16424	−0.16622	−0.16660	−0.17012
F	−0.20982	−0.20472	−0.20641	−0.20684	−0.21123
Ne	−0.25500	−0.12483	−0.25038	−0.25088	−0.25610
H_2	−0.03383	−0.03352	−0.03578	−0.03595	−0.03595
LiH	−0.06056	−0.05968	−0.06226	−0.06263	−0.06351
BeH	−0.07010	−0.06907	−0.07199	−0.07242	−0.07374
BH	−0.11740	−0.11353	−0.12053	−0.12136	−0.12764
CH($^2\Pi$)	−0.14169	−0.13730	−0.14283	−0.14379	−0.14979
CH($^4\Sigma^-$)	−0.11546	−0.11291	−0.11562	−0.11679	−0.11991
NH	−0.16798	−0.16297	−0.16596	−0.16691	−0.17223
OH	−0.21513	−0.20766	−0.21036	−0.21120	−0.21802
FH	−0.26438	−0.02542	−0.25777	−0.25872	−0.26689
BeH_2	−0.09926	−0.09690	−0.10055	−0.10080	−0.10354
BH_2	−0.12179	−0.11842	−0.12218	−0.12271	−0.12677
CH_2 ($^3B_1$)	−0.15339	−0.14851	−0.15168	−0.15257	−0.15800
CH_2 ($^1A_1$)	−0.17559	−0.16846	−0.17355	−0.17443	−0.18343
NH_2	−0.20929	−0.20057	−0.20387	−0.20483	−0.21337
OH_2	−0.26382	−0.25096	−0.25521	−0.25625	−0.26700

In more general terms, the problem may be outlined as done by McQuarrie (2008). If we have a closed shell molecule consisting of 2N electrons and we use a basis set of K basis function, the HF treatment will result in K molecular orbitals and 2K spin orbitals. The fist 2N orbitals of lowest energy will be occupied and the remaining 2(K-N) spin orbitals will remain unoccupied or "*virtual*". Denoting Ψ_0 the Slater determinant formed by the occupied orbitals and using it as reference state, all the excited configurations, formed by replacing a single occupied spin orbital with a virtual one, will be defined in terms of energy difference with this reference point. Using the standard notation and identifying the occupied spin orbitals as χ_i, χ_j, the virtual ones as χ_a,χ_b,... and as Ψ_i^a the "*singly excited*" determinant obtained by promoting an electron from the occupied spin orbital χ_i to the virtual spin orbital χ_a, we may moreover define a single excitation operator C_1, such that (McQuarrie 2008):

$$C_1\Psi_0 = \sum_i \sum_a c_i^a \Psi_i^a \qquad (11.2.1)$$

with the c_i^a constants determined variationally in order to achieve the lowest energy.

With a double substitution, it is obtained a doubly-excited configuration, and the summation of all of them leads to:

$$C_2\Psi_0 = \sum_{i<j} \sum_{a<b} c_{ij}^{ab} \Psi_{ij}^{ab} \qquad (11.2.2)$$

with the summations restricted to $i < j$ and $a < b$ in order to ensure that all possible configuration are accounted only one time.

The "*extended*" or "*full configuration interaction*" wave function Ψ_{CI} is obtained by extending the procedure to all 2N electrons:

$$\begin{aligned} \Psi_{CI} &= c_0\Psi_0 + C_1\Psi_0 + C_2\Psi_0 + \ldots + C_{2N}\Psi_0 = \\ c_0\Psi_0 &+ \sum_i \sum_a c_i^a \Psi_i^a + \sum_{i<j} \sum_{a<b} c_{ij}ab\Psi_{ij}ab + \ldots + \\ &\sum_{i<j<\ldots a<b<\ldots} \sum_{ij} c_{ij}^{ab\ldots} \Psi_{ij}^{ab\ldots} \end{aligned} \qquad (11.2.3)$$

with the last term involving "*the promotion of all 2N electrons from the occupied spin orbitals to all possible spin orbitals*" (McQuarrie 2008). The more the basis set is extended, the more the solution of the Schrödinger equations approach exactness. The computation becomes however the more expensive the larger is the adopted basis set (of the order $O(n^2N^4)$ nonzero elements, for a molecular n-electron system adopting a basis set with N elements).

11.2.1 Configuration Interaction Singles (CIS), Doubles (CID), and Singles-Doubles (CISD)

I will outline here briefly the procedure of Pople et al. (1977).

The definition (11.2.3) is rewritten in the form (Pople et al. 1977):

$$\Psi_{CI} = c_0\Psi_0 + \sum_i^{occ} \sum_a^{virt} c_{ia}\Psi_{i\to a} + \frac{1}{4}\sum_{ij}^{occ} \sum_{ab}^{virt} c_{ijab}\Psi_{ij\to ab} \qquad (11.2.4)$$

antisymmetric in the suffix pairs ij and ab. Moreover, it is denoted ψ_s (s $\geq$ 0) a many-electron basis function, and

$$H = H^{(0)} + V \tag{11.2.5}$$

the perturbed Hamiltonian, subject to the perturbation operator V, expandable in power series. The diagonal element E_s of the unperturbed Hamiltonian is (Pople et al. 1977):

$$E_0 = \sum_{k}^{occ} \epsilon_k \tag{11.2.6}$$

and those of the perturbed one are

$$E_{i\to a} = E_0 + (\epsilon_a - \epsilon_i) \tag{11.2.7.1}$$

$$E_{ij\to ab} = E_0 + \left(\epsilon_a + \epsilon_b - \epsilon_i - \epsilon_j\right) \tag{11.2.7.2}$$

The matrix elements (V_{st}) of the perturbation operator V are translated into a set of related matrix elements ($\overline{V}_{st}$):

$$\overline{V}_{st} = V_{st} - V_{00}\delta_{st} \tag{11.2.8}$$

The lowest energy expectation values of the non-relativistic Hamiltonian H are:

$$\epsilon = \frac{\int \Psi^* H \Psi d\tau}{\int \Psi^* \Psi d\tau} \tag{11.2.9}$$

Using Ψ_0 in the above equation, the initial unrestricted Hartree-Fock energy is:

$$E_{UHF} = E_0 + V_{00} \tag{11.2.10}$$

The CI equations are of type:

$$\sum_t H_{st} c_t = E c_s \quad (s = 0, 1, 2, \ldots) \tag{11.2.11}$$

and must be solved iteratively, posing

$$w_s = \sum_t \overline{V}_{st} c_t \tag{11.2.12}$$

$$(E_s + V_{00} - \epsilon)c_s + w_s = 0 \tag{11.2.13}$$

The iteration proceeds as follows (Pople et al. 1977):

- *Step 1*: at the iteration number m=1, it is defined a vector $c_s^{[m-1}$ that satisfies the normalization condition

$$\sum_s \left|c_s^{[m-1]}\right|^2 = 1 \tag{11.2.14}$$

and fulfills, at the starting point:

$$c_0^{[0]} = 1; \quad c_s^{[0]} = 0 \quad (s > 0) \tag{11.2.15}$$

where the index in square brackets is related to the iteration step

- *Step 2*: It is formed the wave vector w_s for this iteration:

$$w_s^{[m]} = \sum_t \overline{V}_{st} c_t^{[m-1]} \tag{11.2.16}$$

- *Step 3*: Because the wave function returning the expectation value (11.2.9) obeys:

$$\Psi = c_0\Psi_0 + \sum_{s>0} c_s\Psi_s \tag{11.2.17}$$

the variational energy $\mathcal{E}^{[m]}$ is obtained from the vector $c_s^{[m-1]}$:

$$\epsilon^{[m]} = \sum_s c_s^{[m-1]*}\left\{(E_s + V_{00})c_s^{[m-1]} + w_s^{[m]}\right\} \tag{11.2.18}$$

- *Step 4*: the next vector $c^{[m]}$ is found by applying:

$$c^{[m]} = \frac{w_s^{[m]}}{\epsilon^{[m]} - E_s - V_{00}} \tag{11.2.19}$$

- *Step 5*: the vector $c^{[m]}$ is multiplied by a normalization constant in order to fulfill eqn. (11.2.14).

In the first iteration cycle, the first four steps (i.e. before normalization) return:

$$c_0^{[0]} = 1; \quad c_s^{[0]} = 0 \tag{11.2.20.1}$$

$$w_0^{[1]} = 0; \quad w_s^{[1]} = \overline{V}_{s0} \tag{11.2.20.2}$$

$$\epsilon^{[1]} = E_0 + V_{00} \tag{11.2.20.3}$$

$$c_0^{[1]} = 1; \quad c_s^{[1]} = \frac{\overline{V}_{s0}}{E_0 - E_s} \tag{11.2.20.4}$$

We see thus that, at the end of the first cycle the energy is unperturbed and corresponds to the unrestricted HF energy $\mathcal{E}_{UHF}$ (Eq. 11.2.20.3) .

At the second iteration, the variational energy is:

$$\epsilon^{[2]} = E_{UHF} + \frac{\sum\limits_{s>0} \frac{V_{0s}V_{s0}}{E_0 - E_s} + \sum\limits_{s,t>0} \frac{V_{0s}\overline{V}_{st}V_{s0}}{(E_0 - E_s)(E_0 - E_t)}}{1 + \sum\limits_{s,t>0} \frac{V_{0s}V_{s0}}{(E_0 - E_s)^2}} \tag{11.2.21}$$

Table 11.2 resumes the results obtained by Pople et al. (1977) with the CIS, CID, and CISD procedures, compared with the results of a third-order Moller-Plesset expansion (MP3; Pople et al. 1976). The last column provides the results of a further "*size-consistent correction*", operating beyond CISD and based on:

$$\frac{\Delta E_{size-corr}}{\Delta E_{CISD}} = \frac{\sqrt{n^2 + 2n\tan^2 2\theta} - n}{2(\sec 2\theta - 1)} \tag{11.2.22}$$

where

$$\tan\theta = \frac{1}{x\sqrt{m}} \Delta E_{CISD} \tag{11.2.23}$$

$$m = n/2 \tag{11.2.24}$$

$$x = \langle \Psi_0 | H | \Psi_{SD} \rangle \tag{11.2.25}$$

The size-consistent correction is aimed to reducing the error induced by the fact that "*unlinked clusters do not cancel in limited CI expansions*" (Pople et al. 1977).

For small correlation corrections, Eq. (11.2.22) may be expanded in power series of θ:

$$\frac{\Delta E_{size-corr}}{\Delta E_{CISD}} = 1 + \left(1 - 2n^{-1}\right)\theta^2 + \ldots\ldots \tag{11.2.26}$$

The corresponding Langhoff-Davidson formula (Langhoff and Davidson 1974) gives:

$$\frac{\Delta E_{size-corr}}{\Delta E_{CISD}} = 1 + \theta^2 + \ldots\ldots \qquad (11.2.26)$$

which converges to (11.2.26) for large values of n.

In Table 11.3 are listed the total energies obtained at CID and CISD theory level by Pople et al. (1976), [all calculated on MP2 / (841/41) optimized geometries] and compared with the results of UHF calculations.

Table. 11.3 Total energies of the investigated molecules (Pople et al. 1977). Data in *Hartree.*

Molecule	UHF	CID	CISD
H	−0.49928	−0.49928	−0.49928
Li	−7.43178	−7.46454	−7.46458
Be (^{1}S)	−14.57124	−14.64650	−14.64685
Be (^{3}P)	−14.50407	−14.54227	−14.54231
B (^{2}P)	−24.52813	−24.62175	−24.62242
B (^{4}P)	−24.44694	−24.49529	−24.49534
C (^{3}P)	−37.68404	−37.79634	−37.79687
C (^{5}S)	−37.58882	−37.65368	−37.65376
N	−54.38785	−54.51879	−54.51904
O	−74.78785	−74.95407	−74.95445
F	−99.36960	−99.57601	−99.57644
Ne	−128.48170	−128.73208	−128.73258
H_2	−1.13111	−1.16689	−1.16706
LiH	−7.98292	−8.04519	−8.04555
BeH	−15.14378	−15.21577	−15.21620
BH	−25.12457	−25.24510	−25.24592
CH($^2\Pi$)	−38.27096	−38.41380	−38.41475
CH($^4\Sigma^-$)	−38.28024	−38.39586	−38.39703
NH	−54.96450	−55.13046	−55.13141
OH	−75.39109	−75.60145	−75.60229
FH	−100.01597	−100.27374	−100.27469
BeH_2	−15.76145	−15.86200	−15.86225
BH_2	−25.75347	−25.87566	−25.87618
CH_2 ($^3B_1$)	−38.92711	−39.07879	−39.07968
CH_2 ($^1A_1$)	−38.88058	−39.05412	−39.05501
NH_2	−55.56641	−55.77027	−55.77123
OH_2	−76.02567	−76.28088	−76.28192

For a more-in-depth treatment of CI procedures, I remand the interested reader to Foresman et al. (1992), Head-Gordon et al. (1994,1995), Pople et al. (1977), Krishnan et al. (1980b), Raghavachari and Pople (1981).

11.2.2 *Quadratic Configuration Interaction (QCI)*

Pople et al. (1987) introduced the Quadratic Configuration Interaction (QCI) as a new procedure devoted to minimizing the error caused by size inconsistency.

They defined this purpose, single, double, and triple substitution operators of form:

$$\mathsf{T}_1 = \sum_{ia} a_i^a t_i^a \tag{11.2.27.1}$$

$$\mathsf{T}_2 = \frac{1}{4}\sum_{ijab} a_{ij}^{ab} t_{ij}^{ab} \tag{11.2.27.2}$$

$$\mathsf{T}_3 = \frac{1}{36}\sum_{ijkabc} a_{ijk}^{abc} t_{ijk}{}^{abc} \tag{11.2.27.3}$$

where $t_i^a, t_{ij}ab, t_{ijk}^{abc}$ are "*elementary substitution operators*" and $a_i^a, a_{ij}^{ab}, a_{ijk}^{abc}$ are array coefficients antisymmetric in both the occupied (i, j, k,..) and virtual (a, b, c,...) indexes. In this sort of representation, $\mathsf{T}_2\psi_0$ is "*a general linear combination of all determinants obtained by double substitution in* ψ_0" (Pople et al. 1987).

Applying various functions of the T operators to ψ_0:

$$\Psi \equiv f(\mathsf{T}_1, \mathsf{T}_2, \ldots)\Psi_0 \tag{11.2.28}$$

and determining the unknown a coefficients by appropriate projection of the Schrodinger function (H – E)ψ, various types of antisymmetric wave functions can be obtained:

$$\langle \Psi_0 | \mathsf{H} - \mathrm{E} | \Psi \rangle = 0 \tag{11.2.29.1}$$

$$\left\langle \Psi_i^a | \mathsf{H} - \mathrm{E} | \Psi \right\rangle = 0 \tag{11.2.29.2}$$

$$\left\langle \Psi_{ij}^{ab} \mathsf{H} - \mathrm{E} | \Psi \right\rangle = 0 \tag{11.2.29.3}$$

H in the above equations is the full Hamiltonian and E is the energy, Ψ_i^a is the singly substituted determinant $\Psi_i^a = t_i^a\Psi_0$, $\Psi_{ij}^{ab} = t_{ij}^{ab}\Psi_0$, and so on.

Defining

$$\mathsf{H} = \mathsf{F} + \mathsf{V} \tag{11.2.30}$$

with F as Fock Hamiltonian [see also Eq. (11.2.5)], and posing:

$$E_{HF} = \langle \Psi_0 | \mathsf{H} - E | \Psi \rangle \tag{11.2.31.1}$$

$$E = E_{HF} + E_{corr} \tag{11.2.31.2}$$

with E_{corr} as correlation energy,

$$\overline{\mathsf{H}} = \mathsf{H} - E_{HF} \tag{11.2.31.3}$$

$$\overline{\mathsf{V}} = \mathsf{V} - \langle \Psi_0 | \mathsf{V} | \Psi \rangle \tag{11.2.31.4}$$

the QCI projection equations are (Pople et al. 1987):

$$E_{corr} = \langle \Psi_0 | \mathsf{H} | \mathsf{T}_2 \Psi_0 \rangle \tag{11.2.32.1}$$

$$a_i^a E_{corr} = \left\langle \Psi_i^a | \overline{\mathsf{H}} | (\mathsf{T}_1 + \mathsf{T}_2 + \mathsf{T}_1 \mathsf{T}_2) \Psi_0 \right\rangle \tag{11.2.32.2}$$

$$a_{ij}^{ab} E_{corr} = \Psi_{ij}^{ab} | \overline{\mathsf{H}} | \left(1 + \mathsf{T}_1 + \mathsf{T}_2 + \frac{1}{2} \mathsf{T}_2^2 \right) \Psi_0 \Big\rangle \tag{11.2.32.3}$$

These equations may be compared with the analogous CISD equations, in order to understand better the significance of "*size mismatch*":

$$E_{corr} = \langle \Psi_0 | \mathsf{H} | \mathsf{T}_2 \Psi_0 \rangle \tag{11.2.33.1}$$

$$a_i^a E_{corr} = \left\langle \Psi_i^a | \overline{\mathsf{H}} | (\mathsf{T}_1 + \mathsf{T}_2) \Psi_0 \right\rangle \tag{11.2.33.2}$$

$$a_{ij}^{ab} E_{corr} = \left\langle \Psi_{ij}^{ab} | \overline{\mathsf{H}} | (1 + \mathsf{T}_1 + \mathsf{T}_2) \Psi_0 \right\rangle \tag{11.2.33.3}$$

The left-hand sides of Eqs. (11.2.33.2,3) are quadratic in the a vectors and the right-hand sides are linear, while in the QCI Eqs. (11.2.32.2,3) are quadratic on both sides. In the case of two infinitely separated molecules X and Y, operating a double substitution ij→ab on X, the corresponding double equation (11.2.33.3) is:

$$a_{aj}^{ab}(X) E_{corr}(X) + a_{ij}^{ab}(X) E_{corr}(Y) = \left\langle \Psi_{ij}^{ab} | \overline{\mathsf{H}} | (1 + \mathsf{T}_1 + \mathsf{T}_2) \Psi_0 \right\rangle \tag{11.2.34}$$

with the second-term on the left implying that the value of the a_{ij}^{ab} coefficient on X depends on the nature of Y. The addiction of quadratic terms in T_1, T_2

(Eqs. 11.2.33.2,3) eliminates the term on the right involving E_{corr} and is the simplest way of correcting this type of error.

11.3 Complete Basis Set (CBS) Extrapolation

The major source of error in most *ab initio* calculations of molecular energies is the truncation of the one-electron basis set. The CBS procedure correcting for this error adds to molecular SCF energies small corrections based on an asymptotic convergence of the multicenter angular momentum expansion.

11.3.1 Effect of Angular Momentum in He-Like Atoms

Angular momentum provides a natural ordering parameter for a sequence of atomic one-electron basis sets. This has been stressed by Schwartz (1962) who developed formally a relation explaining how the correlation energy evolves when the azimuthal quantum number l increases from 0 to infinity (or, in practice, when successive s functions, p functions, d functions, f functions,…., are added in the formal calculus). For an hypothetical two-electron (i.e. helium-like) atom with a nucleus imposing a strong central charge Z, the ground state energy may be expanded as (Schwartz 1962):

$$E = +\left(\frac{e^2}{a_0}\right)\left(-Z^2 + \frac{5}{3}Z + E_2 + ...\right) \tag{11.3.1}$$

and the wave function as:

$$\psi = \psi_0 + \frac{\psi_1}{Z} + ... \tag{11.3.2}$$

The Coulomb interaction between the electrons is related to their mutual distance r_{12} and their angular position by:

$$\frac{e^2}{r_{12}} = e^2 + \sum_{l=0}^{\infty} P_l(\cos\theta)\frac{r_1^l}{r_2^{l+1}} \quad (r_1 < r_2) \tag{11.3.3}$$

where r_1 and r_2 are the distances of the two electrons from the central charge, θ is the angle between the two position vectors and the factor $P_l(\cos\theta)$ defines the angular structure.

Through the approximation:

$$\psi_1 \approx \left(\frac{r_{12}}{2a_0}\right)\psi_0 \tag{11.3.4}$$

and integrating over angles, Schwartz (1962) obtained:

$$\psi_1(l) = \frac{P_l(\cos\theta)}{2a_0}\psi_0 \frac{r_<^l}{r_>^{l+1}}\left[\frac{r_<^2}{2l+3} - \frac{r_>^2}{2l-1}\right] \tag{11.3.5}$$

For large l and small δ, setting $r_<^2 = r_>^2\,(1-\delta)$ this function is approximated by

$$-\left(\frac{r}{2a_0 l^2}\right)e^{-\delta l}(1+\delta l)\psi_0 P_l(\cos\theta) \tag{11.3.6}$$

which is sharply peaked around r_1=r_2 and falls off rapidly outside a distance $|(r_1 - r_2)/r_1| \approx l^{-1}$.

This implies a strong radial correlation and a sharp cusp in the second derivative (appearing in the variational treatment). Due to these two properties, attempting to represent $\psi_1(l)$ for large l through sums of product functions of type $u(r_1)v(r_2)$ is unpractical, because slowly convergent (Schwartz 1962).

However, computing the variational limit of the second-order energy E_2 through a variational coefficient λ_l, Schwartz (1962) showed that the angular component has for $l >> 1$, an upper bound:

$$-E_2(l) \geq \frac{45}{256}l^{-4} \tag{11.3.7}$$

and that:

$$\lim_{l\to\infty} \lambda_l = 1 \tag{11.3.8}$$

This fact, coupled with the peaking at $r_1 = r_2$ indicates that, besides being an upper bound, Eq. (11.3.7) is the correct asymptotic result, hence, it is also the CBS-limit for a helium-like atom with high nuclear charge. Besides emphasizing this important achievement, it is worth to stress that, according to Schwartz (1962), assuming that the correlation energy in He is simply the difference between the exact energy and the HF approximation is wrong because the HF method actually includes the major part of the radial interaction in an averaged way [i.e. −0.1110 *Hartree*; roughly 88% of the $E_2(0)$ according to Schwartz 1962]. Thus, what is generally referred to as radial correlation is only a relatively small part of the total wave function. To this purpose, it is interesting to note that, at MP2/631G(d) level, it is obtained for He an E_2 energy contribution (EUMP2) of −0.0112 *Hartree*, which is roughly 8.9% of the bulk, based on the estimates of Schwartz (1962).

11.3.2 Extension to Many-Electrons Atoms

For a many-electrons atom, the CBS computation is more complex because it involves pair-wise computations among the various electrons in the system. The computational procedure was assembled by Petersson and coworkers (Petersson 1977; Nyden and Petersson 1981; Petersson and Frish 2000; Petersson and Licht 1981; Petersson and Al-Laham 1991; Petersson et al. 1985, 1988) and involves recurrent MP2 calculations for each pair of electrons and a summation of the MP2 correlation pair energies to define the bulk amount at the investigated CBS level.

Denoting e_{ij} the pair energies between occupied orbitals (a, b) and virtual orbitals (i,j):

$$e_{ij} = \sum_{a,b}^{virt} C_{ij}^{ab} V_{ij}^{ab} \tag{11.3.9}$$

where C_{ij}^{ab} and V_{ij}^{ab} are the Hamiltonian matrix elements for the first-order Configuration Interaction (CI) coefficients:

$$C_{ij}^{ab} = \frac{V_{ij}^{ab}}{e_i + e_j - e_a - e_b} \tag{11.3.10}$$

and

$$V_{ij}^{ab} = \langle ij|\frac{1}{r_{12}}|ab\rangle \tag{11.3.11}$$

The bulk second-order energy is:

$$E_2 = \sum_{ij}^{occ} e_{ij} \tag{11.3.12}$$

Petersson and Nyden (1981) revealed the existence of an "*interference factor*" that reduces the effect of basis set truncation in the approach to the CBS limit. This interference factor is the square of the sum of the coefficients of the configurations that are included in the direct calculation:

$$\lim_{l \to \infty} \Delta e_j^{(\infty)} = \left[C_{SCF} + \sum_{a,b<l} C_{ijab} \right]^2 \Delta e_l \tag{11.3.13}$$

C_{SCF} in the above equation if the coefficient of the SCF configuration, and is approximately equal to 1. Since the other coefficients are small negative numbers, the interference factor varies between 0 and 1 (Nyden and Petersson 1981).

The net effect of the interference factor is a "*destructive interference between the SCF configuration and the low-lying excitations*" (Ochterski et al. 1996) and is a quantitative expression of the well-known observation that full CI energies converge more rapidly with basis set than MP2 energies.

Replacing each pair of occupied orbitals, i j, with Pair Natural Orbitals (PNO), which are the virtual orbitals that transform the matrix C_{ij}^{ab} to diagonal form, it is obtained the expansion for $\alpha\beta$ electron pairs (Petersson et al. 1985; Petersson and Al-Laham 1991):

$$e_{ij}^{\alpha\beta} = \sum_{a}^{PNO} C_{ij}aaV_{ij}^{aa} \tag{11.3.14}$$

Rearranging the terms in Eq. (11.3.14) in a monotonically decreasing order, gives the asymptotic form (Nyden and Petersson 1981):

$$\begin{matrix} \lim e_{ij}^{\alpha\beta}(N) \\ N \to \infty \end{matrix} = e_{ij}^{\alpha\beta}(CBS) + \left(\frac{25}{512}\right)\frac{f_{ij}^{\alpha\beta}}{(N + \delta_{ij})} \tag{11.3.15}$$

with

$$e_{ij}^{\alpha\beta}(N) = \sum_{a=2}^{N} C_{ij}^{aa}V_{ij}{}^{aa} \tag{11.3.16}$$

The overlap factor $f_{ij}^{\alpha\beta}$ in Eq. (11.3.15) provides the exact attenuation of the inter-orbital pairs relative to intra-orbital pairs, and it was developed in a model case study involving two infinitely separated He-like ions (Petersson et al. 1985):

$$f_{ij}^{\alpha\beta} = |S|_{ij}^{2} \tag{11.3.17}$$

$$|S|_{ij} = \int |\varphi_i(r)\varphi_j(r)| d\tau \tag{11.3.18}$$

$|S|_{ij}$ is the integral of the *absolute value* of the product of the molecular orbitals, or "*absolute overlap integral*" (Petersson 1977).

The parameter, δ_{ij} in Eq. (11.3.15) is a necessary modification to impose size consistency for finite N but it has no effect on the limit as N approaches infinity. This parameter is intimately linked to the basis set truncation error, which depends on the portion of the virtual space not included in the direct calculation and converges as $(N+\delta)^{-1}$ (Octherski et al. 1996).

The asymptotic convergence of the infinite-order (i.e. full CI) pair energies is given by

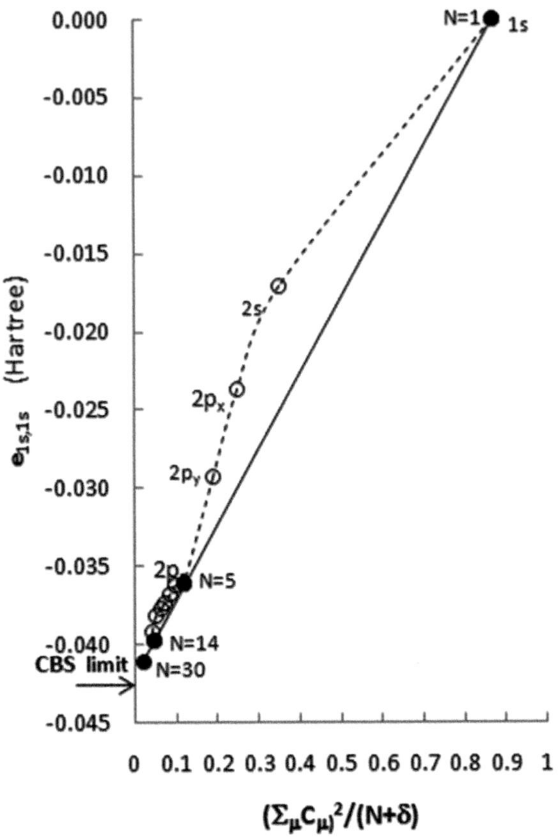

Fig. 11.2 The CBS limit construction in the case of Helium. The dashed curves demonstrate the typical behavior of partially completed shells (open circles). Only completed shells of Helium PNO's (N= 1,5,14,30; filled circles) are useful for extrapolations to the CBS limit. Modified from Ochterski et al. (1986)

$$\lim_{N \to \infty} e_{ij}^{\alpha\beta}(N) = e_{ij}^{\alpha\beta}(CBS) + \left[\sum_{a=1}^{N} C_{ij}^{a_i a_j}\right]^2 \left(\frac{25}{512}\right) \frac{f_{ij}^{\alpha\beta}}{(N + \delta_{ij})} \tag{11.3.19}$$

Drawing in a x-y binary plot $y = e_{ij}^{\alpha\beta}(CBS)$ with respect to $x = \left[\sum_{a=1}^{N} C_{ij}^{a_i a_j}\right] \frac{f_{ij}^{\alpha\beta}}{(N+\delta_{ij})}$ we effectively see in the case of He a straight correlation with slope (25/512) attaining the CBS limit at x=0 (Fig. 11.2)[1]. As noted by Ochterski et al. (1996), only completed shells of helium PNO's (N= 1,5,14,30) are useful for extrapolations to the CBS limit.

[1] It should be noted that in Nyden and Petersson (1981), the same slope (i.e. their eqn. 29) is represented as (225/4608).

Table. 11.4 CBS estimates (Petersson et al.,1988) of the total non-relativistic energy for closed-shell atoms and hydrides of first-row elements compared with experiments [Accad et al. (1975) for He, Viellard and Clementi (1968) for the remaining compounds]

Elements	CBS	Experimental	Δ
He	−2.90363	−2.903724	0.00009
Be	−14.66763	−14.66755	−0.00008
Ne	−128.93625	−128.9366	0.00035
Hydrides	CBS	Experimental	Δ
LiH	−8.06973	−8.07028	0.00055
BH_3	−26.59805	−26.598	−0.00005
CH_4	−40.51170	−40.514	0.00230
NH_3	−56.56590	−56.563	−0.00290
H_2O	−76.43819	−76.437	−0.00119
HF	−100.45614	−100.4563	0.00016

The correction to the second-order (MP2) correlation energy (ΔE_2) may be formally represented as:

$$\Delta E_2 = \sum_{ij} \left[e_{ij}(CBS) - e_{ij}(N) \right] \tag{11.3.20}$$

In Table 11.4 are listed CBS estimates of the total non-relativistic energy for closed-shell atoms and hydrides of first-row elements after Petersson et al. (1988). With the exception of NH_3 where the correlation energy is clearly overestimated (and, in a minor extent of H_2O), the CBS extrapolations agree with the experimental values to within the error limits of the experiments for all these atoms and molecules with more than four electrons.

By virtue of the interference factor correcting the CBS overestimation, an infinite order correction may be proposed for the approximate full CI energies MP4 or QCISD(T) (Pople et al. 1987) as well (Petersson and Al-Laham 1991):

$$\Delta E_\infty = \sum_{ij} \left\{ \left[\sum_{a=1}^{N_{ij}} C_{ij}^{a_i a_j} \right]^2 \left[e_{ij}(CBS) - e_{ij}(N) \right] \right\} \tag{11.3.21}$$

Since the individual pair coupling terms are not available from the QCISD(T) calculation, the interference correction is not applied to the pair coupling terms (Petersson and Al-Laham 1991).

Table. 11.5 CBS-SCF and CBS-QCISD(T) energies of first-row atoms, compared with experimental non-relativistic energies for first-row atoms (data in *Hartree*). From Petersson and Al-Laham (1980)

Element	CBS-SCF	CBS-QCI	Experimental	$\Delta_{CBS\text{-}SCF}$	$\Delta_{CBS\text{-}QCI}$
H	−0.49995	−0.49995	−0.50000	0.00005	0.00005
He	−2.86163	−2.90350	−2.90372	0.04209	0.00022
Li	−7.43257	−7.47759	−7.47785	0.04528	0.00026
Be	−14.57299	−14.66786	−14.66735	0.09436	−0.00051
B	−24.53286	−24.65341	−24.65384	0.12098	0.00043
C	−37.69361	−37.84463	−37.84510	0.15149	0.00047
N	−54.40434	−54.58965	−54.58950	0.18516	−0.00015
0	−74.81894	−75.06643	−75.06720	0.24826	0.00077
F	−99.41609	−99.73230	−99.73130	0.31521	−0.00100
Ne	−128.54643	−128.93322	−128.93670	0.39027	0.00348

The computed CBS-SCF and CBS-QCI energies are compared with experimental non-relativistic energies for first-row atoms in Table 11.5. The listed experimental values given in Table 11.5 are the total non-re1ativistic experimental energies obtained by relativistic energies + mass correction + lambda shift (Jankowski and Malinowski 1980).

Chapter 12
Quantum Mechanical Wave Functions

We have seen in the initial parts of this book that a *wave* may be defined as a disturbance in some physical system which is *periodic* in both space and time. The term is here adopted in its more general quantum–mechanical sense. I.e.: the wave function is a complex-valued probability amplitude. How this wave function may be conformed to attain a satisfactory representation of the probability density ? The first pioneering studies in this sense are due to Slater (1930). The *Slater Type Orbitals* (STO) extended and generalized the procedures already set up by Zener (1930). Differently from Zener, however, the shielding constants and other parameters were not fixed by variational procedures but adjusted merely to get agreement with observables (ionization energy, x-ray energy levels, sizes, etc.). Frank Boys in (1950) promoted the utilization of *Gaussian Type Orbitals* (GTO) in the place of Slater-Type-Orbitals (STO), through an application of the Gaussian Product Theorem. McWeeny (1953) proposed then that each STO may be eventually replaced by a linear combination of a limited number of GTOs, and extensive applications were then proposed by Foster and Boys (1960), Reeves and Fletcher (1965) and O-ohata et al. (1966).

After this necessarily short historical reconstruction we pass to analyze technical details apt to solve in a practical way some computational aspects, and to better conform the wave functions in space. We see thus the significance of a Contracted Gaussian Function, the Split Valence Basis Sets, the Polarized and Diffuse Functions. The nature of the *Basis Set Superposition Error* (BSSE) and the Counterpoise Procedures apt to reduce (or even eliminate) it are then discussed. We encounter then the various forms of *Plane Waves* (PW), originally promoted by Sommerfeld and Runge (1911), following "*an idea of Debye*" (Sommerfeld, 1930) and popular nowadays in Geochemistry because particularly apt to speed-up calculations on periodic systems. We introduce finally the *Effective Core Potentials* (ECPs), beneficial again for simple computational aspects: heavy elements have a great number of inner electrons that do not participate in a sensible way to the

G. A. Ottonello, *Quantum Geochemistry*, Springer Geochemistry,
https://doi.org/10.1007/978-3-031-21837-8_12

constitution of chemical bonds. The effect of inner (or "*core*") electrons may be thus abstracted from the computation, replacing the Coulomb potential of valence electrons with ECPs having fewer nodes.

12.1 Premises

At the light of the de Moivre's theorem:

$$\exp(i\phi) \equiv \cos\phi + i\sin\phi \tag{12.1.1}$$

with ϕ as real number, for a general complex number z such that:

$$r = |z| \tag{12.1.2}$$

$$\theta = \arg(z) \tag{12.1.3}$$

with r = modulus and θ = argument, the complex number may be written in the form:

$$z = r\exp(i\phi) \tag{12.1.4}$$

Since a space (x) and time (t) dependent monodimensional wavefunction ψ(x,t) may be generally written in the form:

$$\psi(x,t) = A\cos(kx - \omega t + \Phi) \tag{12.1.5}$$

with A as amplitude, k as wavenumber, ω as angular frequency and φ as phase angle, the function (12.1.5) corresponds to the real part (Re) of a complex wavefunction:

$$\psi(x,t) = \mathrm{Re}\{\psi_0 \exp[i(kx - \omega t)]\} \tag{12.1.6}$$

with

$$\psi_0 = A\exp(i\phi) \tag{12.1.7}$$

In fact (Fitzpatrich, 2010):

$$\begin{gathered}\mathrm{Re}\{\psi_0 \exp[i(kz - \omega t)]\} = \mathrm{Re}\{A\exp(i\varphi)\exp[i(kx - \omega t]\} \\ = \mathrm{Re}\{A\exp[i(kx - \omega t + \varphi]\} = A\mathrm{Re}\{\exp[i(kx - \omega t + \varphi]\}\end{gathered} \tag{12.1.8}$$

Omitting, for ease of notation Re, Eq. (12.1.6) is usually written in the form:

$$\psi(\mathrm{x,t}) = \psi_0 \exp[\mathrm{i(kx} - \omega \mathrm{t})] \tag{12.1.9}$$

and generalized in three dimensions as:

$$\psi(\mathrm{x,t}) = \psi_0 \exp[\mathrm{i(kr} - \omega \mathrm{t})] \tag{12.1.10}$$

with **k** as wavevector and **r** as space vector.

Disregarding the time-dependence one finally has:

$$\psi(\mathrm{x,t}) = \psi_0 \exp(\mathrm{ikr}) \tag{12.1.11}$$

12.2 Slater Type Orbitals (STO)

Slater (1930) in a pioneering contribution devoted to the comprehension of the shielding effect of inner electrons, i.e. the "*Atomic Shielding Constants*", introduced simple rules aimed at "*giving approximate analytic atomic wave functions for all the atoms, in any stage of ionization*" (Slater, 1930) by extending and generalizing the procedures already set up by Zener (1930). Differently from Zener, however, the shielding constants and other parameters were not fixed by variational procedures but adjusted merely to get agreement with experiments (ionization energy, x-ray energy levels, sizes, etc.). Because the nodes in the wave function were found by Zener to be unimportant, Slater neglected them, and considered as the radial part of the wave function of one electron simply:

$$\psi_{\mathrm{n^*,s}}(\mathrm{r}) = \mathrm{r}^{n^*-1} \exp\left[-\frac{(\mathrm{Z-s})\mathrm{r}}{n^*}\right] \tag{12.2.1}$$

reflecting the asymptotic form at large distances for a hydrogen-like wave function of quantum number n^* in the field of a nuclear charge (Z—s). n^* in Eq. (12.2.1) is the "*effective quantum number*" and (Z-s) is the nuclear charge deprived of some amount by the shielding effect of inner electrons (hereafter "effective nuclear charge" Z_{eff}).

The values of n^* and Z_{eff} are given by the following rules:

(1) for the effective quantum number:

n	1	2	3	4	5	6
n^*	1	2	3	3.7	4	4.2

(12.2.2)

(B) For determining Z-s,

- the electrons are divided into the following groups, each having a different shielding constant: [1s]; [2s, p]; [3s, p]; [3d]; [4s, p]; [4d]; [4f]; [5s, p]; [5d]; etc. That is, the s and p of a given *n* are grouped together, but the d and f are separated.
- the shells are considered to be arranged from inside out in the above outlined order

(C) The shielding constant s is formed, for any group of electrons, from the following contributions:

(a) Nothing from any shell outside the one considered.
(b) An amount 0.35 from each other electron in the group considered (except in the 1s group, where 0.30 is used instead).
(c) If the shell considered is an s, p shell, an amount 0.85 from each electron with total quantum number (*n*-1), and an amount 1.00 from each electron (*n*-2) or less
(d) If the shell is a d or f, an amount 1.00 from every electron closer to the nucleus (including electrons of same *n* but smaller *l*).

We see in Table (12.1) an exemplificative application of the above scheme to carbon and iron.

Table 12.1 Slater's scheme for the assessment of the shielding constant. Examples for Carbon and Iron $\zeta \equiv Z_{eff} = Z\text{-s}$. The f terms are the shielding contributions

CARBON $1s^2 2s^2 2p^2$											
Group	**e(1)**	**f(1)**	**e(2)**	**f(2)**	**e(3)**	**f(3)**	**e(4)**	**f(4)**	**Z**	s	ζ
2s, 2p	3	0.35			2	0.85			6	2.75	3.25
1s	1	0.3							6	0.3	5.7
IRON $1s^2 2s^2 2p^6 3s^2 3p^6 3d^6 4s^2$											
Group	**e(1)**	**f(1)**	**e(2)**	**f(2)**	**e(3)**	**f(3)**	**e(4)**	**f(4)**	**Z**	s	ζ
4s	1	0.35			14	0.85	10	1	26	22.25	3.75
3d	5	0.35	8	1	8	1	2	1	26	19.75	6.25
3s, 3p	7	0.35			8	0.85	2	1	26	11.25	14.75
2s, 2p	7	0.35			2	0.85			26	4.15	21.85
1s	1	0.3							26	0.3	25.7
e(1)	Electrons in the same group										
e(2)	Electrons in groups with same principal quantum number *n* and azimutal quantum number < *l*										
e(3)	Electrons in groups with principal quantum number *n*-1										
e(4)	Electrons in groups with same principal quantum number *n*-2										

It may be easily seen that, based on the simple form of the radial function (12.2.1), one may rapidly derive the atomic energy (and the ionization potentials) and the radial extent at the maximum radial charge density. In fact, if we take as an atomic wave function a product of one-electron wave functions of type (12.2.1) for each electron, then the atomic energy is simply the sum of $-[\varsigma/n^*]^2$ for each electron, times the function coming from the second derivative in the Laplacian operator. For iron, for instance, this gives E = $-2(25.70)^2 - 8(21.85/2)^2 - 8(14.75/3)^2 - 6(6.25/3)^2 - 2(3.75/3.7)^2 = -2497.2$ *Ry* ≡ 1248.6 *Hartree* (an almost identical to the energy is obtained with the STO-3G functions, i.e. −1248.511 *Hartree*). The maximum of the radial charge density for a shells comes, by differentiation, at $(n^*)^2/\zeta$. For example, the radius, as defined in this way, of the outer 4 s shell of Fe is $(3.7)^2/3.\ 75 = 3.65$ *Bohr* = 1.95 *Å*.

Because the probability of finding the electron somewhere in space is unity, i.e. in vectorial notation:

$$\int |\Psi|^2 dr = 1 \tag{12.2.4.1}$$

The normalization of function (12.2.1) yields:

$$\psi_{n^*,s}(\mathrm{r}) = \mathrm{Nr}^{n^*-1} \exp\left[-\frac{(\mathrm{Z}-\mathrm{s})\mathrm{r}}{n^*}\right] \tag{12.2.3}$$

with

$$\mathrm{N} = (2\varsigma)^{n^*}\sqrt{\frac{2\varsigma}{(2\mathrm{n}^*)!}} \tag{12.2.4}$$

The normalization factor coming from the integral:

$$\int_0^\infty x^n \exp(-\mathrm{cx})\mathrm{dx} = \frac{\mathrm{n}!}{\mathrm{c}^{\mathrm{n}+1}} \tag{12.2.5}$$

The angular part of the Slater orbitals is commonly expressed in spherical harmonics satisfying:

$$\int_0^\pi \int_0^{2\pi} \mathrm{Y}(\theta,\phi)^2 \sin\theta \mathrm{d}\theta \mathrm{d}\phi = 1 \tag{12.2.6}$$

Table 12.2 lists STO's up to $n = 3$. The part in square bracket is the angular term with normalized real spherical harmonics expressed in Cartesian coordinates.

Table 12.2 Slater type orbitals

Group	Angular $\psi(\theta, \varphi)$	Radial $\psi(r)$	Explicit form $\psi(\theta, \varphi, r)$
1s	$\sqrt{\frac{1}{4\pi}}$	$2\varsigma^{3/2}\exp(-\varsigma r)$	$\sqrt{\frac{\varsigma^3}{\pi}}\exp(-\varsigma r)$
2s	$\sqrt{\frac{1}{4\pi}}$	$\frac{2}{3}\sqrt{3}\varsigma^{5/2} r\exp(-\varsigma r)$	$\sqrt{\frac{\varsigma^5}{3\pi}} r\exp(-\varsigma r)$
$2p_x$	$\sqrt{\frac{3}{4\pi}\frac{x}{r}}$	$\frac{2}{3}\sqrt{3}\varsigma^{5/2} r\exp(-\varsigma r)$	$\sqrt{\frac{\varsigma^5}{\pi}} x\exp(-\varsigma r)$
$2p_{xy}$	$\sqrt{\frac{3}{4\pi}\frac{y}{r}}$	$\frac{2}{3}\sqrt{3}\varsigma^{5/2} r\exp(-\varsigma r)$	$\sqrt{\frac{\varsigma^5}{\pi}} y\exp(-\varsigma r)$
$2p_{xz}$	$\sqrt{\frac{3}{4\pi}\frac{z}{r}}$	$\frac{2}{3}\sqrt{3}\varsigma^{5/2} r\exp(-\varsigma r)$	$\sqrt{\frac{\varsigma^5}{\pi}} z\exp(-\varsigma r)$
3s	$\sqrt{\frac{1}{4\pi}}$	$\frac{2}{15}\sqrt{10}\varsigma^{7/2} r^2\exp(-\varsigma r)$	$\sqrt{\frac{2\varsigma^7}{45\pi}} r^2\exp(-\varsigma r)$
$3p_x$	$\sqrt{\frac{3}{4\pi}\frac{x}{r}}$	$\frac{2}{15}\sqrt{10}\varsigma^{7/2} r^2\exp(-\varsigma r)$	$\sqrt{\frac{2\varsigma^7}{15\pi}} rx\exp(-\varsigma r)$
$3p_y$	$\sqrt{\frac{3}{4\pi}\frac{y}{r}}$	$\frac{2}{15}\sqrt{10}\varsigma^{7/2} r^2\exp(-\varsigma r)$	$\sqrt{\frac{2\varsigma^7}{15\pi}} ry\exp(-\varsigma r)$
$3p_z$	$\sqrt{\frac{3}{4\pi}\frac{z}{r}}$	$\frac{2}{15}\sqrt{10}\varsigma^{7/2} r^2\exp(-\varsigma r)$	$\sqrt{\frac{2\varsigma^7}{15\pi}} rz\exp(-\varsigma r)$
$3d_{z2}$	$\sqrt{\frac{5}{4\pi}\frac{3z^2-r}{2r^2}}$	$\frac{2}{15}\sqrt{10}\varsigma^{7/2} r^2\exp(-\varsigma r)$	$\frac{1}{3}\sqrt{\frac{\varsigma^7}{2\pi}}(3z^2 - r^2)\exp(-\varsigma r)$
$3d_{xz}$	$\sqrt{\frac{5}{4\pi}}\sqrt{3}\frac{xz}{r^2}$	$\frac{2}{15}\sqrt{10}\varsigma^{7/2} r^2\exp(-\varsigma r)$	$\sqrt{\frac{2\varsigma^7}{3\pi}} xz\exp(-\varsigma r)$
$3d_{yz}$	$\sqrt{\frac{5}{4\pi}}\sqrt{3}\frac{yz}{r^2}$	$\frac{2}{15}\sqrt{10}\varsigma^{7/2} r^2\exp(-\varsigma r)$	$\sqrt{\frac{2\varsigma^7}{3\pi}} yz\exp(-\varsigma r)$
$3d_{xy}$	$\sqrt{\frac{5}{4\pi}}\sqrt{3}\frac{xy}{r^2}$	$\frac{2}{15}\sqrt{10}\varsigma^{7/2} r^2\exp(-\varsigma r)$	$\sqrt{\frac{2\varsigma^7}{3\pi}} xy\exp(-\varsigma r)$
$3d_{x2\text{-}y2}$	$\frac{1}{2}\sqrt{\frac{5}{4\pi}}\sqrt{3}\frac{x^2-y^2}{r^2}$	$\frac{2}{15}\sqrt{10}\varsigma^{7/2} r^2\exp(-\varsigma r)$	$\sqrt{\frac{\varsigma^7}{6\pi}}(x^2 - y^2)\exp(-\varsigma r)$

12.3 Gaussian Type Orbitals (GTO)

The utilization of Gaussian Type Orbitals (GTO) in the place of Slater-Type-Orbitals (STO) was promoted by Frank Boys in (1950) through an application of the '*Gaussian Product Theorem*'. This theorem states that the product of two Gaussian Probability functions centered on two different atoms is a finite sum of Gaussians centered on a point along the axis connecting them. Boys (1950) wrote: "*It is shown that by taking Gaussian functions, and functions derived from these by differentiation with respect to the parameters, complete systems of functions can be constructed appropriate to any molecular problem, and that all the necessary integrals can be explicitly evaluated. These can be used in connexion with the molecular orbital method, or localized bond method, or the general method of treating linear combinations of many Slater determinants by the variational procedure*" (Boys, 1950).

The many-electron Schrödinger equation for stationary states was expressed by Boys (1950) as follows:

$$H\Psi \equiv \left\{\sum_{i} K_i - \sum_{I,i} Z_I V_{Ii} + \frac{1}{2} M_{ij}\right\}\psi = E\Psi \quad (12.3.1)$$

where ψ is a function anti-symmetric with respect to the different sets of (space and spin) variables (x_i, y_i, z_i, v_i) and

$$K_i = -\frac{1}{2}\nabla_i^2 \quad (12.3.2)$$

$$V_{Ii} = \left[(x_i - P_{I_x})^2 + \left(y_i - P_{I_y}\right)^2 + (z_i - P_{I_z})^2\right]^{-1/2} \quad (12.3.3)$$

$$M_{ij} = \left[\left(x_i - x_j\right)^2 + \left(y_i - y_j\right)^2 + \left(z_i - z_j\right)^2\right]^{-1/2} \quad (12.3.4)$$

x_i, y_i, z_i in Eq, (12.3.3) are the space variables corresponding to the total number of electrons and $P_{I_x}, P_{I_y}, P_{I_z}$ are those of nuclei which are to be regarded as fixed and with integral charges Z_I.

The complete class of Gaussian functions consist of all functions, and of all linear combinations of functions, of the type $x^l y^m z^n \exp\left(-ar_A^2\right)$ where *l, m, n,* are integers, a is real positive and r_A obeys the general notation

$$x_A = x - A_x \quad (12.3.5)$$

$$r_A = \sqrt{x_A^2 + y_A^2 + z_A^2} \quad (12.3.6)$$

where A_x (and A_y, A_z) denoting the position of the nuclear center A are real. Extending the notation to other centers, i.e. B_x B_y, B_z,…and posing

$$aA = \exp\left(-ar^2\right) \quad (12.3.7.1)$$

$$AB^2 = \sum_{x,y,z} (A_x - B_x)^2 \quad (12.3.7.2)$$

$$CP^2 = \sum_{x,y,z} (C_x - P_x)^2 \quad (12.3.7.3)$$

$$P_x = \frac{aA_x + bB_x}{a + b} \quad (12.3.7.4)$$

$$Q_x = \frac{cC_x + dD_x}{c + d} \tag{12.3.7.5}$$

$$F(z) = \frac{1}{\sqrt{z}} \int_0^{\sqrt{z}} \exp(-u)du \tag{12.3.7.6}$$

the various integrals composing the bulk electronic energy of the multicenter moiety (overlap – kinetic energy – nuclear interaction – electronic interaction) are:

$$(aA|bB) = \left(\frac{\pi}{a+b}\right) \exp\left(-\frac{AB^2 ab}{a+b}\right) \tag{12.3.8}$$

$$(aA|K|bB) = \left[\frac{3ab}{a+b} - \frac{2AB^2a^2b^2}{(a+b)^2}\right] \left(\frac{\pi}{a+b}\right) \exp\left(-\frac{AB^2ab}{a+b}\right) \tag{12.3.9}$$

$$aA\left|V_c\right|bB = \frac{2\pi}{(a+b)} \exp\left(-\frac{AB^2ab}{a+b}\right) F\{CP^2(a+b)\} \tag{12.3.10}$$

$$aAbB|M|cCdD = \frac{2\pi^{3/2}}{(a+b)(c+d)\sqrt{(a+b+c+d)}} \exp\left\{-\frac{AB^2ab}{a+b} - \frac{CD^2cd}{c+d}\right\} F\left[\frac{PQ^2(a+b)(c+d)}{a+b+c+d}\right] \tag{12.3.11}$$

Applying the Boys' procedure, four-center integrals can be reduced to finite sums of two-center integrals, and in a next step to finite sums of one-center integrals. The speedup in computation achieved in this way is generally of 4–5 orders of magnitude.

12.4 Gaussian Expansions of Slater-Type Atomic Orbitals (STO-NG)

An LCAO-SCF variational calculation involves the construction of molecular orbitals (ψ_i) stemming from the summation of basis sets (ϕ_μ) multiplied by the appropriate molecular orbital expansion coefficients ($c_{\mu i}$):

$$\Psi_i = \sum_\mu c_{\mu i} \phi_\mu \tag{12.4.1}$$

Minimal basis sets of Slater's type (STO) for molecules or crystals with principal quantum number n not higher than 2 have the form:

$$\phi_{1s}(\zeta_1, \mathrm{r}) = \left(\frac{\zeta_1^3}{\pi}\right)^{1/2} \exp(-\zeta_1, \mathrm{r}) \tag{12.4.2.1}$$

$$\phi_{2s}(\zeta_2, \mathrm{r}) = \left(\frac{\zeta_2^5}{3\pi}\right)^{1/2} \mathrm{r} \cdot \exp(-\zeta_2, \mathrm{r}) \tag{12.4.2.2}$$

$$\phi_{2p}(\zeta_2, \mathrm{r}) = \left(\frac{\zeta_2^5}{\pi}\right)^{1/2} \mathrm{r} \cdot \exp(-\zeta_1, \mathrm{r}) \cos\theta \tag{12.4.2.3}$$

(equations similar to 12.4.2.3 hold for other 2p functions).

In the above equations, the ζ exponents may be fixed (e.g. Slater, 1930), or else treated as a variational parameters. Each STO may be moreover eventually replaced by a linear combination of a limited number of Gaussian-type orbitals, as first suggested by McWeeny (1953) to predict X-ray scattering in H_2, then applied by Foster and Boys (1960) to formaldehyde (HCHO) and later generalized by Reeves and Fletcher (1965) and O-ohata et al. (1966) which proposed Gaussian expansions for 1s, 2s, 2p, 3s, 3p and 3d STO's (but see also Sect. 7.2.5 and references therein).

As noted by Pereira Gomes and Custodio (2002), the source of inspiration for the elaboration of STO-NG expansions was the exact representation of the 1s STO function as an integral transform, proposed by Kikuchi (1954):

$$\exp(-\mathrm{r}) = \frac{2}{\sqrt{\pi}} \int_0^\infty \exp\left[-\left(\frac{\mathrm{r}}{2\mathrm{t}}\right)^2 - \mathrm{t}^2\right] \mathrm{dt} \tag{12.4.3}$$

valid for non-negative r with $\mathrm{r} = \sqrt{\mathrm{s}}$ and $4\mathrm{t}^2 = 1/\mathrm{s}$.

According to O-ohata et al. (1966), the Gaussian Type Orbitals (GTO's) have "*the decisive advantage in evaluating the values of molecular integrals*".

The definition of (normalized) STO's adopted by O-ohata et al. (1966) is the following:

$$\varphi_{n,l,m_l}^{\mathrm{STO}}(\zeta) = \mathrm{R}_n^{\mathrm{STO}}(\zeta, \mathrm{r}) \mathrm{Y}_{l,m_l}(\theta, \phi) \tag{12.4.4.1}$$

$$\mathrm{R}_n^{\mathrm{STO}}(\zeta, \mathrm{r}) = \mathrm{N}_n^{\mathrm{STO}}(\zeta) \mathrm{r}^{n-1} \exp(-\zeta \mathrm{r})) \tag{12.4.4.2}$$

$$\mathrm{N}_n^{\mathrm{STO}}(\zeta) = [(2n)!]^{-1/2} (2\zeta) \mathrm{r}^{n+1/2} \tag{12.4.4.3}$$

This form may be easily compared with that one of the (normalized) GTO's:

$$\phi^{\mathrm{GTO}}_{n,l.m_l}(\zeta) = \mathrm{R}^{\mathrm{GTO}}_{n}(\alpha,\mathrm{r})\mathrm{Y}_{l,m_l}(\theta,\phi) \tag{12.4.5.1}$$

$$\mathrm{R}^{\mathrm{GTO}}_{n}(\alpha,\mathrm{r}) = \mathrm{N}^{\mathrm{GTO}}_{\mathrm{n}}(\alpha)\mathrm{r}^{\mathrm{n}-1}\exp\left(-\alpha\mathrm{r}^2\right) \tag{12.4.5.2}$$

$$\mathrm{N}^{\mathrm{STO}}_{n}(\alpha) = 2^{n+1}\left[(2n-1)!\right]^{-1/2}(2\pi)^{-1/4}\alpha^{(2n+1)4} \tag{12.4.5.3}$$

where α is the GTO exponent.

The approximate expansion of STO's in terms of GTO's is expressed as:

$$\phi^{\mathrm{STO}}_{n,l.m_l}(\zeta) \cong \sum_{\mathrm{i}=1}^{\mathrm{N}} \mathrm{c_i}\phi^{\mathrm{GTO}}_{i,n',l.m_l}(\alpha_{\mathrm{i}}) \tag{12.4.6}$$

and the integral

$$\mathrm{V}(\zeta;\mathrm{c_i},\alpha_{\mathrm{i}}) = \int\left[\phi^{\mathrm{STO}}_{n,l.m_l}(\zeta) - \sum_{i=1}^{\mathrm{N}} \mathrm{c_i}\Phi^{\mathrm{GTO}}_{i,n',l.m_l}(\alpha_{\mathrm{i}})\right]^2 \mathrm{dV} \tag{12.4.7}$$

(to be minimized by a least squares fitting procedure) represents the performance of the method. Obviously, the larger is the set of GTO's adopted, the better is the result. It must be noted that *n'* in the right part of Eq. (12.4.6), which replaces *n* in the left part, has a precise significance. In fact, if the power exponent of r is written as r^{n-1} then odd *n* goes with even *l* and even *n* with odd *l*. Thus, to the STO $2\mathrm{s_S}$ corresponds the GTO $1\mathrm{s_G}$ and the GTO's are named as $(1\mathrm{s})_\mathrm{G}$, $(3\mathrm{s})_\mathrm{G}$, $(5\mathrm{s})_\mathrm{G}$,…; $(2\mathrm{p})_\mathrm{G}$, $(4\mathrm{p})_\mathrm{G}$,…; $(3\mathrm{d})_\mathrm{G}$, $(5\mathrm{d})_\mathrm{G}$,…; etc. As already stressed by Huzinaga (1965), it is only under this restriction that various atomic and molecular integrals over GTO's can be performed with great ease.

The values attained by $\mathrm{V}(\zeta;\mathrm{c_i},\alpha_{\mathrm{i}})$ for Gaussian expansions of STO's with the increasing values of N are listed in Table (12.3).

In the procedure of Hehre et al. (1969), each STO (ϕ_μ) is replaced by and atomic orbital ϕ'_μwhich is the sum of K = 2,6 GTO's. The combination is initially obtained for ζ = 1 and then uniformly scaled.[1]

Thus:

$$\phi'_\mu(\zeta,\mathrm{r}) = \zeta^{3/2}\phi'_\mu(1,\zeta\mathrm{r}) \tag{12.4.8}$$

$$\phi'_{1\mathrm{s}}(1,\mathrm{r}) = \sum_{5\mathrm{k}}^{\mathrm{K}} \mathrm{d}_{1\mathrm{s,k}}\mathrm{g}_{1\mathrm{s}}(\alpha_{1\mathrm{k}},\mathrm{r}) \tag{12.4.9.1}$$

[1] The adopted mathematical symbolism differ slightly from what we have seen before in the same paragraph but I prefer to conform to the Authors' style, to avoid misinterpreting.

Table 12.3 Square deviation for different Gaussian expansions of STO's (from O-ohata et al. 1966), slightly modified

Types	N of terms			
	4	6	8	10
$(1s)_S - (1s)_G$	6.88030×10^{-5}	4.00686×10^{-6}	2.39259×10^{-7}	7.66301×10^{-8}
$(2s)_S - (1s)_G$	2.70069×10^{-5}	4.26723×10^{-6}	1.08542×10^{-7}	2.96730×10^{-8}
$(2p)_S - (2p)_G$	4.02689×10^{-5}	2.05513×10^{-6}	4.90599×10^{-7}	
$(3s)_S - (1s)_G$	2.04490×10^{-6}	1.80054×10^{-7}	1.57802×10^{-7}	
$(3p)_S - (2p)_G$	3.07501×10^{-6}	2.37588×10^{-7}	1.16694×10^{-7}	
$(3d)_S - (3d)_G$	5.44367×10^{-5}	1.91591×10^{-6}	3.30641×10^{-7}	

$$\phi'_{1s}(1,r) = \sum_k^K d_{1s,k} g_{1s}(\alpha_{1k}, r) \quad (12.4.9.2)$$

$$\phi'_{2p}(1,r) = \sum_k K d_{2p,k} g_{2p}(\alpha_{2k}, r) \quad (12.4.9.3)$$

where g_{1s} and g_{2p} are the GTO's

$$g_{1s}(\alpha, r) = \left(\frac{2\alpha}{\pi}\right)^{3/4} \exp(-\alpha r^2) \quad (12.4.10.1)$$

$$g_{2p}(\alpha, r) = \left(\frac{128\alpha^5}{\pi^3}\right)^{1/4} r \exp(-\alpha r^2) \cos\theta \quad (12.4.10.2)$$

As already seen in the O-ohata et al. (1966) application, also here the 2s exponential function is replaced by a linear combination of 1s Gaussian functions. However, while in O-ohata et al. (1966) two different sets of α values were adopted for 2s and 2p, here the combinations for ϕ'_{2s} and ϕ'_{2p} share the same set of Gaussian exponents α_{2k} (Table 12.4). The constants d and α in eqns. (12.4.9.1–12.4.9.3) are chosen to minimize the integrals:

$$\epsilon_{1s} = \int \left(\Phi_{1s} - \Phi'_{1s}\right)^2 d\tau \quad (12.4.11.1)$$

$$\epsilon_{2s} + \epsilon_{2p} = \int \left(\Phi_{2s} - \Phi'_{2s}\right)^2 d\tau + \int \left(\Phi_{2p} - \Phi'_{2p}\right)^2 d\tau \quad (12.4.11.2)$$

with $d\tau$ as one-electron volume element (cf. Eq. 9.1.9).

Table 12.4 Coefficients and exponents for Gaussian expansion of Slater orbitals. From Hehre et al. (1963) with modifications

K	α_{1s}	d_{1s}	ε_{1s}	$\alpha_{2s} = \alpha_{2p}$	d_{2s}	ε_{2s}	d_{2p}	ε_{2p}
2	1.51623×10^{-1}	6.78914×10^{-1}	3.16×10^{-3}	9.74545×10^{-2}	9.63782×10^{-1}	5.60×10^{-3}	6.12820×10^{-1}	3.48×10^{-3}
	8.51819×10^{-1}	4.30129×10^{-1}		3.84244×10^{-1}	4.94718×10^{-2}		5.11541×10^{-1}	
3	1.09818×10^{-1}	4.44635×10^{-1}	3.31×10^{-4}	7.51386×10^{-2}	$7.\times10^{-1}$	6.42×10^{-4}	3.91957×10^{-1}	3.60×10^{-4}
	4.05771×10^{-1} 1	5.35328×10^{-1}		2.31031×10^{-1}	3.99513×10^{-1}		6.07684×10^{-1}	
	2.22766	1.54329×10^{-1}		9.94203×10^{-1}	-9.99672×10^{-2}		1.55916×10^{-1}	
4	8.80187×10^{-2}	2.91626×10^{-1}	4.38×10^{-5}	6.28104×10^{-2}	4.97767×10^{-1}	7.95×10^{-5}	2.46313×10^{-1}	4.82×10^{-5}
	2.65204×10^{-1}	5.32846×10^{-1}		$1.\ 63{,}541\times10^{-1}$	5.58855×10^{-1}		5.83575×10^{-1}	
	9.54620×10^{-1}	$2.\times10^{-1}$		$5.\ 02{,}989\times10^{-1}$	2.97680×10^{-5}		2.86379×10^{-1}	
	5.21686	5.67523×10^{-2}		2.32350	-6.22071×10^{-2}		4.36843×10^{-2}	
5	7.44527×10^{-2}	$1.\times10^{-1}$	6.88×10^{-6}	$5.\ 44{,}949\times10^{-2}$	3.46121×10^{-1}	1.17×10^{-6}	1.56828×10^{-1}	7.28×10^{-6}
	$1.\ 97572 \times 10^{-1}$	4.82570×10^{-1}		1.27920×10^{-1}	6.12290×10^{-1}		$5.\ 10{,}240\times10^{-1}$	
	5.78648×10^{-1}	3.31816×10^{-1}		$3.\ 29,\times10^{-1}$	1.28997×10^{-1}		3.73598×10^{-1}	
	2.07173	1.13541×10^{-1}		1.03250	-6.53275×10^{-1}		1.07558×10^{-1}	
	1.13056×10^{1}	2.21406×10^{-2}		5.03629	$-2.\ 94{,}086\times10^{-2}$		1.25561×10^{-2}	
6	6.51095×10^{-2}	1.30334×10^{-1}	1.24×10^{-6}	4.85690×10^{-2}	2.40706×10^{-1}	2.01×10^{-6}	1.01708×10^{-1}	1.22×10^{-6}
	1.58088×10^{-1}	4.16492×10^{-1}		1.05960×10^{-1}	5.95117×10^{-1}		4.25860×10^{-1}	
	4.07099×10^{-1}	3.70563×10^{-1}		2.43977×10^{-1}	$2.\ 50{,}242\times10^{-1}$		4.18036×10^{-1}	
	1.18506	$1.\ 68{,}538\times10^{-1}$		6.34142×10^{-1}	-3.37854×10^{-2}		1.73897×10^{-1}	
	4.23592	4.93615×10^{-2}		2.04036	-4.69917×10^{-2}		3.76794×10^{-2}	
	2.31030×10^{-1}	9.16360×10^{-3}		1.03087×10^{1}	-1.32528×10^{-2}		3.75970×10^{-3}	

Table 12.5 Differences in the HF total energies (*Hartree*) calculated with fixed exponent and various GTO expansions, with respect to STO calculations. From Hehre et al. (1963)

Molecule	STO-2G	STO-3G	STO-4G	STO-5G	STO-6G
HF	3.9213	0.9511	0.2581	0.0816	0.0284
CO	4.4154	1.1139	0.3099	0.0993	0.0350
CH_4	1.5422	0.3988	0.1156	0.0370	0.0130
C_2H_2	3.0235	0.7770	0.2233	0.0717	0.0250
C_2H_4	3.0393	0.7837	0.2254	0.0725	0.0253
C_2H_6	3.0481	0.7894	0.2279	0.0731	0.0255
BH_3	1.0417	0.2730	0.0794	0.0256	0.0089
B_2H_6	2.0718	0.5431	0.1581	0.0514	0.0179
NH_3	2.1700	0.5514	0.1560	0.0497	0.0176
HCN	3.6447	0.9281	0.2628	0.0845	0.0295
H_2CO	4.4459	1.1201	0.3134	0.0999	0.0350

Given the basis functions (12.4.8, 9), the molecular orbital expansion coefficients (c_{vi}) related to the elements of the Fock matrix (F_{uv}) and the elements of the overlap matrix (S_{uv}) (Eq. 9.2.1) can be found by minimizing the total electronic energy calculated from a single-determinant wavefunction. For closed-shell molecules, this leads to the Roothaan-Hall Eqs. 9.2.2 to 9.2.4. For open-shell molecules, one may use unrestricted single-determinant wavefunctions, with different molecular orbitals for α and β electrons. The LCAO-SCF Eq. (9.2.1) then has to be generalized as indicated by Pople and Nesbet (1954). The differences in the total HF energies computed with the various GTO expansions with respect to STO are listed in Table (12.5). These are associated to the inadequate behavior of Gaussian functions near the nuclei (e.g., lack of cusps; cf. also Sect. 7.2.5).

The dissociation energies computed via K-GTO expansions converge more rapidly to STO values than total energies (Hehre et al., 1963). This is due to the fact that the error introduced when STO's are replaced by small GTO expansions is very similar in the molecule and in the separate atoms. The atomization energy is basically overestimated (see Table 12.3 in Hehre et al., 1963), but the biases decrease systematically with the increase of K. For the STO-3G level the overestimation is in the order of $2 \times 10^{-3} \div -3 \times 10^{-3}$ *Hartree* for each bond in the molecule (Hehre et al., 1963).

Three new integral transforms of STO have been presented by Pereira Gomes and Custodio (2002) using Laplace transforms expressed in term of Gaussian functions. The use of any of these three alternatives depends only on appropriate numerical integration techniques. The STO-NG basis set can then be developed using either only 1s Gaussian functions, the same Gaussian exponents for each shell, or using the first Gaussian of each symmetry (Pereira Gomes and Custodio, 2002).

12.5 Gaussian Basis Sets

A basis set is the mathematical form adopted to describe the atomic orbitals used, in their turn, to approximate, through combinatory methods, the total electronic wavefunction in a system. For example, in the LCAO-SCF procedure each molecular orbital is written in the form:

$$\psi_i = \sum_{\mu} c_{\mu i} \phi_{\mu} \tag{12.5.1}$$

where ϕ_μ are basis functions and $c_{\mu i}$ are variational coefficients. The basis set ϕ_μ is usually described as "*minimal*" if it contains only atomic orbitals for the shells occupied or partly occupied in the atomic ground state. For example, to describe H_2O the minimal basis set is composed of two basis for the 1s orbitals of H and five basis for the 1s, the 2s, and the three 2p orbitals of oxygen. This minimal set is valid basically for all organic molecules with only first-row atoms. However, even a minimal basis set representation for a simple diatomic molecule cannot lead to a proper LCAO-SCF computation since the MO expansion coefficients cannot be variationally optimized. For example, a minimal basis set computation for the C_2 molecule is constrained to select the π_u (2p) orbitals as 1.0 $2p_\pi(A)$ + 1.0 $2p_\pi(B)$. (with A and B as carbon centers) to build up the MO orbital.

Larger basis sets are usually called "*extended*" and give extra flexibility to the molecular-orbital description, allowing such features as anisotropic atomic shape and angular distortion.

The adoption of extended basis sets has obviously a very high computational cost. Thus, Clementi (1985) and Clementi and Davis (1966a,b) promoted the utilization of "*contracted*" basis sets which combine linearly the Gaussians. The contracted set reduces in a substantial manner the number of super-matrix elements in the SCF computation, and therefore allows the utilization of a basis set of considerable magnitude.

12.5.1 *Contracted Gaussian Function*

We have already seen that the (primitive) Gaussian functions are of the form (Sect. 7.3.6):

$$g_P(\alpha, r) = N x^l y^m z^n \exp\left(-\alpha r^2\right) \tag{7.3.82.1}$$

where N is a normalization factor; *l, m, n* are integers (0,1,2,3), and α is the orbital exponent. A "*contracted Gaussian*" is a linear combination of primitive or "*uncontracted*" Gaussians where each component Gaussian of the linear combinations has the same value for *l,m,n* as in Eq. (7.3.82.1); i.e., s Gaussians are combined with s Gaussians, $2p_x$ Gaussians with $2p_x$ Gaussians, etc. The coefficients of the linear expansion are taken from the eigenvectors of the atomic computation. The number of linearly combined Gaussians in a contracted Gaussian function, designated as γ, depends from the chemical composition of the investigated substance.

The first compound adopted as test case by Clementi and Davis (1966a) is the ethane molecule (C_2H_6) in its eclipsed *(D_{3h})* and staggered *(D_{ad})* conformers in the aim of determining the magnitude of the rotational barrier (cf. Fig. 12.1).

The first compound adopted as test case by Clementi and Davis (1966a) is the ethane molecule (C_2H_6) in its eclipsed *(D_{3h})* and staggered *(D_{ad})* conformers in the aim of determining the magnitude of the rotational barrier (cf. Fig. 12.1).

In Table 12.6 are listed the 86 primitive Gaussians adopted by Clementi and Davis (1966a) in the contraction procedure. The functions are centered, in general, on atomic nuclei and the coordinate axes at the various centers are parallel to a master coordinate axis, which, in this case, is the z axis (C–C bond direction). In the first column are indicated all the centers having the same type of function (eclipsed configuration), with the value between parentheses referring to the first, second, etc., equivalent atom. The second column specifies the s or p symmetry of the Gaussian function. The p functions appear in the computation as p_x (indicated as x), p_y (indicated as y), and p_z (indicated as z). As indicated in the table by writing all

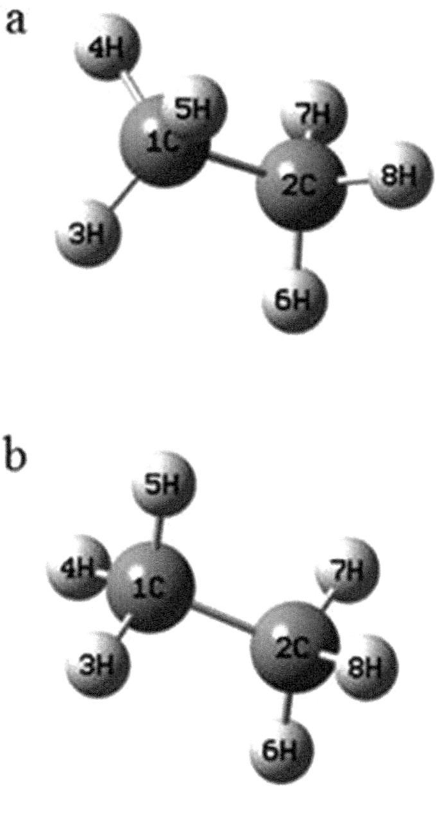

Fig. 12.1 eclipsed *(D_{3h})* and staggered *(D_{ad})* conformers of the ethane molecule (H_3C-CH_3) (parts **a** and **b**, respectively) resolved at HF/ 4-31G theory level. The z-axis coincides with the C–C bond. The hydrogens of HC_3 in the staggered conformer are rotated of 60° with respect to the corresponding hydrogens of CH_3. The staggered conformer at this theory level is more stable than the eclipsed one by about 11.6 *kJ/ mol*. This is the height of the so called "*rotational barrier*"

three designations in the second column, the same values of the orbital exponents hold for x, y, z functions. The index (from 1 to 86) associated to each Gaussian is reported in the third column. If more than one center appears in the first column, more than one index will correspondingly appear in the third column. If more than one symmetry type appears in the second column, and more than one center appears in the first column, then the indices in the third column will correspond to the first type of Gaussian for the first center, second center, etc., to the second type of Gaussian for the first center, second center, etc., and so on. For example, the indices 51, 57, 63, 69, 76, 81 of the table represent a $2p_x$ on C(l), $2p_x$ on C(2), $2p_y$ on C(l), $2P_y$ on C(2), $2p_z$ on C(l), $2p_z$ on C(2) Gaussian function, all with orbital exponent equal to 0.657707. The orbital exponents are given in the last column of the table.

In Table 12.7 are reported the contracted Gaussian sets used by Clementi and Davis (1966a) for C_2H_6. The primitive (uncontracted) Gaussian functions of Table (12.6) (denoted as β in Clementi and Davis, 1966a) are designated via their indices (3rd column of Table 12.6). In the second column is reported the corresponding type (s, x, y, z) and the "*corresponding Slater-type analog*" (Clementi and Davis, 1966a). Thus, γ_1 is a contracted Gaussian of s type, which has analog functions to a 1s Slater-type function on C(1). The third column of Table 12.7 designates the center for a given contracted Gaussian. It must be also noted that:

(a) γ_5, γ_6, γ_7, γ_8 [centered on C(2)] are not reported since there is a one-to-one correspondence between these and γ_1, γ_2, γ_3, γ_4.
(b) γ_{11}, γ_{12} [centered on H(2)], γ_{13}, γ_{14} [centered on H(3)], γ_{15}, γ_{16} [centered on H (4)], γ_{17}, γ_{18} [centered on H(5)], γ_{19}, γ_{20} [centered on H(6)] are not reported since there is a one-to-one correspondence between these and γ_9, γ_{10}.
(c) γ_{23}, γ_{24} [centered on C(2)], γ_{27}, γ_{28} [centered on C(2)], γ_{31}, γ_{32} [centered on C (2)] are not reported since there is a one-to-one correspondence between these three pairs of functions and the three pairs γ_{21}, γ_{22}, γ_{25}, γ_{26} and γ_{29}, γ_{30}.

The last step in the contraction procedure is the generation of Symmetry Adapted Functions (SAF). A SAF can be defined as "*the smallest linear combination of atomic functions, which transform as an irreducible representation of the molecular point group*" (Clementi and Davis, 1966a).

In the case of C_2H_6, for the eclipsed conformer (D_{sh}) the occupied orbitals transform according to the A_1', A_2'', E', and E'' irreducible representations. The ground-state electronic configuration for the eclipsed form is $1a_2''(2)$, $1a_1''(2)$, $2a_2''(2)$, $1e'(4)$, $3a_1(2)$, $1e''(4)$. Since the SAF are real, the $1e'$, $1e''$, le_u, and $1e_g$ orbitals split into degenerate pairs, i.e.: $1e' \rightarrow 1e_a', 1e_b'$; $1e'' \rightarrow 1e_a'', 1e_b''$; $1e_u \rightarrow 1e_{u,a}$, $1e_{u,b}$; $1e_g \rightarrow 1e_{g,a}$, $1e_{g,b}$. The un-normalized SAF for the C_2H_6 eclipsed conformer are listed in Table 12.8.

We have in this sort of construction a "*hierarchy of functions*" (Clementi and Davis, 1966b):

Table 12.6 Primitive Gaussian functions adopted by Clementi and Davis (1966b) in the contraction procedure for the C_2H_6 eclipsed conformer. In Clementi and Davis (1966a) these functions are denoted β

Center	Type	Indices						Orbital exponents
C(l), C(2)	s					1	11	9470.52
C(l), C(2)	s					2	12	1397.56
C(l), C(2)	s					3	13	307.539
C(l), C(2)	s					3	14	85.5419
C(l), C(2)	s					5	15	26.9117
C(l), C(2)	s					6	16	9.409
C(l), C(2)	s					7	17	3.50002
C(l), C(2)	s					8	18	1.06803
C(l), C(2)	s					9	19	0.400166
C(l), C(2)	s					10	20	0.135124
H(l), H(2), H(3), H(4), H(5), H(6)	s	21	26	31	36	41	46	0.12332
H(l), H(2), H(3), H(4), H(5), H(6)	s	22	27	32	37	42	47	0.45376
H(l), H(2), H(3), H(4), H(5), H(6)	s	23	26	33	38	43	48	2.0133
H(l), H(2), H(3), H(4), H(5), H(6)	s	24	29	34	39	44	49	13.3615
H(l), H(2), H(3), H(4), H(5), H(6)	s	25	30	35	40	45	50	0.07983
C(l), C(2)	x, y, z	51	57	63	69	76	81	0.657707
C(l), C(2)	x, y, z	52	58	64	70	76	82	1.78729
C(l), C(2)	x, y, z	53	59	65	71	77	83	5.77636
C(l), C(2)	x, y, z	54	60	66	72	78	84	25.3655
C(l), C(2)	x, y, z	55	61	67	73	79	85	0.0910638
C(l), C(2)	x, y, z	56	62	68	74	80	86	0.248046

(1) standard Gaussian functions, centered in general, but not necessarily, on one of the atomic nuclei (g_p, or else β; 1st-set);
(2) linear combinations of Gaussian, on one center, the "*contracted Gaussian set*" where the linear expansion coefficients are taken from educated guesses, based on the atomic orbital expansion coefficients of atomic computations (γ; 2nd-set);
(3) symmetry functions; these are expanded either in terms of uncontracted Gaussian functions or in terms of the contracted Gaussian set, or in terms of both, free and contracted (symmetry-adapted functions SAF, else χ; 3rd-set);
(4) molecular orbitals, which are variational combinations of the smallest linear combination of atomic functions (Φ; 4th-set).

Table 12.7 Contracted Gaussians set for C_2H_6 (From Clementi and Davis, 1966a)

Contracted Gaussian	Type	Center
$\gamma_1 = 0.42695\beta_6 + 0.35791\beta_7$	s; 1s	C(1)
$\gamma_2 = 0.00045\beta_1 + 0.00358\beta_2 + 0.01934\beta_3 + 0.07736\beta_4 + 0.22779\beta_5$	s; 1s'	C(1)
$\gamma_3 = 0.08502\beta_8 + 0.60689\beta_9$	s; 2s	C(1)
$\gamma_4 = \beta_{10}$	s; 2s'	C(1)
$\gamma_9 = 0.50807\beta_{21} + 0.47449\beta_{22} + 0.\lambda 3424\beta_{23} + 0.01906\beta_{24}$	s; 1s	H(1)
$\gamma_{10} = \beta_{25}$	s; 1s'	H(1)
$\gamma_{21} = 0.35871\beta_{51} + 0.18263\beta_{52} + 0.05479\beta_{53} + 0.00875\beta_{54}$	x; $2p_x$	C(1)
$\gamma_{22} = 0.43276\beta_{55} + 0.20347\beta_{56}$	x; $2p_x$'	C(1)
$\gamma_{25} = 0.35871\beta_{63} + 0.18263\beta_{64} + 0.05479\beta_{65} + 0.00875\beta_{66}$	y; $2p_y$	C(1)
$\gamma_{26} = 0.43276\beta_{67} + 0.20347\beta_{68}$	y; $2p_y$'	C(1)
$\gamma_{29} = 0.35871\beta_{75} + 0.18263\beta_{76} + 0.05479\beta_{77} + 0.00875\beta_{78}$	z; $2p_z$	C(1)
$\gamma_{30} = 0.43276\beta_{79} + 0.20347\beta_{80}$	z; $2p_z$'	C(1)

Table 12.8 Un-normalized Symmetry-adapted functions for C_2H_6 (eclipsed conformer). From Clementi and Davis (1966b)

$\chi_1 = \gamma_1 + \gamma_5$
$\chi_2 = \gamma_2 + \gamma_6$
$\chi_3 = \gamma_3 + \gamma_7$
$\chi_4 = \gamma_4 + \gamma_8$
$\chi_5 = \gamma_9 + \gamma_{11} + \gamma_{13} + \gamma_{15} + \gamma_{17} + \gamma_{19}$
$\chi_6 = \gamma_{10} + \gamma_{12} + \gamma_{14} + \gamma_{16} + \gamma_{18} + \gamma_{20}$
$\chi_7 = \gamma_{29} - \gamma_{31}$
$\chi_8 = \gamma_{30} - \gamma_{32}$
$\chi_9 = \gamma_1 - \gamma_5$
$\chi_{10} = \gamma_2 - \gamma_6$
$\chi_{11} = \gamma_3 - \gamma_7$
$\chi_{12} = \gamma_4 - \gamma_8$
$\chi_{13} = \gamma_9 + \gamma_{11} + \gamma_{13} - \gamma_{15} + \gamma_{17} + \gamma_{19}$
$\chi_{14} = \gamma_{10} + \gamma_{12} + \gamma_{14} - \gamma_{16} - \gamma_{18} - \gamma_{20}$
$\chi_{15} = \gamma_{29} + \gamma_{31}$
$\chi_{16} = \gamma_{30} + \gamma_{32}$
$\chi_{17} = \gamma_9 - l\gamma_{11} - l\gamma_{13} + \gamma_{15} - l\gamma_{17} - l\gamma_{19}$
$\chi_{18} = \gamma_{10} - l\gamma_{12} - l\gamma_{14} + \gamma_{16} - l\gamma_{18} - l\gamma_{20}$
$\chi_{19} = \gamma_{21} + \gamma_{23}$
$\chi_{20} = \gamma_{22} + \gamma_{24}$
$\chi_{21} = \gamma_{11} - \gamma_{13} + \gamma_{17} - \gamma_{19}$
$\chi_{22} = \gamma_{12} - \gamma_{14} + \gamma_{18} - \gamma_{20}$
$\chi_{23} = \gamma_{25} + \gamma_{27}$
$\chi_{24} = \gamma_{26} + \gamma_{28}$
$\chi_{25} = \gamma_9 - l\gamma_{11} - l\gamma_{13} - \gamma_{15} + l\gamma_{17} + l\gamma_{19}$
$\chi_{26} = \gamma_{10} - l\gamma_{12} - l\gamma_{14} - \gamma_{16} + l\gamma_{18} + l\gamma_{20}$
$\chi_{27} = \gamma_{21} - \gamma_{23}$
$\chi_{28} = \gamma_{22} - \gamma_{24}$

Actually, however, the construction proceeds in an opposite sense. To understand why, we must return to the actual problem, which is basically, the determination of the variational energy of the system. The Hamiltonian of the problem, written in atomic units is:

$$\mathsf{H} = \sum_{i} \left(-\frac{1}{2}\nabla_i^2 - \frac{Z_a}{r_i} \right) + \sum_{i,j} \frac{1}{r_{i,j}} - \sum_{a,b} \frac{Z_a Z_b}{r_{ab}} \tag{12.5.2}$$

where the first term is the kinetic operator for the ith electron, the second term is the potential between the ith electron and the nuclear point charge Z_a, the third term is the electron–electron potential between the ith and jth electrons, and the last term is the nuclear potential for two point charges at distance r_{ab}.

Since the hamiltonian (12.5.2) is spin-independent, the spin-orbitals are a simple product of a spin function and an orbital (spatial) function. The wave function ψ for the problem is thus the standard antisymmetrized product of spin-orbitals, based on Pauli exclusion principle.

The SCF procedure gives, at the variational limit, the orbitals, which are approximate solutions to the Hamiltonian (12.5.2). returning an energy:

$$E = \frac{\langle \psi \mid \mathsf{H} \vert \psi \rangle}{\langle \psi \mid \psi \rangle} \tag{12.5.3}$$

In a molecular system the AO's are linearly combined, generating MO's (Φ), with a given geometrical configuration and a given electronic configuration and state. Each Φ, is characterized by an index λ, which specifies the irreducible representation of the point group for the molecule in consideration, and by a second index i which identifies the ith MO of the λth irreducible representation. The MO's are expanded in terms of the smallest linear combination of atomic functions.

Designating by χ the symmetry-adapted functions and by ψ the Gaussian functions, a symmetry function and a molecular orbital are expanded, respectively, as:

$$\chi_{\lambda q} = \sum_{r} d_{\lambda q r} \psi_r \tag{12.5.4}$$

and

$$\Phi_{\lambda i} = \sum_{q} c_{\lambda i q} \chi_{\lambda q} \tag{12.5.5}$$

with the $d_{\lambda qr}$ coefficients selected in order to insure that the ψ_r transform properly, and the $c_{\lambda iq}$ as the expansion coefficients for the molecular orbital $\Phi_{\lambda i}$, variationally obtained in the SCF technique.

In the case of C_2H_6, through the contraction procedure, the two-electrons integrals are reduced to 139,565, with respect to the much larger number of bi-electronic integrals involving primitive Gaussian functions (of the order of several millions). The super-matrix for the SCF computation in terms of the normalized SAF has dimension 112 × 112. For one-electron integrals the reduction is not important, passing from 528 of a given type (i.e., overlap, kinetic, potential) to 250.

The computation with the 28 (normalized) SAF gives a barrier height in reasonable agreement with experiments (15.149 *kJ/mol*, with respect to 12.677 ± 1.255 *kJ/mol* according to Lide, 1958).

Several contributions proposing contracted Gaussians are available in literature, such as, among the others, Ditchfield et al. (1970) for 1st-row atoms (B to F), Mclean and Chandler (1980) for 2nd-row atoms (Na to Ar); Binning and Curtiss (1990) for 3rd-row atoms (Ga to Kr).

12.5.2 Split-Valence Basis Sets

In the HF calculation of C_2H_6 (Fig. 12.1) I adopted the 4-31G basis set of Ditchfield et al. (1971). This is an extended basis set arising from a fixed combination of Gaussian functions, where each inner shell is represented by a single basis function taken as a sum of four Gaussians and each valence orbital is split into inner and outer parts described by three and one Gaussian function, respectively. This basis set derives from the contracted set obtained by Ditchfield et al. (1970) by minimizing the energy of the atomic ground state for hydrogen and for the first row atoms boron to fluorine. The expansion coefficients and Gaussian exponents of the split-valence basis set were again determined by Ditchfield et al. (1971) through a variational approach to the atomic ground state. Optimization of valence-shell scaling factors showed however that considerable rescaling of atomic functions occurs in molecules, the largest effects being observed for hydrogen and carbon.

In their "*Quantum Mechanical Development*", Ditchfield et al. (1971) choose to represent the basis set with one function (ϕ) for inner shells and two sets of ϕ functions for the valence shells. In carbon, for example, the split-valence set is represented by the basis functions:

$$\phi_{1s}(r) = \sum_{k}^{N_1} d_{1s,k} g_s(\alpha_{1k}, r) \tag{12.5.6.1}$$

$$\phi'_{2s}(r) = \sum_{k'=1}^{N'_2} d_{2s,k'} g_s(\alpha_{2k'}, r) \tag{12.5.6.2}$$

$$\phi'_{2p_x}(r) = \sum_{k'=1}^{N'_2} d_{2px,k'} g_{px}(\alpha_{2k'} r) \tag{12.5.6.3}$$

$$\phi''_{2s}(r) = \sum_{k''=1}^{N''_2} d_{2s,k'} g_s(\alpha_{2k''}, r) \tag{12.5.6.4}$$

$$\phi''_{2p_x}(r) = \sum_{k''=1}^{N''_2} d_{2px,k''} g_{px}(\alpha_{2k''}, r) \tag{12.5.6.5}$$

where g_{1s} and g_{2p} are the GTO's already seen at Sect. (12.4; Eqns. 12.4.10.1–12.4.10.2).

It may be noted that 2s and 2p share the same Gaussian exponents $\alpha_{2k'}$ and $\alpha_{2k''}$.

The functions ϕ' and ϕ'' represent the inner and outer parts of the valence shell, because the $\alpha_{2k''}$ exponent is smaller than the $\alpha_{2k'}$.

Since each molecular orbital is obtained in the form

$$\Phi_i = \sum_{\mu} c_{\mu i} \phi_{\mu} \tag{12.5.7}$$

with $c_{\mu i}$ as variational coefficients, the relative weighting of the ϕ', ϕ'' parts in the expansion (12.5.7) "*gives the molecular orbital* Φ_i *some in–out flexibility*" (Ditchfield et al.,1971).

For hydrogen, there is no inner shell, but the valence orbital is still represented by two functions:

$$\phi'_{1s}(r) = \sum_{k'}^{N'_1} d_{k'} g_s(\alpha_{k'}, r) \tag{12.5.8.1}$$

$$\phi''_{1s}(r) = \sum_{k''}^{N''_1} d_{k''} g_s(\alpha_{k''}, r) \tag{12.5.8.2}$$

An expansion 4-31G is meant simply to indicate that for the heavy atoms $N_1 = 4$, $N_2' = 3$, $N_2'' = 1$. As already noted, each atomic orbital is represented thus by four Gaussians, but the valence orbitals are split into two parts consisting of three and one Gaussian, respectively. The choice is dictated by a compromise between the needs for accuracy and efficiency. The three-one splitting of the valence shell results in two parts "*contributing comparable total amounts to the orbitals of the free atom*" (Ditchfield et al., 1971).

As anticipated, the α and d parameters were selected by Ditchfield et al. (1971) in order to minimize the calculated energy of the atomic ground state, through the open shell procedure of Pople and Nesbet (1954) (all the investigated

Table 12.9 The hydrogen atom basis functions of Ditchfiled et al. (1971). The ground state is in Russel-Saunders notation (cf. Sect. 6.3)

Element	gr. state	α'	d'	α"	E (*Hartree*)
Hydrogen	^2S	13.00773	0.0334946	0.1219492	−0.499278
		1. 962,079	0.2 34,727		
		0.444529	0.8137573		

atoms are multiplets). The minimization of the atomic energy at ground state was carried out with the pattern search technique of Hooke and Jeeves (1961). The results are shown for H in Table (12.9) and for C, N, O, F in Table (12.10) and Fig. (12.2) shows the various contributions to the 1s orbital.

Besides the 4-31G, other sets of "*double-zeta*" spit valence basis sets[2] have been proposed in literature, especially from the research group of John Pople. I may quote as examples of double-zeta basis sets the 3-21G and 6-21G of Binkley et al. (1980) for atoms Li to Ne; the set of Gordon et al. (1982) for 2nd-row elements (Na to Ar); and the 6-31G basis set of Francl et al. (1982) (Na to Ar).

Krishnan et al. (1980) promoted the utilization of "*triple zeta*" expansions (6-311G) for first row elements (Li to Ne) and McLean and Chandler (1980) extended the procedure to the neutral second row atoms, Z = 11–18, and for the negative ions P^-, S^-, and Cl^-. Triple zeta basis set expansion were promoted for 3rd-row elements by Binning and Curtiss (1990), McGrath and Radom (1991), Curtiss et al. (1995). Blandeau et al. (1997) proposed a set of all electron 6-311G expansions for K and Ca, stemming from the (14s,9p) → [8s,5p] double-zeta quality basis sets of Schäfer et al. (1992).

The normalized Gaussian functions are of form:

$$\phi_{1s} = \sum_{k=1}^{6} d_{1s,k} g_s(\alpha_{1k}) \tag{12.5.9.1}$$

$$\phi'_{2s} = \sum_{k'=1}^{3} d'_{2s,k'} g_s(\alpha_{2k'}) \tag{12.5.9.2}$$

$$\phi'_{2p} = \sum_{k'=1}^{3} d'_{2p,k'} g_p(\alpha_{2k'}) \tag{12.5.9.3}$$

$$\phi''_{2s} = g_s(\alpha''_2) \tag{12.5.9.4}$$

[2] The term *double-zeta* could appear misleading when referred to the double exponent. In fact the exponent is denoted with a Greek zeta (ζ) in STO's (cf. Eqs. 12.4.4.1-12.4.4.3 for example) but with a Greek a (∝) in GTO's (see for instance Eqs. 12.4.9.1–12.4.9.3).

Table 12.10 The 4-31G first-row atom basis functions of Ditchfiled et al. (1971)

Element	gr. state	α_1	d_{1s}	α_2'	d_{2s}'	d_{2p}'	α_2''	E (*Hartree*)
Carbon	^3P	486.9669	0.0177258	8.673525	−0.123837	0.0635454	0.1697095	−37.636918
		73.37109	0.1234779	2.096619	−0.2273385	0.2982678		
		16.41346	0.4338754	0.6046513	1.185174	0.7621032		
		4.344984	0.565042					
Nitrogen	^4S	671.2795	0.0175983	12.64524	−0.1174893	0.0640203	0.2352813	−54.327925
		101.2017	0.1228462	2.981719	−0.213994	0.3112026		
		22.69997	0.4337821	0.8494318	1.174502	0.7346079		
		6.040609	0.5614182					
Oxygen	^3P	883.2729	0.0175506	16.52326	−0.113401	0.0685453	0.295585	−74.705481
		133.1293	0.1228292	3.856837	−0.1772865	−3.312254		
		29.90641	−0.4348836	1.092729	1.150408	0.7346079		
		7.978677	0.5600108					
Fluorine	^2P	112.6163	0.0174758	21.49537	−0.1110571	0.0698888	0.3730318	−99.265481
		169.7432	0.1225231	4.989778	−0.1683221	0.3393875		
		38.18151	0.4349985	1.403574	1.143626	0.727959		
		10.21204	0.5598122					

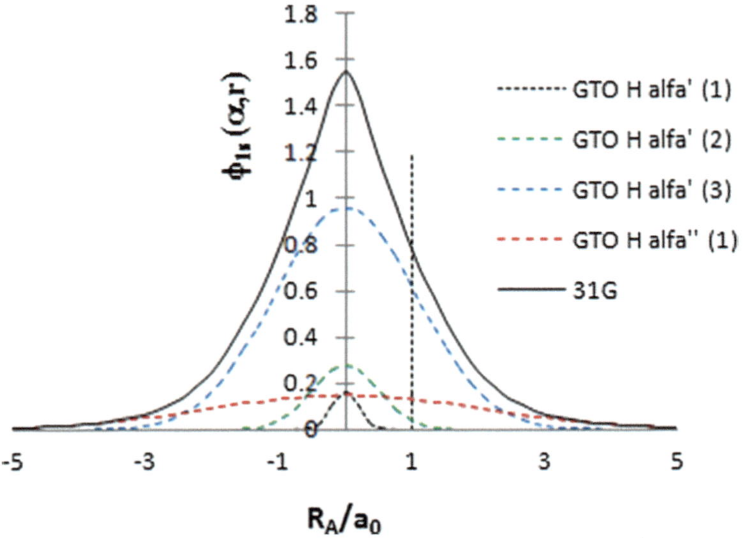

Fig. 12.2 Radial plot of the $\phi_{1s'}$ and $\phi_{1s''}$ functions for hydrogen. In plotting the α'' contribution it is assumed for graphical reasons d″ = 1.0. The vertical dashed bar delineates the Bohr's radius

Table 12.11 The hydrogen atom basis functions of Krishnan et al. (1980)

Element	gr. state	α'	d′	α''	α'''	E (*Hartree*)
Hydrogen	^2S	33.865	0.0254938	0.32584	0.102741	−0.49981
		5.09479	0.190373			
		1.15879	0.852161			

$$\phi''_{2p} = g_p(\alpha''_2) \tag{12.5.9.5}$$

$$\phi'''_{2s} = g_s(\alpha'''_2) \tag{12.5.9.6}$$

$$\phi'''_{2p} = g_p(\alpha'''_2) \tag{12.5.9.7}$$

where, $g_s(\alpha)$ and $g_p(\alpha)$ are normalized s-, p- type primitive Gaussian functions with exponent α. For comparative purposes with the double-zeta set, I report hereafter in Table (12.11) the triple-zeta expansion for hydrogen, according to Krishnan et al. (1980) and, as an example, the 6-311G expansion for C, according to the same Authors Table 12.12.

Table 12.12 The 6-311G basis functions of Krishnan et al. (1980) for carbon

Element	gr. state	α_1	d_{1s}	α_2'	d_{2s}'	d_{2p}'	α_2''	α_2'''	E (*Hartree*)
Carbon	^3P	456.324	0.0019667	20.9642	0.11466	0.0402487	0.483856	0.145585	−37.712868
		682.024	0.0152306	4.80331	0.919999	0.237594			
		154.973	0.0761269	1.45933	−0.0030307	0.815854			
		44.4553	0.260801						
		13.029	0.616462						
		1.82773	0.221006						

12.5.3 Polarized and Diffuse Functions

The adoption of polarized and diffusion additional terms in the basis set expansion is deemed to allow an extra flexibility in size and shape to the molecular orbitals, beyond the normal set.

The addition of polarization functions to split-valence extended Gaussian basis sets was promoted by Harriharan and Pople (1973). These are d-type Gaussians with the same d-exponent (0.8) for the first row atoms C, N, O and F and p-type gaussians with p-exponent 1.1 for hydrogen. The advantage obtained by adding polarization functions is testified by the lowering of the computed LCAO-MO-SCF ground-state energy on some key molecules (Table 12.13). Harriharan and Pople (1973) proceeded in two steps: they first introduced a single (uncontracted) d-gaussian functions to the 6-31G set for heavy atoms (the hydrogen functions being unmodified) and then added the p polarized function on hydrogen.

The additional six (2nd-order, unnormalized, cartesian) d-type functions are Hariharan and Pople, 1973):

$$\left(x^2, y^2, z^2, xy, yz, zx\right) \exp\left(-\alpha_d r^2\right) \tag{12.5.10}$$

and are equivalent to the five pure d-type functions:

$$\left(d_{3z^2-r^2}, d_{xz}, d_{xy}, d_{x^2-y^2}\right) \exp\left(-\alpha_d r^2\right) \tag{12.5.11}$$

plus the s-type function

$$r^2 \exp\left(-\alpha_d r^2\right) \tag{12.5.12}$$

The conversion between the two forms is easily obtained by applying (Krishnan et al. 1980):

$$d_{3z^2-r^2} = z^2 - \frac{1}{2}\left(x^2 - y^2\right) \tag{12.5.13}$$

$$d_{x^2-y^2} = \frac{\sqrt{3}}{2}\left(x^2 - y^2\right) \tag{12.5.14}$$

By adding p-polarization functions on the hydrogen atoms:

$$(x, y, z) \exp\left(-\alpha_p r^2\right) \tag{12.5.15}$$

(with fixed $\alpha_p = 0.8$), the ground energy of the target molecules is effectively lowered of some extent with respect to the simple d-polarization effect (Table 12.13). The energy lowering due to d functions is found to depend on the local symmetry around the heavy atom. The mean absolute deviation between theory and experiment for 6-31G** functions is around 4 *kcal/mole*.

Table 12.13 Total HF/6-31G(d,p) energies (*Hartree*) with standard (*≡d) and (**≡d,p) polarization functions versus estimated Hartree–Fock limits (from Harriharan and Pople, 1973: slightly modified)

Molecule	6-31G	6–31 G*	6–31 G**	Limit
H_2	−1.12676	−1.12676	−1.13129	−1.1336
CH_4	−40.18038	−40.19506	−40.20159	−40.225
NH_3	−56.1632	−56.18374	−56.19499	−56.225
OH_2	−75.98508	−76.00987	−76.02255	−76.065
FH	−99.98343	−100.00281	−100.01122	−100.071
HC ≡ CH	−76.79261	−76.81732	−76.82138	−76.86
HC = N	−92.82763	−92.87317	−92.87515	−92.92
$C^- - O^+$	−112.66722	−112.73718	−112.73718	−112.791
N ≡ N	−108.86762	−108.94234	−108.94234	−108.997
F − F	−198.64605	−198.6729	−198.6729	−198.77

Diffuse functions are Gaussian functions with small orbital exponents, causing them to be larger in radial extent (i.e. expanding the tail). The possibility of occupying larger spatial extensions is particularly useful in describing electrons associated to anions, or lone-pair electrons, or electrons in soft molecules, located far away from the nucleus. Chandrasekar et al. (1981) showed that the addition of a set of diffuse functions to the 4-31G basis set of Ditchfield et al. (1991) results in a dramatic improvement in calculated proton affinities, especially when the Möller-Plesset P2 correction for electron correlation is included. Clark et al. (1983) proposed 4–31+G and 3–21+G basis sets for first row elements (Li-F). The basis set exponents were first optimized for the anions H^-, Li^-, BeH^-, BH_2^-, CH_3^-, NH^-, OH^-, and F^-, in their singlet ground-state configurations. A single diffuse sp shell was then added to the 6-31G basis set of Ditchfield et al. (1971) and to the 6-21G basis sets of Binkley et al. (1980), and the exponents were optimized with respect to anion total energy. The core and inner two-valence shells remained unaltered. The two sets of diffuse exponents obtained in this way are listed in Table 12.14. The first set (4–31+G) was derived from the 6-31G basis set and the second set ("*all other*" in Table 12.14) was used for the 3–21+G, 6–21+G, and 6–31+G* expansions. Clark et al. (1983) recommended replacement of the 4–31+G by the more efficient 3–21+G basis set. A single, standard set of diffuse exponents should be used to augment all basis sets.

In Table 12.15 we see as an example the 6–31+G(d) basis set for carbon. The function 1 is a linear combination of 6 primitive Gaussians built up with the exponent and s-coefficients listed in the table, i.e.:

1 s = 0.002 g(3047,**r**) + 0.014 g(457.4,**r**) + 0.069 g(103.9,**r**) + 0.232 g(29.21,**r**) + 0.468 g(9.287,**r**) + 0.362 g(3.164,**r**).

The function 2 s is given by:

2 s = -0.119 g(7.868,**r**)-0.161 g(1.881,**r**) + 1.143 g(0.544,**r**).

Table 12.14 Diffuse functions for 1st-row elements. p set from Dunning and Hay (1977). sp set (for 4-31G) and "all other basis sets (b. s.)" from Clark et al. (1983)

Element	p	sp	All other b.s
H	–	–	0.036
Li	–	0.012	0.0074
Be	–	0.026	0.0207
B	0.019	0.034	0.0315
C	0.034	0.04	0.0438
0	0.059	0.068	0.0845
N	0.048	0.053	0.0639
F	0.074	0.09	0.1076

Table 12.15 Example of combined (diffused + polarized) basis functions. The 6–31+ G(d) basis set of carbon

Shell 1s				
N	Exponent	s-coefficient		
1	3047.5249	0.0018347		
	457.36952	0.0140373		
	103.94869	0.0688426		
	29.210155	0.2321844		
	9.286663	0.4679413		
	3.163927	0.362312		
Shell 2sp				
N	Exponent	s-coefficient	p-coefficient	
2–5	7.8682724	−0.1193324	0.0689991	
	1.8812885	−0.1608542	0.316424	
	0.5442493	1.1434564	0.7443083	
Shell 3sp				
N	Exponent	s-coefficient	p-coefficient	
6–9	0.1687145	1	1	
Shell 4sp				
N	Exponent	s-coefficient	p-coefficient	
10–13	0.0438	1	1	
Shell 5d				
N	Exponent	s-coefficient	p-coefficient	d-coefficient
14–19	0.8	–	–	1

The functions $2p_x$, $2p_y$, $2p_z$ are:
$2p = 0.069\ g(7.868,\mathbf{r}) + 0.316\ g(1.881,\mathbf{r}) + 0.744\ g(0.544,\mathbf{r})$.
and so on.

It must be also stressed that each primitive Gaussian $g(\alpha,\mathbf{r})$ is of the form (7.3.82.1) and thus involves a normalization coefficient.

12.5.4 Basis Set Superposition Error and Counterpoise Procedure

When two atoms approach each other within a single molecular fragment, or two molecules approach each other, such as two monomers in a dimeric species, the atomic basis functions may partially overlap. This overlap causes an augmented electron density and an artificial stabilization of the complex. The Basis Set Superposition Error (BSSE) so induced may be defined "*intramolecular*" when involves two (or more) atoms within a given molecule or "*intermolecular*" when it involves two molecular clusters.

As noted by Boys and Bernardi (1970) BSSE effects may be particularly problematic in the main region of importance for thermodynamic properties, that is, where electrostatic, overlap and dispersion forces have the same order of magnitude and in which these are required to an accuracy of 10^{-5} to 10^{-7} *Hartree.*

Consider the interaction potential of two molecules A and B at a distance r apart as:

$$\Delta_{AB}E(r) \equiv \overline{E}_{AB}(r) - \overline{E}_A - \overline{E}_B \qquad (12.5.16)$$

where the total energy of the interacting system $\overline{E}_{AB}$ and the energies of the two separate molecules $\overline{E}_A$ and $\overline{E}_B$ are obtained from solutions of the appropriate Schrödinger equations. As anticipated $\Delta_{AB}E(r)$ is a function of distance and orientation and can be reduced by introducing "*some errors in the energy calculations for two separate molecules as nearly equal to the corresponding errors in the bimolecular calculation as possible*" (Boys and Bernardi, 1970), i.e. the "*counterpoise procedure*".

The counterpoise bivariational procedure implies the adoption of the "*transcorrelated wave equation*" of Boys and Handy (1969). Equation (12.5.16) is rewritten in the form:

$$\Delta^C_{AB}E(r) \equiv \overline{E}_{AB}(r) - \overline{E}^C_A - \overline{E}^C_B \qquad (12.5.17)$$

with the superscript C implying that all the counterpoise procedures have been used and the total energy of the two molecules at a distance r has the separate energies subtracted with the latter computed by the same points and functions as for E_{AB}.

The transcorrelated method adopts an approximate wavefunction of form:

$$C\Phi = \prod_{i>j} f(r_i, r_j)\mathcal{A}(\phi_1, \phi_2, \ldots) \tag{12.5.18}$$

where $f(r_i, r_j)$ are correlation functions to be included for all pairs of electrons and $\phi_1, \phi_2, \ldots$ are spin orbitals.

The computed energy must satisfy:

$$\left(C^{-1}HC - E\right)\Phi = 0 \tag{12.5.19}$$

Denoting as Ys a set of adjustable parameters in the functions $f(r_i, r_j)$, their values are constrained by:

$$\left\langle \frac{\partial}{\partial Y_s}(C^{-1}\Phi)|H|C\Phi \right\rangle = 0 \tag{12.5.20}$$

$$\left\langle \left(\frac{\partial}{\partial d_s}\log C\right)|C^{-1}HC - CHC^{-1}|\Phi \right\rangle = 0 \tag{12.5.21}$$

The d_s in Eq. (12.5.21) are optimization parameters which occur in $f(r_i,r_j)$. The optimum CΦ and the optimum E can only be obtained when Eqs. (12.5.20, 21) are simultaneously satisfied.

The computations are carried out in practice by creating "*empty basis sets*" through the adoption of ghost atoms, which have the basis set of the real ones, but no electrons to fill it, nor nuclear charges.

In Geochemistry, the intramolecular BSSE may be particularly relevant in aqueous metalorganic complexes and deserves attention. In Table 12.16 are given as an example the details of the BSSE in the case of aqueous complex Fe(II)–monoacetate (Ottonello and Vetuschi Zuccolini, 2008). The calculations were first carried out on the entire complex $[(CH_3COO)Fe]^+$, then were repeated assigning, in two distinct runs, a zero charge to the ghost cation (Fe^{2+} subtracted in Table 12.16) and to the ghost ligand [$(CH_3COO)^-$ subtracted, in Table 12.16]. The two fragments were then recalculated as single isolated moieties [Fragment Fe^{2+} and Fragment $(CH_3COO)^-$ in the same table]. As we may see, the overlap of basis functions generates an error which is quite relevant for the central metal. The correction is obviously positive (i.e. the bonding energy is overestimated), thus reducing the mismatch with the experiments in terms of Gibbs free energy of formation from the elements, which remains however quite high (i.e. $\Delta G^\circ_f =$ −518.1 *kJ/mole* with respect to -468.8 *kJ/mole* for PCM calculations carried out with UA0 radii.

As an example of intermolecular BSSE correction I report in Fig. 12.3 the results of Hellmann et al. (2008) on the methane dimer, another case study quite important in Geochemistry. The methane dimer has 17 conformers with different angular

Table 12.16 BSSE calculations for $[(CH_3COO)Fe]^+$ at B3LYP/6-31G(d,p) theory level (Ottonello and Vetuschi Zuccolini, 2008)

Moiety	Electron energy (*Hartree*)	Bonding energy (*kJ/mol*)
$[(CH_3COO)Fe]^+$	−1491.82598913	
Fe^{2+} subtracted	−228.50258623	
$(CH_3COO)^-$ subtracted	−1262.65304543	
Fragment Fe^{2+}	−1262.64465360	
Fragment $(CH_3COO)^-$	−228.50219759	
With counterpoise correction		−1760.024
Without counterpoise correction		−1783.078
Counterpoise on Fe^{2+}		22.033
Counterpoise on $(CH_3COO)^-$		1.021
BSSE		23.054

orientations of the methane molecules. The lowest energy is attained with the conformer I (represented in the inset of Fig. 12.3).

The ab initio computations of Hellmann et al. (2008), carried out at CCSD(T)/Aug-cc-pVQZ level and Counterpoise correction, define a pronounced well with the minimum at 3.633 Å, an internuclear C–C distance slightly lower with respect to the experimental minimum of Matthews and Smith (1976) which is at ~ 3.85 Å. Repeating the calculations with a much less extended basis set, in order to emphasize the role of the Counterpoise correction, we see that we are always in a repulsive state (i.e. lack of a minimum) and that an hypothetical minimum (at ~ 4.175 Å) is eventually attained only when disregarding the BSSE. This is obviously connected with the fact that BSSE progressively increases with the shortening of the interatomic distance, and that the adopted functional does not take into consideration the dispersive forces.

In periodic structures the BSSE is evaluated through the "*Geometrical Counterpoise Correction*" (GCP). In the GCP approach, the Boys-Bernardi counterpoise is put in action through a semi-empirical, repulsive pair potential $V_A(r)$, which decays exponentially with the inter-atomic distance r; Brandenburg et al., 2013):

$$V_A(r) = e_A^{miss} \frac{\exp(-\alpha r^{\beta})}{\sqrt{S \cdot N_B^{virt}}} \tag{12.5.22}$$

e_A^{miss} in the above equation is the energy difference between a large basis set and the (generally smaller) target for an atom A inside a weak electric field. It is thus a measure of the energy loss arising from incompleteness of the atomic target basis

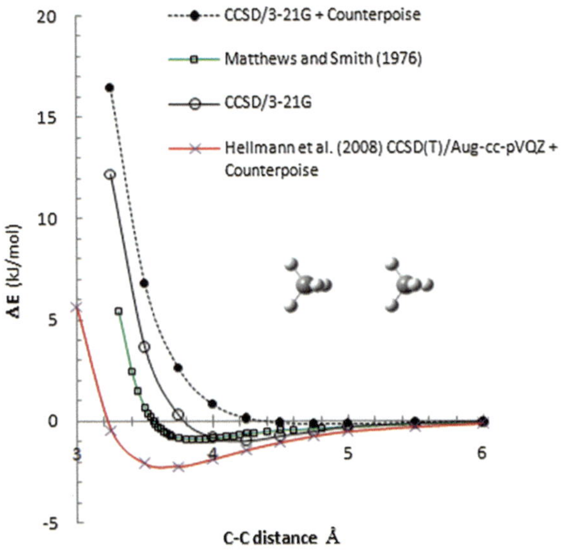

Fig. 12.3 The potential well of the methane dimer (conformer I, shown in the inset)

(Kruse and Grimme, 2012). S is the Slater-overlap, N_B^{virt} is the number of virtual orbitals, and the empirical parameters α and β have the duty of normalizing the potential.

The GCP energy (E_{GCP}) is obtained by summing up the potentials over all atom pairs:

$$E_{GCP} = \frac{\sigma}{2} \sum_{A \neq B}^{pairs} \sum_{T} V_A(\|r_A - r_B + T\|) \tag{12.5.23}$$

where σ is a global scaling parameter and the summation in T runs over all distinct pairs AB inside a large super-cell of certain radius, in order to satisfy the periodic boundary conditions.

The four fit parameters (σ, α, β, η), needed for each basis set, have been obtained by Brandenburg et al. (2012) by least-squares fitting of the GCP-corrected energies of the large (S66 × 8) non-covalent interaction benchmark set of Řezáč et al. (2011).

The GCP procedure is implemented in the recent release (17) of the CRYSTAL package (Civalleri, 2017).

12.6 Plane Waves

The possible utilization of plane waves (PW) in Wave Mechanics was promoted by Sommerfeld and Runge (1911), following "*an idea of Debye*" (Sommerfeld, 1930): *"Der wellenoptische Ansatz fur eine ebene Welle heisst nun* $u = u_0 \exp[ik(\alpha x + \beta y + \gamma z)]$ *wobei* $n(\alpha x + \beta y + \gamma z)$ *der Lichtweg längs der Fortpflanzungsrichtung der Welle ist. Wir setzen daher mit Rücksicht auf die Bedeutung des Eikonals als Lichtweg entsprechend* $u = u_0 \exp\left(i\frac{k}{n}E\right)$ *dabei suchen wir* $\mathbf{E}$ *als Ortsfunktion so zu bestimmen...."* (Sommerfeld and Runge, 1911); the "*Eikonal*" in the above sentence being the characteristic function for ray optics.

The term Plane Wave (*ebene welle*) is somewhat misleading when adopted in quantum mechanics, because it is conceptually associated with a particular choice of amplitude. PW is indeed periodic and continuous (and differentiable), and the "*planes*" arise eventually from the selection of a particular amplitude: the wave maxima, a series of equally spaced parallel planes, normal to the wavevector.

Neglecting this lexical inconsistency we may agree with Pisani (1996a) in asserting that PW's are, in a sense, "*the ideal basis functions for a periodic system*".

Denoting $\phi_k(r)$ a PW of the general wave vector **k** at position **r**:

$$\phi_k(r) = \frac{1}{\sqrt{\Omega}} \exp(ik \cdot r) \tag{12.6.1}$$

where Ω is the volume of the first Brillouin Zone (BZ), it may be shown that, at the light of the Bloch Theorem (cf. Sect. 8.4) it is also valid:

$$\phi_k(r) = \frac{1}{\sqrt{\Omega}} \exp(ik \cdot r) = \frac{1}{\sqrt{\Omega}} \exp[i(\mathcal{K} + K) \cdot r] = \phi_K(r; \mathcal{K}) \tag{12.6.2}$$

where **K** is a vector of the reciprocal lattice and $\mathcal{K}$ is the translationally equivalent vector in the first BZ.

Any function belonging to the realm of continuous differentiable functions can be expanded into an orhtonormal complete PW set, with arbitrary precision. Denoting r_x the radius of a sphere centered at the origin in the reciprocal space, the number of **k** vectors [hence of PW's = $\phi_K(r;\|)$] contained in the sphere is $n_{PW} = \left(\frac{4}{3}\pi r_x^3\right) \times \Omega^{-1}$. For a volume V in the direct space of 1000 $\AA$, $\Omega = (8\pi^3/V) = 0.248\ \AA^{-1}$ and $n_{PW} = 534$. The finest detail that can be revealed in the direct space with such a set is $1/r_x = 0.316\ \AA$, i.e. about half the Bohr's radius. This precision may be satisfactory for valence electrons, but it is definitely inacceptable for core electrons which deserve a separate treatment (i.e. linearized-aughmented PW, or Pseudo Potentials, see later).

As discussed at Sect. (8.6.4), the variational Kohn–Sham procedure to be adopted in determining the ground state energy (E_0) involves few steps:

(1) Find the n/2 lowest eigenvalues (and the corresponding eigenfunctions of a one-electron effective Hamiltonian h_{eff}

$$\mathsf{h}_{eff}\psi_i(r) \equiv \left[-\frac{\nabla^2}{2} + V_{eff}(r)\right]\psi_i(r) = \varepsilon_i\psi_i(r) \quad (12.8.6.44)$$

(2) Calculate the density:

$$\rho(r) = 2\sum_i |\psi_i(r)|^2 \quad (12.8.6.45)$$

(3) Re-compute $V_{eff}(\mathbf{r})$ as a functional of the density $\rho(r)$:

$$V_{eff}(r) = V(r) + \int \frac{\rho(r')}{|r - r'|} dr' + \mu_{exc}(r; [\rho] \quad (12.8.6.46)$$

(4) Replace $V_{eff}(\mathbf{r})$ in (8.6.44) and repeat the calculus, until convergence is attained.

Limiting ourselves to pseudo-valence electrons and denoting as ρ'(**r**') their density, the potential generated at the general point **r** by the distribution of pseudo-valence electrons $v_H(\mathbf{r})$ (2nd term on the right in Eq. 8.6.46) will be:

$$v_H(r) = \int \frac{\rho'(r')}{|r - r'|} dr' \quad (12.6.3)$$

Since ρ'(**r**) is (in direct space) a periodic function, and a PW is periodic as well, Eq. (12.6.3) may be replaced by:

$$v_H(r) = \sum_{k \neq 0} v_k \exp(ik \cdot r) \quad (12.6.4)$$

$$\rho'(r) = \sum_K \rho_k \exp(iK \cdot r) \quad (12.6.5)$$

The knowledge of the ith occupied orbitals allows to determine $\rho_{\mathbf{k}}$:

$$\psi_i(r,\mathcal{K}) = \sum_{\mathcal{K}} c_{i,k,\|}\phi_{\mathcal{K}}(r, \mathcal{K}) \quad (12.6.6.1)$$

$$\rho_k = 2\sum_{\mathcal{K}}\sum_i |c_{i,k,\mathcal{K}}|^2 \quad (12.6.6.2)$$

and the application of the Poisson's relation (Pisani 1996a):

$$\nabla^2 v_H(r) = 4\pi\rho'(r) \tag{12.6.7.1}$$

gives the value of $v_{\mathbf{k}}$:

$$v_k = 4\pi\rho_k(r)/K^2 \tag{12.6.7.2}$$

The number of PW's necessary to obtain the appropriate accuracy depends upon the "*cutoff energy*"[3] $T < r_x^2$. All PW's with $(\mathcal{K} + \mathbf{K}^2) < T$ must be included.

12.6.1 Augmented Plane Waves

The Augmented Plane Wave (APW) procedure was proposed by Slater already in 1937 as "*a new method for approximating the solutions of the problem of the motion of an electron in a periodic potential, as a crystal lattice*" (Slater, 1937). In this procedure the potential is supposed to be spherically symmetrical within spheres surrounding the nuclei, and constant outside.[4] Within the spheres, the wave function is expanded in spherical harmonics and radial solutions of the wave equation. Outside the spheres the wave function is represented by plane waves, joining continuously at the surface.

The spherically symmetrical potential within the nth sphere was assumed by Slater (1937) to be $U_n(|r - r_n|)U$, with $\mathbf{r}$ as radius vector to an arbitrary point and $\mathbf{r_n}$ as radius vector to the nth nucleus. Outside all the spheres, the potential was assumed to be constant. To get continuity Slater assumed that each of the potentials U_n reduces to this constant (zero) value at the radius R^n. The spherical coordinates with respect to the nucleus are $|r\text{-}r_n|, \theta, \phi$. The wave function is:

$$\psi_i = \sum_{l=0}^{\infty} \sum_{m=-l}^{l} A_{lm} P_l^{|m|}(\cos\theta)\exp(im\phi)u_{nl}(|r - r_n|) \tag{12.6.8}$$

with the radial function u_{nl} satisfying the equation:

$$-\frac{1}{r^2}\frac{d}{dr}\left(r^2\frac{du_{nl}}{dr}\right) + \left[\frac{l(l+1)}{r^2} + U_n\right]u_{nl} = Eu_{nl} \tag{12.6.9}$$

where E is energy.

---

[3] The cutoff energy T is usually given in Rydbergs (*Ry*), for historical reasons. 1 *Ry* = 0.5 *Hartree*.

[4] This approximation is generally known as *Muffin Tin*, for the peculiar shape of the potential in the spheres (resembling a *Muffin*) and flat outside (i.e. the *tin*). Slater, however never used this term in its 1937 paper.

Any series of the form (12.6.8), with arbitrary coefficients A_{lm}, satisfies the differential equation within the sphere. Because there are two independent solutions of (12.6.8) for any energy value E (one regular at the origin and the other regular at infinity) the selected energy value is that one regular at the origin because it concerns the interior of the sphere.

Outside the spheres the solution of the problem was set by Slater (1937) in terms of a series of plane waves of form:

$$\psi_i = \sum_k v_k \exp(ik \cdot r) \tag{12.6.10}$$

where each potential at **k** satisfies the difference equation

$$(k_0 + k_i)^2 v(k_0 + k_i) + \sum_{k_j} W(k_j) v(k_0 + k_i + k_j) = Ev(k_0 + k_i) \tag{12.6.11}$$

where the matrix component of the potential energy W($\mathbf{k}_i$) between the wave functions is normalized in such a way that W(0) is the average potential energy through the cell.

For a crystal composed of several unit cells:

$$F_n(k_j) = \int U_n(|r - r_n|) \exp(ik_j \cdot |r - r_n|) dr = 4\pi \int_0^{R_n} r^2 U_n(r) \frac{\sin(|k_j| r)}{|k_j| r} dr \tag{12.6.12}$$

A single unperturbed function consists of a single plane wave outside the spheres, together with the necessary spherical functions within the spheres. The matrix components of energy must be set up between these unperturbed functions. The secular equation however must be solved numerically, because it involves the energy explicitly (and also implicitly through the ratio of the slope of the various radial functions to the functions themselves, at the surfaces of the spheres).

The exponential exp ($i\mathbf{k}_i \cdot \mathbf{r}$) was then expanded by Slater (1937) in spherical coordinates about the nucleus of the nth atom, at r_n, and through various passages involving spherical Bessel functions (see Sect. A.18.2.1), he arrived to determine the wave function inside the sphere as:

$$\psi_i = \exp(ik_i \cdot r_n) \sum_{l=0}^{\infty} \sum_{m=-l}^{l} (2l+1) i^l \frac{j_l(k_i R_n)}{u_{nl}(R_n)} \times u_{nl}(|r - r_n|) \frac{(l - |m|)}{(l + |m|)} P_l^{|m|}(\cos\theta) \times P_l^{|m|}(\cos\theta_i) \exp[im(\phi - \phi_i)] \tag{12.6.13}$$

The value (12.6.13) inside the sphere at r_n, together with the expression exp ($i\mathbf{k} \cdot r$) outside all the spheres, determines the function ψ_i completely.

As noted by Slater (1937) however, the numerical solution of the secular equation was not an easy task due to the energy-dependent radial basis functions: "*It is a more difficult secular equation than one usually meets. In the first place, the functions are not orthogonal, so that the energy E appears explicitly in the non-diagonal terms. But worse than this, the energy appears implicitly in the expressions* $u_{nl}'(R_n)/u_{nl}(R_n)$*, which depend on the energy, and which appear both in diagonal and in nondiagonal terms*" (Slater, 1937).

12.6.2 Linearized Augmented Plane Waves

An important improvement to APW came from Andersen (1975) who introduced a linearization of this energy dependency (LAPW). The obvious computational advantage of using the variational principle for Schrodinger's equation with energy-independent basis functions is that the secular equations become linear in energy, that is, they reduce to the eigenvalue equations

$$\sum_{\mathrm{L}} \left(\mathrm{H}^{\mathrm{k}}_{\mathrm{LL}} - \mathrm{E}^{k}_{\mathrm{j}} \mathrm{O}^{\mathrm{k}}_{\mathrm{LL}} \right) \mathrm{A}^{\mathrm{k}}_{\mathrm{Lj}} = 0 \tag{12.6.14}$$

where the wave functions are strictly orthonormal and all eigenvalues $\mathrm{E}^{\mathrm{k}}_{\mathrm{j}}$ and eigenvectors $\mathrm{A}^{\mathrm{k}}_{\mathrm{Lj}}$ of a given Bloch vector **k** can be found simultaneously by diagonalization procedures.

I resume briefly hereafter part of what Andersen considered "*the essence of the LAPW method*", (i.e. his paragraphs IIA, IIIA, but not VB).

12.6.2.1 Linear Method for Single Sphere (Basic Formalism)

An approximation, valid in the neighborhood of some arbitrary energy E_v to the eigenvalues of a single atomic (or muffin-tin) sphere of radius S, and with the spherically symmetric Hamiltonian $[-\nabla^2 + v(r)]$ (in Rydberg units; *Ry*), is obtained by adopting as radial basis functions, for each value of *l*:

$$\phi_{vl}(\mathrm{r}) = \phi_l(\mathrm{E}_{v,\mathrm{r}}) \tag{12.6.14}$$

and its derivative

$$\dot{\phi}_{vl}(\mathrm{r}) = \left.\frac{\partial \phi_l(\mathrm{E},\mathrm{r})}{\partial \mathrm{E}}\right|_{\mathrm{E}_v} \tag{12.6.15}$$

Omitting the subscript l, the normalization yields:

$$\langle \Phi^2(E) \rangle = \int_0^S \varphi(E, r) r^2 dr \equiv 1 \tag{12.6.16}$$

$\phi(r)$ and $\dot{\phi}(r)$ are orthogonal. The radial logarithmic derivatives at the sphere boundary yield:

$$D_v = {S\phi_v\prime(S)}/{\phi_v(S)} \tag{12.6.17.1}$$

$$D_{\dot{v}} = {S\dot{\phi}_v\prime(S)}/{\dot{\phi}_v(S)} \tag{12.6.17.1}$$

In terms of spherical harmonics the radial basis functions may be represented as:

$$\Phi_{lm}(D, r) = i^l Y_{lm}(r)\Phi_l(D, r) \tag{12.6.18}$$

where Y_{lm} is the "*spherical harmonic with the phase*", i.e., with the phase factor $(-1)^m$, introduced originally by Condon and Shortley (1935):

$$Y_{lm}(\theta, \phi) = (-1)^m \sqrt{\frac{(2l+1)}{4\pi} \frac{(l-m)!}{(l+m)!}} P_l^m(\cos\theta) \exp(im\phi) \tag{12.6.19}$$

$P_l^m(x)$ in the above equation is the associated Legendre Polynomial. The normalization factor in Eq. (12.6.19) is such that the spherical harmonics are normalized to one. These functions are orthonormal and complete. A convenient trial function is represented by the linear combination:

$$\Phi(D, r) = \phi_v(r) + \omega(D)\dot{\phi}_v(r) \tag{12.6.20}$$

where, besides l, also the subscript v is omitted. Differentiating the Schrödinger equation with respect to energy:

$$(H - E_v)\dot{\phi}_v(r) = \phi_v(r) \tag{12.6.21}$$

the Hamiltonian matrix elements in the sphere are given by (Andersen 1975):

$$\langle \Phi_{l'm'}(D')|H - E_v|\Phi_{lm}(D)\rangle = \delta_{l'l}\delta_{m'm}\omega_l(D) \tag{12.6.22}$$

and the normalization integrals are:

$$\langle \Phi_{l'm'}(D')|\Phi_{lm}(D)\rangle = \delta_{ll}\delta_{m'm} \times \left|1 + \langle \phi_{vl}^2 \rangle \omega_l(D')\omega_l(D)\right] \tag{12.6.23}$$

12.6.2.2 LAPW Formalism for a Solid

In the interstitial region the adopted energy-independent APW is the plane wave $\exp[i(k+G)\cdot r]$ specified by the reciprocal-lattice vector **G** and the Bloch vector **k**. Inside any sphere it is adopted the linear combination of the functions (12.6.20) which matches continuously and differentiably onto the plane wave:

$$\psi_K(r) = \sum_{lm} 4\pi Y^*_{lm}(k) j_l(KS) \frac{\Phi_{lm}(\widetilde{D}, r)}{\Phi_l(\widetilde{D})} \tag{12.6.24}$$

where **K** = **k** + **G,** and $\tilde{D} = \tilde{D}_l(K^2)$ is the logarithmic derivative at the Muffin Tin sphere for a free electron of angular momentum l and wave number K (see Eq. 2.29 in Andersen 1975).

The variational freedom, arising from the *over completeness*[5] of the plane-wave set in the interstitial region, is used in the LAPW method to form, inside the sphere and for each value of angular momentum, the linear combination Φ[D(E),r] corresponding to the proper energy. The Hamiltonian and overlap matrices for the Muffin Tin potential are obtained from Eqs. (12.6.22, 12.6.23, 12.6.24) and the LAPW matrix is:

$$H^k_{G'G} - \varepsilon O^k_{G'G} = \left(\delta_{G'G} - \Omega^{-1}\langle \exp[i(G - G']) \cdot r\right)\left(K^2 + V_{mtz} - E\right) + \sum_l W_l(K', K) Y_l(K', K, E) \tag{12.6.25}$$

where $\varepsilon = E\text{-}E_v$.

For a diagonal element, the first term is $(\Omega_I/\Omega)(K^2 - V_{mtz} - E)$, and, in the second term, $\sum_l W_l(K', K) = \Omega_S / \Omega$, with Ω_I and Ω_{SS} as volumes of the interstitial and spherical regions, respectively.

[5] The concept of *over completeness* was introduced by Duffin and Schaeffer (1952) for a class of non-harmonic Fourier series. In this case it is applied to plane waves. A complete system of plane waves is *overcomplete* if removal of an element from the system results in a system that is still complete.

12.6.3 Full Potential Linear Augmented Plane Waves (FPLAPW)

It is basically a generalization of the LAPW procedure obtained removing the remaining shape approximation to the potential inside the atomic spheres through a novel technique for solving Poisson's equation for a general charge density. The acronymic derives from the fact that "*all contributions to the potential are completely taken into account in the Hamiltonian matrix elements*" (Wimmer et al., 1981).

The charge density inside sphere α is given by

$$\rho(\mathrm{r}) = \sum_{v} \rho_v(\mathrm{r}_\alpha)\mathrm{K}_v(\hat{\mathrm{r}}_\alpha) - 2\mathrm{Z}_\alpha\delta(\mathrm{r}_\alpha) \tag{12.6.26}$$

where

$$\mathrm{K}_\mathrm{v}(\widehat{\mathrm{r}}_\alpha) = \sum_{m} \mathrm{c}_m(v)\mathrm{Y}_{l_v m}(\widehat{\mathrm{r}}) \tag{12.6.27}$$

is a lattice harmonic;

$$\mathrm{r}_\alpha = \mathrm{r} - \boldsymbol{\tau}_\alpha \tag{12.6.28}$$

with $\boldsymbol{\tau}_\alpha$ as position of the atomic sphere α.

The charge density in the interstitial is given by:

$$\rho(\mathrm{r}) = \sum_{\mathrm{n,s}} \rho_{\mathrm{n,s}}^{\mathrm{PW}} \cos(\mathrm{k_n z})\Phi_\mathrm{s}(\mathrm{r}) \tag{12.6.29}$$

with

$$\Phi_\mathrm{s}(\mathrm{r}) = \frac{1}{\mathrm{n_0}} \sum_{\mathrm{R}} \exp\left[\mathrm{i}\hat{\mathrm{R}}_\mathrm{s}\mathrm{G}_\mathrm{s} \cdot (\mathrm{r} - \mathrm{t_R})\right] \tag{12.6.30}$$

as two-dimensional (2D) plane-wave star function.

$\hat{\mathrm{R}}$ in Eq. (12.6.30) is the point-group part of the 2D space group operation R; $\mathbf{G}_\mathrm{s}$ is a 2D star representative reciprocal-lattice vector; $\mathrm{t_R}$ is a non-primitive 2D translation vector, and $\mathrm{n_o}$ is the number of 2D space-group operations.

Finally, in the vacuum, the charge density is given by:

$$\rho(\mathrm{r}) = \sum_{\mathrm{s}} \rho_\mathrm{s}(\mathrm{z})\Phi_\mathrm{s}(\mathrm{r}) \tag{12.6.31}$$

The representation of the charge density in the interstitial and vacuum region already has the desired form for application of the Poisson's relation (cf. Equations 12.6.7.1, 12.6.7.2) but inside the spheres, because of the rapid variation of the charge density near the nuclei, a Fourier expansion for the charge density would be extremely slowly convergent. Hence, the true rapidly varying charge inside the MT spheres is replaced by another, smoother charge (i.e. the "*full pseudo-charge-density*" Wimmer et al., 1981) able of creating the correct multipole moments.

The total potential corresponding to the total charge density, after the appropriate substitutions is:

$$V(r) = \sum_{\nu} V_{\nu}(r_{\alpha}) K_{\nu}(\hat{r}_{\alpha}) \tag{12.6.32.1}$$

$$V(r) = \sum_{n,s} V_{n,s}^{PW} \cos(k_n z) \Phi_s(r) \tag{12.6.32.2}$$

$$V(r) = \sum_{s} V_s(z) \Phi_s(r) \tag{12.6.32.3}$$

in the spheres, in the interstitial and in the vacuum, respectively.

In Fig. (12.4) we see an emblematic application of the above concepts in the case of the O_2 molecule (Wimmer et al. 1981).

The total valence density is shown in part (**a**) of the figure. The "plane-wave" density (part **b**) was obtained by Wimmer et al. (1981) using the expansion (12.6.29) not only in the interstitial region but also inside the spheres and in the vacuum region. The continuation of the plane-wave charge into the MT spheres is qualitatively similar to the original charge valence charge density. The total charge density and the total FLAPW potential are shown in parts **c** and **d** of the same figure. The total potential is composed of the Coulomb plus the exchange–correlation terms. The fact that there are no discontinuities across the boundaries in real space between the spheres, the interstitials and the vacuum testifies that Eqs. (12.6.32.1–12.6.32.3) converge in a satisfactory way at the limit. The FLAPW procedure is at the core of the WIEN2k program package, perhaps the most successful software tool for the computation of electronic structure of solids using DFT and Plane Waves. For a more in dept treatment of the FLAPW procedure and its implementation in WIEN2k, I remand the interested reader to Andersen (1975), Weinert (1981), Wimmer et al. (1981) Weinert et al. (1981, 1982), Blaha et al. (2020).

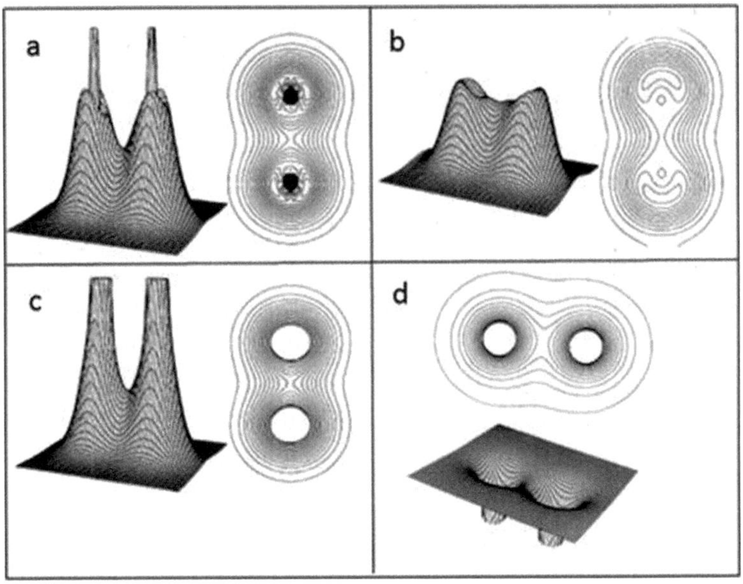

Fig. 12.4 Valence **a**, PW component **b**, total FLAPW charge density **c** and total FLAPW potential **d** for the O_2 molecule. The lowest contour value and the lowest contour spacings in **a**,**b**,**c** are 0.06 electrons/(a.u.) and the outermost contour in **d** is at -0.8 *Ry*, which is also the contour spacing. The cutoff in the figure (**d**) is at -13.6 *Ry* Modified from Wimmer et al. (1981)

12.7 Effective Core Potentials

Heavy elements have a great number of inner electrons that do not participate in a sensible way to the constitution of chemical bonds. The effect of inner (or "*core*") electrons may be "*frozen*" (i.e. abstracted from the computation) and their effect conveniently taken into account by replacing the Coulomb potential of valence electrons with Effective Core Potentials (ECPs) with fewer nodes.

The construction of ECP or "*pseudopotentials*" followed through the years two distinct lines of reasoning. In the first one, smoothed pseudopotentials were empirically fitted to simply reproduce experimental energy bands (in this case the resulting wave functions are only approximated). In the second one, orthogonalized plane-waves (OPW) were used to derive pseudopotentials from atomic calculations. In this case, the resulting wave functions generally exhibit the correct shape outside the core region though differing from the real wave functions by a normalization factor.

The effective reduction of the number of electrons (and their respective basis sets) and the much simpler form of the wave function lead to a substantial reductions in computational effort. Perhaps more importantly, an ECP allows an efficient treatment of relativistic effects. This second aspect is of particular relevance in

Geochemistry, in the investigation of reactivity-transport of heavy elements and ore formation.

To account for relativistic effects in a consistent way for all atoms in the Periodic Table, electron interaction effects are described in the LDA framework. The ground-state energy of a system of interacting electrons in the potential exerted by the nucleus is, in this view, a functional of the electron density ρ(r):

$$E[\rho] = T[\rho] + E_{coulomb}[\rho] + \int V_{ext}(r)\rho(r)dr + E_{xc}[\rho] \quad (12.7.1)$$

where $T[\rho]$ is the kinetic energy of the (non interacting) electrons in the relativistic formulation of Dirac (1957); $E_{coulomb}[\rho]$ is the Coulomb energy of the electrons:

$$E_{coulomb}[\rho] = \frac{1}{2}\int\int \frac{\rho(r)\rho(r')}{|r - r'|}drdr' \quad (12.7.2)$$

$V_{ext}(r)$ is actually the nuclear potential exerted on an electron at distance r:

$$V_{ext}(r) \equiv V_{nuclear}(r) = -\frac{Z}{r} \quad (12.7.3)$$

and $E_{xc}[\rho]$ is the exchange–correlation energy.

The variational solution of Eq. (12.7.1) is obtained by replacing the usual Schrodinger-type equations with two coupled equations for minor and major wave-function components $G_i(r)$ and $F_i(r)$, respectively, whose radial extents fulfill (Bachelet et al. 1982):

$$\frac{dF_i(r)}{dr} - \frac{\mathcal{K}}{r}F_i(r) + \alpha[\varepsilon_i - V(r)]G_i(r) = 0 \quad (12.7.4.1)$$

$$\frac{dG_i(r)}{dr} + \frac{\mathcal{K}}{r}G_i(r) - \alpha\left[\frac{2}{\alpha^2} - \varepsilon_i - V(r)\right]F_i(r) = 0 \quad (12.7.4.2)$$

and return a relativistic electron density:

$$\rho(r) = \sum_{i}^{\substack{\text{occupied}\\ \text{states}}} \left[|F_i(r)|^2 + |G_i(r)|^2\right] \quad (12.7.5)$$

$\mathcal{K}$ in Eqs. (12.7.4.1, 12.7.4.2) is a non-zero integer quantum number obeying:

$$\mathcal{K} = \begin{cases} l, & \text{for} \quad j = l - \frac{1}{2} \\ -(l+1), & \text{for } j \quad j = l + \frac{1}{2} \end{cases} \quad (12.7.6)$$

Moreover, $\alpha = c^{-1} = 1/137.04$, and $\hbar = m = e = 1$, in atomic units (Bachelet et al. 1982).

The decision about how many electrons may be frozen depends from the kind of information one wants to get out from calculus, and also on the amount of data at disposal to conform the potential. Depending on the level of frozen electrons, the ECPs are defined as "*small-core*" or "*large-core*". As an example, in the case of bromine $^{79}_{35}Br$, a small-core ECP has 10 frozen electrons (i.e. 1s, 2s, 2p), while a large-core ECP has 28 frozen electrons (1s, 2s, 2p, 3s, 3p, 3d) and only the remaining 7 electrons (4s and 4p) are treated explicitly.

Excellent lectures about this subject may be found on the web (Carlsson 2007; Dal Corso 2011), here I limit myself to present some notes based mainly on the papers of Hamann et al. (1979), Bachelet et al. (1982), Christiansen et al. (1979), Vanderbilt (1990).

12.7.1 Norm-Conserving Pseudopotentials

The Norm-conserving pseudopotentials (Hamann et al., 1979) were developed in order to fulfill four "*desirable properties*" (Bachelet et al. 1982)::

(1) real and pseudo valence eigenvalues must agree for a chosen "*prototypic*" atomic configuration;
(2) real and pseudo atomic wave functions must agree beyond a chosen "*core radius*" r_c;
(3) the integrals from 0 to r of the real and pseudo charge densities must agree for $r > r_c$, for each valence state (norm conservation);
(4) the logarithmic derivatives of the real and pseudo wave function and their first energy derivatives must agree for $r > r_c$.

As noted by Hamann et al. (1979) properties (3) and (4) are crucial in order to attain optimum transferability among a variety of chemical environments in self-consistent calculations in which the pseudo charge density is treated as a real physical object.

Hamann et al. (1979) proposed the following identity (in atomic units):

$$2\pi\left[(r\phi)^2\frac{d}{d\varepsilon}\frac{d}{d}r\ln\phi\right]_R = 4\pi\int\phi^2r^2dr \tag{12.7.7}$$

where ϕ is the solution of the radial Schrodinger equation at energy ε (not necessarily an eigenvalue) which is regular at $r = 0$, and the radial logarithmic derivative of ϕ is simply related to the scattering phase shift. This identity, showing that any norm-conserving pseudopotential satisfies an important transferability criterion, is related to the Friedel sum rule, and was pointed out in connection with pseudopotentials already by Shaw and Harrison (1967).

The consequence of Eq. (12.7.7) is that if two potentials v_1 and v_2 yield solutions ϕ_1 and ϕ_2 which have the same integrated charge inside a sphere of radius R, the linear energy variation around ε of their scattering phase shifts at R is identical.

The norm-conserving pseudopotentials are constructed in five steps (Bachelet et al., 1982):

(i) Dirac's Eqs. (12.7.4.1, 12.7.4.2) are solved for a chosen atomic reference (ν) configuration. Only the major wavefunction component $G_{\mathcal{K}}(r)$ is adopted for valence electrons outside the core region:

$$\left[\frac{1}{2}\frac{d^2}{dr^2} - \frac{\mathcal{K}(\mathcal{K}+1)}{2r^2} + V^{\nu}(r)\right] G_{\mathcal{K}}(r) = \varepsilon G_{\mathcal{K}}(r) \qquad (12.7.8)$$

(ii) A first-step pseudo-potential $V^{\nu}_{1(lj)}$ is built up for each angular momentum quantum state (lj) by cutting off the singularity near r = 0 in the screened full-core potential V^{ν}:

$$V^{\nu}_{1(lj)}(r) = V^{\nu}(r)\left\{1 - f\left[\frac{r}{r_{c,(lj)}}\right] + c^{\nu}_{(lj)} f\left[\frac{r}{r_{c,(lj)}}\right]\right\} \qquad (12.7.9)$$

where $f\left[\frac{r}{r_{c,(lj)}}\right]$ is a smoothing function with a cutoff at $r \approx r_{c,(lj)}$ and approaching 0 as $x \to \infty$ and 1 as $r \to 0$; $c^{\nu}_{(lj)}$ is a normalization constant. The assumption:

$$f\left[\frac{r}{r_{c,(lj)}}\right] = \exp\left[-\left(\frac{r}{r_{c,(lj)}}\right)^{\lambda}\right] \qquad (12.7.10)$$

with λ = 3.5 yields optimum results in tests on a variety of atoms.

A multiplicative constant $\gamma^{\nu}_{(lj)}$ is then introduced, in order to render the normalized function $\omega^{\nu}_{1,(lj)}(r)$ consistent with the full-core valence wave function beyond $r_{c,(lj)}$:

$$\gamma^{\nu}_{(lj)}\omega^{\nu}_{1,(lj)}(r) \to G_{(lj)}(r) \quad \left(r > r_{c,(lj)}\right) \qquad (12.7.11)$$

(iii) In the short range, the intermediate normalized function is replaced by:

$$\omega^{\nu}_{2,(jj)}(r) = \gamma^{\nu}_{(jj)}\left\{\omega^{\nu}_{1,(j)}(r) + \delta^{\nu}_{(jj)} r^{l+1} f\left[\frac{r}{r_{c,(jj)}}\right]\right\} \qquad (12.7.12)$$

in order to agree with the full-core valence wave function for $r > r_{c,(lj)}$.

$\delta^{\nu}_{(lj)}$ in Eq. (12.7.11) is the smaller solution of the quadratic equation

$$\left[\gamma^{v}_{(jj)}\right]^2 \int \left\{\omega^{v}_{1,(j)}(r) + \delta^{v}_{(j)} r^{l+1} f\left[r/r_{c,(j)}\right]\right\}^2 dr = 1 \tag{12.7.13}$$

(iv) Knowing $V^{v}_{1(lj)}(r)$ and $\omega^{v}_{2,(lj)}(r)$, the final screened pseudopotentials $V^{v}_{2(lj)}(r)$ which produce the nodeless eigenfunctions $\omega^{v}_{2,(lj)}(r)$ at eigenvalues ε_j are obtained analytically inverting the radial Schrödinger equation (Bachelet et al., 1982):

$$V^{v}_{2(lj)}(r) = V^{v}_{1(lj)}(r) + \frac{\delta^{v}_{(lj)} r^{l+1} f\left[r/r_{c,(lj)}\right]}{2\omega^{v}_{2,(lj)}(r)} \times \left\{ \frac{\left[\gamma^{v}_{(lj)}\right]^2 \left[r/r_{c,(lj)}\right]^{2\gamma^{v}_{(j)}} - \left[2l\gamma^{v}_{(jj)} + \gamma^{v}_{(j)}\left(\gamma^{v}_{(j)}+1\right)\right]\left[r/r_{c,(j)}\right]^{\gamma^{v}_{(j)}}}{r^2} + 2\varepsilon_j - 2V^{v}_{1(lj)}(r) \right\} \tag{12.7.14}$$

(v) In the last step, the screened potentials $V^{v}_{2(lj)}(r)$ are unscreened using the nodeless pseudo-wave function $\omega^{v}_{2,(lj)}(r)$:

$$V^{ion}_{l\pm1/2}(r) = V^{v}_{2(lj)}(r) + \int \frac{\rho^{v}(r')}{|r - r'|} dr' - \frac{\delta E_{xc}|\rho^{v}|}{\delta\rho^{v}(r)} \tag{12.7.15}$$

where

$$\rho^{v}(r) = \sum_{\substack{\text{occupied}\\ \text{valence}\\ \text{states}}} \left[\frac{\omega^{v}_{2,(lj)}(r)}{r}\right]^2 \tag{12.7.16}$$

The bare-ion pseudopotentials $V^{ion}_{l\pm1/2}(r)$ so obtained are, to first order, independent of changes in the atomic prototype configuration v (Bachelet et al. 1982).

For practical computations it is defined an average pseudopotential weighted by the different j degeneracies of the $l \pm 1$ states $\left[V^{ion}_{l}(r)\right]$:

$$V^{ion}_{l}(r) = \frac{1}{2l+1}\left[lV^{ion}_{l-1/2}(r) + (1+1)V^{ion}_{l+1/2}\right] \tag{12.7.17}$$

The strength of spin–orbit effects $\left[V^{so}_{l}(x)\right]$ can accordingly be defined in terms of difference between the two spin states:

$$V^{so}_{l}(r) = \frac{2}{2l+1}\left[V^{ion}_{l+1/2}(r) - V^{ion}_{l-1/2}\right] \tag{12.7.18}$$

The pseudopotential is decomposed into a long-range Coulomb part (l independent) and a short-range l-dependent pseudopotential terms:

$$V_l^{ion}(r) = V_{core}(r) + \Delta V_l^{ion} \tag{12.7.19}$$

The core potential originates from Gaussian-type effective core charges:

$$V_{core}(r) = -\frac{Z_v}{r}\left\{\sum_{i=1}^{2} c_i^{core} \operatorname{erf}\left[\left(\alpha_i^{core}\right)^{1/2} r\right]\right\} \tag{12.7.20}$$

with Z_v as valence charges, and approaches smoothly a finite value for $r \rightarrow 0$ (Bachelet et al., 1982).

The potentials ΔV_l^{ion} and $V_l^{so}(r)$ are expanded in Gaussian-type functions, e.g.:

$$\Delta V_l^{ion}(r) = \sum_{i=1}^{3} \left(A_i + r^2 A_{i+3}\right) \exp\left(-\alpha_i r^2\right) \tag{12.7.21}$$

Each atom is thus characterized by (Bachelet et al. 1982):

(i) a valence charge Z_v and two sets of linear coefficients (c) and decay constants (α) for the core $c_i^{core}, \alpha_i^{core}, i = 1, 2$.
(ii) two sets of three linear coefficients each, A_i and A_{i+3} for each l value and corresponding to the decay constants α_i, i = 1,2,3 for the average potential.
(iii) a similar set as (ii) for each spin–orbit difference potential, provided the spin–orbit splitting of the eigenvalues is larger than a chosen threshold value of 0.05 *eV*.

Finally, because too many significant figures must be retained in the A_i's for practical tabulation, the coefficients A_i, A_{i+3}, i = 1,3, are transformed into a set of coefficients C_i, i = 1,6 for an orthonormal basis set:

$$C_i = -\sum_{l=1}^{6} A_l Q_i l \tag{12.7.22}$$

where Q_{il} is the orthogonality matrix

$$Q_{il} = \begin{cases} 0 & \text{for} \quad i = l \\ \left[S_{il} - \sum_{k=1}^{i-1} Q_{ki}^2\right]^{1/2} & \text{for} \quad i = l \\ \left[\dfrac{S_{il} - \sum_{k=1}^{i-1} Q_{ki} Q_{kl}}{Q_{ii}}\right. & \text{for } i < l \end{cases} \tag{12.7.23}$$

S_{il} is the overlap matrix:

$$S_{il} = \int_0^\infty r^2 \Phi_i(r)\Phi_l(r)dr \qquad (12.7.24)$$

and

$$\Phi_i(r) = \begin{cases} \exp(-\alpha_i r^2) & i = 1,2,3 \\ r^2 \exp(-\alpha_i r^2) & i = 4,5,6 \end{cases} \qquad (12.7.25)$$

As an applicative example I report in Table (12.17) the ECP parameters of Bachelet et al. (1982) for Au.

Figure 12.5 shows the ion-core pseudopotentials for Au. The 5d electrons experience a strongly attractive potential, not observed for 6s, 6p and 6f states.

The $j = l$-1/2 spin–orbit components of the Au valence wave functions are shown in Fig. 12.6. The 6s and 5d wave functions are calculated from the atomic ground state ($5d^{10}\ 6s^1$) while the 6p and 5f states are obtained from excited states (see Bachelet et al., 1982; for details).

Similar procedures were envisaged by other authors in the same years (Kahn et al. 1976; Hay et al. 1978; Christiansen et al. 1979). Christiansen and coworkers noted that the normalized valence pseudo-orbitals χ_v are usually defined as linear combinations of valence and core HF orbitals (Φ_v, Φ_c) of a given l quantum number:

$$\chi_v = C_v\Phi_v + \sum_c^{core} C_c\Phi_c \qquad (12.7.26)$$

with expansion coefficients chosen "*such that* χ_v *is smooth and nodeless near the origin*" (Christiansen et al., 1979).

The effective potential U_{EP} is obtained moreover by requiring that it reproduces both the pseudo-orbital χ_v and the true eigenvalue ε_v of the Hartree–Fock orbital,

$$\left(-\frac{1}{2}\nabla^2 - \frac{Z}{r} + U_{EP} + W_{PS,val}\right)\chi_v = \varepsilon_v\chi_v \qquad (12.7.27)$$

The potential $W_{PS,val}$ in Eq. (12.7.27) comprises the usual Coulomb and exchange terms arising from interaction of the electron in the χ_v orbital with the other electrons in the occupied pseudo-orbitals, and the other symbols have the usual significance. From Eq. (12.7.27), the local potential at distance r from the nucleus is:

$$U_{Ep}(r) = \frac{\left[\left(\varepsilon_v + \frac{1}{2}\nabla^2 + \frac{Z}{r} - W_{PS,val}\right)\chi_v\right]}{\chi_v} \qquad (12.7.28)$$

Table 12.17 Pseudopotential parameters for gold. (From Bachelet et al., 1982)

Atom	Z_A	Z_V		L	α_1	α_2	α_3	c_1	c_2	c_3	c_4	c_5	c_6
Au	79	11		core	2.85	1.14		2.3778	−1.3778				
				0	1.09	1.24	1.62	−8.9867	−1.0964	−0.1164	0.2108	−0.0732	−0.0077
				1	1.02	1.12	1.34	−7.8473	−0.5246	−0.6230	0.1483	0.0363	0.2780
				2	3.34	3.78	4.97	−3.6440	0.0850	0.0286	−0.0597	−0.0456	0.0047
				3	1.03	1.17	1.39	−5.6797	−2.0821	−0.0032	0.1156	−0.1733	−0.0507
			so	1	0.26	0.39	0.49	−0.3833	−0.1289	0.0164	−0.0189	0.0144	−0.0110
			so	2	0.89	1.04	1.39	−0.0791	−0.0452	−0.0148	−0.0113	−0.0058	−0.0191

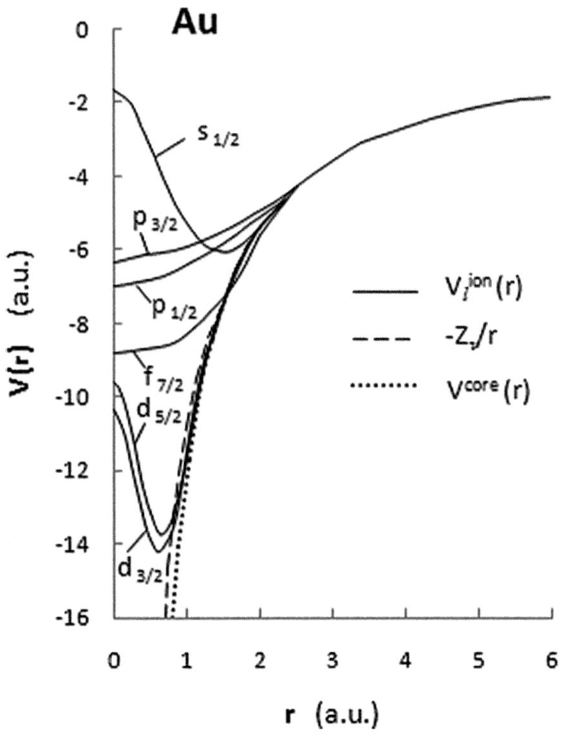

Fig. 12.5 Ion-core pseudopotentials for Au Modified from Bachelet et al. (1982)

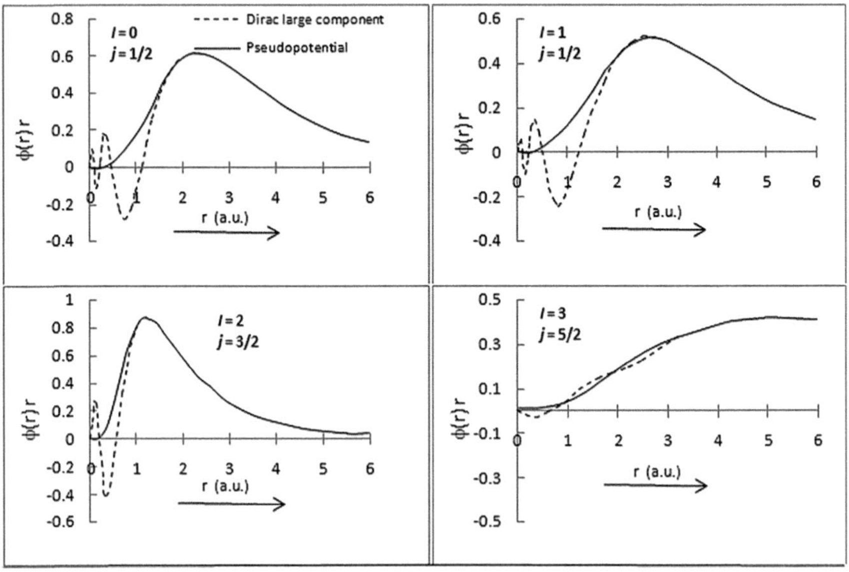

Fig. 12.6 Valence wave functions for gold. Modified From Bachelet et al. (1982)

If r_c is the radius at which the most diffuse core orbital becomes negligibly small, then, for $r > r_c$:

$$\left.\begin{array}{r}\chi_v(r) \approx C_v\phi_v(r)\\ U_{EP} \approx W_{HF,core} + W_{HF,val} - W_{PS,val}\end{array}\right\} \quad r > r_c \tag{12.7.29}$$

where $W_{HF,core}$ and $W_{HF,val}$ are the (coulomb + exchange) potential arising from the HF core and from valence orbitals, respectively. For large values of r, $W_{HF,core}$ decays with the inverse of the distance, i.e.:

$$W_{HF,core} \approx {}^{N_c}/_{r} \tag{12.7.30}$$

where N_c denotes the number of core electrons. Hence, if the EP generates excessive long tails, this must be ascribed to $W_{PS,val}$ and/or $W_{HF,val}$.

Christiansen et al. (1969) showed then that the difference $W_{HF,val}$ - $W_{PS,val}$ in Eq. (12.7.29) may be recast in terms of difference of the spherically averaged Coulomb potentials for corresponding HF orbitals and pseudoorbitals [$J_{HF,v}(r)$ and $J_{PS,v}(r)$, respectively] when these are summed over all other occupied valence orbitals together with corresponding non-spherical Coulomb and Exchange terms:

$$J_{HF,v}(r) - J_{HF,v}(r) \approx \frac{1}{r}\int_r^\infty |\phi_v(t)|^2 (t^2 - rt)dt \quad (r > r_c) \tag{12.7.31}$$

It is clear From Eqs. (12.7.29) and (12.7.31) that the mismatch of the pseudo and HF orbitals in the valence region can be eliminated only if χ_v and ϕ_v are required to be identical in that region. Christiansen et al. (1969) thus abandoned the constraint (12.7.26) and partitioned ϕ_v into core and valence terms posing simply:

$$\chi_v = \phi_v + f_v \tag{12.7.32}$$

with the function f_v attaining zero outside the core region and conformed in such a way to cancel the oscillations of ϕ_v in the core region.

In Table 12.18 are listed the Gaussian expansions of Christiansen et al. (1979) for the chlorine effective potentials. These expansion fulfill:

$$U_{EP}(r) - \frac{N_c}{r} = r^2 \sum_i C_{il}\gamma_{n,il} \exp(-\zeta_{il} r^2) \tag{12.7.33}$$

In Fig. 12.7 we see the amplitude of the wavefunction arising from the EP developed by Christiansen et al. (1979) for chlorine, compared with the HF wavefunction. r-match is the innermost point at which the matching results in a nodeless radial pseudoorbital. The outer part spanning the region from r-match to infinity is just the numerical HF valence orbital in that region. In the inner part (from zero to r-match) it is operative the ECP pseudoorbital (12.7.28).

Table 12.18 Gaussian expansions of Christiansen et al. (1979) for the chlorine Effective Potential (EP)

Expansion coefficients				
n_{il}	ζ_{il}	$U_s - U_d$	$U_p - U_d$	U_d
2	5	−2700.311508	−149.0917485	−56.3781074
2	5.5	6004.363424	−5920.867317	266.4739971
2	6	−12937.73745	−2774.331675	−171.1705515
2	6.72	8235.893211	6517.461179	2.3506943
2	28.89	101.6005388	−35.8188407	2.5515185
2	96.63	199.3749764	286.0375129	5.5724718
1	5.3	6448.538315	9714.154963	−119.0912105
1	6	−6312.698496	−9481.013584	108.1034976
1	33.87	−112.9121148	−211.1595365	0.9877129
0	6.16	13.0730405	26.0712424	
0	67.33	−10.0730405	−21.0712424	

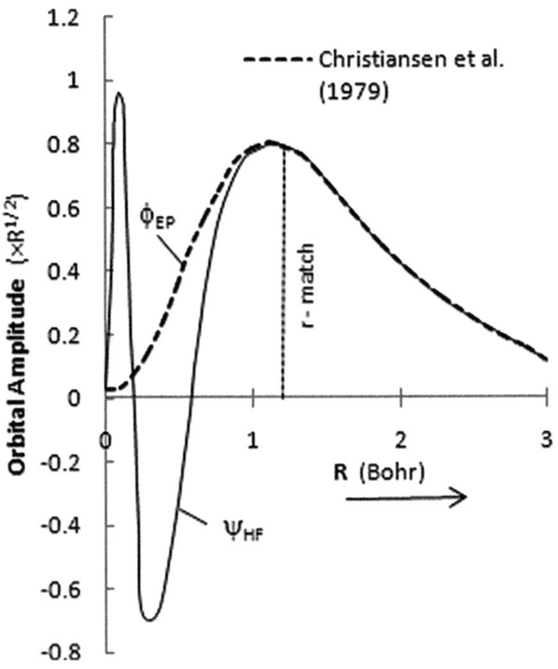

Fig. 12.7 The chlorine 3s pseudoorbital of Christiansen et al. (1979) compared with the Hartree–Fock wavefunction

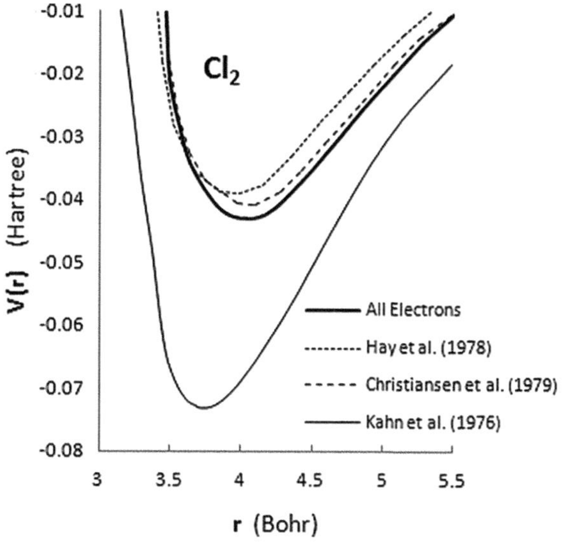

Fig. 12.8 Potential energy curves for the ground state of Cl_2 from all-electron and effective potential calculations according to various authors Modified from Christiansen et al. (1979)

In Fig. 12.8 the EP dissociation curves of Cl_2 of Christiansen et al. (1979) are compared with EP calculations of Kahn et al. (1976) and.Hay et al. (1978) and All Electron (AE) calculations. The agreement between AE and the EP procedure of Christiansen is quite good, as observed also for F_2 and LiCl (see Figs. 12.2 and 12.4 in Christiansen et al. 1979).

12.7.2 Ultrasoft Pseudopotentials (USP)

The norm conservation implies that the valence electrons (i.e. outside the core region) are not affected by changes in the charge distribution inside the core region, provided that the total charge is not changed. This is a consequence of the Gauss Law:

$$V(r) = \int \frac{n(r')dr'}{|r - r'|} \tag{12.7.34}$$

The true charge distribution, concentrating the charge on the nuclear center may be thus replaced by a smoother one inside the core region. However, in the presence of tightly bound orbitals having a substantial fraction of their weight inside the core region of the atom, the cutoff energy for the plane-wave basis set is too high, and the only way to obtain smooth pseudo wave functions is to release the norm conservation criterium.

As shown by Vanderbilt (1990) this can be done by splitting the pseudo wave functions into two parts: an ultrasoft valence wave function $\phi_l^{US}(r)$ that do not fulfill the norm conservation, plus a core augmentation charge counterbalancing the charge deficit in the core region (Carlsson, 2007):

$$Q_{nm}(r) = \psi_n^{AE*}(r)\psi_m^{AE}(r) - \phi_n^{US*}(r)\phi_m^{US(r)} \quad (12.7.35)$$

The ultrasoft pseudopotential takes the form (Carlsson, 2007):

$$V^{US} = V_{local}(r) + \sum_{nml} D_{nm}^0 |\beta_n^l\rangle\langle\beta_m^l| \quad (12.7.36)$$

and is practically obtained by inverting the Schrödinger equation with an opportune choice of the cut-off radius and the angular momentum for the local potential.

D_{nm}^0 is defined by

$$D_{nm}^0 = \langle\phi_n^l|\chi_m^l\rangle + \varepsilon_m q_{nm} \quad (12.7.37)$$

with ε_m as eigenvalue and

$$q_{nn} = \int_0^{r_c} Q_{nn}(r)dr \quad (12.7.38)$$

The β projector functions (restricted inside the cut-off region for the wave functions) define the non-local part of the pseudo potential:

$$|\beta_n\rangle = \sum_m \frac{|\chi_m\rangle}{\langle\chi_m \mid \phi_n\rangle} \quad (12.7.39)$$

The χ-functions are defined through

$$|\chi_n\rangle = (\varepsilon_n - T - V_{local})|\phi_n\rangle \quad (12.7.40)$$

(see Vanderbilt 1990; Carlsson 2007; for a more exhaustive treatment).

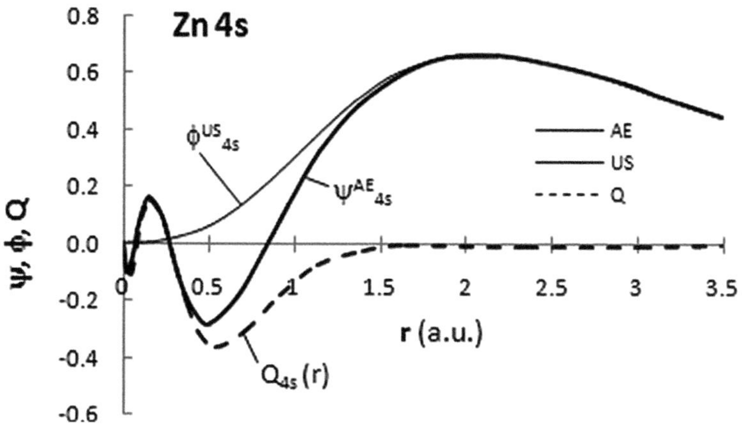

Fig. 12.9 All electron (AE) and Ultra-Soft (US) potentials for the 4 s electrons of zinc. The core augmentation charge counterbalances the charge deficit in the core region delimited by the cut-off radius Modified from Carlsson (2007)

In Fig. 12.9 it is shown the Q-function for the augmentation charge for the 4s orbital of zinc, as the difference between the all-electron and the pseudo wave functions (Eq. 12.7.35; Carlsson 2007).

Chapter 13
Ab Initio Molecular Dynamics

Although in some way promoted by Ehrenfest almost one century ago (Ehrenfest 1927), Ab Initio Molecular Dynamics (AIMD) developed massively only in the eighties of the last century, under the influx of a seminal work of Car and Parrinello (1985). Since then, there has been a continuous blooming of technical developments and applications. Twelve years ago an excellent reference treatise (Marx and Hutter 2010) appeared, which resumes in a lucid and exhaustive way the basic theory and the advanced methods. Here, as done for other procedures and applications, I limit myself to abstract some concepts that may be of help in understanding the various applications to Earth's materials. I will begin resuming concisely the work done by Hans C. Andersen (1980), which enlightens the fundamental relations between Volume (V) and Energy (E) in a microcanonical ensemble of N particles, and the thermodynamic trajectories in macroscopic systems. The scientific paper of Andersen (1980), like the fairy tales written by his much older homonymous, is a real poem. I will introduce progressively two approaches that were in some manner precursor to the Car-Parrinello development. The Car-Parrinello procedure will be briefly outlined in the last paragraph. I will make no mention of the various potentials (mainly force field) adopted in classical MD. A detailed review may be found in Van Gunsteren and Berendsen (1990).

13.1 Molecular Dynamics (MD) Simulations at Constant P and/or T

While in Newtonian mechanics, the time evolution is obtained by computing the total force being exerted on each particle of the system at the light of Newton's second law, in Hamiltonian mechanics, the time evolution is obtained by computing the Hamiltonian of the system (H) in the generalized coordinates and inserting it into Hamilton's equations:

G. A. Ottonello, *Quantum Geochemistry*, Springer Geochemistry,
https://doi.org/10.1007/978-3-031-21837-8_13

$$\frac{d\mathbf{p}}{dt} = -\frac{\partial H}{\partial \mathbf{q}} \tag{13.1.1}$$

$$\frac{d\mathbf{q}}{dt} = -\frac{\partial H}{\partial \mathbf{p}} \tag{13.1.2}$$

where $\mathbf{q}$ are the generalized coordinates and $\mathbf{p}$ their conjugate momenta.

Andersen (1980) developed molecular dynamics methods for simulating a fluid subject to a constant pressure (P), constant temperature (T), or constant T and P. The trajectory averages for these three types of simulations correspond to averages over the isoenthalpic-isobaric (NHP), canonical (NVT), and isothermal-isobaric (NTP) ensembles, respectively. The average of any property over these trajectories is equivalent to an average over a microcanonical (NVE) ensemble if the trajectory passes through all parts of phase space that have the specified energy.

Andersen (1980) considered an atomic fluid composed of N atoms, with coordinates r_1, r_2,…r_N in a cubic volume V with periodic boundary conditions[1] Each component of each coordinate is a number between 0 and $V^{1/3}$. The potential energy of the atoms is

$$U(\mathbf{r}^N) = \sum_{i<j} u(r_{ij}) \tag{13.1.3}$$

The "*minimum image convention*" implies that r_{ij} in the above summation "*is to be interpreted as the distance between r_i and either r_j or the nearest image of particle* j, *whichever is closer*" (Andersen 1980):

$$r_{ij} = |\mathbf{r}_{ij}| \equiv \min_n |\mathbf{r}_i - \mathbf{r}_j + nV^{1/3}| \tag{13.1.4}$$

Given a Lagrangian $\mathcal{L}$ in terms of the r coordinates and $dr/dt \equiv \dot{r}$ velocities, with m as atom mass:

$$\mathcal{L}_1(\mathbf{r}^N, \dot{\mathbf{r}}^N) = \frac{m}{2}\sum_i \dot{\mathbf{r}}_i \cdot \dot{\mathbf{r}}_i - \sum_{i<j} u(r_{ij}) \tag{13.1.5}$$

the momenta are calculated by differentiating the Lagrangian with respect to the (generalized) velocities:

$$\mathbf{p}_i = \frac{\partial \mathcal{L}(\mathbf{r}^N, \mathbf{r}^N)}{\partial \dot{\mathbf{r}}_i} = m\dot{\mathbf{r}}_i \tag{13.1.6}$$

[1] The periodic boundary conditions imply that if particle i is at r_i there is a set of image particles at positions $r_i + \mathbf{n}V^{1/3}$, where $\mathbf{n}$ is a vector with integer components.

The Hamiltonian H is the Legendre transformation of $\mathcal{L}$ with **p** as the dual variable and **q**,t fixed:

$$H_1(\mathbf{r}^N, \mathbf{p}^N; V) = \sum_i \dot{\mathbf{r}}_i \cdot \mathbf{p}_i - \mathcal{L}_1 = \frac{1}{2m}\sum_i \mathbf{p}_i \cdot \mathbf{p}_i + \sum_{i<j} u(r_{ij}) \tag{13.1.7}$$

Denoting $\mathbf{r}_{ij}$ the unit vector in the direction of r_i–r_j and taking into account the minimum image convention, the Hamiltonian equations of motion are (Andersen 1980):

$$\frac{\partial \mathbf{r}_i}{\partial t} = \frac{\partial H_1}{\partial \mathbf{p}_i} = \frac{\mathbf{p}_i}{m} \tag{13.1.8}$$

$$\frac{\partial \mathbf{p}_i}{\partial t} = -\frac{\partial H_1}{\partial \mathbf{r}_i} = -\sum_{j\neq i} \hat{\mathbf{r}}_{ij} u'(r_{ij}) \tag{13.1.9}$$

where u'(r_{ij}) is the time-derivative of u and $\hat{\mathbf{r}}_{ij}$ is a unit vector in the direction of $r_i - r_j$ using the minimum image convention.

Consider now a general measurable thermodynamic property F. According to statistical thermodynamics, the measured F is equal to the ensemble average of the function F over a suitably chosen ensemble of states. Denoting F_{NVE} (N,V,E) the property of F in the NVE microcanonical ensemble with (N,V,E) arguments, the microcanonical ensemble average of the function F is (Andersen 1980):

$$F_{NVE}(N, V, E) = \frac{1}{N!\Omega(N, V, E)} \int_V d\mathbf{r}^N \int d\mathbf{p}^N \times \delta[H_1(\mathbf{r}^N, \mathbf{p}^N; V) - E] F(\mathbf{r}^N, \mathbf{p}^N; V) \tag{13.1.10}$$

where $\Omega(N, V, E)$ is the microcanonical ensemble partition function:

$$\Omega(N, V, E) = \frac{1}{N!} \int_V d\mathbf{r}^N \int d\mathbf{p}^N \times \delta[H_1(\mathbf{r}^N, \mathbf{p}^N; V) - E] \tag{13.1.11}$$

and δ[x] is the Dirac's δ function. The canonical average is

$$F_{NVT}(N, V, T) \equiv \frac{1}{N!Q(N, V, T)} \int_V d\mathbf{r}^N \int d\mathbf{p}^N \times \exp\left[-\frac{H_1(\mathbf{r}^N, \mathbf{p}^N; V)}{k_B T}\right] F(\mathbf{r}^N, \mathbf{p}^N; V) \tag{13.1.12}$$

$$Q(N,V,T) = \frac{1}{N!}\int_V d\mathbf{r}^N \int d\mathbf{p}^N \times \exp\left[-\frac{H_1(\mathbf{r}^N,\mathbf{p}^N;V)}{k_B T}\right] \tag{13.1.13}$$

where Q(N,V,T) is the Laplace transform of $\Omega(N,V,E)$ with respect to E with $(kT)^{-1}$ as transform variable.

The isothermal-isobaric ensemble average is:

$$F_{NPT}(N,P,T) = \frac{1}{N!\Delta(N,P,T)}\int_0^\infty dV \int_V d\mathbf{r}^N \int d\mathbf{p}^N \times \exp\left[-\frac{PV + H_1(\mathbf{r}^N,\mathbf{p}^N;V)}{k_B T}\right] F(\mathbf{r}^N,\mathbf{p}^N;V) \tag{13.1.14}$$

with

$$\Delta(N,P,T) = \frac{1}{N!}\int_0^\infty dV \int_V d\mathbf{r}^N \int d\mathbf{p}^N \times \exp\left[-\frac{H_1(\mathbf{r}^N,\mathbf{p}^N;V)}{k_B T}\right] \tag{13.1.15}$$

To these ensembles, Andersen (1980) associates the NPH ensemble, related to the NPT ensemble "*in the same way as the microcanonical ensemble is related to the canonical ensemble*" (Andersen 1980).

$$F_{NPH}(N,P,H) \equiv \frac{1}{N!\Gamma(N,P,H)}\int_0^\infty dV \int_V d\mathbf{r}^N \int d\mathbf{p}^N \times \delta\left[PV + H_1(\mathbf{r}^N,\mathbf{p}^N;V) - H\right] F(\mathbf{r}^N,\mathbf{p}^N;V) \tag{13.1.16}$$

$$\Gamma(N,P,H) = \frac{1}{N!}\int_0^\infty dV \int_V d\mathbf{r}^N \int d\mathbf{p}^N \times \delta\left[PV + H_1(\mathbf{r}^N,\mathbf{p}^N;V) - H\right] \tag{13.1.17}$$

The Laplace transform of Γ(N,V,H) with respect to H with $(k_B T)^{-1}$ as transform variable gives Δ(N,P,T).

The trajectory average (or "*time average*") of a function $F(\mathbf{r}^N,\mathbf{p}^N;V)$ is given by

$$\overline{F} = \lim_{T\to\infty} T^{-1}\int_0^T F\left[\mathbf{r}^N(t),\mathbf{p}^N(t);V(t)\right]dt \tag{13.1.18}$$

By making an initial choice of $r^N(0)$ and $p^N(0)$ and fixing N and V, the Hamilton's Equation (13.1.8, 13.1.9) may be solved and the trajectory average calculated. Along the trajectory the potential + kinetic energy (i.e. H_1) is conserved, and, assuming that the trajectory spends equal times in all equal volumes with that same value of energy, the trajectory average is equal to the microcanonical ensemble average:

$$\overline{F} = F_{NVE}(N, V, E) \tag{13.1.19}$$

provided that $\overline{F}$ is calculated from Eq. 13.1.18, using the solutions (13.1.8, 13.1.9).

13.1.1 MD at Constant P

V is, in this case, a dynamical variable. The trajectory average of any property corresponds to the NPH ensemble average of that property. The coordinates $\mathbf{r}_i$, i = 1,2,...*N* are replaced by scaled dimensionless coordinates $\boldsymbol{\rho}_i$ (ranging from 0 to 1):

$$\boldsymbol{\rho}_i = \frac{\mathbf{r}_i}{V^{1/3}} \quad (i = 1, 2, ...N) \tag{13.1.20}$$

Adopting the minimum energy convention for the $\boldsymbol{\rho}$ vectors in the cube of constant volume, the new Lagrangian to consider is (Andersen 1980):

$$\mathcal{L}_2(\boldsymbol{\rho}^N, \dot{\rho}^N, Q\dot{Q}) = \frac{m}{2} Q^{2/3} \sum_{i=1}^{N} \dot{\rho}_\imath \cdot \dot{\rho}_\imath - \sum_{i<J=1}^{N} u\left(Q^{1/3} \rho_{ij}\right) + \frac{1}{2} M\dot{Q}^2 - \alpha Q \tag{13.1.21}$$

Comparing Eqs. 13.1.21 and 13.1.5 it is evident that the first two terms on the right of the two equations are equivalent, under the change of coordinates. Hence, the new variable Q may be interpreted as the (variable) volume V. The third term on the right of Eq. 13.1.21 is the kinetic energy for the motion of Q and the fourth term is the potential energy associated to Q (with α and M as constants).

To achieve a physical interpretation of this Lagrangian we must imagine a fluid stored in a container of variable volume and compressed by a piston of mass M acting under the influence of an external pressure inducing a αV potential, analogous to a PV potential for the original system. We must moreover imagine that the piston does not act in one direction, but that the expansion and/or compression is isotropic. The coordinate of the piston is thus not linear but just Q. This representation is however only approximate because assuming that Q $\equiv$ V would imply that

$$\dot{\mathbf{r}}_i = Q^{1/3} \dot{\rho}_i + \frac{1}{3} Q^{-1/3} \dot{Q} \boldsymbol{\rho}_i \tag{13.1.22}$$

Since the kinetic energy factors arising from the second term on the right in Eq. 13.1.22 do not appear in Eq. 13.1.21, the interpretation is only approximate.

Nevertheless, denoting as "*scaled system*" (Andersen 1980) the system that adopts the dimensionless variable ρ, a "*correspondence*" between the scaled system and the original one is given by:

$$V = Q \tag{13.1.23.1}$$

$$\mathbf{r}_i = Q^{1/3}\boldsymbol{\rho}_i \tag{13.1.23.2}$$

$$\mathbf{p}_i = \boldsymbol{\pi}_i / Q^{1/3} \tag{13.1.23.3}$$

where $\boldsymbol{\pi}_i$ is the momentum conjugate to $\boldsymbol{\rho}_i$:

$$\pi_i = \frac{\partial \mathcal{L}_2}{\partial \dot{\rho}_i} = mQ^{2/3}\dot{\rho}_i \tag{13.1.24}$$

At the light of this correspondence, a trajectory for the original system may be generated from the trajectory of the scaled one:

$$V(t) = Q(t) \tag{13.1.25.1}$$

$$\mathbf{r}_i(t) = Q^{1/3}(t)\boldsymbol{\rho}_i(t) \tag{13.1.25.2}$$

$$\mathbf{p}_i(t) = \boldsymbol{\pi}_i(t) / Q^{1/3}(t) \tag{13.1.25.3}$$

and the corresponding equations of motion are (Andersen 1980):

$$\frac{d\mathbf{r}_i}{dt} = \frac{\mathbf{p}_i}{m} + \frac{1}{3}\mathbf{r}_i\frac{d\ln V}{dt} \tag{13.1.26.1}$$

$$\frac{d\mathbf{p}_i}{dt} = -\sum_{i(\neq j)=1}^{N} \hat{\mathbf{r}}_{ij}u'(r_{ij}) - \frac{1}{3}\mathbf{p}_i\frac{d\ln V}{dt} \tag{13.1.26.2}$$

$$\frac{Md^2V}{dt^2} = -\alpha + \frac{\left[\frac{2}{3}\sum_{i=1}^{N}\frac{\mathbf{p}_i\cdot\mathbf{p}_i}{2m} - \frac{1}{3}\sum_{i<j=1}^{N}\hat{\mathbf{r}}_{ij}u'(r_{ij})\right]}{V} \tag{13.1.26.3}$$

At the light of Eq. 13.1.18, the trajectory defined by these equations of motion can be used to calculate time averages of any function $F(r^N, p^N; V)$. Moreover "*the time average of any F calculated from this trajectory* is *equal to the ensemble average of F for an isoenthalpic-isobaric ensemble in which the pressure* is α" (Andersen 1980), i.e.

$$\overline{F} = F_{NPH}(N, \alpha, H) \tag{13.1.27}$$

In other words, the value of P for the ensemble is the value of α in the Lagrangian of the scaled system and "*the value of H for the ensemble is the energy of the trajectory of the scaled system minus the time average kinetic energy associated with the motion of Q*" (Andersen 1980).

13.1.2 MD at Constant T

In the constant temperature molecular dynamics, the energy of a system of N particles fluctuates due to the presence of stochastic forces that act on the atoms of the sample and change their kinetic energy. The equations of motion in the NVT ensemble are the Hamilton Equations 13.1.8, 13.1.9, to which the stochastic collision term $d\mathbf{p}_i/dt$ is added. Fixed the temperature T, it is also selected the mean rate ν at which each particle undergoes a collision. The times at which different particles suffer collisions are statistically uncorrelated, and the probability that a particular particle suffers a stochastic collision, in any small time interval Δt, is νΔt. The times of collision for each particle are thus fixed, generating through random numbers the time intervals between successive collisions, and distributing them according to the probability law

$$P(t) = \nu \times \exp(-\nu t) \tag{13.1.28}$$

Based on this law, the probability that an interval between collisions is between t and t + Δt is P(t)dt.

An initial set of positions $\mathbf{r}^N(0)$ and momenta $\mathbf{p}^N(0)$ is selected and the Hamiltonian equations of motion are integrated from t = 0 to the time of the first stochastic collision. It is then added the momentum of the collision of the ith particle, chosen at random from a Boltzmann distribution at T, while all other particles remain unaffected. The Hamilton equations for all particles are then integrated from the time of the first stochastic collision to the second, and so on. It is in this way generated a $\mathbf{r}^N(t)$, $\mathbf{p}^N(t)$ trajectory for N particles in a volume V with periodic boundary conditions.

Again, as done for the NPV ensemble, the time averages may be obtained at the light of Eq. 13.1.18; i.e.:

$$\overline{F} = F_{NVT}(N, V, T) \tag{13.1.29}$$

As anticipated, to perform a constant temperature molecular dynamics calculation, it is necessary to choose a value of ν. Since the average kinetic energy of a particle before collision is $\frac{3}{2}k_B(T+\Delta T)$ and, after collision, it is $\frac{3}{2}k_BT$, each stochastic collision changes the energy of the system by $-\frac{3}{2}k_B\Delta T$. Since the total rate of occurrence of stochastic collisions is Nν, the rate of energy gain is $-\frac{3}{2}k_B\Delta T$. By dimensional analysis, the rate of heat gain (in energy per unit time) is $-ak\Delta TV^{1/3}$ where a is a dimensionless constant that depends on the shape of the sample and upon the temperature distribution (Andersen 1980), and k is the thermal conductivity. Equating this rate of heat gain to $-\frac{3}{2}k_B\Delta T$ and indicating with $\rho = N/V$ the number density of the particle, it is obtained:

$$\nu = \frac{2}{3}\frac{ak}{k_B\rho^{1/3}N^{2/3}} \tag{13.1.30}$$

For the hypothetical condition of a single particle (i.e. N = 1) it is reasonable to assume that the required stochastic collision frequency should be the actual collision frequency ν_c of the single particle (all the others belonging to the heat bath), hence the above equation reduces to

$$\nu_c = \frac{2}{3}\frac{ak}{k_B\rho^{1/3}} \tag{13.1.31}$$

Thus, we may rewrite Eq. 13.1.30 in the form:

$$\nu = \frac{\nu_c}{N^{2/3}} \tag{13.1.32}$$

This tells us that, for large enough N, the stochastic collision frequency will be much smaller than the intermolecular collision frequency. Therefore, "*for most of the time, most of the molecules will be moving according to the conservative equations of motion for a closed system. The stochastic interruptions will be infrequent, but they will cause the energy of the system to relax to a value appropriate to the temperature T at a rate appropriate for a system of N particles and will cause the energy to fluctuate about its equilibrium value with the magnitude appropriate to a canonical ensemble*" (Andersen 1980).

13.1.3 The NTP Ensemble

The Boltzmann distribution for particle momentum π_i is proportional to $m\exp\left(-\frac{\pi_i\cdot\pi_i}{2mQ^{2/3}k_BT}\right)$ and the effect of each stochastic collision is to replace the momentum of the affected particle by a new value chosen at random from this distribution (for any particular value of Q). At the light of the correspondence (13.1.24), a trajectory for the original system may be generated from the trajectory of the scaled one (Eqs. 13.1.25.1, 13.1.25.2, 13.1.25.3) and this trajectory allows the computation of time averages for any function can be used to calculate time averages of any function $F(r^N, p^N; V)$ at the light of Eq. 13.1.18.

Hence: *the time average of any F calculated from this trajectory* is *equal to the ensemble average of F for an isothermal-isobaric ensemble in which the pressure is* α *and the temperature* is *T*" (Andersen 1980), i.e.

$$\overline{F} = F_{NPT}(N, \alpha, T) \tag{13.1.33}$$

13.2 Ehrenfest Molecular Dynamics

The non-relativistic time-dependent Schrödinger equation may be written in the form (Marx and Hutter 2010):

$$i\hbar\frac{\partial}{\partial t}\Phi(\{\mathbf{r}_i\},\{\mathbf{R}_I\};t) = H(\{\mathbf{r}_i\},\{\mathbf{R}_I\};t) \tag{13.2.1}$$

where H is the Hamiltonian

$$\mathsf{H} = -\sum_I \frac{\hbar^2}{2M_I}\nabla_I^2 - \sum_i \frac{\hbar^2}{2m_e}\nabla_i^2 + \frac{1}{4\pi\varepsilon_0}\sum_{i<j}\frac{e^2}{|\mathbf{r}_i - \mathbf{r}_j|} - \frac{1}{4\pi\varepsilon_0}\sum_{I,i}\frac{e^2 Z_I}{|\mathbf{R}_I - \mathbf{r}_i|} + \frac{1}{4\pi\varepsilon_0}\sum_{I<J}\frac{e^2 Z_I Z_J}{|\mathbf{R}_I - \mathbf{R}_J|} \tag{13.2.2}$$

for the electronic $\{\mathbf{r}_i\}$ and nuclear $\{\mathbf{R}_I\}$ degrees of freedom. M_I, Z_I, m_e, and e in the above equation are respectively mass and charge of the Ith nucleus and mass and charge of the electron. ε_0 is the vacuum permittivity.

This full Hamiltonian may be conveniently rewritten in the form (Marx and Hutter 2010):

$$\mathsf{H} = -\sum_I \frac{\hbar^2}{2M_I}\nabla_I^2 + \mathsf{H}_e(\{\mathbf{r}_i\},\{\mathbf{R}_I\}) \tag{13.2.3}$$

where H_e is the clamped-nuclei part:

$$\mathsf{H}_e = -\sum_i \frac{\hbar^2}{2m_e}\nabla_i^2 + V_{n-e}(\{\mathbf{r}_i\},\{\mathbf{R}_I\}) \tag{13.2.4}$$

If both electrons and nuclei move quantum mechanically in time-dependent effective potentials, their motions may be defined in terms of coupled Schrödinger equations over the electronic (ψ) and nuclear (χ) wave functions:

$$i\hbar\frac{\partial\psi}{\partial t} = -\sum_i \frac{\hbar^2}{2m_e}\nabla_i^2\psi + \left\{\int \chi^*(\{\mathbf{R}_I\};t)V_{n-e}(\{\mathbf{r}_i\},\{\mathbf{R}_I\})\chi(\{\mathbf{R}_I\};t)d\mathbf{R}\right\}\psi \tag{13.2.5}$$

$$i\hbar\frac{\partial\chi}{\partial t} = -\sum_I \frac{\hbar^2}{2M_I}\nabla_I^2\chi + \left\{\int \psi^*(\{\mathbf{r}_I\};t)H_e(\{\mathbf{r}_i\},\{\mathbf{R}_I\})\psi(\{\mathbf{r}_i\};t)d\mathbf{r}\right\}\chi \tag{13.2.6}$$

These functions are at the basis of the Time Dependent Self Consistent Field (TDSCF) method of Dirac (1930). Substituting to the quantum–mechanical nuclear wave function the classical-mechanical nuclear position, the time-dependent wave equation for the electrons becomes (Marx and Hutter 2010):

$$\begin{aligned} i\hbar\frac{\partial\psi}{\partial t} &= -\sum_i \frac{\hbar^2}{2m_e}\nabla_i^2\psi + V_{n-e}(\{\mathbf{r}_i\},\{\mathbf{R}_I(t)\})\psi \\ &= H_e(\{\mathbf{r}_i\},\{\mathbf{R}_I(t)\})\psi(\{\mathbf{r}_i\},\{\mathbf{R}_I\};t) \end{aligned} \tag{13.2.7}$$

with the nuclei moving according to the classical Newtonian equation of motion in the effective (Ehrenfest) potential V_e^E:

$$M_I\dot{\mathbf{R}}_I(t) = -\nabla_I V_e^E(\{\mathbf{R}_I(t)\}) \tag{13.2.8}$$

Solving numerically the coupled set of "*quantum/classical*" (Marx and Hutter 2010) Eqs. (13.2.7, 13.2.8):

$$i\hbar\frac{\partial\psi}{\partial t} = H_e\psi \tag{13.2.9}$$

$$M_I\mathbf{R}_I(t) = -\nabla_I\langle H_e\rangle \tag{13.2.10}$$

the time-dependent electronic Schrödinger equation may be resolved "*on the fly*" as the nuclei are propagated using classical mechanics. This is a concrete way of getting out from the "*dimensionality bottleneck*", without involving approximations in the global potential energy surface or reductions in the number of active degrees of freedom (Marx and Hutter 2010).

Solving the time-independent Schrödinger equation, the electronic function returns "*instantaneous adiabatic electronic states*":

$$H_e(\{\mathbf{r}_i\};\{\mathbf{R}_I\})\psi_k = E_k(\{\mathbf{R}_I\})\psi_k(\{\mathbf{r}_i\};\{\mathbf{R}_I\}) \tag{13.2.11}$$

(where E_k is the adiabatic energy and $\{\mathbf{R}_I\}$ are the instantaneous nuclear positions at time t), and expanding the electronic wave function in a basis of electronic states ψ_l:

$$\psi(\{\mathbf{r}_i\},\{\mathbf{R}_I\};t)=\sum_{l=0}^{\infty}c_l(t)\psi_l(\{\mathbf{r}_i\};\{\mathbf{R}_I\}) \tag{13.2.12}$$

(where $c_l(t)$ are complex time-dependent coefficients), the Ehrenfest equations of motion corresponding to Eqs. 13.2.7 and 13.2.10 are (Doltsinis and Marx 2002; Marx and Hutter 2010):

$$M_I\ddot{\mathbf{R}}_I(t)=\sum_k|c_k(t)|^2E_k=-\sum_k|c_k(t)|^2\nabla_IE_k+\sum_{k,l}c_k^*c_l(E_k-E_l)\mathbf{d}_I^{kl} \tag{13.2.13}$$

$$i\hbar\dot{c}_k(t)=c_k(t)E_k-i\hbar\sum_l c_l(t)D^{kl} \tag{13.2.14}$$

where D^{kl} are non-adiabatic coupling elements given by:

$$D^{kl}=\int\psi_k^*\frac{\partial}{\partial t}\psi_l d\mathbf{r}=\sum_I\dot{\mathbf{R}}_I\int\psi_k^*\nabla_I\psi_l d\mathbf{r}=\sum_I\dot{\mathbf{R}}_I\mathbf{d}_I^{kl} \tag{13.2.15}$$

The restriction to a single electronic state (in most cases the ground state ψ_0) in the expansion (13.2.12) leads to:

$$M_I\ddot{\mathbf{R}}_I(t)=-\nabla_I\langle\psi_0|\mathsf{H}_e|\psi_0\rangle \tag{13.2.16}$$

$$i\hbar\frac{\partial\psi_0}{\partial t}=\mathsf{H}_e\psi_0 \tag{13.2.17}$$

Because it includes rigorously non-adiabatic transitions between different electronic states, the Ehrenfest approach, which is the oldest "*on the fly*" method to solve coupled quantum/classical equations, has been often adopted in the investigation of scattering problems, but its application to systems with many degrees of freedom (typical in the investigation of condensed matter) has not been widespread, though recently it gained rejuvenated attention in the treatment of electronic subsystems (see Marx and Hutter, 2010; for an extended literature reference).

13.3 Born—Oppenheimer (BO) Molecular Dynamics

In this approach, the nuclei propagate under the classical mechanics laws between two successive molecular dynamic steps, and, at each step, the electronic structure part is resolved under a time-independent approach, e.g. by the time-independent Schrödinger equation. This implies that "*the time-dependence of the electronic structure is imposed and dictated by the parametric dependence on the classical dynamics of the nuclei which it just follows*" (Marx and Huttner 2010). In a bare formulation, this is readily defined for the electronic ground state as:

$$-\nabla_I \min_{\psi_0} \{\langle\psi_0|\mathsf{H}_e|\psi_0\rangle\} = M_I\ddot{\mathbf{R}}_I(t) \tag{13.3.1}$$

$$E_0\psi_0 = \mathsf{H}_e\psi_0 \tag{13.3.2}$$

The corresponding equations of motion at the Harthree–Fock (HF) theory level are:

$$-\nabla_I \min_{\{\varphi_i\}} \{\langle\psi_0|\mathsf{H}_e^{HF}|\psi_0\rangle\} = M_I\ddot{\mathbf{R}}_I(t) \tag{13.3.3}$$

$$-\mathsf{H}_e^{HF}\phi_i + \sum_j \Lambda_{ij}\phi_j = 0 \tag{13.3.4}$$

where the variational minimum of the energy expectation value $\langle\psi_0|\mathsf{H}_e^{HF}|\psi_0\rangle$ depends upon the classical term of nuclear motion $M_I\ddot{\mathbf{R}}_I(t)$ at time t, and Λ_{ij} are the associated Lagrange multipliers necessary to impose the Born–Hoppenheimer constraint, i.e.

$$\mathcal{L} = -\langle\psi_0|\mathsf{H}_e^{HF}|\psi_0\rangle + \sum_{i,j} \Lambda_{ij}\left(\langle\phi_i \mid \phi_j\rangle - \delta_{ij}\right) \tag{13.3.5}$$

with ϕ_i, ϕ_j non-orthonormal orbitals and δ_{ij} as deviation from the orthonormality condition

$$\langle\phi_i \mid \phi_j\rangle = \delta_{ij} \tag{13.3.6}$$

The isomorphic set of equations valid at Kohn–Sham DFT theory level is (Marx and Hutter 2010; and references therein):

$$-\nabla_I \min_{\{\phi_i\}} \{\langle\psi_0|\mathsf{H}_e^{KS}|\psi_0\rangle\} = M_I\ddot{\mathbf{R}}_I(t) \tag{13.3.7}$$

$$-\mathsf{H}_e^{KS}\phi_i + \sum_j \Lambda_{ij}\phi_j = 0 \tag{13.3.8}$$

13.4 Car–Parrinello (CP) Molecular Dynamics

Car and Parrinello (1985) define the Kohn–Sham (KS) electron density in terms of occupied single-particle orthonormal orbitals:

$$n(\mathbf{r}) = \sum_i |\phi_i(\mathbf{r})|^2 \tag{13.4.1}$$

and define the BO potential energy surface in terms of the energy functional:

$$E[\{\phi_i\}, \{\mathbf{R}_I\}, \{\alpha_v\}] = \sum_i \int_\Omega d^3r\phi_i^*(\mathbf{r})\left[-\frac{\hbar^2}{2m}\nabla^2\right]\phi_i(\mathbf{r}) + U[n(\mathbf{r}), \{\mathbf{R}_I\}, \{\alpha_v\}] \tag{13.4.2}$$

U in the above equation is a functional collecting of the effective electronic potential energy and the electronic repulsion. $\mathbf{R}_I$ are the nuclear coordinates and α_v are all the possible external constraints imposed on the system (i.e., volume Ω, strain, etc.,)

The KS functional is conventionally minimized with respect to orthonormal orbitals through the self-consistent KS equation:

$$\left\{-\frac{\hbar^2}{2m}\nabla^2 + \frac{\partial U}{\partial n(\mathbf{r})}\right\}\phi_i(\mathbf{r}) = \epsilon_i\ \phi_i(\mathbf{r}) \tag{13.4.3}$$

Car and Parrinello (1985) replace the conventional way of minimizing the KS functional with the "*dynamical simulated annealing*", which is computationally much more efficient and also allows the study of finite temperature properties (i.e. not only at the athermal limit).

In this procedure, the parameters $\{\phi_i\}, \{\mathbf{R}_I\}, \{\alpha_v\}$ in the functional (13.4.2) are considered as time-dependent. It is thus introduced the Lagrangian

$$\mathcal{L} = \sum_i \frac{1}{2}\mu \int_\Omega |\dot{\phi}_i|^2 d^3\mathbf{r} + \int_I \frac{1}{2}M_I\dot{\mathbf{R}}_I^2 + \sum_v \frac{1}{2}\mu_v\dot{\alpha}_v^2 - E[\{\phi_i\}, \{\mathbf{R}_I\}, \{\alpha_v\}] \tag{13.4.4}$$

with μ and μ_v arbitrary parameters of appropriate units, and with the constraints between positional variables and time expressed by

$$\int_\Omega \phi_i^*(\mathbf{r}, t)\varphi_i(\mathbf{r}, t)d^3r = \delta_{ij} \tag{13.4.5}$$

The corresponding Newtonian equations of motion are obtained through the application of the Euler–Lagrange equations

$$\frac{d}{dt}\frac{\partial\mathcal{L}}{\partial\dot{\mathbf{R}}_I} = \frac{\partial\mathcal{L}}{\partial\mathbf{R}_I} \tag{13.4.6.1}$$

$$\frac{d}{dt}\frac{\delta \mathcal{L}}{\delta \dot{\phi}_i^*} = \frac{\delta \mathcal{L}}{\delta \phi_i^*} \tag{13.4.6.2}$$

returning:

$$\mu \ddot{\psi}_i(\mathbf{r}, t) = -\frac{\delta E}{\delta \psi_i^*(\mathbf{r}, t)} + \sum_{ik} \Lambda_{ik} \psi_k(\mathbf{r}, t) \tag{13.4.7.1}$$

$$M_I \ddot{\mathbf{R}}_I = -\nabla_{\mathbf{R}_I} E \tag{13.4.7.2}$$

$$\mu_v \ddot{\alpha}_v = -\frac{\partial E}{\partial \alpha_v} \tag{13.4.7.3}$$

where Λ_{ij} are Lagrangian multipliers introduced to fulfill the holonomic constraints of Eq. 13.4.5.

The first three terms on the right of Eq. 13.4.4 define a "*classical*" kinetic energy whose equilibrium value can be calculated as the temporal average over the trajectories generated by the equations of motion (13.4.7.1, 13.4.7.2, 13.4.7.3). By variation of the velocities $\left\{\dot{\phi}_i\right\}, \{\dot{\mathbf{R}}_I\}, \{\dot{\alpha}_v\}$ in Eq. (13.4.4) the temperature of the system (which is related to them by suitable normalization) can be slowly reduced and for $T \rightarrow 0$ the equilibrium state of minimal E is attained. At equilibrium, the Lagrangian Eq. (13.4.4) "*describes a real physical system whose representative point in configurational space lies on the BO surface*" (Car and Parrinello 1985). In fact, at equilibrium $\ddot{\psi}_i = 0$. In this case, Eq. 13.4.7.1 is identical (within a unitary transformation) to the Kohn–Sham Eq. (13.4.3), and the eigenvalues of the Λ matrix coincide with the occupied KS eigenvalues.

The basic idea of the CP approach, i.e. to introduce a fictitious Newtonian dynamics for the electronic variables, results in a very efficient computational algorithm, because the updating of the electronic degrees of freedom does not involve explicit minimization of the electronic density functional for each ionic configuration. The coupled set of equations of motion for the ionic and electronic variables generates an "*ion dynamics*" that closely approximates a microcanonical evolution. The intrinsic reason why CP AIMD is so successful in practice are however not simple to decipher. Pastore et al. (1991) noted that "*While considerable progress has been made on many computational aspects of the method, little attention has yet been given to developing the underlying theory. Consequently, there have been some misunderstandings about the justification for the method*". Through detailed computations on simplified systems Pastore et al. (1991) showed that "*the CP dynamics of the electronic variables consists of small oscillations whose eigenfrequencies are approximately given by the differences between eigenvalues of the effective Schrodinger equation*". They argued then that "*precisely this oscillating dynamics makes it possible to approximate the desired Born–oppenheimer dynamics of the ionic system*". The critical assessment of Pastore et al. (1991) has been revisited (and the conclusions confirmed) by Marx and Hutter

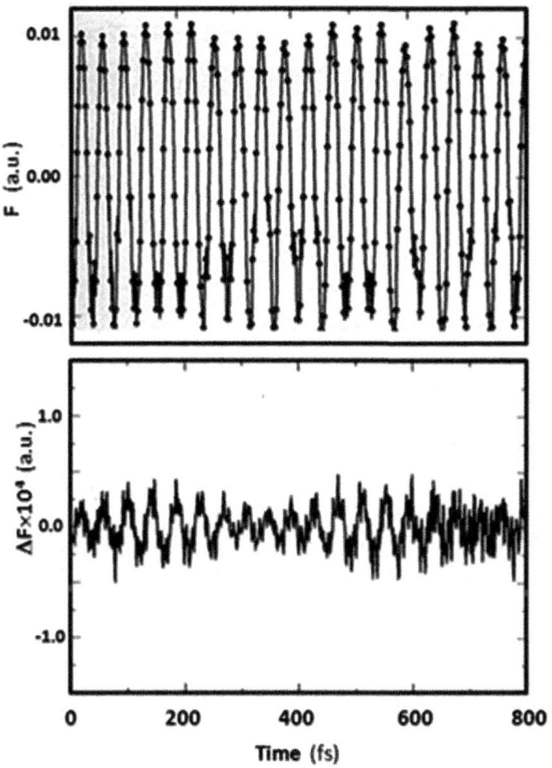

Fig. 13.1 X-component of the force (F) acting on one Si atom in a simplified diamond lattice. In the upper part of the figure the CP (solid line) and BO (dots) are compared in a relatively large time interval (800 *fs*). In the lower part the magnified delta between the two computations (CP-BO; $\Delta F \times 10^4$) reveals a constant oscillatory behavior (negligible under averaging). Modified from Marx and Hutter (2010)

(2010) with an improved sampling procedure. Figure (13.1) shows the comparison of the x-component of the force acting between two silicon atoms in a simplified periodic diamond lattice.

The AIMD computations were carried out by Marx and Hutter (2010) at LDA theory level with norm-conserving pseudopotential for core atoms and plane wave basis for valence orbitals. The adopted time step was 0.3 Femtoseconds ($1\ fs = 10^{-15}\ s$), with $\mu = 300$ a.u. and a propagation time of 6.3 Picoseconds ($1\ ps = 10^{-12}\ s$). The Car-Parrinello forces were shown to match closely the Born-Opperneimer forces obtained from a full self consistent minimization of the electronic energy at each time step of the Car-Parrinello simulation (upper part of Fig. 13.1). Augmenting the resolution of a 10^4 factor (lower part of Fig. 13.1) it is seen that the deviations are oscillatory as the actual forces are, but the oscillations remain practically unvaried over the entire simulation period at the same scale, and actually negligible for two main reasons:

(i) the smallness and boundedness of ΔF;
(ii) *the intrinsic averaging effect of small-amplitude high-frequency oscillations within a few molecular dynamic time steps, i.e. on the sub-Femtosecond time scale, which is irrelevant for nuclear dynamics* (Marx and Hutter, 2010).

I remand the interested reader to Payne (1989), Pastore et al. (1991), Car et al. (1991), Marx and Hutter (2010), and references therein, for an extended and in-depth discussion on this matter.

Chapter 14
Electron Delocalization Indices

Electron delocalization/localization is related to the two-electron, or *pair*, density, which is the simplest quantity that describes the correlated motion of a pair of electrons (Bader and Stephens 1974; Poater et al. 2005). The Lewis model describes the electronic structure of a molecule in terms of electron pairs: bonding, non-bonding, and core. Thus, it is through the pair density that the Lewis model of electronic structure finds physical expression (Fradera et al. 1999). As shown in this chapter, electron descriptors associated to electron pairs and defined in terms of distributions related to the pair density have been proposed.

As a necessary introduction to the concept of "electron correlation" I report hereafter some introductory notes from a seminal work of McWeeny (1960), which is prodromal to a better understanding of the significance of this term in the context of electron localization/delocalization studies

14.1 Spinless Density Matrix and "*Correlation*"

According to McWeeny (1954, 1960), for many (computational) purposes the simultaneous values of particle spins are of secondary importance, and it is convenient to introduce *spinless density matrices.*

Denoting $P_n(1,\ldots n)dv_1\ldots dv_n$ the probability of finding any n particles, in any order, in volume elements $dv_1\ldots dv_n$ at points 1…n, the spinless density matrix of order n- is.[1].

[1] The adoption of the same symbol P_n is justified by the fact that, when $r_1=r_{1'}$, $r_2 = r_{2'}$, …n = n' then $P_n\ (r_1.r_2,\ldots n)dv_1dv_2\ldots dv_n \equiv P_n(r_{1'},r_{2'},\ldots r_{n'})dv_{1'}dv_{2'}\ldots dv_{n'}$

G. A. Ottonello, *Quantum Geochemistry*, Springer Geochemistry,
https://doi.org/10.1007/978-3-031-21837-8_14

$$P_n(1,...n;1',...n') = \int_{\omega'_1\to\omega_1,...\omega'_n\to\omega_n} [\rho_n(1,...n;1',...n']d\omega_1...d\omega_n \tag{14.1.1}$$

where ρ_n is the n-order spin density matrix, $\int d\omega_i$ indicates integration over the ith spin variable and the arguments in P_n are spatial variables. In this view, while $P_1(1)$ is the charge density function (or "*number density*") and integrates to N (the number of electrons in the system), the diagonal element $P_2(1, 2)$ of the second order density matrix $\rho_2(1,2;1',2')$ is the pair function of the 1,2 particle pair (McWeeny 1955).

With this form, the energy E of the system may be expressed as:

$$E = \int_{1'\to 1} \left[-\left(\frac{\hbar}{2m}\right)\nabla^2(1)P_1(1;1')\right]dv_1 + \int V(1)P_1(1)dv_1 + \frac{1}{2}\int g(1,2)P_2(1,2)dv_1dv_2 == T + V_e + V_{ee} \tag{14.1.2}$$

where g(1,2) is a scalar coefficient (representing the classical term $1/r_{12}$ linked to electron repulsion), T is the kinetic energy, V_e, is the total potential energy of the individual electrons in the given potential field [a static "*charge cloud*" of density $P_1(1)$], and V_{ee} is the average interaction energy of electrons distributed with pair function $P_2(1,2)$.

Although the form of $P_1(1)$ is expected to reflect the average effect of N—1 other electrons, to a first approximation it may be assumed that the motion of one electron would be independent of the instantaneous positions of the others. In this case, (quasi-independent particles) $P_2(1,2)$ would be given by the multiplicative law for the simultaneous occurrence of two independent events:

$$P_2(1,2) = P_1(1)P_1(2) \tag{14.1.3}$$

However, "*no properly antisymmetric wave function can give such a result*" (McWeeny 1960). So, it may be posed as general case:

$$P_2(1,2) = P_1(1)P_2(2) + C(1,2) \tag{14.1.4}$$

where C(1,2) describes the departure from quasi-independence due to *correlation* between the motions of the two particles. Moreover, writing

$$P_2(1,2) = P_1(1)P_1(2)[1 + f(1,2)] \tag{14.1.5}$$

with [1+f(1,2)] as correlation factor, the conditional probability of finding a particle at point 2, when one is known to be at point 1, is then

$$\frac{P_2(1,2)}{P_1(1)} = P_1(2)[1 + f(1,2)] \tag{14.1.6}$$

The presence of a particle at point 1 therefore changes the probability of finding a particle at a second point, 2, by a fractional amount f(1,2) and an absolute amount $P_1(2)f(1,2)$. Mutual repulsion between the particles results in a correlation factor which is negative for $2 \rightarrow 1$, and thus a "*hole*" is created in the 1-particle probability function, surrounding any point at which a particle is already supposed to be. This hole is of depth $P_1(2) \times f(1,2)$ (McWeeny, 1960).

14.2 Electron Delocalization Descriptors

Several descriptors, measuring the extent to which electrons are shared between two or more atoms (or other chemical "*entities*") have been introduced through years (Matito et al. 2007). They are usually defined in terms of the Exchange Correlation Density (XCD), which represents the deviation, due to Coulomb and Fermi correlation of electron motions, between the true pair density of a system and that given by the purely classical description of a product of independent electron densities. The various descriptors until now proposed, differ from one another in the way atoms or chemical entities are defined and the XCD is expressed and approximated (Matito et al. 2007). Probably the most frequently used formulation is that of Fradera et al. (1999), whereby atomic domains are derived from the Quantum Theory of Atoms in Molecules (QTAIM) (Bader 1994). These domains are quantum objects defined by their property of being bounded by a surface through which the flux of the gradient of the electron density is zero at any point (zero-flux surface). In Fradera's et al. approach, one of the various possible forms for the XCD is doubly integrated over either the same QTAIM atomic domain, Ω, or over two different domains Ω and Ω' to yield the so-called localization [$\lambda(\Omega,\Omega)$; hereinafter $\lambda(\Omega)$, for the sake of simplicity] or delocalization $\delta(\Omega,\Omega')$ indices, respectively. The number of electrons $N(\Omega)$ on an atom Ω is then given by:

$$N(\Omega) = \lambda(\Omega) + \frac{1}{2}\sum_{\Omega' \neq \Omega} \delta(\Omega', \Omega) \tag{14.2.1}$$

where $\lambda(\Omega)$ is the number of electrons that are fully localized within Ω and $\delta(\Omega,\Omega')$ the number of those that are delocalized (exchanged) between Ω and Ω' (with $\Omega \neq \Omega'$). In practice, $N(\Omega)$ equals the sum of the number of electrons that, on a statistical basis, always stay in Ω and of half the number of those that are exchanged between Ω and all the remaining basins. The λ and δ indices lend themselves to insightful statistical interpretations of electron localization/delocalization effects because they are intimately related to the variances and covariances of the electron populations of their associated domains (Poater et al. 2005). Moreover, DI's provide estimates of the number of electrons delocalized (shared) between different atoms, no matter whether they are bonded or not to one another. Hence, DI's have been largely employed to highlight delocalization effects in complicated chemical bonding situations, like those typically occurring in organometallic systems (Macchi and Sironi 2003), as ingredients of cumulative indices to quantify aromaticity (Bultinck 2007), or as a method to understand polymerization/depolymerization effects that may be ascribed to metal complexation (alkali or alkali-earth metals) of silicate or silico-aluminate rings in molten substances (Gatti et al. 2012). As indicated by Eq. (14.2.1), the change of the electron population on an atom due to a perturbation translates into a change in either the number of electrons localized in its basin, or in the number of electrons exchanged with other basins, or, more generally, in both of these numbers. A perturbation may in some cases also lead to a negligible change in the electron population, which is, however, the result of large, almost compensating, variations of the numbers of the localized and exchanged electrons (we will see later a nice example of this case; i.e.. the analysis of thermal deformation in Na_2SiO_3). Hereafter I resume in an extremely concise fashion the various concepts that lead to the formulation of DI's.

14.2.1 Exchange Correlation Density (XCD)

Similarly to the "*electron density*", which contains information about the position of a single electron regardless the position of the other N-1 electrons, the "*pair density*" accounts for the distribution of an electron pair, irrespective of the position of the remaining N-2 electrons. The XCD enables us to "extract" from the pair density the information on the correlated motion of two electrons, that is of that kind of information that cannot be retrieved from the uncorrelated one electron distributions.

Hence XCD is defined as minus the difference between the pair density of the system and a fictitious pair distribution constructed from two independent one particle distributions. In the formalism of Matito et al. (2007) this is expressed as:

$$\gamma_{\mathrm{xc}}(\mathbf{x}_1, \mathbf{x}_2) = \gamma^{(1)}(\mathbf{x}_1)\gamma^{(1)}(\mathbf{x}_2) - \gamma^{(2)}(\mathbf{x}_1, \mathbf{x}_2) \tag{14.2.2}$$

$\gamma^{(1)}(\mathbf{x}_1)$ and $\gamma^{(2)}(\mathbf{x}_1, \mathbf{x}_2)$ in the above equation are the diagonal parts of the 1st-order Reduced Density Matrix (1-RDM) and of the 2nd-order Reduced Density Matrix (2-RDM); (simply "*electron density*" and "*electron pair density*" in the current literature).

For monodeterminantal wavefunctions (HF or DFT within the Kohn–Sham formalism) the XCD can be expressed in terms of the 1-RDM (Matito et al. 2007):

$$\gamma_{xc}(\mathbf{x}_1, \mathbf{x}_2) = \gamma^{(1)}(\mathbf{x}_1; \mathbf{x}_2)\gamma^{(1)}(\mathbf{x}_2; \mathbf{x}_1) = \left|\gamma^{(1)}(\mathbf{x}_1; \mathbf{x}_2)\right|^2 \tag{14.2.3}$$

This function is close to zero for two points distant in space. When used for correlated wave functions it is only an approximation to the true XCD and is usually referred as "*HF-like exchange– correlation density*", or simply HF-XCD (Matito et al. 2007). The localization and delocalization indices determined through Eq. (14.2.3) for Kohn-Sham DFT wavefunctions are very close to those obtained at HF theory level (Gatti and Lasi 2007), although this similarity does not imply that an electron-pair density matrix may be obtained at the DFT level of theory with the formalism valid for HF wavefunctions. As we will see later in an application section, $\lambda(\Omega)$ and $\delta(\Omega,\Omega')$ indices obtained with the approximate formula (Poater et al. 2002) that makes use of an HF-XCD matrix are of high utility in deciphering the role of ad-atoms in determining stability and conformation of silicate and silico-aluminate ring complex in silicate melts.

14.2.2 Electron Sharing Indexes (ESIs)

The Electron Sharing Index (ESI) or "*bond order*" (Fulton 1993; Fulton and Mixon, 1993) measures the extent the electrons are shared by neighboring atoms. ESIs are classified according to the way atoms are defined. In a LCAO-MO approach the atoms are formally defined "*by the Hilbert-space basis functions assigned to them*" (Matito et al. 2007). Of course, in this case the basis functions must include diffuse terms, in order to fully appreciate the true atomic extension. Other procedures assign to atoms discrete spatial regions, or domains, with sharp boundaries, such as Voronoi cells, muffin-tin spheres, Wigner-Seitz cells, or Bader's atomic basins (Bader 1994), derived from the Quantum Theory of Atoms in Molecules (QTAIM).

Adopting the QTAIM space partitioning, Cioslowski and Mixon (1991) defined the ESI for an *isopycnic* (i.e. equal density) transformation as

$$\delta^C(A, B) = 2\sum_i \lambda_i^2 S_{ii}(A)S_{ii}(B) \tag{14.2.4}$$

S_{ii} (A) in the above equation are the diagonal elements of the atomic overlap matrix (AOM) of thelocalized spin orbitals integrated over the domain of atom A and λ_i is the occupancy.

The ESI of Fulton (1993), which adopts again the QTAIM space partitioning and the 1-RDM may be simplified as follows (Matito et al. 2007):

$$\delta^F(A, B) = 2\sum_{ji} \lambda_i^{1/2}\lambda_j^{1/2} S_{ij}(A)S_{ij}(B) \tag{14.2.5}$$

Finally Fradera et et al. (1999) proposed what it is nowadays known as the delocalization index (DI):

$$\delta(A,B) = 2\int\limits_{A} \gamma^{(1)}(\mathbf{x}_1)d\mathbf{x}_1 \int\limits_{B} \gamma^{(1)}(\mathbf{x}_1)d\mathbf{x}_1 - 2\int\limits_{B}\int\limits_{A} \gamma^{(2)}(\mathbf{x}_1,\mathbf{x}_2)d\mathbf{x}_1 d\mathbf{x}_2 \quad (14.2.6)$$

This expression is founded on the seminal work of Bader and Stephens (1975) which used XCD explicitly as an ESI. In Eq. (14.2.6) the 1-RDM and 2-RDM are integrated over the atomic domains to provide respectively the atomic (1st term on the right) and atomic pair populations (2nd term on the right).

Introducing for XCD the HF-like expression (14.2.3), adopting Bader basins and associating to S_{ii} the MO overlaps over the atomic basins, Eq. (14.2.6) reduces basically to Eqs. (14.2.4, 14.2.5) (Gatti personal communication). A lucid explanation of the ESIs and their relations to XCD may be found in Gatti and Macchi (2012).

According to Bochicchio et al. (2003), in the case of a correlated wavefunction, the DI of Eq. (14.2.6) can be expressed as the sum of two terms, the first one $\delta^A(A,B)$ (exchange term) corresponding to the ESI of Angjan et al. (1994,1999) and the second one corresponding to the second order density matrix (cumulant) term $\delta^{cum}(A,B)$:

$$\delta(A,B) = \delta^A(A,B) + \delta^{cum}(A,B) \quad (14.2.7)$$

This cumulant term, which measures the difference between the true XCD and the HF-XCD, is obviously absent for monodeterminantal wavefunctions.

According to Fradera et al. (1999) the number of electrons localized in the atomic basin, or "*Localization Index*" (LI) is half of its same-atom contribution:

$$\lambda(A) = \frac{\delta(A,A)}{2} \quad (14.2.8)$$

Because the expected number of electrons in the basin is

$$\langle N(A)\rangle = \int_A \gamma^{(1)}\mathbf{r}_1 d\mathbf{r}_1 \quad (14.2.9)$$

its variance,or "*uncertainty*" [δ(A)], is interpreted as the atomic valence for closed shell systems (Mayer 1983; Matito et al. 2007):

$$\begin{aligned}\delta(A) &= \langle N^2(A)\rangle - \langle N(A)\rangle^2 = \int_A \rho(\mathbf{x}_1)d\mathbf{x}_1 - \int_A \gamma_{xc}(\mathbf{x}_1,\mathbf{x}_2)d\mathbf{x}_1 d\mathbf{x}_2 \\ &= N(A) - \lambda(A) \end{aligned} \quad (14.2.10)$$

Similarly, the covariance in a population of different atoms (Matito et al. 2007):

$$V[N(A), N(B)] = \langle N(A)(B)\rangle - \langle N(A)\rangle\langle N(B)\rangle = \iint_{BA}\int_{xc}(\mathbf{x}_1, \mathbf{x}_2)d\mathbf{x}_1 d\mathbf{x}_2 = -\frac{\delta(A, B)}{2} \quad (14.2.11)$$

is proportional to the ESI between the two atoms A and B. Finally, the number of electrons of atom A shared with the rest of the atoms in the molecule (or, in other terms, its "*valence*") is the sum of all the ESI in which it is involved (Matito et al. 2007):

$$\delta(A) = \sum_{A \neq B} \delta(A, B) \equiv -2\sum_{A \neq B} V[N(A), N(B)] \quad (14.2.12)$$

In Table 14.1 I report the QTAIM-ESI calculations of Matito et al. (2007) for some selected molecules of Geochemical interest (H_2, N_2, CO, HF, HCl, NH_3, CH_4, H_2O, CO_2, C_6H_6). The adopted theory at configuration interaction level was CISD/ 6-311++G(2d,2p), with correlation involving only the valence electrons (frozen core). The molecules were also investigated at HF and DFT/B3LYP level on the CISD geometry, to analyze both the effect of the inclusion of electron correlation and of the formal differences in the various indexes (Eqs.14.2.4, 14.2.6). Unrelaxed density matrices were obtained from the GAUSSIAN03 package while the atomic overlap matrices were generated with AIMPAC (Biegler-König et al. 1982).

$\delta^{\sigma\sigma}(A, B)$, $\delta^{\sigma\sigma'}(A, B)$ In Table (14.1) are the contributions accounting for the (α,α) and (α,β) spin cases:

$$\delta(A, B) = \delta^{\sigma\sigma}(A, B) + \delta^{\sigma'\sigma'}(A, B) + 2\delta^{\sigma\sigma'}(A, B) \quad (14.2.13)$$

$$\delta^{\sigma\sigma}(A, B) = 2\int_B\int_A \gamma_{xc}^{\sigma\sigma}(\mathbf{r}_1, \mathbf{r}_2)d\mathbf{r}_1 d\mathbf{r}_2 \quad (14.2.14.1)$$

$$\delta^{\sigma\sigma'}(A, B) = 2\int_B\int_A \gamma_{xc}^{\sigma\sigma'}(\mathbf{r}_1, \mathbf{r}_2)d\mathbf{r}_1 d\mathbf{r}_2 \quad (14.2.14.2)$$

Comparing the various magnitudes the following conclusions may be concisely abstracted (see Matito et al. 2007; for a more extended and in-depth appraisal):

(1) The fact that $\delta_{A,B}^{B3LYP}$ are much closer to $\delta_{A,B}^{HF}$ than the CISD results $\delta_{A,B}$ implies that, in general, the coulomb correlation is not fully reflected in $\delta_{A,B}^{B3LYP}$, in spite of the fact that correlation is present in B3LYP (Matito et al., 2007).

Table 14.1 QTAIM-ESI for various diatomic molecules of peculiar geochemical interest. When the ESI refers to the same atom, half of the value is given instead.Data from Matito et al. (2007)

Molecule	pair	$\delta^{HF}_{A,B}$	$\delta^{B3LYP}_{A,B}$	$\delta_{A,B}$	$\delta^{\sigma\sigma}_{A,B}$	$\delta^{\sigma\sigma\prime}_{A,B}$	$\delta^{A}_{A,B}$	$\delta^{F}_{A,B}$	**Cum.**	$\delta^{\sigma\sigma} - \delta^{A}$
H_2	H–H	0.500	0.500	0.575	0.500	0.075	0.490	0.578	0.085	0.010
	H–H'	1.000	1.000	0.849	1.000	0.151	0.950	0.844	0.100	0.050
N_2	N–N	5.480	5.477	5.896	5.564	0.332	5.408	5.796	0.488	0.156
	N–N'	3.041	3.046	2.209	2.872	0.664	2.833	2.407	0.625	0.039
CO	C–C	3.870	3.944	4.004	3.932	0.072	3.860	4.052	0.144	0.072
	C–O	1.569	1.813	1.377	1.620	0.244	1.617	1.466	0.241	0.003
	O–O	8.561	8.244	8.620	8.448	0.172	8.196	8.481	0.424	0.252
HF	H–H	0.023	0.033	0.045	0.032	0.014	0.030	0.050	0.015	0.001
	H–F	0.403	0.484	0.436	0.463	0.027	0.442	0.426	0.006	0.021
	F–F	9.574	9.483	9.519	9.506	0.014	9.357	9.524	0.162	0.149
HCl	H–H	0.253	0.258	0.333	0.273	0.059	0.261	0.342	0.071	0.012
	H–Cl	1.046	1.062	0.933	1.052	0.119	1.011	0.915	0.078	0.041
	Cl–Cl	16.702	16.680	16.734	16.675	0.059	16.461	16.744	0.273	0.214
NH_3	N–N	6.687	6.581	6.815	6.645	0.170	6.515	6.793	0.300	0.130
	N–H	0.897	0.919	0.784	0.897	0.113	0.870	0.798	0.086	0.027
	H–H	0.188	0.199	0.244	0.199	0.045	0.192	0.253	0.052	0.007
	H–H'	0.019	0.021	0.034	0.022	0.011	0.019	0.018	0.015	0.004
CH_4	C–C	3.868	3.982	4.312	4.022	0.290	3.948	4.238	0.364	0.074
	C–H	0.982	0.984	0.821	0.966	0.145	0.942	0.858	0.120	0.025
	H–H	0.485	0.455	0.520	0.463	0.057	0.449	0.530	0.071	0.014
	H–H'	0.044	0.044	0.054	0.044	0.010	0.039	0.035	0.015	0.004

(continued)

Table 14.1 (continued)

Molecule	pair	$\delta^{HF}_{A,B}$	$\delta^{B3LYP}_{A,B}$	$\delta_{A,B}$	$\delta^{\sigma\sigma}_{A,B}$	$\delta^{\sigma\sigma\prime}_{A,B}$	$\delta^{A}_{A,B}$	$\delta^{F}_{A,B}$	**Cum.**	$\delta^{\sigma\sigma} - \delta^{A}$
H_2O	O–O	8.609	8.418	8.538	8.476	0.061	8.321	8.541	0.217	0.155
	O–H	0.628	0.703	0.615	0.676	0.061	0.651	0.612	0.037	0.025
	H–H	0.064	0.083	0.107	0.080	0.027	0.077	0.114	0.030	0.003
	H–H'	0.007	0.010	0.019	0.011	0.008	0.008	0.008	0.011	0.003
CO_2	C–C	2.234	2.408	2.496	2.368	0.128	2.321	2.467	0.175	0.047
	C–O	1.152	1.408	1.125	1.254	0.128	1.237	1.155	0.112	0.016
	O–O	8.566	8.198	8.484	8.404	0.081	8.245	8.486	0.239	0.159

(2) Both $\delta^{\mathrm{HF}}_{\mathrm{A,B}}$ and $\delta^{\mathrm{B3LYP}}_{\mathrm{A,B}}$ are much closer to $\delta^{\sigma\sigma}_{\mathrm{A,B}}$ or $\delta^{A}_{\mathrm{A,B}}$ rather then to $\delta_{\mathrm{A,B}}$. This again indicates that "*what makes the difference in the calculations of* $\delta_{\mathrm{A,B}}$ *is definitely the Coulomb correlation included in 2-RDM*" (Matito et al. 2007).

(3) Since $\delta^{\mathrm{F}}_{\mathrm{A,B}}$ can be regarded as an approximation to the actual XCD for correlated calculations, it partially includes the Coulomb information of the 2-RDM, hence is nearer to $\delta_{\mathrm{A,B}}$ than $\delta^{\mathrm{A}}_{\mathrm{A,B}}$.

(4) In order to judge the approximation of HF-XCD to recover the actual XCD for correlated calculations, one must focus on the cumulant term (Cum. in Table 14.1), while the difference $\delta^{\sigma\sigma} - \delta^{\mathrm{A}}$ (last column in Table 14.1) isolates the effect of Coulomb correlation from the actual approximation given by HF-XCD.

(5) The inclusion of electron correlation in the calculation of ESI plays a marginal role in non-bonded or weakly bonded species but has a slight influence on ionic species, and a prominent role in equally-shared or polar covalent bonding (Matito et al. 2007).

14.3 Electron Delocalization and its Lewis Model Representation

Perhaps the most immediate perception of the intimate connection between DI's and the Lewis model is through the graphical representation of the *bond order*, which, according to IUPAC is "*a theoretical index of the degree of bonding between two atoms relative to that of a normal single bond, i.e. the bond provided by one localized electron pair*" (Muller, 1994). This is evident, as an example, in the representation of symmetrical *molecular entities* in which a lone pair of electrons or a vacant p-orbital is conjugated with a double bond (Fig. 14.1; parts a and b), or in the representation of aromaticity in a benzene ring conceived as resonating Kekulé structures (part c and d in the same figure).

However, as masterfully outlined by Fradera et al. (1999), the intimate connection between the quantum-mechanical concept of *delocalization* and the Lewis model goes far beyond this visual perception. To appreciate this, one must consider in detail the nature of the "*Fermi Hole*" that is created in the pair density by antisymmetry requirements.[2] According to Fradera et al. (1999) in fact, the Fermi hole may be viewed as a description of how the density of an electron of given spin (the reference electron *e**) is spread out in space, thereby excluding the presence of an identical amount of same-spin density. It follows from this consideration that all physical measures of the localization or delocalization of an electron are determined by the corresponding localization or delocalization of its Fermi hole. Since in a

[2] We will revisit in more detail the concept of Fermi Hole when dealing with the first excited state of He (Sect. 16.3.2).

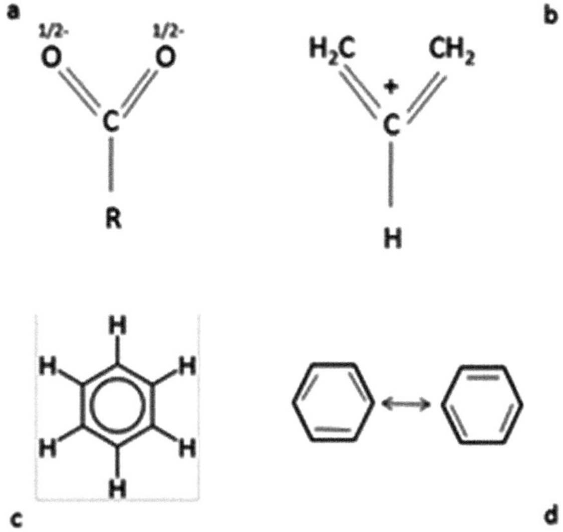

Fig. 14.1 Lewis-type representations of electron delocalization

close shell system this applies separately for electrons with opposite spin (i.e. α and β), the density of the Fermi hole determines the spatial pairing and, consequently, allows to count for the number of electron pairs that are localized to a given atom and the number that are shared between pairs of atoms, which is "*the very information needed to assign a Lewis structure*" Fradera et al., 1999). At the HF theory level each electron moves in the self-consistently determined average field of the remaining electrons and the unique form of correlation is between the motions of same-spin electrons resulting from the antisymmetry requirement (Fermi correlation). Hence, the Lewis model finds its most direct expression at this theory level.

14.4 An Application: Delocalization Indexes in Na_2SiO_3 at Pre-Melting Conditions

Na_2SiO_3 has an unusual melting behaviour most likely associated to an anomalous thermal expansion (Belmonte et al., 2016). Because this anomalous behavior is precursor to melting, melting itself is possibly caused by insufficient electronic delocalization between adjacent basins.

Two of the Raman active modes of the substance exhibit also an anomalous behavior (respectively a non linear shift with temperature above ~900 *K* and a sudden increase in the area ratio with T about 100 *K* below the melting point). In the aim of investigating in detail these two peculiar vibrations Belmonte et al. (1996) abstracted from the variational structure a cluster corresponding to an elongated packing along the C axis. The cluster contains a tetrameric chain surrounded by Na atoms in the positions dictated by the *Cmc2₁* symmetry (Fig. 14.2).

The spectral analysis carried out with the GAUSSIAN package at the same level of theory adopted for the crystal assigns to the band ν_{54} of the cluster, which corresponds to the ν_{29} mode of the crystal, a frequency of 597 cm^{-1}. The mode is effectively a mixed Si−O bending + Si−O−Si stretching motion affecting mainly the tetrameric chain (see the displacement vectors in Fig. 14.2a). The Na atoms are immobile and completely unaffected by the thermal stress. In Fig. 14.2b are shown the displacement vectors of the ν_{64} mode of the cluster (954.2 cm^{-1}), which corresponds to the ν_{34} of the crystal (954.8 cm^{-1}). This vibrational mode is essentially described by a stretching motion mostly affecting the bridging oxygens of the chain. Again, the Na atoms are unaffected by thermal vibration

Figure 14.3 shows the σ—bonding orbital of the$[Na_8Si_4O_{13}]^{2-}$ cluster abstracted by the crystal structure (MO N. 50, out of 273; 125 occupied). The electron cloud (drawn at an isodensity limit of 0.03 $e/Å^3$) encompasses essentially all the atoms of

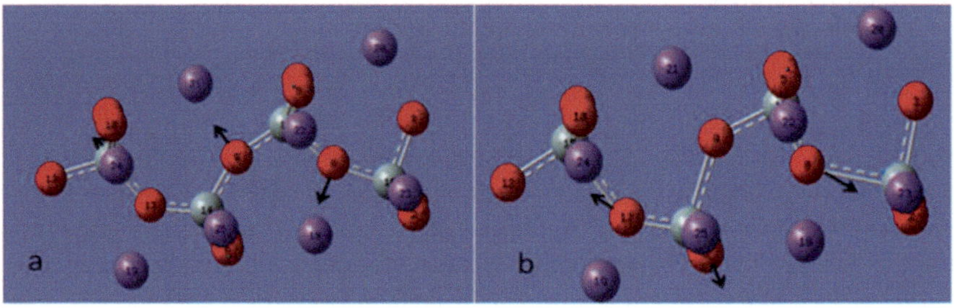

Fig. 14.2 The $[Na_8Si_4O_{13}]^{2-}$ cluster, obtained from $1\times1\times3$ packing of the $Cmc2_1$ variational structure of the crystal. The displacement vectors affecting Si and O atoms in the chain are superimposed: left side: ν_{54} = 597.1 cm^{-1}; right side: ν_{64} = 954.2 cm^{-1}. From Belmonte et al. (2016). With permission from the American chemical society

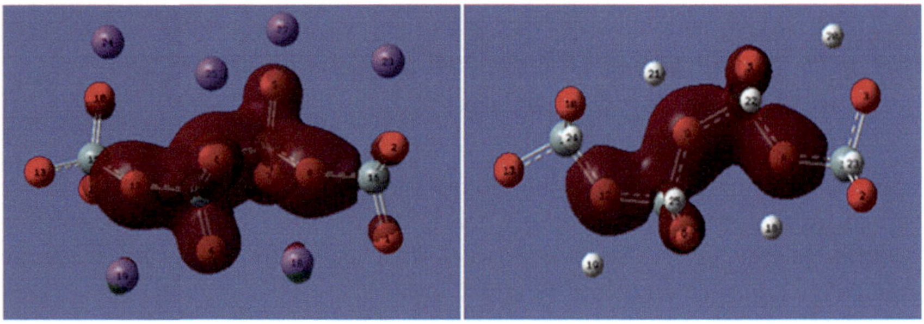

Fig. 14.3 σ-bonding inner orbital of the $[Na_8Si_4O_{13}]^{2-}$ (left side) and $[H_8Si_4O_{13}]^{2-}$ (right side) clusters resolved at the B3LYP theory level with $Cmc2_1$ symmetry and constant atomic cohordinates. The adopted basis sets are the Triple Zeta Valence basis set with Polarization of Peitinger et al. (2013) for Si and O, the 8-511G basis set of Dovesi et al. (1991) for Na, and the 5-11G* set of Dovesi et al. (1983) for H. Electron clouds are drawn for an isodensity limit of 0.03 $e/Å^3$. The eigenvalue for $[Na_8Si_4O_{13}]^{2-}$ is −1.087 *Hartree* and that of $[H_8Si_4O_{13}]^{2-}$ is −1.061 *Hartree*. From Belmonte et al. (2016). With permission from the American Chemical Society

the chain besides the terminal ones. In analogy to what already seen for zeolitic rings, (Gatti et al. 2012) this MO may be interpreted as a Pauling double bond, contributing substantially to the structural stability of the chain (Belmonte et al. 2016). The eigenvalue of MO is −1.087 *Hartree*, quite similar to the energy levels observed in similar MOs of zeolitic rings (Gatti et al. 2012). When Na is replaced by H, while keeping constant the symmetry and structure of the cluster, the electron cloud keeps the same shape, while its eigenvalue rises to −1.061 *Hartree*.

To investigate the role of deformation on structural stability, the cluster was submitted to electron localization/delocalization analysis (Belmonte et al. 2016). The interionic distances of the $[Na_8Si_4O_{13}]^{2-}$cluster in the $Cmc2_1$ground state geometry (GST) of Na_2SiO_3 and those of the vectorially deformed structure (according to the ν_{54} and ν_{64} modes) are listed in Table (14.2). In the same Table are also reported the corresponding bond critical point (bcp) electron density and delocalization indexes values. Significant variations in the electron density value induced by the deformation are observed in some Si-O bonds (bold character in Table 14.2).

Table 14.2 Electron localization/delocalization analysis of the $[Na_8Si_4O_{13}]^{2-}$ at the ground structural state (GS) and subject to ν_{54} and ν_{64} maximum deformation: ρ_{bcp} = bond critical point; δ(X,Y) = delocalization index (Belmonte et al., 2016). Data highlighted in bold refer to the changes upon deformation that are deemed to be more relevant and in most cases also somewhat unexpected

X-Y		GS	ν_{54}	ν_{64}
Si15-O13	R_e (Å)	1.6988	1.8442	1.7358
	ρ_{bcp} , *au*	0.1286	0.1016	0.1198
	δ (X,Y)	0.4171	**0.5394**	0.3682
Si15-O12	R_e (Å)	1.6892	2.0132	1.1994
	ρ_{bcp} , *au*	0.1173	**0.0624**	**0.3907**
	δ (X,Y)	0.3090	**0.2345**	**0.5802**
Si14-O12	R_e (Å)	1.6988	1.9885	2.0182
	ρ_{bcp} , *au*	0.1220	**0.0800**	**0.0698**
	δ (X,Y)	0.3382	**0.4014**	**0.3126**
Si15-O11	R_e (Å)	1.6188	1.7063	1.7916
	ρ_{bcp} , *au*	0.1484	0.1265	0.1071
	δ (X,Y)	0.4718	**0.5993**	0.4050
Si15-O10	R_e (Å)	1.6188	1.7161	1.7916
	ρ_{bcp} , *au*	0.1446	0.1202	0.1027
	δ (X,Y)	0.4302	**0.5204**	0.3591
O11-Na19	R_e (Å)	3.8511	3.8437	3.89478
	ρ_{bcp} , *au*	no bcp	no bcp	no bcp
	δ (X,Y)	0.0012	0.0006	0.0022
O10-Na24	R_e (Å)	2.2875	2.1376	2.23075
	ρ_{bcp} , *au*	0.0293	0.0374	0.0339
	δ (X,Y)	0.2224	0.2225	0.2864
Si15-Si14	R_e (Å)	3.1075	3.5930	2.87209
	ρ_{bcp} , *au*	no bcp	no bcp	no bcp
	δ (X,Y)	0.0030	0.0021	0.0048

Table 14.3 Electron populations (Bader's net atomic charges) of atoms Ω (Belmonte et al. 2016). Data highlighted in bold refer to the most relevant changes upon deformation

Ω	GS	ν_{54}	ν_{64}
Si14	10.847	**11.218**	**11.192**
Si15	10.870	**11.111**	10.867
O10	9.649	9.632	9.613
O11	9.669	9.609	9.631
O12	9.676	9.520	9.695
O13	9.608	9.536	9.546
$\sum(O)$	38.602	**38.297**	38.485

These variations are associated to progressive changes of the bonds as a result of the deformation: with the increase of the distance, the electron sharing, hence the degree of covalence, increases, and less charge separation between Si and O occurs. The atomic population of the involved atoms changes toward a lower ionicity (Si less positive, O less negative) when the structure is deformed according to the two investigated modes (Table 14.3). Nevertheless, the investigated deformations are certainly not responsible of any bond-rupture in the pre-melting T region. There is in fact no significant weakening of the structure, because when the bond length increases, the modification of the bond type from more polar to less polar tends to oppose the decomposition.

Chapter 15
Nuclear Magnetic Resonance

Nuclear Magnetic Resonance (NMR) is an extensively adopted experimental procedure in Geochemistry. What determines the NMR is the torque exerted by an applied magnetic field on the axial component of the angular momentum of the investigated nuclide along the quantization axis. A nuclide is constituted by a fixed proportion of protons and neutrons, which are themselves constituted by quarks and gluons. Quarks, protons, and neutrons, like electrons and baryons, are Fermions, and, as such, they possess an intrinsic angular momentum whose component along the quantization axis is ½. Protons and neutrons, moreover, move along nuclear orbits, centered on the center of mass of the nucleus, and the spin-orbit interaction determines a peculiar angular momentum whose component along the quantization axis is what it is improperly defined as "*spin*". To better understand this, it is appropriate to recall briefly some concepts concerning the nature of the nucleus, as we know it from the plethora of experimental information at our disposal.

15.1 Nuclear Spin

Nuclear mass is always lower than the sum of proton and neutron masses, due to mass-energy conversion which transforms part of the mass into binding energy. For example, ^{4}He has a mass 4.002604, while the resulting mass of two protons + two neutrons is 4.032981. Based on Einstein's equation:

$$\mathrm{E} = \mathrm{m}c^2 \tag{15.1.1}$$

(where c is the speed of light: $2.99792458 \times 10^{8}\ m \times s^{-1}$), the mass differential:

$$\mathrm{dm} = 4.032981 - 4.002604 = 0.030377 \tag{15.1.2}$$

G. A. Ottonello, *Quantum Geochemistry*, Springer Geochemistry,
https://doi.org/10.1007/978-3-031-21837-8_15

corresponds to a binding energy of 28.3 *MeV* (i.e. about 7.1 *MeV* per nuclear constituent). The average binding energy per nucleon is roughly constant (7.4–8.8 *MeV*), except in a few of the lightest nuclei. It increases slightly from light elements to A = 60 (iron and nickel nuclei) and then decreases more or less linearly with increasing A.

Since experimental evidence indicates that the volume and total binding energy of nuclei are proportional to the number of nucleons, it may be deduced that nuclear matter is incompressible and that each nucleon is able to interact only with a limited number of surrounding nucleons. This evidence leads to the formulation of the binding energy and mass of nuclei in terms of a "*charged drop*" model; i.e. the nuclei are assimilated to charged liquid drops with surface tension. Total binding energy E_B is the result of various energy contributions, each with simple functional dependence on mass A and charge Z of the nucleus (Myers and Swiatecki 1966):

$$E_B = c_1 A\left[1 - k\left(\frac{N-Z}{A}\right)^2\right] - c_2 A^{2/3}\left[1 - k\left(\frac{N-Z}{A}\right)^2\right] - c_3 Z^2 A^{-1/3} + c_4 Z^2 A^{-1} + \delta \tag{15.1.3}$$

The first term on the right in Eq. (15.1.3) is the *volume energy*, proportional to the number of protons and neutrons in the nucleus. It contains a symmetry correction term, proportional to $(N - Z)^2/A$, accounting for the fact that nuclear forces are at maximum when N = Z and decrease symmetrically on both sides of N = Z. The second term is the *surface energy*, and accounts for the fact that nucleons at the surface of the nucleus have unsaturated forces which reduce binding energy proportionally to $A^{2/3}$ (hence to the surface, A being proportional to volume). The third term is the electrostatic *Coulomb interaction*, accounting for Coulomb repulsion between protons, which have omologous charges. This term also reduces the binding energy of the nucleus. Since the Coulomb interaction is proportional to Z^2, it becomes increasingly important with the increase of Z and is responsible for the observed progressive relative increase of N/Z. The fourth term in Eq. (15.1.5) is a correction to Coulomb energy, due to the non-uniform charge distribution (or "*diffuse boundary*). Although Eq. (15.1.3) has only six adjustable parameters,[1] it is able to reproduce the energy of the approximately 1200 nuclides of known mass with satisfactory precision. Nevertheless, this theory is somewhat empirical and it is not able to explain the fact that nuclides of peculiar conformation (in terms of number of protons and neutrons) are particularly stable.

As anticipated, what it is usually denoted "spin" of a nuclide (I_z), represents the z-component (along the quantization axis) of the vector **I**, which is, itself, the sum

[1] When binding energy is expressed in *MeV*, the empirically adjusted coefficients have the following values: $c_1 = 15.677$ *MeV*, $c_2 = 18.56$ *MeV*, $c_3 = 0.717$ *MeV*, $c_4 = 1.211$ *MeV* and k = 1.79. δ is associated to A (if A is odd, δ×A = 0; if it is even, δ×A = ±132).

of the orbital and intrinsic contributions of the constituting nucleons. For a single nucleon, we have:

$$j = l + s \tag{15.1.4}$$

$$\langle j^2 \rangle = \hbar^2 j(j+1) \tag{15.1.5}$$

$$\langle j_z \rangle = \langle l_z + s_z \rangle = \hbar m_j \tag{15.1.6}$$

where $m_j = -j, -j + 1, \ldots j - 1$. The coupling follows the rules already seen at Sect. 6.2 (see Fig. 6.3 for a graphical explanation). From the above equation, it is also evident that

$$m_j = m_l + m_s = m_l \pm 1/2 \tag{15.1.7}$$

Since m_l is always an integer, m_j must be half-integer. There are moreover only two possible values for j, namely l + ½ or l − ½. For $l = 1$ (p states), for example, we may have $p_{3/2}$ and $p_{1/2}$.

Considering now a nuclide is composed of A nucleons, we may pose:

$$\mathbf{I} = \sum_{i=1}^{A} (l_i + s_i) = \mathbf{L} + \mathbf{S} = \sum_{i=1}^{A} \boldsymbol{j}_i \tag{15.1.8}$$

$$I_z = \sum_{i=1}^{A} \langle j_z \rangle \tag{15.1.9}$$

The observed values of I_z range from 0 to 8 over the entire periodic table. One may generally state that:

- An even/even parity leads to a zero I_z. Examples of this case are:

$${}^{4}_{2}\text{He}; {}^{12}_{6}\text{C}; {}^{16}_{8}\text{O}; {}^{32}_{16}\text{S}.$$

- An odd/odd parity leads to integer I_z. Examples of this case are

$${}^{2}_{1}\text{H}(I_z = 1); {}^{10}_{5}\text{B}(I_z = 1); {}^{14}_{7}\text{N}(I_z = 3).$$

- Even/odd and odd/even parities lead to half-integer values. As an example

$${}^{1}_{1}\text{H}\left(I_z = \frac{1}{2}\right); {}^{3}_{1}\text{H}\left(I_z = \frac{1}{2}\right); {}^{17}_{8}\text{O}\left(I_z = \frac{5}{2}\right); {}^{19}_{9}\text{F}\left(I_z = \frac{1}{2}\right); {}^{23}_{11}\text{Na}\left(I_z = \frac{3}{2}\right); {}^{31}_{15}\text{P}\left(I_z = \frac{1}{2}\right).$$

There is no simple formula to predict I_z from the number of protons and neutrons, though some hints may be obtained by application of the Wigner–Mayer–Jensen *Nuclear Shell Model*. This model was mainly conceived to explain the nature of the "*magic numbers*" 2, 8, 20, 28, 50, 82, 126; i.e. nuclides whose binding energies are significantly higher than those predicted by a charged drop model. These numbers, by analogy with the electrons behavior, are conceived as representing "*filled shells*".

To understand this, we must recall that the degeneracy of the quantum states in the nucleus obeys the rules of a tridimensional harmonic oscillator subjected to a spherical potential field (cf. Sect. 3.5); i.e. $\frac{(n+1)(n+2)}{2}$. Considering the spin, the degeneracy is doubled, i.e. $(n+1)(n+2)$. For every specific level, the overall number of states is $4l + 2$. The spin-orbit interaction must then be added and the spherical potential, which steadily increases with the distance from the center of the coordinates (Eqs. 3.5.1, 3.5.2), must be replaced with a more realistic potential valid for the nucleus. The Woods–Saxon mean field potential (Woods and Saxon 1954):

$$V(r) = \frac{V_0}{1 + \exp\left(\frac{r-R}{a}\right)} \tag{15.1.10}$$

where $R = r_0A^{1/3}$ is the nuclear radius, taken as proportional to the atomic mass A, $r = 1.25 \times 10^{-15}$ *m*, $a = 5 \times 10^{-16}$ *m* and $V_0 = 50$ *MeV*, accomplishes this duty, because it approaches a constant value at high R. It is worth to stress that the Woods–Saxon potential implicitly adopts the incompressibility concept which is the basis of the charged drop model, but now the nucleons are no more distributed at random. The net result of replacing the spherical potential $V = \frac{1}{2}\left(k_x x^2 + k_y y^2 + k_z z^2\right)$ with the potential (15.1.10) is in fact a reduction in the energy levels of high l orbits. This alters the levels of the shells of the canonical 3D oscillator and conform their degeneracy to what is experimentally observed, i.e. the *magic numbers*. The new degenerate levels are described in terms of the quantum numbers j, m_l, and parity, instead of l, m_l, and m_s (Fig. 15.1).

The energy splitting dictated by spin-orbit interaction, which increases with the orbital quantum number, determines the overlaps shown in the figure and returns the correct degeneracy predicted by the magic numbers. Knowing the total angular momentum of each nucleon we must now combine them to get the overall angular momentum of a given nuclide. Taking as an example $^{17}_{8}O$ with 8 protons and 9 neutrons, we may observe that the 8 protons fill the first shell and their angular momenta cancel out. So do the first 8 neutrons. The 9th neutron occupies the first energy level above the first shell, which is $1d_{5/2}$. Since this neutron is in a d shell ($l = 2$) its parity $p = (-1)^l$ is positive. Hence the nuclide $^{17}_{8}O$ is expected to have positive parity and overall angular momentum $J_z = 5/2$.

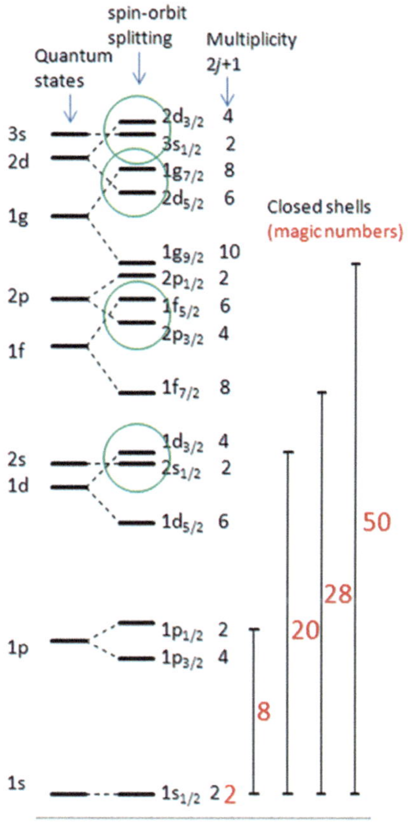

Fig. 15.1 Energy levels of the nuclear shell model. The subscripts indicate the value attained by the total angular momentum J_z. The multiplicity is given by $2j + 1$. Values in red are the first five magic numbers, corresponding to the degeneracy of the closed shells (i.e. 8 = 2 + 6; 20 = 2 + 6 + 12; and so on). Overlapping levels are highlighted by green circles

Table 15.1, taken from Tossel and Vaughan (1992) lists all the possible nuclides of potential use in NMR investigation.

To achieve a minimum energy the spins of protons and neutrons are coupled when possible in parallel. This has to do with their constitution (two up quarks + one down quark + gluons for protons and one up quark + two down quarks + gluons for neutrons).[2]

In deuterium (^2H) for example, proton and neutron are coupled in parallel, giving an overall I_z of 1. In Tritium (^3H) one of the neutrons couple with the same spin proton, the two neutrons are anti-parallel, and the uncoupled neutron assign to ^3H a net spin of ½, just like ^1H. If the spin of all particles is paired there is no net spin and $I_z = 0$.

[2] This does not violate Pauli exclusion principle because protons and neutrons are not indistinguishable particles.

Table 15.1 Some nuclides of potential interest in NMR. % is the percentual isotopic abundance. The quadrupole moment eQ is in $cm^{-1}\ \times 10^{24}$. The frequency is in MHz

Nuclide	I_z	eQ	%	Frequency	Nuclide	I_z	eQ	%	Frequency
${}^{1}_{1}H$	1/2	–	99.985	500	${}^{63}_{29}Cu$	3/2	0.16	69.1	132.5
${}^{2}_{1}H$	1	0.0028	0.015	76.8	${}^{67}_{30}Zn$	5/2	0.15	4.1	31.3
${}^{7}_{3}Li$	3/2	−0.03	92.58	194.3	${}^{71}_{31}Ga$	3/2	0.112	39.6	152.5
${}^{9}_{4}Be$	3/2	0.0512	100	70.3	${}^{73}_{32}Ge$	9/2	−0.2	7.8	17.4
${}^{10}_{5}B$	3	0.074	19.58	53.7	${}^{79}_{35}Br$	3/2	0.33	50.5	125.3
${}^{11}_{5}B$	3/2	0.0355	80.42	160.4	${}^{85}_{37}Rb$	5/2	0.27	71.25	48.3
${}^{13}_{6}C$	1/2	–	1.1	125.7	${}^{89}_{39}Y$	1/2	–	100	24.5
${}^{14}_{7}N$	1	0.016	99.6	36.1	${}^{93}_{41}Nb$	9/2	−0.2	100	122.2
${}^{17}_{8}O$	5/2	−0.026	0.037	67.8	${}^{95}_{42}Mo$	5/2	0.12	15.7	32.6
${}^{23}_{11}Na$	3/2	0.14	100	132.3	${}^{109}_{47}Ag$	1/2	–	48.18	23.3
${}^{25}_{12}Mg$	5/2	ND	10.1	30.6	${}^{113}_{48}Cd$	1/2	–	12.26	110.9
${}^{27}_{13}Al$	5/2	0.149	100	130.3	${}^{115}_{49}In$	9/2	1.14	95.72	109.6
${}^{29}_{14}Si$	1/2	–	4.7	99.3	${}^{119}_{50}Sn$	1/2	–	8.58	186.4
${}^{31}_{15}P$	1/2	–	100	202.4	${}^{133}_{55}Cs$	7/2	−0.003	100	65.6
${}^{35}_{17}Cl$	3/2	−0.0789	75.5	49.0	${}^{139}_{57}La$	7/2	0.21	99.9	70.6
${}^{39}_{19}K$	3/2	0.11	93.1	23.3	${}^{199}_{80}Hg$	1/2	–	16.8	89.1
${}^{45}_{21}Sc$	7/2	−0.22	100	121.5	${}^{205}_{81}Tl$	1/2	–	70.5	288.5
${}^{51}_{23}V$	7/2	−0.04	99.76	131.4	${}^{207}_{82}Pb$	1/2	–	22.6	104.6

15.2 Magnetic Dipole Moment

To a non-zero **I** it is always associated a non-zero magnetic dipole moment **μ** through the relation:

$$\boldsymbol{\mu} = \gamma \mathbf{I} \tag{15.2.1}$$

where γ is the gyromagnetic ratio.

The multipole expansion of the current distribution may be written as follows:

$$\mathbf{A}(\mathrm{r}) = \frac{\mu_0}{4\pi}\left\{\frac{1}{\mathrm{r}}\int \mathbf{j}(\mathbf{r}')\mathrm{dv}' + \frac{1}{r^3}\int \mathbf{j}(\mathbf{r}')(\mathbf{r}\cdot\mathbf{r}')\mathrm{dv}' + \ldots\right\} \tag{15.2.2}$$

where **A**(r) is the vector potential, r′ is the position relative to the origin, and the vector function **j**(**r**′) gives the magnitude and direction of the electric current per unit volume dv′ at that point. Limiting the expansion to the lowest order (dipole), Eq. (15.2.2) may be written in the form:

$$\mathbf{A}(\mathrm{r}) = \frac{\mu_0}{4\pi} \frac{\boldsymbol{\mu} \times \mathbf{r}}{\mathrm{r}^3} \tag{15.2.3}$$

where the leading non-vanishing term is the *magnetic dipole moment* **μ**:

$$\boldsymbol{\mu} = \frac{1}{2} \int \mathbf{r}' \times \mathbf{j}(\mathbf{r}') \mathrm{d}\mathrm{v}' \tag{15.2.4}$$

The integrand in Eq. (15.2.4) includes the charge density and the vector product $\mathbf{r}' \times \mathrm{v}'$, which, in the case of a particle with mass m is l/m (with l as angular momentum). Because the charge density may be written also in the form $e|\psi(\mathbf{r}')|^2$, the quantum-mechanical expression of the magnetic dipole moment is (Krane 1987):

$$\boldsymbol{\mu} = \frac{e}{2\mathrm{m}} \int \psi^*(\mathbf{r}') l \psi(\mathbf{r}') \mathrm{d}v' \tag{15.2.5}$$

If the wave function corresponds to a state of definite l_z, only the z component of the integral is non-vanishing, hence, operating the substitution $l_z = m_l \hbar$ (Krane 1987):

$$\mu_z = \frac{e}{2\mathrm{m}} \int \psi^*(\mathbf{r}') l_z \psi(\mathbf{r}') \mathrm{d}v' = \frac{e\hbar}{2\mathrm{m}} m_l \tag{15.2.6}$$

Substituting in Eq. (15.2.6) the proton mass m_p, we get the nuclear magneton μ_N:

$$\mu_\mathrm{N} = \frac{e\hbar}{2\mathrm{m_p}} = 3.15245 \times 10^{-8} \quad (eV/\mathrm{T}) \tag{15.2.7}$$

Considering the spin, Eq. (15.2.6) may be extended in the form (Krane 1987):

$$\mu = \frac{(\mathrm{g}_l l + \mathrm{g}_s s)\mu_\mathrm{N}}{\hbar} \tag{15.2.8}$$

where the *g factors* g_l and g_s account for the orbital and intrinsic spin contributions (for protons g_l = 1 and g_s = 5.5856912; for neutrons g_l = 0 and g_s = −3.8260837).

In a nuclide, the effect of all nucleons is accounted for through the simple summation:

$$\mu = \sum_{\mathrm{i}=1}^{\mathrm{A}} \frac{(g_{l,i} l_i + g_{s,i} s_i)\mu_\mathrm{N}}{\hbar} \tag{15.2.9}$$

15.3 Electric Quadrupole Moment

Considering the distribution of charges inside the nucleus, the electric potential V(r) is:

$$V(r) = \frac{1}{4\pi \in_0} \int \frac{\rho(r')dv'}{|r - r'|} \tag{15.3.1}$$

where $\in_0$ is the vacuum permittivity. Expanding the factor $|r - r'|$ and assuming that the observation point is far from the nucleus (i.e. $r \gg r'$) it is obtained:

$$V(r) = \frac{1}{4\pi \in_0} \left[\frac{1}{r} \int \rho(r')dv' + \frac{1}{r^2} \int \rho(r')r' \cos\theta dv' + \frac{1}{r^3} \int \rho(r')r'^2 \frac{1}{2} (3\cos^2\theta - 1)dv' + \ldots \right] \tag{15.3.2}$$

The first term gives the total charge *Ze*, which is uninteresting from the point of view of nuclear structure, and the second term vanishes (see Krane 1987; for details). Thus the first interesting term in the expansion is the quadrupole term. The nuclear quadrupole moment is thus defined as:

$$eQ = \int \rho(r')r'^2 (3\cos^2\theta - 1)dv' \tag{15.3.3}$$

The angle θ in Eq. (15.3.3) is measured from the *z* axis, corresponding to the maximum projection of the nuclear spin. Based on the value of Q the nuclei are spherical (Q = 0), or non-spherical ($Q \neq 0$). If $r'^2(3\cos^2\theta - 1) = 3z'^2 - r'^2$ is on the average positive, there is more of the charge density along the z′ axis than within the average radius and the nuclide is *prolate deformed* (Q > 0). If on the contrary $3z'^2 - r'^2$ is negative the z′ axis contain less of the nuclear charge density and there is corresponding flattening (*oblate deformed* nuclide; Q < 0).

A nice representation of the various cases has been proposed by Tossel and Vaughan (1992) and is here reproduced in Fig. 15.2 with minimal modifications.

15.4 Larmor Precession

The magnetic quantum number m_l can take values from $+I_z$ to $-I_z$ in integer steps. For a nucleus of I_z = ½, two degenerate states are accessible, namely $m_l = 1/2$ and $m_l = -1/2$. If the nucleus is placed in a magnetic field the degeneracy is eliminated due to the interaction between the nuclear magnetic dipole moment and the external magnetic field. Denoting $\mathbf{B}_0$ the magnetic field, the energy of the μ_z component of the magnetic dipole moment $\boldsymbol{\mu}$ along the quantization axis *z* is:

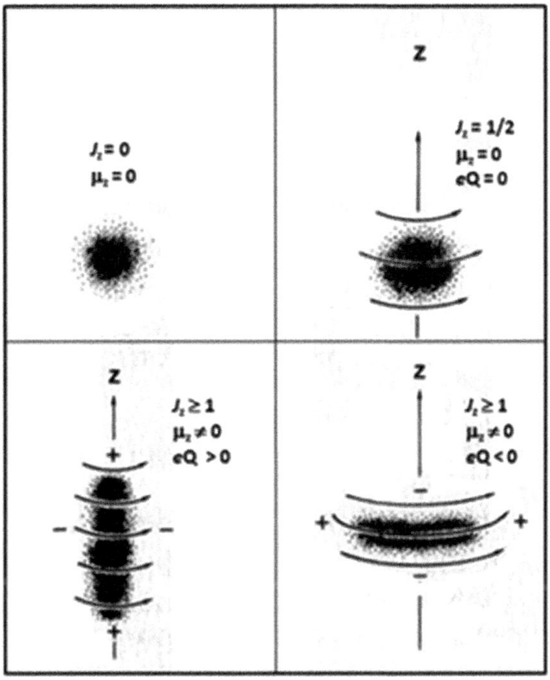

Fig. 15.2 Nuclei with different magnetic properties. Modified from Tossel and Vaughan (1992)

$$E = -\mu_z B_{0,z} = -\gamma m_l \hbar B_{0,z} \tag{15.4.1}$$

and the energy difference between the two degenerate states is simply

$$\Delta E = \gamma \hbar B_{0,z} \tag{15.4.2}$$

Similarly to the effect exerted by the gravitational field on a spinning top, the external magnetic field exerts a torque on the nuclear magnetic moment:

$$\tau = \mu \times \mathbf{B}_0 = \gamma \mathbf{I} \times \mathbf{B}_0 \tag{15.4.3}$$

As a net effect, the angular momentum vector **I** precesses about the external field axis with an angular frequency depending on the magnitude B of the applied field (Larmor frequency):

$$\omega = \gamma B \equiv \frac{\Delta E}{\hbar} \tag{15.4.4}$$

Equation (15.4.4) which is obeyed at "*resonance condition*", determines slight fluctuations in the magnetic field, which are the object of the NMR measurement.

To detect the resonance condition, a small oscillating magnetic field, with a frequency sufficiently close to the Larmor frequency, is superimposed on the constant magnetic field $\mathbf{B}_0$. This perturbation determines a Rabi's oscillation between the two degenerate states aligned along or against $B_{0,z}$. Assuming that the system at time t = 0 is in the state $|\phi_1\rangle$:

$$|\psi(0)\rangle = |\phi_1\rangle \tag{15.4.5}$$

the probability $\mathcal{P}_{12}(\mathrm{t})$ of finding it in the state 2 at time *t* is given by the Rabi's formula (Cohen-Tannoudji et al. 1977)

$$\mathcal{P}_{12}(\mathrm{t}) = \frac{4\left|\mathrm{W}_{12}^2\right|}{4\left|\mathrm{W}_{12}^2\right| + \Delta\mathrm{E}_{12}^2}\sin^2\left[\frac{\mathrm{t}}{2\hbar}\sqrt{4\left|\mathrm{W}_{12}^2\right| + \Delta\mathrm{E}_{12}^2}\right] \tag{15.4.6}$$

where W_{12} is the perturbation or "*coupling*" and $\Delta\mathrm{E}_{12} \equiv \Delta\mathrm{E}$.

The Rabi's oscillations are analyzed in terms of the precession of the spin magnetization around the effective magnetic field in a reference frame rotating with the frequency ω_{RF} (Abragam 1961).

Because the electrons surrounding a given nuclide produce a magnetic field which shields somewhat the nucleus from the applied field, the frequency required to achieve resonance is also reduced. Is indeed this "*chemical shift*", reflecting the electron density distribution around the nuclide, which is the object of investigation: a high shielding (i.e. high electron density) causes an "*upfield*" shift of the resonance and a low shield (low electron density) a "*downfield*" shift. In practice, the chemical shift is expressed in *ppm* with respect to a reference compound:

$$\delta = \frac{\mathcal{F}_{\mathrm{sample}} - \mathcal{F}_{\mathrm{reference}}}{\omega_{\mathrm{s}}} \times 10^6 \tag{15.4.7}$$

with $\mathcal{F}_{\mathrm{sample}}$ and $\mathcal{F}_{\mathrm{reference}}$ as absorption lines in *Hz*, and ω_{s} as operating frequency of the spectrometer.

15.5 Gauge Invariant Atomic Orbitals (GIAO)

The orbital electronic currents induced by the applied magnetic field in molecules with unpaired electrons cause, as anticipated, chemical shifts of the nuclear resonance frequencies (and causes also diamagnetic polarization). The total magnetic moment of the induced currents is $\chi_{\mathrm{m}}\mathbf{B}_0$, with χ_{m} as molecular diamagnetic (negative) susceptibility: The secondary magnetic field generated by these currents at nucleus *A* is $-\sigma^{\mathrm{A}}\mathbf{B}_0$, with $-\sigma^{\mathrm{A}}$ as magnetic screening constant. The total field experienced by the nucleus A is thus:

$$\mathbf{B}^{\mathrm{A}} = \mathbf{B}_0\left(1 - \sigma^{\mathrm{A}}\right) \tag{15.5.1}$$

The magnitudes χ_{m} and σ^{A} are tensors, with three principal axes, and values. Only the mean of these values is experimentally available in fluids.

For a single atom A, the diamagnetic susceptibility can be conveniently split into two parts (Pople 1962):

$$\chi^{\mathrm{A}} = \chi_{\mathrm{d}}^{\mathrm{A}} + \chi_{\mathrm{p}}^{\mathrm{A}} \tag{15.5.2}$$

$\chi_{\mathrm{d}}^{\mathrm{A}}$ is the diamagnetic contribution for an atom whose electrons move freely about the direction of the magnetic field $\mathbf{B}_0$, with the Larmor angular frequency

$$\omega = \frac{e\mathbf{B}_0}{2\mathrm{m}c} \tag{15.5.3}$$

For an isolated spherical atom, the molar atomic susceptibility is given by the Langevin formula

$$\chi_{\mathrm{d}}^{\mathrm{A}} = -\frac{\mathrm{N}e^2}{6\mathrm{m}c^2}\sum_{\mathrm{i}} \bar{\mathrm{r}}_{\mathrm{i}}^2 \tag{15.5.4}$$

with $\bar{\mathrm{r}}_{\mathrm{i}}^2$ as mean square distance of the ith electron from the nucleus.

In a molecule, however, the free circulation of electrons is prevented by the hindering effect of the electric field imposed by the other electrons. This generates the second paramagnetic (positive) contribution $\chi_{\mathrm{p}}^{\mathrm{A}}$. The calculation of the hindering effect requires the adoption of electronic wave functions for electrons in a magnetic field. In 1937 London proposed to impose gauge invariance,[3] in molecular calculations by attaching field dependent complex phase factors to the atomic orbitals. The Gauge Independent Atomic Orbital (GIAO) χ_ν of London (1937) involves the modification of the basis function ϕ_ν through a complex factor which depends on the gauge of the vector potential **A(r)**:

$$\chi_\nu = \exp\left[-\left(\frac{\mathrm{i}}{c}\right)\mathbf{A}_\nu \cdot \mathbf{r}\right]\phi_\nu \tag{15.5.5}$$

As noted by Ditchfield (1974), GIAO is an "*unfortunate term*", because it is quite evident that the factor depends on the gauge. In fact, Pople (1962) proposed to adopt rather the term "*Gauge Dependent Atomic Orbitals*". The London atomic

---

[3] A gauge theory is a type of field theory in which the Lagrangian is invariant under local transformation from a Lie group (a differentiable manifold).

orbitals were employed by Hameka (1958, 1962) for calculating the magnetic susceptibility and shielding constants of H_2.

Ditchfield (1974) adopted London orbitals in ab initio calculations of shielding constants in the framework of SCF perturbation theory.

The treatment of Ditchfield (1974) begins with the Hamiltonian for a closed shell molecule in the total magnetic field arising from an external field $\mathbf{B}_0$ plus the dipole fields arising from nuclear magnetic moments $\boldsymbol{\mu}_A$, $\boldsymbol{\mu}_B$, … at the nuclear positions $\mathbf{R}_A$, $\mathbf{R}_B$, …. In atomic units the Hamiltonian has the form:

$$\mathrm{H}(\mathbf{B}_0,\mu)=\frac{1}{2}\sum_{j}\left\{\left[\mathbf{p}_j+\frac{1}{c}\mathbf{A}'(\mathrm{r}_j)\right]^2-2\sum_{A}\frac{Z_A}{\mathbf{r}_{jA}}\right\}+\sum_{j\neq m}\frac{1}{\mathbf{r}_{jm}}+\sum_{A\neq B}\frac{Z_AZ_B}{\mathbf{R}_{AB}} \tag{15.5.6}$$

where the summation in j is over all electrons and the summation in A is over all nuclei. $\mathbf{A}'(\mathrm{r}_j)$ is the vector potential describing the total magnetic field at the position of electron j:

$$\mathbf{A}'(\mathrm{r}_j)=\frac{1}{2}\mathbf{B}_0\times\mathrm{r}_j+\sum_{A}\frac{\mu_A\times\mathbf{r}_{jA}}{\mathrm{r}_{jA}^3}=\mathbf{A}(\mathrm{r}_j)+\sum_{A}\frac{\mu_A\times\mathbf{r}_{jA}}{\mathrm{r}_{jA}^3} \tag{15.5.7}$$

Z_A, Z_B in Eq. (15.5.6) are the nuclear charges of atoms A and B; $\mathbf{r}_j$ in Eq. (15.5.7) is the distance vector from electron j to some arbitrary origin; $\mathbf{r}_{iA}$, $\mathbf{r}_{jm}$, $\mathbf{R}_{AB}$ are the vectors $(\mathbf{r}_j-\mathbf{R}_A)$, $(\mathbf{r}_j-\mathbf{r}_m)$ and $(\mathbf{R}_A-\mathbf{R}_B)$, respectively.

Imposing the Coulomb gauge (div A = 0) and replacing the quantum mechanical momentum $\mathbf{p}_j$ with $-i\nabla_j$ leads to (Ditchfield 1974)

$$\mathrm{H}(\mathbf{B}_o,\mu_A)=\mathrm{H}^{(0)}++\sum_{\alpha}\mathrm{H}_\alpha\mathrm{H}_\alpha^{(1,0)}+\sum_{\alpha}\mu_{A\alpha}\mathrm{H}_{A\alpha}^{(0,1)}+\frac{1}{2}\sum_{\alpha}\sum_{\beta}\mathrm{H}_\alpha\mathrm{H}_{\alpha\beta}^{(2,0)}+\sum_{\alpha}\sum_{\beta}\mathrm{H}_\alpha\mathrm{H}_{A\alpha\beta}^{(1,1)}\mu_{A\beta} \tag{15.5.8}$$

H_α and H_β in Eq. (15.5.8) are directional components of the external magnetic field (α and β are used to indicate Cartesian coordinates x, y or z), and

$$\mathrm{H}^{(0)}=\sum_{j}\left(-\frac{1}{2}\nabla_j^2-\sum_{A}\frac{Z_A}{\mathbf{r}_{jA}}\right)+\sum_{j\neq m}\frac{1}{\mathbf{r}_{jm}}+\sum_{A\neq B}\frac{Z_AZ_B}{\mathbf{R}_{AB}} \tag{15.5.9.1}$$

$$\mathrm{H}_\alpha^{(1,0)}=-\left(\frac{i}{2c}\right)\sum_{j}\mathrm{L}_{j\alpha} \tag{15.5.9.2}$$

$$\mathrm{H}_{\alpha\beta}^{(2,0)}=\left(\frac{1}{4c^2}\right)\sum_{j}\left(\mathrm{r}_j^2\delta_{\alpha\beta}-\mathrm{r}_{j\alpha}\mathrm{r}_{j\beta}\right) \tag{15.5.9.3}$$

$$H_{A\alpha}^{(0,1)} = -\left(\frac{i}{c}\right)\sum_j \frac{L_{jA\alpha}}{r_{jA}^3} \tag{15.5.9.4}$$

$$H_{A\alpha\beta}^{(1,1)} = \left(\frac{1}{2c^2}\right)\sum_j \frac{\left(r_j \cdot r_{jA}\delta_{\alpha\beta} - r_{j\alpha}r_{A\beta}\right)}{r_{jA}^3} \tag{15.5.9.5}$$

$$L_{j\alpha} = \left(r_j \times \nabla_j\right)_\alpha \tag{15.5.9.6}$$

$$L_{jA\alpha} = \left(r_{jA} \times \nabla_j\right)_\alpha \tag{15.5.9.7}$$

with $\delta_{\alpha\beta}$ as Kronecker delta.

The energy $E(\mathbf{B}_0, \mu_A)$ associated to the Hamiltonian is found by resolving the Schrödinger equation:

$$H(\mathbf{B}_0, \mu_A)\Psi(\mathbf{B}_0, \mu_A) = E(\mathbf{B}_0, \mu_A)\Psi(\mathbf{B}_0, \mu_A) \tag{15.5.10}$$

where $\Psi(\mathbf{B}_0, \mu_A)$ is the wave function describing the molecule in the presence of $\mathbf{B}_0$ and $\boldsymbol{\mu}_A$.

Expanding $\Psi(\mathbf{B}_0, \mu_A)$ and $E(\mathbf{B}_0, \mu_A)$ in a Taylor series about their zero-field values, Ditchfield (1974) found:

$$E(\mathbf{B}_o, \mu_A) = E^0 + \sum_\alpha E_\alpha^{(1,0)} H_\alpha +$$
$$\sum_\alpha E_{A\alpha}^{(0,1)} H_\alpha + \frac{1}{2}\sum_\alpha\sum_\beta H_\alpha E_{\alpha\beta}^{(2,0)} H_\beta + \sum_\alpha\sum_\beta H_\alpha E_{\alpha\beta}^{(1,1)} \mu_{\alpha\beta} + \frac{1}{2}\sum_\alpha\sum_\beta \mu_{A\alpha} E_{A\alpha\beta}^{(0,2)} \mu_{A\beta} + \cdots \tag{15.5.11.1}$$

or, in a simpler form:

$$E(\mathbf{B}_0, \mu_A) = E^0 - \sum_\alpha \gamma_\alpha H_\alpha -$$
$$\sum_\alpha \mu_{A\alpha} H_\alpha - \frac{1}{2}\sum_\alpha\sum_\beta H_\alpha \chi_{\alpha\beta} H_\beta + \sum_\alpha\sum_\beta H_\alpha \sigma_{A\alpha\beta}\mu_{\alpha\beta} + \ldots \tag{15.5.11.2}$$

where γ_α is a component of the permanent magnetic moment of the molecule ($\gamma_\alpha = 0$ for singlet state molecules with zero orbital angular momentum); the third term represents the direct interaction between the nuclear magnetic moment and the external magnetic field; the fourth term describes the diamagnetic polarization of the molecule.

The electronic currents induced by the external magnetic field generate a total magnetic moment which, in direction α is $\sum_\beta \chi_{\alpha\beta} H_\beta$ (where $\chi_{\alpha\beta}$ is a component of the nuclear magnetic susceptibility tensor χ). The same currents generate a secondary magnetic field at nucleus A, with magnitude $\sum_{\beta} \sigma_{A\alpha\beta} H_{\beta}$ where $\sigma_{A\alpha\beta}$ is a component of the magnetic shielding tensor σ^{A}.

The total magnetic field experienced by nucleus A, which determines its NMR frequency is the one given by Eq. (15.5.1), as anticipated.

Following London (1937) and replacing the MO's ϕv with the GIAO MO's χ_{v}, the LCAO procedure becomes (cf. Eq. 15.5.5):

$$\chi_{v}(\mathbf{B}_{0}) = \exp\left[-\left(\frac{i}{c}\right)\mathbf{A}_{v} \cdot \mathbf{r}\right]\phi_{v} = f_{v}\phi_{v} \tag{15.5.12}$$

$$\psi_{j}(\mathbf{B}_{0}, \mu_{A}) = \sum_{v} c_{v,j}(\mathbf{B}_{0}, \mu_{A})\chi_{v}(\mathbf{B}_{0}) \tag{15.5.13}$$

with $c_{v,j}$ as LCAO coefficients. The solution of the Schrödinger equation (15.5.10) was then obtained by Ditchfield, at the light of SCF perturbation, as a single determinant of the GIAO MOs $\psi_{j}(\mathbf{B}_{0}, \mu_{A})$:

$$\Psi(\mathbf{B}_{0}, \mu_{A}) = \left|\psi_{1}(\mathbf{B}_{0}, \mu_{A})\overline{\psi}_{1}(\mathbf{B}_{0}, \mu_{A}).\ldots..\psi_{n}(\mathbf{B}_{0}, \mu_{A})\overline{\psi}_{n}(\mathbf{B}_{0}, \mu_{A})\right| \tag{15.5.14}$$

Differentiating $E(\mathbf{B}_{0}, \mu_{A})$ with respect to $\mu_{A\beta}$ and recalling the relationship between $E^{(1,1)}_{A\alpha\beta}$ and $\sigma_{A\alpha\beta}$ (Eqs. 15.5.11.1 and 15.5.11.2) Ditchfield (1974) showed that $\sigma_{A\alpha\beta}$ can be split into diamagnetic and paramagnetic components:

$$\sigma_{A\alpha\beta} = \sigma^{d}_{A\alpha\beta} + \sigma^{p}_{A\alpha\beta} \tag{15.5.15}$$

$$\sigma^{d}_{A\alpha\beta} = \sum_{v\lambda} P^{(0)}_{v\lambda}\left(H^{(1,1)}_{Av\lambda}\right)^{d}_{\alpha\beta} \tag{15.5.16}$$

$$\sigma^{p}_{A\alpha\beta} = \sum_{v\lambda}\left[P^{(0)}_{v\lambda}\left(H^{(1,1)}_{Av\lambda}\right)^{p}_{\alpha\beta} + \left(P^{(1,0)}_{v\lambda}\right)_{\alpha}\left(H^{(0,1)}_{Av\lambda}\right)^{p}_{\beta}\right] \tag{15.5.17}$$

where the $P_{v\lambda}$ are terms of the density matrix $P(\mathbf{B}_{0}, \mu_{A})$:

.

$$P^{(0)}_{v\lambda} = 2\sum_{j}^{occ} c^{(0)}_{vj} c^{(0)}_{\lambda j} \tag{15.5.18}$$

$$\left(P^{(1,0)}_{v\lambda}\right)_{\alpha} = 2\sum_{j}^{occ}\left[c^{(0)}_{vj}\left(c^{(1,0)}_{\lambda j}\right)_{\alpha} - \left(c^{(1,0)}_{vj}\right)_{\alpha} c^{(0)}_{\lambda j}\right] \tag{15.5.19}$$

Ditchfield stressed that, although Eq. (15.5.15) provides a theoretical value for the magnetic shielding constant, in this particular form it gives no indication of the important atomic (local and non-local) and interatomic mechanisms contributing to $\sigma^{\mathrm{d}}_{\mathrm{A}\alpha\beta}$ and $\sigma^{\mathrm{p}}_{\mathrm{A}\alpha\beta}$ and by which shielding occurs. Through matrix deconvolution techniques operating on the vacant and occupied orbitals, Ditchfield showed then that the diamagnetic term $\sigma^{\mathrm{d}}_{\mathrm{A}\alpha\beta}$ is actually the sum of three contributions: $\left(\sigma^{\mathrm{AAAA}}_{\mathrm{A}\alpha\beta}\right)^{\mathrm{d}}$, $\left(\sigma^{\mathrm{BBBB}}_{\mathrm{A}\alpha\beta}\right)^{\mathrm{d}}$, $\left(\sigma^{\mathrm{interatomic}}_{\mathrm{A}\alpha\beta}\right)^{\mathrm{d}}$, and the paramagnetic term $\sigma^{\mathrm{p}}_{\mathrm{A}\alpha\beta}$ arises from seven distinct contributions: $\left(\sigma^{\mathrm{AAAA}}_{\mathrm{A}\alpha\beta}\right)^{\mathrm{p}}$, $\left(\sigma^{\mathrm{ABBA}}_{\mathrm{A}\alpha\beta}\right)^{\mathrm{p}}$, $\left(\sigma^{\mathrm{ACDA}}_{\mathrm{A}\alpha\beta}\right)^{\mathrm{p}}$, $\left(\sigma^{\mathrm{BCCB}}_{\mathrm{A}\alpha\beta}\right)^{\mathrm{p}}$, $\left(\sigma^{\mathrm{BBBB}}_{\mathrm{A}\alpha\beta}\right)^{\mathrm{p}}$, $\left(\sigma^{\mathrm{BCDB}}_{\mathrm{A}\alpha\beta}\right)^{\mathrm{d}}$, $\left(\sigma^{\mathrm{interatomic}}_{\mathrm{A}\alpha\beta}\right)^{\mathrm{p}}$. The significance that Ditchfield assigned to the various terms is the following:

(1) $\left(\sigma^{\mathrm{AAAA}}_{\mathrm{A}\alpha\beta}\right)^{\mathrm{d}}$ is the contribution to the screening field at nucleus B due to the diamagnetic Langevin-type currents on atom B itself. For an isolated spherical atom, it is the only contribution.

(2) $\left(\sigma^{\mathrm{AAAA}}_{\mathrm{A}\alpha\beta}\right)^{p}$, $\left(\sigma^{\mathrm{ABBA}}_{\mathrm{A}\alpha\beta}\right)^{\mathrm{p}}$ and $\left(\sigma^{\mathrm{ACDA}}_{\mathrm{A}\alpha\beta}\right)^{\mathrm{p}}$ are contributions due to the paramag-netic- type currents on atom B. The first term is a purely intra-atomic contribution. $\left(\sigma^{\mathrm{ABBA}}_{\mathrm{A}\alpha\beta}\right)^{\mathrm{p}}$ arises because the external magnetic field acting on atom B mixes in certain excited electronic states of the molecule and thereby induces a current flow on atom A. Similarly, $\left(\sigma^{\mathrm{ACDA}}_{\mathrm{A}\alpha\beta}\right)^{\mathrm{p}}$ describes the shielding effect of currents on atom A which arise from the external magnetic field acting on the pair of atoms C and D.

(3) $\left(\sigma^{\mathrm{BBBB}}_{\mathrm{A}\alpha\beta}\right)^{\mathrm{d}} + \left(\sigma^{\mathrm{BBBB}}_{\mathrm{A}\alpha\beta}\right)^{\mathrm{p}}$ is the total contribution made to the screening at atom A by the atomic circulations (on B) produced by the magnetic field acting on atom B.

(4) $\left(\sigma^{\mathrm{BCCB}}_{\mathrm{A}\alpha\beta}\right)^{\mathrm{p}}$ and $\left(\sigma^{\mathrm{BCDB}}_{\mathrm{A}\alpha\beta}\right)^{\mathrm{d}}$ are contributions to the secondary magnetic field at A due to the paramagnetic-type currents on atom B which arise from the magnetic field acting on atom C, and on the pair of atoms C and D, respectively.

(5) $\left(\sigma^{\mathrm{interatomic}}_{\mathrm{A}\alpha\beta}\right)^{\mathrm{d}} + \left(\sigma^{\mathrm{interatomic}}_{\mathrm{A}\alpha\beta}\right)^{\mathrm{p}}$ is the contribution to the screening due to currents which cannot be localized on any atom.

The chemical shifts calculated for first-row atoms in various molecules by Ditchfield (1974) with three distinct basis sets (STO-5G, LEMAO-5G, and 4-31G) are compared in Table 15.2 with the corresponding experimental values. The

Table 15.2 Calculated (Ditchfield 1974) versus observed NMR chemical shifts. Units are *ppm*. ^{13}C chemical shifts are given relative to CH_4, ^{14}N shifts relative to NH_3, ^{17}O shifts relative to H_2O and ^{19}F shifts relative to HF. A negative value indicates a downfield shift. The experimental ^{13}C shift of CH_4 relative to Total Magnetic Shielding (TMS) is 2.1 *ppm*. All ^{1}H shifts are given relative to CH_4. The experimental ^{1}H shift of CH_4 relative to TMS is −0.23 *ppm*

Molecule	Nucleus	STO-5G	LEMAO-5G	4-31G	Experimental	Reference
CH_4	C	0.0	0.0	0.0	0.0	[1]
	H	0.00	0.00	0.00	0.00	[2]
C_2H_6	C	−11.8	2.2	−7.4	−8.0	[1]
	H	0.05	0.03	−0.35	−0.75	[2]
CH_3F	C	− 50.5	−75.0	−65.4	−77.5	[1]
	F	49.2	128.7	74.0	76	[3]
	H	−1.18	−3.29	−4.04	−4.00	[2]
C_2H_4	C	−111.6	−152.5	−130.8	−125.6	[4]
	H	−5.21	−6.20	−5.61	−5.18	[2]
H_2CO	C	−147.9	−250.4	−199.6	−197	[5]
	O	−757.6	−860.3	−858.8	−580 to −600	[6]
	H	−9.34	−10.02	−10.39	−9 to −10	[7]
C_2H_2	C	−56.9	−90.0	−75.2	−−76	[4]
	H	−1.41	−2.27	−1.25	−1.35	[2]
HCN	C	−97.5	−147.5	−123.5	−120 ± 10	[8]
	N	−266.5	−360.5	−308.4	−297 ± 20	[9]
	H	−3.10	−3.79	−2.95	−2.83	[2]
NH_3	N	0.0	0.0	0.0	00	
	H	1.01	0.13	0.86	0.05	[2]
OH	O	0.0	0.0	0.0	0.0	
	H	1.37	−1.45	0.06	−0.60	[2]
FH	F	0.0	0.0	0.0	0.0	
	H	2.20	−4.16	−2.12	− 2.50	[2]

[1] Spiesecke and Schneider (1961); [2] Schneider et al. (1958); [3] Frankiss (1963); [4] Ditchfield and Ellis (1972); [5] Lauterbur (private communication), quoted in Newmann and Moskowitz (1969); [6] Emsley et al. (1966); [7] Bow (1969); [8] Appleman and Dayley (private communication to Ditchfield); [9] Gierke and Flygare (1972)

extended 4-31G basis set generally gives more accurate results than the minimal sets STO-5G and LEMAO-5G.[4]

A rather efficient implementation of the London method for calculations of nuclear magnetic shieldings at the Hartree-Fock self-consistent field (SCF) level (perhaps the most widely adopted procedure nowadays) was presented by Wolinski

[4] The valence shell scale factor ζ values of the Least Energy Minimal Atomic Orbitals (LEMAO) (Ditchfield et al. 1970) were treated as additional variational parameters in the unperturbed molecular calculations to allow for changes in the atomic size depending on the molecular environment.

et al. (1990). Using modern techniques borrowed from analytical derivative methods, Wolinski and coworkers were able to improve the efficiency of the GIAO method significantly. The computational effort for the ab initio calculation of the NMR chemical shifts is, in this procedure, only −2.5 times of the energy calculation.

The first-order density matrices needed for paramagnetic shielding are solutions of the appropriate Coupled Hartree–Fock (CHF) equations with the external magnetic field as a perturbation (Stevens et al. 1963). While in the GIAO procedure, Ditchfield (1974) employed explicitly field-dependent basis functions (Eq. 15.5.5), Wolinsky et al. (1990) found more convenient to adopt the Fock-Dirac density matrix (McWeeny 1962) which is invariant with respect to unitary transformation of the orbitals. Besides this major modification, the improvement in efficiency was obtained through:

1. Calculation of the perturbed two-electron integral by the quadrature technique of Dupuis et al (1976);
2. Elimination of the storage for perturbed two-electron integrals;
3. Formulation of the coupled-perturbed Hartree-Fock equations fully in atomic orbital (AO) basis;
4. Solution of the CHF perturbation by a conjugate gradient type method (Stiefel 1958);
5. Screening of the perturbed two-electron integrals and the calculation of only those that contribute significantly to the magnetic shieldings.

In Tables 15.3, 15.4, and 15.5 I report some of the NMR chemical shielding calculations carried out by Wolinski et al. (1990) in order to assess the efficiency of the procedure. Particularly what it is shown in these tables is the convergence of the calculated shielding tensors with the basis set quality in methane (CH_4), methyl fluoride (CH_3F), and hydrogen sulfide (H_2S). Calculations were performed by Wolinsky et al. (1980) using the experimental geometries. Various basis sets were adopted in order to assess the effect of basis set completeness on the calculations. It may be thus seen that convergence is essentially achieved at the triple zeta plus

Table 15.3 Convergence of the calculated NMR shielding with the basis set in the GIAO method for methane (CH_4) (Wolinsky et al. 1990)

Basis set	Energy, a.u.	Isotropic shielding (*ppm*)	
		^{13}C	^{1}H
6-31G	−40.180513	207.36	32.93
6-31G (d, p)	−40.201662	202.62	31.75
6-311G	−40. 188113	203.59	32.83
6-311G (d, p)	−40.209003	197.09	31.88
6-311G+(d,p)	−40.209080	196.97	31.87
6-2111G+(d,p)	−40.210176	194.73	31.62
6-311G+(2d,2p)	−40.212095	196.33	31.67

Table 15.4 Convergence of the calculated NMR shielding with the basis set in the GIAO method for methyl fluoride (CH_3F) (Wolinsky et al. 1990). 'par' and 'perp' denote the component of the shielding parallel and perpendicular to the C, axis; 'iso' is the isotropic average

Basis set	^{13}C shielding (*ppm*)			^{19}F shielding (*ppm*)		
	Iso	Par	Perp	Iso	Par	Perp
6-31G	142.28	198.94	113.95.	489.35	439.93	514.06
6-31G (d,p)	137.00	192.69	109.15	485.48	438.71	508.86
6-311G	136.82	194.61	107.93	487.18	437.71	511.92
6-311G (d,p)	128.36	186.95	99.08	483.06	437.59	505.79
6-311G+(d,p)	127.39	187.53	97.32	484.44	438.95	507.18
6-311IG+(2d,2p)	126.23	187.09	95.79	483.34	438.88	505.57

Table 15.5 Convergence of the calculated NMR shielding with the basis set in the GIAO method for hydrogen sulfide (H_2S) (Wolinsky et al. 1990)

	^1H shielding (*ppm*)				^{33}S shielding (*ppm*)			
	Iso	A	B	C	Iso	A	B	C
6-311G	32.8	26.0	29.9	42.3	767.9	648.9	738.8	916.0
6-2111G	32.8	26.0	29.9	42.3	745.1	614.2	713.0	908.3
6-2111G+	32.8	26.0	30.0	42.3	750.5	617.4	714.4	919.7
6-2111G+(2s)	32.8	26.0	30.0	42.3	750.6	617.6	714.3	919.8
6-2111G+(d,p)	31.0	24.4	26.9	41.8	730.7	573.5	690.2	928.4
6-2111G+(3d,p)	30.9	24.5	26.8	41.6	713.2	542.8	670.6	926.2
6-2111G+(3d,2p)	30.8	24.3	26.6	41.5	715.7	546.1	672.6	928.3
6-2111G+(5d,3p)	30.8	24.3	26.5	41.5 7	12.6	541.7	668.7	927.2

polarization basis set level for ^1H, ^{13}C, and ^{19}F isotropic shielding, as well as for principal components of the carbon and fluorine shielding tensors in methyl fluoride. The influence of the diffuse orbitals is insignificant. In the calculations with diffuse functions, the latter were placed only on heavy atoms

For sulfur in H_2S (Table 15.5), Wolinsky et al. (1980) adopted the (12s, 9p) basis set of McLean and Chandler (1980), contracted to [6s5p], and augmented by one, two, or three sets of d functions, and by a set of diffuse s and p functions. The hydrogen s-type basis set in the 6-2111G expansion was kept at the 6-311G level. Moreover, in the 6-2111G+(2s) expansion, the basis set was augmented by two s-type functions on the sulfur atom. The Authors noted that "*decontracting the 2p region has a large effect on the* ^{33}S *chemical shift, while enlarging the McLean-Chandler basis further with s functions, or diffuse p functions, leads only to minor changes. Adding diffuse functions results in a change of −5 ppm. Polarization functions are, as expected, important, and three sets of d functions are needed to obtain convergence to a few ppm*" (Woolinsky et al. 1980). A, B, and C in Table 15.5 are absolute shieldings and the anisotropy is defined as A − (B + C)/2.

More recently, using London orbitals, Haser et al. (1992) determined the shielding constants of large carbon clusters: the fullerenes C_{60} and C_{70}, by direct calculations. In the same year, Gauss (1992) investigated correlation effects on calculated ^{17}O chemical shift, in the framework of the MP2 theory. One year later, Ruud et al. (1993), through London orbitals, calculated the molecular magnetizabilities of a series of diamagnetic molecules H_20, NH_3, CH_4, PH_3, H_2S, CO_2, CSO, CS_2, and C_3H_4.

15.6 An Application: Isotropic Chemical Shifts of ^{23}Na and ^{29}Si in the $[Na_8Si_4O_{13}]^{2-}$ Cluster

Let me consider again the very same substance already treated as an applicative example for electron delocalization studies, namely the $[Na_8Si_4O_{13}]^{2-}$ cluster abstracted from the Na_2SiO_3 crystal.

Raman and ^{29}Si NMR spectra suggest that the onset of pre-melting in Na_2SiO_3 occurs 160 K below the room-pressure congruent melting point, when rearrangements of the Si_2O_6 chains begin to take place well after sodium mobility has itself set in at a much lower temperature of about 850 K. Na site hopping, based on ^{23}Na NMR spectrum and partial averaging of the low-temperature biaxial chemical shift powder pattern in ^{29}Si NMR spectra, seems to suggest rapid diffusion of Na^+ near the melting point, inducing extensive librational motion of SiO_4 tetrahedra (George et al. 1998).

The Ramsey contributions to the J_{NM} coupling tensor between sodium and oxygen in $Cmc2_1$ cluster $[Na_8Si_4O_{13}]^{2-}$ evaluated adopting the GIAO procedure (Table 15.6) are dominated by the interaction between the magnetic moments

Table 15.6 Ramsey contributions to the J_{NM} coupling tensor between sodium and oxygen and between silicon and oxygen in the $[Na_8Si_4O_{13}]^{2-}$ cluster evaluated with the GIAO procedure at DFT/B3LYP theory level and with the same basis set adopted for the $Cmc2_1$ Na_2SiO_3 crystal (see caption to Fig. 14.3). $d_{M\text{-}O}$ = metal to first neighbor distance; FC = Fermi Contact; SD = Spin Dipolar interaction; PSO = Paramagnetic Spin Orbit term; DSO = Diamagnetic Spin Orbit term. All values are expressed in Hz. (nb) = non-bridging oxygen; (b) = bridging oxygen; (nb'9) = terminal oxygen. (See Fig. 14.2 for atom labeling. From Belmonte et al. (2016))

Couple	Coordination	$d_{M\text{-}O}$ ($Å$)	FC	SD	PSO	DSO	J_{NM}
$Na_{24}O_{10(nb)}$	VI	2.29	−20.9	−0.5	0.5	−0.1	−21.0
$Na_{25}O_{6(nb)}$	VI	2.29	−37.7	−0.1	1.0	−0.1	−36.9
$Na_{22}O_{5(nb)}$	VI	2.29	−33.6	−0.1	0.7	−0.1	−33.0
$Na_{23}O_{2(nb)}$	VI	2.29	−32.3	0.0	0.9	−0.1	−31.4
$Na_{19}O_{12(b)}$	?	2.42	−6.5	−0.1	1.0	−0.1	−5.7
$Na_{21}O_{9(b)}$	?	2.42	−4.1	−0.1	0.8	−0.1	−3.5
$Na_{18}O_{8(b)}$	?	2.42	−3.7	−0.1	0.8	−0.1	−3.1

(continued)

Table 15.6 (continued)

Couple	Coordination	$d_{M\text{-}O}$ (Å)	FC	SD	PSO	DSO	J_{NM}
$Na_{20}O_{7(nb')}$	?	2.37	−15.7	−0.1	1.3	−0.1	−14.2
$Si_{15}O_{10(nb)}$	IV	1.62	19.4	0.0	7.0	−0.1	26.1
$Si_{15}O_{11(nb)}$	IV	1.62	24.8	−0.3	9.5	−0.1	33.9
$Si_{15}O_{12(b)}$	IV	1.69	16.8	−0.4	4.2	−0.1	20.5
$Si_{15}O_{13(nb')}$	IV	1.70	25.6	−1.4	10.5	−0.1	34.5
$Si_{14}O_{12(b)}$	IV	1.70	12.6	−0.5	5.6	−0.1	17.6
$Si_{14}O_{6(nb)}$	IV	1.62	16.5	−0.4	8.6	−0.1	24.5
$Si_{14}O_{9(b)}$	IV	1.69	9.7	−0.4	5.4	−0.2	14.5
$Si_{14}O_{4(nb)}$	IV	1.62	20.6	−0.4	9.3	−0.1	29.4
$Si_{17}O_{9(b)}$	IV	1.70	7.6	−0.4	4.9	−0.2	12.0
$Si_{17}O_{5(nb)}$	IV	1.62	15.5	−0.4	8.5	−0.1	23.5
$Si_{17}O_{8(b)}$	IV	1.69	7.9	−0.4	5.7	−0.1	13.1
$Si_{17}O_{7(nb')}$	IV	1.62	19.4	−0.4	9.0	−0.1	28.2
$Si_{16}O_{8(b)}$	IV	1.70	12.2	−0.5	4.1	−0.1	15.7
$Si_{16}O_{2(nb)}$	IV	1.62	16.8	−0.3	7.9	−0.1	24.3
$Si_{16}O_{3(nb')}$	IV	1.69	22.6	−0.4	8.0	−0.1	30.1
$Si_{16}O_{1(nb)}$	IV	1.62	26.0	−0.4	9.9	−0.1	35.5

associated with the nuclear spins (i.e. the Fermi Contact FC). The J_{NM} coupling tensor between sodium and oxygen is clearly affected by the interionic distance and decidedly more intense when non-bridging oxygen centers (nb in Table 15.6) are involved. The calculated isotropic chemical shifts of ^{23}Na with respect to the isotropic shielding of a one-molal NaCl solution computed at the same theory level through IEFPCM (Tomasi and Persico 1994) [$\sigma_{ref} = 622.8$ *ppm*; $\delta_{iso} = -(\sigma - \sigma_{ref}) = 33 \div 51$] is slightly higher than what observed in the crystalline substance (George et al. 1998) ($\delta_{iso} = 11 \div 25$) although substantially comparable. The calculated isotropic shift of ^{29}Si with respect to the isotropic shielding of tetramethylsilane [$Si(CH_3)_4$] calculated at the same theory level [$\sigma_{ref} = 347.2$ *ppm*; $\delta_{iso} = -(\sigma - \sigma_{ref}) = -51 \div -75$] is again in substantial agreement with what experimentally observed on the crystal (George et al. 1998) ($\delta_{iso} = -79.6$).

Part III
Applications

Chapter 16
Atmospheric Gases

The atmosphere appeared very early on Earth. Its mean age is in fact 4.4 Ga, according to Argon and Xenon isotope dating (Fegley et al., 1986; and references therein). Volcanic gases have been, through the geologic eras, one of the main inputs to the global atmospheric composition. Nevertheless, some volcanic gases are quite peculiar, and do not survive in large amounts in the atmosphere for long lapses of time. In the Archean the atmosphere was mainly composed of nitrogen (N_2; $\sim$ 90% by volume), with minor, but still consistent, amounts of carbon dioxide (CO_2; 10% by volume) and limited amounts of H_2 ($\sim$0.1%). O_2 begun to be elapsed in great amounts in the atmosphere at 2.45 Ga through photosynthetic processes operated by cyanobacteria (then, later, by algae and plants) and is now at 20.946%. On time-scales of 100 to 1000 years this value has been held nearly constant by equal rates of photosynthetic production and respiratory consumption (Broecker, 1970). The increasing concentrations of CO_2, NO_2, SO_2 and non-methane hydrocarbons is mostly associated to anthropogenic pollution. An air free of the above components is often used as a reference sample in the calibration and operation of instruments, under the name of "*zero air*" (McNaught and Wilkinson,1997). According to the IUPAC Goldbook (McNaught and Wilkinson,1997), the mean composition of the atmosphere is now still dominated by nitrogen (78.084%). Disregarding H_2O, which is quite variable (0 to 4% by volume), the third component is Ar (0.934% by volume). All other gases occur at part per million level by volume (*ppmv*), including CO_2 which is now at 407 *ppmv*. Ozone (O_3) is concentrated in a part of the stratosphere called "*ozonosphere*" (15–35 km). Its bulk concentration is extremely low ($\sim$0.04 *ppmv*) but its capability of absorbing the majority of UV solar emission (also causing an upward thermal inversion) is extremely beneficial to life in all forms we know. Very informative review papers have been produced on the evolution of the Earth's atmosphere, to whom I refer the interested reader (Holland, 2006; Kasting and Tazewell Howard, 2006; Kasting and Ono, 2006; Zahnle et al., 2010).

G. A. Ottonello, *Quantum Geochemistry*, Springer Geochemistry,
https://doi.org/10.1007/978-3-031-21837-8_16

16.1 Premises

The atoms composing most part of the Earth's atmosphere are the results of nuclear fusion reactions in 1st and 2nd generation stars. (Table 16.1). Hydrogen atoms and part of ^4He are believed to have been created during the "*Big Bang*" by proton-electron combination.

All naturally occurring radioactive isotopes constituting the atmospheric gaseous molecules are listed in Table 16.2. Nuclides with atomic mass A ≥ 40 are the results of early cosmic events. Lighter radionuclides are continuously produced in the lower atmosphere by interaction of neutrons (produced by cosmic rays and slowed to thermal energies) with atmospheric nitrogen, oxygen, and argon nuclei.

16.2 H_2

Hydrogen has two stable isotopes ^1_1H (0.99985) and ^2_1H (0.00015) and a short lived radioactive isotope (Tritium) ^3_1H produced in the atmosphere by interaction of ^{14}N with cosmic-ray neutrons:

$$^{14}_{7}N + n \rightarrow ^{3}_{1}H + ^{12}_{6}C \tag{16.2.1}$$

The concentration of ^3_1H in the atmosphere (measured in Tritium Units TU, $1\,TU = {}^3_1H/{}^1_1H \times 10^{18}$) rapidly increased after 1950 due to nuclear tests. Tritium decays to stable ^3_2He by β^- emission ($t_{1/2} = 12.26\ a$; see Kamensky et al., 1991, for an account of $^3H - ^3He$ dating of ground waters).

Isotope ^2_1H (Deuterium) discovered by Urey et al. (1932) is usually denoted by symbol D. The large relative mass difference of H and D induces sensible fractionation ascribable to equilibrium, kinetic and diffusional effects. the main difference in the calculation of equilibrium isotopic fractionation effects in hydrogen molecules with respect to oxygen arises from the fact that the rotational partition function of hydrogen is non-classical. Rotational contributions to the isotopic fractionation do not cancel out at high T, as in the classical approximation, and must be accounted for in the estimates of the partition function ratio f.

16.2.1 Thermophysical and Thermodynamic Properties

The H_2 molecule is a Symmetrical Top $D_{\infty h}$ (linear with inversion center). Its ground state is $^1\Sigma_g$ (singly degenerate, even). The two H atoms are connected by a single bond at an equilibrium distance of 0.741 *Å* and vibrate with an harmonic frequency $\nu_1 = 4401\ cm^{-1}$ (Huber and Herzberg, 1979). The ionization energy is 15.426 *eV*

Table 16.1 Nucleosynthesis and the main nuclear reactions taking place in stars. (*): unstable; n: neutron; ν^+: neutrino; β^+: positron; γ: : gamma radiation (from Ottonello, 1997)

NUCLEOSYNTHESIS (3 min after the Big Bang → 1 h after the Big Bang; log T ~ 9; about 13% of nucleons present as neutron)		
$^1H + n \rightarrow {}^2D + \gamma$	$^2D + {}^1H \rightarrow {}^3He + \gamma$	$^3He + {}^3He \rightarrow {}^4He + {}^1H + {}^1H$
HYDROGEN BURNING (first generation stars; log T ~7)		
$^1H + {}^1H \rightarrow {}^2D + \beta^+ + \nu^+$	$^2D + {}^1H \rightarrow {}^3He + \gamma$	$^3He + {}^3He \rightarrow {}^4He + {}^1H + {}^1H$ $^3He + {}^4He \rightarrow {}^7Be + \gamma$
$^7Be + e^- \rightarrow {}^7Li + \nu^+$	$^7Li + {}^1H \rightarrow {}^8Be^* \rightarrow {}^4He + {}^4He$	
$^7Be + {}^1H \rightarrow {}^8B + \gamma$	$^8B \rightarrow {}^8Be^* + \beta^+ + \nu^+$	$^8Be^* \rightarrow {}^4He + {}^4He$
CARBON–NITROGEN CYCLE (second- and later generation stars)		
$^{12}C + {}^1H \rightarrow {}^{13}N + \gamma$	$^{13}N \rightarrow {}^{13}C + \beta^+ + \nu^+$	
$^{13}C + {}^1H \rightarrow {}^{14}N + \gamma$	$^{14}N + {}^1H \rightarrow {}^{15}O + \gamma$	$^{15}O \rightarrow {}^{15}N + \beta^+ + \nu^+$ $^{15}N + {}^1H \rightarrow {}^{12}C + {}^4He$
HELIUM-BURNING (red giants; log T ~8)		
$^4He + {}^4He \rightarrow {}^8Be$	$^8Be + {}^4He \rightarrow {}^{12}C + \gamma$	$^{12}C + {}^4He \rightarrow {}^{16}O + \gamma$ $^{16}O + {}^4He \rightarrow {}^{20}Ne + \gamma$
CARBON-BURNING reactions (log T ~8.8)		
$^{12}C + {}^{12}C \rightarrow {}^{24}Mg + \gamma$		
$^{12}C + {}^{12}C \rightarrow {}^{23}Mg + n$		
$^{12}C + {}^{12}C \rightarrow {}^{23}Na + {}^1H$		
$^{12}C + {}^{12}C \rightarrow {}^{20}Ne + \alpha$		
$^{12}C + {}^{12}C \rightarrow {}^{16}O + 2\alpha$		
OXYGEN-BURNING reactions (log T~ 9.4)		
$^{16}O + {}^{16}O \rightarrow {}^{31}S + n$		
$^{16}O + {}^{16}O \rightarrow {}^{31}P + {}^1H$	$^{31}P + \gamma \rightarrow {}^{30}Si + {}^1H$	$^{30}Si + \gamma \rightarrow {}^{29}Si + n$ $^{29}Si + \gamma \rightarrow {}^{28}Si + n$
$^{16}O + {}^{16}O \rightarrow {}^{28}Si + \alpha$		

and the proton affinity is 4.377 *eV* (NIST, 2020). Based on the NIST-JANAF tabulation for monatomic hydrogen, the dissociation energy is 0.164566 *Hartree*. The rotational constant of H_2 is 60.853 cm^{-1} and the calculated one at B3LYP/6.31G(d,p) level[1] is 60.633 cm^{-1}. The components of the electric dipole moment are zero and those of the quadrupole moment are xx = -0.260 Debye (*D*)×*Å*, yy = −0.260 *D*×*Å*, zz = 0.520 *D*×*Å*. The electrical dipole polarizability is α = 0.787 $Å^3$ (Olney et al., 1997) and the calculated one at B3LYP/6-31G(d,p) level is α = 0.372 $Å^3$.

[1] In the sake of conformity I will limit myself in most circumstances to list in this chapter quantum-mechanical data at simple B3LYP/6-31G(d,p) theory level for all gaseous molecule. Quite extensive tabulations are provided in the CCCBDB database and are easily accessed by anyone on the web. The comparison of the various theories is very informative (https://cccbdb.nist.gov).

Table 16.2 Naturally occurring radionuclides in the Earth's atmosphere. *a* = *anna* (years), *d* = days. EC = electron capture

Radio nuclide	Decay process	Half-life		Isotopic abundance (%)	Stable end-product
^{3}H	β^-	12.26	*a*	1.38×10^{-4}	^{3}He
^{14}C	β^-	5730	*a*	–	^{14}N
^{37}Ar	EC	34.8	*d*	–	^{37}Cl
^{39}Ar	β^-	269	*a*	–	^{39}K
^{81}Kr	EC	2.13×10^5	*a*	–	^{81}Br
^{85}Kr	β^-	10.6	*a*	–	^{85}Rb

At standard state (pure gas, $P = 10^{-5}$ *Pa*, 298.15 *K*) H_2 has an entropy of 114.716 *J/(mol × K)* and an isobaric heat capacity of 20.786 *J/(mol × K)* (Chase, 1998). The Shomate equation for the isobaric heat capacity in the range $298.15 \leq T (K) \leq 1000$ is $C_P = 33.066178 - 11.369417 \times 10^{-3} \times T + 11.432816 \times 10^{-6} \times T^2 - 2.772876 \times 10^{-9} \times T^3 - 0.158558 \times 10^6 \times T^{-2}$ (NIST, 2000; based on refitted data of Chase 1998). The triple point occurs at 13.95 *K*. The critical temperature is $T_c = 33.18$ *K*, the critical pressure is $P_c = 13.00$ *bar* and the critical density is $\rho_c = 15.4$ *mol/l* (NIST, 2020). As anticipated, the rotational behavior of H_2 is non-classical. The classical expression for the rotational partition function (Q_{rot}) is

$$Q^*_{rot} = \frac{hcB}{k_B T} = \frac{\hbar^2}{4Ik_B T} \tag{16.2.2}$$

where B is the rotational constant in cm^{-1}, I is the momentum of inertia, and the remaining symbols have the usual significance. Bigeleisen and Mayer (1947) suggested to account of non-classical behavior of hydrogen molecules by expanding the deviation from the classical rotational partition function in the polynomial form:

$$\frac{Q_{rot}}{Q^*_{rot}} = \left(1 + \frac{\sigma}{3} + \frac{\sigma^2}{15} + \frac{\sigma^3}{315} + \ldots\ldots\right) \tag{16.2.3}$$

where $\sigma = 1/Q^*_{rot}$.

Bottinga (1969) evaluated the non classical rotational partition function of water vapor, molecular hydrogen and methane through the asymptotic expansion of Strip and Kirkwood (1951):

$$Q_{rot} = \left[\pi/\left(s_x s_y s_z\right)\right]^{1/2}\left\{1 + \left[2s_x + 2s_y + 2s_z - \left(s_x s_y/s_z\right) - \left(s_y s_z/s_x\right) - \left(s_z s_x/s_y\right)\right]/12 + \ldots\right. \tag{16.2.4}$$

where s_x is the classical rotational partition function contribution about the *x* axis:

$$s_x = Q^*_{rot,x} = \frac{\hbar^2}{4I_x k_B T} \quad (16.2.5)$$

and I_x is the the *x* axis component of momentum of inertia.

Bottinga (1969) calculated the isotopic fractionation relative to the exchange reactions:

$$\frac{1}{2}H_2O + D \Leftrightarrow \frac{1}{2}D_2O + H \quad (16.2.6.1)$$

$$\frac{1}{4}CH_4 + D \Leftrightarrow \frac{1}{4}CD_4 + H \quad (16.2.6.2)$$

$$\frac{1}{2}H_2O + \frac{1}{4}CD_4 \Leftrightarrow \frac{1}{2}D_2O + \frac{1}{4}CH_4 \quad (16.2.6.3)$$

Results of his calculations are listed in Table 16.3 in terms of fractionation factors α and Δ^*_{AB}.

Note that Bottinga (1969) defines this last term as:

$$\Delta^*_{AB} = (\alpha - 1) \times 1000 \quad (16.2.7)$$

Taking as phase B the water vapor, the parameter Δ^*_{AB} is equal to the deuterium enrichment in phase A with respect to water, hence is equivalent to the δ notation adopted in geochemical literature, under the assumption of equilibrium.

$$\alpha_{AB} = \frac{\delta_A/10^3 + 1}{\delta_B/10^3 + 1} = \frac{\delta_A + 10^3}{\delta_B + 10^3} \quad (16.2.8)$$

$$\Delta_{AB} = \left(\frac{\delta_A + 10^3}{\delta_B + 10^3} - 1\right) \times 10^3 \quad (16.2.9)$$

16.3 Noble Gases

Argon (from the Greek term "*argues*", i.e., inactive) is present in the Earth's atmosphere in a rather conspicuous amount (0.934% in volume). The atmospheric amounts of the other noble gases is decidedly subordinated (Ne ≈18 *ppmv*; He ≈5 *ppmv*; Kr ≈ 1.1 *ppmv*; Xe ≈0.08 *ppmv*). Natural Argon has three isotopes, in the following proportions: ^{36}Ar = 0.003365, ^{38}Ar = 0.000632 and ^{40}Ar = 0.996003 (Rosman and Taylor, 1998). ^{40}Ar is generated by the natural decay of ^{40}K (10.5%), which also decays into ^{40}Ca (89.5%). The ^{40}K to ^{40}Ar decay branch is dominated by electron capture, although a limited amount of positron emission of low energy

Table 16.3 Deuterium fractionation in the system water-hydrogen-methane,after Bottinga (1969) (wv = water vapor, me = methane, hy = hydrogen)

T (°C)	ln α		10^3ln α	Δ^*_{AB}	
	wv-hy	me-hy	wv-me	me-wv	hy-wv
0	1.395	1.387	7.3	−7.3	−752.1
10	1.355	1.322	13.3	−13.3	−736.9
20	1.280	1.261	18.9	−18.7	−721.9
25	1.253	1.232	21.8	−21.2	−714.5
30	1.228	1.204	24.0	−23.7	−707.1
40	1.179	1.151	28.7	−28.2	−692.5
50	1.134	1.101	33.0	−32.4	−678.2
60	1.091	1.054	36.9	−36.3	−664.1
70	1.051	1.010	40.6	−39.8	−650.3
80	1.012	0.968	43.9	−43.0	−636.7
90	0.977	0.930	47.0	−45.9	−623.4
100	0.943	0.893	49.9	−48.6	−610.4
120	0.880	0.825	54.8	−53.4	−585.2
140	0.824	0.765	59.0	−57.3	−561.1
160	0.773	0.710	62.5	−60.6	−538.2
180	0.726	0.661	65.4	−63.3	−516.3
200	0.684	0.616	67.8	−65.5	-495.4
220	0.645	0.576	69.6	−67.2	−475.5
240	0.610	0.539	71.1	−68.6	−456.5
260	0.577	0.505	72.2	−69.7	−438.4
280	0.547	0.474	73.0	−70.4	−421.2
300	0.519	0.445	73.5	−70.9	−404.8
320	0.493	0.419	73.8	−71.2	−389.2
340	0.469	0.395	73.9	−71.3	−374.3
360	0.446	0.372	73.8	−71.2	−360.1
380	0.425	0.352	73.5	−70.9	−346.5
400	0.406	0.333	73.2	−70.5	−333.6
450	0.362	0.291	71.7	−69.2	−303.9
500	0.325	0.255	69.6	−67.3	−277.5
550	0.293	0.226	67.2	−65.0	−253.9
600	0.265	0.201	64.7	−62.3	−232.9
650	0.241	0.179	61.9	−60.0	−214.1
700	0.220	0.161	59.1	−57.4	−197.3

is observed ($\beta^+ \approx 0.001\%$). Neon has three isotopes $^{20}Ne = 0.9048$, $^{21}Ne = 0.0027$, $^{22}Ne = 0.0925$. Helium has two isotopes $^{3}He = 1.37 \times 10^{-6}$, $^{4}He = 0.99999863$. Krypton has six isotopes $^{78}Kr = 0.0035$, $^{80}Kr = 0.0228$, $^{82}Kr = 0.1158$, $^{83}Kr = 0.1149$, $^{84}Kr = 0.5700$, $^{86}Kr = 0.1730$. Xenon has nine isotopes $^{124}Xe = 0.0009$, $^{126}Xe = 0.0009$, $^{128}Xe = 0.0192$, $^{129}Xe = 0.2644$, $^{130}Xe = 0.0408$, $^{131}Xe = 0.2118$, $^{132}Xe = 0.2689$, $^{134}Xe = 0.1044$, $^{136}Xe = 0.0887$ (Rosman and Taylor, 1998).

16.3.1 Thermophysical and Thermodynamic Properties

Noble gases (ground state ^1S), behave as ideal monatomic gases, and, in the obvious absence of vibrational contributions, the integrated heat capacity from the athermal limit to any T is simply 5/2 × RT. The remaining relevant properties (NIST, 2000) are resumed in Table 16.4.

16.3.2 The Spin–Orbit Interaction in He

I already introduced Helium as the simplest many-electrons atom, discussing to some extent the role of spin–orbit interaction (Sect. 6.4.3). In fact, we have seen that the Grotrian diagram of the spectrum of atomic Helium (Morton et al., 2006; Fig. 6.6) tells us that the first excited level $1s^12p^1$ is the triplet $^3P_2$ (ortho-helium).

Table 16.4 Some thermodynamic and thermophysical physical properties of noble gases (from NIST, 2000)

Property	Dimension units	He	Ne	Ar	Kr	Xe
$S_{298.15}$	*J/(mol × K)*	126.15	146,33	154,85	164.10	169.68
$C_{P,298.15}$	*J/(mol × K)*	20.79	20.79	20.79	20.79	20.79
$T_{boiling}$	*K*	–	–	87.5	119.78	165.02
T_{fusion}	*K*	–	–	83.8	–	–
$T_{triple\ point}$	*K*	–	24.56	87.78	115.76	161.38
$P_{triple\ point}$	*bar*	–	0.4337	0.689	0.7315	0.816
$T_{crytical}$	*K*	5.2	–	150.86	209.46	289.74
$P_{crytical}$	*bar*	2.274	–	4.8979	55.2019	–
$\rho_{crytical}$	*mol/l*	17.40	–	13.41	11.0	8.371
K^0_{Henry}	*mol/(kg × bar)*	0.00038	0.00045	0.0014	0.0025	0.0043
Polarizability	*Å*^{3}	0.208	0.381	1.664	2.498	4.005
dlnK/d(1/T)	*K*	92	450	1500	1900	–
Ionization Energy	*eV*	24.857	21.565	15.759	14.000	12.130
Proton affinity	*eV*	1.843	2.060	3.826	4.397	5.117

A triplet of symmetric spin functions is coupled with an antisymmetric function in the positional coordinates of the two electrons. The quantum statistics resulting from the acceptance of only antisymmetric wave functions (Fermi–Dirac statistics) applies to many problems and it has been found that, for protons and electrons, this is the conditions observed in Nature. In the case of Helium, the wave functions obtained by multiplying the symmetric orbital wave function by the single antisymmetric spin function

$$\chi_A = \frac{1}{\sqrt{2}}[\alpha(1)\beta(2) - \beta(1)\alpha(2)] \tag{16.3.1}$$

generates singlet states (empty circles in Fig. 6.7) while triplet states are obtained by multiplying the antisymmetric orbital wave function by the symmetric spin function

$$\chi_S = \frac{1}{\sqrt{2}}[\alpha(1)\beta(2) - \beta(1)\alpha(2)] \tag{16.3.2}$$

with zero component along the quantization z axis (filled circles in Fig. 16.7).

For two electrons 1 and 2 at positions r_1 and r_2, we may denote $\Psi_\pm$ their spatial wave functions determined by the principal quantum numbers n_1, n_2, the azimuthal quantum numbers l_1 and l_2 and the magnetic quantum number $m_{l,1}$, $m_{l,2}$ (with - denoting the antisymmetric and + the symmetric states, respectively):

$$\Psi_\pm(\mathbf{r}_1, \mathbf{r}_2) = \frac{1}{\sqrt{2}}\left[\psi_{n_1,l_1,m_{l,1}}(\mathbf{r}_1)\psi_{n_2,l_2,m_{l,2}}(\mathbf{r}_2) \pm \psi_{n_2,l_2,m_{l,2}}(\mathbf{r}_1)\psi_{n_1,l_1,m_{l,1}}(\mathbf{r}_2)\right] \tag{16.3.3.1}$$

In the case of proto-helium, with electronic configuration $1s^1 2p^1$, replacing in Eq. 16.3.3.1 the general expression with the real form of the wave functions (Table 5.4) we get:

$$\Psi_\pm(\mathbf{r}_1, \mathbf{r}_2) = \frac{1}{\sqrt{2}}\left[\frac{1}{\sqrt{\pi}}(Z/a_0)^{3/2}\exp(-\sigma) \pm \frac{1}{4\sqrt{2\pi}}(Z/a_0)^{3/2}\sigma\exp\left(-\frac{\sigma}{2}\right)\sin\theta\sin\phi\right] \tag{16.3.3.2}$$

where $\psi_{n_2,l_2,m_{l,2}}$ in Eq. 16.3.3.1 is the 2py orbital and $\sigma = \rho\frac{n}{2} = \frac{Zr}{a_0}$

When electrons 1, 2 approach each other in the antisymmetric condition, the probability density [which is the squared modulus of $\Psi_\pm(\mathbf{r}_1, \mathbf{r}_2)$] becomes progressively smaller, and, eventually, cancel out when the two terms in the spatial part of the wavefunction sum to zero. This "*zero probability*" condition may be interpreted as an exclusion zone that cannot be penetrated by the other electron. The result is a correspondingly decreased electron–electron repulsion, and a more energetically stable condition. A "*Fermi hole*" is in this way created. In other terms, because of antisymmetry of the positional coordinates, the two electrons of same spin remain distant from each other and this cause a lower repulsion, while the opposite would be observed with the opposite combination. In fact, for symmetric spatial functions, the probability density may eventually be doubled when the two electrons are in the same position (symmetric in positional coordinates and antisymmetric in spin coordinates; para-helium). This electron clumping (or "*Fermi heap*") results in a correspondingly increased electron–electron repulsion, and a less stable arrangement.

A nice visualization of a He $1s^12p^1$ Fermi hole and a He $1s^12p^1$ Fermi heap caused by an incident light polarized along y may be found in Dill (2006) and is here reproduced in Fig. (16.1).

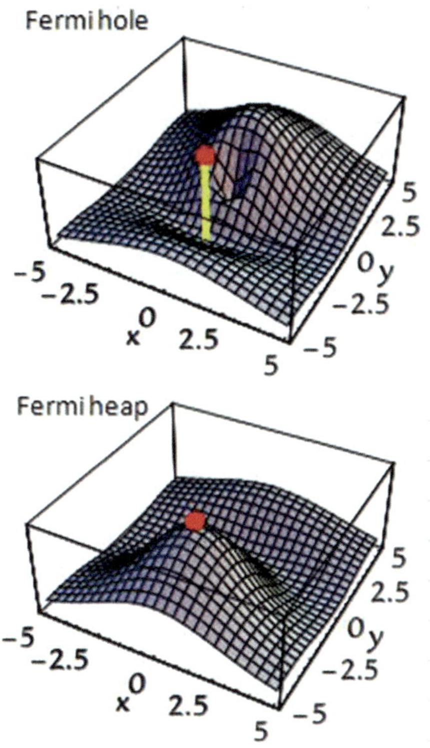

Fig. 16.1 Fermi hole and Fermi heap for He $1s^12p^1$ for one electron at $y = -2a_0$ in the xy plane. The height of the surface is the probability density of the other electron. Modified from Dill (2006)

16.3.3 Solubility of Noble Gases in Water: Application of Scaling Particle Theory (SPT) and Polarized Continuum Model (PCM)

When molecules are embedded in a solvent of discrete permittivity the electrostatic field modifies their conformation and stabilizes them through electrostatic interactions. Dispersive, repulsive and cavitation contributions to the bulk energy in solution are present as well. These factors are important when dealing with Gibbs free energy changes caused by the transfer of the solute from the hypothetical gaseous state (isolated moiety) to the standard state of "*hypothetical one-molal solution referred to infinite dilution*",which is adopted for all solutes in simple or complex solvents. In estimating the parameters necessary to depict the solvent properties one usually tries to reproduce the experimental Henry's Law constant of a nominally non-interacting spherical species (e.g. noble gases), maximizing in this way the role of the (otherwise minimal) energy spent in creating in the solvent a cavity able to host it in the solvent.

Transferring a molecule from the isolated state of a gas phase to that of a solute in the solvent medium involves various energy terms which, for the sake of simplicity, may be grouped to define a solvation energy contribution to the bulk Gibbs free energy of the substance. This Gibbs free energy of solvation (ΔG°_{solv}) is made up of two main terms related to coupling work and molecular motion, which can themselves be further subdivided into several contributions (Tomasi and Persico, 1994):

$$\Delta G^0_{solv} = W(M/S) + \Delta G_{mm} =$$
$$\Delta G_{el} + \Delta G_{cav} + \Delta G_{disp} + \Delta G_{rep} + RTln\left(\frac{q_{rotg} \times q_{vilg}}{q_{rots} \times q_{vils}}\right) - RTln\left(\frac{n_{Mg} \times \Lambda^3_{Mg}}{n_{Ms} \times \Lambda^3_{Ms}}\right) \tag{16.3.4}$$

where $q_{rot,g}$, $q_{vib,g}$, $q_{rot,s}$and $q_{vib,s}$ are the microscopic contributions to the rotational and vibrational partition function in the gaseous state and in the solvent, respectively, $n_{M,g}$ and $n_{M,s}$ the number densities of the molecule M in the gas and in the solvent, and $\Lambda_{M,g}$ and$\Lambda_{M,s}$ the momentum partition functions of M in the two states.

The "*coupling work*" terms are the first four terms on the right-hand part of Eq. 16.3.4. They include electrostatic solute–solvent interaction, cavitation, dispersion and repulsive contribution which are related to the work necessary to transfer the molecule M into the solvent S (Tomasi and Persico, 1994; Tomasi et al., 1999; see also Sect. 7.7.3).

For the solute molecule, interactions with the solvent perturbs its energy as derived from:

$$\left[\widehat{\mathrm{H}}^{0}+\widehat{\mathrm{V}}_{\mathrm{R}}\right]=\mathrm{E}\Psi \tag{16.3.5}$$

where $\widehat{\mathrm{H}}^{0}$ is the Hamiltonian of the solute *in-vacuo*, $\hat{\mathrm{V}}_{\mathrm{R}}$ the electrical potential of the reaction field and Ψ the solute wave function in solution. In the classical approximation (Floris and Tomasi, 1986; Floris et al., 1991) the sum of solute–solvent dispersive and repulsive interactions $\Delta G_{disp}+\Delta G_{rep}$ is computed as double summations of terms extending to the solute atoms placed within the cavity and solvent atoms outside a cavity limited by a tessellated surface (see Sect. 7.7.3; for details).

The electrostatic interactions between the "*nominally*" inert elements He, Ne, Ar, Kr, Xe, Rn and the H_2O solvent are virtually negligible but an accurate description of polarizabilities and permanent moments is crucial to depict the weak interactions taking place. The computational effort on noble gases has been quite intensive and it is recognized that diffuse polarization functions are necessary for a correct appraisal of their static dipole polarizability (α). The augmented (aug) correlation consistent (cc), polarized valence (PV) series of Gaussian basis sets (Dunning, 1989; Woon and Dunning, 1993,1995) are the fastest converging and most efficient series hitherto available for calculating atomic energies of some of the noble gases. A systematic study on static electrical responses has been carried out by Woon and Dunning (1995) on the aug-cc-PV-XZ basis sets (with X = D,T,Q,5 for double, triple, quadruple and quintuple polarized valence, respectively) and it has been shown that both the adopted theory level and the polarization expansion affect the final results. It was also shown that the utilization of diffuse functions (aug) improves dramatically the accuracy. A further series of cc basis sets have been developed for the post-*d* group 16–18 with small-core relativistic pseudo-potentials (Peterson et al., 2003) (PP). Because the outer-core (*n*21) *spd* shells are explicitly treated together with the *nsp* valence shell these basis sets have been expanded into cc-pVXZ-PP and aug-cc-pVXZ-PP and show systematic convergence of both Hartree–Fock and correlated total energies. These sets have been adopted by Ottonello and Richet (2014) for preliminary calculations concerning the heavier elements in the series Xe and Rn. In Fig. 16.2 it is shown how the polarization extent affects the static dipole moment of the various noble gases in vacuum. Figure 16.2 is analogous to Fig. 16.1a in Woon and Dunning (1995) but here the adopted DFT theory with hybrid (B3LYP) functional is preferred for computational economy.

We may see that the XZ dependence of the calculated static dipole polarizability varies significantly with the number X of additional functions in the basis set and with the adopted theory level. While at RHF theory level the calculated value is

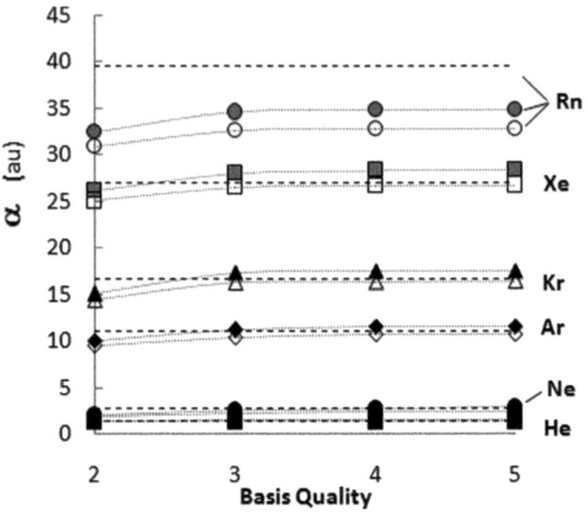

Fig. 16.2 The dipole polarizability of noble gases at RHF (open symbols) and B3LYP (solid symbols) level of theory obtained with a variable (X) quality in aug-cc-PVXZ basis sets. Modified from Ottonello and Richet (2014)

systematically lower with respect to observations (dashed lines in Fig. 16.2), a crossover between computation and experiment is attained at DFT/B3LYP level between the double and triple expansion for all elements but Rn. For this element the adopted PP functions and the ECP results in α values substantially lower with respect to observations. This fact, and the lack of dispersive and repulsive parameters for Xe and Rn prevent any reasonable calculation (though approximate estimates of repulsive + dispersive energy terms could be obtained by adopting Eqs. 133–134 in Tomasi and Persico, 1994).

The most important form of energy involved in solute/solvent interaction is, in most cases, the "*cavitation energy*". The formulation of Pierotti (Pierotti, 1963, 1965, 1976) for the Gibbs free energy of cavitation (ΔG_{cav}) and its enthalpic and entropic components (H_{cav}, S_{cav}, respectively) is based on the Scaled Particle Theory (SPT) of Reiss and coworkers (Reiss et al., 1959, 1960; Reiss, 1965). Denoting by σ_s the diameter of the solvent in atomic units (Å), the reversible work necessary to create in the solvent a cavity of radius $r \leq \sigma_s/2$ is expressed as:

$$W_0(r, \rho) = kT \ \ln \ p_{0,i}, \tag{16.3.6}$$

where $p_{0,i}$ the complement to 1 of the probability of finding a solute molecule i at the place of a solvent molecular center:

$$p_{0,i} = \left(1 - p_{0,r}\right) = \left(1 - \frac{4}{3}\pi r^3 \rho_s\right). \tag{16.3.7}$$

To make room in the solvent for a solute of diameter σ_i the necessary cavity must have a radius $r = \frac{\sigma_s + \sigma_i}{2}$. Denoting by $Y = \frac{\pi \rho_s \sigma_s}{6}$ the reduced number density and $\varsigma = \sigma_i/\sigma_s$, we may write the work done as (Pierotti, 1976):

$$\frac{W(r,\rho)}{kT} = -\ln(1-Y) + \left(\frac{3Y}{1-Y}\right)\varsigma + \left[\left(\frac{3Y}{1-Y}\right) + \frac{9}{2}\left(\frac{Y}{1-Y}\right)^2\right]\varsigma^2 + \frac{YP}{\rho_s kT}\zeta^3 \tag{16.3.8}$$

where P is the (isotropic) pressure. The partial molar Gibbs free Energy of cavitation may be assumed to be equal to the reversible work $\overline{G}_{cav} = W(r,\rho) \equiv \Delta G_{cav}$. The cavitation enthalpy has the following form (Pierotti, 1976):

$$H_{cav} = Y\alpha_V RT^2(1-Y)^{-3}\left[(1-Y)^2 + 3(1-Y)\xi + 3(1+2Y)\varsigma^2\right] + Y\frac{N_0 P}{\rho}\varsigma^3 \tag{16.3.9}$$

whereas the cavitation entropy may be obtained from the thermodynamic relationship

$$S_{cav} = (\Delta G_{cav} - H_{cav})/T. \tag{16.3.10}$$

Inert gases have a spherical shape and little interact electronically with the solvent. Assessment of their solubility in various fluids thus is a simple but stringent test of the reliability of the parameterization adopted to model the Gibbs free energy of cavitation. To make meaningful comparisons, I first recall that the bulk Gibbs free energy of solution is simply related to the Henry's Law constant K_H by:

$$\Delta G_s = RT \ln K_H = \Delta\overline{G}_{cav} + \Delta\overline{G}_{el} + \Delta\overline{G}_{disp} + \Delta\overline{G}_{rep} + RT\ln\left(\frac{RT}{\overline{V}_s}\right) \tag{16.3.11}$$

where $\overline{V}_s$ is the molar volume of the solvent. Adoption of this simplified form implies that the rotational, vibrational and momentum partition functions must be identical in the two states (cf. Eq. 16.3.4).

The enthalpy of solution is given by the following partial derivative of Henrys' law constant (Pierotti, 1976):

$$\Delta H_S = \left(\frac{\partial \ln K_H}{\partial(1/RT)}\right)_P = \overline{H}_{cav} + \overline{H}_{el} + \overline{H}_{disp} + \overline{H}_{rep} - RT + \alpha_s RT^2 \tag{16.3.12}$$

where α_s is the thermal expansion coefficient of the solvent. The entropy of solution is:

$$\Delta S_s = -\left(\frac{\partial \Delta G_s}{\partial T}\right)_P = \overline{S}_{cav} + \overline{S}_{el} + \overline{S}_{disp} + \overline{S}_{rep} - R - R\ln\left(\frac{RT}{\overline{V}_s}\right) + \alpha_s RT \tag{16.3.13}$$

The isobaric heat capacity of solution is (Pierotti, 1963, 1965):

$$\Delta C_{P_s} = \overline{C}_{P_{Cav}} + \overline{C}_{P_{el}} + \overline{C}_{P_{disp}} + \overline{C}_{P_{rep}} - R + 2\alpha_s RT + RT^2(\partial\alpha_s/\partial T)_P^2 \tag{16.3.14}$$

with

$$\overline{C}_{P_{Cav}} = \left[\frac{2}{T} - \alpha_s + \frac{(\partial\alpha_s/\partial T)_P}{\alpha_s}\right] \times \overline{H}'_{cav} - R\left[\frac{Y\alpha_s T}{(1-Y)^2}\right]^2 \times \left[(1-Y)^2 + 6\varsigma(1-Y) + 3\varsigma^2(4Y+5)\right] \tag{16.3.15}$$

where $\overline{H}_{cav}\prime$ is given by Eq. 16.3.9 without the last term on the right-side and

$$\overline{C}_{P_{el}} + \overline{C}_{P_{disp}} + \overline{C}_{P_{rep}} = \alpha_s \times \left(\overline{H}_{el} + \overline{H}_{disp} + \overline{H}_{rep}\right) \tag{16.3.16}$$

The molar volume of the solute $\overline{V}$ is given by the partial derivative of its chemical potential in solution with respect to pressure P and is composed of a cavitation + an interaction contribution to whom the compressional term of the solvent is added (Pierotti, 1976):

$$\overline{V} = \overline{V}_{cav} + \overline{V}_{int} + \beta_s RT \tag{16.3.17}$$

where β, s is the isothermal compressibility of the solvent and

$$\overline{V}_{cav} = 82.5(\beta_s/\alpha_s)\frac{\overline{H}_{cav}}{RT} + (\pi/6)N_0\sigma_s^3 \tag{16.3.18}$$

$$\overline{V}_{int} = \beta_s \overline{G}_{int} \tag{16.3.19}$$

To address properly all the above energy terms one must introduce some simplifying assumptions, namely, that the electronic, dispersive and repulsive contributions to the Gibbs free energy of solution in Eq. 16.3.11 are purely enthalpic (which is implicit in the ab initio assessment). With this proviso, the corresponding entropy terms in Eq. (16.3.13) vanish.

One of the most important parameters necessary to quantitatively evaluate the solute/solvent interaction is the solvation diameter (cf. Sect. 7.7.3), or the, corresponding, "*solvation radius*".

Simple assessments of solvation radii are based on the collisional theory (Clever et al., 1957) or also on spherical shapes. The assumption of a spherical shape for the solvent molecules has the obvious advantage of eliminating solvent orientation around the solute. This assumption ($\sigma_S = \sqrt[3]{6V/\pi}$) results however in a large overestimation of the appropriate value to assign to the real solvent According to Stearn and Eyring (1937), a simple cubic shape (i.e. $\sigma_S = \sqrt[3]{V}$) is able to reproduce the linear dependency of the free energy of activation for N–Br exchanges with the reciprocal of the dielectric constant experimentally observed in organic solvents rich in C_6H_6 component (see their Fig. 1 and their Table II). Likewise, the Henry's law constant of pure benzene is well accounted for in this way, but the approach fails for denser solvents with higher dielectric constants. A mean radius for non-spherical solvent molecules can be formally derived from the scaled particle theory for a mixture of hard spheres (Lebovitz et al., 1965). An extension to hard spheres of unequal diameter has been also proposed (Gogonea et al., 1998). Actually, as anticipated by Reiss (1965), even for a simple non-spherical solvent composed of molecules with identical shape, σ_s is a thermally averaged quantity over various orientations of the solvent with respect to the cavity surface (see infra) of the solute molecule (or atom).

In Table 16.5 are resumed some selected solvent properties necessary to carry out the solute/solvent interaction calculations. The properties of water are compared to those of a non-polar solvent such as benzene (C_6H_6).

The obtained solvation energy terms are listed in Table 16.6.

The Henry's Law constants for He, Ne, Ar and Kr obtained from the Ar-based parameterization of solvent properties are compared with high-T experiments in Table 16.7 and in Table 16.8 are provided the thermodynamic standard state properties of the various series elements in the investigated media.

As shown in Fig. 16.3, the Ar-based parameterization may be safely adopted to depict the properties of the other terms in the noble gases series.

16.3.4 Solubility of Noble Gases in Silicate Melts

The solubility of noble gases in silicate melts gives fundamental clues on topics such as the early history of our planet. Solubility investigations can be fraught with experimental difficulties at the relevant high temperatures or pressures. In addition, they cannot readily address the wide range of silicate composition that has to be dealt with. By contrast, theoretical calculations represent a promising way to address such problems. An ab initio assessment of the energy of neutral and charged species in complex solvents is currently the most rigorous procedure to estimate homogeneous reaction constants among species in solution. Perhaps the noble gas solubility is the simplest case to treat.

Adopting Argon as a probe to depict its high-T solubility in melts through an appropriate choice of the solvent diameter σ_s, along the guidelines outlined for water and benzene, the obtained σ_s is shown to be consistent with a simple functional form based on the molecular volume of the solvent (Ottonello and Richet, 2014). The σ_s appropriate to melts fits nicely on the trend between a non polar

Table 16.5 Selected physical properties of the solvent at T = 298.15 *K*, P = 1 *bar*. Data for H_2O are compared with those of a non-polar solvent such benzene C_6H_6

Solvent	H_2O	C_6H_6
Molar volume (*cc/mol*)	18.070	89.400
Molar weight (*g/mol*)	18.015	78.112
Density (*g/cc*)	0.997	0.873
Molecular volume ($\AA^3$)	30.006	148.452
Numeral density ($\AA^{-3}$)	0.03333	0.00674
Solvation radius ($\AA$)	1.385	2.630
Static polarizability ($\AA^3$)	1.47	10.32
ε	78.39	2.247
ε^{∞}	1.776	2.244
α_V ($K^{-1} \times 10^5$)	25.7	138
α'_V ($K^{-2} \times 10^9$)	–	–
β_V ($Mbar^{-1}$)	44.59	93.8

Table 16.6 Solution energy terms for Ar in water and benzene at T = 25 °*C*, P = 1 *bar*. The interaction Gibbs free energy ΔG_i is the summation of electrostatic + repulsive + dispersive terms computed at B3LYP/aug-cc-PVTZ theory level. Data are in *kJ/mol* (G, H) and *J/(mol × K)* (S, C_P). Calcd = calculated; Obsd = observed. The T limit for the heat capacity coefficients depicting the heat capacity of solution is 353.5 *K* for benzene, 372.8 *K* for water. From Ottonello and Richet (2014)

Solvent		ΔG_{cav}	H_{cav}	S_{cav}	$\Delta G_i \cong H_i$	ΔG_s	ΔH_s	ΔS_s	ΔC_p a	ΔC_p b × 10^3
H_2O	Calcd	23.398	3.759	-65.9	-15.841	25.825	−14.011	−133.6	38.787	−92.388
	Obsd	–	–	–	–	26.250[a]	−11.213[a]	−125.5[a]	–	–
Benzene	Calcd	18.522	20.900	8.0	−13.640	18.827	5.801	−43.7	315.563	−1059.82
	Obsd	–	–	–	–	17.435[b]	1.243[b]	−54.4[b]	–	–

[a] Lide (2002); [b] Clever et al. (1957)

(C_6H_6) and a polar solvent (H_2O) (Fig. 16.4) is thus satisfactorily reproduced by a simple expression of the form:

$$\sigma_s = 2 \times (1.1383 r_{coll} - 0.3836) \qquad (16.3.19)$$

where r_{coll} is the Stearn-Eyring collisional radius (i.e. $r_{coll} = \sqrt[3]{V/8}$).

Solubility calculations for Ar, as already seen for the water and C_6H_6 solvents, were extended by Ottonello and Richet (2014) to He, Ne and Kr whose dispersive and repulsive coefficients are available from theory. The static dielectric constant (ε)

Table 16.7 Result of the SPT–PCM application to noble gases compared with experimental observations (values in brackets) in terms of Henry's Law constant ($\ln K_H$) at T = 298.15 K. From Ottonello and Richet (2014)

Element	H_2O	C_6H_6
He	10.30	8.40
	(11.87)[a]	(9.48)[b]
Ne	11.62	9.03
	(11.72)[a]	(9.18)[b]
Ar	10.42	7.59
	(10.59)[a]	(7.05)[b]
Kr	9.72	6.75
	(10.01)[a]	(5.93)[b]

[a] Lide (2002); [b] Clever et al. (1957)

Table 16.8 Ab initio standard state* thermodynamic parameters for noble gases in the investigated media. T_{lim} is the high-T limit of validity of the heat capacity polynomial $C_P = a + b \times T$. From Ottonello and Richet (2014)

Element		**H°$_f$** (kJ/mol)	**S°** [J/(mol × K)]	**V°** (J/bar)	**C_P**		**T_{lim}** (*K*)
					a	**b ×10^3**	
He	Water	−3.906	27.4	1.323	34.004	−37.769	372.8
Ne		−6.249	28.8	2.475	48.472	−66.971	
Ar		−14.010	21.2	3.393	59.576	−92.388	
Kr		−18.049	23.6	3.880	64.928	−104.793	
He	Benzene	5.656	75.3	2.576	163.228	−451.282	353.3
Ne		8.320	99.2	4.280	259.547	−776.657	
Ar		5.801	111.2	5.476	336.354	−1059.82	
Kr		4.240	123.0	6.077	373.311	−1198.02	

* hypothetical one-molal solution referred to infinite dilution at T = 298.15 *K*, P = 10^5 *Pa*

of the investigated silicate melts and its optical counterpart (ε^{∞}) were resolved through application of a modified form of the Clausius–Mossotti relation:

$$\frac{\varepsilon - 1}{\varepsilon + 2} \cong \frac{4\pi}{3V_s} \sum_j^{\text{oxides}} x_j \alpha_j^{\text{static}} \tag{16.3.20}$$

where the static polarizabilities α_j^{static} are the values calculated by Lasaga and Cygan (1982). At optical frequencies it was applied

$$\frac{\varepsilon^{\infty} - 1}{\varepsilon^{\infty} + 2} = \frac{n^2 - 1}{n^2 + 2}, \tag{16.3.21}$$

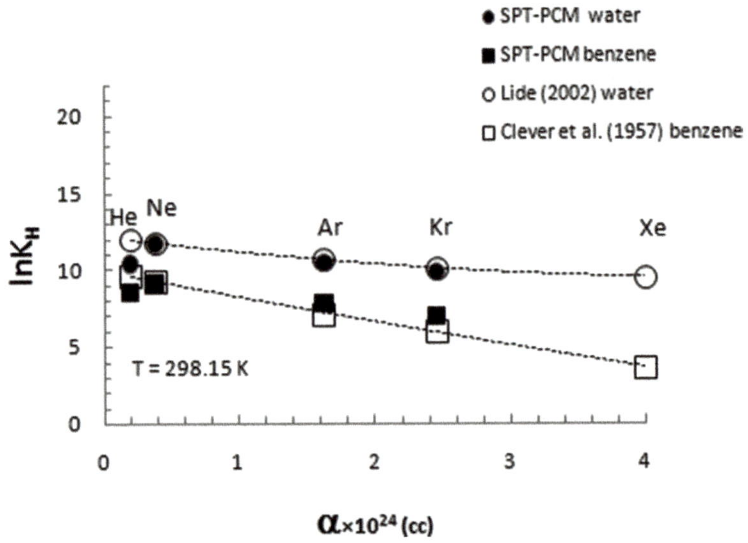

Fig. 16.3 Ln K_H versus polarizability of inert gases in water and benzene at 298.15 *K*. Modified from Ottonello and Richet (2014)

Fig. 16.4 Collisional versus solvation radius in the investigated media. The dotted line is the correlation equation relating a non polar (C_6H_6) and a polar (H_2O) solvent. Modified from Ottonello and Richet (2014)

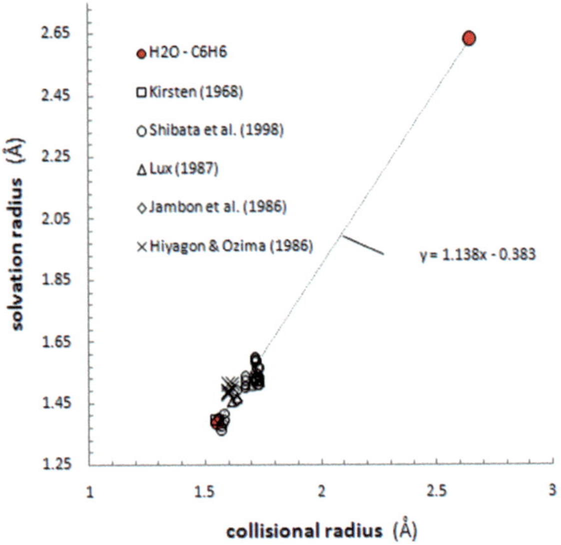

Table 16.9 Gibbs free Energy difference between observation and calculus at the experimental T, P conditions for some of the investigated materials. The error progression from Ar-based solvation radii (a) to radii obtained by the general relation between solvation radius and molecular volume of the solvent (b) is acceptable. Electrostatic, dispersive and repulsive solute/solvent interactions were computed for all elements at the B3LYP/aug-cc-PVTZ theory level with the GAUSSIAN package (Frisch et al., 2004). The high bias at He level could be ascribed to the adopted van der Waals radius

Error	He	Ne	Ar	Kr	Samples	T (°C)	P Kbar	Reference	Notes
ΔG (*kJ/mol*)	−23.5	−2.5	–	6.1	4(He,Ne); 5 (Ar,Kr)	1350	0.001	[1]	(a)
	−20.3	−2.2	–	3.8	1 (reversal)	1500	0.001	[2]	(a)
	−15.7	5.3	10.1	15.1					(b)
	−34.7	−6.3	–	–	Enstatite	1500	0.001	[3]	(a)
	−34.7	−9.1	4	–					(b)
	–	1.5	–	2.4	NS1	1200	1.89	[4]	(a)
	–	4.5	4	6.9					(b)
	–	3.2	–	2.4	NS1	1300	1.97	[4]	(a)
	–	7.3	5.5	8.6					(b)
	–	4.6	–	0.7	NS1	1400	1.88	[4]	(a)
	–	11.5	9.1	11					(b)

(a) = Ar-based; (b) = general relation; [1] Lux (1987); [2] Jambon et al. (1986); [3] Kirsten (1968); [4] Shibata et al. (1998)

where the refractive index n vary within a restricted range (Vo-Thanh et al., 2005), namely, 1.58 to 1.64.

As shown in Table 16.9 and Fig. 16.5b,c,d, adopting the general relation instead of the solubility-based radius (Fig. 16.5a) induces only limited errors. Evidently the non negligible polarization properties of the investigated solvents induce some form of preferred orientation of non-spherical moieties around the solute cavity, the effect being the more intense the higher is the solvent polarizability.

16.4 O_2

Oxygen (O) is present in Nature with three stable isotopes: ^{16}O (0.997630), ^{17}O (0.000375) and ^{18}O (0.001995). Therefore in all quantum mechanical calculations of thermochemical and thermophysical properties involving O atoms, the adopted reference mass is ^{16}O. Molecular oxygen (O_2) is essentially a byproduct of biological activity, and, as such, somewhat ephemeral in Nature.

The isotopic composition of atomic oxygen in marine water and in aerial O_2 are not identical and the latter is enriched in ^{18}O. The deviation of $\delta^{18}O$ of air O_2 from that of seawater oxygen is known as the Dole Effect (DE) (Dole, 1935). According to Kroopnick and Craig (1972) DE amounts to 23.5‰ ± 0.3‰. Although for a

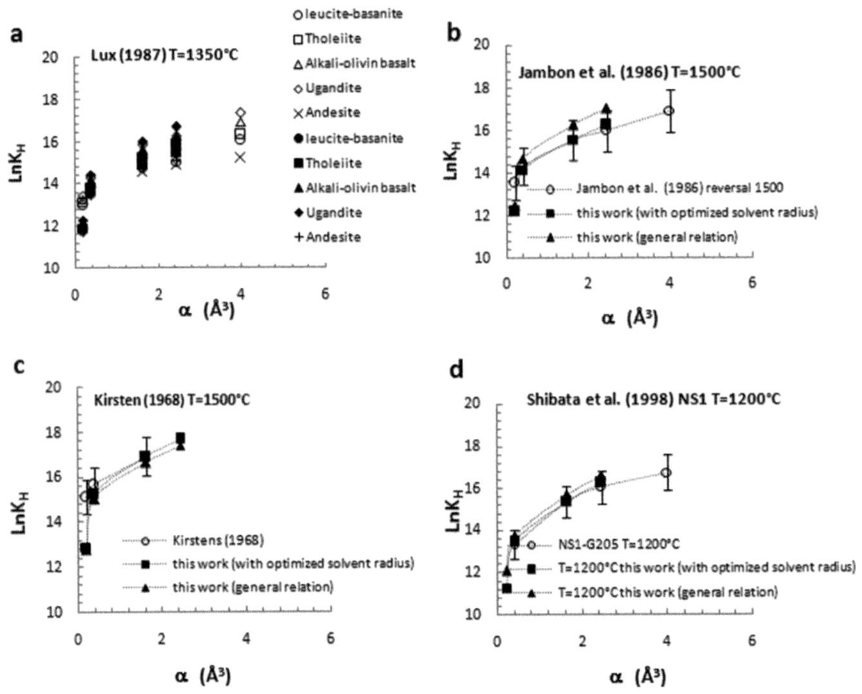

Fig. 16.5 Results of the SPT-PCM application to noble gases in silicate melts at high T: (**a**) experiments of Lux (1987) (open symbols) compared with SPT-PCM calculations (filled symbols). The electrostatic + dispersive + repulsive interaction was computed at B3LYP/aug-cc-PVTZ theory level. The Gibbs free energy of cavitation was split into enthalpic and entropic contributions as indicated by Pierotti (1963, 1965, 1976). The experimental Ar solubility ($LnK_{H,Ar,T}$) was adopted as a target value in an inverse modeling procedure that returned the correct σ_s at target value of $LnK_{H,Ar,T}$. (**b, c, d**) some selected experiments (open symbols) reproduced through inverse modeling (σ_s at target value of $LnK_{H,Ar,T}$; filled squares) and through the general equation relating the molecular volume and the collisional radius r_{coll}; filled triangles). The 5% error bars on experiments are not meant to reproduce the reported experimental accuracy but rather a likely range of uncertainty in the computational procedure. The various plots stress the role of polarizability on the solubility of the inert gases previously evidenced in other theoretical investigations. Modified from Ottonello and Richet (2014)

long time it has been accepted that there is no oxygen isotope fractionation in photosynthesis, recent studies (Luz and Barkan, 2011) show that O_2 released by certain marine phytoplankton primary producers, is significantly enriched (up to 6‰) in ^{18}O with respect to the substrate seawater and that the magnitudes of the global DE should not be sensitive to past changes in the ratio of land-to-sea photosynthetic rates.

16.4.1 Thermophysical and Thermodynamic Properties

As anticipated at Sect. 7.4, O_2 has quite unusual properties. It has symmetry $D_{\infty h}$ (linear with inversion center) but its outer electrons are not in a closed shell. Its ground state is $^3\Sigma_g^-$ (i.e. triply degenerate, even, with sigma symmetry). Due to this fact, the oxygen molecule in its ground state has a magnetic moment arising from the spin. This magnetic moments is observed not only in the spectrum of the molecule but also in large-scale properties in the O_2 gas.

The two O atoms are connected by a double bond at an equilibrium distance of 1.208 *Å* and vibrate with a harmonic frequency $\nu_1 = 1580\ cm^{-1}$ (Huber and Herzberg, 1979). The ionization energy is 12.070 *eV* and the electron affinity 0.451 ± 0.007 *eV* (NIST, 2020). The first excited state is a doubly degenerate $^1\Delta_g$ at + 0.9817 *eV* from the ground state (Huber and Herzberg, 1979). Based on the NIST-JANAF tabulation for monatomic oxygen, the dissociation energy is 0.18799475 *Hartree*.

The rotational constant of O_2 is 1.43768 cm^{-1} and the calculated one at B3LYP/6-31G(d,p) level is 1.429 cm^{-1}. The components of the electric dipole moment are zero and those of the quadrupole moment are xx = 0.155 *D×Å*, yy = 0.155 *D×Å*, zz = −0.310 *D×Å*. The electrical dipole polarizability is $\alpha = 1.562$ $Å^3$ (Olney et al., 1997) and the calculated one at B3LYP/6-31G(d,p) level is $\alpha = 1.031$ $Å^3$.

At standard state (pure gas, P = 10^{-5} *Pa*, 298.15 *K*) O_2 has an entropy of 205.147 *J/(mol × K)* and an isobaric heat capacity of 29.376 *J/(mol × K)* (Chase, 1998). The Shomate equation for the isobaric heat capacity in the range 298.15 ≤ T (*K*) ≤ 700 is $C_P = 31.32234 - 20.23532 \times 10^{-3} \times T + 57.86644 \times 10^{-6} \times T^2 - 36.50624 \times 10^{-9} \times T^3 - 0.007374 \times 10^6 \times T^{-2}$ (NIST, 2000; based on refitted data of Chase 1998). At room pressure O_2 boils at 90.2 *K*. The triple point is at 54.33 *K*. The critical temperature is $T_c = 154.58$ *K*, the critical pressure is $P_c = 50.43$ *bar* and the critical density is $\rho_c = 13.60$ *mol/l* (NIST, 2020).

16.4.2 Photosynthetic Production

Oxygen photosynthesis is "*one of the most important metabolisms to have evolved on Earth as it has enabled complex life forms to emerge*" (Sanchez Baracaldo and Cardona, 2020). The first rise of oxygen in the atmosphere (Great Oxidation Event; GOE) took place at 2.32–2.40 *Ga* (Bekker et al., 2004; Lyons et al., 2014) and is ascribed to the activity of Cyanobacteria, though the fossil record is not conclusive (Rasmussen et al., 2008; Schirrmeister et al., 2016; Sanchez Baracaldo and Cardona, 2020). A second major oxygenation event (Neoproterozoic Oxidation Event; NOE) at 0.8–0.6 *Ga* brought the oxygen concentration level to values similar to present. The progressive accumulation of oxygen experienced over geological times is due to the fact that carbon fixation by oxygenic phototrophs has exceeded by far the respiration of organic matter. The imbalance of these two processes has resulted in the drawdown of atmospheric CO_2, whose depletion (and the concomitant depletion in CH_4 due to oxidative processes) generated eventually long periods of glaciations, associated to lack of greenhouse effect.

The basic photosynthetic reaction is usually formalized as:

$$6CO_2 + 6H_2O \Leftrightarrow C_6H_{12}O_6 + 6O_2 \tag{16.4.1}$$

with glucose ($C_6H_{12}O_6$) representing the organic compound. To proceed rightward, this reaction, which is energetically quite unbalanced ($\Delta G_1 = +2827.38$ *kJ/mol* at room P,T), needs to be promoted by photosynthetic electron transport: H_2O is split by the oxygen-evolving complex to provide protons and electrons to the chloroplastic electron chain. Four electrons are required for every O_2 evolved, so that gross O_2 production (GOP) is related to linear electron transport (J) according to J/4 (Holloway-Phillips, 2018).

To elucidate the charge-transfer mechanisms involved in the process, Zhang and Friesner (1998) carried out ab initio electronic structure calculations of electron transfer couplings between chromophores in the bacterial photosynthetic reaction center and showed that the couplings conform to an explicit intermediate mechanism.

The calculations were carried out at RHF/6-31G(d,p) theory level on six pairs of molecules in the RC structure in the bacterium *Rhodopseudomonas Viridis*, experimentally determined by Deisenhofer et al. (1985): three for the L-side acceptors (P_LB_L, P_MB_L, B_LH_L) and the corresponding three for the M-side acceptors (P_LB_M, P_MB_M, B_MH_M). The P_LB_L L-side acceptor is shown as an example in Fig. 16.6.

The basic mechanism of bioenergetic conversion is the Proton Coupled Electron Transfert [(PCET); Hammes-Shiffer (2009)]. In the co-linear PCET both the proton and electron are transferred from the donor (D in Fig. 16.7) to the acceptor (A) along the same direction while in the orthogonal PCET, proton and electron are transferred to disparate acceptors (A and B).

PCET reactions are described in terms of non-adiabatic transitions between the reactant and product electron-proton vibronic states. The PCET rate increase as the electronic coupling and temperature increase, and as the total reorganization energy and equilibrium proton donor–acceptor distance decrease. Also, the rate constant is predicted to increase as the driving force becomes more negative (Hammes-Shiffer, 2009),

The rate constant expression for fixed proton donor–acceptor distance is:

$$K = \sum_{\mu} P_{\mu} \sum_{\nu} \frac{\left|V^{el} S_{\mu\nu}^{(0)}\right|^2}{\hbar} \sqrt{\frac{\pi}{\lambda_{\mu\nu} k_B T}} \exp\left[-\frac{\left(\Delta G_{\mu\nu}^0 + \lambda_{\mu\nu}\right)^2}{4\lambda_{\mu\nu} k_B T}\right] \tag{16.4.2}$$

The summations in Eq. 16.4.2 are carried out over reactant (μ) and product (ν) vibronic states, P_{μ} is the Boltzmann probability for the reactant state μ, V^{el} is the

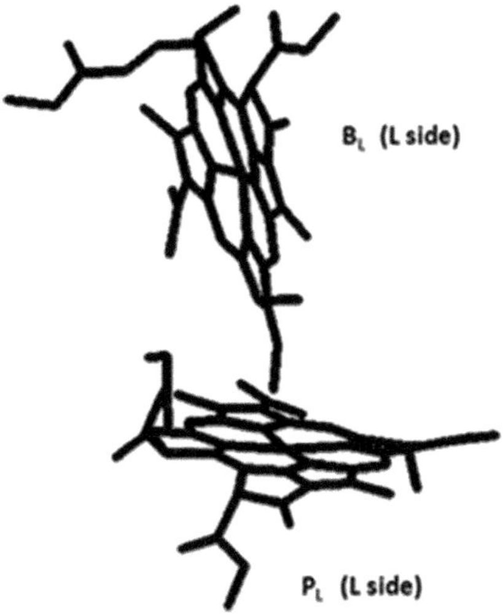

Fig. 16.6 Structure of P_LB_L from the bacterial photosynthetic reaction center. The bottom molecule is the BacterioChlorophyll of the special pair on the L side P_L, the top molecule is the BacterioChlorophyl monomer of the L side B_L. Modified from Zhang et al. (1997)

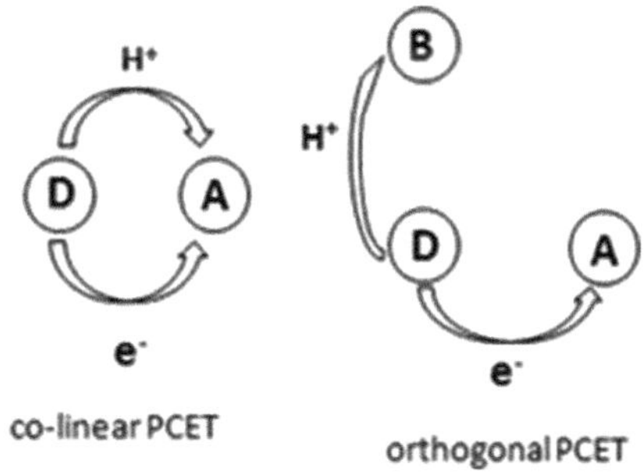

Fig. 16.7 Co-linear and orthogonal proton-coupled electron transfer (schematic)

electronic coupling, $S_{\mu\nu}^{(0)}$ is the proton vibrational wave function overlap, $\lambda_{\mu\nu}$ is the total reorganization energy, and $\Delta G_{\mu\nu}^{0}$ is the free energy of reaction for vibronic states μ and ν (Hammes-Shiffer, 2009).

The initial steps of photosynthesis take place in the "*photosynthetic reaction center*" (RC) constituted by a chlorophyll–protein complex. Under light excitation, an electron is transferred from the excited state of the special pair of bacteriochlorophylls (BChl), P*, to the primary acceptor, a bacteriopheophytin on the L-side of RC designated H_L. This takes place in 3 *ps* over a distance of $\sim$ 20 $\AA$ but the kinetic is not apparently affected by temperature. Because such a direct-transfer kinetics is implausible, the Bacteriochlorophyll protein (B_L) participate most likely either as an explicit intermediate or via a super-exchange mechanism (Zhang and Friesner, 1998; and references therein).

Designating state 1 the donor state, state 2 the intermediate state, and state 3 the acceptor, the off-diagonal electron transfer matrix elements for an explicit intermediate mechanism H_{12} should be of the order of 10–30 cm^{-1}, and H_{23} should be somewhat larger than this. For super-exchange, the effective coupling between states 1 and 3 is obtainable applying (Zhang and Friesner, 1998):

$$J_{eff} = \frac{H_{12}H_{13}}{E_1 - E_2} \quad (16.4.3)$$

with the energy states E_1 and E_2 evaluated at the cross point of the donor and acceptor potential surfaces.

The electron transfer (ET) matrix elements determined adopting the pseudo-spectral (PS) numerical procedure of Zhang et al. (1997) are shown in Table 16.10.

The values for the H_{ij} on the L-side are incompatible with a super-exchange model with a large energy gap, while the value of 122 cm^{-1} for H_{23} is qualitatively consistent with a rapid B → H transfer. The P → B transfer rate was obtained by Zang and Friesner (1998) adopting as exciton representation of the P excited state $\left(P_L^*P_M - P_M^*P_L\right)/\sqrt{2}$, and using the localized P_1B_2 coupling matrix elements of Table 16.10. Zang and Friesner (1998) obtained in this way J_{eff} = 17 cm^{-1} in good agreement with the value of 22 cm^{-1} obtained by Makri et al. (1996) by a rigorous quantum dynamical fitting of the experiments.

Another quite interesting and promising quantum mechanical investigation of the photosynthetic processes at the basis of the massive production of atmospheric O_2 is due to Wawrzyniak et al. (2008), Wawrzyniak (2011). In this case the object of the investigation were the NMR spectra obtainable from an accurate DFT/6–311 ++G(d,p) investigation followed by the GIAO treatment (see Sect. 15.5). The

Table 16.10 Ab initio calculation of electrotransfer coupling matrix elements (Zang and Friesner, 1998)

Reaction	Coupling H_{ij} (cm^{-1})
L-side of RC	
$P_L^-B_L \rightarrow P_LB_L^-$	5.10
$P_L^*B_L \rightarrow P_L^+B_L^-$	7.24
$P_M^-B_L \rightarrow P_MB_L^-$	30.68
$P_M^*B_L \rightarrow P_M^+B_L^-$	32.27
$B_L^-H_L \rightarrow B_LH_L^-$	122.17
P without Mg on L-side	
$P_L'^-B_L \rightarrow P_L'B_L^-$	9.14
$P_M'^-B_L \rightarrow P_M'B_L^-$	35.02
M-side of RC	
$P_L^-B_M \rightarrow P_LB_M^-$	17.20
$P_L^*B_M \rightarrow P_L^+B_M^-$	17.41
$P_M^-B_M \rightarrow P_MB_-^-$	12.64
$P_M^*B_M \rightarrow P_M^+B_M^-$	19.10
$B_M^-H_M \rightarrow B_MH_M^-$	95.45

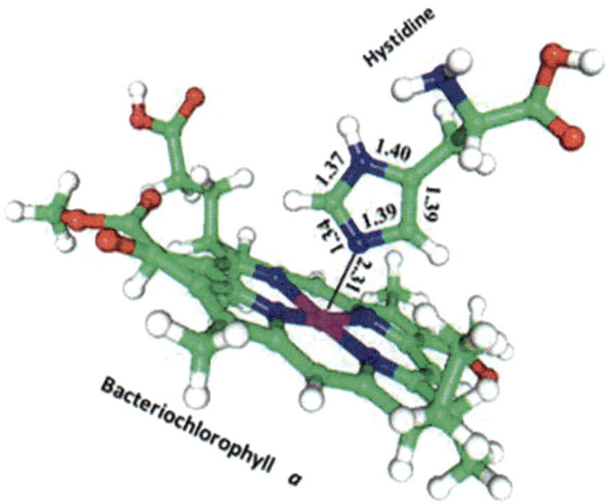

Fig. 16.8 DFT-optimized structure of the neutral BChl *a*–His complex. The phytyl tail is truncated at the esther group and substituted by a hydrogen atom. The bond lengths in the imidazole ring and the Mg–N distance are in Å. Modified from Wawrzyniak (2011)

investigated moiety was the Bacteriochlorophill *a*—Histidine complex extracted from one of the B850 BChl dimer units in the 2FKW PDB crystallographic structure of *Rps. Acidophila* (Cherezov et al., 2006; His 30 and BChl 1601) and shown in Fig. 16.8. According to the crystal structure, the two BChl *a* molecules in the dimer are slightly asymmetric, but the two histidines coordinated to them appear to be equivalent according to the NMR spectra.

A neutral amino acid termination (COOH–NH_2) was used to saturate the broken peptide bond, and the phytyl tail of the bacteriochlorophyll *a* was truncated at the esther group and saturated by a hydrogen atom (see Fig. 16.8). After full DFT optimization, the obtained GIAO ^1H and ^{13}C chemical shieldings were referred to the TMS chemical shifts scale, while the computed ^{15}N chemical shieldings were referred to the value of $^{15}NH_3$.[2]

The obtained DFT chemical shifts (Table 16.11) support the proposed assignment of the histidines protonation state in LH2 and confirm that all the histidines except His 37 participate in a hydrogen bond (Wawrzyniak et al., 2008; Wawrzyniak, 2011). Besides the chemical shifts, also the electron charge difference map between the BChl *a*–His complex and its two fragments, histidine and bacteriochlorophyll *a* was calculated with the ADF program (Velde et al., 2001). In this calculation the 6–311++G(d,p) basis set was replaced by Triple-Zeta with Polarization (TZP) basis expansions. At the light of the computational results, the protein environment in LH2 appears to exerts a stress on the Mg-coordinated histidines that induces a considerable electron charge transfer to the bacteriochlorophylls *a* (Wawrzyniak et al., 2008; Wawrzyniak, 2011). This charge transfer appears in fact to depend in a crucial manner on the relative distance and orientation of the BChl *a*–His complex. The combined chemical-shift + electron-transfer DFT

[2] The experimental chemical shifts referred to $(^{15}NH_4)_2SO_4$ powder were converted to the $^{15}NH_3$ scale by adding 20 *ppm* (Duncan, 1997).

Table 16.11 Calculated histidine and 4-methylimidazole chemical shifts [*ppm*] for different protonation states in comparison with available experimental data for crystal histidine. 'x' denotes an inter-molecular hydrogen bond with a water molecule at the pyrrole nitrogen. The experimental chemical shifts for ^{13}C are from Alia et al. (2001); those for ^{15}N are from Munowitz et al. (1982). The DFT-GIAO results are from Wawrzyniak et al. (2008), Wawrzyniak (2011). "..x" denotes an inter-molecular hydrogen bond with a water molecule at the pyrrole nitrogen

	Histidine (experimental)		Histidine (DFT)					4-Methylimidazole (DFT)		
	Cationic	Neutral_	Cationic	Neutral	Neutral..x	Neutral	Anionic	Neutral	Neutral	Neutral..x
C_γ	128	137	144.3	148.5	147.7	131.6	137.7	147.3	129.8	130.7
C_δ	119.3	113.5	122.6	117.9	118	134.5	133.4	113.9	133.6	133.4
C_ε	136.4	138.6	132.9	136	137.1	137.3	154	136.3	136.8	137
N_π	175	242	212	302	301	190	293	306	183	192
N_τ	172	167	188	175	186	303	295	174	304	302
H_δ			7.2	6.9	6.9	7.0	7.0	6.7	6.9	6.8
H_ε			8.0	7.4	7.5	7.3	7.4	7.3	7.3	7.4

results "*strongly suggest that the protein environment is forcing the histidine to stay close to the BChl a, thereby inducing a combined structural change and a large charge transfer. When the constraints from the protein are removed, the structure relaxes and histidine moves away from its close position to the bacteriochlorophyll a*" (Wawrzniak, 2011).

16.4.3 Solubility of O_2 in Water

The solubility of oxygen in water has been the subject of a dedicated effort, since the early measurements by Bunsen. Battino et al. (1983) provided an extensive review of basically all experimental data at $P_{O2,g} = 1.01325\times10^5$ *Pa* in the thermal ranges 273.15 ≤ T (*K*) ≤ 348.15 and 348.15 ≤ T (*K*) ≤ 573.15.

Table 16.12 resumes the recommended solubility values determined by Battino et al. (1983) in the low T range from a critical survey of over 40 papers. After the exclusion of 25 papers whose standard deviations from the smoothing curve

$$x_l = \left[A_0 + A_1\tau^{-1} + A_2 \ln \tau + A_3\tau + A_4\tau^2\right] \tag{16.4.4}$$

with $\tau = T(\mathrm{K})/100$ and x_l as mole fraction in solution, was exceedingly high, the 149 selected data points in the remaining 15 papers gave, after regression, the listed

Table 16.12 O_2 solubility in water the thermal range 273.15 ≤ T (*K*) ≤ 348.15 for a gaseous pressure $P_{O2,g} = 1.01325\times10^5$ *Pa*. (From Battino et al., 1983). (a) Battino et al. (1983); (b) Benson et al. (1979); (c) SPT-PCM procedure. x_l = moles × 10^5

T (K)	x_l (a)	x_l (b)	x_l (c)
273.15	3.912	3.958	4.158
278.15	3.460	3.465	3.666
283.15	3.070	3.075	3.248
288.15	2.756	2.761	2.895
293.15	2.501	2.508	2.591
298.15	2.293	2.301	2.328
303.15	2.122	2.131	2.102
308.15	1.982	1.991	1.904
313.15	1.867	1.874	1.732
318.15	1.773	1.718	1.578
323.15	1.697	1.697	1.442
328.15	1.635	1.631	1.321
333.15	1.586	1.575	1.212
338.15	1.549	1.530	1.114
343.15	1.521	1.493	1.026
348.15	1.502	1.463	0.947

values, which may compared with the high precision measurements of Benson et al. (1979) and with the result of the SPT-PCM procedure in the last column of the table.

Battino et al. (1983) recommended the system of oxygen dissolved in water as a reference system, since it is the most commonly measured and most precisely known gas solubility. It is therefore quite appropriate to see how a first principles investigation may reproduce the data. To this purpose I performed some SPT-PCM calculations along the guidelines previously discussed for noble gases. When resolved at DFT/6–31+G* SCRF-IEFPCM level, with UAHF radii, the UATM assigns to the each oxygen a radius of 1.545 *Å*, and an electrostatic scaling factor γ_{el} = 1.20. The computed cavity volume, limited by a tessellated surface (composed of tesserae of an average area of 0.2 $Å^2$; see the inset in Fig. 16.9) is 39.397 $Å^3$. The solvation radius is 1.385 *Å* and the returned solute/solvent interaction energy is H_{int} = −17.071 *kJ/mol*, arising from a polarized solute —solvent electrostatic interaction of −0.418 *kJ/mol*, a dispersive energy contribution of −20.627 *kJ/mol* and a repulsive energy of 3.974 *kJ/mol*. The Pierotti-Reiss energy terms are resumed in Table 16.13. The upper part of the Table returns the adopted solvent–solute parameters at T = 298.15 *K* and P = 10^5 Pa, and the lower part resumes the results for some temperatures in the range from 0 to 75 °*C* at the same pressure. The adopted solute radius (which gives the best agreement with the experiments) is 2.045 *Å*, somewhat lower than a radius of a spherical moiety (2.111 *Å*).

In Fig. 16.9, and Table 16.12, the SPT-PCM calculations (solid line) are compared with the experimental observations. The inset in the same figure shows the adopted SCFR tessellation of the O_2 molecule (see Sect. 7.7.2). The solute radius is the unique variable adopted in the computational approach and its value (2.045 *Å*)

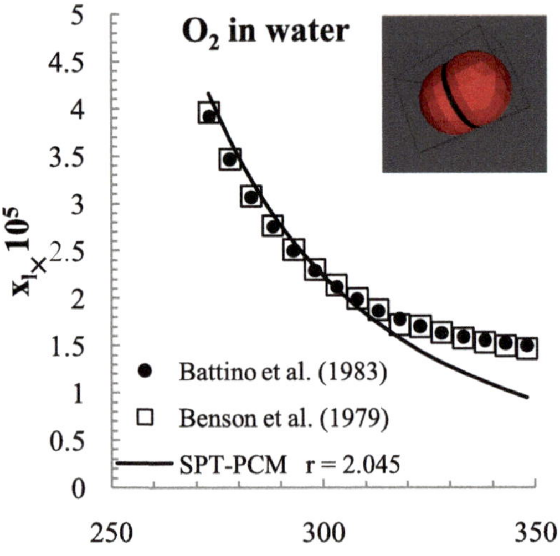

Fig. 16.9 Experimental versus computed SPT-PCM abundances of dissolved oxygen in water at P = 1 *bar* and various T. The inset shows the adopted SCRF tessellation

Table 16.13 O_2 solubility in water. SPT-PCM calculations. Thermal expansion obeys $\alpha_T = \alpha_0 + \alpha_1 \times T + \alpha_2 \times T^2$

Solvent		Solute		
$V°$ *(cc/mol)*	18.0686	R *(Å)*	2.045	
ρ *(g/cc)*	0.9970	H_{int} *(kJ/mol)*	−17.071	
ρ_{num} *(Å^{-3})*	0.0333	G_{cav} *(kJ/mol)*	25.639	
R_{solv} *(Å)*	1.385	H_{cav} *(kJ/mol)*	3.934	
$\alpha_{p,static}$ *(Å^3)*	6.8963	S_{cav} *[J/(mol × K)]*	−72.7987	
ε	78.3900	V_{cav} *(cc/mol)*	45.607	
ε^{∞}	1.776	V_{int} *(cc/mol)*	−7.724	
α_0 *(K^{-1})*	0.006394	V_{solute} *(cc/mol)*	39.005	
$\alpha_1 \times 10^5$	3.490			
$\alpha_2 \times 10^8$ *(K)*	−4.237			
β *($Mbar^{-1}$)*	45.248			
T *(°C)*	0.0	25.0	50.0	75
G_{sol} *(kJ/mol)*	22.941	26.477	29.987	33.524
H_{sol} *(kJ/mol)*	−15.941	−15.434	−15.329	−15.725
S_{sol} *[J/(molxK)]*	−142.347	−140.572	−140.230	−141.457
LnK_H (adimensional)	10.1011	10.6804	11.1603	11.5808
LnK_{sol} (adimensional)	−10.1011	−10.6884	−11.1603	−11.5808
LnK_{sol} (cc STP/g)	−2.9749	−3.5542	−4.0341	−4.4545
LnK_{sol} (m^3 STP/g)	−16.7904	−17.3697	−17.8496	−18.2700

is quite reasonable due to the non perfectly spherical conformation of the solute molecule. The thermal effect on O_2 solubility in water is reproduced only in a qualitative way. At T = 348.5 *K* the SPT-PCM calculation underestimates the effective amount of O_2 in solution by roughly 36%.

16.4.4 Solubility of O_2 in Silicate Melts: The Normal Oxygen Electrode

SPT-PCM computations on the very same substances investigated to depict the solubility of noble gases (Ottonello and Richet, 2014) show that the solubility of molecular oxygen in silicate melts does not differ much from the solubility observed in water. The amount of dissolved O_2 increases regularly with the increase of numeral density of the solvent from ~ 7 *ppm* for the less dense to ~ 26 *ppm* (molar fraction) for the most dense fictive silicate liquid at equilibrium with a pure O_2 atmosphere at T = 298.15 *K* and P = 10^5 *Pa.*

Besides this simple consideration it is more interesting to evaluate the solvent effect on the *normal oxygen electrode* reaction

$$\frac{1}{2}O_{2,g} + 2e^- \rightarrow O^{2-}_{sln} \tag{16.4.5}$$

which, in a Lux-Flood notation replaces the more familiar *normal hydrogen electrode* reaction:

$$\frac{1}{2}H_{2,g} + e^- \rightarrow H^+_{sln} \tag{16.4.6}$$

that assigns a zero potential to the redox scale but is inapplicable in aprotic system such as silicate melts.

As far as I know there are no experimental data concerning the thermodynamic properties of O^{2-} in solution and also the gas-phase properties are scanty. The calculations of Moretti and Ottonello (2022) at B3LYP/6–31+G(d,p) theory level assign to O^{2-}_g a strong electron affinity resulting in a quite unstable ion ($G^0{}_f$ = 778.6 *kJ/mol*). However, when in solution O^{2-}_{sln} is stabilized by a strong electrostatic interaction with the solvent molecules. Dispersive and repulsive interactions counterbalance each other and the net result is that the standard potential of the redox couple (16.4.5) is little affected by the solvent composition (Fig. 16.10).

Because in computing the standard redox potential one encounters the difficulty of assigning the correct thermodynamic magnitudes to electrons in solution, Moretti and Ottonello (2022) circumvented the problem by adopting the procedure of Larson et al. (1968): instead of writing the redox reaction as usual (Eq. 16.4.5), the redox balance was obtained as

$$\frac{1}{2}O_{2,g} + H_{2,g} \Leftrightarrow O^{2-}_{sln} + 2H^+_{sln} \tag{16.4.7}$$

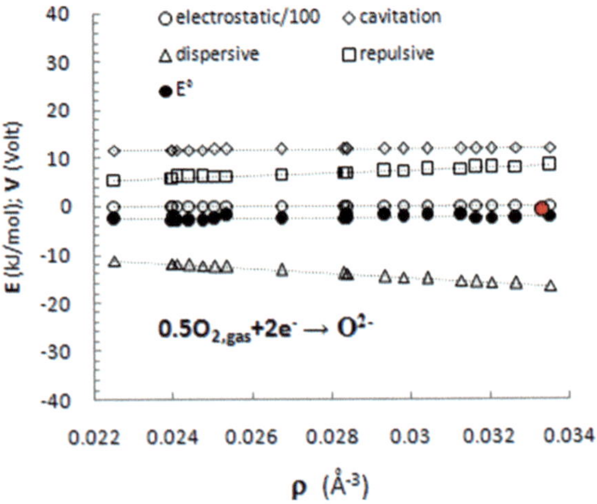

Fig. 16.10 The normal oxygen electrode reaction in various silicate melts, compared to water (red dot). Modified from Moretti and Ottonello (2022)

The identity of the two formulation can be easily understood when subtracting (16.4.5) from (16.4.7). In fact this brings to

$$H_{2,g} \Leftrightarrow 2H^{+}_{sln} + 2e^{-} \tag{16.4.8}$$

which is the standard hydrogen redox potential, implying a (conventional) zero Gibbs free energy difference. The standard redox potential computed for fictive silicate melt solvents is not dissimilar from what computed for water and is virtually unaffected by composition. Table 16.14 lists the properties of O^{2-}_{sln} in various substances after Moretti and Ottonello (2022). These were obtained by application of the SPT-PCM theory. Calculations were carried out at the DFT/B3LYP theory level with a 6–31+G(d,p) basis set. The adopted solute radius for O^{2-} was 1.26 Å. The adopted scaling factor to obtain conventional magnitudes from ab initio absolute values was $H^0_{f,abs,i} - H^0_{f,conv,i} = 386.2 \times Z_i$ (*kJ/mol*) for the enthalpy of formation from the elements of a ionic solute of charge Z_i and $S^0_{abs,i,i} - S^0_{conv,i} = -152.9 \times Z_i[J/(mol \times K)]$ for the entropy (Vetuschi Zuccolini et al., 2011). The entropy of formation from the elements $S^0_{f,conv,i}$ was based on Eq. 16.4.7 and the Gibbs free energy of formation from the elements was obtained by applying $G^0_{f,conv,i} = H^0_{f,conv,i} - T \times S^0_{f,conv,i}$. The redox potential arises simply from application of the Faraday law:

$$E^{\circ} = -G^0_f/nF \tag{16.4.9}$$

where F is the Faraday constant and n the number of flowing electrons.

16.5 O_3

Ozone (trioxidane) forms from O_2 under the action of UV radiation and electrical discharges within the Earth's atmosphere. It is concentrated in a tiny belt of roughly 10 *km* (from ~ 20 *km* to ~30 *km*). The cycle of ozone in the atmosphere was already described in the thirties of the past century by Chapman (1930) as:

$$O_2 + h\nu \rightarrow O + O \tag{16.5.1.1}$$

$$O + O + M \rightarrow O_2 + M \tag{16.5.1.2}$$

$$O + O_2 + M \rightarrow O_3 + M \tag{16.5.1.3}$$

$$O_3 + h\nu \rightarrow O + O_2 \tag{16.5.1.4}$$

Table 16.14 Solution energy terms and thermodynamic properties of O^{2-} in various media at T = 25 °*C*, P = 1 *bar*. The interaction Gibbs free energy ΔG_i is the summation of electrostatic + repulsive + dispersive terms computed at B3LYP/6–31+G(d,p) theory level. Data are in *kJ/mol* (G, H), *J/(mol × K)* (S), *J/bar* (V) and *V* (E°). From Moretti and Ottonello (2022)

Solvent	ΔG_{cav}	H_{cav}	S_{cav}	$\Delta G_i \cong H_i$	ΔG_s	ΔH_s	ΔS_s	$G^0{}_f$	$H^0{}_f$	S^0	V	E^0
Andesite(1)	11.650	0.347	−37.913	−1115.078	−1086.339	−1117.175	−103.425	451.463	449.657	227.196	−0.347	−2.340
Leuc-bas(1)	11.701	0.583	−37.289	−1167.629	−1138.717	−1169.466	−103.130	399.085	397.366	227.491	−0.345	−2.068
Tholeiite(1)	11.694	0.493	−37.568	−1069.305	−1040.415	−1071.240	−103.389	497.387	495.592	227.232	−0.367	−2.578
AOB(1)	11.816	0.616	−37.564	−1185.202	−1155.890	−1187.002	−104.349	381.912	379.830	226.272	−0.380	−1.979
Ugandite(1)	11.864	0.753	−37.267	−1097.045	−1067.560	−1098.693	−104.420	470.242	468.139	226.201	−0.383	−2.437
Entatite(2)	11.982	0.436	−38.723	−1153.905	−1123.999	−1155.903	−107.007	413.803	410.929	223.614	−0.437	−2.144
Mean comp(3)	11.817	0.559	−37.758	−1089.806	−1060.490	−1091.669	−104.574	477.312	475.163	226.047	−0.381	−2.473
BH-257(4)	11.854	0.552	−37.907	−1167.043	−1137.594	−1168.913	−105.044	400.208	397.919	225.577	−0.396	−2.074
BH-266(4)	11.888	0.567	−37.972	−1175.327	−1145.758	−1177.181	−105.394	392.044	389.651	225.227	−0.405	−2.032
NS1-G205(5)	11.685	0.659	−36.981	−985.416	−956.558	−987.169	−102.669	581.244	579.663	227.952	−0.303	−3.012
NS2-G205(5)	11.671	0.566	−37.247	−991.943	−963.133	−993.799	−102.852	574.669	573.033	227.769	−0.306	−2.978
NS3-G205(5)	11.653	0.441	−37.603	−1000.436	−971.689	−1002.429	−103.103	566.113	564.403	227.518	−0.311	−2.934
CMS1-G212(5)	11.916	0.547	−38.130	−1206.958	−1177.291	−1208.834	−105.795	360.511	357.998	224.826	−0.411	−1.868
CMS2-G212(5)	11.927	0.504	−38.314	−1085.330	−1055.619	−1087.252	−106.097	482.183	479.580	224.524	−0.417	−2.499
CMS3-G212(5)	11.939	0.458	−38.506	−1086.543	−1056.791	−1088.516	−106.405	481.011	478.316	224.216	−0.423	−2.493
NCS1-G205(5)	11.658	0.508	−37.395	−995.457	−966.692	−997.376	−102.915	571.110	569.456	227.706	−0.308	−2.960
NCS2-G205(5)	11.762	0.599	−37.443	−1023.657	−994.532	−1025.477	−103.788	543.270	541.355	226.833	−0.346	−2.815
NCS3-G211(5)	11.821	0.599	−37.639	−1047.213	−1017.881	−1049.032	−104.479	519.921	517.801	226.142	−0.371	−2.694
MS1-G156(5)	11.957	0.416	−38.709	−1024.745	−994.927	−1026.765	−106.785	542.875	540.067	223.836	−0.431	−2.813
Molten silica(6)	11.588	0.008	−38.840	−1024.745	−996.218	−1027.216	−103.967	541.584	539.616	226.654	−0.326	−2.807
Water	11.977	1.737	−34.344	−1434.192	−1404.306	−1434.752	−102.117	133.496	132.080	228.504	0.678	−0.692

[1] Lux G. (1987); [2] Kirsten T. (1968); [3] Janbon et al. (1986); [4] Hiyagon H. and Ozima M. (1986); [5] Shibata et al. (1998); [6] Ottonello and Richet (2014).

$$O_3 + O \rightarrow 2O_2 \quad (16.5.1.5)$$

Oxygen molecules are dissociated into oxygen atoms by photolysis with a slow kinetic due to partial recombination (Eqs. 16.5.1.2). Ozone is produced by reaction of O_2 with atomic oxygen through collisional processes involving a neutral molecule M (16.5.1.3). This process is kinetically faster than (16.5.1.2) but the produced ozone interconverts to O_2 under the action of UV rays (16.5.1.4). In the interconversion, the UV radiation transforms into thermal energy, heating the stratosphere.

16.5.1 Thermophysical and Thermodynamic Properties

Ozone ground state is $^1A_1$. The structure is quite similar to that of H_2O, with C_{2V} symmetry (but see also later on). According to Tanaka et al. (1970) the interatomic distance is 1.272 *Å* (1.278 for Herzberg, 1966), and the bent angle is 116.78°. The central oxygen is sp^2 hybridized with one lone pair. The experimental dipole moment is 0.53 *Debye*. The molecule, which is weakly paramagnetic, can be represented as a resonance Lewis structure with bond order 1.5 (Fig. 16.11).

O_3 is an asymmetric top (rotational symmetry number 2). The rotational constants of O_3 are A = 3.55348 cm^{-1}, B = 0.44525 cm^{-1}, C = 0.39479 cm^{-1} (Herzberg, 1966) and the calculated ones at B3LYP/6.31G(d,p) level are A = 3.720 cm^{-1}, B = 0.449 cm^{-1}, C = 0.401 cm^{-1}. The total electric dipole moment is 0.530 *Debye* and the components of the quadrupole moment are xx = 9.200 $D \times Å$, yy = −17.100 $D \times Å$, zz = 7.900 $D \times Å$ (NIST, 2000). The electrical dipole polarizability is $\alpha = 3.079\ Å^3$ (Gussoni et al., 1998) and the calculated one at B3LYP/6-31G(d,p) level is $\alpha = 0.372\ Å^3$.

The experimental harmonic frequencies listed by NIST are $\nu_1 = A_1 = 716\ cm^{-1}$; $\nu_2 = A_1 = 1085\ cm^{-1} \nu_3 = A_1 = 1135\ cm^{-1}$.

The ionization energy is 12.530 *eV* and the electron affinity 2.103 *eV*. The dissociation energy is 0.227042 *Hartree* (NIST, 2000). At standard state (pure gas, P = 10^5 *Pa*, 298.15 *K*) O_3 has an entropy of 238.932 *J/(mol × K)* and an isobaric heat capacity of 39.238 *J/(mol × K)*. The enthalpy of formation from the elements is 142.647 *kJ/mol*. The Shomate equation for the isobaric heat capacity in the range $298.15 \leq T(K) \leq 1200$ is $C_P = 31.32234 - 79.86001 \times 10^{-3} \times T - 66.02603 \times$

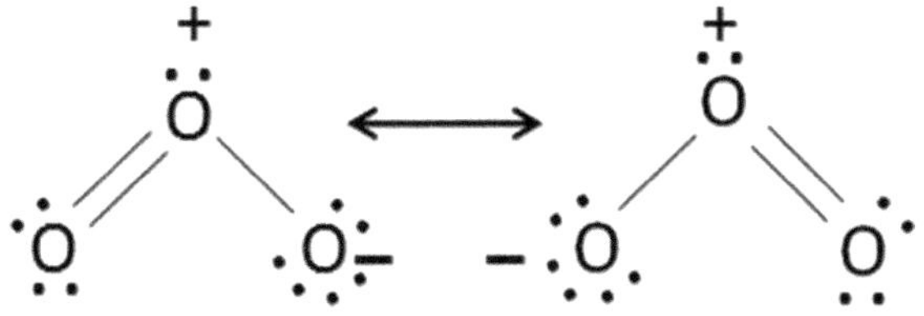

Fig. 16.11 Resonating Lewis structures of the O_3 molecule

$10^{-6} \times T^2 + 19.58363 \times 10^{-9} \times T^3$ $-7.9251 \times 10^4 \times T^{-2}$ (Chase, 1998). The melting point is at 80.7 *K*. Its boiling point is at 161.3 *K*. The critical temperature is $T_c = 261.15$ *K* (NIST, 2020).

16.5.2 Ground State Potential Energy, Excited States and Hartley Continuum

The electronic structure of ozone in the ground state and the first excited levels has been extensively investigated (Hay et al., 1982; and references therein; Lohr and Helman, 1987; Yamashita et al., 1992; Müller et al., 1998; Tyuterev et al., 1999; Siebert et al., 2002). The Hartley band of UV absorption (220 to 300 *nm*) arises from the $^1B_2 \leftarrow X^1A_1$ electronic transition. The ground state (X^1A) of O_3 dissociates to $O_2(^3\Sigma_g-)$ + $O(^3P)$. The $^1B_2$ state of O_3 correlates with the O_2 ($^1\Delta_g$) + O (^1D) and is dissociative (Hay et al., 1982). The 2^1A_1 state lies below the $^1B_2$ state and is also dissociative, leading to $O_2(^1\Delta_g)$ + O(1D). The absorption cross section of the 2^1A_1 dissociation is however several orders of magnitude smaller than the $^1B_2$ state, hence it is expected to play a minor role in the photochemical process. Nevertheless, the low energy part of the Hartley continuum (Huggins band) has been attributed by Brand et al. (1978) to the 2^1A_1 state (and to portions of the $^1B_2$ surface by Simons et al., 1973). The dominant aspect of the ozone ultraviolet spectrum is a continuum having its distribution centered near 39,200 cm^{-1}. The Hartley bands near the center of the spectrum appear as a relatively minor perturbation of the underlying continuum. The Huggins band structure in the long wavelength tail (28,000–32,000 cm^{-1}) is more pronounced but much weaker in an absolute sense (Simons et al., 1973).

The spectral analysis of Simons et al. (1973) is extremely accurate. The real absorption coefficients at any T were written as a sum of a band component $k^b_{\nu T}$ plus continuum components arising from each (appreciably populated) vibrational level of the ground state:

$$k_{\nu T} = k^b_{\nu T} + \sum k^{ijk}_{\nu T} \quad (16.5.2)$$

The fraction of ozone in a particular vibrational state at temperature T is:

$$P_{ijk} = \frac{\exp\left(-hc\nu_{ijk}/K_B T\right)}{Q_{vib}(T)} \quad (16.5.3)$$

In Table 16.15 are listed the vibrational partition function and the population of the lower electronic state at three distinct temperatures adopted by the Authors in their deconvolution investigation.

The small fluctuations in the spectrum at low T were replaced by the "*continuum component*": an empirical function averaging the real fluctuations within ten distinct

Table 16.15 Distribution of vibrational populations of O_3 (Simons et al., 1973)

T	Q_{vib}	P_{000}	P_{010}	P_{001}	P_{100}
195	1.0066	0.993	0.0058	0.0005	0.0003
300	1.0477	0.954	0.0327	0.0063	0.0047
333	1.0719	0.933	0.0451	0.0104	0.0080

v ranging from v = 32,000 cm^{-1} to 48,000 cm^{-1}. These are ten exponential functions of the type:

$$k_\nu^0 = 17.7 \exp\left[1.311 \times 10^{-3}(\nu - 32000)\right] \qquad (16.5.4)$$

valid in the range 32,000 ≤ v ≤ 33,000.

The spectrum at high T was divided into two components $k_{\nu T}^{000}$ and $k_{\nu T}^{\oplus}$, which is the component due to all vibrationally excited ozone. In Fig. 16.12 we see the results obtained by Simons et al. (1973) by subtracting $k_{\nu T}^{000} \equiv P^{000} k_\nu^0$ from the real spectrum at T = 300 *K*. The dotted curve in figure is a component $\left(1 - P^{000}\right) k_\nu^0$, the continuum component normalized to the expected magnitude of $k_{\nu T}^{\oplus}$. The shift between the two continuum spectra is approximately given by the weighted average vibrational excitation $\langle \Delta G \rangle$ (Simons et al., 1973):

$$\langle \Delta G \rangle = \sum \frac{\left[\nu_{ijk} \exp\left(-hc\nu_{ijk}/kT\right)\right]}{\left(Q_{vib} - 1\right)} \qquad (16.5.5)$$

In Fig. 16.13 are shown the Potential Energy Surfaces (PES) for the four $^1A'$ states of ozone involved in photodissociation by UV radiation computed by Hay

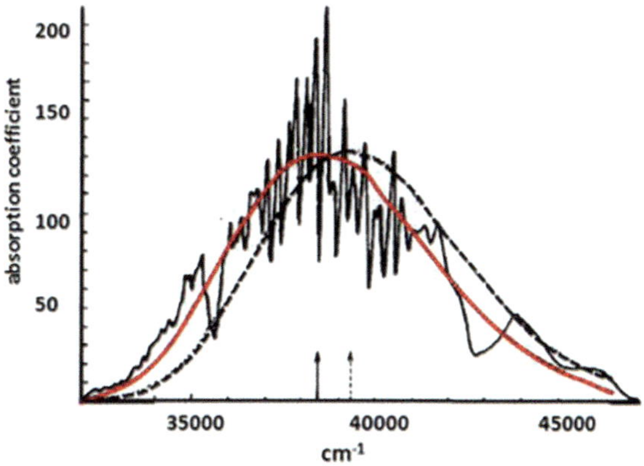

Fig. 16.12 Ozone UV spectrum at T = 300 *K* with the $P^{000} k_\nu^0$ continuum component subtracted. The smooth red solid curve is the continuum component arising from the sum of excited vibrational levels of the electronic ground state. The smooth dotted curve is the continuum component normalized to the expected magnitude of $k_{\nu T}^{\oplus}$. Modified from Simons et al. (1973)

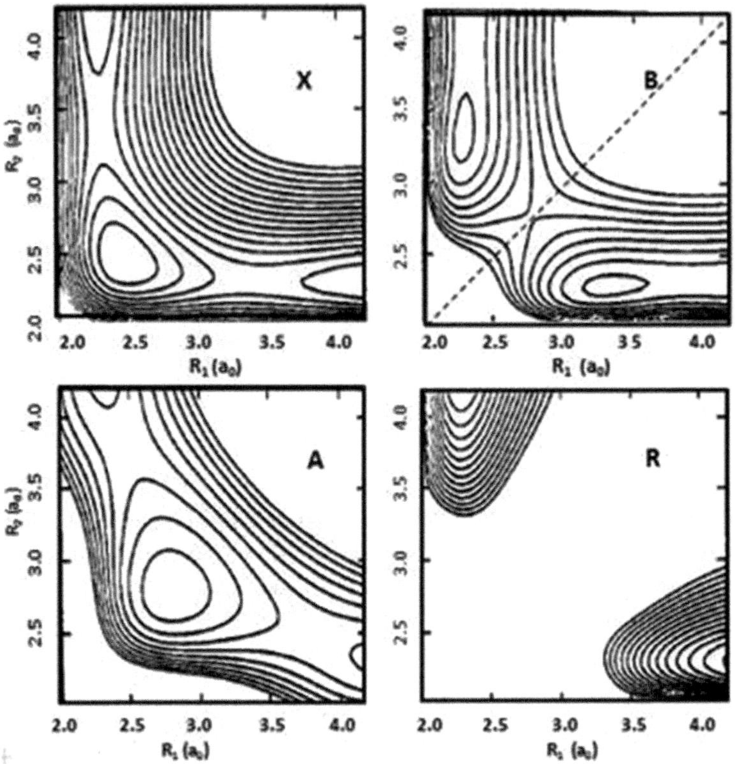

Fig. 16.13 Contour maps of the X, A, B, and R potential energy surfaces with the bond angle fixed at 116°. R_1, and R_2, are the O–O bond lengths. The dashed line on the B PES has been added to better identify the saddle structure described by Hay et al. (1982). Modified from Hay et al. (1982)

et al. (1982) by a generalized valence bond (GVB-PP) multi-configuration SCF technique (Wilson and Hopper, 1981) with a $^3s^2p^1d$ contracted Gaussian basis set.

Many investigations followed through the years but the essence of the message remains unchanged. The four states are designated as X,A, B, and R and are shown as a function of the interatomic distances R_1 and R_2 for a fixed bond angle $\theta = 116°$. The ground-state (X) surface shows the characteristic well corresponding to the stable O_3 species at $R_1 = R_2 = 2.42\ a_0$. The B surface, corresponding to the $^1B_2$ state of the Hartley band in the Franck–Condon region, is generally characterized by a barrier along the $R_1 = R_2$ diagonal relative to its dissociation limit $O_2(^1\Delta_g) + O(^1D)$. Excursions perpendicular to the diagonal are monotonically downhill toward dissociation. The minimum barrier height is found to be 1.14 *eV* relative to $O_2(^1\Delta_g) + O(^1D)$, at a geometry of $R_1 = R_2 = 2.75\ a_0$ and $\theta = 111°$. At this point the $^1B_2$ state is 5.01 *eV* above the calculated ground-state minimum (Hay et al., 1982). Varying one bond length (R_2) and fixing the other geometrical parameters at $R_1 = 2.5\ a_0$ and $\theta = 110°$ the calculated potential energy curves for the $O_3 \rightarrow O_2 + O$ reaction converge at $R_2 \approx 3.3\ a_0$ for the A, B and R surfaces (see to this purpose Fig. 1 in Hay et al., 1982).

Besides the "*open structure*" visible in Fig. 16.13, there is also another minimum, corresponding to the D_{3h} symmetry: the "*ring minimum*" with the three oxygen atoms forming an equilateral triangle. The ring minimum lies significantly above the open minima. The characteristics of the various minima, after the compilation of Siebert et al. (2002) are reported in Table 16.16.

Electronic structure calculations performed by Siebert et al. (2002) at the multi-reference configuration interaction level, with complete active space SCF reference functions and correlation-consistent polarized quadruple zeta atomic basis functions, provided a global PES for the ground electronic state of ozone, covering the three identical C_{2v} open minima, the D_{3h} ring minimum and the $O(^3P)-O_2(^3\Sigma_g^-)$ dissociation threshold. The shape of the PES along the dissociation path is very important for the reaction dynamics. In Fig. 16.14 it is shown the potential energy along the minimum energy dissociation path for removal of an end atom. A limited barrier (0.006 *eV* above the dissociation limit) is observed at intermediate separation ($R_1 = 3.82\ a_0$, $R_2 = 2.28\ a_0$, $\theta = 115.3°$). The inset in figure shows the minimum energy path for isomerization from the ring to the open minimum. The energy is normalized in such a way to return E = 0 for the $O + O_2$ asymptote.

Table 16.16 Characteristic features of the O_3 potential energy surface (From Siebert et al., 2002)

	R_1	R_2	a	Energy	D_0
Open minimum	2.4094 [a]	2.4094	116.79	1.027 [b]	0.944
Expt	2.4052 [c]	2.4052	116.75	1.132 [b,d]	1.052 [e]
Theory (Xie et al., 2000)	2.4081	2.4081	116.86	1.049	
Theory (Muller et al., 1998)	2.4026	2.4026	116.9	1.076	
Ring minimum	2.7249	2.7249	60.00	1.34 [f]	
Theory (Muller et al., 1998)	2.717	2.717	60.00	1.27 [f]	
Theory (Lee, 1990)	2.715	2.715	60.00	1.26 [f]	
TS open/ring	2.68	2.68	83.1	2.37 [f]	
Diss. barrier	3.82	2.28	115.3	0.006 [g]	
Van der Waals minimum	4.81	2.28	116.4	0.020 [h]	

[a] Distances in a_0, angles in degrees, and energies in *eV*
[b] Dissociation energy D_e.
[c] Equilibrium bond distances and bond angle from Tyuterev et al. (1999.2000).
[d] Cited in Xie et al. (2000).
[e] Gole and Zare (1972).
[f] Energy above open minimum.
[g] Barrier height, with respect to the O_1O_2 asymptote.
[h] Depth of van der Waals minimum.

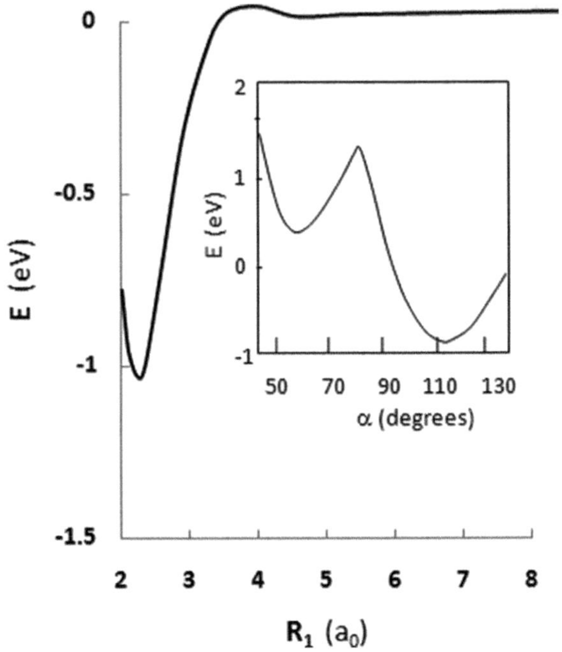

Fig. 16.14 Potential energy along the minimum energy path as one of the end atoms of O_3 is removed. The inset shows the minimum energy path for isomerization from the open to the ring minimum. Modified from Siebert et al. (2002)

16.5.3 The Ozone Isotope Effect

The isotopic composition of ozone in the ozonosphere has been measured directly in the stratosphere by mass spectrometry (Mausberger, 1981) and it has been found a maximum enhancement of 40% over the statistical expectation for the $^{50}O_3$ isotopologue at 32 *km*. The relative abundance of $^{50}O_3$ decreases toward higher and lower altitudes, and attains the standard value below 24 *km*. The relative abundances of ^{16}O, ^{17}O and ^{18}O in ozone apparently do not strictly obey a simple mass dependent fractionation of the type Δm/m and the heavy isotopologues ^{49}O, ^{50}O are above the statistical expectation (Mauersberger 1981, 1987). The origin of this anomaly are still under debate (Mauersberger et al., 2003; Tuzson, 2005). It has been suggested by Heidenreich and Thiemens (1986) that the anomalous enrichment of heavy isotopologues of O_3 is due to a doubling of the lifetime of the collision complex for the asymmetric isomers, associated to a different number of rotational levels (i.e. a factor of 2 difference in the number of rotational levels for the symmetric and asymmetric isomers). According to Yang and Epstein (1987a,b), the anomaly is rather due to the isotopomeric patterns of O_2 generated by pre-dissociation of vibrationally hot ozone, with the pre-dissociation rate depending on both mass and symmetry configuration of O_3.

Bates (1987) severely criticized both models based on ozone symmetry, and suggested that the anomaly is rather linked to difference in the lifetimes of the collision complex due to lack of randomization of the activated complex before

Table 16.17 Calculated and experimental ozone geometries and harmonic vibrational Frequencies

	$^1A_1$		Exper.	$^1B_2$
	(1)	**(2)**	**(3)**	**(1)**
$r(\AA)$	1.275	1.264	1.272	1.463
$\alpha(°)$	116.9	117.9	116.8	111.05
$v_1(cm^{-1})$	1152	1217	1135	822.8
$v_2(cm^{-1})$	717	706	716	341.3
$v_3(cm^{-1})$	1057	1192	1089	1282.6

(1) Lee and Scuseria (1990) CCSD(T)/ [4 s 3p 2d 1f]ANO; (2) this work. DFT frequencies scaled by factor 0.9612; (3) Barbe et al. (1974).

dissociation. According to Valentini (1987) the anomaly is generated in the non-adiabatic transition from the Π or Δ electronic states of O_2 to Σ. According to Morton et al. (1990) the anomaly is due to the $O(^3P) + O_2(^3Z_g)$ recombination reaction.

Calculated and experimental ozone geometries and harmonic vibrational frequencies of the ^{48}O isotopologue are listed in Table 16.17. In this table are also shown the results of ab initio calculation carried out at G3/B3LYP theory level. G3/B3LYP is a variation of G3, promoted by Baboul et al. (1999)., with the geometries and zero-point energies obtained from DFT B3LYP with 6-31G* basis set expansions, instead of geometries from MP2 second-order perturbation theory, and zero-point energies from HF theory. The ground state energy of the DFT procedure returns for $^{48}O_3$ –225.399069 *Hartree*. Proceeding as done at Sect. 11.1 it may be seen that the error at this theory level is ~ 0.029 *Hartree*, much lower than the error associated to the MP2/cc-PVQZ procedure (cf. Table 11.1). The performance of the various computational procedures has been carefully investigated by Lee and Scuseria (1990) to whom I remand for reference. Apparently, the best agreement with experimental observation in geometry and vibrational properties is obtained adopting the CCSD(T} theory in conjunction with a [4s 3p 2d 1f] expansion of Atomic Natural Orbitals (ANO).

Low T (150–180 *K*) infrared measurements carried out by Barbe et al. (1974) on $^{16}O_3$ and $^{18}O_3$ revealed the existence of thirty-three vibrorotational bands. Deconvolution of these bands allowed the computation of harmonic frequencies and the observation of Darling-Dennison resonance[3] (Darling and Dennison, 1940). The frequencies were shown to obey the rule

$$\frac{^{18}\nu_i}{^{16}\nu_i} = \left(\frac{m_{16}}{m_{18}}\right)^{1/2} \tag{16.5.6}$$

3 A resonance interaction between certain of the overtone bands which arises from the near equality of v_1 and v_3 (observed in the water molecule).

It was thus possible to calculate harmonic frequencies for partially isotopic substituted species as well. The values obtained by Barbe et al. (1974) for species: $^{18}O^{16}O^{18}O$ and $^{16}O^{18}O^{16}O$ for C_{2v} symmetry and $^{18}O^{18}O^{16}O$, $^{18}O^{16}O^{18}O$, $^{16}O^{18}O^{16}O$ and $^{16}O^{16}O^{18}O$ for C_s symmetry are compared in Table 16.18 with the values obtained at DFT/B3LYP level with the basis set 6-31G*.

In comparing the various frequencies it must be noted that, while these are of symmetry $2A_1 + 1B_2$ in the first two isotopologues, the introduction of mass asymmetry implies the transformation of the symmetric stretching and symmetric bending into asymmetric motions of type B_2 in the four remaining isotopologues.

The (vibrationally-dependent) thermodynamic properties of the six investigated isotopologues are listed in Table 16.19. I remand the reader to Sects. 7.10.3, 7.10.4 for the essential details on the adopted procedures.

We may note that, while entropy agrees fairly with the tabulated values, the enthalpy of formation is underestimated by ~ 25 *kJ/mol*. Notwithstanding this bias the relative values among the various isotopologues are certainly accurate, so that we see that the stability of the complex is determined by the presence of the heavy isotope (i.e. −2.009 *kJ/mol* with a triple substitution, −1.387 *kJ/mol* with a double substitution and −0.620 *kJ/mole* with a single substitution).

Anderson et al. (1997) investigated the rate coefficients of the four ozone formation channels:

$$^{16}O + {}^{16}O^{16}O + M \rightarrow {}^{16}O^{16}O^{16}O + M \tag{16.5.7.1}$$

$$^{18}O + {}^{16}O^{16}O + M \rightarrow {}^{18}O^{16}O^{16}O + M \tag{16.5.7.2}$$

$$^{16}O + {}^{18}O^{18}O + M \rightarrow {}^{16}O^{18}O^{18}O + M \tag{16.5.7.3}$$

$$^{18}O + {}^{18}O^{18}O + M \rightarrow {}^{18}O^{18}O^{18}O + M \tag{16.5.7.4}$$

Based on the experiments, reaction (16.5.7.2) had a rate coefficient of 0.92 with respect to the standard reaction (16.5.7.1), while the rate coefficient of reaction (16.5.7.3) was 1.53. The Authors' conclusion was that the mass asymmetry of the molecule does not guarantee a kinetic advantage with respect to mass-symmetric isotopomers.

In Table 16.20 are listed the coefficients of of the Taylor expansion promoted by Polyakov and Kharlashina (1995) to depict the separative effect of each moiety in an investigated couple exchange of the type already seen at Sects. 7.9.1, 7.9.2:

$$f_A = 1000 \ln \beta_A = \sum_{i=1}^{3} a_{i,A} X_i^i \tag{16.5.8}$$

with $X_i = 10^6/T^2$.

As the symmetry factors and relative masses cancel each other out in an isotopic exchange reaction, isotopic fractionation factors are readily obtained by applying:

Table 16.18 Calculated harmonic frequencies for isotopic substitution in ozone

	ν_1	ν_2	ν_3	References
$^{16}O^{16}O^{16}O$	1134.9	716.0	1089.2	[1]
	1216.8	706.6	1192.1	[2]
$^{18}O^{18}O^{18}O$	1070.0	674.8	1026.5	[1]
	1147.1	666.1	1123.7	[2]
$^{18}O^{18}O^{16}O$	1091.9	691.1	1035.4	[1]
	1173.8	682.0	1130.6	[2]
$^{18}O^{16}O^{18}O$	1100.9	682.8	1064.2	[1]
	1173.8	682.0	1130.6	[2]
$^{16}O^{18}O^{16}O$	1105.8	707.1	1052.6	[1]
	1206.9	690.4	1170.0	[2]
$^{16}O^{16}O^{18}O$	1122.6	699.6	1071.6	[1]
	1206.9	690.4	1170.0	[2]

[1] Barbe et al.,1974; [2] this work. DFT frequencies scaled by 0.9612.

Table 16.19 Ab initio vibrational properties of O_3 isotopologues resolved at G3B3LYP level

Magnitude	$^{16}O^{16}O^{16}O$	$^{18}O^{18}O^{18}O$	$^{18}O^{18}O^{16}O$	$^{18}O^{16}O^{18}O$	$^{16}O^{18}O^{16}O$	$^{16}O^{16}O^{18}O$	Dimension
E_{DFT}	−225.406453	−225.406453	−225.406453	−225.406453	−225.40645	−225.40645	*Hartree*
E_{ZPC}	0.007384	0.006961	0.007078	0.007078	0.00727	0.00727	*Hartree*
Therm corr U	0.010342	0.009945	0.010053	0.010053	0.010237	0.010237	*Hartree*
Therm corr H	0.011286	0.010889	0.010998	0.010998	0.011181	0.011181	*Hartree*
Therm corr G	−0.015749	−0.016514	−0.016277	−0.016277	−0.015985	−0.015985	*Hartree*
U	27.15416	26.112344	26.396856	26.396856	26.878016	26.878016	*kJ/mol*
$U_{vibrational}$	19.719192	18.673192	18.957704	18.957704	19.438864	19.438864	*kJ/mol*
C_V	29.595064	30.29216	30.06204	30.06204	29.794264	29.794264	*J/(mol × K)*
$C_{V,vibrational}$	4.640056	5.347152	5.121216	5.121216	4.85344	4.85344	*J/(mol × K)*
S	238.07797	241.31638	240.17415	240.17415	239.22857	239.22857	*J/(mol × K)*
$S_{vibrational}$	1.397456	1.690336	1.585736	306.085	1.489504	1.489504	*J/(mol × K)*
H^0_f	167.312	166.27	166.556	166.556	167.036	167.036	*kJ/mol*
S^0	238.069	241.31	240.183	240.183	239.223	239.223	*J/(mol × K)*
G^0_f	188.08	186.071	186.693	186.693	187.46	187.46	*kJ/mol*

Table 16.20 Polynomial coefficients for calculating the separative effects of O_3 and O_2 isotopologues

Isotopologue	a_1	a_2	a_3
667 and 676	4.480530	−0.155878	0.004262
677	8.961169	−0.311340	0.008962
767	12.676388	−0.467585	0.013012
777	17.159462	−0.620294	0.017140
778 and 787	21.128951	−0.752232	0.020658
788	25.098564	−0.883841	0.024166
878	28.391042	−1.016395	0.027893
668 and 686	8.447823	−0.290671	0.007903
688	16.896075	−0.579851	0.015760
868	23.905787	−0.866547	0.023890
888	32.362457	−1.145758	0.031302
67	6.258357	−0.363959	0.012632
77	12.528039	−0.722104	0.024954
68	11.806093	−0.681199	0.023552
78	18.085722	−1.034025	0.035593
88	23.652071	−1.341074	0.045987

$$1000 \ln \alpha = \sum_{i=1}^{3} \left(a_{i,A} - a_{i,B}\right) \cdot X_i^i \tag{16.5.9}$$

The separative effects reproduced by the coefficients in Table 16.19 were generated by ab initio vibrational frequencies scaled by the factor 0.9612 (Table 16.20).

Based on the above listed coefficients one may easily compute, at all T of interest, the fractionation factors for the 30 possible formation channels, based on the ab initio vibrational properties of O_2 and O_3 (with O_2 calculated at the same theory level of O_3) (Table 16.21).

Mauersberger et al. (1993) produced various isotopologues from mass 48 to 54 from two isotopically enriched O_2 gas mixtures at P = 0.09 bar and room temperature, and observed a sort of Gaussian shaped distribution centered on mass 51 when arranging the isotopic fractionation of the ozone isotopomers against their molecular mass. When compared with the values of Table 16.21 the agreement is poor, possibly due to the fact that kinetic effects are quite sensible at room T (Fig. 16.15).

It must be also stressed that Mauersberger et al. (2003) define the percent enrichment of a given mass M as:

Table 16.21 Separative effects and isotopic fractionation for all the possible formation channels of ozone at T = 298.15 *K*. Values in bold highlight the maximum computed fractionation for the various O_3 isotopologues

N	Isotopologue	Exchange reaction	1000lnβ(O_3)	1000lnβ (O_2)	1000lnα	α	Δ×100
1	$^{16}O^{16}O^{16}O$	$^{16}O+^{32}O_2+^{48}O_3 \leftrightarrow ^{16}O+^{32}O_2+^{48}O_3$	0	0	0	**1**	0
2	$^{16}O^{16}O^{17}O$	$^{16}O+^{33}O_2+^{48}O_3 \leftrightarrow ^{16}O+^{32}O_2+^{49}O_3$	36.74521	42.32659	-5.58138	**0.99443**	-0.55658
3	$^{16}O^{16}O^{17}O$	$^{17}O+^{32}O_2+^{48}O_3 \leftrightarrow ^{16}O+^{32}O_2+^{49}O_3$	36.74521	0	36.74521	**1.03743**	3.74287
4	$^{16}O^{17}O^{16}O$	$^{16}O+^{33}O_2+^{48}O_3 \leftrightarrow ^{16}O+^{32}O_2+^{49}O_3$	36.74521	42.32659	-5.58138	0.99443	-0.55658
5	$^{16}O^{17}O^{16}O$	$^{17}O+^{32}O_2+^{48}O_3 \leftrightarrow ^{16}O+^{32}O_2+^{49}O_3$	36.74521	0	36.74521	**1.03743**	3.74287
6	$^{16}O^{17}O^{17}O$	$^{16}O+^{34}O_2+^{48}O_3 \leftrightarrow ^{16}O+^{32}O_2+^{50}O_3$	73.52575	85.07566	-11.54992	0.98852	-1.14835
7	$^{16}O^{17}O^{17}O$	$^{17}O+^{33}O_2+^{48}O_3 \leftrightarrow ^{16}O+^{32}O_2+^{50}O_3$	73.52575	42.32659	31.19916	**1.03169**	3.1691
8	$^{17}O^{16}O^{17}O$	$^{16}O+^{34}O_2+^{48}O_3 \leftrightarrow ^{16}O+^{32}O_2+^{50}O_3$	101.95668	85.07566	16.88101	1.01702	1.70243
9	$^{17}O^{16}O^{17}O$	$^{17}O+^{33}O_2+^{48}O_3 \leftrightarrow ^{16}O+^{32}O_2+^{50}O_3$	101.95668	42.32659	59.63009	**1.06144**	6.14438
10	$^{16}O^{16}O^{18}O$	$^{16}O+^{34}O_2+^{48}O_3 \leftrightarrow ^{16}O+^{32}O_2+^{50}O_3$	69.49919	80.13514	-10.63594	0.98942	-1.05796
11	$^{16}O^{16}O^{18}O$	$^{18}O+^{32}O_2+^{48}O_3 \leftrightarrow ^{16}O+^{32}O_2+^{50}O_3$	69.49919	0	69.49919	**1.07197**	7.19712
12	$^{16}O^{18}O^{16}O$	$^{16}O+^{34}O_2+^{48}O_3 \leftrightarrow ^{16}O+^{32}O_2+^{50}O_3$	69.49919	80.13514	-10.63594	0.98942	-1.05796
13	$^{16}O^{18}O^{16}O$	$^{18}O+^{32}O_2+^{48}O_3 \leftrightarrow ^{16}O+^{32}O_2+^{50}O_3$	69.49919	0	69.49919	**1.07197**	7.19712
14	$^{16}O^{17}O^{18}O$	$^{16}O+^{35}O_2+^{48}O_3 \leftrightarrow ^{16}O+^{32}O_2+^{51}O_3$	106.31189	123.269404	-16.95751	0.98319	-1.68145
15	$^{16}O^{17}O^{18}O$	$^{17}O+^{34}O_2+^{48}O_3 \leftrightarrow ^{16}O+^{32}O_2+^{51}O_3$	106.31189	80.13514	26.17676	1.02652	2.65224
16	$^{16}O^{17}O^{18}O$	$^{18}O+^{33}O_2+^{48}O_3 \leftrightarrow ^{16}O+^{32}O_2+^{51}O_3$	106.31189	42.32659	63.9853	**1.06608**	6.60767
17	$^{17}O^{17}O^{17}O$	$^{17}O+^{34}O_2+^{48}O_3 \leftrightarrow ^{16}O+^{32}O_2+^{51}O_3$	138.93655	85.07566	53.86089	**1.05534**	5.53378
18	$^{16}O^{18}O^{18}O$	$^{16}O+^{36}O_2+^{48}O_3 \leftrightarrow ^{16}O+^{32}O_2+^{52}O_3$	139.12664	161.8141	-22.68745	0.97757	-2.2432
19	$^{16}O^{18}O^{18}O$	$^{18}O+^{34}O_2+^{48}O_3 \leftrightarrow ^{16}O+^{32}O_2+^{52}O_3$	139.12664	80.13514	58.99151	**1.06077**	6.07662
20	$^{18}O^{16}O^{18}O$	$^{16}O+^{36}O_2+^{48}O_3 \leftrightarrow ^{16}O+^{32}O_2+^{52}O_3$	193.27489	161.8141	31.4608	1.03196	3.19609
21	$^{18}O^{16}O^{18}O$	$^{18}O+^{34}O_2+^{48}O_3 \leftrightarrow ^{16}O+^{32}O_2+^{52}O_3$	193.27489	80.13514	113.13976	**1.11979**	11.97884
22	$^{17}O^{17}O^{18}O$	$^{17}O+^{35}O_2+^{48}O_3 \leftrightarrow ^{16}O+^{32}O_2+^{52}O_3$	171.902569	123.269404	48.63317	1.04984	4.98352
23	$^{17}O^{17}O^{18}O$	$^{18}O+^{34}O_2+^{48}O_3 \leftrightarrow ^{16}O+^{32}O_2+^{52}O_3$	171.902569	85.07566	86.82691	**1.09071**	9.07079

(continued)

Table 16.21 (continued)

N	Isotopologue	Exchange reaction	1000lnβ(O_3)	1000lnβ (O_2)	1000lnα	α	Δ×100
24	$^{17}O^{18}O^{17}O$	$^{17}O+^{35}O_2+^{48}O_3\leftrightarrow^{16}O+^{32}O_2+^{52}O_3$	171.902569	123.269404	48.63317	1.04984	4.98352
25	$^{17}O^{18}O^{17}O$	$^{18}O+^{34}O_2+^{48}O_3\leftrightarrow^{16}O+^{32}O_2+^{52}O_3$	171.902569	85.07566	86.82691	**1.09071**	9.07079
26	$^{17}O^{18}O^{18}O$	$^{17}O+^{36}O_2+^{48}O_3\leftrightarrow^{16}O+^{32}O_2+^{53}O_3$	204.897415	161.8141	43.08332	1.04402	4.40249
27	$^{17}O^{18}O^{18}O$	$^{18}O+^{35}O_2+^{48}O_3\leftrightarrow^{16}O+^{32}O_2+^{53}O_3$	204.897415	123.269404	81.62801	**1.08505**	8.50521
28	$^{18}O^{17}O^{18}O$	$^{17}O+^{36}O_2+^{48}O_3\leftrightarrow^{16}O+^{32}O_2+^{53}O_3$	230.46709	161.8141	68.653	1.07106	7.10645
29	$^{18}O^{17}O^{18}O$	$^{18}O+^{35}O_2+^{48}O_3\leftrightarrow^{16}O+^{32}O_2+^{53}O_3$	230.46709	123.269404	107.19769	**1.11315**	11.31543
30	$^{18}O^{18}O^{18}O$	$^{18}O+^{36}O_2+^{48}O_3\leftrightarrow^{16}O+^{32}O_2+^{54}O_3$	263.6254	161.8141	101.8113	**1.10717**	10.71745

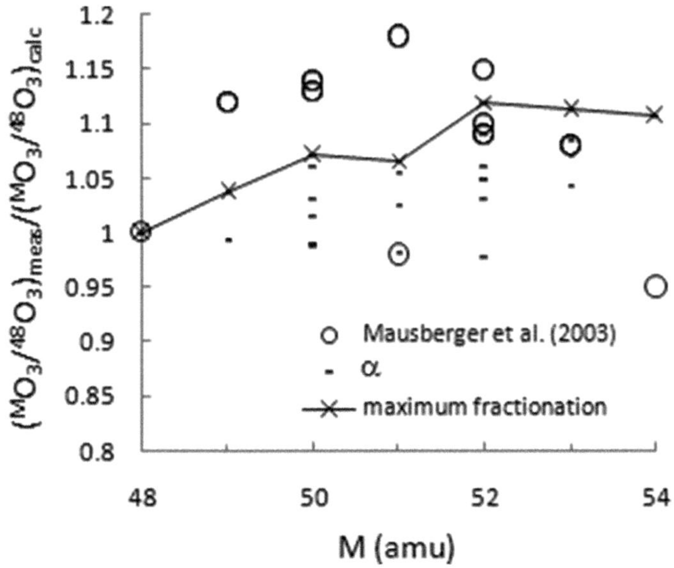

Fig. 16.15 Measured enrichment or depletion of all possible ozone isotopologues with respect to $^{16}O^{16}O^{16}O$ at room temperature, compared with ab initio fractionation factors (α)

$$E(^{M}O_3)\% = \left[\frac{\left(^{M}O_3/^{48}O_3\right)_{\text{measured}}}{\left(^{M}O_3/^{48}O_3\right)_{\text{calculated}}} - 1\right] \times 100 \tag{16.5.10}$$

where the $\left(^{M}O_3/^{48}O_3\right)_{\text{calculated}}$ values are based on the natural proportions of ^{16}O, ^{17}O and ^{18}O (see Sect. 16.4).

The isotopic anomalies in the relative abundances of the various isotopologues depend primarily on their vibrational behavior, but also on the relative time scales for odd oxygen production, destruction, and interconversion. Stemming from the observation of Cicerone and McCrumb (1980) that more $^{34}O_2$ is photodissociated than $^{32}O_2$ in the stratosphere (with a consequent enhancement of the $^{18}O/^{16}O$ density ratio beyond the elemental ratio), Kaye and Strobel (1983) investigated the process:

$$^{34}O_2 + h\nu \rightarrow {}^{16}O + {}^{18}O \tag{16.5.11}$$

$$^{18}O + {}^{32}O_2 + M \rightarrow {}^{50}O_3 + M \tag{16.5.12}$$

A detailed analysis of the vibrational–rotational-traslational and kinetic effects (see their A,B appendixes, respectively for details) lead them to define the concentration ratio R of heavy ozone to normal ozone as:

$$R = \frac{[^{50}O_3]}{[^{48}O_3]} = \frac{2f(1+K_{eq})}{K_{eq}} \quad (16.5.13)$$

$$K_{eq} = 1.94 \times \exp\left(\frac{32}{T}\right) \quad (16.5.14)$$

where K_{eq} is the equilibrium constant of the reaction

$$^{18}O + {}^{32}O_2 \Leftrightarrow {}^{16}O + {}^{34}O_2 \quad (16.5.15)$$

and f is the natural $^{18}O/^{16}O$ ratio (i.e. $f \approx 0.204$).

The predicted R values in the stratosphere and mesosphere according to Eq. 16.5.13 are compared in Table 16.22 with the results of our exchange reaction 10 in Table 16.21:

$$^{16}O^{34}O_2 + {}^{48}O_3 \Leftrightarrow {}^{16}O + {}^{32}O_2 + {}^{50}O_3 \quad (16.5.16)$$

Although the two procedures are quite dissimilar, the results are in good agreement, with a mean bias on R of 1×10^{-4}. This suggests that the kinetic effects have a limited influence on the overall net balance.

Comparing the computational results with the experimental observations of Morton et al. (1990) (Fig. 16.16) we may see that, most likely, Eqs. 10 to 13 in Table 16.21 concur on an equal basis to the formation of the $^{50}O_3$ isotopomers, and Eqs. 2 to 5 to the formation of $^{49}O_3$. Here, being the T range more extended, a possible kinetic effect may be envisaged as well.

Table 16.22 Predicted isotopic ratio R and normalized isotopic fractionation $\alpha \times 3f$ among the $^{50}O_3$ and $^{48}O_3$ in Stratosphere and Mesosphere. (R,f from Kaye and Strobel, 1983)

z (*km*)	T (*K*)	*R*	$\alpha \times 3$	bias	$(R/3f\text{-}1)\times 100$	$(\alpha\text{-}1)\times 100$
15	216.6	0.00589	0.00597	−8.4E-05	−3.78440	−4.46693
20	216.6	0.00589	0.00597	−8.4E-05	−3.78440	−4.46693
25	221.7	0.00589	0.00600	−1.0E-04	−3.68160	−3.77333
30	230.7	0.00591	0.00603	−1.2E-04	−3.51049	−2.86147
35	241.5	0.00592	0.00605	−1.3E-04	−3.32085	−2.14206
40	255.3	0.00593	0.00607	−1.4E-04	−3.10038	−1.59845
45	267.7	0.00594	0.00608	−1.4E-04	−2.92048	−1.32506
50	271.6	0.00594	0.00608	−1.3E-04	−2.86708	−1.26624
55	263.9	0.00594	0.00607	−1.4E-04	−2.97392	−1.39335
60	249.3	0.00592	0.00606	−1.4E-04	−3.19343	−1.79544
65	232.7	0.00591	0.00603	−1.2E-04	−3.47414	−2.70202
70	216.2	0.00589	0.00597	−8.2E-05	−3.79265	−4.52790
75	205	0.00587	0.00590	−2.3E-05	−4.03575	−6.72629
80	195	0.00586	0.00579	6.6E-05	−4.27443	−9.77404

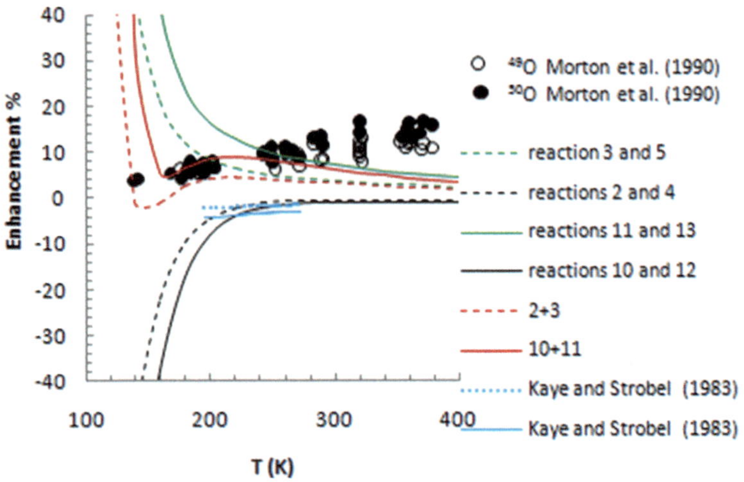

Fig. 16.16 Thermal effect on the $^{49}O_3$ and $^{50}O_3$ enrichment factors. The experimental observations of Morton et al. (1990) and the Kay and Strobel (1983) model are compared with the vibrational isotopological exchanges of Table 16.21. The adopted Kay-Strobel equilibrium constant for isotopomer $^{49}O_3$ (solid blue line) is $K_{eq} = 1.96 \times \exp(16.8/T)$

16.6 N_2

After the first atmosphere, composed essentially of gases of the solar nebula (primarily hydrogen and simple hydrides), the late heavy bombardment by asteroids contributed substantially to the increase of N_2 (plus CH_4,H_2,H_2O, and NH_3; Zahnle et al., 2010; and references therein) which is now the major atmospheric constituent (78.084% by volume).

16.6.1 Thermophysical and Thermodynamic Properties

In N_2 the two N atoms are triply bonded, sharing 6 electrons. The corresponding Lewis structure is shown in the upper part of Fig. 16.17, while in the lower part it is shown the HOMO computed at B3LYP/6-31G* theory level for an isodensity of 0.10 $e/Å^3$. The electronic structure of N_2 is $\sigma 2s^2$, $\sigma^* 2s^2$, $\sigma 2p_x^2$, $\pi 2p_y^2 \approx \pi 2p_z^2$. There are $N_b = 8$ bonding orbitals and $N_a = 2$ antibonding orbitals, hence the bond order is $½(N_b - N_a) = 3$.

N_2 does not possess a permanent electric dipole moment and thus the pure rotation and vibrorotation spectra of this molecule are forbidden. The ground state $^1\Sigma_g^+$ has been derived from the few Raman observations and quadrupole transitions, combined with UV spectra (see Edwards et al., 1993; and references therein). According to Huber and Herzberg (1979) the interatomic distance in N_2 is 1.098 $Å$.

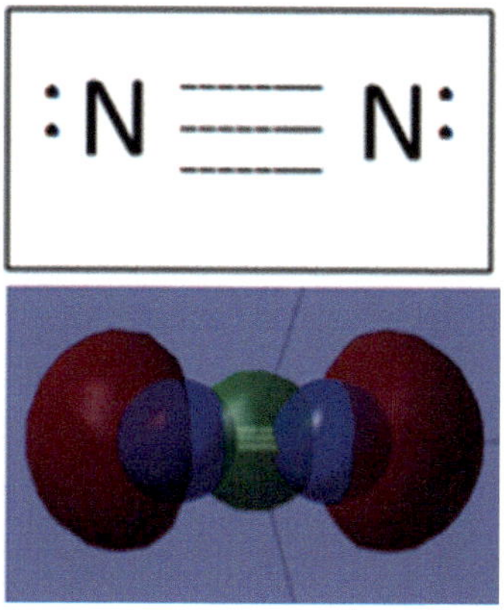

Fig. 16.17 Lewis structure of N_2 (upper part) and B3LYP/6-31G* HOMO at 0.10 *e/Å*^{3} isodensity elongated along the z axis (lower part)

The experimental dipole polarizability is 1.710 *Å*^{3}. The molecule is a prolate symmetric top with rotational symmetry number 2.

The rotational constant of N_2 is B = 1.99824 *cm*$^{-1}$ and the calculated one at B3LYP/6.31G(d,p) level is B = 1.970 *cm*$^{-1}$. The components of the electric dipole moment are zero and those of the quadrupole moment are xx = -0.967 $D \times Å$, yy = 0.697 $D \times Å$, zz = -1.394 $D \times Å$. The electrical dipole polarizability is α = 1.710 *Å*^{3} (Olney et al., 1997) and the calculated one at B3LYP/6-31G(d,p) level is α = 1.265 *Å*^{3}.

The experimental harmonic frequency listed by NIST is ν_1 = 2359 *cm*$^{-1}$. A quite similar frequency (2362 *cm*$^{-1}$) is obtained at simple B3LYP/6-31G* theory level, with the nominal scaling factor (0.9612). The zero point energy at this theory level is 0.005598 *Hartree*. The ionization energy according to NIST (2000) is 15.581 *eV* and the proton affinity is 493.8 *kJ/mol*. The dissociation energy computed by Ochtersky et al. (1996) at G2(MP2) theory level is 936.8 *kJ/mol*, against a reported experimental energy of 941.8 *kJ/mol* (see their Table III). At standard state (pure gas, P = 10^{-5} *Pa*, 298.15 *K*) N_2 has an entropy of 191.61 $J/(mol \times K)$ and an isobaric heat capacity of 29.124 $J/(mol \times K)$ (Chase, 1998). The Shomate equation for the isobaric heat capacity in the range 298.15 ≤ T(*K*) ≤ 500 is $C_P = 28.98641 + 1.853978 \times 10^{-3} \times T \; -9.647459 \times 10^{-6} \times T^2 + 16.63537 \times 10^{-9} \times T^3 + 0.0117 \times 10^4 \times T^{-2}$ (NIST, 2000; based on refitted data of Chase 1998). The melting point is at 63.3 *K*. The boiling point is at 77.4 *K*. The critical temperature is T_c = 126.2 K and the critical pressure is P_c = 34 *bar*. The triple point occurs at T_t = 63.3 *K* and P_t = 0.126 *bar* (NIST, 2020).

16.6.2 The Role of N_2 in Aurorae and Airglows

Because N_2 is the main atmospheric gas, the N_2 molecule plays a significant role in atmospheric phenomena including aurorae and airglows (Chamberlain, 1961;

Table 16.23 Relative brightness of auroral band systems (*kR*) (From Vallance Jones, 1971). Numbers in brackets include allowance for deactivation and cascading. Values of Green et al. (1969) normalized to I (1N N_2^+ = 100)

	Observed	Predicted
N_2 1P	1000 ± 500	575 (850)
N_2 2P	100 to 200	160
N_2 LBH	40 or 125	190 (350)
N_2 VK	45 o 1400	(57)
N_2^+ 1N	120	(120)
N_2^+ M	700 ± 300	410 (640)
O_2^+ 1N	50 ± 25	30
O_2 Atm	200–500	6 (25)
O_2 IR Atm	500–1000	13 (9)

McCormac, 1971). For an extensive critical review and compilation of the observed and predicted spectroscopic data on the molecules N_2 and N_3 and the ions N_2^-, N_2^+, N_2^{2+}. I remand the interested reader to Lofthus and Krupenie (1977). The N_2^+, A–X Meinel bands were first identified in the aurora at wavelengths longer than 9000 *Å*. UV emissions below 3100 *Å* were registered by rocket or satellite observations while IR bands beyond 20,000 *Å* were observed by equipped balloons and rockets. Aircraft observations proved feasible in the region 10,000–20,000 *Å*. Direct observation from ground of comet tails showed the spectrum of the N_2^+ B-X (1-) bands (Lofthus and Krupenie, 1977). R. J. Strutt (Lord Rayleigh) in 1911 described a "*chemically active modification*" of nitrogen, produced by an electrical discharge in flowing nitrogen. The yellow-green afterglow following such discharge is known as the "*Lewis-Rayleigh afterglow*". Other afterglows have also been studied but the spectroscopic features of active nitrogen in terms of simultaneous excitation mechanisms has not yet a comprehensive explanation.

In Table 16.23 are listed the observed N_2 bands and their intensities normalized to IBC3 aurora. The auroral intensity is given in terms of the International Brightness Coefficient (IBC) which is scaled to the intensity of the 557.7 *nm* forbidden oxygen [OI] emission line, on a logarithmic scale from IBC1 ($I_{557.7}$ *nm* = 1 *kR*) to IBC4 ($I_{557.7}$ *nm* = 1000 *kR*) (Vallance Jones 1971). The IBC3 intensity corresponds to an intensity of 100 *kR* for the 557.7 *nm* [OI] line. An IBC3 aurora is exceptionally bright and occurs very rarely. The predicted intensities in Table 16.23 are those calculated by Green et al. (1969). Normalizing the sum of the predicted band intensities by the observed intensities of one or more of the stronger unblended bands are obtained the "*total system intensities*" (Vallance Jones, 1971).

Potential energy curves and other electronic transition dipole moments for the low-lying states of N_2 have been investigated ab initio by Yu and Hogbin (2014) using CASSCF) and an aug-cc-pVqZ basis set. Their calculated cross-sections provide a database for studying the elementary processes and the properties in N_2 plasma, fulfilling the lack of an available database of the cross-sections of electron-impact excitations and radiations. The geometry optimization was carried out by the Authors by keeping the ground state with symmetry of $D_{\infty h}$. Besides the ground state $X^1\Sigma_g^+$, also the potential energy curves of the singlet states $a\Pi_g^1$,

$a'^1\Sigma_u^-$, $w^1\Delta_u$ and the triplet states $A_3\Sigma_u^+$, $B_3\Pi_g$, $C^3\Pi_u$, $W^3\Delta_u$, $B'^3\Sigma_u^-$ were investigated in the range of internuclear separation 0.7 – 2.65 Å, with 10 orbitals and 10 valence electrons in the active space. Bond equilibrium distances R_e, harmonic vibrational frequencies ν_e and dissociation energies D_e of the ground state and excited states were obtained through fitting of a Morse potential of type:

$$V(R) = D_e\{1 - \exp[-a(R - R_e)]\}^2 \tag{16.5.17}$$

with the width of the potential energy curve related to the vibrational frequency ν_e through

$$a = \sqrt{\frac{2\pi^2 c\mu}{D_e h}}\nu_e \tag{16.5.18}$$

and μ as reduced mass.

Equilibrium bond distances, vibrational frequencies and dissociation energies obtained by Yu and Hogbin (2014) for the investigated states are compared in Table 16.24 with experimental observations.

Transition dipole moments in the range R = 0.7–2.5 Å for the triplet–triplet and the singlet–singlet transitions were obtained through the CASSCF procedure by averaging the wavefuntions of two states. The electron-impact vibrational excitation cross sections of the transition bands

$$a'\Pi_g(v=0) \rightarrow a'\Sigma_u^-(v'=0-6) \tag{16.5.19.1}$$

$$a'\Pi_g(v=0) \rightarrow w'\Delta_u(v'=0-6) \tag{16.5.19.2}$$

$$B^3\Pi_g(v=0) \rightarrow W^3\Delta_u(v'=0-6) \tag{16.5.19.3}$$

$$B^3\Pi_g(v=0) \rightarrow B'^3\Sigma_u^-(v'=0-6) \tag{16.5.19.4}$$

where then obtained from the calculated spectroscopic constants and the transition dipole moments by a "*similarity function method*" (Yu and Hogbin, 2014). The

Table 16.24 Bond distance (*Å*), vibrational frequency (cm^{-1}) and dissociation energy (*eV*) of ground and excited states of N_2 after Yu and Hogbin (2014). Values in brackets are experimental observations (Müller et al., 2001)

	R_e	ν_e	D_e
$X^1\Sigma_g^+$	1.092 (1.097)	2358 (2359)	9.82 (9.90)
$a^1\Pi_g$	1.230 (1.220)	1676 (1694)	6.30 (6.08)
$a'^1\Sigma_u^-$	1.277 (1.275)	1572 (1530)	5.81 (6.22)
$w^1\Delta_u$	1.270 (1.269)	1607 (1559)	6.14 (5.73)
$A_3\Sigma_u^+$	1.336 (1.286)	1196 (1461)	3.19 (3.68)
$B_3\Pi_g$	1.225 (1.212)	1586 (1733)	4.63 (4.90)
$W^3\Delta_u$	1.287 (1.279)	1461 (1507)	4.51 (4.87)
$B'^3\Sigma_u^-$	1.284 (1.278)	1485 (1517)	4.74 (5.26)

Table 16.25 Excitation energy (*eV*) of the v,v' bands (From Yu and Hogbin, 2014)

v,v'	a-a'	a-w	B-W	B-B'
(0,0)	0.2764	0.4271	0.0302	0.7035
(0,1)	0.4774	0.6281	0.2060	0.9045
(0,2)	0.6784	0.8040	0.3819	1.0804
(0,3)	0.8543	1.0050	0.5528	1.2563
(0,4)	1.0302	1.1809	0.7186	1.4070
(0,5)	1.2060	1.3819	0.8844	1.5829
(0,6)	1.3819	1.5578	1.0452	1.7337

calculated cross sections from lower energy states with vibrational band v = 0 to higher energy states with vibrational bands v' = 0–6 allowed to define the excitation threshold energy of each transition band (Table 16.25).

16.6.3 Solubility of N_2 in Water

Within the Eh–pH stability range of water (thin solid lines in Fig. 16.18) N_2 occupy a large part of the stability field (thick solid lines in Fig. 16.18 denoting the limits of relative predominance of neighboring species; i.e. "*isoactivity lines*"). It must be noted however that both NO_3^- and NH_4^+ persist metastably in zones where the thermodynamic activity of N_2 would be predominant.

N_2 has received considerably less attention with respect to O_2 concerning the homogeneous solubility in water. With the measurements of Rettich et al. (1983), performed adopting the high precision equilibration procedure of Benson and Krause (see Rettich et al., 1983; and references therein) and the fitting procedure of Battino et al. (1984), a reasonable agreement was found. The retrieved data of Battino et al. (1984), based on the fitting relation (Eq. 16.4.4), limited to the first three coefficients, are plotted in Fig. 16.19, superimposed to the experimental data of Rettich et al. (1983) and to the results of SPT-PCM calculations. All in all, the solubility of N_2 is quite similar to that of O_2. When resolved at DFT/6-31G(d) SCRF-IEFPCM level, with UAHF radii, the UATM assigns to the each nitrogen a radius of 1.590 *Å*. The computed cavity volume is 41.429 $Å^3$. The returned solute/solvent interaction energy is $H_{int} = -17.656$ *kJ/mol*, and arises from a polarized solute – solvent electrostatic interaction of −1.381 kJ/mol, a dispersive energy contribution of −20.878 *kJ/mol* and a repulsive energy of 4.602 *kJ/mol.* To reproduce the experimental observations at room temperature we must adopt a solute radius of 2.145 *Å*. As observed for O_2, also in the case of N_2 the solubility in water at high T is underestimated (Fig. 16.19).

The failure of the SPT procedure in mimicking the observed gas solubility (N_2 and O_2) in water at high T is difficult to justify. It is of interest to recall, to this purpose, that the SPT equation adopted to describe the reversible work necessary to

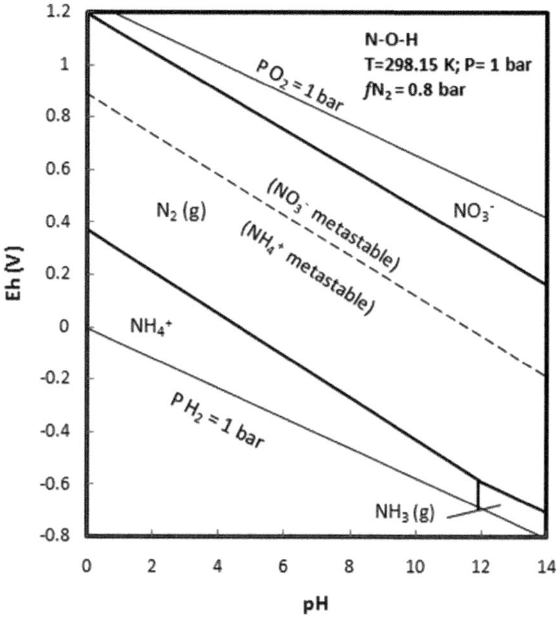

Fig. 16.18 Eh–pH diagram of the N–O-H system at T = 298.15 K and P = 1 bar. Modified from Brookins (1988)

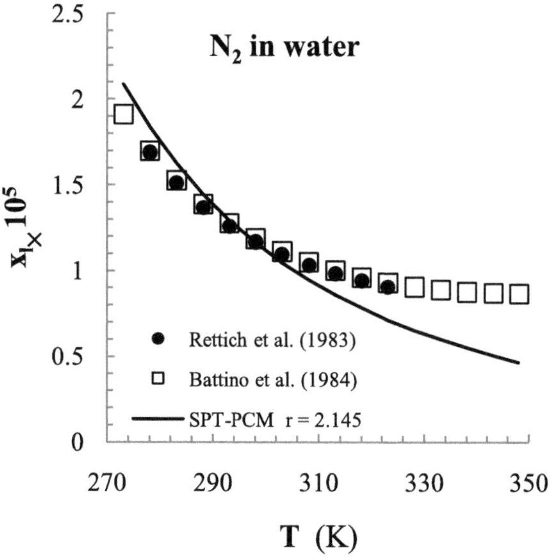

Fig. 16.19 Experimental versus computed SPT-PCM abundances of dissolved nitrogen in water at P = 1 bar and various T

create in the solvent an appropriate cavity (Eq. 16.3.8) was obtained by Pierotti (1976) through an asymptotic expansion of type:

$$W(r, \rho) = K_0 + K_1 r + K_2 r^2 + K_3 r^3 \tag{16.6.1}$$

and this, as noted by Pierotti (1976), is almost the same form as required in classical termodynamics, i.e.

$$W(r) = \frac{4}{3}\pi r^3 P + 4\pi r^2 \gamma \left(1 - \frac{4\delta}{r}\right) \tag{16.6.2}$$

the only difference being the absence of the constant term K_0 in the thermodynamic equation. The term involving pressure (P) in Eq. 16.6.2) is the volume work. The term involving γ (the surface tension) is the surface work. The $4\delta/r$ term corrects the surface tension for the surface curvature. K_0 (which is negligible in macroscopic cavities), has, in SPT, form

$$K_0 \equiv W_0 = kT \ln\left(1 - \frac{\pi \sigma_1^3 \rho}{6}\right) \tag{16.6.3}$$

with σ_1 as solvent diameter and ρ as numeral density.

16.7 NH_3

NH_3 (Ammonia) has been an essential precursor for the terrestrial amino acid synthesis in the pre-biotic age on Earth (Brach, 2007). Still now, NH_3 is a fundamental nitrogen source for life on Earth. An enormous amount of NH_3 is in fact synthesized daily from chemically inert N_2 by Haber–Bosch process and by enzymatic catalysis of nitrogenase (Wagner, 2011). As anticipated, NH_3 was present from the very beginning on our planet's atmosphere under the effect of bombardment by asteroids during the Late Heavy Bombardment age (LHB; 3.9–4.0 Ga; Fegley et al., 1986; Zahnle et al., 2010). It must be stressed however that NH_3 may form as well in alkaline aqueous solutions by deprotonation of NH_4^+ (Fig. 16.18). We will see in some detail later on the LHB production, that has been recently reinvestigated by AIMD procedures.

16.7.1 Thermophysical and Thermodynamic Properties

In NH_3 (Ammonia) all atoms are covalently bonded, with lone pair electrons localized on N. The corresponding Lewis structure is shown in the upper part of

Fig. 16.20 Lewis structure of NH_3 (upper part) and B3LYP/6-31G** HOMO at 0.10 *e/Å*^{3} isodensity, elongated along the z axis (lower part). Van der Waals radii adopted

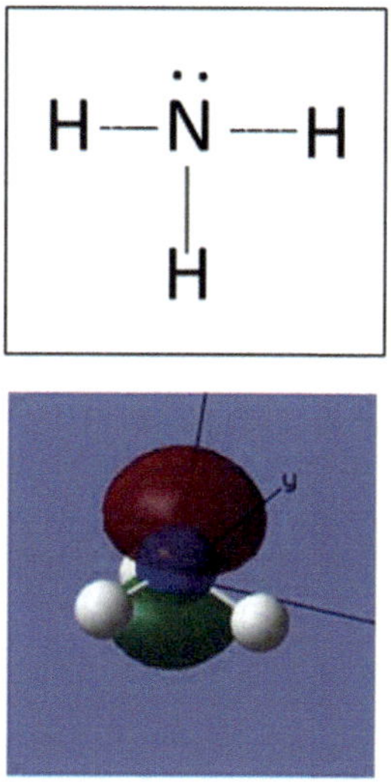

Fig. 16.20, while in the lower part it is shown the HOMO computed at B3LYP/6-31G(d,p) theory level for an isodensity of 0.10 *e/Å*^{3}.

The ground state of ammonia is $^1A_1$ with C_{3V} point group. The N–H interatomic distance is 1.012 *Å* and the H–N–H angle is 106.67° (Herzberg, 1966). The experimental dipole polarizability is 2.103 *Å*^{3} (NIST, 2000). The molecule is an asymmetric top with rotational symmetry number 1. When resolved at B3LYP/6-31G(d,p) theory level, the N–H distance is 1.018 *Å* and the H–N–H angle is 105.7°.

The rotational constants of NH_3 are A = B = 9.44430 cm^{-1}, C = 6.19600 cm^{-1} (Herzberg, 1966) and the calculated one at B3LYP/6.31G(d,p) level are A = B = 9.798 cm^{-1}, C = 6.348 cm^{-1}. The total electric dipole moment is 1.470 *Debye* and those of the components of the quadrupole moment are xx = yy = 1.160 $D \times Å$, yy = -2.320 $D \times Å$. The electrical dipole polarizability is α = 2.103 *Å*^{3} (Olney et al., 1997) and the calculated one at B3LYP/6-31G(d,p) level is α = 1.270 *Å*^{3}.

The lowest B3LYP/6-31G** vibrational frequency is a bending mode of symmetry A_1 (IR and Raman active; 1047 cm^{-1} scaled by the 0.9612 factor) there is then a symmetric stretching mode (Raman active) of same symmetry at 3326 cm^{-1} (scaled), two asymmetric stretching modes of symmetry E (IR and Raman active at 3449 cm^{-1}, scaled) and two asymmetric bending modes of same symmetry (IR and Raman active, at 1628 cm^{-1}, scaled). The zero point energy at this theory level is 0.0034442 *Hartree* (unscaled). The best agreement with the experimental dissociation energy (1157.7 *kJ/mol*; Chase et al., 1985) is obtained at CBS-Q theory level (1156.9 *kJ/mol*, Ochtersky et al., 1996; see Tables 7.32; 10.10 and 10.19). In terms of ionization potential (10.18 *eV*), B3LYP and B98 functionals give almost the same results [i.e. 10.12 and 10.11 eV (Becke, 1993b; Schmider and Becke, 1989;

Table 10.11]. For the proton affinity the agreement is poorer (855.2 *kJ/mol* and 856.9 *kJ/mol*, respectively, against 847.3; Table 10.12).

At standard state (pure gas, $P = 10^{-5}$ *Pa*, 298.15 *K*) the NIST-JANAF Tables list an enthalpy of formation from the elements of NH_3 of -45.898 *kJ/mol* (Chase, 1998), the entropy is 192.734 *J(mol × K)* and the isobaric heat capacity is 35.652 *J/(mol × K)*. The Shomate equation for the isobaric heat capacity in the range $298.15 \leq T(K) \leq 1400$ is $C_P = 19.99563 + 49.7719 \times 10^{-3} \times T - 15.37599 \times 10^{-6} \times T^2 + 1.921168 \times 10^{-9} \times T^3 + 1.89174 \times 10^5 \times T^{-2}$. (NIST, 2000; based on refitted data of Chase 1998). At G2 theory level the obtained enthalpy of formation is −41.2 *kJ/mol* and the entropy is 201.1 *J/(mol × K)* (see Table 7.33). The melting point is at 194.95 *K*. The critical temperature is $T_c = 405.4$ *K* and the critical pressure is $P_c = 113$ *bar*. The triple point occurs at $T_t = 63.3$ *K* and $P_t = 0.126$ *bar* (NIST, 2000).

16.7.2 LHB Production

Oxygen isotope data from zircon crystals in Archean suggest that oceans existed on Earth as early as 4.3 *Ga*. The fact that much of the Earth surface was covered by ocean from the late Hadean pushed Nakazawa et al. (2005) to investigate through high yield shock synthesis the direct production of ammonia from iron, water and nitrogen by impacting asteroids. The leading hypothesis was that these objects would have had to traverse a water-saturated atmosphere, raising a post-impact plume containing large amounts of H_2O and N_2, as well as vaporized, molten, and hot fragments of the extraterrestrial object (composed of silicates, sulfides and Fe–Ni alloys, and responsible of an overall reducing system-Eh). In the experiment, powder mixtures of Cu_3N (as source of nitrogen), H_2O and Fe were used as starting material sample. The sample (and the atmospheric air) were encapsulated in the cavity (0.5 cm^3) of a steel container (30 *mm* long and 30 *mm* in diameter). The impact wall was 3 *mm* thick. Shock waves were generated by a single-stage propellant gun. The impacting material was a stainless steel flyer of 2 *mm* thickness. Using Hugoniot relations and the impedance match method, the Authors estimated an impact pressure of 2.9 *GPa* on iron and 6.3 *GPa* on Cu_3N. SEM observations indicating vapor growth of copper crystals suggested temperatures over the boiling point of Cu, due to exothermic chemical reactions induced by the shock wave in the cavity. The results showed clearly the formation of NH_3, with yields depending on the availability of iron, water, or nitrogen.

Since the impact velocity in the experiment of Nakagawa et al. (2005) was substantially lower than typical impact velocities of meteorites on the early Earth, and the specific production mechanisms were unclear, Shimamura et al. (2016) carried out a computational experiment consisting of a multiscale shock technique-based on ab initio molecular dynamics simulations (MSST-AIMD; Reed et al., 2003). This method allows the molecular dynamics simulation of the system under dynamical shock conditions for orders of magnitude longer time periods than is possible using the popular non-equilibrium molecular dynamics approach (Reed et al., 2003). While the AIMD follows the trajectories of all atoms while computing interatomic interactions quantum mechanically, based on the Hellmann–Feynman theorem and the Navier–Stokes equations, MSST models the propagation of steady shock waves for compressible flow. The propagation of the shock wave is modeled using the 1D Euler equations for compressible flow, which neglect thermal transport. Nevertheless, the conservation of mass, momentum, and energy, respectively, everywhere in the wave is obeyed. Solutions of these equations (steady in the frame of the shock wave moving at speed v_s) are obtained by indicating the position at time t as:

$$(x, t) \rightarrow x - v_s t \tag{16.7.1}$$

Integration over x yields a modification of the Hugoniot relations,

$$u = v_s\left(1 - \frac{\rho_0}{\rho}\right) \tag{16.7.2}$$

$$p - p_0 = v_s^2 \rho_0 \left(1 - \frac{\rho_0}{\rho}\right) \tag{16.7.3}$$

$$e - e_0 = p_0\left(\frac{1}{\rho_0} - \frac{1}{\rho}\right) + \frac{v_s^2}{2}\left(1 - \frac{\rho_0}{\rho}\right)^2 \tag{16.7.4}$$

u in the above equations is the particle velocity (i.e. the local speed of the material in the laboratory frame (particle velocity), ρ is the density (the reciprocal of the specific volume $\boldsymbol{v}$), e is the energy per unit mass, and p is the negative component of the stress tensor in the direction of shock propagation. The subscript 0 denotes values before the shock wave. Choosing $u_0 = 0$, the material is thought as initially at rest in the laboratory frame.

The molecular dynamics simulation is carried out by employing the Lagrangian:

$$\mathcal{L} = T\left(\left\{\dot{\mathbf{r}}_i\right\}\right) - V(\{\mathbf{r}_i\}) + \frac{1}{2}Q\dot{\boldsymbol{v}}^2 + \frac{1}{2}\frac{v_s^2}{2\boldsymbol{v}_0^2}(\boldsymbol{v}_0 - \boldsymbol{v}) \tag{16.7.5}$$

with T and V as kinetic and potential energies per unit mass, and Q as mass-like parameter for the simulation of the cell size (Reed et al., 2003). The equation of motion for the system specific volume $\boldsymbol{v}$ is

$$Q\ddot{\boldsymbol{v}} = \frac{\partial T}{\partial v} - \frac{\partial T}{\partial v} - p_0 - \frac{v_s^2}{v_0^2}(v_0 - v) \tag{16.7.6}$$

When $\ddot{\boldsymbol{v}} = 0$, Eq. 16.7.6 reduces to Eq. 16.7.3.

The system investigated by Shimamura et al. (2016) was composed of a Fe_{36} slab, 16 N_2, and 38 H_2O molecules. The atomic configuration of the supercell (a total of 182 atoms in a 29.72 *Å* × 8.580 *Å* × 8.580 *Å* volume) was prepared by placing a 2×3×3 bcc Fe slab at the center with the two surfaces perpendicular to the x direction, surrounded by 38 H_2O molecules randomly replaced by N_2 molecules. Figure 16.21 shows the volume ratio (V/V_0), pressure (P), particle velocity (u_p), and temperature (T) attained as a function of time for the 5 *km/s* shock-wave simulation. For all the four quantities, there are rapid changes at around 0.3 *ps*. Within 4 *ps* the four parameters converge to 0.58, 27.6 *GPa*, and 2.30 *km/s*. Temperature increases rapidly at around 0.3 *ps*, and, subsequently, more gradually from ∼ 1500 to ∼ 2100 *K*. This gradual increase was ascribed by the Authors to exothermic reactions, in agreement to what anticipated by Nakazawa et al. (2005). The MSST-AIMD simulations showed rapid NH_3 production through an associative mechanism, as seen in the catalyst of nitrogenase enzyme. A first NH_3 molecule formed at ∼ 1.4 *ps*, followed by a second one at ∼ 2 *ps* and a third one at 3.6 *ps*. Scaling to real time, the annual production amount of NH_3 during shock compression is estimated to be about 4.3×10^7 *tons*, somewhat larger than the estimated annual production amount from the shock experiment by Nakazawa et al. (1.08×10^7 *tons*).

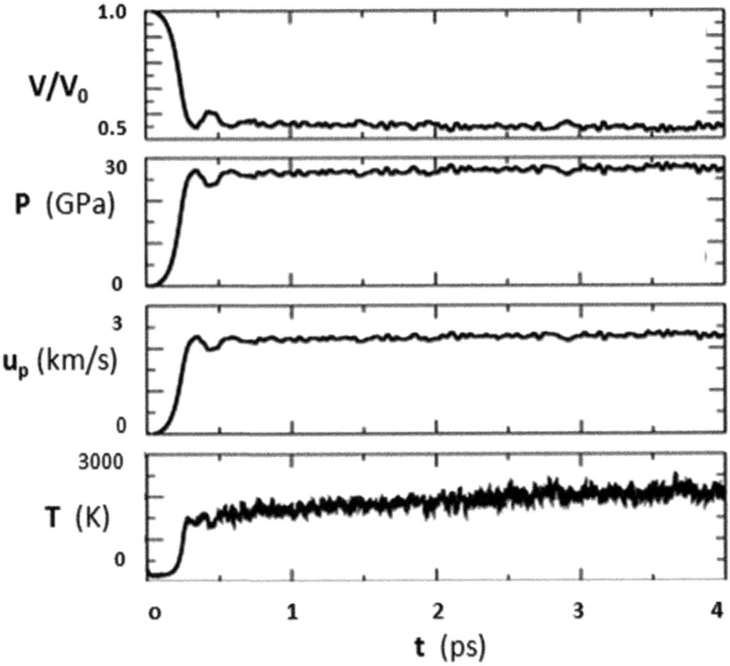

Fig. 16.21 Time evolution of volume ratio, pressure, particle velocity, and temperature in the 5 *km/s* shock-wave simulation experiment of NH_3 production. Modified from Shimamura et al. (2016)

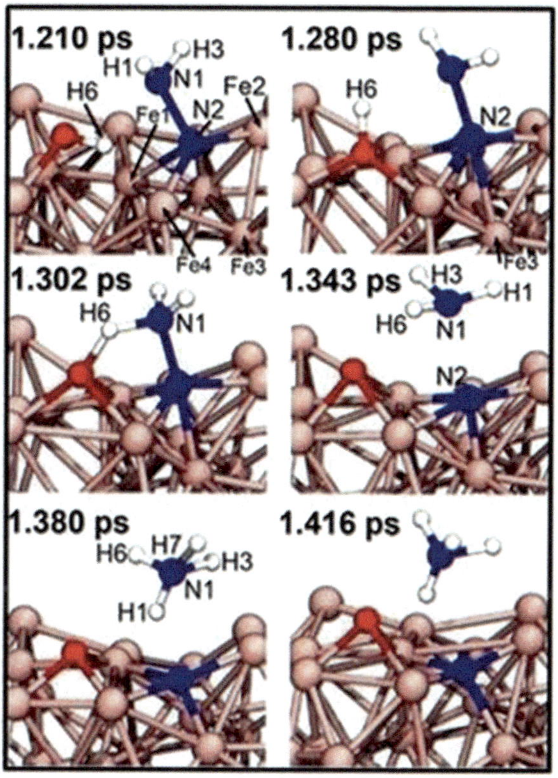

Fig. 16.22 Atomistic configurations at 1.210, 1.280, 1.302, 1.343, 1.380, and 1.416 *ps* during the first observed productions of an NH_3 molecule on the Fe slab and the subsequent NH_4^+. White, blue, red, and pink spheres represent H, N, O, and Fe atoms, respectively. Modified from Shimamura et al. (2016)

The atomistic configurations at 1.210, 1.280, 1.302, 1.343, 1.380, and 1.416 *ps* during the first observed productions of an NH_3 molecule on the Fe slab and the subsequent NH_4^+ abstraction is shown in Fig. 16.22.

16.8 CO_2

We have already seen (Sect. 7.5) the CO_2 molecule treated as an example of linear hetero-nuclear closed shell configuration. CO_2 has symmetry $D_{\infty h}$ (linear with inversion center). There are in CO_2 twenty two electrons in a closed shell ${}^1\Sigma_g^+$ configuration. The LCAO SCF results for the $1\sigma_g$ and $1\sigma_u$ orbitals indicate rather large contributions of the carbon orbitals to the 2s orbitals of oxygen, and, on the basis of the criterion of maximum overlap, they suggest a bonding character for $1\sigma_g$ (see Fig. 7.23). The first quantum mechanical characterization of CO_2 is due to Mulliken (1935) and Moffitt (1949). Both approaches were semi-empirical but gave a correct appraisal of the ground state, the bond order, and the ionization potential. The first non-empirical approach, based on LCAO and Slater-type AOs is due to Mulligan (1951).

16.8.1 Thermophysical and Thermodynamic Properties

The C-O interatomic distance is 1.162 *Å* and the O-C-O angle is 180° (Herzberg, 1966. When resolved at B3LYP/6-31G(d) theory level, the C-O distance is 1.169 *Å*. The molecule is a prolate symmetric top with rotational symmetry number 2.

The rotational constants of CO_2 is B = 0.39021 cm^{-1} (Herzberg, 1966) and the calculated one at B3LYP/6.31G(d,p) level is B = 0.386 cm^{-1}. The total electric dipole moment is zero and the components of the quadrupole moment are xx = yy = 2.139 $D \times Å$, yy = −4.278 $D \times Å$. The electrical dipole polarizability is $\alpha = 2.507\ Å^3$ (Olney et al., 1997) and the calculated one at B3LYP/6-31G(d,p) level is $\alpha = 1.850\ Å^3$.

The IR and Raman spectra of CO_2 arising from ab initio calculation (Fig. 7.32) show that the most IR intense mode is the σ_u asymmetric stretching (B3LYP/6-31G* scaled = 2341 cm^{-1}, against 2349 cm^{-1} observed; NIST, 2000), while the intensity of the (doubly degenerate) π_u bending mode is quite limited (calculated = 615 cm^{-1}, observed = 672 cm^{-1}). The σ_g Raman active mode (calculated = 2341 cm^{-1}, observed = 2349 cm^{-1}) is IR-inactive. The zero point energy at this theory level is 0.011592 *Hartree* (unscaled).

The best agreement with the experimental dissociation energy (1597.9 *kJ/mol*; Chase et al., 1985) is obtained at G2 and MP2 theory level (1607.1 *kJ/mol*, Ochtersky et al., 1996; 1594.1 *kJ/mol*, Johnson et al., 1992; see Tables 7.32; 10.10 and 10.19). Also the Becke's 3-parameters hybrid functional B3-PW91 gives an accurate result (calculated = 1594.1 *kJ/mol*; Foresman and Frish, 1993). The experimental ionization potential is 13.777 *eV*; the electron affinity is −0.600 *eV* and the proton affinity is 540.5 *kJ/mol* (NIST, 2000).

At standard state (pure gas, P = 10^{-5} *Pa*, 298.15 *K*) the NIST-JANAF Tables list an enthalpy of formation from the elements of CO_2 $\Delta_f H° = -393.522$ *kJ/mol* (Chase, 1998), the entropy is S° = 213.795 *J(mol × K)* and the isobaric heat capacity is $C_{P,298.15} = 37.129$ *J/(mol × K)*. The Shomate equation for the isobaric heat capacity in the range $298.15 \le T(K) \le 1200$ is $C_P = 24.99735 + 55.18696 \times 10^{-3} \times T - 33.69137 \times 10^{-6} \times T^2 + 7.948387 \times 10^{-9} \times T^3 - 1.36684 \times 10^5 \times T^{-2}$ (NIST, 2000; based on refitted data of Chase 1998). The Heat of formation computed via the B98 functional is −400.12 *kJ/mol* ((Schmider and Becke, 1998; Table 10.14), while at G2 theory level one obtains a better agreement with what experimentally observed (−392.5 *kJ/mol*; Vetuschi Zuccolini et al., 2011; Table 7.33). Also the estimated entropy at this theory level is sufficiently accurate [213.4 *J(mol × K)*; Table 7.33].

The critical temperature is $T_c = 304.2$ *K* and the critical pressure is $P_c = 73.8$ *bar*. The triple point occurs at $T_t = 216.58$ *K* and $P_t = 5.185$ *bar* (NIST, 2000).

16.8.2 IR Absorption and Role in the Greenhouse Effect

The Moon formed by the collision of two planets, one roughly the size of Mars and the other roughly the size of Earth (Zahnle et al., 2010; and references therein). The collision took place 30 – 100 *Ma* after the formation of the Solar system. Since the Hadean time (Moon-forming impact) CO_2 was the major atmospheric gas apart from water. A strong greenhouse effect kept the early Earth warm while CO_2 remained in the atmosphere. Progressively, however, CO_2 reacted with the seafloor and was subducted into the mantle through "*vigorous hydrothermal circulation through the oceanic crust and rapid mantle turnover*" (Zahnle et al., 2010). It is estimated that this process could have moved 100 *bar* of CO_2 from the atmosphere into the mantle in less than 10 million years (Sleep 2010). However, the rest of the CO_2 cycle remains obscure, "*which makes it difficult to predict how quickly this would have happened or what the asymptotic atmospheric CO_2 level would have been*" (Sleep et al. 2001). As noted by Zahnle et al. (2007) "*prior to the origin of life, in the absence of an abundant potent greenhouse gas, the surface should have been ice covered and very cold*". Nowadays the concern goes in the opposite direction: the massive release of CO_2 in the atmosphere due to combustion of fossil fuels, initiated in the industrial era, entails a progressive raise of temperature.

Figure 16.23 shows the relative transmittance of 200 *mmHg* of a CO_2 gas diluted to a total pressure of 600 *mmHg* with N_2 and measured by a DOW KBr Foreprism instrument in transmission mode. The spectrum results superimposing 5 different data sets in the range 460 to 2000 cm^{-1},with a resolution of 4 cm^{-1}, which are available on the NIST Chemistry WebBook. The spectrum exhibit a strong absorption centered at $\sim$669 cm^{-1} and various side bands of progressively lower intensity moving away from the central peak.

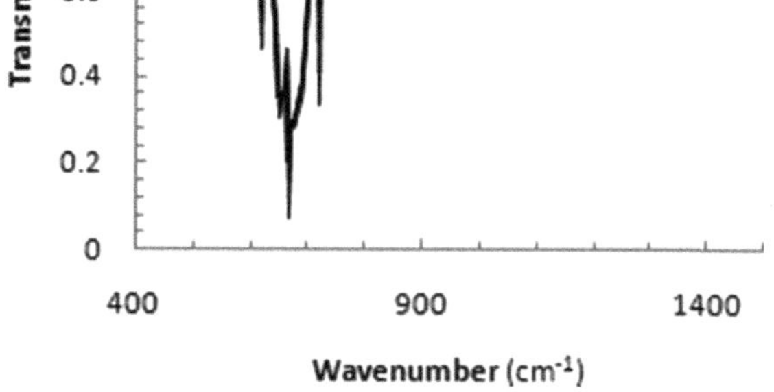

Fig. 16.23 Transmittance of a N_2-diluted CO_2 gas (from the NIST Chemistry WebBook collection of 5 data sets in the spectral range 460 to 2000 cm^{-1})

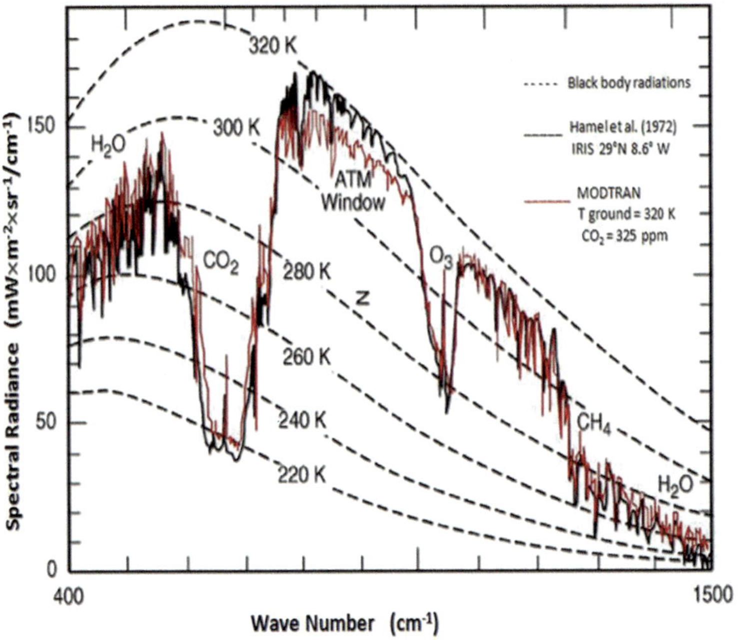

Fig. 16.24 Spectral Radiance ($mW \times m^{-2} \times sr^{-1}/cm^{-1}$) measured at 29°N, 8.6°W by the IRIS instrument carried on the Nimbus 4 meteorological satellite flying over Sahara. On the same figure are superimposed the black body radiations at various T and the results of the MODTRAN software simulation (The University of Chicago) for a ground temperature of 320 *K*, and a CO_2 concentration of 325 *ppm* (by volume)

In Fig. 16.24 we see the IR spectroscopic measurements obtained by Hamel et al. (1972) with an IRIS Instrument carried on the Nimbus 4 meteorological satellite flying over Sahara at 29° latitude N and 8.6° longitude W. On the same figure are superimposed the black body radiations at various T and the results of the MODTRAN software simulation[4] for a ground temperature of 320 *K*, a CO_2 concentration of 325 *ppm* (by volume). The total energy flux from all IR light (upward IR heat flux) is in W/m^2. The Spectral Radiance in wavelength (not to be confused with Radiance[5]) is, in SI units, expressed in Watts per square metre per steradian per metre ($W{\cdot}sr^{-1}{\cdot}m^{-3}$), but commonly converted in Watts per steradian per square metre per nanometre ($W{\cdot}sr^{-1}{\cdot}m^{-2}{\cdot}nm^{-1}$). A steradian is the solid angle subtended at the center of a unit sphere by a unit area on its surface. A complete spherical body has a solid angle of 4π *sr*. In the original paper of Hanel et al.

[4] MODTRAN is a freeware code released by The Chicago University. Available at http://climatemodels.uchicago.edu.

[5] The SI unit of radiance is in Watts per steradian per square metre ($W{\cdot}sr^{-1}{\cdot}m^{-2}$).

(1972) the (spectral) radiance is expressed in Ergs per second per steradian per cm^3 and has been here converted in the usual notation for comparative purposes.

The zones of absorption operated by CO_2 and O_3 at the investigated latitude and longitude are clearly visible in Fig. 16.24, as well as the atmospheric window between the two.

To understand the features implicit in Figs. 16.23, 16.24, we must recall that the spectral distribution of energy, in a quantized electromagnetic field in complete thermal equilibrium with the radiating object, is obtained multiplying the energy of each photon $E = h\nu$ by the number of states available at ν and by the probability that the state is occupied (Planck Law):

$$u(\nu, T) = \frac{2h\nu^3}{c^2} \frac{1}{\exp(h\nu/k_B T) - 1} \tag{16.8.1}$$

u(ν,T) is then the thermal radiation power of a black body per unit area of radiating surface per unit of solid angle and per unit frequency. Remembering that the wavelength λ is related to frequency by $\lambda = c/\nu$, the Planck equation can be re-expressed as:

$$u(\lambda, T) = \frac{2hc^2}{\lambda^5} \frac{1}{\exp(hc/k_B \lambda T) - 1} \tag{16.8.2}$$

The power output is obtained by integrating Eq. 16.8.1 over the field of spectral emission (Stefan-Boltzmann Law):

$$P = \sigma_{SB} A T^4 \tag{16.8.3}$$

where σ_{SB} is the Stefan-Boltzmann constant (σ_{SB} = 5.670 × 10–8 $W{\cdot}m^{-2}{\cdot}K^{-4}$) and A is the radiating surface area. Based on the Wien's displacement Law, the black body radiation peaks at different temperatures, with maxima inversely proportional to T (see the various maxima in the black body radiations of Fig. 16.24):

$$\lambda_{peak} = b_w / T \tag{16.8.4}$$

with b_W = Wien's displacement constant ($b_W = 2.898 \times 10^{-3}$ $m{\cdot}K$).

When the black body is replaced by a planetary "*grey body*", the black body radiation is attenuated by a frequency dependent emissivity factor F(ν) caused by the presence on Green House Gases (GHG), each one with a peculiar infrared photon scattering cross section $IR_{cs,i}$ and with a partial pressure P_i (Rhodes, 2017):

$$F(\nu) = \left(1 + \sum_{i=1}^{n} IR_{cs,i} \times P_i\right)^{-1} \tag{16.8.5}$$

With the increase of CO_2 partial pressure in the atmosphere, the emissivity factor decreases at all the CO_2 absorption lines of Fig. 16.23, decreasing the radiative emission to the outer space. This is seen by an outer observatory as an apparent decrease in the temperature of the black body. Where the GHG are non influent, an "*atmospheric window*" opens (Fig. 16.24).

The potential effects of GHG on the climate was hypothesized already in the nineteenth century by Fourier (1827) and Arrhenius (1896) but the present day concern originated in the seventies of the twentieth century when Broeker (1975) coined the term "*global warming*" and anticipated that a doubling of the CO_2 concentration with respect to preindustrial levels would have raised the global temperature of 2.4 *K*. The "*Radiative Forcing*" (RF)[6] superimposes on the orbital-induced insolation variability that is known to affect the planet. Hays et al. (1976) showed that "*For 500,000 years, major climatic changes have followed variations in obliquity and precession*". Their analysis assumed that "*the obliquity of the earth's axis (with a period of about 41,000 years) and the precession of the equinoxes (period of about 21,000 years) are the underlying, controlling variables that influence climate through their impact on planetary insolation*". With their work Hays et al. (1976) gave new emphasis to the "*Milankovitch cycles*" (Milankovitch, 1913) implying that the insolation varies with time as a result of complex gravitational interactions that cause periodic changes in the parameters of the Earth's motion around the Sun.

The geological data of Hays et al. (1976) were represented by:

(i) the oxygen isotopic composition $\delta^{18}O$ of planktonic foraminifera;
(ii) an estimate of summer sea-surface temperatures (T_s) at the core site, derived from a statistical analysis of radiolarian assemblages;
(iii) the percentage of *Cycladophora davisiana*, the relative abundance of a radiolarian species not used in the estimation of T_s.

Identical samples were analyzed for the three variables at 10-*cm* intervals through two cores of the CLIMAP project (RC1 1–120 and E49-18).

For any particular latitude the intensity of solar radiation received at the top of the atmosphere varies quasi-periodically with three elements of the earth's orbit: eccentricity (e, with an average period of 93 *ka*), obliquity ($\in$, with an average period of about 41 *ka*) and the longitude of perihelion (Π, moving equinoxes). Specifically, the intensity of incoming solar radiation at a particular latitude and season varies as $e \times \sin\Pi$. To analyze the quasi-periodic secular variations in $\in$, e, $\sin\Pi$, and $\Delta e \times \sin\Pi$ (expressing with this term the difference between $e \times \sin\Pi$ calculated at any particular time in the past with respect to the value for June 1950 A.D.), Hays et al. (1976) adopted spectral techniques, specifying the interval over which frequencies were to be analyzed, and a gain function for band-pass Tukey

[6] RF is the net change in the energy balance of the Earth system due to some imposed perturbation (Myhre et al., 2013).

filters (Goodman, 1960) (a 40 K filter centered on a frequency of 0.025 cycle per *ka*; and a 23 K filter centered on a frequency of 0.043 cycle per *ka*).

Their results were astonishingly precise in mimicking the observed seasonality changes (see Figs. 16.8 and 16.9 in Hays et al., 1976 and their simplified counterparts in Broeker, 1985). This "*clever mathematical scheme*" (Broecker, 1985), comparing the amplitude history of tilt-induced and the precession induced seasonality changes with the amplitude history of the $\delta^{18}O$ of planktonic foraminifera, provided a dramatic evidence that Earth climate (at least in the last 600 *ka*) was controlled by orbitalic details (see also Fig. 7.14 in Broecker, 1985).

From the analysis of Hays et al. (1976) is follows that:

1. "*The 42,000-year climatic component has the same period as variations in the obliquity of the earth's axis and retains a constant phase relationship with it*"
2. "*The 23,000-year portion of the variance displays the same periods (about 23,000 and 19,000 years) as the quasi-periodic precession index*"
3. "*The dominant, 100,000-year climatic component has an average period close to, and is in phase with, orbital eccentricity*".

Hays et al. (1976) also noted that the correlation induced by the dominant climatic component, associated to eccentricity, requires an assumption of non-linearity, differing in this aspect from the higher-frequency orbital variations "*which can be explained on the assumption that the climate system responds linearly to orbital forcing*". Indeed, the eccentricity of the Earth's orbit varied in an apparent complex way in the last 0.5 Gyr. However, when examining its time derivative we may note that the flexi behave almost harmonically and control locally the amplitude of the climatic excursion which can be tentatively approximated by an Hermite orthogonal function (n = 10) of type (Fig. 16.25; see also Eqn. 3.3.12, and Sect. 3.3, A.14):

$$\psi_n(x) = \frac{e'}{2} \times \left[N_n \cdot H_n(\xi) \cdot \exp\left(-\frac{\xi^2}{2} \right) \right] - 0.025 \qquad (16.8.6)$$

$$e' = 0.025 - 0.009 \times \sin(2\pi t') \qquad (16.8.7)$$

$$t' = -0.25 + 0.0025t \quad (ka) \qquad (16.8.8)$$

where x = [0.04 × t(*ka*)-4 × ncycle + 1] is the scaled time and "ncycle" is an integer defined at the beginning of each cycle (with ncycle 1 = −25 → −125 *ka*; ncycle 2 = −125 → −225 *ka*; ncycle 3 = −225 → −325 *ka*; and so on).

The "*climatic response*" of the Earth seems in this way to follow, at the *ka* scale, harmonic oscillations with Hermitian simmetry, by the entire solar system.

At the end of the twentieth century the anthropic "*global warming*" received an apparent confirmation through the geochemical investigation of ice cores (Lorius et al., 1990): past temperature increases are accompanied by marked increases in atmospheric CO_2 concentrations. Extensive geochemical analyses of ices cores in Antarctica (Petit et al., 1999; Lüthi et al., 2008), besides the isotopic deltas of

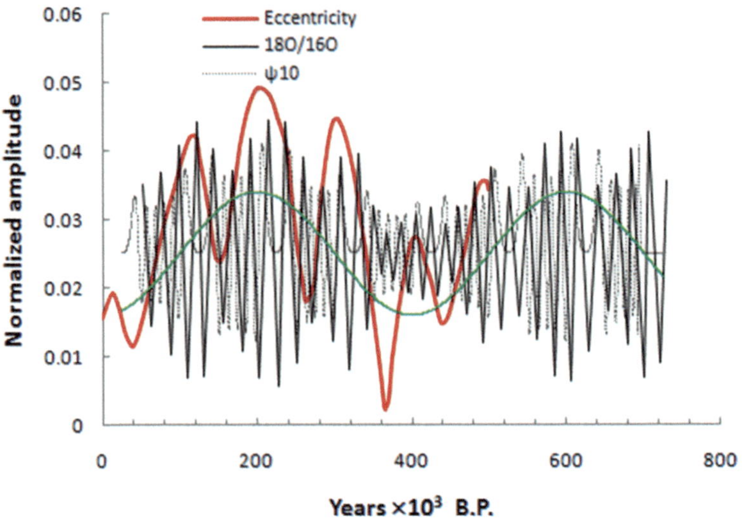

Fig. 16.25 Periodicity of the orbital forcing. Minima and maxima in $\delta^{18}O$ associated to precession frequency (corresponding respectively to conditions of minimum and maximum extension of ice volume) are shown to be markedly influenced by orbitalic eccentricity. The Ψ10(x) Hermitian function (dotted line) is superimposed only to stress that the system behaves as an harmonic oscillator

Deuterium in ice (δD_{ice}; a proxy of local temperature change) and of $\delta^{18}O$ in O_2, report also atmospheric CO_2 and CH_4 concentrations at the time of snow deposition, and sodium and dust contents over a time interval of 800 *ka*. The analyses at the Vostok site (spanning 420 *ka*; Petit et al., 1999), while confirming through a smart spectral analysis the leading role of eccentricity (100 *ka*) followed by obliquity (41 *ka*) and precession [23 and 19 *ka* (almost negligible); see their Fig. 16.4] provided further evidence of a strong correlation between T and GHG. The ancillary role of GHG was expressed with the statement "*This high correlation indicates that CO_2 and CH_4 may have contributed to the glacial-interglacial changes over this entire period by amplifying the orbital forcing*...".[7] They estimated this contribution to be ~ 3 W/m^2, corresponding to a glacial-interglacial ΔT_{GHG} of ~ 0.75 *K*. According to Berger (1978), the long-term variations of daily insolation in the last 400 *ka* at a constant latitude [65°N; i.e. "*the latitude around which the ice caps of glacial times were centered*" (Broecker, 1985)] and at mid-june has a lag of roughly 100 W/m^2 (i.e. from + 50 to -50 W/m^2) and the interglacial T-lag in the same lapse of time is roughly 10 *K* (see also Fig. 16.3 in Petit et al., 1999). Few calculations carried out with MODTRAN, show that the

[7] According to Richet (2021) there are epistemological weaknesses in the Vostok site investigation, residing in the inability of the model to account for the features of the glacial–interglacial transitions, the differing durations of interglacial periods and the constant 7 *ka* time lag between temperature and CO_2 decreases following deglaciation.

decrease of upward IR heat flux to the outer space induced by an increase of atmospheric CO_2 from 300 to 400 *ppm* is around 0.64 W/m^2 with a surface T of 250 *K* and rises to 1.42 W/m^2 when the surface T is 300 *K*. The T effect is almost linear. The calculations indicate that the possibility of a runaway effect (increasing T the forcing increases, and so on) is not remote. Another warning concerns the T effect on the solubility of CO_2 in water (which decreases with the increase of T) and the effect of CO_2 on the acidity of Earth's oceans (which increases with the partial pressure of CO_2 in the atmosphere). I will treat this aspect later on, when dealing with aqueous solutions.

16.8.3 Solubility of CO_2 in Water

The solubility of CO_2 in water is partly reactive. Aqueous solutions of carbon dioxide contain in fact both $CO_{2(a)}$ and $H_2CO_{3(a)}$ neutral molecules.

Conversion from the linear form $CO_{2(a)}$ to the trigonal-planar form $H_2CO_{3(a)}$, i.e.:

$$CO_{2(a)} + H_2O_{(a)} \rightarrow H_2CO_{3(a)} \tag{16.8.9}$$

is relatively slow (about 3% of $CO_{2(a)}$ is converted into $H_2CO_{3(a)}$ per second at room conditions; cf. Kern, 1960). The dehydration process:

$$H_2CO_{3(a)} \rightarrow CO_{2(a)} + H_2O_{(a)} \tag{16.8.10}$$

is relatively faster (net rate constant $K_r = 2.0 \times 10^{-3}\ s^{-1}$ at 0 °*C*, according to Jones et al., 1964).

The equilibrium kinetics of reactions (16.8.9) and (16.8.10) show that, at room P, T and in stationary conditions, the $CO_{2(a)}/H_2CO_{3(a)}$ ratio is near 600 (Kern, 1960). Notwithstanding this fact, in Geochemistry it is customary to represent aqueous CO_2 as if it were entirely in the form of carbonic acid $H_2CO_{3(a)}$, not distinguishing planar from trigonal forms.

The solubility of nominal $CO_{2(a)}$ in the T-range $273 \leq T\ (K) \leq 353$ at P = 1 *bar* has been examined by Wilhelm et al. (1976). Experimental observation after Morrison and Billett (1952), Murray and Riley (1971) and Weiss (1975) were retrieved in the form

$$R \ln(x_1) = [A_0 + A_1/T + A_2 \ln T + A_3 T] \tag{16.8.11}$$

The best fit over 17 data gave $A_0 = -317.658$ [*cal/(mol × K)*], $A_1 = 17{,}371.2$ [*cal/mol*], $A_2 = 43.0607$ [*cal/(mol × K)*], $A_3 = -0.00219107$ [$cal/(mol \times K^2)$], with a percentage standard deviation of 0.54.

When resolved at DFT/6–31+G(d) SCRF-IEFPCM level, with UAHF radii, the UATM assigns to the carbon atom a radius of 1.770 *Å*. The computed cavity volume is 59.372 $Å^3$. The returned solute/solvent interaction energy is

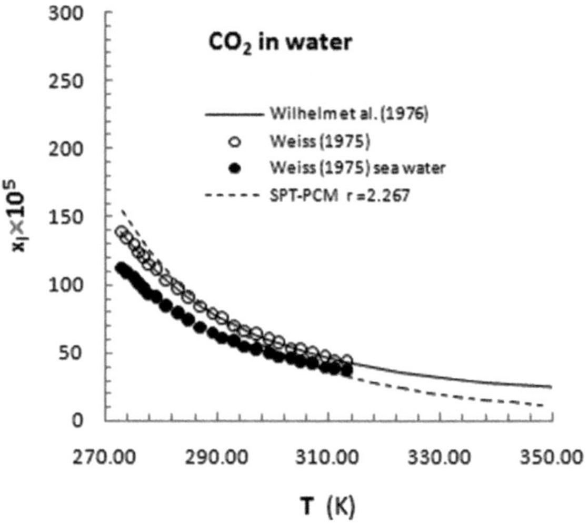

Fig. 16.26 Experimental versus computed SPT-PCM abundances of dissolved CO_2 in water at P = 1 *bar* and various T

H_{int} = −29.790 *kJ/mol,* and arises from a polarized solute—solvent electrostatic interaction of -7.615 *kJ/mol*, a dispersive energy contribution of −26.359 *kJ/mol* and a repulsive energy of 4.184 *kJ/mol.* To reproduce the experimental observations at room temperature we must adopt a solute radius of 2.267 *Å*. As already seen for O_2 and N_2 also in the case of CO_2, the solubility is underestimated at high T, although less severely (Fig. 16.26). Being a non polar species, CO_2 is subject to the "*salting out*" phenomenon (i.e. the activity coefficient of the species increases with the ionic strength of the solution). As a result, the molar fraction of dissolved species decreases with the increase of salinity. In Fig. 16.26 we may see that, based on the experimental observations of Weiss (1975), this effect is quite visible in sea water, especially at low T.[8]

At T = 298.15 *K* and P = 1 *bar* the SPT-PCM procedure predicts for CO_2 a cavitation Gibbs free energy $G°_{cav}$ = 30.466 *kJ/mol*, a cavitation enthalpy $H°_{cav}$ = 4.721 *kJ/mol* and a cavitation entropy $S°_{cav} = -86.352 J/(mol \times K)$. The volume of the solute is V_{solute} = 45.865 *cc/mol* and arises from a cavitation volume V_{cav} = 58.223 *cc/mol* and an interaction volume V_{int} = −13.479 *cc/mol*. Summing up the solute–solvent interaction enthalpy ($H°_{int}$ = −29.790 *kJ/mol*) with the cavitation enthalpy, the enthalpy of solution is $H°_{solution}$ = −27.368 *kJ/mol*. The solution entropy is $S°_{solution}$ = −154.125 *kJ/mol* and the Gibbs free energy of solution is $G°_{solution}$ 18.586 *kJ/mol*. The isobaric heat capacity of solution is $C_{P,solution}$ = 13.809. Adding the solution terms to the thermodynamic parameters of

[8] It must be stressed however that, at the pH conditions ruling in sea water, the Eh-pH predominance field is firmly occupied by bicarbonate ion, hence the nominal CO_2 molar fraction in solution does not represent the full amount of dissolved inorganic carbon.

the gaseous species we obtain the corresponding values for the solute, i.e. $\Delta_f G°_{solute} = -376.743$ *kJ/mol*; $\Delta_f H°_{solute} = -420.890$ *kJ/mol*; $S°_{solute} = 59.670$ *J (mol × K)* and $C_{P,298.15,solute} = 50.703$ *J/(mol × K)*.

16.9 CH_4

CH_4 (Methane) is a GHG, with absorption bands located between 1200 and 1400 cm^{-1} and between 2700 and 3200 cm^{-1} (NIST, 2000; see also Fig. 16.24). Urey (1951) was the first Scientist to note that "*The primitive atmosphere of the earth contained water, ammonia, and methane as well as large quantities of hydrogen which escaped*". More than half century later Schaefer and Fegley (2007) found that methane is one of the major gases produced by heating H-type chondrites up to their solidus temperature (1225 K). Although the experiment of Schaefer and Fegley (2007) was mainly focused on asteroid-sized bodies, the prominence of methane led them to extend the work to Earth (Schaefer and Fegley, 2010). They concluded that Earth's early atmosphere contained CH_4, H_2, H_2O, N_2, and NH_3, similar to the gas mixtures used in the Miller–Urey synthesis of organic compounds (Miller and Urey, 1955). Figure 16.27 resumes their results. The temperature axis can be interpreted as quench temperatures in the cooling plume of impact ejecta. When melting the High iron ordinary chondrites (H-type), the gases are very reduced and at 100 *bar* pressure CH_4 is dominant. The disappearance of CH_4 at very low temperatures indicates that graphite has become the stable form of carbon. At higher T, CO replaces CH_4 as the major carbon gas.

With the increase of atmospheric oxygen through the geological eras, CH_4 decreased dramatically, but (as seen for CO_2) the anthropogenic activity is working now in the opposite direction. Globally averaged surface CH_4 concentrations have risen from 722 ± 25 *ppb* in 1750 to 1803 ± 2 *ppb* by 2011 (Myhre et al., 2013), while the Vostok records indicate for the last 150 kyr maximum seasonal fluctuations from roughly 300 *ppbv* to 700 *ppbv* (Petit et al., 1999). The RF for CH_4 from 1750 to 2011 is estimated to be 0.48 ± 0.05 W/m^2, with an uncertainty dominated by the radiative transfer calculation (Myhre et al., 2013).

16.9.1 Thermophysical and Thermodynamic Properties

CH_4 is (besides H_2O) the most cited example of sp^3 hybridization. Perhaps it is less known that the proposal of a tridimensional CH_4 is due to Van 't Hoff in 1874, when he was a lecturer in chemistry and physics at the Veterinary College in Utrecht, and, as such, his proposal of a "*non-planar*" CH_4 was not readily accepted by the Academia. The first detailed ab initio calculation of the orbital energies of CH_4 is due to Nesbet (1960) who obtained approximate Hartree–Fock molecular orbitals as linear combinations of a set of basis orbitals of the form

Fig. 16.27 Gas compositions in equilibrium with ordinary H-type (high iron) chondrites at 100 *bar*. Modified from Schaefer and Fegley (2010)

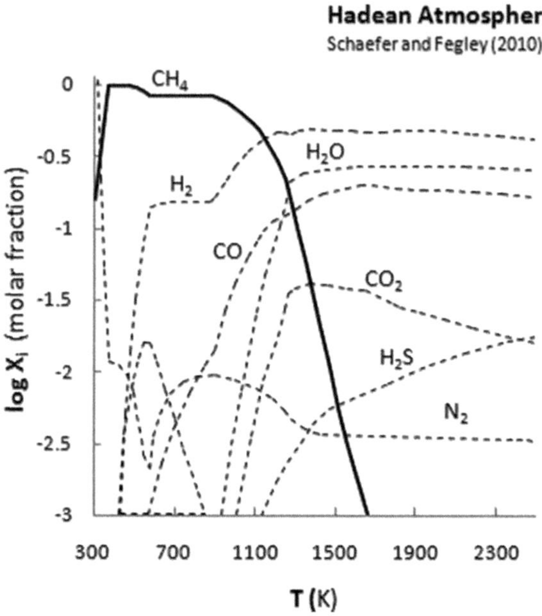

f(x,y,z) × exp($-\alpha r^2$) centered on carbon and hydrogen. Five independent s orbitals and two p orbitals on carbon were combined with one simple Gaussian orbital on each hydrogen nucleus. Correlation effects were also accounted for by a perturbation method.

Figure 16.28 show the conformation of HOMO resolved at DFT/B3LYP-6-31G (d) theory level and Table 16.26 reports the molecular orbital coefficients for a simple calculus carried out at RHF/STO-3G level. I recall that a MO may be seen as a linear combination of applicable functions in the adopted basis set (see Sect. 7.3.7 and Eq. 7.3.82.3). The triply degenerate HOMO is essentially composed of the 1s of the hydrogens and the $2p_x$, $2p_y$, $2p_z$ of C as anticipated by Nesbet (1960). The eigenvalue of this orbital at G2 theory level (−0.544 *Hartree*) is also in substantial agreement with the eigenvalue (−0.535 *Hartree*) of Nesbet (1960).

Ab initio calculations carried out at B3LYP/6-31G(d,p) theory level depict a triply degenerate asymmetric stretching mode of t_2 symmetry, IR and Raman active, at 1303.8 cm^{-1} (scaled). It is then observed a Raman active doubly degenerate libration mode of e symmetry at 1518 cm^{-1} (scaled). There is then a Raman active symmetric stretching mode of symmetry a_1 at 2929.2 cm^{-1} (scaled) and a triply degenerate mode IR and Raman active of t_2 symmetry at 3041 cm^{-1} (scaled). The zero point energy at this theory level is 0.045032 *Hartree* (unscaled).

The best agreement with the experimental dissociation energy (1642.2 *kJ/mol*; Chase et al., 1985) is obtained at G2 theory level (1641.4 *kJ/mol*, Ochtersky et al., 1996; see Tables 7.32, 10.10 and 10.19). The experimental ionization potential is 12.61 *eV* and the proton affinity is 543.15 *kJ/mol* (NIST, 2000).

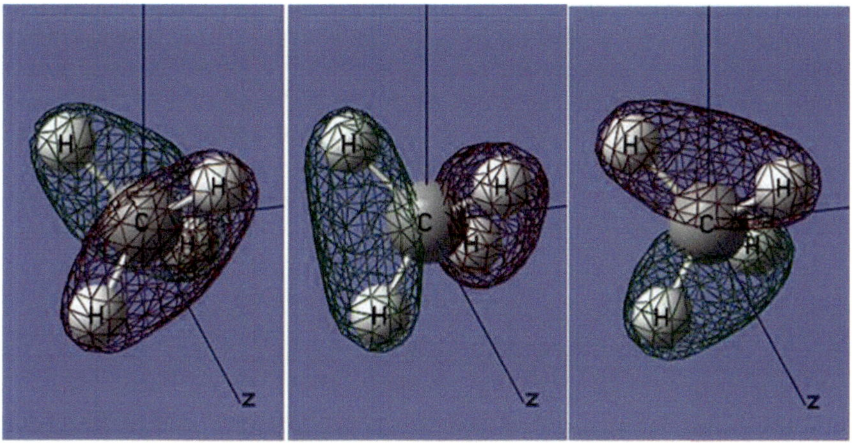

Fig. 16.28 G2 HOMO of CH_4 at 0.10 *e/Å*^{3} isodensity (eigenvalue = −0.544 *Hartree*). Van der Waals radii adopted

Table 16.26 Molecular orbital coefficients of HOMO in CH_4, resolved at RHF/STO-3G level. The MO eigenvalue is, at this simple level −0.308 *Hartree*

AO	Label	Atom	AO	MO (3)	MO (4)	MO (5)
1	1	C	1s	0	0	0
2			2s	0	0	0
3			$2p_x$	0	0.57197	0
4			$2p_y$	0	0	0.57197
5			$2p_z$	0.57197	0	0
6	2	H	1s	0.30082	0.30082	0.30082
7	3	H	1s	0.30082	−0.30082	−0.30082
8	4	H	1s	−0.30082	0.30082	−0.30082
9	5	H	1s	−0.30082	−0.30082	0.30082

At standard state (pure gas, P = 10^{-5} *Pa*, 298.15 *K*) the NIST-JANAF Tables list an enthalpy of formation from the elements of CH_4 $\Delta_f H°$ = −74.873 *kJ/mol* (Chase, 1998), the entropy is S° = 186.251 *J(mol × K)* and the isobaric heat capacity is $C_{P,298.15}$ = 35.639 *J/(mol × K)*. The Shomate equation for the isobaric heat capacity in the range 298.15 ≤ T(*K*) ≤ 1300 is $C_P = -0.703029 + 108.4773 \times 10^{-3} \times T - 42.52157 \times 10^{-6} \times T^2 + 5.862788 \times 10^{-9} \times T^3 + 6.78565 \times 10^5 \times T^{-2}$ (NIST, 2000; based on refitted data of Chase 1998). The Heat of formation computed via the B98 functional is −74.14 *kJ/mol* (Schmider and Becke, 1998; Table 10.14), while at G2 theory level one obtains a better agreement with what experimentally observed (−74.5 *kJ/mol*; Vetuschi Zuccolini et al., 2011; Table 7.33). Also the estimated entropy at this theory level is sufficiently accurate [185.9 *J(mol × K)*; Table 7.33].

The critical temperature is $T_c = 190.6\ K$ and the critical pressure is $P_c = 46.1\ bar$. The triple point occurs at $T_t = 90.67\ K$ and $P_t = 0.1169\ bar$ (NIST, 2000).

The rotational constants of CH_4 are A = B = C = 5.2412 cm^{-1} (Herzberg, 1966) and the calculated one at B3LYP/6.31G(d,p) level is 5.262 cm^{-1}. The electric dipole and quadrupole moments are zero. The electrical dipole polarizability is $\alpha = 2.448\ Å^3$ (Olney et al., 1997) and the calculated one at B3LYP/6-31G(d,p) level is $\alpha = 1.879\ Å^3$.

16.9.2 The High-T Spectrum of CH_4

Investigating the "*opacity*" of CH_4 is important not only for a better understanding of the Earth's climate but also to investigate the atmosphere of other planetary or extra solar bodies. In fact, a careful spectroscopic analysis of the light coming from distant objects is the only way to get clues about their composition and thermal history. Among these objects the "*failed stars*" deserve a special attention because accounting for part of the missing mass in our galaxy. The brown dwarfs (T-dwarfs, or also "*failed stars*") are sub-stellar bodies having masses which are intermediate between the most massive gaseous giant planets and the least massive stars (i.e. from 13 to 80 M_J with M_J as Mass of Jupiter $\approx 1.9 \times 10^{27}\ kg$). Their mass is insufficient to trigger the nuclear fusion of Hydrogen into Helium and their surface temperature is relatively low, emitting thus the majority of their light in the infrared region. The Near Infra-Red (NIR) spectrum of Gliese 229B (the prototypical T dwarf) is dominated by absorption bands from methane (CH_4) (Fig. 16.29), features that were found only in the giant planets of the Solar System and Titan[9]. (see Allard et al., 1997; for an accurate account on the astronomical importance of brown dwarfs).

Transitions involving changes in both vibrational and rotational states (i.e. "*rovibrational*" or "*ro-vibrational*") may emit or absorb photons. The difference in energy levels associated to the transition is in most cases IR and/or Raman active and is peculiar to a given molecule[10]. For a given vibrational transition, the same theoretical treatment as for pure rotational spectroscopy gives the rotational quantum numbers, energy levels, and selection rules. As anticipated, CH_4 is a spherical top. In linear and spherical top molecules, rotational lines are found as simple progressions at both higher and lower frequencies relative to the pure vibration frequency.

---

[9] Titan's methane cycle bears a striking similarity to Earth's water cycle, and apparently conforms its planetary surface.

[10] Since changes in rotational energy levels are typically much smaller than changes in vibrational energy levels, changes in rotational state are said to give the "fine structure" to the vibrational spectrum.

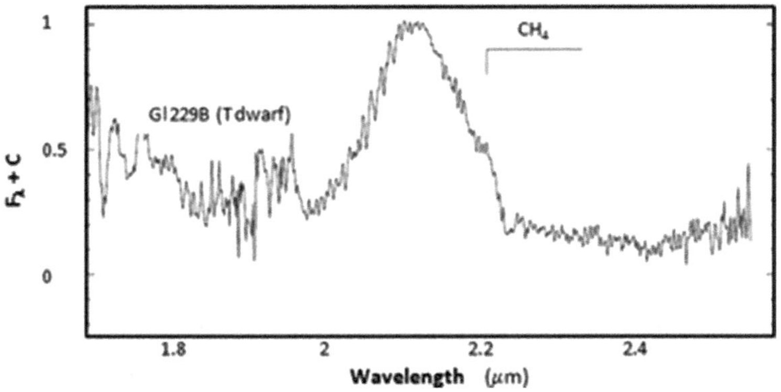

Fig. 16.29 Near-infrared spectrum of the T dwarf Gl 229B. The flux scale is in units of F_λ normalized to one at 2.20 *km* through a constant C. Modified from Kirkpatrick et al. (1999)

Fig. 16.30 Structure of the Methane molecule. Carbon (not shown) is at the center of the tetrahedron. Modified from Slater (1968)

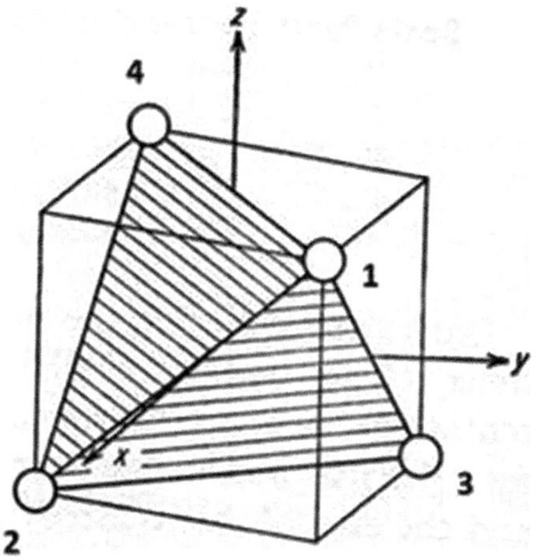

Figure 16.30 is the representation of the Methane molecule after Slater (1968). It is introduced here for its historical importance in the irreducible representations of the group T_d.

There are five classes of operations belonging to T_d which transform the molecule into itself. The irreducible representation consist of two one-dimensional representations, one two-dimensional representation, and two three-dimensional representations. The simple basis functions for the five irreducible representations are given in Table 16.27.

The carbon 1 s and 2 s orbitals have the symmetry of the irreducible representation Γ_1 while the three 2p functions ($2p_x$, $2p_y$ and $2p_z$) have the symmetry of the irreducible representation Γ_4. In other terms "*the degeneracy of the p states is not removed by the tetrahedral field*" (Slater, 1968).

Table 16.27 Basis functions for the irreducible representation of the group T_d (Slater, 1968)

Γ_1	A_1	One-dimensional:	1		
Γ_2	A_2	One-dimensional.	$x^4(y^2 - z^2) + y^4(z^2 - x^2) + z^x(x^2 - y^2)$		
Γ_3	E	Two-dimensional.	$3x^2 - r^2$	$\sqrt{3}(y^2 - z^2)$	
Γ_4	T_1	Three-dimensional:	x	y	z
Γ_5	T_2	Three-dimensional:	$x(y^2 - z^2)$	$y(z^2 - x^2)$	$z(x^2 - y^2)$

The "*opacity*" of CH_4 was not well known until the beginning of this century. The existing CH_4 database consisting of room temperature measurements of the strongest bands (Rothman et al., 1992, 1998) was in fact incomplete. This fact prompted Schwenke and Partridge (2001) to perform a detailed ab initio investigation of the molecule in order to reproduce the missing lines. Their work builds upon the anharmonic force field refinement of Gray and Robiette (1978) and the ab initio quartic force field investigation of Lee et al. (1995).

The construction of an accurate Potential Energy Function (PES) and of the associated Dipole Moment Surface (DMS) is prodromal to the comprehension of the nuclear motions. This can be done quantum mechanically in the framework of the Born–Oppenheimer approximation, solving for the lowest electronic eigenvalue and associated dipole moment for each nuclear geometry of interest, or, alternatively, stemming from experimental vibrational spectra (possibly of various isotopologues of the same molecule). In this case, the force constants are estimated through fitting procedures with a less computational demanding effort. Of course, the two procedures may be combined, ensuring that the appropriate Hamiltonian identifies correctly eventual resonances and anharmonic interactions as well. With the PES at hand, the rovibrational energy levels associated to nuclear motion generate the line positions and the rovibrational wave functions which, in conjunction with DMS provide the intensities.

Gray and Robiette (1978) calculated cubic and quartic anharmonic force field constants by combining a re-fitting procedure of spectroscopic data on the isotopologues $^{12}CH_4$, $^{13}CH_4$, $^{12}CD_4$, $^{12}CH_3D$, $^{12}CHD_3$ and $^{12}CH_2D_2$ and ab initio calculations. Six of the thirteen cubic force constants were determined experimentally and the remaining seven were obtained ab initio. For the quartic force field only one force constant was refined. The obtained equilibrium C-H bond length (r_e), on which the force constant are based (r_e = 1.0858 *Å*) is not far from what experimentally observed (r_e = 1.087 *Å*; Herzberg, 1966). Lee et al. (1995) obtained a very accurate ab initio quartic force field for CH_4 and its isotopologues with the CCSD(T) procedure and the correlation consistent basis set of Dunning (1989) (cc-pVDZ, cc-pVTZ, and cc-pVQZ). Their results, compared with the spectroscopically based parameters of Gray and Robiette (1978) are resumed in Table 16.28. The force constants are based on the same coordinate system adopted by Gray and Robiette (1978): i.e. a non-redundant set of curvilinear symmetry coordinates obtained as linear combinations of the ten geometrically-defined

Table 16.28 Equilibrium bond distance (*Å*), rotational constant (cm^{-1}), harmonic Frequencies (cm^{-1}), and quadratic force constants in units of $aJ/Å^2$ (stretch-stretch), $aJ/(Å \times rad)$ (stretch-bend) and aJ/rad^2 (bend-bend) obtained by Lee et al. (1995), compared with the experimental observations of Gray and Robiette (1978). The total (electronic + zero point) CCSD (T) energies for the cc-pVDZ, cc-pVTZ and cc-pVQZ basis sets obtained by Lee et al. (1995) are −40.387627, −40.438099, and −40.450888 *Hartree*, respectively. Note that 1 attaJoule (*aJ*) corresponds to 10^{-18} *J*)

	cc-pVDZ	**cc-pVTZ**	**cc-pVQZ**	**Expt** (Gray and Robiette, 1978)
r_e	1.1037	1.0890	1.0879	1.0858 ± 0.001
B_e	5.1492	5.2892	5.2999	
ν_1 (a_1)	3040.4	3034.7	3036.2	3025.5
ν_2 (e)	1550.5	1570.8	1570.4	1582.7
ν_3 (t_2)	3175.0	3153.9	3157.1	3156.8
ν_4(t_2)	1333.8	1344.0	1345.3	1367.4
F_{11}	5.48924	5.46865	5.47384	5.4358
F_{22}	0.57964	0.57919	0.57770	0.5841
F_{33}	5.43475	5.36602	5.37696	5.3788
F_{43}	20.20217	20.21036	20.21057	20.2213
F_{44}	0.53801	0.53227	0.53225	0.5481

curvilinear internal displacement coordinates δr_1, δr_2, δr_3, δr_4, $\delta\alpha_{12}$, $\delta\alpha_{13}$, $\delta\alpha_{14}$, $\delta\alpha_{23}$, $\delta\alpha_{24}$, $\delta\alpha_{34}$. These symmetry-reduced coordinates are

$$S_1(a_1) = \frac{1}{2}(\delta r_1 + \delta r_2 + \delta r_3 + \delta r_4)$$

$$S_{2a}(e) = \frac{1}{\sqrt{12}}(2\delta\alpha_{12} - \delta\alpha_{13} - \delta\alpha_{14} - \delta\alpha_{23} - \delta\alpha_{24} + 2\delta\alpha_{34})$$

$$S_{2b}(e) = \frac{1}{2}(\delta\alpha_{13} - \delta\alpha_{14} - \delta\alpha_{23} + \delta\alpha_{24}) \quad S_{3x}(t_2) = \frac{1}{2}(\delta r_1 - \delta r_2 + \delta r_3 - \delta r_4)$$

$$S_{3y}(t_2) = \frac{1}{2}(\delta r_1 - \delta r_2 - \delta r_3 + \delta r_4) \qquad S_{3z}(t_2) = \frac{1}{2}(\delta r_1 + \delta r_2 - \delta r_3 - \delta r_4)$$

$$S_{4x}(t_2) = \frac{1}{\sqrt{2}}(\delta\alpha_{24} - \delta\alpha_{13}) \qquad S_{4y}(t_2) = \frac{1}{\sqrt{2}}(\delta\alpha_{23} - \delta\alpha_{14})$$

$$S_{4z}(t_2) = \frac{1}{\sqrt{2}}(\delta\alpha_{34} - \chi\alpha_{12})$$

It should be noted that the atom numbering and the Cartesian coordinates adopted in defining the curvilinear symmetry coordinates of Table 16.28 do not correspond to the notation of Slater (1968). While in Slater (1968) the sides of the cube are two units in length and hydrogen 1 has coordinates (1,1,1), hydrogen 2 (1,−1,−1), hydrogen 3 (−1,1,−1) and hydrogen 4 (−1,−1,1), the coordinates of Gray and Robiette (1978) correspond to a cube of dimension $a = \frac{1}{\sqrt{3}} r_e$ and hydrogen 1 has coordinates (a,a,a), hydrogen 2 (−a,−a,a), hydrogen 3 (a,−a,−a) and hydrogen 4 (−a,a,−a).[11] As noted by Lee et al. (1995), there is generally good agreement between the CCSD(T)/cc-pVQZ and experimental harmonic frequencies, with the largest difference being 22.1 cm^{-1}.

Stemming from the symmetry relations of Table 16.28 and the potential expansion

$$V = E_0 + \frac{1}{2}\sum_{ij} F_{ij}\Delta S_i \Delta S_j + \frac{1}{6}\sum_{ijk} F_{ijk}\Delta S_i \Delta S_j \Delta S_k + \frac{1}{24}\sum_{jjkl} F_{ijkl}\Delta S_i \Delta S_j \Delta S_k \Delta S_l \tag{16.9.1}$$

Lee et al. (1995) determined the unique cubic and quartic force constants, based on the symmetry relationships of Gray and Robiette (1978) and Raynes et al. (1987), for the cubic and quartic force fields, respectively. Fundamental[12] vibrational frequencies were then determined by second-order perturbation theory anharmonic analyses. Their results for CH_4 are resumed in Table 16.29.

A PES of optic expansion was generated by Schwenke and Partridge (2001) working essentially at the same theory level of Lee et al. (1985). The aim of this study was to attain an accuracy better than a few cm^{-1}. In fact, the quartic expansion is apparently (qualitatively) incorrect from about 5500 cm^{-1} above the zero point energy. The functional form adopted by Schwenke and Partridge (2001) to represent the PES of CH_4 involves a simple zero-order function V^0 and a correction function V^c:

$$V = V^0 + V^c \tag{16.9.2}$$

The zero-order function is in the form:

$$V^0 = \sum_i V^s(R_i) + \sum_i V^a(R_i, R_j, \alpha_{ij}) \tag{16.9.3}$$

where α_{ij} is the angle between Ri and R_j. V^s is the "*stretching function*" (Schwenke and Partridge, 2001), given by:

---

[11] Basically atom 2 replaces atom 4 of Slater, atom 4 replaces atom 3 and atom 3 replaces atom 2. Moreover the unit length is replaced by a.

[12] The therm "fundamental" is deserved to vibrational therms associated to transitions from the ground vibrational quantum number v=0 to v=1 while "overtones" are generated by transitions from v=0 to v=2, etc.

Table 16.29 Vibrationally averaged bond distance, fundamentals, and anharmonic constants for CH_4, (Lee et al., 1995) The fundamental constants are indicated as ω_i. The experimental anharmonic constants of Gray and Robiette (1978) are also listed for comparative purposes. The listed experimental fundamentals differ slightly from the observed ones which are: 2916.5, 1533.3, 3019.2, and 1310.8 cm^{-1}. All magnitudes are in cm^{-1}, besides r_e and r_z which are expressed in Å

	cc-pVTZ	cc-pVQZ	Expt.		cc-pVTZ	cc-pVQZ	Expt.
r_e	1.0890	1.0879	1.0858	X_{23}	212.843	212.847	210.69
r_z	1.1021	1.1010	1.0985	X_{24}	25.231	25.207	212.09
ν_1	3034.7	3036.2	3025.5	X_{33}	228.469	228.336	229.18
ν_2	1570.8	1570.4	1582.7	X_{34}	28.246	28.243	23.98
ν_3	3153.9	3157.1	3156.8	X_{44}	25.831	25.854	215.63
ν_4	1344.0	1345.3	1367.4	G_{22}	20.390	20.426	0.92
ω_1	2911.1	2912.9	2920.9	G_{33}	11.606	11.544	12.13
ω_2	1533.1	1532.7	1532.4	G_{34}	20.747	20.742	21.25
ω_3	3011.6	3015.4	3022.5	G_{44}	4.999	5.020	9.27
ω_4	1312.7	1314.0	1308.4	T_{23}	0.408	0.406	0.41
X_{11}	212.513	212.476	211.00	T_{24}	20.906	20.907	20.1
X_{12}	218.349	218.485	216.33	T_{33}	3.266	3.246	3.40
X_{13}	252.746	252.487	250.32	T_{34}	0.189	0.182	0.20
X_{14}	20.764	20.706	6.12	T_{44}	0.535	0.535	2.00
X_{22}	20.332	20.295	22.98	S_{34}	0.019	0.016	20.11

$$V^s = \sum_{i=1}^{4} C_i^s Y_i(R) \tag{16.9.4}$$

with

$$Y = 1 - \exp -\beta_s(R - R_0)] \tag{16.9.5}$$

and V^a is the "*bending function*", given by

$$V^a = \exp\left\{-\beta_a\left[(R_i - R_0)^2 + (R_i - R_0)^2\right]\right\} \times \sum_{n=2}^{4} C_n^a\left(\cos\alpha_{ij} + \frac{1}{3}\right)^n \tag{16.9.6}$$

The parameters in V^0 were determined by (Schwenke and Partridge, 2001) performing an equally weighted least squares fit to the calculated points, which range in energy from 0 to about 26,000 cm^{-1} above the minimum. The zero-order function was designed in order to be qualitatively correct for all geometries. The PES can be represented in a "*relatively compact*" way (Schwenke and Partridge, 2001) as:

$$V = \sum_{i_A=1}^{N_A} F_{i_A}\left(\chi_1, \chi_2, \tilde{\alpha}\tilde{\beta}\tilde{\gamma}\right) G_{i_A}(R_1, R_2, R_3, R_4) \qquad (16.9.7)$$

with the matrix elements over stretching degrees of freedom (G) evaluated by numerical quadrature and the matrix elements of the angular degrees of freedom (F) determined by expanding the PES in terms of basis functions depending on the angles $\chi_1, \chi_2, \tilde{\alpha}\tilde{\beta}\tilde{\gamma}$. The optic expansion is obtained in this way with $N_A = 881$, instead of the 5687 terms necessary to define V^c and the 31,653 basis functions used to expand the angular dependence V^a. The maximum error below 13,000 cm^{-1} above the minimum is about 8 cm^{-1}, while the maximum error below 10,000 cm^{-1} is about 3 cm^{-1}. More important, the experimental observations of Venuti et al. (1999) were in this way integrated by 52 new bands and assignments (see Table 16.1 in Schwenke and Partridge, 2001). The computed harmonic frequencies agree with those of Lee et al. (1985), namely $\nu_1 = 3035.11$, $\nu_2 = 1571.28$, $\nu_3 = 3154.47$, and $\nu_4 = 1344.14$ (cf. Table 16.29).

16.10 H_2O

H_2O is nowadays almost entirely trapped in the oceans in its liquid aggregation state, but it was not so in the past. The Hadean atmosphere was in fact most likely dominated by gaseous H_2O. Assuming that the planetesimal which impacted the Earth was about the size of Mars and with a composition similar to CI carbonaceous chondrites (with 3.5% of reduced carbon and up to 10% by weight of H_2O; Hashimoto et al. 2007) the impact velocity would have exceeded 5 *km/s* and, under these conditions, most of the volatiles in the impactor would have been delivered directly into the growing atmosphere (Lange et al., 1985; Zahnle et al., 2010). Also in the case of an atmosphere slowly degassed by a solid Earth, H_2O would have been one of the major components (together with CO_2 and N_2; Holland, 1962). The predominance of oxidized components would have been dictated in this case by the fact that the gas/melt equilibria was buffered by the QFM (Quartz-Fayalite-Magnetite) assemblage. There are evidences (whole-rock abundances of Cr and V and the composition of Cr-rich spinels in modern and ancient volcanics) that this redox state has not changed over the past 3.8 *Ga* (Delano, 2001; Zahnle et al., 2010).

16.10.1 Thermophysical and Thermodynamic Properties

Has we have already seen in Chap. 7, the isolated di-hydrogen monoxide (H_2O) molecule has been extensively investigated as a case study from the very beginning of Quantum Mechanics (see Sect. 7.6). The various treatments include

Valence-Bond (VB) approaches (Coolidge, 1932; Van Vleck and Cross, 1934), the Slater-Pauling Directed Valence theory (Slater, 1931), LCAO-SCF methods (Mulliken, 1934; Duncan and Pople, 1953; Ellison and Shull, 1954) and Lennard–Jones perturbation procedures (Lennard–Jones,1949; Pople, 1950).

Figure 16.31 show the conformation and the molecular orbital coefficients of the occupied orbitals in H_2O resolved at simple RHF/STO-3G theory level and Table 16.30 reports the molecular orbital coefficients. We may see that the "*form*" of di-hydrogen oxide is not dictated by the HOMO, which is indeed the AO $2p_z$ of

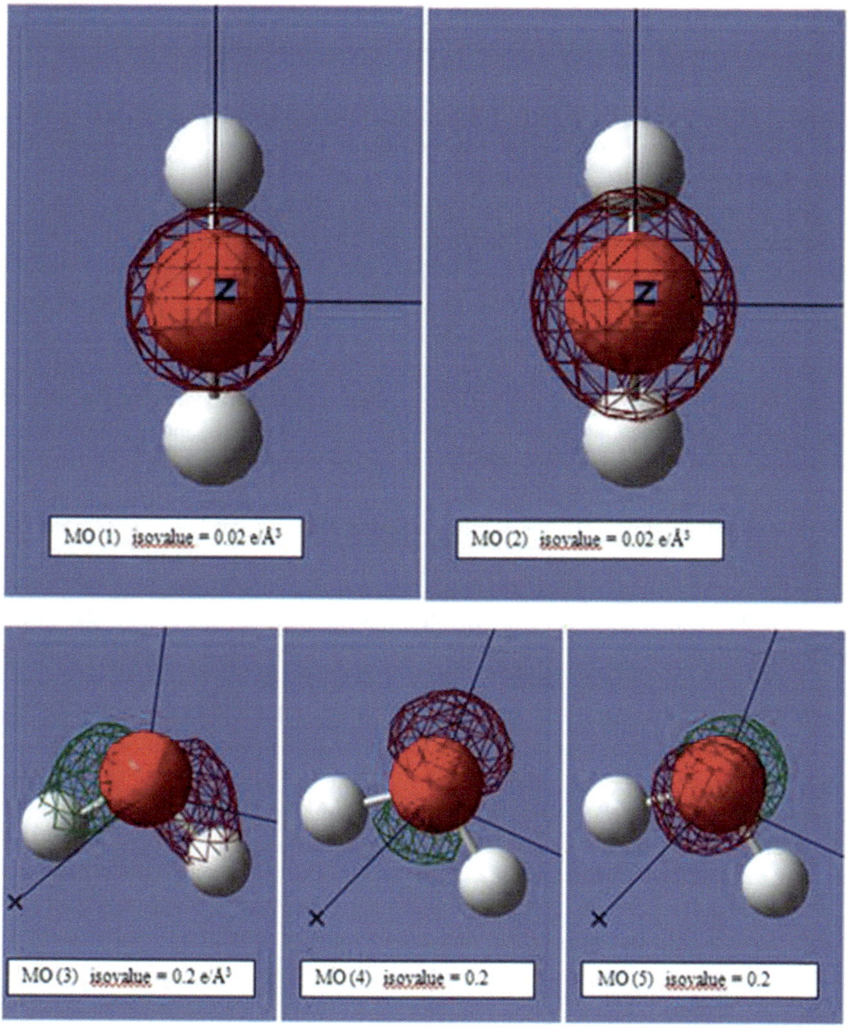

Fig. 16.31 The occupied orbitals of H_2O resolved at simple RHF/STO-3G level. Van der Waals radii adopted

Table 16.30 Molecular orbital coefficients of occupied orbitals in H_2O, resolved at RHF/STO-3G level. The MO eigenvalues are listed in brackets

AO	Label	Atom		MO (1) (–20.235)	MO (2) (−1.261)	MO (3) (−0.624)	MO (4) (0.441)	MO (5) (−0.387)
1	1	O	1s	0.99412	−0.23325	−0.00001	−0.09980	0.00000
2			2s	0.02657	0.83537	0.00008	0.52067	0.00000
3			$2p_x$	0.00000	−0.00003	0.59973	0.00003	0.00000
4			$2p_y$	−0.00415	−0.12329	0.00000	0.79169	0.00000
5			$2p_z$	0.00000	0.00000	0.00000	0.00000	1.00000
6	2	H	1s	−0.00597	0.15953	0.44237	−0.27320	0.00000
7	3	H	1s	−0.00597	0.15957	−0.44237	−0.27314	0.00000

oxygen but rather by MO (3) which arises from the $2p_x$ of oxygen and the 1s of the two hydrogens. The strong dipolar properties of the molecule arise mainly from MOs 3 and 4.

The ground state of H_2O is $^1A_1$ with C_{2v} configuration. The O–H interatomic distance is 0.958 *Å* and the H–O-H angle is 104.4776° (NIST,2000; and references therein). The molecule is an asymmetric top with rotational symmetry number 1. When resolved at B3LYP/6-31G(d,p) theory level, the O–H distance is 0.969 *Å* and the H-C-H angle is 103.098°. The lowest B3LYP/6-31G** harmonic vibrational frequency is a bending mode of symmetry A_1 (IR and Raman active; 1600 cm^{-1}, when scaled by the 0.9612 factor). There is then a symmetric stretching mode (IR and Raman active) of symmetry A_1 at 3653 cm^{-1} (scaled), and an asymmetric stretching mode of symmetry B_2 (IR and Raman active) at 3762 cm^{-1} (scaled). The zero point energy at this theory level is 0.021370 *Hartree* (unscaled). For a comparative appraisal of frequencies obtained at the various theory levels see Table 7.26 . The rotational constants of H_2O (A,B,C) are, respectively 27.877, 14.512 and 9.285 cm^{-1} (Herzberg, 1966). The total electric dipole moment is 1.855 *Debye* (*D*) and the components of the quadrupole moment are xx = −2.500 $D\times Å$, yy = 2.630 $D\times Å$, zz = 0.150 $D\times Å$. The electrical dipole polarizability is $\alpha = 1.501\ Å^3$ (Olney et al., 1997).

The ionization energy is 12.621 ± 9,992 *eV*, and the proton affinity is 691 *kJ/mol* (NIST, 2000). The best agreement with the experimental dissociation energy (917.6 *kJ/mol*; Chase et al., 1985) is obtained at G2 theory level (918.4 *kJ/mol*, Ochtersky et al., 1996; see Table 7.32). In terms of ionization potential (12.62 *eV*), the B3LYP and B98 functionals give the same result (i.e. 12.54 *eV*; Schmider and Becke, 1989; Table 10.11).

At standard state (pure ideal gas, $P = 10^{-5}$ *Pa*, 298.15 *K*) the NIST-JANAF Tables list an enthalpy of formation from the elements of H_2O of −241.826 *kJ/mol* (Chase, 1998), the entropy is 188.834 *J(mol × K)* (Chase, 1998) and the isobaric heat capacity is 33.590 *J/(mol × K)* (NIST, 2000). The Shomate equation for the isobaric heat capacity in the range $500 \le T(K) \le 1700$ is $C_P = 30.092 + 6.832514 \times 10^{-3} \times T + 6.793435 \times 10^{-6} \times T^2 + -2.534480 \times 10^{-9} \times T^3 + 0.82139 \times 10^4 \times T^{-2}$ (NIST, 2000; based on refitted data of Chase 1998).

At G2 theory level the obtained enthalpy of formation from the elements is −237.6 *kJ*/mol and the entropy is 188.2 *J/(mol × K)* (see Table 7.33). Detailed calculations of the isochoric heat capacity and entropy in the T range 298.15 ≤ T (*K*) ≤ 1700 are shown in Table 7.34.

16.10.2 The Real H_2O Gas: Virial Expansions and Empirical Models

Di-hydrogen oxide does not obey the equation of state of perfect gases, due mainly to the dipolar properties of the di-hydrogen monoxide. Deviations from the equation of state of perfect gases may be expressed either in terms of pressure or of volume. It is obvious that specific interactions among the molecules of a real gas are translated into modifications of volume with respect to the volume occupied by n moles of an ideal perfect gas at the same temperature T or, T and volume being equal, into modifications of pressure P exercised by the gaseous system towards the exterior. Analytically, this may be expressed through the concept of "*fugacity*". Fugacity f_i of a real pure gas i is obtained through:

$$RT \ln f_i = RT \ln P + \int_{P_r}^{P} \left(\overline{V}_i - \overline{V}_{perfect}\right) dP \tag{16.10.1}$$

where $\overline{V}_i$ is the molar volume of the real gas and $\overline{V}_{perfect}$ is the volume occupied by one mole of perfect gas in the same P, T conditions. The relationship between fugacity f_i and pressure P_i of a real gas i is given by:

$$f_i = P_i \Gamma_i \tag{16.10.2}$$

In Eq. 16.10.2, Γ_i is the "*fugacity coefficient*":

$$\Gamma_i = \exp\left[\int_{P_r}^{p} \frac{\left(\overline{V}_i - \overline{V}_{perfect}\right) dP}{RT}\right] \tag{16.10.3}$$

Values of Γ_i for gases of geochemical interest for which experimental data are not available in literature may be obtained through the "*universal chart of gases*". Construction of this chart is based on the fact that the thermodynamic properties of all real gases are similar functions of P and T, when the parameters of the equation of state are reduced with respect to the values of critical temperature T_c, critical pressure P_c and critical volume V_c of the compound of interest. We can thus introduce the concepts of "*reduced pressure*" P_r, "*reduced temperature*" T_r and "*reduced volume*" V_r:

$$P_{r,i} = \frac{P_i}{P_{c,i}} \tag{16.10.4.1}$$

$$T_{r,i} = \frac{T_i}{T_{c,i}} \tag{16.10.4.2}$$

$$V_{r,i} = \frac{V_i}{V_{c,i}} \tag{16.10.4.3}$$

The Γ_i values allow the calculation of fugacity and, thence the chemical potential of the gaseous component at the P, T of interest, applying:

$$\mu_{i,g} = \mu_{i,g}^0 + RT \ln\left(\frac{f_i}{f_i^0}\right) = \mu_{i,g}^0 + RT \ln a_i \tag{16.10.5}$$

with a_i as *thermodynamic activity.*

The "*universal chart of gases*" assumes that all gaseous species exhibit the same sort of deviation from ideal behaviour at the same values of T_r, P_r and V_r. This fact, known as the "*principle of corresponding states*", is analytically expressed by "*deviation parameter*" (or "*compressibility factor*") Z_g. For n = 1:

$$Z_g = \frac{PV}{RT} = f(P_r, T_r) \tag{16.10.6}$$

Deviation parameter Z_g is operationally expressed as a virial expansion on P or V:

$$Z_{g(P,T)} = 1 + BP + CP^2 + DP^3 + \ldots \tag{16.10.7.1}$$

$$Z_{g(V,T)} = 1 + B'V + C'V^{-2} + D'V^{-3} + .. \tag{16.10.7.2}$$

Virial expansion coefficients B, C, D and B', C', D' in Eqs. 16.10.7.1, 16.10.7.2 are simple functions of T and are specific to each gaseous species. Virial expansion equations like (16.10.7.1, 16.10.7.2) do not reproduce the behaviour of real gases in the vicinity of the critical point and in high P, T conditions with satisfactory precision.

Deviations from ideality of real gases may also be expressed through the van der Waals equation:

$$\left(P + aV^{-2}\right)(V - b) = RT \tag{16.10.8}$$

term aV^{-2} is defined as "*internal pressure*" or "*cohesive pressure*" and term b as "*co-volume*".

An alternative (and more accurate) form of equation of state for real gases is the Redlich-Kwong equation:

$$\left[P + \frac{a}{V(V+b)T^{1/2}}\right](V - b) = RT \qquad (16.10.9)$$

Parameters a and b in Eqs. 16.10.8, 16.10.9 have the same significance. As noted by Ganguly and Saxena (1987), the Redlich-Kwong (RK) equation for H_2O is quite accurate (but see also chapter 6 in Ganguly, 2008, for an outline of modified RK forms).

The "*universal chart of gases*" is not accurate in the proximity of the critical point. The PVT properties of H_2O within the critical region are well described by the *non-classical* (asymptotic scaling) equation of state of Levelt Sengers et al. (1983). Outside the critical region and up to 1000 °C, 15 *kbar*, PVT properties of H_2O are accurately reproduced by the *classical* equation of state of Haar et al. (1984). An appropriate description of the two equations of state is beyond the purposes of this textbook and I refer readers to the excellent revision of Johnson and Norton (1991) for an appropriate treatment.

A comprehensive review of the various forms of potentials adopted for H_2O in its various aggregation states has been presented by Reimers et al. (1982).

The various models reviewed by Reimers et al. (1982) adopt different geometrical schemes which may be resumed as shown in Fig. 16.32. Points 1 and 2 in this Figure correspond to hydrogen atoms, points 3 and 4 to negative charges, point 5 to the oxygen atom, and point 6 to the centre of charge for one molecule. Ab initio (and semi-empirical) models adopt the left hand scheme. There are +q point charges at positions 1 and 2 and −2q at position 3. Empirical models adopt the Bjerrum scheme shown at the right hand of Fig. 16.32: there are +q charges at positions 1 and 2 -q charges and at positions 3 and 4.

Every potential may be written as a sum of nine different types of pair interaction between one point on each molecule. The nine types of interaction and the acronyms adopted by Reimers et al. (1982) are resumed in Table 16.32. For the sake of

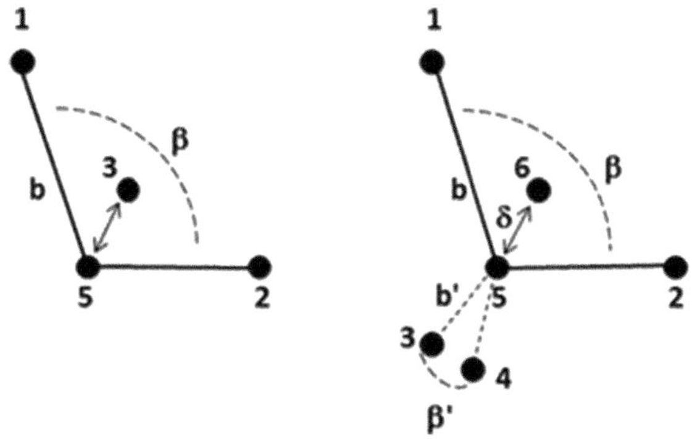

Fig. 16.32 Definition of molecular points in ab initio (left) and empirical (right) models

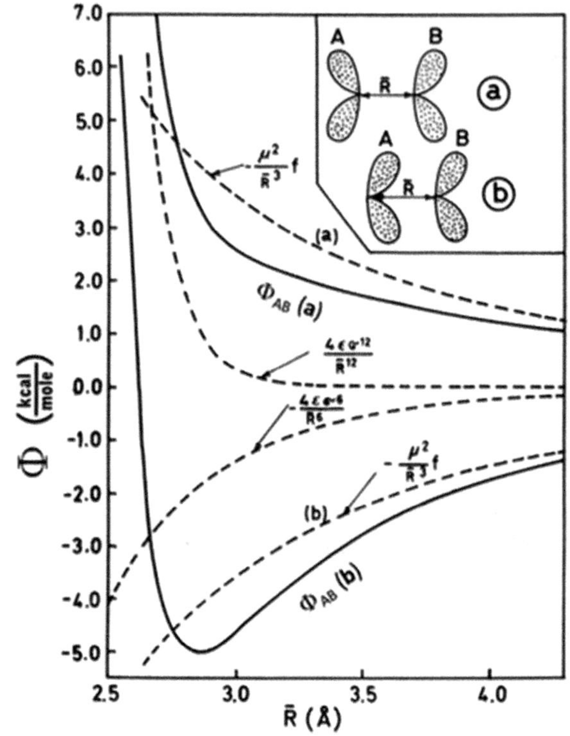

Fig. 16.33 Molecular interaction potentials in Stockmayer's (1941) model for H_2O vapor. (a): antiparallel dipolar moments; (b): parallel dipolar moments. Modified from Eisemberg and Kauzmann (1969)

brevity I recall here only few empirical models among the extensive compilation of Reimers et al. (1982).

Figure 16.33 shows the energy potential of two gaseous molecules of H_2O (A-B), according to the classical model of Stockmayer (1941).

Bulk potential Φ_{AB} obeys the equation:

$$\Phi_{AB} = -\frac{\mu^2 f}{\overline{R}^3} - \frac{4\varepsilon\sigma^6}{\overline{R}^6} + \frac{4\varepsilon\sigma^{12}}{\overline{R}^{12}} \qquad (16.10.19)$$

The first term on the right in the sum of potentials in Eq. 16.10.19 is the electrostatic term (represented here as a simple dipole moment). Factor f is the angular function of dipole orientation:

$$f = \sin\theta_A \sin\theta_B \cos(\phi_A - \phi_B) - 2\cos\phi_A \cos\phi_B \qquad (16.10.20)$$

Angles θ, φ in Eq. 16.10.20 define the dipole orientation with respect to a co-ordinate system centred on the oxygen nuclei. The second term on the right represents the dispersive energy (independent of molecular orientation), and the third one the repulsive interactions (also independent of orientation). σ is the "*collision diameter of the gaseous molecule*" and represents the distance at which attractive dipole moments and repulsive energy terms mutually compensate (i.e., when $\overline{R} = \sigma$, the series of potentials of Eq. 16.10.29 reduces to the first term), and ε represents the "*characteristic energy*" of the gaseous molecule. It is of interest to recall that the σ value observed by Stockmayer (1941) for gaseous H_2O is identical to the hydrogen-bond distance in ice I (Eisemberg and Kauzmann, 1969). Note in Fig. 16.33 that bulk potential Φ_{AB} depends greatly on the mutual orientation of dipoles: molecules with parallel dipoles result in more stable bonds than molecules with antiparallel dipoles. Note moreover, as already stated, that repulsive and dispersive forces are not affected by the state of dipole orientation.

Rowlinson (1949, 1951), promoted a revisitation of the Stockmayer potential consisting of a Lennard–Jones interaction between centers of charge plus two point multipole functions. The simplest form, (denoted RWL/D by Reimers et al., 1982) is essentially a Stockmayer potential form with $\varepsilon = 3.18$ *kJ/mol*, $\sigma = 2.648$ *Å*, $\mu = 1.8242$. Adding the dipole-quadrupole interaction (RWL/DQ in Reimers et al., 1982), the characteristic energy ε is reduced to 2.958 *kJ/mol*, the collision diameter σ is enhanced to 2.725 *Å*, the dipole–dipole moment μ_{dd} is 1.8367 and the dipole-quadrupole moment is $\mu_{dq} = 1.8185$. The two bulk potentials are here represented as (see also Table 16.31 and Reimers et al., 1982]:

$$\Phi_{AB}^{(RWL/D)} = LJ(6,6') + DIP(6,6') \tag{16.10.21}$$

$$\Phi_{AB}^{(RWL/DQ)} = LJ(6,6') + DIP(6,6') + DQ(6,6') \tag{16.10.22}$$

where 6 and 6' denote, respectively the centers of charge of the A and B molecules. In the model of Ben-Naim and Stillinger (1972) the Lennard–Jones interaction is developed on the oxygen centers and the bulk potential is represented as:

$$\Phi_{AB}^{(BNS)} = LJ(5,5') + \sum_{i=1}^{4}\sum_{j=1'}^{4'} C(i,j) + SW(5,5') \tag{16.10.23}$$

The potential hitherto taken into consideration consider H_2O as a rigid un-deformable molecule. The potentials proposed by Watts (1972) for water dimers are valid for deformable molecules and have been developed through fitting to spectroscopic data for each isolated water molecule, and to the second virial coefficent of steam. The RWK1 Watt's potentials is in the form:

$$\begin{aligned}\Phi_{AB}^{(RWK1)} = {} & D1(5,5') + ER(3,3') + ER(1,1') + ER(1,2') + ER(2,1') + ER(2,2') + \\ & M(3,1)' + M(3,2') + M(1,3') + M(2,3') + \sum_{i=1}^{3}\sum_{j=1'}^{3'} C(i,j)\end{aligned} \tag{16.10.24}$$

In RWK2, D2 replaces D1.

Table 16.31 Functional forms of the potential energy between A and B H_2O molecules (from Reimers et al. (1982). I in (16.10.16.4) is the ionization energy

N	description	Acronym	Condition	Formula	Number
1	Coulomb	C(A,B)		$q_A q_B / r_{AB}$	(16.10.10)
2	Exponential Repulsion	ER(A,B)		$A_{AB} \exp(-r_{AB}\alpha_{AB})$	(16.10.11)
3	Exponential Attraction	EA(A,B)		$-A'_{AB} \exp(-r_{AB}\alpha'_{AB})$	(16.10.12)
4	Lennard–Jones	LJ(A,B)		$4\varepsilon\left[(\sigma/r_{AB})^{12} - (\sigma/r_{AB})^{6}\right]$	(16.10.13)
5	Switch function	SW(A,B)	$R_{AB} \leq R_L$	0	(16.10.14.1)
			$R_{AB} \geq R_U$	1	(16.10.14.2)
			$R_L \leq R_{AB} \leq R_U$	$\frac{(R_{AB}-R_L)^2(3R_U-R_L-2R_{AB})}{(R_U-R_L)^3}$	(16.10.14.3)
6	Morse	*M*(A,B)		$A_{AB} \exp[-\alpha_{AB}(R_{AB} - R_{min})]\{\exp -\alpha_{AB}(R_{AB} - R_{min}) - 2\}$	(16.10.15)
7	Dispersion terms	DSW1	$R_{AB} \geq R_{SW}$	1	(16.10.16.1)
			$R_{AB} \leq R_U$	$\exp\left[-(R_{SW}/R_{AB} - 1)^2\right]$	(16.10.16.1)
		D1(A,B)		$-\mathrm{DSW1}\left(C_6/R_{AB}^6 + C_8/R_{AB}^8 + C_{10}/R_{AB}^{10}\right)$	(16.10.16.3)
		R_{SC}		$R_{AB} \times 0.94834673 = R_{AB}(I_{H_2O}/I_H)^{2/3}$	(16.10.16.4)
		$R_{au.}$		$R_{SC}/0.52977 \quad (a.u.)$	(16.10.16.5)
		F.		$1 - R_{au}^{2.326} \exp(-R_{au})$	(16.10.16.6)
		g_n		$1 - \exp\left[(-2.1/n)R_{AB} + (-0.109n^{-1/2})R_{AB}^2\right]$	(16.10.16.7)
		D2(A,B)		$-F\left[C_6\left(\frac{g_6}{R_{SC}}\right)^6 + C_8\left(\frac{g_8}{R_{SC}}\right)^8 + C_{10}\left(\frac{g_{10}}{R_{SC}}\right)^{10}\right]$	(16.10.16.8)
8	Dipole–dipole	DIP(A,B)		$(3\mu^2/R_{AB}^5)(a_A \cdot R_{AB})(a_B \cdot R_{AB}) - (\mu^2/R_{AB}^3)(a_A \cdot a_B)$	(16.10.17)
9	Dipole-quadrupole	DQ(A,B)		$\left(\frac{15\mu Q}{2R_{AB}^7}\right)\left[(a_A \cdot R_{AB})(b_B \cdot R_{AB})^2 - (a_B \cdot R_{AB})(b_A \cdot R_{AB})^2\right]$	(16.10.18)

16.10.3 The Real H_2O Gas: Ab Initio Potentials

Already more than half a century ago, McCarty and Babu (1970) promoted translational and rotational quantum corrections to the Stockmayer potential of steam. They observed that the best fit of the experimental data is obtained with values of ε sufficiently large that the ratio k_BT/ε attains values lower than 1.0 in the range of temperatures where the second virial coefficient B(T) can be measured. According to McCarty and Babu (1970) the quantum corrections to the second virial coefficient are particularly relevant for H_2O and NH_3, attaining an order of roughly 10% at the lowest temperature for which B(T) data are available. A quick glance at their Table II shows that the corrections are not negligible even at high T. The observations of McCarty and Babu (1970) were confirmed four years later by MacRury and Steele (1974a,b) which noted that rotational quantum effects are appreciable even at 300 K and higher, but also that the dipole-induced dipole energy (i.e. a "*polarizable dipole moment*") plays an important role in the calculation.

Popkie et al. (1973) calculated the Hartree–Fock energy of the water dimer for 216 different nuclear configurations using the basis set of Clementi and Popkie (1972) and optimizing further the set near the equilibrium configuration.[13] The study confirmed the importance of polarization (internal charge transfer). The computed energies were used to derive a simple analytical expression that reproduces the Hartree–Fock potential energy surface to a high degree of accuracy. The adopted charge distribution is that one proposed long ago by Bernal and Fowler (1933) and here reproduced in Fig. 16.34 This form combines at best numerical accuracy with mathematical simplicity.

The analytical potential of Popkie et al. (1973) in atomic units is

$$\Delta E = q^2\left(\frac{1}{r_{13}}+\frac{1}{r_{14}}+\frac{1}{r_{23}}+\frac{1}{r_{24}}\right)+\frac{4q^2}{r_{78}}-2q^2\left(\frac{1}{r_{18}}+\frac{1}{r_{28}}+\frac{1}{r_{37}}+\frac{1}{r_{47}}\right)+$$
$$+a_1\exp(-b_1r_{56})+a_2[\exp(-b_2r_{13})+\exp(-b_2r_{14})+\exp(-b_2r_{23})+\exp(-b_2r_{24})]+$$
$$+a_3[\exp(-b_3r_{16})+\exp(-c_3r_{26})+\exp(-b_3r_{35})+\exp(-b_3r_{45})] \tag{16.10.29}$$

The O_5–M_7 (and O_6–M_8) distance is 0.436 a.u. and $a_1 = 582.\ 277{,}054$, $a_2 = 0.143789$, $a_3 = 5.470184$, $b_2 = 1.\ 221{,}756$, $q^2 = O.\ 449{,}387$, $b_1 = 2.\ 520{,}593$. The accuracy of this fit is indicated by the low values of the standard deviation. The fit is very good for attractive regions but somewhat less accurate for the repulsive regions. The standard deviation is 0.0018 a.u. for all 216 points while is restricted to 0.0002 a.u. for the 80 binding points (i,e, those with negative ΔE).

[13] This resulted in a large Gaussian basis set with optimized orbital exponents for the oxygen 3d- and 4f-type and hydrogen 2p- and 3d-type functions, in order to get results close to the Hartree-Fock limit. In Matsuoka et al. (1974) this basis set is defined as of "double-zeta plus polarization quality".

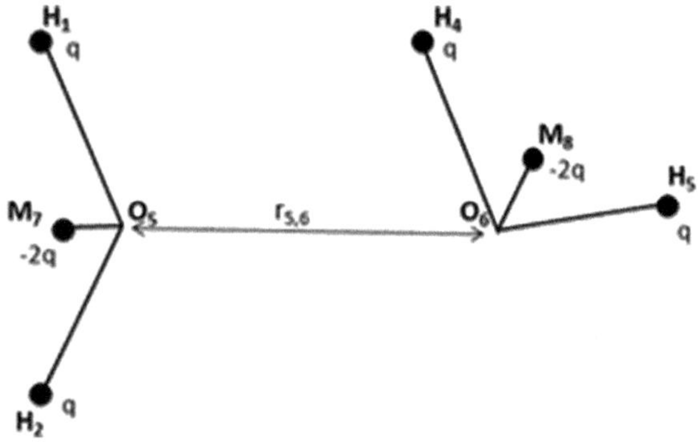

Fig. 16.34 Point charge model adopted by Popkie et al. (1973) in the derivation of the analytical $H_2O - H_2O$ interaction potential

In the notation of Reimers et al. (1982) the potentials of Popkie et al. (1973) takes the form

$$U_{AB}^{(HF)} = \sum_{i=1}^{3} \sum_{j=1'}^{3'} [C(i,j) + ER(i,j)] \qquad (16.10.30)$$

The Hartree- Fock potential of Popkie et al. (1973) is compared in Fig. 16.35 with the empirical potentials of Rowlinson (1951) and Ben-Nairn and Stillinger (1972). The energy surfaces correspond to configurations where the plane of molecule 1 is perpendicular to the plane of molecule 2 and the symmetry axis of molecule 2 points toward the oxygen nucleus of molecule 1. The variables are the oxygen–oxygen distance and the angle between the symmetry axes of the two molecules. From the comparative representation it may be seen that both the HF potential of Popkie et al. (1973) (**a** in Fig. 16.35) and the potential of Rowlinson (1951) (**b** in the same Figure) depict nearly linear hydrogen bonds, with the Rowlinson potential yielding a much smaller O–O equilibrium separation. The Ben-Naim-Stillinger potential (Ben-Naim and Stillinger, 1972; **c** in Fig. 16.35) corresponds to a not very pronounced bent hydrogen bond, more structured in the lone pair – lone pair repulsive region.

The iso-energy contour map **a** in Fig. 16.35 obeys *two geometrical constraints.* First, the geometry of each molecule is kept rigid (and chosen to be the known experimental geometry of equilibrium internuclear distances). Second, both molecular planes are constrained to be perpendicular to each other, with the oxygen of the molecule B lying in the plane of the molecule A. In a further contribution, Clementi and coauthors (Kirstenmaker et al., 1974) relaxed some of the above-mentioned constraints, namely, for each position of the molecule B, its Hydrogens were left free to be either *in* or *out* of the plane. This corresponds to letting the rotational axis which passes through the oxygen nucleus of the molecule B, not to be perpendicular to the molecular plane of A.

A "*more faithful*" analytical representation of the Hartree–Fock potential was in this way obtained (Kirstenmaker et al., 1974). The new numerical fitting constants

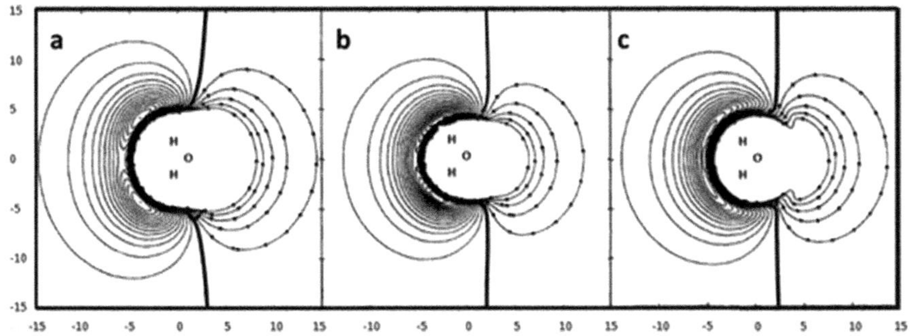

Fig. 16.35 Comparison of the Hartree–Fock potential (**a**) with the Rowlinson (**b**) and Ben Naim – Stillinger (**c**) potentials. The contour interval is 0.0005 a.u. Solid lines represent negative (attractive) energies, lines with dots represent positive (repulsive) energies, and the heavy line with crosses is the nodal contour of zero energy (relative to the energy of the separated monomers). The scale is in atomic units with the oxygen atom at the origin. Modified from Popkie et al. (1973)

were: $a_1 = 113.$ 996,966, $a_2 = 1.242839$, $a_3 = 6.$ 508,362, b1 = 2.100760, $b_2 = 1.653882$, b3 = 2.071387, $q^2 = 0.419427$. The $O_5{-}M_7$ (and $O_6{-}M_8$) distance M was reduced to 0.427 a.u.

With the relaxed constraints the standard deviation was 0.0018 a.u. for all 230 HF computations points while it remained at 0.0002 a.u. for the binding points. Stemming from this new potential Kistenmaker et al. (1974) investigated the magnitude of two-body contributions in a trimeric assemblage $(H_2O)_3$. This kind of study is not relevant for the gaseous state but it is of interest as a prodromal approach to the structure-energy of "*water*" (i.e. the liquid aggregation state). For the $(H_2O)_3$ complex, the best configuration's optimization yielded a cyclic geometry with nearly C_3 symmetry, The energy of stabilization for this structure obtained through an analytical fit of the HF sampling was −0.0196 *Hartree*, corresponding to −51 *kJ/mol*. To understand the significance of this number it is necessary to anticipate that if we designate with the numerals 1, 2, and 3 the first, second, and third molecules of H_2O in the $(H_2O)_3$ complex, the energy for anyone of the three molecules at infinite separation will be $V_1 = V_2 = V_3$ (zero interaction). Designating V_{ij} the interaction energy for the molecules i and j

$$V_{ij} = E_{ij} - V_i - V_j \qquad (16.10.31)$$

with E_{ij} as the total energy of a dimer of water molecules, having the geometry of the ith and jth molecules in the $(H_2O)_3$ complex, the total energy E_{123} for the $(H_2O)_3$ complex, is given by:

$$E_{123} = V_1 + V_2 + V_3 + V_{12} + V_{13} + V_{23} + V_{123} \qquad (16.10.32)$$

Since the binding is obtained relative to $V_1 + V_2 + V_3$, these quantities are, by definition, equal to zero in the study of the stabilization energy and V_{123} is defined, by the above equation, as the energy needed to make the equality numerically correct. The two-body contributions to the stabilization energy of the trimer obtained by Kistenmaker et al. (1974) are $V_{12} = -0.00667$ a.u., $V_{13} = -0.00671$ a.u., $V_{23} = -0.00610$ a.u. The three-body contribution to the stabilization energy of the trimer is the difference between the total stabilization energy, relative to three molecules of water infinitely separated (−0.02128 a.u.) and the sum of the two-body contributions (−0.01948 a.u.): $V_{123} = -0.00180$ a.u.

In the notation of Reimers et al. (1982) the potential of Kistenmaker et al. (1974) takes the form

$$U_{AB}^{(HF/D)} = U_{AB}^{(HF)} + D2(5, 5') \quad (16.10.33)$$

with $U_{AB}^{(HF)}$ as in Eq. 16.10.30. The form of the dispersion interaction D2 is due to Ahlricks et al. (1976).

Adopting the same basis set of Popkie et al. (1973), Matsuoka et al. (1974) calculated the electronic wavefunctions and energies for the ground state of the H_2O dimer, using the configuration-interaction (CI) approach.

The ground state wavefunction was expanded in the form

$$\Psi = C_{SCF}\Phi_{SCF} + \sum_i C_i\Phi_i^{inter} + \sum_i C_i\Phi_i^{intra} \quad (16.10.34)$$

where Φ's are orthonormal configuration state functions (CSF) and C's are variational coefficients. The Slater Determinants (SD) were built from an orthonormal one-particle basis set of spatial orbitals, expanded in terms of a set of elementary basis functions (EBF) and constrained to be each a singlet state function. The first one of the three classes of functions in Eq. 16.10.34 is the SCF state function, taken as reference. The second and third classes are meant to represent the main contribution to inter- and intra-molecular correlation energies.

Designating as A and B, respectively the two interacting H_2O monomers, the SCF state function was expanded in the form:

$$\Phi_{SCF} = 1\phi_A^2 2\phi_B^2 3\phi_A^2 4\phi_A^2 5\phi_A^2 6\phi_A^2 7\phi_B^2 8\phi_B^2 9\phi_B^2 10\phi_B^2 \quad (16.10.35)$$

The first two orbitals in Eq. 16.10.35 correspond essentially to the 1s orbitals of the two oxygens and were designated by Matsuoka et al. (1974) "*core orbitals*". The other eight orbitals were designated as "*internal orbitals*".

By exciting a pair of electrons from i and j internal orbitals into *k* and *l* SCF virtual orbitals, two classes of α *inter* and *intra* correlation functions were generated. Considering all the possible states arising from an orbital configuration which interact with the SCF there are initially 36,864 *inter* and 37,057 *intra* virtual configurations ($\overline{\Phi}_{ij,\alpha}^{kl}$) to be evaluated. However, adopting the first order Rayleigh-Schrodinger perturbation method the $\overline{\Phi}_{ij,\alpha}^{kl}$ SCF virtual orbitals with SCF as the 0th order wavefunction were converted into two sets of approximate *inter* and *intra* natural orbitals (NO).[14] The orbitals in these sets were then ordered according to their occupation number and the first 10 of each set were excluded because equivalent to SCF occupied orbitals. Then an "*ordered orbital set*" (indexed as ϕ_{11} to ϕ_{26}) was obtained by combining the core and internal orbitals with the eight leading orbitals each from $\{\phi^{\mathrm{inter}}\}$ and $\{\phi^{\mathrm{intra}}\}$ and orthonormalizing them. A more limited number of SCF was in this way generated:

$$\{\Phi^{\mathrm{inter}}\} = \left\{\Phi_{\mathrm{ij},\alpha}^{kl}\right\} \quad \mathrm{i} = 3, \ldots, 6; \mathrm{j} = 7, \ldots, 10; k, l = 11, \ldots, 26 \quad (16.10.36.1)$$

$$\{\Phi^{\mathrm{intra}}\} = \left\{\Phi_{\mathrm{ij},\alpha}^{kl}\right\} \text{ and } \left\{\Phi_{\mathrm{p}}^{\mathrm{q}}\right\} \begin{matrix} \mathrm{i,j} = 3, \ldots, 6; \text{ or } = 7, \ldots, 10; \\ k, l = 11, \ldots, 18; p = 3, \ldots, 10; q = 11, \ldots, 26 \end{matrix} \quad (16.10.36.2)$$

with $\left\{\Phi_p^q\right\}$ SCF's generated by single excitation. The number of SCF so obtained was 5280 [4096 $\{\Phi^{\mathrm{inter}}\}$ and 1184 $\{\Phi^{\mathrm{intra}}\}$.respectively]. The final CI wavefunction was:

$$\Psi = \mathrm{C_{SCF}}\Phi_{\mathrm{SCF}} + \sum_{\mathrm{i}=1}^{4096} \mathrm{C_i}\Phi_{\mathrm{i}}^{\mathrm{inter}} + \sum_{\mathrm{i}=1}^{1084} \mathrm{C_i}\Phi_{\mathrm{i}}^{\mathrm{intra}} \quad (16.10.37)$$

The potential energies for the water dimer computed by Matsuoka et al. (1976) shows that the maximum dimerization binding energy is achieved in a linear configuration (−23.4 *kJ/mol*) and, more important, that a relevant part of the energy arises from correlation effects (−4.6 *kJ/mol*).

In the notation of Reimers et al. (1982) the potential of Matsuoka et al. (1976) has form

$$\mathrm{U_{AB}^{(CI)}} = \mathrm{U_{AB}^{(HF)}} + \mathrm{EA}(1, 3') + \mathrm{EA}(2, 3') + \mathrm{EA}(3, 1') + \mathrm{EA}(3, 2') \quad (16.10.38)$$

This form is valid also for the potential developed by Lie et al. (1976) for water.

[14] NO's are defined as the eigenfunctions of the one-electron reduced density matrix, the eigenvalues of which are referred to as occupation numbers (Hay, 1973).

Chapter 17
Volcanic Gases

The atomic composition of volcanic gases is conveniently described by the C–O–H–S system, since these four atoms concur to form almost all the gaseous molecules associated with magmas. The major molecular species present in volcanic gases are H_2O, CO_2, SO_2.[1] The amount of SO_2 released by volcanoes in the Earth's atmosphere is impressive. In the eruption of Mount *Pinatubo* (Philippines; June 1991), the second-largest terrestrial *eruption* of the twentieth century, the volcano released into the atmosphere 1.5×10^{10} *kg* of SO_2. The sulfuric acid aerosols rapidly formed by oxidation in the presence of water cooled the Earth's surface by 1–2 °*C*. The more recent eruption of Manaru Voi (Ambae Island, Vanuatu, South Pacific Ocean; July 2018) injected in the atmosphere 6×10^8 *kg* of SO_2. Apart from these impressive events, the bulk volcanic eruptions on Earth release, yearly, roughly 2×10^8 *kg* of SO_2. The second sulfur species in order of relative abundance is H_2S, followed by S_2, whose partial pressure is two orders of magnitude lower. OCS, SO_3, CS_2, and S_8 occur at level of traces and ultra-traces. Besides the above species, chlorine follows in order of abundance, in the form of hydrochloric acid HCl, conspicuous in low-T (i.e. below 600 °*C*) fumaroles. The amount of HCl in high-*T* volcanic gases does not generally exceed 1% in moles. Other atoms forming gaseous molecules in volcanic emissions are F, N, and B. Fluorine is present in the form of hydrofluoric acid HF, typically in a ratio of 1/10 with respect to HCl. Nitrogen is present as N_2 and NH_3. Bulk amounts of N and B are, however, extremely low, and the presence of N_2 is indicative of probable atmospheric contamination. In this chapter, besides discussing some classical Geochemical aspects of volcanic gases, I introduce the results of some quantum mechanical investigation on the compositional effect of the melt on sulfur abundance and speciation.

[1] I already introduced H_2O and CO_2 as "*atmospheric gases*", and this emphasizes how ephemeral is the definition.

G. A. Ottonello, *Quantum Geochemistry*, Springer Geochemistry,
https://doi.org/10.1007/978-3-031-21837-8_17

17.1 Stoichiometric Constraints: The Role of Oxygen

Most gaseous emissions associated with basaltic lavas have compositions lying along a tie-line connecting stoichiometry:

$$H:O = 2:1 \tag{17.1.1}$$

with stoichiometry:

$$C + S:O = 2:1 \tag{17.1.2}$$

This confirms that the major molecular species present in volcanic gases are H_2O, CO_2, and SO_2. Figure 17.1 shows the O–H–S–C compositional tetrahedron after Gerlach and Nordlie (1975a).

In this representation, the compositional limits of volcanic gases are composed of vertical planes with atomic ratios:

$$C:S = 1:2 \tag{17.1.3}$$

and:

$$C:H = 1:1 \tag{17.1.4}$$

respectively, by the H–O–C plane and the surfaces established by the hematite–magnetite (HM) and quartz–fayalite– magnetite (QFM) oxygen-buffers:

$$3Fe_2O_3 \Leftrightarrow 2Fe_3O_4 + \frac{1}{2}O_2 \tag{17.1.5}$$

$$3SiO_2 + 2Fe_3O_4 \Leftrightarrow 3Fe_2SiO_4 + O_2 \tag{17.1.6}$$

The upper surface in the compositional tetrahedron of Fig. 17.1 corresponds to f_{O_2} conditions determined by buffer Eq. (17.1.5) at T = 250 °*C* and represents the highest oxidation limit observable in terrestrial volcanic gases. The surface immediately below this one is defined by the same buffering equilibrium at T = 800 °*C*, and the lower one (buffer 17.1.6) is attained at maximum reduction. According to Gerlach and Nordlie (1975a, b, c), this compositional space is representative of the chemistry of all terrestrial volcanic gases.

Although, as already stated, the major molecular species present in volcanic gases are H_2O, CO_2, and SO_2, several other molecules also occur, although in subordinate amounts. Figure 17.2 shows the complete speciation in gaseous phases with different C/H, C/S relative amounts at T = 1250 °*C*, as a function of f_{O_2}. As we can see, molecules S_2, H_2, H_2S, COS, S, SO_3, CS_2 and O_2 are also present in molar proportions dependent on f_{O_2}.

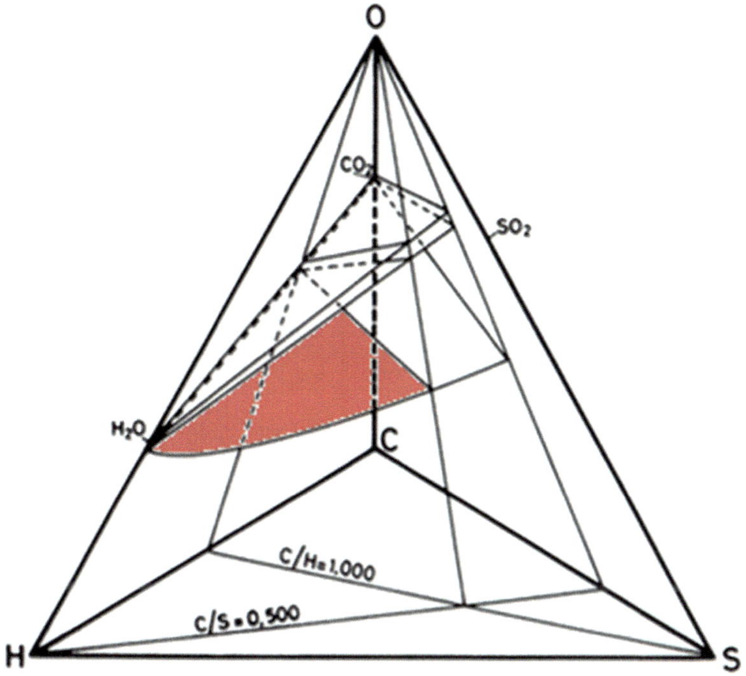

Fig. 17.1 O–H–S–C compositional tetrahedron. The red surface delineates the maximum reduction. Modified from Gerlach and Nordlie (1975a)

Besides buffers (17.1.5, 17.1.6), Fig. 17.2 shows the location of nickel–nickel oxide (NNO) and hematite–wüstite (HW) f_{O_2}-buffering equilibria:

$$Ni + \frac{1}{2}O_2 \Leftrightarrow NiO \tag{17.1.7}$$

$$2FeO + \frac{1}{2}O_2 \Leftrightarrow Fe_2O_3 \tag{17.1.8}$$

Another equilibrium that controls the chemistry and speciation of volcanic gases is the multiple buffer involving gaseous species S_2 and O_2:

$$Fe_3O_4 + 3S_2 \Leftrightarrow 3FeS_2 + 2O_2 \tag{17.1.9}$$

Application of the buffering equilibrium of Eq. (17.1.9) to the atomic composition of the gaseous emissions of Kilawea (Hawaii) (Gerlach and Nordlie 1975c) results in a speciation state rather similar to that actually observed by analysis (Table 17.1). In calculating mass balance through equilibrium constant $K_{17.1.9}$, the authors assumed O_2 and S_2 to behave as perfect gases, i.e.: $f_{O_2} = P_{O_2}$ and $f_{S_2} = P_{S_2}$. This assumption is plausible, since T is high and bulk pressure is low.

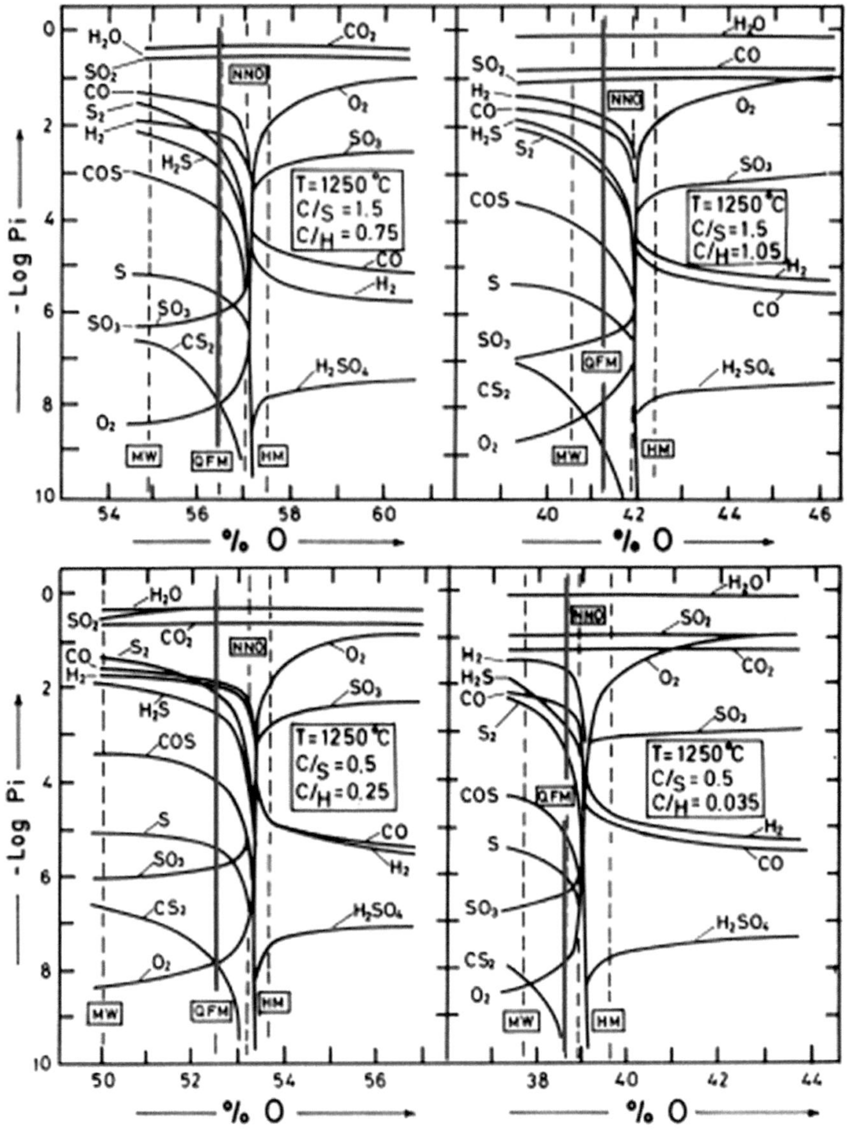

Fig. 17.2 Speciation of volcanic gases with different C/H and C/S relative amounts as a function of f_{O_2} at T = 1250 °*C*. QFM buffer marked in red. Modified from Gerlach and Nordlie (1975b)

Table 17.1 Calculated speciation for gases of Alae Crater (Hawaii), compared with analytical results of gaseous emissions of Kilawea (Hawaii). From Gerlach and Nordlie 1975c)

Gaseous species	T = 800 °*C*, P = 1 *bar*, C/S = 1.5)	
	Alae crater (calculated)	Kilawea emissions (measured; Nordlie 1971)
H_2O	4.8×10^{-1}	3.1×10^{-1}
CO_2	3.1×10^{-1}	4.1×10^{-1}
SO_2	2.0×10^{-1}	2.7×10^{-1}
H_2S	3.6×10^{-3}	2.3×10^{-3}
H_2	1.2×10^{-3}	6.9×10^{-4}
S_2	2.0×10^{-3}	2.3×10^{-3}
CO	7.2×10^{-4}	8.6×10^{-4}
OCS	6.9×10^{-5}	8.9×10^{-5}
SO_3	4.3×10^{-8}	6.6×10^{-8}
CS_2	3.5×10^{-9}	4.4×10^{-9}
O_2	7.0×10^{-14}	8.6×10^{-14}
S_8	1.3×10^{-15}	2.5×10^{-15}

17.2 Sulfur Species

17.2.1 Thermophysical and Thermodynamic Properties

Sulfur dioxide, SO_2, has a resonance structure. By moving an electron pair in one of the double bonds towards oxygen, it is originated a lone pair on one of the two oxygen atoms, which develops a charge of −1 (Fig. 17.3: upper part).

With 1 π bond and 2 σ bonds, the bond order ($1 + \pi/\sigma$) is 1.5. The ground state of SO_2 is $^1A_1$. The structure is quite similar to that of H_2O and O_3, with C_{2V} symmetry. According to Herzberg (1966), the interatomic distance is 1.432 *Å*, and the bent angle is 119.5°. The molecule is an asymmetric top with rotational symmetry number 2. In the lower part of Fig. 17.3, it is shown the HOMO of SO_2 computed at B3LYP/6-31G(d,p) theory level for an isodensity of 0.02 $e/Å^3$. When resolved at B3LYP/6-31G(d,p) theory level, the S–O distance is 1.463 *Å* and the O–S–O angle is 119.05°.

The total electric dipole moment is 1.630 *Debye* and the components of the electric quadrupole moment are $xx = 3.860\ D \times Å$, $yy = -4.910\ D \times Å$, $zz = 1.020\ D \times Å$. The electrical dipole polarizability is $\alpha = 3.882\ Å^3$ (NIST 2000), and the calculated one at B3LYP/6-31G(d,p) level is $\alpha = 2.872\ Å^3$.

The experimental harmonic frequencies listed by NIST (2000) are $\nu_1 = 1168\ cm^{-1}$ (a_1), 516 cm^{-1} (a_1), 1381 cm^{-1} (b_2). The frequencies obtained at B3LYP/6-31G(d,p) level with scaling (0.9612) are $\nu_1 = 1097.4\ cm^{-1}$ (a_1), 483.3 cm^{-1} (a_1), 1286.8 cm^{-1} (b_2). All frequencies are IR and Raman active. The (unscaled) zero point energy at this theory level is 0.006797 *Hartree*. The ionization

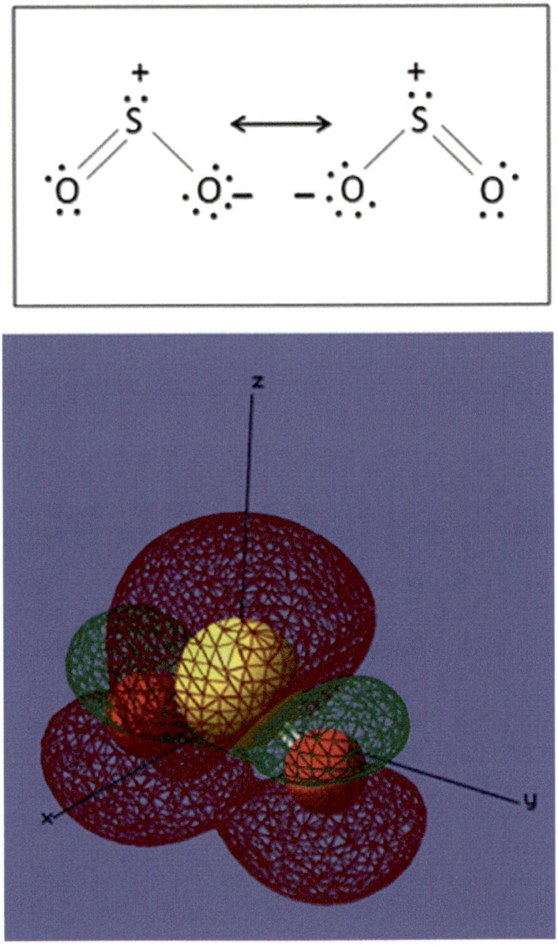

Fig. 17.3 Lewis structure of SO_2 (upper part) and B3LYP/6-31G(d,p) HOMO at 0.02 $e/Å^3$ isodensity. Van der Waals radii adopted

energy according to NIST (2000) is 12.394 *eV*, and the proton affinity is 672.3 *kJ/mol*. The dissociation energy computed by Ochtersky et al. (1996) at G2(MP2) theory level is 1060.1 *kJ/mol*, against a reported experimental energy of 1062.7 *kJ/mol* (see Table 7.32). At standard state (pure ideal gas, $P = 10^{-5}$ *Pa*, 298.15 *K*), SO_2 has an entropy of 248.212 $J/(mol \times K)$ and an isobaric heat capacity of 39.878 $J/(mol \times K)$ (Chase 1998). The enthalpy of formation from the elements at standard state is −296.842 *kJ/mol* (Chase 1998). When resolved at B3LYP/6-31G (d,p) theory level, the returned entropy is 248.786 $J/(mol \times K)$ and the enthalpy of formation from the elements at standard state ($H°_f$) is −165.208 *kJ/mol*. Carrying out the calculations at G2 theory level $H°_f$ lowers to −275.7 *kJ/mol* (Curtiss et al. 1995), indicating a substantial role of correlation effects in stabilizing the substance. The Shomate equation for the isobaric heat capacity in the range $298.15 \leq T\ (K) \leq 1200$ is $C_P = 21.43049 + 74.35094 \times 10^{-3} \times T - 57.75217 \times 10^{-6} \times$

$T^2 + 16.35534 \times 10^{-9} \times T^3 + 8.6731 \times 10^4 \times T^{-2}$ (NIST 2000; based on refitted data of Chase 1998). The melting point is at 200.75 *K*. The critical temperature is $T_c = 430.34$ *K*. The triple point occurs at $T_t = 197.64$ *K* and $P_t = 0.0167$ *bar* (NIST 2020).

Dihydrogen sulfide, H_2S, has again C_{2V} symmetry and its ground state is $^1A_1$. The structure is quite similar to that of SO_2 with a lone pair localized on S. The bond is purely covalent since the sulfur octet is obtained by sharing one electron with each of the two hydrogen atoms. According to Cook et al. (1975), the interatomic distance is 1.336 *Å*, and the bent angle is 92.11°. The molecule is an asymmetric top with rotational symmetry number 2. When resolved at B3LYP/6-31G(d,p) theory level, the S–O distance is 1.348 *Å* and the H–S–H angle is 92.70°.

The experimental harmonic frequencies listed by NIST are $\nu_1 = 2722$ cm^{-1}, $\nu_2 = 1215$ cm^{-1}, $\nu_3 = 2733$ cm^{-1}. The frequencies obtained at B3LYP/6-31G(d,p) level with scaling (0.9612) are $\nu_1 = 2582$ cm^{-1} (A_1), $\nu_2 = 1177$ cm^{-1} (A_1), $\nu_3 = 2601$ cm^{-1} (B_2). The dissociation energy computed by Ochtersky et al. (1996) at CBS-Q theory level is 723.8 *kJ/mol*, against a reported experimental energy of 724.7 *kJ/mol* (see Table 7.32).

Diatomic sulfur, S_2, (or, more commonly, "*disulfur*") has $D_{\infty h}$ symmetry and $^3\Sigma_g$ ground state. Like O_2, it is an open shell (triplet) with two unpaired electrons. According to Huber and Herzberg (1979), the interatomic distance is 1.889 *Å*. The molecule is a prolate symmetric top with rotational symmetry number 2. When resolved at B3LYP/6-31G(d,p) theory level, the S–S distance is 1.927 *Å*. The experimental harmonic frequency is $\nu_1 = 720$ cm^{-1} (Huber and Herzberg 1979), while the frequency obtained at B3LYP/6-31G(d,p) level with scaling (0.9612) is $\nu_1 = 781.9$ cm^{-1} (SGG; Raman active). The dissociation energy computed by Ochtersky et al. (1996) at G2 theory level is 427.6 *kJ/mol*, against a reported experimental energy of 421.3 kJ/mol (see Table 7.32).

Carbonyl sulfide, OCS, has $C_{\infty v}$ point group symmetry and $^1\Sigma$ ground state. It is linear with an O–C interatomic distance of 1.160 *Å* and a C–S distance of 1.560 *Å* (Herzberg 1966). The molecule is a prolate symmetric top with rotational symmetry number 2. When resolved at B3LYP/6-31G(d,p) theory level, the S–S distance is 1.927 *Å*. The harmonic frequencies obtained at B3LYP/6-31G(d,p) level with scaling (0.9612) are $\nu_{1,2} = 489.8$ (PI), $\nu_3 = 837.7$ cm^{-1} (SG) and $\nu_4 = 2047.2$ cm^{-1} (SG). All frequencies are IR and Raman active.

Sulfur trioxide, SO_3 has a trigonal planar structure, belonging to the D_{3h} point group and its ground state is $^1A_1$. The substance undergoes mesomerism with double resonance as schematically shown in Fig. 17.4.

According to Meyer et al. (1991), the substance is nonpolar with S–O–S angles of 120° and S–O interatomic distances of 1.418 *Å*. The molecule is an asymmetric top with rotational symmetry number 2. When resolved at B3LYP/6-31G(d,p) theory level, the S–O distance is 1.453 *Å*. The harmonic frequencies obtained at the same theory level with scaling (0.9612) are $\nu_1 = 489.8$ cm^{-1} (IR active), $\nu_{2,3} = 478.4$ cm^{-1} (IR and Raman active), $\nu_4 = 983.7$ cm^{-1} (Raman active) and $\nu_{5,6} = 1299.5$ cm^{-1} (IR and Raman active).

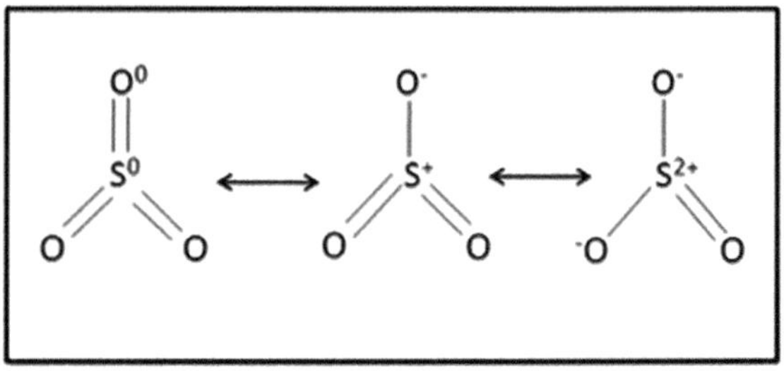

Fig. 17.4 Mesomerism in SO_3

Carbon difsulfide, CS_2, has $D_{\infty h}$ symmetry and ${}^1\Sigma_g$ ground state. At standard state, it occurs at the liquid aggregation state but its 1-*bar* boiling point is at 319.19 *K*, hence it is eminently gaseous in volcanic regimes. In the gaseous state, it is a linear molecule with a C–S interatomic distance of 1.554 *Å* (Herzberg 1966). The molecule is a prolate symmetric top with rotational symmetry number 2. When resolved at B3LYP/6-31G(d,p) theory level, the C-S distance is 1.563 *Å*. The harmonic frequencies obtained at the same theory level with scaling (0.9612) are $\nu_{1,2} = 388.9\ cm^{-1}$ (PIU), $\nu_3 = 647.9\ cm^{-1}$ (SGG), $\nu_4 = 1497.7\ cm^{-1}$ (SGU). All frequencies are IR active but none is Raman active.

Octasulfur, S_8, has a "*crown*" conformation with D_{4d} symmetry and its ground state is ${}^1A_1$. Its boiling point is at 717.8 *K*. According to Cotton and Wilkinson (1972), all S–S bond lengths are almost equal, at 2.059 ± 0.002 *Å*, the S–S–S-angles are 107.9° ± 0.6, and the S–S–S–S dihedral angles are 98.9°. The molecule is an asymmetric top with rotational symmetry number 1. When resolved at B3LYP/6-31G(d,p) theory level, the S–S distance is 2.0996 ± 0.0003 *Å*; the S–S-S angles are 109.158° ± 0.002, and the S–S–S–S dihedral angles are 97.35° ± 0.06. As we may see in Fig. 17.5, the "*crown*" conformation arises from the HOMO polarities with alternate positive wave functions along the *z*-axis, while the negative wavefunction forms a continuous distorted ring on the *xy* plane.

The thermophysical properties of H_2S and of the other minor sulfur species in volcanic gases are resumed in Table 17.2.

Care must be exercised when comparing theoretical and experimental polarizabilities. In fact, while the calculated polarizabilities are usually zero-frequency (static) electronic polarizabilities (α), the experimental ones (α_{exp}) contain, besides the electronic term (α_{el}) also vibrational (α_{vib}) and orientation (α_{or}) contributions:

$$\alpha_{exp} = \alpha_{el} + \alpha_{vib} + \alpha_{or} \tag{17.2.1}$$

In Table 17.3 I report the results of MP2/def2-QZVPD calculations carried out by Thakkar and Wu (2015) on various sulfur species of geochemical interest. Polarizability data have been converted from a.u. into $Å^3$ by scalar multiplication with factor 0.1481847. With the adopted basis set, developed by Rappoport and Furke (2010), stemming from the original def2-QZVP basis set of Weigend and Ahlrichs (2005), by adding a small number of diffuse basis functions chosen to

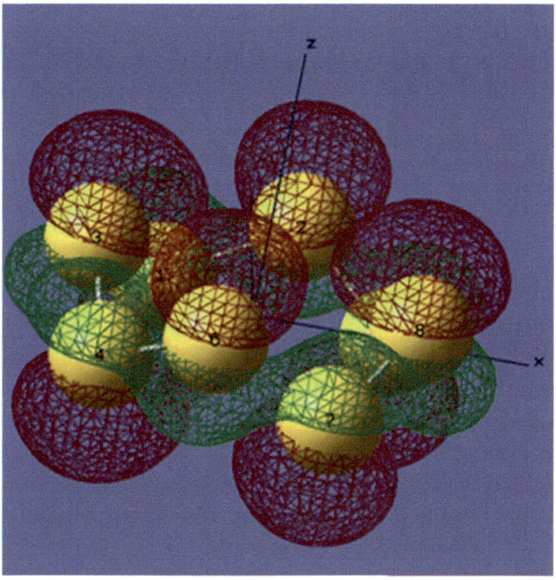

Fig. 17.5 B3LYP/6-31G (d,p) HOMO of S_8 at 0.02 *e/Å*^{3} isodensity. Van der Waals radii adopted

Table 17.2 Thermophysical properties of minor sulfur species

Property	H_2S	OCS	SO_3	S_2	CS_2	S_8
State	$^1A_1$	$^1\Sigma$	$^1A_1$	$^3\Sigma_g$	$^1\Sigma_g$	$^1A_1$
Configuration	C_{2v}	$C_{\infty v}$	D_{3h}	$D_{\infty h}$	$D_{\infty h}$	D_{4d}
$T_{boiling}$ *(K)*	212.87[a]	–	–	–	319.2[c]	–
T_{fusion} *(K)*	190.85[b]	–	290.1[l]	–	161[c]	–
T_{triple} *(K)*	187.66[a]	134.31[f]	–	–	161.59[n]	–
P_{triple} *(bar)*	0.232[a]	–	–	–	–	–
$T_{crytical}$ *(K)*	373.5[a]	378.8[g]	–	–	552[o]	–
$P_{crytical}$ *(bar)*	89.70[a]	63.49[g]	–	–	–	–
I_e *(eV)*	10.457[c]	11.18[c]	12.80[c]	9.356[c]	10.073[c]	9.3[p]
P_a *(kJ/mol)*	705.0[d]	628.5[d]	588.3[d]	–	681.9[d]	–
El. Dip. Mom. *(D)*	0.970[c]	0.715[c]	0	0	0	–
Quad. Mom. xx $(D\times Å)$	–	0.292[h]	–	–	-1.724	–
Quad. Mom. yy $(D\times Å)$	–	0.292	–	–	-1.724	–
Quad. Mom. zz $(D\times Å)$	–	-0.584	–	–	3.448[p]	–
Dipole polar. $(Å^3)$	3.631[e]	5.090[i]	4.297[m]	–	8.749[c]	–

(a) Goodwin (1983); (b) Beckmann and Waentig (1910); (c) NIST (2000); (d) Hunter and Lias (1998); (e) Olney et al. (1997); (f) Clusius and Weigand (1940); (g) Robinson and Senturk (1979); (h) Benkova and Sadlej (2004); (i) Gussoni et al. (1998); (l) Kondrat'ev and Ztrizhov (1985); (m) Thakkar and Wu (2015); (n) Stull (1937); (o) Majer and Svoboda (1985); (p) Watson et al. (1997)

Table 17.3 MP2/def2-QZVPD static electronic dipole polarizabilities α (in $Å^3$) (Thakkar and Wu 2015) compared with the experimental values (α_{exp}) recommended by Hohm (2013)

Molecule	α	α_{exp}
SO_2	3.79	3.80 ± 0.07
H_2S	3.63	3.66 ± 0.01
COS	5.14	5.00 ± 0.09
CS_2	8.38	8.19 ± 0.15
SO_3	4.25	4.24 ± 0.22

maximize atomic Hartree–Fock static polarizabilities, the results are extremely satisfactory.

Table 17.4 resumes the results of few ab initio calculations I carried out on the minor sulfur species at B3LYP/6-31G(d,p) and G2 theory levels, with the procedures outlined at Sects. (7.10.3, 7.10.4). When compared with experimental evidences and assessments, we may note that the computations return entropy values in reasonable agreement with reported data, while serious biases are evident in the values of the enthalpy of formation from the elements at standard state obtained with the simple DFT procedure.

17.2.2 Redox-Controlled Speciation of Sulfur in Magmas

A simple glance to Fig. 17.2 is sufficient to appreciate how dramatic is the role of oxygen fugacity in determining the speciation of sulfur-bearing gases. In a "*magma*", i.e. a system where crystals, melts, and gases may be conceived to coexist stably at equilibrium, the redox state is controlled by the presence of iron oxides, and, subordinately, by other components of multivalent elements. The mass fraction of melt and crystals usually dominates, hence the f_{O_2} buffered by homogeneous equilibria in the liquid is immediately transferred to the gas phase as well. It is worth to stress that in many experiments this operative condition is not observed and, rather, it is the gaseous mixture that imposes its intrinsic redox state to melt equilibria. The case of sulfur is particularly illustrative of this dual approach. In a seminal paper appeared more than half century ago Fincham and Richardson (1954) wrote "*As part of a general programme of research on the nature of inorganic melts, a study has been made of the manner in which sulphur is held in liquid silicates and aluminates. Mixtures of CaO–SiO*$_2$*, MgO–SiO*$_2$*, FeO–SiO*$_2$*, CaO–Al*$_2$*O*$_3$ *and CaO–SiO*$_2$*–Al*$_2$*O*$_3$ *have been brought into equilibrium, at temperatures between 1350 and 1650 °C, with gas phases made by mixing H*$_2$*, CO*$_2$ *and SO*$_2$ *at room temperature. It has been shown that, when* P_{O_2} *is less than about 10*$^{-5}$ *atm, a sulphur atom can only enter the melt by displacing a suitable oxygen atom. The sulphur is then held in the slag entirely as sulphide, its concentration being controlled by the general equilibrium,* $\frac{1}{2}S_2 + (O)melt = \frac{1}{2}O_2 + (S)melt$.

Table 17.4 Thermodynamic properties of minor sulfur species obtained at B3LYP/6-31G(d,p) theory level and with the G2 procedure (values in italics), compared with various assessments. $C_p = A + B \times T + C \times T^2 + D \times T^3 + E \times T^{-2}$

Property	Dimension	H_2S	OCS	SO_3	S_2	CS_2	S_8	Refs.
Mult		1	1	1	3	1	1	1
E_{el}	*Hartree*	−399.3916241	−511.5355463	−623.7592096	−796.3628899	−834.4888683	−3185.5505036	1
		−398.9240740	*−510.9395470*	*−623.1146170*	*−795.4521770*	*−833.5216510*		1
ZPC	*Hartree*	0.015075	0.009159	0.011795	0.001598	0.006929	0.011797	1
		0.014685	*0.008804*	*0.012303*	*0.001655*	*0.006508*		1
Therm U	*Hartree*	0.017923	0.012015	0.015391	0.004071	0.010027	0.022971	1
		0.002848	0.002881	*0.003516*	*0.002463*	*0.003127*		1
Therm H	*Hartree*	0.018867	0.012959	0.016336	0.005015	0.010971	0.023915	1
		0.003793	*0.003825*	*0.00446*	*0.003408*	*0.004071*		1
Therm G	*Hartree*	−0.004490	−0.013365	−0.01468	−0.02092	−0.016003	−0.026984	1
		−0.019517	*−0.022512*	*−0.026333*	*−0.022465*	*−0.022926*		1
U (therm.)	*kJ/mol*	47.057	31.543	40.409	10.686	26.326	60.308	1
S°	*J/(mol ×K)*	205.685	231.806	273.119	228.379	237.526	448.211	1
		205.267	*231.923*	*270.282*	*227.837*	*237.735*		1
		205.757	231.57	256.769	228.165	237.977	430.311	2
$C_{V,298.15}$	*J/(mol ×K)*	25.744	33.359	43.614	24.242	36.941	149.950	1
$C_{P,298.15}$	*J/(mol ×K)*	34.192	41.497	50.661	32.490	45.664	156.043	2
$H°_f$	*kJ/mol*	−8.035	−118.302	−199.99	151.603	143.027	356.759	1
		−23.967	*−158.818*	*−386.696*	*133.509*	*95.574*		1
		−20.502	−138.41	−395.765	128.600	116.949	100.416	2
							381.09	3
A	*J/(mol ×K)*	26.884120	34.538920	24.025030	33.513130	35.853910	180.669700	4
$B \times 10^3$	*J/mol*	18.678090	43.053780	119.460700	5.065360	52.491210	1.918988	4

(continued)

Table 17.4 (continued)

Property	Dimension	H_2S	OCS	SO_3	S_2	CS_2	S_8	Refs.
$C \times 10^6$	$J \times K/mol$	3.434203	−26.617730	−94.386860	−1.059670	−40.837430	−0.520130	4
$D \times 10^9$	$J \times K^2/mol$	−3.378702	6.338844	26.962370	0.089905	12.001550	0.044394	4
$E \times 10^{-6}$	$J/(mol \times K^3)$	0.135882	−0.327515	−0.117517	−0.211911	−0.224831	−2.270155	4
T-limit	K	1400	1200	1200	6000	1000	6000	4

(1) this work; (2) Chase (1998); (3) Joback and Reid (1987); (4) NIST (2000)

When P_{O_2} is greater than about 10^{-3} atm, the sulphur is held as sulphate and the corresponding equilibrium is ½S_2 +3/2O_2 + (O)melt = (SO_4)melt" (Fincham and Richardson 1954).

Let me consider now the dominant sulfur species of volcanic gases, i.e. SO_2. We may write two distinct heterogeneous equilibria relating SO_2 with the main sulfur species in the liquid:

$$SO_{2,g} + \frac{1}{2}O_{2,g} + O_m^{2-} \Leftrightarrow SO_{4,m}^{2-} \quad (17.2.2)$$

$$SO_{2,g} + O_m^{2-} \Leftrightarrow S_m^{2-} + \frac{3}{2}O_{2,g} \quad (17.2.3)$$

Equilibria (17.2.2, 17.2.3) are similar to those ones suggested by Fincham e Richardson (1954) and involving native sulfur. Because molecular oxygen appears in the reactant side of equilibrium (17.2.2) and in the product side of equilibrium (17.2.3), the mass action effects demand that the amount of reduces species S^{2-} in the melt decreases with the increase of fO_2 in the gas phase and the amount of the oxidized species SO_4^{2-} in the melt increases with the increase of f_{O_2}. Moreover, the stoichiometric coefficients of O_2 (−1/2 and 3/2) are different as well. We may thus identify two distinct redox regimes where one of the two equilibria is dominant. Katsura and Nagashima (1974) showed that, due to this fact, the sulfur solubility in natural basaltic melts equilibrated with a H_2–CO_2–SO_2 gaseous mixture exhibits a minimum in the fugacity range $0 < \Delta_{QFM} < +1$ (with Δ_{QFM} representing the deviation from the f_{O_2} imposed by the Quartz–Fayalite–Magnetite assemblage). The buffering gaseous mixture contained 2.1% of SO_2 in weight. The relevant data are resumed in Table 17.5. Based on these data, for oxygen partial pressures (P_{O_2}) higher than 10^{-7}, all sulfur was present in the melt as SO_4^{2-} while, for log $P_{O_2} < -7$, all sulfur dissolved as S^{2-}.

Besides the oxygen fugacity, also the melt composition affects sensibly the bulk solubility and speciation of sulfur (Carroll and Rutherford 1988; Carroll and Webster 1994). The compositional effects superimpose on the simple redox control imposed by the gaseous assemblage and the simple logarithmic relation a priori envisageable between bulk sulfur amount and oxygen fugacity is apparently observed only in the immediate neighborhood of the solubility minimum (Fig. 17.6).

Table 17.5 Sulfur contents (weight percent) in the experiment of Katsura and Nagashima (1974) on an Hawaiian tholeiite melt at 1250 °*C* and 1 *bar*, with 2.1% of SO_2 in the gaseous phase

Log P_{O_2}	S_{total}	S^{2-}	SO_4^{2-}
−3.65	0.006	<0.001	0.006
−5.38	0.003	<0.001	0.003
−6.79	0.001	<0.001	0.001
−8.97	0.009	0.008	0.001
−10.56	0.022	0.022	<0.001
−12.2	0.11	0.11	<0.001

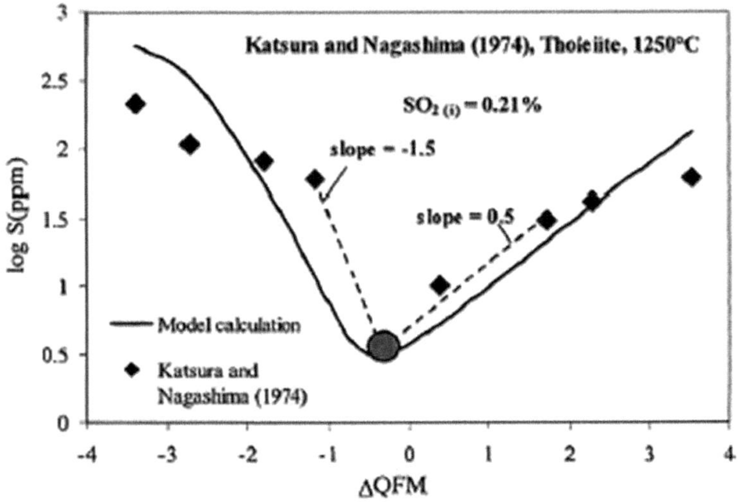

Fig. 17.6 Bulk sulfur solubility in tholeiite according to the experimental results of Katsura and Nagashima (1974). The circle shows the position of the expected minimum, based on theoretical slopes. The solid line returns model calculations based on a polymeric approach to silicate melts coupled with a Flood-Grjotheim treatment of ionic quasi-chemical species. From Moretti and Ottonello (2005). With permission from Geochimica et Cosmochimica Acta

17.2.3 *Compositional Effect of the Melt: The Classical Approach*

To understand the compositional effect of the melt, it is convenient to introduce the concepts of "*sulfide capacity*" (Fincham and Richardson 1954) and "*sulfate capacity*" (Moretti and Ottonello 2003a, b):

$$C_{S^{2-}} = [S]_{wt\%}\left(\frac{f_{O_2}}{f_{S_2}}\right)^{1/2} \tag{17.2.4}$$

$$C_{S^{6+}} = [S]_{wt\%} f_{O_2}^{-3/2} f_{S_2}^{-1/2} \tag{17.2.5}$$

The compositional control operated by the two above equilibria is easily understood if one admits that every molten sulfide and sulfate contributes to the bulk solubility of sulfur (Moretti and Ottonello 2003b):

$$M_{2/\nu}S \Leftrightarrow M^{\nu+} + S^{2-} \tag{17.2.6}$$

$$M_{2/\nu}SO_4 \Leftrightarrow M^{\nu+} + SO_4^{2-} \tag{17.2.7}$$

and that the amount of free oxygen in the melt depends upon the Lux-Flood character of the molten oxides:

$$M_{2/\nu}O \Leftrightarrow M^{\nu+} + O^{2-} \tag{17.2.8}$$

Equilibria (17.2.6, 17.2.7, 17.2.8) involve "*ionic quasi-chemical species*" whose nature is conveniently interpreted in terms of a Flood-Grjotheim thermochemical cycle (Flood and Grjotheim 1952; see also Gaskell 2000; Moretti and Ottonello 2005; for a detailed treatment).

In short, the disproportionation of oxide, sulfate, and sulfide components, usually described in terms of equilibria with an externally buffered gaseous phase:

$$Ms_{2/\nu}O_{(m)} + \frac{1}{2}S_{2(g)} \Leftrightarrow M_{2/\nu}S_{(m)} + \frac{1}{2}O_{2(g)} \tag{17.2.9}$$

$$M_{2/\nu}O_{(m)} + \frac{1}{2}S_{2(g)} + \frac{3}{2}O_{2(g)} \Leftrightarrow M_{2/\nu}SO_{4(m)} \tag{17.2.10}$$

$$K_{O-S,M} = \frac{a_{M_{\nu/2}S}}{a_{M_{\nu/2}O}}\left(\frac{a_{O_2}}{a_{S_2}}\right)^{1/2} \tag{17.2.11}$$

$$K_{O-SO_4,M} = \frac{a_{M_{\nu/2}SO_4}}{a_{M_{\nu/2}O}}\left(a_{O_2}^{-3/2}a_{S_2}^{-1/2}\right) \tag{17.2.12}$$

are more conveniently generalized in Flood–Grjotheim terms:

$$\ln K'_{O-S} = \sum_{i=1}^{n} N'_{A_i^{\nu+}} \ln C_{A_i} K_{O-S,A_i} \tag{17.2.13}$$

$$\ln K'_{O-SO_4} = \sum_{i=1}^{n} N'_{A_i^{\nu+}} \ln C_{A_i} K_{O-SO_4,A_i} \tag{17.2.14}$$

where N′ are "*electrically-equivalent fractions*":

$$N'_{A^{\nu+}} = \frac{\nu_A^+ n_A}{\nu_A^+ n_A + \nu_B^+ n_B + \ldots} \tag{17.2.15}$$

and C_A, C_B take into account the energy gaps connected to the different standard state notations (see Moretti and Ottonello 2005; for details). The Temkin activity (Temkin 1945) of a given component in a multicomponent melt involves mixing over two distinct matrixes, i.e. a "*cation matrix*" and an "*anion matrix*". The anion matrix contains, besides free ions such as O^{2-}, S^{2-} SO_4^{2-}, also polymeric units SiO_4^{4-}, $Si_2O_6^{4-}$ etc. The Lux–Flood acid–base behavior of each molten oxide causes a disproportionation between "*network formers*" and "*network modifiers*. The activity of a generic oxide M_2O is thus in the presence of H_2O and sulfur species:

$$a_{M_2O} = \left[\frac{(M^{\nu+})}{\sum \text{cations}}\right]^{\frac{2}{\nu}} \times \frac{(O^{2-})}{\left[(O^{2-}) + (S^{2-}) + (SO_4^{2-}) + (OH^{-}) + \sum \text{structons}\right]} \tag{17.2.16}$$

and the relative molar abundances of sulfide and sulfate with respect to free oxygen may be shown, through few passages, to obey (Moretti and Ottonello 2003a, b, 2005):

$$\begin{aligned} \frac{(S^{2-})}{(O^{2-})} &= K_{O-S,M} \\ &\times \exp\left(\frac{\mu^0_{M_{2/\nu}S,\text{melt}} - \mu^0_{M_{2/\nu}O,\text{melt}} + \mu^*_{M_{2/\nu}O,\text{melt}} - \mu^*_{M_{2/\nu}S,\text{melt}}}{RT}\right)\left(\frac{a_{S_2}}{a_{O_2}}\right)^{1/2} \\ &= K'_{O-S,M}\left(\frac{a_{S_2}}{a_{O_2}}\right)^{1/2} \end{aligned} \tag{17.2.17}$$

$$\begin{aligned} \frac{(SO_4^{2-})}{(O^{2-})} &= K_{O-SO_4,M} \times \exp\left(\frac{\mu^0_{M_{2/\nu}SO_4,\text{melt}} - \mu^0_{M_{2/\nu}O,\text{melt}} + \mu^*_{M_{2/\nu}O,\text{melt}} - \mu^*_{M_{2/\nu}SO_4,\text{melt}}}{RT}\right) \times \\ a_{S_2}^{1/2} a_{O_2}^{3/2} &= K'_{O-SO_4,M} \times a_{S_2}^{1/2} \times a_{O_2}^{3/2} \end{aligned} \tag{17.2.18}$$

where the starred chemical potentials define fully-dissociated Temkin-model salts, not to be confused with the usual chemical potentials (μ^0 terms) defined on an energy basis.

Tables 17.6 and 17.7 resume the Arrhenius coefficients of the various equilibrium constants (lnK = A +B/T), according to Moretti and Ottonello (2003a, b, 2005).

The model may be applied equally well to natural silicate melts and metallurgical slags. As an example, we may see in the upper part of Fig. 17.7 how the model reproduces the marked negative correlation between sulfur solubility and acidity of the system, and, in the lower part, the effect of iron on sulfur solubility in reducing conditions.

The satisfactory model performance is evident in the statistical quantile–quantile representation of Fig. 17.8. Consistent deviations are observed only at very low

Table 17.6 Arrhenius coefficients of $K_{O\text{-}S,M}$ and $K'_{O\text{-}S,M}$ (italics)

M	$A_{O-S,M}$	$B_{O-S,M}$	Data source	Remarks
P^{5+}	2.1987	−24,786	a,c	
	1.732	*−24,786*	f	
	2.157	*−24,786*	g	
Si^{4+}	2.0492	−36,418	a,d	
	12.752	*−36,418*	f	
	12.9577	*−36,418*	g	
Ti^{4+}	*−4.7934*	*−16,575.9*	f	
	−4.8018	*−16,575.9*	g	
Al^{3+}	2.1838	−28,926	a	
	7.7087	*−28,926*	f	
	7.8105	*−28,926*	g	
Fe^{3+}	*15.0328*	*−20,000*	f	
	6.6613	*−20,000*	g	
Cr^{3+}	*0*	*0*		
Fe^{2+}	1.7787	−15,265	e	
	−5.91	*−8897.9*	b	
	5.4414	*−15,265*	f	
	8.4132	*−15,265*	g	
Mn^{2+}	−4.4868	−9420.8	b	
	−1.1437	−8391.7	a,d	
	−0.917	*−8391.7*	f	
	−1.0202	*−8391.7*	g	
Ca^{2+}	−5.0645	−6804.2	b	
	−4.0239	*−6804.2*	f	
	−4.229	*−6804.2*	g	
Mg^{2+}	−3.4254	−15,181	b	
	−4.1698	*−15,181*	f	
	−5.0762	*−15,181*	g	
Na^{+}	14.731	−16,396	a	
	14.53	*−16,396*	f	
	11.2533	*−16,396*	g	
K^{+}	3.91	−16,432	a,c	
	11.266	*−16,432*	f	
	10.6606	*−16,432*	g	
H^{+}	0.427	−18,764	d	
	14.7606	*−18,764*	g	With Papale model (Papale 1997)
	11.7851	*−18,764*	g	With Burnham model (Burnham 1975)

(a) Barin and Knacke (1973); (b) Pelton et al. (1993); (c) recalculated from Ghiorso et al. (1983); (d) Barin et al. (1977); (e) Robie et al. (1978); (f) Moretti and Ottonello (2003a); (g) Moretti and Ottonello (2005)

Table 17.7 Arrhenius coefficients of $K_{O-SO_4,M}$ and $K'_{O-SO_4,M}$ (italics). References as in Table 17.6

M	$A_{O-SO_4,M}$	$B_{O-SO_4,M}$	Data source	Remarks
P^{5+}	0	0	g	
Si^{4+}	*−26.8836*	*77,291.1*	g	
Ti^{4+}	*−22.6614*	*98,230*	g	
Al^{3+}	*−31.0955*	*85,879.8*	g	
Fe^{3+}	*−31.9273*	*85,879.8*	g	
Cr^{3+}	*0*	*0*	g	
Fe^{2+}	*−42.0516*	*99,540.8*	g	
Mn^{2+}	*−51.4008*	106,414.8	g	
Ca^{2+}	*−43.5173*	107,754.9	g	
Mg^{2+}	−42.881	95,831	a	
	−36.3719	*95,831*	g	
Na^{+}	−21.812	100,934	a	
	−21.6124	*100,934*	g	
K^{+}	−31.642	100,741	a	
	14.814	100,741	g	
H^{+}	−37.695	67,292	d	
	−60.3994	*67,292*	g	With Papale model (Papale 1997)
	−116.805	*67,292*	g	With Papale model and volumetric annealing
	56.344	*67,292*	g	With Burnham model (Burnham 1975)
	−101.959	*67,292*	g	With Burnham model and volumetric annealing

values of C_S. Most likely these may be ascribed to analytical problems, which become more and more serious with the decrease of sulfur amount in the system and with the increase of the "*sulfurizing power*" (f_{S_2}/f_{O_2}) of the gaseous mixture at equilibrium.

17.2.4 Compositional Effect of the Melt: The Quantum–Mechanical View

I begin with the simplest check, which is the comparison of the solubility of SO_2 in water at standard P, T conditions (i.e. $P = 10^5\ Pa$; $T = 298.15\ K$) predicted by Tomasi's PCM and the experimental observations. As noted by Battino, however (in Young 1983), also this simple comparison is complicated by the fact that reported solubilities are bulk solubilities, "*i.e., they are taken as if all of the SO_2-containing species in solution were simply $SO_{2(aq)}$*". Moreover, due to the high

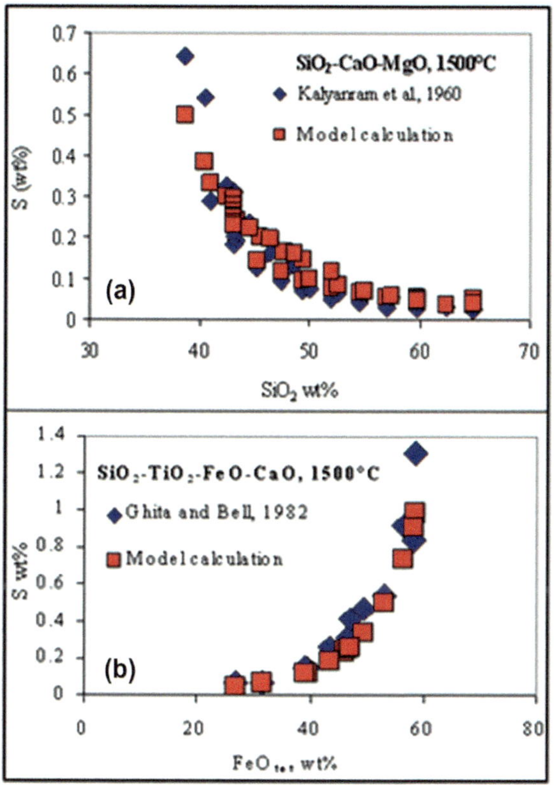

Fig. 17.7 Experiments versus model results. **a** Sulfur solubility as a function of the system acidity (represented by the percentual amount in weight of silica in the system. **b** Effect of the bulk iron amount (expressed as weight percent of total iron as FeO). Experimental data after Kalyanram et al. (1960) (**a**) and Ghita and Bell (1982) (**b**). Model calculations after Moretti and Ottonello (2005)

solubilities exhibited by this system and the chemical interactions involved, the system does not obey Henry's law even over relatively small extrapolations.

Notwithstanding these complexities, in a restricted T-range, SO_2 behaves similarly to other gaseous species investigated so far, namely a marked solubility decrease with the increase of T, ascribable to the entropic part of the cavitation energy (Fig. 17.9). In the investigated T-range ($278.15 \leq T(K) \leq 328.15$) the smoothed data obey the simple relation (Young 1983):

$$\ln x_1 = \left[-25.26286 + 45.7552\tau^{-1} + 5.685450\ln\tau\right] \quad (17.2.19)$$

with $\tau = T(K)/100$ and x_1 as mole fraction in solution.

When resolved at DFT/631G(d,p) SCRF-IEFPCM level, with UAHF radii, the UATM assigns to each oxygen in sp^2 hybridization a radius of 1.590 Å, and to S in sp^3 hybridization a radius of 1.980 Å. The electrostatic scaling factor is $\gamma_{el} = 1.20$.

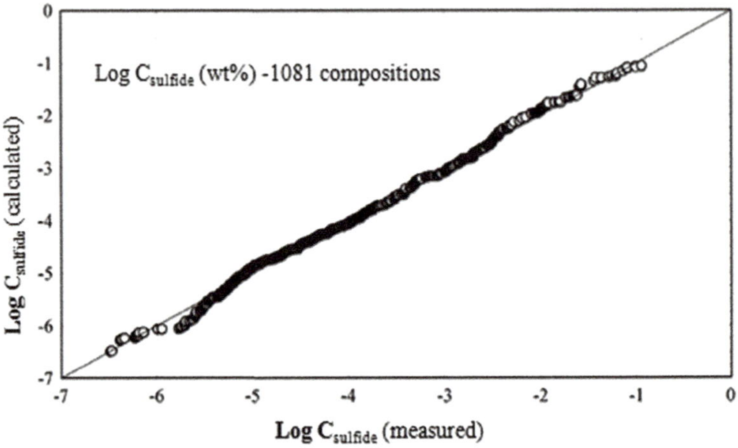

Fig. 17.8 Quantile–quantile representation of the measure versus observed populations of sulfide capacity in anhydrous systems. From Moretti and Ottonello (2005). With permission from Geochimica et Cosmochimica Acta

Fig. 17.9 Experimental versus computed IEFPCM and SPT-PCM abundances of dissolved SO_2 in water at P = 1 bar and various T

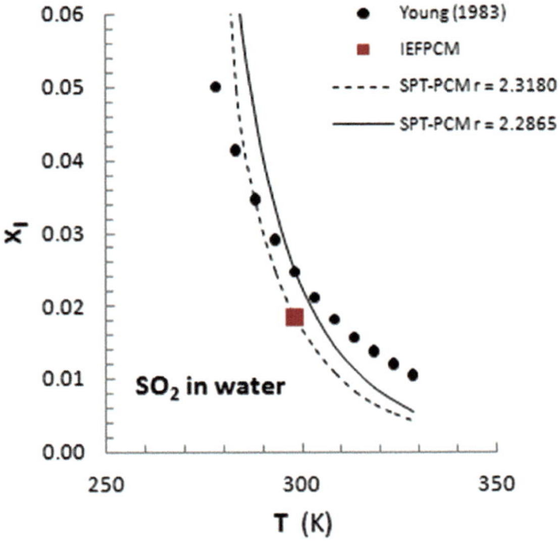

The computed cavity volume, limited by a tessellated surface is 74.843 $Å^3$. The solute/solvent interaction energy is $H_{int} = -44.560$ *kJ/mol*, arising from a polarized solute–solvent electrostatic interaction of -12.678 *kJ/mol*, a dispersive energy contribution of -36.945 *kJ/mol* and a repulsive energy of 5.063 *kJ/mol*. The computed cavitation energy is 34.644 *kJ/mol*. The bulk solvation energy is $\Delta G_{solv} = -9.916$ *kJ/mol*, corresponding to a nominal Hernry's Law constant $\ln K_H = 3.99996$ and to an amount of $SO_{2,aq}$ $x_l = 0.0183$. The nominal (smoothed) amount of $SO_{2,aq}$ calculated through

Eq. (17.2.19) at the same P, T conditions is $x_l = 0.0246$. The performance of PCM is excellent, if one considers that actually the true $SO_{2,aq}$ is somewhat lower than the declared one, due to aqueous speciation equilibria involving other sulfur species. To achieve a perfect match between observed and computed values, one should adopt a radius of 2.2685 *Å*, lower to some extent with respect to the SPT-PCM radius necessary to reproduce the IEFPCM results (i.e. r = 2.318 *Å*). The cavitation entropy is in this way would be reduced from $S_{cav} = -89.619$ *J/(mol × K)* to $S_{cav} = -87.586$ *J/(mol × K)*. In the following, I will adopt nevertheless the solute radius that reproduces the IEFPCM, assuming that the minor biases with respect to experiments may be conceived as an uncertainty on T (the bias at T = 298.15 corresponds to an hypothetic lowering of T of roughly 10 *°C*; see Fig. 17.9). As a next step, I decided to run the *ab initio* computations on the tholeiite composition of Katsura and Nagashima (1974), for its historical importance. At standard state (P = 10^5 *Pa*; T = 298.15 *K*), the fictive liquid has a molar volume V° = 21.5595 *cc/mol* and a corresponding numeral density of 0.02793 $Å^{-3}$. The computed solvation radius is 1.492 *Å*. The Causius-Mossotti polarization is 4.421 $Å^3$, the dielectric constant is ε = 4.215, and the dielectric constant at infinite frequency is $\varepsilon^{\infty} = 2.594$. With these input values, the SCRF-IEFPCM routine of the GAUSSIAN package (Frish et al. 2004) returns, at the athermal limit, a solute/solvent interaction energy $H_{int} = -44.476$ *kJ/mol*, arising from a polarized solute–solvent electrostatic interaction of −12.552 *kJ/mol*, a dispersive energy contribution of −36.987 *kJ/mol* and a repulsive energy of 5.063 *kJ/mol*. If we were to compute the expected amount of SO_2 in the fictive melt of tholeiitic composition at room conditions it would be sufficient to adopt the computed Gibbs free energy of solvation[2] (see Table 17.8), and we would obtain a molar fraction of solute quite similar to that one expected for the water solvent (i.e. $x_{SO_2,melt} = 0.0193$). However, because the T condition of interest is much higher, we must undergo the SPT-PCM computations anticipated at Sect. 16.3.3 to have a correct appraisal of the entropic and enthalpic components of the cavitation energy (see Eqs. 16.3.9–16.3.10).

We see in this way that the entropic part is dominant and negative, and this is basically why the solubility of the SO_2 molecule in the tholeiite solvent decreases markedly with the increase of T. Namely, the Gibbs free energy of solvation rises from 9.791 to 210.214 *kJ/mol*, and the Henry's Law constant rises correspondingly from 51.9 to $10^{7.208}$. Hence, SO_2 is practically insoluble in a melt at high T. Things are quite different for the species SO_4^{2-} and S^{2-} which display a highly negative solvation energy, due basically to strong electrostatic interactions with the solvent.

The simplest way to quantitatively estimate the amount of SO_4^{2-} and S^{2-} in a silicate melt at equilibrium with an atmosphere of known f_{O_2} fugacity is to write down a heterogeneous equilibrium of form

[2] I adopt here and in the following the term "*solvation*" in a rather extensive way for whatever solvent and not specifically water.

Table 17.8 Thermochemical properties of sulfur species. Values in italics are relative to the solute state. Neutral species resolved at B3LYP/6-31G(d,p) theory level. The basis set adopted for anions includes diffuse functions [(6-31G+(d,p)]

Property	Dimension	SO_2	SO_4^{2-}	S^{2-}
E_{el}	*Hartree*	−548.58749	−699.01039	−398.03897
		−548.59119	*−699.28829*	*−398.3367*
ZPC	*Hartree*	0.006797	0.014396	0
		0.006715	*0.01451*	*0*
Therm U	*Hartree*	0.009882	0.018591	0.001416
		0.009806	*0.018655*	*0.001416*
Therm H	*Hartree*	0.010826	0.019535	0.00236
		0.01075	*0.019599*	*0.00236*
Therm G	*Hartree*	−0.017426	−0.013219	−0.013915
		−0.017518	*−0.010711*	*−0.013915*
G_{cav}	*kJ/mol*	34.644	42.426	13.472
E_{elect}	*kJ/mol*	−12.552	−736.761	−782.533
E_{disp}	*kJ/mol*	−36.987	−43.472	−36.903
E_{rep}	*kJ/mol*	5.062	7.657	24.267

$$S^{2-}_{melt} + 2O_{2,gas} \Leftrightarrow SO4^{2-}_{4,melt} \qquad (17.2.20)$$

For the sulfate anion there are, as far as I know, no thermodynamic data. SO_4^{2-} at ground state is $1A_1$ with T_D symmetry. The molecule is a prolate symmetric top with rotational symmetry number 1. When resolved at B3LYP/6-31+G(d,p) theory level, the S–O distances are 1.532 $\AA$, the O–S–O angles are 109.471° and the dihedral angles are 35.264°. In Fig. 17.10, it is shown the HOMO of SO_4^{2-} computed at B3LYP/6–31+G(d,p) theory level for an isodensity of 0.02 $e/\AA^3$. The MO is composed essentially by the 1s orbital of S and 2p orbitals of O.

The harmonic frequencies obtained at the same theory level with scaling (0.9612) are $\nu_{1,2} = 381.9\ cm^{-1}$(e, Raman active), $\nu_{3,4,5} = 545.69\ cm^{-1}$ (IR and Raman active), $\nu_6 = 970.3\ cm^{-1}$ (a_1, Raman active), $\nu_{6,7,8} = 1119.1\ cm^{-1}$ (IR and Raman active).

Stemming from the computed ab initio thermochemical properties (Table 17.8), the Electron Convention (EC) enthalpy of SO_4^{2-} at the gaseous state and room P, T conditions is $H_{SO^{2-}_{4,gas,298}} = -384.008 kJ/mol$ and the entropy is $S_{SO^{2-}_{4,gas,298}} = 288.431\ J/(mol \times K)$. The isochoric heat capacity has, besides translational and rotational terms, also a conspicuous vibrational part. From the computed harmonic frequencies, the isochoric heat capacity C_V is easily computed at various T. The isobaric heat capacity is obtained adding to C_V the term $\frac{\alpha^2}{\beta}TV$, which, for an ideal gas, reduces to:

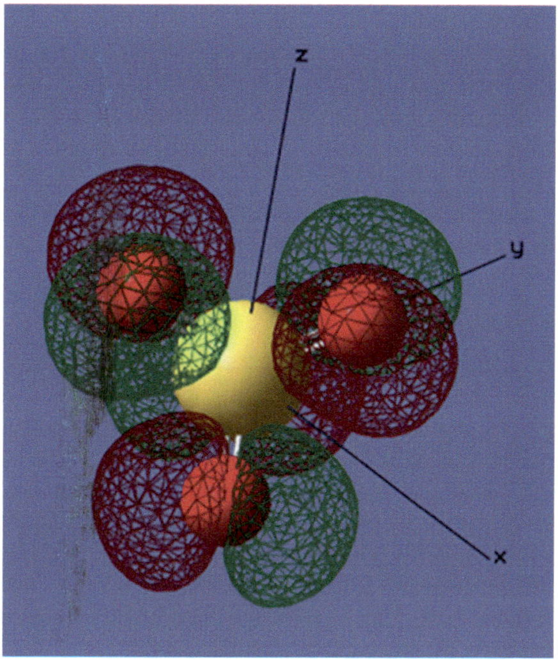

Fig. 17.10 B3LYP/6–31+G** HOMO of SO_4^{2-} at 0.02 $e/Å^3$ isodensity. Van der Waals radii adopted

$$C_P - C_v = \frac{P}{T}V = R \tag{17.2.21}$$

The C_P values, thus, obtained at the various T are easily retrieved in the form of a simple three terms heat capacity polynomial $C_P = f(T)$, valid in the T range $298.15 \leq T(K) \leq 2000$ [i.e. $C_{P,SO^{2-}_{4,gas}} = A + B \times T + C \times T^{-2}$; with $A = 99.891$ *J/(mol × K)*, $B = 4.6067 \times 10^{-3}$ *J/mol*, $C = -2.4122 \times 10^6$ *J/(mol × K³)*]. The EC enthalpy at T, P obtained by integration is $H_{SO^{2-}_{4,gas,1523}} = -263.010$ *kJ/mol*, and the entropy is $S_{SO^{2-}_{4,gas,1523}} = 443.942$ *J/(mol × K)*. The ensuing Gibbs free energy is $G_{SO^{2-}_{4,gas,1523}} = -939.200$ *kJ/mol*. Adding to the Gibbs free energy of the gas the solvation term obtained by SPT-PCM at T = 1523.15 ($G_{solv} = -393.528$ *kJ/mol*) the Gibbs free energy of SO_4^{2-} in the tholeiitic solvent at T = 1523.15 *K* and $P = 10^5$ *Pa* is $G_{SO^{2-}_{4,melt,1523}} = -1332.728$ *kJ/mol*.

The treatment for S^{2-} is quite simple. Stemming from the computed ab initio thermochemical properties (Table 17.8), the EC enthalpy of S^{2-} at the gaseous state and room P, T conditions is $H_{S=,gas,298} = 441.606$ *kJ/mol*, and the entropy is $S_{S=,gas,298} = 33.036$ *J/(mol × K)*. The isochoric heat capacity is simply translational ($C_V = 1.5R$), and the isobaric heat capacity is $C_P = 2.5R$. The enthalpy at T,P obtained by integration is $H_{S=,gas,1523} = 467.070$ *kJ/mol* and the entropy is $S_{S=,gas,1523} = 66.938$ *J/(mol × K)*. The ensuing Gibbs free energy is $G_{S=,gas,1523} = 365.113$ *kJ/mol*. By adding to the Gibbs free energy of the gas, the solvation term obtained by SPT-PCM at T = 1523.15 ($G_{solv} = -545.211$ *kJ/mol*) the resulting Gibbs free energy of S^{2-} in the tholeiitic solvent at T = 1523 *K* and P = 1 *bar* is $G_{S=,melt,1523} = -180.098$ *kJ/mol.*

Adopting the NIST-JANAF thermodynamic parameters I got, for $O_{2,gas}$, $G_{O_2,gas,1523} = -352.409$ *kJ/mol.* The ensuing Gibbs free energy of reaction (17.2.20) at T = 1523.15 *K* and P = 1 *bar* is $\Delta G_{17.2.20} = -447.812$ *kJ/mol.*

Assuming ideal mixing of the sulfur species in the anion matrix of the melt, their molar ratio is easily obtained by applying:

$$\frac{x_{SO_4 melt}^{2-}}{x_{S^{2-}_{melt}}} \frac{\left[SO^{2-}_{4,sln}\right]}{\left[S^{2-}_{sln}\right]} = K_{17.2.20} \times fO^2_{2,gas} \tag{17.2.22}$$

The results of this computational exercise are compared in Fig. 17.11 with the unique datum of Katsura and Nagashima (1974).

The comparison is encouraging, if one remembers that the computation is purely theoretical with no ad-hoc assumptions. Repeating the calculations in the absence of solvation terms (i.e. gas phase; $\Delta G_{17.2.20} = -594.088$ *kJ/mol*), we may deduce that the effect of the solvent is a stabilization of the sulfide species at the expense of sulfate anion. The slope of the relations in the logarithmic scale is obviously 2 and imposed by the reaction stoichiometry.

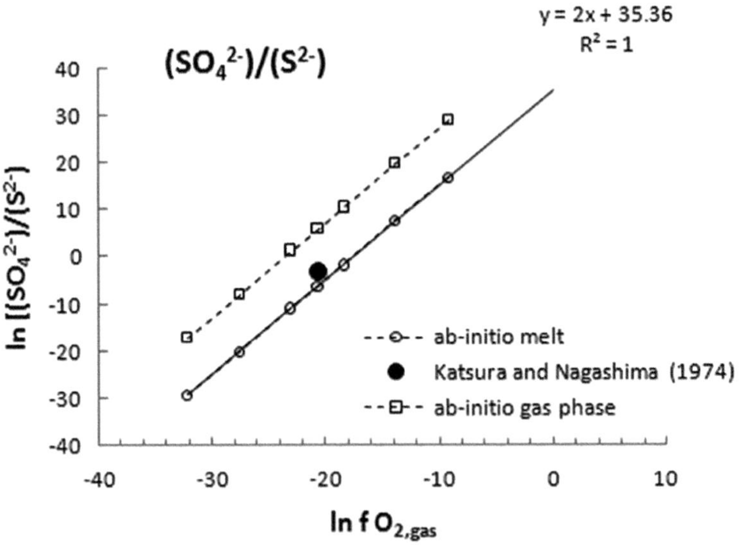

Fig. 17.11 Ab initio sulfur speciation in a tholeiitic melt at T = 1523.15 *K*, P = 10^5 *Pa*, compared with experiments. The computed speciation in the gas phase is also shown

17.3 CO

I treat this species at the end of the chapter because it is stable only under very reducing conditions, attained only when the system develops in closed conditions without interaction with the atmosphere. The CO partial pressure decreases of about two orders of magnitude when the amount of atomic oxygen in the system exceeds ~57% at high T (see Fig. 17.2). According to Giggenbach (1996), the presence of a "*rapidly responding species*" like CO in volcanic emission is ascribable to the buffering control exerted by the H_2S–SO_2 redox couple, whose presence reflects evaporation of elemental sulfur at shallow levels and hydrothermal regimes.

Carbon monoxide (CO) has $D_{\infty v}$ symmetry and $^1\Sigma_g$ ground state. It is a linear molecule with a C–O interatomic distance of 1.128 *Å* (NIST 2000). The molecule is a prolate symmetric top with rotational symmetry number 1. When resolved at B3LYP/6-31G(d,p) theory level, the C–O distance is 1.138 *Å*. The observed total electric dipole moment is 0.112 *Debye*, and the components of the electric quadrupole moment are $xx = 1.420\ D \times Å$, $yy = 1.420\ D \times Å$, $-2.839\ D \times Å$ (NIST 2000). The electrical dipole polarizability is $\alpha = 1.953\ Å^3$ (Olney et al. 1997) and the calculated one at B3LYP/6-31G(d,p) level is $\alpha = -2.069\ Å^3$. The experimental harmonic frequency is $v_1 = 2170\ cm^{-1}$ (Huber and Herzberg 1979), and the calculated one at B3LYP/6-31G(d,p) theory level with scaling is $v_1 = 2123\ cm^{-1}$ (sg, IR and Raman active). The (unscaled) zero point energy at this theory level is 0.005533 *Hartree*. The ionization energy according to NIST (2000) is 14.014 *eV*, and the proton affinity is 594.0 *kJ/mol* at C level and 426.3 *kJ/mol* at O level (Huntner and Lias 1998). The dissociation energy computed by Ochtersky et al. (1996) at G2(MP2) theory level is 1074.9 *kJ/mol*, against a reported experimental energy of 1070.3 *kJ/mol* (see Table 7.32). At standard state (pure ideal gas, $P = 10^{-5}\ Pa$, 298.15 K) CO has an entropy of 197.66 *J/(mol × K)* and an enthalpy of formation from the elements of −110.53 *kJ/mol* (Cox et al. 1984). When resolved at G2 theory level, the returned entropy is 197.3 *J/(mol × K)*, and the enthalpy of formation from the elements at standard state ($H°_f$) is −112.7 *kJ/mol* (Table 7.32). The Shomate equation for the isobaric heat capacity in the range $298.15 \leq T\ (K) \leq 1300$ is $C_P = 25.56759 + 6.096130 \times 10^{-3} \times T + 4.054656 \times 10^{-6} \times T^2 - 2.671301 \times 10^{-9} \times T^3 + 1.31021 \times 10^5 \times T^{-2}$ (NIST 2000; based on refitted data of Chase 1998). The boiling point is at 81.63 *K*. The critical temperature is $T_c = 134.45\ K$. The triple point occurs at $T_t = 67.95\ K$ and $P_t = 34.988\ bar$ (NIST 2020).

17.4 Geothermal Emissions

17.4.1 Composition of Geothermal Gases

Analysis of temperature profiles in deep geothermal fields, between two and a half *km* depth, almost invariantly reveals P, T conditions along the boiling curve of water (i.e. at all depths, the temperature of the fluid is that of the aqueous solution at the P of interest; cf. Table 17.9).

Table 17.9 Thermal profile of water along boiling curve (from Haas 1971)

T (°C)	P (*bar*)	Depth (*m*)	Density (*g*/*cm*^{3})
100	1	0.0	0.958
110	1.4	4.5	0.951
120	2	10.4	0.943
130	2.7	18.2	0.935
140	3.6	28.2	0.926
150	4.8	40.9	0.917
160	6.2	56.8	0.907
170	7.9	76.4	0.897
180	10	100.5	0.887
190	12.6	129.7	0.876
200	15.6	164.7	0.865
210	19.1	206.8	0.853
220	23.2	256.4	0.84
130	28	314.9	0.827
240	33.5	383.3	0.814
250	39.8	462.9	0.799
260	46.9	555.2	0.784
270	55.1	661.8	0.768
280	64.2	784.6	0.751
290	74.4	925.6	0.732
300	85.9	1088	0.712
310	98.7	1273	0.691
320	122.9	1487	0.667
330	128.6	1732	0.64
340	146.1	2017	0.609
350	165.4	2350	0.573
360	186.7	2746	0.525
370	210.5	3243	0.446

Table 17.10 Gas composition in geothermal fluids (from Ellis 1967, 1970; and Henley 1984). P_{sep} = pressure of separation of gaseous phase

Locality	P_{sep} (*bar*)	$F_{H_2O,f}$	X_g	‰ Concentration in gas					
				CO_2	H_2S	CH_4	H_2	N_2	NH_3
(1)	2	0.3	0.2	917	44	9	8	15	6
(2)	9.8	0.2	1.2	936	64	nd	nd	nd	nd
(3)	11	0.19	10.04	956	18.4	11.8	1.01	8.89	4.65
(4)	2.87	0.19	24.8	945	11.7	28.1	3	2.1	10.2
(5)	7.6	0.289	5.88	822	79.1	39.8	28.6	5.1	23.1
(6)	8.6	0.414	2.95	932	55	4.1	3.6	1.2	4.3
(7)	20	0.135	0.248	962	29	1	2	6	nd
(8)	1	0.38	0.507	957	43.9	nd	nd	nd	nd
(9)	–	1	20	941	16	12	23	8	nd
(10)	–	1	5.9	550	48	95	150	125	nd

(1) Wairakei—New Zealand; (2) Tauhara—New Zealand; (3) Broadlands—New Zealand; (4) Ngawha—New Zealand; (5) Cerro Prieto—Mexico; (6) Mahio-Tongonan—Philippines; (7) Reykjanes—Iceland; (8) Salton Sea—California (9) Geysers—Wyoming; (10) Larderello—Italy

There are two main reasons for this:

(1) rocks are extremely efficient heat insulators;
(2) the liquid–vapor transition for H_2O involves quite high evaporation enthalpy, which controls the thermal state of the system.

At shallower depths, the vapor phase separates from the liquid one and migrates toward the surface, sometimes heating superficial aquifers and finally giving rise to *fumarole* activity. The gaseous phase is dominated by H_2O mixed with subordinate amounts of CO_2, H_2S, CH_4, H_2, N_2, and NH_3. Table 17.10 lists representative analyses of geothermal gases from the main geothermal fields of the world. It also lists the pressure of separation of the gaseous phase from the liquid, the fraction of H_2O vapor in the fluid ($F_{H_2O,f}$) and the "*fraction of gas in the vapor phase*" (X_g). The latter definition is purely operative, since H_2O vapor is a gas to all intents and purposes, and the gaseous phase is not unmixed at the various P, T of interest. The need to distinguish the total amount of non-H_2O gaseous components from gaseous H_2O (i.e. "*vapor*") is dictated by the fact that as the gaseous phase flows through superficial aquifers, further enrichment in H_2O may take place by simple evaporation.

17.4.2 Solubility of Geothermal Gases in the Boiling Aquifer

Because the development of a gaseous phase in geothermal fields is associated with boiling aquifers and is generally induced by decompression during convective transport toward the surface or to heat transfer from reservoir rocks exceeding the

Table 17.11 Coefficients for vapor/liquid gas distribution coefficients (Giggenbach's Eq. 17.4.1) for various species in T range 100 < T (°*C*) < 340 (from Giggenbach 1980). Values in brackets represent the SPT-PCM regression

Species	A	B
NH_3	1.4113 (1.781)	−0.00292 (−0.0024)
H_2S	4.0547	−0.00981
CO_2	4.7593	−0.01092
CH_4	6.0783	−0.01383
H_2	6.2283	−0.01403
N_2	6.4426 (6.056)	−0.01416 (−0.0064)

T-limit of the boiling curve at all P, it is extremely important to ascertain the thermodynamics of the heterogeneous equilibria between the gas phase and the aqueous solutions. Giggenbach (1980) proposed simple linear equations relating the mass distribution constants of the main reactive gases to T expressed in Celsius:

$$\log K_i' \equiv \log\left(\frac{X_{i,g}}{X_{i,1}}\right) = A + BT \tag{17.4.1}$$

These constants, listed in Table 17.11 and valid for T in centigrades, were derived by Giggenbach (1980) from the respective Henry's constant by applying:

$$K_i' = K_{h,i} \times \frac{Z_{g,H_2O}}{P_{H_2O}} \tag{17.4.2}$$

where Z_{g,H_2O} is the "*compressibility factor*" of gaseous H_2O.

As we have already seen (Eq. 16.10.6), the compressibility factor is given by:

$$Z_{g,i} = \frac{P_i V_i}{n_i RT} \tag{17.4.3}$$

The molar fraction of gaseous species i in gaseous mixture (i + H_2O_g) is given by:

$$X_{i,g} = \frac{n_i}{n_i + n_{H_2O_{(g)}}} \tag{17.4.4}$$

If gaseous component i is at trace concentration level, we may also write:

$$X_{i,g} = \frac{n_i}{n_i + n_{H_2O_{(g)}}} \approx \frac{n_i}{n_{H_2O_{(g)}}} \tag{17.4.5}$$

Re-expressing n_i and $n_{H_2O_{(g)}}$ through their respective compressibility factors, we obtain:

$$X_{i,g} = \frac{P_i}{P_{H_2O}} \times \frac{Z_{g,H_2O}}{Z_{g,i}} \tag{17.4.6}$$

Moreover, dividing by the molar fraction of species i in the liquid, we have:

$$K_i' = \frac{X_{i,g}}{X_{i,l}} = \frac{P_i}{X_{i,l}} \times \frac{Z_{g,H_2O}}{Z_{g,i}} \times \frac{1}{P_{H_2O}} = K_{h,i} \times \frac{Z_{g,H_2O}}{Z_{g,i}} \times \frac{1}{P_{H_2O}} \tag{17.4.7}$$

In the transition $K' \rightarrow K_{h,i}$ the compressibility factor of the trace component in the gaseous phase is usually neglected since, as the concentration is low, the partial pressure of the gaseous species, is virtually one.

As we may see in Fig. 17.12, Giggenbach's procedure supersedes all geotechnical treatments, like the Benson–Krause or the Potter–Klinne equations (Potter and Clinne 1978), and testifies for a perfect Arrhenian behavior of the gaseous species.

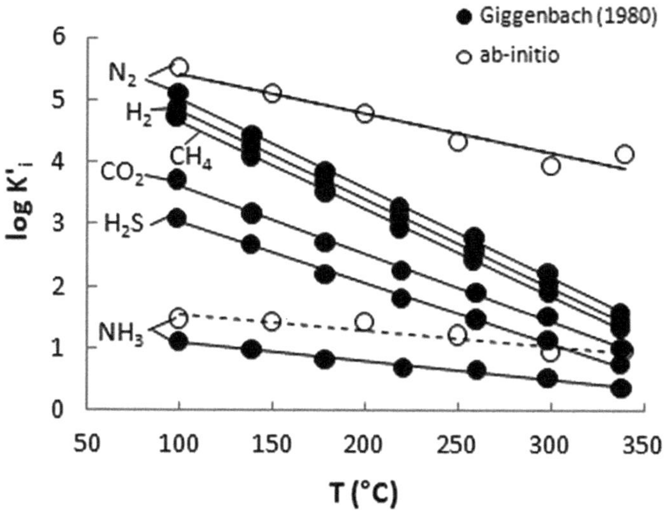

Fig. 17.12 Mass distribution constants of various gaseous species as a function of T after Giggenbach (1980) (solid symbols). The open symbols are the results of SPT-PCM calculations for N_2 and NH_3 (Table 17.2)

Table 17.12 SPT-PCM calculations for NH_3 along the boiling curve. The adopted solute radius for NH_3 is 1.77 *Å*. The B3LYP/6-31G(d,p) IEFPCM solute–solvent interaction is −37.321 *kJ/mol* (−18.702 electrostatic, −25.690 dispersive, +7.071 repulsive). The cavitation energy at T = 25 °*C*, P = 1 *bar* is +20.251 *kJ/mol*. For N_2 see Sect. 16.6.2

T (°*C*)	P (*bar*)	V_s, H_2O	Z_{g,H_2O}	N_2		NH_3	
				ln K_H	log K′	ln K_H	log K′
100	1.0	1.694	0.997	12.683	5.507	3.377	1.465
150	4.8	0.374	0.931	13.375	5.097	4.915	1.422
200	15.6	0.132	0.955	13.814	4.786	6.046	1.413
250	39.8	0.049	0.818	13.881	4.341	6.672	1.211
300	85.9	0.020	0.658	13.979	3.955	7.034	0.939
340	146.1	0.010	0.523	15.168	4.141	7.858	0.967

Few calculations for N_2 and ammonia, based on the SPT-PCM theory and no ad-hoc assumptions (Fig. 17.12 and Table 17.12), although underestimating the solubility of both gases in boiling aquifers, confirm substantially the intuition of Giggenbach (1980).

Chapter 18
Aqueous Solutions

The importance of aqueous solutions in geochemistry may be appreciated if we recall that two-thirds of our planet is covered by water and that waters of various salinities help to determine the volcanic and metamorphic processes in the Earth's upper mantle and crust. But why H_2O is mainly concentrated on the planetary surface in the liquid aggregation state, and why it is "salty"? The H_2O "*component*" is advectively transported to the Earth's surface from the deep planetary interiors and progressively released thanks essentially to the effects of the T, P intensive variables on its "*cavitation energy*"[1] When the bulk H_2O amount exceeds the solubility product by nearly adiabatic decompression, it is released in the *gaseous* state. When the temperature of the atmosphere is sufficiently low, the H_2O gas condenses and forms a "*liquid*" phase, essentially thanks to weak Hydrogen Bonding (HB) effects. Nevertheless, the conformation of the H_2O molecule, under all aggregation states (*i.e.* gaseous, liquid, and the various crystalline forms) is virtually the same: a strange dipole possessing the capability of neutralizing the charges in solution through the "*solvation-hydrolysis*" process.

18.1 The Structure of the Solvent

Despite extensive studies, the microscopic nature of the liquid aggregation state of H_2O remains elusive (see Guillot, 2002; for an extensive review). Computer simulations are extremely important because somewhat complementary to experiments and provide important information in a wide P,T range of interest, however, the parameterization against experimental data, calls into question their heuristic

[1] H_2O has in silicate melts a complex reactivity, and good part of it forms neutral hydroxyl terminations $[OH]^0$ on the polymeric units present in the melt. Hence the cavitation energy applies only to the undissociated moiety.

G. A. Ottonello, *Quantum Geochemistry*, Springer Geochemistry,
https://doi.org/10.1007/978-3-031-21837-8_18

capability when P,T conditions different from those used in the parameterization itself must be afforded. This is especially true for empirical force fields based on two-body interactions. Many-body (cooperative) effects may be eventually taken into account by parameterizing many-body potentials against *ab initio* potential energy surfaces of small water clusters, but, as noted by Lin et al. (2009) "*the inability to describe reactions remains*".

18.1.1 Models of Distortion of the Hydrogen Bond

The models of distortion of the hydrogen bond, for their immediacy, do not require extensive explanations. I introduce first these models for their historical valence. Such models consider that transition from the crystalline to the liquid state mainly involves distortion of the hydrogen bond, but no bond rupture.

The model of Pople (1950) assumes that, in the liquid state, each H_2O molecule is linked to other three molecules; distance $\overline{R}_0$ between neighboring oxygen nuclei is constant, but the distance between next neighbors varies according to the degree of distortion of the hydrogen bond. As shown in Fig. 18.1A, in Pople's model, the hydrogen bond is considered undistorted when the O–H bond direction of a molecule falls along the line connecting neighboring oxygen nuclei. As the degree of distortion increases (ϕ angles in Fig. 18.1A), the internal energy of the liquid also increases. This may be quantified by

$$\Delta U = K_{\phi}(1 - \cos \phi) \tag{18.1.1}$$

K_{ϕ} in Eq. 18.1.1 is the constant of distortion of the hydrogen bond.

The degree of distortion of the hydrogen bond in water varies with temperature (from 26° to 30° for $0 < T(°C) < 100$) and is zero in ice, where H_2O molecules are linked to form perfect tetrahedra.

The simple electrostatic model of Bernal and Fowler (1933), which assumed a displacement of negative charges along the bisector of the H–O-H angle, coupled with a repulsion-dispersion term centered on oxygen, was prophetic in predicting the possibility of tetrahedral arrangements of water molecules and found confirmation fifty years later (Jorgensen et al., 1983).

The "*random network model*" of Bernal (1964) is a development of Pople's model. In Bernal's version, the water molecules are conceived as forming an irregular network of rings (not open chains, as in Pople's model). Most of the rings have five members, since the O–H bond in a single H_2O molecule is near the angle 108° characteristic of a five-membered ring, but four, six, seven, or even more H_2O molecules may be present in some rings.

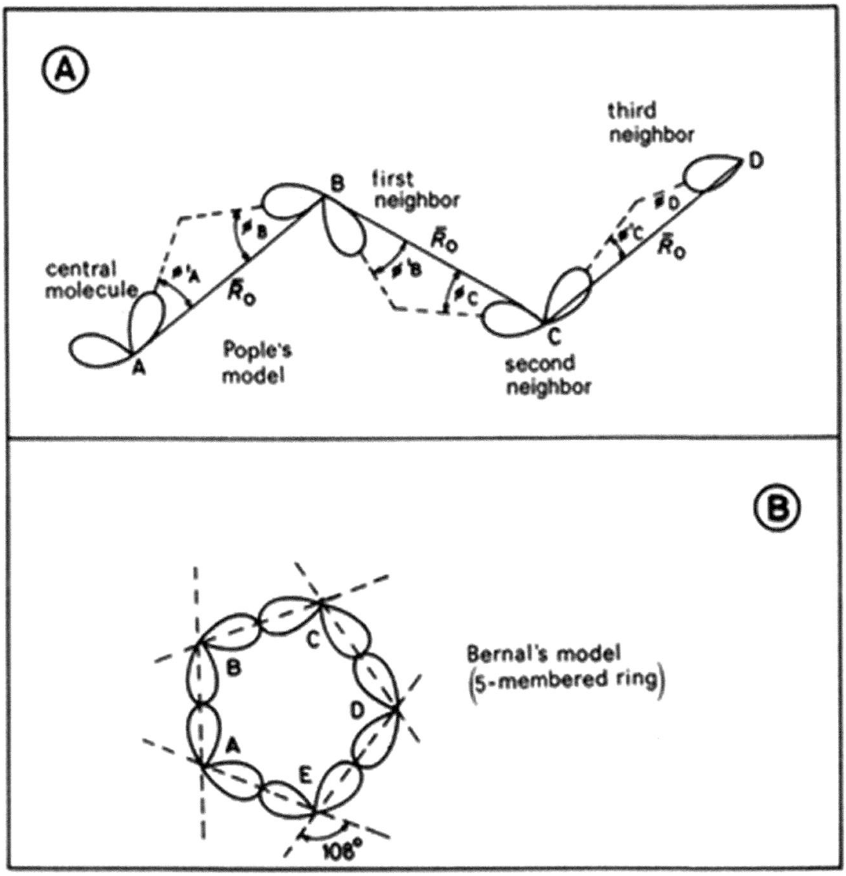

Fig. 18.1 Models of distortion of hydrogen bond for water. Note that bond angle between two O–H bonds of a single molecule is 105°; formation of a five-membered ring (part B) thus implies distortion of 3°. From Ottonello (1997). With permission from Columbia University Press

18.1.2 Ab Initio Water Dimers and Trimers

The old models of water structure cannot be disregarded and, as we will see later on, their validity is partially confirmed by recent ab initio Molecular Dynamics (AIMD) investigations, but, to achieve an appropriate understanding of intermolecular interactions in large water clusters, it is necessary to unravel first what happens in small clusters such as dimers and trimers. As an example, I may quote the difficulty in the interpretation of spectroscopic data associated with the hydrogen-bond tunneling motions that occur both in small water clusters as well as in the bulk liquid (Fraser, 1991; Pugliano and Saykally, 1992). The microwave spectroscopic studies of partially deuterated species carried out by Dyke et al.

(1977) and Odutola and Dyke (1980) gave a first neighbors O–O interionic distance of 2.976 *Å*. As noted by Xantheas and Dunning (1993), in ab initio computations, this bond length is very sensitive to the level of correlation and to the extension of the adopted basis set. It ranges from 2.971 *Å* for HF/6-31G(d) calculations to 3.026 *Å* for HF/6–311 ++G(3df,3pd) basis set. Introducing correlation, the variation is much smaller, i.e. from 2.913 *Å* at the MP2/6-31G(d) level to 2.911 *Å* at the MP2/ 6–31++g(3df,3pd) level (see references in Xantheas and Dunning, 1993). O–O equilibrium separation distances (R_e) comparable to those obtained with the MP2 theory were observed at the CISD level of theory (*ibid*). The MP2/aug-cc-pVDZ calculation carried out by Xantheas and Dunning (1993) gave an equilibrium O–O separation distance $R_e = 2.920$ *Å*. Computations at MP4 theory level carried out by the same Authors, keeping frozen all internal MP2 coordinates but the separation distance, gave $R_e = 2.926$ *Å*, and, by adding the experimentally estimated anharmonic correction of 0.03 *Å* (Odutola and Dyke, 1980) to the computed MP4 R_e, they predicted $R_e = 2.956$ *Å*, which is only 0.02 *Å* shorter with respect to the measured value. The dissociation energy (D_e) of the water dimer calculated by Xantheas and Dunning (1993) at the MP2/aug-cc-pVDZ level ($D_e = 22.13$ *kJ/mol*) is in excellent agreement with the experiments of Curtiss et al. (1979) ($D_e = 22.59 \pm 2.92$ *kJ/mol*) and Reimers et al. (1982) ($D_e = 22.59 \pm 0.84$ *kJ/mol*). The MP2/ aug-cc-pVDZ calculations of the dipole moment and polarizability of the monomer are, respectively, within 1% and 3.5% of experiment. Table 18.1 lists the rotational constants calculated for various deuterated species by Xantheas and Dunning (1993) adopting the MP2 optimal structure constrained to obey $R_e = 2.956$ *Å*. This yields rotational constants of B = 6214 *MHz* and C = 6213 *MHz*. The geometries of the partially deuterated species ($D_xH_{4-x}O_2$, x = 0–4), were obtained by taking the computed structure at the (constrained) MP4 level and substituting deuterium atoms at different positions, increasing the O-D bond lengths by 0.002 *Å*, and decreasing the O-D-O angles by 0.75°. The calculated rotational constants are within 0.5% of the measured ones. The notation adopted in the first column of Table 18.1 corresponds to proton acceptor–proton donor dimers.

Accurate measurements of an intermolecular vibration of the per-deuterated water dimer $(D_2O)_2$ by tunable far infrared laser spectroscopy (FIR-VRT; Pugliano and Saykally, 1992) revealed the existence of five vibration–rotation-tunneling bands of the isotope, located near 84 cm^{-1}. These bands were interpreted as due to the ν_8 acceptor wag (Pugliano and Saykally, 1992) and involve large amplitude motion of the hydrogen bond. The spectra indicated moreover donor tunneling

Table 18.1 Experimental (Expt., Odutola and Dyke, 1980) and calculated (Calc., Xantheas and Dunning, 1993) rotational constants for various isotopic species of the water dimer

½(B + C) (*MHz*)			
Species	Expt.	Calc.	Expt.- Calc.
H_2O–H_2O	6214	6161	53
D_2O–DOH	5675	5674	1
HOD -D_2O	5648	5663	15
H_2O–D_2O	5893	5928	35

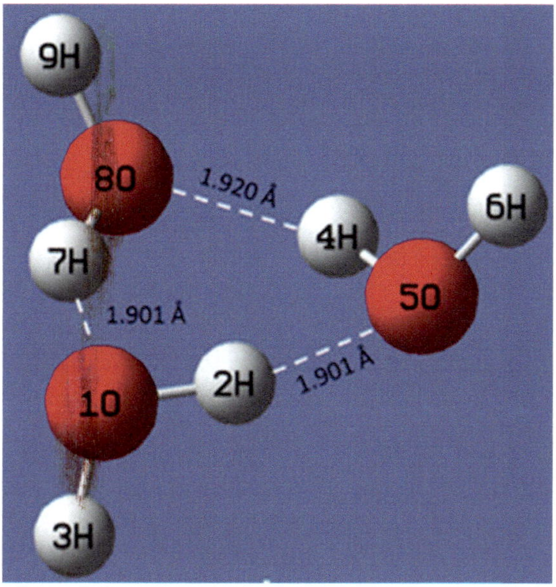

Fig. 18.2 The MP2/Aug-cc-pVDZ water trimer according to Xantheas and Dunning (1992). The adopted atom labeling does not correspond to the original one of the Authors

motions to the excited vibrational coordinate and strong coupling of the donor–acceptor interconversion (Pugliano and Saykally, 1992).

Figure 18.2 shows schematically the structure of the water trimer $(H_2O)_3$ resolved at MP2/Aug-cc-pVDZ level by Xantheas and Dunning (1992).

It is a cyclic structure with three nonlinear hydrogen bonds. The deviation of the hydrogen bonds from nonlinearity is δ (1O-2H-5O) = 28.9°, δ (5O-4H-8O) = 31.6°, and δ (8O-7H-1O) = 28.7°. The three hydrogens that participate in the hydrogen bonds (namely 2H, 4H, and 7H in Fig. 18.2) lie very close to the plane of the three oxygens, the corresponding dihedral angles being 0.6°, 3.1°, and 0.1°, respectively. Two of the three terminal hydrogens (6H,9H) are arranged on the same side of the plane of the oxygens, but the third one (3H) is on the other side. The torsional angles of these three hydrogens with respect to the oxygen plane are 118.4° (3H), 237.3° (6H), and 230.9° (9H). This arrangement gives rise to a chiral structure, i.e. a non-superposable mirror image. The structural parameters are listed in Table 18.2 and are in qualitative agreement with the experimental observations of Pugliano and Saykally (1992). In comparing the two sets of data, it must be noted that, while the experimental parameters are vibrationally averaged value, the calculated ones represent equilibrium distances in the absence of vibrational motions (i.e. athermal limit).

In Table 18.3 are listed the rotational constants for $(H_2O)_3$ and $(D_2O)_3$ computed by Xantheas and Dunning (1992) with the optimum structure obtained from the MP2/aug–cc–pVDZ calculations. They differ by 83, 177, and 243 *MHz* from the measured constants. By adding an anharmonic correction of +0.03 *Å* to the O–O bond length, as done for the water dimer, the error is reduced to 40, 62, and 180 *MHz*. With O–O bond lengths equal to R_e(MP4) + 0.03 = 2.837 *Å* and the rest of

Table 18.2 Calculated (Calc.; Xantheas and Dunning, 1993) and experimental (Expt.; Pugliano and Saykally, 1992) structural parameters for the water trimer. Distances are in *Å* and angles in degrees (From Xantheas and Dunning, 1992; with modifications in labeling and extension)

	Calc	Expt
R (O–O)	2.8	2.94, 2.97, 2.97
R (1O–3H)	0.965	
R (5O–6H)	0.964	
R (8O–9H)	0.964	
R (1O–2H)	0.978	
R (5O–4H)	0.977	
R (8O–7H)	0.978	
δ (1O–2H–5O)	151.1	152
δ (5O–4H–8O)	148.4	150
δ (8O–7H–1O)	151.3	153
2H–(1O5O8O)	178.2	180
4H–(5O8O1O)	196	180
7H–(8O1O5O)	180.6	180
3H–(1O5O8O)	118.4	
6H–(5O8O1O)	237.3	
9H–(8O1O5O)	230.9	

Table 18.3 Calculated (Xantheas and Dunning, 1992) and experimental (Pugliano and Saykally, 1992) rotational constants for $(H_2O)_3$ and $(D_2O)_3$. Interatomic distances are in *Å* and the rotational constants are in *MHz*. From Xantheas and Dunning (1992)

Species	Geometry	R (O–O)	A”	B”	C”
$(H_2O)_3$	MP2	2.800[a]	6861	6789	3476
(D_2O_3)	MP2	2.800[a]	5968	5880	3049
	MP2	2.830[b]	5845	5765	2986
	MP2	2.830[c]	5843	5763	2986
	MP4	2.837[d]	5823	5744	2976
	Experimental	2.97, 2.97, 2.94	5885	5703	2806

(a) R_e MP2/Aug–cc–pVDZ
(b) $R_o = R_e$ MP2/Aug–cc–pVDZ + 0.03*Å*
(c) as (b) plus corrections for different O–D bond lengths (+0.002 *Å*) and D–O–D angles (–0.75°)
(d) $R_o = R_e$ MP4/Aug–cc–pVDZ + 0.03*Å*

the internal coordinates equal to their MP2 optimal values, the difference from experimental results is reduced to 62, 41, and 170 *MHz* (see Xantheas and Dunning, 1992; for an extended discussion).

18.1.3 AIMD Structures

Ab initio molecular dynamics (AIMD) simulations are particularly apt to investigate a compositionally simple fluid like H_2O. The first AIMD simulation for liquid water employing a Car-Parrinello quantum–mechanical description of the molecular interactions (Car and Parrinello, 1985; see Sect. 13.4) is due to Laasonen et al. (1993). The simulation was carried out on a small system size (32 molecules) and short simulation lengths (1.5 and 2 *ps* for equilibration and production, respectively), due to the limited computer resources available at that time. Various investigations were then carried out through the years on bigger system's sizes. The outcome of AIMD water simulations appeared to be very sensitive to the particular choice of functional and basis set. Accurate investigations of the role of various functional in the AIMD simulation of the thermophysical properties of water have been carried out by Lin et al. (2009, 2012). The General Gradient Approximation (GGA) leads to an "*overstructured*" water, which diffuses one order of magnitude more slowly than experiment. Hybrid functionals improve slightly the prediction, but the key toward a better comprehension of the true nature of the liquid seems to be provided by a sort of "*compromise*": the BLYP exchange–correlation functional which, for its nature, is completely unable to describe dispersion forces, may be coupled with an empirical vdW correction and dispersion-corrected atom-centered potentials (DCACPs; Lin et al., 2009). This provides a sort of balancing between the role of H-bonding and vdW interactions.

In the Dispersion-Corrected Atom-Centered Potential (DCACP) models, the long-range electron density correlation energy (attractive) is represented as a sort of atom–electron interaction:

$$\hat{v}_l^{\mathrm{DCACP}}(r, r') = \sum_{l=0}^{l_{\max}} \sum_{m=-l}^{+l} \mathrm{Y}_{lm}(\hat{\boldsymbol{r}})\hat{p}_l(\mathrm{r})\sigma_1\hat{p}_l(r')\mathrm{Y}_{lm}^*(\hat{\boldsymbol{r}}') \tag{18.1.2}$$

where Y_{lm} are spherical harmonics, $\hat{\mathrm{p}}_l(\mathrm{r})$ are normalized projectors, represented as Gaussian functions of the form

$$\hat{\mathrm{p}}_l(\mathrm{r}) = \frac{\sqrt{2}\mathrm{r}^l \exp\left[-\mathrm{r}^2/(2\sigma_2^2)\right]}{\sigma_2^{l+3/2}\sqrt{\Gamma(l+3/2)}} \tag{18.1.3}$$

$\mathrm{r} = |\mathbf{r} - \mathbf{R}_l|$ is the distance from the position of nucleus l, $\mathbf{r}$ is the unit vector in the direction of $\mathbf{r} - \mathbf{R}_l$, and σ_1, σ_2 are fitting parameters obtained by minimizing a penalty functional $\mathcal{P}$ composed of two energy and one nuclear force-dependent terms (Lin et al., 2009):

$$\min_{\{\sigma_\mathrm{i}\}} \mathcal{P} = \min_{\{\sigma_\mathrm{i}\}} \left[\left|\mathrm{E}^{\mathrm{ref}}(\mathrm{r}_{\min}) - \mathrm{E}(\mathrm{r}_{\min}, \{\sigma_\mathrm{i}\})\right|^2 + \left|\mathrm{E}^{\mathrm{ref}}(\mathrm{r}_{\mathrm{mid}}) - \mathrm{E}(\mathrm{r}_{\mathrm{mid}}, \{\sigma_\mathrm{i}\})\right|^2 + \sum_l \mathrm{w}_l |\mathrm{F}_l(\mathrm{r}_{\mathrm{mid}}, \{\sigma_\mathrm{i}\})|^2 \right] \tag{18.1.4}$$

r_{min} and r_{mid} in Eq. 18.1.4 are the equilibrium and midpoint distances[2] of the reference calculations. $E^{ref}(r)$ is the reference interaction energy at r and w_l is a weighting factor. $\mathcal{P}$ is minimized by tuning the amplitude σ_1 and the width σ_2.

The inclusion of the midpoint energy term in the penalty functional operated by Lin et al. (2008) was the first attempt to improve the long-range behavior of the resulting interaction energy curves, which, in older versions of DCACPs, weren't able of reproducing the asymptotic dipolar nature of vdW interactions. The analytical form of the adopted DCACP model is that one of a separable dual-space Gaussian pseudopotential (GTH; Goedecker et al., 1996) and, as such, it is straightforward to be applied it in plane wave DFT calculations.

Lin et al. (2009) carried out the AIMD computations with the version 3.11 of the freeware code CPMD, released jointly by the IBM Corporation and the Max Planck Institute of Stuttgart (Germany; http://www.cpmd.org). The adopted plane wave cutoff was 125 *Ry*. The parameters for DCACPs were taken from Lin et al. (2007), and those for the empirical vdW corrections were taken from Williams and Malhotra (2006). The volume of the microcanonical ensemble was 20^3 $Å^3$. The gradient tolerances were set to 10^{-5} a.u. and 10^{-7} a.u. for the nuclear and electronic degrees of freedom, respectively. I remand the interested reader to Lin et al. (2009) for all the remaining computational details. I only add here that the dimers (and trimers) were obtained stemming from fixed monomer geometries obtained previously at the same theory level (i.e. the unique degree of freedom is the O–O distance). The initial structure for the liquid simulation (64 water molecules in a cubic 12.42^3 $Å^3$ periodic box, corresponding to a light water density of 1 g/cm^3) was taken from Kuo et al. (2004).

The equilibrium interaction energies of the water dimers and trimers calculated by Lin et al. (2009) through the various procedures are compared in Table 18.4 to "*high-level reference values*" (*sic*, Lin et al., 2009). DCACP and BLYP-D (empirical-vdW-correction-augmented BLYP) calculations show the smallest deviations from the reference values. The effects of having well-described vdW interactions are evident especially in the interaction energy.

Three years later, Lin and coauthors (Lin et al., 2012) extended the computation to the PBE functional (see Sect. 10.5.4) and its revised form revPBE (Zhang and Yang, 1998). They also introduced damped atom-pairwise dispersion corrections of the C_6R^{-6} form. By comparing the GGA density functionals (BLYP, PBE, and revPBE) with data from their van der Waals (vdW) corrected counterparts, they observed that the vdW correction to BLYP simulations provides softer and more mobile structures, while no practical effects were observed on PBE. Augmenting the revPBE functional with the damped atom-pairwise dispersion corrections, softens the already under-structured revPBE water, while including DCACP renders the structure more "*glassy*" (*sic*, Lin et al., 2012). At low T, the isothermal compressibility, the heat capacity, and the thermal expansivity of liquid water

[2] The midpoint distance is where the interaction energy is half of the equilibrium energy in the reference curve.

Table 18.4 AIMD Structural Parameters and interaction energies of optimized water dimer and cyclic trimer obtained by Lin et al. (2009) with various procedures, compared with the best *ab initio* results of Klopper et al. (2000) (dimer) and the extrapolated MP2 results of Nielsen et al. (1999) (trimer). The interionic distances (r_{OO}) are in *pm*, the O–H-O angles (∠OHO) in degrees, and the interaction energies (E_{int},) is in kJ/mol

Method	$(H_2O)_2$			$(H_2O)_3$	
	r_{OO}	∠OHO	E_{int}	r_{OO}	E_{int}
BLYP	292	171	−17.11	283	−54.85
DCACP	291	170	−21.71	281	−69.50
BLYP-D	291	170	−22.84	283	−71.00
B3LYP	291	172	−18.54	281	−58.16
Reference	291	175	−21.00	278	−66.53

diverge. These anomalous properties have been related to the local presence of two metastable glassy ice forms, the low-density (LDA) and high-density amorphous (HDA) ices, experimentally observed and separated by a first-order-like phase boundary. At low T ($\sim 80\ K$) and room P, the densities of the two polymorphs are quite different (i.e., $\rho_{LDA} = 0.93\ g/cm^3$ and $\rho_{HDA} = 1.17\ g/cm^3$; Finney et al., 2002). The molecular arrangement in LDA resembles cubic or hexagonal crystalline ice, while that of HDA is similar to liquid water at standard T (more disordered). Shiratani and Sasai (1998) introduced a Local Structure Index (LSI), in order to quantify the degree of inhomogeneity in the local molecular environments of liquid water (and other aggregation states as well). The LSI order parameter is essentially, "*the mean-squared deviation among the radial distances corresponding to the set of molecules that surround a given water molecule*" (Santra et al., 2015). LSI finds its quantitative expression in the inhomogeneity index I of the distribution of radial distances. To compute I one proceeds as follows:

(1) a cut-off distance is selected (for liquid water typically 3.7 Å);
(2) the set of radial oxygen–oxygen distances $\{r_j\}$ corresponding to the N neighboring molecules that are within the cut-off distance are ordered in a decreasing sequence: $r_1 < r_2 < \cdots < r_j < r_{j+1} < \cdots < r_N$;
(3) the mean (Δ) overall relative distances $\Delta_{j+1,\ j} \equiv r_{j\ +\ 1} - r_j$. is calculated;
(4) I is then given by $I = \frac{1}{N}\sum_{j=1}^{N}\left[\Delta_{j+1,j} - \langle\Delta\rangle\right]^2$

The different local molecular environments that can be distinguished and quantified by the LSI order parameter are shown in the schematic diagram of Santra et al. (2015), here reproduced in Fig. 18.3. A low I value implies that the molecule is in a locally disordered environment, with a relatively high packing of neighboring molecules in the interstitial region and hence an increase in the local atomic number density (left part of Fig. 18.3). A high value of I is associated to a more ordered local environment. In the case shown in the right part of Fig. 18.3, the neighboring water molecules are concentrated in the region between 2.7 and 3.2 Å

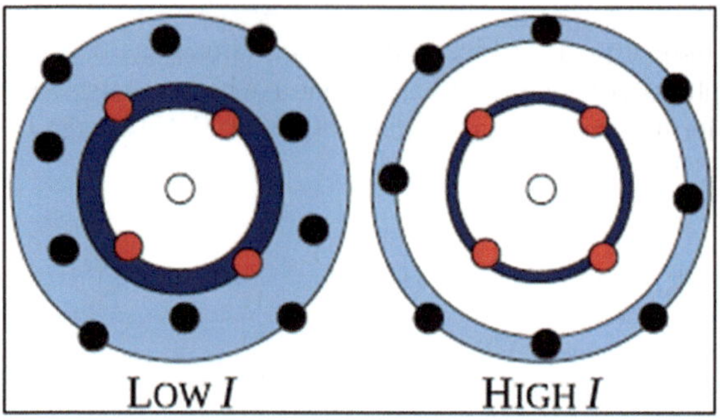

Fig. 18.3 High-density-like and locally disordered (left) versus low-density-like and locally ordered (right) environments, corresponding respectively to low and high values of the LSI order parameter, I. The red circles are first neighbors in the 1st coordination shell (dark blue) and the black circles are next-nearest neighbors in the 2nd coordination shell (light blue). From Santra et al. (2015). Reproduced with permission from Taylor & Francis Ltd, www.tandfonline.com

and sparse in the region between 3.2 and 3.8 *Å*. This leads to a relatively low local atomic number density and to a more prominent separation between the first and second coordination shells (Santra et al., 2015).

The first attempt to classify the structures that characterize the Low-Density Liquid (LDL) and High-Density Liquid (HDL) molecular environments through AIMD in the framework of DFT is again due to Santra et al. (2015). Previous investigations (essentially based on classical force fields in conjunction with several order parameters) showed that the LSI order parameter is bimodal in the corresponding inherent potential energy surface (IPES) of both supercooled and ambient liquid water. The procedure adopted by Santra et al. (2015) consisted in exploring the results of a systematic hierarchy of DFT exchange–correlation functional at standard temperature and density conditions. The AIMD investigation showed that, for all the adopted exchange–correlation functional, the probability distributions of the LSI order parameter P(I) are unimodal in shape and rapidly decaying from maxima located in the low-LSI range. This was interpreted by the authors as indicative of the fact that the majority of water molecules at ambient conditions occur in disordered HDL environments. However, the net HDL/LDL ratio was shown to depend upon the adopted exchange–correlation functional. This complexity was ascribed by the authors to thermal effects. In fact, by removing thermal excitations and obtaining the IPES corresponding to ambient liquid water generated at the PBE0 + TS-vdW(SC) level of theory, the resultant P(I) appeared to be bimodal in shape and exhibited a HDL/LDL ratio near to 3:1. The molecules in LDL were shown to form small connected clusters amid the predominant population of molecules in HDL. Two snapshots of typical liquid water structures (configurations) found in the IPES of ambient liquid water computed with the PBE0 + TS-vdW(SC) exchange–correlation potential at 330 *K* are shown in Fig. 18.4.

As a final approach to the complexity of liquid water, through AIMD, I may quote the recent investigation of Chen et al. (2017), which applied the Strongly

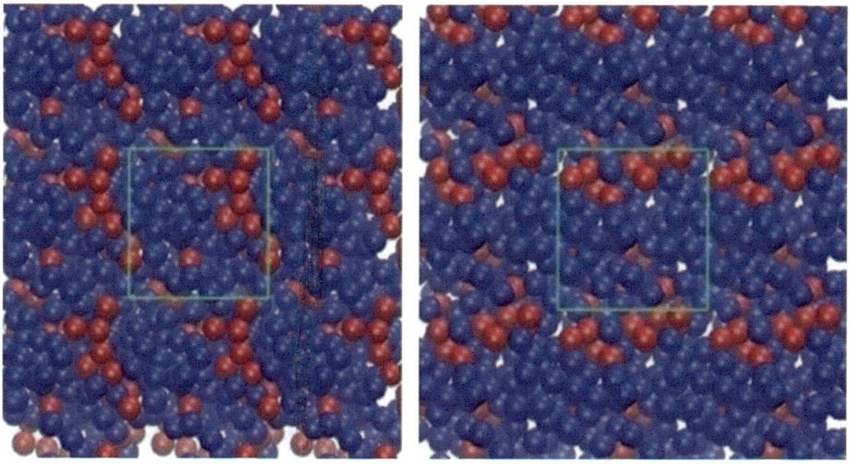

Fig. 18.4 Snapshots of typical liquid water structures (configurations) found in the IPES of ambient liquid water computed with the PBE0 + TS-vdW(SC) potential at 330 *K* (only the oxygen atoms shown for clarity). The blue spheres represent water molecules characterized by I < Imin (i.e. HDL water molecules). The red spheres represent water molecules characterized by I > Imin (i.e. LDL water molecules). The green boxes refer to the simulation cell containing 128 water molecules. From Santra et al. (2015). Reproduced with permission from Taylor & Francis Ltd, www.tandfonline.com

Constrained and Appropriately Normed (SCAN) meta-GGA functional (Sun et al., 2015). SCAN was shown by Sun et al. (2016) to predict the energy of gas-phase water hexamers and ice phases with quantitative accuracy, while other exchange–correlation functional (even with vdW corrections) failed. Based on the investigation of Chen et al. (2017), SCAN predicts structural, electronic, and dynamic properties of liquid water in excellent agreement with experimental measurements. The fact that SCAN is inherently nonempirical, adds value to its heuristic properties. Due to its ability to describe vdW interactions on intermediate length scales, the dynamics of liquid water obtained through AIMD-SCAN are in near quantitative agreement with experiments, and the correct density ordering between liquid water and ice is predicted.

The Radial Distribution Functions (RDFs) obtained by Chen et al. (2017) with the SCAN and the PBE functional are compared in Fig. 18.5 with the results of X-ray diffraction (Soper and Benmore, 2008; Skinner et al., 2013) and Neutron diffraction experiments (Soper and Benmore, 2008).

As noted by Chen et al. (2015), the near perfect agreement between the SCAN $g_{OO}(r)$ and experiment, evident in part A of the figure, is nontrivial, due to the complex interplay among hydrogen bonds, covalent bonds and vdW dispersive interactions. The region between the first and second peaks of $g_{OO}(r)$ predominantly consists of non-H-bonded water molecules that occupy the interstitial space between H-bonded neighbors. The shorter O–H bond length of the SCAN prediction, with respect to PBE, in better agreement with experimental observation, is

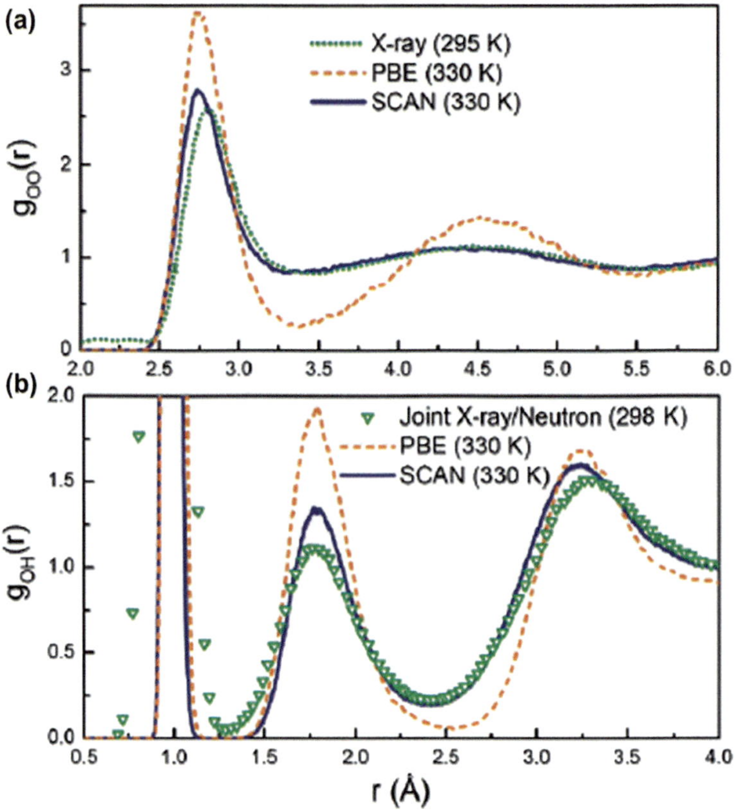

Fig. 18.5 PBE and SCAN oxygen–oxygen [g_{OO}(r), part A] and oxygen-hydrogen [g_{OH}(r); part B] Radial Density Functions of liquid water at T = 330 *K*, compared with experimental RDF evidences. From Chen et al. (2017). With permission from PNAS

interpreted by the Authors as due to an enhancement of the covalency, which strengthens the O–H bond, reducing the interionic distance. The good agreement in the SCAN-predicted versus observed second and third coordination shells is evident in part B of the figure. Other fundamental properties of water (and ice) that appear to be better reproduced by the AIMD-SCAN procedure are resumed in Table 18.5.

The original experimental water density at 300 *K* listed by the Authors (0.99656 *g/cc*) has been replaced here by the value 330 K, i.e. the value that can be retrieved from the NIST Chemistry WebBook (i.e. 0.98479 *g/cc*; NIST, 2020). Notwithstanding these slight modifications, the achievement of AIMD-SCAN is impressive if one recalls that almost all DFT-based approaches predict a solid phase that is denser than the liquid (i.e. $\rho_w/\rho_{Ih} > 1$). The dipole moments of liquid water and ice (μ_w and μ_{Ih} in Table 18.5) were computed by Chen et al. (2017) through Maximum Localized Wannier Functions (MLWFs; Marzari et al., 2012), generated from a unitary transformation of the occupied Kohn–Sham eigenstates. The

Table 18.5 AIMD properties of water (w, 330 *K*) and ice (I_h, 273 *K*) predicted by SCAN and PBE functionals in the isobaric-isothermal ensemble, compared with experimental evidences (From Chen et al., 2017; with modifications)

Property	SCAN	PBE	Experimental
ρ_w *(g/cc)*	1.050 ± 0.027	0.850 ± 0.016	0.98479[a]
ρ_{Ih} *(g/cc)*	0.964 ± 0.023	0.936 ± 0.013	0.9167[b]
$\Delta\rho$ *(g/cc)*	0.086 ± 0.035	-0.086 ± 0.021	0.068
ρ_w/ρ_{Ih}	1.089 ± 0.038	0.908 ± 0.021	1.074
μ_w *(D)*	2.97 ± 0.29	3.12 ± 0.28	2.9 ± 0.6[c]
μ_{Ih} *(D)*	3.29 ± 0.21	3.35 ± 0.21	–
E_g *(eV)*	4.92 ± 0.14	4.43 ± 0.19	8.7 ± 0.6[d]
q	0.68 ± 0.18	0.83 ± 0.11	0.593[e]
D *(Å/ps)*	0.190 ± 0.025	0.018 ± 0.002	0.187[f]
τ_2 *(ps)*	2.9 ± 0.4	7.1 ± 0.5	2.4[g]

[a]NIST (2020); [b]Haynes (2005); [c]Badyal et al. (2000); [d]Bernas et al. (1997); [e]Soper and Benmore (2008); [f]Mills (1973); [g]Ropp et al. (2001)

computed μ_w is in excellent agreement with the observations of Baydal et al. (2000). The band gap E_g, calculated by Chen et al. (2017) through averaging over eight randomly selected configurations from the trajectories differs significantly from what experimentally observed (Haynes 2005). The underestimation of E_g is a recurrent feature of General Gradient Approximation functionals. The tetrahedral order parameter q listed in Table 18.5 quantifies the amount of tetrahedral order (DiStasio et al., 2014). A perfect tetrahedral local environment has q = 1 and lowering of q below this value indicates progressive distortion. The value calculated through AIMD-SCAN agrees, within uncertainty, with what experimentally observed by Soper and Benmore (2008). The diffusion coefficient (D) and the rotational correlation time (τ_2) are again in good agreement when the SCAN procedure is adopted, while the PBE predictions are rather poor.

18.2 Thermophysical and Thermodynamic Properties of the Solvent

18.2.1 P,T, and Redox Stability Limits

The liquid aggregation state, at a 10^5 *Pa* pressure and in the absence of solutes, occurs only within a very narrow T range (i.e. the 100 *K* limiting the centigrade scale, from 0° to 100°)) and redox conditions. Rising the pressure, the maximum T of stability attains the critical point (T_c = 373.917 *°C*, P_c = 220.46 *bar*), beyond which, the liquid aggregation does not survive.

Concerning the redox state of the system, at room P,T, the maximum oxidation state is defined by the dissociation of the H_2O molecule, with the formation of hydrogen ions and gaseous oxygen, i.e.:

$$O_{2,g} + 4H^+_{aq} + 4e^- \Leftrightarrow 2H_2O_{aq} \tag{18.2.1}$$

The standard potential of Eq. 18.2.1 is $E^0_1 = 1.228\ V$. At standard state, the activity of gaseous oxygen is one by definition, and standard potential E^0_1 thus refers to H_2O in equilibrium with an atmosphere of pure O_2 at $T = 25\ °C$, $P = 1\ bar$. Combining the Nernst equation with Eq. 18.2.1, and transforming natural logarithms into 10-based logarithms, we obtain:

$$\begin{aligned} Eh = 1.228 + \frac{2.303RT}{nF} 4 \log a_{H^+} + \frac{2.303RT}{nF} \log f_{O_2} = \\ = 1.228 + 0.0591 \log a_{H^+} + 0.0148 \log f_{O_2} \end{aligned} \tag{18.2.2}$$

The maximum reduction limit for the thermodynamic stability of water is defined by the reduction of hydrogen ions in solution with the formation of diatomic gaseous molecules H_2 (normal hydrogen electrode):

$$2H^+_{aq} + 2e^- \Leftrightarrow H_{2,g} \tag{18.2.3}$$

The redox potential dictated by Eq. 18.2.3 is related to standard potential E^0_1 through:

$$Eh = 0.000 + 0.0591 \log a_{H^+} - 0.0295 \log f_{H_2} \tag{18.2.4}$$

18.2.2 Thermochemical Cycles and the "Hydronium Ion" $[H_3O]^+$

According to the NIST-JANAF tables, the Gibbs free energy of formation of the gaseous proton H^+ at T= 298.15 *K*, P = 1 bar is 1516.990 *kJ/mol*. According to Tissandier et al. (1998), the proton's Gibbs free energy of solvation at the same T,P conditions is –1104.500 *kJ/mol*. Summing up the two terms we get an "*absolute*" molar Gibbs free energy of formation of the aqueous proton from the stable diatomic gas of 412.490 *kJ/mol*. From this energy term, we must subtract the "*liberation free energy*" of the aqueous proton (Ben Naim, 1987). Although the momentum partition function $\Lambda_M{}^3$ of M in the gas phase and in the aqueous phase could be assumed to be identical at first approximation, the numeral densities n_M of M in the gas and in water are substantially different, given the fact that the state of *hypothetical 1-molal solution referred to infinite dilution* is adopted for M as standard state in the aqueous phase. Adopting an ideal gas as reference state, $n_{M,g} =$

4.034 10^{-5} *mol/cc* ≡4.034 10^{-4} *bar/J*. In solution, however, with the adopted standard state of reference $n_{M,s}$ = 1 10^{-3} *mol/cc* ≡1×10^{-2} *bar/J*. The energy contribution of the last term on the right in Eq. 18.3.3, i.e. the *liberation free energy* changed of sign (see later), is +7.959 *kJ/mol*. It should be noted that being the standard state of the solvent different from that of solutes (i.e. *pure solvent*) also the scaling is different. H_2O molecules that, in the ideal gaseous state have a numeral density $n_{M,g}$ = 4.034 × 10^{-5} *mol/cc* (≡4.034 × 10^{-4} *bar/J*), are constrained in an even more restricted space in solution ($n_{M,s}$ = 0.0555 *bar/J*) and the ensuing scaling energy for the solvent attains +17.916 *kJ/mol*. When writing deprotonation equilibria, the liberation free energy of solute species in reaction cancels out, but their individual values must be accounted for when determining their individual bulk Gibbs free energy.

In the ab initio calculation of pK_a, there are two possible ways of approaching the Gibbs free energy change involved in deprotonation of the acid (HA), namely:

$$\mathrm{HA} \Leftrightarrow \mathrm{A}^- + \mathrm{H}^+ \tag{18.2.5}$$

$$\mathrm{HA} + \mathrm{H_2O} \Leftrightarrow \mathrm{A}^- + \mathrm{H_3O}^+ \tag{18.2.6}$$

The species H_3O^+ or "*hydronium ion*" is not clearly defined (Klamt et al., 2003). The adoption of H_3O^+ in the place of H^+, in the calculation of pKa requires the knowledge of its solvation-free energy $G_{solv,H3O+}$. According to Pliego and Riveros (2002), the value of $G_{solv,H3O+}$ is −461.1 *kJ/mol*, while Pearson (1986) assigns to it –435 *kJ/mol* and a PCM-UAHF computation (Barone et al., 1997) restitutes −440.6 *kJ/mol*. Pliego (2003) argued for the correctness of the experimental value of Pliego and Riveros (2002) against the computational one, but his arguments were based upon a debatable choice of standard state in the gas phase and are here disregarded.

We have already seen that an "*absolute*" molar Gibbs free energy of formation of the aqueous proton from the stable diatomic gas of 412.490 *kJ/mol* is obtained by summing up the experimental Gibbs free energy of solvation of Tissandier et al. (1998) with the Gibbs free energy of formation of the gaseous proton from the stable diatomic molecule. The Gibbs free energy of formation of the "*aqueous*" H_2O is –237.1 *kJ/mol*. Because the "experimental" pK_a is based on the measured (galvanic cell) activity of H^+ in solution, to have identical pK_a values with the two formalisms (18.2.5, 18.2.6) we must admit that no energy transfer is involved in the reaction

$$\mathrm{H}^+ + \mathrm{H_2O} \Leftrightarrow \mathrm{H_3O}^+ \tag{18.2.7}$$

Hence, the (conventional) Gibbs free energy of formation of the hydronium ion would be the same of water (−237.1 *kJ/mol*). Subtracting to it the experimental hydration energy, we get for the gaseous hydronium a (conventional) Gibbs free energy of +198 *kJ/mol*. The conventional value of the Gibbs free energy of formation from the elements that we get at G2-MP2 level for the gaseous hydronium is

+633.5 *kJ/mol* (Vetuschi Zuccolini et al., 2011). Adding to it the solvation energy of Pliego and Riveros (2002), scaling and adding the liberation free energy we have –232.1 *kJ/mol*, which is 5 *kJ* higher than the value assigned to H_2O. Given the fact that the uncertainty in the experimental hydration energies is five times higher than this bias, we can confidently consider the notations (18.2.5) and (18.2.6) as energetically equivalent.

Background information: The pK$_a$ scale of organic and inorganic acids

Calculating ab initio accurate pK_a values of organic and inorganic acids is perhaps the most demanding task because an error of 5.69 *kJ/mol* in the G of reaction for equilibria (18.2.5) or (18.2.6) results in an error of 1 unit in the pK_a scale. We have already seen that the two formulations are energetically equivalent (within uncertainty) but we also stressed that the likely errors in the evaluation of gas-phase energy and solvation energy may be quite high. In the attempt to reduce systematic errors, several authors proposed different thermochemical cycles. Saracino et al. (2003) adopted equilibrium (18.2.5) with the cycle of Fig. 18.6a.

They performed gas-phase calculations with the PBE0 functional (Adamo and Barone, 1999) and computed solvation energies with the PCM (Cossi et al., 1998). They adopted for the gaseous proton the Gibbs free energy of formation obtained by Topol et al. (1999) (−26.28 *kJ/mol*) and for the proton solvation the datum of Tissandier et al. (1998). The pK_a values obtained in this way by Saracino et al. (2003) for carboxylic acids are in remarkably good agreement with observations, yielding pK_a values, which are less than half a pK_a unit from experiments (Sergeant and Dempsey, 1979; Table 18.6).

Da Silva et al. (1999) adopted Eq. 18.2.6 and split the bulk hydration energy in "*solvation*" and "*relaxation*" terms, with the latter (E_{relax}) representing the gain in energy when the molecular geometry is relaxed in the gas phase (Fig. 18.6b).

To compute energies in vacuo, Namazian and Halvani (2006) adopted the same basis set with the hybrid functional B3LYP and the PCM for calculating the solvation energy. To calculate the pK_a of 66 carboxylic acids, they explored two different cycles. The first cycle (column a in Table 18.6) combines the isodesmic reaction (18.2.8) and the dissociation reaction (18.2.9) of acetic acid as a reference molecule for the calculation of the Gibbs free energy change involved in equilibrium (18.2.6)[3]:

$$HA_{aq} - CH_3COO^-_{aq} \Leftrightarrow A^-_{aq} + CH_3COOH_{aq} \tag{18.2.8}$$

$$CH_3COOH_{aq} + H_2O_{aq} \Leftrightarrow CH_3COO^-_{aq} + H_3O^+_{aq} \tag{18.2.9}$$

[3] In fact, summing up the two reactions, one gets equilibrium (18.2.6).

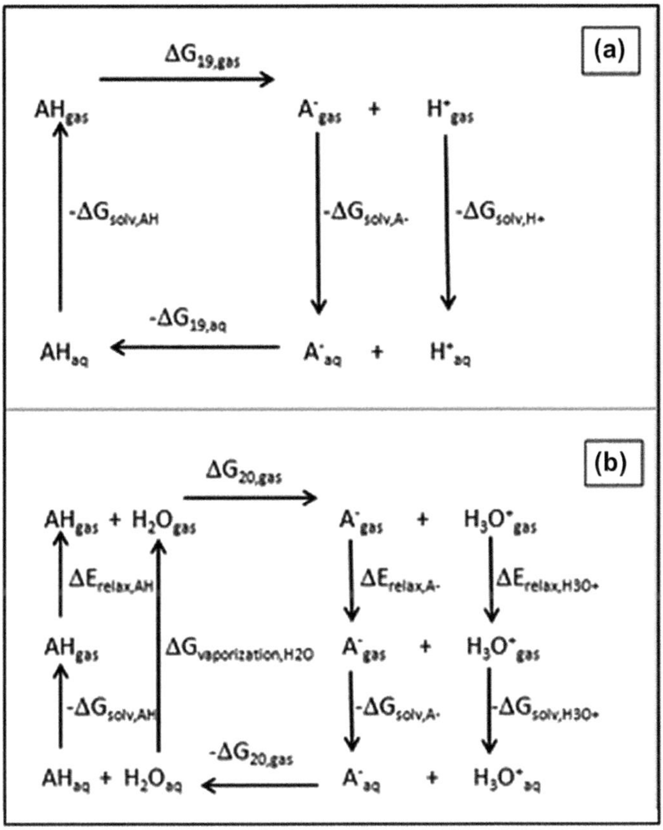

Fig. 18.6 Thermochemical cycles adopted for the calculation of pK_a

Table 18.6 Ab initio pK_a of carboxylic acids

Molecule	Saracino et al. (2003)	Liptak and Shields (2001)		Namazian and Halvani (2006)		da Silva et al. (1999)	Klamt et al. (2003)	Sergeant and Dempsey (1979)
		CBS-QB3	CBS-APNO	a	b			
HCOOH	4.37	3.53	3.93	2.87	2.84	2.92	3.57	3.77
CH_3COOH	4.62	5.19	4.91	4.76	4.73	4.32	4.88	4.76
FCH_2COOH	2.83	–	–	1.18	1.15	1.14	3.14	2.59
$ClCH_2COOH$	2.96	3.36	3.13	2.25	2.23	1.57	3.06	2.85
$CNCH_2COOH$	1.61	2.34	2.31	–	–	–	2.82	2.45
HOOCCOOH	0.7	1.20	1.51	–	–	-	1.69	1.23
$CH3CH_2COOH$	4.67	–	–	4.6	4.57	4.76	–	4.87
$(CH_3)_3CCOOH$	5.64	5.19	6.10	5.1	5.07	5.42	–	5.03

The second cycle adopted by Namazian and Halvani (2006) is nothing else than cycle of Fig. 18.6b without the distinction of relaxation + solvation terms. The average error obtained by these authors on 66 carboxylic acids is 0.5 pK_a units.

Klamt et al. (2003) explored a large variety of organic and inorganic acids, adopting in solvation calculations the Conductor Like Screening Model for real solvents (COSMO-RS; Klamt, 1998). Calculations, based on equilibrium (18.2.5), revealed the existence of a systematic inconsistency in the slope of the pK_a scale, which is only roughly one half of the expected one:

$$pK_{a,experim} = 0.58(\pm 0.01)\frac{\Delta G_{diss,calc}}{RT\ln(10)} + 1.66(\pm 0.10) \qquad (18.2.10)$$

The reason of this systematic deviation is unknown but it is probably already present in gas-phase calculations.

18.3 The Energy of Ionic Solutes

18.3.1 Solvation Energy

Basically, we have two distinct conceptual representations apt to describe the status of a "*solute*" in a solvent like water: an "*explicit*" and an "*implicit*" solvation. An explicit solvation is obtained by arranging the water molecules around the solute with the (energetically) appropriate geometry in order to maximize the coulombic attraction generated by charge predominance. We see as an example in Fig. 18.7 how a simple Fe^{2+} cation dissolved in water "*explicitly*" appears, when surrounded by two solvation spheres (6 water molecules octahedrally arranged in the inner sphere and 12 in the second).

With the increase of the charge of the central cation also the number of solvation spheres basically increases as well. The trick adopted in thermodynamics to avoid this sort of uncertainty (*i.e.* number and position of solvating molecules) is simply to sum up to the energy of the central cation those of the H_2O molecules in the appropriate amount. When writing a balanced equilibrium (for example with a solid at saturation) the H_2O molecules appear explicitly at the product side and are thus virtually non-influent in the equilibrium. The solute thus reduces (energetically) to the simple ion (and this is the notation usually adopted in the thermo-chemical tables). The standard state is that of "*hypothetical one molal solution referred to infinite dilution*". We have already seen extensively how to treat the solvation "*implicitly*". This is the most promising way to proceed, and, as we have already seen at Sect. 7.7.3 it is at the core of the Polarized Continuum Model (PCM) approach.

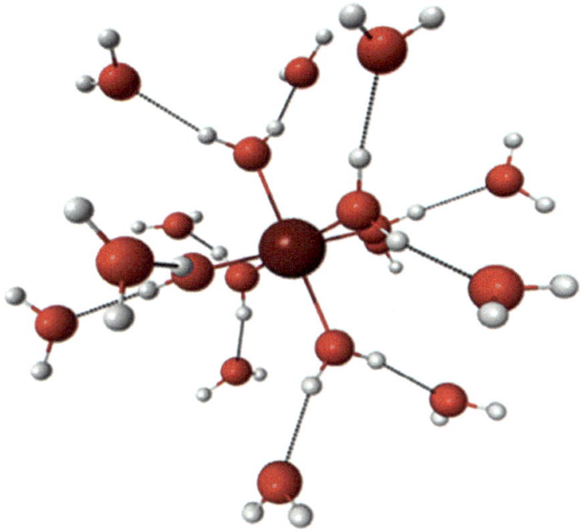

Fig. 18.7 The "*aqueous*" Fe^{2+}. Actually a $[Fe(H_2O)_{18}]^{2+}$ charged cluster of 55 atoms. From Ottonello (2018)

18.3.2 Ab Initio Enthalpy of Formation from the Elements, Some Examples

The computation begins with a detailed investigation of the dissociation energy of the molecule in the gaseous state, in the terms already discussed at Sect. 7.10.2. Then, the energy of the solute is obtained by adding the appropriate solution energy terms. In some cases (transition elements, heavy nuclei), it is appropriate to include, already in gas-phase calculations, corrective terms (relativistic and non-relativistic) to achieve an appropriate appraisal of the true energy of the isolated molecule. In fact, when comparing DFT energy difference obtained with a non-relativistic Hamiltonian in an all-electron computation with experimental energy differences between different ionization states, one has to reduce the errors arising from both neglecting relativistic effects and approximating electron–electron correlation in differently charged systems. Although this sort of uncertainty is partly overcome when recurring to energy-adjusted effective core potentials (Hay and Wadt, 1985a, b; Wadt and Hay, 1985; Dolg et al., 1987), in the case of transition elements there are ambiguities discussed in some detail by Wadt and Hay (1984). Particularly, when comparing computed versus experimental Gibbs free energies of formation of aqueous complexes of chromium arranged in deprotonation or chlorination series, one observes systematic shifts, which are complex function of the first ligands of the central transition metal and of its formal ionization state (Ottonello and Vetuschi Zuccolini. 2005b). Although this shift is apparently linear, it turns out to be apparently periodic when molecules are arranged according to their net charge. Moreover, if we analyze the amplitude of the periodic function (i.e. the overall recurrent "*bias*" in observed versus computed energies (Fig. 18.8) we may see that it has to do with the charge-dependent error in the all electron non-relativistic

Fig. 18.8 Energy bias versus net charge of the aqueous species. Open circles are experimental-computed (correction free) energies and filled circles are energy correction terms arising from Eq. (18.3.2). The continuous periodic function is drawn only to put in evidence the recurrent character of the bias

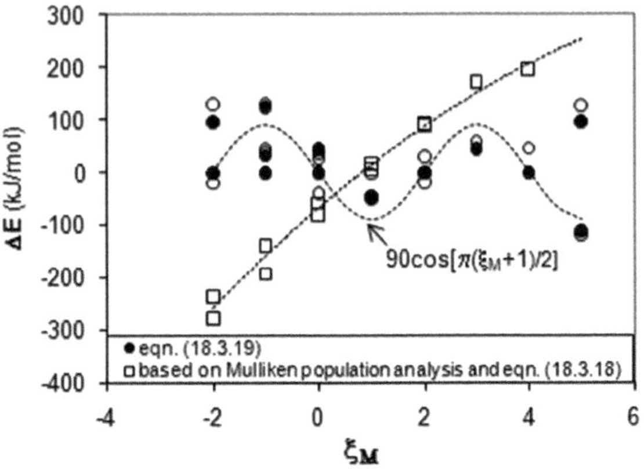

computation of ionization energies in the gaseous state (i.e. it corresponds roughly to the first derivative of the error with respect to ion charge; cf. Table 18.7 and Fig. 18.9.

This observation rules out any possible error associated to the solvation process. Moreover, the charge-dependent periodicity of the error suggests that we cannot account of this error in terms of a simple polynomial E = f(q) dependence associated to the Mulliken population analysis, as done for instance in the case of Na in the CBS-Q procedure of Ochterski et al. (1996). In fact, although a continuous (3rd order) function may be envisaged also for chromium, based on the discrete computed values of ionic potential and electron affinity:

$$\Delta E = 90.145q - 24.376q^2 + 0.931q^3 \quad (18.3.1)$$

the molecular correction obtained when substituting to q the net charge obtained from Mulliken analysis of the population in the actual molecule gives rise to an odd monotonic correction (Fig. 18.8), despite the fact that the depicted function has the advantage of overlapping with satisfactory approximation to the relativistic and correlation corrections necessary for Cl and O ions as well (Fig. 18.9).

This observation led (Ottonello and Vetuschi Zuccolini, 1995b) to hypothesize that the observed gas-phase single ion DFT all-electron biases are actually no more than values corresponding to local minima of two periodic functions (valid respectively for ionization and electron affinity) which may be combined appropriately in the form of a complementary function of general validity for chemically complex molecules of the central cation M:

$$\Delta E' = -\left(\frac{\partial \Delta E_{ea}}{\partial q^-} + \frac{\partial \Delta E_{ip}}{\partial q^+} n_M\right) \cos\left(\frac{3\pi n^-}{2 q_M n_M}\right) \quad (18.3.2)$$

where $\frac{\partial \Delta E_{ea}}{\partial q^-}, \frac{\partial \Delta E_{ip}}{\partial q^+}$ are the slopes of the linear interpolants in the anionic and cationic region, respectively, q_M and n_M are charge and number of metal cations in the

Table 18.7 Ab initio gas-phase energetic of Cr molecules (Ottonello and Vetuschi Zuccolini, 1995b) compared with experimental observations. (1) = JANAF Tables (Chase, 1998); (2) NBS Tables Wagman et al., 1982). (3) based on electron affinity of CrO_2 after Wenthold et al. (1997)

Molecule	E_e (*Hartree*)	E_0 (*Hartree*)	H_{corr} (*Hartree*)	E' (*Hartree*)	$H°_{f,0}$ (*kJ/mol*)	$H°_{f,298.15}$(*) (*kJ/mol*)	$H°_{f,0}$ (*kJ/mol*) Experim	$H°_{f,298.15}$ (*kJ/mol*) Experim	Ref
H	−0.50027	−0.50027	0.00236	0.00000	216.0	218.0	216.0	218.0	1
H^+	0.00000	0.00000	0.00236	0.00000	1529.5	1537.7	1528.0	1536.4	1
H_2	−1.17854	−1.16836	0.01348	0.00000	−8.5	−8.3	0.0	0.0	1
Cl^-	−460.25223	−460.25223	0.00236	−0.04397[a]	−300.4	−305.0	−229.3	−233.9	1
Cl	−460.13624	−460.13624	0.00236	0.0000	119.6	121.2	119.7	121.3	1
Cl_2	−920.34988	−920.34869	0.00470	0.0000	39.1	39.1	0.0	0.0	1
O	−75.06062	−75.06062	0.00236	0.0000	246.8	248.6	246.9	249.4	1
O^-	−75.05273	−75.05273	0.00236	−0.04397[a]	152.1	147.7	105.9	101.7	1
O^{2-}	−74.56925	−74.56925	0.00236	−0.10864[a]	1251.6	1241.1			1
O_2	−150.32004	−150.31626	0.00709	0.00000	−18.4	−18.4	0.0	0.0	1
H_2O	−76.41974	−76.39919	0.02515	0.00000	−206.5	−209.3	−238.9	−241.8	1
Cr^-	−1044.27154	−1044.27154	0.00236	−0.04397[a]	304.7	300.7	331.1	327.0	1
Cr	−1044.28103	−1044.28103	0.00236	0.00000	395.3	397.4	395.3 394.6	397.5 396.6	1 2
Cr^+	−1044.05338	−1044.05338	0.00236	0.02541[a]	1059.7	1068.0	1048.1 1047.3	1056.5 1055.6	1 2
Cr^{2+}	−1043.47299	−1043.47299	0.00236	0.03437[a]	2607.0	2621.5	2639.3	2655.6	2
Cr^{3+}	−1042.32162	−1042.32162	0.00236	0.02902[a]	5615.9	5636.6	5626.2	5648.4	2
Cr^{4+}	−1040.49502	−1040.49502	0.00236	0.01148[a]	10,365.5	10,392.5	10,364.2	10,388.9	2
Cr^{5+}	−1037.91484	−1037.91484	0.00236	−0.01611[a]	17,067.4	17,100.5	17,067.0	17,100.0	2
Cr^{6+}	−1034.54608	−1034.54608	0.00236	−0.05164[a]	25,818.8	25,858.1	25,811.1	25,849.2	2
Cr^{7+}	−1028.61679	−1028.61679	0.00236	−0.09296[a]	41,277.6	41,323.2	41,263.0	41,309.1	2

(continued)

Table 18.7 (continued)

Molecule	E_e (*Hartree*)	E_0 (*Hartree*)	H_{corr} (*Hartree*)	E' (*Hartree*)	$H°_{f,0}$ (*kJ/mol*)	$H°_{f,298.15}$(*) (*kJ/mol*)	$H°_{f,0}$ (*kJ/mol*) Experim	$H°_{f,298.15}$ (*kJ/mol*) Experim	Ref
Cr^{8+}	−1021.79367	−1021.79367	0.00236	−0.13797[a]	59,073.6	59,125.3	59,086.0	59,140.8	2
CrO	−1119.48053	−1119.47794	0.00592	−0.01818[b]	230.5	230.8	187.9 222.2	188.3	1 2
CrO_2	−1194.77663	−1194.77140	0.00973	0.01730[c]	−41.0	−41.9	−73.8 −59.0	−75.3	1 2
CrO_2^-	−1194.89508	−1194.89017	0.00928	0.00000[c]	−392.1	−399.6	−306.2		3
$Cr(H_2O)_6^{3+}$	−1502.01391	−1501.85950	0.16968	0.01730[c]	1333.2	1310.9			
$Cr(H_2O)_5(OH)^{2+}$	−1501.94745	−1501.80579	0.15734	0.00000[c]	−100.7	−123.9			
$Cr(H_2O)_5Cl^{2+}$	−1886.32696	−1886.19718	0.14524	0.00000[c]	39.0	19.3			
$Cr(H_2O)_4(OH)_2^+$	−1501.70598	−1501.57662	0.14407	−0.01730[c]	−1073.9	−1101.6			
$Cr(H_2O)_4Cl_2^+$	−2270.50177	−2270.39561	0.12047	−0.01730[c]	−889.5	−910.2			
$Cr(H_2O)_3(OH)_3$	−1501.32913	−1501.21969	0.12294	0.00000[c]	−1620.9	−1653.7			
$Cr(H_2O)_3Cl_3$	−2654.52114	−2654.43985	0.09520	0.00000[c]	−1322.4	−1342.2			
$Cr_2(H_2O)_8(OH)_2^{4+}$	−2850.65832	−2850.42354	0.26054	0.00000[c]	1238.8	1203.5			
$Cr_2(H_2O)_{10}(OH)^{5+}$	−2927.00974	−2926.73707	0.30076	0.03646[c]	2880.3	2840.3			
$Cr_3(H_2O)_9(OH)_4^{5+}$	−4122.65887	−4122.36672	0.32406	−0.04291[c]	1867.8	1820.8			
$Cr_2O_7^{2-}$	−2615.59061	−2615.56566	0.03567	0.03646[c]	−1532.5	−1555.3			
$HCr_2O_7^-$	−2616.24480	−2616.20904	0.04726	0.04764[c]	−1662.8	−1681.6			
CrO_4^{2-}	−1345.29566	−1345.28291	0.01840	0.00000[c]	−611.3	−630.3			
$HCrO_4^-$	−1346.05260	−1346.02904	0.03010	0.01223[c]	−1008.7	−1023.4			
H_2CrO_4	−1346.58125	−1346.54627	0.04183	0.01730[c]	−823.9	−835.8			
CrO_3Cl^-	−1730.46217	−1730.45054	0.01791	0.01223[c]	−948.0	−959.4			

(*) Integrated electronic heat capacity added (net charge × 6.197 *kJ/mol*). a: based on Eq. 18.3.1 in tex. b: based on Eq. 18.3.1 in text and a cationic slope coefficient obtained by the local partial derivative of Eq. 18.3.1; c: based on Eq. 18.3.2

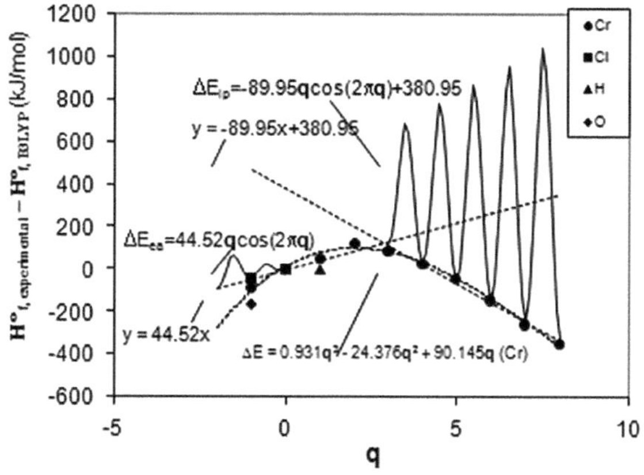

Fig. 18.9 Ab initio experimental values of ionization potential and electron affinity of atomic constituents. The third order equation in figure is the continuous function of charge (solid line) passing through the obtained discrete differences. The two periodic functions whose local minima correspond to discrete differences in computed—observed electron affinity and computed—observed ionization potentials (solid lines) have constant terms corresponding to the obtained linear interpolation over negative and positive charges, respectively (dotted lines). Modified from Ottonello and Vetuschi Zuccolini (1995)

molecule, n^- are the net negative charges counterbalancing metal charges in the molecule and the 3/2 term arises from phase opposition.

Having at hand the correct enthalpy of formation from the elements of the gaseous species, we proceed to evaluate the solvation energy terms. Instead of building up the actual solute cluster in solution, as previously seen for Fe^{2+} (Fig. 18.7), we may exploit the fundaments of Tomasi's Polarized Continuum Model (see Sect. 7.7.3) and conceive the Gibbs free energy of solvation ($G^0{}_{solv}$) as composed of two main terms (coupling work and molecular motion).

$$\begin{aligned}\Delta G^0_{solv} &= W(M/S) + \Delta G_{mm} = \\ &\Delta G_{el} + \Delta G_{cav} + \Delta G_{disp} + \Delta G_{rep} \\ &+ RT\ln\left(\frac{q_{rot,g} \times q_{vib,g}}{q_{rot,s} \times q_{vib,s}}\right) - RT\ln\left(\frac{n_{M,g} \times \Lambda^3_{M,g}}{n_{M,s} \times \Lambda^3_{M,s}}\right)\end{aligned} \tag{18.3.3}$$

where $n_{M,g}$ and $n_{M,S}$ are the numeral densities of the molecule M in the gas and in the solvent and $\Lambda_{M,g}$, $\Lambda_{M,s}$ are the momentum partition functions of M in the two states. The "*coupling work*" terms (first four terms on the right-end part of Eq. 18.3.3, i.e. electrostatic solute–solvent interaction, cavitation, dispersion and repulsive contribution, are related to the work necessary to "*build up*" the molecule M in the solvent S (Tomasi and Persico, 1994).

The electrostatic perturbation induced by the reaction field is:

$$[\hat{H}^0 + \hat{V}_R] = E\Psi \tag{18.3.4}$$

where $\hat{H}^0$ is the Hamiltonian of the solute *in-vacuo*, $\hat{V}_R$ the electrical potential of the continuum and Ψ the solute wave function in solution. the dispersive and repulsive contributions are expressed in terms of perturbing operators which act on the Atomic Orbital (AO) basis set (Amovilli and Mennucci, 1997; Cossi and Barone, 2000):

$$\Delta G_{rep} = \left\langle \Psi \middle| \hat{h}_{rep} \middle| \Psi \right\rangle \tag{18.3.5}$$

$$\Delta G_{dis} = \left\langle \Psi \middle| \hat{h}_{dis} + \frac{1}{2}\hat{X}_{dis} \middle| \Psi \right\rangle \tag{18.3.6}$$

In practical computations, however, the sum of solute–solvent dispersive and repulsive interactions $\Delta G_{dis} + \Delta G_{rep}$ are obtained as double summations of terms extending to the solute atoms placed within the cavity and solvent atoms outside a cavity limited by a tessellated surface:

$$\Delta G_{dis} + \Delta G_{rep} = \sum_{a}^{\text{solute}} \sum_{s}^{\text{solvent}} \rho_s \sum_{i}^{\text{tesserae}} a_i \Gamma_{as} \tag{18.3.7}$$

where ρ_s is the number density of the solvent in atomic units ($\rho_s = N_0/V_s$) ($Å^{-3}$), a_i the area of the ith tessera composing the surface and Γ_{as} a set of dispersive and repulsive parameters obtained from crystallographic data (Caillet and Claverie, 1978).

The molecular motion effects are defined in terms of microscopic contributions to the rotational and vibrational partition function q_{rot} and q_{vib} (subscript g for the gaseous state and s for solution) and of the numeral densities of the molecule M in the gas and in the solvent ($n_{M,g}$ $n_{M,S}$) while $\Lambda_{M,g}$, $\Lambda_{M,s}$ are the momentum partition functions of M in the two states. The *liberation free energy* is the amount of energy arising from the last term on the right in Eq. 18.3.3, and intimately linked to substantially different numeral densities in the gas and in the liquid.

Absolute solvation-free energy is usually computed by Boltzmann averaging over all minima (Ben Naim, 1987; Barone et al., 1997), i.e.[4]:

[4] Assuming that only a single conformer (c) exists in the gas phase is equivalent to stating that its equilibrium mole fraction P_c is equal to 1 in the gas phase and that eventual minor structural modifications taking place during solvation and related to the reaction field are fully accounted for by the variational principle. This is an oversimplification of reality but the errors generated by this assumption are by far less important than those arising from the neglect of correlation effects and in the empiricism adopted in the assessment of non-electrostatic entropy contributions.

$$\exp\left[\frac{-\Delta G_{solv}}{RT}\right] = \sum_c P_c \left[\frac{-\Delta G_{solv,c}}{RT}\right] \quad (18.3.8)$$

Neglect of electron–electron correlation affects the results in principle in an underestimate of the stabilizing effect of the reaction field. According to Foresman et al. (1996), DFT or Møller-Plesset Perturbation theory should be adopted in place of HF whenever electron–electron correlation effects are large compared to solvent effects but are less important when computing quantities with large solvent effect components. According to Miertuš and Tomasi (1982) and Cammi and Tomasi (1994), being G_{solv} a function of the dielectric constant of the solvent (ε) and of the cavity size, which are both T-dependent magnitudes, the correct approach is to combine the partial derivatives as follows:

$$\frac{\partial \Delta G_{solv}}{\partial T} = \frac{\partial \Delta G_{solv}}{\partial \varepsilon}\frac{\partial \varepsilon}{\partial T} + \frac{\partial \Delta G_{solv}}{\partial n}\frac{\partial n}{\partial T} \quad (18.3.9)$$

The GAUSSIAN code (Frish et al., 2004), which embodies the IEFPCM procedure, returns the "*coupling work*" terms (first four terms on the right-end part of Eq. 18.3.3 but, while the electrostatic solute–solvent interaction, the dispersion and the repulsive contribution are essentially enthalpic, splitting the Gibbs free energy of cavitation into entropic and (subordinated) enthalpic terms requires application of the Scaled Particle Theory (SPT) in the terms already seen at Sect. 16.3.3, and, notably Eqs. 16.3.9–16.3.10. As noted by Pierotti (1976) "*The essence of the scaled particle theory is that work is required to exclude the centers of molecules from any specified region of space in a fluid*". Hence, the number density ρ_s appearing in the expression of the Gibbs free energy of cavitation:

$$\Delta G_{cav} = \left\{ -\ln(1-Y) + \left(\frac{3Y}{1-Y}\right)\varsigma + \left[\left(\frac{3Y}{1-Y}\right) + \frac{9}{2}\left(\frac{Y}{1-Y}\right)^2\right]\varsigma^2 + \frac{YP}{\rho_s}\varsigma^3 \right\} \quad (18.3.10)$$

and in the cavitation enthalpy H_{cav}:

$$H_{cav} = Y\alpha_V RT^2(1-Y)^{-3}\left[(1-Y)^2 + 3(1-Y)\varsigma + 3(1+2Y)\varsigma^2\right] + Y\frac{N_0 P}{\rho_s}\varsigma^3 \quad (16.3.9)$$

has a non-negligible role. As we have already seen at Sect. (16.3.3), the enthalpy of solution was obtained by Pierotti (1976) as the following partial derivative of Henrys' law constant:

$$\Delta H_S = \left(\frac{\partial \ln K_H}{\partial(1/RT)}\right)_P = \overline{H}_{cav} + \overline{H}_{el} + \overline{H}_{disp} + \overline{H}_{rep} - RT + \alpha_s RT^2 \tag{16.3.12}$$

where α_s is the thermal expansion coefficient of the solvent.

Table 18.8 lists the results obtained by Ottonello and Vetuschi Zuccolini (1995b) on various solute species of chromium. The absolute ab initio enthalpy of formation from the elements at standard state was obtained simply by adding to the enthalpy

Table 18.8 Computed absolute and conventional enthalpy of formation from the elements (*kJ/mol*), compared with experimental evidences and guess values. Values in italics obtained by adding cavitation and solution terms. Notes: **a**: Wagman et al. (1982); **b**: Dellien et al. (1976);. **c**: consistent with calculated G°f and S° in Ottonello and Vetuschi Zuccolini (1995b); **d**: obtained from Ziemniak et al. (1998) by adding the contribution of inner shell water molecules. **e**. obtained from the reaction $Cr(H_2O)_6^{3+} + nCl^- \Leftrightarrow Cr(H_2O)_{6-n}^{(3-n)+} + nH_2O$ which, based on experimental data, gives $H^0_{f,Cr(H_2O)_{6-n}^{(3-n)+}} = -1993.0 + 161.4n \quad (R^2 = 0.993)$; **f**: calculated with the method of Baes and Mesmer (1981); **g**: consistent with calculated G°f and S° of Baes and Mesmer (1981); **h**: Ball and Nordstrom (1998); **i**.: Latimer (1952); **l**: based on the observed $H°_f$ of $Cr_2O_7^{2-}$ and Table 3 in Brito et al. (1997); **m**: Tissandier et al. (1998); **n**: Martinez et al. (2000) with HIW pair potentials

Aqueous species	$H°_{f,aq,}$ abs, ab initio	$H°_{f,aq,}$ conv, ab initio	$H°_{f,aq,\ conv,}$ exp	$H°_{soln,}$ ab initio	$H°_{solv,}$ Born	$H°_{solv,exp}$
H^+	368.1 *431.0*	−18.1 *7.3*	0	*−1106.7*	−218.8	−1150.1[m]
$Cr(H_2O)_6^{3+}$	−689.1 *−693.0*	−1847.7 *−1964.1*	−1954.0[b]; −1970.9[b]; −1999.1[a]	−1954.6 *−2003.9*	−1919.9	2397 ± 188[n]
$Cr(H_2O)_5(OH)^{2+}$	−1099.6 *−1044.1*	−1872.0 *−1891.5*	−1903.9[i]; −1912.0[b]; −1939.7[a]	−975.7 *−920.2*	−855.3	
$Cr(H_2O)_4(OH)_2^+$	−1431.1 *−1438.5*	−1817.3 *−1862.2*	−1879.0[b]; −1907.1[a]	−374.9 *−336.9*	−285.5	
$Cr(H_2O)_3(OH)_3^0$	−1780.7 *−1811.2*	−1780.7 *−1811.2*	−1824.0[c]; −1864.6[d]	−127.0 *−157.5*	−111.7	
$Cr(H_2O)_5Cl^{2+}$	−1073.9 *−991.5*	−1846.3 *−1838.9*	−1816.0[b]; −1859.8[a]; −1991.7[i]	−1093.2 *−1010.8*	−942.9	
$Cr(H_2O)_4Cl_2^+$	−1322.3 *−1311.0*	−1708.5 *−1734.7*	−1678.0[b]; −1687.8[i]; −1720.5[a]	−457.5 *−400.8*	−344.9	
$Cr(H_2O)_3Cl_3^0$	−1482.1 *−1523.8*	−1482.1 *−1523.8*	−1508.0[e]	−139.9 *−181.6*	−130.5	
CrO_2^-	−634.5 *−687.2*	−248.3 *−263.5*	–	−234.9	−269.2	

(continued)

Table 18.8 (continued)

Aqueous species	$H^\circ_{f,aq,}$ abs, ab initio	$H^\circ_{f,aq,}$ conv, ab initio	$H^\circ_{f,aq, conv,}$ exp	$H^\circ_{soln,}$ ab initio	$H^\circ_{solv,}$ Born	$H^\circ_{solv,exp}$
$Cr_2(H_2O)_8(OH)_2^{4+}$	−1785.9 *−1622.8*	−3330.6 *−3317.6*	−3260.7[f]; −3280.7[g]	−2989.4 *−2826.3*	−2707.6	
$Cr_2(H_2O)_{10}(OH)^{5+}$	−1668.6 *−1434.1*	−3599.5 *−3552.6*	−3597.9[f]	−4413.2 *−4274.4*	−4121.3	
$Cr_3(H_2O)_9(OH)_4^{5+}$	−2395.0 *−2212.0*	−4325.9 *−4330.5*	−4169.7[f]	−4328.5 *−4032.8*	−3875.9	
CrO_4^{2-}	−1596.3 *−1671.8*	−824.0 *−824.4*	−879.0[h]; −894.3[i]	−966.0 *−1041.5*	−999.6	
$HCrO_4^-$	−1242.6 *−1310.0*	−856.4 *−886.3*	−873.0[h]; −878.3[b]; −923.1[i]	−187.1 *−286.6*	−255.7	
H_2CrO_4	−880.0 *−902.0*	−880.0 *−902.0*	−841.0[b]; −905.0[h]	1.2 *−66.2*	−40.6	
$Cr_2O_7^{2-}$	−2245.6 *−2377.6*	−1473.2 *−1530.2*	−1476.2[h]; −1523.0[i]	−594.6 *−822.3*	−777.1	
$HCr_2O_7^-$	−1851.7 −1949.1	−1465.5 −1525.4	−1487.4[l]; −1534.2[i]	−45.0 −267.5	−234.0	
CrO_3Cl^-	−1212.2 −1210.6	−826.0 −786.9	−754.8[b]	−220.7 −251.2	−220.6	

of the isolated gaseous molecule the solute–solvent interaction in terms of electrostatic + dispersive + repulsive contributions. The conventional counterparts were obtained by applying the nominal scaling factor $H^0_{f,abs} - H^0_{f,conv} = 386.1 \times \xi_M$. I repeated here the calculations in terms of the Pierotti (1976) Eq. (16.3.12), i.e. adding the cavitation enthalpy (H_{cav} in Eq. 16.3.12) and the last two solution terms on the right of the same equation.

As we may see in Fig. 18.10, the nominal scaling factor obtained by coupling the solvation energy of proton of Tissandier et al. (1998) with the NIST-JANAF enthalpy of formation of the gaseous proton (Chase, 1998) is sufficiently consistent with the population regression in terms of solvation (i.e. 387.8 *kJ/mol* instead of 386.1 kJ/mol) but it cannot be applied when cavitation and solution terms are taken into account. In fact, the solution energy of the proton is higher than the simple electrostatic + repulsive + dispersive interaction of 63.6 *kJ/mol* (43.4 *kJ/mol* higher with respect to the estimate of Tissandier et al., 1998; see also Table 18.8), and also the scaling factor of the whole population is higher, attaining 423.7 *kJ/mol.* Applying the appropriate scaling factor and including the cavitation and solution terms reduces the mean absolute error from 53.8 to 36.7 *kJ/mol.*

Disregarding the minor corrections on the zero point electronic energy of the isolated species and operating with insufficiently extended basis sets affects the accuracy of the procedure. As an example, we may see in Table 18.9 and

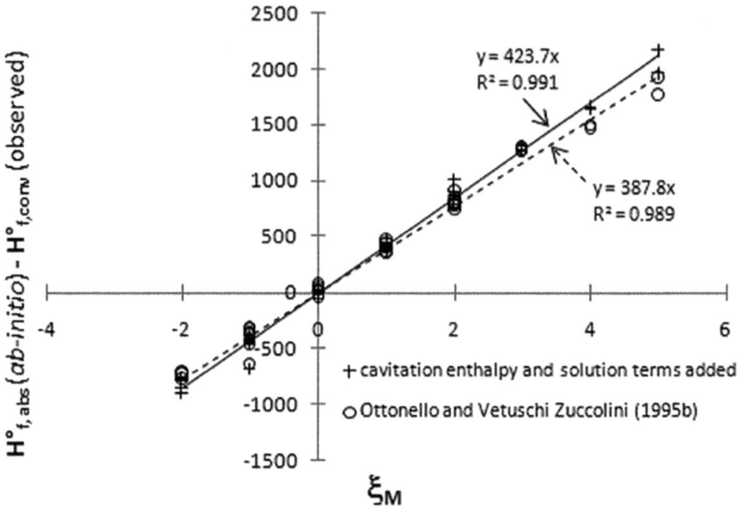

Fig. 18.10 Significance of the scaling factor. The regressions are obtained by imposing a zero intercept (the liberation free energy term is translational, hence essentially entropic and absent in the computation of enthalpy). The nominal scaling (386.1 *kJ/mol*) is applicable only whenever the simple electrostatic + dispersive + repulsive contributions to the solvation energy are taken into account

Fig. 18.11, the results obtained by our research group on inorganic and metal–organic Fe species (Ottonello and Vetuschi Zuccolini 2005, 2009).

18.3.3 Ab Initio Gibbs Free Energy of Formation from the Elements

In retrieving the conventional Gibbs free energy of formation from the elements from the absolute one, we may proceed as anticipated at Sect. (18.2.2), i.e. adopt a conventional "*absolute*" molar Gibbs free energy of formation of the aqueous proton from the stable diatomic gas of 412.490 *kJ/mol* and subtract the "*liberation free energy*" of the aqueous proton (Ben Naim, 1987). Alternatively, one may provisionally adopt a scaling factor appropriate to the applied theory level. In Fig. 18.12, I we may see that indeed the regression line for the Cr population $G^0_{f,abs} - G^0_{f,conv} = 423.1 \times \xi_M - 11.75$ (kJ/mol) is not far from the expected one:

$$G^0_{f,abs} - G^0_{f,conv} = \left(\overline{G}^0_{f,gas,H^+} + \Delta G^0_{solv,H^+}\right) \times \xi_M + RT\ln\left(RT \times n_{M,s}\right) \quad (18.3.11)$$

Table 18.9 Computed absolute and conventional enthalpy of formation from the elements of Fe-hydrates and Fe-acetates, compared with literature estimates

Aqueous species	$H^0_{f,g\ ab\text{-}initio}$	G_{el}	G_{cav}	G_{disp}	G_{rep}	$H^{\circ}_{soln,ab\text{-}initio}$	$H^{\circ}_{f,\ aq,\ abs,ab\text{-}initio}$	$H^{\circ}_{f,aq,conv,exp}$	IEFPCM procedure
$[Fe(H_2O)_6]^{2+}$	−20.209	−812.8	75.8	−86.6	26.9	−862.6	−882.8	−1807.2[a]	1
		−839.8	82.6	−64.3	8.7	−884.4	−904.6		2
		−824.4	84.7	−85.7	25.5	−873.2	−893.4		3
		−847.2	81.8	−85.4	26.2	−895.5	−915.7		4
		−895.9	86	−62.8	8.1	−939.0	−959.2		5
$[Fe(H_2O)_5(OH)]^{+}$	−901.987	−328	86.9	−81.3	27.7	−369.8	−1271.8	−1755.8[a]	3
$[Fe(H_2O)_4(OH)]^{+}$	−592.982	−346.1	74.2	−72.3	21	−387.8	−980.7	−1470.0[a]	3
$[Fe(H_2O)_4(OH)_2]^{0}$	−1428.248	−117.6	88.3	−76.3	19.2	−162.7	−1591.0	−1692.5[a]	3
$[Fe(H_2O)_3(OH)_3]^{-}$	−1539.253	−292.5	92.1	−72.3	19.2	−333.0	−1872.2	−1669.2[a]	3
		−277.7	89	−69.5	17.3	−317.8	−1857.1		3
		−299.9	91.1	−50	4.8	−332.7	−1871.9		4
$[Fe(OH)_3]^{-}$	−534.495	−306.8	56	−31.6	4.8	−327.0	−861.5	−811.8[a]	5
$[Fe(H_2O)_6]^{3+}$	1533.662	−1929.4	72.6	−99	42.6	−1976.4	−442.8	−1764.6[a]	1
		−1883.4	79.2	−64	8.5	−1928.4	−394.8		2
		−1944.4	75	−98.6	41.6	−1991.6	−458.0		3
		−1959.4	74.3	−99.5	44.6	−2004.6	−471.0		4
		−1890.6	80.5	−64	8.6	−1935.3	−401.7		5
$[Fe(H_2O)_5(OH)]^{2+}$	107.448	−906.9	76.4	−82.3	24.3	−954.9	−847.4	−1722.0[a]	3
$[Fe(H_2O)_4(OH)]^{2+}$	452.608	−939.7	68.7	−72.4	21.5	−981.9	−529.3	−1436.2[a]	3
$[Fe(H_2O)_4(OH)_2]^{+}$	−820.597	−107.6	83.8	−76.1	19.4	−153.1	−973.7	−1684.3[a]	4
$[Fe(H_2O)_3(OH)_2]^{+}$	−557.286	−326.8	68.7	−69.1	21.3	−365.9	−923.2	−1398.5[a]	4
$[Fe(H_2O)_3(OH)_3]^{0}$	−1412.188	−115.1	80.6	−72.2	18.5	−158.1	−1570.3	−1646.7[a]	3
		−103.1	78.7	−71.8	18.9	−145.6	−1557.8		4
		−125.9	84.2	−53.7	6.2	−162.1	−1574.3		5

(continued)

Table 18.9 (continued)

Aqueous species	$H^0_{f,g\ ab\text{-}initio}$	G_{el}	G_{cav}	G_{disp}	G_{rep}	$H^{\circ}_{soln,ab\text{-}initio}$	$H^{\circ}_{f,\ aq,\ abs,ab\text{-}initio}$	$H^{\circ}_{f,aq,conv,exp}$	IEFPCM procedure
$[Fe(OH)_3]^0$	−546.826	−200.3	53.5	−31.1	4.6	−220.6	−767.4	−789.2[a]	5
$[Fe(H_2O)_2(OH)_4]^-$	−1568.018	−265.6	82	−12.6	18.7	−248.6	−1816.6	−1669.2[a]	4
		−291.6	87.1	−72.4	21.7	−330.5	−1898.5		5
$[Fe(OH)_4]^-$	−973.89	−312.5	65.4	−38	5.6	−336.7	−1310.6	−1034.8[a]	5
$[(CH_3COO)Fe]^+$	505.9	−546.7	47.9	−39.7	7	−574.1	−68.2	−581.8[b]	3
		−688	56.9	−27.1	1.5	−706.8	−200.9		5
		−495.7	47.7	−39.8	7	−523.3	−17.4		4
$[(CH_3COO)_2Fe]^0$	−844.4	−110.2	81	−75.6	13.6	−161.4	−1005.8	−1084.1[b]	3
		−92.8	99	−49.7	2.6	−126.1	−970.5		5
$[(CH_3COO)Fe]^{2+}$	1993.8	−1526.9	44.2	−46.9	14.9	−1554.2	439.6	-	3
		−1766.9	56.5	−27.2	1.5	−1785.9	207.9		5
		−1331.6	44.4	−47.2	15.4	−1358.7	635.1		4
$[(CH_3COO)_2Fe]^+$	−55	−390.2	79.7	−75.3	13.5	−441.5	−496.5	-	3
		−436.6	98.2	−49.5	2.6	−469.9	−524.9		5
		−299.9	79.9	−75.4	13.9	−350.8	−405.8		3
$[(CH_3COO)_3Fe]^0$	−1255.2	−59.9	113.1	−99.2	17.2	−125.8	−1381.0	-	3
		−39.1	132.8	−62	2.6	−79.1	−1334.3		5
		−48.2	113.1	−99.2	17.5	−113.8	−1369.0		4

1. UHF/STO-3G with UAHF radii 2. UHF/STO-3G with UA0 radii. 3. UHF/6-31G(d,p) with UAHF radii. 4. B3LYP/6-31G(d,p) with UAHF radii. 5. B3LYP/6-31G(d,p) with UA0 radii. [a] HKF estimates of Shock et al. (1997) coupled with the appropriate number of salvation molecules; i.e. $[Fe(OH)_4]^- \equiv [FeO_2]^- + 2[H_2O]^0$; b Shock and Koretsky (1993)

Table 18.10 Ab initio Gibbs free energy of formation from the elements of aqueous molecules of the Cr and Fe transition elements (Ottonello and Vetuschi Zuccolini 2005, 2008, 2009). Gas phase ab initio Gibbs free energies based on stationary state structures were resolved at the B3LYP/6-31G(d,p) theory level. All values are in *kJ/mol*

Aqueous species	$G^0_{f,g}$ ab initio	G_{el}	G_{cav}	G_{disp}	G_{rep}	G_{abs} = > G_{conv}	$G°_{f,aq}$, ab initio
$Cr(H_2O)_6^{3+}$	1581.6[1]	−1956.2	69.1	−98.7	42.2	−1282.8	−1636.8[3]
		−1975.7	71.3	−99.7	45.4		−1651.9[4]
		−1974.1	71.4	−99.6	45.1		−1650.4[2]
$Cr(H_2O)_5(OH)^{2+}$	123.8[1]	−871.5	70.3	−83.6	25.9	−855.2	−1582.2[2]
$Cr(H_2O)_4(OH)_2^{+}$	−862.8[1]	−290.9	70.5	−76.2	21.2	−427.6	−1557.8[2]
$Cr(H_2O)_3(OH)_3$	−1422.2[1]	−113.8	70.3	−76.0	23.3	0.0	−1510.4[2]
$Cr(H_2O)_5Cl^{2+}$	248.7[1]	−960.7	75.2	−82.4	22.5	−855.2	−1543.9[2]
$Cr(H_2O)_4Cl_2^{+}$	−705.4[1]	−351.4	81.5	−79.7	19.5	−427.6	−1455.3[2]
$Cr(H_2O)_3Cl_3$	−1168.5[1]	−132.9	86.6	−75.8	15.4	0.0	−1267.8[2]
CrO_2^{-}	−414.2[1]	−274.3	31.7	−20.8	4.9	427.6	−237.8[2]
$Cr_2(H_2O)_8(OH)_2^{4+}$	1648.6[1]	−2758.8	111.8	−124.2	40.8	−1710.4	−2784.2[2]
$Cr_2(H_2O)_{10}(OH)^{5+}$	3355.6[1]	−4199.3	118.4	−132.2	40.1	−2138.0	−2947.3[2]
$Cr_3(H_2O)_9(OH)_4^{5+}$	2405.0[1]	−3949.3	141.8	−147.2	42.8	−2138.0	−3636.9[2]
CrO_4^{2-}	−571.9[1]	−1019.3	45.2	−34.8	7.8	855.2	−709.9[2]
$HCrO_4^{-}$	−962.5[1]	−263.5	47.3	−38.2	9.9	427.6	−771.5[3]
		−277.3	49.2	−38.3	10.2		−783.2[2]
H_2CrO_4	−760.7[1]	−41.4	48.9	−41.0	10.8	0.0	−775.4[2]
$Cr_2O_7^{2-}$	−1444.2[1]	−791.8	77.9	−50.0	9.3	855.2	−1335.6[2]
$HCr_2O_7^{-}$	−1560.5[1]	−236.0	79.9	−53.7	11.6	427.6	−1323.2[2]
CrO_3Cl^{-}	−913.4[1]	−224.8	53.6	−39.2	6.6	427.6	−681.6[2]
$[Fe(H_2O)_6]^{2+}$	224.5[1]	−812.8	75.8	−86.6	26.9	−840.9	−1405.2[5]
		−839.8*	82.6	−64.3	8.7	−840.9	−1421.2[2]
		−824.4	84.7	−85.7	25.5	−840.9	−1408.4[7]
		−847.2*	81.8	−85.4	26.2	−840.9	−1433.0[6]
		−895.9*	86	−62.8	8.1	−840.9	−1473.0[3]
$[Fe(H_2O)_5(OH)]^{+}$	−650.8[1]	−328.0**	86.9	−81.3	27.7	−420.4	−1357.9[7]
$[Fe(H_2O)_4(OH)]^{+}$	−416[1]	−346.1	74.2	−72.3	21	−420.4	−1151.6[7]
$[Fe(H_2O)_4(OH)_2]°$	−1182.7[1]	−117.6**	88.3	−76.3	19.2	0.0	−1261.1[7]
$[Fe(H_2O)_2(OH)_2]°$	−748.7[1]	−161.9	63.3	−58.2	17.7	0.0	−879.7[7]
$[Fe(H_2O)_3(OH)_3]^{-}$	−1307.4[1]	−292.5	92.1	−72.3	19.2	420.4	−1132.5[7]
		−277.7	89	−69.5	17.3	420.4	−1119.9[6]
		−299.9	91.1	−50	4.8	420.4	−1133.0[3]
$[Fe(OH)_3]^{-}$	−634.7[1]	−306.8	56	−31.6	4.8	420.4	−483.9[3]
$[Fe(H_2O)_6]^{3+}$	1794.3[1]	−1929.4	72.6	−99	42.6	−1261.3	−1372.3[5]
		−1883.4	79.2	−64	8.5	−1261.3	−1318.8[2]
		−1944.4	75	−98.6	41.6	−1261.3	−1385.5[7]
		−1959.4*	74.3	−99.5	44.6	−1261.3	−1399.1[6]
		−1890.6	80.5	−64	8.6	−1261.3	−1324.6[3]

(continued)

Table 18.10 (continued)

Aqueous species	$G^0_{f,g}$ ab initio	G_{el}	G_{cav}	G_{disp}	G_{rep}	G_{abs} => G_{conv}	$G^\circ_{f,aq,}$ ab initio
$[Fe(H_2O)_5(OH)]^{2+}$	351.6[1]	−906.9*	76.4	−82.3	24.3	−840.9	−1369.9[7]
$[Fe(H_2O)_4(OH)]^{2+}$	635.3[1]	−939.7	68.7	−72.4	21.5	−840.9	−1119.5[7]
$[Fe(H_2O)_4(OH)_2]^+$	−592.2[1]	−107.6**	83.8	−76.1	19.4	−420.4	−1085.1[6]
$[Fe(H_2O)_3(OH)_2]^+$	−391.8[1]	−326.8	68.7	−69.1	21.3	−420.4	−1110.1[6]
$[Fe(H_2O)_3(OH)_3]^\circ$	−1177.6[1]	−115.1*	80.6	−72.2	18.5	0.0	−1257.9[7]
		−103.1	78.7	−71.8	18.9	0.0	−1246.9[6]
		−125.9	84.2	−53.7	6.2	0.0	−1258.9[3]
$[Fe(OH)_3]^\circ$	−492.4[1]	−200.3	53.5	−31.1	4.6	0.0	−657.7[3]
$[Fe(H_2O)_2(OH)_4]^-$	−1348.8[1]	−265.6	82	−12.6	18.7	420.4	−1097.9[6]
		−291.6	87.1	−72.4	21.7	420.4	−1175.6[3]
$[Fe(OH)_4]^-$	−877.9[1]	−312.5	65.4	−38	5.6	420.4	−729.1[3]
$[(CH_3COO)Fe]^+$	528.0[1]	−546.7	47.9	−39.7	7.0	−420.4	−415.9[7]
		−688.0	56.9	−27.1	1.5		−541.2[3]
		−495.7	47.7	−39.8	7.0		−365.3[6]
$[(CH_3COO)_2Fe]^\circ$	−724.4[1]	−110.2	81.0	−75.6	13.6	0	−807.3[7]
		−92.8	99.0	−49.7	2.6		−747.0[3]
$[(CH_3COO)Fe]^{2+}$	2013.6[1]	−1526.9	44.2	−46.9	14.9	−840.9	−334.0[7]
		−1766.9	56.5	−27.2	1.5		−541.2[3]
		−1331.6	44.4	−47.2	15.4		−148.3[6]
$[(CH_3COO)_2Fe]^+$	30.4[1]	−390.2	79.7	−75.3	13.5	−420.4	−754.4[7]
		−436.6	98.2	−49.5	2.6		−759.2[3]
		−299.9	79.9	−75.4	13.9		−663.6[6]
$[(CH_3COO)_3Fe]^\circ$	−1117.2[1]	−59.9	113.1	−99.2	17.2	0	−1138.1[7]
		−39.1	132.8	−62.0	2.6		−1065.0[3]
		−48.2	113.1	−99.2	17.5		−1116.1[6]

1. B3LYP/6-31G(d,p). 2. IEFPCM UHF/STO-3G with UA0 radii. 3. IEFPCM B3LYP/6-31G(d,p) with UA0 radii. 4. IEFPCM B3LYP/3-21G with UA0 radii. 5. IEFPCM UHF/STO-3G with UAHF radii. 6. IEFPCM B3LYP/6-31G(d,p) with UAHF radii. 7. IEFPCM UHF/6-31G(d,p) with UAHF radii

when the computation is carried out in the terms of the IEFPCM theory (i.e. a coupling work given by the cavitation + electrostatic + repulsive + dispersive terms in Eq. (18.3.3) and identical rotational and vibrational partition functions in the gas phase and in solution (fifth term on the right in Eq. 18.3.3). Concerning the last term of molecular motion, it is obvious that we may pose:

$$\mathrm{RT}\ln\left(\frac{n_{M,g}\times\Lambda^3_{M,g}}{n_{M,s}\times\Lambda^3_{M,s}}\right)\equiv \mathrm{RT}\ln\left(\frac{\mathrm{RT}\times n_{M,g}\times\Lambda^3_{M,g}}{\mathrm{RT}\times n_{M,s}\times\Lambda^3_{M,s}}\right)$$
$$\cong \mathrm{RT}\ln\left(\frac{\mathrm{RT}\times n_{M,g}}{\mathrm{RT}\times n_{M,s}}\right) = -\mathrm{RT}(\mathrm{RT}\times n_{M,S}) \tag{18.3.12}$$

Table 18.11 Ab initio activity coefficient of Na^+ in diluted aqueous solutions of various ionic strengths (logarithmic values) obtained with the IEFPCM procedure, compared with the Poisson–Boltzmann and the Debye–Hückel approximations

Log M	Log I	Debye-Hückel	Poisson-Boltzmann[a]	[a]	[b]
−7	−7.30103	−0.000262	−0.00037	−0.00038	
−6.3	−6.60206	−0.000586	−0.00053	−0.00054	
−6	−6.30103	−0.000828	−0.00117	−0.00119	
−5.3	−5.60206	−0.00185	−0.00262	−0.00265	
−5	−5.30103	−0.002615	−0.00370	−0.00375	
−4.3	−4.60206	−0.005825	−0.00824	−0.00837	
−4	−4.30103	−0.008216	−0.01162	−0.01182	
−3.3	−3.60206	−0.018165	−0.02565	−0.02622	
−3	−3.30103	−0.025474			−0.03227
−2.3	−2.60206	−0.055022			−0.06454
−2	−2.30103	−0.075876			−0.08471
−1.3	−1.60206	−0.153538			−0.15329
−1	−1.30103	−0.202700			−0.19362
−0.3	−0.60206	−0.353944			−0.28640
0	−0.30103	−0.429963			−0.32674

[a]Numerical integration of the full (nonlinearized) equation [31]
[a] Cancés et al. (1997); Cossi et al. (1998); [b] Vetuschi Zuccolini et al. (2011)

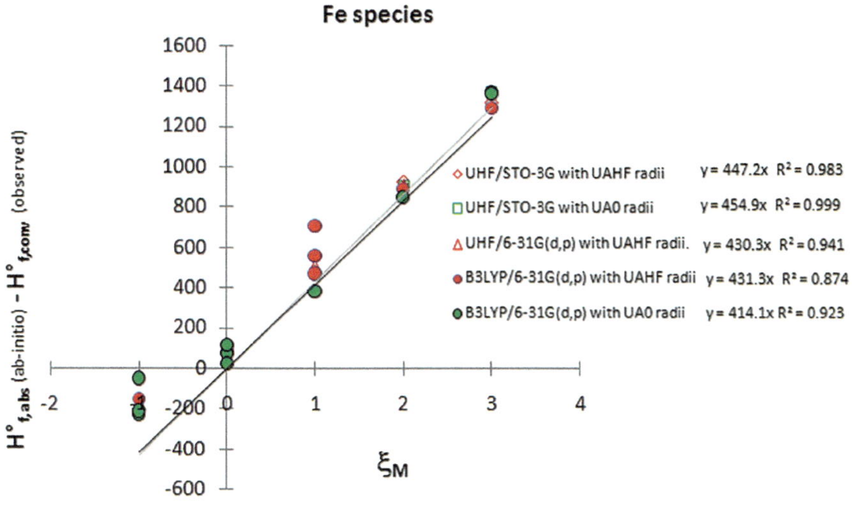

Fig. 18.11 Apparent enthalpy scaling factors for Fe hydroxides and Fe-acetates. All gas phase calculations were carried out at B3LYP/6-31G(d,p) theory level. Adopting simple STO-3G expansions in IEFPCM calculations rises significantly the nominal scaling factor of the population. The R^2 correlation is not much significant due to the limited sample population

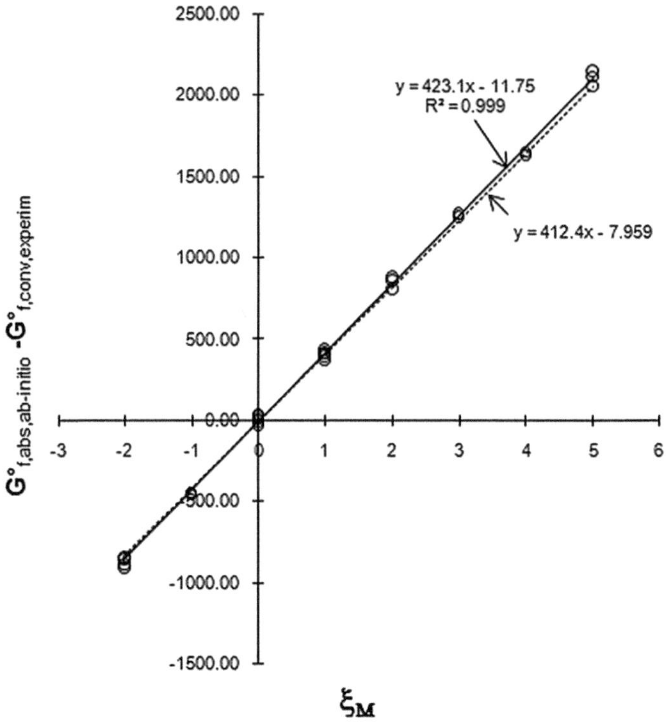

Fig. 18.12 Apparent Gibbs free energy scaling factors for Cr hydroxides. All gas phase calculations were carried out at B3LYP/6-31G(d,p) theory level. The nominal regression based on Eq. 18.3.11 is superimposed for comparative purposes

Hence, we see that the liberation free energy is correctly embodied in the IEFPCM theory. I stress, in passing, that this molecular motion contribution is absent in the SPT, where the liberation energy of the solvent is present in the computation of ΔG_s (last term on the right in Eq. 18.3.12), but not that of the solute.

In Table (18.10) are resumed the results of ab initio calculations carried out at the Geochemical Laboratory of the University of Genova (LdG) on inorganic and organic complexes of the transition elements Fe and Cr under various oxidation states (Ottonello and Vetuschi Zuccolini 2005, 2008, 2009). Gas phase calculations were carried out at the B3LYP/6-31G(d,p) theory level. Corrective terms to the electronic energy of the gaseous molecule were applied to Cr species, in the terms discussed at Sect. 18.3.2, while no corrections were provided for Fe species.

When comparing the ab initio values with the tabulated ones, the population regression for Cr-species is not far from the theoretical one and non-dispersed (R^2 = 0.998). Moreover, the theoretical calculation identifies the chromito ion $[CrO_2]^-$ as a dubious species (Fig. 18.13). Indeed, this compound has been simply hypothesized but its existence has never been confirmed by experiments.

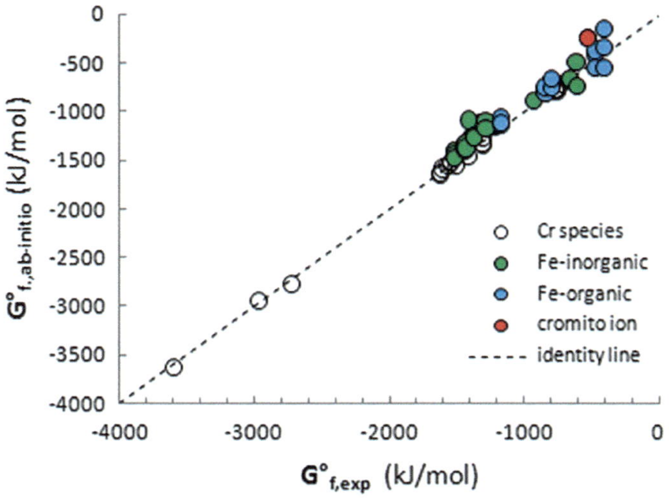

Fig. 18.13 Ab initio versus experimental values of the Gibbs free energy of formation from the elements of aqueous complexes of the transition elements Cr and Fe (Ottonello and Vetuschi Zuccolini 2005, 2008, 2009). The existence of the aqueous chromito ion $[CrO_2]^-$ is dubious. From Ottonello (2018)

In the absence of corrective terms, the ab initio computation reproduces the experimental data of iron species with more difficulty (Fig. 18.13). This is especially true for the organic species, where the adopted theory level is determinant. Despite this unsatisfactory result, the computed de-protonation equilibria in solution of iron species are not far from what experimentally observed in terms of energy. This is simply due to error cancellation effects encountered when one writes homogeneous mass-balanced reactions.

18.3.4 Activity Coefficients of Aqueous Solutes

Because non-electrostatic interactions in solution (i.e. cavitation, repulsion, and dispersion terms) are not appreciably affected by the solvent salinity and variations in the rotational partition function and momentum may be neglected, the activity coefficient of the ith ion in solution at different ionic strengths (I) may be obtained through application of the Integral Equation Formalism of the Polarized Continuum Model (IEFPCM) by comparing the solute–solvent electrostatic interaction at a ionic strength I ($G_{el,I}$) with the value obtained at the standard state of infinite dilution ($G_{el,0}$):

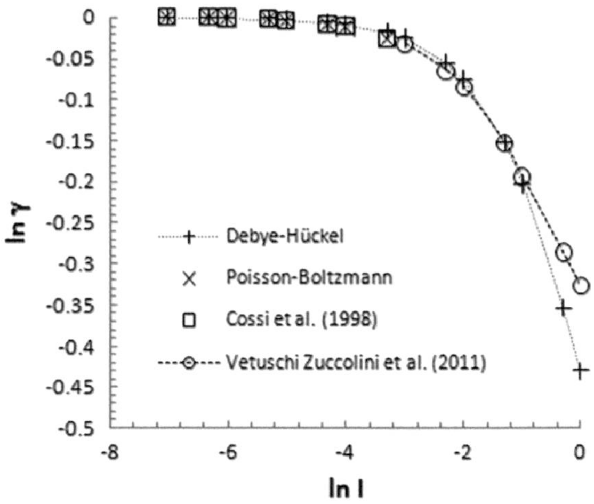

Fig. 18.14 Natural Logarithm of the activity coefficient of Na^+ in aqueous solution calculated with the IEFPCM procedure at the HF/6-31G(d) (Cossi et al., 1998) and B3LYP/6-31G(d) level (Vetuschi Zuccolini et al., 2011) compared to average values obtained by IPBECP (numerical integration of the Poisson–Boltzmann equation) and to the Debye–Hückel predictions. Modified from Vetuschi Zuccolini et al. (2011)

$$\ln \gamma_i = \frac{G_{el,I} - G_{el,0}}{RT} \tag{18.3.13}$$

In Table (18.11), values of the activity coefficient of Na^+ obtained with the IEFPCM procedure at various ionic strengths are compared with the results of the Poisson–Boltzmann and of the Debye–Hückel equations.

In Fig. 18.14, we see that the IEFPCM calculation is so precise to allow establishing the concentration limit at which the Debye–Hückel prediction fails.

Extensive calculations carried out in the framework of the IEFPCM model will allow in near future precise estimates of the activity coefficients in solutions at high salinity conditions, rendering obsolete the empirical procedures popularly adopted.

Chapter 19
Silicate Melts

We have already seen that, like other fused salts, silicate melts too are ionic liquids (Sect. 16.3.4). In an ionic melt, coulombic forces between charges of opposite sign lead to relative short-distance ordering of ions, with anions surrounded by cations and vice versa. The probability of finding a cation replacing an anion in such ordering is effectively zero and, from a statistical point of view, the melt may be considered as a *quasi-lattice*, with two distinct reticular sites which may be defined as "*anion matrix*" and "*cation matrix*". Applications of this "*quasi-lattice*" formulation in terms of Temkin's model (Temkin 1945; Richardson 1956) is, however, inadequate for the compositional complexity of natural melts, mainly because, in a compositionally complex melt, the types of anions and consequently the entity of the anion matrix vary in a complicated way with the composition. Nevertheless, Polymer Chemistry shows that, the larger the various polymers become, the more their reactivity becomes independent of the length of the polymer chains. This fact, known as the "*principle of equal reactivity of co-condensing functional groups*" has many practical consequences that could be beneficial in limiting the extension of the real configurons in ab initio investigations. We will see in fact in this chapter that the structure of compositionally simple melts like pure SiO_2 or pure Al_2O_3 may be explored with simple quantum chemical simulations. The exploration of chemically complex melts is however still at an embryonic stage, though extremely important conclusions may be attained through AIMD techniques. I begin this chapter with some hint on the fundamental building block $[SiO_4]^{4-}$ showing that it is indeed unstable at the gaseous state, and that only the dielectric continuum prevents its decomposition. Stemming from this observation, I will stress then the beneficial role of PCM in the investigation of the reactive properties of the liquid.

G. A. Ottonello, *Quantum Geochemistry*, Springer Geochemistry,
https://doi.org/10.1007/978-3-031-21837-8_19

19.1 The $[SiO_4]^{4-}$ Monomer and the Role of sp^3 Hybridization

In Pauling's words (Pauling 1949) "*if quantum theory had been developed by the chemist rather than the spectroscopist it is probable that the tetrahedral orbitals [...] would play the fundamental role in the theory, in place of the s and p orbitals*". As we may see in the historical representation of Pauling (1949) (Fig. 19.1), apart from the principal quantum number *n*, the analogy between bond strength/bond length relation in the s-p combination for carbon and silicon is indeed almost complete and the maximum bond strength (i.e. the magnitude of the bond orbital in its angular dependence) is observed for both elements not far from the 1:3 = s:p proportion. The tetrahedral coordination of the $[SiO_4]^{4-}$ group is, in this sense, regarded as a direct consequence of the capability of the silicon atom to form four hybrid orbitals sp^3, (with the p shell of silicon filled by the four oxygen donors) each directed toward one of the apexes of a regular tetrahedron.

Though the sp^3 terminology implies that the weight of the respective s and p components of a hybrid sp^3 orbital are in a 1:3 proportion and this proportion find its counterpart in quantum chemistry[1] this is not a strict nor unique prerequisite to form a bond.[2] There are in fact, besides the appropriate directionality of the AOs, other two conditions to satisfy:

- overlapping of discrete AOs along the opportune direction;
- delocalization of electrons between adjacent basins.

Collins et al. (1972), with a limited basis set, observed a considerable participation of the 3d orbital of silicon with the valence orbitals of oxygen to form σ-bonding e orbitals and π-bonding t_2 orbitals in the valence region. Tossell et al. (1973) with MS-SCF $X\alpha$ indicated as highest occupied valence the orbital $1t_1$, which is essentially the nonbonding AO 2p of oxygen, followed in order of decreasing energy by $5t_2$ (triply degenerate) MOs with π symmetry ($\sim 5\ eV$) formed by combination of the AOs 3s and 3p of silicon with the AO 2p of oxygen, then by 1e (doubly degenerate) and $4t_2$ MO ($\sim 7\ eV$) also formed by combination of the AO 3p of silicon with the AO 2p of oxygen but with σ symmetry.

In Fig. 19.2 are shown the occupied valence orbitals of $[SiO_4]^{4-}$ according to Tossell et al. (1973).

---

[1] The sp^3 wave function has form $N(s + \sqrt{3} \times p_\sigma)$ where N is a normalization constant, $\lambda = \sqrt{3}$ is the ratio of the orbital coefficients and p_σ is one of the four sigma bonds directed toward the apexes of the tetrahedron. Because the electron density is proportional to the square of the wave function, the squared ratio of the coefficients is 3 and hence the s to p proportion is maintained.

[2] Because symmetry operators commute in the Hamiltonian, once the structure is selected by imposing the symmetry of nuclear centers, whatever licit operation combining lines and rows does not alter the determinant and returns (at the variational limit) MOs of appropriate (though often unconsciously imposed) symmetry, but not necessarily an sp^3 combination among 3s and 3p AOs of silicon and not necessarily in the 1:3 proportion.

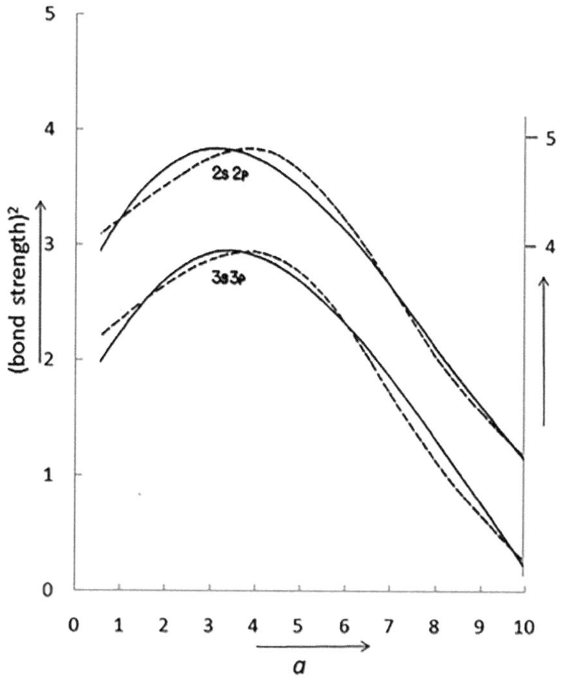

Fig. 19.1 Bond strength/bond length for hybrid s-p orbitals with varying s to p proportions ($a = 0$ pure p orbitals; $a = 10$ pure s orbitals). Dashed curves = square of bond strength; solid curves = bond length. The upper pair of curves are for L orbitals (2s and 2p of carbon). The lower pair, with shifted vertical scale, is for M orbitals (3s and 3p of silicon). From Pauling (1949); redrawn, with modifications

The solid portion of the central blocks is the electron density in the Si sphere, the empty section that in the O sphere and the hatched one that within the inter-atomic region. The adopted Si–O distance (1.607 Å) is that of α-Quartz. The moiety is stabilized with a Watson sphere[3] of charge +4.

Background Information: *X*α-SCF procedure
The Johnson-Slater *X*α-SCF procedure is a practical method to calculate with good precision (and computational economy) electronic energy levels and eigenfunctions in solids (and polyatomic clusters as well). It is adopted the "*muffin-tin*" approximation, i.e.: (1) a spherically symmetrical potential within spheres surrounding the various nuclei; (2) a constant potential in the region between the spheres; (3) a spherically symmetrical potential outside a sphere surrounding the molecule. The one-electron Schrödinger equation is solved with the "*multiple-scattering procedure*". One begins the SCF calculation with an assumed potential, solves the one-electron equation, and finds various spin orbitals u_i with eigenvalues $\in_i$, then, through "*suitable criteria*" (Slater and Johnson 1972) decides which spin orbitals are occupied; the charge

[3] The Watson sphere is a spherically symmetric potential imposed around the ion of interest to simulate the electrostatic potential of the surrounding ions. It was first suggested by Watson (1958) in order to stabilize free oxygen ions O^{2-}, in order to calculate their wave function.

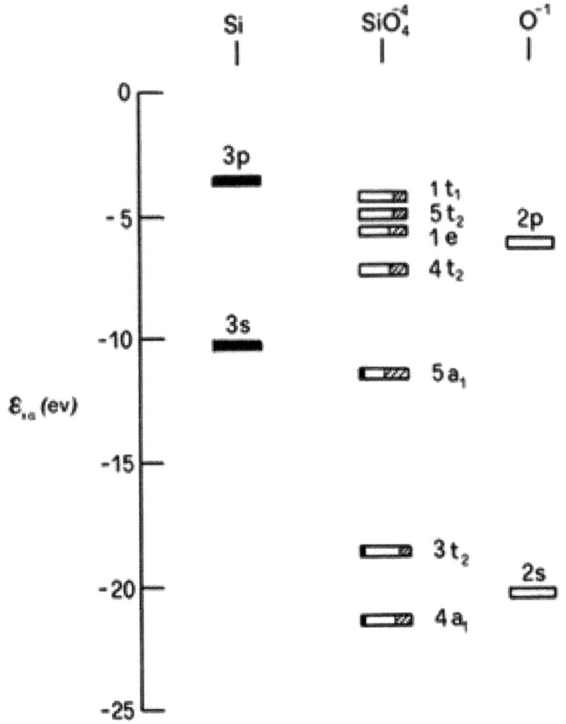

Fig. 19.2 Occupied valence orbitals of $[SiO_4]^{4-}$ according to the SCF-Xa scattered method. From Tossell et al. (1973). With permission from Chemical Physics Letters

density arising from the selected orbitals (and the nuclei) determines the potential. After carrying out the averaging required by the muffin-tin method, the resulting potential is adopted as the starting point of the next iteration. In the $X\alpha$-SCF procedure, the "suitable criterium" consists in assuming that occupied orbitals are those of lowest eigenvalue, up to the Fermi level. Total energy is computed from the resulting charge densities with a statistical approximation for the exchange correlation associated to an undetermined factor α. The spin orbitals and occupation numbers are varied to minimize this total energy. The value of α for an isolated atom is that one determining a statistical total energy identical to the precise Hartree–Fock value. The α's of the various atoms within the atomic spheres (and a suitable average in the region between them) determine the bulk energy of the substance. The spin orbitals arising in this way are quite accurate. In practice, the calculations proceed as follows (Slater and Johnson 1972):

- The charge density at point (1) is written in the form

$$\rho(1) = \sum_i n_i u_i^*(1) u_i(1) \tag{19.1.1}$$

where the occupation number n_i is zero for an empty spin orbital and one for an occupied spin orbital;
- the charge density is separated in spin up $\rho_\uparrow(1)$ and spin down $\rho_\downarrow(1)$.
- Assuming the internuclear repulsion to be constant (i.e. nuclei in fixed position), the expectation value of the total energy E is expressed in the form

$$\begin{aligned}\langle E\rangle = {} & \sum_i n_i \int u_i^*(1) f_1 u_i(1) dv_1 \\ & + \frac{1}{2}\int \rho_\uparrow(1)\left[\int \rho(2) g_{12} dv_2 + U_\uparrow(1)\right] dv_1 \\ & + \frac{1}{2}\int \rho_\downarrow(1)\left[\int \rho(2) g_{12} dv_2 + U_\downarrow(1)\right] dv_1\end{aligned} \tag{19.1.2}$$

with the first term representing the average of the one-electron operators and the other terms representing the interactions of the electrons with the electron clouds. $\int \rho(2) g_{12} dv_2$ is the coulomb interaction including the self-interaction of the electron with itself and $U_\uparrow(1)$ (the exchange–correlation potential) is a negative term removing the self-interaction.

- $U_\uparrow(1)$ is now replaced with a statistical approximation in the form

$$U_{X\alpha}(1) = -9\alpha\left[(3/4\pi)\rho_\uparrow\right]^{1/3} \tag{19.1.3}$$

This functional dependence of U on the density (constant times the 1/3 power of the density) may be associated to the concept of the Fermi hole [Slater and Johnson 1972; but see also Sects. (14.3) and (16.3.2)]

- The expectation value of the energy is now a function of the occupation numbers n_i and a functional of the spin orbitals u_i. Varying the u_i in order to minimize the total energy at fixed occupation numbers, the resulting one electron Schrödinger equations are of the type

$$\left[-\nabla_i^2 + V_C(1) + V_{X\alpha}(1)\right] u_i(1) = \epsilon_i\, u_i(1) \tag{19.1.4}$$

with $-\nabla_i^2$ representing the kinetic energy in the appropriate units, $V_C(1)$ as the electrostatic potential energy (electronic + nuclear) at position (1), and $V_{X\alpha}(1)$, different for spin up and spin down conditions, is represented by

$$V_{X\alpha\uparrow}(1) = \frac{2}{3}U_{X\alpha\uparrow}(1) = -6\alpha\left[(3/4\pi)\rho_{\uparrow}\right]^{1/3} \quad (19.1.5)$$

- The eigenvalues of the one electron Schrödinger equations are then (Slater and Johnson 1972):

$$\in_{iX\alpha} = \frac{\partial \langle E_{X\alpha} \rangle}{\partial n_i} \quad (19.1.6)$$

Various seminal contributions on the utilization of Xα procedures in Theoretical Geochemistry are due to Tossell to whom I remand the interested reader (Tossell 1975, 1976, 1977, 1978).

The importance of these two pioneering investigations stems on the fact that the obtained orbital energies of the various MOs were shown to be perfectly consistent with experimental observables (X-ray emission spectra) not associated (in an explicit way) with the structure of the group. Collins et al. (1972) also noted that because (according to the MOs assignation) there are 24 electrons in unbound molecular orbitals, the $[SiO_4]^{4-}$ ion is formally non-existent in the gaseous state. A DFT calculation carried out at the B3LYP/6-31+G(d,p) theory level on the $[SiO_4]^{4-}$ ion (Richet and Ottonello 2014) confirms what observed long ago concerning the inner orbital calculations [orbital coefficients and ensuing character assignation, and energy; Table 19.1]. If we immerge however the $[SiO_4]^{4-}$ ion in a reaction field of appropriate permittivity ($\varepsilon = 3.8$, taken as representative of the dielectric property of molten silica) we may see that important change are observed, at the same theory level, in the orbital population (and energy) of both valence and inner region (Table 19.1) and the cluster become "stable" to all extents.

Concerning the fractional ionic character of the bond, electron density maps for the α-quartz polymorph indicate that the charge transfer between silicon and oxygen is not complete and that a residual charge of +1.0 (±0.1) electron units (*e.u.*) remains localized on silicon, whereas a charge of −0.5 (±0.1) *e.u.* is localized on each oxygen. The interpretation of this fact in terms of the bond ionicity is again not as univocal as it may appear at first glance. According to Pauling (1980), electron density maps confirm the fractional ionic character (f_i = 51%) deduced from electronegativity. However, starting from the same experimental evidence, and adopting as fractional ionic character: f_i = formal charge/nominal charge, Stewart et al. (1980) obtain $f_i = 1.0/4.0 = 0.25$. The assumption of Stewart et al. (1980) is more in line with the electronegativity theory for molecular orbitals. If we perform a Mulliken population analysis on our gaseous ionic cluster we obtain a charge of 0.543*e* localized on the central metal and charges of −1.136*e* on each of the four bonded oxygens, while the charge localized on Si rises to 1.917 and those of the four bonded oxygens rise to −1.479 when the ion is placed in the dielectric

Table 19.1 Energy and orbital assignments in $[SiO_4]^{4-}$. (a) Collins et al. (1972); (b) Richet and Ottonello (2014)

Reference	(a)		(b)		(b)		
Theory level	HF/3-21G(d)		DFT/6-1+G(d,p)		DFT/6-31+G(d,p)-PCM		
Total energy	−15,632.40		−16,033.53		−16,068.68		
Valence orbitals	$1t_1$	20.54	$1t_1$	18.01		$3t_1$	0.406
	$5t_2$	18.01	1e	16.81			
	1e	14.91	$4t_2$	16.68			
	$4t_2$	12.73	$3t_2$	15.04			
	$5a_1$	10.28	$4a_1$	12.76			
Mainly O(2s)	$3t_2$	−13.19			Mainly O(2p)	$3t_2$	−0.621
	$4a_1$	−14.15				2e	−0.877
Si(2p)	$2t_2$	−81.59	$2t_2$	−74.41	Si(3p)	$3t_2$	−2.404
Si(2s)	$3a_1$	−134.37	$3a_1$	−119.05	Si(3s)	$1a_1$	−4.526
O(1s)	$2a_1$	−547.12	$2a_1$	−494.67	O(2s,3s)	$3t_2$	−16.306
O(1s)	$1t_2$	−547.12	$1t_2$	−494.67	Si(4s)	$1a_1$	−17.575
Si(1s)	$1a_1$	−1818.63	$1a_1$	−1775.08	Si(2p)	$3t_2$	−91.935
					Si(2s)	$1a_1$	−136.554
					O(1s)	$1a_1$	−512.376
					O(1s)	$3t_2$	−512.376
					Si(1s)	$1a_1$	−1792.53

medium. Applying the definition of Stewart et al. (1980), we thus see that for the isolated cluster the fractional ionic character of the bond is $f_i = 0.136$ (in simple terms, a ionic cluster $[SiO_4]^{4-}$ is essentially covalent) while $f_i = 0.479$ in the dielectric medium. Notwithstanding the decrease of covalence, the cluster is, however, more stable, due essentially to the electrostatic solute/solvent interaction.

19.2 Polymerization Extent and Donor Pressure of the Medium

Monomers exist in silicate liquids only at low acidity conditions and are progressively replaced by polymeric structures with the progressive increase of acidic oxides (such as SiO_2, Al_2O_3, TiO_2, CO_2) in the system. Understanding of the structure of aluminosilicate melts relies on an accurate assessment of binding energies between the network-modifier cations and the anionic framework, which is nominally made up of TO_4^{4-} (T = Si, Al) tetrahedra interconnected in a great many ways. Moreover, regardless of any possible structural complexity, the existence of Al^{3+} in tetrahedral coordination with oxygen exerts special constraints on short-range order because charge-compensating cations are required to ensure local electroneutrality. The interpretation of the complex interactions and the ensuing short and medium range order in silicate, aluminate and borate melts, passes thus through a precise assessment of binding energies between Network Modifying (NM) ad-atoms, free oxide ions (O^{2-}) and ionic clusters built up by the Network Formers (NF).

In the framework of the Lux-Flood theory, the oxides of low-field-strength ions decompose through a basic dissociation scheme liberating free oxide ions O^{2-} which increases the bulk donor pressure of the system, while high-field-strength ions attract free oxygen, lowering the bulk donor pressure. The free oxide ions de-polymerize the framework structure built up by the NFs; on the other hand, free cations, and especially either alkalis or alkaline earths owing to their high abundance, act as Lewis acidic centers and, as such, they affect the basicity of the framework oxygen centers and partially recombine with them by electrostatic interaction. While the energy amount involved in a simple dissociation scheme may be readily evaluated through a quantum–mechanical computation of the electronic energy and thermal corrections of neutral isolated molecules of appropriate stoichiometry, the second and third (subordinated) energy terms require more sophisticated ab initio computations. In fact, the electrical field imposed on the surroundings of an atom affects its orbital energy via changes of the external potential acting on the core electrons. External localized charges affect, besides orbital energies, also other parameters that characterize the electronic state of the perturbed atom, such as its atomic net charge, orbital population and hybridization. In their turn, these changes act in such a direction to screen the external influence. To resume in simple terms what above stated, denoting $M_xO_{xZM/2}$ a generic basic oxide of the NM with charge Z_M and assuming it interacts with a prototypical NF oxide such as SiO_2 the energy steps that need to be attentively evaluated are the following:

$$M_xO_{xZ_M/2} \Leftrightarrow xM^{Z_M+} + \frac{xZ_M}{2}O^{2-} \tag{19.2.1}$$

$$\frac{xZ_M}{2}O^{2-} + SiO_2 \Leftrightarrow SiO_{(4+xZ_M)/2}^{-xZ_M} \tag{19.2.2}$$

$$xM^{Z_M+} + SiO_{(4+xZ_M)/2}^{-xZ_M} \Leftrightarrow \left[M_x \cdot \left(SiO_{(4+xZ_M)/2}\right)\right] \tag{19.2.3}$$

In the above simplified reaction scheme, Eq. (19.2.1), which is the energetically predominant one, represents a basic dissociation; Eq. (19.2.2) is a de-polymerization process acting on a formal tectosilicate and Eq. (19.2.3) is an electrostatic recombination between a partially de-polymerized network and the nominally free NM that, in the case of Z_M = 2+, reconstitutes a formal meta-silicate complex.

19.2.1 Nephelauxetic Effect

The amount of energy associated to equilibria (19.2.1 and 19.2.2) is easily detected through spectroscopic observations. The Lewis basicity of a NM donor is usually expressed by its h-Jørgensen function (Jørgensen 1969), which is, in its turn, connected to the frequency shift of the $^3P_1 \rightarrow {}^1S_0$ transition when passing from a free-oxygen donor medium (ν_f) to the state of interest (ν) through scalar operations:

$$h = \frac{1 - {}^{\nu}/_{\nu_f}}{\mu} \tag{19.2.4}$$

with μ as scaling parameter for the probe ion. In a complex medium composed of basic donor centers and acidic acceptors, the net charge transfer is intimately linked to the difference in the Jørgensen's h functions of the various atoms, which may be eventually translated into an optical basicity scale in terms of oxide components. In fact, in a series of stimulating papers Duffy and coworkers (Duffy 1989, 1990, 1993; Duffy and Ingram 1970, 1971, 1973, 1974, 1976) promoted the utilization of the nephelauxetic effect (Eq. 19.2.4) as an optical scale measuring the extent of electron donation by oxide centers to p-block elements.

Metal ions such as Tl^+ (group III), Pb^{2+} (group IV), Bi^{3+} (group V) have an electron pair in the outermost 6s orbital. When trace concentrations of the metal are dissolved in melts and glasses, coordination with the ligand field anions results in formation of MOs, which increase the electron density of the inner shells. The consequent shielding of nuclear charges affects the energy involved in the outermost 6s $\rightarrow$ 6p transitions, which becomes smaller when the inner shell electron density is increased. Lowering of the 6s $\rightarrow$ 6p transition energy is experimentally observed as a dramatic red-ward shift of the 6s $\rightarrow$ 6p UV absorption band when the p-block free ion is immersed in a ligand field. Figure 19.3 is a schematic representation of what above stated according to Duffy and Ingram (1976).

The spectroscopic shift of the $^1S_0 \rightarrow {}^3P_1$ absorption band experienced by Pb when passing from a free ion (Pb^{2+}) condition to Pb^{II} in an O^{2-} ligand field is for instance 60,700 − 29,700 = 31,000 cm^{-1} (Duffy and Ingram 1971, 1976). Figure 19.4 shows a typical absorption spectrum for Pb^{2+} in a borate glass, compared with the corresponding spectrum for the chloro-complex of Pb(ll) in hydrochloric acid. The absorption band is more broad in the glass, but this feature is unimportant in view of the enormous frequency shift generated by the donor pressure.

For Bi^{III}, the analogous red-ward shift is 28.8 kK (1 kK = 1000 cm^{-1}), and it is 18.3 kK for T^{II}. This phenomenon is quantitatively understood in terms of ligand field theory by analogy with the behavior of 3d, 4d, and 5f transition ions (Jørgensen 1962, 1969; and references therein). In octahedral 3d chromophores, for example, the energy splitting between anti-bonding $\Psi^*_{z^2}, \Psi^*_{x^2-y^2}$ MOs and the d_{xy}, d_{xz}, and d_{yz} AOs of the central atom $\Delta_{cov-\sigma}$ is linearly related to the position of the

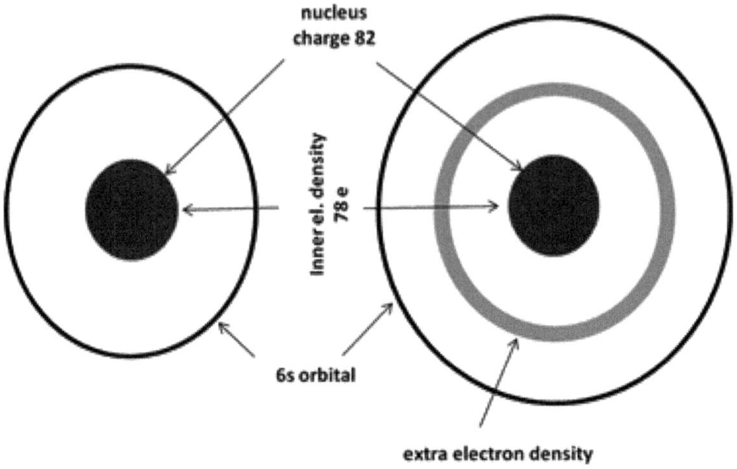

Fig. 19.3 Schematic diagram of a free Pb^{2+} ion before (left) and after (right) receiving negative charge from neighboring oxide ions. Modified from Duffy and Ingram (1976)

Fig. 19.4 Absorption spectrum of Pb^{2+} in sodium borate glass 20% Na_2O) [Pb^{2+}] = 0.00113 M (**a**), path length 1.50 *mm*) and in 11 M HC1 (**b**). Modified from Duffy and Ingram (1976)

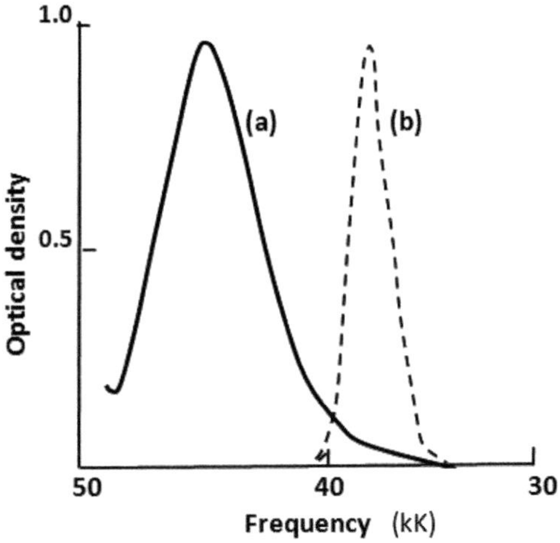

ligand in the spectrochemical series (represented by parameter *f*) and to the representative parameter of the central cation (*g*), according to the simple relationship (Jørgensen 1969):

$$\Delta_{cov-\sigma} = f_{(ligand)} \cdot g_{(cation)} \tag{19.2.5}$$

The precision achieved by this simple equation in describing $\Delta_{cov-\sigma}$ in $3d^3$, $3d^6$, and $3d^8$ chromophores is remarkable (Table 5.8 in Jørgensen 1969). However, the energy shift induced by changes in *f* is not large enough to allow a basicity scale to be proposed on the basis of Eq. (19.2.5) (Table 5.5 in Jørgensen 1969).

Nevertheless, as already discussed, the spectroscopic shift induced by the nephelauxetic effect on p-block metals by the ligand field is much more pronounced.

In a chemically complex melt or glass, the capability of transferring fractional electronic charges from the ligands to the central cation depends in a complex fashion on the melt or glass structure, which affects the polarization state of the ligand itself. Nevertheless, the mean polarization state of the various ligands (mainly the oxide ions O^- and O^{2-} in natural silicate melts) and their ability to transfer fractional electronic charges to the central cation are conveniently represented by the "*optical basicity*" of the medium, i.e. the ratio h/h*, where h is the Jørgensen's (1962) function of the ligand in the polarization state of interest and h* is the same function relative to the ligand in an unpolarized state (mainly free O^{2-} ions in an oxidic medium; Duffy and Ingram 1971):

$$\Lambda = \frac{h}{h^*} = \frac{1-\beta}{1-\beta^*} = \frac{\nu_{free} - \nu_{glass}}{\nu_{free} - \nu^*} \quad (19.2.6)$$

with $\nu_{free} = {}^1S_0 \rightarrow {}^3P_1$ absorption band of the free p-block cation; $\nu_{glass} = {}^1S_0 \rightarrow {}^3P_1$ absorption band measured in the glass; $\nu* = {}^1S_0 \rightarrow {}^3P_1$ absorption band in a free O^{2-} medium.

The reciprocal of optical basicity Λ of a cation, i.e. the "*basicity moderating parameter*" γ of Duffy and Ingram (1973) represents the tendency of an oxide forming metal M to reduce the localized donor properties of oxide ions, and is related to the optical basicity of the medium by:

$$\gamma_M = \frac{Z_M \times r_M}{|Z_O| \times \Lambda_{MO}} \quad (19.2.7)$$

where Z_M = formal oxidation number of cation in MO; Z_O = formal oxidation number of oxide ion in MO and r_M = stoichiometric ratio between number of cations and number of total oxide ions in the medium.

Although γ_M reduces to Λ_{MO}^{-1} in a simple single oxide medium, it has the property of describing the Jorgensen's h function in complex systems according to Duffy and Ingram (1976):

$$h = h^* \left[1 - \frac{Z_A \times r_A}{|Z_O|}\left(1 - \frac{1}{\gamma_A}\right) - \frac{Z_B \times r_B}{|Z_O|}\left(1 - \frac{1}{\gamma_B}\right) - \cdots \right] \quad (19.2.8)$$

with A, B,... oxide forming cations. As an example, in a $0.5CaO–0.5Al_2O_3$ medium:

$$
\begin{aligned}
\gamma_{0.5(\mathrm{CaO})0.5(\mathrm{Al_2O_3})} &= \left[1-\frac{Z_A\times r_A}{|Z_O|}\left(1-\frac{1}{\gamma_A}\right)-1-\frac{Z_B\times r_B}{|Z_O|}\left(1-\frac{1}{\gamma_{\mathit{textB}}}\right)\right]\\
&= \left[1-\frac{2\times 0.5/2}{2}\left(1-\frac{1}{\gamma_A}\right)-\frac{3\times 0.5/2}{2}\left(1-\frac{1}{\gamma_B}\right)\right]\\
&= \left[1-\frac{1}{4}\left(1-\frac{1}{\gamma_A}\right)-\frac{3}{4}\left(1-\frac{1}{\gamma_B}\right)\right]\\
&= \left[1-\frac{1}{4}-\frac{3}{4}+\frac{1}{4\gamma_A}+\frac{3}{4\gamma_B}\right]=\frac{1}{4}\gamma_A+\frac{3}{4}\gamma_B
\end{aligned}
\tag{19.2.9}
$$

As we will see later on, this has a strict analogy with Flood-Grjotheim electric equivalent fractions in a thermochemical cycle $\nu_{Ca}=\frac{Z_A\times n_A}{\sum Z}=\frac{1}{4};\ \nu_{Al}=\frac{Z_B\times n_B}{\sum Z}=\frac{3}{4}.$

Empirical relationships between observed (or inferred) optical basicity of simple oxides (Table 19.2) and atomistic properties of the intervening cations, such as the Pauling's and Sanderson's electronegativities (χ_P and χ_S, respectively, in the same table) or the free ion polarizability have been proposed long ago (Duffy and Ingram 1974; Young et al. 1992).

A strict connection between optical basicity and bond ionicity may be envisaged by equating the spectroscopic definition of fractional ionic character of a bond (Phillips 1970) with Pauling's relation:

$$
f_i=\frac{E_i^2}{\left(E_i^2+E_c^2\right)}=1-\exp\left[-\frac{1}{4}(\chi_O-\chi_M)^2\right] \tag{19.2.10}
$$

where χ_O and χ_M are the Pauling electronegativity of oxygen and metal, respectively, and

$$
E_i=\left(E_g^2-E_c^2\right)^{1/2} \tag{19.2.11}
$$

E_i in Eq. (19.2.7) is the "*ionic energy gap*" and E_g is the total energy gap between bonding and anti-bonding orbitals. E_c corresponds to E_g for the non-polar covalent bond in the same row of the periodic table, with a correction for inter-atomic spacing:

$$
E_g=\frac{\hbar\omega_p}{(\varepsilon_\infty-1)^{1/2}} \tag{19.2.12}
$$

with $\hbar\omega_p$ as plasma frequency for valence electrons and ε_∞ as optic dielectric constant.

Based on the Λ values of the various oxides listed in Table 19.2 and the corresponding f_i values calculated from Pauling's electronegativities, the following operational relationship may be proposed:

Table 19.2 Optical basicity Λ and basicity moderating parameter of the central cation γ according to various sources. Pauling's and Sanderson's electronegativities (Pauling 1932, 1960, 1980; Sanderson 1967) are also listed. Λ, γ, χ_S: a dimensional; χ_P: *eV*

Oxide	Λ						γ	χ_p	χ_s
	(1)	(2)	(3)	(4)	(5)	(6)	(6)		
H_2O			0.40			0.39	2.56	2.15	3.55
Li_2O						1.00	1.00	1.0	0.74
B_2O_3			0.42			0.42	2.38	2.0	2.84
Na_2O		1.15	1.15	1.15	1.15	1.15	0.87	0.9	0.70
MgO	0.78	0.78	0.78	0.78	0.78	0.78	1.28	1.2	1.99
Al_2O_3	0.60	0.60	0.60	0.61	0.59	0.59	1.69	1.5	2.25
SiO_2	0.48	0.46	0.48	0.48	0.48	0.48	2.09	1.8	2.62
P_2O_5	0.40	0.40	0.33	0.8	0.40	0.40	2.50	2.1	3.34
SO_3	0.33		0.25			0.33	3.03	2.5	4.11
K_2O		1.40	1.4	1.40	1.40	1.36	0.74	0.8	0.41
CaO	1.00	1.00	1.00	1.00	1.00	0.99	1.00	1.0	1.22
TiO_2		0.65		0.61	0.61	0.58	1.72	1.6	1.60
Cr_2O_3	0.70					0.58	1.72	1.6	1.88
MnO	0.94–1.03	0.98		0.90	0.59	0.59	1.69	1.5	2.07
FeO	0.86–1.08	1.03	1.00	1.03	0.51	0.48	2.09	1.8	2.10
Fe_2O_3	0.73–0.81	0.77		1.21	0.48	0.48	2.09	1.8	2.10
CoO						0.51	1.96	1.7	2.10
NiO						0.48	2.09	1.8	2.10
Cu_2O						0.43	2.30	1.9	2.60
ZnO		0.82–0.98				0.58	1.72	1.6	2.84
SrO	1.10					1.03	0.97	1.0	1.00
SnO						0.48	2.09	1.8	3.10
BaO	1.15	1.15		1.15	1.15	1.12	0.89	0.9	0.78
PbO						0.48	2.09	1.8	3.08

(1) Duffy (1989, 1990); (2) Young et al. (1992); (3) Duffy and Ingram (1974); (4) Sosinsky and Sommerville (1986); (5) Gaskell (1982); (6) from Pauling's electronegativity; see eqn. A-7 in Ottonello et al. (2001), note that $\Lambda = \gamma^{-1}$

$$\Lambda = (4.6242 - 4.6702 \times f_i)^{-1} \quad \left(R^2 = 0.95\right) \tag{19.2.13}$$

On the basis of the above equality, the following approximation holds:

$$\Lambda \approx \frac{1}{4.6\left(1 - \frac{E_i^2}{E_i^2 + E_c^2}\right)} \tag{19.2.14}$$

which suggests an optical basicity of 0.225 for a purely covalent nonpolar bond.

Direct estimates of the basicity moderating parameter of the central cation may be also obtained from Pauling's or Sanderson's electronegativities by application of the second-order polynomials

$$\gamma = 0.8969 - 0.6703\chi_P + 0.7404\chi_P^2 \quad (R^2 = 0.97) \tag{19.2.15.1}$$

$$\gamma = 0.7398 - 0.60193\chi_S + 0.1880\chi_S^2 \quad (R^2 = 0.96) \tag{19.2.15.2}$$

These equations may be adopted whenever literature values are controversial or lacking (see an exemplificative application in Ottonello et al. 2001).

19.2.2 Optical Basicity and Polymerization Constant

The chemical interaction between network formers and network modifiers constitutes by far the main form of energy gained (or lost) by the liquid in the mixing process.

The charge balance of a polymerization reaction involving SiO_4^{4-} monomers in a melt (m) may be formally described by a homogeneous reaction involving three forms of oxygen: singly bonded O^-, doubly bonded (or "*bridging oxygen*") O^0 and *free* oxygen O^{2-} Fincham and Richardson (1954):

$$2O_m^- \Leftrightarrow O_m^0 + O_m^{2-} \tag{19.2.16}$$

Equation (19.2.16) is representative of a reaction between SiO_4^{4-} monomers:

$$SiO_4^{4-} + SiO_4^{4-} \Leftrightarrow Si_2O_7^{6-} + O^{2-} \tag{19.2.17}$$

Polymer chemistry shows that, the larger the various polymers become, the more their reactivity becomes independent of the length of the polymer chains. This fact, known as the "*principle of equal reactivity of co-condensing functional groups*", has been verified in fused polyphosphate systems, which, for several properties, may be considered as analogous to silicate melts; cf. Fraser (1977) with polymeric chains longer than 3 PO_4^{4-} units (Meadowcroft and Richardson 1965; Cripps-Clark et al. 1974). Assuming this principle to be valid, the equilibrium polymerization constant K_P:

$$K_P = \frac{(O^0)(O^{2-})}{(O^-)^2} \tag{19.2.18}$$

(in which the terms in brackets represent the number of moles per unit mole of melt) is thus always representative of the polymerization process, independent of the effective length of the polymer chains.

The Gibbs free energy change involved in Eq. (19.2.16) is:

$$\Delta G_P^0 = -RT \ln K_P \tag{19.2.19}$$

Since two moles of O^- produce one mole of O^0 and one of O^{2-}, the Gibbs free energy of mixing per mole of silicate melt is given by:

$$\Delta G_{mixing} = \frac{(O^-)}{2} RT \ln K_P \tag{19.2.20}$$

$(O^-)/2$ in Eq. (19.2.10) are the number of contacts among functionals of the polymeric units of the Toop-Samis model (Toop and Samis 1962a, b).

Toop and Samis (1962a, b) showed that, in a binary melt $MO–SiO_2$, in which MO is the oxide of a basic cation completely dissociated in the melt, the total number of bonds per mole of melt is given by:

$$2(O^0) + (O^-) = 4N_{SiO_2} \tag{19.2.21}$$

where N_{SiO_2} are the moles of SiO_2 in the $MO–SiO_2$ melt. The number of bridging oxygens in the melt is thus:

$$(O^0) = \frac{4N_{SiO_2} - (O^-)}{2} \tag{19.2.22}$$

Mass balance gives the number of moles of free oxygen per mole of melt:

$$(O^{2-}) = (1 - N_{SiO_2}) - \frac{(O^-)}{2} \tag{19.2.23}$$

where $(1 - N_{SiO_2})$ are the moles of basic oxide in the melt. Equations (19.2.20 and 19.2.23) yield:

$$K_P = \frac{[4N_{SiO_2} - (O^-)][2 - 2N_{SiO_2} - (O^-)]}{4(O^-)^2} \tag{19.2.24}$$

(O^-) is thus given by the quadratic equation:

$$(O^-)^2(4K_P - 1) + (O^-)(2 + 2N_{SiO_2}) + 8N_{SiO_2}(N_{SiO_2} - 1) = 0 \tag{19.2.25}$$

which may be solved for discrete values of K_P and N_{SiO_2}.

As anticipated, in the light of the principle of equal reactivity of co-condensing units, K_P cannot be expected to be affected by the length and conformation of the structons and, as we may see in Fig. 19.5, it is simply related to the basicity moderating parameter of the network former γ_A (Table 19.3).

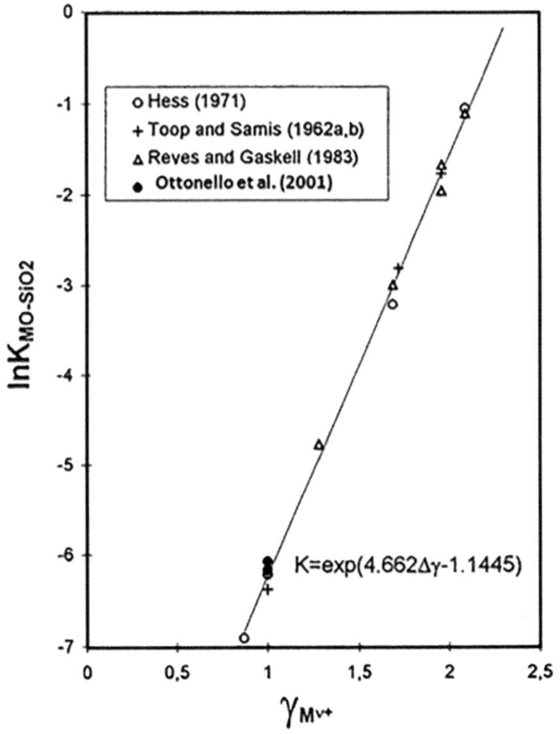

Fig. 19.5 Relationship between polymerization constant of the Toop–Samis model in MO–SiO_2 melts and basicity moderating parameter of the network former (γ_A). Modified from Ottonello et al. (2001)

Although in Fig. 19.5 a single value of K_P pertains to each MO, a more strict analysis, based on the liquidus topology in pseudo-binary Network Modifier (NM)–Network Former (NF) phase diagrams shows that ΔG_{mixing} can be split into its enthalpic and entropic components, and hence that the polymerization constant is more conveniently represented in the form (Table 19.4):

$$-\frac{\Delta G_P^0}{RT} = -\frac{\Delta H_P^0}{RT} + \frac{\Delta S_P^0}{R} = \ln K_P = \frac{A}{T} + B \qquad (19.2.26)$$

with the slopes of the function clearly individualized in semi-logarithmic plots of the constant with respect to the optical basicity difference (hence electronegativity gap) between network former and network modifier ($\Delta\Lambda$). In Fig. 19.6, we see in fact that the electronegativity gap between network-former and network modifier dictates, at all T, the extension of polymerization.

This is not only observed in silicate systems (Fig. 19.6; Table 19.4) but in titanate, aluminate, and borate systems as well (Fig. 19.7).

Pushing further the analysis, and assuming that slopes and intercepts of Eq. (19.2.26) are simple linear functions of reciprocal T, allows to extend further the generalization and to propose a quite complete set of values for all NM-NF binaries of interest in petrology, glass industry and material science (Table 19.5).

Table 19.3 Toop–Samis polymerization constants in MO–SiO_2 melts at various T conditions, as inferred from Gibbs free energy of mixing curves and thermodynamic activity data. Values of the basicity moderating parameter of the network modifier are also listed. Bold values adopted in Fig. 19.5

MO	γ_A	K_P	T (°C)	Refs.	MO	γ_A	K_P	T (°C)	Refs.
NaO	0.87	0.001		(1)	MnO	1.69[a]	0.049		(1)
		0.001		**(3)**			**0.04**		(5,3)
							0.05	1600	(6)
K_2O	0.71	0.0005		(1)					
		0.001		(2,3)	CoO	1.96[a]	0.174		(1)
							0.1		(5,3)
							0.141	1450	(6)
CaO	1	0.002		(1)					
		0.0017		**(4)**	FeO	1.96	0.174		(1)
		0.002		(5,3)			0.05		(5,3)
		0.0016	~1600	(7,8)			**0.17**	1600	(4)
		0.0021	1500	(11)			**0.188**	1600	(6)
		0.0023	1600	(12)			0.1	1325	(9)
							0.18	1960	(10)
MgO	1.28	0.0073		(1)					
		0.0085	1600	(6)	NiO	2.09[a]	0.318		(1)
							0.35		(5,3)
SnO	2.09[a]	0.318		(1)	PbO	2.09[a]	0.1237		(1)
		0.331	1100	(6)			0.002–0.01		(3)
							0.04	1100	(4)
ZnO	1.72[a]	0.057		(1)			0.032	1000	(6)
		0.06	1300	(4)			0.138	727	(7)
		0.017	1560	(6)					

(1) Ottonello et al. (2001); (2) Charles (1969); (3) Hess (1971); (4) Toop and Samis (1962a, b); (5) Masson (1965,1968); (6) Reyes and Gaskell (1983); (7) Navrotsky (1994); (8) Masson et al. (1970); (9) based on activity data of Schumann and Ensio (1951); (10) based on activity data of Distin et al. (1971); (11) based on activity data of Carter and MacFarlane (1957) and Sharma and Richardson (1962); (12) based on activity data of Zou et al. (1982). [a]Based on Pauling electronegativity.

Preliminary investigations of liquidus relations in REE_2O_3–Al_2O_3 systems (with REE = generic lanthanidic element) show that the role of electronegativity is evident also within a single transition series. These regularities (present also in cationic sub-lattice interactions) are extremely useful whenever the lack of accurate phase diagrams prevent predictions based on liquidus topology.

Table 19.4 Chemical interaction of the binary silicate melts (from Ottonello and Moretti 2004)

Join	$\ln Kp = A/T + B$		R^2	T range (°C)	Notes
	A	B			
K_2O–SiO_2	−31,708	0	–	1127–1527	−1
	−29,540	0	–	1223	−2
	−36,967	0	0.9922	1500–1800	−3
Na_2O– SiO_2	−23,336	0	0.9995	1000–1800	−3
CaO–SiO_2	−15,372	0	0.9958	1000–2000	−4
	−14,807	0	0.9943	1000–2000	−2
MgO–SiO_2	−9809.5	0	0.9955	1400–2000	−3
ZnO–SiO_2	−6460.1	0	–	1400–2000	−3
MnO–SiO_2	−6183.8	0	0.9596	1000–2000	−3
	−5649.1	0	–	1600	−1
PbO– SiO_2	−5330	0	0.5825	1000–1800	−3
	−4098.1	0	–	1273	−1
FeO– SiO_2	−3600	0	0.9973	1000–2000	−4
Fe_2O_3– SiO_2	7569.5	−7.2752	0.935	1000–2000	−5
TiO_2– SiO_2	4667.3	−3.2092	0.9107	1500–1900	−3
ZrO_2– SiO_2	2685.9	−6.382	0.9658	1400–2000	−3
NiO– SiO_2	1507.7	−1.7772	0.9825	1500–2000	−3
Al_2O_3– SiO_2	0	−1.4059	–	1000–2000	−3
B_2O_3– SiO_2	0	−1.066	–	1000–2000	−3

(1) Ottonello (2003); (2) Ottonello and Moretti (2004), based on experiments of Eliezer et al. (1978); (3) Ottonello and Moretti (2004), based on the Modified Quasi Chemical parameterization; (4) Ottonello (2001); (5) Ottonello et al. (2001)

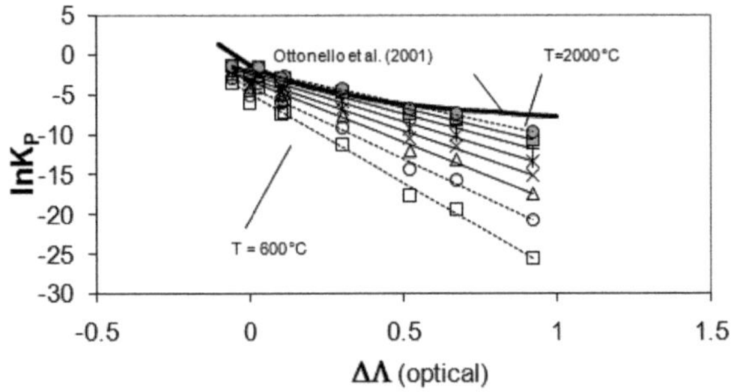

Fig. 19.6 Polymerization constant in MO–SiO_2 melts at various T (with MO = basic oxide) plotted against the difference in the optical basicity of MO and SiO_2. The adopted basicity values are those of column 6 in Table 19.4. Selection of other basicity values within uncertainty ranges (columns 1–6 in the same table) do not alter sensibly the observed relations. Dashed lines are extrapolations at T beyond the T limits investigated by Ottonello (2005) for most part of basic oxides. The solid heavy line is the functional form adopted by Ottonello et al. (2001) and Moretti and Ottonello (2003) when the role of T was still unclear. From Ottonello and Moretti (2004). With permission from Journal of Physics and Chemistry of Solids

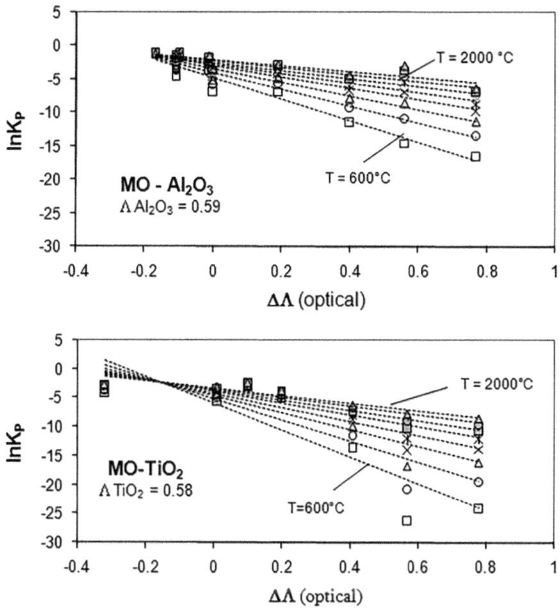

Fig. 19.7 Polymerization constant in MO–Al_2O_3 and MO–TiO_2 melts at various T (with MO = basic oxide) plotted against the difference in the optical basicity of MO and the network former

The Extended Polymeric Approach (EPA)

Although the original Toop-Samis model is a powerful tool in deciphering the complex relations existing between chemical interaction and configurational disorder, there are minor points that must be arranged, namely:

- assessment of the low polymerization limit;
- extension to the multicomponent field.

From the usual expression of the thermodynamic activity of a component in mixture, we get:

$$a_{MO} = \exp\left[\frac{\frac{(O^-)}{2}RT\ln K_p + (1 - X_{MO})\frac{\partial G_{mixing}}{\partial X_{MO}}}{RT}\right] \quad (19.2.27)$$

Being the mean number of monomers in the polymer chain defined as:

$$\overline{\nu}_{Si} = (SiO_2)\Big/\sum \text{anions} \quad (19.2.28)$$

adopting Temkin model activities of the fused salts it is quite evident that along any binary join we may pose:

Table 19.5 NM-NF parameters for a Toop-Samis description of chemical interaction in binary melts. Bold values obtained by topological analysis of liquidus relations at P = 10^5 *Pa*. All the listed values must be taken as provisional and subject to refinement

		MO–SiO_2		MO–Al_2O_3		MO–B_2O_3		MO–TiO_2	
Oxide	Λ	A	B	A	B	A	B	A	B
$\overline{}$ K_2O	1.36	**−31,708**	**0**	**−14,521**	**0**	−19,096	0.6	**−22,061**	**1.1244**
Na_2O	1.15	**−23,336**	**0**	**−8217.7**	**0**	−14,590	0.26	**−26,314**	**3.8359**
BaO	1.12	−16,751	0	−10,881.8	−0.143	−13,946	0.211	−16,539	0.5063
SrO	1.03	−14,492	0	−9605.9	−0.216	−12,015	0.065	−14,387	0.1356
Li_2O	1	−14,406	0	−9180.6	−0.240	−11,372	0.016	-3865.8	−1.6771
CaO	0.99	**−14,807**	**0**	**−10,081**	**0**	**−11,157**	**0**	**-10,528**	**−1.6101**
MgO	0.78	**−9809.5**	**0**	**−5948.4**	**−0.0853**	−6651.3	−0.341	**-1689.5**	**−2.973**
Al_2O_3	0.59	**0**	**−1.406**	–	–	**0**	**−1.083**	−3865.8	−1.6771
MnO	0.59	**−6183.8**	**0**	**−5013.3**	**−1.1943**	−2574.7	−0.649	**−3572.4**	**−1.5715**
Cr_2O_3	0.58	−6201.2	0	−3226.2	−0.583	−2360.1	−0.665	−3626.6	−1.7183
TiO_2	0.58	**4667.3**	**−3.209**	**639.61**	**−1.923**	−2360.1	−0.665	–	–
ZnO	0.58	**−6460.1**	**0**	**−3239.9**	**0**	−2360.1	−0.665	−3626.6	−1.7183
CoO	0.51	−4833.6	0	−2233.8	−0.640	−858.25	−0.779	−1952.8	−2.0067
Fe_2O_3	0.48	−4247.57	0	−1808.5	−0.665	−214.57	−0.828	−1235.5	−2.1303
SiO_2	0.48	–	–	**0**	**−1.079**	**0**	**−0.844**	−1235.5	−2.1303
FeO	0.48	**−3600**	**0**	**−3066**	**0**	−214.57	−0.828	**−987.33**	**−1.8905**
NiO	0.48	**1507.7**	**−1.777**	−1808.5	−0.665	−214.57	−0.828	−1235.5	−2.1303
PbO	0.48	**−5330**	**0**	**−3983**	**0**	−214.57	−0.828	−1235.5	−2.1303
SnO	0.48	−4247.56	0	−1808.5	−0.665	−214.57	−0.828	−1235.5	−2.1303
Cu_2O	0.43	−3270.8	0	−1099.7	−0.705	858.21	−0.909	−39.883	−2.3362
B_2O_3	0.42	**0**	**−1.066**	**0**	**−1.083**	–	–	199.23	−2.3774

(continued)

Table 19.5 (continued)

		MO–SiO_2		MO–Al_2O_3		MO–B_2O_3		MO–TiO_2	
Oxide	Λ	A	B	A	B	A	B	A	B
P_2O5	0.4	−2684.7	0	−674.4	−0.730	1501.9	−0.958	677.47	−2.4598
H_2O	0.39	−2489.3	0	−532.6	−0.738	1716.4	−0.974	916.59	−2.501
SO_3	0.33	−1317.2	0	318.0	−0.787	3003.8	−1.071	2351.3	−2.7482
CO_2	0.26	50.331	0	1310.4	−0.844	4505.7	−1.185	−2080.1	−1.7337
ZrO_2	0.26	**2551.7**	**−2.288**	**36.039**	**−1.942**	4505.7	−1.185	**−2080.1**	**−1.7337**

$$\overline{\nu}_{Si} = \frac{(SiO_2)}{(O^{2-})/a_{MO,binary} - (O^{2-})}. \tag{19.2.29}$$

Though the domain of Eq. (19.2.29) spans the entire compositional range, it is obvious that $\overline{\nu}_{Si}$ can be never lower than unity (i.e. monomer) and, consequently, the limit of maximum depolymerization defines a limiting activity represented by:

$$a_{MO} = \frac{(O^{2-})}{(O^{2-}) + (SiO_2)} \qquad \in \left[(SiO_2) \leq (SiO_2)_{min}\right]. \tag{19.2.30}$$

The molar abundances of the various oxygen species cannot be thus expected to be those of the original Toop-Samis model, especially when strongly basic oxides are present in the liquid phase.

In solving thermodynamic activity on a Temkin model basis, Toop and Samis 1962a, b) observed that the mean extension of the polymer chains is univocally defined by a "*polymerization path*" depicted in terms of mean number of silicon atoms per polymer unit $\overline{\nu}_{Si} = N_{SiO_2}/\sum$ anions) versus the stoichiometric ratio $(O^-)/\left[(O^-) + (O^0) + Si^{IV}\right]$ cf. Figures 2, 3 in Toop and Samis 1962a). Their further assumption that a single polymerization path of general validity in a ternary system may be proposed on the basis of viscosity data cannot, however, be shared. In fact, each MO–SiO_2 system pertains a different reaction constant K_P, and, being the activity of the basic dissolved oxide MO implicitly defined by the partial derivative of the Gibbs free energy of mixing at any point of the compositional space of interest, we have:

$$\sum \text{anions} = (O^{2-}) \Big/ \exp\left\{\left[G_{mixing} + (1 - X_{MO})\frac{\partial G_{mixing}}{\partial X_{MO}}\right] \Big/ RT\right\} \tag{19.2.31}$$

$$\frac{\partial G_{mixing}}{\partial X_{MO}} = \frac{1}{2} RT \ln K_{18} \frac{\partial (O^-)}{\partial X_{MO}} \tag{19.2.32}$$

$$(O^-) = \frac{2X_{MO} - 4 + \left[(4 - 2X_{MO})^2 - 32(4K_p - 1)\left(X_{MO}^2 - X_{MO}\right)\right]^{1/2}}{8K_{18} - 2} \tag{19.2.33}$$

$$(O^0) = \frac{4 - 4X_{MO} - (O^-)}{2} \tag{19.2.34}$$

$$(O^{2-}) = X_{MO} - \frac{(O^-)}{2} \tag{19.2.35}$$

$$Si^{IV} = (SiO_2) \tag{19.2.36}$$

where terms in brackets denote number of moles and X_{MO} is the molar fraction of a generic basic oxide MO in a MO–SiO_2 binary join.

Indeed, distinct "*polymerization paths*" relating the mean number of silicon atoms in the polymeric units ($\overline{\nu}_{Si}$) to the relative proportion of singly bonded oxygen in the unit $(O^-)/\left[(O^-) + \left(O^0\right) + Si^{IV}\right]$ may be calculated for the two limiting NM-NF binaries on the basis of the system of Eqs. (19.2.31 and 19.2.36). These paths differ sensibly from each other.

The apparent complexity introduced by recognizing that each oxide imposes its own imprinting on the extension of polymeric units turns out to be a powerful tool in deciphering the reactive properties of chemically complex melts. This is due o the fact that chemical interactions within the anion matrix of a n-component system are completely fixed by the interaction properties observed along the n − 1 limiting binaries. The intrinsic thermodynamic significance of this evidence may be recast to the Flood-Grjotheim thermodynamic cycle (Flood and Grjotheim 1952, 1953). Denoting $K_{P,A\text{-}F}$, $K_{P,B\text{-}F}$ the polymerization constants valid along the limiting A–F, B–F binaries, the polymerization constant of the extended Toop-Samis model turns out to be (Moretti and Ottonello 2003):

$$\ln K_P = N'_{A^{\nu+}} + \ln K_{P,A-F} + N'_{B^{\nu+}} + \ln K_{P,B-F} + \cdots \tag{19.2.37}$$

The above equation states that in a compositionally complex melt NM cations A, B of charge ν_A^+, ν_B^+ interact with free oxygen O^{2-} and hence with the polymeric units built up by the network former NF , with a simple proportionality described in terms of $N'_{A^{\nu+}}, N'_{B^{\nu+}}$ "*electrically equivalent fractions*"):

$$N'_{A^{\nu+}} = \frac{\nu_A^+ n_A}{\nu_A^+ n_A + \nu_B^+ n_B + \cdots} \tag{19.2.38}$$

Application of the Flood-Grjotheim thermodynamic cycle to the system CaO–FeO–SiO_2 reproduces satisfactorily the bulge in the thermodynamic activity of FeO experimentally observed by Elliot (1955) long ago (Fig. 19.8). No high order terms are necessary to reproduce the observed complex thermodynamic behavior, besides a topologically negligible interaction in the cation sublattice.

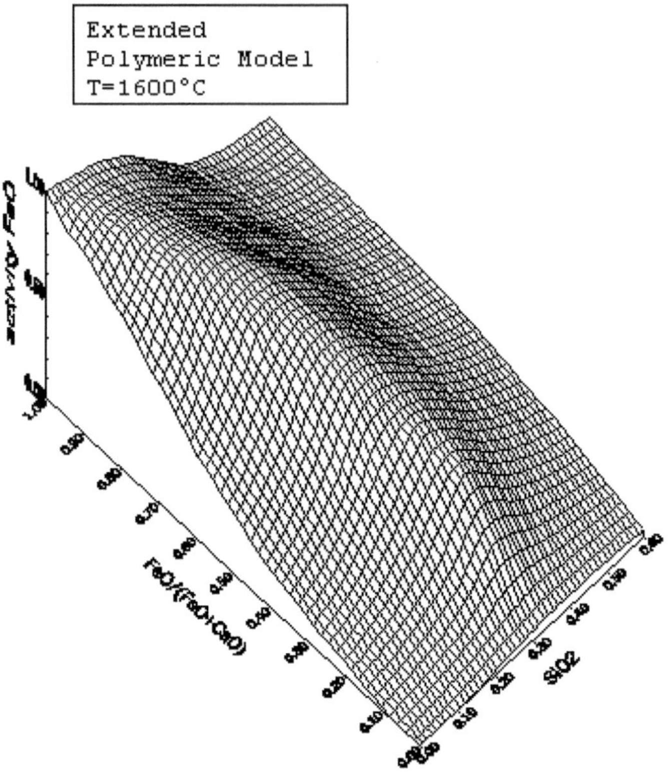

Fig. 19.8 FeO activity surface at 1600 °*C* for the liquid FeO–SiO_2–CaO ternary system. Graphical representation obtained by inverse-distance contouring of 9900 activity values generated by the extended polymeric model, adopting, as limiting K_p, 0.21 for the FeO–SiO_2 join and 0.002 for the CaO–SiO_2 join, respectively. Assumed cationic interaction $W_{Ca\text{-}Fe} = -33$ *kJ/mol*. Strain energy contributions included. From Ottonello (2001). With permission from Journal of Non Crystalline Solids

19.2.3 Quantum–Mechanical Appraisal

Resuming briefly what we have seen until now, we may state that the donor pressure affects sensibly the polymerization capability of a NF and, judging from the inferred relations between optical basicity and polymerization constant, this effect is simply related to the donor pressure of basic oxides, in an additive fashion. The best way to understand why it is so, we must investigate how the donor pressure affects the energy of the bonding MOs leading to stable Si–O–Si linkages and formation of "*bridging oxygens*" (O^0).

As already seen at Sect. 19.1, the permittivity of a silica melt has a strong stabilizing effect on $[SiO_4]^{4-}$ monomers but, what about the effect on Si–O–Si bonding? A seminal investigation on this matter was carried out through Extended Hückel Molecular Orbital calculations by Louisnathan and Gibbs (1972) on both the silicate ion $[SiO_4]^{4-}$ and the orthosilicic acid group $[H_4SiO_4]°$: the calculations showed a negligible effect of angular distortion on bond overlap. Here, we are more concerned with the energy of polymers and how these are affected by the reaction field. It is thus useful to repeat the calculations already done on the monomer $[SiO_4]^{4-}$ (Table 19.1) on a dimeric cluster $[Si_2O_7]^{6-}$. In doing this, I will consider not only the permittivity of the medium, as previously done for $[SiO_4]^{4-}$, but also the solvation radius, that, as already seen, sensibly affects the solute–solvent interaction in the terms accounted for by IEFPCM. Table 19.6 resumes the results of calculations carried out at simple B3LYP/3-21G theory level with UAHF radii.

The computation is quite informative even at this simple theory level. First of all, we may see that the nephelauxetic effect is quite evident on $[SiO_4]^{4-}$ where the Si–O- distance changes from 1.672 *Å* in vacuum to 1.703 *Å* in a CaO medium. It is not so in $[Si_2O_7]^{6-}$ where the Si–O^0 distance shrinks from 1.923 *Å* in vacuum to 1.728 *Å* in the CaO solvent. Also, the Si–O^- distance changes from 1.723 *Å* in vacuum to 1.685 *Å* in CaO solvent. The Mulliken charges localized on Si increase substantially in $[SiO_4]^{4-}$ but are virtually unaffected in $[Si_2O_7]^{6-}$. More important is the effect of the polarized continuum that stabilizes both solvated clusters. The amount of ionic character of the bond (f_i), calculated dividing the Mulliken charges on the terminal oxygens by the formal charge (−2) does not change much with the medium and it is in line with the Pauling's relation (19.2.10) with $\chi_O = 3.5$ and $\chi_{Si} = 1.8$ (Pauling 1949), which brings to $f_i = 0.514$.

Figure 19.9 shows a peculiar α MO of solvated $[Si_2O_7]^{6-}$ that I interpret as a Pauling's double bond, and is probably the main contributor to the effective dimeric stability in solution. Its orbitalic energy is −1.024 *Hartree*. All occupied orbitals in this solvated complex have negative energies with the HOMO at −0.358 *Hartree*. In gas phase, (isolated moiety) only 17 MOs out of the 45 occupied have negative energies with HOMO at +0.844 Hartree, which is indicative of intrinsic instability.

The virtual effect of the polarzed continuum on the polymerization extent is easily obtained by summing up the solvation terms to the gas phase energies (Table 19.6). I assumed in the calculations that the solvation entropy arises from electrostatic + cavitation terms. I obtained the electrostatic contribution to solvation entropy from the Born model and the cavitation entropy from the scaled particle theory (Pierotti 1976). The solvation enthalpy is the summation of electrostatic + repulsive + dispersive contributions and the (subordinated) cavitation enthalpy (Eq. 16.3.9). For the polymerization reaction (19.2.17) I obtained for the CaO medium A = −37,078 and B = 27.32 against A = −3807, B = 0 of the topological estimates (cf. Eq. 19.2.26 and Table 19.5). This confirms that the reaction proceeds in the depolymerization sense and that the entropic contributions are quite subordinated. At T = 1000 *K* topological estimates and *ab-initio* computations are virtually coincident. However, at high T the ab initio values of $\ln K_P$ are exceedingly negative with respect to topological estimates while, at room conditions, the extrapolated value of the topological constant is $\ln K_{P,topological} = -41.7$, against an ab initio predicted value $\ln K_{P,ab\ initio} = -92.7$. These biases could arise primarily from entropic effects but we could tentatively ascribe this failure also to the effect of the ionic cloud. We know in fact that the ionic atmosphere affects (only) the

Table 19.6 IEFPCM calculations of $[Si_2O_7]^{6-}$, $[SiO_4]^{4-}$ and $[O]^{2-}$ in molten silica and lime media. The solvation parameters adopted for the lime medium are $r_{solv,CaO}$ = 1.209 Å; ε_{CaO} = 4.791. Results of gas phase calculations are also listed. $\Delta G_{solvation}$ is the summation of cavitation + interaction energy terms. The electronic energy (E_{el}) and the thermal corrections to U, H, and G are in *Hartree*. Inter-atomic distances are in *Å*. Cavitation energy ΔG_{cav} interaction energies (electrostatic, dispersive, repulsive), solvation Gibbs free energy and enthalpy (ΔG_{solv}, ΔH_{solv}) and Gibbs free energy of formation from the elements and enthalpy of formation from the elements ($G^0_{f,298}$, $H^0_{f,298}$) are in *kJ/mol*. Solvation entropy (ΔS_{solv}) and entropy at standard state (S^0_{298}) are in *J/mol × K*). Values in italics obtained accounting for the iconicity of the solvent and assuming full CaO dissociation

Moiety	$[Si_2O_7]^{6-}$	$[Si_2O_7]^{6-}$	$[Si_2O_7]^{6-}$	$[SiO_4]^{4-}$	$[SiO_4]^{4-}$	$[SiO_4]^{4-}$	$[O]^{2-}$	$[O]^{2-}$	$[O]^{2-}$
Medium	Gas	SiO_2	CaO	Gas	SiO_2	CaO	Gas	SiO_2	CaO
E_{el}	−1097.553670	−1099.452725	−1106.243439	−587.500564	−587.083199	−590.814109	−74.569748	−74.577336	−74.640585
ZPE	0.022859	0.024782	0.021891	0.013192	0.012595	0.010957	0	0	0
Therm U	0.032408	0.034085	0.030841	0.017668	0.017175	0.015825	0.001416	0.001416	0.001416
Therm H	0.033353	0.035029	0.031785	0.018612	0.01812	0.016769	0.00236	0.00236	0.00236
Therm G	−0.012612	−0.011662	−0.013846	−0.012428	−0.013123	−0.014941	−0.013915	−0.013915	−0.013915
Si–O^0	1.923	1.757	1.728	–	–	–	–	–	–
Si–O^-	1.723	1.673	1.685	1.672	1.692	1.703	–	–	–
Mull. Si	0.551	0.619	0.619	0.234	0.371	0.375	–	–	–
Mull. O^-	−1.048	−1.057	−1.057	−1.058	−1.093	−1.094	–	–	–
Mull. O^0	−0.811	−0.890	−0.898	–	–	–	–	–	–
f^i	0.524	0.529	0.529	0.529	0.546	0.547	–	–	–
ΔG_{cav}	–	70.1	75.7	–	40.7	44.0	–	7.4	7.6
E_{electr}	–	−5031.1	−5576.1	–	−2838.2	−3136.7	–	−1585.0	−1751.0
	–	–	*−7009.0*	–	*−3928.2*	*−3944.1*	–	*−2184.7*	*−2190.3*
E_{disp}	–	−37.1	−74.9	–	−25.2	−51.0	–	−17.5	−35.5
E_{rep}	–	9.5	19.2	–	7.4	14.9	–	20.3	41.2
ΔG_{solv}	–	−4983.6	−5556.3	–	−2815.4	−3129.1	–	−1574.9	−1737.8
ΔH_{solv}	–	−5087.8	−5653.5	–	−2872.5	−3185.0	–	−1591.3	−1753.8
ΔS_{solv}	–	−332.8	−326.1	–	−191.5	−187.8	–	−55.2	−53.6

(continued)

Table 19.6 (continued)

Moiety	$[Si_2O_7]^{6-}$	$[Si_2O_7]^{6-}$	$[Si_2O_7]^{6-}$	$[SiO_4]^{4-}$	$[SiO_{4]}{}^{4-}$	$[SiO_4]^{4-}$	$[O]^{2-}$	$[O]^{2-}$	$[O]^{2-}$
$G^0{}_{f,298}$	4605.7	−382.9	−950.6	2670.2	−145.2	−458.8	2000.5	425.6	262.7
		−2237.0	*−22383.5*		*−1235.1*	*−1266.2*		−174.1	*−176.5*
$H^0{}_{f,298}$	4466.2	−621.6	−950.6	2599.7	−272.8	−585.3	2000.1	408.8	246.4
		−2475.7	*−2620.3*		*−1362.8*	*−1392.7*		*−190.2*	*−192.9*
$S^0{}_{f,298}$	413.6	80.8	87.5	276.5	85.0	88.7	143.3	88.1	89.7

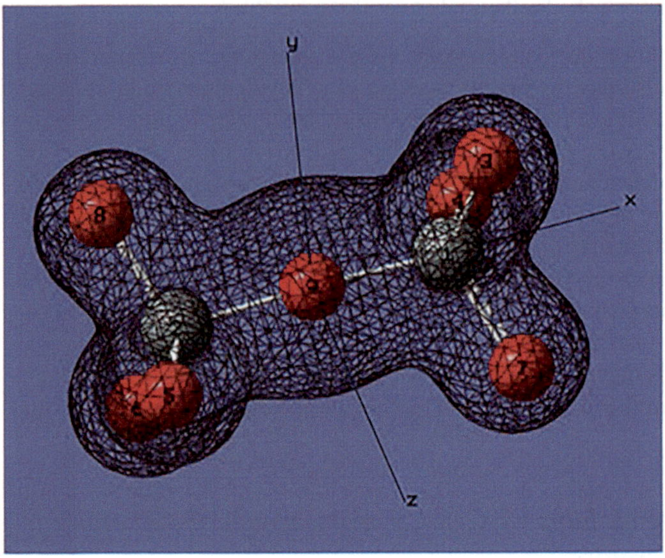

Fig. 19.9 B3LYP/3-21G bonding MO of solvated $[Si_2O_7]^{6-}$ at a 2×10^{-2} *e/Å*^{3} isodensity, elongated along the *x* axis. The eigenvalue is −1.024 *Hartree*

electrostatic solute–solvent interactions (Barone et al. 1997). The value one must assign to the ionic strength of the two media is debatable because it involves an a-priori decision about the dissociation extent of the molten oxide. In Fig. 19.10 we see XPS measurements of oxygen speciation in quenched CaO–SiO_2 melts after Park and Rhee (2001) compared with predictions of the Extended Polymeric Approach (EPA).

No doubt that the Toop-Samis model reproduces fairly well the experimental observations about the oxygen speciation at high T and that the SiO_2 medium is fully polymerized even at this high T (hence zero ionicity) while the CaO medium is fully dissociated. In a fully dissociated CaO medium, there are 17.832 mol of Ca^{2+} and 17.832 mol of O^{2-} per 10^3 *g* of solvent, resulting in a ionic strength I = 71.328. The effect of the ionic atmosphere may be quantitatively evaluated by application of the IEFPCM formalism (Barone et al. 1997). We may thus see that the solute–solvent electrostatic interaction in the CaO medium increases substantially (Table 19.6) leading to an overall decrease of roughly 250 *kJ/mol* in ΔG_P (from 219.7 to 27.7 *kJ/mol*). More important, the apparent polymerization enthalpy × goes practtically to zero and the apparent entropy lowers to –7.9 *J/*(*mol*×*K*) per unit mole of melt. We could also consider that the ionicity of the medium does not vary linearly with composition. At $0.33X_{SiO2}$, there are roughly 0.78 × 2 mol of O^- and 0.1 × 2 mol of O^{2-} per unit mole of melt. The molar weight of $Ca_{0.667}Si_{0.333}O_{1.333}$ is 57.411 g/mol. There are thus 17.418 mol of $Ca_{0.667}Si_{0.333}O_{1.333}$ per 10^3 g of solvent and the ionic strength is:

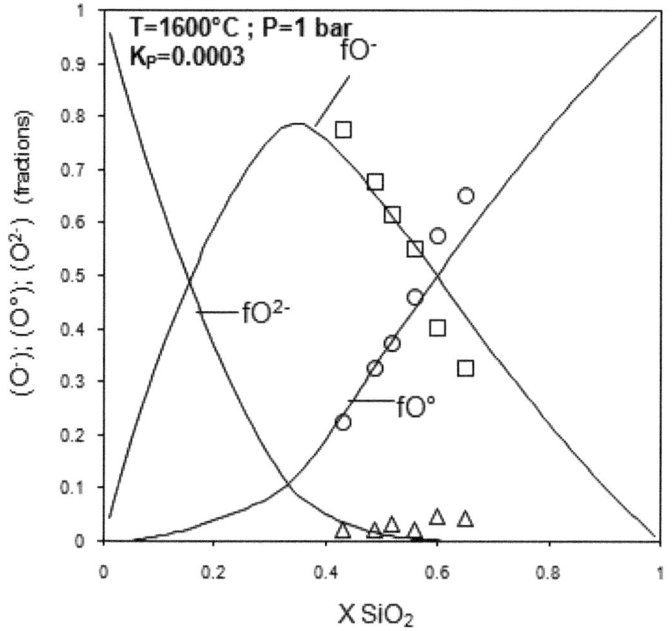

Fig. 19.10 XPS measurements of oxygen speciation in quenched CaO–SiO_2 melts after Park and Rhee (2001) compared with Toop-Samis predictions from Ottonello (2005)

$$I = \frac{1}{2} \cdot \sum_i m_i Z_i^2 = \frac{1}{2}(11.618 \times 4 + 27.172 \times 1 + 3.484 \times 4) = 43.79 \quad (19.2.39)$$

The decrease of ionic strength with the increase of SiO_2 is thus almost linear,[4] and the limited deviations from linearity would not have implications about the actual value of K_P, which could be effectively constant for a simple pseudo-binary join like CaO–SiO_2. In summary, the ab initio calculations are promising, but, undoubtedly, in the next future, it will be beneficial a comprehensive ab initio investigation of the various joins, considering other polymeric aggregates, with sufficiently extended basis sets at high theory level, in order to achieve a critical appraisal of the procedure.

[4] A simple linearity would imply I = 47.790.

19.3 Polymer Structures: The Role of Ad-Ions

The energy terms associated to Eq. (19.2.3) are rather complex to depict. In computational investigations carried out on zeolitic frameworks, Vayssilov and coworkers (Vayssilov and Rösch 1999, 2000; Vayssilov et al. 1999, 2000) have convincingly shown that the stabilizing effect of an external cation on a NF network may be precisely estimated in terms of Kohn–Sham core-level (O1s) energy shifts, ΔE_b, with respect to a given reference state. Vayssilov and Rösch (1999, 2000) in addition determined to what extent these energy shifts are related (in a simple scalar way) to the electronic properties of the oxygen atoms and, in particular, to the changes in their basicity. For zeolite cages, for example, the proton affinity (PA) of framework oxygens increases by about 25% from 700 to 930 *kJ/mol* as a net result of the external electrical field, giving rise to a linear correlation between PA and ΔE_b ($R^2 = 0.986$). In a more recent investigation, Gatti et al. (2012) extended the zeolite investigations of Vayssilov and coworkers for alkali ad-atoms to alkaline earth in alumino-silicate ring complexes.

As done in previous investigations (Pacchioni et al. 1999; and references therein), Gatti et al. (2012) assumed the core-level binding energy shift (ΔE_b O1s)) to be sufficiently approximated by the difference in the Kohn–Sham orbital energies of the oxygen donors ε^{KS} in the complex with respect to a state of reference:

$$\Delta E_b(O1s) = -\left[\varepsilon^{KS}(O1s) - \varepsilon^{KS}_{ref}(O1s)\right] \tag{19.3.1}$$

This reference state was represented by four-membered all-Si rings (4MR) for the M-Si_3Al rings and six-membered all-Si rings (6MR) for the M-Si_5Al/M-Si_4Al_2 cages Fig. 19.11.

The same analysis was extended to the eigenvalues of the first σ-bonding and σ-antibonding MOs of the oxygen atoms of the investigated ring structures. The study largely confirmed the results of Vayssilov and Rösch (1999, 2000) about the stabilizing effect of ad-atoms on the donor centers (Fig. 19.12). Notwithstanding some scatter in the data (probably due to the fact that the T-cations of the rings were not held in fixed, crystallographic positions), the observed difference in stabilization energies between alkali and alkaline-earth cations is nonetheless systematic. The insertion of Al atoms in the ring differentiates the core level energies of the oxygen donors abscissa axis in Fig. 19.12.

The various ad-atoms exert a stabilizing effect, shifting the core level energies of all donor centers in the ring in a sub-parallel way independently of their relative position. The magnitude of the shift depends upon the nominal charge of the ad-atom and upon the row of the periodic table to which it belongs. The oxygen donor centers in 6MR are stabilized by -3.6 ± 0.4, -3.9 ± 0.5, -7.3 ± 0.1, and -7.6 ± 0.2 *eV* by K^+, Na^+, Ca^{2+}, and Mg^{2+} ad-ions, respectively. The energy shifts are even greater for 4MR where the stabilization effects attain -3.7 ± 0.1, -4.1 ± 0.1, -8.1 ± 0.1 and -9.0 ± 0.1 *eV*, respectively. These effects are also observed on the low-lying σ-bonding and antibonding MOs of the oxygen

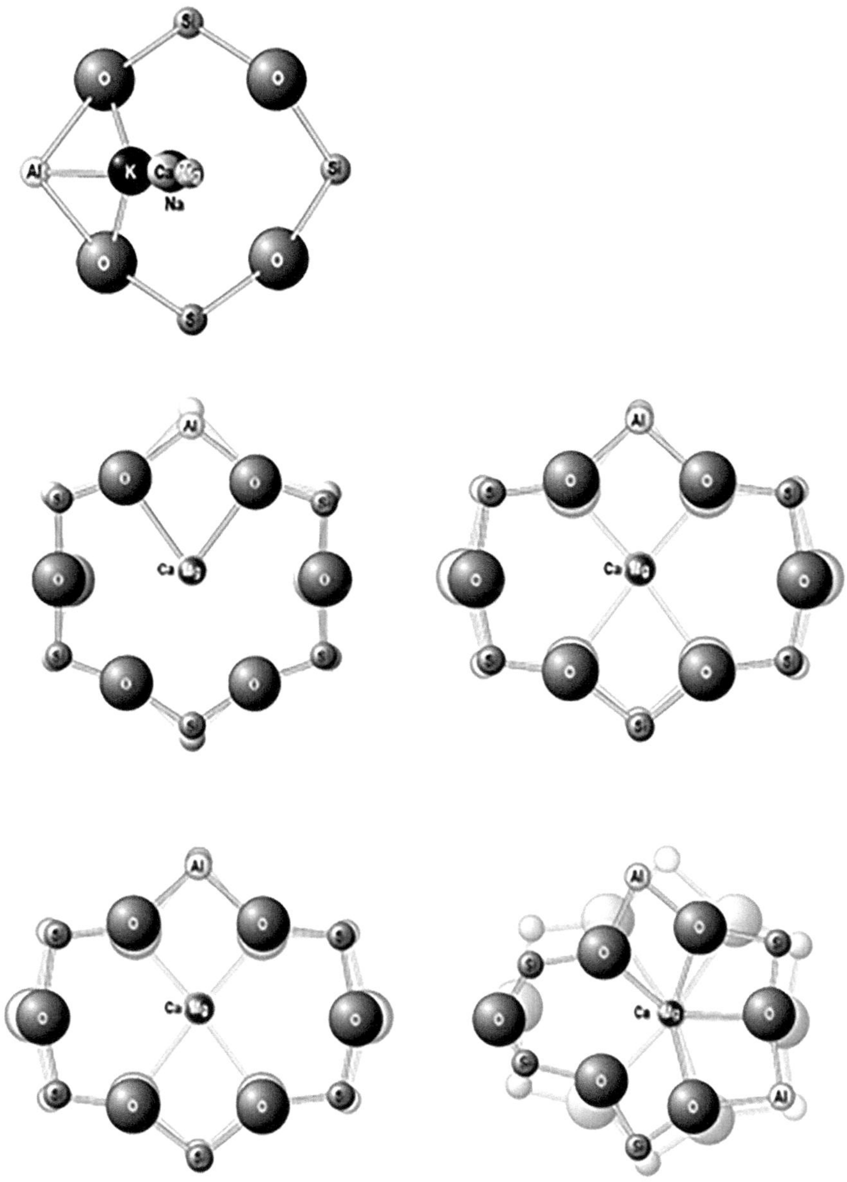

Fig. 19.11 Planar views of the calculated four- and six-membered ring geometries and positions of ad-atoms investigated by Gatti et al. (2012). For clarity, the dangling hydrogens bonded to silicons and aluminums are not shown. In the four-membered ring, the variations in ring geometries when the ad-atom is changed are slight enough not to be shown. Modified from Gatti et al. (2012)

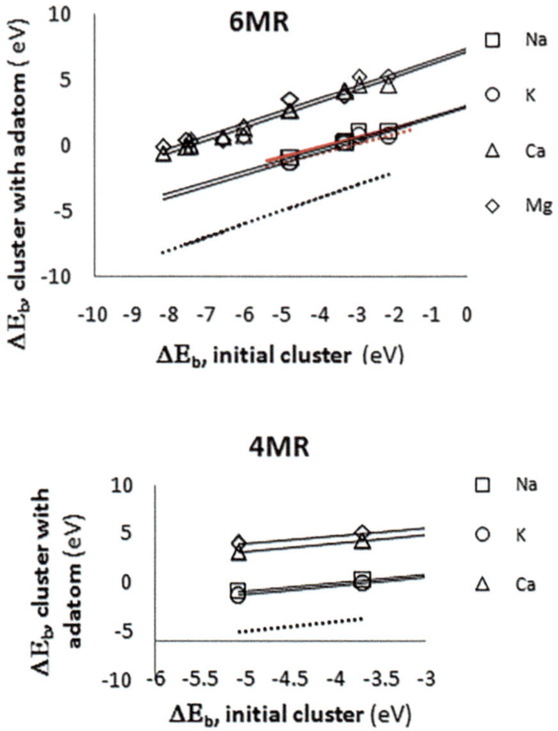

Fig. 19.12 Binding-energy shifts ΔE_b of oxygen atoms of cation-containing 6MR (upper part) and 4MR (lower part) clusters versus the corresponding values in the initial all-Si rings without central cation (Gatti et al. 2012). Thin solid lines are the best interpolant. The interpolations of Vayssilov and Rösch (1999, 2000) for six rings are also shown for comparative purposes. Al-substitution in the ring differentiates the core-level energies of the oxygen donors. The effect of the ad-atom is seen as an upward shift of the sub-parallel interpolating lines. Modified from Gatti et al. (2012)

framework, but in a less systematic fashion. Clear relationships with the core-level shifts are found when the effects of alkali complexation are evaluated through electron localization/delocalization indices, which are defined in terms of the whole wave function and not just of the individual orbitals. Complexation with cations does not only involve a small but significant electron sharing of the cation with the oxygen atoms in the ring, but also enhances electron exchange among oxygen atoms, while reducing that between the O atoms and the Si or Al atoms bonded to them. The enhanced electron delocalization among O atoms upon cation complexation is also manifest in the Pauling's double-bond nature of the bent bonding MO between non-adjacent oxygen centers in O-based ring structures such as those ones represented in Fig. 19.13.

The Electron Localization/Delocalization Analysis was carried out by Gatti et al. (2012) adopting the descriptor of Fradera et al. (1999) (see Sect. 14.2). In Fradera's et al. approach, the exchange–correlation density (XCD) is doubly integrated over either the same QTAIM atomic domain, Ω, or over two different domains Ω and Ω' to yield the so-called localization [$\lambda(\Omega, \Omega)$; hereinafter $\lambda(\Omega)$, for the sake of simplicity] or delocalization $\delta(\Omega, \Omega')$ indices, respectively. The number of electrons $N(\Omega)$ on an atom Ω is then given by:

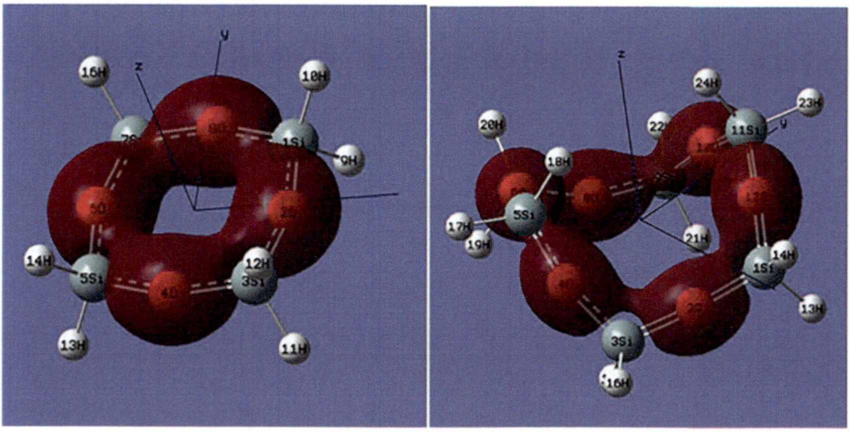

Fig. 19.13 The σ-bonding orbitals of the [Si_4 $(OH_2)_4$] and [Si_6 $(OH_2)_6$] rings resolved at the B3LYP/TZVP theory level (Gatti et al. 2012). The isodensity limit is 0.03 *e/Å*^{3} for both structures. The eigenvalue of the 4MR is −1.019 *Hartree* and that of the 6MR is −1.016 *Hartree*. The electron clouds essentially occupy the xy plane in a standard arrangement although, apparently, the co-planarity of the donor centers is severely perturbed in the 6MR. From Gatti et al. (2012). With permission from The American Chemical Society

$$N(\Omega) = \lambda(\Omega) + \frac{1}{2}\sum_{\Omega' \neq \Omega} \delta(\Omega', \Omega) \tag{19.3.2}$$

where $\lambda(\Omega)$ is the number of electrons that are fully localized within Ω and δ (Ω, Ω′) the number of those that are delocalized (exchanged) between Ω and Ω′ (with $\Omega \neq \Omega'$). In practice, N(Ω) equals the sum of the number of electrons that, on a statistical basis, always stay in Ω and of half the number of those that are exchanged between Ω and all the remaining basins. The λ and δ indices lend themselves to insightful statistical interpretations of electron localization/delocalization effects because they are intimately related to the variances and covariances of the electron populations of their associated domains (Poater et al. 2005). As indicated by Eq. (19.3.1), the change in the electron population on an atom due to a perturbation translates into a change in either the number of electrons localized in its basin, or in the number of electrons exchanged with other basins, or, more generally, in both of these events. A perturbation may, in some cases, also lead to a negligible change in the electron population, which is, however, the result of large, almost compensating, variations in the numbers of the localized and exchanged electrons. In summary, Eq. (19.3.1) leads to valuable insights into the changes in the bonding capacity of an atom in response to perturbations, such as those due to the alkali or alkaline-earth metal complexation investigated by Gatti et al. (2012). Table 19.7 lists, as an example, the electron localization/delocalization analysis of 4MRs with ad-atoms. The table reports, for a given system and for each of its unique atoms by symmetry Ω, the number of electrons that are fully localized in the basin of this

Table 19.7 Delocalization analysis (Gatti et al. 2012) in 4MR with Ad-atom. *Italics:* localization index. (*) the two H give very different contributions, H15: 0.437, H16: 0.267; (**) the two H give quite different contributions H15: 0.068, H16: 0.124; (***) this is for Ca-H16; Ca-H15 is just 0.003

$NaSi_3AlO_4H_8$	Si1	Si3	Al7	O2	O8	O4	O6	$(H9, H10)_{av}$	$(H11, 12)_{av}$	$(H15, H16)_{av}$	Na
Si1	*10.106*	0.005	0.004	0.364	0.470	–	–	0.483	–	–	0.002
Si3	0.005	*10.103*	–	0.401	–	0.401	–	–	0.484	–	0.001
Al7	0.004	–	*10.003*	–	0.221	–	0.221	–	–	0.362	0.001
O2	0.364	0.401	–	*8.835*	0.168	0.160	–	0.114	0.115	–	0.040
O8	0.470	–	0.221	0.168	*8.901*	–	0.121	0.118	–	0.092	0.088
Na	0.002	0.001	0.001	0.040	0.088	0.040	0.088	–	–	–	*9.940*
$KSi_3AlO_4H_8$											**K**
Si1	*10.100*	0.005	0.004	0.366	0.470	–	–	0.474	–	–	0.001
Si3	0.005	*10.101*	–	0.407	–	0.407	–	–	0.480	–	0.001
Al7	0.004	–	*9.997*	–	0.228	–	0.228	–	–	0.345	0.005
O2	0.366	0.407	–	*8.835*	0.158	0.154	–	0.116	0.117	–	0.034
O8	0.470	–	0.228	0.158	*8.911*	–	0.116	0.121	–	0.093	0.072
K	0.001	0.001	0.005	0.034	0.072	0.034	0.072	–	–	–	17.909
$MgSi_3AlO_4H_8$											**Mg**
Si1	*10.128*	0.004	0.003	0.334	0.430	–	–	0.520	–	–	0.003
Si3	0.004	*10.128*	–	0.371	–	0.371	–	–	0.523	–	0.002
Al7	0.003	–	*10.022*	–	0.189	–	0.189	–	–	0.400	0.002
O2	0.334	0.371	–	*8.818*	0.206	0.198	–	0.102	0.104	–	0.114
O8	0.430	–	0.189	0.206	*8.886*	–	0.144	0.107	–	0.086	0.171
Mg	0.003	0.002	0.002	0.114	0.171	0.114	0.171	–	–	–	*9.935*

(continued)

Table 19.7 (continued)

$NaSi_3AlO_4H_8$	Si1	Si3	Al7	O2	O8	O4	O6	$(H9, H10)_{av}$	$(H11, 12)_{av}$	$(H15, H16)_{av}$	Na
$CaSi_3AlO_4H_8$											**Ca**
Si1	*10.116*	0.004	0.004	0.345	0.438	–	–	0.501	–	–	0.002
Si3	0.004	*10.120*	–	0.386	–	0.386	–	–	0.510	–	0.001
Al7	0.004	–	*10.006*	–	0.219	–	0.219	–	–	0.352 (*)	0.010
O2	0.364	0.401	–	*8.829*	0.180	0.180	–	0.105	0.109	–	0.091
O8	0.438	–	0.219	0.180	*8.912*	–	0.133	0.113	–	0.096 (**)	0.112
Ca	0.002	0.001	0.010	0.091	0.112	0.091	0.112	–	–	0.116 (***)	*17.880*

atom, $\lambda(\Omega)$, and the number of electrons that are exchanged between the other system's atoms Ω', $\delta(\Omega, \Omega')$. In practice, to each system, a N × M matrix is associated where the number of rows N equals the number of unique atoms by symmetry in the system (excluding H atoms) while the number of columns M equals the number of atoms having unique exchange interactions with the unique atoms listed in the first column of the Table. The $\lambda(\Omega)$ values (in italic) are the diagonal elements of each matrix, while the $\delta(\Omega, \Omega')$ values (in roman type) correspond to the out-of diagonal elements. Only the electron exchanges for 1–2 and 1–3 interactions are reported, where 1–2 refers to two directly bonded atoms, i.e. Si–O, Al–O, Si–H, and 1–3 to two atoms that are only indirectly bonded and lie in relative position 1–3 (i.e. the underlined atoms in Si–O–Si, O–Si–O, Si–O–Al, O–Al–O). Electron exchanges with the ad-atoms are reported for all M atoms (but H).

In these "small" rings, the optimized cation location is in general highly displaced from the center of the ring and the most perturbed atoms are those that turn out to be closer to the complexing cations. As a rule, the adsorbed metals sit significantly closer to the oxygen atoms bound to one Al (Fig. 19.11), because the valence electrons of these O, due to the higher ionicity of the Al–O with respect to the Si–O bond, are less engaged in bonding and thus more prone to bond to the adatom and exchange electrons with their nearest neighbor O atoms bonded to Si atoms only. Indeed, the adatom-O bond is not purely ionic but retains a small, yet non-negligible, electron exchange, about twice as great as for the oxygen atoms bonded to Al, which are, obviously, those closer to the adatom.

Another general feature is that the O atoms largely enhance their ability to exchange electrons with their nearest neighbor O atoms, whereas the bonds to Si and Al become less shared and more ionic, as suggested by generally lower $\delta(\mathrm{Si,O})$ and $\delta(\mathrm{Al,O})$ values. In $NaSi_3AlO_4H_8$ rings, $\delta(0,0)$ increases from 0.106 to 0.121 for the O atoms bonded to Al and from 0.149 to 0.168 for those bonding only with Si whereas the latter oxygen atoms, O_2/O_4, increase their mutual exchange from 0.136 to 0.160. Because the Na atom retains a 10% fraction of its valence charge, it has the potential to delocalize and exchange electrons with the atoms in the ring. Indeed, whereas its exchange with Si and Al atoms is practically null, it exhibits a small, but non-negligible, electron exchange with O atoms. This difference testifies to the special effect of interactions between the oxygens and metal cations on electron localization/delocalization within the ring chain (Gatti et al. 2012). Complexation with K instead of Na leads to similar (though slightly weakened) effects in agreement with the trends found for the core energy shifts. Insertion of an alkaline-earth metal instead of an alkali metal does not qualitatively alter the picture described for Na/K complexation, but strongly enhances the effects. Like alkali metals, the third-row alkaline-earth metal (Mg) has a larger impact than the fourth-row element (Ca), again in agreement with the trends of core-energy shifts. For Mg, the exchange is remarkably strong as shown by $\delta(08,\mathrm{Mg}) = 0.171$ and $\delta(02,\mathrm{Mg}) = 0.114$, which indicate that the Mg···O ring interaction is far from being purely ionic (Gatti et al. 2012).

19.4 Configurational Disorder

In a simple Temkin's fused salts model, cation matrix and anion matrix are treated as independent sub-systems, due to electrostatic reasons (i.e. the probability of finding a cation, at the place of an anion is virtually zero) and the thermodynamic activity of solutes has no configurational entropic contributions, besides matrix ideality. In silicate melts and slags the treatment is more complex, especially in high acidity regions were conformational ordering takes place. This conformational ordering is prodromal to immiscibility. The existence of miscibility gaps in silicate melts has been experimentally known for decades, both in terrestrial vulcanites, such as those present in the Archean greenstone belt of Brasil (Coltorti et al. 1987), or in lunar basalts (Roedder and Wieblen 1971). Immiscibility zones are also quite evident along simple binary joins, like CaO–SiO_2, MgO–SiO_2, FeO–SiO_2 etc., and, apparently, they are an intrinsic property of the polymeric network that cannot be reproduced by a simple Fincham-Richardson reactivity [see Sect. 19.2.2].

The branch of material science that was most successful in the comprehension of un-mixing phenomena concerns the rubberlike materials, and, more particularly, the role of block-copolymers in microphase separation. Although there are no experimental evidences about the eventual formation of local block-copolymeric structures in silicate melts or glasses, the recent findings in this branch of material science are relevant in the comprehension of un-mixing phenomena that take place in SiO_2-rich melts as well.

The physics of microphase separation in block copolymers is understood in terms of an expanded Flory–Huggins model. The Flory–Huggins model belongs to the class of Quasi-chemical approaches. Flory itself defined his method as "*a restricted variation of the quasi-chemical method used by Guggenheim*" (cf. page 507 in Flory 1953). Basically, the main merit of the Flory–Huggins approach is to acknowledge that molar fractions are unsuitable chemical parameters whenever molecules largely different in size are interacting among themselves and with much smaller stoichiometric units in the melt phase. Volume fractions are thus substituted to molar fractions based on the identity:

$$\phi_i = \frac{n_i \nu_i}{\sum_{i=1}^{k} n_i \nu_i} \tag{19.4.1}$$

where ϕ_i is the volume fraction of the ith component in mixture, n_i its molar amount, ν_i its partial molar volume, and the summation at denominator is extended to all k components in mixture.

Block copolymers are constituted by two or more distinct blocks interconnected by linear or branched sequencing. The relative arrangements of the various blocks give rise to an astonishing number of distinct configurations and are responsible of the many useful and desirable properties exploited by the industry of rubberlike materials. The simplest architecture is the linear AB diblock: a long sequence of type A monomers covalently bonded to a chain of type B monomers. The relative arrangements of the two blocks are characterized by "*a fluidlike disorder on the molecular scale and a high degree of order at longer length scales*" (Bates and Fredrickson 1999).

In interpreting structural and reactive properties of diblock copolymers, the following working assumptions are commonly made:

- Each molecule in the melt is composed of N segments of which a fraction f forms the A block and fraction 1 − f forms the B block;
- A and B segments each represent a sufficient length of polymers, such that they can be treated as Gaussian, and where internal conformational states of the segments produce a Hookian entropic penalty of stretching;
- The statistical length of a segment is its average RMS end-to-end length when no tension is applied, and is related to the effective spring constant;
- Average segment concentration is forced to be uniform (incompressibility constraint).

The interaction between blocks A and B is described in terms of a χ_{AB} Flory–Huggins interaction parameter which, in units of the thermal energy k_BT, is expressed as follows:

$$\chi AB = \frac{Z}{k_BT}\left[\varepsilon_{AB} - \frac{1}{2}(\varepsilon_{AA} + \varepsilon_{BB})\right] \qquad (19.4.2)$$

with Z = number of nearest-neighbor monomers to a copolymer configuration cell. If χ_{AB} is large chemical interaction leads to a macroscopic segregation of A and B blocks (binodal or spinodal decomposition). If, on the contrary, χ_{AB} is sufficiently low, the thermodynamic forces driving separation are counterbalanced by entropic restoring contributions i.e. "*chain elasticity*"; intimately connected to the covalent character of the bond but (Hookian in nature) reflecting the requirement that, to keep the dissimilar A and B portions of each molecule apart, copolymers must adopt extended configurations (Bates and Fredrickson 1999).

19.4.1 Mean Field Approximation Theories

The Helmholtz free energy F (incompressibility constraint) is (Fredrickson et al., 1994; Fredrickson and Liu, 1995):

$$F = \frac{k_BT}{4\pi^2}\int_0^\infty q^2 \ln[\phi_A S_A(q) + \phi_B S_B(q)]dq \qquad (19.4.3)$$

where q is the reciprocal of the distance in the search space and S_A, S_B are form factors of the A and B block copolymers (Fredrickson et al. 1994; Fredrickson and Liu 1995; Chen et al. 2000, 2002). These "*form factors*" assume precise values depending upon the extension and the geometry of the block. For a linear chain L of monomers of statistical Kuhn length b the form factor is simply $S_L = N_L g(x)$ with N_L = degree of polymerization and:

$$x = q^2 b^2 N_L / 6 \tag{19.4.4}$$

$$g(x) = \frac{2(e^{-x} + x - 1)}{x^2} \tag{19.4.5}$$

For branched polymers, the form factors are more complex and are obtained by combination of arm and backbone contributions (see Fredrickson et al. 1994, 1995; for an appropriate treatment).

19.4.2 Matsen-Schick Self-consistent Mean Field Method

The spectral analysis developed by Matsen and Shick in a series of articles (Matsen and Schick 1994a, b; Matsen and Bates 1996; Matsen 1998) is a powerful predictive tool in deciphering the complex microphase separation processes taking place in rubberlike materials. The theory stems from the Self Consistent Mean Field (SCMF) treatment of Hong and Noolandi (1981). The form factor functionals are here substituted by space-occupation functions $\mathbf{r}_\alpha(s)$ describing the space curve occupied by the αth copolymer and s is the parameterizatin variable.

In a system of *n* starblock copolymers composed of M AB diblock arms joined together at a central core, s = 0 at the core, s = f at the AB junction and s = 1 at the end of the first arm. In the second arm s will be 1, 1 + f, 2 and so on, until the end of the last arm where s = M is reached. The space occupation function is thus piecewise continuous with discontinuities at integer values (Fig. 19.14).

Background Information: Matsen-Schick Analysis

The partition function for a system of p polymers of polymerization N and density ρ_0 is:

$$Q = \int D r_P P[\mathbf{r}_p; M + f] \delta\left[1 - \hat{\phi}_A - \hat{\phi}_B\right] \times \exp\left\{-\frac{\rho_0}{N} \int d\mathbf{r} \chi_{AB} N \hat{\phi}_A(\mathbf{r}) \hat{\phi}_B(\mathbf{r})\right\} \tag{19.4.6}$$

P is the probability density functional for a given curve (see Eq. 2 in Matsen and Schick 1994a, for the approximate form adopted to describe P); $\hat{\phi}_A(\mathbf{r}); \hat{\phi}_B(\mathbf{r})$ are (dimensionless) A and B monomer density operators:

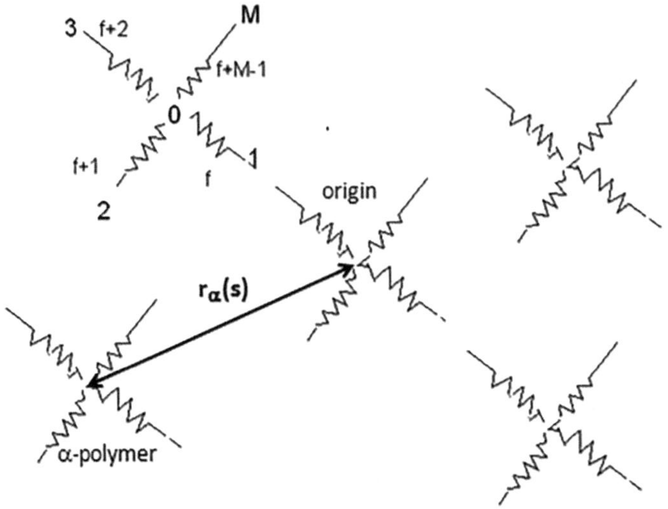

Fig. 19.14 Positional parametrization of starblock copolymers. The copolymer melt is composed of starblock copolymers made up of alternating A and B blocks consisting of fN and (1 − f)N monomers. The positional vector defining the relative position of the various branches of the nth copolymer is defined in terms of polymer number and branch number. The space function is piecewise continuous with discontinuities at integer values

$$\hat{\phi}_A(\mathbf{r}) = \frac{N}{\rho_0} \int_0^{M+f} ds\gamma(s) \cdot \delta[\mathbf{r} - \mathbf{r}_P(s)] \quad (19.4.7)$$

$$\hat{\phi}_B(\mathbf{r}) = \frac{N^M}{\rho_0} \int_0^{M+f} ds(1 - \gamma(s)) \cdot \delta[\mathbf{r} - \mathbf{r}_P(s)] \quad (19.4.8)$$

$\gamma(s)$ is a discontinuity function that attains the value 1 when s corresponds to an A-monomer region and the value 0 for a B-monomer region. The product of density operators in Eq. (19.4.6) represents the various space-dependent interactions in the copolymer melt.

To render the problem mathematically tractable the following steps are performed:

(1) a Delta functional integral $1 = \int D\Phi_A \delta\left[\Phi_A - \hat{\phi}_A\right]$ permits the replacement of the chain density operator by segment volume fractions ϕ_A
(2) Segment volume fractions are expanded as linear combinations of basis functions:

$$\phi_A(r) = \phi_{A,1}f_1(r) + \phi_{A,2}f_2(r) + \phi_{A,3}f_3(r) + \cdots \tag{19.4.9}$$

(3) Basis functions (Fourier transforms) are selected such that they possess the symmetry of the phase being considered; i.e. for the gyroid phase with symmetry $Q_{Ia\bar{3}d}$ (Matsen and Schick 1994b):

$$\begin{aligned}
f_1(r) &= 1 \\
f_2(r) &= \sqrt{8/3}[\cos(X)\sin(Y)\sin(2Z) + \cos(Y)\sin(Z)\sin(2X) + \cos(Z)\sin(X)\sin(2Y)] \\
f_3(r) &= \sqrt{4/3}[\cos(2X)\cos(2Y) + \cos(2Y)\cos(2Z) + \cos(2Z)\cos(2X)] \\
f_4(r) &= \ldots\ldots\ldots \\
f_5(r) &= \ldots\ldots\ldots
\end{aligned} \tag{19.4.10}$$

with $X = 2\pi x/D$ and D = size of the cubic gyroid) unit cell.

Through several passages one arrives to SCMF equations reducing the problem to that one of average densities of A, B monomers at **r** in the ensemble of non-interacting polymers subject to self-consisten external fields (cf. Eqs. 11–15 in Matsen and Schick 1994a).

19.4.3 Chemical Interaction and Long-Range Disorder

In silicate melts, as we have already seen, chemical interactions are dominant, but strain energy contribution to the Gibbs free energy of mixing do appear as well in the high SiO_2 portion of the system. Our appraisal is still insufficient to appreciate the complex phenomenology taking place within these instability regions, although various "*operative*" functionals have been proposed through the years (Ottonello 2001, 2005; Belmonte et al. 2017a, b).

Since the Toop-Samis model embodies ideal and excess Gibbs free energy contributions to the Gibbs free energy of the liquid $MO–SiO_2$ mixture in a single term (cf. Eq. 19.2.20), it is conceptually appropriate to depict the Gibbs free energy amount necessary to the opening of a liquid solvus at high SiO_2 content as an elastic energy contribution, that, in analogy to what we have seen for rubberlike materials for a chain of $\bar{v}_{Si}$ monomers extended to a distance x can be expressed as:

$$G_{strain} = \frac{3RT}{2\bar{v}_{Si}}\left(\frac{x}{a}\right)^2 \tag{19.4.11}$$

Table 19.8 Parameters of the hybrid polymeric model for the $FeO–SiO_2$ join (Ottonello 2001)

$\ln K_{p},FeO–SiO_2 = -3600/T$
$\chi_1,FeO–SiO_2 = -4.7469 + 0.002328 \times T$
$\chi_2,FeO–SiO_2 = 29.316 - 0.01426 \times T$
$\chi_3,FeO–SiO_2 = -45.828 + 0.021348 \times T$
$\chi_4,FeO–SiO_2 = 32.129 - 0.013545 \times T$
$\ln K_{p},CaO–SiO_2 = -15{,}372/T$
$\chi_1,CaO–SiO_2 = -3.075 + 0.0011465 \times T$
$\chi_2,CaO–SiO_2 = 29.490 - 0.01147 \times T$
$\chi_3,CaO–SiO_2 = -81.311 + 0.032785 \times T$
$\chi_4,CaO–SiO_2 = 67.579 - 0.028095 \times T$

x/a in Eq. (19.4.11) is a bending term taking into account the relative arrangement of monomers in the chain. This term may be arbitrarily expanded in a polynomial of type:

$$\frac{x}{a} = \chi_1 \cdot N_{SiO_2} + \chi_2 \cdot N^2_{SiO_2} + \chi_3 \cdot N^3_{SiO_2} + \chi_4 \cdot N^4_{SiO_2} + \cdots \quad (19.4.12)$$

The strain coefficients $\chi_1 \ldots \chi_n$, calculated at the various T conditions with a non-linear minimization routine, depict a linear T-dependence. Since the polymerization path along any join is defined in terms of $\overline{\nu}_{Si}$ versus N_{SiO_2} the strain energy calculated along a given binary at various T conditions is intimately related to the polymerization extent along any particular compositional path. Table 19.8 lists for example the parameterization developed long ago with this method for the $FeO–SiO_2$ and $CaO–SiO_2$ joins.

The computed bending terms indicate increased departure from linear arrangement with the increase of silica concentration in the system [note that being the x/a term squared in Eq. (19.4.11), both positive and negative x/a values are appropriate, which is typical of angular deformation terms]. In Fig. 19.15 are shown the computed bending factors and computed elastic strain energy versus composition for the $FeO–SiO_2$ join, at the various T of interest. The computed strain energy is slightly asymmetrical in the compositional space of interest.

The procedure may be easily extended to multicomponent melts, such as the $CaO–FeO–SiO_2$ ternary system:

$$\left(\frac{x}{a}\right)_{CaO-FeO-SiO_2} = \chi_{1,CaO-FeO-SiO_2} \cdot N_{SiO_2} + \chi_{2,CaO-FeO-SiO_2} \cdot N^2_{SiO_2} + \cdots \quad (19.4.13)$$

and

$$\chi_{1,CaO-FeO-SiO_2} = \chi_{1,CaO-SiO_2} \cdot \frac{X_{CaO}}{X_{CaO} + X_{FeO}} + \chi_{1,FeO-SiO_2} \cdot \frac{X_{FeO}}{X_{CaO} + X_{FeO}} \quad (19.4.14)$$

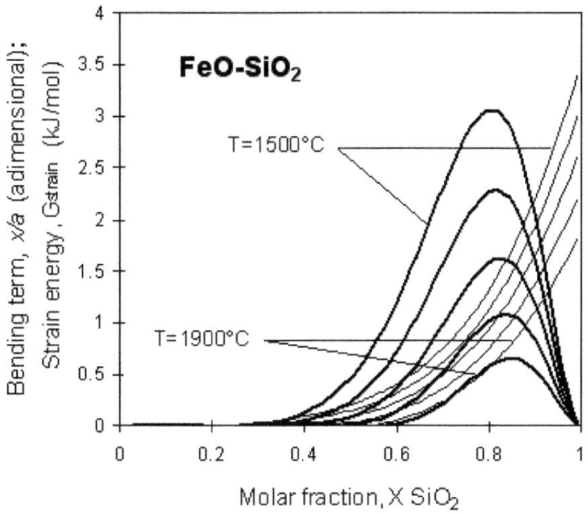

Fig. 19.15 Elastic strain energy contribution to the bulk Gibbs free energy of mixing heavy lines and bending terms thin lines) computed for the join FeO–SiO_2 at various T conditions. From Ottonello (2001). With permission from the Journal of Non Crystalline Solids

The observed conformation of the Gibbs free energy of mixing surface in the ternary CaO–FeO–SiO_2 is obtained in this way with reasonable approximation. This fact may be appreciated when comparing the computed excess Gibbs free energy of mixing in the ternary at T = 1600 °C with the assessment of Elliott (1955) (Fig. 19.16).

19.4.4 A Quantum-Chemical View

Although the model is operatively successful, it is not quite satisfactory from the theoretical point of view, and perhaps also in this case, quantum–mechanical arguments may provide precious hints.

There is general consensus on the fact that the strain energy is largely entropic in nature. In copolymer blends, such as rubberlike materials, as we have already seen, this energy is associated to conformational adjustments of polymer chains (see Fredrickson et al. 1994; Bates and Fredrickson 1999; and references therein for an extended discussion). This form of energy is usually adopted in the theory of microphase separation to limit (and not to enhance) decomposition. This energy (differently from the contribution represented by Eq. (19.4.11)) is in fact usually conceived as "*an entropic restoring force that serves to limit the phase separation to mesoscopic dimensions*" (Bates and Fredrickson 1999). We think (Belmonte and Ottonello, in prep.) that, in silicate liquids, this amount of entropy is not configurational in nature but is rather due to a limited increase in the vibrational frequency of Rigid Unit Modes (RUM) affecting adjacent monomeric groups in polymer chains and induced by the presence of NM cations coulombically

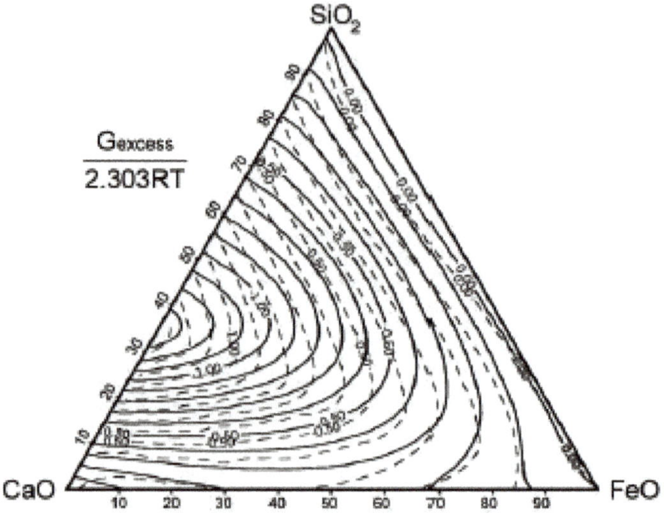

Fig. 19.16 FeO activity surface at 1600 °C for the liquid FeO–SiO_2–CaO ternary system. Graphical representation obtained by inverse-distance contouring of 9900 activity values. Model isoactivity curves (solid lines) are compared with the estimates of Elliott (1955) (dashed lines). Assumed cationic interaction $W_{Ca-Fe} = -33$ *kJ/mol*. Strain energy contributions included. From Ottonello (2001). With permission from the Journal of Non Crystalline Solids

contouring the structons. As such, entropy has not as a "*restoring*" effect, such as in rubberlike materials, but, rather, favors decomposition. Consider for instance a simple dimer, such as that one already represented in Fig. 19.9. The first three optical modes of this linear dimer are RUMs and the second one with $\nu_2 = 75.265\ cm^{-1}$ is particularly a bending motion (Fig. 19.17).

Treating this RUM as a simple Einstein's oscillator we may obtain its contribution to the vibrational entropy of the liquid by applying:

Fig. 19.17 Bending RUM of the $[Si_2O_7]^{6-}$ dimeric unit in a Polarized Continuum with $\varepsilon_0 = 3.8$. The force constant of the RUM is 0.0534 *mDyne/Å* and the reduced mass m^* is 16.0034 *a.m.u.*. Displacement vectors are superimposed to visualize the kind of motion

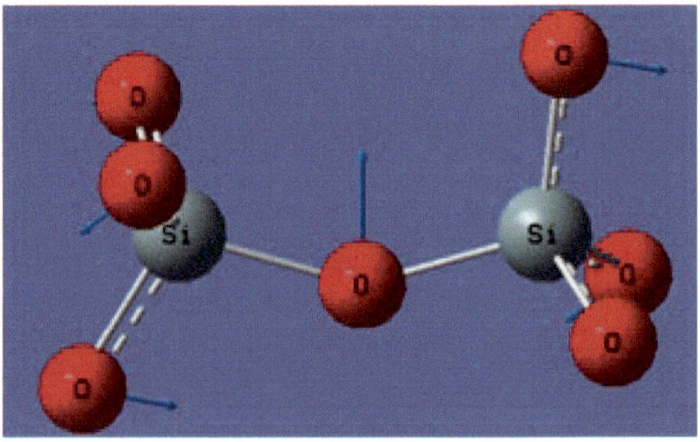

$$S_{\nu_1} = R\left[\frac{X_1}{e^{X_1} - 1} - \ln\left(1 - e^{X_1}\right)\right] \quad (19.4.15)$$

where $X = \frac{2\pi ch}{k_B}$, with c = speed of light, is the a-dimensional frequency. A simple increase of 10 cm^{-1} in its vibrational frequency causes a decrease of −0.52 $J/(mol \times K)$ virtually unaffected by T in the T-range of interest (i.e. from 1000 to 3000 K). Because there is a direct relation between linear frequency ν and molecular force constant; i.e.

$$\Delta F_K = 4\pi^2 m^* \times \left[(\nu + \Delta\nu)^2 - \nu^2\right] \quad (19.4.16)$$

with m* = reduced mass of the molecule, this increase of vibrational frequency may be ascribed to a slight increase in the force constant of the bond F_K (~0.015 $mDyne/\AA$) caused by the stabilizing effect that NM cations have on the low-lying σ-bonding MOs of the oxygen framework, such as the Pauling's double bond of oxygen acting on next-neighbor oxygens in silica and silico-aluminate rings (Gatti et al. 2012), and, most likely, on linear chains as well.

Consider now for instance the $MgO–SiO_2$ join. Figure 19.18 shows the phase diagram obtained for P = 1 bar, through a full ab initio optimization of all the phases nucleating in the system. There is clearly a solvus in the high acidity region

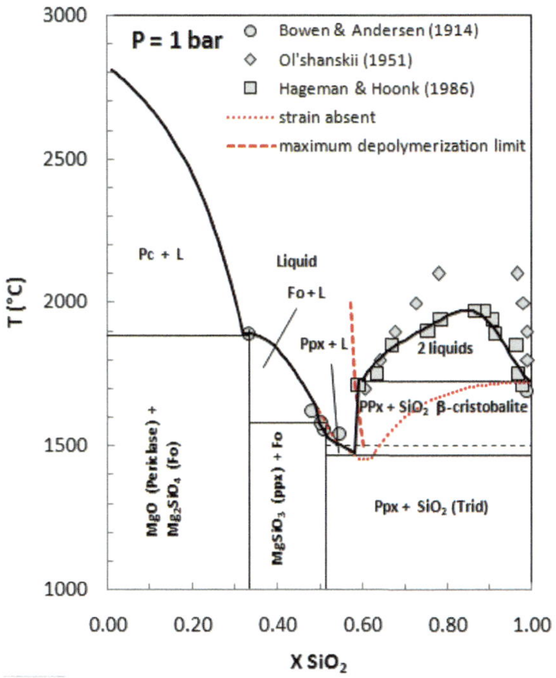

Fig. 19.18 $MgO–SiO_2$. HPA model predictions at room pressure (heavy line) are compared with the experimental observations of Bowen and Andersen (1914), Ol'shanskii (1951) and Hageman and Hoonk (1986). The solvus occurs only in the presence of polymeric species. In the absence of strain energy (red dotted line) the solvus at high SiO_2 content does not occur even at room pressure (Belmonte and Ottonello, in prep.)

(although the experimental evidences are somewhat contrasting). According to our calculations, the solvus extends from $X_{SiO2} \cong 0.59$ to pure X_{SiO2}. The consolute point is at T = 1971 °*C*, $X_{SiO2} \cong 0.84$.

The fact that the decomposition takes place only when polymeric species are present is evident when superimposing the maximum depolymerization limit on the isobaric Schreinemaker's projection at all P (i.e. not only at room pressure), confirming thus that the tendency to un-mixing is a peculiar character of polymer blends also in inorganic systems. It is also evident that, in the absence of strain energy contributions, there is no immiscibility.

The opening of the solvus in Fig. 19.18 is reproduced by a simple Redlich–Kister expansion on X_{NF} weighted on the amount of effective contacts:

$$\Delta G_{strain} = \frac{(O^{-})}{2}\eta \tag{19.4.17}$$

with the η parameter conveniently expanded as a function of the molar fraction of Network Formers in the system (X_{NF}) and intensive variables P, T.

$$\eta = \sum_{k=0} n\eta_k \cdot (2X_{NF} - 1)^n \tag{19.4.18}$$

$$\eta_0 = \eta_{0,H} - T \times \eta_{0,S} + P \times \eta_{0,V} \tag{19.4.19}$$

Two parameters are sufficient to reproduce the solvus: $\eta_{1,S} = 3$ *J/mol*, $\eta_{6,S} = -10$ *J/mol*. Because the increase in Gibbs free energy associated to the vibrational entropy loss is quite limited, the substance remains metastably homogeneous until an intrinsic instability region is attained. The loci of intrinsic instability of the MgO–SiO_2 join are depicted by unitary activity of the SiO_2 component in the liquid with solid state as reference, as schematically shown in Fig. 19.19. This is not surprising, because the energy properties and vibrational features of the silicate liquid indicate that this can be seen as an aggregate of a D_{6h} (α-cristobalite) network and monomers locally ordered in the short-medium range and present in a mutual arrangement lacking of spatial continuity (Ottonello et al. 2010b). A slight increase in energy may thus favor nucleation and growth of crystal seeds.

This simple example tells us that the tendency to un-mixing is a peculiar character of polymer blends also in inorganic systems, whenever there is a high acid–base contrast between the two components in mixture, hence, a high donor pressure. This donor pressure causes an increase in the force constant of RUM bending motions between adjacent monomeric groups, and, consequently, a slight increase in frequency which lowers the vibrational entropy of the substance. Because the ensuing increase in Gibbs free energy is quite limited the substance remains metastably homogeneous until a critical instability region is attained.

Although the simple parameterization adopted to conform the solvus is operatively sound, it is not yet satisfactory from the theoretical point of view, because largely empirical. A better appraisal could be achieved by coupling SCMF

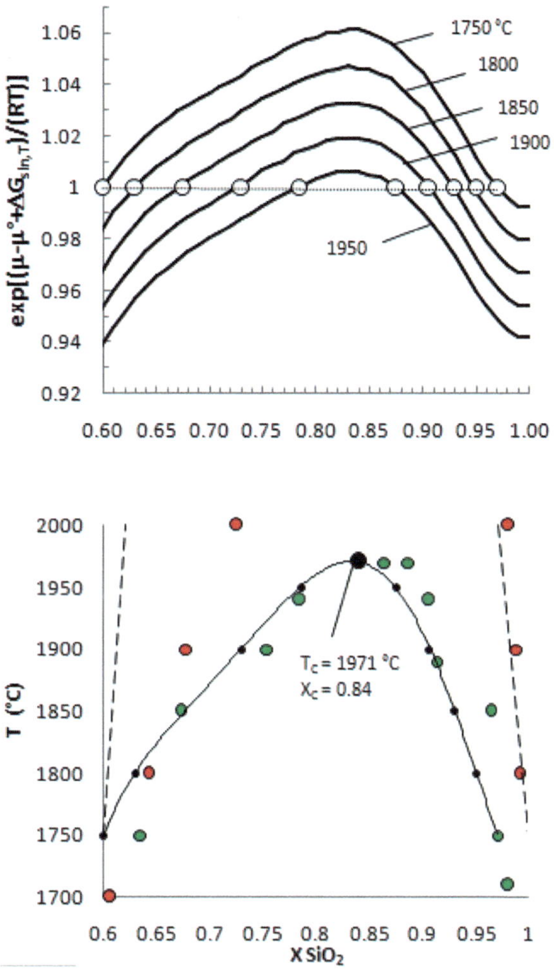

Fig. 19.19 Intrinsic instability range depicted by the loci of unitary activity of SiO_2 in the liquid referred to solid at standard state (upper part of the figure) and projected onto the X-T space (lower part of the figure). The superposition of the experimental observations (Ol'shanskii 1951; red dots; and Hageman and Hoonk 1986; green dots) on the lower part of the figure explains the large biases between the two experimental sets as due to different cooling rates and degree of attainment of the intrinsic instability limit. From Belmonte and Ottonello, in prep.)

procedures with a quantum chemical assessment for this topologically complex region. In fact, the short- and medium-range periodicity in silicate melts and glasses could locally resemble those ones of all-Si zeolites, with cationic "*cages*" of variable dimensions, depending upon silicate network connectivity (Fig. 19.20). In this sort of arrangement, it is still possible to isolate polymeric subunits (inset in Fig. 19.20) of quasi-crystalline short- and medium-range order, but whose relative positioning perturbs the long-range periodicity.

The following working assumptions could be then proposed to render the problem feasible:

- A simple irreducible A-A starblock-like representation of the network is selected as responsible of the short- and medium-range order observed in silicate melts and glasses Fig. 19.14

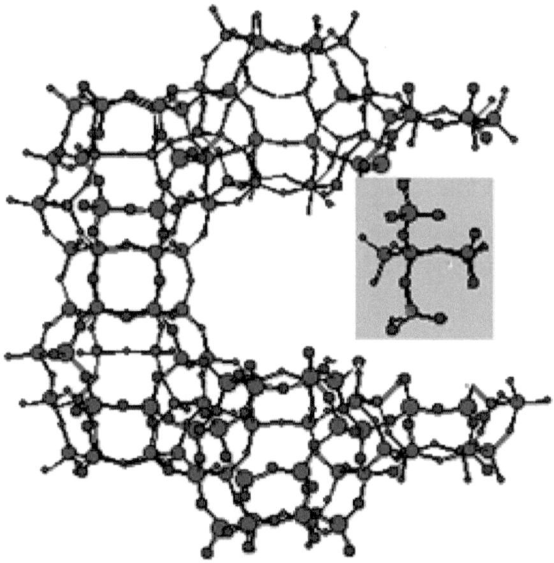

Fig. 19.20 Structure of an all-Si zeolite with "*cages*" of variable dimension and coordination number determined by the SiO_4^{4-} network connectivity. In the inset, the irreducible representation of a "*starblock-like*" copolymer composed of identical interacting A-A monomers of the same Kuhn length is proposed. The χ_{AA} interaction parameter is that one arising from the HPA treatment. The bending between adjacent polymers depends upon the value attained by χ_{AA} and determines the medium range order. Long-range periodicity is locally interrupted by depolymerization and the relative arrangement of starblock-like copolymers contribute eventually to structural instability (and consequent unmixing) through strain energy contributions. From Ottonello (2005)

- It is assumed that in the two-phase region, coexisting liquids mimic locally the structures of crystalline phases observed at the monotectic;
- Basis functions representative of the geometry of fictive intermixing A-A starblocks (or eventually A-B starblocks, in the case of aluminosilicate melts) are assumed to be represented by the Fourier expansions of the spatial groups of the two solid phases at monotectic.

19.5 The SiO_2 and Al_2O_3 Liquids and Their Vitrification

Despite half century of sustained effort, the melting process remains phenomenologically elusive (Cotterill 1980). Concerning more specifically the solids of interest in Earth Sciences, although this is not clearly perceived by the scientific community, even the melting temperatures (T_m) of the most simple substances and the role of pressure in modifying them are poorly known. Even for a substance like MgO, a single cubic solid polymorph at all P conditions, with no solid state transitions, the results found in the literature are controversial, and different theoretical calculations and high-pressure experiments provide melting curves, which can differ from each other up to several hundred *K* (Belmonte 2017; Belmonte et al. 2017b). Basically, although the "*lattice instability*" criterion established by Lindemann more than one century ago (Lindemann 1910; based on a critical amplitude of the thermal motions) is cogent, it cannot bring to concrete quantitative assessments. Better to assume that the thermodynamic and vibrational properties of a liquid (and/or a glass) arise from a mixture of sub-structural entities locally ordered but lacking of spatial continuity (polyamorphism). This idea is not new. A brilliant application of the theory to molten silica has been offered by Kalampounias et al. (2006) on the basis of Raman-mode assignments. To reproduce the observed thermodynamic properties of such a liquid aggregate at the melting point, and to model it through an accurate ab initio investigation of the static and vibrational properties of the substance there are three constraints to obey: mass balance (1), charge balance (2) and energy balance (3):

(1) the overall stoichiometry of the investigated substance must be respected;
(2) the overall charge of the liquid, when composed of the charged "*configurons*" (in the sense of Ojovan 2008), is zero;
(3) the isochoric and isobaric heat capacities of the substance, its enthalpy of formation from the elements and its entropy at T_m must be consistent with the experimental observations.

19.5.1 SiO_2 Melting

As we will see later on when dealing with crystals, silica undergoes, in the P-T range relevant to planetary studies, many solid-state transitions. When the temperature is sufficiently high to attain the liquid state, the intersection of the liquidus with two adjacent primary phase fields generates an invariant point characterized by a "*cusp*". This cusp is theoretically required at the invariant point (Schreinemakers 1915a, b, c), though, in many cases, it is not easily visualized in the experimental data. As an example, we may see in Fig. 19.21 the cusp at the invariant point where liquid, Cohesite and Stishovite coexist stably. Similar cusps, though less marked, appear, at lower P, at the β-Quartz–Coesite limit ($P \approx 4.8$ *GPa*, $T \approx 2500$ *K*) and at the β-Crystobalite–β-Quartz limit ($P \approx 0.5$ *GPa*, $T \approx 2000$ *K*). Cusps at phase boundaries are also observed in chemically complex systems. As shown by Presnall et al. (1979) in a seminal contribution of high importance in Petrology, the cusp at ~0.9 GPa in the CaO–MgO–Al_2O_2–SiO_2 (CMAS) system is most likely responsible of the extensive and continuous melting at ocean ridges (tholeiite melts), thus, it practically shaped a large part of the Earth's surface. Obviously, any sound thermodynamic assessment must generate this feature, and the adoption of first principles in determining the properties of the various aggregation states is

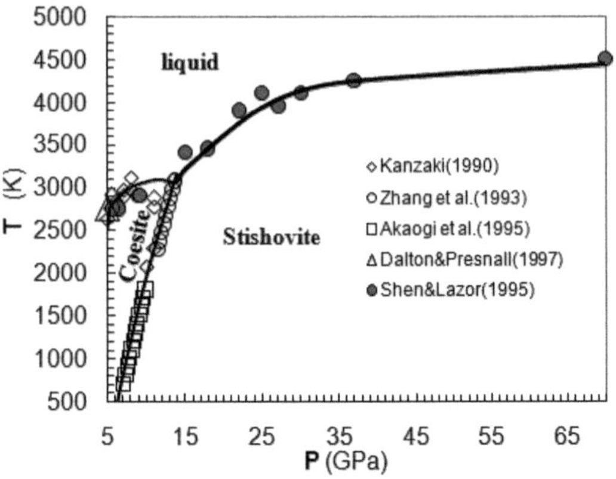

Fig. 19.21 SiO_2 phase diagram between 5 and 70 *GPa*. Melting curves (solid lines) and extrinsic stability limits of Coesite and Stishovite obtained by Belmonte et al. (2017a) are compared with various experimental observations (solid symbols)

almost compulsory. Silica melting at all P conditions is satisfactorily reproduced by the thermophysical-thermodynamic parameters listed in Tables 19.9 and 19.10 (see this purpose Figs. 10, 11 and 12 in Belmonte et al. 2017a).

The assessment resumed in the two tables was largely based on experimental observations but it benefited also substantially from the fact that many of the properties of the nucleating solids were obtained ab initio, and also the vibrational properties of the liquid were obtained from first principles. It is thus of some interest to resume the conceptual framework behind this construction.

19.5.2 SiO_2 Vitrification

The "*structural arrest*" that takes place with decreasing temperatures in the glass transition range implies loss of translational freedom. More precisely, the temperature of this sharp transition is in reality a mode-coupling temperature T_{MC} above which liquid diffusion is no longer ruled by activated jumps between inherent structures, like in solids, but by molecular motion along the unstable directions separating domains at variational state from activated structures, i.e. saddle structures in the interpretation of Grigera et al. (2003). The change from the glassy to the liquid state has obviously profound thermodynamic consequences because it implies cessation of coherent displacement of all atoms in the disordered structure and the onset of translational (and rotational) motions of discrete fragments ("*configurons*" in the sense of Ojovan 2008). If one assumes that the glass transition is the temperature at which translational and rotational motions are frozen in, the heat capacity gap between the liquid and glassy states at T_g should depend in a simple way upon the level of fragmentation, on the adjustments of centers of atomic vibration associated to local strain and on the onset of coherent displacements (i.e. acoustic modes). Fragments of particular relevance are, in this sense, besides the fundamental $[SiO_4]^{4-}$ block and oxygen in various ionization states, the $[SiO_2]^{\circ}$

Table 19.9 Thermodynamic properties of SiO_2 liquid. $H^0_{f,298}$ = enthalpy of formation from the elements at $T_r = 298.15$ *K*, $P_r = 1$ *bar*; S^0_{298} = standard state entropy ($T_r = 298.15$ *K*, $P_r = 1$ *bar*); C_P = isobaric heat capacity ($C_p = a + bT + cT^{-2} + dT^{-1/2} + eT^{-3} + fT^2 + gT^3 + hT^{-1}$). Belmonte et al. (2017b)

Component	$H^0_{f,298}$ (*J/mol*)	S^0_{298} (*J/mol × K*)	a	b × 10^3	c × 10^{-5}	d	e × 10^{-8}	f × 10^6	g × 10^9	h × 10^{-4}
SiO_2	−911,746.2	33.887	88.455	−3.00137	−48.527	−114.33	7.2829	0.71332	0.0059239	0.0

Table 19.10 Thermophysical properties of liquid SiO_2. V^0_{298} = standard state molar volume (T_r = 298.15 *K*, P_r = 1 *bar*); K_0 = bulk modulus at the athermal limit (i.e. T = 0 *K*; P = 0 *GPa*); $K'_0 = (dK/dP)_0$ = pressure derivative of the bulk modulus (assumed independent on T); $(dK/dT)_P$ = temperature derivative of the bulk modulus; α_V = volume thermal expansion coefficient ($\alpha_V = \alpha_0 T + \alpha_1 + \alpha_2 T^{-1}$). Belmonte et al. (2017a)

Component	V^0_{298} (*cc/mol*)	K_0 (*GPa*)	K'_0	$(dK/dT)_P$ (*bar/K*)	$\alpha_0 \times 10^7$ (K^{-2})	$\alpha_1 \times 10^7$ (K^{-1})	$\alpha_2 \times 10^3$
SiO_2	26.0	5.0	12.8	0.0	−0.3611	843.1	0

molecule, $[Si_nO_{3n}]^{(2n)-}$ polymeric rings up to n = 6, $[Si_nO3_{n+1}]^{(2n+1)-}$ chains up to n = 3 and the $[(SiO_2)_4]^\circ$ D_{6h} arrangement, which yields the tetragonal α-cristobalite polymorph by symmetry operations (Ottonello et al. 2010). Calculations were thus carried out at DFT/B3LYP theory level in order to achieve an appropriate knowledge of their vibrational behavior. The adopted basis set was 6-31G+(d,p) with polarized and diffuse functions for anions and 6-31G(d,p) for neutral solvated complexes. The $[(SiO_2)_4]^\circ$ D_{6h} unit was resolved, within the LCAO approximation and in the framework of the Bloch's theorem (see Sect. 8.4), with the same basis set previously adopted for another crystalline polymorph of SiO_2 [Ottonello et al., 2009; i.e. a 88-31G(d) contraction for silicon and a 8-411(d) set for oxygen, as proposed by Nada et al. (1996)]. Solvation energies were obtained at IEFPCM theory level. The solute molecules were embedded in cavities of molecular shape surrounded by a continuum dielectric medium (with dielectric constant $\varepsilon = 3.8$ and solvation radius b_{solv} = 1.385 *Å*), The solvent accessible surface (SAS) was built up adopting UA0 atomic radii (see Sect. 7.7.3). The vibrational contributions of all optical modes to the isochoric heat capacity of the substance were assumed to be due to simple Einstein oscillators. The contribution of acoustic modes was estimated from longitudinal and transversal sound velocities through the sine-wave dispersion relation anticipated by Kieffer (1979a, b, c). It was finally assumed that the collective modes observed by Benassi et al. (1996), which propagates through the glass with a velocity of 5.8 *km/s*, represent an excess phonon mode travelling through the Brillouin zone with a sine-wave dispersion similar to those of the other acoustic modes and attaining a maximum frequency (ν_{excs}) at the limit of the Brillouin zone (K_{max}).

Background Information: Structure Energy Details

Structural and energetic details of the investigated clusters are resumed in Table 19.11 Among the various molecules investigated, the most stable in the reaction field are the fundamental block $[SiO_4]^{4-}$ and the dimeric ring $[Si_2O_6]^{4-}$ (Table 19.11). This latter has a wider Si–O°–Si angle than expected (87°) due to attractive terms acting between neighboring Si atoms, which contribute substantially to stabilize the molecule in the reaction field. Disregarding $[Si_2O_6]^{4-}$, the relative stability follows the trend already

Table 19.11 All electron DFT geometry and energy (electronic, zero point correction and thermal corrections to internal energy U, enthalpy H and Gibbs free energy (G) of the unit cell $[(SiO_2)_4]^0$ D_{6h} (B3LYP/LCAO) and of monomeric and chain species in a reaction field of permittivity $\varepsilon = 3.8$ (B3LYP/IEFPCM). Solvent cavities generated with the Unified Atom Topological Model. NBO = Non-Bridging Oxygen; O° = bridging oxygen. Values of the vibrational, rotational, and translational contributions to the isochoric heat capacity and entropy of the solute are also listed. The numeral density is the number of moles of solute per unit volume (n_s). The liberation free energy is $RT\ln(n_s/n_g)$ with $n_g = 4.034 \times 10^{-4}$ (*bar/J*) numeral density in the ideal gas. The Gibbs free energy of formation of the solute species is obtained by summing up the various solvation terms with the Gibbs free energy of the cluster at variational state in the reaction field. (*) periodic structure (Ottonello et al. 2010b)

Cluster	$[SiO_2]^0$	$[SiO_4]^{4-}$	$[Si_2O_7]^{6-}$	$[Si_2O_4]^0$	$[Si_2O_6]^{4-}$	$[Si_3O_6]^0$	$[Si_3O_9]^{6-}$	$[Si_4O_{12}]^{8-}$	$[Si_5O_{15}]^{10-}$	$[(SiO_2)_4]^0$ D_{6h}
Basis set	6-31G(d,p)	6-31+G(d,p)	6-31+G(d,p)	6-31G(d,p)	6-31+G(d,p)	6-31G(d,p)	6-31+G(d,p)	6-31+G(d,p)	6-31+G(d,p)	88-31G(d)[Si]–8-411(d)[O]
Molecular mass (a.m.u.)	59.967	91.957	167.918	119.925	151.923	179.9	227.885	303.847	379.808	239.851
SAS-delimited volume (Å^3)	60.001	85.106	149.949	102.175	128.044	147.2	187.9	253.885	323.626	–
numeral density (*bar/J*)	0.273	0.024	0.013	0.018	0.014	0.012	0.01	0.007	0.006	0.009
I_x (a.u.)	0	437.7	863.6	1596.3	595.6	1561.5	2631.6	4901.9	8097.3	–
I_z	264.2	437.7	2788.5	1123.6	1538.4	1561.7	2632.7	4905.3	8102.6	–
I_z	264.2	437.7	2789.6	1283.2	1804.9	3123.2	4005.2	8113.4	–	–
Si-NBO	1.521	1.695	1.677	1.517	1.626	1.518	1.624	1.65	1.655	–
Si-O°	–	–	1.726	1.68	1.75	1.641	1.719	1.699	1.699	1.625
NBO-Si-O°	–	–	109	135	113	127	110	110	110	–
NBO-Si-NBO	–	109	110	180	115	–	113	111	110	–
O°-Si-O°	–	–	–	89	87	107	103	107	106	108–146
Si-O°-Si	–	–	177	91	93	134	137	163	178	94–125
Si-Si-Si	–	–	–	–	–	60	60	90	108	109–112
Dihedral Si-O°-Si-O°	–	–	–	0	0	0	0	0	0	51

(continued)

Table 19.11 (continued)

Cluster	$[SiO_2]^0$	$[SiO_4]^{4-}$	$[Si_2O_7]^{6-}$	$[Si_2O_4]^0$	$[Si_2O_6]^{4-}$	$[Si_3O_6]^0$	$[Si_3O_9]^{6-}$	$[Si_4O_{12}]^{8-}$	$[Si_5O_{15}]^{10-}$	$[(SiO_2)_4]^0$ D_{6h}
E_{el} *(Hartree)*	−439.92713	−589.6279	−1104.0939	−880.01114	−1030.0048	−1320.0773	−1544.1622	−2058.6543	−2572.618	−1760.50056
ZPC *(Hartree)*	0.006098	0.011	0.021923	0.01737	0.021907	0.02762	0.0356	0.044222	0.055179	0.045091
Therm U *(Hartree)*	0.009459	0.015861	0.031678	0.023198	0.029479	0.036848	0.047151	0.061468	0.077204	0.0085386
Therm H	0.010403	0.016805	0.032622	0.024142	0.030423	0.037792	0.048096	0.062412	0.078148	0.0085426
Therm G	−0.013696	−0.014878	−0.014	−0.012862	−0.010439	−0.007964	−0.002096	−0.001631	0.003485	−0.0061903
$S_{vib,298.15}$ $[J/(mol \times K)]$	7.6	26	115.4	44.7	67.5	102.2	132.9	243.2	327.5	129.7
$S_{rot,298.15}$	44.8	87.8	126.7	112.7	120.9	127.2	132.6	140.7	147.2	–
$S_{tra,298.15}$	159.8	165.1	172.6	168.4	171.4	173.5	176.5	180	182.8	–
$C_{V,vib,298.15}$ $[J/(mol \times K)]$	9.4	44.4	115.8	53.6	89.8	103.4	153.1	231.8	303.3	152.7
$C_{V,rot,298.15}$	12.5	12.5	12.5	12.5	12.5	12.5	12.5	12.5	12.5	–
$C_{V,tra,298.15}$	12.5	12.5	12.5	12.5	12.5	12.5	12.5	12.5	12.5	–
$H°_{f\ gas\ 298.15}$ *(kJ/mol)*	−185.2	1472	2958.8	−773.6	38.2	−1315.2	2456	3870.3	6657.9	−3765.7*
$S°_{gas\ 298.15}$ $[J/(mol \times K)]$	212.2	279	414.8	325.9	359.8	402.9	442	564	657.5	129.7*
$G°_{f\ gas\ 298.15}$ *(kJ/mol)*	−181.7	1541.7	3098	−737.2	150.6	−1235	2653.9	4141.7	7029.2	−3537.3*
ΔG_{electr} (kJ/mol)	−29.7	−2718.9	−4935.2	−49.4	−2374.9	−86.4	−4831.2	−7352.8	−10,495.3	–
ΔG_{cav}	39.2	54.6	89	59.3	74.4	80.4	103.5	136.9	171.5	–
ΔG_{rep}	1.5	2.7	3.9	2.4	3.4	3.1	4.4	5.4	6.4	–
ΔG_{disp}	−14.7	−26.1	53.6	−25.1	−34.5	−33.8	−46.2	−57.3	−67.9	–
ΔG_{lib}	16.2	10.1	8.6	9.5	8.9	8.5	7.9	7.2	6.6	7.7
$G°_{f,298.15}$ *(kJ/mol)*	−185.4	−1145.9	−1690.6	−749.9	−2180.9	−1263.2	−2115.7	−3126	−3356.2	−3537.3

envisaged by Galeener (1982). Puckering is observed only for the five-membered ring whereas the remaining forms are virtually planar. Increasing the number of Si atoms in the chains results into a substantial instability in the reaction field because for Si > 2 the chain breaks during the Berny optimization procedure and a variational state is never achieved. The relative stability of the various forms is substantially enhanced by electrostatic interaction with the continuum which is obviously the more severe the higher the number of unsaturated bonds. The energy spent in forming the cavity and the repulsion term are highly subordinated in magnitude and virtually counterbalanced by dispersive contributions. The liberation free energy (i.e. the term arising from reduction of the space available to the moiety in the continuum with respect to the gaseous state) is almost negligible. The contributions of the internal modes of the various clusters to the isochoric heat capacity between 298.15 and 3000 *K* are listed in Table 19.12 together with the acoustic terms calculated with a sine-wave dispersion relation by assuming for the glass transverse velocities $v_{s1} = v_{s2} = 3.74$ *km/s*, and a longitudinal velocity $v_p = 5.5$ km/s[21–23] at an internal freedom 3n − 3 + 1 = 85. The contribution of the additional acoustic-like mode traveling at $v_{exc} = 5.8$ *km/s* (Benassi et al. 1996) is also listed, together with translational and rotational terms for non-linear clusters. At all temperatures of interest, the difference between the isobaric and isochoric heat capacities is negligible for either the glass or the liquid state, owing to small isobaric thermal expansion coefficient α and low isothermal bulk modulus K in both aggregation states (Bucaro and Dardy 1976; Krol et al. 1986; Kress and Carmichael 1991; Meade and Jeanloz 1987).

The computed isochoric heat capacity arising from internal modes, when normalized on a 1 Si atom basis (Fig. 19.22), increases regularly with the number of Non-Bridging Oxygens (NBO) in the molecule, due to the substantial contributions of high-frequency stretching motions. Disregarding the contributions of acoustic terms and anharmonicity one could hypothesize that the fundamental block $[SiO_4]^{4-}$ and the two-membered ring $[Si_2O_6]^{4-}$ prevail in the glass simply because they almost overlap to the observed calorimetric heat capacity (Fig. 19.22). Considering, however, the Gibbs free energy of formation of these two moieties from the elements in their standard state, it is readily seen that these are excessively stable in the reaction field, due, among various contributions, to a conspicuous electrostatic interaction with the continuum, so that the cluster $[(SiO_2)_4]^0$, representative of a D_{6h} network has instead the most appropriate energy.

The highest frequency modes observed in glass are a Raman-active mode (R) at 1200 cm^{-1} followed by two infrared-active (IR) modes at 1184 and 1084 cm^{-1} and a R mode at 1055 cm^{-1} (Flubacker et al. 1959). The low-frequency range is covered by an intense Raman continuum extending from 560 cm^{-1} down to 8 cm^{-1} with intensity maxima at 490, 440, and 48 cm^{-1}. In the intermediate frequency range, there is a coupled IR, R mode

Table 19.12 Optical (opt), acoustic (ac), excess (exc), translational (trasl), rotational (rot), and anharmonic (C_P-C_V) contributions to the heat capacity of liquid and glass at some selected temperatures (see Ottonello et al. 2010b; for a complete outline). The excess term present at the glass state arises from a sine wave dispersion treatment of the 5.8 *km/s* traveling wave (Benassi et al. 1996). The corresponding cutoff frequency at the limit of the Brillouin zone depends upon the actual extension of coherent blocks. The listed rotational contributions apply to all clusters but the linear form $[SiO_2]^\circ$ for which a 2R value C_V = 8.315 [*J/(mol* × *K)*] is appropriate. The anharmonicity contribution α^2KVT corresponding to the difference between the isobaric and isochoric heat capacities of the substances is negligible and is listed only for completeness

Cluster	$[SiO_2]^0$	$[SiO_4]^{4-}$	$[Si_2O_4]^\circ$	$[Si_2O_6]^{4-}$	$[Si_2O_7]^{6-}$	$[Si_3O_6]^0$	$[Si_3O_9]^{6-}$	$[(SiO_2)_4]^0$	$[Si_4O_{12}]^{8-}$	$[Si_5O_{15}]^{10-}$	Glass	Glass	Liquid	Liquid	Glass	Liquid
T	$C_{V,opt}$	$C_{V,opt}$	$C_{V,opt}$	$C_{V,opt}$	$C_{V,opt}$	$C_{V,opt}$	$C_{V,opt}$	$C_{V,opt}$	$C_{V,opt}$	$C_{V,opt}$	$C_{V,ac}$	$C_{V,exc}$	$C_{V,trasl}$	$C_{V,rot}$	$(C_P-C_V) \times 10^5$	$(C_P-C_V) \times 10^5$
10	0	0	0	0	1.268	0.006	0.008	0.101	3.572	4.663	0.098	0.009	12.472	12.472	0.000	0.000
100	3.049	5.395	16.553	22.236	40.07	40.075	47.915	45.732	89.956	123.179	3.289	1.147	12.472	12.472	0.065	0.196
200	6.734	27.501	37.79	61.335	84.817	79.159	109.826	107.594	175.413	232.152	3.492	1.169	12.472	12.472	0.13	0.393
298.15	9.394	44.397	53.56	89.792	115.826	103.406	153.147	152.733	231.768	303.306	3.531	1.169	12.472	12.472	0.194	0.585
500	14.731	60.981	73.978	120.6	147.188	134.26	201.566	208.908	292.093	378.533	3.552	1.183	12.472	12.472	0.326	0.981
1000	21.163	70.929	91.259	141.118	166.757	161.16	235.135	253.58	332.393	428.308	3.56	1.186	12.472	12.472	0.652	1.963
1500	23.108	73.056	95.75	145.739	171.023	168.241	242.846	264.653	341.487	439.495	3.562	1.187	12.472	12.472	0.979	2.944
2000	23.877	73.825	97.46	147.43	172.572	170.943	245.683	268.809	344.817	443.588	3.563	1.187	12.472	12.472	1.305	3.925
2500	24.251	74.185	98.278	148.227	173.3	172.237	247.021	270.785	346.385	445.515	3.563	1.187	12.472	12.472	1.631	4.907
3000	24.459	74.382	98.73	148.663	173.697	172.952	247.755	271.873	347.245	446.571	3.563	1.188	12.472	12.472	1.957	5.888

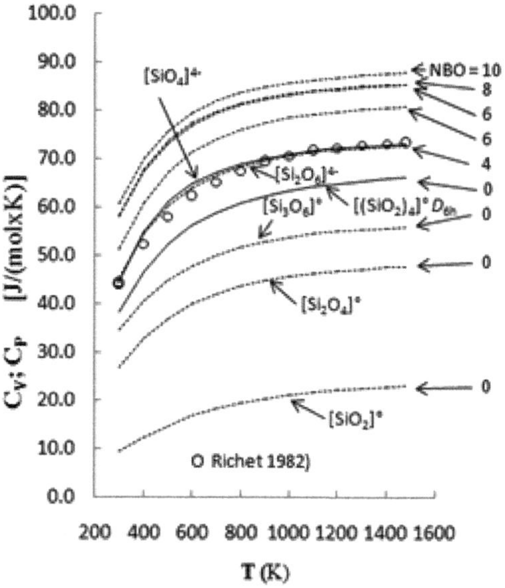

Fig. 19.22 Isochoric-optic heat capacity of the investigated clusters on 1 Si atom basis, compared with the isobaric heat capacity of the glass. NBO is the amount of Non-Bridging Oxygens in the molecule. Modified from Ottonello et al. (2010b)

at 804 cm^{-1}, an IR mode at 695 cm^{-1} and a R mode at 605 cm^{-1}. Table 19.13 lists the vibrational frequencies obtained by Ottonello et al. (2010b) on the various clusters.

Although high-frequency Si–O stretching modes are present in all clusters, they attain a frequency compatible with what observed only for the rigid D_{6h} network, and in agreement with experimental (Kalampounias et al. 2006) and computational investigations (Pasquarello and Car 1998; Umari et al. 2003), for three- and four-membered rings. The close similarity of the high-frequency mode distribution of cristobalite and glass was recognized long ago, although its significance has been largely debated (see Kieffer 1979a; Flubacher et al. 1959; and references therein). Concerning the low-frequency zone it may be noted that modes capable of generating substantial deviations from the Debye T^3 law in the phonon dispersion spectrum are present in D_{6h} with two R-active modes at 60.8 and 123 cm^{-1}. Low frequency (predominantly IR-active) modes associated with Rigid Unit Modes (RUM; i.e. virtually coherent motions of the tetrahedra, evidenced in bold in Table 19.13) or to tilting of the NBOs with respect to the plane of the ring are present in the $[Si_2O_7]^{6-}$ dimeric chain, as anticipated by Buchenau et al. (1986) on the basis of elastic and inelastic neutron scattering experiments, and in all ring structures with Si > 3. Their frequency tends to decrease with an increase of the number of Si atoms in the ring, due to progressive lowering of the relevant force constant and attains values as low as 23.5 and 26 cm^{-1} in the four- and five-membered ring, respectively.

Although some features of the vibrational spectrum of glass lead several authors to adopt the $[SiO_4]^{4-}$ cluster as the fundamental vibrating unit, at least

Table 19.13 Calculated B3LYP/PCM wavenumbers (cm^{-1}) of the various clusters, arranged in order of increasing frequency. Modes listed in roman are IR-active, those in italics are Raman-active, those in roman and underlined are predominantly IR-active, those in italics and underlined are predominantly Raman-active. Frequencies in bold are those ones that may be referred to Rigid Unit Modes (RUM). (Ottonello et al. 2010b)

Mode	$[SiO_2]^0$	$[SiO_4]^{4-}$	$[Si_2O_4]^0$	$[Si_2O_6]^{4-}$	$[Si_2O_7]^{6-}$	$[Si_3O_6]^0$	$[Si_3O_9]^{6-}$	$[(SiO_2)_4]^0$	$[Si_4O_{12}]^{8-}$	$[Si_5O_{15}]^{10-}$
v_1	253.1	*323.0*	115.3	**127.8**	**36.2**	90.3	**85.4**	*60.8*	**23.5**	**26.0**
v_2	*983.8*	*323.0*	230.2	178.7	**75.3**	*92.6*	**95.9**	*123.0*	**65.5**	**33.2**
v_3	1440.0	441.9	*251.8*	*272.6*	**97.6**	101.2	**104.4**	155.7	*71.8*	**59.8**
v_4		441.9	*300.4*	305.4	*213.2*	231.2	*221.6*	155.7	***78.0***	*77.2*
v_5		441.9	427.5	*312.6*	*251.4*	231.6	239.6	*220.5*	***88.1***	***80.8***
v_6		*651.8*	*481.0*	328.2	*259.1*	294.6	*240.9*	275.0	**103.7**	***98.2***
v_7		734.9	*687.2*	391.1	320.7	295.6	275.7	275.0	*140.0*	***102.0***
v_8		734.9	767.0	*409.7*	325.3	*316.5*	278.7	288.2	211.5	**116.0**
v_9		734.9	*850.9*	473.8	*413.3*	*317.1*	*287.6*	296.8	215.5	***117.2***
v_{10}			886.2	*514.9*	*415.5*	320.0	*290.5*	*367.8*	216.4	*150.4*
v_{11}			1290.8	*570.4*	434.1	391.9	*292.2*	*381.1*	*224.1*	*178.2*
v_{12}			*1336.2*	695.9	435.1	453.8	*414.4*	386.2	252.2	*179.1*
v_{13}				*701.4*	439.1	595.4	*416.4*	386.2	*257.4*	190.6
v_{14}				714.4	*510.8*	720.8	437.9	*417.9*	*268.6*	190.9
v_{15}				827.0	679.9	720.9	470.7	*452.3*	296.3	219.8
v_{16}				*918.5*	*761.6*	991.4	471.3	477.6	297.0	*247.4*
v_{17}				*929.1*	772.0	1028.4	477.2	477.6	331.6	*248.7*
v_{18}				944.6	773.5	1028.5	*495.4*	510.5	353.3	297.2
v_{19}					782.1	1298.5	*553.4*	619.9	*372.8*	298.2
v_{20}					784.0	1298.9	651.7	619.9	*408.8*	314.6
v_{21}					843.2	*1304.6*	653.2	*777.4*	*409.4*	315.8
v_{22}							*805.5*	784.0	415.4	322.5

(continued)

Table 19.13 (continued)

Mode	$[SiO_2]^0$	$[SiO_4]^{4-}$	$[Si_2O_4]^0$	$[Si_2O_6]^{4-}$	$[Si_2O_7]^{6-}$	$[Si_3O_6]^0$	$[Si_3O_9]^{6-}$	$[(SiO_2)_4]^0$	$[Si_4O_{12}]^{8-}$	$[Si_5O_{15}]^{10-}$
ν_{23}							856.0	*784.3*	419.5	324.6
ν_{24}							856.5	789.3	433.7	*352.7*
ν_{25}							907.8	789.3	437.8	*381.1*
ν_{26}							908.2	*1073.0*	438.5	381.7
ν_{27}							*915.4*	1076.8	*491.5*	408.1
ν_{28}							972.0	1076.8	549.0	*418.6*
ν_{29}							972.7	*1077.2*	551.4	*419.0*
ν_{30}							978.7	1082.0	*648.7*	*423.7*
ν_{31}								*1133.5*	*763.3*	423.9
ν_{32}								1200.8	810.8	433.6
ν_{33}								1200.8	810.9	434.0
ν_{34}									*843.3*	434.3
ν_{35}									847.7	470.5
ν_{36}									*848.0*	516.0
ν_{37}									*859.5*	516.7
ν_{38}									862.1	*618.2*
ν_{39}									864.4	*619.0*
ν_{40}									938.7	*762.4*
ν_{41}									939.0	*762.6*
ν_{42}									952.5	809.5
ν_{43}										809.8
ν_{44}										824.6
ν_{45}										825.1
ν_{46}										*831.8*

(continued)

Table 19.13 (continued)

Mode	$[SiO_2]^0$	$[SiO_4]^{4-}$	$[Si_2O_4]^0$	$[Si_2O_6]^{4-}$	$[Si_2O_7]^{6-}$	$[Si_3O_6]^0$	$[Si_3O_9]^{6-}$	$[(SiO_2)_4]^0$	$[Si_4O_{12}]^{8-}$	$[Si_5O_{15}]^{10-}$
v_{47}										*831.9*
v_{48}										846.1
v_{49}										847.8
v_{50}										860.4
v_{51}										926.6
v_{52}										927.0
v_{53}										*957.2*
v_{54}										958.5

in the intermediate-high frequency range (Gaskell 1970; Bell et al. 1968; Bell and Dean 1970), some of these features, such as the high-frequency modes at 1200 cm^{-1}, were observed only in the D_{6h} network and only the triply degenerate IR modes of the $[SiO_4]^{4-}$ cluster at 442 and 735 cm^{-1} may be tentatively assigned to the 468 cm^{-1} IR mode experimentally seen in glass, and to the 695 cm^{-1} IR band, respectively. The outlined energy properties and vibrational features suggested that none of the investigated clusters, when taken alone, justifies the observed energy and vibrational properties of the silica glass, whereas an aggregate of the D_{6h} network and the monomer locally ordered in the short-medium range is rather present in the glass, in a mutual arrangement lacking of spatial continuity.

19.5.3 The SiO_2 Glass at Low T

Based on the above outlined thermodynamic assessment, the glass inherits the vibrational properties of the fundamental block $[SiO_4]^{4-}$ and of the tridimensional network represented by the $[(SiO_2)_4]^0$ D_{6h} unit. If the two moieties are present in a 1:2 proportion, the bulk internal freedom is 85, corresponding to 81 internal modes, with four additional coherent motions represented by the two degenerate transversal acoustic waves traveling at 3.74 *km/s*, a longitudinal acoustic wave propagating at 5.5 *km/s* and the coherent (excess) mode at 5.8 *km/s*. At low T, the bulk heat capacity is dominated by the acoustic terms while optical contributions are less determinant. The analysis of acoustic modes in this T-region is thus particularly stringent. The resulting heat capacity at low T fits rather well with the measurements of Flubacher et al. (1959) and Westrum (1956) and almost overlaps with the cubic-spline refinement of Richet (1982) (Fig. 19.23).

The calorimetric Debye temperature indicates, however, that the acoustic phonons are only partly responsible for the low-T heat capacity. A comparison with analogous calculations made by Kieffer (1979c) adopting the paracrystalline model (PD) of Robinson (1965),[5] where the contribution of acoustic modes is by far subordinated with respect to the approach of Ottonello et al. (2010b),[6] indicates in fact that the calorimetric behavior is substantially controlled by internal modes at very low temperatures as well and that the underestimation of PD at low T arises most likely from the adopted conformation of the optical branches. The intense continuum observed by Flubacher et al. (1959) in the Raman spectra of vitreous

[5] A structure composed of 225 puckered pentagonal dodecahedra corresponding to 1125 SiO_2 molecules.

[6] In the PD model, the contribution of acoustic modes to the total freedom is 8.9×10^{-4}, while in the approach of Ottonello et al. (2010b) is 4.8×10^{-2}.

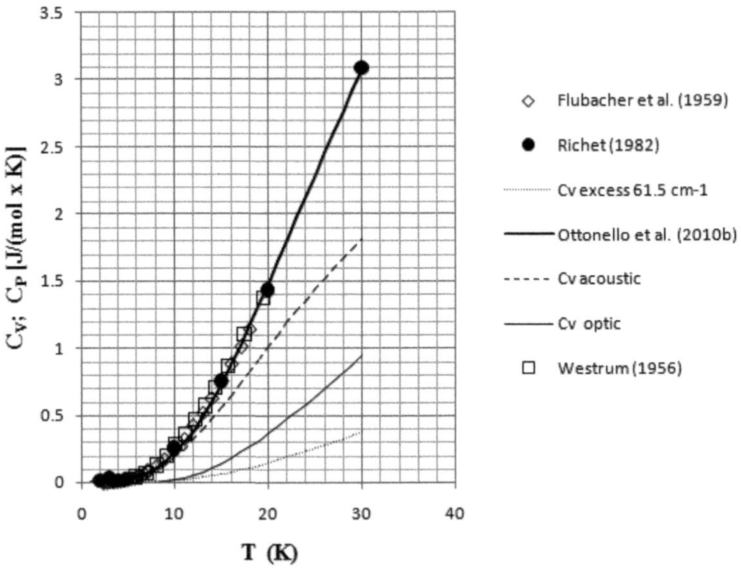

Fig. 19.23 Ab initio low-*T* heat capacity of SiO_2 (Ottonello et al. 2010b) compared to experimental observations. Acoustic and excess phonon contributions are also shown. The data of Richet (1982) were sampled from the cubic-spline refinements of all existing low-T data (Table 5.16 in Richet 1982)

silica (from 560 to 8 cm^{-1}) was interpreted by them as due to optical modes of very low frequency. On the other hand, inelastic neutron scattering measurements led Leadbetter (1969) to assign the modes below 40 cm^{-1} to broadened acoustic branches. This observation, and the fact that the differential cross section for neutron scattering is almost identical in cristobalite and silica glass (Leadbetter 1969), suggested to Kieffer (1979c) that above 50 cm^{-1} "*the optic modes merge with and are indistinguishable from the acoustic modes*". The lower limit of the optical continuum was thus placed by Kieffer (1979c) at 13.5 cm^{-1}, while the lowest optical branch observed by Ottonello et al. (2010b) for the D_{6h} cluster is at 60.8 cm^{-1}.

In the calculations of Ottonello et al. (2010), the overlap of acoustic and optical branches is preserved. Assuming the range of coherence to be coincident with the assemblage composed by two units D_{6h} and one $[SiO_4]^{4-}$ in a variable bonding state, the cutoff frequency of transversal branches at the Brillouin zone boundary was placed by the Authors at 34.1 cm^{-1}, the longitudinal branch at 50.1 cm^{-1} and the excess phonon at 52.8 cm^{-1}. To explain the downward extension of the spectrum to 8 cm^{-1} one must admit, however that, at low T, the coherence zone extends to islands composed of at least 6000 atoms, thus almost duplicating the primitive cell of the PD model. This conceptual model implies, however, partial elimination of the mode degeneracy (eventually associated with local strain). Under these conditions, the contribution of acoustic branches to the bulk isochoric heat capacity of the substance would be negligible at all T.

19.5.4 The Boson Peak

A plot of the low-T heat capacity normalized to T-cubed (Fig. 19.24) reveals with great detail the deviations from the Debye T^3 law (Richet 2009; Richet et al. 2010). This excess heat capacity, usually denoted as "*boson peak*" (Buchenau et al. 1986; Zotov et al. 1999) because of an intensity distribution reproduced approximately by Bose–Einstein statistics, has long been known (Anderson 1959). The form of the Boson peak is highly dependent upon minimal variations in the measured calorimetric C_P. Based on the experimental observations, the maximum in silica glass is placed at about 10 K (Richet 2009; Richet et al. 2010) and the maximum deviation attains 2.4×10^{-4} $J/(mol \times K)$. The Boson peak is confined toward the athermal limit by a sudden rise of the excess heat capacity, as evident when coupling the low-T data refinement of Richet (1982) with the observations of Hornung et al. (1969). According to Kieffer (1979c), the linear term one obtains when deconvoluting the excess heat capacity as a summation of two terms (the first one linear with T and the second one cubic with T) can be derived from a DOS spectrum of phonon excitations, which is independent of energy. According to Sette et al. (1998), the phonon peak is actually due to the excess collective mode, propagating as a sound wave traveling at 5.8 *km/s*, which can be treated as an acoustic term with a sine-wave dispersion relation. The obtained acoustic excess heat capacity matches quite well the position of the maximum (Fig. 19.24) and, more importantly, justifies, when coupled with the other acoustic modes, the sudden rise-up experimentally observed at low T and corresponding to the linear part of the T-dependent excess. The contribution of this mode, when taken alone, is insufficient, however, to explain the observed intensity of the deviation.

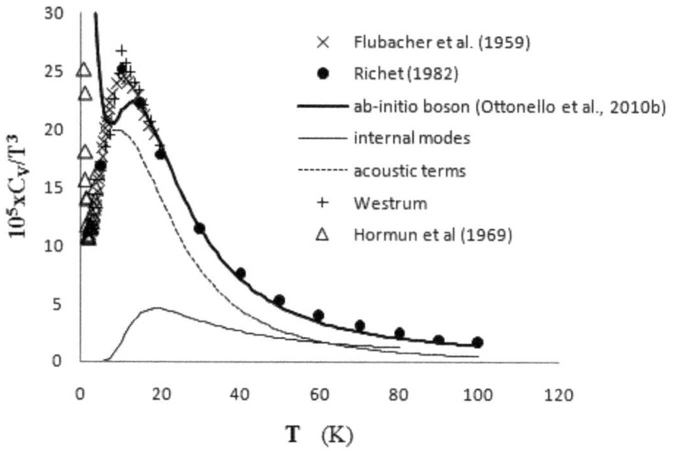

Fig. 19.24 Ab initio C_V/T^3 model values compared with low-T C_P/T^3 observations. In the T-range of interest, the difference between isochoric and isobaric terms is practically negligible. Modified from Ottonello et al. (2010b)

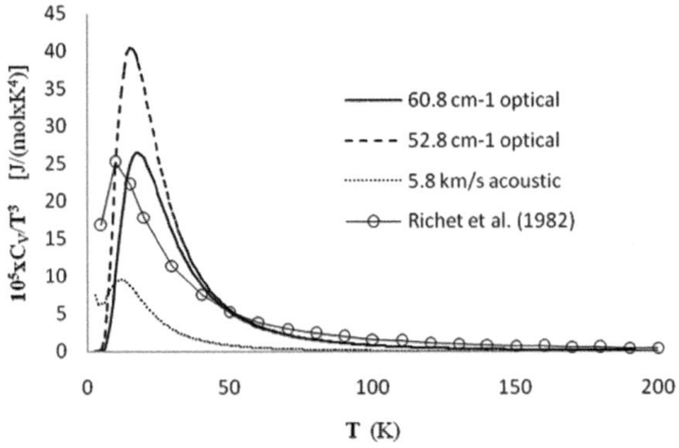

Fig. 19.25 Individual contributions to the excess heat capacity. The Einstein oscillators give an excess contribution with a maximum shifted at higher T with respect to the observed boson peak. A sine-wave dispersed acoustic mode fits the position but explains only 1/3 of the observed intensity. The excess heat capacity arising from the cubic-spline refinement of experimental data made by Richet (1982) is shown for comparative purposes. From Ottonello et al. (2010b). With Permission from the American Institute of Physics

In Fig. 19.25, we see that, on the contrary, the lowest optical mode of the D_{6h} structure explains, when taken alone, the whole intensity of the boson peak, but not the position.[7]

19.5.5 The Glass Transition

Richet et al. (1982) performed an accurate investigation of amorphous silica by drop calorimetry between 900 and 1880 *K*. A functional form of the isobaric heat capacity was later obtained by Richet and Bottinga (1984) with a simultaneous fit of C_P and relative enthalpy ($H_T - H_{Tref}$) (glass) or by partial differentiation of relative enthalpy (liquid). The glass transition was identified by a sudden change in the slope of the fitted enthalpy curve (Fig. 1 in Richet and Bottinga 1984) and occurs at T_g = 1480 *K*. At this T one observes a heat capacity gap corresponding to 0.959R. This gap (0.925R in the ab initio investigation; Fig. 19.26) is due to cessation of translational and rotational motions below T_g, partly compensated by coherent displacements of all atoms in the disordered structure (acoustic modes). If one assumes relaxation of the centers of motion to be negligible at T_g, then the heat capacity of the stable liquid above T_g would correspond simply to the isochoric contributions due to internal modes plus the rotational and translational

[7] This low frequency mode is not far from the mode hypothesized by Richet (2009) through a deconvolution of experimental heat capacity of glasses of various composition, based on the inversion procedure of Grayson-Smyth and Stanley (1950) (96 cm^{-1}).

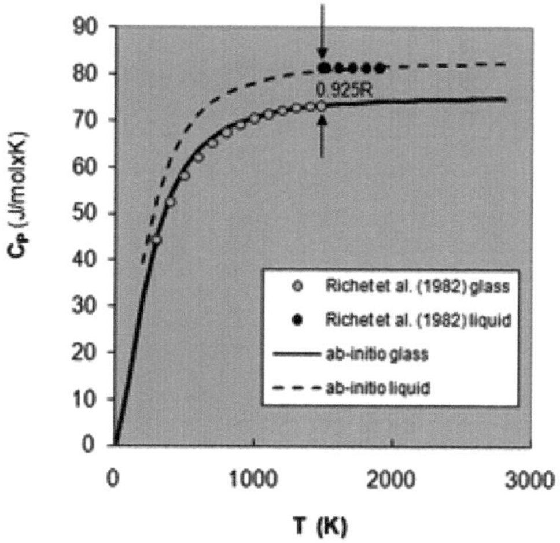

Fig. 19.26 Ab initio versus observed heat capacity of glass and liquid. The heat capacity gap at T_g (1480 *K*) is 0.925R, not far from the value observed by Richet et al. (1982) $\Delta C_P = 0.959R$

contributions of the various units. Below T_g rotational and translational movements cease out and are substituted by coherent motions of all atoms in the mixed clusters (Table 19.14). As we may see in Fig. 19.26, this conceptual framework reproduces satisfactorily both the heat capacity curves of the glass and the liquid within their respective stability ranges and the heat capacity gap observed at T_g. The computed heat capacity gap is 0.925R, which, when compared to the experimentally observed one, assesses a reproducibility of 0.3 *J/mol*, well within experimental uncertainty.

According to Richet et al. (1982), the enthalpy of fusion and entropy of fusion at the 1-bar melting point of the high-T β-cristobalite polymorph of SiO_2 (T_{fus} = 1996 ± 5 K) are $\Delta H_{fus} = 8.9 \pm 1.0\,kJ/mol$; $\Delta S_{fus} = 4.5 \pm 0.5\,J/(mol \times K)$. These values are, within errors, quite similar to the values listed on the NIST-JANAF Tables [Chase 1998; $\Delta H_f = 9.581 \pm 2.1\,kJ/mol$; $\Delta S_{fus} = \Delta H_{fus}/T_{fus} = 4.8 \pm 1$ $J/(mol \times K)$]. The enthalpy of formation and entropy of the SiO_2 liquid at T_{fus} [$H_{f,T_{fus},liquid}$ = −781.606 *kJ/mol* and $S_{T_{fus},liquid}$ = 172.696 *J/(mol × K)*, respectively] may be obtained by subtracting ΔH_f and ΔS_f from the data of Berman et al. (1985) for the high-T crystalline polymorph. The corresponding values at T_g and provisional values for the fictive liquid at T_{ref} = 298.15 *K* [$H°_{f,liquid}$ = −919.759 *kJ/mol*; $S°_{f,liquid}$ = 17.987 *J/(mol × K)*] are obtainable by application of the heat capacity coefficients of Richet and Bottinga (1985). Adopting for the SiO_2 glass the heat capacity equation of Richet and Bottinga (1985), deduced from $H_{f,Tg}$ and S_{Tg}, and assuming provisionally the glass/liquid transition to be reversible (at the limit of infinite relaxation), the enthalpy of formation of the glass at 298.15 *K* and 1 *bar* is $H°_{f,glass}$ = −901.486 *kJ/mol* and its entropy is $S°_{glass}$ = 48.066 *J/(mol × K)*. We may now compare the resulting $G°_{f,298.15}$ of the glass (−915.817 *kJ/mol*) with the DFT/PCM assessment. If the $[SiO_4]^{4-}$ and $[(SiO_2)_4]^0$ moieties are assumed to make up the glass at 298.15 *K*, their proportions should be 36% and 64%, respectively, to

Table 19.14 Isobaric heat capacity of the various species (C_P; [*J/(mol × K)*]) on a 1 Si atom basis in the glass (italics) and liquid state (Ottonello et al. 2010b) compared with experimental observations (Richet et al. 1982). At the glass state, C_P is obtained by summing up the optic and the acoustic (3 branches + 1 excess phonon) terms assumed to correspond to ~4.5% of all modes (anharmonicity contribution negligible). At the liquid state, C_P is obtained by adding to the optic contributions the rotational and translational components (acoustic terms absent; anharmonicity negligible)

T	$[SiO_2]^0$	$[SiO_4]^{4-}$	$[Si_2O_6]^{4-}$	$[Si_2O_7]^{6-}$	$[Si_3O_6]^0$	$[Si_3O_9]^{6-}$	$[(SiO_2)_4]^0$	$[Si_4O_{12}]^8$	$[Si_5O_{15}]^{10}$	Experimental
298.15	*14.9*	*49.9*	*50.4*	*63.4*	*40.0*	*56.6*	*43.7*	*63.5*	*66.2*	*44.1*
500	*20.3*	*66.5*	*65.8*	*79.1*	*50.3*	*72.7*	*57.8*	*78.6*	*81.2*	*58.1*
1000	*26.7*	*76.5*	*76.1*	*88.9*	*59.3*	*83.9*	*68.9*	*88.6*	*91.2*	*70.6*
1100	*27.3*	*77.1*	*76.8*	*89.6*	*60.0*	*84.7*	*69.8*	*89.3*	*91.9*	*71.6*
1200	*27.7*	*77.6*	*77.4*	*90.1*	*60.5*	*85.3*	*70.4*	*89.9*	*92.4*	*72.3*
1300	*28.1*	*78.0*	*77.8*	*90.5*	*61.0*	*85.8*	*70.9*	*90.3*	*92.8*	*72.8*
1400	*28.4*	*78.3*	*78.1*	*90.8*	*61.3*	*86.2*	*71.4*	*90.6*	*93.2*	*73.2*
1480	*28.6*	*78.6*	*78.4*	*91.0*	*61.6*	*86.4*	*71.6*	*90.9*	*93.4*	*74.2*
1480	43.8	98.0	85.3	97.9	64.3	89.2	72.3	91.6	91.9	81.4
1500	43.9	98.0	85.3	98.0	64.4	89.3	72.4	91.6	92.3	81.4
2000	44.7	98.8	86.2	98.8	65.3	90.2	73.4	92.4	93.3	81.4
2500	45.0	99.1	86.6	99.1	65.7	90.7	73.9	92.8	93.9	81.4
3000	45.2	99.3	86.8	99.3	66.0	90.9	74.2	93.0	94.2	81.4

reproduce the $G°_{f,298.15}$ derived from the thermochemical cycle, once configurational disorder has been accounted for, on the basis of permutability. These values are close to 33% and 67%, which are the values obtained by Ottonello et al. (2010b) from the vibrational model.

There is above T_g a marked decrease of the excitation energy in the liquid with the increase of T, interpreted as due to the onset of diffusion at a molecular scale (Sette et al. 1998). According to Grigera et al. (2003), this is the mode-coupling temperature T_{MC} (which in glycerol, the prototypical glass-forming material, occurs at $\sim 100\ K$ above T_g) where there is hybridization of acoustic modes with high-energy modes that soften upon approaching the saddle-phonon transition. The transition from glass to liquid, which involves rupture of bonds to achieve transitional freedom, may be seen as the passage from a variational (minimum energy) to an activated state along the direction of the saddle. Moreover, the transition direction may be tentatively identified on the basis of the generally accepted view that the motion of the first imaginary mode furnishes the direction of reaction toward the activated state (locating thus the two neighboring atoms subjected to bond rupture). Simple electrostatic considerations suggest that hybridization occurs when high-frequency stretching modes associated with terminal Si atoms of the D_{6h} units are replaced by RUM-like vibrations between two rigid $[Si_4O_7]^{2+}$ blocks which interact coulombically with an interposed monomer $[SiO_4]^{4-}$. In this view, the glass at T_{MC} is simply a "wrong" accretionary form of α-cristobalite attained by rotation of a terminal accretionary monomeric group (Fig. 19.27B, C; Ottonello et al. 2010b).

The transition from the T_{MC} glass to the liquid corresponds to a depolymerization step of the type:

$$\left[(SiO_2)_9\right]^0 + 2O^{2-} \Leftrightarrow \left[(SiO_2)_8\right]^0 + [SiO_4]^{4-} \tag{19.5.1}$$

which most likely involves actually in the liquid two charged species $[Si_4O_7]^{2+}$ and the $[SiO_4]^{4-}$ unit in a coulomb-interacting ensemble (cf. Fig. 19.27A, C):

$$\left[(SiO_2)_9\right]^0 \Leftrightarrow 2[Si_4O_7]^{2+} + [SiO_4]^{4-} \tag{19.5.2}$$

While the glass transition is associated with a sharp modification of the heat capacity, which occurs when the rotational and translational freedom is replaced by coherent motions of all atoms in the ensemble, at higher T, the process is most likely kinetically mediated by the number of bonds, which are effectively broken at the various T above T_g. Moving along a cooling path we may tentatively see T_{MC} as the T at which the transformation process begins and T_g the temperature at which it ends (all bonds reconstituted) and build up a schematic TTT diagram of the process where the "*fictive*" temperatures represent the T conditions of attainment of a specific glass configuration for a given value of ordering parameter X_T (Fig. 19.28) with T_{MC} and T_g as boundary conditions at the limiting values of X. For a given cooling path, the ordering state is given by the contour, which appears as a tangent

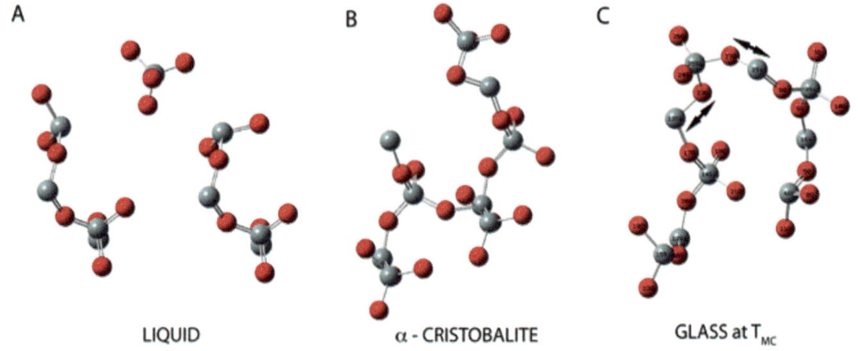

Fig. 19.27 Conceptual schematic representation of liquid. (**A**), crystalline (**B**), and glassy (**C**) silica cluster aggregates. In (**A**), the liquid is seen as an aggregate of un-bonded $[SiO_4]^{4-}$ and $[Si_4O_7]^{2+}$ polymers. In (**B**), the correct positioning of the $[SiO_4]^{4-}$ monomer at the Si termination of one of the $[Si_4O_7]^{2+}$ chains and Si–O bonding in the intermediate region generate the correct accretionary stage of α-cristobalite. In (**C**), a $[SiO_4]^{4-}$ group bonded to two Si terminations prevents the long-range tetragonal symmetry; two high frequency Si–O stretching modes are replaced by one RUM-like stretching mode along the Si–O bond directions indicated by the arrows. This mode propagates coherently throughout the whole ensemble. This is probably the hybridization condition of mode-coupling described by Sette et al. (1998). From Ottonello et al. (2010b): With permission from The American Institute of Physics

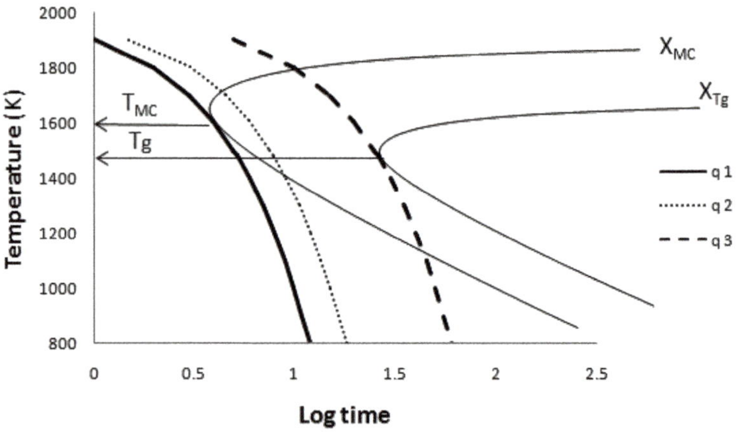

Fig. 19.28 Schematic TTT diagram illustrating the role of the cooling rate in determining fictive temperatures of transition: at a cooling rate of −100 *K/s* (representative of rapid quenching) the Mode Coupling temperature T_{MC} is attained. This T is here envisaged as the T of beginning of the transition process, defined by the tangency point of the cooling rate q_1 with the X_{MC} ordering curve. The transition ends at T_g where the cooling rate $q_3 = -20$ *K/s* is tangent to the ordering state X_{Tg}. The enthalpy of transition with the above limiting rates is 250 *kJ/mol*, much higher than the value expected from simple bonding energy considerations (dotted cooling path) and implies a high energy of activation of the transition process. From Ottonello et al. (2010b). With permission from The American Institute of Physics

to the cooling path. Interestingly, adopting as enthalpy of activation of the process the value involved in the equilibria (19.5.1 and 19.5.2) and assuming as valid the Arrhenian dependency law of the various fictive T on the logarithm of the cooling rate q envisaged by Moynihan et al. (1974), to have a T_g coincident with that one experimentally observed by Richet (1982), Richet et al. (1982), we would assume a cooling rate of q = −66 *K/min*, about 10 times higher than the cooling rates normally adopted. On the other hand, assuming that a (somewhat fast) cooling rate q = −20 *K/s* is appropriate to detect the true T_g of pure amorphous silica, the enthalpy of activation would be ~250 *kJ/mol*, implying a substantial kinetic hindrance to the process.

The "*microscopic*" definition of fictive temperature implicitly adopted by Ottonello et al. (2010b) (i.e. according to Mauro et al. 2009, "*a mapping of the non equilibrium glassy structure to comparable equilibrium liquid structures*") is only one among the number of different definitions being currently practiced in the field, but if has the advantage to recast the structural state of the glass in a kinetic framework. Though the ab initio vibrational model is able to reproduce satisfactorily the heat capacity of SiO_2 glass and liquid, identifying the main processes acting around the glass transition and responsible for the heat capacity gap between liquid and glassy states (and justifying their observed Gibbs free energies as well), the structural details of the glass below T_g remain partly elusive. The "*acoustic-like*" nature of the excess phonon experimentally observed by IXS investigations (Benassi et al. 1996) and ascribed by Ottonello et al. (2010b) to a RUM-like motion between the building block $[SiO_4]^{4-}$ and the D_{6h} aggregates, requires coherency over the entire ensemble. On the other hand, the broad downward extension of the vibrational continuum (i.e. the "*broadened acoustic branches*" of Leadbetter 1969) requires partial elimination of the mode degeneracy, possibly induced by local strain. The 2:1 proportion between D_{6h} accretionary islands and $[SiO_4]^{4-}$ has thus most likely only a statistical significance with the actual extension of paracrystalline unit variable with T. Under the above provisos, an appropriate conceptual framework for the glass is that one of an energy landscape with a continuously broken ergodicity developed by Gupta and Mauro (2007) on the framework of Palmer's theory (Palmer 1982).

19.5.6 Al_2O_3 Melting and Vitrification

At the melting point, α-alumina (the stable form of crystalline Al_2O_3; see Sect. 20.2.2) undergoes a dramatic structural reorganization that does not involve simple rupture of bonds subjected to thermal stress, but deep rearrangements in short-range ordering with a substantial change in the coordination of the central cation. Based on neutron scattering experiments on aerodynamically-levitated laser-heated drops of molten alumina at 2500 *K*, about 62% of the former VI-fold coordinated Al switches to IV-fold coordination, 24% is in V-fold coordination and only 2% remains in the former VI-fold coordination of the solid state (Landron et al. 2001). The density deficit between crystalline and amorphous states attains in this way 27% (Landron et al. 2001). Although the heat capacity of α-alumina has been the object of intensive investigation (Castanet 1984), only few experimental studies are available on the liquid (Shpil'rain et al. 1972; and references therein). These data are nevertheless internally consistent and allow to establish that the isobaric heat capacity of the liquid is roughly constant above liquidus ($C_P = 184 \pm 15$ *J/mol × K*). The enthalpy of fusion at $T_f = 2325$ *K*, as determined from the difference between the heat capacity delta of α-alumina from the melting point to the athermal limit $\left(H_{\alpha,2325} - H_{\alpha,0}\right)$ and the corresponding value for the liquid $(H_{L,2325} - H_{L,0})$, is 107 ± 5 *kJ/mol* (Castanet 1984; Shpil'rain et al. 1972).

The NIST-JANAF tables (Chase 1998) assume the glass transition to take place at $T_g = 1350$ *K* and to be high order (i.e. no appreciable discontinuity in the enthalpy and entropy at the glass transition). Accepting this T_g as sufficiently representative of the thermal state at the glass transition, the heat capacity gap at T_g appears to be roughly consistent with what expected from the simple contributions of the translational and rotational partition functions (Q_{trans}, Q_{rot}) of the monomeric groups (Belmonte et al. 2013). Inelastic X-ray scattering measurements carried out at T_f on a levitated sample have shown persistence of an (excess) collective excitation corresponding to a longitudinal sound velocity of 7350 ± 40 *m/s* (Sinn et al. 2003). This additional phonon mode bears analogy with what observed for silica (see Sects. 19.5.2 and 19.5.5, Benassi et al. 1996; Grigera et al. 2003) and may be interpreted as a Rigid Unit Mode (RUM) motion between configurons. Indeed, in the frame of the Mode Coupling Theory, it has been hypothesized that the alumina excess phonon is associated to breaking and reforming of the bonds involving the $[AlO_4]^{5-}$ structural unit (Sinn et al. 2003). The above-recalled observations pushed Belmonte et al. (2003) to attempt an ab initio investigation of the processes which are likely to occur in alumina at the two transition points, along the guidelines previously followed to investigate SiO_2 (Ottonello et al. 2010b). To this purpose, besides the $[(Al_2O_3)_6]^\circ$ *D3d6* (*R-3c*) structure, resolved within the LCAO approximation and in the framework of the Bloch's theorem, also the monomeric units $[AlO_4]^{5-}$ and $[AlO_6]^{9-}$ were investigated at DFT/B3LYP theory level. The adopted basis set was the 85-11G(d) and 8-411G(d) contractions for Al and O adopted by Montanari et al. (2006) in their ab initio investigation of the vibrational partition function of corundum. This basis set was implemented with a 5-11G* contraction for the hydrogen centers placed on the oxygen terminations of the $[AlO_6]^{9-}$ unit to saturate the dangling bonds of the $[Al_2O_9H_5]^{7-}$ cluster (investigated only in the aim of estimating the force constant of the stretching mode of the Al^{VI}–Al^{IV} coupling). Solvation energies were obtained by application of IEFPCM (with a dielectric constant $\varepsilon = 4.575$ and a solvation radius $b_{solv} = 1.641$ *Å*). Some energy features of the investigated clusters are listed in Table 19.15.

The VI-fold coordinated unit $[AlO_6]^{9-}$ was abstracted from the refined $\alpha - Al_2O_3$ structure and subjected to single-point calculations in the aim of assessing the effect of short-range ordering. The geometry of the fundamental block $[AlO_4]^{5-}$ is virtually unaffected when passing from the isolated state to the state of solute, but inter-ionic

Table 19.15 All-electron DFT/B3LYP IEFPCM geometry and energy (electronic; zero point correction; electrostatic, dispersive and repulsive solute–solvent interaction; cavitation energy) of Al monomers in a reaction field of permittivity $\varepsilon = 4.6$. Solvent cavities generated with the United Atom Topological Model dimensions. $I_{x,y,z}$ are the axial components of the momentum of inertia expressed in atomic units. Values of the vibrational, rotational, and translational entropy at 298.15 *K* are also listed. (‡) single point calculation; (**) activated state. Values in brackets are relative to the gaseous state (isolated molecule)

Cluster	$[AlO_4]^{5-}$	$[AlO_5]^{7-\ (‡,**)}$	$[AlO_6]^{9-\ (‡)}$
Molecular mass (a.m.u.)	90.916	106.956	122.951
SAS-delimited volume ($Å^3$)	94.407	108.221	120.225
I_x (a.u.)	543.1 (658.8)	573.5	822.9
I_z	543.1 (658.8)	581.1	823.0
I_z	543.1 (658.8)	826.0	833.6
$Al\text{-}O^-$	1.86 (2.08)	1.86–1.97	1.86–1.97
$O^-\text{-}Al\text{-}O^-$	109	90.8–101.2	79.6–101.2
E_{el} (polarized) (*Hartree*)	−543.443583	−617.973304	−692.666926
ZPC (*Hartree*)	0.008659	0.010811	0.013966
$S_{vib,298.15}$ [*J/(mol × K)*]	36.4 (54.9)	(44.0)	74.3
$S_{rot,298.15}$	90.5 (92.9)	113.4	116.4
$S_{tra,298.15}$	165.0 (165.0)	167.0	168.8
ΔG_{electr} (kJ/mol)	−4542.3	−8782.1	−13,895.1
ΔG_{cav}	39.0	42.7	47.1
ΔG_{rep}	1.5	1.5	1.8
ΔG_{disp}	−14.1	−15.9	−18.3

distances shrink considerably and the energy is substantially stabilized in the reaction field by electrostatic interaction with the continuum (Table 19.15). The energy spent in forming the cavity and the repulsion term are highly subordinated in magnitude and virtually counterbalanced by dispersive contributions (Table 19.15). In spite of all efforts, Belmonte et al. (2013) were unable to achieve a stationary state for the $[AlO_5]^{7-}$ cluster in the dielectric continuum and limit themselves to present the results for an activated state with a single imaginary frequency. Besides the clusters of Table 19.15 also a much bigger configuron $[Al_{30}O_{45}]^0$ was also resolved (at simple HF/STO-3G level). The optical contributions to the isochoric heat capacity of the various moieties between 298.15 and 3000 *K* are listed in Table 19.16 together with the acoustic terms calculated with the sine-wave dispersion relation by assuming for the glass transverse and longitudinal velocities identical to those of corundum ($v_{s1} = 6.12$ *km/s*; $v_{s2} = 6.59$ *km/s*; $v_p = 10.855$ *km/s*; Kieffer 1973a). The contribution of the additional acoustic-like mode traveling at $v_{exc} = 7.32$ *km/s* (Sinn et al. 2003) is also listed, together with translational and rotational terms and the anharmonicity contribution $C_P - C_V = T\alpha^2KV$ obtained adopting the parameters listed in Table 19.17.

The computed isochoric heat capacities in the low-T range 0–300 K are compared in Figs. 19.29 and 19.30 with the partial molar heat capacities obtained by

Table 19.16 Components of the bulk heat capacity of Al_2O_3 in the various aggregation states at selected temperatures in the range 298.15 $\leq$ T(*K*) $\leq$ 3000 (Belmonte et al. 2013). All data are in *J/(mol × K)*. The optical terms of $[AlO_4]^{5-}$ were obtained at DFT/B3LYP level of theory in a polarized continuum having the properties listed in Table 19.17 (IEFPCM) and adopting a 85-11G(d) basis set for Al and 8-411G(d) for O. The optical modes of α-Al_2O_3were obtained with the CRYSTAL code at DFT/B3LYP level in the LCAO approximation. The optical modes of the $[Al_{30}O_{45}]^0$ cluster are relative to the gaseous (isolated) state and were resolved at simple HF/STO-3G level

T(*K*)	Anharmonic	Acoustic (coherent)	Excess (coherent)	Trans + Rot	Optical $[AlO_4]^{5-}$	Optical α-Al_2O_3	Optical $[Al_{30}O_{45}]^0$	Liquid	Glass
298.15	0.048	9.595	3.244	37.861	53.249	68.444	1017.784	134.350	80.739
300	0.049	9.604	3.246	37.861	53.455	68.818	1022.447	134.754	81.062
400	0.065	9.936	3.339	37.861	61.536	84.116	1236.267	150.688	95.758
500	0.081	10.097	3.384	37.861	65.892	92.908	1385.684	159.421	105.940
1000	0.159	10.318	3.444	37.861	72.438	106.890	1681.750	172.746	126.038
1500	0.235	10.359	3.455	37.861	73.754	109.819	1755.688	175.456	131.095
2000	0.309	10.374	3.459	37.861	74.223	110.872	1783.511	176.425	133.043
2500	0.381	10.381	3.461	37.861	74.441	111.365	1796.759	176.876	134.007
3000	0.450	10.385	3.462	37.861	74.560	111.634	1804.058	177.123	134.567

Table 19.17 Physical properties of amorphous Al_2O_3 adopted in the calculations by Belmonte et al. (2013)

Solvent	Al_2O_3	References
Nominal Density at $T = 298.15$ K (g/cc)	$3.049^{(*)}$	Glorieux et al. (1999)
Molar weight (g/mol)	101.961	
Fictive Molar volume (cc/mol)	$33.438^{(*)}$	
Numeral density $(Å^{-3})$	0.01801	
Solvation radius $(Å)$	1.641	
ε	$4.575^{(\&)}$	
K_0 (GPa)	$5.485^{(\$)}$	
$\alpha_V \left(K^{-1} \times 10^5\right)$	$2.831^{(£)}$	Stebbins et al. (1984)
$\alpha'_V \left(K^{-2} \times 10^9\right)$	$-0.736^{(£)}$	Stebbins et al. (1984)
v_{s1} (km/s)	$6.12^{(‡)}$	Kieffer (1979a)
v_{s2} (km/s)	$6.590^{(‡)}$	Kieffer (1979a)
v_P (km/s)	$10.855^{(‡)}$	Kieffer (1979a)
v_{exc} (km/s)	7.350	Sinn et al. (2003)

(*) Obtained extending arbitrarily the observed T-ρ relationship
(&) Based on the Clausius-Mosotti relation with a static polarizability of 7.7 $Å^3$
($) Based on the melting curve of the pure component
(£) Obtained from the listed volume coefficients
(‡) Assumed to be identical to α-Al_2O_3 (corundum)

Richet et al. (2009) by deconvolution of calorimetric data in the system CaO–Al_2O_3–SiO_2. At such low T, the anharmonicity effect is negligible and $C_P \cong C_V$. We may note in Fig. 19.29 that the bulk isochoric heat capacity obtained by summing up the optical terms of the $[AlO_4]^{5-}$ vibrating unit with the acoustic contributions is in fairly good agreement with the observations of Richet and coworkers for the tetrahedrally coordinated $^{IV}Al_2O_3$ component. This means that the heat capacity of $^{IV}Al_2O_3$, arises entirely from one mole of $[AlO_4]^{5-}$ units vibrating in a reaction field with charges compensated by atomic centers that do not contribute to the optical terms. The fact that the contribution of acoustic terms brings almost precisely to the correct bulk value means that the computation is also correct in terms of vibrational freedom [$3N - 3 = 12$ optical contributions +4 (longitudinal + transverse + excess coherent phonon) acoustic contributions].

When comparing the exa-coordinated forms α-Al_2O_3 and $[AlO_6]^{9-}$ with $^{VI}Al_2O_3$ partial molar properties (Fig. 19.30), we see that the observed partial molar property is satisfactorily reproduced only when one adopts the optical terms of the crystalline polymorph and neglects the acoustic contributions. This basically would mean that the vibrational modes of the VI-coordinated amorphous substance lose the degeneracy due to long-range disorder and that the actual contributions of coherent modes (in terms of vibrational freedom) are negligible. The virtual inexistence of a “free” $[AlO_6]^{9-}$ vibrating units is indicated by the excessively high optical contribution they would impose on the amorphous medium.

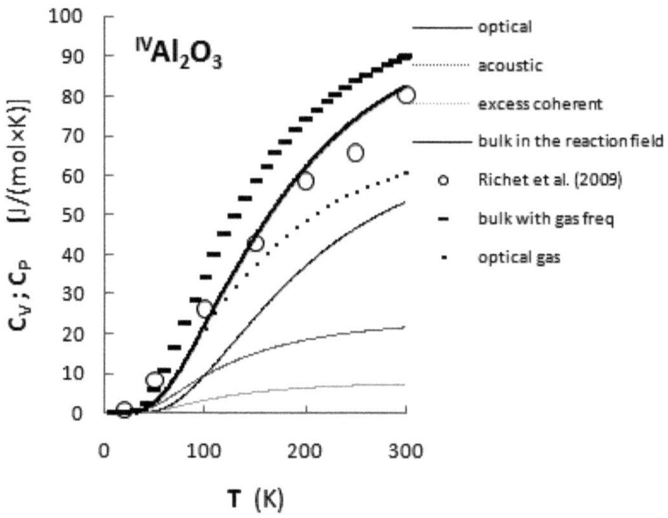

Fig. 19.29 Calculated (Belmonte et al. 2013) versus observed (Richet et al. 2009) isobaric heat capacity of an amorphous Al_2O_3 substance composed of $[AlO_4]^{5-}$ vibrating units in a dielectric medium of low permittivity. The "bulk" isobaric heat capacity was obtained by summing up to the optical terms of the three acoustic modes and the excess coherent phonon. The effect of the reaction field on the frequency of stretching motions is quite evident: neglecting this effect an odd result is obtained (bulk with gas frequencies in figure). Modified from Belmonte et al. (2013)

19.5.7 The Al_2O_3 Liquid

Accepting that the alumina excess phonon is associated to breaking and reforming of the bonds involving the $[AlO_4]^{5-}$ structural unit (Sinn et al. 2009) and that the liquid inherits the vibrational properties of both tetra-coordinated and VI-fold coordinated units (Figs. 19.19, 19.29 and 19.30) it is rather consequential to hypothesize that the alumina "*liquid*" is actually a mixture of structural "*islands*" corresponding to minimal extensions of the *D3d*6 framework and $[AlO_4]^{5-}$ monomers (Belmonte et al. 2013). A half unit-cell of $[(Al_2O_3)_6]^\circ$ *D3d*6 (*R-3c*) corresponds in fact to a positively-charged cluster of formula $[Al_{12}O_{11}]^{14+}$ which may be contoured by $[AlO_4]^{5-}$ units in an approximate 1:3 proportion to achieve neutrality (Fig. 19.31).

The isobaric heat capacity C_P of the cluster (Fig. 19.32), obtained when summing up to the vibrational effect of the optical modes the contributions of the translational and rotational terms (appropriately scaled to the adopted molar notation):

$$C_{V,trasl+rot} = 2 \times \left(0.263 \times \frac{3R}{12} + 0.737 \times 3R\right) = 4.5535R \qquad (19.5.3)$$

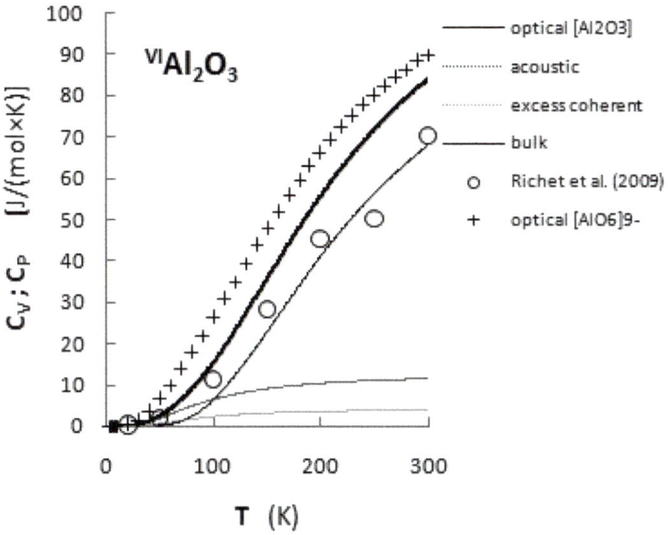

Fig. 19.30 Calculated (Belmonte et al. 2013) versus observed (Richet et al. 2009) isobaric heat capacity of an amorphous Al_2O_3 substance composed of exa-coordinated vibrating units in a dielectric medium of low permittivity. The "bulk" isobaric heat capacity is obtained by summing up to the optical terms the three acoustic modes and the excess coherent phonon and is evidently too high with respect to observations. The effect of the reaction field on the frequency of stretching motions is quite evident also in this case but not sufficient to achieve a good consistency with observations that correspond rather to the simple optical terms of α-Al_2O_3. Modified from Belmonte et al. (2013)

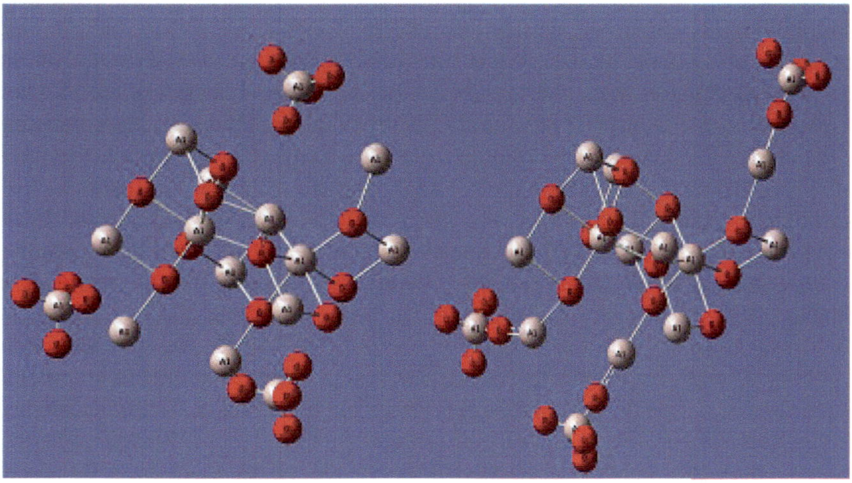

Fig. 19.31 Schematic representation of the mode-coupling excitation. The $[AlO_4]^{5-}$ monomeric units surround the corundum-like islands of exa-coordinated aluminum (left) and create pioneering bonds continuously breaking and reforming at appropriate structural sites (right). From Belmonte et al. (2013). With permission from The American Institute of Physics

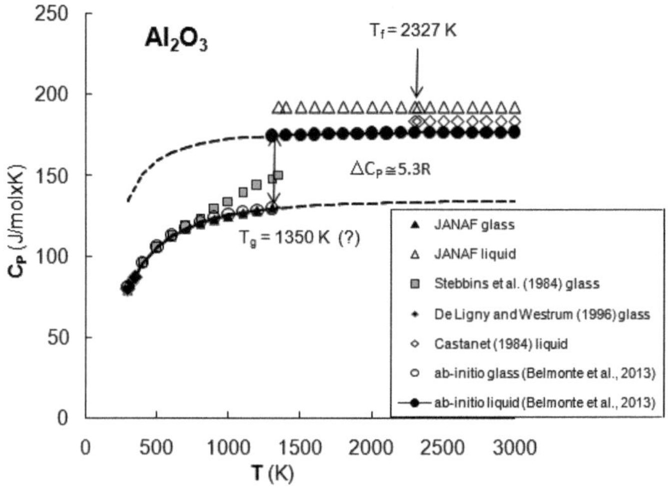

Fig. 19.32 Ab initio vs observed isobaric heat capacity of amorphous Al_2O_3 at high T. The "*liquid*" is seen as an aggregate of $[Al_{12}O_{11}]^{14+}$ islands coulombically contoured by $[AlO_4]^{5-}$ units in an approximate 1:3 proportion (actually 0.263:0.737). Coherent motions are missing and anharmonicity is almost negligible. The effect of the translational and rotational partition function corresponding to 4.55R is added to the vibrational part (the structural subunits are free to rotate and translate in the dielectric medium). The glass heat capacity is obtained by summing to the 222 optical modes of the $[Al_{30}O_{45}]^0$ cluster (resolved at simple HF/STO-3G level of theory) the acoustic phonons and the coherent excess mode, after scaling to the Al_2O_3 stoichiometry. The glass transition is not known with precision but the heat capacity gap between super-heated glass and liquid is almost constant (from 5.39R at T = 1350 *K* to 5.18 at T = 2300 *K*)

and the, almost negligible, anharmonic effects (Table 19.16), is not far from the estimates of Castanet (1984), which are based on a critical assessment of the existing calorimetric data.

Assuming the internal modes of all the structural islands to be degenerate, the bulk internal freedom of such a "liquid" assemblage would be 36. With the decrease of T, at the mode-coupling temperature, a further (coherent) excitation corresponding to a longitudinal sound velocity $v_{excess} = 7350 \pm 40$ *m/s* (Sinn et al. 2003) would be active. Assuming our statistical ensemble to have a density corresponding to experimental observations (i.e. $\rho = 3.986$ *g/cc*), its primitive unit cell would contain $Z_{PRIM} = 2.4$ atoms with a radius at the limit of the Brillouin zone defined as:

$$K_{MAX} = 6\pi N_0 / \sqrt[3]{\frac{4\pi Z_{PRIM} M}{\rho}} = 8.344\ \mathring{A} \tag{19.5.4}$$

where M is the molar mass. The corresponding wavenumber ν_{excess} is then obtained by applying (Belmonte et al. 2013):

$$\nu_{excess} = \frac{2 \times \mathrm{v}_{excess} \times \mathrm{K}_{MAX}}{\pi \times c} = 206.420 \quad (cm^{-1}) \tag{19.5.5}$$

where c is the speed of light.

19.5.8 The Al_2O_3 Glass

Assuming for simplicity that all atoms in the glass at $T \leq T_g$ are subjected to the normal acoustic excitations observed in the α-Al_2O_3 polymorph and proceeding as done for the excess phonon, Belmonte et al. (2013) obtained $\nu_{S1} = 172.580\ cm^{-1}$, $\nu_{S2} = 185.834\ cm^{-1}$, ν_P was then obtained by summing up the effects of the acoustic + excess coherent modes to the optical contributions. In doing this the Authors took into account that the 219 internal degrees of freedom of the glass would render the effect of coherent modes ineffective if not scaled to the appropriate stoichiometry. Also, they noted that at the glassy state the structural sub-units are unable to rotate and translate, so that the two terms are absent. The obtained heat capacity (Table 19.16) is nearly coincident with the few data at disposal: namely, the three calorimetric assessments of De Ligny and Westrum (1996), the assessment of Stebbins et al. (1984) when limited to $T \leq 800$ K, and the assessment of the JANAF tables (Chase 1998) over the entire stability range of the glassy polymorph (Fig. 19.32). Figure 19.33 shows the variational "*glass*" $[Al_{30}O_{45}]^0$ obtained by Belmonte et al. (2013) at simple HF/STO-3G level from two $[Al_{12}O_{11}]^{14+}$ units and six $[AlO_4]^{5-}$ monomer units:

$$2[Al_{12}O_{11}]^{14+} + 6[AlO_4]^{5-} \Leftrightarrow [Al_{30}O_{45}]^0 + O^{2-} \tag{19.5.6}$$

Lowering the T, the various $[AlO_4]^{5-}$ units rotate around the pioneer bonds and become clamped by the structural islands. Further displacive readjustments of the structure dictated by steric forces in the absence of accretional structural constraints change the coordination state of Al centers about the first oxygen neighbors. The proportion of IV-fold to V-fold to VI-fold coordinated centers attains in this way 66.6–27.8–5.6%, quite similarly to the 70.4–27.3–2.3% proportion experimentally observed by Landron et al. (2001) in the liquid.

The molar volume of the model glass at the athermal limit is 424.394 *cc/mol*, which, when reduced to the usual stoichiometry (Al_2O_3) returns a density of 3.603 *g/cc*, substantially lower than the crystalline counterpart (3.894 *g/cc*). Further structural details about the model glass may be obtained by computational investigation of its magnetic properties. The Ramsey contributions to the J_{NM} coupling tensor between aluminum and oxygen under various coordination states in the $[Al_{30}O_{45}]^{\circ}$ glass evaluated at simple HF/STO-3G level with the Gauge-Independent Atomic Orbital (GIAO) method (see Sect. 15.5) are highly variable within a given coordination state and dominated by the interaction between the magnetic moments associated with the nuclear spins (i.e. the Fermi Contact FC) (Table 19.18).

Fig. 19.33 The HF/STO-3G variational "*glass*" $[Al_{30}O_{45}]^0$ (Belmonte et al. 2013). The Al central atoms are represented with different colors according to their coordination states. Displacive movements dictated by steric forces in the absence of crystalline constraints produce 14 four-fold coordinated centers (green), 5 five-fold coordinated centers (light blue) while the exa-coordinated centers are reduced to 1 (violet); oxygen centers are shown in red and low-coordinated Al centers in pink. The proportions IV:V:VI (67%–28%–6%) attained by displacive reorganization are not much dissimilar to what was experimentally observed by Landron et al. (2001) in the liquid near the melting point

The calculated isotropic chemical shifts of ^{27}Al in the various coordination states with respect to the isotropic shielding of a one-molar $Al(NO_3)_3$ solution computed at the same theory level ($\sigma_{ref} = 537.36\ ppm$) follows the same trend observed in crystalline substances although substantially reduced in magnitude due to the lowering of structural constraints and the magnetic asymmetry is almost negligible in all coordination states. For IV-fold coordinated ^{27}Al, Belmonte et al. (2013) observed a mean value $\delta_{iso} = -(\sigma - \sigma_{ref}) = 4 \pm 20\ \ ppm$; for V-fold coordinated ^{27}Al they observed $\delta_{iso} = -13 \pm 11\ ppm$ and for VI coordinated ^{27}Al $\delta_{iso} = -80\ ppm$. The corresponding experimental values in crystalline counterparts are, respectively, 64, 36, and 16 *ppm* in sillimanite (Massiot 1996), andalusite (Rocha 1998), and alumina (Jakobsen et al. 1988).

Table 19.18 Ramsey contributions to the J_{NM} coupling tensor between aluminum and oxygen under various coordination states in the $[Al_{30}O_{45}]^{\circ}$ glass evaluated at simple HF/STO-3G level with the GIAO procedure (Belmonte et al. 2013). FC = Fermi Contact; SD = Spin Dipolar interaction; PSO = Paramagnetic Spin Orbit term; DSO = Diamagnetic Spin Orbit term. All values are expressed in *Hz*. See Fig. 19.33 for atom labeling

Couple	Coordination	FC	SD	PSO	DSO	J_{NM}
$Al_{17}O_{33}$	IV	−6.2	0.0	3.1	−0.3	−3.7
$Al_{17}O_{45}$	IV	−9.1	−0.6	3.0	−0.3	−7.0
$Al_{17}O_{64}$	IV	−17.8	−0.1	3.6	−0.3	−14.6
$Al_{17}O_{74}$	IV	−25.0	0.2	8.4	−0.2	−16.6
$Al_{2}O_{26}$	V	−12.2	−0.6	3.4	−0.3	−9.6
$Al_{2}O_{28}$	V	−17.6	−0.6	3.7	−0.3	−14.8
$Al_{2}O_{31}$	V	−1.4	−0.4	1.9	−0.3	−0.2
$Al_{2}O_{38}$	V	−13.3	−0.3	4.3	−0.3	−9.9
$Al_{2}O_{44}$	V	−9.1	−0.4	2.9	−0.3	−6.9
$Al_{12}O_{27}$	VI	−3.4	−0.5	2.3	−0.3	−2.0
$Al_{12}O_{28}$	VI	−18.1	−0.4	2.8	−0.3	−16.1
$Al_{12}O_{36}$	VI	−0.8	−0.5	3.1	−0.3	1.5
$Al_{12}O_{39}$	VI	−2.0	−0.7	5.1	−0.2	2.1
$Al_{12}O_{40}$	VI	1.5	−0.5	2.5	−0.3	3.3
$Al_{12}O_{46}$	VI	3.7	−0.5	1.0	−0.3	0.0

19.6 Silicate Melts as Solvents

The application of IEFPCM to solutions with organic and inorganic solvents is so successful that in the most powerful ab initio computational packages, such as GAUSSIAN for example (Frisch et al. 2004), the solvent parameters are conformed in such a way to reproduce, through PCM calculations, the dielectric properties leading to the observed solubility, i.e. the "*coupling work*" terms in Eq. (16.3.4). It is worth to stress here, in passing, that also to depict macroscopic thermodynamic properties such as the Gibbs free energy of mixing in a binary join we treated implicitly the melt as a "*solvent*" and investigated its interaction with a single molecule hosted in it (see Sect. 16.3.3). The dielectric properties of "*silicate liquid solvents*" (ε, ε^{∞}) were opportunely parameterized through the Clausius–Mossotti relation (Eq. 16.3.20) and the refractive index (n in Eq. 16.3.21). Argon was then adopted as a probe to obtain the solvent diameter σ_s, based on the plethora of experimental observations on noble gases solubility at various *P*, *T* conditions. The obtained σ_s proved to be consistent with a simple functional form based on the molecular volume of the solvent, fitting nicely the trend between a non-polar (C_6H_6) and a polar solvent (H_2O) [Fig. 16.4 and Eq. (16.3.19)].

The obtained Ar solution energies in few selected melts are resumed in Table 19.19. The interplay between cavitation and the other forms of energy is complex, and the solubility trend for inert gases in melts at high T is the opposite of

Table 19.19 Solution energy terms for Ar in various media at T = 298.15 *K*, P = 10^5 *Pa*. The interaction Gibbs free energy ΔG_i is the summation of electrostatic + repulsive + dispersive terms computed at B3LYP/aug-cc-PVTZ theory level. Data are in *kJ/mol* (G, H) and *J/(mol × K)* (S, C_P). Calcd = calculated; Obsd = observed; Extrp = extrapolated. The T limit for the heat capacity coefficients depicting the heat capacity of solution is 353.5 *K* for benzene, 372.8 *K* for water and 1623.15 *K* for the silicate liquids (Ottonello and Richet 2014)

Solvent		ΔG_{cav}	H_{cav}	S_{cav}	$\Delta G_i \cong H_i$	ΔG_s	ΔH_s	ΔS_s	ΔC_p (a)	ΔC_p (b × 10^3)
H_2O	Calcd	23.398	3.759	−65.9	−15.841	25.825	−14.011	−133.6	38.787	−92.388
	Obsd	–	–	–	–	26.250[a]	−11.213[a]	−125.5[a]	–	–
Benzene	Calcd	18.522	20.900	8.0	−13.640	18.827	5.801	−43.7	315.563	−1059.82
	Obsd	–	–	–	–	17.435[b]	1.243[b]	−54.4[b]	–	–
Leucite-bas	Calcd	19.613	1.024	−62.3	−11.757	25.090	−13.149	−128.3	−0.791	−5.672
	Extrp	–	–	–	–	52.133[c]	18.828[c]	−111.7[c]	–	–
Tholeiite	Calcd	20.231	0.906	−64.9	−11.632	25.802	−13.152	−130.7	−1.659	−4.798
	Extrp	–	–	–	–	60.886[c]	28.451[c]	−108.8[c]	–	–
AOB	Calcd	21.675	1.239	−68.5	−13.180	26.007	−14.353	−135.4	0.885	−7.407
	Extrp	–	–	–	–	87.107[c]	59.413[c]	−92.9[c]	–	–
Andesite	Calcd	18.294	0.549	−59.5	−11.088	24.299	−12.981	−125.0	−4.465	−2.323
Ugandite	Calcd	23.241	1.689	−72.3	−13.807	27.064	−14.514	−139.5	4.556	−11.523

[a]Lide (2002); [b]Clever et al. (1957); [c]Lux (1987)

that observed in water and benzene at low temperature. This is basically due to the fact that the cavitation energy is largely entropic in nature and the $-TS_{cav}$ energy term becomes progressively dominant with the increase of T.

Solubility calculations were then extended to He, Ne, and Kr whose dispersive and repulsive coefficients are available from theory, and it was shown that their ab initio Henry's constants at high T reproduce with discrete accuracy the observed increase with the static polarizability of the series element (Fig. 16.5). It was then shown that conforming exactly the solvent radius on the observed effective Ar solubility or adopting the simple molar-volume-based expression bring essentially to the same results (Table 16.9) and that the Ar-based parameterization may be safely adopted to depict the properties of the other terms in the inert gases series (Sect. 16.3.4).

Having shown that the solubility of non-reactive species in silicate melts may be accurately reproduced by ab initio procedures, the next obvious step was to extend the calculus to reactive components such as H_2O.

There is a general consensus that the reactive solubility of H_2O in silicate melts is well represented by a homogeneous equilibrium among quasi-species of type:

$$[OH_2]_{melt} + [O]^0_{melt} \Leftrightarrow 2[OH]^0_{melt} \quad (19.6.1)$$

where the molecule $[OH_2]$ is assumed to react with the doubly bonded oxygen of the network $[O]^\circ$ forming the $[OH]^\circ$ functional.

As a case study, it was selected the wet solvus of silica. The preliminary investigation of the SiO_2–H_2O system indicated that, in the light of the existing experiments, the electrostatic interaction, reproduced through a scaling factor applied to the van der Waals UAHF radius, conforms an apparent potential well at low H_2O content in the system (Fig. 19.34).

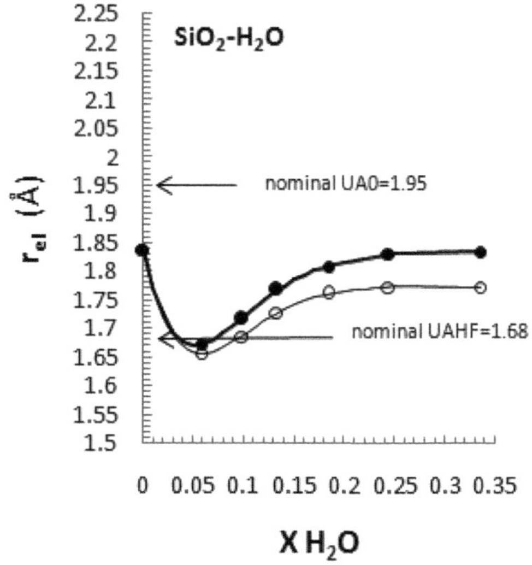

Fig. 19.34 Electrostatic radius of OH_2 in the H_2O–SiO_2 mixed solvent. The shrinking of the electrostatic radius with the increase of H_2O component in the solvent can be ascribed to the formation of hydrogen bonding between H_2O and hydroxyl functionals of polymer clusters in solution when the latter are in sufficient amount to surround water molecule in an appropriate position. From Ottonello et al. (2015). With permission from The American Institute of Physics

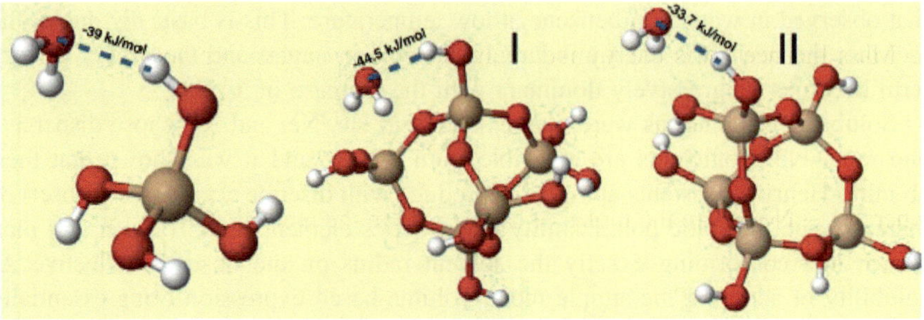

Fig. 19.35 HB effects in a polymerized melt. Hydrogen bonding between discrete OH_2 molecules and hydroxyl terminations of silica polymers. In position I, molecular water approaches the $[Si_5O_6(OH)_8]$ neutral cluster at an OH functional termination pertaining to a singly hydroxylated monomeric group. In position II, molecular water approaches a doubly hydroxylated unit. In both cases (and also in the case of HB with the orthosilicic acid unit), the HB distance is ~1.98 Å. The net gain in stabilization at the athermal limit is $E_{HB} \cong -44.5$ *kJ/mol* when molecular water $[OH_2]$ approaches a hydroxyl functional $[OH]^{\circ}$ of a singly-hydroxylated monomeric group, while it reduces to $E_{HB} \cong -33.7$ *kJ/mol* for a doubly-hydroxylated monomeric group. From Ottonello et al. (2015). With permission from The American Institute of Physics

The apparent shrinking of the electrostatic radius in Fig. 19.34, corresponding to a net stabilization of the solute, is actually due to Hydrogen Bonding (HB) between molecular water $[OH_2]$ and the neutral $[OH]^{\circ}$ hydroxyl functional of the liquid structons. To strengthen this hypothesis, detailed HB calculations performed adopting the SSB-D Functional as implemented in NWCHEM (Valiev et al. 2010) were carried out on the $[Si_5O_6\,OH)_8]^0$ cluster (Fig. 19.35; Table 19.20). This cluster is a good proxy of the tridimensional fully polymerized network in the presence of DHMO in the system, a structure of intermediate Lewis acidity corresponding to the OH terminated analog of the structure 4 of Civalleri et al. (1998) with three four-membered rings. This structure that most likely forms in the SiO_2-rich portion of a H_2O–SiO_2 binary system has eight hydroxyl terminations, two of them pertaining to the two singly hydroxylated monomers.

In the intermediate to low acidity range, the system is most likely completely depolymerized and orthosilicic acid units $[H_4SiO_4]^0$ are present. The SSB-D calculations return zero point ΔH_{HB} energies of −44.5 *kJ/mol* when one OH_2 molecule approaches the OH functional of the singly hydroxylated monomer and of −33.7 *kJ/mol* when it approaches the OH functional of the doubly hydroxilated monomer. The ΔH_{HB} energy is −39 *kJ/mol* when OH_2 approaches the orthosilicic acid unit. These energies are quite consistent with the estimates of the ΔH_{HB} energy in BLYP-D geometrically optimized structures of water (Wendler et al. 2010). Taking into account the effects of HB, the bulk H_2O solubility was reproduced by the simple speciation model of Eq. 19.6.1, with the $[OH]^{\circ}$ functional partly re-bound to $[OH_2]$ through HB. The computational results were then recast in terms of an asymmetric Redlich–Kister mixing model (Table 19.21).

Table 19.20 SSB-D functional calculations for hydroxyl-terminated clusters bonded to a water molecule. E_{el} = electronic energy; ZPC = zero point correction; Therm-U, H, G = thermal corrections to internal energy, enthalpy and Gibbs free energy at T = 298.15 *K*. S = entropy. All magnitudes are expressed in *Hartree* if not otherwise specified. Single and double asterisks identify positions I and II, respectively, in Fig. 19.23

Cluster	E_{el}	ZPC	Therm-U	Therm-H	Therm-G	S $[J/(mol \times K)]$
$[Si_5O_6\ (OH)_8]$	−2519.54504048	0.15020900	0.17479200	0.17573600	0.09734061	690.3
$[OH_2]$	−76.95233356	0.02167100	0.02450500	0.02544800	0.00338239	194.3
$[Si_5O_6\ (OH)_8\ (OH_2)]$*	−2596.52239820	0.17632700	0.20342800	0.20437200	0.12099580	734.2
$[Si_5O_6\ (OH)_8\ (OH_2)]$**	−2596.51816044	0.17660700	0.20347000	0.20441400	0.12150486	730.1
$[Si\ (OH)_4]$	−596.25156713	0.05827514	0.06538668	0.06633087	0.02893090	329.3
$[Si\ (OH)_4\ (OH_2)]$	−673.22632926	0.08434200	0.09401300	0.09495700	0.05038105	392.5
HB Reaction (T = 298.15 *K*, P = 10^5 *Pa*)			ΔU (kJ/mol)	ΔH (kJ/mol)	ΔS $[J/(mol \times K)]$	ΔG (kJ/mol)
$[Si_5O_6\ (OH)_8] + [OH_2] <=> [Si_5O_6\ (OH)_8\ (OH_2)]^{*}$			−43.2	−45.7	−150.4	−0.8
$[Si_5O_6\ (OH)_8] + [OH_2] <=> [Si_5O_6\ (OH)_8\ (OH_2)]^{**}$			−31.2	−33.7	−154.6	12.4
$[Si\ (OH)_4] + [OH_2] <=> [Si\ (OH)_4\ (OH_2)]$			−36.5	−39.0	−131.1	0.1

Table 19.21 Redlich–Kister parameterization of the SPT-PCM enthalpy of mixing in the H_2O–SiO_2 liquid Ottonello et al. (2015)

Coefficient	Model 1	Model 2
A0	−132.860	−19.703
A1	−898.290	−204.120
A2	−1142.000	−740.470
A3	1235.400	17.163
A4	1745.200	3188.700
A5	−3641.200	3124.100
A6	−4882.100	−539.320
Mean error (*kJ*)	0.084	0.141
Standard error (*kJ*)	0.120	0.224

Model (1) $$G_{mixing} = X_A X_B \sum_{i=0}^{6} A_i \times (X_A - X_B)^i + RT[X_A \ln(X_A) + X_B \ln(X_B)]$$

Model (2) $$G_{mixing} = X_A X_B \sum_{i=0}^{6} A_i \times (X_A - X_B)^i$$

The model, strictly valid only for the SiO_2–H_2O binary system, was later extended to the realm of all natural silicate melt compositions by operating through inverse methods on an extended database (Papale 1997; Papale et al. 2006).

19.7 AIMD Investigations

19.7.1 The H_2O–SiO_2 System

Pöhlmann et al. (2004) investigated by AIMD procedures a silica liquid containing 3.84 wt% H_2O at T = 3000 and 3500 *K*, in the frame of DFT-GGA approximation, with the PBE functional (Perdew et al. 1996; see Sect. 10.5.4). Troullier-Martins pseudopotentials (Troullier and Martins 1991) were adopted for the core electrons of oxygen and silicon. The energy cutoff was set at 50 *Ry*. The simulation box contained 30 Si, 64 O, and 8 H atoms and was prepared stemming from an equilibrated liquid silica simulation, by substitution of two randomly chosen Si with 8 H (4 H for each Si). The computation was carried out with the CPMD code (Car and Parrinello 1985; Hutter et al. 1995; see Sect. 13.4). A first run involved equilibration at 5000 *K*, followed by cooling down at 3500 and 3000 *K* and annealing for 5 *ps*. The equilibration at the two temperatures involved the utilization of the Nose–Hoover thermostat (Hoover 1985) and an electronic mass of 600 a.u. (energy × time^2) at a time step of 4.5 a.u. (0.1088 *fs*).

The short-range correlations of the liquid were analyzed in terms of radial distribution functions (RDFs; see also Sect. 18.1.3) of the type (Pöhlmann et al. 2004):

$$g_{\alpha\beta}(r) = \frac{D_{\alpha\beta}(r)}{4\pi r^2 \rho_\alpha A_\beta dr} \tag{19.7.1}$$

where $D_{\alpha\beta}(r)$ is the number of interatomic distances between α and β atoms found between r and r + dr; ρ_α is the numeral density of α atoms in the microcanonical ensemble:

$$\rho_\alpha = \frac{N_\alpha}{V} \tag{19.7.2}$$

and

$$A_\beta = N_\beta - \delta_{\alpha\beta} \tag{19.7.3}$$

The corresponding integrated coordination numbers (ICNs) are given by (Pöhlmann et al. 2004):

$$ICN_{\alpha\beta}(r) = \frac{1}{A_\alpha} \int_0^r D_{\alpha\beta}(r')dr' \tag{19.7.4}$$

$$A_\alpha = N_\alpha - \delta_{\alpha\beta} \tag{19.7.5}$$

RDF and ICN for the SiO_2–H_2O liquids at the two investigated temperatures are shown in Fig. 19.36.

The first striking observation is that "*the distributions become broader as the temperature is increased*" (Pöhlmann et al. 2004). The RDFs involving the matrix atoms Si and O (left panel) show in fact a first peak followed by a well-defined first minimum which becomes more pronounced if the temperature is decreased from 3500 to 3000 *K*.

From the position of the Si–O first peak, it can be deduced that the most probable Si–O distance is around 1.65 ± 0.02 *Å* for 3500 *K* and around 1.63 ± 0.02 *Å* for 3000 *K* (Pöhlmann et al. 2004). The first maximum in the H–Si distribution (right panel), is found around 2.3 *Å*, excluding the presence of stable molecular Si–H units. The first well-defined peak in the H–O distribution functions, located at 0.99 ± 0.01 *Å*, followed by a well-defined minimum at 1.48 ± 0.03 *Å* at 3500 *K* and at 1.43 ± 0.03 *Å* at 3000 *K*, reveals that the O–H bond are the dominant stable configuration for the hydrogen atoms. The existence of stable H_2 molecules can be excluded due to the lack of a well-defined first peak in the RDF for H–H. At T = 3500 *K*, on the contrary, it is observed a weak peak at $r \approx 2.0$ *Å*, a distance very close to the H–H distance in a free water molecule. This was taken by Pöhlmann et al. (2004) as an evidence that water molecules do exist at this temperature. The ICN for Si–O exhibits a plateau at a value of 4, indicating that SiO_4 tetrahedra are the main units, even in the presence of H_2O. However, the inflection point in the ICN for Si–Si at around 3.6 indicates that the tetrahedral network is partially broken (Pohlmann et al. 2004).

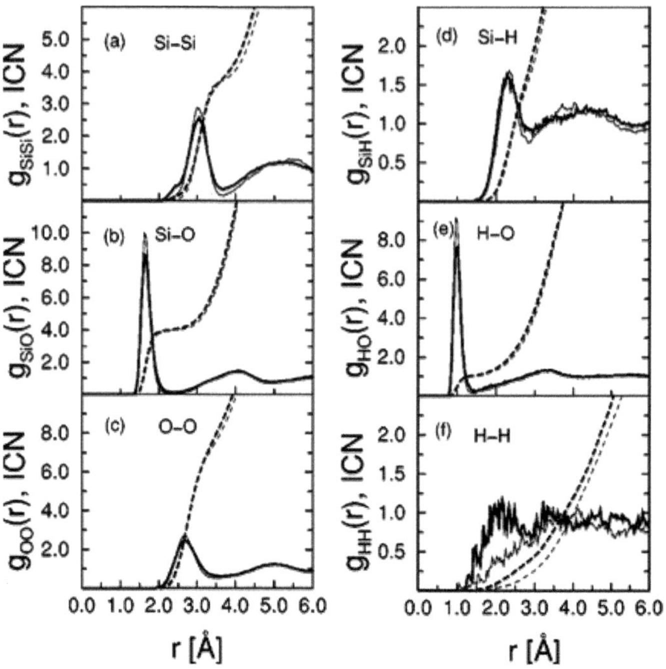

Fig. 19.36 RDF (solid lines) and ICN (dashed lines) of Si–Si (**a**), Si–O (**b**), O–O (**c**), Si–H (**d**), H–O (**e**), and H–H (**f**) for the SiO_2–H_2O liquids at 3500 *K* (bold lines) and 3000 *K* (thin lines). From Pöhlmann et al. (2004). With permission from The American Physical Society

19.7.2 The MgO–SiO_2 System

One year later, Stixrude and Karki (2005) provided another important application of AIMD to Earth's sciences, investigating the structure and the freezing point of $MgSiO_3$ Liquid in Earth's Lower Mantle. Computations were carried out at the Brillouin zone center Γ under local density approximation level (LDA; Ceperley and Alder 1980; see also Sect. 10.4). In their investigation, they adopted the ultrasoft pseudopotentials of Kresse et al. (1992), implemented in the Vienna ab initio simulation code package (Kresse and Hafner 1993a, b) and explored a range of volumes $V = V_x$ to $V = V_{x/2}$, where $V_x = 38.9$ cm^3/mol is the experimental value for the liquid at the ambient-pressure melting point (1830 *K* according to Lange and Carmichael 1987). Simulations were performed in the canonical ensemble [constant NVT; see Sect. 13.1] through the use of the Nosé thermostat (Nosé 1984). The electronic structure at the Brillouin zone center was computed at an energy cutoff of 400 *eV*. Since the Pulay finite basis set corrections to the pressure (P_{Pulay}) were found to vary linearly with volume V from 2 *GPa* at $V = V_x$ to 5 *GPa* at $V = V_{x/2}$, a uniform correction of 2 *GPa*, for over-binding of the LDA (P_{emp}), was added to the pressure on top of the Pulay term (as previously done by

Oganov et al. 2001; and Wentzcovitch et al. 2004; on the Perowskite polymorph of $MgSiO_3$). This further readjustment involves, for thermodynamic self-consistency, an adjustment to the energy of the type:

$$E_{emp}(V) = -P_{emp}(V - V_0) \tag{19.7.6}$$

with V_0 as zero pressure reference volume and $E_{emp}(V) = 0$ (de Koker and Stixrude 2009).

The simulations included the influence of finite temperature on the electronic structure through the Mermin functional (see Sect. 8.8.3) and were confined to the Born–Oppenheimer surface. Stixrude and Karki (2005) found that, over the pressure regime of Earth's mantle, the mean silicon-oxygen coordination number of magnesium metasilicate liquid changes progressively from 4 at low P, to 6, attained at the pressure experienced by the system at the base of the Earth's mantle. The increase with P is smooth (i.e. not discontinuous or limited to narrow P ranges) and no discernible influence of temperature was found.

Because the thermal pressure increases markedly on compression, to reproduce the computational observations, the authors adopted the Mie-Grüneisen form

$$P(V,T) = P_c(V,T_0) + \frac{\gamma}{V}C_V(T - T_0) \tag{19.7.7}$$

P_c in Eq. (19.7.7) is the reference term at $T_0 = 3000\ K$, and γ is the Grüneisen parameter (See also Sect. 8.9.4, for a more extended appraisal of ab initio EOSs). The Mie-Grüneisen equation of state of the liquid involves a third-order Eulerian finite strain expression for P_c and $T_0 = 3000\ K$, $V_0 = 1.13\ V_x$, $K_0 = 10.1\ GPa$, $K_0' = 7.6$, $C_V = 3.7\ Nk_B$, and $\gamma = 0.21\text{–}1.75\ (V\text{–}V_0)/V_x$ (Fig. 19.37).

Since the internal energy U and pressure P vary linearly with temperature along isochors, γ and Cv may be obtained (within computational uncertainty) from the simulations, by applying (Stixrude and Karki 2005):

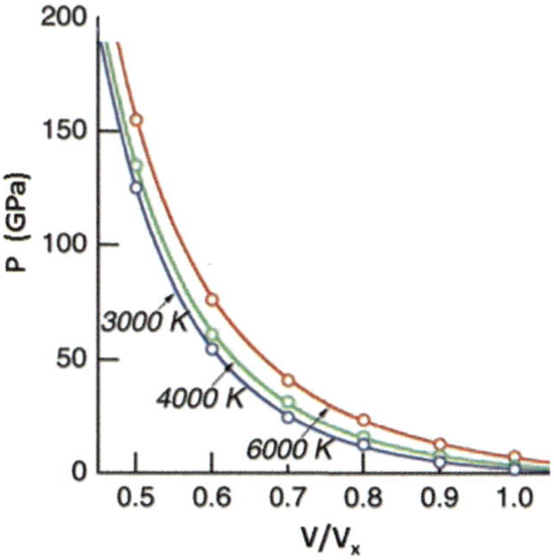

Fig. 19.37 Equation of state of $MgSiO_3$ liquid. FPMD results (open symbols) and Mie-Grüneisen equation of state fit (lines) to the FPMD results (uncertainties smaller than the size of the symbols). Modified from Stixrude and Karki (2005)

$$C_V = \left(\frac{\partial U}{\partial T}\right)_V \tag{19.7.8}$$

$$\gamma = \frac{V}{C_V}\left(\frac{\partial P}{\partial T}\right)_V \tag{19.7.9}$$

As the coordination number of the liquid increases on compression, the liquid adopts values of γ characteristic of the more highly coordinated state. This unusual behavior is consistent with theoretical analyses of the influence of pressure-induced coordination changes on γ (Jeanloz and Roufosse 1982) and it is consistent with previous observations (Boehler and Kennedy 1977; Wasserman et al. 1993; Vočadlo et al. 2003). The density contrast between liquid and crystal decreases by a factor of nearly 5 over the mantle pressure regime and is 4% at the core-mantle boundary.

By integrating the Clausius–Clapeyron equation, Stixrude and Karki (2005) obtained the melting curve $\left(\frac{\partial T_M}{\partial P}\right)$ of $MgSiO_3$:

$$\frac{\partial T_M}{\partial P} = \frac{\Delta V}{\Delta H/T} \tag{19.7.10}$$

The magnitudes on the right-hand side of Eq. (19.7.10) were obtained from AIMD over the entire lower mantle regime. The integration constant was set by assuming a single fixed point along the melting curve (P = 25 *GPa*, T = 2900 *K*). The ab initio melting curve yielded a melting temperature at the core-mantle boundary of 5400–5600 *K*.

Five years later, the same authors (Karki and Stixrude 2010) investigated through AIMD the transport properties of $MgSiO_3$ and found that the shear viscosity (η) of $MgSiO_3$ liquid varies by two orders of magnitude over the mantle pressure regime. The shear viscosity was calculated with the Green–Kubo relation:

$$\eta = \frac{V}{3k_B T}\int_0 \left\langle \sum_{i<j} \sigma_{ij}(t+t_0)\cdot\sigma_{ij}(t_0)\right\rangle \tag{19.7.11}$$

σ_{ij} in Eq. (19.7.11) is the stress tensor over the parametric coordinates (i and j = x, y, z), t is time, with t_0 as origin, and the remaining symbols have their usual significance. The activation volume was obtained as the partial derivative of the natural logarithm of shear viscosity with respect to pressure at constant T:

$$V_\eta^* = \left(\frac{\partial \ln \eta}{\partial P}\right)_T \tag{19.7.12}$$

The calculated viscosity varies systematically with P by a factor of ~ 140 over the entire mantle pressure regime at 4000 *K*. The increase is rather regular over most part of the investigated P–T range but for the lowest P, T conditions, where is observed an inversion near 5 *GPa* at 3000 *K*. This inversion is ascribed by the Authors to the presence of V-fold coordinated silicon. The activation volume also varies systematically, with a tendency to decrease with increasing P and with increasing T. The contrasting effects of T and P result in modest increases of viscosity along temperature profiles characteristic of Earth's interior (i.e. a factor of 10 from the surface to the core-mantle boundary when moving along a slightly super-liquidus magma ocean isentrope such as that one proposed by Stixrude et al. 2009).

More recently, de Koker et al. (2013) extended the investigation to other intermediate components in the MgO–SiO_2 system: Mg_5SiO_7, $Mg_3Si_2O_7$, $MgSi_2O_5$, $MgSi_3O_7$, and $MgSi_5O_{11}$.

This investigation, combined with previous observations on pure liquid MgO (de Koker and Stixrude 2009; Karki et al. 2006), pure liquid SiO_2 (Karki et al. 2007), Mg_2SiO_4 (de Koker et al. 2008), $MgSiO_3$ (de Koker and Stixrude 2009; Stixrude and Karki 2005) and an internally consistent thermodynamic treatment of the AIMD observables (de Koker and Stixrude 2009) allowed the authors to formulate a quite complete and convincing picture of the thermodynamics of the entire MgO–SiO_2 system in the liquid state, as resumed in Table 19.22.

Due to the seminal importance of this treatment in Geochemistry, I resume concisely the computational strategy adopted by the authors, remanding the interested reader to the original works of de Koker and Stixrude (2009) and Stixrude et al. (2009), de Koker et al. (2013) for a more extensive and in depth treatment.

It is assumed as framing relation that the Helmhotz free energy of the liquid is composed of three separable terms (Mc Quarrie 1984; de Koker and Stixrude 2009):

$$F(V, T) = F_{ig}(V, T) + F_{xs}(V, T) + F_{el}(V, T) \qquad (19.7.13)$$

where $F_{ig}(V, T)$ is an ideal gas term arising from momenta of the atoms (McQuarrie 1984):

$$F_{ig} = \sum_M N_M F_{ig,M} + RT \sum_M N_M \ln X_M \qquad (19.7.14)$$

$$F_{ig,M} = -RT\left[\ln V_M + \frac{3}{2}\ln T + \frac{3}{2}\ln\left(\frac{m_M k_B}{2\pi\hbar^2}\right) + 1\right] \qquad (19.7.15)$$

with m_M, V_M and X_M as mass, unit volume and fraction of particles of type M and N_M as number of atoms of each type per formula unit.

$F_{xs}(V, T)$ is an excess "*ideal gas*" term expanded in f and θ about the origin, in a 2-D Taylor series (de Koker and Stixrude 2009):

$$f = f(V); \quad f(V_0) = 0; \quad \theta(T_0) = 0 \qquad (19.7.16)$$

Table 19.22 Thermodynamic properties of liquids in the $MgO–SiO_2$ binary join at 0 *GPa* and 3000 *K* after de Koker et al. (2013)

Compound	MgO	Mg_5SiO_7	Mg_2SiO_4	$Mg_3Si_2O_7$	$MgSiO_3$	$MgSi_2O_5$	$MgSi_3O_7$	$MgSi_5O_{11}$	SiO_2
X_{SiO2}	0	0.167	0.333	0.4	0.5	0.667	0.75	0.833	1
V/V_0	16.46	107.50	58.4	100.5	41.8	67.5	93.5	146	27.8
K_T (*GPa*)	33.49	24.33	19.5	18.35	17.2	12.92	12.53	13.06	6.2
K_S (*GPa*)	41.75	29.49	20.7	19.55	18.1	13.1	12.96	13.2	6.2
K_T'	4.81	5.61	17.97	5.97	6.17	6.9	6.72	5.161	4.94
$\alpha \times 10^{-6}$ (K^{-1})	95	107	195	71	65	40	63	35	39
S [$J/(mol \times K)$]	174.1	1125.4	1785.3	1010.8	415.3	641.1	859.3	1283.2	205.5
C_V (Nk_B)	3.6	3.9	12.3	4.2	4.2	4.5	4.4	4.5	4.6
C_P (Nk_B)	4.5	4.8	13.2	4.5	4.4	4.5	4.6	4.6	4.6
γ	0.87	0.66	0.93	0.31	0.27	0.12	0.18	0.1	0.06

$$F_{xs}(V,T) = \sum_{i=0}^{\vartheta'_f} \sum_{j=0}^{\vartheta'_\theta} \frac{a_{ij}}{i!j!} f^i \theta^j \tag{19.7.17}$$

where $\vartheta'_f, \vartheta'_\theta$ denote the respective orders of expansion in f and θ and the primes indicate that the summations extend to $i + j < \vartheta'_f, \vartheta'_\theta$.

$F_{el}(V, T)$ is the electronic part of the Helmholtz free energy:

$$F_{el}(V,T) = -\varsigma \left[\frac{1}{2} T - T_{el} - T T_{el} \ln \frac{T}{T_{el}} \right] \tag{19.7.18}$$

with ζ as thermo-electronic heat capacity coefficient depending only on V. This form is obtained by integration, by assuming that the electronic heat capacity is zero at $T < T_{el}$. Under the same proviso, the electronic entropy is

$$S_{el} = \varsigma \left[T - T_{el} - T_{el} \ln \frac{T}{T_{el}} \right] \tag{19.7.19}$$

The AIMD electronic entropy is obtained from the electronic eigenvalues ($\in_i$) (Kresse and Furthmüller 1996):

$$S_{el}^{DFT} = -2k_B \sum_i \{ \phi(\in_i) \ln(\phi \in_i) + [1 - \phi(\in_i)] \ln[1 - \phi(\in_i)] \} \tag{19.7.20}$$

where $\phi(\in_i)$ is the Fermi–Dirac distribution function

$$\phi(\in_i) = \frac{1}{\exp\left(\frac{\in_i - \in_f}{k_B T}\right) + 1} \tag{19.7.21}$$

$\in_f$ is the Fermi energy.

Knowing the electronic entropy, the V-dependent thermo-electronic heat capacity coefficient ζ is easily obtained through Eq. (19.7.19). This allows to obtain the electronic component of the Helmholtz free energy (Eq. 19.7.18), the electronic pressure (P_{el}), and the electronic internal energy (U_{el}) by applying:

$$P_{el} = \frac{\partial \varsigma}{\partial V} \left[\frac{1}{2} \left(T^2 - T_{el}^2 \right) - T T_{el} \ln \frac{T}{T_{el}} \right] + \varsigma \frac{\partial T_{el}}{\partial V} \left[\left(T - T_{el} \right) - T \ln \frac{T}{T_{el}} \right] \tag{19.7.22}$$

$$U_{el} = \frac{1}{2} \varsigma (T - T_{el})^2 \tag{19.7.23}$$

The ideal gas contributions so pressure and internal energy (P_{ig}, U_{ig}) are obtained from the fundamental relations:

$$P_{ig} = -\left(\frac{\partial F_{ig}}{\partial V}\right)_T \tag{19.7.23}$$

$$U_{ig} = F_{ig} - T\left(\frac{\partial F_{ig}}{\partial T}\right)_V = F_{ig} + TS_{ig} \tag{19.7.24}$$

Removing the relevant U_{ig}, P_{ig}, U_{el}, P_{el} from the AIMD internal energy and pressure (E_{MD}, P_{MD}) allows the parameterization of the excess terms U_{xs}, P_{xs}:

$$U_{xs}(V,T) = U_{MD} - U_{el} - U_{ig} = \sum_{i=0}^{\vartheta'_f}\sum_{j=0}^{\vartheta'_\theta} \frac{a_{ij}}{i!j!} f^i \theta^{j-1}\left(\theta - ij\frac{\partial T}{\partial T}\right) \tag{19.7.25}$$

$$P_{xs}(V,T) = P_{MD} - P_{el} - P_{ig} = \sum_{i=0}^{\vartheta'_f}\sum_{j=0}^{\vartheta'_\theta} \frac{ia_{ij}}{i!j!} f^{i-1}\theta^j \tag{19.7.26}$$

A single least squares inversion applied to Eqs. (19.7.25 and 19.7.26) allows to constrain coefficients a_{ij}, which in turn are related to physical properties at the reference pressure (P_0) and temperature (T_0) (see Appendix A in de Koker and Stixrude 2009).

The remaining thermodynamic parameters follow by fundamental relations:

$$K_T = V\left(\frac{\partial^2 F}{\partial V^2}\right)_T \tag{19.7.27}$$

$$\alpha = -\frac{1}{K_T}\left(\frac{\partial^2 F}{\partial V \partial T}\right) \tag{19.7.28}$$

$$C_V = -T\left(\frac{\partial^2 F}{\partial T^2}\right)_V \tag{19.7.29}$$

$$\gamma = \frac{\alpha K_T V}{C_V} \tag{19.7.30}$$

$$C_P = C_V(1 + T\alpha\gamma) \tag{19.7.31}$$

$$K_S = K_T(1 + T\alpha\gamma) \tag{19.7.32}$$

$$K'_T = \left(\frac{\partial K}{\partial P}\right)_T \tag{19.7.33}$$

$$G = F - V\left(\frac{\partial F}{\partial V}\right)_T = F + PV \qquad (19.7.34)$$

The investigation of de Koker et al. (2013) shows that the progressive increase of Z_{Si-O} coordination number from 4 to 6 upon twofold compression, from $V/V_x = 1$ to $V/V_x = 0.5$, already observed by de Koker and Stixrude (2005) in $MgSiO_3$, is common to all considered compounds in the join but SiO_2, in which the increase with P is somewhat delayed. This behavior is paralleled by the Grüneisen parameter (see Fig. 4 in de Koker et al. 2013). The excess volume along the $MgO–SiO_2$ join is quite asymmetric, especially at low P, and becomes progressively less marked with the increase of P, attaining virtual ideality at P = 135 *GPa*. The investigation, coupled with a similar approach to the thermodynamics of solids nucleating in the system at high P, has deep geophysical implications in quantifying the density contrast between co-existing liquid and crystals and providing a plausible explanation for the origin of ultra-low velocity zones at the base of the mantle, ascribed to the presence of a liquid phase.

19.7.3 The $MgO–SiO_2–H_2O$ System

An AIMD investigation of an hydrous $MgSiO_3–H_2O$ liquid, with 10 wt% water, was carried out by Mookherjee et al. (2008) in the framework of LDA with pseudo-potentials. They found that pressure has a deep influence on speciation of the water component, which changes from being dominated by hydroxyls and water molecules at low P to extended structures at high P.

The isochoric heat capacity, C_V, obtained with the same procedure previously described for $MgSiO_3$, decreases by about 8% upon twofold compression, from 3.63 ± 0.21 to 3.34 ± 0.11 Nk_B, where N is the number of atoms in the micro-canonical ensemble. The Grüneisen parameter, γ, increases by a factor of three over the same compression range. The density of the hydrous melt is less than that of the anhydrous melt under all explored P, T conditions. However, according to Mookherjee et al. (2008), the density contrast between anhydrous and hydrous system is "*small enough that hydrous melts may be neutrally buoyant in the mantle*", providing a possible explanation of a low-velocity layer on top of the 410 *km* discontinuity (Song et al. 2004). With the increase of pressure, the partial molar volume of water in the silicate melt, $\overline{V}_{H_2O}$ approaches asymptotically that of pure water, and becomes virtually identical to it at the highest investigated P condition (see Fig. 4b in Mookerjee et al. 2008).

According to Karki and Stixrude (2010), a silicate melt with 10% in weight of H_2O is two to four times less viscous than the anhydrous melt at all P, T conditions of their AIMD simulation, and the regions of anomalous pressure dependence of the viscosity and diffusion seen for $MgSiO_3$ are weak or absent in the case of hydrous silicate liquid. The dynamical enhancement induced by H_2O (smaller viscosity and

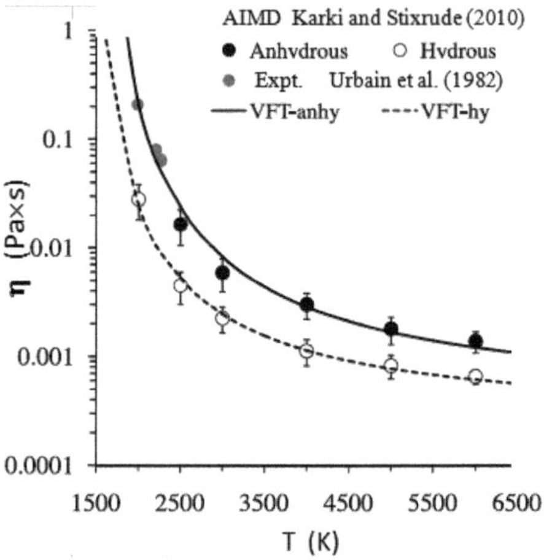

Fig. 19.38 AIMD shear viscosity (η) of anhydrous (solid symbols) and hydrous (open symbols) $MgSiO_3$ melts, compared with experimental evidences at ambient P and high T. The solid and dashed lines are the loci of VFT equations. Modified from Karki and Stixrude (2010)

larger diffusivity) was taken by the Authors as an evidence that "*water systematically depolymerizes the melt*" (Karki and Stixrude 2010).

As shown in Fig. 19.38, the calculated viscosities in both anhydrous $MgSiO_3$ and hydrous system show large and systematic deviations from an Arrhenian behavior and are consistent with a Vogel-Fulcher-Tammann (VFT) equation of type:

$$\eta(\mathrm{T}) = \mathrm{A}\exp\left(\frac{\mathrm{B}}{\mathrm{T}-1000}\right) \tag{19.7.12}$$

with A = 0.00033 $Pa \times s$ and B = 6400 K for anhydrous liquid, and A = 0.00024 $Pa \times s$ and B = 4600 K for hydrous liquid.

19.7.4 More Complex Systems

Through the utilization of maximally localized Wannier orbitals (see Sect. A-21 and 18.1.3), Vuilleumier et al. (2011) were able to quantify the redistribution of electronic density around oxygen atoms as a function of the cationic environment in compositionally complex melts such as Rhyolite and $An_{36}Di_{64}$ (besides pure silica and $MgSiO_3$). Computations were carried out with the CPMD code at DFT-BLYP theory level. Trouiller–Martins pseudopotentials were used for the core electrons of all atoms but Al, for which it was employed a Bachelet–Hamann–Schlüter pseudopotential (Bachelet et al. 1982). For Mg, it was adopted a non-local core correction while Na and K were described by semicore pseudopotentials. The non-local form for all atoms was then achieved through the Kleinman–Bylander procedure (Kleinman and Bylander 1982). The kinetic energy cutoff adopted in the plane wave expansions was 70 Ry. Composition and extension of the microcanonical ensembles are specified in Table 19.23. To reduce the computational

Table 19.23 Chemical composition (weight %) of the melts investigated by Vuilleumier et al. (2011). Values in brackets are the atoms of each species in the microcanonical ensemble

	Silica	Rhyolite	$An_{36}Di_{64}$	Enstatite
SiO_2	100 (33)	77.9 (26)	50.7 (18)	59.9 (20)
Al_2O_3	0	12.7 (5)	14.3 (6)	0
MgO	0	0	11.3 (6)	40.1 (20)
CaO	0	0	23.7 (9)	0
Na_2O	0	4.6 (3)	0	0
K_2O	0	4.7 (2)	0	0

time, the initial configurations for AIMD simulations were prepared from fully equilibrated runs generated by classical MD simulations based upon previously developed empirical potentials.

Table 19.24 lists the detailed bonding conditions in the four investigated media in terms of oxygen species: bridging oxygen $[O]^0$, oxygen functionals $[O]^-$, and unbound (or "*free*") oxygen ions $[O]^{2-}$. The AIMD runs were performed at the experimental density (measured or extrapolated) of the corresponding melt at atmospheric pressure. The temperature of the simulation runs fluctuated around 2360 K for SiO_2, 2420 K for rhyolite, 2500 K for $An_{36}Di_{64}$, and 2550 K for molten enstatite. Noteworthy features are the presence of a limited amount of triclusters in molten SiO_2 and Rhyolite, and the high amount of oxygen functionals in the strongly depolymerized $MgSiO_3$ melt. In this compound it is also observed the presence of a slight amount of free oxygen $[O]^{2-}$, consistent with Toop-Samis model calculations based on the procedures outlined at Sects. 19.2.2 and 19.2.3. (Fig. 19.39).

For each oxygen atom, there are four doubly occupied Wannier orbitals (WO). The four WOs are on average tetrahedrally distributed around each oxygen atom.

Table 19.24 Oxygen speciation in the investigated media, after Vuilleumier et al. (2011)

	Silica	Rhyolite	$An_{36}Di_{64}$	Enstatite
$[O]^0$	**100.0**	**97.4**	**60.7**	**34.0**
Si–O–Si	99.5	67.1	24.0	34.0
Si–O–Al	–	27.8	33.0	–
Al–O–Al	–	1.5	2.5	–
$O–Si_3$	0.5	0.3	0.0	0.0
$Si_2–O–Al$	–	0.4	0.9	–
$Al_2–O–Si$	–	0.2	0.3	–
$O–Al_3$	–	0.1	0.0	–
$[O]^-$	**0.0**	**2.5**	**39.1**	**65.9**
Si–O	–	2.5	36.1	65.5
Al–O	–	0.0	3.0	–
$[O]^{2-}$	**–**	**0.0**	**0.0**	**0.4**

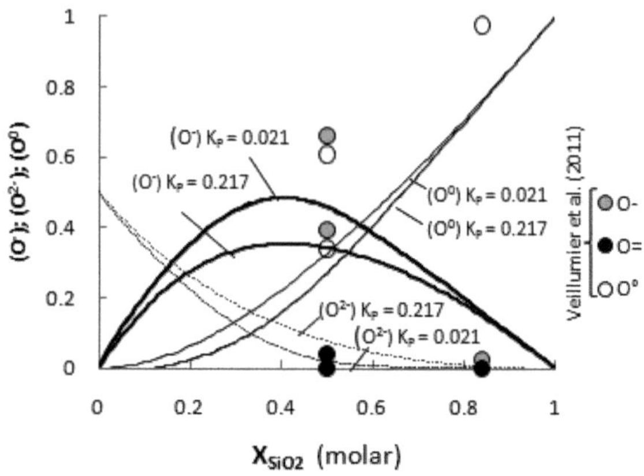

Fig. 19.39 AIMD oxygen speciation after Vuilleumier et al. (2011) compared with Toop-Samis model calculations (EPA); (see Table 19.5 for the adopted parameters)

Each WO, being localized, possesses a well-defined center (WOC), which occurs at a $R_{C\text{-}WOC}$ distance from the core center of the oxygen on which is localized. Figure 19.40 shows the distribution function of core-Wannier orbital center distances, $R_{C\text{-}WOC}$, associated with oxygen atoms in the four simulated melts, after Vuilleumier et al. (2011).

This distribution is bimodal, with the maximum centered at about 0.29 Å in Silica and Rhyolite and at a slightly larger distance ($\sim$0.34 Å) in $An_{36}Di_{64}$ and Enstatite, corresponding to lone pair 1p orbitals interacting with the NMs. The (common) maximum at 0.44 Å corresponds to an "*electronic displacement induced by Si–O–Si*"; Vuilleumier et al. (2011). Because in silica there are virtually only doubly-bonded (*bridging*) oxygens, the two distances 0.29 and 0.44 Å can be used as reference. Their evolution with composition (in terms of NMs) is interpreted by

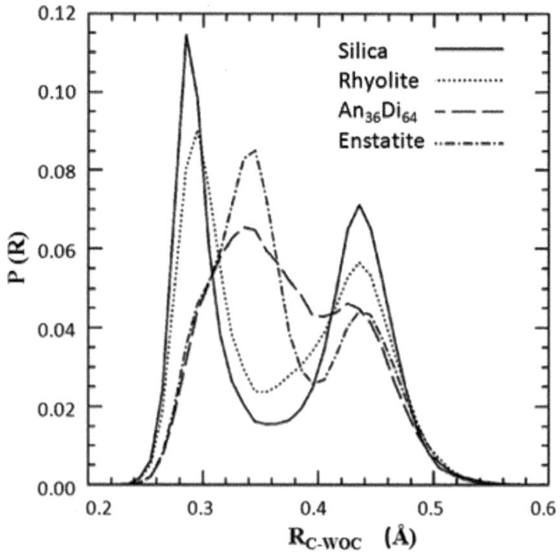

Fig. 19.40 Averaged distribution functions of the intra-atomic distances between the oxygen core and the four WOCs. From Villeumier et al. (2011). With permission from the Journal of Non Crystalline Solids

Table 19.25 Composition of the pyrolite melt after 0%, 30%, and 50% crystallization (Caracas et al. 2019). MO = Mantle Ocean

MO				30% crystallization		50% crystallization	
Atoms	AIMD	Oxides	wt%	AIMD	wt%	AIMD	wt%
O	80			86		81	
Si	24	SiO_2	44.58	21	38.96	18	34.38
Al	3	Al_2O_3	4.73	3	4.72	3	4.86
Fe	4	FeO	8.88	6	13.31	7	15.99
Mg	30	MgO	37.38	31	38.58	30	38.44
Ca	2	CaO	3.47	2	3.46	3	5.35
Na	1	Na_2O	0.96	1	0.96	1	0.99
Total	153	Total	100	150	100	143	100

Villeumier et al. (2013) as an evidence that the Al–O bond in silicate melts is less covalent than the Si–O bond. Also, the covalent character of the M–O bond appears to decrease progressively for the series Mg–O, Ca–O, Na–O, and K–O.

Quite recently Caracas and coworkers (Caracas et al. 2019; Solomatova and Caracas 2021) provided insightful AIMD investigations on "*pyrolite*".[8] After more than half a century, geologists still assign to this peculiar composition a particular relevance in deep mantle investigations. In the first investigation, Caracas et al. (2019) coupled diamond-anvil cell experiments and AIMD procedures to investigate the primordial Mantle Ocean (MO). Their primary interest was focused on the density contrast between primary phase crystals formed upon cooling of Mantle Ocean and the residual melt. The initial liquid composition of the pyrolite Mantle Ocean and the residual composition after extensive crystallization are listed in Table 19.25.

The AIMD investigation was carried out at the Γ point of the Brillouin zone with Projector Aughmented Wavefunctions (PAW; cf. Sect. 8.8.3), under the GGA approximation, with the PBE96 functional (Sect. 10.5.4). The adopted kinetic energy cutoff was 550 *eV* for the plane-wave basis set and of 800 *eV* for the augmentation charges. Due to the presence of iron in the microcanonical ensembles (Table 19.15), all simulations, at all volumes and temperatures, were spin-polarized. The local magnetic moments were set initially to 1 Bohr magnetons per Fe atom. This proved sufficient to break the symmetry of the magnetic wave-functions, that converged to their state within a few steps. All simulations were "*isokinetic*", i.e. fixed T, V, and number of particles (Caracas et al. 2019). These non-microcanonical

[8] The term "pyrolite" was coined long ago by Ringwood (1962): "*It is assumed that the mantle immediately below the Mohorovicic discontinuity consists dominantly of dunire and peridotire. This passes downward into more primitive material, which is chemically equivalent to a mixture of 1 part of basalt to 4 parts of dunite. For convenience, this hypothetical primitive material is called 'pyrolite.'*" (Ringwood, 1962).

simulations may be seen as "*an extreme version of the Nose–Hoover thermostat where the oscillation period of the temperature is set to the time step*" (Caracas et al. 2019). Because the kinetic energy does not fluctuate, the isochoric heat capacity could not be predicted, but this is the unique limitation of this much faster procedure, since the equations of state and diffusion properties are similar to the NVT ensembles, within error bars (see Hernandez and Caracas 2016; for an extended discussion). Cations were initially placed in the cubic boxes on a crystalline-like arrangement of ions and equilibrated at very high T (5000 *K*) followed by long thermalization runs (several picoseconds). Calculations were then performed at 3000, 4000, and 5000 *K*. Pressure was progressively increased by decreasing the volume of the simulation cell, attaining values exceeding 150 *GPa*. The computed density–pressure points were fitted with a third-order Birch-Murnagham EOS at each isotherm [see Sect. 8.9.2]. The thermal expansion was fitted to the modified expansion proposed by Holland and Powell (1998):

$$V_T = V_0\left[1 + \alpha_0(T - T_0) - 2\alpha_1\left(T^{1/2} - T_0^{1/2}\right)\right] \tag{19.7.12}$$

with α_0 and α_1 as thermal expansion coefficients and T_0 = 3000 *K* as reference T at all pressures. The densities of the high-P polymorph perovskite ($MgSiO_3$) and bridgmanite ($Mg_{0.75}Fe_{0.25}SiO_3$) at 3000 and 4000 *K* were calculated as well. Table 19.26 resumes the computational results in terms of computed pressure for various densities at the three investigated temperatures.

The results indicate that, upon crystallization, the density of the residual melt increases progressively. The increase at 30% crystallization is around 0.05–0.1 *g/cm*^{3}, and about 0.15–0.2 *g/cm*^{3} at 50% crystallization. Moreover, these increases are systematically larger at higher pressures and temperatures. With the progressive increase of FeO in the residual liquid, Bridgmanite replaces Perovskite as crystallizing phase, and the density of the crystals increase by about 0.2 *g/cm*^{3}.

Concerning the short-range order, the observations of Caracas et al. (2019) are in line to what previously seen by other researchers: while at low P the IV-fold coordination dominates (SiO_4 and AlO_4 tetrahedra), with the increase of P, the coordination number passes progressively to 5, 6 and 7. At the bottom of Mantle Ocean, most of Fe is surrounded by 7 or 8 O atoms, and Mg by 8 or 9. The increased coordination number of Si and Al is compensated by progressive polymerization.

The iron atoms, which at low pressure have a local magnetic moment close to 4 Bohr magnetons (cf. Sect. 6.1.2), attain at 4000 *K* and 85 *GPa*, an intermediate spin configuration (close to 2). But, even at the pressures and temperatures characteristic to the bottom of the Mantle Ocean, the nature of the melt is still "*metallic*" since the individual magnetic moment on the iron atoms is non-null.

The viscosity of the melt does not evolve much during crystallization (at least up to 50% solidification) and the bulk transport properties of the crystal mush are rather mainly affected by the configuration of solid crystals and their accumulation according to the buoyancy relations.

Table 19.26 Computed pressure (*GPa*) at 3000, 4000, and 5000 *K* for the initial Mantle Ocean and after 30 and 50% crystallization. (Caracas et al. 2019)

Density (g/cm^3)	T = 3000 *K*	T = 4000 *K*	T = 5000 *K*
0% crystals			
5.377	138.6 ± 10.9	147.3 ± 3.2	157.9 ± 3.2
4.645	73.1 ± 9.4	83.8 ± 2.7	91.7 ± 3.3
4.04	41.4 ± 2.6	48.0 ± 2.8	53.6 ± 2.8
3.535	22.7 ± 1.7	27.0 ± 2.4	31.2 ± 2.4
3.111	11.8 ± 1.6	14.5 ± 1.9	17.5 ± 2.0
2.753	4.8 ± 1.4	7.1 ± 1.7	9.3 ± 1.9
2.447	1.3 ± 1.3	2.9 ± 1.5	4.6 ± 1.6
2.185	−0.6 ± 1.3	0.6 ± 1.4	
1.947		−0.2 ± 1.3	0.5 ± 1.2
30% crystals			
6.278			255.1 ± 3.8
5.7191	69.0 ± 1.9		
5.383	132.1 ± 2.1	138.9 ± 2.5	149.1 ± 3.1
4.65	70.3 ± 2.1	78.1 ± 4.1	87.0 ± 2.9
4.044	38.4 ± 2.3	45.0 ± 2.9	50.7 ± 2.6
3.539	21.1 ± 1.8		
3.115	10.7 ± 1.6	13.0 ± 1.8	16.4 ± 2.0
2.738	4.4 ± 1.5	6.2 ± 1.7	6.8 ± 1.6
2.434	1.1 ± 1.5	2.5 ± 1.3	4.1 ± 2.0
2.174	−0.7 ± 1.1	0.5 ± 1.3	
1.949			−0.3 ± 1.3
50% crystals			
6.29	203.4 ± 2.4	212.8 ± 2.9	223.9 ± 3.3
5.393	111.9 ± 2.2	120.6 ± 2.7	129.7 ± 3.0
4.658	60.4 ± 2.1	68.0 ± 2.6	75.8 ± 3.1
4.052	32.9 ± 2.7	38.4 ± 2.4	43.8 ± 2.5
3.121	8.0 ± 1.4	10.6 ± 2.1	13.9 ± 2.0
2.455	0.2 ± 1.3	1.6 ± 1.5	3.3 ± 1.9
2.124	−1.0 ± 1.3	0.0 ± 1.2	2.7 ± 1.5
1.904		−0.5 ± 1.1	0.3 ± 1.2

Based on their observations, Caracas et al. (2019) defined a Density Crossover Depth (DCD) as the depth at which the densities of the residual liquid and the crystallizing phases are identical (i.e. zero buoyancy). The first crystals forming from a fully molten pyrolite at the bottom of Mantle Ocean [$110 \leq$ P (*GPa*) $\leq$ 120; $4000 \leq$ T (*K*) $\leq$ 500] are neutrally buoyant. The progressive removal of Si and Fe enrichment upon crystallization moves the neutral buoyancy region to ~80 *GPa*. At 50% crystallization [$3000 \leq$ T (*K*) $\leq$ 3500] neutral buoyancy occurs at around 50 *GPa* (DCD ≈ 1250 *km*).

Table 19.27 Volatile Concentrations in Pyrolite Melt (py) wt% = weight percent; mol% = mole percent. Also listed are the numbers of atoms in the microcanonical ensembles (Solomatova and Caracas 2021)

Compound	H (mol%)	H (wt%)	H_2O (wt%)	C (mol%)	C (wt%)	CO_2 (wt%)	Atoms
Pure pyropite (py)	0	0	0	0	0	0	155
py-2Mg+4H	0.13	2.58	1.13	0	0	0	157
py-4Mg+8H	0.26	5.1	2.29	0	0	0	161
py-8Mg+16H	0.53	9.94	4.72	0	0	0	165
py+4H_2O	0.24	4.85	2.18	0	0	0	177
py+8H_2O	0.48	9.04	4.27	0	0	0	161
py+4CO	0	0	0	1.44	2.48	5.26	165
py+4CO_2	0	0	0	1.41	2.42	5.16	169
py+8CO	0	0	0	2.78	4.73	10.2	177
py+8CO_2	0	0	0	2.68	4.52	9.82	166
py-2Mg-Si+8H+4CO	0.25	4.82	2.2	1.47	2.41	5.37	155

Two years after this study, Solomatova and Caracas (2021) with the same AIMD procedures studied the role of added volatiles to the pyrolite melt. Table 19.27 resumes the investigated compositions.

The CO, CO_2, and H_2O molecules were added in the interstitial voids of pyrolite melt at low pressure. After thermalization, the melts were compressed in about 6–10 steps along the same isotherms previously investigated by Caracas et al. (2019) (3000, 4000, 5000 *K*). Additional simulations at T = 2000 *K*, on volatile-free and volatile-bearing pyrolite melts with 4.85 wt% H and 2.42 wt% C, were also performed over a pressure range of about 0 to 40 *GPa*, in order to have a direct comparison with experimental observations.

The introduction of hydrogen in the ensemble involved the Mg ⇔ 2H substitution, in three concentrations. A more complex substitution (2 Mg + Si ⇔ 8H) was also conceived in order to minimize the effect on the Mg:Si ratio. The addition of four CO molecules on top of this substitution was imagined to mimic a carbonated-hydrous melt (last row in Table 19.27).

The addition of volatiles renders all the melts more buoyant compared to the reference volatile-free pyrolite (Caracas et al. 2019). The effect is pressure-dependent, largest at the surface, decreasing to about 20 *GPa*, and remaining roughly constant to 135 *GPa*. At all pressures, the polymerization of the silicate melt is strongly affected by the amount of oxygen added to the system while being only weakly affected by the specific type of added volatile. The computational results in terms of ambient pressure density (ρ_0), bulk modulus (K_0), and its isothermal derivative $\left[K' = \left(\partial K/\partial P\right)_T\right]$ are shown in Table 19.28. Also shown, for comparative purposes, is the density at 135 *GPa* (ρ_{135}) and the bulk modulus at the same pressure (K_{135}), along the various isotherms.

Table 19.28 Results of the third-order Birch Murnaghan EOS fitting on AIMD observations along various isotherm (Solomatova and Caracas 2021). Bulk moduli are in *GPa* and densities are in *g/cm*^{3}

	ρ_0	K_0	K'	ρ_{135}	K_{135}
T = 2000 K					
Pure pyrolite (py)	2.49 ± 2	24 ± 2	4.7 ± 2	–	–
py+4H_2O	2.32 ± 3	15 ± 2	5.5 ± 3	–	–
py+4CO_2	2.37 ± 5	18 ± 3	5.0 ± 4	–	–
T = 3000 K					
pure pyrolite (py)	2.26 ± 2	14 ± 1	5.7 ± 2	5.32 ± 1	536 ± 4
py-2Mg+4H	2.17 ± 3	11 ± 2	6.1 ± 3	5.32 ± 2	544 ± 10
py-4Mg+8H	2.13 ± 4	11 ± 1	6.0 ± 3	5.26 ± 1	534 ± 5
py-8Mg+16H	1.98 ± 5	8 ± 1	6.4 ± 4	5.20 ± 1	533 ± 3
py+4H_2O	2.11 ± 6	10 ± 2	6.3 ± 4	5.24 ± 1	540 ± 4
py+4CO	2.11 ± 7	9 ± 2	6.5 ± 5	5.29 ± 2	541 ± 7
py+4CO_2	2.12 ± 7	10 ± 2	6.2 ± 4	5.27 ± 1	538 ± 5
T = 4000 K					
Pure pyrolite (py)	2.08 ± 3	12 ± 1	5.4 ± 1	5.24 ± 1	512 ± 2
py-2Mg+4H	2.01 ± 2	10.6 ± 8	5.4 ± 1	5.21 ± 1	510 ± 2
py-4Mg+8H	1.94 ± 2	8.9 ± 6	5.6 ± 1	5.18 ± 1	512 ± 2
py-8Mg+16H	1.74 ± 4	5.2 ± 8	6.2 ± 2	5.11 ± 1	512 ± 2
py+4H_2O	1.94 ± 6	9 ± 1	5.6 ± 2	5.16 ± 1	513 ± 3
py+8H_2O	1.81 ± 6	7 ± 1	5.6 ± 2	5.09 ± 1	505.7 ± 5
py+4CO	1.85 ± 5	6 ± 1	6.2 ± 3	5.20 ± 1	517 ± 4
py+4CO_2	1.92 ± 3	8.2 ± 9	5.8 ± 2	5.18 ± 1	515 ± 2
py+8CO	1.69 ± 6	4 ± 1	6.7 ± 5	5.16 ± 1	515 ± 2
py+8CO_2	1.86 ± 4	7 ± 1	5.8 ± 3	5.15 ± 1	512 ± 3
py-2Mg-Si+8H+4CO	1.78 ± 3	6.1 ± 6	5.9 ± 1	5.13 ± 1	509 ± 1
T = 5000 K					
pure pyrolite (py)	1.80 ± 4	7.2 ± 9	5.5 ± 2	5.15 ± 2	499 ± 3
py-2Mg+4H	1.73 ± 6	6 ± 1	5.6 ± 2	5.11 ± 1	497 ± 3
py-4Mg+8H	1.53 ± 7	3 ± 1	6.0 ± 4	5.08 ± 4	495 ± 5
py-8Mg+16H	1.47 ± 7	2.8 ± 9	6.4 ± 5	5.00 ± 2	498 ± 3
py+4H_2O	1.66 ± 7	6 ± 1	5.6 ± 3	5.06 ± 2	498.8 ± 4
py+4CO	1.60 ± 5	4.1 ± 8	6.0 ± 3	5.11 ± 1	498 ± 3
py+4CO_2	1.65 ± 7	5 ± 1	5.8 ± 3	5.10 ± 2	497 ± 3

The data concerning pure pyrolite along the 2000 *K* isotherm are in good agreement with experimental observations, and, more generally, the comprehensive set of data is quite important for the comprehension of deep mantle rheology and evolution of the early Mantle Ocean.

Chapter 20
Crystals

The treatment of natural crystals cannot be very comprehensive, for obvious reasons, and I limit myself to some rock-forming minerals and to some phases of peculiar interest in Geochemistry and Cosmochemistry, that received the dedicated attention of Quantum Chemists. Excellent textbooks appeared on this matter in the past. I may quote, for instance, especially for what concerns carbonates and sulfides, the theoretical compendium of Tossel and Waughan (1992). The Native Element Carbon offers the opportunity to report here what has been done, from the Quantum Mechanical point of view, on the Band Theory, with the goal of understanding the properties of the various allotropes and of the relations between *Graphite* and monolayer *Graphene* (so important for industrial applications). Copper offers the opportunity of exploring the electronic arrangements responsible of its high electrical and thermal conductivities, and the role of pressure in enhancing the stability of the crystalline state with respect to the liquid. In the case of oxides, multiple oxides, and silicates, emphasis is given to the simplest thermodynamic and thermophysical observables, i.e. those necessary to attain a sufficient capability of reproducing observed intrinsic and extrinsic stability limits. Kinetic aspects are treated only in an ancillary way, and I refer the interested reader to Volume 3 of the Springer Series "*Advances in physical Geochemistry*" edited by S.K. Saxena" and Volume 8 of the MSA Series "*Reviews in Mineralogy*" edited by P.P. Ribbe and to Lasaga (1998) for a more in-dept treatment. In passing, the interesting properties of Magnetite "*The simplest compound for which interesting relations appear to occur between electronic conductivity and the arrangement of cations*" (Verwey et al. 1947), and so important in Earth Sciences for the development of the Vine–Matthews theory of sea floor spreading (Vine and Matthews 1963), are also briefly outlined.

G. A. Ottonello, *Quantum Geochemistry*, Springer Geochemistry,
https://doi.org/10.1007/978-3-031-21837-8_20

20.1 Native Elements

Among the 3D-periodic substances that form in nature as pure elements belong *metals* and *metalloids*. Some of these occurrences have practical importance. Among *metals*, Cu, Ag, and Au in the group of Gold, Pt, and Fe in the Platinum group. Among *metalloids*, only C and S have practical importance, while As, Sb, and Bi (all isomorphous) occur rarely in nature as primary ores. I make here a further selection, restricting the discussion to elements whose quantum chemical treatment promoted significant seminal investigation. I treat first the carbon allotropes for applications to metalloids and then the copper allotropes for applications to metals.

20.1.1 Carbon

Although at present are known roughly 500 tridimensional allotropes[1] of carbon (either original or "*duplicate*" observations, see Fig. 3 in Hoffmann et al. 2016), we are here interested only in the main naturally occurring forms of graphite and diamond and the relations between graphite and graphene.

20.1.1.1 Graphite and Graphene

The natural occurrences of graphite are mainly of metamorphic origin, by recrystallization of carbonaceous substances. Subordinated amounts of graphite are found in pegmatites and other intrusive rocks. The "normal" structure of graphite is hexagonal holohedral (space group $D_{6h}^{4} - P6_3/mmc$). This structural form was first proposed by Hull (1917) and then confirmed by Bernal (1924) and Hassel and Mark (1924). Lukesh (1950), however, stressed that, according to X-rays observations, the symmetry of graphite can be no higher than twofold, due to the presence of unequal C–C bond lengths. Lukesh and Pauling (1950) proposed thus that the structure has actually space-group $D_{2h}^{23} - Fmmm$ and constants $a_0 = 2.456$, $b_0 = 4.254$, and $c_0 = 6.696$. In the Bernal structure, the carbon atoms form a regular hexagonal ring-structure (Fig. 20.1) with axes $a_0 = 2.46$ and $c_0 = 6.7$ (Laves and Baskin 1956). Two carbon atoms have coordinates $00\frac{1}{4}$; $00\frac{3}{4}$ and two carbon atoms have coordinates $\frac{1}{3}\frac{2}{3}\frac{3}{4}$; $\frac{2}{3}\frac{1}{3}\frac{1}{4}$. The distance separating the two layers is so large that each layer may be considered as a giant molecule. The superimposed layer molecules are held together by weak van der Waal forces (and overlapping between the $2p_z$ orbitals of carbon atoms on neighboring layers; see later). The layer molecule "*resonates many valence-bond structures*" (Pauling 1949).

[1] The adjective term "*allotrope*" is deserved to elements, the corresponding term "*polymorph*" is more appropriate for chemically complex substances.

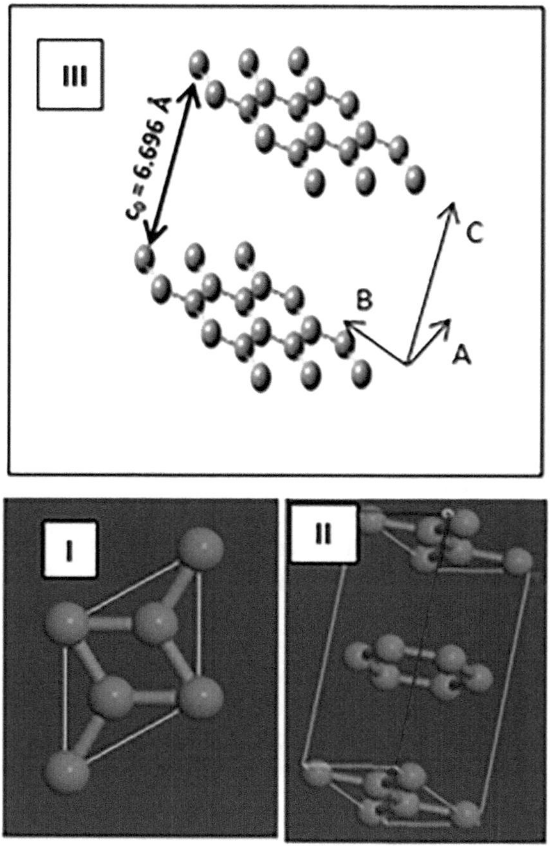

Fig. 20.1 Structure of graphite. (I) $D_{6h}^{4} - P6_3/mmc$; $a_0 = b_0 = 2.456$ Å, $c_0 = 6.696$ Å, $\alpha = \beta = 90°$, $\gamma = 120°$ (Wyckoff 1963). (II) $D_{2h}^{23} - Fmmm$; $a_0 = 2.456$ Å, $b_0 = 4.254$ Å, $c_0 = 6.696$ Å, $\alpha = \beta = \gamma = 90°$ (Lukesh and Pauling 1950). (III) hexagonal layers along the C-axis. Parts (I) and (II) generated by the AMCSD software (Downs and Hall-Wallace 2003)

Natural graphite normally occurs with the hexagonal allotrope, but, in some occurrences, up to 5% of the structure is rhombohedral (Debye and Scherrer 1916; Laves and Baskin 1956) ($a_0 = 3.635$ Å; $\alpha = 39.49°$; space group $D_{3d}^{5} - \overline{R}3m$) with the six C atoms at ± (u,u,u), and u = 1/6 (Laves and Baskin 1956). All graphite allotropes are made up of interconnected planes of carbon atoms stacked with an inter-planar spacing along the C-axis (~3.35 Å; see Fig. 20.1). The various allotropes exhibit, however, different stacking sequences of the carbon planes. In the AAA stacking (simple hexagonal graphite), all carbon atoms belonging to consecutive layers are located directly on top of each other. In the Bernal allotrope, half of the atoms are placed directly above each other in adjacent planes, and the other half is placed above the center of the hexagonal ring in the adjacent plane (ABAB stacking). In the rhombohedral allotrope, half of the atoms occur below atoms in the adjacent plane and half above hexagonal ring centers (ABC stacking; Charlier et al. 1994).

The two-dimensional honeycomb structure of the carbon monolayers is known as "*graphene*" (with the suffix "*ene*" reflecting the presence of numerous Pauling's double bonds in the structure). Because each carbon atom in a single graphene sheet contributes one electron to a conduction band extending over the whole sheet, its unusual "*semimetallic*" properties stimulated an intensive research through the years. The most exciting property is undoubtedly superconductivity, associated to bands close to the Fermi level, and observed in twisted bilayer graphene (Oh et al. 2021; and references therein). From the initial mono- and bi-layers, attention progressively turned to graphene few-layer allotropes (FLG; see Lui et al. 2011; and references therein). The crystallographic stacking of the individual graphene sheets in FLG, provides an additional degree of freedom, with the electronic properties strongly influenced by the stacking order.

Charlier et al. (1994) investigated ab initio the AAA, ABAB, and ABC allotropes of graphite in the DFT framework under LDA. Their attention was mainly focused on the effect of stacking on the overlap between conduction and valence bands. The LDA exchange–correlation energy was parameterized through the Perdew–Zunger functional (Perdew and Zunger 1981). Norm-conserving pseudopotentials for the core electrons of carbon, and an extended plane wave basis set (up to 1700 plane waves, depending on the required accuracy), were adopted, in the framework of the momentum-space formalism (Ihm et al. 1979). The Brillouin Zone (BZ) was sampled with an appropriate selection of Special **k** points and an appropriate number of plane waves, in order to get maximum truncation error on energies on the order of 5 *meV*.

The calculations showed that the bonding between graphene sheets is not uniquely dominated by van der Waals dispersive interactions, because a sensible overlap between the $2p_z$ orbitals perpendicular to the graphene sheets is evident.

The valence charge densities of the three allotropes obtained within a maximum kinetic energy of 64 *Ry* and 28 special **k** points are shown in Fig. 20.2.

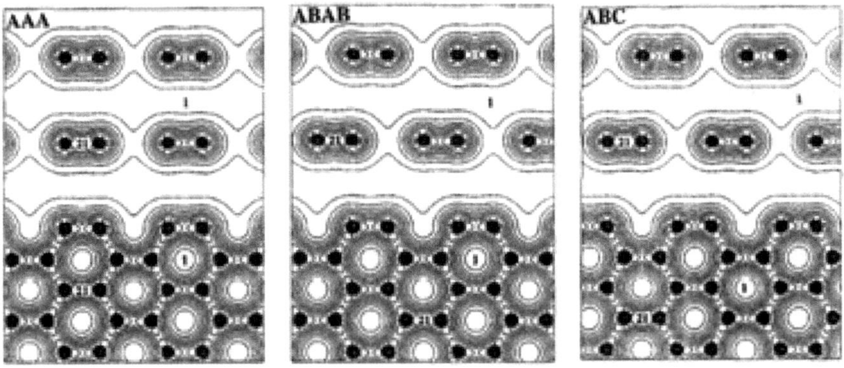

Fig. 20.2 Ab initio valence electron pseudo-charge densities of simple hexagonal (AAA), Bernal (ABAB), and rhombohedral (ABC) allotropes of graphites. The contour intervals are in units of 0.1 $e/Å^3$, spaced by 0.2 $e/Å^3$. Upper part: density along (110), (perpendicular to graphitic planes). Lower part: density in the graphitic planes (001). From Charlier et al. (1994). With permission from Carbon (Elsevier Science Ltd.)

For the AAA allotrope were adopted by Charlier et al. (1994) 989 plane waves, for the ABAB allotrope 2 × 995 plane waves, and for the ABC allotrope 992 plane waves. The isodensity lines in Fig. 20.2 display the main features of binding: the quite weak bonds in the plane (110) perpendicular to the graphitic layers (upper part in Fig. 20.2), responsible for the easy cleavage of these materials (independently of the stacking sequence), and the strong covalent bonds between carbon atoms in the plane (001) (lower part). The density along the c-axis reflects the absence (AAA) or the presence (ABAB & ABC) of shift between neighboring planes in the stacking. The hexagonal symmetry arises thus quite evidently in the in-plane density.

A pioneering investigation on the structure of the electronic energy bands in graphite was carried out by Wallace (1947) using the "*tight binding*" approximation. The approximation concerns the wave function appropriate for two neighboring (A and B) carbon atoms on a single graphite sheet:

$$\psi = \varphi_1 + \lambda\varphi_2 \tag{20.1.1}$$

with

$$\varphi_1 = \sum_{\mathrm{A}} \exp[2\pi i\mathbf{k}\cdot\mathbf{r}_{\mathrm{A}}]X(\mathbf{r}-\mathbf{r}_{\mathrm{A}}) \tag{20.1.2.1}$$

$$\varphi_2 = \sum_{\mathrm{B}} \exp[2\pi i\mathbf{k}\cdot\mathbf{r}_{\mathrm{B}}]\mathrm{X}(\mathbf{r}-\mathbf{r}_{\mathrm{B}}) \tag{20.1.2.2}$$

X(**r**) represents the normalized orbital $2p_z$ wave function for an isolated atom. The two summations are taken over A (or B) and over all the lattice points generated from them, by primitive lattice translations. The energy is obtained through the application of the variational principle. Wallace found in this way (i.e. neglecting the effect of the $2p_z$ overlap) that Graphite is a *semi-conductor* with zero activation energy, i.e. there are no free electrons at zero temperature, but they are created at higher temperatures by excitation to a band contiguous to the highest one which is normally filled. The electrical conductivity, calculated with the *mean-free-path* procedure was shown by Wallace to be one hundred times greater along a single bi-dimensional graphene sheet with respect to the conductivity across neighboring planes.

Charlier et al. (1994) found that the weak interactions between graphitic planes modify the "*zero-gap semiconductor*" situation depicted by Wallace (1947). Being the band structures of the three investigated allotropes dominated by two-dimensional in-plane interactions, the main features of the energy bands of graphite were obtained with a good accuracy [except for the energy range close to the Fermi energy ($\in_F$)] using a two-dimensional calculation. Figure 20.3 shows the band structures of the three graphitic stackings investigated by Charlier et al. (1994) in different special directions of the BZ (shown in the upper part of the figure; see to this purpose also Sect. 8.3.2).

The hexagonal Brillouin space adopted for every allotrope is relative to a six-atom unit cell for rhombohedral graphite. The dispersion of the σ bands

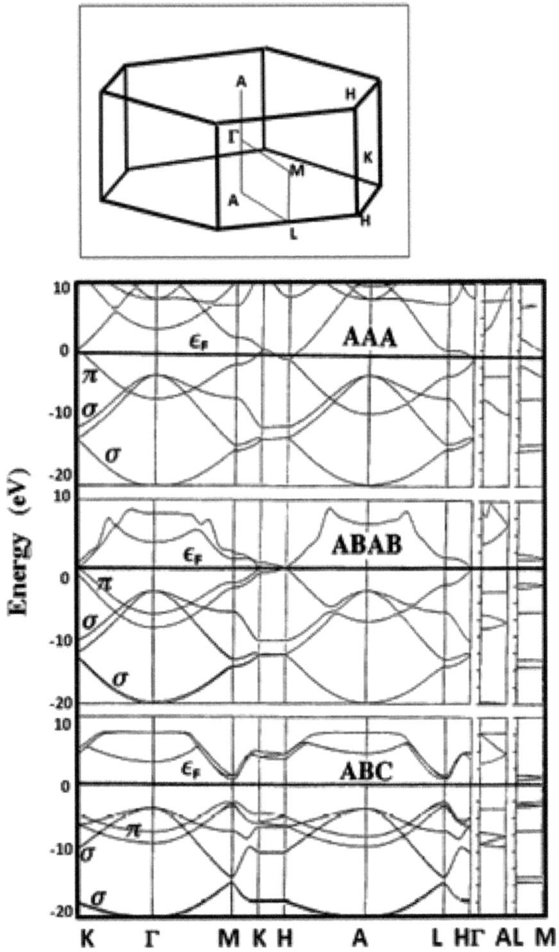

Fig. 20.3 Ab initio band structures of simple hexagonal, Bernal, and rhombohedral graphites along different lines in the hexagonal BZ imposed on the rhombohedral structures using a six-atom unit cell (upper part of the figure). The length of the three band structures has been scaled in order to make the comparison easier. Modified from Charlier et al. (1994)

perpendicular to the basal planes is quite limited but it is not so for the π bands along the c-axis. In the AHL plane of the BZ (see the upper part of the figure for the location of the special points) the tridimensional structure of the AAA and ABAB forms, is very similar to the monolayer band structure observed by Zunger (1978). There is, moreover, a strong dispersion of the π bands at the Γ point.

Summarizing the various observations, it may be stated that the strong intra-layer bonding in graphite arises from sp^2 hybridization between the 2s and $2p_x$, $2p_y$ AOs of carbon (σ bands in Fig. 20.3), while the weak inter-layer bonding arises from the overlap between $2p_z$ AOs of carbon atoms in distinct planes (π bands in the same figure). The resulting band structure is constituted by bonding (σ, π) and antibonding (σ^*, π^*) states (valence and conduction bands, respectively). The weak interactions between graphitic planes modify the simple bi-dimensional situation described by Wallace (1947), and create a small or wide

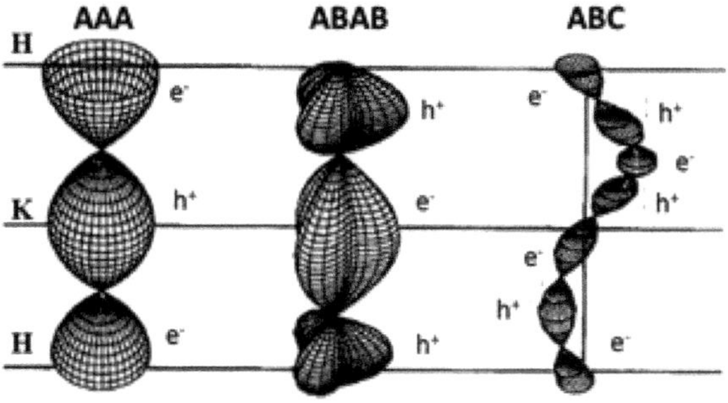

Fig. 20.4 Tridimensional tight-binding Fermi surface of simple hexagonal (AAA; Charlier et al. 1991; magnified), Bernal (ABAB; Slonczewski and Weiss 1958), and rhombohedral (ABC; Haering 1958; McClure 1969; magnified) graphites in arbitrary units. All the three Fermi surfaces are centered on one of the six vertical HKH edges of the hexagonal BZ (lengths adjusted). e^- and h^+ labels, respectively, electron and hole pockets. From Charlier et al. (1994). With permission from Carbon (Elsevier Science Ltd.)

Table 20.1 Bulk energy difference $E_{RG} - E_{BG}$ for different functionals at different electronic temperatures (Nery et al. 2021)

Functional	$E_{RG} - E_{BG}$ *(meV/atom)*		
	T = 284 *K*	T = 1010 *K*	T = 3158 *K*
LDA	−0.025	+0.020	+0.138
Grimme	−0.025	+0.020	+0.138
GGA-PBE	−0.040	+0.002	+0.116

overlap semimetal, depending on the allotropic structure. This is clearly reflected in the tight binding model Fermi surfaces of the three allotropes (Fig. 20.4). Any modification of the stacking sequence introduces a variation of the Fermi surface volume. This significantly affects the electrical conductivity, due to the different number of carriers (Charlier et al. 1991).

Quite recently, Nery et al. (2021) performed a careful DFT-LDA investigation of the energy difference between the ABAB allotrope (denoted by them "*multilayer rhombohedral stacked graphene*, RG) and the ABC allotrope ("*multilayer stacked Bernal graphene*", BG) in order to clarify the origin of literature discrepancies. Energies of all stacking sequences (SSs) were calculated for N = 12 layers, and an Ising model was used to fit them with few parameters.

In their investigation the authors where initially concerned with the bulk energy differences between the two polymorphs at various electronic temperatures. Their results are resumed in Table 20.1.

Calculations were carried out adopting the experimental lattice parameters $a_0 = 2.461$ *Å*, $c_0 = 3.347$ *Å*. We may immediately appreciate the fact that LDA and the Grimme functional (see Sect. 10.5.6) give identical results. This fact was interpreted by Nery et al. (2021) as an evidence of a minimal dispersive energy in the structure. Both functionals return a higher stability of the RG allotrope at low T.

A slightly higher differential stability at low T is predicted by the Perdew–Burke–Ernzerhof functional under General Gradient Approximation (GGA-PBE; see Sect. 10.5.4).

Introducing the electronic temperature through "*Fermi–Dirac smearing*" the energy difference, which is negative at low T (284 *K*) becomes progressively more and more positive with the increase of T (1010 and 3158 *K*; see Table 20.1).

To appreciate the significance of this observation, we must recall that the electron density [n(**r**)] is obtained from the Kohn–Sham eigenvectors:

$$n(\mathbf{r}) = \sum_i f_i |\phi_i(\mathbf{r})|^2 \tag{20.1.3}$$

Integration over the BZ of the system gives the electron density in the 3D crystal:

$$n(\mathbf{r}) = \sum_i \int_{BZ} f_{i\mathbf{k}} |\phi_{i\mathbf{k}}(\mathbf{r})|^2 d\mathbf{k} \tag{20.1.4}$$

The index i runs over all states and f_i, f_{jk} are the occupation numbers. These numbers are 0 or 1 for all bands far from the Fermi surface $\in_F$. However, for the bands crossing $\in_F$ the occupancies jump suddenly from 1 to 0. For these bands, the integer occupation numbers are conveniently replaced by a function that varies smoothly between the two limiting values. The Fermi–Dirac distribution assumes

$$f_{ik} = \frac{1}{\exp[(\in_{ik} - \mu)/\sigma] + 1} \tag{20.1.5}$$

where μ is the chemical potential and $\sigma = k_B T$ is the broadening.

Introducing the Fermi–Dirac distribution implies however that the variational internal energy U(n) must be replaced by the (variational) Helmholtz free energy functional (Mermin 1965; see also Sect. 8.8.3):

$$F(n) = U(n) - TS_{el} \tag{20.1.6}$$

where S_{el} is the electronic entropy.

Nery et al. (2021) investigated then how energy changes at the various electronic temperatures passing from bulk 3D polymorphs to slabs with a definite amount of layers N. Their results are returned in the form of a "*phase diagram*" (Fig. 20.5). At high T (i.e. above ~700 *K*), the BG allotrope is the stable form, independently of the number of grapheme layers. The preferential stability of BG extends to lower T when the layer sequence is reduced to few sheets. The fact that natural graphite occurs mainly with the BG structure is interpreted by the authors as due to "persistent metastability" (i.e. graphite forms at relatively high T and there is no structural inversion with the lowering of T).

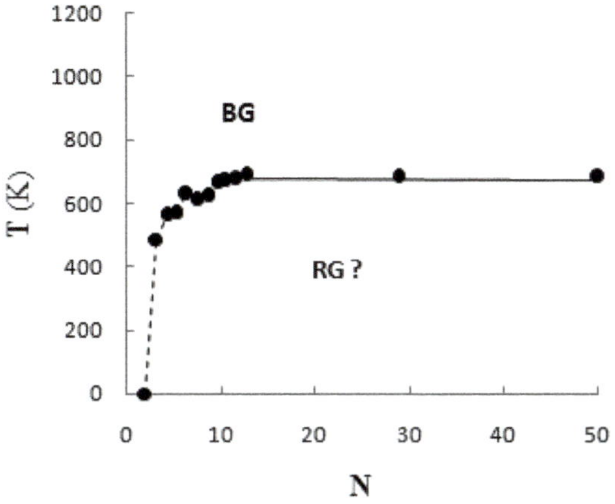

Fig. 20.5 Thermal stability of multilayer graphene. The solid line for N $\geq$ 12 drawn by Nery et al. (2021) as a linear fit on reciprocal T is here replaced by a straight-line regression. Modified from Nery et al. (2021)

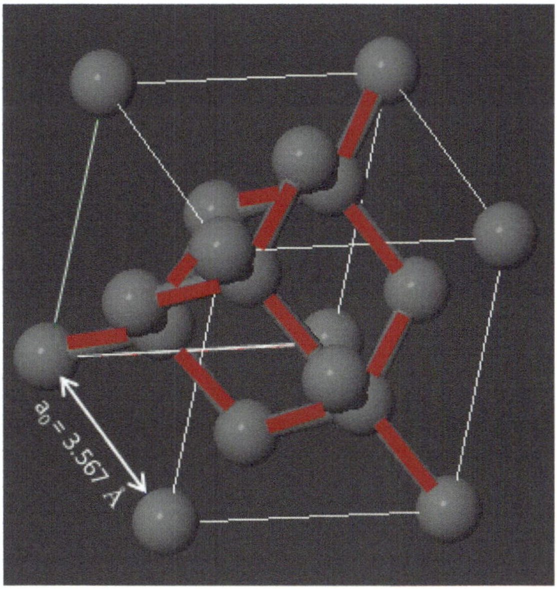

Fig. 20.6 The face-centered cubic allotrope of diamond. Red bonds put in evidence the sp_3 hybridization of carbon AOs leading to quite stable tetrahedral groups. $a_0 = b_0 = c_0 = 3.567$ *Å*, $\alpha = \beta = \gamma = 90°$ (Wyckoff 1963)

20.1.1.2 Diamond

Diamond of "*primary*" origin forms in the Earth's Mantle at depth (150–250 *km*) and occurs as accessory phase in micaceous peridotites rapidly transported to the Earth's surface under low oxygen fugacity conditions (kimberlites, lamproites). *Secondary* diamond is observed in alluvial placers.

Figure 20.6 shows the cubic face-centered (Fcc) structure of diamond (space group $Fd\bar{3}m$), which is the unique high-density 3D form of carbon.

The structure (Bragg and Bragg 1913) arises from interconnected tetrahedra generated by sp^3 hybridization of 2s and 2p carbon AOs (see to this purpose also Sect. 19.1). The hexagonal form "*Lonsdaleite*" found half-century ago at the Meteor Crater in Arizona is actually a "*faulted-twinned*" cubic diamond occurrence (Nemeth et al. 2014) and does not exist as a discrete allotrope.

The lattice dynamical properties of diamond were first investigated ab initio by Pavone et al. (1993) in light of DFT with pseudopotentials and plane waves. Harmonic phonon dispersion curves and phonon eigenvectors were evaluated within the *linear-response* framework, obtaining the parameters of internal strain. Thermal expansion and mode Grüneisen parameters were then obtained under the Quasi-Harmonic Approximation (QHA).

According to the static linear response theory, the lattice distortion corresponding to a given phonon can be considered as a static perturbation acting on the electrons of the crystal. From the Feynman–Hellmann theorem (Hellmann 1937; Feynman 1939), in fact, the linear electron-density response due to an external static perturbation yields the correct energy up to third order (see Sects. 7.2.3 and 8.8.1). The energy cutoff (E_{cut}) adopted in the plane waves pseudopotential method was $55 \leq E_{cut} \leq 85$ *Ry*. The exchange–correlation energy in the LDA framework was parameterized as indicated by Perdew and Zunger (1981). Norm-conserving pseudopotentials for carbon were generated through a fitting minimization of the squared differences between the atomic all-electron and pseudo-eigenvalues and eigenfunctions outside the core sphere (von Barth and Car; unpublished).

Figure 20.7 shows the valence charge density calculated by Pavone et al. (1993) in the (110) plane of diamond. This plot emphasizes the role of 2p orbitals in the generation of a double-hump in the electron density along the bonds.

The equilibrium lattice structure at the athermal limit was obtained by Pavone et al. (1993) with the symmetry-preserving procedure discussed at Sect.. 8.9.1, repeated for kinetic energy cutoff up to 85 *Ry*. Bulk modulus ($K_0 = 4.73$ *Mbar*) and its baric derivative ($K' = 3.5$) were obtained by fitting the zero point energy at constant energy cutoff and variable cell edge to a first-order Birch–Murnaghan

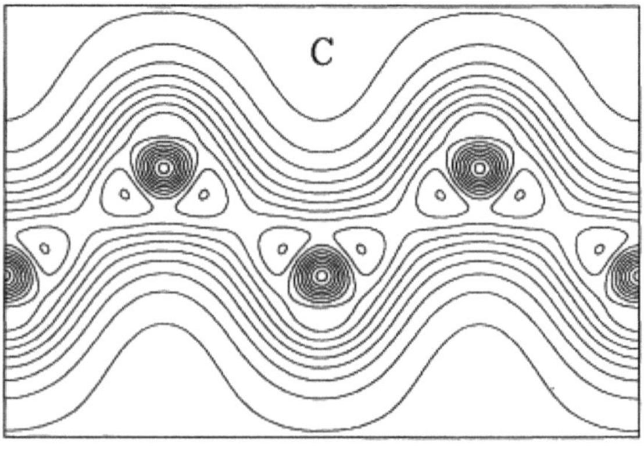

Fig. 20.7 DFT/LDA valence charge density in the (110) plane of diamond. Contour intervals are set at 0.2 *e/Å* The maximum charge along the bonds is 2.0 *e/Å*. From Pavone et al. (1993). With permission from The American Physical Society

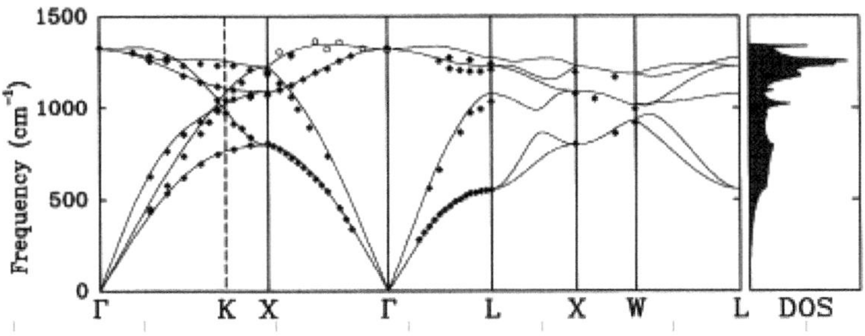

Fig. 20.8 Ab initio Phonon dispersion curves and density of states (DOS) of diamond (solid lines) compared with results of experimental neutron scattering (diamonds; Warren et al. 1967) and synchrotron scattering (open circles; Burkel 1991). The transept X−Γ corresponds to the (100) direction in the bulk crystal. From Pavone et al. (1993). With permission from The American Physical Society

equation. The obtained values are in reasonable agreement with experimental observations.

Figure 20.8 shows the bulk phonon dispersion curves obtained by Pavone et al. (1993) at a kinetic energy cutoff of 55 *Ry*, which ensures an accuracy of 1–2% in the calculated frequencies. The agreement with experimental observations (neutron and synchrotron scattering) is quite good. The rightward part of the figure is the density histogram of the various vibrational modes (Density of States, DOS). Interesting observations may be done when comparing the dispersion curves of diamond with those of another semiconductor like Si (see Sects. 8.7.3, and Fig. 8.11). Contrary to silicon, the transversal branches are not flat and the corresponding sharp feature in the DOS histogram is lost. Also, the maximum of the optical branches is not at the zone center (Γ point) as it is the case of silicon and germanium. Moreover, the minimum of the uppermost branch at the Γ point in diamond is ascribable to over-bending, which is absent in silicon.

The T-dependent thermal expansion [α(T)] was obtained by Pavone et al. (1993) in the framework of QHA (see also Sect. 8.9.3):

$$\alpha(\mathrm{T}) = \frac{1}{3\mathrm{V}}\frac{\partial \mathrm{V}}{\partial \mathrm{T}} \cong \frac{1}{\mathrm{K}_0}\sum_{\mathrm{q,j}} \gamma_{\mathrm{j}}(\mathbf{q})\mathrm{C}_{\mathrm{V,j}}(\mathbf{q},\mathrm{T}) \tag{20.1.7}$$

$$\gamma_{\mathrm{j}}(\mathbf{q}) = -\frac{\mathrm{V}_0}{\omega_{\mathrm{j}}(\mathbf{q},\mathrm{V}_0)}\left[\frac{\partial \omega_{\mathrm{j}}(\mathbf{q},\mathrm{V})}{\partial \mathrm{V}}\right]_{\mathrm{V}=\mathrm{V}_0} \tag{20.1.8}$$

where $\gamma_{\mathrm{j}}(q)$ is the mode Grüneisen parameter and $\mathrm{C}_{\mathrm{V,j}}(\mathbf{q},\mathrm{T})$ is the contribution of the mode $\mathbf{q}_{\mathrm{j}}$ to the isochoric heat capacity of the substance. The calculated α(T) is consistent with experimental observations (Slack and Bartram 1975) for T not exceeding $\sim 500\ K$. Deviations from observed values at higher T are due to high-order anharmonic contributions not accounted for by theory.

Diamond has the highest thermal conductivity (k_L) of any known bulk material (more than an order of magnitude higher than common semiconductors such as silicon and germanium; $k_L \geq 3 \times 10^3\ W \cdot m^{-1} \cdot K^{-1}$). The lattice thermal conductivity of diamond has been calculated by Ward et al. (2009) based on an exact solution of the Boltzmann Transport Equation (BTE).

$$\vec{\mathcal{V}}_\lambda \cdot \nabla T \frac{\partial n_\lambda^0}{\partial T} = \left.\frac{\partial n}{\partial t}\right| c \tag{20.1.9}$$

where $\vec{\mathcal{V}}_\lambda$ is the phonon velocity, ∇T is the (small) temperature gradient that perturbs the phonon distribution function, $n_\lambda = n_\lambda^0 + n_\lambda^1$ (with n_λ^0 here as Bose distribution and n_λ^1 as non-equilibrium part proportional to the small ∇T). The collision term $\partial n_\lambda^0 / \partial T$ describes the scattering into and out of the state. The anharmonicity of the interatomic potential causes phonons to scatter inelastically from one another. The scattering may be of two types: Normal and Umklapp. In the normal scattering, two phonons k_1 and k_2 entering the BZ generate a phonon k_3 that remains inside the BZ, while, in the Umklapp process, the phonon k_3 exceeds the Brillouin space. The strong bond stiffness (and light atomic mass) in diamond creates extremely high phonon frequencies and acoustic velocities. In these conditions the phonon–phonon Umklapp scattering at room temperature is quite weak, and, as a result, k_L is extremely large. To determine the phase space of all possible phonon scattering events that conserve both energy and crystal momentum, it is necessary an accurate knowledge of the entire phonon dispersion. Interatomic Force Constants (IFCs) were calculated by Ward et al. (2009) (as in Pavone et al. 1993), in the framework of DFT under LDA (Perdew and Zunger 1981) and BHS (Bachelet et al. 1982; see Sect. 12.7.1). The calculated phonon dispersion has the same accuracy previously appreciated in the investigation of Pavone et al. (1993). With the calculated phonon dispersions, the phase space of all possible three phonon scattering events that conserve both energy and crystal was determined by applying

$$\omega_j(\vec{q}) \pm \omega_j'\left(\vec{q}'\right) = \omega_j''\left(\vec{q}''\right) \tag{20.1.10}$$

$$\vec{q} \pm \vec{q}' = \vec{q}'' + \vec{K} \tag{20.1.11}$$

In the normal process $\vec{K} = 0$, while, in the Umklapp process, $\vec{K} \neq 0$. In Table 20.2, the calculated zone center frequency $[\omega_{LO}(\Gamma)]$, the transverse (TA) and longitudinal (LA) acoustic velocities, and the three-phonon-phase space (P_3) of diamond are compared with the same magnitudes in Si and Ge.

The ab initio lattice thermal conductivity of isotopically enriched and naturally occurring diamond as a function of temperature obtained by Ward et al. (2009) are compared with experimental observations in Fig. 20.9. The open diamonds (Berman et al. 1975), open circles (Onn et al. 1992), plusses (Wey et al. 1993), and

Table 20.2 Zone center frequencies [$\omega_{LO}(\Gamma)$, (*THz*)], Transverse and longitudinal acoustic velocities [v_{TA}, v_{LA}, (*m/s*)] and three-phonon-phase space [P_3, (10^{-2})] in diamond , compared to other semiconductors. From Ward et al. (2009)

	$\omega_{LO}(\Gamma)$	v_{TA}	v_{LA}	P_3
Diamond	39.17	12,567	17,326	0.0796
Si	15.49	5413	8216	0.3536
Ge	8.92	3251	4844	0.5795

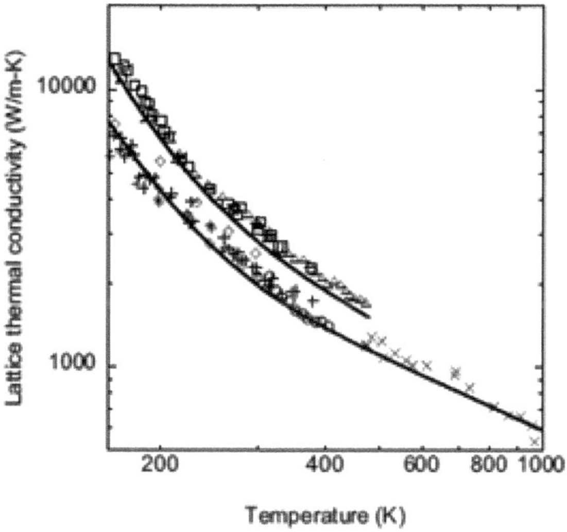

Fig. 20.9 Ab initio lattice thermal conductivity (k_L) of isotopically enriched and naturally occurring diamond as a function of temperature (solid lines), compared with measured values (symbols). From Ward et al. (2009). With permission from The American Physical Society

crosses (Olson et al. 1993) are for naturally occurring type IIa diamond with 0.989 ^{12}C and 0.011 ^{13}C. The open squares (Olson et al. 1993) are measured values for isotopically enriched diamond with 0.9993 ^{12}C, 0.0007 ^{13}C. Triangles (Wey et al. 1993) are measured values for 0.9999 ^{12}C, and 0.0001 ^{13}C.

The calculated curves for both the isotopically enriched(mass variance = 4.86×10^{-6}) and naturally occurring diamonds (mass variance = 7.54×10^{-5}) are in very good overall agreement with the experimental data at all T. As noted by Ward et al. (2009) this agreement is particularly impressive since there are no adjustable parameters.

Fractionation factors for the two stable carbon isotopes ^{12}C and ^{13}C for graphite–CO_2, graphite–CH_4, and diamond–CO_2 equilibria have been calculated by Bottinga (1969a, b). Some selected results of vibrational calculations of Bottinga (1969a, b) in the T range 0–1000 °C are resumed in Table 20.3.

The seminal importance of Bottinga's investigations in the field of Isotope Geochemistry is widely recognized and his results are still valid now after more than an half-century.

Table 20.3 Calculated fractionation factors (1000lnα) for ^{13}C exchange among carbon dioxide, calcite, graphite, diamond, and methane (Bottinga 1969a, b). Gr = graphite; Dm = diamond

	$1000 \ln \alpha_{AB}$			
T (°C)	Gr-CH_4	CO_2.Gr	CO_2-Dm	Dm-Gr
0	63.4	14.3	2.8	11.5
50	45.8	14.9	7	7.8
100	33.8	15.1	9.6	5.9
150	–	–	10.9	3.9
200	19.5	14.4	11.5	2.9
250	–	–	11.6	2.1
300	11.8	13.1	11.4	1.7
350	–	–	11	1.3
400	7.4	11.5	10.5	1.1
450	6	10.8	9.9	0.9
500	4.9	10.1	9.3	0.8
550	4	9.4	8.7	0.7
600	3.3	8.8	8.2	0.6
650	2.7	8.2	7.7	0.6
700	2.3	7.7	7.2	0.5
750	–	–	6.7	0.5
800	–	–	6.3	0.5
850	–	–	5.9	0.5
900	–	–	5.5	0.5
1000	–	–	4.4	0.4

Bottinga's (1969a, b) calculations of the carbon fractionation factors between carbon polymorphs and coexisting gaseous phases were carried out along the guidelines outlined at Sect. 7.9. The phonon frequency distribution in the solid phases was obtained by application of lattice dynamics to crystalline polymorphs. After an opportune selection of static potentials describing the interaction between neighboring carbon atoms in the structure, as a function of interatomic distances and angles, the equations of motions were obtained for all atoms in the unit cell and the normal modes of vibration of all r particles in the crystal composed of N unit cells:

$$\int_0^{\omega_{max}} g(\omega)d\omega = 3Nr \tag{20.1.12}$$

were calculated by solving the 3r × 3r dynamical matrix obtained by the general solution of wave vectors. The vibrational partition function was then obtained by applying

$$\ln Q_{vib} = \frac{k_B T}{hc} \sum_n \frac{g_n}{\omega_{n+1} - \omega_n} \times \int_{X_n}^{X_{n+1}} \ln\left[\frac{\exp(-X/2)}{1 - \exp(-X)}\right] dX \qquad (20.1.13)$$

were the various g_n (i.e. the eigenvalues of the dynamical matrix) are the number of modes in the vibrational frequency interval $\omega_{n+1} - \omega_n$. The vibrational spectrum of graphite was based on the interionic potential model of Yoshimori and Kitano (1956) and adopts the force constants of Young and Koppel (1965) (assumed identical for ^{12}C and ^{13}C). The normalized phonon spectrum of diamond was based on the interionic potential model of Aggarwal (1967) and the dynamical matrix equations of Herman (1959) (1/48 of the BZ; 3.46×10^6 frequencies).

According to the calculations of Bottinga (1962b) for the C–O system, the high-density form diamond is preferentially enriched in ^{13}C with respect to the low-density form graphite (although the enrichment is quite limited). Moreover, the CO_2–diamond carbon fractionation curve exhibits a broad maximum around 300 *°C*.

20.1.2 Copper

As a prototypical example of metal, I introduce hereafter some observations on copper. The electron configuration of Cu is $1s^2 2s^2 2p^6 3s^2 3p^6 3d^{10} 4s^1$ (see Table 6.3). Unlike other metals with unfilled d-shells, the bond is eminently metallic (arising only from delocalized 4*s* electrons). Metallic bonding is responsible of the high electrical (σ_{el} = 59.6 *S/m)* and thermal [k_L = 401 *W/(m · K)*] conductivity.

Crystalline Copper occurs in nature as in two allotropic forms: a *Fm3m* face-centered (fcc) cubic closed packet structure (Fig. 20.10) stable at low T up to *TPa* P conditions (Fratanduono et al. 2020) and a *Fm3m* body-centered (bcc) form, detected for the first time by Bolesta and Fomin (2014) in an MD simulation of shock compression. In nature, the fcc allotrope (cubic and, occasionally, ottaedric crystals) form dendritic aggregates by reduction of organic substances. More rarely, native *Fm3m* fcc copper precipitates from hydrothermal solutions in the late pneumatolytic stages of magma evolution.

Melting of the fcc allotrope has been investigated ab initio by Vočadlo et al. (2004) in the P range $0 \leq P \leq 100$ *GPa* by the "*phase-coexistence approach*". The ab initio total energies (and ionic forces of systems representing solid and liquid copper) were calculated with the PAW implementation of DFT, and the GGA approximation (see Sect. 8.8.3). The PAW calculations were performed using a cutoff energy of 273 *eV* and a core radius of 2.3 a.u. The $3d^{10}4s^1$ states were treated as valence, with an Ar-core. Approximated melting loci were obtained with the aid

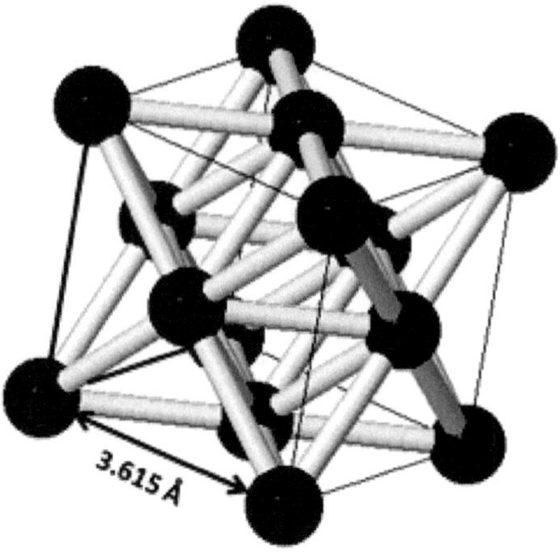

Fig. 20.10 The *Fm3m* cubic face-centered structure of Copper (Wyckoff 1963). The structure generated by the AMCSD software (Downs and Hall-Wallace 2003) and partly redrawn

of an empirical reference system (based, in this case, on the Embedded-Atom Model, EAM). The approximated melting curve was corrected using calculated free energy differences between the reference and the ab initio system (Vočadlo et al. 2004).

In practice, the coexistence simulation proceeds as follows:

(1) It is created a supercell containing the perfect fcc crystal;
(2) The ensemble is thermalized at a temperature slightly below the expected melting curve (i.e. in the solid state at limit);
(3) The simulation is halted, and the positions of the atoms in one half of the cell are held fixed, while the other half is heated to a very high T (i.e. inside the stability range of the liquid);
(4) The molten half is re-thermalized to the expected melting temperature;
(5) The fixed atoms are released and the whole system is allowed to evolve freely at constant (NVE) for a long time, so that solid and liquid come into equilibrium.

Denoting $U_{AI}(r_1 \ldots r_N)$ and $U_{ref}(r_1 \ldots r_N)$ the total energy functions of the ab initio and reference systems as a function of ionic positions r_i and $T_{m,AI}$, $T_{m,ref}$ the melting temperatures at P, the lowest order correction formula (with ℓ = liquid and s = solid) is sufficient (in the case of Cu) to mimic the ab initio solid and liquid very precisely (Vočadlo et al. 2004).

$$\Delta T_m \equiv T_m^{AI} - T_m^{ref} \approx \frac{\Delta G^{\ell s}\left(T_m^{ref}\right)}{S_{ref}^{\ell s}} \tag{20.1.14.1}$$

$$\Delta G^{\ell s} = G_{AI}^{\ell s} - G_{ref}^{\ell s} \tag{20.1.14.2}$$

$$S_{ref}^{\ell s} = S_{ref}^{\ell} - S_{ref}^{s} \tag{20.1.14.3}$$

Because the simulation is carried out at constant V and T, the Gibbs free energy difference ΔG is replaced by the corresponding Helmholtz free energy difference ΔF:

$$\Delta G(P,T) \cong \Delta F(V,T) - \frac{1}{2}\frac{V\Delta P^2}{K_T^{ref}} \tag{20.1.15}$$

where K_T^{ref} is the bulk modulus at T of the reference model and ΔP is the change of pressure when U_{ref} is replaced by U_{AI} at constant V and T, and

$$\Delta F(V,T) = \langle \Delta U \rangle_{ref} - k_B T \ln\left[\left\langle \exp\left(-\frac{\delta \Delta U}{k_B T}\right)\right\rangle_{ref}\right] \tag{20.1.16.1}$$

$$\Delta U = U_{AI} - U_{ref} \tag{20.1.16.2}$$

$$\delta \Delta U = \Delta U - \langle \Delta U \rangle_{ref} \tag{20.1.16.3}$$

Expanding the second term on the right-hand side of Eq. (20.1.16.1) in a Taylor series, it is finally obtained (Vočadlo et al. 2004)

$$\Delta F(V,T) = \langle \Delta U \rangle_{ref} - \frac{1}{k_B T}\langle \delta \Delta U^2 \rangle_{ref} \tag{20.1.17}$$

The EAM assumes the internal energy to be composed essentially of two terms:

$$U_{ref} = \frac{1}{2}\sum_{i \neq j} \in \left(\frac{a}{r_{ij}}\right)^n - \in C \sum_i \left[\sum_{j \neq i}\left(\frac{a}{r_{ij}}\right)^m\right]^{1/2} \tag{20.1.18}$$

The first term on the right of Eq. (20.1.18) represents an inverse-power repulsive pair potential and the second term on the right of it is the sum of embedding energies of the individual ions in the sea of conduction electrons. Vočadlo et al. (2004) adopted in the initial calculations the parameters a, e, C, m, and n proposed by Belonoshko et al. (2000). These parameters where then further tuned up by minimizing the fluctuation strength

$$\sigma^2 \equiv \frac{\left\langle (\delta \Delta U)^2 \right\rangle_{ref}}{N} \tag{20.1.19}$$

Table 20.4 lists the zero point volume (V_0) of the fcc crystal, the bulk modulus at the athermal limit (K_0), and its first and second P-derivative (K′, K″) obtained by Vočadlo et al. (2004) under General Gradient (GGA) and Local Density (LDA) approximation. The results of additional GGA calculations carried out by

Table 20.4 Ab initio Volume (V_0, *Å/atom*), Bulk modulus (K_0, *GPa*) and first and second P-derivatives of *Fm3m* fcc copper after Vočadlo et al. (2004) (K′ adimensional, K″ GPa^{-1})

	V_0	K_0	K′	K″
GGA Ar-core	12.043	134.721	4.9	−2.7
GGA $Ne3s^2$-core	12.041	139.190	4.55	0.5
LDA	10.938	182.674	5.10	−3

treating explicitly the 2p states ($Ne3s^2$-core) are also reported in the table. The difference in P at a volume of 8.7 $Å^3$*/atom* between Ar-core and $Ne3s^2$-core calculations is <0.5 *GPa*. All calculations were performed adopting a $15 \times 15 \times 15$ Monkhorst–Pack, with 120 K points in the irreducible Brillouin sampling zone, returning a precision better than 1 *meV/atom*. The energies, calculated with a set of volumes in the range $7 \leq V \leq 14$ $Å^3$ were fitted by Vočadlo et al. (2004) to a fourth-order logarithmic EOS (Poirier and Tarantola 1998). The experimental cell volume of Cu is 47.24×10^{-24} cm^3 (Robie and Bethke 1962), which, with 4 atoms per unit cell, brings to 11.81 $Å^3$*/atom*, in reasonable agreement with the GGA results. The bulk modulus at the athermal limit is 142 *GPa* (Ledbetter and Naimon 1974) again in reasonable agreement with the GGA results, especially for the $Ne3s^2$-core calculations.

Phonon frequencies were calculated by Vočadlo et al. (2004) using the small displacements method (see Sect. 8.8.3) with a 64-atom supercell. The BZ sampling was performed with a $5 \times 5 \times 5$ Monkhorst–Pack grid equivalent to 18 k-points in the irreducible zone. The phonon dispersion curves obtained under GGA at the experimental volume are in good agreement with the phonon frequencies measured by Nilsson and Rolandson (1973) at 80 *K*. The melting relations of the fcc allotrope calculated by Vočadlo et al. (2004) are in reasonable agreement with the results of Belonoshko et al. (2000) up to a confining pressure of ~50 *GPa* (see later).

Recent shock compression investigations on copper (Sharma et al. 2020), showing structural transformation from the face-centered cubic lattice to the body-centered one at subsolidus, induced Smirnov (2021) to perform a quite comprehensive ab initio investigation of the relative stability of fcc/bcc, proposing a P–T relative stability diagram of the two substances in a wide P–T range (Fig. 20.11).

Smirnov (2021) adopted for DFT-GGA calculations the all-electron full-potential linear muffin-tin orbital method (P-LMTO) of Savrasov (1996). This way of representing first-order corrections to the one-electron wave functions greatly facilitates the treatment of the materials with localized orbitals. The internal parameters were further adjusted in order to ensure an accuracy of ~0.1 *mRy/atom*. Pseudopotentials were adopted in the core electrons of a $Ne3s^2$ configuration with 3s valence states. Exchange–correlation terms were accounted for with the PBE functional. Integration in the Brillouin zone was carried out with the "*improved tetrahedron method*" of Blöchl et al. (1994). The adopted integration mesh was $30 \times 30 \times 30$.

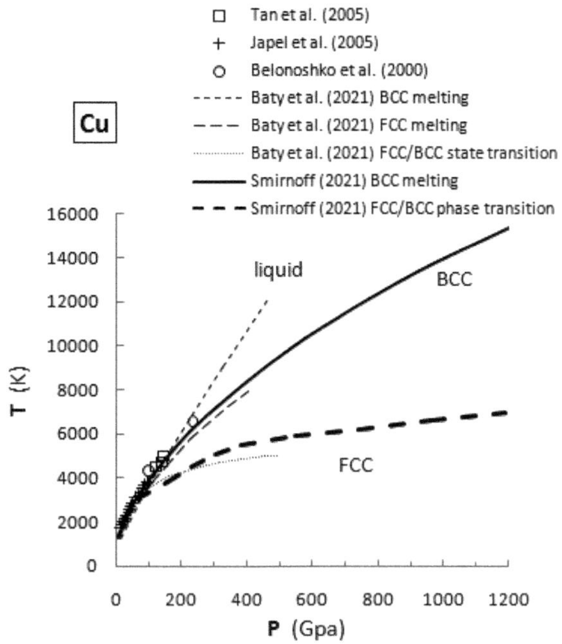

Fig. 20.11 Melting and subsolidus relations of the Cu allotropes. Ab initio computations are compared with the few existing experimental observations

The phonon spectra, based on the linear response theory where obtained from the FP-LMTO code (Savrasov 1996) with a Monkhorst–Pack grid 10×10×10. The EOS was calculated with the "*universal equation of state for compressed solids*" of Parsafar and Mason (1994). In this approach, the thermal pressure is included from the very beginning; E(ρ, T), P(ρ,T) are expanded over the density ρ in the virial form

$$P(V, V_0) = A_0 + A_1\left(\frac{\rho}{\rho_0}\right) + A_2\left(\frac{\rho}{\rho_0}\right)^2 \tag{20.1.20}$$

and the thermal dependence of A_0, A_1, A_2 is regressed for T > θD in the form (Parsafar and Mason 1994):

$$A_i(T) + A_i + b_i T - c_i T \ln T \tag{20.1.21}$$

The loci of the melting curve were localized by Smirnoff (2021) at the light of the Lindemann criterion (see however also Sect. 19.5, to this purpose). Concerning the phase stability relations at subsolidus, the QHA calculations of Smirnov (2021) show that the higher stability of the bcc structure at high T, compared to the fcc allotrope, is due mainly to softer low-frequency phonon modes, and to smaller contribution of lattice vibrations to free energy at high T.

Recent AIMD investigations carried out by Baty et al. (2021) confirm essentially the results of Smirnoff (2021) about the subsolidus transition, although the depicted loci of melting of the two allotropes are quite different (see Fig. 20.11). The calculations of Baty et al. (2021) were carried out with the VASP package under GGA with PBE exchange correlation. The adopted electron core-valence representation was [Mg]$3p^6 3d^{10} 4s^1$. For the core electrons were adopted the PAW pseudopotentials, while the 17 valence electrons were represented with a plane-wave basis set with a cutoff energy of 460 *eV*. The EoS of the two allotropes at the athermal limit were computed with very dense Monkhorst–Pack grids (60 × 60 × 60 for fcc and 75 × 75 × 75 for bcc) in order to achieve an appropriate convergence to full energy (< 0.1 *meV/atom* in the whole P range) and E-V were regressed along a Birch–Murnaghan third-order finite elasticity isothermal EoS:

$$P_{st} = -\left(\frac{\partial E_{ai}}{\partial V}\right)_{T=0} = \frac{3}{2}K_0\left[\eta^{-7/3} - \eta^{-5/3}\right]\left\{1 - \frac{3}{4}(4 - K')\left[\eta^{-2/3} - 1\right]\right\} + P_0 \tag{8.9.1}$$

where η is the dimensionless variable "*dilaton*" ($\eta = V/V_0$) and P_0 is the reference pressure (usually $P_0 = 0$; see Sect. 8.9.2) In Baty et al. (2021) however the dilaton was replaced by ρ/ρ_0. The regressed parameters were $\rho_0 = 9.02$ *g/cm*^{3}, $K_0 = 133.0$ *GPa*, $K' = 5.1$ for fcc, in excellent agreement with experiments (Wallace 2002; Dewaele et al. 2004), and $\rho_0 = 8.98$ *g/cm*^{3}, $K_0 = 131.7$ *GPa*, $K' = 5.1$ for bcc.

Six discrete points along the melting curves of the two allotropes of Cu were computed by Baty et al. (2004) with the *Z* method implemented in VASP (Belonoshko et al. 2006; Burakovsky et al. 2015). Simulations were carried out at Γ point with supercells of ~500 atoms (an average of five computer runs per point, with a time step of 1 *fs* and a total length of 1.5×10^4–2.5×10^4 time steps per run).

In Fig. 20.12 is shown, as an example, the time evolution of T at a density of 8.2627 *g/cm*^{3} for the fcc allotrope. Melting occurs at T = 2750 *K*. It is detected by the decrease of average T from ~1400 to 1280 *K* and the corresponding increase[2] of average P from ~ −2 to −1.2 *GPa*. At T = 2500 *K* the system remains a superheated solid. This is quite evident because the average T (and P) is roughly constant during the 20*ps* of running time. At T = 3000 *K*, on the contrary, melting starts after only 2*ps* of running time, and the melting process takes only 1*ps*, thus the system melts virtually immediately.

As a conclusive note it is worth to stress that the steep slope of the melting curve of copper, first observed by Belonoshko et al. (2000) and then confirmed by Japel et al. (2005) and Baty et al. (2021) is most likely due to the 3d filled shell of the metal. This hypothesis was first stressed by Ross et al. (2004) and then reproposed by Japel et al. (2005). Laser-heated diamond cell experiments carried out for Cu up to a confining pressure of 97 *GPa* and for Ni to 60 *GPa*, showed in fact that the melting slope for Cu is about 2.5 times steeper than for Ni (Japel et al. 2005). This

[2] Since V is fixed and the total energy $E = k_B T + PV$, a decrease of T corresponds to an increase of P.

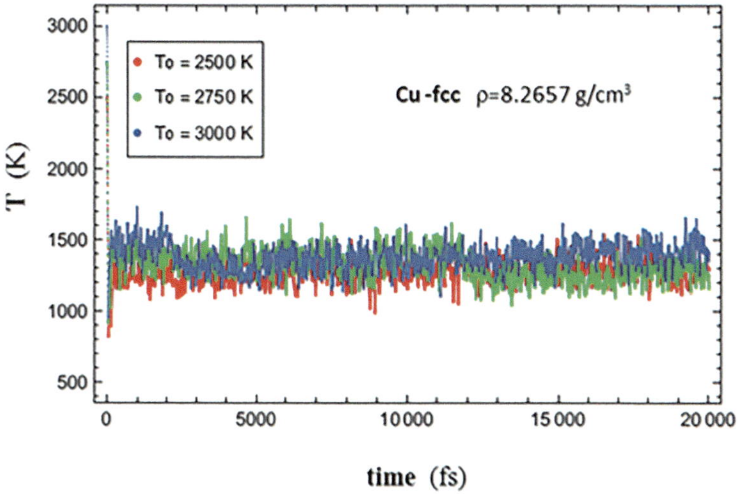

Fig. 20.12 The melting point of fcc-Cu at a density of 8.2627 *g/cc* (Z AIMD procedure). Modified from Baty et al. (2021)

was interpreted by Japel et al. (2005) in light of a semiempirical model developed by Ross et al. (2004). In this model, the slower melting slope of Ni (with respect to Cu) is due to loss of structural periodicity associated to s-d electron transfer (absent in Cu, which has a filled 3d shell).

20.2 Oxides

Again, for space reasons, I will introduce hereafter only some quantum mechanical applications to the three major rock-forming oxides, SiO_2, Al_2O_3 and MgO.

20.2.1 SiO_2

Silica has 22 polymorphs, although only some of them are of geochemical interest, namely, the crystalline polymorphs Quartz, Tridymite, Cristobalite, Coesite, and Stishovite (in their structural modifications of low and high T, usually designated, respectively, as α and β forms) and the amorphous phases Chalcedony and Opal (hydrated amorphous silica).[3] The crystalline polymorphs of silica are tectosilicates (dimensionality = 3). Table 20.5 reports their structural properties, after the

[3] Generated by hydrothermal deposition from SiO_2-saturated solutions.

Table 20.5 Structural properties of SiO_2 polymorphs after Smyth and Bish (1988)

Polymorph	Quartz	Coesite	Stishovite	Cristobalite	Tridymite
Density (*g/cm*^{3})	2.648	2.909	4.287	2.318	2.269
Molar volume (*cm*^{3})	22.688	20.657	14.017	25.925	26.478
Z	3	16	2	4	48
System	Trigonal	Monoclinic	Tetragonal	Tetragonal	Monoclinic
Class	*32*	*2/m*	*4/mmm*	*422*	*m*
Group	$P3_221$	*C2/c*	$P4_2/mnm$	$P4_12_12$	*Cc*
Cell parameters					
a	4.9134	7.1464	4.1790	4.978	18.494
b	4.9134	12.3796	4.1790	4.978	4.991
c	5.4052	7.1829	2.6651	6.948	25.832
β		120.283			117.75
Volume (*Å*^{3})	113.01	548.76	46.54	172.17	2110.2
References	Le Page et al. (1980)	Smyth et al. (1987)	Baur and Kahn (1971)	Peacor (1973)	Kato and Nukui (1976)

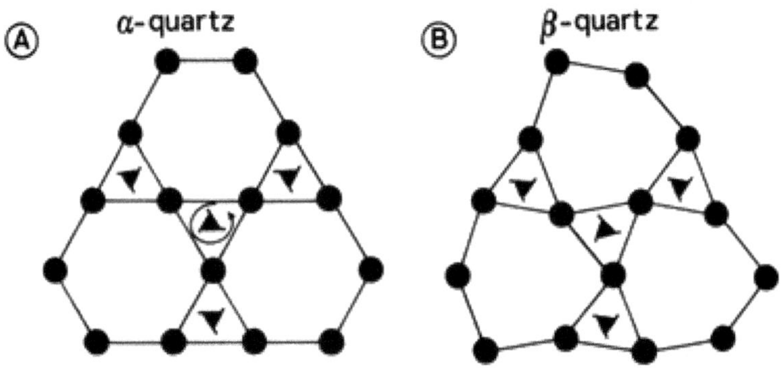

Fig. 20.13 Quartz structure projected along axis z. Filled dots are Si atoms (oxygen atoms not shown). α-quartz (**A**) – β-quartz (**B**) transition takes place by simple rotation of tetrahedra on ternary helicogyre. Modified from Gottardi (1972)

synthesis of Smyth and Bish (1988). As we will see later on some parameters are still subject to refinements. The formula unit for all polymorphs is SiO_2 and the molar weight is 60.085 *g/mol*. Note the conspicuous variation of the number of formula units per unit cell from phase to phase.

In part A of Fig. 20.13 are shown the relative positions of Si atoms in the α-quartz structure projected along axis z. This simplified representation puts in evidence the hexagonal arrangement of Si atoms, which can be internally occupied by univalent cations through coupled substitution involving Al^{3+} in tetrahedral position (the hexagonal ring is actually a spiral with atoms positioned on separate planes).

In part B of the same figure it is shown the structure of high-T β-quartz. The α-quartz – β-quartz transition, which may be described as the overlap of a first-order and a λ transition, takes place by simple rotation of tetrahedra on the ternary helicogyre.

Belmonte et al. (2017a) proposed an assessment of the thermophysical parameters for various SiO_2 polymorphs, largely based, in the absence of experiments, on first-principles (Demuth et al. 1999; Ottonello et al. 2009, 2010). Because the rheology of a substance is quite important in establishing its extrinsic stability limits at the high thermobaric regimes ruling in the planetary interiors, the properties of the most dense crystalline polymorph (Stishovite) assumed in this assessment a key role and are here outlined first. The thermochemical parameters adopted by Belmonte et al. (2017a) for most part of silica polymorphs are those of Berman (1988). This choice was dictated by the fact that Berman (1988) adopted α-Quartz as a reference phase for the enthalpy of formation from the elements for a large number of solids, and his model-depicted extrinsic stability limits for the low-density polymorphs of silica are still unmatched in their accuracy with respect to experiments. The effect of second-order (lambda) polymorphic transitions on the heat capacity of SiO_2 phases have been quantified by Belmonte et al. (2017a) according to the procedure of Berman (1988) by the following equation (applied for $T_{ref,\lambda} < T < T_\lambda$ at 1-bar pressure):

$$C_{P,\lambda} = T \cdot (X_{\lambda 1} + X_{\lambda 2} T)^2 = b_\lambda T + f_\lambda T^2 + g_\lambda T^3 \qquad (20.2.1)$$

where the parameters b_λ, f_λ and g_λ are those of the Shomate's expansion (see Sect. 8.10.7) for $C_{P,\lambda}$.

20.2.1.1 Stishovite

Stishovite, the rutile-type *P4₂/mnm* tetragonal aggregation form of silica is found exclusively in "*impactites*" (i.e. rocks formed by meteorite impact). This polymorph of silica has been investigated ab initio at the DFT/B3LYP theory level under the General Gradient Approximation (GGA) by Ottonello et al. (2009b). All-electron calculations were performed with the CRYSTAL package (Roetti 1996), within the LCAO approach within the Kohn–Sham formalism by using the hybrid B3LYP functional. The adopted basis set was a 88-31G(d) contraction for silicon and a 8–411(d) set for oxygen, as proposed by Nada et al. (1996). The DFT exchange_-correlation contribution was evaluated by numerical integration over the cell volume (Pascale et al. 2004). Radial and angular points of the atomic grid were generated through Gauss–Legendre and Lebedev quadrature schemes with a grid pruning (75,434)p for all calculations. This pruned grid spans the radial range as proposed by Gill et al. (1993) with five shells with different angular points. The reciprocal space was sampled according to a regular sublattice with a shrinking factor leading to 18 k points in the irreducible BZ. The gradient with respect to the atomic coordinates and lattice parameters was evaluated analytically (Doll 2004). The equilibrium structure was determined by using a modified conjugate gradient algorithm as proposed by Schlegel (1982).

There are two independent atoms in the conventional cell: one (Si) in special position and one (O) in general position. The compressibility of stishovite is anisotropic, with cell edge a being approximately twice as compressible as the cell edge c. Provisional values of the cell edges and oxygen positional parameters (based on the data of Ross et al. 1990) were assigned for increasing nominal pressures from 0 to 30 *GPa*. At each nominal pressure the c/a ratio was varied maintaining constant the nominal volume and the best c/a ratio (in terms of electron energy) was selected as representative of the dimensionally relaxed substance. A geometry optimization on each selected (c; a) couple allowed to relax further the internal coordinates and to shape the potential well of the substance (Fig. 20.14). The conformation of the potential well so obtained is consistent with experimental evidence. The bulk modulus of the substance at zero pressure K_0 and its P-derivative K′ were obtained by Gaussian integration of the Birch–Murnaghan (BM) third-order finite strain isothermal EoS.

In the harmonic approximation, the Born–Oppenheimer potential energy surface of a system with N nuclei, V(**x**), which is a function of vector **x** of the 3N coordinates of the atoms, has form:

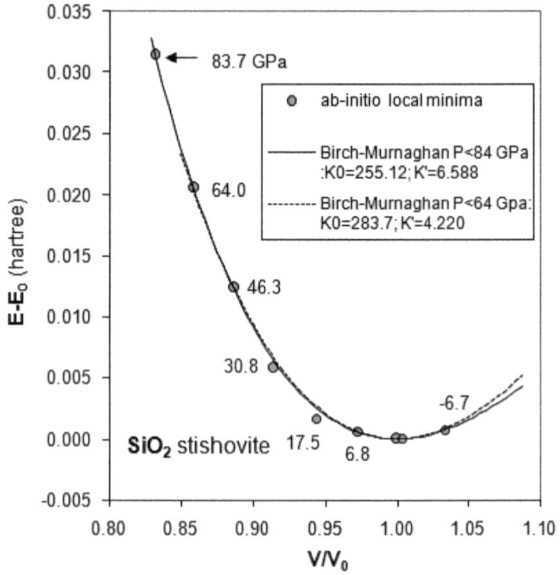

Fig. 20.14 Ab initio potential well of Stishovite at the athermal limit. Modified from Ottonello et al. (2009b)

$$V_{(0)} = \frac{1}{2}\sum_{ij} u_i \mathsf{H}_{ij} u_j = \frac{1}{2}\langle u|\mathsf{H}|u\rangle \tag{20.2.2}$$

where u_i represents a displacement of the ith Cartesian coordinate from its equilibrium value, H is the Hessian matrix of the second derivatives of V(x), evaluated at equilibrium, with respect to the displacement coordinates

$$\mathsf{H}_{ij} = \left[\frac{\partial^2 V}{\partial u_i \partial u_j}\right]_0 \tag{20.2.3}$$

In CRYSTAL, the first derivatives of energy with respect to the atomic positions are calculated analytically (Doll 2001; Doll et al. 2001) for all u_j coordinates, while second derivatives at **u** = 0 (where all first derivatives are zero) are calculated numerically using a 2-point or 3-point formula (see Pascale et al. 2004; for a more explicit formulation of the method).

In the ditetragonal unit cell of Stishovite there are 6 atoms corresponding to 15 true vibrational modes. The irreducible representation of vibrational modes, with a group theoretical treatment of the optical zone-center (gamma point; **k** = 0) yields

$$\Gamma = 1B_{1g} + 1B_{1u} + 6Eu + 2Eg + 1A_{2g} + 1A_{2u} + 1B_{2g} \tag{20.2.4}$$

In Table 20.6 are listed the ab initio vibrational frequencies and their Gruneisen mode gamma $\gamma_i = -\partial \ln \nu_i / \partial \ln V$. The calculated frequencies are very similar to previous theoretical results obtained by Lee and Gonze (1994) at the LDA level of theory with a plane-wave/pseudopotential method.

Table 20.6 Calculated vibrational frequencies of Stishovite (Ottonello et al. 2009b), and their gamma modes in stishovite. Acoustic frequencies represent the upper limit at $\mathbf{k}_{max}$

N	$\nu\ (cm^{-1})$	Mode	γ
1	252.9	Acoustic	–
2	252.9	Acoustic	–
3	413.9	Acoustic	–
4	217.5	B_{1G}	−1.4
5	399.2	B_{1U}	0.7
6_7	459.8	E_U	2.5
8_9	571.9	E_U	2.1
10_11	587.1	E_G	1.5
12	602.5	A_{2G}	1.0
13	664.7	A_{2U}	1.1
14	758.6	A_{1G}	1.2
15	759.4	B_{1U}	1.3
16_17	834.4	E_U	0.9
18	954.8	B_{2G}	1.3

At the long-wavelength limit corresponding to the center of the BZ ($\mathbf{k} \to 0$), the first three solutions of the dispersion relation

$$\omega = \omega_i(k) \quad 1 \le i \le 3n \tag{20.2.5}$$

where ω is the angular frequency and n is the number of atoms in the unit cell, vanish (Born and Huang 1954) and must be obtained by solving the non-zero components of the strain matrix acoustic modes at Γ point.

Background information: computing the acoustic modes
Introducing Cauchy relations and accounting for the internal dependencies imposed by the symmetry of the crystal, the Christoffel determinant for the tetragonal crystals reduces to (Authier and Zarembovitch 2006):

$$\begin{vmatrix} c_{33} - \rho v^2 & 0 & 0 \\ 0 & c_{44} - \rho v^2 & 0 \\ 0 & 0 & c_{44} - \rho v^2 \end{vmatrix} \tag{20.2.6}$$

for an incident wavevector parallel to [001] and to

$$\begin{vmatrix} c_{11} - \rho v^2 & 0 & 0 \\ 0 & c_{66} - \rho v^2 & 0 \\ 0 & 0 & c_{44} - \rho v^2 \end{vmatrix} \tag{20.2.7}$$

for an incident vector parallel to [111].

The resolutions of the first determinant are: $v_L = \sqrt{c_{33}/\rho}$; $v_{S_1} = \sqrt{c_{44}/\rho}$; $v_{S_2} = \sqrt{c_{44}/\rho}$ and the solutions of the second determinant are $v_L = \sqrt{c_{11}/\rho}$; $v_{S_1} = \sqrt{c_{66}/\rho}$; $v_{S_2} = \sqrt{c_{44}/\rho}$.

Because the elastic constants are the second derivative of the energy density with respect to the strain components:

$$c_{ij} = \frac{1}{V} \frac{\partial^2 E}{\partial \epsilon_i \partial \epsilon_j}\bigg|_0 \tag{20.2.8}$$

one may evaluate their discrete values by imposing a certain amount of strain ε along the crystallographic direction corresponding to the component of the dynamical matrix. The non-zero elements of the fourth rank strain tensor for a tetragonal crystal are:

$$A = c_{11} = c_{22}; B = c_{12}; C = c_{13} = c_{23}; D = c_{66}; E = c_{44} = c_{55}; F = c_{33} \tag{20.2.9}$$

where

$$c_{11} \Rightarrow \begin{vmatrix} \epsilon & 0 & 0 \\ 0 & 0 & 0 \\ 0 & 0 & 0 \end{vmatrix} \quad A = \frac{2b_{11}}{V} \tag{20.2.10}$$

and b_{11} is the second coefficient of a fourth-order polynomial relating energy and strain. Other elastic constants may be obtained in a similar way as

$$c_{12} \Rightarrow \begin{vmatrix} \epsilon & 0 & 0 \\ 0 & \epsilon & 0 \\ 0 & 0 & 0 \end{vmatrix} \quad B = \frac{2b_{12}}{V} - c_{11} \tag{20.2.11}$$

$$c_{13} \Rightarrow \begin{vmatrix} \epsilon & 0 & 0 \\ 0 & 0 & 0 \\ 0 & 0 & \epsilon \end{vmatrix} \quad C = \frac{b_{13}}{V} - \frac{c_{11}}{2} - \frac{c_{33}}{2} \tag{20.2.12}$$

$$c_{66} \Rightarrow \begin{vmatrix} 0 & \epsilon & 0 \\ \epsilon & 0 & 0 \\ 0 & 0 & 0 \end{vmatrix} \quad D = \frac{b_{66}}{2V} \tag{20.2.13}$$

$$c_{44} \Rightarrow \begin{vmatrix} 0 & 0 & 0 \\ 0 & 0 & \epsilon \\ 0 & \epsilon & 0 \end{vmatrix} \quad E = \frac{b_{44}}{2V} \tag{20.2.14}$$

$$c_{33} \Rightarrow \begin{vmatrix} 0 & 0 & 0 \\ 0 & 0 & 0 \\ 0 & 0 & \epsilon \end{vmatrix} \quad F = \frac{2b_{33}}{V} \tag{20.2.15}$$

Calculations on Stishovite were carried out by Ottonello et al. (2009b) sampling the energy E with respect to different values of the strain variable. A range of $-0.035 < \epsilon < 0.035$ with a step 0.005 was spanned. The 15 energy points were then fitted with a fourth-order polynomial, as anticipated. Results are summarized in Table 20.7.

The computed elastic constants are in reasonable agreement with the results of Brillouin scattering experiments (Weidner et al. 1982; Brazhkin et al. 2005) and generally lower than those calculated ab initio by using plane-wave/pseudopotential methods at the LDA level of theory (Cohen 1991; Karki et al. 1997; Holm and Ahuja 1999). The shear constant c_s, which is given by

$$c_s = \frac{1}{9} \times (c_{11} + c_{12} + 2c_{33} - 4c_{13}), \tag{20.2.16}$$

Table 20.7 Ab initio strain tensor components in Stishovite, compared with experimental evidences and previous computations. c_s is the shear constant (Ottonello et al. 2009b)

Strain	Ottonello et al. (2009b)	Weidner et al. (1982)	Brazhkin et al. (2005)	Cohen (1991)	Holm and Ahuja (1999)
c_{11}	439	453 ± 4	466 ± 3		480
c_{12}	196	211 ± 5	207 ± 8		245
c_{13}	188	203 ± 4	204 ± 4	221	220
c_{33}	754	776 ± 5	775 ± 4	807	735
c_{44}	243	252 ± 2	258 ± 2		260
c_{66}	316	302 ± 3	310 ± 6		340
c_s	155	156	156	158	146

is in excellent agreement with experimental data. The average shear modulus is found from the Hill average (Hill 1952) of the Voigt (upper, μ_V) and Reuss (lower, μ_R) bounds:

$$\mu_V = \frac{1}{5} \times (2c_s + 3c_{44}) = 207.8\ GPa \tag{20.2.17}$$

$$\mu_R = \left[\frac{1}{5} \times \left(\frac{2}{c_s} + \frac{3}{c_{44}}\right)\right]^{-1} = 198.0\ GPa \tag{20.2.18}$$

$$\overline{\mu}_{VRH} = \frac{1}{2}(G_V + G_R) = 202.9\ GPa \tag{20.2.19}$$

Adoption of a geometric mean, as suggested by Kumazawa (1969) yields essentially the same result:

$$\overline{\mu} = \sqrt{\mu_V \mu_R} = 202.8\ GPa \tag{20.2.20}$$

Young's modulus E and Poisson's ratio ν_P are, respectively,

$$E = \frac{9K_0\overline{\mu}_{VRH}}{3K_0 + \overline{\mu}_{VRH}} = 377.0\ \mathrm{GPa} \tag{20.2.21}$$

$$\nu_P = \frac{3K_0 - 2\overline{\mu}_{VRH}}{2 \times (3K_0 + \overline{\mu}_{VRH})} = 0.202 \tag{20.2.22}$$

By applying the general relation $v_{ij} = \sqrt{c_{ij}/\rho}$ it may be deduced that the lower and upper limits of longitudinal sound velocities are 10.236 and 13.416 *km/s*, respectively, while shear velocities span a more restricted range (7.616–8.685 *km/s*). The averaged shear and longitudinal wave velocities are given by

$$\overline{v}_s = \sqrt{\overline{\mu}_{VRH}/\rho} = 6.959\left(\frac{km}{s}\right) \quad (20.2.23)$$

$$\overline{v}_L = \sqrt{\frac{K_0 + \frac{4}{3}\times\overline{\mu}_{VRH}}{\rho}} = 11.389\left(\frac{km}{s}\right) \quad (20.2.24)$$

The maximum frequency of each acoustic branch is calculated according to the relation:

$$\omega_i = v_i\left(\frac{6\pi N_0}{ZV}\right) \quad (20.2.25)$$

where the term in brackets ($\mathbf{k}_{max}$) defines the BZ boundary (i.e. a radius af a sphere with the same volume of the BZ). Conversion to linear frequencies yielded the following results:

$$\nu_1 = \nu_2 = \overline{v}_s = 252.9\ cm^{-1};\quad \nu_3 = 413.9\ cm^{-1} \quad (20.2.26)$$

The first two acoustic modes at $\mathbf{k}_{max}$ are somewhat higher than the estimates of Kieffer (1979c) (186 and 227 cm^{-1}, respectively) while the longitudinal mode is in good agreement with her datum of 405 cm^{-1}.

The quasi-harmonic (QHA) gamma-mode analysis of the $\alpha_T K_T$ product gave:

$$\alpha_T K_T = \frac{R}{ZV}\sum_{i=4}^{3n}\gamma_i e^{-X_i}\left(\frac{X_i}{e^{X_i}-1}\right)^2 \quad (20.2.27)$$

(with X_i = ith a-dimensional frequency and γ_i = Gruneisen gamma of the ith normal mode ν_i; i.e. $\gamma_i = \partial \ln \nu_i / \partial \ln V$).

The quasi-harmonic treatment of vibrational frequencies allowed to establish the thermal dependency of the bulk modulus (see Anderson 1995; for a detailed treatment):

$$\left(\frac{\partial K}{\partial T}\right)_P = -\overline{\alpha}\,\overline{K} \times \left[K' - q^{ht} + 1\right] \quad (20.2.28)$$

where $\overline{\alpha}\,\overline{K}$ is the limiting αK product at $T \rightarrow \infty$, and

$$\gamma^{ht} = \gamma_{(T\rightarrow\infty)} = \left(\frac{1}{3n-3}\right)\sum_{i=4}^{3n}\gamma_i \quad (20.2.29)$$

$$q^{ht} = \frac{\partial \ln \gamma^{ht}}{\partial \ln V} \tag{20.2.30}$$

For the thermal analysis the first three (acoustic) vibrational modes were treated by Ottonello et al. (2009b) at the light of the Kieffer's model sound wave dispersion (see Sect. 8.10.4) and coupled with the calculated ab initio vibrational frequencies of the remaining 3n−3 modes (Table 20.6). This QHA gamma-mode analysis was later extended by Belmonte et al. (2017a) to 5000 *K*; a value more consistent with the high-T regimes that may be attained in the planetary interior. The analysis confirmed that $\alpha_T K_T = f(T)$ has at high T a constant slope attaining 0.0059 *GPa/K* at T = 5000 *K*. Adopting the previous $\alpha_T = f(T)$, the thermal derivative dK/dT was remodeled in order to achieve consistency over the entire T range of interest. The bulk modulus at the athermal limit obtained in this way ($K_0 = 270$ *GPa*) is somewhat higher than the value obtained from the simple static analysis of the potential well ($K_0 = 250$ *GPa*) and intermediate between the values derived, respectively, from General Gradient Approximation (GGA) and Local Density Approximation (LDA) analyses (see Demuth et al. 1999; and references therein). Because the anharmonicity at high T is slightly affected by the introduced modifications, also the parameters of the isobaric heat capacity obtained by Ottonello et al. (2009a) were re-fitted by Belmonte et al. (2017a) with non-linear minimization procedures. The compressional behavior of Stishovite and the other silica polymorphs, in a wide baric range consistent with the pressure attained in the planetary interior, is shown in Fig. 20.15. Experimental observations in the low-intermediate P range are superimposed for comparative purposes.

In Fig. 20.16a are shown the $P = 10^5$ *Pa* ab initio isochoric (C_V) and isobaric (C_P) heat capacities of Stishovite, compared with low-T C_P experimental observations.

In the harmonic approximation, the isochoric heat capacity, C_v, of one mole of a crystalline solid is given by (see also Sect. 8.10.3 and 8.10.4):

$$C_V = \frac{3R}{Z}\left(\frac{2}{\pi}\right)^3 \sum_{i=1}^{3} \int_0^{X_i} \frac{\left[\arcsin(X/X_i)^2\right] X^2 e^X dX}{\left(X_i^2 - X^2\right)^{1/2} \left(e^{X-1}\right)^2} + \frac{R}{Z} \sum_4^{3n} e^{X_i} \left(\frac{X_i}{e^{-X_i} - 1}\right)^2 \tag{20.2.31}$$

where n is the number of atoms in the unit cell and Z is the number of formula units per unit cell. The first term on the right of Eq. (20.2.31) is the acoustic contribution at $\mathbf{k}_{max}$ represented in terms of a sine-wave dispersion relation (Kieffer 1979a, b). The second term is the contribution of all remaining (optic) modes. Using the standard relation between C_V and the isobaric heat capacity, C_P is readily obtained, i.e.:

$$C_P = C_V + T\alpha_T^2 K_{P,T} V_{P,T} \tag{20.2.32}$$

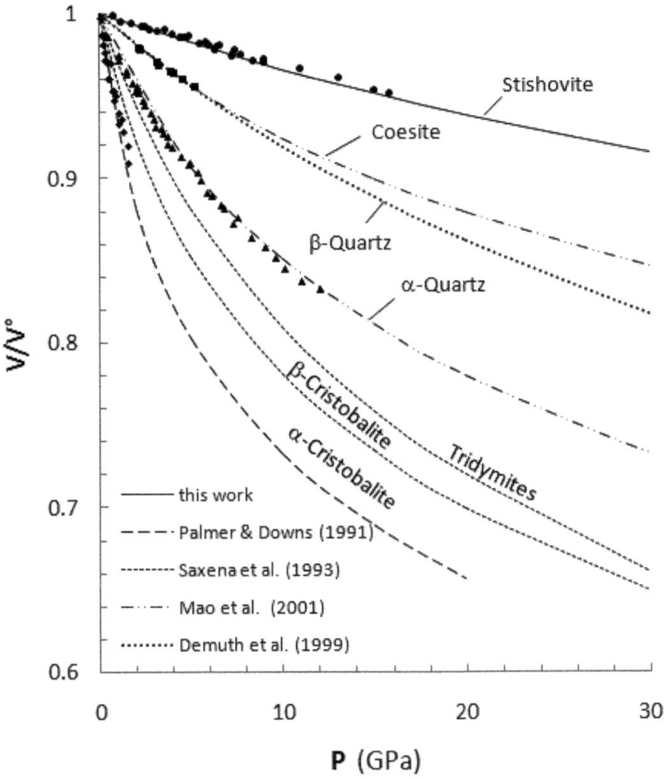

Fig. 20.15 Compressional behavior of Silica polymorphs, experiments versus assessments. Stishovite: Olinger (1976), Ross et al. (1990), Liu et al. (1999). Coesite: Levien and Prewitt (1981), α-Quartz: Vaidya et al. (1973), Olinger and Halleck (1976), D'Amour et al. (1979), Levien et al. (1980), Cartz and Jorgensen (1981), Jorgensen (1978). α-Cristobalite: Palmer and Downs (1991). From Belmonte et al. (2017a). With permission from CALPHAD (Elsevier Science Ltd.)

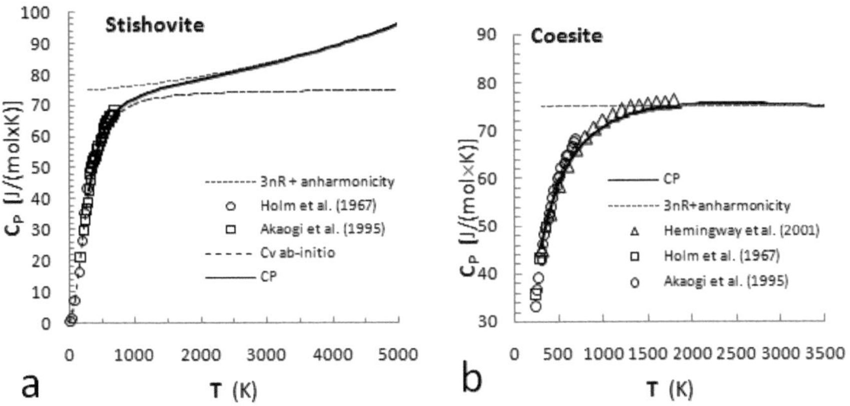

Fig. 20.16 Isobaric and isochoric heat capacity of Stishovite (**a**) and Coesite (**b**), compared with low-T observations. The solid curve is the continuous function generate by the Shomate's heat capacity expansion [see Table 20.6]. From Belmonte et al. (2017a). With permission from CALPHAD (Elsevier Science Ltd.)

The isochoric heat capacity C_V obeys obviously the Dulong-Petit limit at high temperature. The ab initio C_P data nicely overlap the results of cryogenic experiments by Holm et al. (1967) at low T and of differential scanning calorimetry experiments of Akaogi et al. (1995) at 183–703 *K*, At higher T, C_P converges toward the 3nR + ($TV\alpha^2$ K) condition dictated by anharmonicity. The anharmonicity is definitely non-negligible and attains ~14.7 *J/(mol × K)* at the maximum T of investigation. The calculated isobaric heat capacity at 298.15 *K* is 42.807 *[J/(mol × K)]*, which is close to the value tabulated by Robie et al. (1995) (42.99 *[J/(mol × K)]*, but the minor inflection envisaged at T ≈ 1300 *K* by these authors is not reproduced by the ab initio harmonic frequencies.

Tables 20.8 and 20.9 resume, respectively, the thermodynamic and thermophysical properties of all crystalline polymorphs of Silica in the assessment of Belmonte et al. (2017a) (see Sect. 8.10.7 for an outline of the adopted procedures). The entropy of Stishovite at standard state [S^0_{298} = 27.809 *J/(mol × K)*] is that one obtained ab initio by Ottonello et al. (2009b) while the enthalpy of formation from the elements at standard state was selected by Belmonte et al. (2017a) in order to reproduce the invariant point of coexistence with liquid and Coesite and the melting curve of the substance. It is somewhat lower than current estimates (i.e. −874 *kJ/mol* against −861 *kJ/mol* indicated by Robie and Hemingway (1995) and in line with the assessment of Mao et al. (2001).

Background information: scaling from absolute to conventional magnitudes

Comparison of the ab initio enthalpy of formation from the elements of the substance with experimental observations requires the assessment of a thermochemical cycle in the terms discussed at Sect. 8.10.7, i.e.

$$\overline{H}^0_{f,298.15} = \left(\Delta U_{298.15-0} + P\Delta V_{298.15-0}\right) - D_0 + \sum_{i=1}^{n/Z} n_i H_{f,A_i,0} - \sum_{i=1}^{n/Z} n_i \Delta H_{element_i,0\rightarrow 298.15} \quad (8.10.56)$$

where the term in brackets is the thermal correction to the enthalpy at 298.15 $H_{f,A_i,0}$ is the enthalpy of formation of the ith gaseous atom from the stable element at T = 0 *K*, P = 1 *bar* (with the summation extended to the n/Z atoms in the molecule; cf. Table 20.10), $\Delta H_{element_i,0\rightarrow 298.15}$ is the enthalpy difference between T = 298.15 and T = 0 for the monatomic elements (Table 20.11) and D_0 is the zero point dissociation energy of the gaseous molecule into gaseous atoms at 0 *K*:

$$D_0 = \sum_{i=1}^{n/Z} n_i E_{A_i} - \left(E_{HF,crystal} + E_{ZPE,crystal}\right) \quad (8.10.57)$$

Table 20.8 Thermodynamic properties of silica polymorphs in the assessment of Belmonte et al. (2017a). $H^0_{f,298}$ = enthalpy of formation from the elements at $T_r = 298.15$ *K*, $P_r = 10^5$ *Pa (kJ/mol)*; S^0_{298} = entropy of the substance at the same T, P conditions [*J/(mol × K)*]; C_P = isobaric heat capacity (Shomate equation: $C_p = a + bT + cT^{-2} + dT^{-1/2} + eT^{-3} + fT^2 + gT^3 + hT^{-1}$)

Polymorph	$H^0_{f,298}$	S^0_{298}	Aa	b × 10^3	c × 10^{-5}	d	e × 10^{-8}	f × 10^6	g × 10^9	h × 10^{-4}
α-quartz	−910.700	41.460	80.01	0	−35.467	−240.3	4.9157	0	0	0
β-quartz	−908.627	44.207	80.01	0	−35.467	−240.3	4.9157	0	0	0
α-cristobalite	−907.753	43.394	83.51	0	−24.554	−374.7	2.8007	0	0	0
β-cristobalite	−906.377	46.029	83.51	0	−24.554	−374.7	2.8007	0	0	0
Low- tridymite	−907.610	44.029	75.37	0	−59.581	0	9.5825	0	0	0
Mean tridymite	−907.137	45.316	75.37	0	−59.581	0	9.5825	0	0	0
High tridymite	−907.045	45.524	75.37	0	−59.581	0	9.5825	0	0	0
Coesite	−906.662	40.668	97.552	−2.3932	−17.960	−634.24	3.8529	−0.65017	0.12384	−0.26112
Stishovite	−874.361	27.809	73.768	1.0205	−58.832	56.402	8.3752	0.65429	0	0

Table 20.9 Thermoelastic properties of silica polymorphs in the assessment of Belmonte et al. (2017a). V^0_{298} = standard state molar volume (T_r = 298.15 *K*, $P_r = 10^5$ *Pa*). K_0 = bulk modulus at the athermal limit; K′ = (dK/dP) (assumed not dependent on T). $K_{0,T} = K_0 + k_1T + k_2T^2$ gives the thermal dependence of the bulk modulus. The volumetric thermal expansion at all *T* and P_r is reproduced by the following polynomial expansion on *T*: $\alpha_V = \alpha_0 T + \alpha_1 + \alpha_2 T^{-1} + \alpha_3 T^{-2} + \alpha_4 T^{-3}$

Polymorph	V^0_{298} (*cc/mol*)	K_0 (*GPa*)	K′	k_1 (*bar/K*)	k_2 (*bar/K²*)	$\alpha_0 \times 10^7$ (K^{-2})	$\alpha_1 \times 10^7$ (K^{-1})	$\alpha_2 \times 10^3$	α_3 (*K*)	α_4 (K^2)
α-quartz	22.69	40.07	6	−53.678	0.0039	−0.6756	699.08	0	0	0
β-quartz	23.9	132.6	3.9	−1153.8	0.3218	−0.07	0	0	0	0
α-cristobalite	25.76	11.5	9	0	0	2.55	−2424.5	0	0	0
β-cristobalite	27.242	14.0	7.1	−20	0	−0.173593	235.599	0	0	0
Low tridymite	27.170	30.8	4.1	0	0	−0.27655	313.204	0	0	0
Mean tridymite	27.170	30.8	4.1	0	0	−0.27655	313.204	0	0	0
High tridymite	27.306	30.8	4.1	0	0	−0.27655	313.204	0	0	0
Coesite	20.640	100.3	8.4	−249.584	0.0252	−0.004534	101.67	0	0	0
Stishovite	14.000	270.1	6.59	−474.75	0.03514	0.05904	167.77	2.2872	−2.1518	237.0245

Table 20.10 Thermochemical reference data for the various atoms of interest. $H_{f,0}$ = enthalpy of formation of the gaseous atom from the stable element at T = 0 *K*, P = 1 *bar* (Chase et al. 1998). $\Delta H_{element_i,0\rightarrow 298.15}$ is the enthalpy difference between T = 298. 15 *K* and T = 0 for the monoatomic element (Chase et al. 1998). $S^0_{298.15}$ is the standard state entropy of the monoatomic element. E_{Ai} and E^*_{Ai} are electronic energies of the gaseous atoms computed, respectively, with the same basis set of the bulk and with the basis set supplemented by two more diffuse sp shells, respectively. E^{**}_{Ai} are the values returning a correct heat content of stable elements at standard state

Species	$H_{f,0}$ (*kJ/mol*)	$\Delta H_{element_i,0\rightarrow 298.15}$ (*kJ/mol*)	$S^0_{298.15}$ (*J/mol × K*)	E_{Ai} –B3LYP (*Hartree*)	E^*_{Ai}-B3LYP (*Hartree*)	E^{**}_{Ai} (*Hartree*)
Si	445.668	4.3320	18.820	−280.011087	–	−289.295420578
O	246.790	4.3415	102.5735	−75.055332	−75.05984193	−75.057906111

Table 20.11 Ab initio heat content of stishovite and α-quartz compared to assessments

Substance	D_0 (*kJ/mol*)	$H^0_{f,0}$ (*kJ/mol*)	$H^0_{f,298.15}$ (*kJ/mol*)	$H^0_{f,0}$ (*kJ/mol*) experim.	$H^0_{f,298.15}$ (*kJ/mol*) experim.	References
O_{gas}	– –	246.8	248.6	246.9 246.8	249.4 249.2	JANAF Tables NBS Tables
Si_{gas}	–	445.7	448.6	445.7 451.3	450.0 455.6	JANAF Tables NBS Tables
O_2 gas	487.4	6.2	0.0	–	0.0	–
Si cubic	444.8	0.8	0.0	–	0.0	–
SiO_2 tetragonal	1774.1	−834.8	−845.8	–	−861.3	Robie and Hemingway (1995)
SiO_2 trigonal	1840.4	−901.2	−906.2	–	−910.7	Robie and Hemingway (1995)

This energy term corresponds to the sum of the electronic energies of the gaseous atoms at 0 K less the electronic + zero point energy of the crystal.

As already noted at Sect. 8.10.7, the estimate of appropriate electronic energies for an isolated gaseous atom is still a matter of debate. One may eventually adopt the theoretical Hartree–Fock limit or add diffuse functions with respect to the basis set adopted in the bulk. A more practical way is to assign to the gaseous atoms electronic energies returning a correct dissociation energy (D_0) of the element at its stable state (i.e. the gaseous diatomic molecule O_2 and the cubic hexakisoctahedral crystalline polymorph of Si). This requires however evaluating the variational structure of both molecules and to carry out their vibrational analysis. The B3LYP energies of gaseous O_2 and crystalline Si are listed in Table 20.11. In evaluating the thermal corrections of Si Ottonello et al. (2009b) adopted the acoustic phonons of Brazhkin et al. (2000) and the experimentally observed entropy ($S^0 = 18.86$ *J/(mol × K)*; Robie and Hemingway 1995).[4] This procedure returns for the tetragonal form of SiO_2 an enthalpy of formation from the elements at standard state of –845.8 *kJ/mol*; i.e. roughly 15 *kJ/mol*, higher than current estimates. To verify the internal consistency of this datum the authors carried out an additional investigation on the trigonal form of SiO_2 (α-quartz). The ab initio electron energy of Stishovite and α-quartz and thermal corrections to U, H and G are listed in Table 20.12. In computing

[4] The anharmonicity of Si is quite pronounced (cf. Brazhkin et al. 2000; and references therein), and the anharmonicity contribution to entropy is definitely non-negligible even at low T.

thermal corrections acoustic contributions were accounted for as well. To this purpose Ottonello et al. (2009b) utilized the elastic properties of Mc Skimmin et al. (1956) and their treatment according to Kieffer (1979c). The resulting ab initio enthalpy of formation of α-quartz was –906.2 *kJ/mol.*; 4.5 *kJ/mol* higher than current estimates (Table 20.11).

Although the apparently cumbersome procedure of estimating the B3LYP energies of the elements at their stable state is not compulsory (the enthalpy of solid state transition at T = 298.15 *K*, P = 1 *bar* may be eventually obtained by simply summing up the electronic energy and the thermal correction for each polymorph listed in Table 20.12 and then subtract the results) it has the advantage of returning additional information concerning the intrinsic stability of each substance, such as for instance its zero-point dissociation energy (Table 20.11). The ab initio enthalpy of formation of the two SiO_2 polymorphs imply a metastable enthalpy of transition α-quartz = > stishovite of +60 *kJ/mol*, 11–23 *kJ/mol* higher than observed (i.e. 49 ± 2 *kJ/mol*, Holm et al. 1967; 52 ± 2 *kJ/mol*, Akaogi and Navrotsky 1984; 37 ± 2 *kJ/mol* Akaogi et al. 1995). The computed relative stability of Stishovite with respect to α-quartz is +67 *kJ/mol*, which is in agreement with previous GGA calculations ranging between 49 and 61 *kJ/mol* (Hamann 1996; Zupan et al. 1998; Demuth et al. 1999) and, as expected from previous studies (Hamann 1996; Zupan et al. 1998), markedly larger than LDA calculations (Keskar and Chelikowsky 1992; Liu et al. 1994; Lee and Gonze 1994, 1995; Teter et al. 1995; Hamann 1996; Zupan et al. 1998; Demuth et al. 1999).

20.2.1.2 Coesite

This polymorph has been found coexisting with garnet in the Dora-Maira Massif (Western Alps) and in various other high-P metamorphic terrains in the former Soviet Union and in China. In melting experiments, with the decrease of P at subsolidus temperatures, Stishovite transforms reversibly into this less dense phase.

Table 20.12 DFT/B3LYP electronic energy, zero point energy of optical modes (ZPC) and thermal corrections to U, H and G (P = 1 *bar*; T = 298.15 *K*; inclusive of acoustic branches) of the investigate phases. Data are in *Hartree* and on a gram formula weight basis

Phase	E_{el}	ZPC	Therm U	Therm H	Therm G
SiO_2-stishovite	−440.097777091	0.010551533	0.012600267	0.012600821	0.009441827
SiO_2-quartz	−440.123389195	0.011187465	0.013822510	0.013823421	0.009091540

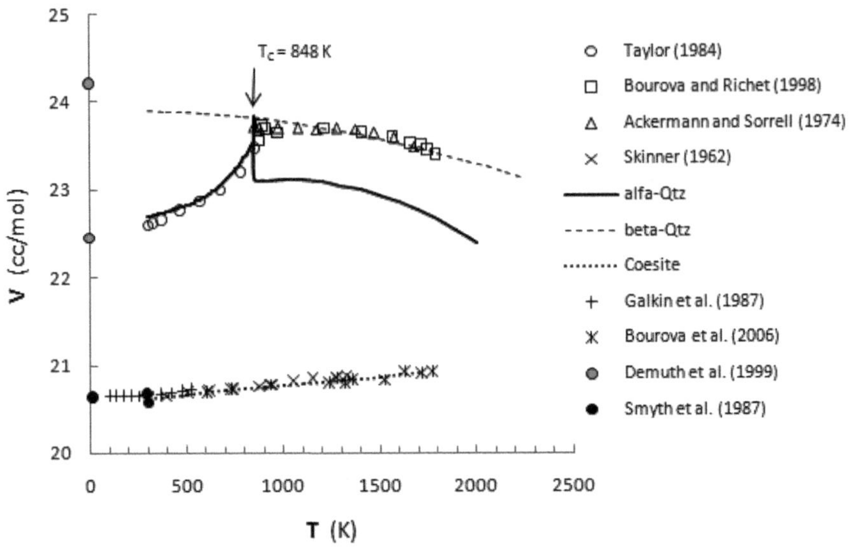

Fig. 20.17 Model volumes of Quartz and Coesite, compared to experimental observations and first-principles computations. Modified from Belmonte et al. (2017a)

The adopted space group of monoclinic Coesite is *C2/c* (Smyth et al. 1987; Geisinger et al. 1987), though the space group $P2_1/c$ has also been proposed (Kirfe et al. 1979). The DFT-LDA bulk modulus of Coesite is $K_0 = 94.3$ *GPa*, with a P derivative $K' = 4.8$ (Demuth et al. 1999). These compressibility values are only approximately consistent with the experimental observations of Levien and Prewitt (1981) and the assessment of Mao et al. (2001) returning a slightly higher bulk modulus and baric derivative (i.e. $K_0 = 100.3$ *GPa*, $K' = 8.4$; Table 20.9) coupled with a marked thermal depencence of the compressibility ($K_1 = -249.584$ *bar/K*; $K_2 = 0.0504$ bar/K^2). The standard state molar volume and the thermal expansion in Table 20.9 are based on the experiments of Bourova et al. (2006) (Fig. 20.17). The isobaric heat capacity (Fig. 20.16b) is essentially based on the experiments of Hemingway et al. (2001) refitted by Belmonte et al. (2017a) in the form of a Shomate's equation consistent with the Dulong-Petit limit expectations at high T. Based on the selected thermophysical parameters (Belmonte et al. 2017a) and differently from Stishovite, the anharmonicity contribution to the isochoric heat capacity is almost negligible and the assessed values of S^0 and H_f^0 are nearly similar to those of Mao et al. (2001).

20.2.1.3 Quartz

With the decrease of P, Coesite transforms reversibly into β-Quartz for temperatures above ~1550 *K* and to α-Quartz at lower T. The assessment of the thermophysical properties of the two Quartz polymorphs is complicated by the existence of a lambda transition coupled to a slight first-order contribution. The λ transition of quartz has been treated in many ways with somewhat contradictory results in terms of second-order contributions to the bulk thermodynamic properties of the phase (Helgeson et al. 1978; Hosieni et al. 1985; Salje et al. 1991; Berman and Brown 1985, 1986; Berman 1988). Belmonte et al. (2017a) adopted the procedure of Berman and coworkers: a common background (or "*lattice*" cf. Berman 1988) heat capacity with superimposed λ. Based on the experiments of Ackermann and Sorrell (1974) and Bourova and Richet (1998) the standard state molar volume of β-Quartz was set by Belmonte et al. (2017a) to 23.9 *cc/mol*, and it was assumed to be affected by a small negative thermal expansion represented through a single coefficient (i.e. $\alpha = -0.07 \times T$; Table 20.9). The simple first-order treatment of the α-Quartz to β-Quartz transition gives a constant Clapeyron slope $k = dT/dP = 0.0237$ *K/bar* (see Table 20.13). The λ contribution to entropy at T_C is $S_{\lambda,C} = 2.158$ *J/(molK)*. The resulting $V_{\lambda,C}$ at the critical temperature ($T_C \cong T_\lambda = 848$ *K*) is $V_{\lambda,C} = 0.55$ *cc/mol*, which, when subtracted to the first-order volume of β-Quartz, returns $V^0 = 22.95$ *cc/mol*. This value is still somewhat too high with respect to the observation of Taylor (1984) ($V^0 = 22.69$ *cc/mol*). Notwithstanding this fact, as we may see in Fig. 20.17, the evolution of the volume with T for the α-Quartz polymorph below the phase transition is sufficiently consistent with experiments. The potential well of the idealized $P6_222$ structure of β-Quartz according to the DFT-LDA computations of Demuth et al. (1999) denotes a marked stiffness of the substance at the athermal limit (i.e. $K_0 = 132.6$ *GPa*) while the unique experimental observation at 873 *K* (i.e. 56.4 *GPa* at T = 873 *K*; Pies et al. 1979) testifies for a marked T dependency, interpreted by Demuth et al. (1999) as due to the activation of low-frequency Rigid Unit Modes (RUM). For the $P3_221$ α-polymorph the same authors declare at the athermal limit a DFT/LDA bulk modulus of 35.4 *GPa*, while, with the same procedure, Hamann (1996) obtained $K_0 = 45.0$ *GPa*. Belmonte et al. (2017a) adopted the assessment of Mao et al. (2001) which, based on several sets of internally consistent experiments assign $K_0 = 40.07$ *GPa* and $K' = 6$. Concerning the thermal dependence of the bulk modulus on temperature, Belmonte et al. (2017a) adopted again the assessment of Mao et al. (2001) while, the thermal dependence of the compressibility of β-quartz was based on the experimental observations of Cohen (1979) and Thompson and Wennemer (1979) of the Clapeyron slope between the two polymorphs (Table 20.9). The ensuing anharmonicity is definitely non-negligible and attains ~2.3 *J/mol* at T = 3000 *K*. The isobaric heat capacity remains below the Dulong-Petit asymptotic limit at all T of interest.

20.2.1.4 Cristobalite

Cristobalite is rarely found in volcanic rocks and sometimes occurs in opal gems. The V^0, K_0, K′ values for the α-polymorph (space group $P4_12_12$) listed in Table 20.9 are those of Downs and Palmer (1994) (i.e. $V^0 = 25.76$ *cc/mol*; $K_0 = 11.5$ *GPa*; K′ = 9). The experimental bulk modulus is well in line with the DFT/LDA predictions of Teter et al. (1995) and Demuth et al. (1999) (i.e. 11.9 and 12.8 *GPa*, respectively) though the baric derivative is much higher than the computed ones (3.0 and 6.3, respectively). For the β-polymorph the adopted bulk modulus and its baric derivative are those obtained ab initio by Demuth et al. (1999) for the $P2_13$ space group (i.e. $K_0 = 14.0$ *GPa*; K′ = 7.1). The standard state volume ($V^0 = 27.242$ *cc/mol*) and thermal expansion of β-Cristobalite adopted by Belmonte et al. (2017a) are consistent with the observations of Swainson and Dove (1995) and Bourova and Richet (1998) (Fig. 20.18 and Table 20.9).

To reproduce the experimental Clapeyron slope of the α/β transition, Belmonte et al. (2017a) assumed that the α-polymorph has a complex thermal expansion (negative from room conditions to T_λ and positive at higher T). This complex behavior is consistent with the computations of Teter et al. (1995) at the athermal limit and with the displacive contributions to the volume at T_λ ($\Delta V = 2.77$ *cc/mol* based on the Clapeyron slope; cf. Figure 20.18 and Table 20.13). The λ transition

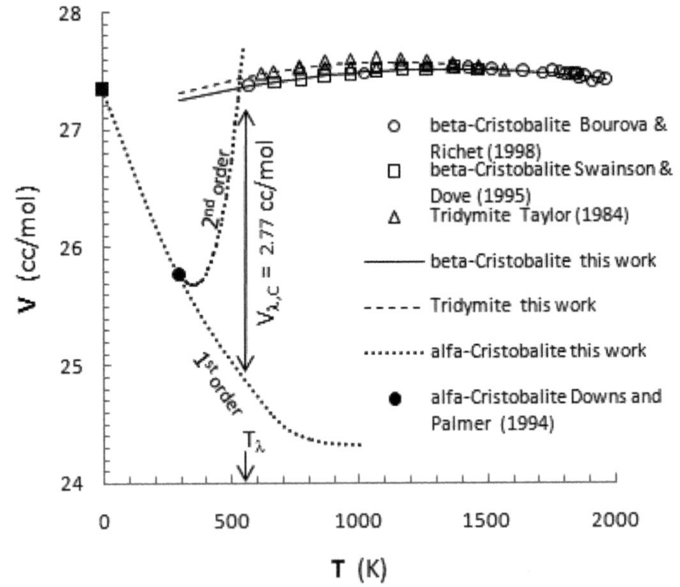

Fig. 20.18 Model volumes of Cristobalite and Tridymite in the assessment of Belmonte et al. (2017a), compared to experimental observations. The effect of λ displacive contributions to the bulk volume of the α polymorph are sketched. From Belmonte et al. (2017a). With permission from CALPHAD (Elsevier Science Ltd.)

Table 20.13 Lambda transitions of SiO_2 polymorphs

Low-T polymorph	T_λ	T_{ref}	$X_{\lambda 1}$	$X_{\lambda 2}$	b_λ	f_λ	g_λ	k	Δ H-1st	References
α-quartz	848	373	−0.09187	0.00024607	0.00844	−4.521E-05	6.06E-08	0.0256	499	Belmonte et al. (2017a)
								(0.0237)		Berman (1988)
									(499)	Berman and Brown (1986)
α-cristobalite	550	298.15	−0.1	0.0004465	0.01000	−8.000E-05	1.60E-07	0.1217	304	Belmonte et al. (2017a)
Low-trydimite	388	308	−0.83	0.0027311	0.70560	−4.586E-03	7.45E-06	0	–	Belmonte et al. (2017a)
Mean-trydimite	440	388	−0.52	0.0014	0.27040	−1.456E-03	1.96E-06	0.0655	–	Belmonte et al. (2017a)

from the α to the β polymorph induces a substantial increase in heat capacity up to a critical limit that, based on the observations of Mosesman and Pitzer (1941) would attain ~23 *J/(mol × K)* at $T_C = 550$ K. Based on the modeled transition at this temperature the delta in the bulk entropy of the two polymorphs is 2.276 *J/(mol × K)* and the *β* polymorph attains at standard state an entropy of 46.029 *J/(mol × K)*, consistent with Berman's (1988) assessment. The lattice heat capacity function of Berman (1988) remains below the Dulong-Petit asymptotic limit in the T-range of extrinsic stability of the substance.

20.2.1.5 Tridymite

In Nature Tridymite is found in porosities of rapidly quenched effusive rocks (mainly andesites and trachites). Actually, under this name are grouped various low-density polymorphs of silica, and numerous phase transitions among them have been proposed (Mosesman and Pitzer 1941; Austin 1954; Cohen and Clement 1980; Raz 1983). The high-T polymorph is closely related to the β-Cristobalite structure (Dmitriev et al. 1997). The ideal $C222_1$ orthorhombic structure of β-Tridymite has been investigated by DFT/LDA procedures by Demuth et al. (1993); the bulk modulus is $K_0 = 30.8$ *GPa* and its baric derivative is 4.1. The experiments of Taylor (1984) on the thermal expansion of Tridymite may be reproduced with a simple two parameters thermal expansion (i.e. $\alpha_0 = -0.276E\text{-}7, \alpha_1 = 313.2E\text{-}7$ with a standard state molar volume $V^0 = 27.306$ *cc/mol*; cf. Figure 20.18 and Table 20.9). Following Berman (1988) it may be assumed that only two phase transitions occur in Tridymites, namely those ones for wich "*acceptable consensus*" has been obtained: a low-to-mean transition occurring at $T_\lambda = 388$ K and a mean-to-high transition with $T_\lambda = 440$ K (Cohen and Klement 1980; Raz 1983). It may be also assumed, for simplicity, that the thermophysical parameters of the "*low*" and "*mean*" polymorphs are identical to those ones of the "*high*" polymorph identified by Belmonte etal. (2017a) with the ideal $C222_1$ orthorhombic structure. The common lattice heat capacity function of Tridymite (Berman 1988) is that one listed in Table 20.8. The second-order contributions of the two transitions (Table 20.13) at T_λ are the following: $\Delta H_\lambda = 473.4$ *J/mol*; $\Delta S_\lambda = 1.287$ *J/(mol × K)*; $\Delta V_\lambda = 0.0$ *J/bar* for the low/mean transition and $\Delta H_\lambda = 91.5$ *J/mol*; $\Delta S_\lambda = 0.208$ *J/(mol × K)*; $\Delta V_\lambda = 0.0136$ *J/bar* for the mean/high transition.

20.2.2 Al_2O_3

Aluminum oxide nucleates under various metastable polymorphic forms in addition to the naturally occurring thermodynamically stable α-Al_2O_3 structure. Pure α-alumina (corundum) has been promoted in 1948 as a heat capacity standard because it does not experience any solid state transformation from the athermal limit to the melting point which, at room pressure, takes place at 2327 ± 6 *K*. The metastable alumina structures (also denoted "*transition aluminas*"), may be divided into two broad categories which have either an fcc or an hcp arrangement of anions (Levin et al. 1993; and references therein). The relative stability of transition (θ-) and α-alumina was the object of the DFT investigation of Lodziana and Parlinsky (2003). Computing the phonon dispersion relations, the phonon density of states, and the free energy of θ and α phases they showed that entropic effects promote potential destabilization of the α phase at high T, but are insufficient to drive transformations between the two polymorphs.

Corundum (α-Al_2O_3) is found as primary occurrence in Nature in micaceous pegmatites, originated in the latest stages of magmatic (pneumatholitic-hydrothermal) differentiation, and in alluvial deposits. The gem qualities *Ruby* and *Sapphire* inherit their color tanks to isomorphic substitution of Al^{3+} with trivalent transition elements. Corundum crystallizes in the trigonal system (space group $D_{3d}^{6}\left(R\bar{3}c\right)$). In Fig. 20.19, we may see the structure generated by the AMCSD software, based on the 1 *bar* investigation of Finger and Hazen (1978). There are 30 atoms in the conventional unit cell, corresponding to six formula units.

The first ab initio investigation of the lattice dynamics of α-Al_2O_3 is due to Heid et al. (2000) which combined the mixed-basis pseudopotential approach of Louie et al. (1979) with density-functional perturbation theory (see Sect. 8.8.1). LDA calculations were performed using the Hedin–Lundqvist exchange–correlation

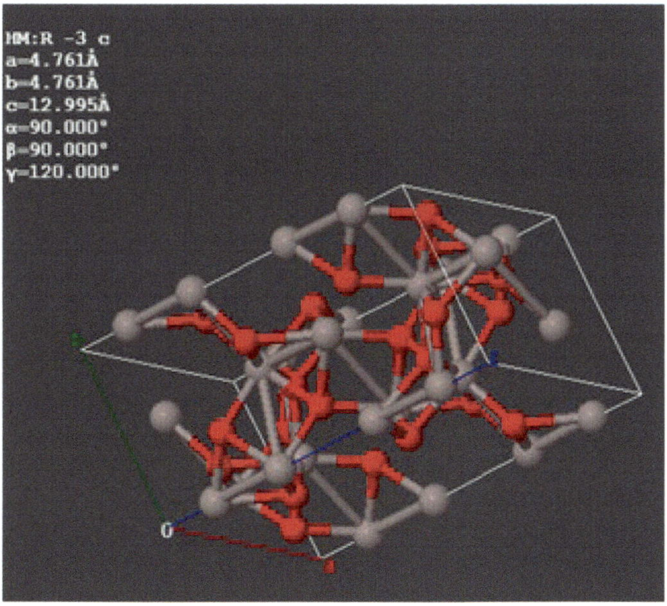

Fig. 20.19 The $D_{3d}^{6}\left(R\bar{3}c\right)$ structure of Al_2O_3 (Finger and Hazen 1978) generated by the AMCSD software (Downs and Hall-Wallace 2003)

Table 20.14 Theoretical values for the screened Born-effective charge tensors $Z_{\alpha\beta}/\sqrt{\epsilon_{zz}^{\infty}}$. For all Al ions charge tensors are equal by symmetry. Values given for O correspond to the ion with rhombohedral coordinates $\left(\frac{1}{4}-u, \frac{1}{4}, -\frac{1}{4}+u\right)$ (From Heid et al. 2000)

	Al			O		
	x	y	z	x	y	z
x	1.65	0	0	−1.15	0	0
y	0	1.65	0	0	−1.04	−0.14
z	0	0	1.63	0	−0.19	−1.09

functional (Hedin and Lundqvist 1971). For core electrons were adopted Hamann–Schlüter–Chiang norm-conserving pseudopotentials (see Sect. 12.7.1). Four local functions of s and p type at each oxygen site were adopted to represent valence states, with cut off at a radius of 1.3 a.u., and plane waves up to a kinetic energy of 20 *Ry*. The sampling of the Brillouin-zone was performed on a 43,434 simple rhombohedral **k**-point mesh corresponding to 13 special points in the irreducible wedge.

Since α-Al_2O_3 is an insulator, a complete characterization of its lattice dynamics requires knowledge of the dielectric tensor $\epsilon_{\alpha\beta}^{\infty}$ and Born effective-charge tensors $Z_{\alpha\beta}^{*}(k)$ for each k determining the non-analytic contribution to the dynamical matrix in the limit $\mathbf{q} \rightarrow 0$. This non-analytic part was extracted by Heid et al. (2000) from calculations for **q** points close to the center of the BZ along three different directions in reciprocal space. All matrix elements of the dielectric and Born effective-charge tensors were thus determined, except for a single scale factor. Since in trigonal symmetry, the dielectric tensor is diagonal, the independent elements are only two: $\epsilon_{xx}=\epsilon_{yy}$ and ϵ_{zz}. These independent elements are shown in Table 20.14 for both ion species. Only small anisotropies are present, the largest for the O ions which occupy sites of low C_2 symmetry. The small anisotropy $(\epsilon_{xx}\epsilon_{zz}= 1.02)$ found by Heid et al. (2000) is in agreement with experimental observations (Barker 1963).

The phonon dispersion calculated by Heid et al. (2000) along the Γ–Z direction (rhombohedral axis), the Γ–A direction [A = (0, 0.5, 0)] and the Γ–D direction [D = (0.5, 0, 0.5)] are in perfect agreement with the neutron scattering experiments of Schober et al. (1993). To obtain the complete phonon spectrum, the dynamical matrix, determined on the 43,434 mesh, was extended to the whole BZ by Fourier interpolation. The phonon DOS of α-Al_2O_3 according to Heid et al. (2000) is shown in Fig. 20.20, where the shaded area denotes the partial density of states of the oxygen ions. The prominent features of the DOS are the pseudogap structures near 43 and 87 *meV*, and the strong peak near 92 *meV*, ascribed by the Authors to a nearly dispersionless oxygen mode.

The first *all-electron* LCAO-DFT ab initio investigation of vibrational frequencies at Γ point of α-Al_2O_3 was carried out by Montanari et al. (2006). The force constant matrix was calculated by the Authors with the finite displacement procedure of Pascale et al. (2004) implemented in the CRYSTAL code. The study

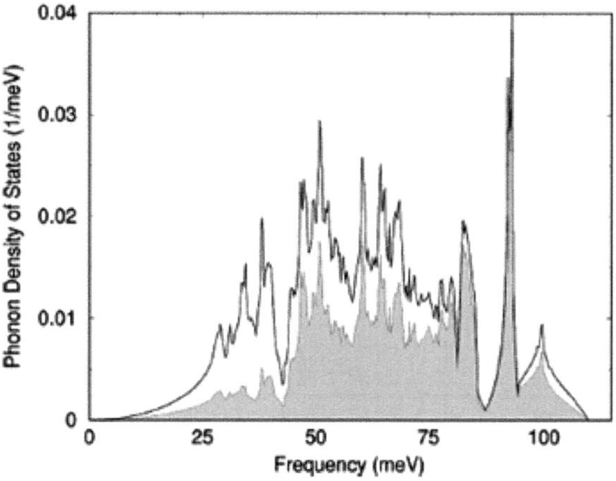

Fig. 20.20 Phonon DOS of α-Al_2O_3. From Heid et al. (2000). With permission from The American Physical Society

was mainly focused on the performance of various exchange–correlation functionals: the Vosko–Wilk–Nusair functional (Sect. 10.4), the Perdew–Wang functional (Sect. 10.5.2), and the hybrid functional B3LYP (Sect. 10.5.5). Also the effect of the completeness of the adopted basis set was the object of investigation. Four all-electron basis sets were tested. The simplest one of these (BS1) was an 86-11G(d) contraction for Al and 8-411G(d) for O (already adopted by Catti et al. 1994; in the investigation of spinel but with $\alpha_{sp} = 0.35$, $\alpha_d = 0.51$ for Al and $\alpha_{sp} = 0.20$, $\alpha_d = 0.57$ for O as most diffuse functions, in $Bohr^{-2}$). The second set (BS2) adopted an 88-11G(d) contractions for Al (with $\alpha_{sp} = 0.20$ for the most diffuse function). In the third basis set (BS3), a double set of diffuse d functions were added to BS1 (with $\alpha_d = 2.0$ and 0.40 for Al and $\alpha_d = 2.0$ and 0.45 for O). The fourth basis set was BS2 with added f functions (with $\alpha_f = 0.8$ for both Al and O). The main results of the comparative investigation are resumed in Table 20.15.

The comparative appraisal shows that the basis set BS1 is sufficient to provide an accurate description of this structure because an extension of the polarization functions produces a quite limited (though systematic) decrease in the lattice parameters. As already observed many times for various compounds, also in α-Al_2O_3 the B3LYP functional overestimates the interionic distances, while LDA underestimates them.

The optical bandgap computed at LCAO-DFT/B3LYP theory level with the BS1 basis set (8.37 *eV*) is consistent with the value measured by French (1990) (8.8 *eV*). This observation is important because the longitudinal frequencies depend on the dielectric properties of the substance. Much lower values are observed with the LDA and PW91 functionals (6.51 and 6.02 *eV*, respectively).

The vibrational properties of α-alumina were later reinvestigated by Belmonte et al. (2013) at the LCAO-DFT/B3LYP theory level, integrating the optical modes with acoustic modes obtained with the analysis of the strain tensor and assuming a Kieffer model sine-wave dispersion (see. Sects. 8.9.2 and 8.10.4). The optical modes obtained by Belmonte et al. (2013) are compared in Table 20.16 with the results of previous investigations and experimental observations.

Table 20.15 *Ab initio* equilibrium geometry of α-Al_2O_3 at DFT/B3LYP theory level obtained with different basis set expansions and different functional (Montanari et al. 2006). The experimental data are from d'Amour et al. (1978)

	BS1	BS2	BS3	BS4	Functional	Experimental
a (Å)	4.799	4.800	4.791	4.780	B3LYP	4.766
	4.748				LDA	
	4.818				PW91	
c (Å)	13.075	13.084	13.069	13.067	B3LYP	13.010
	12.902				LDA	
	13.114				PW91	
c/a	2.725	2.726	2.728	2.734	B3LYP	2.730
	2.717				LDA	
	2.722				PW91	
V_0 ($Å^3$)	43.471	43.520	43.295	43.094	B3LYP	42.649
	41.982	43.520	43.295	43.094	LDA	
	43.944	43.520	43.295	43.094	PW91	
Al_z	0.353	0.353	0.352	0.352	B3LYP	0.352
	0.352				LDA	
	0.353				PW91	
O_x	0.305	0.305	0.306	0.306	B3LYP	0.306
O_x	0.305				LDA	
O_x	0.305				PW91	

We have already seen at Sect. 8.10.7 that this investigation was quite successful in reproducing the isobaric heat capacity of the substance, which is in excellent agreement with the calorimetric assessment of Ditmars and Douglas (1971).

At the melting point, α-alumina undergoes a dramatic structural reorganization, which does not involve simple rupture of bonds subjected to thermal stress, but deep rearrangements in short-range ordering with a substantial change in the coordination of the central cation. We have seen some first-principles interpretations of this complex phenomenon at Sects. 19.5.7 and 19.5.8. The melting behavior of corundum at high-pressure conditions is poorly constrained. The only experimental determination made so far (Shen and Lazor 1995) indicates a melting temperature $T_{fusion} = 3525 \pm 150\ K$ at $P = 25\ GPa$, in agreement with the molecular dynamics simulation of Ahuja et al. (1998).

20.2.3 MgO

Magnesium oxide (*magnesia*), nucleates in its crystalline form in the cubic system (space group *Fm3m*) as an fcc (or B1-type) structure named *Periclase* (Fig. 20.21).

Table 20.16 Optical wavenumbers of α-Al_2O_3 corundum (in cm^{-1}) calculated by Belmonte et al. (2013) at the B3LYP level of theory (1). The adopted basis set for DFT/B3LYP calculations is 85-11G(d) for Al and 8-411G(d) for O. Other theoretical (2) and experimental (3, 4) results are reported for comparison

Mode	ν (cm^{-1})	ν(cm^{-1})	ν (cm^{-1})	ν(cm^{-1})
	(1)	(2)	(3)	(4)
Infrared				
A_{2u}	397.0	391.4	397.5	
A_{2u}	577.6	567.5	582.4	
E_u	384.0	375.4	385.0	
E_u	439.2	434.9	439.1	
E_u	564.9	557.6	569.0	
E_u	639.1	631.1	633.6	
Raman				
A_{1g}	415.1	408.9		418
A_{1g}	646.4	638.9		645
E_g	380.3	374.6		378
E_g	424.8	416.9		432
E_g	454.2	449.7		451
E_g	578.6	574.4		578
E_g	757.7	749.0		751
Silent				
A_{2g}	304.1	298.2		
A_{2g}	551.2	544.9		
A_{2g}	763.7	758.5		
A_{1u}	597.4	593.6		
A_{1u}	704.0	700.5		

(2) Montanari et al. (2006)
(3) Schubert et al. (2000)
(4) Porto and Krishnan (1967)

In nature, Periclase is one of the main constituents of the Earth's lower mantle. At shallower depth, Periclase occurs in *contact metamorphic rocks* and it is found ubiquitously in marbles as a metamorphic product of *dolomitic limestones*.[5] Due to its simple structure and large stability field and the plethora of experimental data in a broad range of P–T conditions Periclase "*has always been considered to be a milestone for computational investigation*" (Belmonte 2017).

As far as I know, the first ab initio computation of the elastic constants and the high symmetry point (Γ, X, and L) phonon frequencies of MgO has been carried out by McCarthy and Harrison (1994) at the HF theory level. In that study, correlation-only corrections to the ground-state periodic HF energies (Causà et al. 1987,1988) were added *a-posteriori*. The most comprehensive ab initio investigation hitherto appeared of thermophysical and thermodynamic properties of Periclase is due to Belmonte (2017). Computations were carried out at the DFT/B3LYP theory level under GGA and LCAO approximation. Magnesium and oxygen were

[5] Occasionally found as well in calcareous ejecta of the Monte Somma-Vesuvius.

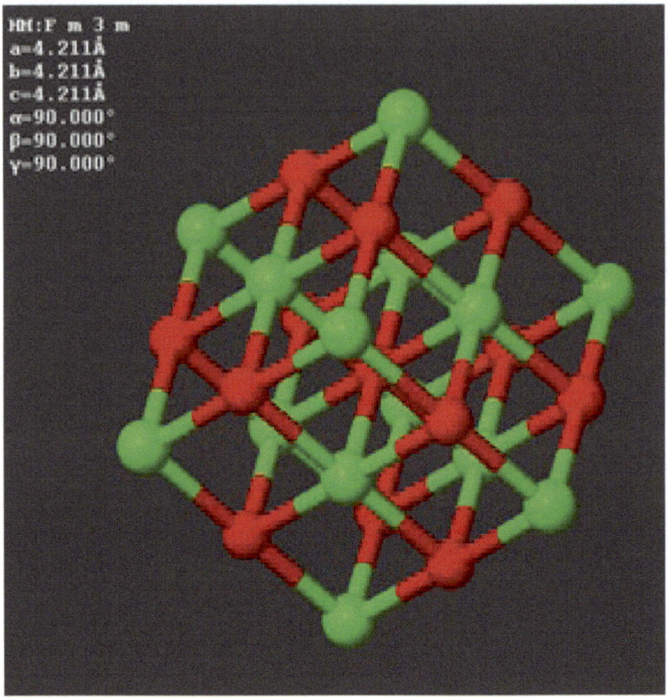

Fig. 20.21 The *Fm3m* structure of MgO (Hazen 1976) generated by the AMCSD software (Downs and Hall-Wallace 2003)

described by (8s)-(511sp)-(1d) and (8s)-(411sp)-(1d) contractions for Mg and O, respectively, with $\alpha_{sp} = 0.2245$ and $\alpha_d = 0.50$ exponents for the most diffuse functions of Mg (in $Bohr^{-2}$) and $\alpha_{sp} = 0.181$ and $\alpha_d = 0.60$ for O. The performance of this set was widely tested in previous investigations (Ottonello et al. 2010; De la Pierre et al. 2011; Belmonte et al. 2013, 2014). MgO displays, at high P–high T conditions, a strong single-crystal elastic anisotropy (Karki et al. 1997; Tsuchiya and Nakamura 2001). Moreover, some aspects related to its thermodynamic and thermophysical properties under those conditions are still poorly defined (Wentzcovitch et al. 2010). These evidences pushed Belmonte to extend the first-principle investigation to very high P and T (i.e. 0–160 *GPa*; 0–3000 *K*).

The geometry optimization, performed through a quasi Newtonian scheme based on analytical energy gradients with respect to atomic coordinates and unit cell parameters (Broyden 1970; Doll 2001), combined with the Broyden–Fletcher–Goldfarb–Shanno (BFGS) algorithm for Hessian updating (Civalleri et al. 2001) was particularly stringent, with thresholds for RMS forces and nuclear displacements set to 0.00015 a.u. and 0.0001 a.u.

The threshold on the self-consistent field (SCF) energy was set to 10^{-8} *Hartree* for geometry optimization and EOS calculations and to 10^{-10} *Hartree* for computation of vibrational frequencies. The Monkhorst–Pack sampling scheme of the reciprocal space with a (75,974)p pruned grid (i.e. 75 radial and 974 angular points) lead to 29 independent **k** vectors in the irreducible part of the BZ. The estimated Δe on the integrated electronic charge density in the unit cell was $\Delta e = 4 \times 10^{-5}$ on a total of 20 electrons per unit cell. Besides the usual vibrational frequencies at Γ point (i.e. $\mathbf{k} = 0$), Belmonte (2017) performed Phonon dispersion calculations at many other special points with a direct-space approach (Parlinski et al. 1997; Alfé 2009) on a $5 \times 5 \times 5$ supercell with 250 atoms and 750 normal modes sampled at 125 k-points. This ensured convergence on harmonic thermodynamic functions (e.g. isochoric heat capacity and vibrational entropy) in the whole P–T range. Instead of defining (as commonly done) a single Mode-gamma Grüneisen parameter (γ_i), for each ith normal mode, the volume dependence of the QHA vibrational frequencies was obtained by third-order least-square fitting of the vibrational frequencies at eight different volume conditions (i.e. V/V_0 = 0.65, 0.70, 0.75, 0.80. 0.85, 0.90, 0.95, and 1.00). The same approach was adopted in determining the thermal pressure, in order to achieve the maximum accuracy in the determination of EoS in a wide P–T range (cf. Sect. 8.9.4). We have also already seen, at Sect. 8.10.8, how successful was the full-phonon dispersion calculation in determining the isobaric heat capacity of the substance.

Here, I limit myself to recall that, having at hand the full phonon dispersion, Belmonte (2017) adopted Eq. (8.9.29) for the $\alpha_T K_T$ product and the Mie–Gruneisen EoS (Eq. 8.9.31) to explore the PVT space.

Figure 20.22 shows, as an example, the isothermal section at T = 3000 *K* of the PVT hyperplane of Periclase. Ab initio PVT data are presented here without any pressure correction.[6] (see Fig. 7 and Table 4 in Belmonte 2017; for a more extensive appraisal). We may see that even at these high-T conditions (relevant for lower mantle investigations) the PVT relations of MgO are perfectly reproduced by the First-Principles Mie–Grüneisen (FPMG) formalism, whether including the zero-point correction term or not (dashed and bold curves in Fig. 20.22, respectively). The effect of the zero-point motions is to slightly raise the PV curve. The FPMG results of Belmonte (2017) are in good agreement with observations and MD investigations at all high T, up to the melting point of the substance.

In Table 20.17 are listed the electronic energies and the thermal corrections obtained by Belmonte (2017) on MgO and component elements at stable state (Belmonte, personal communication). Adopting the thermochemical procedures outlined at Sect. 7.10, the corresponding conventional thermodynamic properties are readily obtained by application of the appropriate thermochemical cycle. As we may readily see, the extension of the supercell does not affect appreciably the static contributions to energy, while sensible variations are observed in the vibrational terms.

---

[6] In Oganov and Dorogupets (2003), a pressure shift of about − 7.7 *GPa* to the calculated static pressures is introduced to account for the systematic overestimation of lattice parameters by GGA (~1–2%) and to match the experimental unit cell volume in ambient conditions. In Wu et al. (2008), a more limited pressure correction is applied to account for the systematic underestimation of volume by LDA (~0.6%) and to reproduce better the experimental results at low pressure.

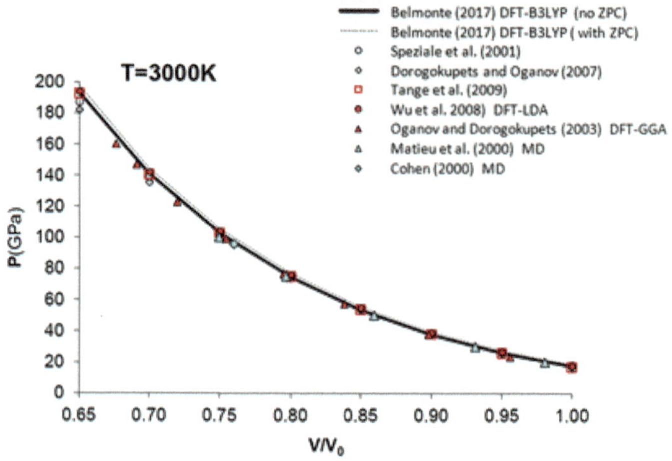

Fig. 20.22 Isothermal section of the PVT space at T = 3000 *K*. The First-Principles Mie–Grüneisen Equations of State of Belmonte (2017) (with and without ZPC, dashed and bold lines, respectively)) are compared with selected experimental data obtained by hydrostatic or quasi-hydrostatic compression in the diamond-anvil (Speziale et al. 2001), semi-empirical assessments (Dorogupets and Oganov 2007), other ab initio LDA (Wu et al. 2008), GGA (Oganov and Dorogupets 2003) and MD simulations (Cohen 2000; Matsui et al. 2000). Modified from Belmonte (2017)

The bulk modulus at the athermal limit obtained by Belmonte (2017) in the absence of ZPC was $K_0 = 167.01$ *GPa*, and its P derivative was $K_0' = (dK/dP)_0 = 3.95$. K_0 reduces to 161.5 when ZPC is accounted for. The thermal effect on the bulk modulus was accounted for by placing $K_T = K_0 + K_1 \times T$, with $K_1 = -308.0$ *bar/K*; the thermal expansion was $\alpha(T) = \alpha_0 T + \alpha_1 + \alpha_2 T^{-1} + \alpha_3 T^{-2} + \alpha_4 T^{-3}$ ($\alpha_0 = 2.608\text{E-}8$, $\alpha_1 = -104.77954\text{E-}7$, $\alpha_2 = 42.9578\text{E-}3$, $\alpha_3 = -17.04307$, $\alpha_4 = 2122.59$); the isobaric heat capacity was $C_P = a + bT + cT^{-2} + dT^2 + eT^{-0.5}$) ($a = 44.202$, $b = 0.39852\text{E-}2$, $c = -1.2908\text{E} + 6$, $d = 0.12101\text{E-}5$, $e = 93.607$).

Other relevant thermophysical and thermodynamic properties obtained by Belmonte (2017) at P = 1 *bar* and T = 298.15–3000 *K* are resumed in Tables 20.18 and 20.19.

20.3 Multiple Oxides (Spinels)

Under this class of substances are grouped oxides with general stoichiometry AB_2O_4, where A and B are the nonequivalent metal sites. In the "*spinel*", sub-group the A and B cations are, respectively, sixfold and fourfold coordinated with an oxygen network arranged in approximated cubic closest packing. In a normal spinel of stoichiometry AB_2O_4, cation A occupies 8a tetrahedral positions; cation B 16d octahedral positions, and oxygen occupies 32e positions with (u, u, u) independent

Table 20.17 Electronic energy and thermal corrections for gaseous O_2, metallic Mg, and Periclase (with various supercell extensions). E_{el}, ZPC, and the other thermal corrections are in *Hartree*, H_f^0 and G_f^0 are in *kJ/mol*. S^0 is in *J/(mol × K)*. (Belmonte, personal communication)

	O_2	Mg	MgO	MgO (2x2x2)	MgO (4x4x4)	MgO (5x5x5)
E_{el}	−150.3050717	−199.9651927	−275.4548366	−275.4548279	−275.4548280	−275.4548280
ZPC	0.0036331	0.0005314	0.0059286	0.0052555	0.0055313	0.0055436
Therm U	0.0036364	0.0009693	0.0078964	0.0070291	0.0074313	0.0074580
Therm H	0.0045806	0.0009698	0.0078968	0.0070296	0.0074317	0.0074584
Therm G	−0.0176924	−0.0027402	0.0047929	0.0043222	0.0045010	0.0044968
H_f^0	0.023	269.927	−602.965	−605.219	−604.163	−604.093
G_f^0	2.711	269.928	−570.79	−572.003	−571.533	−571.545
S^0	196.135	32.67	27.333	23.841	25.808	26.08

Table 20.18 Ab initio B3LYP thermophysical and thermodynamic functions of MgO at P = 1 *bar*, T = 298.15–3000 *K* (Belmonte 2017)

T (*K*)	K_T (*GPa*)	$\alpha \times 10^5$ (K^{-1})	V_T[(a)] (cm^3/mol)	C_V (*J/mol · K*)	C_P (*J/mol · K*)	S (*J/mol · K*)	γ_{th}[(b)]
0	161.5	0.00	11.406	0.000	0.000	0.000	–
100	161.4	0.61	11.412	7.532	7.539	2.290	1.496
298.15	157.7	2.97	11.444	36.142	36.616	26.251	1.483
500	152.5	3.70	11.522	44.069	45.269	47.640	1.474
1000	138.4	4.42	11.760	48.180	51.361	81.314	1.493
1500	123.4	5.03	12.039	49.009	54.657	102.744	1.526
2000	107.8	5.80	12.372	49.304	58.265	118.988	1.568
2500	91.4	6.85	12.775	49.442	63.147	132.583	1.618
3000	74.3	8.45	13.262	49.517	70.594	144.608	1.680

[a]Calculated as $V_T = V_0 \times \exp\left[\int \alpha(T)dT\right]$ by fitting ab initio values of α with a polynomial function in *T* (i.e. $\alpha = \alpha_0 T + \alpha_1 + \alpha_2 T^{-1} + \alpha_3 T^{-2} + \alpha_4 T^{-3}$)
[b]Grüneisen parameter calculated as $\gamma_{th} = \alpha K_T V_T C_V$

Table 20.19 Ab initio B3LYP vibrational contribution to thermodynamic functions of MgO at P = 1 *bar*, T = 298.15–3000 *K* (Belmonte 2017)

T (*K*)	U_{vib} (*kJ/mol*)	H_{vib} (*kJ/mol*)	S_{vib} (*J/mol × K*)	F_{vib} (*kJ/mol*)	G_{vib} (*kJ/mol*)
0	14.555	14.556	0.000	14.555	14.556
100	14.735	14.736	2.290	14.506	14.507
298.15	19.581	19.582	26.080	11.805	11.806
500	27.839	27.840	47.059	4.309	4.310
1000	51.222	51.223	79.318	−28.096	−28.095
1500	75.560	75.561	99.043	−73.004	−73.003
2000	100.149	100.150	113.188	−126.228	−126.226
2500	124.839	124.840	124.207	−185.678	−185.677
3000	149.581	149.582	133.228	−250.104	−250.103

fractional coordinates (space group $Fd\bar{3}m$; 07 number 227). The cell unit comprises 8 units and has edge a_0. The eight cations lying in tetrahedral sites (T) with $\bar{4}3m$ symmetry have coordinates (1/8, 1/8, 1/8), and the sixteen cations in octahedral sites (M) have coordinates (1/2, 1/2, 1/2). The M site has $\bar{3}m$ symmetry and may be slightly distorted. along the threefold axis following the (u, u, u) oxygen coordinates. M is a perfect octahedron for u = 0.25. The oxygen atom is surrounded by three M and one T cations, forming a triangular pyramid with $\bar{3}m$ symmetry.

Adopting the centre of symmetry as origin, the value of u spans the range $0.24 \leq u \leq 0.275$ and is 0.25 for a perfect cubic close packing. When B ions occupy part of the 8a tetrahedral position the structure is defined as partially inverse. The degree of inversion x is then defined as the fraction of tetrahedral sites occupied by B ions; x thus varies from 0 for normal spinel to 1 for perfectly inverse spinel and x = 0.666 corresponds to random distribution.

Spinels are of great importance in Geochemistry and Industry and have been the object of intense investigation. A vast number of thermodynamic data on spinel end-members (Robie et al. 1978; Barin 1995; Chase 1998) and solid solutions (Driessens 1968a–c; O'Neill and Navrotsky 1983; Della Giusta and Ottonello 1993; and references therein) exist together with detailed crystallographic descriptions (Lindsley 1976; Hill et al. 1979). Here I limit myself to report some Quantum Mechanical applications for three end-members of particular importance in Geochemistry (and Industry as well): the "*Spinel*" end member ($MgAl_2O_4$) which gives the name to the sub-group; Mg-chromite ($MgCr_2O_4$) and Magnetite (Fe_3O_4).

20.3.1 $MgAl_2O_4$

It is one of the major end-member components of the solid mixture (Mg,Fe)(Al, Cr)$_2$O$_4$, nucleating in the Earth's upper mantle between ~20 and ~120 *km* (*sp-peridotite* paragenesis). It forms as a pure phase in alkali-deficient magmas or as a metamorphic mineral in metamorphosed limestones. When pure, it is colorless. Trace elements induce blue, greenish, or ruby-red colors. In these varieties, it is used as ornamental stone (*Ceylon's Noble Spinel*, *Badakhshan's Ruby*).

In Fig. 20.23, we see the structure of the "*normal*" Spinel (i.e. x = 0.0) as generated by the AMCSD software (Downs and Hall-Wallace 2003) on the basis of the structural refinement of Peterson et al. (1991) at T = 293 *K*.

As far as I know, the first *ab initio* investigation of $MgAl_2O_4$ is due to Warren et al. (1999) and it was mainly devoted to quantify the influence of the internal disorder on the energy of the substance and its thermal dependence. The potentials associated to 10 different configurations were obtained at the DFT level in both the local density approximation (LDA) and the general gradient approximation (GGA), with plane waves and norm-conserving pseudopotentials. Besides the *normal* $[(Mg)_T(AlAl)_MO_4]$ and *inverse* $[(Al)_T(AlMg)_MO_4]$ configuration, eight other intermediate configurations $[(Mg_{(9-N)/9}Al_{N/9})_T\ (Al_{(9-N)/9}Mg_{N/9})_M]$, at regular interchange extents, where investigated (i.e. x = N/9, with N, the integer number of "*swapped*" Mg, ranging from 1 to 8). Exchanges were performed at random on a supercell of 56 atoms. Calculations were carried out with the CETEP code (Clarke et al. 1992). Integration in the Brillouin zone involved a 2×2×2 Monkhorst–Pack set of **k** points. Results of the ab initio investigation were presented in terms of one-, two-, and three-body interactions between neighboring sites (Table 20.20).

The first term, denoted by the authors as the "*chemical potential*" is associated to the difference in energy between a fourfold and a sixfold coordinated site. In the author's words, it "*represents the energy difference between a Mg atom in a tetrahedron and in an octahedron, and must have its origins in complex electronic*

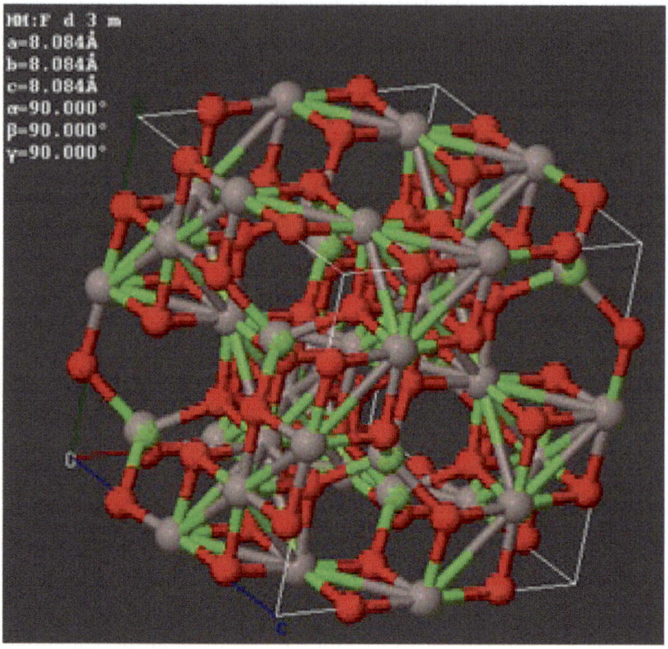

Fig. 20.23 The *Fd3m* structure of $MgAl_2O_4$ (Peterson et al. 1991) as generated by the AMCSD software (Downs and Hall-Wallace 2003)

Table 20.20 Ab initio DFT/LDA and DFT/GGA model parameters (*kJ/mol*) for the internal disorder in $MgAl_2O_4$ after Warren et al. (1999)

Parameter	LDA	GGA
M	41.6625	53.7232
J_{46}	2.7595	6.8215
K_{466}	−1.4762	−2.7402

differences between these two environments" (Warren et al. 1999). The presence of this first-order energy renders the transition non-convergent "*with the order parameter never falling to zero*" (Warren et al. 1999). J_{46} is the traditional pair interaction for first neighbor tetrahedron–octahedron (4–6) pairs and K_{466} is a three-body term for interaction between one tetrahedron and two octahedra sharing a common oxygen. The ab initio derived model parameters were then adopted in a Monte Carlo simulations (10^7 attempted pair exchanges with $8 \times 8 \times 8$ unit cells, each of 24 cations) to take into account thermal fluctuations, in order to predict the disordering over a wide range of temperatures. The fact that simulations started at different temperatures bring to identical results (within the level of noise) testifies to the robustness of the procedure. Vibrational entropy was not included because it was "*not expected to be the dominant effect*" (Warren et al. 1999).

Background information: internal disorder in multiple oxides

The current interpretation of the complex disordering phenomena in multiple oxides is largely based on the theoretical approach of Callen et al. (1956) which discussed the functional dependency of thermodynamic internal energy on ion configuration under the Debye approximation (Debye 1912). Callen et al. (1956) assumed vibrational entropy (S_{vib}) to be explicitly dependent upon vibrational energy (U_{vib}) and on "*one or more parameters analogous to the Debye temperature θ*" (Callen et al. 1956):

$$S_{vib} = S_{vib}(U_{vib}, \theta_D) \tag{20.3.1}$$

Because the bulk internal energy U is composed of a static (U_{st}) and a vibrational part (U_{vib})

$$U = U_{st} + U_{vib}(S_{vib}, \theta_D) \tag{20.3.2}$$

one may anticipate that the bulk entropy of a multiple oxide is

$$S = S_{conf} + S_{vib}(U - U_{st}, \theta_D) \tag{20.3.3}$$

where S_{conf} is the configurational part of entropy arising from internal disorder. Assuming that the functional dependency of θ_D on volume (V) is due exclusively to anharmonicity, Eqs. (20.3.2) and (20.3.3) may be rewritten in the more complete form (Della Giusta and Ottonello 1993):

$$U = U_{st}(S_{conf}) + U_{vib}(S_{vib}, \theta_D) \tag{20.3.4}$$

$$S = S_{conf}(U_{st}) + S_{vib}(U_{vib}, \theta_D) \tag{20.3.5}$$

In this view, the internal disorder arises not only from vibrational terms but from the static potentials of the substance as well. Indeed, a Born-model parameterization of the static potentials in spinel end members and solid mixtures has been attempted in the past on the basis of this conceptual framework (Ottonello 1986; Della Giusta and Ottonello 1993; Della Giusta et al. 1996).

In Fig. 20.24, the model predictions of Warren et al. (1999) are compared with the results of neutron scattering experiments (Redfern et al. 1999) and analytic model fits, in terms of the order parameter Q.

The predicted order parameter is not far from what is experimentally observed at high T, while sensible deviations are observed in the low thermal range. It must be noted however to this purpose that experiments are kinetically limited below 1000 *K* (Warren et al. 1999). Comparing the ab initio results with the static

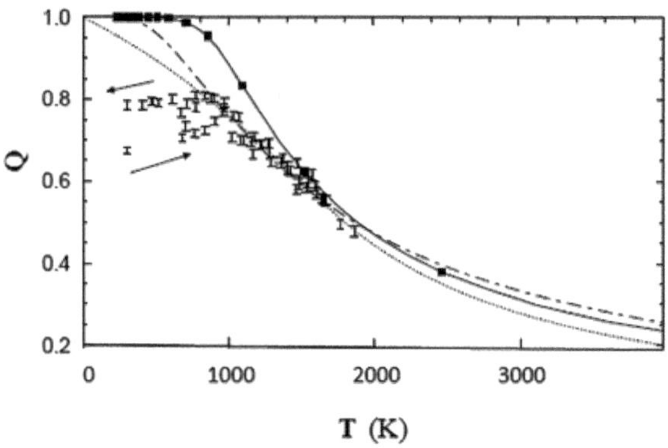

Fig. 20.24 Results of the Monte Carlo simulations of Warren et al. (1999) (filled squares), compared with neutron diffraction experiments (vertical bars; Redfern et al. 1999). The dotted line is the fitting to a Landau model (Redfern et al. 1999) and the dashed line represents the model of O'Neill and Navrotsky (1983). Modified from Warren et al. (1999)

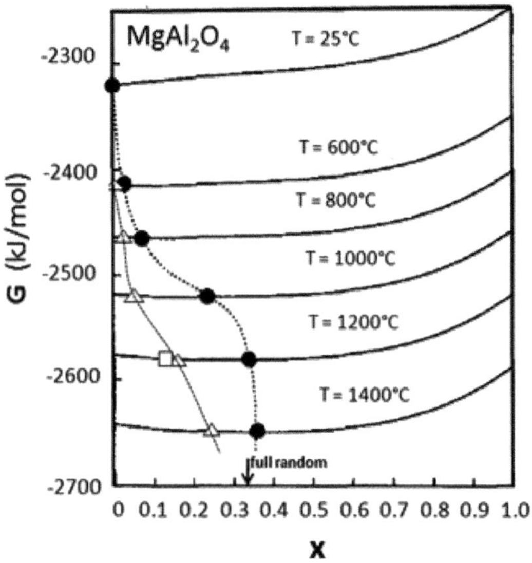

Fig. 20.25 Gibbs free energy values arising from the static interionic potential model at various T conditions and P = 1 *bar* (Della Giusta and Ottonello 1993). Black dots on each curve mark the minimum energy configuration, The superimposed open triangles delineate the minima according to the ab initio procedure of Warren et al. (1999), and the open square is minimum at 1200 °*C* according to Wei and Zhang (2001) The dotted lines connecting the solid symbols are drawn only as an eye-guide to perceive the thermal dependence

potential model (Fig. 20.25), we find again a substantial agreement. Both kinds of calculations converge toward the full random disorder at sufficiently low T. The relative flatness of the Gibbs free energy curves tells us that identifying the minimum requires a high intrinsic precision.

Background information: details on static potentials
In the static potentials approach, compounds are assumed to be essentially ionic solids, and the static lattice energy is represented as composed of coulombic, repulsive, and dispersive terms and of crystal field contributions:

$$U_{st} = E_C + E_R + E_D + E_{CFS} \tag{20.3.6}$$

The Madelung constant is considered as arising from six electrostatic potential contributions (Thompson and Grimes 1977):

$$\mathcal{M} = \alpha_1 Z_{oct}^2 + \alpha_2 Z_{oct} Z_{tet} + \alpha_3 Z_{tet}^4 + \alpha_4 Z_{oct} Z_{anions} + \alpha_5 Z_{anions}^2 + \alpha_6 Z_{tet} Z_{anions} \tag{20.3.7}$$

with Z as ionic charges, α_1, α_2, α_3 constants (Thompson and Grimes 1977) and α_4, α_5, α_6 depending on the shifted oxygen positional parameter $u' = u + 0.125$ (Lindsley 1976; Ottonello 1986). The Coulomb energy E_C is related to the Madelung constant by

$$E_C = \frac{N_0 e 2\mathcal{M}}{a_0 \cdot 4\pi\varepsilon_0} \tag{20.3.8}$$

with a_0 as cell edge and ε_0 as vacuum permittivity. The repulsive energy E_R is in the Huggins–Mayer expanded form (Ottonello 1986):

$$E_R = 4N_0 \sum_i D_{i,tet} X_{i,tet} \times \exp\left(\frac{-\overline{R}_{tet}}{\rho}\right) + 12N_0 \sum_j D_{j,oct} X_{j,oct} \times \exp\left(\frac{-\overline{R}_{oct}}{\rho}\right) \tag{20.3.9}$$

with $D_{i,tet}$ as the repulsive coefficient of the ith ion occurring in a tetrahedral site with atomic proportion $X_{i,tet}$ and ρ as common "*hardness factor*" of the substance. For the dispersive energy E_D, see Appendix I in Ottonello (1987) and Sect. 8.5.3 and for the crystal field stabilization energy E_{CFS} see Sect. 8.10.6. The enthalpy of formation from the elements of the substance at standard state is readily derived by application of the Born–Haber–Fayans (BHF) thermochemical cycle (see Sect. 8.5.5):

$$H^0_{f,T} = \sum_M (I_M + E_M) + 2Ed_{O_2} + 4Ea_O + 7RT - U_{st} \tag{20.3.10}$$

I_M and E_M in Eq. (20.3.10) are, respectively, the ionization and sublimation energies of metals, Ed_{O_2} is the dissociation energy of molecular oxygen and Ea_O is the electron affinity of atomic oxygen. The EOS term 7RT recasts the energy from the athermal limit to T of reference (T = 298.15 *K*). The zero-point vibrational energy is ignored being quite small with respect to all the other energy terms involved in the cycle. The internal disorder at equilibrium and P, T of interest obeys Gibbs free energy minimization. If the vibrational part of the Gibbs free energy does not depend explicitly on configuration, and the kinetic energy of the atoms is sufficiently high to achieve the migration enthalpy at T of interest, then the Gibbs free energy change associated with disorder is purely static, i.e.

$$\Delta G_{conf} = \Delta U_{st} - T\Delta S_{conf} + P\Delta V_{st} \tag{20.3.11}$$

ΔV_{st} in Eq. (20.3.11) is the change in molar volume consequent upon modified disorder (thermal expansion neglected). The minimization principle is then conveniently represented as a minimum on the internal ordering parameter, i.e. $\partial \Delta G_{conf}/\partial x = 0$. This condition translates into (Della Giusta and Ottonello 1993)

$$P\frac{\partial \Delta V_{st}}{\partial x} + \frac{\partial \Delta U_{st}}{\partial x} = T\frac{\partial \Delta S_{conf}}{\partial x} = -k_B T\frac{\partial \ln Q}{\partial x} = -RT\ln\left[\frac{x^2}{(1-x)(2-x)}\right] \tag{20.3.12}$$

where Q is permutability.

The system may be resolved through non-linear hierarchical minimization procedures (James and Roos 1977), independently of the chemical complexity provided that a sufficient number of (G, T, x) observables are at disposal. For a common hardness factor (ρ) of the solid mixture the unknowns reduce to the repulsive coefficients of the intervening cations (and O^{2-}), all the other terms being fixed by theory. As an example, we see in Table 20.21 the static energy terms for the three multiple oxides considered in this book.

The choice of the appropriate E_R to be parameterized depends to some extent on the amount of end-member components in the investigated system[7] and on the accuracy of energy terms at disposal. In the case of $MgAl_2O_4$ (Table 20.22), five parameters concur to determine the internal disorder. Selecting a common value for O^{2-} and a common hardness factor ρ in the family of substances the degree of freedom reduces to 4.

Wei and Zhang (2001) performed band-structure and total-energy calculations of *normal* and *inverse* configuration under LDA with a Ceperley–Alder exchange–correlation potential (see Sect. 10.4). The reciprocal space integration in the Brillouin fcc zone was carried out with two **k** points for the normal structure and six

[7] In this case, the end member components were seven: $MgAl_2O_4$, $FeAl_2O_4$, $NiAl_2O_4$, $MgFe_2O_4$, $FeFe_2O_4$, $MgCr_2O_4$, $FeCr_2O_4$ (Della Giusta and Ottonello 1993).

Table 20.21 Enthalpy of formation at room conditions and static energy terms (athermal limit) for the investigated spinel compounds (Della Giusta and Ottonello 1993), All data are in *kJ/mole*. Asterisks mark the values adopted by the Authors in their in the solid mixture model

Compound	BHF	H_f^0	U	E_C	E_R	E_{DD}	E_{DQ}	E_{CFS}
$MgA1_2O_4$	16638.5	−2310.4[a]	−18,934.3	−22,916.5	4618.8	−471.1	−180.1	0.0
		−2299.3[b]			4629.9			
		−2288.0[c]			4641.2			
		−2300.9[d,*]			4628.3			
		−2295.8[e]			4633.4			
$FeFe_2O_4$	17537.2	−1118.4[a]	−18,656.8	−21,572.7	3811.6	−603.0	−234.4	−56.9
		−1115.7[b]			3814.3			
		−1118.2[c]			3811.8			
		−1111.2[d,*]			3818.8			
		−1113.3[e]			3818.7			
$MgCr_2O_4$	16997.5	−1771.1[a]	−18,769.9	−22,037.2	4363.1	−471.6	−173.6	−449.4
		−1783.6[b]			4350.6			
		−1772.3[d,e,*]			4361.9			

[a]Barin and Knacke (1973), Barin et al. (1977)
[b]Robie et al. (1978)
[c]Helgeson et al. (1978)
[d]Internally consistent set of values
[e]Values obtained at convergency

Table 20.22 Huggins–Mayer parameters (*J/mol*) for the repulsive energy in $MgAl_2O_4$. ρ is in *Å*. χ2 is the mean quadratic error. a = assigned value

ρ	Mg_T^{2+}	Al_T^{3+}	Mg_M^{2+}	Al_M^{3+}	O^{2-}	$\chi^2 \times 10^5$
0.20	182.22	151.73	78.437	144.63	414.31	0.16
0.22	107.87	114.84	56.217	110.21	237.62	0.19
0.24	68.295	104.00	66.120	115.60	118.33	0.11
0.26	61.745	92.435	35.183	87.252	81.809	0.12
0.26	21.443	59.082	45.425	74.949	103.4[a]	0.11
0.28	37.480	82.013	35.966	85.195	51.411	0.10
0.30	13.349	65.416	33.363	76.640	37.586	0.09

k points for the inverse one. Although, as noted by the authors, LDA underestimates the band gap, the errors cancel out when comparing band gap differences arising from contrasting internal atomic distributions but identical composition.[8] The effective oxygen (u) parameter for the inverse spinel was obtained by fitting the averaged tetrahedral (R_{tet}) and octahedral (R_{oct}) bond lengths:

$$R_{tet} = (u - 0.25)a\sqrt{3} \tag{20.3.13}$$

$$R_{oct} = \sqrt{(u - 0.625)^2 + 2\left[(u - 0.375)^2 a\right]} \tag{20.3.14}$$

The averaged bond lengths of Wei and Zhang (2001) where $R_{tet} = 1.96 \pm 0.02$ *Å*; $R_{oct} = 2.05 \pm 0.03$ *Å* for Mg and $R_{tet} = 1.82 \pm 0.02$ *Å*; $R_{oct} = 1.93 \pm 0.02$ *Å* for Al. The inverse band-gap reduction obtained by Wei and Zhang (2001) was −0.66 *eV* (i.e. a higher band gap in the inverse structure) while the inverse energy was 0.50 *eV* (i.e. a lower energy in the *normal* structure). The calculated degree of inversion at equilibrium was x = 0.12 at T = 1200 °*C*.

The computational experiment of Warren et al. (1999) was basically repeated by Da Rocha and Thibaudeau (2003) on five 56-atom cubic supercells with five different degrees of internal disorder. DFT-LDA plane wave calculations were performed with the ABINIT code (http://www.pcpm.ac.be/abinit). Hamann pseudopotentials (Hamann 1989) were adopted for cations, while, for oxygen, it was adopted the Troullier–Martin scheme (Troullier and Martins 1991; Thibaudeu and Gervais 2002). In their study, Da Rocha and Thibaudeau (2003) investigated systematically (for the first time) also the effect of pressure on the internal disorder, and generated infrared phonon spectra through perturbation procedures in order to detect the eventual imprinting of cationic disorder in experimental spectra. To detect the effect of the intensive variable P, the authors assumed that the internal

[8] The lattice vectors for both the normal and inverse spinel were kept to be cubic while relaxing the internal structural parameters.

Table 20.23 Ab initio calculated α, β, μ, and v parameters for different pressures (Da Rocha and Thibodeau 2003): errors on α', β', μ, and v are, respectively, $\pm$ 7.72, $\pm$ 5.79, $\pm$ 0.012, and $\pm$ 0.005 for all pressures

Pressure (kbar)	α	β	μ	v	α'	β'
0	46.31	−24.12	−0.03	−0.04	46.31	−24.12
50	46.31	−24.12	−0.03	−0.03	44.38	−26.05
100	46.31	−24.12	−0.03	−0.03	43.42	−27.98
200	47.28	−26.05	−0.03	−0.02	40.52	−29.91

energy of the investigated substances bears a quadratic dependence on internal disorder (Néel 1950; O'Neill and Navrotsky 1983):

$$\Delta U_{st} = \alpha x + \beta x^2 \tag{20.3.15}$$

Because the cell volume varies with disorder, the (computationally observed) disorder-induced volumetric change was reproduced by two ad hoc parameters (μ and v)

$$\Delta V_{st} = \mu x + v x^2 \tag{20.3.16}$$

with α, β replaced by the effective coefficients α', β':

$$\alpha' = \alpha + \mu P \tag{20.3.17}$$

$$\beta' = \beta + v P \tag{20.3.18}$$

A target stress was then imposed during simulation on the five generated configurations at P = 5, 10, and 20 *GPa* and the average variations of internal energy with respect to P = 0 were evaluated determining in this way the α and β evolution with P. Table 20.23 and Fig. 20.26 resume the obtained results.

For temperature lower than 1000 *K*, P effects on site disorder are almost negligible, while, for higher temperatures, effects are more substantial. For 20 *GPa* the maximum disorder is already attained at T $\approx$ 1400 *K*.

The investigation of Thibaudeau and Gervais (2002) at the same theory level gave ground-state properties consistent with experiments. The (six) generated E–V couples were retrieved in terms of various EOS equations. The bulk modulus at the athermal limit (K_0 = 198 $\div$ 199 *GPa*; depending on the adopted EOS) was consistent with the experimental observations (K_0 = 196 *GPa*; Chang and Barsch 1973; Yoneda 1990; Kruger et al. 1997). The P derivative of the bulk modulus (K' = 3.58 $\div$ 3.63) was somewhat lower with respect to observations (K' = 4.7 $\div$ 5.3). The lowest residual in the fitting procedure was obtained with the Brannan-Stacey EOS (Brennan and Stacey 1979).

The vibrational properties of $MgAl_2O_4$ were calculated by Thibaudeau and Gervais (2002) through application of the Density Functional Perturbation Theory

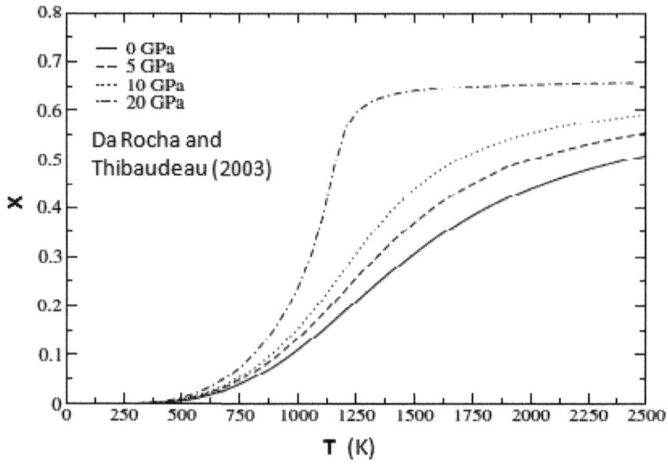

Fig. 20.26 Effect of pressure on the internal disorder in $MgA_{12}O_4$. Modified from Da Rocha and Thibaudeau (2003)

(DFPT) (Baroni et al. 1987; see Sect. 8.8.1). In the primitive unit cell of $MgAl_2O_4$ there are two formula units, leading to 42 normal modes. The 39 optic modes (in addition to three acoustic modes of T1u symmetry) at the Brillouin zone center Γ are distributed among the following symmetries (Thibaudeau and Gervais 2002):

$$\Gamma = \mathrm{A1g}(R) + \mathrm{Eg}(R) + 3\mathrm{T2g}(R) + 4\mathrm{T1u}(IR) + \mathrm{T1g} + 2\mathrm{A2u} + 2\mathrm{Eu} + 2\mathrm{T2u}, \tag{20.2.3.19}$$

where *R* and *IR* denote, respectively, Raman and Infrared activities, and the remaining modes are silent. Because $MgAl_2O_4$ spinel is a polar compound, the DFPT theory allows the calculation of the low- and high-frequency dielectric constants and the Born effective charge tensors directly on all non-equivalent atoms. The results of Thibaudeau and Gervais (2002) in terms of static and optical dielectric constants, and Born effective charges, are in excellent agreement with observations (see Table 20.4). The Born tensors are fully diagonal on cations and mainly diagonal on oxygen atoms up to 2×10^{-2}. The agreement between calculated and experimentally derived diagonal parts of the Born effective tensors is excellent. The optical vibrational frequencies obtained by Thibaudeau and Gervais (2002) are listed in Table 20.24 and are in good agreement with what was experimentally observed (see Table 20.5 for a comparative appraisal). The LO–TO splitting was obtained from the Born effective charge tensors.

Defect formation energies in $MgAl_2O_4$ were more recently investigated by Borges et al. (2016) and Evarestov et al. (2018). The investigation of Borges et al. (2016) was carried out at DFT/GGA theory level with the VASP package, through Projector-Augmented Wave (PAW) procedures (see Sect. 8.8.3) and the Perdew–Burke–Ernzerhof functional (Sect. 10.5.4) with the simple Becke–Johnson

Table 20.24 DFT-LDA-NCPP vibrational frequencies in $MgAl_2O_4$ (Thibaudeau and Gervais 2002)

Symmetry	Frequencies (cm^{-1})			
A1g	760			
Eg	380			
T2g	309	553	661	
T1u (TO)	307	478	565	666
T1u (LO)	309	612	564	854
T2u	191	495		
T1g	315			
Eu	392	602		
A2u	669	742		

exchange potential (Becke and Johnson 2006; and Sect. 10.5.1). The valence electrons considered in the PAWs where $3s^2$ for Mg, $3s^2 3p^1$ for Al and $2s^2 2p^4$ for O. The adopted cutoff energy was 500 *eV*. Defects were imposed on a sufficiently extended supercell (i.e. 2×2×2, corresponding to 112 atoms: 16 Mg, 32 Al, 64 O). Integration in the Brillouin zone involved a 2×2×2 Monkhorst–Pack mesh.

The investigated defects where (in the appropriate Kröger–Vink notation, see Sect. 8.12) $V_O^{\cdot\cdot}$, V_{Mg}'', V_{Al}''', Mg_{Al}', $Al_{Mg}^{\cdot}$, O_i'', $\{V_O^{\cdot\cdot}Al^{\cdot}{}_{Mg}\}^{\cdot\cdot\cdot}$, $\{V_O^{\cdot\cdot}Mg_{Al}'\}^{\cdot}$, $\{V_O^{\cdot\cdot}O_i''\}^{\times}$, $\left\{Mg_{Al}'Al_{Mg}^{\cdot}\right\}^{\times}$

According to Borges et al. (2016), the formation energy of a defect D with charge q [$E_f(D^q)$] is given by

$$Ef(D^q) = [E_{tot}(Dq) - E_{tot}(bulk)] - \sum_i n_i \mu_i + (\epsilon_F + E_V) \tag{20.3.20}$$

E_{tot} (D^q) and E_{tot} (bulk) in Eq. (20.3.20) represent, respectively, the total energy of the supercell with defect in charge state q and the total energy of the perfect crystal (bulk) in the same supercell. ε_F is the Fermi energy, E_V is the valence band maximum, and μ_i is the chemical potential of the added or removed atom i in the structure.

The values of μ_i for Al, Mg, and O must satisfy

$$H^0_{f,MgAl_2O_4} = H^0_{MgAl_2O_4} - H^0_{Mg} - 2H^0_{Al} - 2H^0_{O_2} \tag{20.3.21}$$

where H^0 represents the enthalpy of the stable compound at the standard state. The further constraints that the defect is actually stable in the defective phase and that the formation of MgO and Al_2O_3 is prevented (Eqs. 2–5 in Borges et al. 2016; but see also to this purpose Sect. 8.12.2 and 8.12.3), bring to $\mu_{Mg} = -1.963$ *eV*, $\mu_{Al} = -3.743$ *eV*, $\mu_O = -10.1$ *eV*, for O-poor conditions and $\mu_{Mg} = -7.238$ *eV*, $\mu_{Al} = -11.44$ *eV*, $\mu_O = -4.93$ *eV*, for O-rich conditions (Borges et al. 2016).

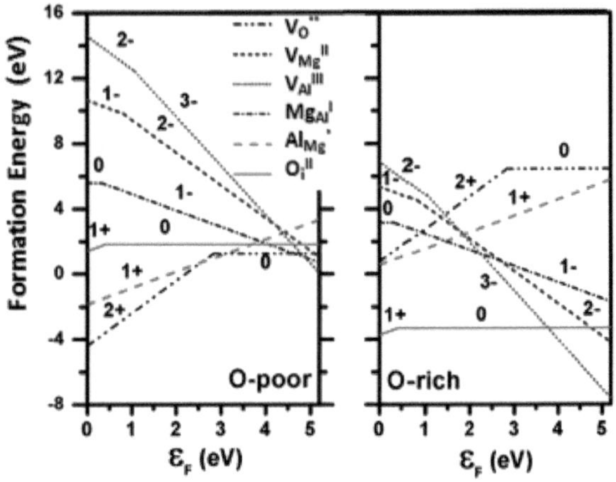

Fig. 20.27 Formation energies of simple (i.e. non-associated) defects in $MgAl_2O_4$ against the energy of the Fermi level. Modified from Borges et al. (2016)

The calculated formation energy for the simple defects in the GGA-calculated band gap region is plotted against the energy of the Fermi level in Fig. 20.27.

Evarestov et al. (2018) investigated the defect energy of interstitial oxygen O_i'' in five vacant Wyckoff positions generated by group-theoretical analysis:

- two positions with site symmetries $S24$ (T_d) and $S12$ (D_{2d}), respectively (parameter-free);
- two positions with symmetries $S4$ (C_{3v}) and $S2$ (C_2), respectively (one parameter-dependent);
- one position with site symmetry $S2$ (C_s) (two parameters-dependent) and one position with site symmetry S_1 (C_1) (three parameter-dependent).

The investigation was carried out at the DFT-LCAO theory level with the most recent release of the CRYSTAL package. The adopted hybrid functional was HSE06, based on a screened coulomb potential (Heyd et al. 2003). The adopted basis set for O was a 6s-211sp-1d expansion as in Baima et al. (2013), and for Mg atoms, the expansion 8s-511sp-1d of McCarthy et al. (1994), already adopted for the bulk properties of MgO. For Al, it was adopted an Effective Core Pseudopotential (ECP) with $3s^2 3p^1 d$ external shell (Causà et al. 1991). The Brillouin zone was sampled with a 4×4×4 Monkhorst–Pack mesh. The ground-state structure and its phonons are in good agreement with observations. The defect energy was defined as

$$E_f\left(O_i''\right]) = [E_{tot}(\text{defective}) - E_{tot}(\text{bulk})] - \frac{1}{2}E_{O_2} \tag{20.3.22}$$

and was found to be minimal for the interstitial site of lowest symmetry (C_1) ($E_f\left(O_i''\right) = 2.31 \div 2.51 \quad eV$) and, as "*intuitively obvious*" (Evarestov et al. 2016) maximal at the highest symmetry $S24$ ($E_f\left(O_i''\right) = 13.0 \div 13.7\ eV$).

20.3.2 $MgCr_2O_4$

Together with *Noble Spinel* ($MgAl_2O_4$), *picrochromite* ($MgCr_2O_4$) is one of the major end member components of the "*spinel group*". In $MgCr_2O_4$ all Mg is in T and Cr in M (i.e. x = 0). Multiple oxides particularly enriched in $MgCr_2O_4$ are found in layered stratiform orthocumulates of ultramafic intrusions. $MgCr_2O_4$ is the major component of the primary ore mineral from which Cr-based industrial reactants are derived. The first ab initio investigation of $MgCr_2O_4$ is due to Catti et al. (1999) which determined its binding energies and the formation enthalpy, with the aim of predicting the decomposition pressure. The study was carried out at the UHF-LCAO theory level and the exchange–correlation effects were added a-posteriori with the GGA-II functional of Perdew et al. (1992) (see Sect. 10.5.2). A later investigation, mainly devoted to the vibrational properties (of central importance in the isotopic investigation of water–rock interaction and of the production process) appeared in 2007. All-electron calculations were performed at the same theory level of the previous study but the correlation effects were obtained through the hybrid B3LYP functional (see Sect. 10.5.5). Further details on the calculations have been already given at Sect. 8.11.

Actually, picrochromite is not the unique $MgCr_2O_4$ polymorph. Erhemberg et al. (2002) showed in fact that below $10 \div 16\ K$ there is a phase transition to a tetragonally distorted polymorph (space group $I4_1/amd$; similarly to $CuCr_2O_4$). Finally, according to Lutz et al. (1991), for all chromites of general formula MCr_2O_4, the true space group is $F\bar{4}3m$ with a lowering of symmetry of the octahedrally coordinated Cr metal (site symmetry C_{3v} instead of D_{3d}), which, particularly for $MgCr_2O_4$, seems confirmed by X-ray diffraction studies (Hilleard 1975).

Notwithstanding this observation, *picrochromite* is usually resolved in the space group $Fd\bar{3}mZ$. There is not a good agreement in the literature also on the cell edge at room conditions [$a_0 = 8.307$ Å (Passerini 1929), $a_0 = 8.312$ Å (Verwey and Heilman 1947), $a_0 = 8.334$ Å (O'Neill and Dollase 1994) and $a_0 = 8.338$ Å (Sawada 1996)]. The structure of picrochromite after Verwey and Heilman (1947) is shown in Fig. 20.28.

Cubic $MgCr_2O_4$ has the "*normal*" spinel structure with Cr atoms located at $\left(\frac{1}{2}, \frac{1}{2}, \frac{1}{2}\right)$ in octahedral co-ordination with oxygen (M), Mg atoms located at $\left(\frac{1}{4}, \frac{1}{4}, \frac{1}{4}\right)$ in tetrahedral co-ordination with oxygen (T), and O in general position (u,u,u). There are three atoms in the rhombohedral primitive cell and 14 atoms in the cubic unit cell corresponding to 39 true vibrational modes. The irreducible representation of vibrational modes, with a group theoretical treatment of the optical zone-center (gamma point; **k** = 0) yields for the cubic system, class $Fd\bar{3}mZ$ (Himmrich and Lutz 1991):

$$\Gamma = A_{1g} + E_g + F_{1g} + 3F_{2g} + 2A_{2u} + 2E_u + 4F_{1u} + 2F_{2u} \tag{20.3.23}$$

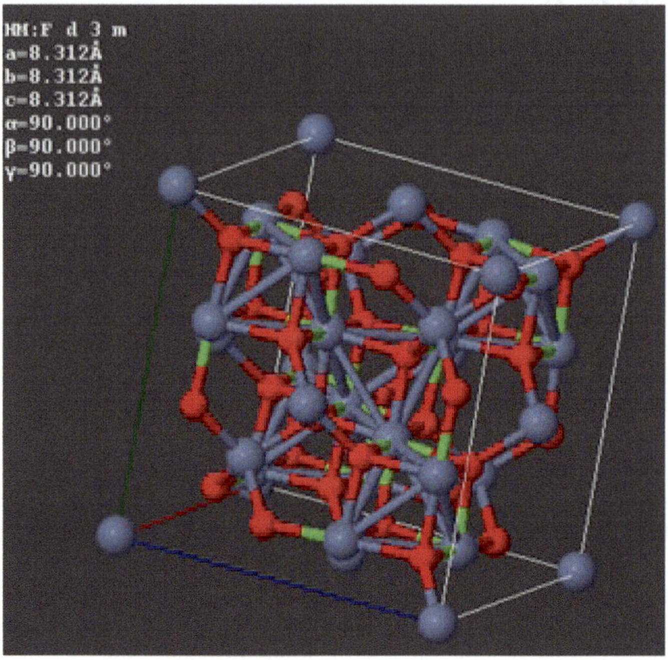

Fig. 20.28 The *Fd3m* structure of $MgCr_2O_4$ (Verwey and Heilman 1947) as generated by the AMCSD software (Downs and Hall-Wallace 2003)

For the ditetragonal–dipyramidal class $I4_1/amd$, the irreducible representation of active modes is

$$\Gamma = 2A_{1g} + 4E_g + 3F_{1g} + F_{2g} + 2A_{1u} + 4A_{2u} + 6E_u + 2F_{1u} + 4F_{2u} \qquad (20.3.24)$$

The first important contributions to vibrational spectra of spinels are due to Waldron (1955), Hafner (1961), and White and De Angelis (1967). Due to new IR data on both normal II–III spinels and isotopically doped substances, their work was later critically revisited by Preudhomme and Tarte (1971). More recent experiments by Basak et al. (1994) and Lutz et al. (1991) were adopted by Gupta et al. (1990, 1992, 1993) to attain a complete assignment on the basis of the de Launay angular force model (DAF; de Launay 1956).

We have already seen at Sect. 8.11, the calculated (UHF and B3LYP) versus observed (and model) wave numbers for the $Fd\bar{3}m$ *Z* cubic polymorph (Table 8.16). Hereafter, in Table 20.25 are listed the DFT/B3LYP results for the $I4_1/amd$ tetragonal isotopologues.

Figure 20.29 shows the potential well of the cubic polymorph of $MgCr_2O_4$ obtained with the procedures outlined at Sects. 8.9.1, 8.9.2. Performing a Gaussian

Table 20.25 B3LYP wave numbers of the $I4_1/amd$ tetragonal phase. Species in bold are IR active (Ottonello et al. 2007)

^{50}Cr	^{52}Cr	^{53}Cr	^{54}Cr	Species
221.6826	218.1320	216.4145	214.7298	$\mathbf{E_U}$
221.7060	218.1409	216.4167	214.7340	F_{1U}
277.7304	275.6935	274.6782	273.6651	$\mathbf{E_U}$
278.0836	276.0382	275.0187	274.0017	$\mathbf{A_{2U}}$
309.5417	309.5417	309.5417	309.5417	F_{1G}
337.9383	331.3811	328.2424	325.1912	A_{1U}
339.2501	332.6638	329.5112	326.4467	F_{2U}
431.0335	431.0335	431.0335	431.0335	A_{2G}
432.0907	432.0907	432.0907	431.7178	E_G
441.3308	436.3861	434.0183	432.0907	$\mathbf{E_U}$
441.8952	436.9557	434.5907	432.2930	$\mathbf{A_{2U}}$
450.4615	450.4615	450.4615	450.4615	F_{2G}
451.8250	451.8250	451.8250	451.8250	A_{1G}
461.2848	459.6743	458.9238	458.2066	$\mathbf{E_U}$
461.5279	459.9617	459.2315	458.5336	F_{1U}
472.2526	469.3757	468.0049	466.6767	$\mathbf{A_{2U}}$
472.5260	469.6546	468.2869	466.9620	$\mathbf{E_U}$
520.9824	511.0067	506.2118	501.5441	F_{2U}
531.3450	531.3363	531.3324	531.3287	A_{1U}
534.5051	534.4447	534.4302	534.4200	F_{2U}
550.0736	550.0736	550.0736	550.0736	F_{1G}
550.6753	550.6753	550.6753	550.6753	E_G
617.1979	614.7568	613.6403	612.5861	$\mathbf{E_U}$
617.2467	614.7810	613.6533	612.5887	$\mathbf{A_{2U}}$
628.1491	628.1491	628.1491	628.1491	E_G
629.5653	629.5653	629.5653	629.5653	F_{1G}
644.8455	644.7366	644.6901	644.6480	F_{2U}
700.5874	700.5874	700.5874	700.5874	A_{1G}

integration over the various V/V_0 states represented by the dilaton ($\eta = V/V_0$; $P_0 = 1\ bar$) K_0, K' were obtained by non-linear minimization, with negligible errors in reproducibility ($K_0 = 192.560\ GPa$ and $\partial K/\partial P = K' = 4.066$; Ottonello et al. 2007),

The obtained K_0 and K' are in reasonable agreement with the literature data when $(\partial K/\partial P)_T$ is fixed to 4 ($K_0 = 189.6$, $K' = 4$; Nestola et al. 2014; $K_0 = 206.9$, $K' = 4$; Yong et al. 2012).[9]

[9] When K′ is set to 4, the third-order Birch–Murnaghan EoS reduces to the second degree.

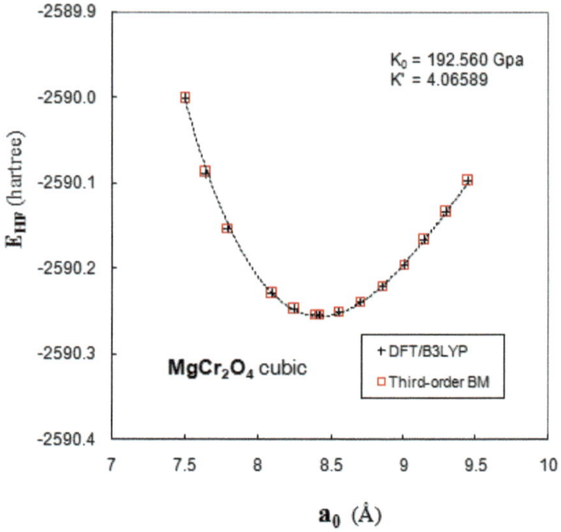

Fig. 20.29 Potential well for the cubic polymorph of $MgCr_2O_4$ resolved at B3LYP level. The third-order Birch–Murnaghan (BM) isothermal finite strain EoS values are compared with the computed values. Modified from Ottonello et al. (2007)

The compressional behavior of the $MgCr_2O_4$ spinel has been also investigated in the 0 ÷ 19 *GPa* range at DFT level by Zhang et al. (2016), with plane waves and pseudopotentials under both GGA and LDA and with the Perdew–Burker–Ernzerhof (PBE) functional. The Authors obtained, at GGA and LDA level, respectively, K_0 = 181.46–216.1 *GPa* and K′ = 4.41–4.5. Zhang et al. (2016) performed an accurate investigation of the effect of P on the oxygen parameter u, showing that it is "*negatively correlated with P*" decreasing by about 0.5–0.6% from 0 to 19 *GPa*. Moreover, the MgO_4 tetrahedron was shown to be much more compressible than the CrO_6 octahedron.
A correct evaluation of the thermal pressure P_{th} requires knowledge of both the isobaric thermal expansion α_T and of the intrinsic temperature derivative $(\partial K_T/\partial T)_V$. Fei (1995), stemming from the isobaric thermal expansion of Skinner (1966), proposed

$$\alpha_{T,MgCr_2O_4} = 0.143 \times 10^{-4} + 1.1191 \times 10^{-8} \times T - 0.1063 \times T^{-2} \quad (T \geq T_{ref}) \tag{20.3.25}$$

It should be noted however that the above function is not well behaved in the vicinity of the athermal limit, since it leads to infinite thermal expansion at this condition, and its practical application is restricted to T > 298.15 K. It is thus appropriate, in the lack of observables, to split the definite integral in Eq. (8.9.30) into two portions, i.e. $0 \rightarrow 298.15$ and $298.15 \rightarrow T$ and to assume the thermal expansion to vary linearly from zero (at the athermal limit) to the value predicted by Eq. (20.3.25) at T = 298.15 (Ottonello et al. 2007):

$$\alpha_{T,MgCr_2O_4} = \frac{\alpha_{T_{ref},MgCr_2O_4}}{T_{ref}} \times T \quad (T \leq T_{ref}) \tag{20.3.26}$$

Indirect estimates of the effect of T on compressibility may be obtained from a quasi-harmonic gamma-mode analysis of the $\alpha_T K_T$ product (see Sect. 8.9.3). Gamma Gruneisen (γ_i) values for all optical modes in cubic $MgCr_2O_4$ were obtained by Ottonello et al. (2007) at the B3LYP level by changing opportunely the cell edge of the polymorph and sampling the obtained frequencies (Table 8.17). The obtained $\alpha_T K_T$ attains at high T a quite constant value of 0.0082 *GPa/K*, not far from the $\alpha_T K_T$ value measured by Anderson et al. (1992) in MgO (0.0064 *GPa/K*). The $\alpha_T K_T$ product was then divided by the computed bulk modulus and the resulting $\alpha_T = f(T)$ curve was compared with the regression of Fei (1995) (Eq. 20.3.25; Fig. 20.30). The two curves have a crossover at high T. The comparison of the two, allowed to establish the T-dependence of compressibility (Ottonello et al. 2007):

$$\frac{\partial K_T}{\partial T} = -0.076 \left(\frac{GPa}{K}\right) \tag{20.3.27}$$

The Kieffer model analysis of the optical and acoustic phonons (see Sect. 8.10.7), coupled with the knowledge of the anharmonicity terms, allowed to estimate the

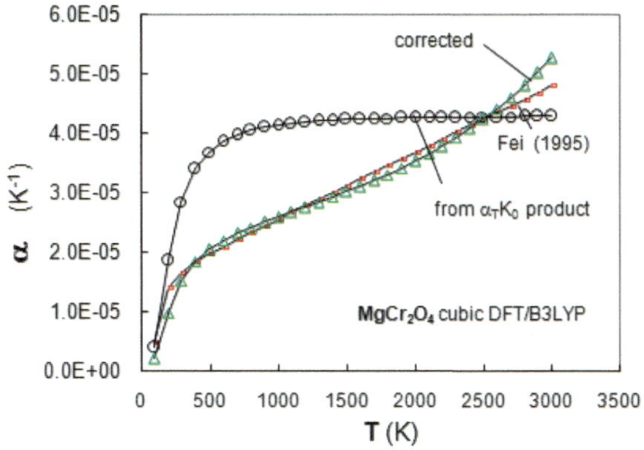

Fig. 20.30 Ab initio thermal expansion of the $MgCr_2O_4$ cubic polymorph, compared with assessments (Eq. 20.3.33). Green open triangles are the results of the thermal correction on bulk modulus. Modified from Ottonello et al. (2007)

Table 20.26 Coefficients of the Haas–Fisher isobaric heat capacity function for the cubic $MgCr_2O_4$ isotopologues. The C_p values [*J/(mol × K)*] are based on DFT/B3LYP calculations fitted by the Haas–Fisher polynomial function $CP = a + b \times T + c \times T^{-2} + d \times T^2 + e \times T^{-1/2}$ (from Ottonello et al. 2007)

Isotopologue	$a \times 10^{-2}$	$b \times 10^{2}$	$c \times 10^{-6}$	$d \times 10^{5}$	$e \times 10^{-2}$
$Mg^{50}Cr_2O_4$	1.5334	0.5516	−5.0906	0.66358	3.9580
$Mg^{52}Cr_2O_4$	1.5320	0.55481	−5.0708	0.66322	3.9997
$Mg^{53}Cr_2O_4$	1.5323	0.55352	−5.0549	0.66344	3.9952
$Mg^{54}Cr_2O_4$	1.5306	0.55817	−5.0528	0.66283	4.0411

discrete values of the isobaric heat capacity for the various isotopologues and to translate them into Haas–Fisher polynomials (Table 20.26).

Background information: details on acoustic modes

The acoustic cut-off frequencies ω_1, ω_2, ω_3 were obtained from transverse and longitudinal velocities by applying

$$\omega_i = \left(\frac{2}{\pi}\right) v_i |\mathbf{k}_{max}| = \left(\frac{2}{\pi}\right) v_i \left(\frac{6\pi^2 N_0}{ZV}\right)^{1/3} \tag{20.3.28}$$

where $\mathbf{k}_{max}$ is the maximum wave vector, the modulus of which corresponds to the radius of a sphere of the same volume as that of the Brillouin zone in reciprocal space and V is the molar volume of the substance (Kieffer 1979a, b). Based on B3LYP computed bulk modulus (K_0 = 192.560 *GPa*), the two transverse shear velocities are $v_{s1} = v_{s2} = 5.192$ *km/s* and the longitudinal velocity v_p is 8.993 *km/s*. The corresponding acoustic wave numbers are then $\nu_1 = \nu_2 = 129.0\ cm^{-1}$ and $\nu_3 = 223.5\ cm^{-1}$.

The first optical mode at $\mathbf{k} \to 0$ was then transposed at $\mathbf{k}_{max}$ by applying (Kieffer 1979a, b)

$$\omega_{1,\mathbf{k}_{max}} = \omega_{1,\mathbf{k}=0}\left(1 - \frac{m_2}{m_1}\right)^{-1/2} \tag{20.3.29}$$

where m_1 and m_2 are the reduced masses of the lightest and heaviest ions, i.e. O and Cr, respectively. The contribution of the acoustic modes is dominant at low T but not negligible even at high T, where, in $MgCr_2O_4$, it approaches 7% of the entire heat capacity [i.e. 100 × (3/3n) = 7.14]. Since this limit is already approached at low T (i.e. T ≥ 400 *K*), high-T isochoric heat capacity estimates could be obtained directly from optic modes by simple normalization (Ottonello et al. 2007).

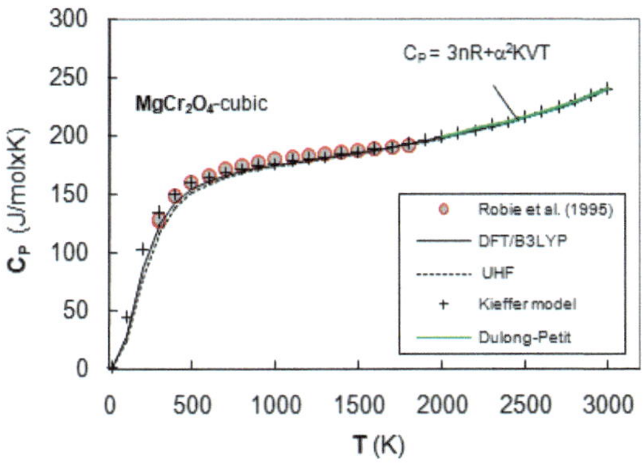

Fig. 20.31 Isobaric heat capacity of the cubic polymorph compared with experimental data. Modified from Ottonello et al. (2007)

The different isotopic masses of ^{50}Cr, ^{52}Cr, ^{53}Cr, and ^{54}Cr affect the thermodynamic properties of *Picrocromite* at T = 298.15 *K* and P = 1 *bar*, through the different thermal corrections. Because ^{52}Cr is the most abundant stable isotope of Cr (83.789%) the values tabulated for $Mg^{52}Cr_2O_4$ may be taken as representative of the natural *Picrocromite* compound.

In Fig. 20.31, the computed B3LYP isobaric heat capacity of $MgCr_2O_4$ is compared with the assessment of Robie et al. (1995). The comparison is overall satisfactory in the T range 298.15–1800 *K*; moreover, the computed heat capacity extends at higher T, attaining the limit imposed by the Dulong–Petit rule and the anharmonic correction. Also in this plot, we may perceive that at the simple UHF level (optical terms) the computation is less satisfactory, while reasonable estimates are achieved by the Kieffer model (optical + acoustic).

Entropy and enthalpy of formation from the elements for the two polymorphs of $MgCr_2O_4$ were evaluated by Ottonello et al. (2007) with the procedures outlined at Sect. 8.10.7.

We have already seen at Sect. 8.11, the thermodynamic properties of the various isotopologues of *Picrocromite* at T = 298.15 *K* and P = 1 *bar*, slightly differing for the different thermal corrections. In Table 20.27 are listed the same properties for the ^{52}Cr tetragonal polymorph at room conditions, and in Table 20.28 is given a selection of the thermodynamic and thermophysical properties of the ^{52}Cr *picrochromite* in the T range from 298.15 to 3000 *K* (see Table 9 in Ottonello et al. 2007; for a more complete outline).

Table 20.27 Ab initio (B3LYP) standard state (T = 298.15 *K*; P = 1 *bar*; pure component) thermodynamic functions of the tetragonal form of the $Mg^{52}Cr_2O_4$ isotopologue (Ottonello et al. 2007)

Magnitude	$Mg^{52}Cr_2O_4$ $I4_1/amd$ B3LYP	Units
C_V	121.839	*J/(mol × K)*
C_P	122.537	*J/(mol × K)*
S_{vib}	91.3095	*J/(mol × K)*
S°	114.595	*J/(mol × K)*
U_{vib}	69.860	*kJ/mol*
U	−2590.240	*Hartree*
ΔU_{corr}	0.026608	*Hartree*
H	−5180.481	*Hartree*
ΔH_{corr}	0.026610	*Hartree*
H_f^0	−1650.553	*kJ/mol*
F_{vib}	33.412	*kJ/mol*
F	−2590.246	*Hartree*
G	−2590.234	*Hartree*
ΔG_{corr}	0.013597	*Hartree*
G_f^0	−1538.563	*kJ/mol*

20.3.3 Fe_3O_4

Fe_3O_4 (*Magnetite*) is structurally quite simple: an fcc inversely ordered multiple oxides where, at the imaginary limit of an instantaneous hopping within (and between) sites, the cations are virtually undistinguishable (Fig. 20.32).

Main magnetite ores on Earth are represented by "*black sands*" originated by depositional enrichment under the action of rivers and sea currents. Its primary origin is, however, either magmatic (primary differentiation or pneumatholytic stages) or by contact metamorphism. The pervasive presence of magnetite in Archean Banded Iron Formations (BIF) is most likely due to dehydration of Fe $(OH)_3$ in anoxic conditions (Klein 2005). The main industrial utilization is of course the iron production.

Magnetite[10] besides being of overwhelming importance in Quantum Chemistry for the quantum–mechanical approach to magnetism (van Vleck 1978) and for some aspects of the band theory (see later), it is important in Geochemistry as well, for many reasons:

[10] From the Greek *μαγνήτης λίθος* (*magnétes líthos*), (Magnesia stone) from the name of an old Asia Minor locality where magnetite deposits were abundant in the antiquity (now Manisa, or Manissa, Turkey).

Table 20.28 Ab initio thermodynamic functions of the cubic $Mg^{52}Cr_2O_4$ isotopologue (DFT/B3LYP). C_V, C_P, and S are expressed in *J/(mol × K)*; U_{vib}, H_{vib}, and F_{vib}, are in *kJ/mol*. V is in *cc/mol*. P_{Th} is thermal pressure, expressed in *GPa*, γ_{th} is the thermal Grüneisen parameter. Θ_{cal} is the calorimetric Debye temperature (Ottonello et al. 2007)

T(*K*)	C_V	C_P	S	U_{vib}	H_{vib}	F_{vib}	V	P_{Th}	γ_{th}	Θ_{cal}
298.15	121.959	122.656	114.814	69.808	69.813	52.772	45.015	0	1.168	829.3
500	152.234	153.882	187.132	98.078	98.083	42.519	45.179	0.7	1.112	837.2
1000	168.534	174.201	301.642	179.527	179.532	−21.765	45.690	2.9	1.325	832.3
1500	171.865	184.752	374.278	264.79	264.795	−154.325	46.339	5.6	1.611	817.5
2000	173.057	197.425	429.072	351.062	351.067	−323.327	47.130	8.8	1.921	795.3
2500	173.612	214.898	474.898	437.745	437.750	−518.048	48.069	12.6	2.251	765.9
3000	173.915	238.895	516.088	524.633	524.638	−732.630	49.164	17	2.602	728.2

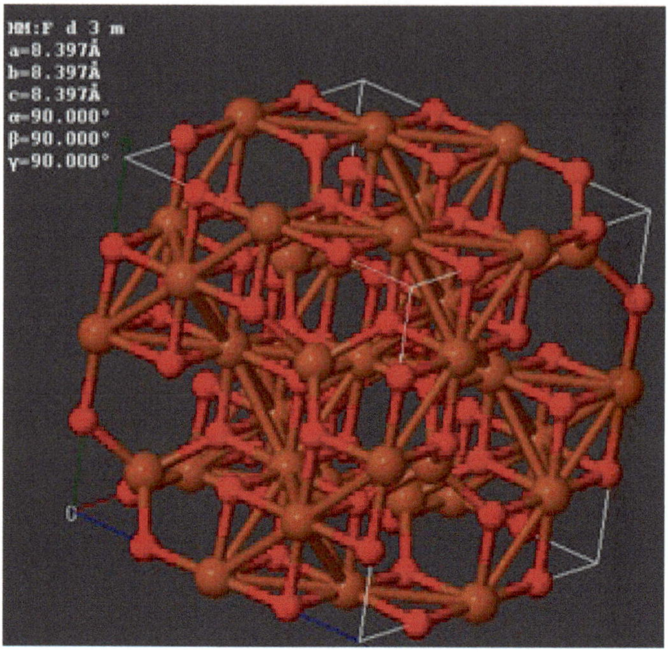

Fig. 20.32 The $Fd\bar{3}m$ structure of Fe_3O_4 (Haavik et al. 2000) as generated by the AMCSD software (Downs and Hall-Wallace 2003)

- it is the most "*magnetic*" of all the Earth minerals, and actually "*Nature's only permanent magnet*" (Lodestone; Wassilewski and Kletetschka 1999)[11];
- together with Quartz and Fayalite, it is a buffer of the oxygen fugacity in volcanic gases (QFM buffer, see Sect. 17.1);
- it is a marker of magnetic inversions on Earth, and, as such, a keystone in the comprehension of the recent lithological evolution of our planet (Vine–Morley–Matthews theory; Morley, unpublished; Vine and Matthews 1963; Morley and Larochelle 1964).[12]

At low P and sufficiently low T, half of Fe^{3+} ions ($5\mu_B$) are in T sites and half in M sites. Fe^{2+} ($4\mu_B$) occupies the other half of M. The bulk moment is thus $4\mu_B$ per formula unit. This gives the bulk crystal an "*antiferrimagnetic*" character.

---

[11] Actually, a *Lodestone* is Fe_3O_4 with inclusions of cubic Fe_2O_3 (*Maghemite*).

[12] Magnetite, present as accessory phase in progressively cooling Mid-Ocean Ridge Basalts (MORB), becomes antiferrimagnetic when passing through the Curie temperature and orients itself parallel to the background geomagnetic field vector, whose direction becomes thus permanently frozen. See to this purpose the pioneering work of Koenigsberger (1938a, b).

Background information: details on magnetism

According to Néel's theory of ferrimagnetism (Néel 1948), the magnetic properties of Fe_3O_4 are due to (collinear) antiferromagnetic ordering of spins in T and M sites.

Stemming from the Weiss theory (Weiss 1907) Néel assumed that the magnetic field acting on a substructure i (i.e. a T or M site of the cubic lattice in our case) is

$$H_i = H_0 + \sum_{k=1}^{n} \gamma_{jk} M_k \qquad (20.3.30)$$

where H_0 is the applied magnetic field, γ_{jk} is the molecular field coefficient of the field exerted on an atom in the ith substructure by its neighbor in the kth substructure, M_k is the average magnetization of the kth substructure and the summation is extended over all the different substructures. If the substructures remain identical in the whole substance (such as in the case of crystalline sites) the spontaneous magnetization on the ith structure is given by (Smart 1955)

$$M_i = Ng\mu_B S B_g \left(\frac{3S}{S+1} \frac{T_c}{T} \frac{M_i}{Ng\,\mu_B S} \right) \qquad (20.3.31)$$

with

N = Number of atoms per unit volume on one substructure;
g = Landé g-factor
S = Spin quantum number
B_g = Brillouin function
μ_B = Bohr magneton

In a simpler form, Eq. (20.3.31) reduces to

$$\sigma_i = B_g \left(\frac{3S}{S+1} \frac{\sigma}{\tau} \right) \qquad (20.3.32)$$

with

σ_i = reduced magnetization of the ith substructure;

$\tau = T/T_c$ = reduced temperature.

This generalized theory predicts that, above the Curie temperature, the magnetic susceptibility χ has form:

$$\chi = \frac{C}{T - \theta} \qquad (20.3.33)$$

with C as Curie constant. For an antiferromagnet, θ is negative and of the same order of T_c (Smart 1955).

For an inverted ferrite such as Fe_3O_4, the magnetic moment is given by

$$\mu_{\pm} = (\mu_{Fe^{3+}})_M + (\mu_{Fe^{2+}})_M \pm (\mu_{Fe^{3+}})_T \qquad (20.3.34)$$

depending on whether the T–M interaction is "*positive*" or "*negative*" (Smart 1955).

The orbital angular momentum of Fe^{3+} is zero (see Sect. 6.1.1) and little affected by the crystal field, thus $\mu_{Fe^{3+}} = 5\mu_B$ for both T and M sites. The "spin only" value of Fe^{2+} is 2S (see Sect. 6.1.2 and Table 6.3), hence $\mu_{Fe^{2+}} = 4\mu_B$. Writing all moments in Bohr magnetons Eq. (20.3.34) leads to $\mu+ = 10 + 4 = 14$ and $\mu- = 4$. The experimental value for Magnetite is ~ 4 (see Table 1 in Smart 1955), which is indicative of a *negative* interaction.

With the increase in T, Fe_3O_4 undergoes various transitions. A first "*Verwey*" transition (Verwey 1939) is observed at 119 *K* and ambient P. Below this T, the phase is monoclinic (Iizumi et al. 1982) and insulating. Above this T, fast hopping in the M site is observed. With the increase of T the degree of inversion progressively changes and, as a result, Fe_3O_4 besides highly conducting, becomes paramagnetic at $T_c = 848.5$ *K* (Grønvold and Sveen 1974; Wu and Mason 1981). Concerning the effect of pressure, the $Fd\bar{3}m$ spinel structure is stable only below 21.8 *GPa* where a "*sluggish*" transition to h-Fe_3O_4 is observed (Fei et al. 199; Haavik et al. 2000). h-Fe_3O_4 (space group *Pbcm*) is paramagnetic at room T (Pasternak et al. 1994).

The first ab initio band structure for Fe_3O_4 is by Yanase and Siratori (1984) which adopted the Augmented Plane Wave method (APW) of Slater (1937) (see Sect. 12.6.1), and the Local Spin Density Approximation (LSDA) of Gunnarson and Lundquist (1976, 1977) in the evaluation of exchange–correlation effects (see also to this purpose Sect. 8.6.4). In this approximation, the starting charge density of the compound is obtained by superposing the self-consistent charge densities of the spin-polarized neutral atoms: (Ar, $4s^2$, $3d_\downarrow^5$) for Fe in T; (Ar, $4s^2$, $3d_\uparrow^5$, $3d_\downarrow^{0.5}$) for Fe in *M*; (He, $2s^2$, $2p^4$) for O (with Ar and He frozen core states, and with the 2s state of oxygen calculated self-consistently). The adopted APW sphere radii of O and Fe in T were taken as one-half of the T–O distance. The radius of Fe in M was obtained by difference (once known as the O radius) from the M–O distance. The special points Γ and X were first selected in the self-consistent iterative procedure. The L point was added when energy attained a 0.003 *Ry* stability. The accuracy of the 3 points was estimated to be at the 0.01 *Ry* level (Yanase and Siratori 1984). The band structure calculations showed a strong hybridization between the 3d states of Fe and the 2p states of O in all bands (see Table I in Yanase and Siratori 1984). Focusing the attention on the main component of the majority spin states

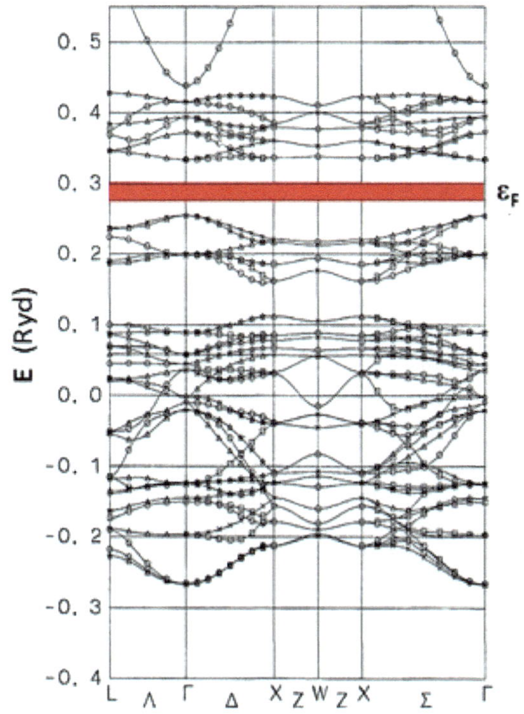

Fig. 20.33 Self-consistent APW energy band structure of majority spin (up spin) electrons in Fe_3O_4, Ryd = Rydbergs. Modified from Yanase and Siratori (1984)

(Fig. 20.33) one sees the 24 2p bands of O, located between −0.27 *Ry* and 0.05 *Ry*. From −0.01 and 0.11 *Ry* occur 12 dε bands of Fe in M. The next eight bands moving upward are dγ bands of Fe in M. Above the band gap, where resides the Fermi level (ε_F), are found the d bands of Fe in T. Concerning the minority spin electrons (spin down), when moving up in the energy scale, the 24p bands of O are followed by 10 d bands of Fe in T. The dε bands of Fe in M occur between 0.23 and 0.33 *Ry*. A small energy gap separates them from the eight dγ bands of Fe in M.

While the delineated band structure is somewhat in line with a simple ionic model, the self-consistent spin and charge distribution are not (i.e. a charge of 23.26 *e* and a spin moment of −3.67 μ_B for Fe in T; a charge of 24.17 *e* and a spin moment of +3.67μ_B for Fe in M; a charge of 7.70 *e* and a spin moment of 0.09μ_B for O; see Table II in Yanase and Siratori 1984).

Two years later, De Groot and Buschow (1986), in a study devoted to the comprehension of the Magneto-Optical Kerr Effect (MOKE) in PtMnSh (a half-metallic ferromagnet), provided ancillary results on Fe_3O_4, obtained with self-consistent Augmented Spherical Wave (ASW) calculations (see later). These publications gave results nearly identical to the previous study, in supporting an "*itinerant electron*" picture (opposed to a "*localized electron*" view, see later) for Fe_3O_4 at high T.

Penicaud et al. (1992) noted that the magnetic properties of Fe_3O_4 at low T are somewhat controversial: "*From a localized electron point of view, the high conductivity of Fe_3O_4 is due to electron hopping between the Fe^{2+} and Fe^{3+} ions occupying the B-sites. The low temperature transition is believed to be an order–disorder transition on these Fe ions. However, from an itinerant electron point of view, the high conductivity is a consequence of a partially filled d-band of the B-site Fe ions, the transition to the insulating state being presumably brought about by a band splitting due to electron correlations or the electron–phonon interaction*" (Penicaud et al. 1992).[13]

They proceeded then to the DFT calculation of the band structure of Fe_3O_4 (and also of the other Co–, Ni–, Mn–, and Sn– ferrites). Effective single particle equations were obtained with the Augmented Spherical Wave (ASW) method of Williams et al. (1979) and the Atomic Sphere Approximation (ASA). In this approximation, the volume of the unit cell equals the sum of the volumes of the touching (muffin-tin) atomic spheres. Atoms are then replaced by spheres of appropriate volume and the empty spaces (60.5% in the case of Fe_3O_4) are replaced by "*Empty Spheres*" (ES; Z = 0). For the Fe_3O_4 cell with $a_0 = 15.8624$ a.u., the radii adopted by Penicaud et al. (1992) were 2.3 a.u. for the two atoms in T (Wyckoff position *a*); 2.5 a.u. for the four atoms in M (Wyckoff position *d*); 1.94 a.u. for the eight oxygens (position *e*); 1.1 a.u. for 4 ES in *c*; 1.8648 a.u. for 12 ES in *f*; and 1.7 a.u. for 2 ES in *b*. The integration in the BZ was carried out initially on 8 special points, extended then to 60 after convergence was attained. Relativistic effects were accounted for, but not those ones relative to spin–orbit coupling.

In Fig. 20.34, we see the band structure of Fe_3O_4 obtained by Penicaud et al. (1992) and subdivided into minority spin (a) and majority spin (b). Comparing Figs. 20.33 and 20.34b, we may see that the band plots are quite similar, besides major differences in the actual position of the Fermi level. This has possibly to do with the fact that, in the investigation of Yanase and Siratori (1984), the interstitial region, which occupied 60.5% of the whole space, was not taken into account by empty spheres.

As noted by Penicaud et al. (1992) (and Yanase and Siratori 1984; as well), for the majority (up spin) electrons, one observes a band gap at the Fermi energy (ε_F); therefore, $N_\uparrow$ is an integer. For the minority spin (down spin) electrons, no band gap is observed, nevertheless, since the unit cell has zero net charges and $N = N_\uparrow + N_\downarrow$, also $N_\downarrow$ is an integer. Hence, in Bohr magnetons:

$$M_S = N_\uparrow + N_\downarrow = 4\mu_B \tag{20.3.35}$$

The magnetic moments for each sphere are listed in Table 20.29. They have opposite signs on the T and M sites, hence the coupling is antiferromagnetic, in

[13] The observation of Penicaud et al. (1992) concerning the "*localized electron*" finds its roots in the pioneering work of Mott and Twose (1961) and Mott (1967). The B site of Penicaud et al. (1992) is the M site in our terminology, and the A site is the T site.

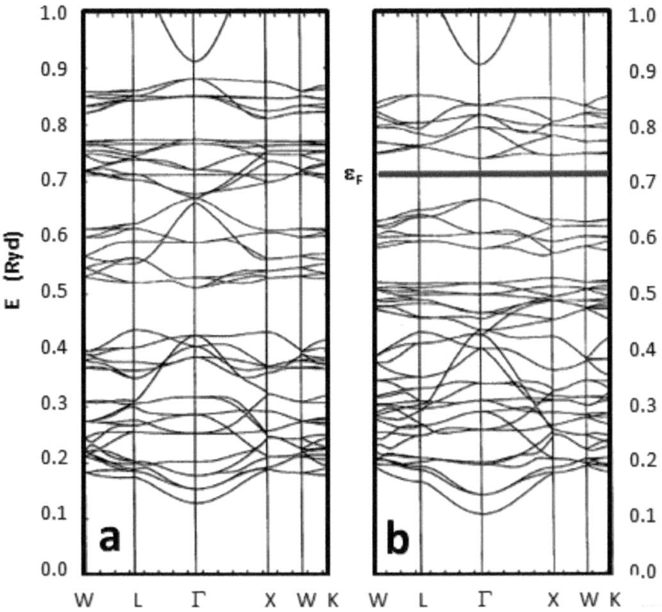

Fig. 20.34 Band structure of Fe_3O_4 according to Penicaud et al. (1992). **a** spin down electrons; **b** spin up electrons; ε_F Fermi energy, Ryd= Rydbergs. Modified from Penicaud et al. (1992)

agreement with the simple ionic model (although the two modes are conceptually very different). Because the ES spheres do carry some charge and are weakly polarized (Penicaud et al. 1992) the sum of charges does not add exactly to 4.

After this preliminary investigation Uhl and Siberchicot (1995) performed an accurate ab initio investigation of the exchange integrals in Fe_3O_4 in the attempt to get insights into the magnetic properties of the high-temperature phase and to calculate magnetic exchange constants, spin excitations, and the Curie temperature within the Heisenberg model. The calculations, which used incommensurate non-collinear spin spiral configurations (Kübler et al. 1992) succeeded in predicting the metallic behavior of the high-temperature phase but not the insulating ground state. The Curie T was, within the Heisenberg model, too high and the exchange constants and spin-wave Energies were "*a little too large*" (Uhl and Siberchicot 1995) with respect to observations.

The exchange constants of the next nearest neighbors were obtained by Uhl and Siberchicot (1995) by resolving the magnetic exchange energy (see their equations 3.3 to 3.5) and are compared in Table 20.30 with experimental observations.

In Fig. 20.35, the acoustic (ω_1, ω_2, ω_3) and optic (ω_4, ω_5, ω_6) spin-wave dispersion curves of magnetite calculated by Uhl and Siberchicot (1995) for **k** vectors along the symmetry line Γ-Δ-*X*-Σ-Γ-Λ-L are compared with the experimental observations. of Watanabe and Crockhouse (1962) and Bourdonnay et al. (1971).

Table 20.29 Lattice constant (a_0), spin electrons ($N_\downarrow$, $N_\uparrow$); valence electrons ($N = N_\downarrow + N_\uparrow$); effective number of valence electrons per formula unit ($Z_V^{eff} = Z_{Fe,T} + 2Z_{Fe,M} - 8$); magnetic moments per formula unit (M_S); distribution of magnetic moments over sites (From Penicaud et al. 1992)

Magnitude	Dimension	Value
a_0	(a.u.)	15.86
$N\downarrow$		22
$N\uparrow$		26
N		48
Z_V^{eff}	(*e*)	16
M_S	μ_B	4
Fe_T	μ_B	−3.37
Fe_M	μ_B	3.63
O	μ_B	0.01

Table 20.30 Calculated and experimental values of nearest-neighbor exchange constants in Fe_3O_4. Data in *MeV*. From Uhl and Siberchicot (1995)

J_{TT}	J_{MM}	J_{TM}	References
−0.18	4.83	−2.88	Uhl and Siberchicot (1995)
−0.11	4.63	−2.92	Uhl and Siberchicot (1995), NN approx
−1.50	4.04	−2.00	Néel (1948)
0.00	4.24	−2.40	Glasser and Milford (1963)
−1.56	0.26	−2.38	Bourdonnay et al. (1971)
−1.52	4,31	−2.42	Möglestue (1968)

Background information: Quantum Mechanical aspects of spin interaction

Assuming the magnetic moment to be localized, the spin system can be described by the Heisenberg Hamiltonian (Bader and Schilling 1979):

$$H = -\sum_{i\mu,j\nu} J_{\mu\nu}(t_{i\mu} - t_{j\nu}) \cdot \hat{S}^z(t_{i\mu}) \cdot \hat{S}^z(t_{j\nu}) \quad (20.3.36)$$

where $\hat{S}z(t_{j\nu})$ is the spin located at the magnetic site $t_{j\nu}$ and $J_{\mu\nu}(t_{i\mu} - t_{j\nu})$ is the exchange constant that describes the interaction between the spins at the sites $t_{i\mu}$ and $t_{j\nu}$. Considering the z component to be invariant (see Sect. 6.1.1) and considering only the linear terms in the transverse components, the Hamiltonian can be rewritten in the form (Uhl and Siberchicot 1995):

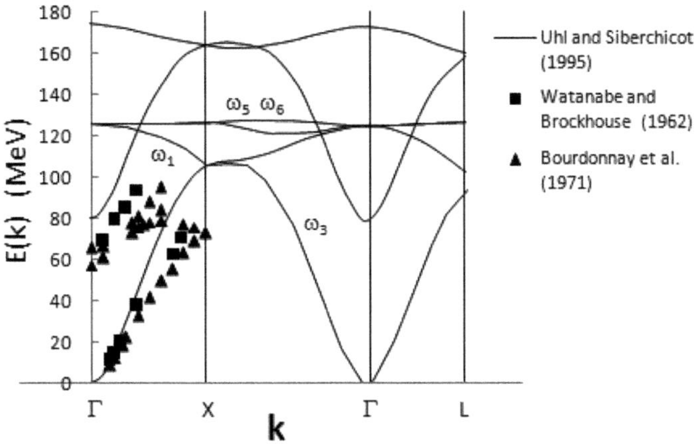

Fig. 20.35 Acoustic and optic spin-wave dispersion curves of magnetite. The results of Uhl and Siberchicot (1995) (solid lines) are compared with the data of Watanabe and Crockhouse (1962) and Bourdonnay et al. (1971). Modified from Uhl and Siberchicot (1995)

$$H = -\sum_{i\mu,j\nu} J_{\mu\nu}(t_{i\mu} - t_{j\nu}) \cdot \left[\hat{S}^z(t_{i\mu}) \cdot \hat{S}^z(t_{j\nu}) + \hat{S}^+(t_{i\mu}) \cdot \hat{S}^-(t_{j\nu})\right] \quad (20.3.37)$$

In Fe_3O_4, the ground state can be seen as "*a collinear ferromagnetic eigenstate of the antiparallel ordered sublattices* $\{T\}$ *and* $\{M\}$" (Uhl and Siberchicot 1995). The corresponding ground state energy is

$$\epsilon_0 = -\sum_{i\mu,j\nu} J_{\mu\nu}(t_{i\mu} - t_{j\nu})\lambda_\mu\lambda_\nu S_\mu(t_{i\mu})S_\nu(t_{j\nu}) \quad (20.3.38)$$

where $S_\nu(t_{j\nu})$ is the spin located at the magnetic site t_j and λ_ν is the direction of magnetization ($\lambda_\nu = -1$ for $\nu \in T$ and $\lambda_\nu = +1$ for $\nu \in M$). Based on the Block theorem (Sect. 8.4) the spin wave energies are defined by

$$H|\mathbf{k}\rangle = (\epsilon_0 + \epsilon(\mathbf{k}))|k\rangle \quad (20.3.39)$$

$$|\mathbf{k}\rangle = \frac{1}{\sqrt{2N}}\sum_\nu C_\nu(\mathbf{k})\sum_j \exp(i\mathbf{k} \cdot t_{j\nu})\hat{S}^-(t_{j\nu})|0\rangle \quad (20.3.40)$$

Fourier transformation of the exchange constants between ν and μ leads finally to (see Uhl and Siberchicot 1995):

$$E_{\nu\mu}(\mathbf{k}) = 2\sum_j \exp[i\mathbf{k} \cdot (t_{j\nu} - t_{0\mu})]J_{\mu\nu}(t_{j\nu} - t_{0\mu}) \quad (20.3.41)$$

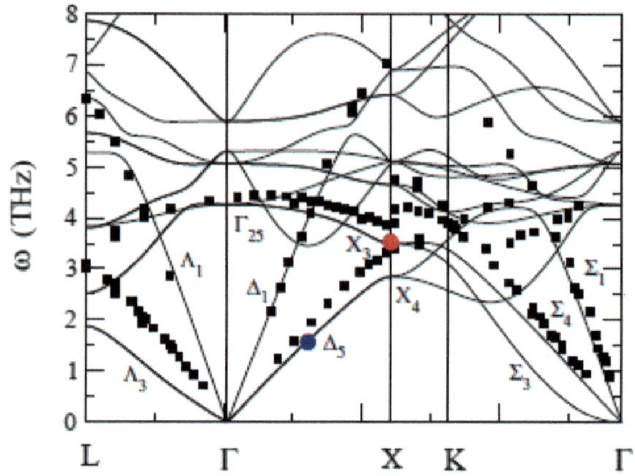

Fig. 20.36 Low-frequency phonons of cubic Fe_3O_4 calculated by Piekarz et al. (2006) at GGA $+U$ theory level with $U = 4$ eV and $J = 0.8$ eV (solid lines), compared with the results of inelastic neutron scattering at room conditions (Samuelsen and Steinsvoll 1974; solid symbols). The two primary order parameters (OP) related to Δ_5 (blue circle) and X_3 (red circle) are superimposed. From Piekarz et al. (2006). With permission from The American Physical Society

The origin of the Verwey transition in Fe_3O_4 was investigated by Piekarz et al. (2006). The ab initio spin-polarized GGA + U study involved an "*on-site*" 3d electron interactions described by the Coulomb element U and the Hund's exchange J (Liechtenstein et al. 1995). Both structures above and below the transition were investigated. The cubic $Fd\overline{3}m$ a × a × a and the monoclinic $P2/c$ $a/\sqrt{2}$, $a/\sqrt{2}, 2 \int$ a supercells contained both 56 atoms. The valence electrons were six for oxygen ($2s^22p^4$) and eight for iron ($3d^74s^1$). Frozen-core electrons were treated by the Projector Augmented Wave (PAW) procedure of Blöckl (1994). Brillouin zone integration involved 4 × 4 × 4 and 4 × 4 × 2 k-point grids for $Fd\overline{3}m$ and $P2/c$ symmetries, respectively. The local interaction was reproduced with $U = 4$ eV and $J = 0.8$ eV (Zaanen and Sawatzky 1990). According to Piekarz et al. (2006), the transition from the high-T cubic to the low-T monoclinic structure is due to an on-site Coulomb interaction U between 3d electrons at Fe ions which amplifies the coupling of phonons to conduction electrons, opening a gap at the Fermi energy level. Phonon waves were calculated only for the cubic polymorph and are plotted in Fig. 20.36. In this representation, low-frequency phonons are classified according to their irreducible representation, and the notation of points and directions in the reciprocal space follows the Bouckaert–Smoluchowslri–Wigner (BSW) scheme, with Λ for the (111) direction, Δ for the (100) direction and Σ for the (110) direction.

Because there is no single irreducible representation of $Fd\overline{3}m$ which lowers the space group directly to $P2/c$, the Verwey transition must involve at least two operations. The two operations (Δ_5 and X_3,) acting simultaneously to reduce the space group from $Fd\overline{3}m$ to $P2/c$ are (Piekarz et al. 2006)

$$Fd\overline{3}m \rightarrow \Delta_5 \rightarrow Pbcm(4) \qquad (20.3.42.1)$$

$$Fd\overline{3}m \rightarrow X_3 \rightarrow Pmna(2) \qquad (20.3.42.2)$$

(with terms in brackets representing the increase in the unit cells).

The space group *P2/c* (4) is formed by the common symmetry elements of *Pbcm* (4) and *Pmna* (2) (Piekarz et al. 2006).

Few years later, Senn et al. (2011) showed that the monoclinic polymorph below the Verwey transition is a superstructure composed of Cc domains, 1 mm in size with up to 24 orientations within magnetite crystals. The structure has 168 variable (x, y, z) atomic coordinates, of which 166 are independent, and only two constrained when fixing the cell origin. The two M–M distances perpendicular to the local orbital ordering axis at each of the eight Fe^{2+} positions tend to be shortened. The minority-spin t_{2g} electron density is significantly delocalized from an Fe^{2+} donor onto two adjacent M-site Fe^{3+} acceptors. This single charge, delocalized over three M sites with associated displacements of the two acceptor M sites and surrounding oxygen atoms (Fig. 20.37) corresponds to "*a highly structured small polaron*" or "*trimeron*" (i.e. linear three-Fe-site units; Senn et al. 2012).

Because the complex structure of the low-T polymorph (with 24 crystallographic iron sites) is a considerable obstruction for spectrum interpretation, Řezníček et al. (2017) carried out an ab initio investigation to obtain a complete set of hyperfine parameters of all the sites. DFT-GGA+*U* calculations (U_{eff} = 4.5 *eV*, R_{MT}(Fe) = 2 a.u., and R_{MT}(O) = 1.5 a.u.) were carried out with 9500 (R_{Kmax} = 6.0) basis functions and 9 **k** points in the irreducible part of the Brillouin zone. The density of charge was Fourier expanded to G_{max} = 16 $Ry^{1/2}$. Spin–orbit coupling was included in the calculations. The ab initio hyperfine parameters allowed for a simulation of a

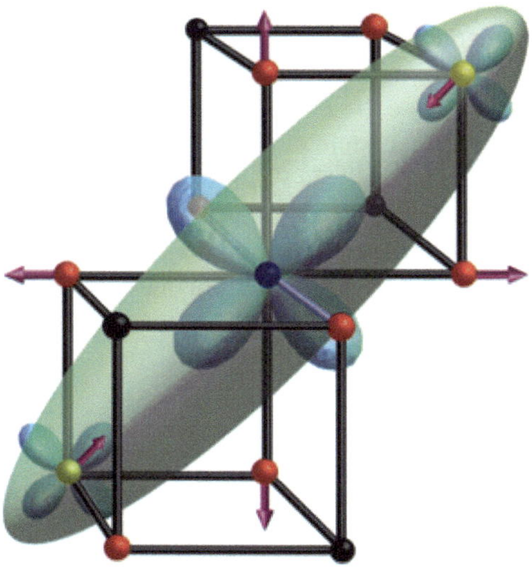

Fig. 20.37 Coupled distribution of a minority-spin electron (blue t_{2g} orbitals with dimension proportional to site population) and associated atomic displacements (purple arrows) within a linear three-Fe-site (*trimeron*) unit. Yellow centers are Fe^{3+} ions at neighboring M sites. Black centers are non-participating Fe^{3+} neighbors. The scalene ellipsoid encompassing the three Fe sites represents the minority-spin electron density. Modified from Senn et al. (2012)

Table 20.31 Average hyperfine parameters assigned to the sextets representing the four groups of iron ions after an ab initio-based refinement of the experimental Mössbauer spectrum (Řezníček et al. 2017)

	Groups			
	$8\times Fe^{3+}$ (T)	$8\times Fe^{3+}$-like (M)	$5\times Fe^{2+}$-like (M)	$3\times Fe^{2+}$-like (M)
Parameter	(T_1, T_2, T_3, T_4, T_5, T_6, T_7, T_8)	(M_{12}, M_{15}, M_8, M_5, M_{10}, M_{11}, M_6, M_9)	(M_4, M_{14}, M_3, M_1, M_2)	(M_7, M_{16}, M_{13})
Isomer shift $\overline{\delta}$ ($mm \times s^{-1}$)	0.386	0.577	0.985	1.018
Hyperfine magnetic field $\overline{B}_{hf}$(T)	50.88	52.58	50.32	36.23
EFG $\overline{V}_{zz}$ ($\times 10^{21}$ $V \times m^{-2}$)	−0.316	−0.489	−4.34	11.90
EFG $\overline{\eta}$	0.392	0.348	0.610	0.039
Line broadening (Γ_N)	0.88	3.31	2.48	2.13

zero-field ^{57}Fe Mössbauer spectrum of the *Cc* phase of magnetite. The 24 iron sites naturally break into four groups: $8 \times Fe^{3+}$ (T), $8 \times Fe^{3+}$-like (M), $5 \times Fe^{2+}$-like (M), and $3 \times Fe^{2+}$-like (M) ions. The hyperfine parameters of the ions within each group share common characteristics (Table 20.31).

20.4 Orthosilicates

I will limit myself to two representative components: Mg_2SiO_4 and Fe_2SiO_4. These components, completely miscible at solid state, form at sufficiently low-P *Olivine*, the main mineral, in terms of mass and volume, of the Earth's upper mantle (*Peridotites,* from the Medieval Latin name *Peridotus* = olivine). Olivine is a solid-state mixture of the polymorphs *Forsterite* and *Fayalite*, in molar proportions of 85–95% and 15–5%, respectively. The intrinsic stability limit of the mixture is assumed to coincide with the seismic discontinuity observed at a depth of about 400 *km*, where olivine undergoes transition to the β-(modified spinel) and, at greater depth, to γ-spinel polymorphs.

20.4.1 Mg_2SiO_4

Mg_2SiO_4 undergoes two main polymorphic transitions. The α polymorph (Forsterite; *Pbnm*), is stable at low P. According to Bowen and Andersen (1914), at room pressure, Forsterite melts congruently at $T = 2163 \pm 20\ K$. At $P \geq 11\ GPa$, it is observed that a solid-state transition to the orthorhombic diphyramidal (*Imma*) form *Wadsleyite*. The volume of transition is slightly negative, hence the stability of the low-P polymorph increases slightly with the increase of pressure. As we will see later on, transition loci have been precisely identified by the experiments of Katsura and Ito (1989), Boehler and Chopelas (1991), and Fei and Bertkha (1999). At higher P a further transition from β to γ (*Ringwoodite*; $Fd\bar{3}m$) is observed. As we will see, these transitions may be reproduced ab initio, through slight adjustments of the volumes of the two polymorphs, if internally consistent functional and basis sets are adopted.

20.4.1.1 Forsterite (α-Mg_2SiO_4)

Forsterite, the α polymorph of Mg-orthosilicate, crystallizes in the orthorhombic bipyramidal class (space group *Pbnm*; non-conventional setting of *Pnma* n. 62). There are six irreducible atoms in the conventional cell and 28 atoms in the unit-cell (four formula units per unit cell). Since it is the most abundant mineral in the Earth's upper mantle, Forsterite has been the subject of numerous investigations by mineralogists and crystallographers. The first structural refinement of forsterite was carried out by Bragg and Brown (1926). In this "*nesosilicate*", isolated $[SiO_4]^{4-}$ groups (*nesos* is the Greek term for "*island*") are mutually linked by Mg^{2+} ions. This arrangement gives rise to two nonequivalent octahedral sites, usually designated M1 and M2, in which Mg^{2+} ions are located, respectively, between the bases of tetrahedra (M1 sites) and between a basis and an edge of two neighboring tetrahedra (M2 sites) (Fig. 20.38). As shown in part A of the figure, tetrahedral apexes are alternate in opposite directions (the dashed line in the figure contours the elementary cell). In part B of the figure are shown details of the two octahedral sites. The M1 octahedron shares six of its 12 edges with the neighboring polyhedra (four with octahedra and two with tetrahedra), whereas the M2 octahedron shares only three edges (two with octahedra and one with a tetrahedron). Because shared edges are usually shorter than non-shared edges, to minimize repulsion between neighboring cations (Pauling's third rule), the octahedral symmetry in both sites is generally distorted (although the degree of distortion is more evident for the M2 site).

First-principles investigations of Forsterite dealt mainly with the elastic and structural properties at high pressure (Brodholt et al. 1996; Da Silva et al. 1997; Jochym et al. 2004), and, more recently also with its vibrational behavior (Noel et al. 2006; Li et al. 2007; Ottonello et al. 2009a).

The investigation of Haiber et al. (1997) was mainly devoted to the comprehension of H^+ absorption in α- and β-Mg_2SiO_4 by ab initio molecular dynamics and has a high seminal importance. The AIMD investigation was preceded by a DFT treatment of static energy in the three Mg_2SiO_4 polymorphs. The lowest energy absorption site for a proton in α- and β-Mg_2SiO_4 was then established by a combination of structural relaxation and molecular dynamics. The DFT investigation

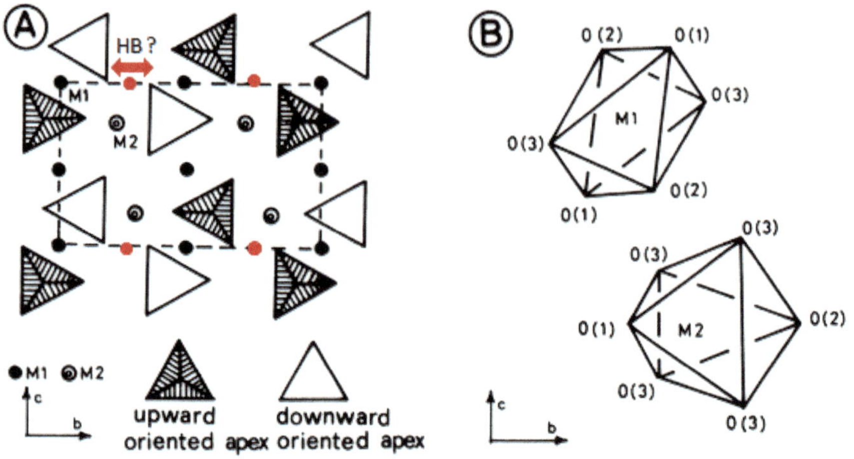

Fig. 20.38 The *Pbnm* structure of α-Mg_2SiO_4. The dashed line in part A of figure contours elementary cell (axes c and b). Mg occupies M1 and M2 positions. In part B are shown details of coordination state and distortion of M1-M2 sites; O(1), O(2), O(3) oxygens occupy nonequivalent positions; distortion from perfect octahedral symmetry is more marked for M2 site. Red dots locate the most likely position for H^+ absorbed in α-Mg_2SiO_4, i.e. between adjacent in-plane O3, through hydrogen bonding (Haiber et al. 1997). Modified from Ottonello (1997)

was carried out at LDA level with plane waves and norm-conserving non-local pseudopotentials for core electrons. For Mg and Si were adopted the pseudopotentials of Stumpf et al. (1990), and for H and O (besides Stumpf's pseudopotentials for benchmark computations) were adopted Troullier–Martins pseudopotentials (Troullier and Martins 1991). Exchange–correlation energies were evaluated with the Perdew functional (see Sect. 10.5.2). The kinetic energy cutoff was set to 110 *Ry*. The sampling of the BZ for α-Mg_2SiO_4 was carried out only at the Γ point. The error associated to the **k** point sampling was comparable to that due to kinetic energy cut off (Haiber et al. 1997). Calculations were carried out with the PWSCF code (see Sect. 8.12.5). MD calculations for pure and H^+ doped Mg_2SiO_4 involved soft pseudopotentials generated with the procedure of Troullier and Martins (1991), and the cutoff energy was set to 70 *Ry*. The adopted supercells contained 56 and 112 atoms. The AIMD simulations were performed using the CPMD code (see Sect. 13.4). The temperature was set to ~700 *K*. For each investigated polymorph the system was equilibrated for ~3000 steps and accumulated statistics for 9000 steps equal to 1 *ps*. The volume was set, for each polymorph, at a value near to the minimum attained by the static calculations (i.e. 580 $Å^3$ for α-, 538 $Å^3$ for β-, and 527 $Å^3$ for γ-Mg_2SiO_4). The structures were characterized by the radial distribution functions (RDF) and by the mean-square displacements of the different atoms. Though the main features of the RDF were similar for the three polymorphs, a noticeable difference was the shape of the first peak of the O–O correlation function (see Fig. 1 in Appendix II of Haiber et al. 1997). Because the equilibrium position

Table 20.32 AIMD H^+ absorption energy in Mg_2SiO_4 polymorphs (Haiber et al. 1997)

Polymorph	α-Mg_2SiO_4			β-Mg_2SiO_4				γ-Mg_2SiO_4
Neighbor	O_1	O_2	O_3	O_1	O_2	O_3	O_4	O_1
$-E_a$ (*eV*)	10.3	11.5	11.8	13.5	12.5	12.8	13.0	11.6
d_{OH} (*Å*)	0.99	1.05	1.10	1.01	1.09	1.10	1.06	1.07

of H^+ and their dynamics are not directly influenced by the V''_{Mg} vacancies "*which act like a homogeneous background of negative charge, ensuring global system neutrality*" (Haiber et al. 1997), the situation was simulated by simply adding one proton in an appropriate position (i.e. near a O^- functional at a distance close to that observed in OH^- hydroxyls, ~ 0.97 *Å*). Absorption sites were identified by the non-equivalent O atom that binds H^+. Computations, initiated with H^+ close to the chosen O atom, were followed by a short MD annealing in order to optimize the geometry. The energy of proton absorption (E_a) was obtained for different configurations as[14]:

$$E_a = [E(H^+|Mg_2SiO_4) - E(Mg_2SiO_4) - E(H^+)] \quad (20.4.1)$$

and is reported in Table 20.32 for the three polymorphs in different positions.

In α-Mg_2SiO_4 the most stable absorption site is close to O_3. This preferential stability may be ascribed to hydrogen bond involving two adjacent O_3 positions. The position of preferential bonding is tentatively delineated in Fig. 20.38. Hydrogen bonding is prevented at both the O_1 and O_2 sites, because the H^+ attached to O^- is separated from any other O atom by a plane of positive Mg^{2+} ions. The absorption energy of the O_3 site in α-Mg_2SiO_4 is 1.7 *eV* less favorable than for β-Mg_2SiO_4. According to Haiber et al. (1997), this large E_a difference between the two polymorphs would imply a ratio for partitioning H^+ between β and α polymorphs much larger than what experimentally observed (i.e. a factor 10 according to (Bell and Rossman 1992; Young et al. 1993; Inoue 1994). It should be noted however that, to the listed energies, it should be added the energy of formation of V''_{Mg}, which, according to static potential calculations, is 1.82 *eV* in M1 sites and 4.37 *eV* in M2 (see Table 8.19). Most likely these defect energies are somewhat larger in β-Mg_2SiO_4 due to the higher structural density.

Fourier Transform Infrared Spectroscopy (FTIR) measurements on hydrothermally annealed olivine samples (Bai and Kohlstedt 1993) revealed that the amount of dissolved H at fixed P, T conditions depends exponentially on fO_2 with logarithmic slopes dictated by the buffering conditions imposed in a first dry annealing at T = 1330 °*C* and room pressure (Fig. 20.39).

[14] Note that I changed sign to the original equation.

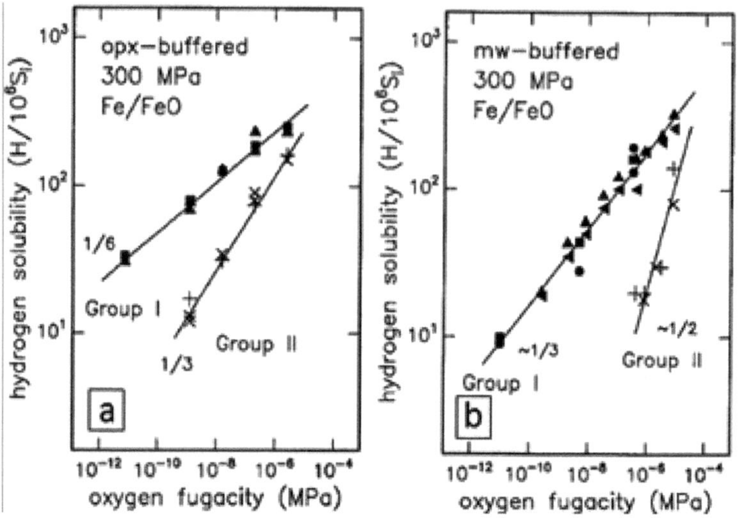

Fig. 20.39 Plot of hydrogen solubility versus oxygen fugacity (Bai and Kohlstedt 1993). **a** samples preannealed under orthopyroxene-buffered conditions and hydrothermally annealed at 300 *MPa* total pressure in Fe capsules (f_{H_2} = 0.158 *bar*); experiments #p9005 (filled boxes and pluses) and #p9016 (filled triangles and crosses) are also shown. **b** samples preannealed under magnesiowustite-buffered conditions and hydrothermally annealed at 300 *MPa* pressure in Fe capsules; experiments #m9007 (filled boxes), #m9008 (dots), #m9013 (filled triangles and pluses) and #m9014 (filled triangles pointing left and crosses) are shown. Modified from Bai and Kohlstedt (1993)

The molar concentration, C_H, of dissolved hydrogen was obtained by Bai and Kohlstedt (1993) by applying

$$C_H = \frac{1}{150 \times \gamma} \int \frac{K(\nu)}{3780 - \nu} d\nu \tag{20.4.2}$$

where K(ν) is the absorption coefficient (cm^{-1}) at wavenumber ν (cm^{-1}) and γ is the orientation factor which, for an unpolarized radiation with the beam parallel to [010] and the O–H bond nearly parallel to [100] assumes the value γ = 0.5 (Paterson 1982; Mackwell et al. 1985). Two distinct IR absorption band groups (absent in dry specimens) were observed after wet annealing. The Group I bands, ($3450 < \nu < 3650\ cm^{-1}$), included 7 well defined peaks (ν = 3483, 3529, 3542, 3576, 3598, 3612 and 3637 cm^{-1}) previously observed by other authors in natural specimens and in wet annealed olivines). The Group II bands, ($3200 < \sim < 3450\ cm^{-1}$) included at least five well defined peaks (ν = 3236, 3329, 3354, 3389 and 3443 cm^{-1}) never observed before.

The (dry) preannealing stage was carried out at room pressure and T = 1300 °*C* at f_{O_2} in the range 10^{-10}–10^{-3} *bar* for times >12 *h*. The second (hydrothermal) annealing stage was performed at 300 *MPa* and 900–1050 °*C* under a hydrogen fugacity f_{H_2} = 0.158 *bar* for 1–5 *h*. The hydrogen solubility associated with Group I bands is proportional to f_{O_2} to the 1/6 power for samples preannealed in contact with orthopyroxene, to the 1/3 power for samples preannealed in contact with magnesiowustite. The hydrogen concentration for Group II bands varies with f_{O_2} to the 1/3 power for opx buffered samples, to the 1/2 power for mw-buffered samples.

An amount of dissolved H proportional to $f_{O_2}^{1/6}$ may be expected for f_{O_2} conditions higher than the intrinsic fugacity of the substance (see Sect. 8.12.3 and Fig. 8.30) if the charge balance is assured by Mg vacancies. The energetically most plausible defect process responsible for proton entrapment in the orthosilicate lattice may be depicted as

$$2H_{2,gas} + Si_{Si}^{\times} + 4O_{O}^{\times} \Leftrightarrow 4H_{i}^{\cdot} + Si_{Si}^{\times} + 2V_{Mg}'' + 2O_{O}^{\times} + 2MgO_{(surf)} \quad (20.4.3)$$

or, in a simplified form:

$$2H_{2,gas} + 2O_{O}^{\times} \Leftrightarrow 4H_{i}^{\cdot} + 2V_{Mg}'' + 2MgO_{(surf)} \quad (20.4.4)$$

Denoting as orthosilicic acid (H_4SiO_4) the H component in mixture and assuming ideal mixing, the activity of H_4SiO_4 in mixture at ~1000 °*C* ranges from 2.5×10^{-6} at $f_{O_2} = 10^{-11}$ *bar* to 1.25×10^{-4} at $f_{O_2} = 10^{-4}$. This implies an Henry's law constant 6.32×10^{4} to 1.26×10^{3}, corresponding to a bulk defect energy of 0.078–0.121 *eV*. It is worth to stress, in passing, that, talking of "*water solubility*", as commonly observed in geological literature, is completely misleading, for several reasons:

(1) water (the liquid aggregate) is not stable at supercritical conditions;
(2) although the equilibrium constant of the gas-phase homogeneous reaction $H_{2,gas} + \frac{1}{2}O_{2,gas} \Leftrightarrow H_2O_{gas}$ is intrinsically displaced toward the product, it decreases progressively with the increase of T (from lnK ≅ 23 at 1000 *K* and room P, to lnK ≅ 8 at 2000 *K* to lnK ≅ 3 at T = 3000);
(3) the alleged amounts of H_2O in orthosilicate crystals are actually obtained by an arbitrary combination of $H_i^{\cdot}$ defects and $O_O^{\times}$.

Forsterite (and also the β- and γ-polymorphs of Mg_2SiO_4; see later) were investigated at DFT/B3LYP theory level under GGA (Ottonello et al. 2009a). For Mg it was adopted the 8-511G(d) expansion of Pascale et al. (2005) with re-optimized coefficients and exponents of the various GTO. For Si and O were adopted the 88-311G(d) set and the 8-411G(d) set of Nada et al. (1996). Other details of the procedure were anticipated at Sect. 8.9.1 to 8.9.3. The computational results (ab initio cell edges vs. ab initio static pressure) are substantially consistent with the experimental evidences of Hazen (1976) at high P (Fig. 20.40). The substance compresses anisotropically and preferentially along the b-axis.

The partial derivative of a third-order polynomial fitted over the molar electronic energy (E_{ai}, *J/mole*)–molar volume (V, *J/bar*) couples allowed to establish the ab initio potential value of static pressure P_{st} corresponding to a given

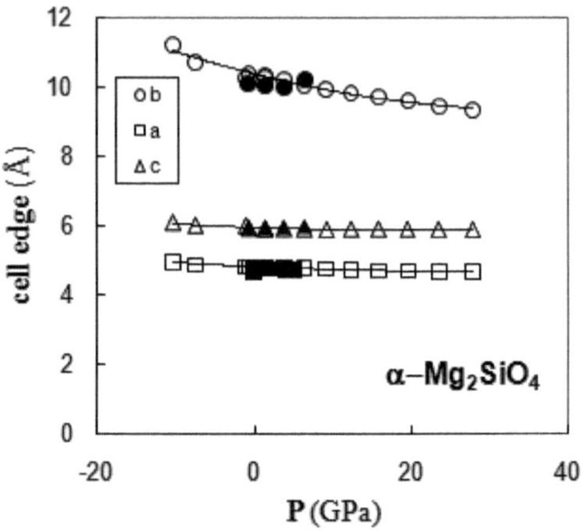

Fig. 20.40 Cell edges of α-Mg_2SiO_4 (*Å*) versus pressure (*GPa*). Ab initio results (Ottonello et al. 2009a; open symbols) are compared with the experiments of Hazen (1976) (filled symbols)

compressional state (see also Fig. 8.16). Data where then retrieved in terms of a third-order Birch–Murnaghan EoS. The results of this exercise are compared in Table 20.33 with other ab initio results and experiments. In passing I recall that K_0 should be always the isothermal compressibility at the athermal limit though in literature K_0 is often meant as the bulk modulus at room conditions (that's basically why I specify with "*static*" the true K_0 in Table 20.33).

There are 28 atoms in the orthorhombic bipyramidal unit cell of α-Mg_2SiO_4. The irreducible representation of vibrational modes, with a group theoretical treatment of the optical zone-center (gamma point; **k** = 0) yields

$$\Gamma = 7B_{2g} + 14B_{2u} + 7B_{3g} + 14B_{3u} + 10A_u + 11B_{1g} + 10B_{1u} + 11A_g \quad (20.4.5)$$

The 81 + 3 vibrational frequencies of α-Mg_2SiO_4 and their mode-gamma analysis are listed in Table 20.34.

The harmonic vibrational frequencies at Γ point listed in Table 20.34 do not differ much from the results of Noel et al. (2006). The need of a scaling factor variable within 0.985 and 0.989 between the experimental wavenumbers (Servoin and Piriou 1973; Iishi 1978; Hofmeister 1987; Chopelas 1991; Reynard 1991; Kolesov and Geiger 2004) and the eigenvalues of the Hessian matrix is nevertheless evident. The correlations between sets of data have in all cases a high internal consistency (R^2 factors variable within 0.993 and 0.995).

The quasi-harmonic gamma-mode analysis of the $\alpha_T K_T$ product showed that $\alpha_T K_T$ has, at high T, a quite constant value, averaging 4.47×10^{-3} *GPa/K* in the T range $1000 \leq T \leq 3000$ and attaining 0.0046 *GPa/K* at the $T \rightarrow \infty$ limit (Fig. 20.41a). According to Anderson (1995) the αK product of Forsterite increases from 3.46×10^{-3} *GPa/K* at 300 *K* to 4.40×10^{-3} *GPa/K* at 1700 *K* with roughly constant values above the Debye temperature of the substance (see Fig. 20.41a, and Table A-7.5 in Anderson 1995).

Table 20.33 Equation of State parameters of forsterite

V_0 (*Å*)	K_0 (*GPa*)	K′	References
Ab-initio			
281.7	134.4	3.88	Brodholt et al. (1996) (static)
293.3	137.0	3.75	Haiber et al. (1997) (static)
298.8	–	–	Jochym et al. (2004) (static)
296.7	130.8	4.001	Ottonello et al. (2009a) (static)
297.5	125.6	4.001	Ottonello et al. (2009a) (298.15 K)
289.5	126.4	4.2	Li et al. (2007) (300 K)
Experimental			
–	127.4	4.8	Meng et al. (1993)
–	–	4.4	Li et al. (1996)
–	–	5.2	Zha et al. (1997)
290.3	–	–	Smyth and Hazen (1973) (296 K)
289.8	–	–	Hazen (1976b)
289.6	–	–	Fujino et al. (1981)
285.5	–	–	Baur (1972)
290.0	–	–	Lager et al. (1981)
292.3	–	–	Bragg and Brown (1926), Belov et al. (1951)
290.2	–	–	van der Wal et al. (1987)
290.3	–	–	Cernik et al. (1950); Della Giusta et al. (1990)
128	–	–	Isaak (1991) (static, extrapolated)
290.1	125	4.0	Downs et al. (1996) (300 K)

The quasi-harmonic treatment of vibrational frequencies allowed to establish the thermal dependency of the bulk modulus in the terms anticipated at Sect. 8.9.3. For forsterite it was obtained $q_{ht} = 1.257$, $K_0 = 4.001$, and an isobaric temperature derivative of bulk modulus $(\partial K/\partial T)$, of -174.592 *bar/K* (i.e. -0.017 *GPa/K*). (see Table 8.5). This procedure disregards the fact that, at low T, the bulk modulus would presumably vary non-linearly with temperature. However, as it may be seen in Fig. 20.41b, non-linearity is virtually absent in the experimental data. The calculated thermal effect on K is lower than that obtained by the best fit to the experimental data, but lies within the uncertainty range obtained by Meng et al. (1993) from a least squares fit of the measured thermal pressure (i.e. $(\partial K/\partial T = -0.0214 \pm 0.03$ *GPa/K*). The ab initio thermal expansion obtained by coupling the αK product with the isobaric temperature derivative is in good agreement with the experimental data of Matsui and Manghnani (1985) (Fig. 20.41c).

Within the lattice dynamics theory, the calculation of thermodynamic properties requires the knowledge of the full phonon dispersion relation (Sect. 8.10.8) and the corresponding vibrational density of states (VDOS). Thermodynamic functions are then expressed as averages over the VDOS. Because of this averaging, however, thermodynamic properties are insensitive to details of the VDOS and simplified approaches such as the Kieffer's model (Sect. 8.10.4) can be successfully applied to

Table 20.34 Calculated vibrational frequencies for Forsterite at Γ point and their gamma modes (Ottonello et al. 2009a). Acoustic frequencies at $\mathbf{K}_{max}$ and their mode gammas are from Chopelas (1991)

N	Mode	ν (cm^{-1})	γ_i	N	Mode	ν (cm^{-1})	γ_i	N	Mode	ν (cm^{-1})	γ_i
1	Acoustic	98	1.1	29	B_{2g}	327.484	1.4	57	B_{1u}	516.014	0.4
2	Acoustic	99	1.1	30	B_{3g}	336.185	1.7	58	A_u	522.529	0.5
3	Acoustic	171	1.6	31	B_{2u}	345.114	1.7	59	B_{3u}	534.356	0.6
4	A_u	109.616	4.5	32	B_{3u}	350.806	1.5	60	B_{2u}	540.598	0.7
5	B_{3u}	149.402	2.4	33	A_g	352.707	1.0	61	A_g	561.630	0.4
6	B_{3g}	195.130	2.9	34	A_u	352.729	2.2	62	B_{1g}	597.161	0.6
7	A_g	195.612	2.1	35	B_{1g}	355.435	1.4	63	B_{3g}	610.826	0.4
8	B_{2g}	203.599	2.5	36	B_{3g}	379.105	1.3	64	B_{2g}	615.208	0.5
9	B_{2u}	204.165	1.9	37	B_{2g}	383.190	0.9	65	A_g	618.363	0.5
10	A_u	210.897	0.6	38	B_{2u}	388.904	1.1	66	B_{2u}	618.696	0.5
11	B_{1g}	230.122	2.0	39	A_u	391.369	1.1	67	B_{3u}	642.857	0.6
12	B_{1u}	238.952	0.9	40	B_{1g}	394.793	1.5	68	B_{1g}	645.894	0.7
13	A_g	241.080	1.0	41	B_{2u}	414.423	1.1	69	A_g	824.218	0.6
14	B_{3g}	258.613	2.1	42	B_{3u}	414.592	1.3	70	B_{3u}	837.277	0.5
15	B_{1g}	264.290	1.9	43	B_{1u}	420.229	0.9	71	B_{1g}	839.895	0.6
16	A_u	273.593	1.9	44	B_{2g}	424.149	1.4	72	B_{2u}	840.668	0.5
17	B_{2u}	285.616	2.3	45	A_g	424.806	0.9	73	A_g	861.568	0.4
18	B_{3u}	292.173	2.7	46	B_{1u}	430.408	0.5	74	B_{1g}	870.299	0.4
19	B_{1u}	296.070	2.9	47	B_{3u}	431.710	1.2	75	B_{3u}	877.930	0.4
20	A_u	301.740	3.0	48	A_u	436.703	0.5	76	B_{1u}	879.689	0.5
21	B_{2g}	304.773	2.3	49	B_{1g}	442.008	0.7	77	B_{3g}	888.507	0.5
22	A_g	305.605	2.2	50	B_{3g}	449.971	0.7	78	A_u	918.916	0.4
23	B_{1u}	306.647	2.0	51	B_{3u}	456.838	1.1	79	B_{2g}	931.329	0.4
24	B_{2u}	307.964	2.1	52	B_{2u}	470.104	1.7	80	B_{2u}	962.855	0.5
25	B_{3u}	309.992	1.7	53	B_{1u}	492.531	0.8	81	A_g	963.245	0.5
26	B_{1g}	318.733	1.8	54	A_u	498.627	0.5	82	B_{1g}	976.713	0.5
27	B_{1u}	321.322	1.9	55	B_{2u}	512.494	2.0	83	B_{2u}	985.226	0.5
28	A_g	323.972	1.7	56	B_{3u}	516.004	0.5	84	B_{3u}	985.288	0.4

solids (see for example Jacobs and De Jong. 2003; Jacobs et al. 2006). The isochoric heat capacity of Forsterite was obtained by Ottonello et al. (2009a) through a *semi-classical* approach, combining the Kieffer's model (Kieffer 1979a, b) for acoustic modes and the Einstein model for optic modes, in the terms outlined at Sect. 8.10.7.

As a demonstration of the validity of such an approach, in Table 20.35 I report a comparison of computed semi-classical isobaric heat capacity and entropy at 298 K for the Mg_2SiO_4 polymorphs with ab initio results obtained at the LDA level by means of a full calculation of the VDOS (Li et al. 2007; Wu and Wentzcovitch 2007; Yu and Wentzcovitch 2006) and calorimetric measurements (Akaogi et al.

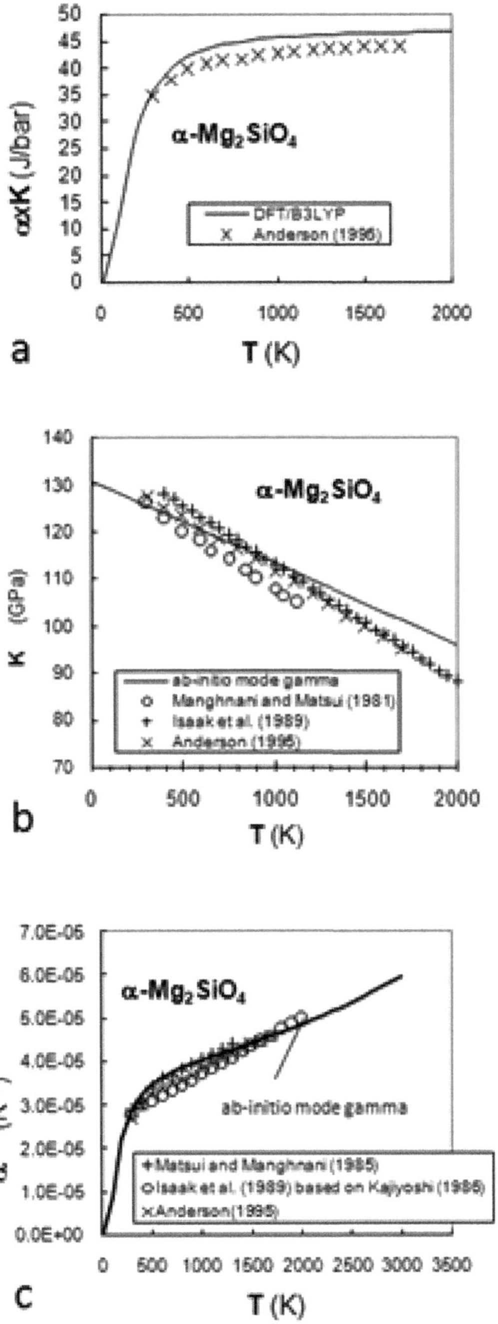

Fig. 20.41 **a** Ab initio α × K product of Forsterite compared with the indications of Anderson (1995). **b** the Ab initio thermal dependency of the bulk modulus compared with experiments. **c** semi-classical isobaric thermal expansion of Forsterite, compared with experiments and estimates. Modified from Ottonello et al. (2009a)

Table 20.35 Comparison of computed isobaric heat capacity and entropy at 298 *K* for Mg_2SiO_4 polymorphs. Results of the semi-classical approach of Ottonello et al. (2009a) are compared with ab initio results obtained at the LDA level by means of a full calculation of the VDOS and calorimetric measurements

	Semi-classical[a]		LDA (full VDOS)[b]	Exp.[c]
	Unscaled	Scaled		
C_P (*J/(mole × K)*)				
α-Mg_2SiO_4	116.8	117.9	119.3	117.9
β-Mg_2SiO_4	116.2	117.3	118.1	116.3
γ-Mg_2SiO_4	113.2	114.5	116.9	114.0
S (*J/(mole × K)*)				
α-Mg_2SiO_4	92.5	94.1	95.9	94.1
β-Mg_2SiO_4	85.7	87.3	88.7	86.4
γ-Mg_2SiO_4	80.2	81.7	85.0	82.7

[a]Ottonello et al. (2009a)
[b]Li et al. (2007), Wu and Wentzcovich (2007), Yu and Wentzcovich (2006)
[c]Akaogi et al. (2007)

2007. The computed B3LYP vibrational frequencies at Γ point (as Einstein oscillators), coupled with the sine-wave dispersed acoustic contributions (Table 20.34) provide evidently a good compromise (Table 20.35).

Coupling the isochoric heat capacity with the anharmonic contributions (in the terms outlined at Sect. 8.10.7) the isobaric heat capacity is readily obtained. The *semi-classical* isobaric heat capacity of forsterite at T = 298.15 *K* is in substantial agreement with the calorimetric datum of Robie et al. (1978) [computed = 116.8 *J/(mol × K)*, measured = 117.9 *J/(mol × K)*] and with the high-T values as well (Fig. 20.42). Applying the scaling factor 0.9875 to vibrational frequencies it is obtained an isochoric heat capacity at standard state and an entropy virtually identical to the tabulated and estimated ones [i.e. $C_{P,298.15}$ = 117.9 *J/(mol × K)*; $S^0_{298.15}$ = 94.114 *J/(mol × K)*]. A non-linear fit to a Haas–Fisher expansion reproduces the computed heat capacities with a mean error of 0.189 *J/(mol × K)*. The obtained coefficients are listed in Table 20.36 together with other relevant thermodynamic magnitudes.

20.4.1.2 Wadsleyite (*β*-Mg_2SiO_4)

Wadsleyite, which is also referred to as "*beta form*" or "*modified spinel*" in the literature, is an orthorhombic bipyramidal (*Ibmm*) polymorph of the Mg-orthosilicate. There are 6 irreducible atoms in the conventional cell. Canonical structure was refined long ago by Moore and Smith (1970) and Baur (1972). The structural refinement of Baur (1972) is shown in Fig. 20.43.

Protonated Wadsleyite was investigated with AIMD by Haiber et al. (1997). Micro-canonical trajectories, covering 1 *ps* after equilibrating the system at several

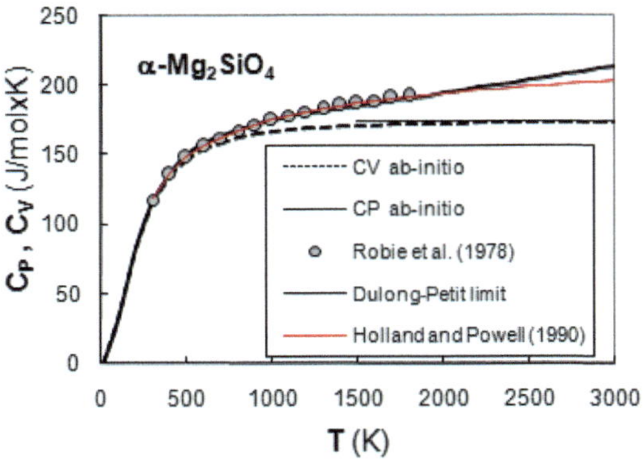

Fig. 20.42 Ab initio isobaric heat capacities of forsterite compared with tabulated values. The isochoric heat capacity curve (dashed line) is also shown to attain as expected the Dulong-Petit limit

Table 20.36 Ab initio thermochemical properties of Mg_2SiO_4 polymorphs at 1 *bar*, 298 *K*. $C_P = a + b \times T + c \times T^{-2} + d \times T^2 + e \times T^{-1/2}$. $\alpha T = \alpha_0 T + \alpha_1 + \alpha_2 T^{-1} + \alpha_3 T^{-2} + \alpha_4 T^{-3}$. Thermal expansion expression valid in the T range $200 \leq T\,(K) \leq 3000$ (Ottonello et al. 2009a)

Magnitude		Forsterite (α)	Wadsleyite (β)	Ringwoodite (γ)
V^0 (*J/bar*)		4.437	4.087	3.963
S^0 (*J/[mol×K]*)		94.1	87.3	81.7
C_P (*J/[mol×K]*)	$a \times 10^{-2}$	1.5513	1.6139	1.5346
	$b \times 10^{2}$	2.1307	2.2270	2.1405
	$c \times 10^{-6}$	−4.4060	−4.1112	−4.7490
	$d \times 10^{5}$	−0.0792	−0.2473	−0.3080
	$e \times 10^{-2}$	0.87536	−0.86514	1.2708
Thermal expansion (K^{-1})	$\alpha_0 \times 10^{7}$	0.10031	0	0
	$\alpha_1 \times 10^{7}$	269.31	464.00	323.00
	$\alpha_2 \times 10^{3}$	5.6082	−13.8005	−7.3084
	α_3	−2.6162	3.6196	1.3745
	α_4	243.828	−395.094	-150.406
K_0 (*GPa*)		130.8	161.8	196.4
K'		4.001	4.415	4.322
$(dK/dT)_P$ (*bar/K*)		−174.592	−117.997	−104.108

temperatures from 400 to 1600 *K*, showed that the stretching of O_1-H in β-Mg_2SiO_4 at the lowest T is remarkably harmonic (Haiber et al. 1997). This allowed a precise determination of the stretching frequency, which resulted to be $3180 \pm 30\ cm^{-1}$, in close agreement with the experimental observations of McMillan et al. (1991) ($3322\ cm^{-1}$). The suggestion that hydrogen bonding with O_1 (or O_2) is the main mechanism for H^+ solution in β-Mg_2SiO_4 was already anticipated by Smyth (1987, 1994) and Downs (1989). According to Smyth (1987, 1994)

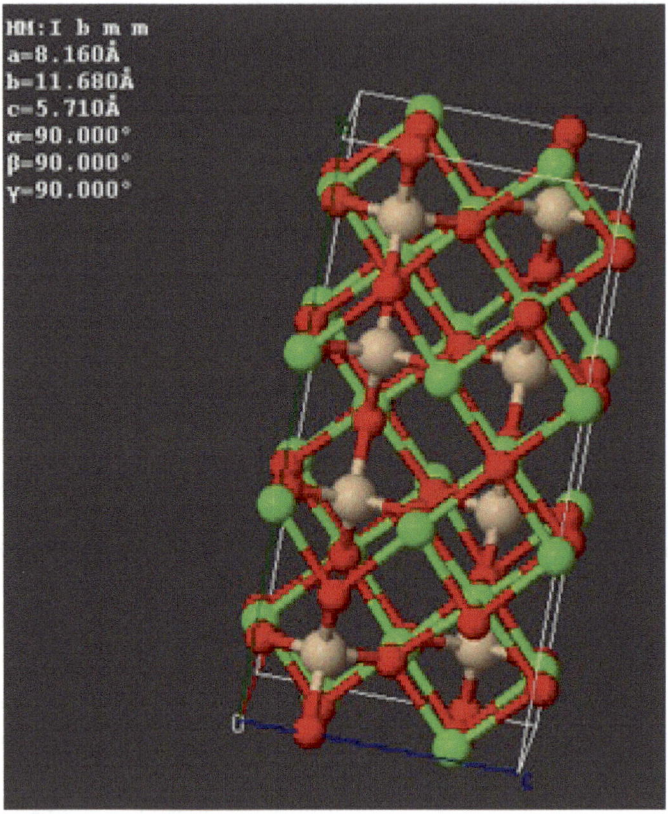

Fig. 20.43 The *Ibmm* structure of Wadsleyite (Baur 1972) as generated by the AMCSD software (Downs and Hall-Wallace 2003)

preferential binding with O_1 (with respect to O_2) is favored by the longer distance from the Si ions and by the greater electronic charge. Downs (1989) basically got the same advice on the basis of optimization of geometry and electronic structure of β-Mg_2SiO_4.

Wadsleyite was investigated by Ottonello et al. (2009a) with the same procedures (theory level, functional, basis sets) adopted for the α-polymorph (see Sects. 8.9.1–8.9.3 and 20.4.1.1). As a guide for the construction of the potential well of the substance were adopted the results of detailed experimental compressional studies (Hazen et al. 2000; Li et al. 1998), in-situ high P–T measurements (Meng et al. 1993) and previous first-principles investigations (Haiber et al. 1997; Kiefer et al. 2001; Wu and Wentzcovitch 2007). Provisional cell edges were assigned to each compressional state and the minimum energy c/b and a/b ratios were refined at the athermal limit over a wide pressure range, as previously described for the α polymorph. The partial derivative of a third-order polynomial fitted to the calculated E-V couples allowed to establish the ab initio potential value

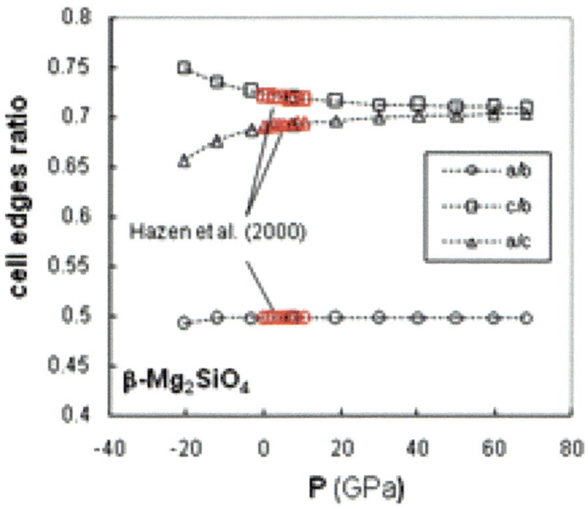

Fig. 20.44 Cell edge ratios of β-Mg_2SiO_4 (adimensional) versus pressure (*GPa*). *Ab initio* results (Ottonello et al. 2009a) are compared with the experiments of Hazen et al. (2000)

of static pressure, P_{st}. As shown in Fig. 20.44, the computed results are consistent with the experimental data of Hazen et al. (2000). The mineral compresses anisotropically and preferentially along the c-axis. The c/b and a/b ratios gradually converge at high pressure while the a/b ratio remains fairly constant.

Setting the nominal static pressure to zero at the athermal limit, a cell volume of 550.0 $Å^3$ (a_0 = 5.731 Å; b_0 = 11.516 Å; c_0 = 8.333 Å) was estimated. A complete optimization of the crystalline structure gave V^0 = 550.4 $Å^3$ (a_0 = 5.734 Å; b_0 = 11.521 Å; c_0 = 8.332 Å) and a B3LYP total energy (gfw basis) E_{B3LYP} = −991.0427538 *Hartree*. As for the α polymorph, also for the β phase the predicted volumes are slightly larger than the one reported from calculations at the LDA level of theory (Haiber et al. 1997; Keifer et al. 2001; Wu and Wentzcovich 2007).

The bulk modulus of the substance at zero pressure (K_0 = 161.826 *GPa*) and its pressure derivative ($\partial K/\partial P = K' = 4.415$) were obtained by Gaussian integration of the Birch–Murnaghan third-order finite strain isothermal EoS and are consistent with the room-temperature estimates of Akaogi et al. (1989) based on the data of Sawamoto et al. (1984) and Tanaka et al. (1987) (i.e. K_T = 173; K′ = 4.3). As shown in Fig. 20.45, the third-order Birch–Murnaghan equation provides also for β-Mg_2SiO_4 a good fit of the computed energies. The results of the static calculations of Ottonello et al. (2009a) are compared in Table 20.37 with those of other static calculations and experimental data.

As for α-Mg_2SiO_4, also in the β polymorph there are 28 atoms in the unit cell, but in this case the irreducible representation of vibrational modes, with a group theoretical treatment of the optical zone-center (Γ point; **k** = 0) yields:

$$\Gamma = 12B_{2g} + 10B_{2u} + 9B_{3g} + 12B_{3u} + 7A_u + 7B_{1g} + 13B_{1u} + 11A_g \quad (20.4.6)$$

The 81 + 3 vibrational frequencies of β-Mg_2SiO_4 and their mode-gamma analysis are listed in Table 20.38.

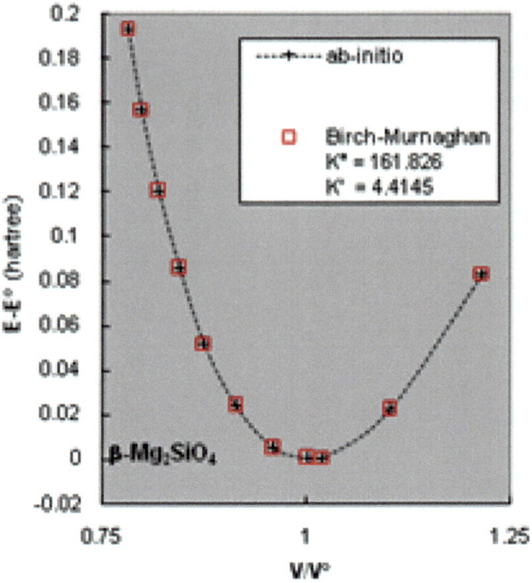

Fig. 20.45 Potential well of Wadsleyite (β-Mg_2SiO_4). The *ab initio* raw data (Ottonello et al. 2009a) are compared with a third-order Birch–Murnaghan fitting

Table 20.37 Equation of State parameters of Wadsleyite

V_0 (*Å*)	K_0 (*GPa*)	K′	References
Calculated			
545.2	179.0	4.23	Haiber et al. (1997)
527.4	180.8	4.34	Kiefer et al. (2001) (static)
535.1	169.2	4.53	Kiefer et al. (2001) (300 K)
528.1	179.2	4.20	Jacobs et al. (2006) (static)
533.5	–	–	Jacobs et al. (2006) (298.15 K)
541.4	165.7	4.44	Wu and Wentzcovitch (2007) (298.15 K)
550.4	161.8	4.415	Ottonello et al. (2009a) (static)
551.7	158.6	4.415	Ottonello et al. (2009a) (298.15 K)
Experimental			
535.3	160 ± 3	4	Hazen et al. (1990)
539.3	172 ± 3	6.3	Hazen et al. (2000)
	172.6	4.8	Meng et al. (1993)
538.1	–	–	Horiuchi and Sawamoto (1981)
–	173	4.3	Akaogi et al. (1989)
–	174	4.3	Fei et al. (1992)

Table 20.38 Ab initio optical frequencies for β-Mg_2SiO_4 at Γ point (Ottonello et al. 2009a) and their mode gamma. Acoustic frequencies at $\mathbf{K}_{max}$ and their mode gammas are from Chopelas (1991)

N	Mode	ν (cm^{-1})	γ_i	N	Mode	ν (cm^{-1})	γ_i	N	Mode	ν (cm^{-1})	γ_i
1	Acoustic	118	0.95	29	B_{2g}	368.2	1.4	57	B_{3u}	526.1	0.9
2	Acoustic	118	0.95	30	B_{3u}	368.2	1.7	58	B_{2u}	535.1	1.0
3	Acoustic	198	1.45	31	B_{1u}	369.1	1.7	59	B_{2u}	548.4	0.7
4	B_{2u}	202.3	3.5	32	B_{1g}	370.0	1.5	60	A_g	561.0	1.2
5	B_{3g}	210.4	2.0	33	B_{1u}	375.6	1.0	61	B_{1u}	563.1	0.9
6	B_{3u}	214.4	5.0	34	B_{3g}	379.1	2.2	62	B_{3u}	577.6	1.2
7	A_u	216.5	1.9	35	A_u	379.6	1.4	63	A_u	578.1	0.6
8	B_{2g}	233.0	2.7	36	B_{1g}	393.2	1.3	64	B_{3g}	593.5	0.9
9	B_{1g}	234.0	4.6	37	B_{3g}	394.0	0.9	65	B_{2g}	607.4	0.6
10	B_{2u}	257.2	1.6	38	A_u	395.8	1.1	66	B_{1u}	617.0	0.5
11	B_{1g}	259.4	1.3	39	B_{2g}	406.2	1.1	67	B_{3g}	618.1	0.8
12	B_{1u}	261.9	3.5	40	A_g	408.2	1.5	68	A_g	641.9	0.7
13	A_g	263.3	3.2	41	B_{2u}	408.9	1.1	69	B_{1u}	705.6	1.0
14	B_{3g}	268.9	0.8	42	B_{1u}	418.4	1.3	70	A_g	734.0	0.7
15	A_g	274. 7	1.0	43	A_g	431.6	0.9	71	B_{2g}	769.1	1.1
16	A_u	276.1	2.0	44	B_{3u}	431.9	1.4	72	B_{3u}	779.4	1.0
17	B_{2g}	284.0	2.7	45	B_{2g}	432.4	0.9	73	B_{2g}	834.1	0.8
18	B_{3u}	296.4	1.8	46	A_g	441.4	0.5	74	B_{3u}	835.2	0.8
19	B_{2g}	299.8	0.5	47	B_{3u}	458.4	1.2	75	B_{1u}	866.7	0.9
20	B_{2u}	302.2	1.5	48	B_{1g}	460.7	0.5	76	B_{1g}	874.0	1.0
21	B_{1g}	302.9	0.8	49	B_{1u}	463.9	0.7	77	A_g	874.0	0.9
22	B_{1u}	307.3	3.1	50	B_{3u}	467.7	0.7	78	B_{3g}	897.4	1.0
23	B_{3u}	325.3	1.4	51	A_u	468.7	1.1	79	A_u	898.4	0.9
24	B_{3g}	328.3	1.9	52	B_{2u}	490.7	1.7	80	B_{1u}	898.6	0.8
25	B_{2u}	335.4	1.8	53	B_{3g}	501.5	0.8	81	B_{2u}	900.0	0.9
26	B_{1u}	353.7	2.6	54	B_{1u}	505.4	0.5	82	A_g	913.7	0.7
27	A_g	353.7	1.1	55	B_{2g}	512. 1	2.0	83	B_{2g}	925.5	0.9
28	B_{2u}	356.2	1.2	56	B_{2g}	516.0	0.5	84	B_{3u}	942.0	0.8

The B3LYP-GGA computed data of Ottonello et al. (2009a) are in line with the results of Wu and Wentzcovitch (2007) obtained at the LDA level of theory.

According to the experimental data of Meng et al. (1993), the αK product of Wadsleyite increases from 3.65×10^{-3} *GPa/K* at 300 *K* to a constant value of 5.16×10^{-3} *GPa/K* above its Debye temperature. The gamma mode analysis of ab initio frequencies yields an average value of αK of 5.48×10^{-3} *GPa/K* at high temperature. As illustrated in Fig. 20.46a, the semi-classical model results agree with Meng et al. (1993) within the experimental uncertainty of the data.

Proceeding in the same manner as done for the α polymorph, the obtained isobaric temperature derivative of the bulk modulus of the β phase is –0.012 *GPa/K*,

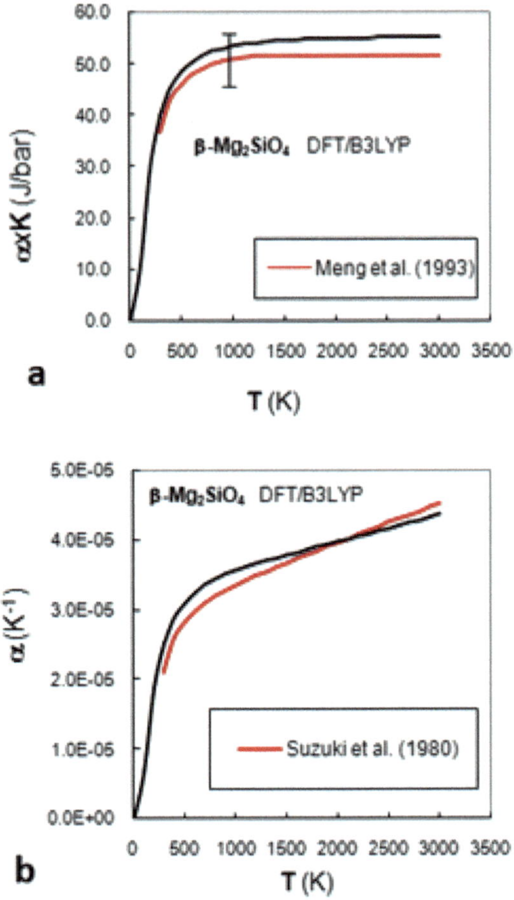

Fig. 20.46 **a** Semi-classical $\alpha \times K$ product of Forsterite compared with the experimental observations of Meng et al. (1993). **b** Semi-classical isobaric thermal expansion of Wadsleyite compared with experiments and estimates. Modified from Ottonello et al. (2009a)

which is substantially smaller in magnitude than the experimental data of Fei et al. (1992) on β-$(Mg_{0.84}Fe_{0.16})_2SiO_4$; −0.027 *GPa/K*. However, when coupling the thermal derivative of the bulk modulus with the αK product the resulting thermal expansion is roughly consistent with the data of Suzuki et al. (1980) (Fig. 20.46b).

The low-temperature ab initio C_P values are in good agreement with the calorimetric measurements of Akaogi et al. (2007) (Fig. 20.47a). The computed high-temperature C_P values are also in good agreement with the thermodynamic assessment of Jacobs and de Jong (2005) up to 1500 *K* (Fig. 20.47b). For higher temperature, the DFT/B3LYP computation gives values intermediate between those of Jacobs and de Jong (2005) and Watanabe (1982) (refitted by Fei and Saxena 1986). The Haas-Fisher polynomial expansion (Table 20.49) reproduces computed heat capacities with a mean error of 0.6 $J/(mol \times K)$. The summation of χ-squared residuals over 28 values is 4.63×10^{-4}. The B3LYP predicted entropy at 1 *bar*, 298.15 *K* is 85.7 $J/(mol \times K)$, which is close to the value of 86.4 ± 0.4 $J/(mol \times K)$ determined by Akaogi et al. (2007), and somewhat lower than the

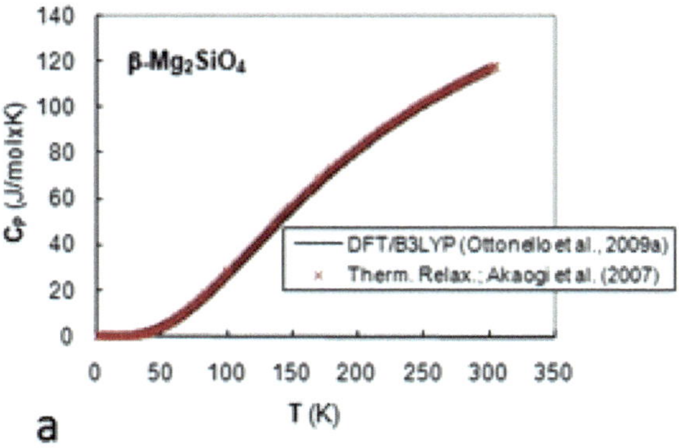

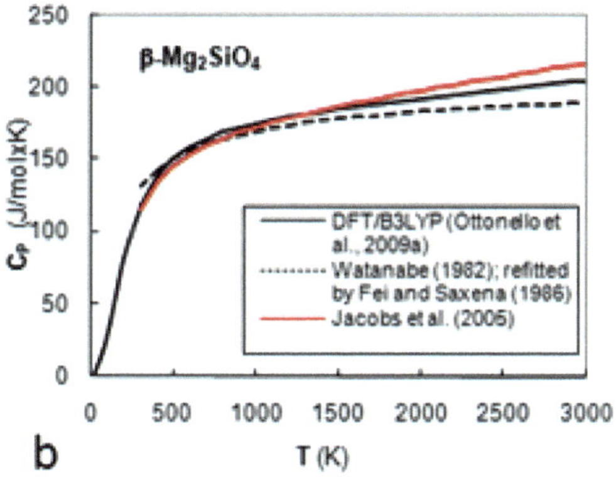

Fig. 20.47 **a** Ab initio low-T isobaric heat capacities of Wadsleyite compared with the thermal relaxation experiments of Akaogi et al. (2007). **b** Ab initio high-T isobaric heat capacity of Wadsleyite compared with the experiments of Jacobs and de Jong (2005) and Watanabe (1982), as refitted by Fei and Saxena (1986)

88.7 $J/(mol \times K)$ predicted by Wu and Wentzcovitch (2007). The third law entropy values determined by Ottonello et al. (2009a) and by Akaogi et al. (2007) are significantly lower than those of Jacobs and de Jong (2005) [91.3 $J/(mol \times K)$] and Saxena et al. (1993) [93.0 $J/(mol \times K)$].

20.4.1.3 Ringwoodite (γ-Mg_2SiO_4)

The γ-polymorph, which is also known as *Ringwoodite*, crystallizes in the cubic system (crystal class Cubic Hexakisoctahedral, Space Group $Fd\overline{3}m$). This polymorph occurs most likely in the Earth's mantle transition zone, delimited by the two seismic discontinuities at 520 and 660 *km* depth. It is practically an *inverse spinel*, with Si occupying the tetrahedral site. There are 14 atoms in the unit cell and 3 irreducible atoms in the conventional cell. Only oxygen is in general position. Based on the ab initio molecular dynamics investigation of Haiber et al. (1997), Ottonello et al. (2009) assigned provisional cell edge (a_0) and oxygen fractional coordinate (u) in a wide pressure range, from a nominally negative pressure of –40 *GPa* (corresponding to a virtual expanded state) to 110 *GPa*. For each pressure condition

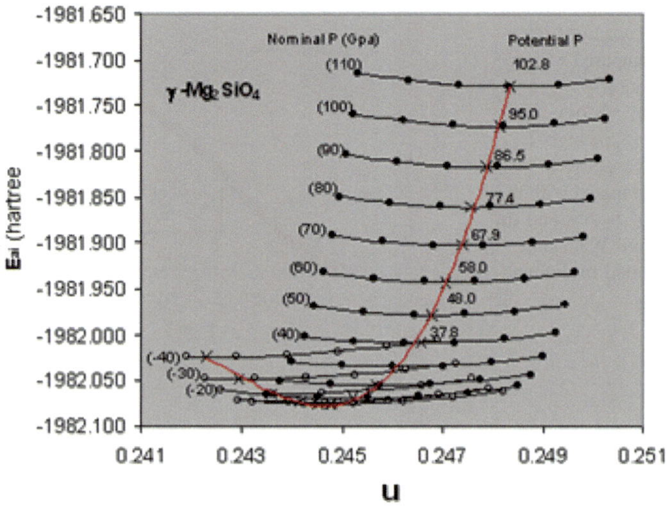

Fig. 20.48 Ab initio electron energy (E_{ai}) of Ringwoodite as a function of internal coordinates. Nominal pressure (values in brackets) and ab initio static potential pressure are indicated on some of the potential curves. Solid circles are virtually compressed states and open circles are virtually expanded states. Crosses locate the results of Bernie optimization. The red line conform an E_{ai}−u potential well

(and corresponding cell edge), the fractional coordinates of oxygen (u) were varied within the range ±0.003 of the initially assigned value to account for the effect of internal strain on the electronic energy of the substance (Fig. 20.48). A geometry optimization on each selected (a_0, u) couple allowed them to further relax the internal coordinates and to determine the potential well of the substance.

On the basis of the partial derivative of a third-order polynomial fitted over 17 molar B3LYP total energy (E_{B3LYP}, *J/mole*)–molar volume (V, *J/bar*) couples, it was established the ab initio potential value of static pressure, P_{st}, corresponding to each selected (a_0, u) (see also Fig. 20.48) and it was located the zero pressure condition. The resulting molar volume V_0 at the a thermal limit was 3.9967 (*J/bar*) (a_0 = 8.0975 *Å*). A final geometry optimization on the conformed minimum (a_0 = 8.097477 *Å*, u = 0.244678) resulted in a B3LYP total energy (gfw basis) E_{B3LYP} = −991.037853 *Hartree*. The obtained B3LYP volumes are intermediate between LDA (Haiber et al. 1997; Kiefer et al. 1999; Yu and Wentzcovich 2006) and GGA (Piekarz et al. 2002) data. However, when compared with the available experimental data (Hazen et al. 1993; Weidner et al. 1984; Ridgen and Jackson 1991), ab initio computations seem to overestimate the cell edge of Ringwoodite for pressures up to ~60 *GPa* and to underestimate it at higher pressures (Fig. 20.49a). Also the oxygen positional parameter is basically overestimated with respect to observations (Fig. 20.49b).

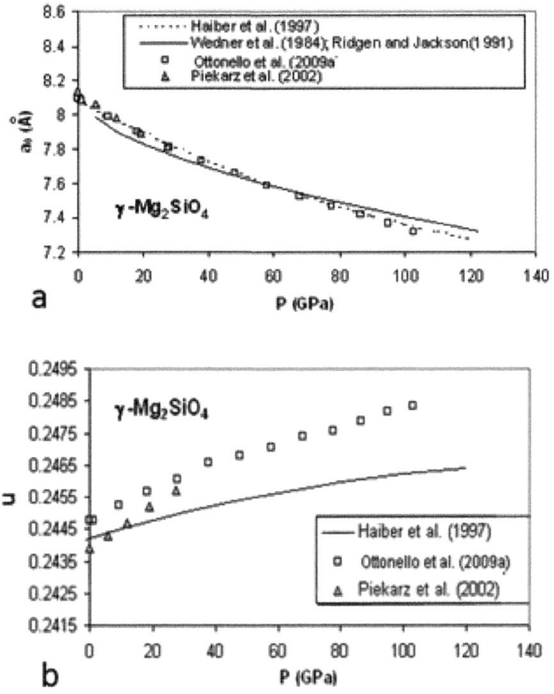

Fig. 20.49 *Ab initio* results for γ-Mg_2SiO_4 compared with literature theoretical estimates and experiments. **a** cell edge (*Å*) versus pressure (*GPa*). **b** Oxygen positional parameter (adimensional) versus pressure (*GPa*)

Application of the Birch–Murnaghan equation through non-linear minimization returned a zero point bulk modulus K_0 = 196.389 *GPa* roughly consistent with the extrapolation to the athermal limit of the experimental values measured at room temperature by Meng et al. (1994) (K_0 = 182 *GPa*), Weidner et al. (1984) (K_0 = 183 *GPa*) and Jackson et al. (2000) (K_0 = 185 *GPa*) (Table 20.39), and a pressure derivative $\partial K/\partial P = K' = 4.322$, which is close to the value measured by Meng et al. (1994); K′ = 4.2.

In the Cubic Hexakisoctahedral unit cell of γ-Mg_2SiO_4 there are 14 atoms corresponding to 39 true vibrational modes. The irreducible representation of vibrational modes, with a group theoretical treatment of the optical zone-center (gamma point; **k** = 0) yields

$$\Gamma = 6E_u + 1B_u + 2E_g + 1A_g + 3F_{1g} + 9F_{2g} + 12F_{1u} + 6F_{2u} \qquad (20.4.7)$$

In Table 20.40 are listed the vibrational frequencies obtained by Ottonello et al. (2009a) and their mode-gamma analysis. The results compare well with the prediction by Yu and Wentzcovitch (2006) at the LDA level of theory and available experimental data (Chopelas et al. 1994).

The mode-gamma analysis of ab initio frequencies yields an average value of αK of 5.01×10^{-3} *GPa/K* at high temperature, which is somewhat larger than the αK value of 4.54×10^{-3} *GPa/K* above the Debye temperature that was determined by

Table 20.39 Equation of State parameters of Ringwoodite

V_0 (*Å*)	K_0 (*GPa*)	K′	References
Calculated			
533.5	189.0	4.08	Haiber et al. (1997) (static)
497.0	206.8	4.0	Kiefer et al. (1999) (static, *K*′ assumed)
539.1	176.0	–	Piekarz et al. (2002) (static)
	189.4	4.33	Jacobs et al. (2006) (static)
527.5	184.6	4.5	Yu and Wentzcovitch (2006) (300 K)
530.9	196.4	4.322	Ottonello et al. (2009a) (static)
531.7	193.3	4.322	Ottonello et al. (2009a) (298.15 K)
Experimental			
–	182.6	5.0	Meng et al. (1993) (static)
523.9	190.3	4.1	Meng et al. (1994) (static, extrapolated)
526.7	182.0	4.2	Meng et al. (1994) (300 K)
526.5	–	–	Katsura et al. (2004) (300 K)

Table 20.40 Ab initio optical frequencies for ringwoodite at Γ point and their mode-gamma (Ottonello et al. 2009a). Acoustic frequencies at $\mathbf{k}_{max}$ and their mode gammas are from Chopelas et al. (1994)

N	Mode	*v* (*cm*)	γ_i
1	Acoustic	138	0.5
2	Acoustic	155	0.5
3	Acoustic	253	1.08
4–6	F_{2u}	264.5	1.8
7–9	F_{2g}	317.7	1.5
10–12	F_{1u}	341.8	1.7
13–15	F_{1g}	362.3	1.0
16–17	E_g	385.1	0.9
18–19	E_u	424.6	2.0
20–22	F_{1u}	438.2	1.1
23–25	F_{2u}	459.6	1.4
26–28	F_{1u}	555.0	1.3
29	E_u	580.5	1.0
30–31	B_u	581.4	0.7
32–34	F_{2g}	633. 8	0.6
35–37	F_{2g}	798.3	1.3
38–40	F_{1u}	810.8	1.2
41	B_u	814.6	1.1
42	A_g	835.2	1.1

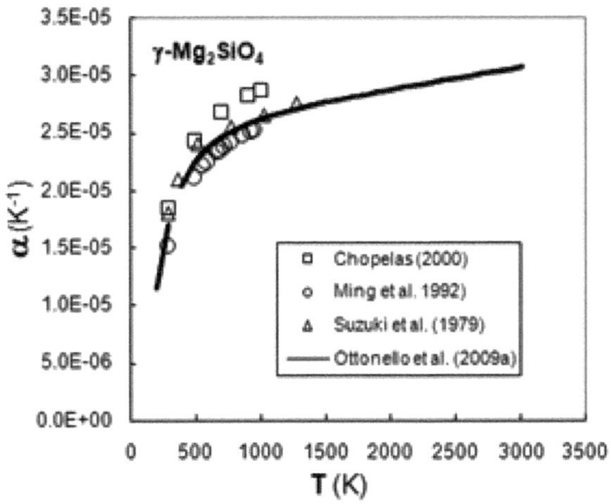

Fig. 20.50 Thermal expansion of Ringwoodite. Results of semi-classical calculations (solid line) are compared with the experimental data of various authors (symbols)

Meng et al. (1993) on the basis of in situ high P–T–X diffraction studies. Using the limiting value of the αK product for T → ∞ (51.052 *bar/K*) the ab initio bulk modulus at the athermal limit (196.389 *GPa*) and its isothermal pressure derivative (K′ = 4.322) it is obtained a value of –0.010 *GPa/K* for the isobaric thermal derivative of the bulk modulus of γ-Mg_2SiO_4 much lower than the value –0.028 ± 3 given by Meng et al. (1993). However, as shown in Fig. 20.50, the thermal expansion obtained by coupling the αK product and the thermal derivative of K is in close agreement with the experiments of Suzuki et al. (1979) and Ming et al. (1992).

The isobaric heat capacities of ringwoodite, calculated following the same procedure adopted for the α and β polymorphs is shown in Fig. 20.51. The computed low-temperature C_P values are again in good agreement with the calorimetric data of Akaogi et al. (2007) (Fig. 20.51a), while the high-T values are in good agreement with those of Chopelas et al. (1994), and Watanabe (1982) (Fig. 20.51b). Haas–Fisher coefficients of polynomial fitting of high-T ab initio heat capacity are listed in Table 20.49. The polynomial function reproduces computed heat capacities with a mean error of 0.6 *J/(mol × K)*. The summation of χ-squared residuals over 28 values is 5.71×10^{-5}.

The DFT/B3LYP third law entropy at 1 *bar*, 298 *K* is 80.2 *J/(mol × K)*, which is close to the recent third law value of 82.7 ± 0.5 *J/(mol × K)* determined by Akaogi et al. (2007) from calorimetric C_P data, but is significantly lower than the estimates of Fei et al. (1990) [89.0 *J/(mol × K)*] Jacobs and de Jong (2005) [88.1–91.8 *J/(mol × K)*, depending on the adopted volume estimates] and Yu and Wentzcovitch (2006) [85 *J/(mol × K)*].

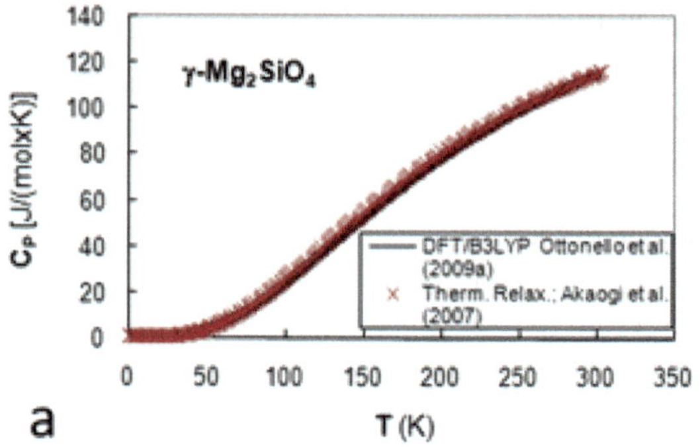

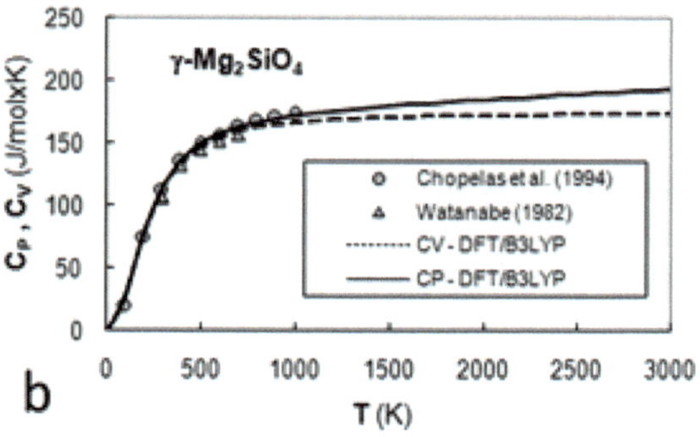

Fig. 20.51 **a** Ab initio low-T isobaric heat capacities of Ringwoodite compared with the thermal relaxation experiments of Akaogi et al. (2007). **b** Ab initio high-T isochoric and isobaric heat capacities of Wadsleyite compared with the experiments of Watanabe (1982) and Chopelas et al. (1994)

20.4.1.4 Phase Transitions

The univariant equilibria governing the relative stabilities of forsterite (α), wadselyite (β) and ringwoodite (γ) in the P–T space:

$$\underset{\text{forsterite}}{Mg_2SiO_4} \Leftrightarrow \underset{\text{wadsleyite}}{Mg_2SiO_4} \tag{20.4.8.1}$$

$$\underset{\text{wadsleyite}}{Mg_2SiO_4} \Leftrightarrow \underset{\text{ringwoodite}}{Mg_2SiO_4} \tag{20.4.8.2}$$

are established in terms of all the thermodynamic and thermophysical parameters of the intervening phases, i.e. besides the changes of Gibbs free energies (ΔG), enthalpies (ΔH), entropies (ΔS) and volumes (ΔV) of these reactions at the standard state (pure end-member phase at T = 298.15 *K* and P = 1 *bar*) also the isobaric heat capacity, the thermal expansion, the bulk modulus and its thermal and baric derivatives for all phases.

Table 20.41 lists electronic energy, zero point energy (ZPE) and thermal corrections of the three Mg_2SiO_4 polymorphs obtained at DFT/B3LYP theory level. Adding the PV product to the thermal correction (Therm-U) one obtains the thermal correction to enthalpy of the substance (Therm-H, same table). Finally, subtracting

Table 20.41 DFT/B3LYP electronic energy, zero point energy of optical modes (ZPE) and thermal corrections (P = 1 *bar*; T = 298.15 *K*; inclusive of acoustic branches) of the investigate phases. Data are in *Hartree* and on a gram formula weight basis. Values in brackets obtained with scaled frequencies (Ottonello et al. 2009a)

Phase	E_{ai}	ZPE	Therm U	Therm H	Therm G
γ-Mg_2SiO_4	−991.037852777	0.022732965	0.028510379 (0.028269282)	0.028511923 (0.028270826)	0.019406396 (0.018991326)
β-Mg_2SiO_4	−991.04275389	0.022582204	0.028672634 (0.028436488)	0.028674236 (0.028438091)	0.018964233 (0.018522545)
α-Mg_2SiO_4	−991.55719925	0.022503083	0.028845553 (0.028607884)	0.028847277 (0.028609613)	0.018366887 (0.01792209)

the TS product at the same P, T to Therm-H one has the thermal correction to the Gibbs free energy (Therm-G, same table). Adding thermal corrections to electronic energies ($E_{ai,crystal}$) and summing up algebraically the various terms in reaction (according to reaction stoichiometry) one obtains directly enthalpy and Gibbs free energy of reaction at standard state of reference (pure phases at P = 1 *bar* and T = 298.15 *K*) without the need of assessing enthalpy and Gibbs free energy of formation from the elements, which are subjected to great uncertainty, due to difficulties in estimating the appropriate electronic energy for isolated gaseous atoms (see Ottonello et al. 2007 for a more detailed account of the problem). The procedure is particularly apt for polymorphic transitions where reactants and products differ only in the structural state.

Some thermodynamic and thermophysical parameters obtained semi-classically on the three polymorphs are listed in Tables 20.42 and 8.5, respectively. Values in brackets in Table 20.42 are relative to "*scaled*" frequencies. The need of a scaling factor between ab initio harmonic frequencies and experimental observation is widely recognized (Scott and Radom 1996; see also Sect. 7.8). Application of the scaling affects to some extent the ab initio vibrationally dependent thermodynamic parameters, restituting discrete values more consistent with direct observations and/or thermodynamic inferences, but is virtually ineffective on relative stabilities and univariant P–T loci.

Table 20.42 Ab initio volume (*J/bar*), entropy [*J/(mol × K)*] and Haas-Fisher isobaric (1 *bar*) heat capacity coefficients for the Mg_2SiO_4 polymorphs (Ottonello et al. 2009a). Valus in parentheses obtained with scaled frequencies

Phase	$V^0_{298.15}$	$S^0_{298.15}$	$a \times 10^{-2}$	$b \times 10^{2}$	$c \times 10^{-6}$	$d \times 10^{5}$	$e \times 10^{-2}$
γ-Mg_2SiO_4	4.0026 (3.925)	80.183 (81.715)	1.5363 (1.5346)	2.1642 (2.1405)	−4.7897 (−4.7490)	−0.31483 (−0.30796)	1.0957 (1.2708)
β-Mg_2SiO_4	4.1530 (4.0495)	85.662 (87.316)	1.5854 (1.6139)	2.3753 (2.2270)	−4.2454 (−4.1112)	−0.27587 (−0.24731)	−0.39472 (−0.86514)
α-Mg_2SiO_4	4.4794 (4.395)	92.452 (94.114)	1.5555 (1.5513)	2.1449 (2.1307)	−4.4249 (−4.4060)	−0.08423 (−0.07918)	0.63793 (0.87536)

In estimating the molar volume at T = 298.15 *K*, P = 1 *bar* one usually starts from the static volume at the athermal limit, neglecting the contribution of vibrational pressure which arises from thermal energy at zero point. This vibrational contribution to the zero point volume can be estimated through application of the approximate Born-Huang relation (Born and Huang 1954):

$$\frac{V_0 - V_{0,st}}{V_{0,st}} = \frac{9}{8} \times \frac{\gamma n R\, \Theta_D}{V_0 K_0} \tag{20.4.9}$$

where Θ_D is the elastic or "*characteristic*" Debye temperature obtained from acoustic frequencies and the other symbols have their usual significance.

As we may see in Table 20.43 the contribution of the zero-point vibrational terms to the zero point volume is non-negligible. In the case of Wadsleyite the increase of 0.040 *J/bar* computed by Ottonello et al. (2009a) is in substantial agreement with the estimates of Jacobs et al. (2006) (i.e. 0.045 *J/bar*). Unfortunately the predicted volume shifts further worsen the bias between ab initio DFT/B3LYP all-electron predictions with respect to experimental evidences. This fact, and the fact that Eq. (20.4.9) obtained factoring out the discrete mode gammas of the individual acoustic branches is only a crude approximation of a more complex reality (the electronic energy arising from the acoustic modes is roughly 1% of the bulk zero point energy of the substance; cf. Tables 20.41 and 20.43) lead Ottonello et al. (2009a) to disregard the vibrational contribution to the zero-point volume.

Gibbs free energy and enthalpy of reaction are readily obtained by summing up algebraically the electronic energies and the thermal corrections of the two polymorphs at equilibrium. The Clapeyron (P–T) slope of a reaction at any state is simply the ratio $\Delta S/\Delta V$ at that state.

The data summarized in Table 20.44 show that the computed enthalpy changes of the phase transitions are comparable to the experimental data and estimates found in the recent literature. In the case of β/γ transition the agreement with the high-temperature drop-solution calorimetric datum of Akaogi et al. (2007) is quite good, especially considering the fact that the transition enthalpy is ~4.8 *ppm* of the bulk energy of the two polymorphs at equilibrium. We also note from the data in Table 20.44 that the frequency scaling does not significantly affect the standard

Table 20.43 Contribution of acoustic branches to the zero point volume of the investigated substances. Elastic Debye temperatures and thermal energies are also listed. From Ottonello et al. (2009a)

Phase	Θ_D	E^0_{ac} (*Hartree*)	V^0 (*J/bar*)	ΔV (*J/bar*)
Forsterite	750	0.00028566	4.5154	0.0472
Wadsleyite	896	0.00034127	4.1837	0.0402
Ringwoodite	880	0.00033517	4.0168	0.0201

Table 20.44 Thermodynamics of α/β and β/γ univariant equilibria. All-electron ab initio predictions (Ottonello et al. 2009a) are compared with. Values in brackets retrieved by graphics from the original referenced work. The Clapeyron slope at standard state is nothing else than the ratio between entropy of reaction and volume of reaction

Magnitude	α/β transition	β/γ transtition	Notes
ΔV (*J/bar*)	−0.313	−0.089	Saxena et al. (1993)
	−0.332	−0.080	Jacobs et al. (2006)
	−0.326	−0.150	Ab-initio
	−0.345	−0.125	Ab-initio; V optimized
ΔS [*J/(mol × K)*]	−1.110	−4.000	Saxena et al. (1993)
	−1.040	−4.800	Jacobs et al. (2006)
	−6.790	−5.479	Ab-initio
	−6.798	−5.601	Ab-initio; scaled ω
ΔH (*kJ/mol*)	32.1	8.8	Saxena et al. (1993)
	(29.1)	(7.6)	Jacobs et al. (2006)
	33.649	12.442	Ab-initio
	33.592	12.429	Ab-initio scaled ω
ΔG (*kJ/mol*)	32.4	9.992	Saxena et al. (1993)
	(29.4)	(9.03)	Jacobs et al. (2006)
	35.611	14.029	Ab-initio
	35.619	14.099	Ab-initio; scaled ω
Slope (*MPa/K*)	3.55	4.494	Saxena et al. (1993)
	3.2	6	Jacobs et al. (2006)
	2.080	3.643	Ab-initio
	1.965	4.401	Ab-initio; *V* optimized
	1.968	4.499	Ab-initio; optimized *V* and scaled ω

state Gibbs free energy and enthalpy changes for the β/γ transition, and its Clayperon slope, which are in good agreement with the estimates of Akaogi et al. (2007). For the α/β transition the ab initio standard state ΔG is close to the optimized value of Saxena et al. (1993), but the Clapeyron slope is more consistent with the data of Akaogi et al. (2007).

A direct comparison with experiments in Fig. 20.52 shows that the bias between the results of ab initio computations and direct experiments is around 2*GPa* for the α/β equilibrium. Scaling on volumes restitutes a Clapeyron slope consistent with observations at T, P of experiments.

For the β/γ transitions raw ab initio predictions are offset of roughly 4 *GPa* with respect to the most recent experiments (Fig. 20.53). Volumes optimization restitutes a Clapeyron slope consistent with the experiments of Kuroda et al. (2000) and with the recent reversal experiments of Inoue et al. (2006).

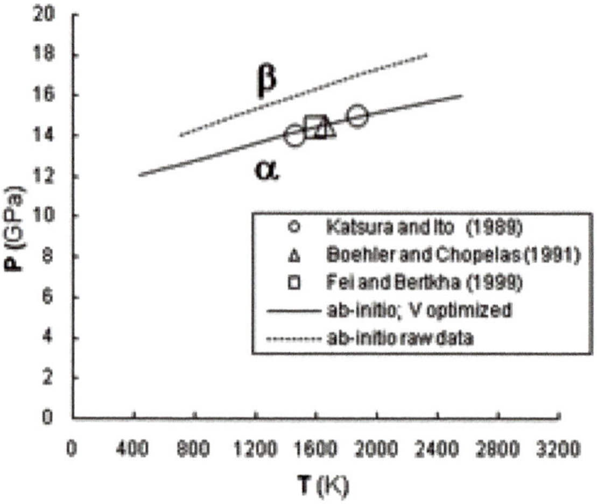

Fig. 20.52 The univariant equilibrium between α and β polymorphts in the P–T space. Results of all-electrons ab initio calculations (Ottonello et al. 2009a) are compared with experimental observations. Agreement with experiments is obtained by scaling the standard state volumes of the α and β polymorphs, respectively, to 4.395 and 4.0495 *J/bar*

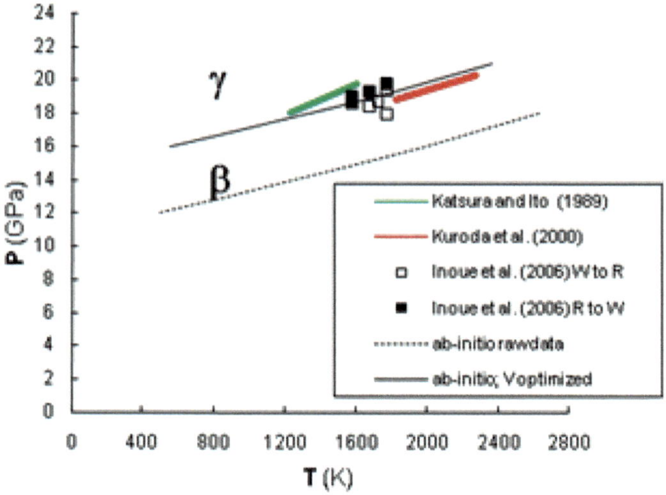

Fig. 20.53 The univariant equilibrium between β and γ polymorphts in the P–T space. Results of all-electrons ab initio calculations (Ottonello et al. 2009a) are compared with experimental observations. Agreement with experiments is obtained by scaling the standard state volumes of the β and γ polymorphs, respectively, to 4.0495 and 3.925 *J/bar*

20.4.2 Fe_2SiO_4

Fe_2SiO_4 is the second most abundant component in natural olivines and, as seen for Mg_2SiO_4, also for Fe_2SiO_4 are observed polymorphic transitions taking place with the increase of P. These transitions however do not include a β-polymorph analogous to β-Mg_2SiO_4 but rather a direct passage $\alpha \rightarrow \gamma$ (*Fayalite* → *Ahrensite*). In Situ (Diamond Anvil) Observation of the $\alpha \rightarrow \gamma$ transition by Synchrotron Radiation (Yagi et al. 1987) locate the univariant equilibrium along the boundary P (*GPa*) = 2.75 + 2.5×10^{-3} T (°*C*). According to Greenberger et al. (2011) at room T and P > 30 *GPa*, γ-Fe_2SiO_4 transform into a pseudo-rhombohedral polymorph. According to Yamanaka et al. (2015), the transition at room T takes place at ~34 *GPa* and the high-P polymorph is a body centered orthorhombic phase (*I*-Fe_2SiO_4) with space group *Imma* and Z = 4. At low T, both α and γ polymorps undergo Para-Magnetic (PM) to Anti-Ferro-Magnetic (AFM) lambda transitions, nicely registered in their isobaric heat capacity spectra (Fig. 20.54).

20.4.2.1 Fayalite (α-Fe_2SiO_4)

Fayalite crystallizes in the orthorhombic system (space group *Pbnm*). A peculiarity of α-Fe_2SiO_4 is pleochroysm: the mineral is pale yellow when the incoming light is parallel to a and b, and orange yellow when it is parallel to c. This peculiarity was

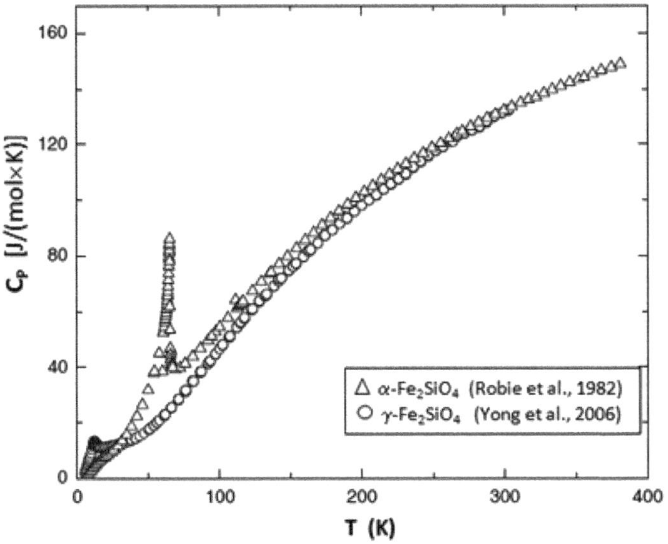

Fig. 20.54 Low-T isobaric heat capacity of α- and γ-Fe_2SiO_4 polymorphs. The Neel transition takes place at 65 *K* in Fayalite (Santoro et al. 1966; Ehremberg and Fuess 1993) and at 11.8 *K* in Ahrensite (Yong et al. 2007). Modified from Yong et al.(2007)

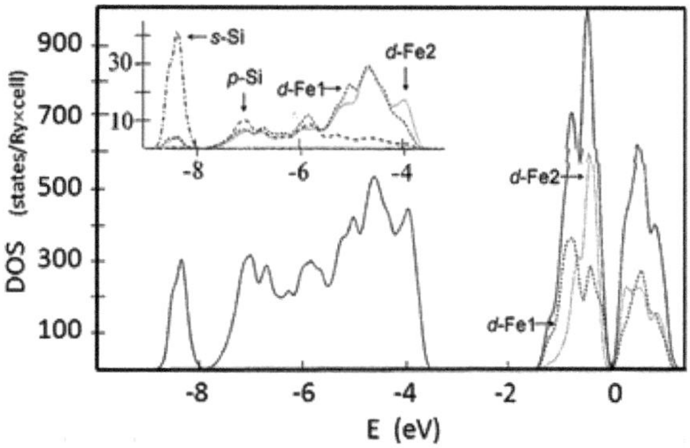

Fig. 20.55 Total and the partial *l*-projected DOS curves for α-Fe_2SiO_4 after Krasovska et al. (1997). The solid line is the total DOS; the dotted dashed line represents the s states of Si, the long-dashed line the p states of Si, the dashed line the d states of Fe1 (Fe_{M1}) and the dotted line the d states of Fe2 (Fe_{M2}). On the right of the figure are shown the *l*-projected partial DOS curves. Modified from Krasovska et al. (1997)

presumably at the origin of the first modern ab initio investigation of the substance (Krasovska et al. 1997). Calculations were carried out at the DFT/LDA theory level with LAPW (see Sect. 12.6.2) and the Hedin–Lundqvist exchange correlation potential (Hedin and Lundqvist 1971). The muffin-tin radii were 0.79 *Å* for O, 0.87 *Å* for Si and 1.22 *Å* for Fe. The (touching) spheres comprised 35% of the unit cell. The electron wave function (generated by a homemade Fortran code) was a linear combination of 1397 energy-independent APWs and 68 localized functions of angular momenta up to $l_{max} = 2$ for Fe, and up to $l_{max} = 1$ for O and Si. The longest reciprocal lattice vector used in the APW set was 5.11 times the adopted muffin-tin radius. The energy band structure of α-Fe_2SiO_4 was calculated over the energy region from the O 2s states to 13 *eV* above the valence band. All eigenenergies in the interval of interest were converged to within 5 *mRy*. Integration over the BZ was carried out with the tetrahedron method of Lehmann and Taut (1972), with a mesh of 115 **k** points (320 tetrahedra) in the irreducible part of BZ (corresponding to 1/8 of the BZ). The results in terms of total and *l*-projected partial DOS curves are shown in Fig. 20.55.

The energy zero in the figure is taken at the valence-band maximum. The calculation of Krasovska et al. (1997) predicts Fe_2SiO_4 to be an insulator with an indirect forbidden gap of 0.25 *eV* that separates occupied Fe t_{2g} states from unoccupied Fe e_g states. The valence band maximum was found at the special point S $[S = 2\pi(0.5/a, 0.5/b, 0)]$ and the conduction band minimum was seen at one-half the distance between Γ and X $[X = 2\pi(0.5/a, 0, 0)]$. The top of the valence band in reciprocal space is an almost dispersionless band in the direction X-S and the bottom of the conduction band is also a flat band in the direction Γ–X. The different symmetries of the M1 and M2 sites translates into different shapes of the partial 3d DOSs of Fe1 and Fe2.

Table 20.45 Energies (*eV*) of the Fe^{2+} 3d levels calculated by the Xa-SW method (Krasovska et al.,1997) and with the model of Burns (1993)

	Fe in M1 (C_i)		Fe in M2 (C_s)	
	Xα-SW	Burns (1993)	Xα-SW	Burns (1993)
e_g	1.568 (a_g)	1.185	1.334 (a′)	0.942
	1.171 (a_g)	0.813	1.097 (a″)	0.889
t_{2g}	0.000 (a_g)	0.000	0.000 (a′)	0.000
	−0.122 (a_g)	−0.099	−0.043(a″)	–
	−0.286 (a_g)	−0.186	−0.174 (a’)	−0.207

Johnson-Slater Xα-SW calculations (Johnson 1966; Slater amd Johnson 1972; see also Sect. 19.1) were then carried out for two $[FeO_6]^{10-}$ clusters to investigate with a better precision the effect of the ligand environment on the energy levels of the central cation. The calculations were carried out adopting the experimental atomic coordinates of Fujino et al. (1981). The effective potential was obtained self-consistently as a muffin-tin potential with non-overlapping muffin-tin spheres. The Hedin–Lundqvist exchange correlation potential was replaced with the potential of von Barth and Hedin (1972) (see Sect. 10.4). The radii of Fe, O and of the Watson's sphere were set, respectively, to 1.19, 0.98, and 3.12 *Å*. The potential inside Watson's sphere was shifted by ~0.5 *Ry* with respect to the potential outside the sphere in order to avoid the unphysical hybridization with electronic states localized outside the sphere. The symmetry of the M1 site was lowered from O_h to C_i, with an ensujing splitting of the t_{2g} and e_g states into non-degenerate {a_g, a_g, a_g} and {a_g, a_g} states, respectively. The M2 symmetry was lowered to C_s, with a consequent splitting of the t_{2g} and e_g states into the singly degenerate {a′, a″, a′} and {a″, a′} states, respectively. The calculations gave results consistent with the predictions of Burns (1993), namely: (1) the Fel t_{2g} levels of Fe in M1 split equidistantly; (2) the upper t_{2g} levels of Fe in M2 (a″ and a′) are very close; (3) the splitting of e_g levels is larger for Fe in M1 with respect to Fe in M2 (see Table 20.45).

Further detailed calculations for clusters of O_h symmetry, varying the Fe–O distance, R, from 2.0 to 2.55 *Å*, showed that the splitting ΔE (*eV*) decreases regularly as a power function of the interionic distance R (*Å*):

$$\Delta E = \frac{C}{R^n} \tag{20.4.10}$$

with n = 4.35, C = 41 and an internal reproducibility of 0.015 *eV* (see Fig. 3 in Krasovska et al. 1997). As stressed by the authors, this relationship between interionic distance and crystal field splitting energy, besides confirming the anticipations of McClure (1965) (but see also Sect. 8.10.6), "*is of fundamental importance in transition-metal Geochemistry and in the derivation of the crystal-field–related thermodynamic properties at high pressures*" (Krasovska et al. 1997).

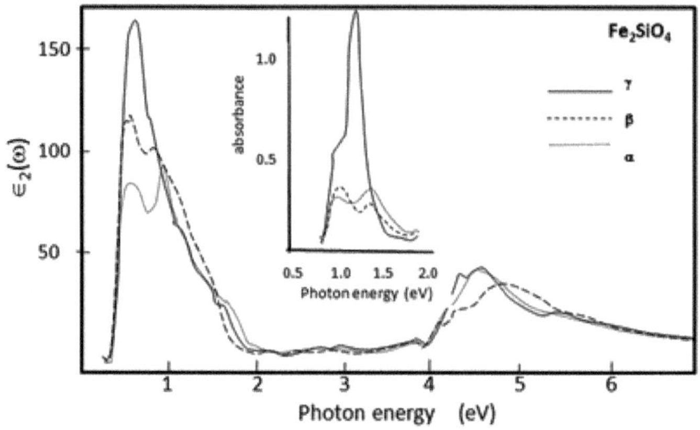

Fig. 20.56 Imaginary part of the dielectric function of fayalite for three light polarizations (Krasovska et al. 1997) compared with the experimental absorbance spectra of Burns (1970) (inset). From Krasovska et al. (1997). With permission from the Mineralogical Society of America

Real $[\epsilon_1(\omega)]$ and imaginary $[\epsilon_2(\omega)]$ parts of the frequency-dependent dielectric function (DF) which determine the absorption coefficient:

$$\alpha(\omega) = \frac{2\omega}{c}\sqrt{\frac{\left[\epsilon_1^2(\omega) + \epsilon_2^2(\omega)\right]^{1/2} - \epsilon_1(\omega)}{2}} \tag{20.4.11}$$

where then calculated by Krasovska et al. (1997), respectively, with the Kramer–Kronig analysis $[\epsilon_1(\omega)$; see references in Krasovska et al. 1997] and by the SCF one-particle approach of Ehrenreich and Cohen (1959) (see Eq. 1 in Krasovska et al. 1997). The theoretical $\epsilon_2(\omega)$ curves of Krasovska et al. (1997) are in good agreement[15] with the experimental observations of Burns (1970) (Fig. 20.56). Namely: (1) the most intense γ spectrum, arising from transitions at the M2 site, has a single well-defined maximum; (2) the α and β spectra (mainly arising from transitions at the M1 site) both have two maxima, with the low-energy maximum more intense in β; (3) the observed downward energy shift (0.4 *eV*) is a typical LDA failure. The only notable difference between theoretical and obsterved spectra is the low-energy shoulder in the γ spectrum which is not resolved in the theoretical curve.

A detailed DFT investigation of the structural, electronic and magnetic properties of α-Fe_2SiO_4 in a plane-wave pseudopotential framework was later presented by Cococcioni et al. (2003) with Local Spin Density (LSDA), and spin-polarized General Gradient Approximation (σ-GGA) procedures. For Fe and O were adopted the Vanderbilt pseudopotentials (Vanderbilt 1990) and for Si was adopted a

[15] I consider the adjective "*qualitative*" adopted by the Authors as too crude, for the complexity of the investigated phenomena.

norm-conserving pseudopotential. Plane waves were expanded to a 31 *Ry* kinetic energy cutoff limit for the electronic states and to 248 *Ry* for the augmented charge density. For GGA it was adopted the PBE exchange correlation functional (see Sect. 10.5.4). The irreducible part of BZ was sampled with 16 special points. Due to the metallic nature of the substance, the Fermi distribution was smoothed with a smearing technique (Methfessel and Paxton 1989) and a broadening width of 10 *mRy*. The structure relaxation at the athermal limit gave lattice parameters and fractional coordinates consistent with the experimental observations for both functionals (see Tables II, III in Cococcioni et al. 2003). According to the investigation of Cococcioni et al. (2003) the Anti Ferro Magnetic (AFM) ordering between corner sharing octahedral involves a superexchange mechanism through oxygen p orbitals. AFM coupling is effective between corner sharing octahedral (Fe–O–Fe angle close to 180 degrees) while a weak FM interaction is observed between edge sharing octahedral (Fe–O–Fe angle close to 90°).

In Fig. 20.57 are shown the σ-GGA band structure and the projected DOS of α-Fe_2SiO_4, after Cococcioni et al. (2003). All the electronic bands originate from iron d level. Two iron d-bands appear to cross the Fermi level, in spite of the fact that experimentally the structure is insulating. As a result, the DOS is not zero at E_F.

According to Cococcioni et al. (2003) the insulating behavior of Fayalite has a Mott–Hubbard origin, i.e. the average on-site repulsion among electrons around the Fermi level is much larger than the relevant band width impeding thus conduction.

Background information: Fayalite as *a Mott–Hubbard insulators*
As already seen at Sect. 10.7, a "*Mott insulator*" (de Boer and Verwey 1937; Mott and Peierls 1937; Mott 1949) is a substance that, based on its band structure is expected to conduct electricity but conduction is prevented by the strong (Coulomb) repulsion between electrons at the limit between valence and conduction band. The insulating state is attained usually at low T and the passage from conducting to insulating state is known as "*Mott transition*". The simplest model able to describe quantitatively the Coulomb energy necessary to open an insulating gap is the Hubbard model (Hubbard 1963). In the Hubbbard model (Sect. 10.7) the Hamiltonian H is composed of two terms: a kinetic term associated to the electron hopping between sites and a potential energy term describing the on-site interaction:

$$\mathsf{H} = -\mathsf{t}\sum_{\mathrm{i},\sigma}\left(c^{+}_{\mathrm{i},\sigma}c_{\mathrm{i}+1,\sigma} + c^{+}_{\mathrm{i},\sigma}c_{\mathrm{i}+1,\sigma}\right) + \mathsf{U}\sum_{\mathrm{i}}\mathbf{n}_{\mathrm{i}\uparrow}\mathbf{n}_{\mathrm{i}\downarrow} \tag{20.4.12}$$

t in Eq. (20.4.12) is the hopping integral, U is the electron repulsion in canonical quantization terms,[16] and $n_{i\sigma} = c^{+}_{i\sigma} c_{i\sigma}$ is the spin-density operator

[16] With wave functions treated as field operators (Dirac 1957).

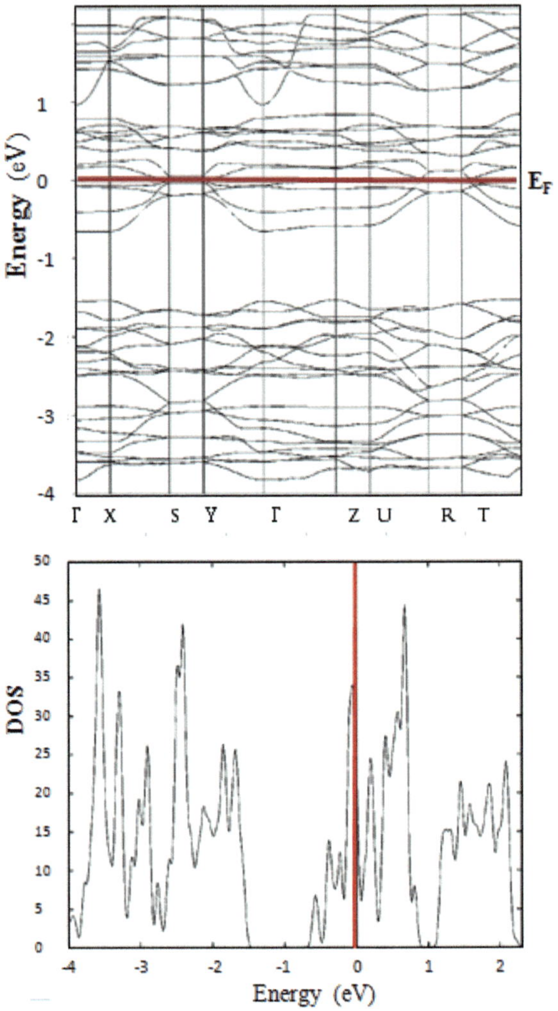

Fig. 20.57 The electronic band structure originated from the iron d bands (upper part) and the projected DOS (lower part). The Fermi level is marked in red. Modified from Cococcioni et al. (2003)

for spin σ on the ith site. In the formalism of Cococcioni et al. (2003), the Hubbard model for the Fayalite Mott insulator is

$$\mathsf{H} = -\sum_{\langle i,j\rangle}\sum_{l,l'} t_{ij} c^{+}_{i,l} c_{i,l'} + \frac{U}{2}\sum_{i}\sum_{l}\sum_{l'\neq l} n_{i,l} n_{i,l'} \tag{20.4.13}$$

with t_{ij} as hopping amplitude between two nearest iron atoms along the same hopping chain and U as average on-site repulsion. $c^{+}_{i,l}$ and $c_{i,l'}$ in Eq. (20.4.13) are, respectively, creation and annihilation fermionic operators on the lth orbital of the ith site, and $n_{il} = c^{+}_{il} c_{il}$ are the corresponding occupation numbers. It is worth to stress that in this formalism the spin degrees of freedom does not appear because only the three t_{2g} minority spin levels (of identical spin) are considered, and the physics of the electrons around the Fermi level is dictated by only two competing parameters: the hopping term and the on-site repulsion.

At the DFT mean-field level of approximation it may be further assumed that (Cococcioni et al. 2003).

$$E_{DFT} \approx \langle H \rangle_{MF} = T[\langle n \rangle] + \frac{N_s U}{2} \frac{(m-1)}{m} \langle n \rangle^2 + E_x[\langle n \rangle] + E_{bg} \quad (20.4.14)$$

with T = average kinetic energy;
m = 3 = number of states per site (t_{2g} levels on each Fe ion);
$\langle n \rangle$ = average electronic occupation of each iron site;
N_s = 8 = number of iron sites in the unit cell;
E_x = exchange term;
E_{bg} = background energy.

Assuming E_x to be approximately contained in the Hartree term, the effective repulsive potential is Cococcioni et al. (2003):

$$U_{eff} = U_{Coulomb} - J \quad (20.4.15)$$

The electronic response to long-wavelength external perturbing potential is however almost insensitive to the background E_{bg}. Comparing thus the electronic compressibility obtained from the DFT calculation (i.e. the curvature of the total energy with respect to long wavelength density fluctuations), since for a charged Fermi liquid the compressibility only depends on the short-range part of the screened electron–electron interaction (which can be identified with the localized U parameter). Cococcioni et al. (2003) finally posed:

$$U = \frac{m}{(m-1)} \left[\frac{\partial \mu}{\partial \langle n \rangle} - \frac{1}{\nu(0)} \right] \quad (20.4.16)$$

where $\frac{\partial \mu}{\partial \langle n \rangle}$ is the linear variation of the Fermi energy, when the number of electron in the unit cell is varied and a uniform compensating background is implicitly added in order to keep the system neutral, and $\frac{1}{\nu(0)}$ is the inverse of the electronic DOS at the Fermi level. This quantity is "*nothing but the same quantity for the noninteracting case that must be subtracted from the interacting result, in order to isolate the contribution originating from short-range interaction*" (Cococcioni et al. (2003). The value $U = 4.5\ eV$ obtained computing the variation of $\langle n \rangle$ defined as the average d-level occupancy per iron sites (as obtained from the projected density of states) is much larger than the calculated electronic band width around the Fermi level ($\sim 1.5\ eV$; see also Fig. 20.57). This implies that the electrons in the minority

t_{2g} bands "*do not have enough kinetic energy to overcome the repulsion they experience on each iron site and thus a Mott localization occurs giving rise to the observed insulating behavior*" (Cococcioni et al. 2003). We will see later on another practical application of DFT + *U*, in the case of Ahrensite.

More recently, DOS, DF, $\in_2(\omega)$ and $\alpha(\omega)$ of α-Fe_2SiO_4 were reinvestigated by Xiao et al. (2018) focusing the attention on the effects of pressure up to 30 *GPa*. The study was carried out at GGA and GGA + *U* theory level with the PBE functional for exchange–correlation. The initial structure adopted a = 4:819 *Å*, b = 10:478 *Å*, and c = 6:087 *Å*, as internal coordinates, with *Pbnm* space group. Coulomb Integrals and Hund's exchange were set to 4.5 and 0.9 *eV*, respectively. In the PAW procedure, PW were expanded to a cut-off limit of 500 *eV* and the convergence criterion was set at 10^{-4} *eV*. The point sampling in BZ was carried out with a 4 × 4 × 4 mesh. The path of the band structure in the irreducible part of BZ was along the **k** points Γ, Z, T, Y, S, X, U, and R.

Xiao et al. (2018) found an anisotropic compressibility of α-Fe_2SiO_4, with the linear compressibility along the b-axis significantly higher that compressibility along a and b (in agreement with the experimental observations of Zhang et al. 2017). Fitting to a third-order Birch–Murnaghan EoS gave K = 129.89 *GPa* (at P = 0 and T = 298.15 *K*) and K′ = 4.18 (K = 135 *GPa*, K′ = 4.0 according to Zhang et al. 2017).

In Fig. 20.58a, b are shown the band structures of α-Fe_2SiO_4 calculated by Xiao et al. (2018) for room pressure and P = 30 *GPa*, respectively. In agreement with the results of Krasovska et al. (1997), the substance is seen to behave as an insulator (at both pressures), though with a higher indirect band gap decreasing progressively from ~1.5 *eV* at P = 0 to ~1.0 at P = 30 *GPa*. The valence band maximum is located in this study (at both pressures) at the Γ point (0, 0, 0), while the minimum in the valence band is observed at the special point T (−0.5, 0, 0.5). Figure 20.58c, d show the (partial and total) DOSs of α-Fe_2SiO_4 obtained by Xiao et al. (2018) at the limiting pressures 0 and 30 *GPa*, respectively. There is a close correspondence in the two investigations in identifying the band energies near the Fermi level as due mainly to the 3d states of iron (with no apparent distinction between M1 and M2 sites in Xiao et al. 2018; cf. Figures 20.58c and 20.55).

The imaginary part of the complex Dielectric Function (DF) of α-Fe_2SiO_4 was computed by Xiao et al. (2018) with basically the same procedures of Krasovska et al. (1997) but exhibits two Gaussian-shaped curves with maxima at 2.81 and 8.05 *eV* differing significantly form the previous investigation. The reason of these discrepancies is to me not clear. According to Xiao et al. (2018) these peaks shift progressively to lower energy as pressure increases.

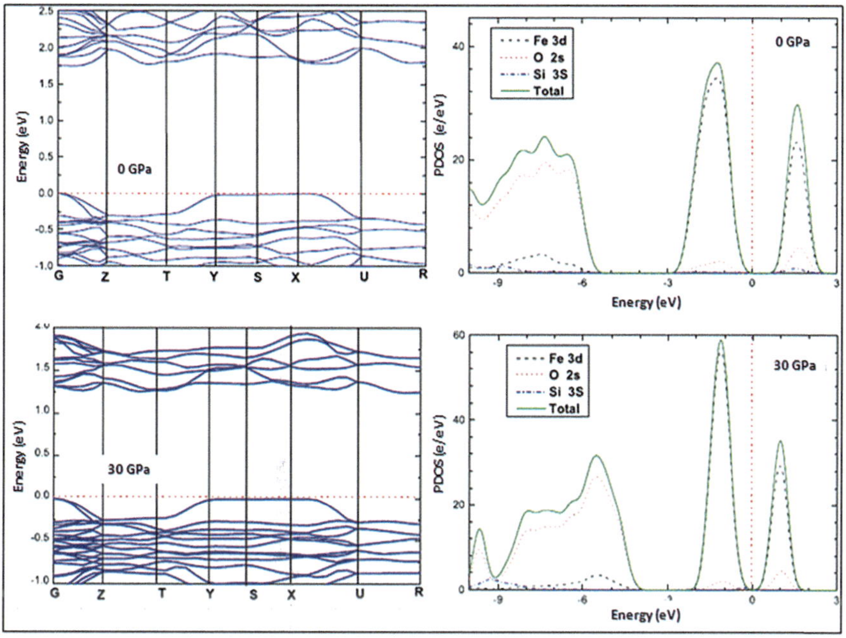

Fig. 20.58 **a**, **b** Band structures of α-Fe_2SiO_4 at 0 and 30 *GPa*, respectively, after Xiao et al. (2018). The special points are, in this study, denominated as follows: G = (0. 0. 0), Z = (0, 0, 0.5), T = −0.5, 0, 0.5), Y = (−0.5, 0.5, 0), S = (0, 0.5, 0), X = (0, 0.5, 0), U = (0, 0.5, 0.5) and R = (−0.5, 0.5, 0.5). **c**, **d** total and partial DOSs at 0 and 30 *GPa*, respectively. The Fermi level is identified by the red dashed line. Assembled from various figures in Xiao et al. (2018)

20.4.2.2 Ahrensite (γ-Fe_2SiO_4)

As anticipated, α-Fe_2SiO_4 trasforms directly into the spinel γ (*Ahrensite*), without the intermediate β-form observed in Mg_2SiO_4. γ-Fe_2SiO_4 crystallizes in space group $Fd\overline{3}m$. It was first synthesized by Ringwood (1958) at T = 600 °*C* and P = 3.8 *GPa*. Natural occurrences of pure Ahrensite were then found in the Unbarger L6 chondrite (Xie et al. 2002) and in the Tissint Martian meteorite (Ma et al. 2016) where they formed by shock melting on fayalite-rich rims of olivine megacrysts. γ-Fe_2SiO_4 has been the object of accurate ab initio inversigations by Derszi et al. (2009, 2011). In the first study, focused on its electronic, magnetic and structural properties and lattice dynamics at the athermal limit and P = 20 *GPa* it was shown that Ahrensite is an Anti Ferro Magnetic (AFM) Mott insulator. In Fig. 20.59 it is shown the structure of Ahrensite, as represented by Derszi et al. (2009), to put in evidence the magnetic details of the substance. The dark and light-red spheres are, respectively, spin-down ($Fe_\downarrow$) and spin-up ($Fe_\uparrow$) iron atoms. These atoms form a tetrahedron. In case of AFM ordering two spin-down Fe atoms are associated with a spin-up atom (or vice-versa).

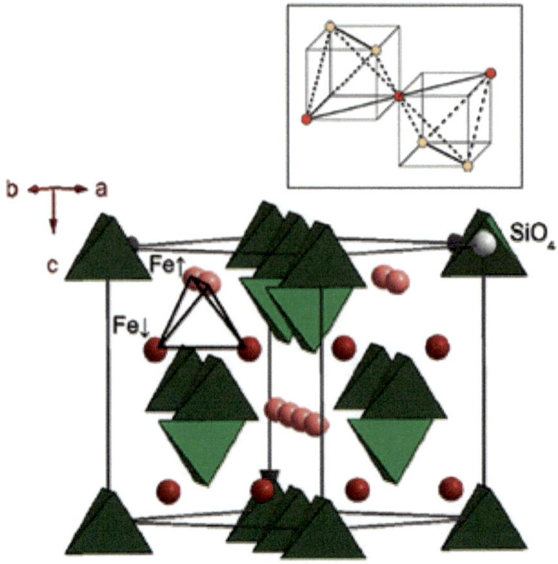

Fig. 20.59 Crystal structure of γ-Fe_2SiO_4 in the AFM configuration. The unit cell is contoured by lines. Chained structures of Fe atoms form in the ab planes. Spin-up and spin-doww Fe atom are identified by red and pink color, respectively. In the inset is shown a schematic representation of nearest-neighbor Fe–Fe spin interactions. Dashed lines represent the AFM bonds and full lines represent the FM bonds. The Fe atom common for both cubes is involved in NN spin interactions along six bonds. From Derszi et al. (2009). With permission from The American Physial Society

Because the electronic interactions between Fe atoms results in a rather strong coupling between spin, charge, orbital, and lattice degrees of freedom, the standard DFT procedures fail to reproduce the observed electronic structure and magnetic properties (Derszi et al. 2009). Spin-polarized DFT-PBE calculations (Sect. 10.5.4) were thus carried out with the PAW method under both GGA (see Sect. 8.8.3) and GGA + U. Practical computing was carried out with the Vienna package (Kresse and Furthmuller 1996). The plane waves kinetic cutoff was set at 500 eV. Sampling of **k** points in the irreducible part of BZ was built on a $4\times4\times4$ Monkhorst–Pack grid. Calculations were performed in the $1\times1\times1$ supercell (56 Atoms: 8 Si, 16 Fe and 32 O). The Coulomb integral U was set to 4.5 eV as in Fayalite (Cococcioni et al. 2003), in order to take into account the on-site Coulomb interactions between 3d electrons (Oleś et al. 2005; and references therein). The Hund's exchange was set to $J_H = 0.90\ eV$. The phonon frequencies for the AFM state were determined using the direct method (cf. Sect. 8.8). Forces on the individual atoms were constrained to be lower than $7 \times 10^{-6}\ eV/\AA$ before stopping the optimization. LO-TO splitting at the Γ point was determined non-analytically (Pick et al. 1970). Effective charges were approximatel by the nominal ionic values for all atoms. The static dielectric constant was set to $\varepsilon = 3.5$ (as experimentally observed in Fayalite; Suto et al. 2002).

Table 20.46 Effect of the Coulomb integral on the structural parameters (a, b, V), enthalpy (H) and magnetic moment (μ) of the FM and AFM states of γ-Fe_2SiO_4 (Derzsi et al. 2009). The AFM structure is tetragonal, hence, also the c edge is shown

Ferromagnetic (FM)					Antiferromagnetic (AFM)				
U (*eV*)	a (Å)	V (Å³)	H (*eV*)	μ (μ_B)	a (Å)	c (Å)	V (Å³)	H (*eV*)	μ (μ_B)
0.0	8.0266	517.124	−363.222	3.571	8.0350	8.0383	518.963	−360.213	3.522
2.0	8.0343	518.614	−354.679	3.603	8.0895	8.0094	524.135	−354.672	3.556
3.0	8.0729	526.125	−348.920	3.632	8.1041	8.0004	525.438	−349.140	3.615
4.5	8.0783	527.181	−343.130	3.684	8.1120	7.9968	526.222	−343.468	3.679
5.5	8.0779	527.103	−339.969	3.716	8.1113	7.9984	526.240	−340.305	3.714

In the AFM configuration the magnetic moments are all parallels in the ab plane and antiparallel along the c direction (Fig. 20.59). This cause reduction from the cubic fcc to the tetragonal bct symmetry (*I*4_1/*amd*). The distortion, which is small under GGA (0.04%), becomes sensible at GGA-*U* level where a direct relation with the imposed value of *U* may be seen in the values of cell edges, unit cell volume, enthalpy and magnetic moment (Table 20.46).

Inclusion of the *U* parameter renders the AFM state energetically more stable than FM at $U > 2\ eV$ and at the expected *U* (4.5 *eV*; Derzsi et al. 2009). The increase in the lattice parameter is smooth in the AFM phase, as, in this case, electron localization is gradual, in contrast to the FM phase. The enegy gap between AFM and FM is 0.338 *eV*. The magnetic interaction is however rather weak and the two states appear almost degenerate at all *U*, which explains the limited λ-feature at the Néel temperature (Derzsi et al. 2009).

The main qualitative change in the crystal structure (and in the magnetic ground state) takes place in the range $2 \leq U\ (eV) \leq 3$. Both FM and AF states behave as metals for $U \leq 2\ eV$, and as insulators for $U \geq 3\ eV$. In the metallic regime the finite DOS around the Fermi level is due entirely to the 3d Fe orbitals. There are four distinct bands, shaped by crystal-field and exchange interactions and extending from −4 *eV* below the Fermi level up to about 2 *eV* above it (Fig. 20.60a). With the introduction of the Coulomb repulsion between electrons ($U = 4.5\ eV$) a gap of ~2 *eV* opens up in the t_{2g} bands and the structure becomes insulating (Fig. 20.60a). The effects of crystal-field splitting and exchange–correlation are schematically depicted in Fig. 20.60b. The splitting dictated by the $[FeO_6]^{10-}$octahedra eliminates the degeneracy of 3d orbitals (Sect. 8.10.6) into t_{2g} and e_g. Both groups are further splitted by the exchange term (E_{EX}) into spin-up (stabilized) and spin-down states (unstabilized). Due to large Hund's exchange interaction, all five lower 3d states [i.e. (t_{2g} + e_g), spin up] are occupied by electrons with majority spins, and only one higher t_{2g} state is occupied by opposite (spin down) minority spin. In the metallic state ($U = 0$) the Fermi level crosses the upper energy minority spin t_{2g} states.

As anticipated, the Coulomb repulsion between electrons removes the degeneracy of t_{2g} orbitals already at $U \sim 2.5\ eV$ (i.e. well below the realistic value $U = 4.5\ eV$) and orbital order occurs. As a consequence a gap opens and γ-Fe_2SiO_4 becomes a (Mott) insulator.

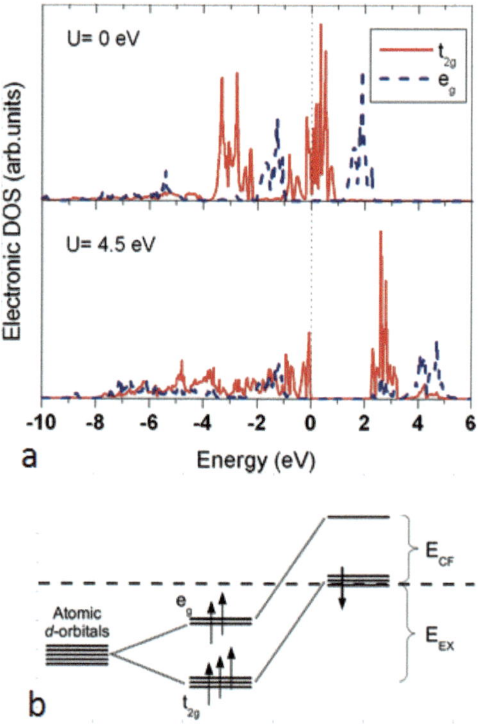

Fig. 20.60 **a** Partial DOS of 3d orbitals (red = t_{2g}; dashed blu = e_g) in γ-Fe_2SiO_4 in the absence ($U = 0$) and presence ($U = 4.5$ *eV*) of Coulomb repulsion. Dashed line marks the Fermi level. **b** Schematic splitting of the atomic 3d orbitals. E_{EX} represents the splitting due to exchange interactions and E_{CF} the effect of the crystal-field. From Derszi et al. (2009). With permission from The American Physial Society

The irreducible representation of vibrational modes, with a group theoretical treatment of the optical zone-center (gamma point; **k** = 0) yielded (Derszi et al. 2009)

$$\Gamma = 1A_g + 2A_{2u} + 4E_u + 2E_g + 12T_{1u} + 6T_{2u} + 3T_{1g} + 9T_{2g} \tag{20.4.17}$$

All E modes are doubly degenerate and all T modes are triply degenerate. A_{1g}, E_g, and T_{2g} modes are Raman active, T_{1u} modes are infrared active, and A_{2u}, E_u, T_{2u}, and T_{1g} are silent (Table 20.47).

In a later investigation Derszi et al. (2011) studied the effect of pressure on the structure and magnetic properties of Ahrensite. The structure was relaxed at various pressures in the range $-12 \leq P$ (*GPa*) ≤ 20. From the phonon spectra Derszi et al. (2011) derived (within the quasiharmonic approximation) thermodynamic properties in good agreement with the experimental observations. The AFM ordering was shown to induce a tetragonal distortion on the structure progressively less marked with the increase of P.

Table 20.47 Ab initio optical frequencies (cm^{-1}) for γ-Fe_2SiO_4 at Γ point for two values of *U* (Derszi et al., 2009)

d	*U* = 0	*U* = 4.5 *eV*	Symmetry	Activity
3	127.04	211.25 (213.43)	T_{1u}	Infrared
3	167.53	166.72	T_{2u}	Silent
1	180.68	315.54	A_{2u}	Silent
2	183.66	234.16	E_u	Silent
3	262.31	214.72	T_{1g}	Silent
3	292.55	351.92	T_{2g}	Raman
3	336.43	345.63 (412.90)	T_{1u}	Infrared
2	372.81	221.49	E_g	Raman
3	412.01	421.05	T_{2u}	Silent
3	492.75	506.63 (583.66)	T_{1u}	Infrared
2	519.13	534.13	E_u	Silent
3	551.23	555.99	T_{2g}	Raman
1	838.14	840.80	A_{2u}	Silent
3	856.37	872.66	T_{2g}	Raman
1	869.84	817.89	A_{1g}	Raman
3	872.98	876.77 (997.44)	T_{1u}	Infrared

Source of Data Derszi et al. (2009). d is the mode degeneracy. Values in parentheses are LO modes obtained from LO-TO splitting

20.5 Metasilicates

I will limit myself again to two representative components: $MgSiO_3$, $CaSiO_3$. These components, partially miscible at solid state, form, at sufficiently low P, *Pyroxenes* [from *pyro* (fire) and *xenos* (alloctonous)]. Pyroxenes occupy a position that is "*chemically central in the realm of rock compositions*" (Robinson 1982). They are therefore found ubiquitously and virtually in any kind of paragenesis developing in the P,T,X conditions typical of Earth's mantle and crust. Pyroxenes are polysilicates (*inosilicates* in the mineralogical classification), crystallizing in two classes: respectively, in the monoclinic (*clinopyroxenes*; spatial groups *C2/c*, $P2_1/c$, *P2/n*) and orthorhombic systems (*orthopyroxenes*; spatial groups *Pbca, Pbcn*). Chain periodicity *p* is 2 and defines the length of the c edge of the unit cell (about 5.3 Å). Infinite silicate chains of composition $[SiO_3]_n$ are formed due to bonding of a tetrahedral group $[SiO_4]$ with two neighbouring groups, through bridging oxygens (Fig. 20.61). In part A of the figure is shown the geometry of these chains, projected along the various axes and in perspective. Part B shows more generally the relative position of the chains in the structure: each chain has an apical and a basal front. Crystallographic sites located between two neighbouring basal fronts are larger (coordination VI or VIII with oxygen) and are generally defined as M2 positions. Sites between two neighbouring apical fronts are smaller (sixfold coordination) and are usually defined as M1 positions. Neighbouring chains may be

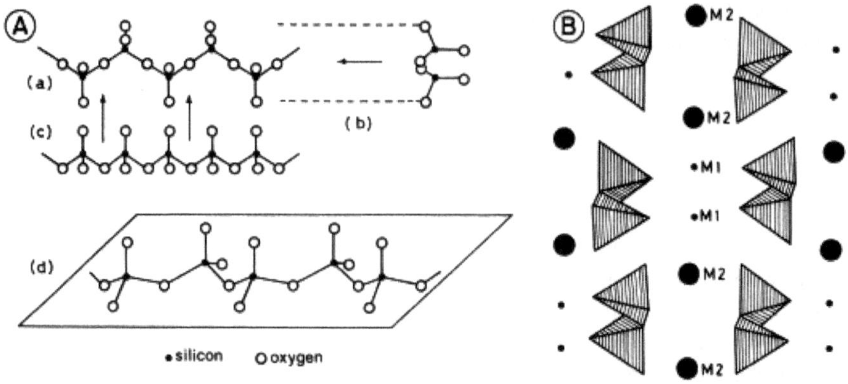

Fig. 20.61 **A** Schematic representation of $[SiO_3]_n$ chains of pyroxenes: (a) projection on plane (100): (b) projection along axis Z; (c) projection along axis Y; (d) perspective representation. **B** Relative arrangement of tetrahedral chains in pyroxene structure (seen in their terminal parts with intercalation of M1 and M2 positions. Modified from Putnis and McConnell (1980)

displaced relative to one other by tilting of the tetrahedra on the bridging oxygens, thus allowing incorporation of different kinds of metal cations in M2 sites, as a function of the degree of distortion.

20.5.1 $MgSiO_3$

$MgSiO_3$ undergoes various polymorphic transitions in the Earth's Mantle, dictated by the thermobaric regimes encountered at depth (i.e. Enstatite, Clinoenstatite, Protoenstatite, High-P clinoenstatite, Majorite, Akimotoite, Perovskite). In Fig. 20.62 I report the results of an ongoing *Ab initio-assisted* investigation (Belmonte and Ottonello, in prep) on the univariant loci of $MgSiO_3$ compared with experiments and thermodynamic assessments. The thermophysical and thermodynamic properties leading to the depicted phase diagrams are listed in Tables 20.48 and 20.49. As we may see, at low-intermediate P the agreement is satisfactory due to an ad hoc optimization of the thermophysical properties of the intervening substances, while at high P, there is a misfit in the triple point Majorite–Akimotoite–Perovskite. This misfit is more severe at DFT/LDA (Yu et al. 2011) with respect to DFT/GGA based assessment (Belmonte and Ottonello, in prep). The bias between assessment (Gasparik 2003) and ab initio predictions at high P could be however partly due to the fact that Majorite is difficult to isolate as a single crystal in high P–T experiments (due to systematic twinning).

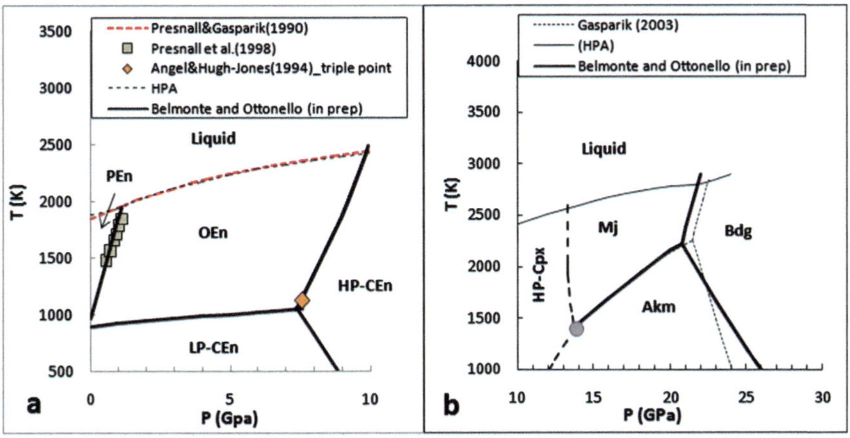

Fig. 20.62 Univariant equilibria and invariant points of $MgSiO_3$ polymorphs at low-intermediate (**a**) and high pressure (**b**). PE = Protoenstatite, LP-CEn = Low-P Clinoenstatite, HP-Cen = High-P clinoenstatite, Mj = Majorite, Akm = Akimotoite, Bdg = Bridgmanite, Filled circle = new invariant point

20.5.1.1 Low-P Clinoenstatite

Monoclinic, space group *P2₁/c*, the first structural refinement is due to Kuno and Hess (1953) (a = 9.618 Å, b = 8.828 Å, c = 5.186 Å, β = 71.38°; Z = 4), essentially confirmed few years later by Morimoto et al. (1959).

Duan et al. (2001) performed an AIMD Full structural optimization of low-P Clinoenstatite at various P adopting an innovative procedure proposed by Wentzcovitch (1991), Wentzcovitch and Martins (1991), i.e.the Damped Molecular Dynamics with Variable Cell Shape (VCS).

Background information: The Variable Cell Shape (VCS) procedure
In the VCS procedure the Bravais lattice "*deformations*" are included among the dynamical variables. The ionic coordinates ($\mathbf{r}_i$, i = 1,….,N) are rescaled as a function of the strain tensor

$$\mathbf{r}_i(\in, \mathbf{q}_i) = (1 + \in)\mathbf{q}_i \qquad (20.5.1)$$

The rescaled coordinates ($\mathbf{q}_i$) and the strain components are "*promoted to dinamical variables*" (Wentzcovitch 1991; Wentzcovitch et al. 1993) through the extended Lagrangian:

$$L = \sum_{i=1}^{N} \frac{m_i}{2} \mathbf{q}_i g(\in_i), \mathbf{q}_i + \frac{W}{2} \mathrm{Tr}\left(\in\in^T\right) - U(\in, \mathbf{q}_i) - P\Omega(\in) \qquad (20.5.2)$$

Table 20.48 Optimized thermophysical properties of crystalline polymorphs of $MgSiO_3$. V^0_{298} = standard state molar volume (T_r = 298.15 *K*, $P_r = 10^5$ Pa). K_0 = bulk modulus; $K' = (dK/dP)_{T=0}$. The volumetric thermal expansion at all T and P_r is reproduced by the following polynomial expansion on T: $\alpha_V = \alpha_0 T + \alpha_1 + \alpha_2 T^{-1} + \alpha_3 T^{-2} + \alpha_4 T^{-3}$

Composition (phase)	V^0_{298} (*cc/mol*)	K_0 (*GPa*)	K′	$(dK/dT)_P$ (*bar/K*)	$\alpha_0 \times 10^7$	$\alpha_1 \times 10^7$	$\alpha_2 \times 10^3$	α_3	α_4	References
$MgSiO_3$ (LP-clinoenstatite)	31.255	121.4	5.5	−210.0	0.07956	300.1	1.599	0	0	(1)
$MgSiO_3$ (Orthoenstatite)	31.33	105.8	8.5[a]	−206.0	0.02694	294.7	0	0	0.5588	(1)
$MgSiO_3$ (Protoenstatite)	32.42	107.0	6.6	−200.0	0.030	266.0	0	0	0	(1)
$MgSiO_3$ (HP-clinoenstatite)	30.337	126.3	5.5	−184.0	0.06784	303.3	1.782	0	0	(1)
$MgSiO_3$ (Akimotoite)	26.354	207.7	4.54	−279	0.10997	82.956456	17.990566	−8.0792	1019.1	(1,2)
$MgSiO_3$ (Majorite)	28.573	160.1	4.82	−237	0.1361365	24.239955	23.562897	−10.0127	1287.32	(1,2)
$MgSiO_3$ (Bridgmanite)	24.318	249.2	4.2	−280	0.0818	198.2	0	−0.474	0	(1,2)

References (1) Belmonte and Ottonello (in prep) (assessed data to reproduce LP-clinoenstatite/orthoenstatite, LP-clinoenstatite/HP- Clinoenstatite, orthoenstatite/protoenstatite and orthoenstatite/HP-clinoenstatite phase transition boundaries)
(2) Belmonte (2013)
Notes [a]$dK'/dT = 0.0015\ K^{-1}$ is assumed for Orthoenstatite

Table 20.49 Optimized thermodynamic properties of $MgSiO_3$ polymorphs. $H^0_{f,298}$ = enthalpy of formation from the elements at $T_r = 298.15$ K, $P_r = 10^5$ *Pa*; S^0_{298} = entropy of the substance at the same T, P conditions; C_P = isobaric heat capacity ($C_p = a + bT + cT^{-2} + dT^{-1/2} + eT^{-3} + fT^2 + gT^3 + hT^{-1}$). References as in Table 20.48

Composition (phase)	$H^0_{f,298}$ (*J/mol*)	S^0_{298} (*J/mol × K*)	a	b × 10^3	c × 10^{-5}	d	e × 10^{-8}	f × 10^6	g × 10^9	h × 10^{-4}	Referecnes
$MgSiO_3$ (LP-clinoenstatite)	−1,5488,081.0	63.400	184.91	−7.4034	29.272	−1808.7	−7.8593	0	0	0	(1)
$MgSiO_3$ (Orthoenstatite)	−1,545,552.0	66.170	133.75	−0.89325	−106.29	38.415	17.483	0	0	0	(1)
$MgSiO_3$ (Protoenstatite)	−1,544,308.0	67.438	132.86	0.7405	−104.36	50.46	16.941	0	0	0	(1)
$MgSiO_3$ (HP-clinoenstatite)	−1,509,046.0	65.800	172.290	5.2079	8.7667	1429.3	4.5571	0	0	0	(1)
$MgSiO_3$ (Akimotoite)	−1,509,046.0	54.013	112.54	12.291	−36.255	59.56	0	−0.35901	0	0	(1,2)
$MgSiO_3$ (Majorite)	−1,516,134.0	68.506	183.7	−17.173	−9.1451	−1507.3	0	4.6114	0	0	(1,2)
$MgSiO_3$ (Bridgmanite)	−1,457,192.0	58.698	110.12	9.5903	−38.879	192.59	0	0.99651	0	0	(1,2)

where $g(\in_i)$ is the metric tensor

$$g(\in) = (1+\in)^T(1+\in) \tag{20.5.3}$$

The first two terms on the right of Eq. (20.5.2) compose the kinetic energy $K(\in, \mathbf{q}_i)$ of particles of mass m_i measured in a frame which "*breathes*" with the cell (Wentzcovitch 1991) and is associated to internal and strain variables; w, in the same equation, is a fictitious mass (adjusted to optimize the dynamical coupling with the other variables); $U(\in, \mathbf{q}_i)$ is the Kohn–Sham (internal) energy, P is the applied pressure (external and constant), and $\Omega(\in)$ is the (variable) cell volume. The (combined) third and fourth terms correspond to the enthalpy.

At any imposed external pressure (P) the structure readjusts in order to achieve the minimum Energy (enthalpy). The trajectories are dictated by the Equation of Motion (EM; Wentzcovich et al. 1993):

$$\ddot{\mathbf{q}}_i = \frac{1}{m_i}(1+\in)^{-1}\mathbf{f}_i - \frac{\dot{g}}{g}\mathbf{q}_j \tag{20.5.4}$$

$$\ddot{\in} = \frac{\Omega\ (\Pi - P)}{w\,(1+\in^T)} \tag{20.5.5}$$

where $\mathbf{f}_i$'s are the Hellman–Feynman forces, and Π the total stress:

$$\Pi = \sum_{i=1}^{N} \frac{1}{\Omega} m_i \mathbf{v}_i \mathbf{v}_i^T + \Sigma \tag{20.5.6}$$

where

$$\mathbf{v}_i(\in, \dot{\mathbf{q}}_i) = (1+\in)\dot{\mathbf{q}}_i \tag{20.5.7}$$

are the (scaled) velocities and Σ is the quantum mechanical stress tensor (Nielsen and Martin 1985)

The first term on the right of Eq. (20.5.6) is the dynamical part of the stress tensor. In the absence of thermal fluctuations (i.e. $\mathbf{v}_i = 0$ for all i's) $\Pi = P$, implying conservation of symmetry and invariance of the Irreducible Brillouin Zone (IBZ) (and, consequently, of the **k** points along the trajectory). Self-consistent solutions of the Kohn–Sham equations are computed at each MD time step and the number of basis functions is variable, to complain with simulations involving a cell with variable shape and volume (Wetzcovitch and Martins 1991).

The LDA exchange–correlation energy was parameterized by Duan et al. (2001) through the Perdew-Zunger functional (Perdew and Zunger 1981). Norm conserving pseudo potentials with partial core corrections were generated with the Troullier–Martin scheme (Troullier and Martins 1991) and an appropriate number of plane waves, in order to get maximum truncation error on energies on the order of 80 *Ry*, was selected. The IBZ was sampled with a zone summation over two special **k**-points generated by a 2 × 2 × 2 Monkhorst–Pack grid. Calculations, carried out for $0 \leq P$ *(GPa)* ≤ 20, showed that the M2 octahedra remain most distorted over the investigated P regime, while M1 octahedra and the tetrahedral sites showed similar levels of distortions. A third-order Birch–Murnaghan fit of the generated volumes gave $V_0 = 30.08$ *cc/mol*, $K_0 = 135$ *GPa*, and $K' = 4.9$. The molar volume is slightly underestimated with respect to experimental observations ($V_0 = 31.28$ *cc/mol*; Angel and Hugh-Jones 1994) due to the usual LDA overbinding, and K_0 is overestimated, for the same reason ($K_0 = 111$ *GPa*, and $K' = 6.6 \pm 1.1$; Angel and Hugh-Jones 1994).

Yu et al. (2010) repeated the calculations of Duan et al. (2001) at DFT/LDA theory level, again with plane waves, and included also a DFT/GGA procedure with the PBE functional. The Troullier–Martins type pseudopotentials for Mg were replaced with von Barth–Car type pseudopotentials (Dal Corso et al. 1993). The plane-wave kinetic energy cutoff was set to 80 *Ry*, the Internal BZ sampling was based on a 4×4×4 Monkhors–Pack grid (shifted from the origin by 1/2, 1/2, 1/2). For high-pressure structural relaxations, it was adopted the VCS MD procedure. Phonon frequencies at Γ-point were obtained through density functional perturbation theory (DFPT; Baroni et al. 2001) on a 2×2×2 q point grid, later interpolated on a denser (8×8×8) q point mesh.

In Table 20.50 the (static) unit cell parameters of $P2_1/c$ clinoenstatite obtained by Yu et al. (2010) at DFT/LDA theory level are compared with previous computational and experimental results.

The cell parameters obtained by Yu et al. (2010) are in better agreement with experiments with respect to the previous calculations of Duan et al. (2001). The bulk modulus at T = 300 *K* is 121.4 *GPa* and its P derivative K′ is 5.5. The QHA thermal expansion at the same T conditions is $\alpha = 2.5 \times 10^{-5}\ K^{-1}$ and the QHA isobaric heat capacity is $C_P = 81.67$ *J/(mol × K)*.

Table 20.50 Unit Cell Parameters of $P2_1/c$ clinoenstatite at room conditions. XRD = X-Ray diffraction; MD = Molecular Dynamics

a (Å)	b (Å)	c (Å)	β (°)	V(Å³)	Method	References
9.606	8.8131	5.170	108.35	415.5	XRD	Ohashi (1984)
9.487	8.668	5.105	107.90	399.5	DFT/LDA	Duan et al. (2001)
9.544	8.755	5.140	107.78	409.0	DFT/LDA	Yu et al. (2010)
9.600	8.672	5.244	108.60	413.8	MD	Matsui and Price (1992)
9.45	8.68	5.09	108.0	397.1	MD	Jahn (2008)

Source of Data Yu et al. (2010)

Table 20.51 Raman-active frequencies (cm^{-1}) in $P2_1/c$ Clinoenstatite. Ab initio results of Yu et al. (2010) (1) are compared with the experimental evidences of Lin (2004) (2) The abbreviations for the experimental spectrum are the following: s, sharp; w, wide; vs, very sharp; m, medium

ν	ν	Mode	I	ν	ν		Mode	ν	ν	i	Mode
(1)	(2)			(1)	(2)			(1)	(2)		
125.8	121	m	A_g	340.5	344.0	s	B_g	535.4			A_g
148.8	143	m	A_g	342.0			A_g	547.4	557	w	A_g
153.2			B_g	349.3			B_g	570.2			B_g
154.0			A_g	363.4	371.0	ms	A_g	582.1	583	mw	B_g
157.8	158	m	B_g	371.3			B_g	653.8	666	s	A_g
191.1			B_g	371.7	388.0	w	A_g	675.0	689	vs	A_g
197.0	195	m	A_g	390.2	405.0	m	B_g	729.4			B_g
202.8	206	m	B_g	395.3	418.0	m	B_g	747.0	755	mw	B_g
202.9			A_g	414.7			B_g	833.4	848	mw	A_g
210.1			A_g	415.2			A_g	888.7			B_g
229.9			A_g	422.5	432.0	m	A_g	899.0			B_g
234.4	233	w	B_g	433.4			A_g	920.7			A_g
246.4	245	ms	B_g	440.6	453.0	w	B_g	922.8	927	mw	B_g
250.8			A_g	452.9			A_g	932.7			A_g
279.7	279	w	A_g	463.7	480.0	w	B_g	1007.6			A_g
292.0			B_g	485.7			B_g	1015.7	1012	vs	B_g
297.7			B_g	500.7			A_g	1035.7	1034	s	A_g
307.6	304	w	A_g	517.8	523.0	m	A_g	1046.3			A_g
327.0	324	mw	B_g	520.6			B_g	1050.6			B_g
329.6			A_g	527.5			B_g	1060.0			B_g

In the unit cell of $P2_1/c$ clinoenstatite there are 40 atoms (8 formula units), hence 120 optical modes. The symmetry assignment at Γ point gives:

$$\Gamma = 30A_g + 30B_g + 30A_u + 30B_u \qquad (20.5.8)$$

The (60) symmetric (g) modes are Raman-active and the 60 asymmetric (u) modes are IR-active. One A_u and two B_u are pure translations. The Raman-active and IR-active frequencies obtained by Yu et al. (2010) for room conditions are compared in Table 20.51 with the experimental results of Lin (2004). Table 20.52 lists the IR-active frequencies. The phonon dispersion and the projected DOS are shown in Fig. 20.63.

There are two clear gaps in the dispersion curves. The modes up to 600 cm^{-1} result from oxygen displacement of the shared corner in the tetrahedral chains, from rotation modes around the c-axis, from translation modes along the *c*-axis and from translation modes of Mg atoms (Yu et al. 2010). The modes between ~600 and ~800 cm^{-1} comprise O–Si–O bond-bending displacements. Above 1000 cm^{-1} stretching modes dominate (Yu et al. 2010).

Table 20.52 Ab initio IR-active frequencies (cm^{-1}) in $P2_1/c$ Clinoenstatite (Yu et al. 2010)

ν	Mode	ν	Mode	ν	Mode
126.5	A_u	351.6	A_u	529.1	B_u
155.6	B_u	360.2	A_u	549.0	B_u
203.0	A_u	374.0	B_u	557.6	A_u
204.8	B_u	377.3	A_u	636.7	B_u
216.4	B_u	390.2	A_u	669.4	B_u
218.5	A_u	416.5	B_u	714.6	A_u
224.4	A_u	418.7	A_u	728.7	A_u
240.4	B_u	419.8	B_u	832.8	B_u
244.4	A_u	425.0	B_u	872.7	A_u
262.4	B_u	429.0	A_u	894.4	A_u
284.0	B_u	453.4	A_u	914.9	A_u
285.2	A_u	462.8	B_u	923.8	B_u
292.2	A_u	464.8	A_u	935.1	B_u
297.9	B_u	471.9	A_u	995.6	A_u
302.7	B_u	476.8	B_u	1003.9	B_u
316.7	A_u	484.1	A_u	1024.9	A_u
323.8	B_u	502.0	B_u	1054.3	B_u
337.1	B_u	511.4	A_u	1073.2	A_u
350.6	B_u	528.1	A_u	1147.8	B_u

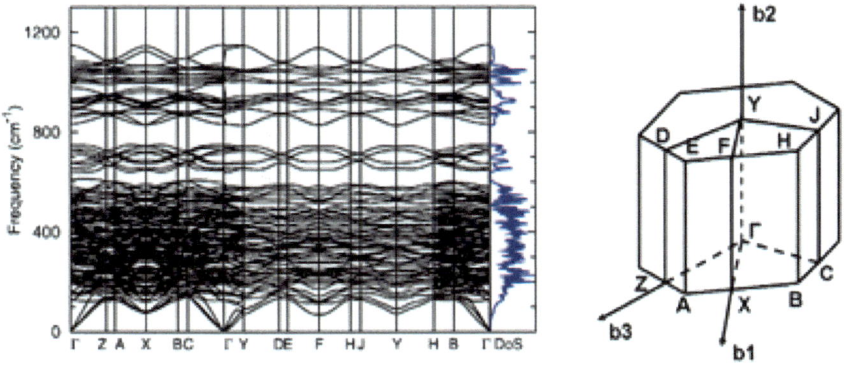

Fig. 20.63 Phonon dispersions and projected DOS of $P2_1/c$ Clinoenstatite at 0 GPa after Yu et al. (2010). In the rightward part of the figure is shown the indicized BZ. From Yu et al. (2010). With permission of The Mineralogical Society of America

20.5.1.2 Orthoenstatite

Orthorhombic, space group *Pbca* (a = 18.233 Å, b = 8.8191 Å, c = 5.1802 Å; Hugh-Jones and Angel 1994). The *Pbca* symmetry is achieved by twinning of the $P2_1/c$ unit cell, leading to a doubled unit cell volume. Orthoenstatite was investigated ab initio by Stangarone (2016), Stangarone et al. (2016). Computations were carried out with the CRYSTAL code. The adopted contracted basis set expansions were 8-511d1G for Mg, 88-31G* for Si and 8-411d11G for O. The adopted functional was the hybrid WC1LYP consisting in a 16% of the Wu-Cohen Exchange functional (Wu and Cohen 2006) and 84% of LYP for correlation (see Sect. 10.3.2). The threshold on the total energy for convergence of the self-consistent field was set to 10^{-8} Hartree for the orbital optimizations and to 10^{-10} Hartree for the construction of the Hessian. Reciprocal space was sampled using a regular sublattice with shrinking factor of 4, corresponding to 24 independent **k** vectors in the irreducible part of the Brillouin zone. The exchange–correlation contribution to the Fock matrix was evaluated by numerical integration over the unit cell volume. Radial and angular points for the integration grid were generated through Gauss–Legendre radial quadrature and Lebedev two-dimensional angular point distributions. A pruned grid with 75 radial and 974 angular points was used (Stangarone 2016). The relaxed volume at the athermal limit was V_0 = 31.90 *cc/mol*, the bulk modulus K_0 = 90.91 *GPa* and its baric derivative K′ = 10.86. Orthoenstatite has 16 formula units per unit cell, hence 240 optical modes. The symmetry assignment at Γ point gives (Stangarone 2016; Stangarone et al. 2016):

$$\Gamma = 30A_g + 30B_{1g} + 30B_{2g} + 30B_{3g} + 30A_u + 30B_{1u} + 30B_{2u} + 30B_{3u} \quad (20.5.9)$$

The (120) A_g, B_{1g}, B_{2g}, B_{3g} modes are Raman active. One B_{1u}, one B_{2u}, one B_{3u} are pure translations and the remaining (29 × 3) B_{1u}, B_{2u}, B_{3u} are IR active. The (30) A_u modes are silent. In Table 20.53 the Raman active mode frequencies obtained by Stangarone (2016) are compared with the experimental observations of Lin (2003). The mean absolute difference between experiment and computational data is $\langle|\Delta\nu|\rangle = 5\ cm^{-1}$. Similar mean absolute differences are obsterved when comparing the computational results with the experimental data of Stalder et al. (2009) ($\langle|\Delta\nu|\rangle = 4\ cm^{-1}$), Chopelas (1999) ($\langle|\Delta\nu|\rangle = 5\ cm^{-1}$), and Zhang et al. (2013) ($\langle|\Delta\nu|\rangle = 5\ cm^{-1}$).

In Fig. 20.64 the Raman spectrum obtained by Stangarone (2016), Stangarone et al. (2016) is compared with the experimental observations of Stalder et al. (2009). The unpolarized spectrum consists almost completely of A_g modes. In order to attain a better comparative appraisal the Raman pattern has been divided by the authors into five ranges (R1 to R5) and six subranges (a and b) (see Stangarone et al. 2016; for an estended discussion). The A_g symmetry modes are the most intense and useful for the purpose of mineral identification and spectroscopic

Table 20.53 Ab initio Raman active frequencies (cm^{-1}) for enstatite (Stangarone 2016) compared to observations (Lin 2003). The ab initio frequencies are rounded at integer values, although four digits are present in the original tables (see Appendix in Stangarone (2016))

(1)		(2)		(1)		(2)		(1)		(2)	
N	Sym	*N*	*i*	*v*	Sym	*v*	*i*	*v*	Sym	*v*	*i*
79	A_g	83	s	327	B_{3g}	327.7	–	550	B_{3g}	–	–
81	B_{2g}	106.8	w	336	A_g	–	–	552	A_g	551.5	m
108	B_{3g}	115.4	w	341	B_{2g}	340.3	–	553	B_{2g}	–	–
127	A_g	134.3	ms	341	B_{1g}	342.9	–	561	A_g	–	–
145	B_{3g}	–	–	343	A_g	343.5	svs	566	B_{3g}	580.3	m
153	B_{1g}	–	–	348	B_{1g}	–	–	594	B_{1g}	594.5	vw
158	B_{3g}	–	–	359	B_{2g}	–	–	610	B_{1g}	–	–
159	A_g	155.3	wm	364	B_{3g}	–	–	610	B_{2g}	–	–
160	B_{2g}	–	–	379	B_{1g}	374	–	659	B_{3g}	–	–
161	B_{1g}	161.5	vw	386	B_{3g}	–	–	668	A_g	663.8	svs
164	B_{2g}	–	–	387	A_g	383.8	wm	688	B_{3g}	-	-
166	B_{3g}	166.7	w	389	B_{2g}	–	–	693	A_g	686.1	vs
184	B_{1g}	190.1	–	395	B_{1g}	–	–	735	B_{1g}	–	–
193	A_g	–	–	401	B_{2g}	–	–	749	B_{2g}	–	–
195	B_{1g}	196.9	wm	406	A_g	402.1	m	759	B_{1g}	–	–
196	B_{3g}	–	–	406	B_{1g}	–	–	760	B_{2g}	750.7	w
201	A_g	201.1	–	412	B_{3g}	–	–	866	A_g	851.1	m
206	A_g	206.5	wsh	415	B_{2g}	–	–	866	B_{3g}	–	–
212	B_{2g}	–	–	422	B_{1g}	–	–	886	B_{1g}	–	–
224	B_{1g}	–	–	422	B_{3g}	–	–	894	B_{2g}	–	–
225	B_{3g}	–	–	426	A_g	421.7	m	906	B_{2g}	–	–
233	B_{3g}	–	–	426	B_{2g}	–	–	919	B_{1g}	–	–
236	A_g	238.2	ms	449	B_{2g}	–	–	924	A_g	926.6	m
238	B_{2g}	239	–	450	B_{3g}	–	–	932	B_{3g}	–	–
245	B_{1g}	–	–	451	A_g	445.5	w	933	A_g	935.2	w
248	B_{2g}	244.8	msh	458	B_{1g}	448	–	943	B_{2g}	–	–
263	B_{3g}	–	–	458	A_g	458.8	vw	945	B_{1g}	–	–
270	A_g	261	–	468	B_{1g}	–	–	957	B_{3g}	–	–
277	A_g	278.2	w	470	B_{3g}	–	–	1012	A_g	1011.3	vs
283	B_{1g}	–	–	474	B_{2g}	472.9	vw	1017	B_{2g}	–	–
285	B_{2g}	–	–	477	B_{2g}	–	–	1023	B_{1g}	–	–
296	B_{3g}	286.5	–	481	A_g	487	–	1024	B_{3g}	–	–
299	B_{1g}	–	–	489	B_{1g}	–	–	1026	B_{3g}	–	–
299	B_{3g}	–	–	498	B_{2g}	–	–	1025	B_{2g}	–	–
304	B_{2g}	–	–	522	B_{1g}	–	–	1027	A_g	1032.9	Svs
305	A_g	301.5	w	523	B_{3g}	–	–	1039	A_g	1034.7	–
311	B_{1g}	–	–	527	B_{2g}	–	–	1043	B_{1g}	–	–

(continued)

Table 20.53 (continued)

(1)		(2)		(1)		(2)		(1)		(2)	
N	Sym	*N*	*i*	*v*	Sym	*v*	*i*	*v*	Sym	*v*	*i*
311	B_{2g}	–	–	531	B_{1g}	–	–	1063	B_{2g}	–	–
318	B_{3g}	–	–	536	A_g	524.9	m	1067	B_{1g}	–	–
325	A_g	323.3	wsh	542	B_{3g}	540.8	–	1153	B_{3g}	–	–

(1) Lin (2003)
(2) The intensity in the spectrum after Lin (2003) is classified as strong (s), medium (m), weak (w) and shoulder (h)

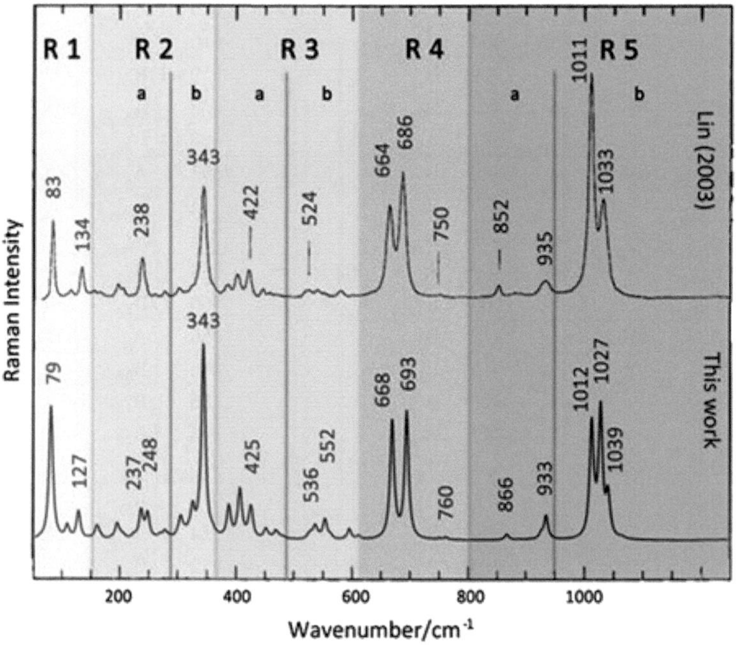

Fig. 20.64 Raman spectrum of Orthoenstatite. In the upper part is shown the experimental spectrum of Stalder et al. (2016) and in the lower part the ab initio derived spectrum of Stangarone (2016), Stangarone et al. (2016). From Stangarone et al. (2016). With permission from The Journal of Raman Spectroscopy

analysis. The assignment is extremely accurate and the comparison between experiment and calculus reveal all the potentiality of first principles in assisting spectral interpretation. For completeness I report also the IR active modes obtained by Stangarone (2016) in Table 20.54.

Table 20.54 Ab initio IR active frequencies (cm^{-1}) of Enstatite (Stangarone 2016)

N	v	Sym	N	v	Sym	N	N	Sym
1	87.48	B_{2u}	30	344.61	B_{3u}	59	549.38	B_{2u}
2	111.58	B_{3u}	31	344.81	B_{2u}	60	552.03	B_{3u}
3	138.18	B_{1u}	32	347.05	B_{3u}	61	565.56	B_{2u}
4	141.70	B_{1u}	33	349.41	B_{1u}	62	572.43	B_{3u}
5	150.53	B_{2u}	34	352.98	B_{1u}	63	580.71	B_{1u}
6	175.03	B_{1u}	35	369.10	B_{2u}	64	645.60	B_{2u}
7	182.97	B_{2u}	36	382.88	B_{3u}	65	654.35	B_{1u}
8	195.57	B_{3u}	37	388.78	B_{1u}	66	681.49	B_{2u}
9	196.17	B_{1u}	38	402.95	B_{3u}	67	699.55	B_{1u}
10	204.95	B_{2u}	39	407.43	B_{2u}	68	730.44	B_{3u}
11	216.71	B_{3u}	40	414.50	B_{1u}	69	752.58	B_{3u}
12	226.44	B_{3u}	41	417.18	B_{3u}	70	865.25	B_{2u}
13	228.04	B_{1u}	42	426.94	B_{2u}	71	865.30	B_{1u}
14	233.82	B_{2u}	43	430.98	B_{1u}	72	884.52	B_{3u}
15	241.63	B_{1u}	44	435.92	B_{3u}	73	911.69	B_{3u}
16	245.59	B_{3u}	45	436.98	B_{2u}	74	918.66	B_{3u}
17	262.68	B_{2u}	46	453.67	B_{2u}	75	928.90	B_{1u}
18	272.27	B_{3u}	47	459.67	B_{1u}	76	930.89	B_{2u}
19	278.85	B_{2u}	48	462.99	B_{1u}	77	934.28	B_{2u}
20	280.69	B_{1u}	49	463.94	B_{3u}	78	969.86	B_{1u}
21	293.44	B_{2u}	50	475.96	B_{3u}	79	1006.55	B_{1u}
22	293.74	B_{1u}	51	478.93	B_{1u}	80	1008.76	B_{3u}
23	297.37	B_{3u}	52	482.80	B_{3u}	81	1011.36	B_{2u}
24	302.67	B_{2u}	53	492.12	B_{3u}	82	1018.50	B_{3u}
25	305.55	B_{1u}	54	506.16	B_{2u}	83	1038.43	B_{1u}
26	320.63	B_{2u}	55	507.57	B_{1u}	84	1050.77	B_{2u}
27	324.06	B_{1u}	56	511.25	B_{3u}	85	1068.17	B_{3u}
28	324.86	B_{2u}	57	532.84	B_{2u}	86	1137.14	B_{1u}
29	330.55	B_{3u}	58	542.86	B_{1u}	87	1156.04	B_{2u}

20.5.1.3 Protoenstatite

Orthorhombic, space group *Pbcn*, Protoenstatite was investigated ab initio at the DFT/LDA theory level by Jahn and Martoňak (2009) with planewaves (energy cutoff = 1000 *eV*) and optimized norm-conserving pseudopotentials (Rappe et al. 1990) in the aim of depicting the relative stabilities of the various $MgSiO_3$ polymorphs at the athermal limit, and their compressibility as (see Figs. 4 and 5 in Jahn and Martoňak 2009). The relaxed structure at the athermal limit gave a = 9.171 *Å*, b = 8.513 *Å*, c = 5.254 *Å*, V_0 = 30.88 *cc/mol* (Jahn and Martoňak 2009). Protoenstatite was also investigated ab initio at the DFT/GGA theory level by Chibisov (2013) in the aim of clearifying the effect of P on the electronic properties, with particular enphasis on the bandgap. Pseudopotentials for the intervening atoms were built with the fhi98PP code (Fuchs and Scheffler 1999). Special points were inspected with a 2×2×2 Monkhorst–Pack grid and a cutoff of 816.34 *eV*. The lattice parameters at the athermal limit were a = 9.2947 *Å*, b = 8.8752 *Å*, c = 5.3553, leading to a volume substantially higher with respect to the results of Jahn and Martoňak (2009), (i.e. V_0 = 33.24 *cc/mol*). The obtained Birch–Murnaghan third-order bulk modulus was 106 *GPa* (presumably setting K′ to 4). According to Chibisov (2013), pressure influences the electronic structure of enstatite, and specifically, its band gap which increases progressively from 4.67 *eV* (at 0 *GPa*) to 4.74 *eV* (at 2 *GPa*). Unfortunately no details are given on the band structure, the number of inspected **k** points and the projected DOS.

20.5.1.4 High-P Clinoenstatite

Monoclinic, space group *C2/c*, (a = 9.201 *Å*, b = 8.621 *Å*, c = 4.908 *Å*, β = 78.5°, V_0 = 28.72 *cc/mol*; Algel and Chopelas 1992) High-P Clinoenstatite (hereafter HP-Cpx) differs essentially from the Low-P $P2_1/c$ polymorph (LP-Cpx) in having a larger β angle. In both polymorphs corner sharing $[SiO_4]^{4-}$ tetrahedra form chains along the c-axis, and silicate layers alternating along the b-axis. Figure 20.65 shows the structure of HP-Cpx, compared to LP-Cpx in the appraisal of Yu and Wentzcovitch (2009).

In this representation it is put in evidence the way the chains of tetrahedra are disposed along the *c* direction: while in LP-Cpx two sorts of O_3–O_3–O_3 counter-clockwise rotations occur (O-chain ≈ 205°; S-chain ≈ 132°), in HP-Cpx only one rotation is present (O-chain ≈ 134°). At the LP-Cpx to HP-Cpx phase transition (which occurs at P $\geq$ 7 *GPa*; see Fig. 20.62a) the S-chain in LP-CPx transforms to an O-chain in HP-Cpx. In their DFT/LDA investigation. Yu and Wentzcovitch (2009) adopted for Mg von Barth–Car type pseudopotentials (Dal Corso et al. 1993), while Troullier–Martins pseudopotentials were adopted for Si and O. The plane-wave kinetic energy cutoff was set to 80 *Ry*, the Internal BZ sampling was based on a 4×4×4 Monkhorst–Pack grid. For high-pressure structural relaxations, it was adopted the VCS MD procedure. Phonon frequencies at Γ-point were obtained through density functional perturbation theory (Baroni et al. 2001). The evaluation of the structural parameters at room conditions involved the utilization of the Quasi Harmonic Approximation (QHA).

According to Yu and Chopelas (2009), the LP-Cpx to HP-Cpx transition is induced by a slight softening of the lowest A_g Raman mode in LP-CPx. This mode is the only one whose frequency decrease with the increase of P (from 125.8 to 124.1 to 119.7 cm^{-1} when pressure is increased from 0 to 5 to 10 *GPa*, respectively). Closer inspection of the atomic displacements (eigenvector) associated with this mode revealed that it folds the S-chain away from the O-chain and reduces the O_3–O_3–O_3 S-chain angle. Enthalpy versus S-chain curves (Fig. 20.65) generated through the application of the A_g mode displacement indicated that "*the chain rotation leading LP-En to HP-En is a discontinuous, first order, enthalpically*

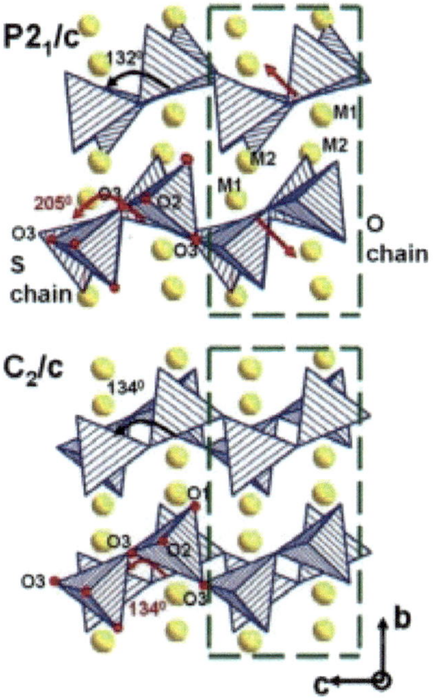

Fig. 20.65 The *P2₁/c* and *C2/c* polymorphs of $MgSiO_3$ in the representation of Yu and Wentzcovitch (2009). With permission from The Mineralogical Society of America

driven, event that could take place along the path provided by the A_g *displacement mode*" (Yu and Wetzcovitch 2009).

C2/c Clinoenstatite was investigated by Yu et al. (2010) with the same procedures previously seen for the low-P polymorph. The obtained cell parameters at P = 7.9 *GPa* are shown in Table 20.55.

The calculated volume at T = 300 *K* was 30.34 *cc/mol*, the bulk modulus at the same T was K = 126.3 *GPa* and its P derivative K′ was 5.5. The QHA thermal expansion at the same T conditions was $\alpha = 2.46 \times 10^{-5}\ K^{-1}$ and the QHA isobaric heat capacity was $C_P = 81.1$ *J/(mol × K).*

In the unit cell of *C2/c* Clinoenstatite there are 20 atoms (4 formula units), hence 60 optical modes. The symmetry assignment at Γ point gives:

Table 20.55 Unit cell parameters of *C2/c* clinoenstatite at P = 7.9 *GPa*. XRD = X-Ray diffraction; MD = Molecular Dynamics

a (*Å*)	b (*Å*)	c (*Å*)	*β* (°)	V (*Å*^{3})	Method	References
9.201	8.621	4.908	101.50	381.5	XRD	Angel et al. (1992)
9.12	8.19	4.904	101.3	359.2	DFT/LDA	Wentzcovitch et al. (1995)
9.167	8.60	4.90	101.0	379.1	DFT/LDA	Yu et al. (2010)

Source of Data Yu et al. (2010)

Table 20.56 Raman-active frequencies (cm^{-1}) in *C2/c* Clinoenstatite. Ab initio results of Yu et al. (2010) (1) are compared with the experimental evidences of Lin (2004) (2) and Chopelas (1999) (3)

ν	ν	ν	Mode	ν	ν	ν	Mode
(1)	(2)	(3)		(1)	(2)	(3)	
162.7	161.1	170	B_g	481.5			B_g
184.3	185.3	185.5	A_g	488.2	490	487.5	A_g
231.9	–	–	B_g	504.8	–	–	A_g
234.6	237.8	238.4	A_g	542.6	542	541.5	B_g
249.6	248.9	249.4	B_g	575.6	563	571	B_g
257.5	264.2	263.9	A_g	593.2	576	–	A_g
290.5	279	–	B_g	614.7	626		B_g
337.7	–	–	B_g	710.7	723	720.9	A_g
341.7	341.2	341.6	A_g	783.5	787	787.4	B_g
363.2	375.1	368.8	B_g	862	863	865.8	A_g
388.7	–	–	A_g	920.5	917	–	B_g
406.6	403.6	407.4	A_g	951.2	944	942.4	B_g
416	410.2	–	B_g	1054.8	1060	1062	A_g
452.3	433.2	447.6	A_g	1090.1	1092	1089	A_g
462.3	450.3	–	B_g	1105.9	–	–	B_g

Table 20.57 Ab initio IR-active frequencies (cm^{-1}) in *C2/c* Clinoenstatite (Yu et al. 2010)

ν	Mode	ν	Mode	ν	Mode
225.0	B_u	418.2	A_u	579.1	A_u
245.6	A_u	425.6	B_u	698.4	B_u
255.1	B_u	442.7	A_u	767.1	A_u
280.7	B_u	464.8	B_u	862.8	B_u
320.6	A_u	501.7	A_u	909.2	A_u
344.4	B_u	530.2	B_u	953.9	A_u
378.2	A_u	533.5	A_u	987.5	B_u
380.0	B_u	533.4	B_u	1103.0	B_u
384.6	A_u	574.3	B_u	1127.8	A_u

$$\Gamma = 14A_g + 16B_g + 14A_u + 16B_u \tag{20.5.10}$$

The (30) symmetric (g) modes are Raman-active and the 30 asymmetric (u) modes are IR-active. One A_u and two B_u are pure translations. The Raman-active frequencies obtained by Yu et al. (2010) for P = 12.5 *GPa* are compared in Table 20.56 with the experimental results of Chopelas (1999) and Lin (2004). Table 20.57 lists the IR-active frequencies. The phonon dispersion and the projected DOS are shown in Fig. 20.66.

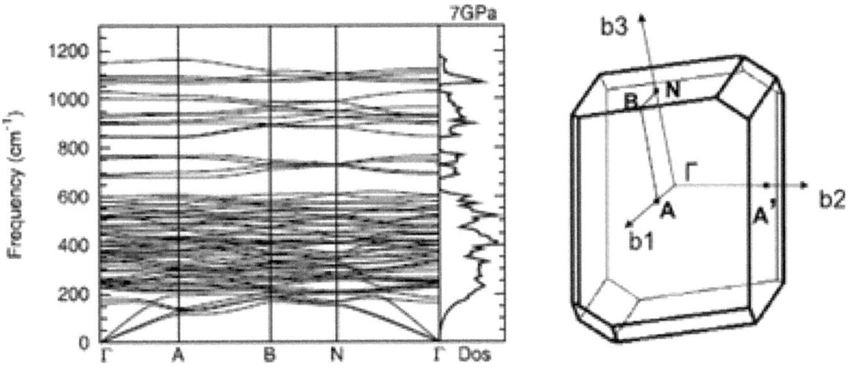

Fig. 20.66 Phonon dispersions and projected DOS of *C2/c* Clinoenstatite at 7 *GPa* after Yu et al. (2010). In the rightward part of the figure is shown the indicized BZ. From Yu et al. (2010). With permission from The Mineralogical Society of America

The phonon dispersion of the of *C2/c* polymorph shows three gaps instead of the two seen for $P2_1/c$. From 0 to $\sim 600\ cm^{-1}$ occur the translational modes of Mg and the translation and rocking modes of silicate chains. Between ~ 800 and $\sim 1000\ cm^{-1}$ occur the O–Si–O bond bending and the Si–O bond stretching motions. Above 1000 cm^{-1} occur the pure Si–O bond stretching modes (Yu et al. 2010).

The *C2/c* clinoenstatite polymorph was also investigated by Belmonte (2013) at DFT/GGA theory level with the hybrid B3LYP exchange–correlation functional and the same basis sets and computational parameters previously adopted for the Mg_2SiO_4 polymorphs (see Sect. 20.4.1.1). The thermoelastic properties of this polymorph (and majorite as well; see later on) were described by Belmonte (2013) adopting the Suzuki equation (Suzuki 1975).

Background information: Suzuki equation

The Suzuki equation (Suzuki 1975) is

$$\frac{V(T, P_0)}{V_0(T_0, P_0)} = \frac{\left[1 + 2k - \left(1 - \frac{4kE_{th}}{Q_0}\right)^{1/2}\right]}{2k} \qquad (20.5.11)$$

where: $V(T, P_0)$ is the volume of the crystal at P = 0 and T of interest; $V_0(T_0, P_0)$ is the volume of the crystal at P = 0 and T = 0 and E_{th} is the thermal energy. Moreover

$$Q_0 = \frac{K_0 V_0}{\gamma_{th}} \tag{20.5.12}$$

$$k = 1/2\,(K' - 1) \tag{20.5.13}$$

This equation is obtained by including the second-order term in the Taylor expansion of the potential energy of the crystal with respect to $\Delta = (V\text{–}V_0)$, i.e.:

$$E(V) = E_0(V_0) + E_0'\Delta + E_0''\frac{\Delta^2}{2} + E_0'''\frac{\Delta^3}{6} + \cdots \tag{20.5.14}$$

with one of the real roots of the resulting quadratic equation adopted in Eq. (20.5.14) (see Anderson 1995; Helffrich 1999; Belmonte 2013; for the mathematical details of the derivation).

The functional form of the thermal expansion $\alpha = f(T)$ obtained by Belmonte (2013) is compared in Fig. 20.67 with the DFT/LDA results of Yu et al. (2010), the MD results of Chopelas (1999) and the XRD results of Shinmei et al. (1999). The results of this last investigation, carried out in situ with synchrotron radiation and a multi-anvil high-pressure apparatus at pressures to 12 *GPa* and temperatures to 1473 *K* did not allow to extend the fitting after the first term of the usual functional form of $\alpha = f(T)$, hence Shinmei et al. (1999) were obliged to adopt a simple linear fit. Nevertheless, their thermophysical parameterization (third-order Birch–Murnaghan) are among the few experimentally-based data for this polymorph ($V_0 = 30.49 \pm 0.13$ *cc/mol*; $K_0 = 106.4 \pm 17.4$ *GPa*, and $K' = 5.4 \pm 2.7$; Shinmei et al. 1999).

20.5.1.5 High-P Polymorphs: Akimotoite–Majorite–Bridgmanite

Akimotoite (ilmenite-type) has a trigonal structure (Space group $R\overline{3}$). Both Si and Mg are in sixfold coordination with O. The $[SiO_6]^{8-}$ and $[MgO_6]^{10-}$ octahedra form discrete layers alternated along the c-axis. The structural refinement of Horiuchi et al. (1982) (Fig. 20.68) assign to cell edges the following values: a = b = 4.7284 *Å*; c = 13.5591 *Å*). The ensuying molar volume is $V_0 = 26.39$ *cc/mol*.

Majorite has a tetragonal structure (Space group $I4_1/a$), there are 16 formula units per unit cell. Denoting the general formula unit as $X_{12}Y_8Z_{12}O_{48}$ we have 12 X cations in dodecahedral sites, 8 Y cations in octahedral sites and 12 Z cations in tetrahedral sites. As in olivine, the $[TO_4]^{4-}$ groups are isolated from one other (*nesosilicates*). The octahedra are also isolated, and are linked by single apexes to tetrahedra. Dodecahedra share edges either with tetrahedra and octahedra or with other dodecahedra. Tetrahedra are fully occupied by Si^{4+}, while both Si^{4+} and Mg^{2+}

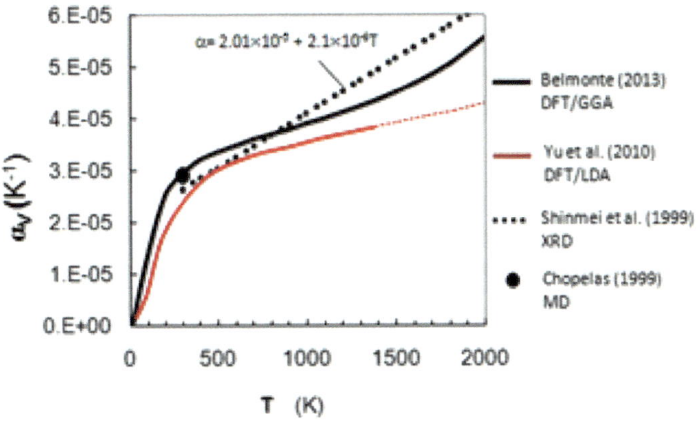

Fig. 20.67 Thermal expansion of *C2/c* Clinoenstatite. Ab initio results of Belmonte (2013) and Yu et al. (2010) are compared with the MD results of Chopelas (1999) and the XRD-multianvil experiment of Shinmei et al. (1999)

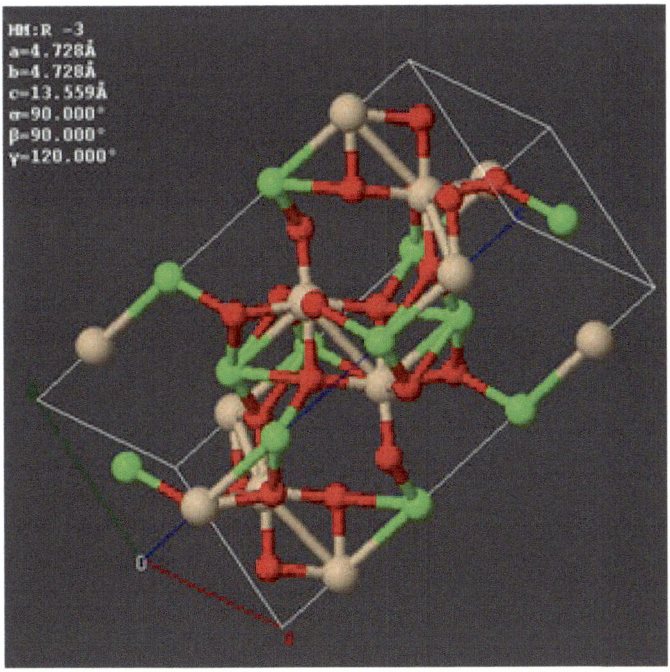

Fig. 20.68 The $R\overline{3}$ structure of Akimotoite (Horiuchi et al. 1982) as generated by the AMCSD software (Downs and Hall-Wallace 2003)

are present in octrahedral sites. Majorite differs from the common cubic garnets in having an almost complete ordering of Mg and Si cations (which is at the origin of the non-cubic symmetry; Angel et al. 1989). The single-crystal XRD refinement of Majorite synthesized at 17 *GPa* and 1800 °*C* in a uniaxial split-sphere anvil device gave a = b = 11.501 *Å*, c = 11.480 *Å*, and V_{cell} = 1518.6 $Å^3$ (Angel et al. 1989), which, once reduced to the usual $MgSiO_3$ formula unit, returns a molar volume V_0 = 28.58 *cc/mol*.

Bridgmanite crystallizes in the orthorhombic system (Space group *Pbnm*) and deviates significantly from the ideal cubic perovskites mainly due to tilting of the relatively regular $[SiO_6]^{8-}$ octahedra and the large distortion of the Mg dodecahedra. A single-crystal XRD refinement of a specimen synthesized at 27 *GPa* and 1830 °*C* gave a = 4.7754, b = 4.9292, c = 6.8969 (Horiuchi et al. 1987). The resulting cell volume (V_{cell} = 162.345 $Å^3$, Z = 4) when scaled to the usual formula unit returns V_0 = 24.442 *cc/mol*.

The three high-P polymorph have been the object of very accurate first-principles investigations, due to their importance in mantle rheology. The most complete investigations recently appeared are those of Belmonte (2013), Hernández et al. (2015), De la Pierre and Belmonte (2016); all carried out at DFT/GGA theory level. In his DFT/GGA investigation of the three high-P polymorphs, Belmonte (2013) adopted (as in *C2/c* clinoenstatite) the same basis sets and computational parameters previously adopted for the Mg_2SiO_4 polymorphs (see Sect. 20.4.1.1). The same basis set was also adopted by De la Pierre and Belmonte (2016) in the new calculations of Majorite. The reciprocal space was sampled by De la Pierre and Belmonte (2016) along the 3 lattice vectors, leading to 6 independent **k** vectors in the irreducible part of the BZ. Hernández et al. (2015) employed in their DFT simulation of Akimotoite-Majorite-Bridgmanite a plane-wave basis set, combined with the Projector augmented Wave (PAW) method (cf. Sect. 8.8.3) and the Perdew–Wang (PW91) functional (see Sect. 10.5.5). A kinetic energy planewave cutoff of 500 *eV* was considered sufficiently accurate for Energy convergence. For Akimotoite an 8×8×8 supercell (2 f.u.) corresponding to 88 special points was adopted. For Majorite the adopted supercell was 4×4×4 (16 f.u.) spanning 14 **k** points and for Bridgmanite the supercell was 6×6×6 (4 f.u.) with 18 **k** points.

The relaxed Majorite structure of De la Pierre and Belmonte (2016) at the athermal limit gave a = b = 11.611 *Å*, c = 11.542 *Å*, V_{cell} = 1556.19 $Å^3$ (V_0 = 29.286 *cc/mol*). The thermophysical parameters obtained by Belmonte (2013) for the same phase are listed in Table 20.48. A nearly similar cell volume was obtained by Hernández et al. (2015) (V_{cell} = 1559.4 $Å^3$; V_0 = 29.347 *cc/mol*). The obtained bulk modulus was K_0 = 146.8 *GPa* and its baric derivative was K′ = 4.6 (Hernández et al. 2015).

For Akimotoite Hernández et al. (2015) obtained a cell volume V_{cell} = 270.65 (V_0 = 27.164 *cc/mol*), a bulk modulus K_0 = 188.4 *GPa* and a baric derivative K′ = dK/dP = 4.6. The parameters obtained for Bridgmanite were V_{cell} = 167.2 (V_0 = 25.173 *cc/mol*), K_0 = 229.7 *GPa*, K′ = 4.4.

Table 20.58 Raman-active frequencies (cm^{-1}) in $R\bar{3}$ Akimotoite. Ab initio results of Hernández et al. (2015) (1) are compared with the Ab initio DFPT/LDA results of Karki (2002) (2) and with the experimental observations of Reinard and Rubie (1996) (3) $\nu(V_{exp})$ are the frequencies interpolated at the experimental volume

Mode	(1)			(2)		(3)	
	ν	$\nu(V_{exp})$	γ	ν	γ	ν	γ
A_g	268	286	2.40	294	1.92	294	1.60
E_g	326	339	1.74	343	1.15	351	1.10
E_g	383	400	2.22	400	1.32	402	1.24
A_g	391	406	1.96	414	1.02	413	1.26
E_g	458	478	2.73	488	1.25	484	1.24
A_g	476	496	2.67	490	1.27	499	1.40
E_g	583	601	2.47	614	0.85	620	0.88
A_g	644	672	3.62	680	1.18	680	1.10
A_g	756	784	3.74	783	1.09	798	1.04
E_i	762	794	4.32	795	1.24	–	–

A group analysis of the Akimotoite polymorph [primitive structure C_{3i} (-3)] gives at Γ point the symmetry assignment:

$$\Gamma = 5A_g + 5A_u + 5E_g + 5E_u \qquad (20.5.15)$$

There are 18 optical modes, the g modes are Raman active and the u modes are IR-active. $5E_g$ and $4E_u$ are doubly degenerate. The Raman-active frequencies obtained by Hernández et al. (2015) and their γ-mode are compared with the results of Karki and Wentzcovitch (2002) obtained at DFPT/LDA theory level and with the experiments of Reynard and Rubie 1996) in Table 20.58. Table 20.59 lists the IR frequencies based on the same authors for computation and Hofmeister and Ito (1992) for experiments. The phonon dispersion and the projected DOS of Akimotoite at 0 *GPa* according to Hernández et al. (2015) are shown in Fig. 20.69. Apparently no gaps are observed between valence and conduction bands, hence this polymorph should have a metallic behavior.

In the primitive cell of Majorite [point group symmetry C_{4h} $(4/m)$] there are 80 atoms. The symmetry assignment of the vibrational modes gives:

$$\Gamma = 25A_g + 33A_u + 27B_g + 31B_u + 28E_g + 34E_u \qquad (20.5.16)$$

The $28E_g$ and $34E_u$ modes are doubly degenerate, the $31B_u$ modes are silent, all the remaining u modes (besides the three purely transaltional modes) are IR active. All g modes are Raman active.

In Tables 20.60 and 20.61 are listed, respectively, the Raman-active and IR-active + silent frequencies at Γ point obtained by De La Pierre and Belmonte (2016) at the DFT/GGA theory level. The computed Raman active modes are in satisfactory agreement with observations. The mean absolute deviation from

Table 20.59 IR-active frequencies (cm^{-1}) in $R\overline{3}$ Akimotoite. Ab initio results of Hernández et al. (2015) compared with the ab initio DFPT/LDA results of Karki (2002) and the experimental observations of Reinard and Rubie (1996)

Mode	(1)		(2)		(3)	
	ν	ν (V_{exp})	γ	ν	γ	ν (range)
A_u	304	325	2	330	1.93	384–421
E_u	317	340	2.03	346	2.07	337–364
E_u	419	435	1.11	439	1.06	450–482
A_u	485	514	1.68	509	1.74	526–552
E_u	554	585	1.59	595	1.54	623–642
E_u	621	651	1.31	648	1.33	670–820
A_u	683	703	0.84	702	0.79	735–820
A_u	759	783	0.9	768	1.00	–

(1) Hernandez et al. (2015)
(2) Karki (2002)
(3) Reinard and Rubie (1996)

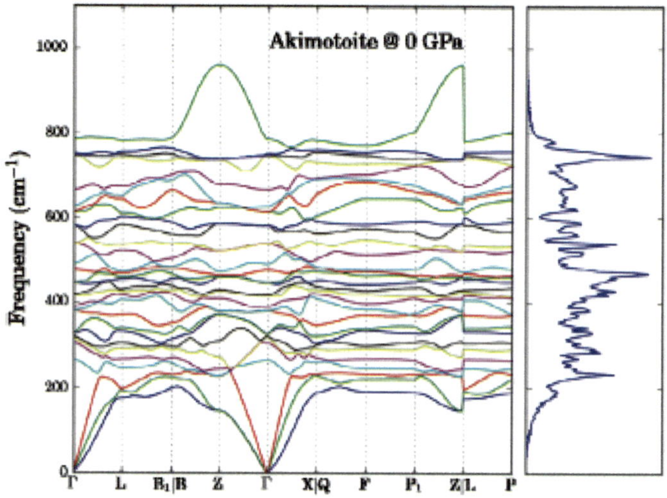

Fig. 20.69 Phonon dispersions and projected DOS of $R\overline{3}$ Akimotoite at 0 *GPa* after Hernández et al. (2015). From Hernández et al. (2015)

experiments is $|\overline{\Delta}| = 3.6; 4.4; 4.9; 4.3\ cm^{-1}$, respectively, for the datasets of McMillan et al. (1989), Rauch et al. (1996), Manghnani et al. (1998) and Chopelas (1999). The computed IR-active modes deviate from the observations of Kato and Kumazawa (1985) and McMillan et al. (1989) by $|\overline{\Delta}| = 5.7;\ 4.6\ cm^{-1}$, respectively.

The phonon dispersion and the projected DOS of Majorite at 0 *GPa* according to Hernández et al. (2015) are shown in Fig. 20.70. There is a large gap between valence and conduction bands ($\sim 5.8\ eV$ at room pressures; Hernandez et al. 2015).

Table 20.60 Raman-active frequencies (cm^{-1}) and their mode-γ in $I4_1/a$ Majorite (De La Pierre and Belmonte 2016)

Mode	v	γ	Mode	v	γ	Mode	v	γ	Mode	v	γ
E_g	116.4	3.36	B_g	299.0	0.98	A_g	429.8	0.86	E_g	653.2	0.54
E_g	138.0	1.94	E_g	308.3	1.25	A_g	458.0	0.45	B_g	670.7	0.55
B_g	154.2	0.22	B_g	315.7	1.59	E_g	462.2	0.72	A_g	795.9	0.83
A_g	154.5	2.08	A_g	315.9	1.93	B_g	466.5	0.61	E_g	802.5	0.82
E_g	162.9	1.43	A_g	317.1	0.9	E_g	481.2	0.5	B_g	812.9	0.69
B_g	164.6	2.54	E_g	320.5	0.71	B_g	482.9	0.73	E_g	850.0	1.09
A_g	177.5	1.38	A_g	338.8	1.07	B_g	500.4	0.49	A_g	867.2	0.59
E_g	189.6	1.54	E_g	344.1	1.05	A_g	506.0	0.72	B_g	878.6	1.03
B_g	200.1	0.94	B_g	346.7	2.08	E_g	509.7	0.73	A_g	886.2	0.67
A_g	210.3	0.36	B_g	352.8	1.04	A_g	532.3	0.73	B_g	886.9	0.84
E_g	211.8	1.67	E_g	357.3	1.09	E_g	540.6	0.66	E_g	888.4	0.76
B_g	212.1	1.53	E_g	363.9	0.99	B_g	559.4	0.63	E_g	909.8	0.83
A_g	229.7	1.37	B_g	367.4	1.38	A_g	567.7	0.72	A_g	926.1	0.77
E_g	237.2	1.27	A_g	368.9	1.15	B_g	585.7	0.43	B_g	957.0	0.79
B_g	242.4	1.26	B_g	382.5	1.16	E_g	593.0	0.69	A_g	960.2	0.86
A_g	256.9	0.9	E_g	393.1	1.1	A_g	602.0	0.83	E_g	984.5	0.86
E_g	257.4	1.81	A_g	396.9	1.04	E_g	611.3	0.73	A_g	986.4	0.88
A_g	276.4	0.86	A_g	402.3	0.64	B_g	622.2	0.67	B_g	1029.6	0.77
B_g	281.6	0.81	B_g	407.1	0.99	A_g	636.5	0.5	E_g	1069.6	0.75
E_g	298.0	0.92	E_g	413.4	0.63	B_g	639.1	0.37	B_g	1073.0	0.74

Table 20.61 IR-active and silent (Bu) modes frequencies (cm^{-1}) and their mode-γ in $I4_1/a$ Majorite (De La Pierre and Belmonte 2016)

Mode	v	γ	Mode	v	γ	Mode	v	γ	Mode	v	γ
E_u	101.3	3.51	B_u	302.2	2.05	B_u	439.2	0.51	B_u	676.2	0.51
E_u	130.9	2.78	A_u	312.1	1.11	E_u	440.5	0.98	E_u	684.8	0.78
A_u	142.4	2.33	B_u	314.7	0.54	A_u	452.7	0.51	A_u	701.6	0.68
B_u	145.7	2.72	A_u	322.6	0.94	E_u	457.6	0.79	E_u	706.8	0.95
A_u	156.4	1.04	E_u	325.8	0.79	B_u	458.8	0.92	B_u	725.7	1.05
B_u	171.9	0.55	B_u	327.5	1.5	A_u	464.0	0.64	A_u	730.4	0.82
E_u	172.3	1.58	E_u	341.4	1.16	A_u	479.8	0.75	B_u	821.2	0.60
A_u	191.8	0.73	A_u	345.1	2.12	B_u	479.8	0.99	E_u	826.3	0.66
B_u	196.9	0.89	A_u	360.5	0.43	E_u	491.1	0.67	A_u	845.1	0.67
E_u	197.7	1.23	B_u	361.3	1.53	B_u	492.4	0.84	E_u	876.1	1.06
E_u	206.8	1.04	E_u	361.7	1.03	E_u	504.3	0.67	B_u	878.5	0.55
A_u	213.4	1.26	B_u	364.8	1.11	A_u	510.3	0.81	A_u	886.4	0.96
B_u	213.7	0.79	E_u	368.8	0.89	B_u	516.1	0.81	E_u	901.4	0.80
E_u	222.5	1.87	A_u	372.4	1.06	E_u	522.5	0.96	E_u	915.1	0.78

(continued)

Table 20.61 (continued)

Mode	ν	γ	Mode	ν	γ	Mode	ν	γ	Mode	ν	γ
E_u	240.2	1.02	A_u	379.6	0.84	B_u	542.0	0.87	B_u	925.5	0.97
A_u	247.1	1.41	B_u	384.4	0.77	E_u	550.1	1.02	A_u	928.4	0.84
B_u	247.6	0.61	E_u	384.9	1.18	A_u	563.2	0.95	B_u	937.7	0.89
E_u	257.3	0.81	B_u	391.3	1.07	B_u	576.4	0.66	A_u	950.4	0.84
A_u	258.8	0.66	A_u	392.1	1.1	A_u	579.7	0.72	B_u	951.9	0.83
B_u	258.8	0.58	E_u	396.5	1.01	E_u	583.2	0.59	E_u	954.2	1.00
B_u	276.9	0.84	A_u	399.0	0.99	A_u	599.6	0.88	A_u	974.5	0.87
E_u	277.0	0.73	E_u	417.1	0.86	E_u	636.8	0.99	E_u	989.1	0.91
A_u	289.8	0.97	A_u	425.5	1.3	B_u	648.9	0.8	B_u	980.4	0.89
E_u	300.3	0.96	B_u	424.7	0.91	A_u	656.7	0.64	A_u	1056	0.77

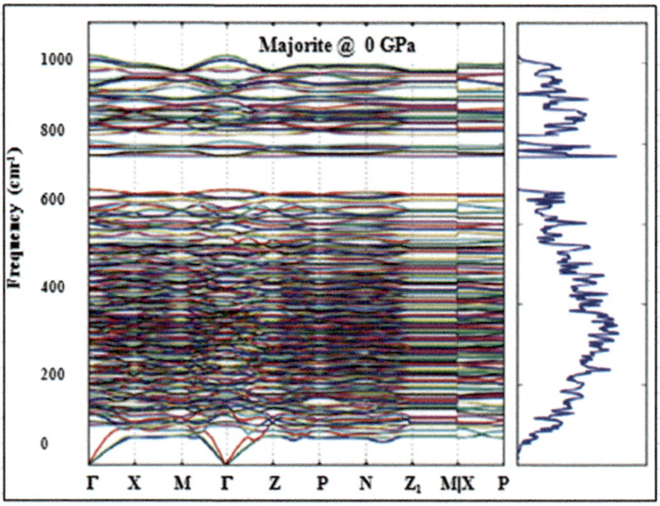

Fig. 20.70 Phonon dispersions and projected DOS of *I4₁/a*Majorite at 0 *GPa* after Hernández et al. (2015). Modified from Hernández et al. (2015)

This gap "*shows a tendency to increase with pressure*" (Hernandez et al. 2015). This polymorph should thus behave as an insulator at all pressures.

In the primitive cell of Bridgmanite [point group symmetry D_{2h} *(mmm)*] there are 20 atoms. The symmetry assignment of the 60 vibrational modes gives (Hernandez et al. 2015):

$$\Gamma = 7A_g + 8A_u + 5B_{1g} + 10B_{1u} + 7B_{2g} + 8B_{2u} + 5B_{3g} + 10B_{3u} \qquad (20.5.17)$$

The three pure translational modes are B_{1u}, B_{2u}, B_{3u}; the 8 A_u modes are silent; the remaining u modes are Raman-active, while all the g modes are IR-active.

Table 20.62 Raman-active frequencies (cm^{-1}) in *Pbnm* Bridgmanite. Ab initio results of Hernández et al. (2015), compared with the Ab initio results of Belousov (2011) and with the experimental oservations of Chopelas (1996). $\nu(V_{exp})$ are the frequencies interpolated at the experimenal volume

Mode	(1)			(2)	(3)	
	ν	$\nu(V_{exp})$	γ	ν	ν	γ
A_g	209	219	3.01	234	245	2.14
B_{3g}	233	250	1.93	242	251	2.06
B_{1g}	258	275	2.15	283	–	–
A_g	265	277	1.58	288	279	1.39
B_{2g}	269	280	1.32	291	–	–
B_{2g}	315	327	1.09	336	327	1.41
B_{1g}	323	337	1.71	340	334	1.63
B_{2g}	352	367	1.35	369	370	1.06
A_g	365	381	1.13	383	379	1.21
A_g	376	394	1.76	396	387	1.34
B_{1g}	414	439	1.9	433	–	–
B_{3g}	417	434	0.84	435	–	–
A_g	475	501	1.72	498	499	1.8
B_{2g}	490	510	1.63	507	–	–
B_{3g}	490	525	2.05	509	–	–
B_{1g}	497	533	2.2	513	–	–
A_g	521	552	2.11	534	–	–
B_{2g}	599	622	1.44	627	–	–
B_{3g}	602	623	1.25	632	–	–
A_g	628	651	1.46	649	–	–
B_{2g}	637	660	1.37	684	–	–
B_{1g}	753	789	0.95	798	–	–
B_{2g}	790	817	1.28	834	–	–
B_{3g}	808	837	1.29	853	–	–

(1) Hernandez et al. (2015)
(2) Belousov (2011)
(3) Chopelas (1996)

The Raman-active frequencies of Bridgmanite obtained by Hernández et al. (2015) for P = 0 *GPa* and their γ-mode are compared with the DFT/GGA results of Belousov (2011) and with the experiments of Chopelas (1996) in Table 20.62. Table 20.63 lists the IR frequencies based on the same authors for computation and Lu et al. (1994) for experiments. The mean absolute deviation from experiments for Raman-active modes and IR-active modes is $|\overline{\Delta}| = 6; 26\ cm^{-1}$, respectively, for both computations. The phonon dispersion and the projected DOS of Bridgmanite at 0 *GPa* according to Hernández et al. (2015) are shown in Fig. 20.71. As in Akimotoite, no gaps are observed between valence and conduction bands, hence also this polymorph should have a metallic behavior.

Table 20.63 IR-active frequencies (cm^{-1}) in *Pbnm* Bridgmanite. Ab initio results of Hernández et al. (2015) compared with the Ab initio results of Belousov (2011) and with the experiments of Lu et al. (1994)

	(1)			(2)	(3)	
Mode	ν	$\nu(V_{exp})$	γ	ν	ν	γ
B_{2u}	158	176	2.16	159	180	1.4[a]
B_{3u}	230	243	1.78	241	222	1.76[a]
B_{1u}	247	262	1.8	258	247	1.29[a]
B_{2u}	288	304	1.66	305	262	1.49[a]
B_{1u}	301	319	1.91	317	386	1.17[a]
B_{3u}	327	348	1.9	341	317	1.28[a]
B_{3u}	364	385	1.72	381	343	–
B_{3u}	408	425	0.99	424	388	–
B_{1u}	418	434	0.75	441	415	–
B_{2u}	419	448	0.59	444	430	–
B_{1u}	448	470	1.78	471	444	–
B_{2u}	459	488	1.76	477	466	–
B_{3u}	464	489	1.78	487	466	–
B_{1u}	487	515	2.04	498	484	–
B_{3u}	508	533	1.86	534	534	1.44
B_{1u}	544	566	1.25	571	534	–
B_{2u}	559	576	1.24	596	614	–
B_{1u}	564	584	1.71	605	597	1.39
B_{3u}	631	652	1.26	669	679	1.15
B_{2u}	671	705	1.92	701	679	–
B_{1u}	692	710	0.97	722	705	–
B_{2u}	699	731	1.79	731	721	–
B_{3u}	716	742	1.78	772	683	–
B_{1u}	749	779	1.6	788	–	–
B_{3u}	762	783	1.03	804	780	1.21
A_u	181	187	1.1	194	–	–
A_u	252	267	2.02	266	–	–
A_u	341	355	0.6	367	–	–
A_u	369	382	0.86	387	–	–
A_u	465	493	1.65	488	–	–
A_u	582	604	1.03	609	–	–
A_u	630	667	2.31	660	–	–
A_u	712	744	1.74	748	–	–

(1) Hernandez et al. (2015)
(2) Belousov (2011)
(3) Lu et al. (1994)
[a] = assignment "*non unambiguous*" according to Hernandez et al. (2015)

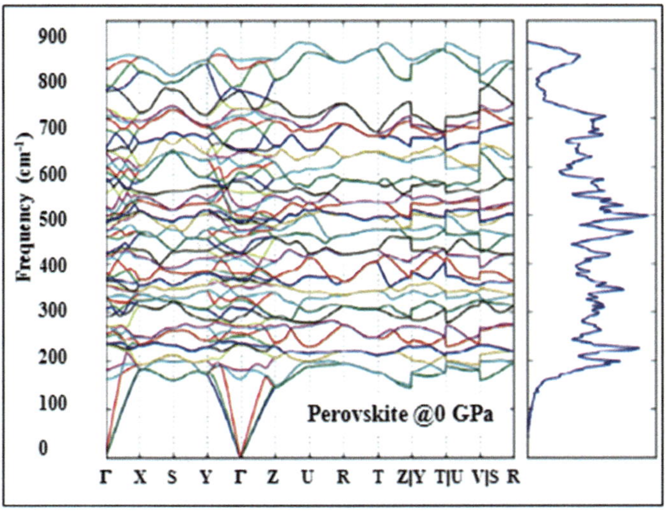

Fig. 20.71 Phonon dispersions and projected DOS of *Pbnm* Bridgmanite at 0 *GPa* after Hernández et al. (2015). Modified from Hernández et al. (2015)

As noted by Belmonte (2013), the thermoelastic properties of Akimotoite are extremely helpful in deciphering the significance of the high seimsmic velocity gradients observed in subduction environments where relatively cold slabs may reach the mantle transition zone (Wang et al. 2004). It is important thus to spend few words in estimating the performance of ab initio procedures in quantifying two important parameters such as thermal expansion and isobaric heat capacity.

Background information: Details on QHA
The theoretical Quasi Harmonic Approximation (QHA) frameworks adopted by Belmonte (2013) and Hernandez et al. (2016) are nearly similar but differ substantially in the hierarchical approach to the thermodynamic magnitudes.

In Belmonte (2013) the thermal expansion derives from the statistical $\alpha_T K_T$ product:

$$\alpha_T K_T(V,T) = \frac{R}{ZV}\sum_{k,i=1}^{3n} \gamma_i(\mathbf{k},V)e^{x_i}\left(\frac{x_i(\mathbf{k},V)}{e^{x_i(\mathbf{k},V)}-1}\right)^2 \qquad (8.9.20)$$

with the summation extended to the acoustic branches. The notation emphasizes that the adimensional frequencies are subjected to phonon dispersion (besides being V-dependent). A mode-γ treatment of the $\alpha_T K_T$ product allows to define the functional form $\alpha = f(T)$ of thermal expansion and the thermal derivative of the bulk modulus. Alternatively, the isobaric thermal expansion is obtained directly from the Suzuki equation 20.5.11.

The isobaric heat capacity is obtained stemming from the isochoric counterpart:

$$C_V = \frac{R}{Z}\sum_{i=1}^{3n} e^{x_i(\mathbf{k},V)}\left(\frac{x_i(\mathbf{k},V)}{e^{x_i(\mathbf{k},V)}-1}\right)^2 \tag{8.10.58}$$

and adding the anharmonicity term:

$$C_P = C_V + T\alpha_T^2 K_T V_T \tag{20.8.10.52}$$

In Hernandez et al. (2016) the procedure begins from the Helmholtz free energy splitted into lattice and vibrational components:

$$F(T,V) = F_{latt}(T,V) + F_{vib}(T,V) \tag{20.5.18}$$

In the QHA the vibrational part of the Helmholtz free energy is

$$F_{vib}(T,V) = k_B T \sum_{n,\mathbf{k}} \ln\left[2\sinh\left(\frac{\hbar\omega_{n,\mathbf{k}}}{k_B T}\right)\right] \tag{20.5.19}$$

An F-T-V grid is then created and a fifth-order Chebyshev polynomial in V is built along the isotherms in the T-V grid. S(T, V), P(T, V) are then obtained by numerical derivation:

$$S(T,V) = -\left(\frac{\partial F}{\partial T}\right)V \tag{20.5.20}$$

$$P(T,V) = -\left(\frac{\partial F}{\partial V}\right)T \tag{20.5.21}$$

The Gibbs free energy is then:

$$G(T,P) = F(T,V) + VP(T,V) \tag{20.5.22}$$

and the enthalpy is:

$$H(T,P) = G(T,P) + TS(T,V) \tag{20.5.23}$$

The isobaric heat capacity is

$$C_P = \left(\frac{\partial H(T,P)}{\partial T}\right)_P \tag{20.5.24}$$

and the isobaric thermal expansion is

$$\alpha_V = \left(\frac{\partial V(T,P)}{\partial T}\right)_P \tag{20.5.25}$$

In Fig. 20.72 we may see a comparative appraisal of these magnitudes.

We may appreciate from Fig. 20.72 that the two approaches are mutually (and strictly) consistent. This is another evidence of the intrinsic beauty of Thermodynamics and of the soundness of the adopted first principles. The model performances for the other two high-P polymorphs are quite similar and do not deserve further comments.

20.5.2 $CaSiO_3$

As for $MgSiO_3$ also $CaSiO_3$ exhibit extensive polymorphism which has been the object of various experimental investigations.

The polymorph stable at room condition is Wollastonite (Triclinic, space group $P\overline{1}$). The structure of Wollastonite contains infinite chains of $[SiO_4]$ tetrahedra sharing common vertices, running parallel to the b-axis (Fig. 20.73).

According to Richet et al. (1991) there is at 995 ± 10 *K* a reversible transition from low-Wollastonite to high-Wollastonite clearly detectable in the mean heat capacity values (see Fig. 1 in Richet et al. 1991; and references therein). The transition "*can be considered calorimetrically as first order*" (Richet et al. 1991) with a very small enthalpy of transition (i.e. ~200 *J/mol*). Wollastonite (Wo) exhibits also a solid state transition to the Cyclosilicate Pseudo-Wollastonite (PsWo). At room pressure the transition takes place at ~1400 *K* (Kushiro 1954). The crystal structure of PsWo is characterized by alternate stacking of layers of ternary (Si_3O_9) tetrahedral rings and *distorted-bicapped* Ca octahedra along [001] (Yang and Prewitt 1999a).[17] PsWo exhibits polytypism due to variable stacking

[17] A *distorted-bicapped* octahedron is formed by eight oxygens, with six of them in an octahedral arrangement plus two from bridging oxygens of the Si_3O_9 ternary rings (one above and one below the octahedron) (Yang and Prewitt 1999a).

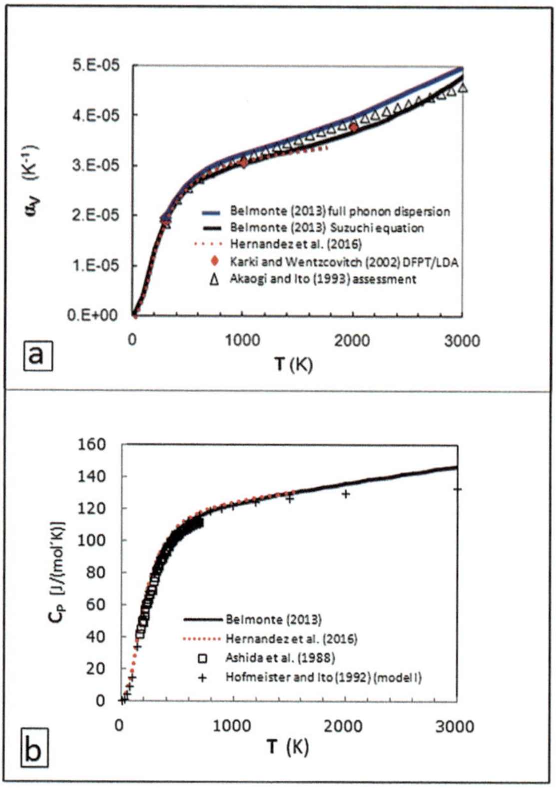

Fig. 20.72 Comparative appraisal of ab initio derived thermophysical magnitudes in Akimotoite. **a** Isobaric (P = 0 *GPa*) thermal expansion. DFT/GGA assessments of Belmonte (2013) and Hernández et al. (2016) are compared with the DFPT/LDA results of Karki and Wentzcovitch (2002) and with the assessment of Akaogi and Ito (1993). **b** Isobaric (P = 0 *GPa*) heat capacity of Belmonte (2013) and Hernández et al. (2016) compared with the experimental results of Ashida et al. (1988) and the model values of Hofmeister and Ito (1992). Modified from Belmonte (2013)

sequence of octahedral layers and ternary rings. According to Yamanaka and Mori (1981) three polytipes exist, the dominant one being a four-multilayer *C*1 phase (with one multilayer composed of one layer of ternary rings plus one layer of CaO_6 octahedra (Yamanaka and Mori 1981). According to Yang and Prewitt (1999b) this polytipe is monoclinic *C2/c*, rather than triclinic *C*1, with a = 6.8394(5), b = 11.870, c = 19.631 *Å*, β = 90.667°, and V_{cell} = 1593.7 $Å^3$. According to Osborn (1942) PsWo melts congruently at 1817 *K*. Transposition on the 1968 T scale places the T of fusion at 1821 *K* (Richet et al. 1991). The enthalpy of fusion is ~56 *kJ/mol* (Barin 1995; Adamkokikova et al. 1980). At high pressure (i.e. above roughly 2.5 *GPa*), PsWo transforms into another triclinic polymorph: Wollastonite II (Essene 1974; Wo-II). Wo-II is structurally quite similar to $CaSiO_3$-walstromite where "*the axial ring Si_3O_9 is isomorphously replaced by the same ring with the inverse axiality*" (Joswig et al. 2003). The Ca positions in Wollastonite-II and $CaSiO_3$-walstromite are almost the same. According to Essene (1974) the univariant boundary has a slightly negative P–T slope boundary [P(*bar*) = −4.7 × T(°*C*) + 32.418]. Above ~10 *GPa* (Ringwood and Major 1967) Wollastonite-II transforms into "*Ca-Perovskite*" (Ca-Pvsk; see also Fig. 20.75). Ca-Pvsk was the object of an extensive investigation because it could constitute a relevant part of the lower

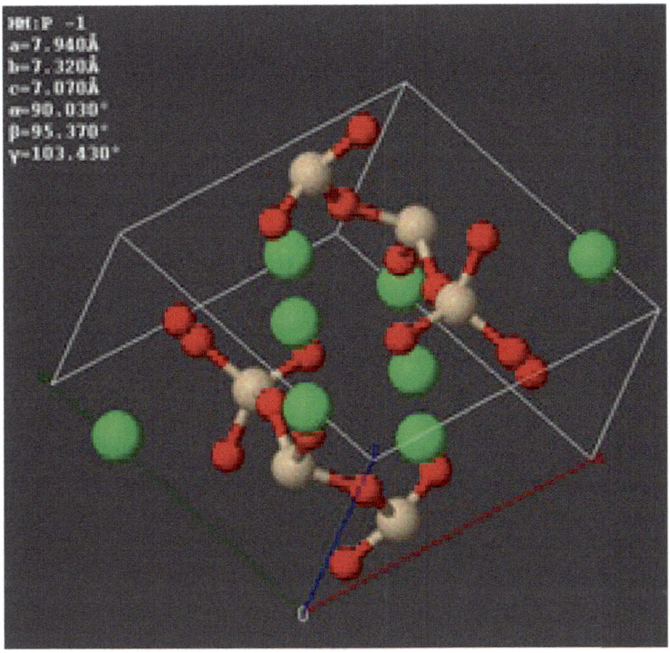

Fig. 20.73 $P\bar{1}$ structure of Wollastonite 1A (Ito et al. 1969), as generated by the AMCSD software (Downs and Hall-Wallace 2003)

transition zone and lower Earth's mantle (6–12 wt%). As kept in evidence by Adams and Oganov (2006) there is however no consensus on the stable cristalline form of this polymorph in the wide P–T regimes of interest:

- cubic, spatial group $Pm\bar{3}m$ at low T and P $\leq$ 134 *GPa* (Liu and Ringwood 1975; Mao et al. 1989; Wang et al. 1996);
- tetragonal, spatial group *I4/mcm* at low T (Stixrude et al. 1996);
- orthorhombic *Pbnm* at room T (Akber-Knutson et al. 2002; Magyari-Köpe et al. 2002);
- tetragonal at room T and 20 $\leq$ P (*GPa*) $\leq$ 46 *GPa* (Ono et al. 2004; Kurashima et al. 2004).

According to Ono et al. (2004) and Kurashina et al. (2004) there should be a (displacive?; see later) transition from a low-T to a high-T polymorph at P $\sim$ 50 *GPa* and T comprised between 500 and 700 *K* (with a Clapeyron slope of 0.18 *GPa/K*).

Joswig et al. (2003) performed a DFT/GGA investigation of Wollastonite II and $CaSiO_3$-walstromite. The computation involved the PBE Exchange–correlation functional (see Sect. 10.5.4) and adopted ultrasoft pseudopotentials with a maximum cutoff energy of the plane waves of 340 *eV*. The irreducibile part of BZ was sampled on the basis of a Monkhorst–Pack grid such that the distances between grid points were of the order of 5×10^{-2} *Å* (Joswig et al. 2003). After relaxation the remaining forces on the atoms were <0.02 *eV/Å*, and the remaining stress was less than 0.05 *GPa* (Joswig et al. 2003). The calculation returned identical total energies for the two structures (within the precision of calculations, i.e. 0.2 *eV*/unit cell, corresponding to 3 *kJ/mol*). Also the compressional behavior of the two polymorphs was identical: $K_0 = 76 \pm 2$ *GPa* and $K' = 3.8$. Comparing the unit cell volume of wollastonite-II at ambient conditions according to Trojer (1969) ($V_0 = 377.69$ $Å^3$) with the in-situ determination of the volume of a $CaSiO_3$-walstromite crystal included into a diamond ($V = 365.39$ $Å^3$; Joswig et al. 1999) and with the released unit cell volume ($V_0 = 375.16$ $Å^3$; Joswig et al. 2003) the Authors deduced that a relative compression of 3.6%, corresponding to a remnant pressure of about 3 *GPa*, affected the included specimen (Joswig et al. 2003).

Adams and Oganov (2006) performed an AIMD investigation of Ca-Pwsk in the aim of resolving the previously mentioned controversies. The investigation was carried out with the Vienna code under GGA with PAW and the PBE functional. Core region cutoffs were set to 2.3 a.u. for Ca ($1s^22s^22p^6$), 1.5 a.u. for Si ($1s^22s^22p^6$), and 1.52 a.u. for O ($1s^2$). The remaining 10 (Ca), 4 (Si) and 6 (O) electrons were treated explicitly. Partial core corrections were done for Ca (2.0 *a.u.*) and Si (1.30 a.u.). The kinetic cutoff in MD calculations was set to 400 *eV*. The unit cell (Z = 4; 20 atoms) was first optimized (relaxed) at the athermal limit and P = 0, 50, 100, and 150 *GPa*. The relaxation proceeded downward to total energy changes below 10^{-6} *eV* per cell. The optimized $2\times1\times1$ tetragonal, orthorhombic, and cubic supercells (80 atoms) constituted the NVT ensembles for MD simulations at T = 500, 1500, 2500, and 3500 *K*. Simulations were run for at least 8 *ps* with 1 *fs* time step. The threshold for attainment of the ground state at each time step was set at 2×10^{-5} *eV*. Equilibration lasted typically several hundred *fs*. The Brillouin zone was sampled only at the Γ point.

At each T the difference in the static and thermal system gave (on the same NTV ensemble) the thermal pressure P_{th} (see Sect. 8.9.4):

$$P_{th}(V,T) = P(V,T) - P_{st}(V) \quad (20.5.26)$$

where $P_{st}(V)$ is the pressure at the athermal limit and P(V, T) embodies the kinetic gas correction $(N-1)k_BTV^{-1}$, with N as number of atoms in the microcanonical ensemble. Following Oganov and Dorogokupets (2003) the Authors defined then an "*effective Grüneisen parameter*" (γ_{eff}) in the form:

$$\gamma_{eff}(V,T) = \frac{P_{th}(V,T)}{(N-3)k_BT} \quad (20.5.27)$$

γ_{eff} has "*a small temperature dependence from which intrinsic anharmonic effects can be estracted*" (Adams and Oganov 2006):

$$\frac{d\ln a(V)}{d\ln V}\frac{T}{2}a(V) = \gamma_{eff}(V,T) - \gamma(V) \quad (20.5.28)$$

where $\gamma(V)$ is the classical Grüneisen parameter and

$$a(V) = a_0 \left(\frac{V}{V_0}\right)^m \tag{20.5.29}$$

Representing the V-dependent part of γ_{eff} as suggested by Al'tshuler et al. (1987):

$$\gamma_{eff}(V) = \gamma_\infty - (\gamma_\infty - \gamma_0)\left(\frac{V}{V_0}\right)^\beta \tag{20.5.30}$$

where γ_∞ if the Grüneisen parameter at infinite compression and γ_0 the same parameter at room conditions (V_0 = 187.31 $Å^3$; $V/V_0 = 1$), Adams and Oganov (2006) obtained the five parameters necessary to conform the *intrinsic anharmonicity* of the cubic structure on the basis of 14 MD runs ($\gamma_\infty = 0.9185$, $\gamma_0 = 1.873$, $a_0 = 1.29 \times 10^{-5}\ K^{-1}$, m = 5.63, $\beta = 4.580$). The computations showed that γ_{eff} depends significantly on T at low pressure, but this dependance is suppressed at high pressure (see Fig. 8 in Adams and Oganov 2006). The positive value of the intrinsic anharmonicity parameter a(V) implies an overall increase of frequencies with temperature if the volume is kept fixed:

$$a(V) = \left\langle \left(\frac{\partial \ln \omega_i}{\partial T}\right) V \right\rangle \tag{20.5.31}$$

According to Dove (1993), intrinsic anharmonic effects suppress dynamical instabilities at high T, hence the cubic $CaSiO_3$ polymorph, dynamically unstable at low temperatures, is stabilized at high temperatures (Adams and Oganov 2006). Besides this inportant conclusion Adams and Oganov (2006) found that the enthalpy of the tetragonal *I4/mcm* structure of $CaSiO_3$ perovskite at 0 *K* is lower than for any other polymorph, hence this is the most probable low-temperature symmetry.

Background information: intrinsic anharmonicity
Accepting the concept of "*intrinsic anharmonicity*" implies acceptance of the idea that phonons are "*quasiparticles*". Denoting as *k* the quantum number of a phonon quasiparticle and as n_k the number of quasiparticles (QP) in the quantum state *k*, the total QP's energy is

$$E_{QP} = \sum_k \hbar\omega_k n_k \tag{20.5.32}$$

their distribution is given by the Bose–Einstein function

$$n_k = \frac{1}{\exp\left(\frac{\hbar\omega_k}{k_B T}\right) - 1} \tag{20.5.33}$$

and the (maximized) entropy is (Landau and Lifshitz 1980):

$$S_{QP} = k_B \sum_k [(n_k + 1)\ln(n_k + 1) - n_k \ln n_k] \tag{20.5.34}$$

Hui and Allen (1975), adopting the self-consistent phonon method of Werthamer (1969) proposed that entropy may be given by the (harmonic) formula

$$S = -k_B \sum_q \ln 2\sin\left(\frac{\hbar\widetilde{\omega}_q}{2k_B T}\right) - \left(\frac{\hbar\widetilde{\omega}_q}{2k_B T}\right)\coth\left(\frac{\hbar\widetilde{\omega}_q}{2k_B T}\right) \tag{20.5.35}$$

where $\widetilde{\omega}_q(T)$ are *T*-dependent renormalized frequencies. Entropy is the only thermodynamic function for which the harmonic formula holds (Hui and Allen 1975). Cochran and Cowley (1967) and Wallace (1973) have shown that the entropy is given correctly by the harmonic formula with the harmonic phonon frequencies replaced by the measurable frequencies, and that this holds to leading order in the usual perturbation theory of anharmonicity.

Using perturbation theory, Balian and De Dominicis (1964) showed that, entropy (at any T) can be formally written in terms of a distribution function $n_{\overline{p}\sigma}$:

$$S = -k_B \sum_{\overline{p}\sigma} n_{\overline{p}\sigma} \ln n_{\overline{p}\sigma} \pm (1 \mp n_{\overline{p}\sigma}) \ln(1 \mp n_{\overline{p}\sigma}) \tag{20.5.36}$$

where the upper sign is for fermions and the lower for bosons. As noted by Hui and Allen (1975), for bosons, this immediately gives Eq. (20.5.35). Moreover the '*statistical*' quasiparticle energy which occurs in $n_{\overline{p}}$ is not necessarily the same as the '*dynamical*' quasiparticle energy defined through the real part of the poles of the single-particle Green function (Balian and De Dominicis; Hui and Allen 1975). For the case of anharmonic effects in crystal dynamics, Cowley (1963) has shown that within the quasi-harmonic approximation, the effective energies in $n_{\overline{p}}$ are those determined by neutron spectrometry.

Ca-Pvsk has been the object of a quite innovative investigation carried out by Sun et al. (2014) which combined AIMD with lattice dynamics:

(1) ionic velocities obtained from AIMD trajectories are projected onto vibrational normal modes;
(2) renormalized phonon frequencies and lifetimes are entracted;
(3) dynamical matrices are built using the renormalized phonon frequencies and the phonon polarization vectors;
(4) the obtained phonon dispersion and the projected DOS are adopted to determine entropy and Helmholtz free energy in the framework of the phonon gas model (PGM; Sun and Allen 2010).

Computations were carried out with the QUANTUM ESPRESSO computational code (Giannozzi et al. 2009) in the framework of the Density Functional Perturbation Theory (DFPT; Baroni et al. 2001; see Sect. 8.8.1) and LDA with Vanderbilt pseudopotentials (see Sect. 12.7.2). Core region cutoffs were set to 2.0 a.u. for Ca ($1s^22s^22p^63s^23p^64s^14p^03d^0$), 1.6 a.u. for Si ($1s^22s^22p^63s^23p^13d^0$), and 1.4 a.u. for O ($1s^22s^22p^4$). Ca-Pvsk was simulated with a 2×2×2 cubic supercell (40 atoms). The plane-wave energy cutoff was set to 36 *Ry*. Because a simple Γ point analysis "*would cause errors of more than 1 GPa in the calculated stresses*" (Sun et al. 2014), a shifted 2×2×2 point mesh was adopted (4 **k** points). Born–Oppenheimer molecular dynamics on the NTV ensembles adopted a Nose–Hoover thermostat with T ranging from 10 to 4500 *K*. The edge of the cubic cell was set to a_0 = 3.45 *Å*, corresponding to a static pressure of 23 *GPa*. The time-step was set to 0.97 *fs* (about 1/30 of the period of the highest-frequency phonon in the system). Fairly long simulations (>50 *ps*) were carried out at each T, in order to minimize the statistical error. The power spectra of velocity autocorrelation functions were computed with the maximum entropy method (Press et al. 1986).

With the phonon frequencies of the 2×2×2 supercell, the renormalized phonon dispersion in the whole BZ was obtained stemming from the power spectrum of the velocity autocorrelation function:

$$\langle \mathbf{v}_{ls} \cdot \mathbf{v}_{ls} \rangle_{\omega} = \sum_{\alpha} \int 0^{\infty} \langle \mathrm{v}_{ls\alpha}(0) \mathrm{v}_{ls\alpha}(\mathrm{t}) \rangle \exp(\mathrm{i}\omega \mathrm{t}) \mathrm{dt} \tag{20.5.37}$$

where $\mathbf{v}_{ls}$ denotes the mass-weighted velocity of particle *s* in the unit cell *l* and $\mathbf{v}_{ls\alpha}$ is its α-Cartesian component.

Individual phonons were resolved projecting first $\mathbf{v}_{ls}$ onto different wave vectors through translational symmetry operations:

$$\mathbf{v}_{\mathbf{q}s} = \sum_{l} \mathbf{v}_{ls} \exp(-\mathbf{i}\mathbf{q}\mathbf{R}_l) \tag{20.5.38}$$

with **q** as wave vector, and then, because the total Vibrational Density of States (VDOS) in a material like Ca-Pvsk is quite complex (i.e. with several vibrational modes of similar frequency and overlapping spectral profiles in the power spectrum):

$$\text{VDOS} = \sum_{ls} \langle \mathbf{v}_{ls} \cdot \mathbf{v}_{ls} \rangle_{\omega} \equiv \sum_{s} \langle \mathbf{v}_{\mathbf{q}s} \cdot \mathbf{v}_{\mathbf{q}s} \rangle_{\omega} \tag{20.5.39}$$

a further projection was then adopted (Sun et al. 2014):

$$\mathrm{v}_{\mathbf{q}j} = \sum_{s\alpha} \mathrm{v}_{\mathbf{q}s\alpha} \hat{\mathrm{e}}_{a}^{*}(\mathrm{s}|\mathbf{q}\mathrm{j}) \tag{20.5.40}$$

where $\hat{\mathrm{e}}_{\alpha}^{*}(\mathrm{s}|\mathbf{q}\mathrm{j})$ is the polarization vector of the harmonic phonon with wave vector $\mathbf{q}$ and branch j (Sun et al. 2014). The resulting power spectrum of the autocorrelation function for $\mathrm{v}_{\mathbf{q},\mathrm{j}}$ is

$$\left\langle \mathrm{v}_{\mathbf{q}j}^{*} \cdot \mathrm{v}_{\mathbf{q}j} \right\rangle_{\omega} = \int_{0}^{\infty} \left\langle \mathrm{v}_{\mathbf{q}j}^{*}(0) \mathrm{v}_{\mathbf{q}j}(\mathrm{t}) \right\rangle \exp(\mathrm{i}\omega \mathrm{t}) \mathrm{dt} \tag{20.5.41}$$

The "*effective dynamical matrix*" is Sun et al. (2014):

$$\tilde{\mathbf{D}}_{\alpha\beta}(ss'|\mathbf{q}) = \sum_{j} \widetilde{\omega}_{\mathbf{q}j}^{2} \hat{e}_{\alpha}(s|\mathbf{q}j) \hat{\mathrm{e}}_{\beta}^{*}(\mathrm{s}'|\mathbf{q}\mathrm{j}) \tag{20.5.42}$$

and differs from the harmonic counterpart $\mathbf{D}$, in having renormalized eigenvalues (i.e. from $\omega_{\mathbf{q}j}^{2}$ to $\widetilde{\omega}_{\mathbf{q}j}^{2}$).

The renormalized phonon dispersions obtained by Sun et al. (2014) for Ca-Pvsk at T = 1000 k is shown in Fig. 20.74. The frequencies extracted from the normal-mode correlation functions are represented by the circles. The solid black lines are the results of Fourier interpolation. The dashed lines are the phonon dispersions determined by fitting an effective harmonic force constant matrix (Hellman et al. 2011), while harmonic phonon dispersions calculated with DFPT are shown in orange. The LO-TO splitting effect, absent in supercell calculations, was taken into account by adding a nonanalytic contribution to the dynamical matrix in the long-wavelength limit (Baroni et al. 2001).

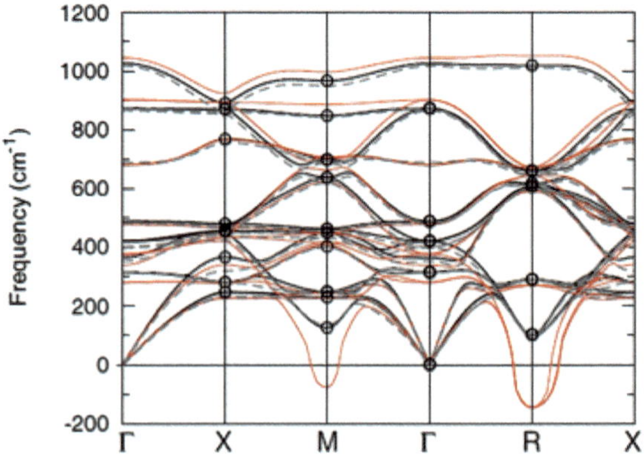

Fig. 20.74 Phonon dispersion of Ca-Pvsk at T = 1000 K. From Sun et al. (2014). With permission from The American Physical Society

Sun et al. (2014) adopted then Eq. (20.5.34) and expressed the vibrational entropy in vectorial notation and with renormalized phonon frequencies as

$$S_{vib} = k_B \sum_{\mathbf{q}j} \left[(n_{\mathbf{q}j} + 1) \ln (n_{\mathbf{q}j} + 1) - n_{\mathbf{q}j} \ln n_{\mathbf{q}j} \right] \tag{20.5.43}$$

$$n_{\mathbf{q}j} = \frac{1}{\exp\left(\frac{\hbar \widetilde{\omega}_{\mathbf{q}j}}{k_B T}\right) - 1} \tag{20.5.44}$$

and estimated the anharmonic entropy S_A in cubic Ca-Pvsk using renormalized frequencies at 600 *K*:

$$S_A = S_{vib}[\widetilde{\omega}(T)] - S_{vib}[\widetilde{\omega}(600)] \tag{20.5.45}$$

where $S_{vib}[\widetilde{\omega}(T)]$ is the vibrational entropy calculated from Eqs. (20.5.43) to (20.5.44) and $S_{vib}[\widetilde{\omega}(600)]$ is the reference calculus with phonon frequencies fixed at those of 600 *K*. The anharmonic part of the Helmholtz free Energy was then evaluated by applying

$$F_A(T) = -k_B T(N-1)) \int_0 T \left[\frac{E(T') - E_0}{(N-1) k_B T'} - 3 \right] \tag{20.5.46}$$

where N is the number of atoms in the supercell, E(T′) is the statistical average of the total Energy for the MD simulation carried out at temperature T′ and E_0 is the static energy of the relaxed structure. Practical calculations were carried out replacing the integrand with the exponential function

$$f(x) = a\left[\exp\left(-\frac{x}{r} - 1\right)\right] \tag{20.5.47}$$

with a and r as fitting parameter. $F_A(T)$ was then obtained by integration, and $S_A(T)$ as $S_A(T) = -(\partial F_A(T)/\partial T$. The anharmonic part of the vibrational entropy calculated in this way for Ca-Pvsk is quite small (i.e. about 3% of the total entropy).

In Fig. 20.75, I reproduce finally a simplified version of the $CaSiO_3$ phase diagram obtained by Belmonte et al. (2017a) through an "*ab initio assisted*" procedure.

For Wo and PsWo Belmonte et al. (2017a) adopted the heat capacity coefficients of Berman (1988) and disregarded the low-Wo to high-Wo transition. Below ~1000 *K* the low-Wo data of Richet et al. (1991) are consistent with the selected heat capacity and a good consistency is observed also for ps-Wo up to ~2000 *K*. Also for the enthalpy of formation from the elements at standard state and the entropy of the two polymorphs the assessment of Berman (1988) was initially selected. In the following optimization, H_f^0 of both polymorphs was lowered by ~0.6 *kJ/mol*. Belmonte et al. (2017) did not make any distinction between

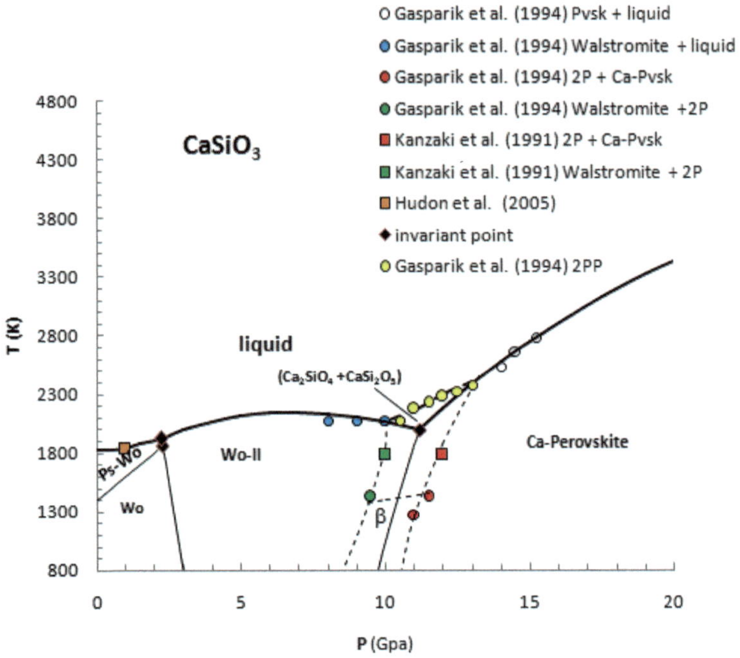

Fig. 20.75 Phase equilibria in the system $CaSiO_3$ after Belmonte et al. (2017a). Thin solid lines delineate the primary phase fields at the subsolidus. The heavy solid line is the melting curve of the substance. Selected experimental data from Gasparik et al. (1994), Kanzaki et al. (1991), and Hudon et al. (2005) are also shown. The low-P subsolidus ($P \leq 3\ GPa$) is consistent with the data of Osborn and Schairert (1941), Kushiro (1964), and Essene (1974). The phase diagram is nearly similar to what proposed by Gasparik et al. (1994). Belmonte et al. (2017) however identified the true invariant point Wo-II/Ca-Pvsk/Liquid which is localized at $P = 11.18\ GPa$ and $T = 1994\ K$. The dotted lines at subsolidus locate the two phases field (β-Ca_2SiO_4 and $CaSi_2O_5$) where Walstromite is replaced with Wo-II

Wollastonite-II (Wo-II) and $CaSiO_3$-walstromite Initially they adopted for the two polymorphs the heat capacity equation provided by Swamy and Dubrovinsky (1997). When coupled with the thermophysical parameters selected or refined in the investigation, a slight (Wo-II) or marked (Ca-Pvsk) discrepancy with the asymptotic Dulong-Petit limit was evident and a partial re-fitting was deemed as necessary. For Wo-II the adopted standard state molar volume was that one of Chatterjee et al. (1994) and the thermal expansion was that one provided by Swamy and Dubrovinsky (1994). The bulk modulus and its thermal derivatives were refined on the basis of the Wo/Wo-II subsolidus univariant curve. For Ca-Pvsk the standard state molar volume is from Wang et al. (1996). Isobaric Thermal expansion and isothermal compressibility were also refined. The optimized (internally consistent) thermophysical and thermodynamic parameters of $CaSiO_3$ according to Belmonte et al. (2017a), which lead to the P–T phase diagram of Fig. 20.75, are listed in Tables 20.64 and 20.65.

Table 20.64 Optimized thermophysical properties $CaSiO_3$ (Belmonte et al. (2017a). V^0_{298} = standard state molar volume (T_r = 298.15 *K*, P_r = 1 *bar*). K_0 = bulk modulus at the athermal limit (i.e. T = 0 *K*; P = 0 *GPa*); K'_0 = $(dK/dP)_0$ = pressure derivative of the bulk modulus (assumed independent on T). $K_T = K_0 + k_1 \times T + k \times T^2$ gives the temperature dependence of the bulk modulus. α_V = volume thermal expansion coefficient ($\alpha_V = \alpha_0 T + \alpha_1 + \alpha_2 T^{-1} + \alpha_3 T^{-2} + \alpha_4 T^{-3}$)

Composition (phase)	V^0_{298} (*cc/mol*)	K_0 (*GPa*)	K′	k_1 (*bar/k*)	k_2(*bar/K*^{2})	$\alpha_0 \times 10^7$ (K^{-2})	$\alpha_1 \times 10^7$ (K^{-1})	$\alpha_2 \times 10^3$	α_3 (*K*)	α_4 (K^2)
$CaSi0_3$ (Wollastonite-I)	39.62	116.6	4	−330	0.027	0.04664	339.404	1.4588	−1.8634	0
$CaSiO_3$ (Pseudowollastonite)	40.08	113	4	−300	0.029	−0.013	466	−8.3137	0	0
$CaSiO_3$ (Wollastonite-II)	37.94	93.9	5	0	0	0.01998	320	0	0	0
$CaSiO_3$ (Ca-perovskite)	27.45	239	4.8	−300	0.01	0.01	300	0	0	0

Table 20.65 Optimized thermodynamic properties of $CaSiO_3$ (Belmonte et al. 2017a). $H^0_{f,298}$ = enthalpy of formation from the elements at $T_r = 298.15\ K$, $P_r = 1\ bar$; S^0_{298} = standard state entropy ($T_r = 298.15\ K$, $P_r = 1\ bar$); C_P = isobaric heat capacity ($C_p = a + bT + cT^{-2} + dT^{-1/2} + eT^{-3} + fT^2 + gT^3 + hT^{-1}$)

Composition (phase)	$H^0_{f,298}$ (*J/mol*)	S^0_{298} (*J/mol × K*)	a	b × 10^3	c × 10^{-5}	d	e × 10^{-8}	f × 10^6	g × 10^9	h × 10^{-4}
$CaSi0_3$ (Wollastonite-I)	−1,632,094	81.81	149.07	0	−36.593	−690.3	4.8435	0	0	0
$CaSiO_3$ (Pseudowollastonite)	−1,628,026	85.279	141.16	0	−58.576	−417.2	9.4074	0	0	0
$CaSiO_3$ (Wollastonite-II)	−1,626,762	82.1	127.954	5.1129	−15.17495	0	0	0	0	0
$CaSiO_3$ (Ca-perovskite)	−1,541,000	74.0	114.66	9.8918	−19.527	−13.052	1.518	−0.2516	−0.15381	−0,476,251

Appendix
Mathematics

A.1 Generalized Coordinates

Whenever the motion of a particle is not free but it is subjected to external forces or the particle is required to move on a given surface, or along a given trajectory, it is preferable to use a coordinate system such that one or more of the coordinates remains constant. Consider as an example Newton's law of motion. In classical mechanics, they are written in the form

$$f_x = m\frac{d^2x}{dt^2} \quad f_y = m\frac{d^2y}{dt^2} \quad f_z = m\frac{d^2z}{dt^2} \tag{A.1.1}$$

but if one wants to describe a planetary motion the forces are simple expressions of polar coordinates, but are quite complex if Cartesian coordinates are adopted. The simplest example of motion subject to constraints is the motion of a rigid body. Considering a rigid body as an ensemble of particles moving in such a way to keep unvaried their relative positions, and fixing the position of one of them, the position of all the other particles in the ensemble is determined. If there are N particles, their positions are specified by 3N. If for simplicity we reduce the ensemble to three non-co-linear particles, we will have nine Cartesian coordinates, but between these nine coordinates there are three relations corresponding to the distances between them, so that the position of the whole ensemble is fixed by only six variables and the ensemble is said to have *six degrees of freedom* (five in the case of co-linearity). In general, a system of N coordinates with F degrees of freedom is determined by F coordinates ($q_1, q_2, \ldots q_F$), the conditions of constraint being expressed by the 3N-F remaining variables ($q_{F+1}, q_{F+2}, \ldots q_{3N}$). In terms of the q's, the rectangular coordinates are given by the 3N functions

G. A. Ottonello, *Quantum Geochemistry*, Springer Geochemistry,
https://doi.org/10.1007/978-3-031-21837-8

$$\begin{array}{lll} x_1 = \varphi_1(q_1 \ldots q_{3N}) & y_1 = \varphi_2(q_1 \ldots q_{3N}) & z_1 = \varphi_3(q_1 \ldots q_{3N}) \\ \cdot & \cdot & \cdot \\ \cdot & \cdot & \cdot \\ \cdot & \cdot & \cdot \\ x_N = \varphi_{3N-2}(q_1 \ldots q_{3N}) & y_N = \varphi_{3N-1}(q_1 \ldots q_{3N}) & z_N = \varphi_{3N}(q_1 \ldots q_{3N}) \end{array} \tag{A.1.2}$$

A.2 Polar Coordinates

The polar coordinates are defined in terms of Cartesian coordinates by

$$\left.\begin{array}{l} x = r\cos\theta \\ y = r\sin\theta \end{array}\right\} \tag{A.2.1}$$

where r is the *radial distance* from the origin, and θ is the polar angle measured counterclockwise from the x-axis (Fig. A.1). The radial distance is related to the Cartesian coordinates by

$$r = \sqrt{x^2 + y^2} \tag{A.2.2}$$

and the polar angle θ by

$$\theta = \arctan\left(\frac{y}{x}\right) \tag{A.2.3}$$

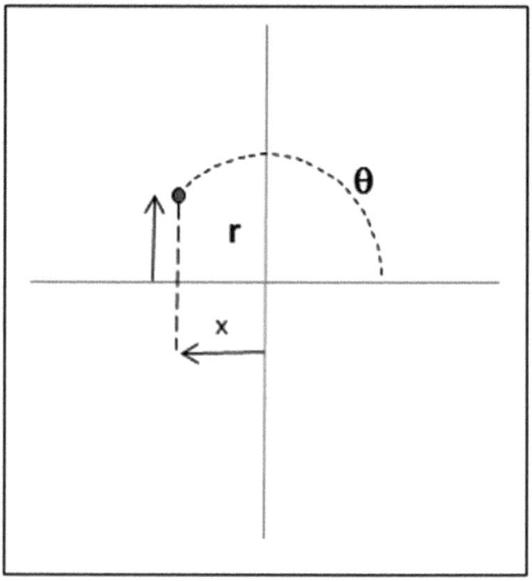

Fig. A.1 Polar coordinates (standard notation)

A.2.1 Cylindrical Polar Coordinates

Superimposing a weight (z)-axis the polar coordinates are generalized to three dimensions. Following the physical representation (Arfken 1985) in terms of the Cartesian coordinates, we have

$$\left.\begin{array}{c} x = r\cos\phi \\ y = r\sin\phi \\ z = z \end{array}\right\} \tag{A.2.4}$$

and in terms of Cylindrical polar coordinates (Fig. A.2)

$$\left.\begin{array}{c} r = \sqrt{x^2+y^2} \\ \phi = \arctan\left(\frac{y}{x}\right) \\ z = z \end{array}\right\} \tag{A.2.5}$$

It should be noted to this purpose that different notational conventions exist in literature (see Weisstein 2018) for an extended discussion.

The Cartesian vector in cylindrical coordinates is implicitly defined by notation (A.2.4) while the unit vector components are

$$\hat{\boldsymbol{\rho}} = \begin{vmatrix} \cos\phi \\ \sin\phi \\ 0 \end{vmatrix}, \quad \hat{\boldsymbol{\phi}} = \begin{vmatrix} -\sin\phi \\ \cos\phi \\ 0 \end{vmatrix}, \quad \hat{\mathbf{z}} = \begin{vmatrix} 0 \\ 0 \\ 1 \end{vmatrix} \tag{A.2.6}$$

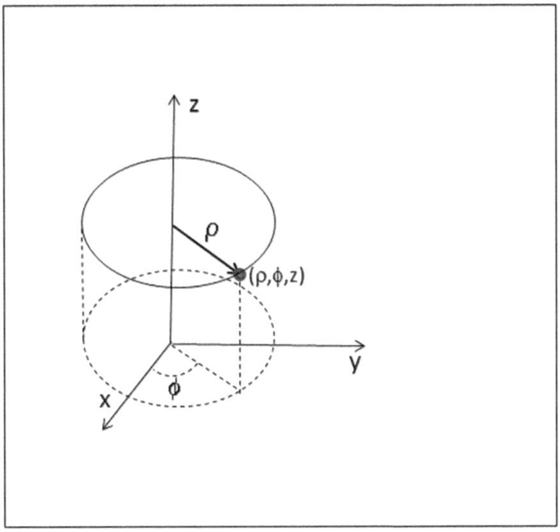

Fig. A.2 Cylindrical polar coordinates in the physical notation (Arfken 1985)

A.2.2 Spherical Polar Coordinates

It is the three-dimensional equivalent of the polar system. The position of the point is specified by its *radial distance* r (distance from a fixed origin, corresponding to zero in Cartesian coordinates), the *polar angle* (measured from a fixed zenith direction), and the *azimuthal angle* (on the orthogonal projection on a reference plane). It should be stressed from the very beginning that the ISO notation adopted in physics does not correspond to the notation adopted in mathematics (Fig. A.3).

Adopting the ISO notation in terms of the Cartesian coordinates, we have

$$\left.\begin{array}{c} \mathrm{x} = \mathrm{r}\sin\theta\cos\phi \\ \mathrm{y} = \mathrm{r}\sin\theta\sin\phi \\ \mathrm{z} = \mathrm{r}\cos\theta \end{array}\right\} \tag{A.2.7}$$

and in terms of spherical polar coordinates

$$\left.\begin{array}{c} \mathrm{r} = \sqrt{\mathrm{x}^2+\mathrm{y}^2+\mathrm{z}^2} \\ \phi = \arctan\left(\frac{\mathrm{y}}{\mathrm{x}}\right) \\ \theta = \arccos\frac{\mathrm{z}}{\sqrt{\mathrm{x}^2+\mathrm{y}^2+\mathrm{z}^2}} \equiv \arccos\frac{\mathrm{z}}{\mathrm{r}} \end{array}\right\} \tag{A.2.8}$$

The conversion into cylindrical coordinates implies

$$\left.\begin{array}{c} \rho = \mathrm{r}\sin\phi \\ \phi = \phi \\ \mathrm{z} = \mathrm{r}\cos\phi \end{array}\right\} \tag{A.2.9}$$

and the reciprocal conversion from cylindrical coordinates to spherical ones implies

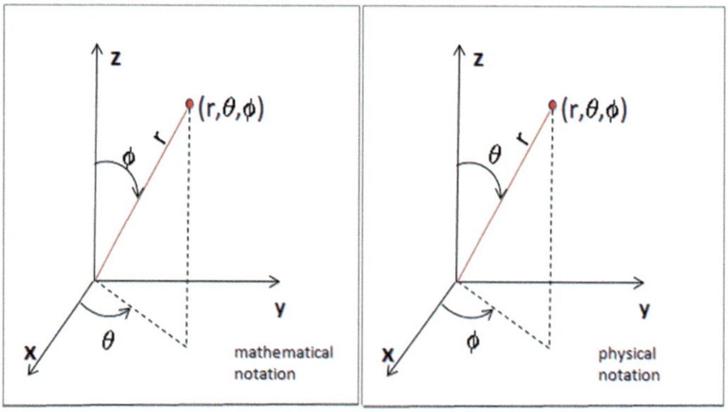

Fig. A.3 Spherical polar coordinates in mathematical and physical (ISO) standard notation

$$\left.\begin{aligned} r &= \sqrt{\rho^2 + z^2} \\ \phi &= \phi \\ \phi &= \arctan\tfrac{\rho}{z} \equiv \arccos\frac{z}{\sqrt{\rho^2 + z^2}} \end{aligned}\right\} \tag{A.2.10}$$

A.3 Spheroidal Coordinates

Spheroidal coordinates are three-dimensional orthogonal systems resulting from the rotation of the two-dimensional elliptic coordinate system about the focal (oblate) or the non-focal (prolate) axes of the ellipses.

The elliptical coordinates are defined as follows:

$$\xi = \frac{r_1 + r_2}{2f} \quad \eta = \frac{r_1 - r_2}{2f} \tag{A.3.1}$$

where r_1 and r_2 are the distances to the foci of a family of confocal ellipses and hyperbolas and 2f is the distance between the foci.

The third set of coordinates consists of planes passing through the new z-axis.

The elliptical coordinates are related to Cartesian coordinates through

$$x = f\xi\eta \quad y = f\sqrt{(\xi^2 - 1)(1 - \eta^2)} \tag{A.3.2}$$

The equations of family of confocal ellipses and confocal hyperbolas are as follows:

$$\begin{aligned} &\frac{x^2}{\xi^2} + \frac{y^2}{\xi^2 - 1} = f^2 \quad (1 < \xi < \infty) \\ &\frac{x^2}{\xi^2} - \frac{y^2}{1 - \eta^2} = f^2 \quad (-1 < \eta < 1) \end{aligned} \tag{A.3.3}$$

A.3.1 Oblate Spheroidal Coordinates

The oblate spheroidal coordinates are obtained by revolving the curves of the elliptic cylindrical coordinates ξ, η about the focal y-axis which is re-labeled the z-axis.

If the system of confocal ellipses and hyperbolas revolves around the minor axis, then

$$\frac{r^2}{\xi^2} + \frac{y^2}{\xi^2 - 1} = f^2 \qquad \frac{r^2}{\xi^2} - \frac{y^2}{1 - \eta^2} = f^2 \tag{A.3.4}$$

With ξ, η, and ϕ as oblate spheroidal coordinates, we have

$$x = r \sin \phi \tag{A.3.5}$$

$$z = r \cos \phi \tag{A.3.6}$$

$$0 \leq \phi \leq 2\pi \tag{A.3.7}$$

The Laplacian is

$$\begin{aligned} \nabla^2 f = \frac{1}{\sin^2 \eta + \sinh^2 \xi} \bigg[& \left(\operatorname{sech}^2 \xi + \tan^2 \eta\right) \frac{\partial^2 f}{\partial \phi^2} + \tanh \xi \frac{\partial f}{\partial \xi} \\ & + \frac{\partial^2 f}{\partial \xi^2} - \tan \eta \frac{\partial f}{\partial \eta} + \frac{\partial^2 f}{\partial \eta^2} \bigg] \end{aligned} \tag{A.3.8}$$

A wave equation expressed in oblate spheroidal coordinates takes the form (Abramovitz and Stegun 1972):

$$\begin{aligned} \nabla^2 \Psi + k^2 \Psi = & \frac{\partial}{\partial \xi} \left[\left(\xi^2 + 1\right) \frac{\partial \Psi}{\partial \xi} \right] + \frac{\partial}{\partial \eta} \left[\left(1 - \eta^2\right) \frac{\partial \Psi}{\partial \eta} \right] \\ & + \frac{\xi^2 + \eta^2}{(\xi^2 + 1)(1 - \eta^2)} \frac{\partial^2 \Psi}{\partial \phi^2} + c^2 (\xi^2 + \eta^2) \Psi = 0 \end{aligned} \tag{A.3.9}$$

where $c = \frac{1}{2} fk$.

A.3.2 Prolate Spheroidal Coordinates

The prolate spheroidal coordinates are obtained by revolving the curves of the elliptic cylindrical coordinates ξ, η about the x-axis which is re-labeled the z-axis.

The equations of family of confocal ellipses and confocal hyperbolas are as follows:

$$\frac{x^2}{\xi^2} + \frac{r^2}{\xi^2 - 1} = f^2 \qquad \frac{x^2}{\xi^2} - \frac{r^2}{1 - \eta^2} = f^2 \tag{A.3.10}$$

The prolate spheroidal coordinates are

$$z = r \sin \phi \tag{A.3.11}$$

$$y = r \cos \phi \tag{A.3.12}$$

$$0 \leq \phi \leq 2\pi \tag{A.3.13}$$

In terms of the distance between the two foci, the ξ and η variables are replaced, respectively, by

$$\lambda = \frac{r_1 + r_2}{R_{12}} \equiv \frac{r_A + r_B}{R_{AB}} \qquad \mu = \frac{r_1 - r_2}{R_{12}} \equiv \frac{r_A - r_B}{R_{AB}} \tag{A.3.14}$$

The Laplacian is

$$\begin{aligned} \nabla^2 f = & \frac{1}{\sin \eta + \sinh \xi} \left[\left(\csc^2 \eta + \operatorname{csch}^2 \xi \right) \frac{\partial^2 f}{\partial \phi^2} + \cot \eta \frac{\partial f}{\partial \eta} \right. \\ & \left. + \frac{\partial^2 f}{\partial \eta^2} + \coth \xi \frac{\partial f}{\partial \xi} + \frac{\partial^2 f}{\partial \xi^2} \right] \end{aligned} \tag{A.3.15}$$

A wave equation expressed in prolate spheroidal coordinates takes the form (Abramovitz and Stegun 1972):

$$\begin{aligned} \nabla^2 \Psi + k^2 \Psi = & \frac{\partial}{\partial \xi} \left[(\xi^2 - 1) \frac{\partial \Psi}{\partial \xi} \right] + \frac{\partial}{\partial \eta} \left[(1 - \eta^2) \frac{\partial \Psi}{\partial \eta} \right] \\ & + \frac{\xi^2 - \eta^2}{(\xi^2 - 1)(1 - \eta^2)} \frac{\partial^2 \Psi}{\partial \phi^2} + c^2 (\xi^2 - \eta^2) \Psi = 0 \end{aligned} \tag{A.3.16}$$

A.4 The Complex Conjugate

The Complex conjugate of a complex number z = a + ib is a complex number z* having the same real part a and an imaginary part of opposed sign, i.e. z* = a – ib.

For a trigonometric function, we have

$$\begin{aligned} & z = r[\cos \theta + i \sin \theta] \rightarrow \quad z^* = r[\cos \theta - i \sin \theta] \rightarrow \\ & z^* = r[\cos(-\theta) + i \sin(-\theta)] \end{aligned} \tag{A.4.1}$$

For an exponential form

$$z = r \exp(i\theta) \rightarrow \quad z^* = r \exp(-i\theta) \tag{A.4.2}$$

Also

$$|z| = |z^*| \tag{A.4.3}$$

$$zz^* = |z|^2 \tag{A.4.4}$$

$$z + z* = 2\ \mathrm{Re}(z) \tag{A.4.5}$$

$$z - z* = 2i\ \mathrm{Im}(z) \tag{A.4.6}$$

In the Argand–Gauss plane z* is symmetric to z with respect to the real axis (Fig. A.4).

A.5 Matrix and Vector Operations

Let $\Re$ be the set of real numbers and $\Re^{m\times n}$ the vector space. We denote in capital letter the matrix and in the corresponding lower case with subscripts i, j the entry, i.e. a_{ij} is a in the ith row and the jth column of matrix A. The matrix elements may be alternatively designated as A(i, j) or $|A|_{ij}$.

$$A \in \Re^{m\times n} \Leftrightarrow A = \left(a_{i,j}\right) = \begin{bmatrix} a_{11} & \ldots & a_{1n} \\ . & & . \\ . & & . \\ . & & . \\ a_{m1} & & a_{mn} \end{bmatrix} \quad a_{ij} \in \Re \tag{A.5.1}$$

The n-by-n matrixes are said to be *square*. A n-by-n matrix of the type

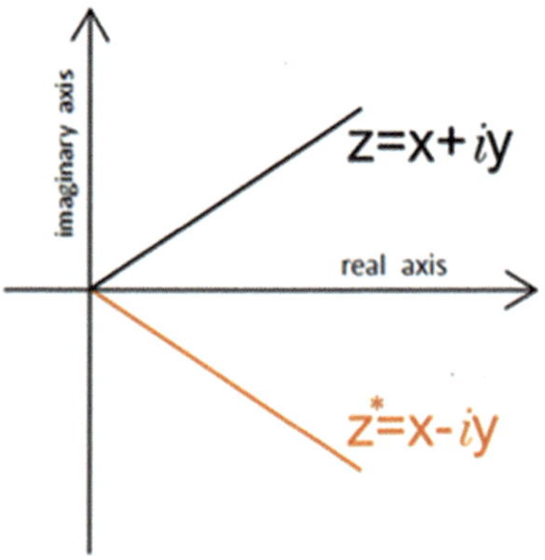

Fig. A.4 The complex number z and its complex conjugate in the Argand–Gauss plane

$$A = \begin{bmatrix} 1 & \dots & 0 \\ \cdot & & \cdot \\ \cdot & & \cdot \\ \cdot & & \cdot \\ 0 & & 1 \end{bmatrix} = I_n \quad e_k^{(n)} = \left(0, \dots, 0, \tfrac{1}{k}, 0, \dots, 0\right) \tag{A.5.2}$$

is said to be an *identity matrix*, it is denoted by I_n and its kth column by $e_k^{(n)}$.

Introducing the Kronecker delta δ_{ij},

$$\delta_{ij} = \begin{cases} 0 & \text{if } i \neq j \\ 1 & \text{if } i = j \end{cases} \tag{A.5.3}$$

we may also write $I = [\delta_{ij}]$.

The following basic operations are possible:

Transposition $\Re^{m\times n} \rightarrow \Re^{n\times m}$:

$$C = A^T \Rightarrow \quad c_{ij} = a_{ji} \tag{A.5.4}$$

Addition $\Re^{m\times n} + \Re^{m\times n} \rightarrow \Re^{m\times n}$:

$$C = A + B \Rightarrow \quad c_{ij} = a_{ji} + b_{ij} \tag{A.5.5}$$

Scalar-matrix multiplication $\Re \times \Re^{m\times n} \rightarrow \Re^{m\times n}$:

$$C = \alpha A \Rightarrow \quad c_{ij} = \alpha a_{ji} \tag{A.5.6}$$

Matrix–matrix multiplication $\Re^{m\times p} \times \Re^{p\times n} \rightarrow \Re^{m\times n}$:

$$C = A \cdot B \Rightarrow \quad c_{ij} = \sum_{k=1}^{r} a_{ik} b_{kj} \tag{A.5.7}$$

Hadamard product $\Re^{m\times n} \times \Re^{m\times n} \rightarrow \Re^{m\times n}$.

A simpler (and commutative) type of matrix multiplication (Horadam 2006; Caro-Lopera et al. 2012) is:

$$C = A \circ B \Rightarrow \quad c_{ij} = \left(a_{ij} b_{ij}\right) \tag{A.5.8}$$

In the case of two 3 × 3 square matrices A and B, for example,

$$\begin{bmatrix} a_{11} & a_{12} & a_{13} \\ a_{21} & a_{22} & a_{23} \\ a_{31} & a_{32} & a_{33} \end{bmatrix} \circ \begin{bmatrix} b_{11} & b_{12} & b_{13} \\ b_{21} & b_{22} & b_{23} \\ b_{31} & b_{32} & b_{33} \end{bmatrix} = \begin{bmatrix} a_{11}b_{11} & a_{12}b_{12} & a_{13}b_{13} \\ a_{21}b_{21} & a_{22}b_{22} & a_{23}b_{23} \\ a_{31}b_{31} & a_{32}b_{32} & a_{33}b_{33} \end{bmatrix} \tag{A.5.9}$$

Kronecker product $\Re^{m\times n} \times \Re^{p\times q} \rightarrow \Re^{m\times n-by-p\times q}$

$$C = A \otimes B \Rightarrow \quad c_{ij} = \left(a_{ij}B\right) \tag{A.5.10}$$

$$A \otimes B \quad = \begin{bmatrix} a_{11}B & \dots & a_{1n}B \\ \cdot & \cdot & \cdot \\ \cdot & \cdot & \cdot \\ \cdot & \cdot & \cdot \\ a_{m1}B & \dots & a_{mn}B \end{bmatrix} \tag{A.5.11}$$

The Kronecker product is not commutative and the Hadamard product is one of its sub-matrices (Caro-Lopera et al. 2012).

Differentiation $\Re^{m\times n} \rightarrow \Re^{m\times n}$:

$$C = \left(c_{ij}(\alpha)\right) \quad \dot{C} = \frac{d}{d\alpha}C = \left[\dot{c}_{ij}(\alpha)\right] \tag{A.5.12}$$

If A and B in $\Re^{n\times n}$ satisfy AB = I_n then B is said to be the *inverse* of A and is denoted as A^{-1}. If A^{-1} exists, then A is said to be *nonsingular*. If A^{-1} does not exist then *A* is *singular*. The transpose of the inverse of A is $(A^{-1})^T = (AT)^{-1} = A^{-T}$.

Given a square matrix A of order n, the determinant of A is the summation of the n! products containing each one a single element of a row or of a column, taken with their own sign or with the opposite sign depending on the class of permutation of the first and second indices (same class or different class).

Hence, if n = 1

$$\det A = a_{11} \tag{A.5.13}$$

if n = 2

$$\det A \begin{vmatrix} a_{11} & a_{12} \\ a_{21} & a_{22} \end{vmatrix} = a_{11}a_{22} - a_{12}a_{21} \tag{A.5.14}$$

if n = 3

$$\det A \begin{vmatrix} a_{11} & a_{12} & a_{13} \\ a_{21} & a_{22} & a_{23} \\ a_{31} & a_{32} & a_{33} \end{vmatrix} = a_{11}a_{22}a_{33} + a_{12}a_{23}a_{31} + a_{13}a_{21}a_{32} - a_{13}a_{22}a_{31} - a_{11}a_{23}a_{32} - a_{12}a_{21}a_{33} \tag{A.5.15}$$

A.5.1 Sarrus Rule

For a rank 3 matrix add to the right of the third column the first 2, then multiply between them the elements located on the three main diagonals and the three secondary ones. Finally, add algebraically the six products, with the last three changed of sign (Fig. A.5).

The following properties of determinants are particularly useful:

(1) For $A, B \in \Re^{n\times n}$ $\det(AB) = \det(A)\det(B)$
(2) For $A \in \Re^{n\times n}$ $\det(A^T) = \det(A)$
(3) For $c \in \Re, A \in \Re^{n\times n}$ $\det(cA) = c^n \det(A)$
(4) For $A \in \Re^{n\times n}$ $\det(A) \neq 0 \rightarrow A =$ *nonsingular*.

A.5.2 Laplace Theorem

The determinant of a square matrix A of order n is equal to the sum of the elements of whatever row or column of A multiplied by their respective algebraic complements:

$$\det(A) = \sum_{i=1}^{n} a_{ij}A_{ij} \quad (i = 1, 2, \ldots, n) \tag{A.5.16}$$

For $A \in \Re^{n\times n}$ the determinant of A is

$$\det(A) = \sum_{j=1}^{n} (-1)^{j+1} \det(A_{1j}) \tag{A.5.17}$$

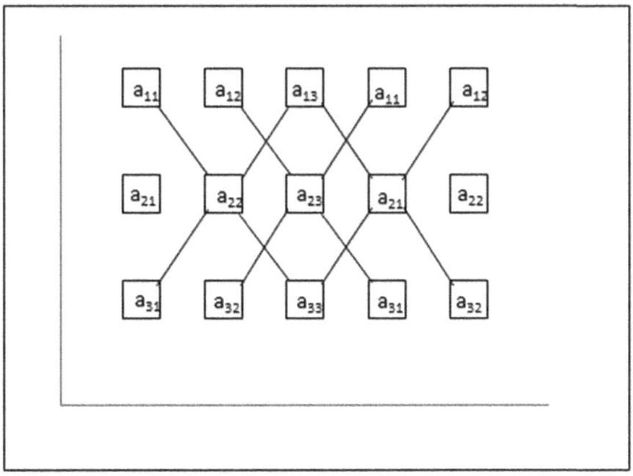

Fig. A.5 Sarrus rule for a third-rank matrix

where $A_{1,j}$ is an (n − 1)-by-(n − 1) matrix obtained by deleting the first row and the jth column of A.

A.5.3 Vector Operations

In this book, I denote as usual all vectors in bold character to distinguish them immediately from scalars. This distinction is however more subtle when adopting a matrix identification.

Denoting the vector space of real n-vectors as $\Re^n$, and the scalar space as a, a_i is the ith component of **a**. The notations a(i) or $|a|_i$ may be alternatively adopted.

Identifying $\Re$ with $\Re^{n\times 1}$ (column vector identification), we have

$$\mathbf{a} \in \Re^n \Leftrightarrow \mathbf{a} = \begin{bmatrix} a_1 \\ \cdot \\ \cdot \\ \cdot \\ a_n \end{bmatrix} \quad a_i \in \Re^{n\times 1} \tag{A.5.18}$$

Identifying $\Re$ with $\Re^{1\times n}$ (row vector identification), we have

$$\mathbf{a} \in \Re^{1\times n} \rightarrow \quad \mathbf{a} = (a_1, \ldots a_n) \tag{A.5.19}$$

With $a \in \Re$ the following basic operations are possible.

A.5.3.1 Scalar-Vector Multiplication

$$\mathbf{b} = \mathbf{ca} \Rightarrow \quad b_i = ca_i \tag{A.5.20}$$

A.5.3.2 Vector Addition

$$\mathbf{z} = \mathbf{a} + \mathbf{b} \Rightarrow \quad z_i = a_i + b_i \tag{A.5.21}$$

A.5.3.3 Dot Product (or Inner Product, if M=n)

A dot product $\mathbf{a} \cdot \mathbf{b}$ involves multiplications between corresponding components of **a** and **b** (Zill and Cullen 2006). The result of a dot product is a scalar, i.e. $c = \mathbf{a} \cdot \mathbf{b}$.

The following notations are also valid:

$$c = \mathbf{a}^T\mathbf{b} \Rightarrow \quad c = \sum_{i=1}^{n} a_i b_i \tag{A.5.22}$$

A.5.3.4 Hadamard Product (or Vector Multiplication)

$$\mathbf{z} = \mathbf{a}^* \circ \mathbf{b} \Rightarrow \quad z_i = a_i b_i \tag{A.5.23}$$

where $\mathbf{a}^*$ is the conjugate transpose of **a**.

For **a** and **b** vectors and corresponding A and B square matrices having Hadamard product $A \circ B$ and D_a and D_b diagonals, the following identity exists:

$$\mathbf{a}^*(A \circ B)\mathbf{b} \Rightarrow \mathrm{tr}\left(D_a^* A D_b B^T\right) \tag{A.5.24}$$

For square matrices A and B, the diagonal elements of AB^T correspond to the row-sums of their Hadamard product

$$\sum_i (A \circ B)_{i,j} = \left(B^T A\right)_{j,j^*} \tag{A.5.25}$$

$$\sum_j (A \circ B)_{i,j} = \left(AB^T\right)_{i,i^*} \tag{A.5.26}$$

A.5.3.5 Cross Product

$$\mathbf{z} = \mathbf{a} \times \mathbf{b} \tag{A.5.27}$$

Each scalar component of $\mathbf{a} \times \mathbf{b}$ is computed by multiplying non-corresponding components of **a** and **b** (Zill and Cullen 2006). The cross product $\mathbf{a} \times \mathbf{b}$ is defined as a vector **z** that is orthogonal to both **a** and **b**, with a direction given by the right-hand rule and a magnitude equal to the area that the vectors span.

$$\mathbf{z} = \mathbf{a} \times \mathbf{b} = |\mathbf{a}||\mathbf{b}| \sin(\theta)\zeta \tag{A.5.28}$$

θ in Eq. A.5.28 is the angle between **a** and **b** and ς is a unit vector perpendicular to the plane containing **a** and **b** in the direction given by the right-hand rule.

The cross product can be expressed in the form of a determinant of a 3×3 matrix which may be easily resolved by application of the Sarrus rule.

A.5.3.6 Outer Product of $x \in \Re^m$ and $y \in \Re^n$

$$xy^T = \begin{bmatrix} x_1y_1 & \dots & x_1y_n \\ & & \\ x_my_1 & & x_my_n \end{bmatrix} \in \Re^{m\times n} \tag{A.5.29}$$

A.6 The Antisymmetrizing Operator

We have seen explicitly the application of the antisymmetrizing operator *u* in the creation of the Slater determinant (Sect. 7.3).

Consider a transposition operator P_{ij} operating on a wave function of N fermions and associate it with the permutation of π coordinates for the ith and jth fermion:

$$\begin{aligned} &\mathsf{P}_{ij}\Psi(1,2,\dots i,\dots j,\dots N) \\ &\equiv \Psi(\pi(1),\pi(2),\dots,\pi(i),\dots\pi(j),\dots\pi(N)) \\ &\equiv \Psi(1,2,\dots j,\dots i,\dots N) \\ &\equiv -\Psi(1,2,\dots i,\dots j,\dots N) \end{aligned} \tag{A.6.1}$$

Imagine that we apply the transposition in sequence (i.e. a "*composed transposition*"). The product of transposition is associative. If the number of transposition is even, the permutation is called "*even*" and has +1 parity. If the number of transposition is odd, the permutation is "*odd*" and the parity is −1.

Let me associate now the linear operator P_{ij} with the permutation π and let me denote the parity of the arbitrary permutation as $(-1)^\pi$. If we apply sequentially all permutations one after the other to the set of all N! permutations (i.e. to the "*Symmetric group*" S_N) we apply the antisymmetrizer $\mathcal{A}$:

$$\mathcal{A} \equiv \frac{1}{N!}\sum_{P\in S_N}(-1)^\pi \mathsf{P} \tag{A.6.2}$$

Consider now a Slater determinant D of wave functions φ_a, φ_b, φ_c with orbital indices a, b, c operating on three positions with coordinates labeled 1, 2, and 3:

$$D = \frac{1}{\sqrt{6}}\begin{vmatrix} \varphi_{a,1} & \varphi_{a,2} & \varphi_{a,3} \\ \varphi_{b,1} & \varphi_{b,2} & \varphi_{b,3} \\ \varphi_{c,1} & \varphi_{c,2} & \varphi_{c,3} \end{vmatrix} \tag{A.6.3}$$

Solving the determinant by Laplace expansion along the first row, we have

$$D = \frac{\varphi_{a,1}}{\sqrt{6}}\begin{vmatrix} \varphi_{b,2} & \varphi_{b,3} \\ \varphi_{c,2} & \varphi_{c,3} \end{vmatrix} - \frac{\varphi_{a,2}}{\sqrt{6}}\begin{vmatrix} \varphi_{b,1} & \varphi_{b,3} \\ \varphi_{c,1} & \varphi_{c,3} \end{vmatrix} + \frac{\varphi_{a,3}}{\sqrt{6}}\begin{vmatrix} \varphi_{b,1} & \varphi_{b,2} \\ \varphi_{c,2} & \varphi_{c,2} \end{vmatrix}$$
$$= \frac{\varphi_{a,1}}{\sqrt{6}}(\varphi_{b,2}\varphi_{c,3} - \varphi_{b,3}\varphi_{c,2}) - \frac{\varphi_{a,2}}{\sqrt{6}}(\varphi_{b,1}\varphi_{c,3} - \varphi_{b,3}\varphi_{c,1}) + \frac{\varphi_{a,3}}{\sqrt{6}}(\varphi_{b,1}\varphi_{c,2} - \varphi_{b,2}\varphi_{c,1})$$
$$= \frac{1}{\sqrt{6}}(\varphi_{a,1}\varphi_{b,2}\varphi_{c,3} - \varphi_{a,1}\varphi_{b,3}\varphi_{c,2} - \varphi_{a,2}\varphi_{b,1}\varphi_{c,3} + \varphi_{a,2}\varphi_{b,3}\varphi_{c,1} + \varphi_{a,3}\varphi_{b,1}\varphi_{c,2} - \varphi_{a,3}\varphi_{b,2}\varphi_{c,1}) \tag{A.6.4}$$

By definition of $\mathcal{A}$, we have

$$\mathcal{A}\varphi_{a,\mathbf{1}}\varphi_{b,2}\varphi_{c,3} = \frac{1}{6}(\varphi_{a,1}\varphi_{b,2}\varphi_{c,3} - \varphi_{a,2}\varphi_{b,1}\varphi_{c,3} + \varphi_{a,3}\varphi_{b,1}\varphi_{c,2} - \varphi_{a,3}\varphi_{b,2}\varphi_{c,1} + \varphi_{a,2}\varphi_{b,3}\varphi_{c,1} - \varphi_{a,1}\varphi_{b,3}\varphi_{c,2}) \tag{A.6.5}$$

hence

$$D = \frac{1}{\sqrt{6}}\mathcal{A}\varphi_{a,\mathbf{1}}\varphi_{b,2}\varphi_{c,3} \tag{A.6.6}$$

A.7 Dirac's Delta Function

The "*delta function*" $\delta(x)$ was promoted by Paul Dirac in 1930 in the aim of getting a precise notation for dealing with "*quantities involving a certain kind of infinity*" (Dirac 1957). The function satisfies the conditions

$$\left.\begin{array}{l} \int_{-\infty}^{+\infty} \delta(x)dx = 1 \\ \delta(x) = 0 \quad \text{for} \quad x \neq 0 \end{array}\right\} \tag{A.7.1}$$

For a function of the real variable x which vanishes everywhere except inside a small domain ϵ surrounding the origin (x = 0) but large enough inside the domain that its integral over the domain is unit, in the limit $\epsilon \to 0$ this function "*will go over into δ(x)*" (ibid.). The exact shape of the function inside the domain does not matter "*provided there are no unnecessary wild variation*" (ibid.).

The $\delta(x)$ was defined by Dirac an "*improper function*", in the sense that it does not have a definite value for each point in its domain and its use is confined to certain simple types of expressions for which "*it is obvious that no inconsistencies can arise*" (ibid.).

The most important property of $\delta(x)$ according to Dirac is exemplified by the equation

$$\int_{-\infty}^{+\infty} f(x)\delta(x)dx = f(0) \tag{A.7.2}$$

In Eq. A.7.2, f(x) is a continuous function of x. The left-hand side of this equation can depend only on the values of f(x) very close to the origin. Replacing f(x) by its value on the origin f(0) does not induce thus essential errors.

Applying then the upper definition in Eq. A.7.1, and making a change of origin in Eq. A.7.2, one obtains

$$\int_{-\infty}^{+\infty} f(x)\delta(x-a)dx = f(a) \tag{A.7.3}$$

where a is any real number. Thus "*the process of multiplying a function of x by δ(x − a) and integrating over all x is equivalent to the process of substituting a for x*" (ibid.). This is true also if the function of x is not simply numerical but also if it is a vector or a linear operator depending on x.

The property of the delta function represented by Eq. A.7.3 is also known as "*sifting property*" (Bracewell 1999).

Equations (A.7.2) and (A.7.3) tell us that, "*although an improper function does not itself have a well-defined value, when it occurs as a factor in an integrand the integral has a well-defined value*" (Dirac 1957).

Consider now a discontinuous function such as, for example,

$$\left.\begin{array}{ll} \varepsilon(x) = 0 & x < 0 \\ \varepsilon(x) = 1 & x > 0 \end{array}\right\} \tag{A.7.4}$$

Replacing the differential coefficient $\dot{\varepsilon}(x)$ of the function $\varepsilon(x)$ with $\delta(x)$ in the left-hand side of Eq. A.7.2 and integrating by parts, one finds for two positive numbers g_1 and g_2:

$$\begin{aligned} \int_{-g_2}^{g_1} f(x)\dot{\varepsilon}(x)dx &= [f(x)\varepsilon(x)]_{-g_2}^{g_1} \\ &- \int_{-g_2}^{g_1} \dot{f}(x)\varepsilon(x)dx = f(g_1) - \int_{0}^{g_1} \dot{f}(x)dx = f(0) \end{aligned} \tag{A.7.5}$$

The delta function thus appears also whenever one differentiates a discontinuous function.

Essential rules for algebraic work involving δ(x) are represented by the following elementary equations (Dirac 1957):

$$\delta(-x) = \delta(x) \tag{A.7.6}$$

$$x\delta(x) = 0) \tag{A.7.7}$$

$$\delta(ax) = a^{-1}\delta(x) \quad (a > 0) \tag{A.7.8}$$

$$\delta(x^2 - a^2) = \frac{1}{2}a^{-1}[\delta(x - a) + \delta(x + a)] \quad (a > 0) \tag{A.7.9}$$

$$\int \delta(a - x)dx\delta(x - b) = \delta(a - b) \tag{A.7.10}$$

$$f(x)\delta(x - a) = f(a)\delta(x - a) \tag{A.7.11}$$

The meaning of any of the above equations is that their two sides give equivalent results as factors in an integrand.

A Fourier series expansion of δ(x − a) gives (Weinstein 2020)

$$a_{\{n\}} = \frac{1}{\pi}\int_{-\pi}^{\pi} \delta(x - a)\cos(nx)dx = \frac{1}{\pi}\cos(na) \tag{A.7.12}$$

$$b_n = \frac{1}{\pi}\int_{-\pi}^{\pi} \delta(x - a)\sin(nx)dx = \frac{1}{\pi}\sin(na) \tag{A.7.13}$$

So that

$$\begin{aligned}\delta(x - a) &= \frac{1}{2\pi} + \frac{1}{\pi}\sum_{n=1}^{\infty}[\cos(na)\cos(nx) + \sin(na)\sin(nx)] \\ &= \frac{1}{2\pi} + \frac{1}{\pi}\sum_{n=1}^{\infty}\cos[n(x - a)]\end{aligned} \tag{A.7.14}$$

A.8 Ket and Bra

The *Ket* notation was introduced by Dirac already in 1930 to convey in a concise way the mathematical relations between vectors representing quantum mechanical states connected through the quantum superposition principle.

Dirac wrote: "*The superposition process is a kind of additive process and implies that states can in some way be added to give new states. The states must therefore be connected with mathematical quantities of a kind which can be added together to give other quantities of the same kind. The most obvious of such quantities are vectors. Ordinary vectors, existing in a space of finite number of dimensions, are not sufficiently general for most of the dynamical systems in quantum mechanics. We have to make a generalization to vectors in a space of an infinite number of dimensions, and the mathematical treatment becomes complicated by questions of convergence….[omissis] …It is desirable to have a special name for describing the vectors which are connected with the states of the system in quantum mechanics, whether they are in a space of a finite or an infinite number of dimensions. We shall call them ket vectors, or simply kets, and denote a general one of them by a special symbol* $|\,\rangle$. *If we want to specify a particular one of them by a label, A say, we insert it in the middle, thus* $|A\rangle$" (Dirac 1957).

When two or more states are superposed, the order in which they occur in the superposition process is unimportant, so that the superposition process is symmetric between the states that are superposed.

Consider now the ket vectors $|A\rangle$ and $|B\rangle$. They may be multiplied by complex numbers c_1 and c_2 and added together, to give another ket vector $|R\rangle$:

$$c_1|A\rangle + c_2|B\rangle = |R\rangle \tag{A.8.1}$$

Excluding the case when c_1 or c_2 is zero, the above equation implies that the state A can be formed by superposition of B and R, and B can be formed by superposition of A and R. The superposition relation is thus symmetrical between all the three states A, B, and R.

If a ket vector is expressible linearly in terms of certain others, it is said to be dependent on them.

If we have a ket vector $|x\rangle$ with x as a continuous variable within a certain range, we may integrate it with respect to x to get another ket vector, i.e.:

$$\int |x\rangle dx = |Q\rangle \tag{A.8.2}$$

Superposing a state with itself we cannot form any new state, but the original state over again. We may write

$$c_1|A\rangle + c_2|B\rangle = (c_1 + c_2)|A\rangle \tag{A.8.3}$$

When $c_1 + c_2 = 0$ the superposition process is nothing at all because the two components cancel each other by an interference process. Apart from this special case, the result must be the same as the original one. We must thus state "*if the ket vector corresponding to a state is multiplied by any complex number, not zero, the resulting ket vector will correspond to the same state*" (Dirac 1957).

In other terms, a "*state*" is specified by the direction of a ket vector and any length one may assign to the ket vector is irrelevant. This implies as well that no distinction can be made between the directions of the ket vectors $|A\rangle$ and $-|A\rangle$.

Dirac developed the *ket* concept in the aim of finding a better rationale in the polarization and the interference of photons. In both cases, one has to deal with the combination (i.e. superposition of translational states) of the two components into which the incident beam is split by passing through a polarization device. Because the Schrödinger equation is linear, the application of Eq. (A.8.1) is particularly apt in quantum mechanics. If we generally expand a wave function ψ in terms of state vectors f_1, f_2, f_3, each one capable of solving a linear equation on ψ, then also the equation

$$\psi = c_1 f_1 + c_2 f_2 + c_3 f_3 \tag{A.8.4}$$

is a solution.

Assume now that we have a number ϕ which is a function of a ket vector $|A\rangle$, or, in other terms, that to each ket vector $|A\rangle$ corresponds one number ϕ. If the function is linear then

1. a number corresponding to $|A\rangle$ + $|A'\rangle$ is the sum of the numbers corresponding to $|A\rangle$ and to $|A'\rangle$;
2. the number corresponding to $c|A\rangle$ is c times the number corresponding to $|A\rangle$.

Hence, the number ϕ corresponding to any $|A\rangle$ may be seen as a scalar product of $|A\rangle$ with some new vector (just one for each linear function of the ket vector $|A\rangle$). These new vectors are denoted "*bra*" and their symbol $\langle\ |$ is the mirror image of the symbol of the ket vector $|\ \rangle$. If we want to specify a particular one of them by a label, B say, we insert it in the middle, thus $\langle B|$.

For the sake of conciseness, the scalar product of a *bra* vector $\langle B|$ and a *ket* vector $|A\rangle$ is represented as $\langle B|A\rangle$ with the two vertical bars contracted to one (note that the bra vectors is always on the left). The symbols $\langle$ and $\rangle$ may be seen as "*a distinctive kind of bracket*" (Dirac 1957). A scalar product appears as a complete bracket expression and a bra vector or a ket vector as an incomplete bracket expression. Hence "*any complete bracket expression denotes a number and any incomplete bracket expression denotes a vector, of the bra or ket kind according to whether it contains the first or second part of the bracket*" (Dirac 1957).

The conditions dictated by points 1. and 2 are symbolically expressed as

$$\langle B|\{|A\rangle + |A'\rangle\} = \langle B|A\rangle + \langle B|A'\rangle \tag{A.8.5}$$

$$\langle B|\{c|A\} = c\langle B|A\rangle \tag{A.8.6}$$

with c being any number.

For a complete definition of the bra vector, one must also consider the limiting condition of a vanishing vector, as a scalar product of a vanishing ket. Symbolically this is represented as

$$\langle \mathrm{P}|\mathrm{A}\rangle = 0, \quad \text{all } |\mathrm{A}\rangle \tag{A.8.7.1}$$

$$\langle \mathrm{P}| = 0 \tag{A.8.7.2}$$

The sum of two bra vectors $\langle \mathrm{B}|$ and $\langle \mathrm{B}'|$ implies that its scalar product with any ket vector $|\mathrm{A}\rangle$ is the sum of the scalar products of vectors $\langle \mathrm{B}|$ and $\langle \mathrm{B}'|$ with vector $|\mathrm{A}\rangle$:

$$\{\langle \mathrm{B}| + \langle \mathrm{B}'|\}|\mathrm{A}\rangle = \langle \mathrm{B}|\mathrm{A}\rangle + \langle \mathrm{B}'|\mathrm{A}\rangle \tag{A.8.8}$$

In an analogous fashion, the product of a bra vector $\langle \mathrm{B}|$ and a number c imply that its scalar product with any ket vector $|\mathrm{A}\rangle$ is c times the scalar product of vector $\langle \mathrm{B}|$ with vector $|\mathrm{A}\rangle$:

$$\{\mathrm{c}\langle \mathrm{B}|\}|\mathrm{A}\rangle = \mathrm{c}\langle \mathrm{B}|\rangle \tag{A.8.9}$$

Until now it has been established by Dirac that between a *bra* and a *ket* there is no connection besides the fact that a scalar product of bra and ket do exist. Now Dirac introduces a new assumption, namely, "*there is a one–one correspondence between the bras and the kets, such that the bra corresponding to* $|\mathrm{A}\rangle + |\mathrm{A}'\rangle$ *is the sum of the bras corresponding to* $|\mathrm{A}\rangle$ *and to* $|\mathrm{A}'\rangle$, *and the bra corresponding to c* $|\mathrm{A}\rangle$ *is* $\bar{c}$ *times the bra corresponding to* $|\mathrm{A}\rangle$, $\bar{c}$ *being the conjugate complex number to c*". This relationship implies that we may call one of them the conjugate imaginary of the other. The bra corresponding to the ket $|\mathrm{A}\rangle$ will be written $\langle \mathrm{A}|$. This however does not imply that bra and ket vectors can be added together. Moreover, they cannot be split up into real and pure imaginary parts. Consider in fact two ket vectors $|\mathrm{A}\rangle$ and $|\mathrm{B}\rangle$. By taking the scalar product of the first with the conjugate imaginary of the second, one obtains a number $\langle \mathrm{B}|\mathrm{A}\rangle$ which depends linearly on $|\mathrm{A}\rangle$ and antilinearly on $|\mathrm{B}\rangle$. Also, if we take the scalar product of $|\mathrm{B}\rangle$ and the conjugate imaginary of $|\mathrm{A}\rangle$ and make the conjugate complex of this scalar product, we obtain the same number. Dirac wrote this symbolically by putting an horizontal bar above the $\langle \mathrm{A}|\mathrm{B}\rangle$.

A.9 Linear Operators

Consider now that, in place of a number ϕ, we have a ket vector $|\mathrm{F}\rangle$ which is a linear function of a ket $|\mathrm{A}\rangle$ (i.e. to each ket $|\mathrm{A}\rangle$ corresponds one ket $|\mathrm{F}\rangle$). If the function is linear the two conditions delineated at points 1. and 2. in Sect. A.8 hold. Under these conditions the passage from $|\mathrm{A}\rangle$ to $|\mathrm{F}\rangle$ corresponds to the application of a *linear operator* α such that

$$|F\rangle = \alpha|A\rangle \tag{A.9.1}$$

In the above products, the ket vector must always be put on the right of the linear operator. The linearity condition is expressed as

$$\alpha\{|A\rangle + |A'\rangle\} = \alpha|A\rangle + \alpha|A'\rangle \tag{A.9.2.1}$$

$$\alpha\{c|A\rangle\} = \alpha|A\rangle + c\alpha|A\rangle \tag{A.9.2.2}$$

Linear operators can be added together:

$$\{\alpha + \beta\}|A\rangle = \alpha|A\rangle + \beta|A\rangle \tag{A.9.3}$$

The product $\alpha\beta$ of a ket vector $|A\rangle$ is "*the linear operator which, operating on* $|A\rangle$ *changes it into that ket which one would get by operating first on* $|A\rangle$ *with* β, *and then on the result of the first operation with* α" (Dirac 1957). This is expressed symbolically as

$$\{\alpha\beta\}|A\rangle = \alpha\{\beta|A\rangle\} \tag{A.9.4}$$

A.10 Commuting Operators

Operators are often characterized by a symbol $\wedge$ superimposed over the letter identifying it, as a sort of hat. In this book, I selected, for simplicity, to distinguish them by a different graphical style. For example, the Hamiltonian Operator has been identified as H. I kept conform to this choice throughout the entire text.

In general terms, if g(x) is the result of an operator O acting on a function of variable x, say f(x) then we will identify this operator as

$$\mathsf{O}f(x) = g(x) \tag{A.10.1}$$

If the operation is a simple partial derivative d/dx, then $g(x) = \dot{f}(x)$, and so on.

If we have two operators acting on the same function, say A and B, and the two operators do not commute, the order of operation dictates the result, i.e.

$$\mathsf{AB}f(x) \neq \mathsf{BA}f(x) \tag{A.10.2}$$

However, if the two operators are "*commuting operators*" then

$$\mathsf{AB}f(x) = \mathsf{BA}f(x) \tag{A.10.3}$$

Hence, we may also write

$$[\mathsf{A}, \mathsf{B}] = \mathsf{AB} - \mathsf{BA} = 0 \quad (A.10.4)$$

Consider now an observable A, if A has a definite value in the system, then the system is in an eigenstate of A. If we can assign two definite values to A and B, then the system is simultaneously an eigenstate of both A and B. Suppose now the observables are A_i and B_j and the function is a wave function $\psi_{A,B}$. The relations between operators and observables require

$$\mathsf{A}\psi_{A_i,B_j} = A_i\psi_{A_i,B_j} \quad (A.10.5.1)$$

$$\mathsf{B}\psi_{A_i,B_j} = B_j\psi_{A_i,B_j} \quad (A.10.5.2)$$

Multiplying, respectively, the above equations by B and A, one obtains

$$\mathsf{BA}\psi_{A_i,B_j} = \mathsf{B}A_i\psi_{A_i,B_j} \quad (A.10.6.1)$$

$$\mathsf{AB}\psi_{A_i,B_j} = \mathsf{A}B_j\psi_{A_i,B_j} \quad (A.10.6.2)$$

Since $\psi_{Ai,Bj}$ is an eigenfunction of A and B we may further transform the above equations into

$$\mathsf{BA}\psi_{A_i,B_j} = A_iB_j\psi_{A_i,B_j} \quad (A.10.7.1)$$

$$\mathsf{AB}\psi_{A_i,B_j} = B_jA_i\psi_{A_i,B_j} \quad (A.10.7.2)$$

Finally, subtracting Eq. (A.10.7.2) from Eq. (A.10.7.1), we get

$$(\mathsf{AB} - \mathsf{BA})\psi_{A_i,B_j} = 0 \quad (A.10.8)$$

This tells us that for two physical quantities to be simultaneously observable, their operators must commute. We may thus say alternatively that.

1. the operators A and B are commuting
2. A and B are compatible observables.
3. A and B have a common eigenbasis.

An useful example of commuting operators in quantum mechanics involves the nature of the quantum numbers in hydrogen-like atoms.

The Hamiltonian operator of an hydrogen-like atom has rotational invariance and it is only function of the electron–nucleus distance r:

$$\mathsf{H} = -\frac{\hbar}{2\mu}\nabla^2 - \frac{Ze^2}{r} \quad (A.10.9)$$

with Z as nuclear charge and μ as reduced mass of the system.

Since the components of the angular momentum operator L do not commute, the rotational invariance implies

$$[\mathsf{L}, \mathsf{H}] = 0 \quad \text{(A.10.10)}$$

The components of the angular momentum (L) do not commute but satisfy the commutation relation:

$$\left[L_i, L_j\right] = i\hbar\varepsilon_{ijk}L_k \quad \text{(A.10.11)}$$

They commute however with the square of the modulus of the angular momentum. Choosing the quantization axis z to exemplify it, we may write

$$\left[L_z, L^2\right] = 0 \quad \text{(A.10.12)}$$

A commuting set of observables will be thus represented by $\{\mathsf{H}, L^2, L_z\}$. We may then specify them as basis set of the Hilbert space:

$$\mathsf{H}\left|E_{n,l,m_l}\right\rangle = E_n\left|E_{n,l,m_l}\right\rangle \quad \text{(A.10.13.1)}$$

$$L^2\left|E_{n,l,m_l}\right\rangle = l(l+1)\hbar^2\left|E_{n,l,m_l}\right\rangle \quad \text{(A.10.13.2)}$$

$$L_z\left|E_{n,l,m_l}\right\rangle = m_l\hbar\left|E_{n,l,m_l}\right\rangle \quad \text{(A.10.13.3)}$$

The eigenstate of the hydrogen-like atom is then completely specified by the $\{E_{n,l,m_l}\}$ set of eigenvalues, hence by the (n, l, m_l) quantum numbers.

A.11 Taylor's Expansions

These procedures are a direct application of Taylor's theorem (Taylor 1715; Strujk 1969). Consider a continuous function (x) in the domain D, containing a and a + h, and derivable as many orders as it is needed. Its Taylor expansion in series is

$$f(a+x) = f(a) + \frac{h}{1!}f'(a) + \frac{h^2}{2!}f''(a) + ::: + \frac{h^n}{n!}f^{(n)}(a) + R_{n+1} \quad \text{(A.11.1)}$$

where R_{n+1} is the remainder for an exact solution:

$$R_{n+1} = \frac{f^{(n+1)}(\xi)}{(n+1)!}h^{n+1} \quad \text{(A.11.2)}$$

with a $\leq \xi \leq$ (a + h).

R_{n+1} tends toward zero when n tends to ∞. For a continuous function, derivable an infinity of times, Taylor's development becomes a MacLaurin series. As it is almost impossible to know the exact value of ξ, usually one searches for a majorant of R_{n+1} in the range (a, a + h).

A.11.1 Approximation

Consider f(x) as a continuous function derivable within the canonical range (−1, +1). The Taylor–MacLaurin development gives

$$\begin{aligned} f(x) = f(0) + \frac{x}{1!}f'(0) + \frac{x^2}{2!}f''(0) \\ + \dots + \frac{x^{2n+1}}{(2n+1)!}f^{(2n+1)}(0) + \frac{x^{2n+2}}{(2n+2)!}f^{(2n+2)}(\xi) \end{aligned} \quad (A.11.3)$$

with x and ξ belonging to the canonical range (−1, +1).

Integrating by parts Eq. A.11.3, all derivatives of odd order disappear and one obtains (Guilpin 1979)

$$\begin{aligned} I = \int_{-1}^{+1} f(x)dx = 2\left[f(0) + \frac{1}{3}f''(0) + \cdots + \frac{1}{(2m+1)}f^{(2m)}(0)\right. \\ \left. + \cdots + \frac{1}{(2n+3)}f^{(2n+2)}(\xi)\right] \end{aligned} \quad (A.11.4)$$

As in the Gauss procedure for the numerical evaluation of the integral, one may pose (Guilpin, 1999)

$$\begin{aligned} J = \sum_{k=0}^{n} H_x f(x_k) = f(0)\sum_{k=0}^{n} H_x \\ + \frac{f'(0)}{1!}\sum_{k=0}^{n} H_x x_k + \frac{f''(0)}{2!}\sum_{k=0}^{n} H_x x_k^2 \\ + \dots + \frac{f^{(2m)}(0)}{m!}\sum_{k=0}^{n} H_x x_k^m + \dots \\ + \frac{f^{(2n+1)}(0)}{(2n+1)!}\sum_{k=0}^{n} H_x x_k^{2n+1} + \frac{f^{(2n+2)}(0)}{(2n+2)!}\sum_{k=0}^{n} H_x x_k^{2n+2} \end{aligned} \quad (A.11.5)$$

with H_x as Gauss coefficient (see Eqn. 7.11 in Guilpin, 1999). If f(x) is a polynomial of degree (2n + 1), the derivative of order (2n + 2) is zero and I = J.

A.11.2 Application I: Differential Equation of the First Order

Consider the differential equation

$$\frac{dy}{dx} = f(x, y) \tag{A.11.6}$$

whose solution for x_0 is y_0.

Starting from x_0 and y_0 one must evaluate y_1 at x_1 then y_2 at x_2, and so on until $x_n = b$. The procedure implies a development in series of $y(x_k + 1)$ in the neighborhood of $y(x_k)$:

$$y_{k+1} = y_k + \frac{h}{1!} y'_k + \frac{h^2}{2!} y''_k + \frac{h^3}{3!} y'''_k \cdots \tag{A.11.7}$$

y'_k is obtained by the differential equation and attains $f(x_k, y_k)$. y''_k and y'''_k are then obtained by the derivatives (Guilpin 1999):

$$y''_k = \left(\frac{\partial f}{\partial x} \cdot \frac{dx}{dx} + \frac{\partial f}{\partial y} \cdot \frac{dy}{dx}\right)_{x_k} = \left(\frac{\partial f}{\partial x} + \frac{\partial f}{\partial y} \cdot y'\right)_{x_k} = g(x_k, y_k) \tag{A.11.8}$$

$$y'''_k = \left(\frac{\partial^2 f}{\partial x^2} + 2y' \frac{\partial^2 f}{\partial x \partial y} \cdot \frac{dy}{dx} + (y')^2 \frac{\partial^2 f}{\partial x^2} + y'' \frac{\partial f}{\partial y}\right)_{x_k} = h(x_k, y_k) \tag{A.11.9}$$

and so on.

Limiting Taylor's development to the first order corresponds to the Euler method. The C program of the procedure may be found at http://www.edpsciences.com/guilpin/.

A.11.3 Application II: Differential Equation of the Second Order

Consider the general case:

$$\frac{d^2y}{dx^2} = g(x, y, y') \tag{A.11.10}$$

One may transform it into:

$$\frac{dy}{dx} = z \tag{A.11.11}$$

$$\frac{dz}{dx} = g(x, y, z) \tag{A.11.12}$$

The following Taylor's series may be then developed:

$$y_{k+1} = y_k + \frac{h}{1!}y'_k + \frac{h^2}{2!}y''_k + \frac{h^3}{3!}y'''_k \cdots \tag{A.11.13}$$

$$z_{k+1} = z_k + \frac{h}{1!}z'_k + \frac{h^2}{2!}z''_k + \frac{h^3}{3!}z'''_k \cdots \tag{A.11.14}$$

with

$$y'_k = f(x_k, y_k, z_k) \tag{A.11.15}$$

$$z'_k = g(x_k, y_k, z_k) \tag{A.11.16}$$

Operating as previously seen for a first-order differential equation, one obtains by derivation (Guilpin 1999):

$$\begin{aligned} y''_k &= \left(\frac{\partial f}{\partial x}\cdot\frac{dx}{dx} + \frac{\partial f}{\partial y}\cdot\frac{dy}{dx} + \frac{\partial f}{\partial z}\cdot\frac{dz}{dx}\right)_{x_k} \\ &= \left(\frac{\partial f}{\partial x} + \frac{\partial f}{\partial y}\cdot y' + \frac{\partial f}{\partial z}\cdot z'\right)_{x_k} = k(x_k, y_k, z_k) \end{aligned} \tag{A.11.17}$$

$$\begin{aligned} z''_k &= \left(\frac{\partial g}{\partial x}\cdot\frac{dx}{dx} + \frac{\partial g}{\partial y}\cdot\frac{dy}{dx} + \frac{\partial g}{\partial z}\cdot\frac{dz}{dx}\right)_{x_k} \\ &= \left(\frac{\partial g}{\partial x} + \frac{\partial g}{\partial y}\cdot y' + \frac{\partial g}{\partial z}\cdot z'\right)_{x_k} = h(x_k, y_k, z_k) \end{aligned} \tag{A.11.18}$$

and so on.

A.12 Fourier Series and Fourier Transforms

Fourier (1822), based on the Newton's law of cooling, showed that the heat flow between two adjacent molecules is proportional to the extremely small difference of their temperatures. In the aim of solving the "*problème du mur*" (wall problem), Fourier showed that a continuous periodic function is conveniently replaced by an infinite sum of a series of sine or cosine terms of appropriate amplitudes and phase coefficients (Fourier series).

A.12.1 Fourier Series

Consider a function f(x) periodic in x with period 2π:

$$f(x) = \frac{a_0}{2} + \sum_{n=1}^{\infty} a_n \cos(nx) + \sum_{n=1}^{\infty} b_n \sin(nx) \tag{A.12.1}$$

To find the values of the coefficients a_n and b_n, we must multiply the two members by $\cos(k\pi)$ and integrate over the period $(0, 2\pi)$ (Guilpin 1999):

$$\int_0^{2\pi} f(x)\cos(kx)dx = \int_0^{2\pi} \frac{a_0}{2} + \sum_{n=1}^{\infty} [a_n \cos(nx) + b_n \sin(nx)]dx \tag{A.12.2}$$

Taking into account the orthogonality relations, we obtain

$$a_k = \frac{1}{\pi}\int_0^{2\pi} f(x)\cos(kx)dx \tag{A.12.3}$$

Multiplying the two members of Eq. A.12.1 by $\sin(k\pi)$ and accounting for the orthogonality relation, we obtain

$$b_k = \frac{1}{\pi}\int_0^{2\pi} f(x)\sin(kx)dx \tag{A.12.4}$$

All k are easily obtained by this method. The initial coefficient $a_0/2$ is due to the fact that

$$a_0 = \frac{1}{2\pi}\int_0^{2\pi} f(x)dx \tag{A.12.5}$$

A.12.2 Fourier Transform

A Fourier transform is a mathematical procedure that decomposes a function (often a "*signal*", i.e. a function of time) into its constituent frequencies.

Consider the function f(t) in the finite range (−T/2, +T/2). Posing $\Omega = 2\pi/\mathrm{T}$ and recalling Eqs. A.12.1, A.12.3, A.12.4, we have

$$\mathrm{f(t)} = \frac{a_0}{2} + \sum_{n=1}^{\infty} a_n \cos(n\Omega t) + \sum_{n=1}^{\infty} b_n \sin(n\Omega t) \tag{A.12.6}$$

$$a_n = \frac{2}{T}\int_{-T/2}^{+T/2} f(t)\cos(n\Omega t)dt \tag{A.12.7}$$

$$b_n = \frac{2}{T}\int_{-T/2}^{+T/2} f(t)\sin(n\Omega t)dt \tag{A.12.8}$$

Posing then $\omega = n\Omega$ we may further develop f(t) as

$$f(t) = \frac{a_0}{2} + \frac{1}{\Omega}\sum_{\omega=\Omega}^{\infty} \Omega[a_n \cos(\omega t) + b_n \sin(\omega t)] \tag{A.12.9}$$

Considering Ω as the $\delta\omega$ increase of ω between two consecutive values, Eq. A.12.9 may be rewritten in the form (Guilpin 1979):

$$f(t) = \frac{1}{T}\int_{-T/2}^{+T/2} f(x)dx + \frac{1}{\pi}\sum_{\omega=\Omega}^{\infty} \delta\omega \int_{-T/2}^{+T/2} f(x)\cos[\omega(x-t)]dx \tag{A.12.10}$$

Assuming that, for $T \to \infty$:

$$\lim \frac{1}{T}\int_{-T/2}^{+T/2} f(x)dx = 0 \tag{A.12.11}$$

$$\delta\omega \to d\omega \tag{A.12.12}$$

$$\sum_{\omega=\Omega}^{\infty} \to \int_{0}^{\infty} \tag{A.12.13}$$

one obtains the Fourier transform of the function f(t) :

$$f(t) = \frac{1}{\pi} \int_{0}^{\infty} d\omega \int_{-\infty}^{+\infty} f(x) \cos[\omega(x - t)] dx \tag{A.12.14}$$

In the case of complex functions, the equivalent of Eq. A.12.14 is (Guilpin 1979)

$$f(t) = \frac{1}{\pi} \int_{-\infty}^{\infty} d\omega \int_{-\infty}^{+\infty} f(x) \exp[i\omega(t - x)] dx \tag{A.12.15}$$

In the common practice, the Fourier transform of f(x) is written in the form:

$$\hat{f}(\xi) = \int_{-\infty}^{+\infty} f(x) \exp(-2\pi i x \xi) dx \tag{A.12.16}$$

and the "*inverse transform*" as

$$f(x) = \int_{-\infty}^{+\infty} \hat{f}(\xi) \exp(2\pi i x \xi) d\xi \tag{A.12.17}$$

A.13 Lagrange's Equations

Let me make a slight change of notation, representing the Cartesian coordinate of the first particle as x_1, x_2, x_3 instead of x_1, y_1, z_1, then x_4, x_5, x_6 will be the Cartesian coordinates of the second particle instead of x_2, y_2, z_2, and so on. If we number the forces and the masses in the same way, Newton's laws turn out to be expressed by a single equation:

$$f_i = m_i \frac{d^2 x_i}{dt^2} \tag{A.13.1}$$

the system in generalized coordinates will be replaced by

$$x_i = \varphi_i(q_1 \ldots q_{3N}) \tag{A.13.2}$$

and the variation in the x's will be related to the variation in the generalized coordinates q's by

$$dx_i = \sum_j \frac{\partial x_i}{\partial q_j} dq_j \tag{A.13.3}$$

If the system is subjected to external forces leading to an infinitesimal displacement of coordinates dx_i, the work done by the forces is

$$dW = \sum_i f_i dx_i = \sum_i \sum_j f_i \frac{\partial x_i}{\partial q_j} dq_j = \sum_j Q_j dq_j \tag{A.13.4}$$

where $Q_j = \sum_i f_i \frac{\partial x_i}{\partial q_j}$ is the generalized force associated with the generalized coordinate q_j.

Replacing f_i in Eq. A.13.4 with the right part of Eq. A.13.1, we get

$$dW = \sum_i m_i \frac{d^2 x_i}{dt^2} dx_i = \sum_i \sum_j \frac{d^2 x_i}{dt^2} \frac{\partial x_i}{\partial q_j} dq_j \tag{A.13.5}$$

hence

$$\sum_i \sum_j \frac{d^2 x_i}{dt^2} \frac{\partial x_i}{\partial q_j} dq_j = \sum_j Q_j dq_j \tag{A.13.6}$$

The above equality is however valid only if the coefficients of each dq_j on both sides are equal (note that the q’s are independent variables and there is thus no implicit relation between the dq’s). The equality constraint gives rise to a set of j equations of type

$$\sum_i m_i \frac{d^2 x_i}{dt^2} \frac{\partial x_i}{\partial q_j} = Q_j \tag{A.13.7}$$

Applying to Eq. A.13.7 the rules of differentiation of implicit functions, we get

$$\sum_i m_i \frac{d^2 x_i}{dt^2} \frac{\partial x_i}{\partial q_j} = \sum_i m_i \frac{d}{dt}\left(\frac{dx_i}{dt} \frac{\partial x_i}{\partial q_j}\right) - \sum_i m_i \frac{dx_i}{dt} \frac{d}{dt} \frac{\partial x_i}{\partial q_j} \tag{A.13.8}$$

Differentiating the q,s in dt: $\frac{dq_k}{dt} = \dot{q}_k$, and posing $\frac{\partial \dot{x}_i}{\partial \dot{q}_k} = \frac{\partial x_i}{\partial q_k} \rightarrow \frac{d}{dt} \frac{\partial x_i}{\partial q_k} = \frac{\partial \dot{x}_i}{\partial q_k}$ we have (Eyring et al., 1944):

$$\begin{aligned} \sum_i m_i \frac{d^2 x_i}{dt^2} \frac{\partial x_i}{\partial q_j} = & \frac{d}{dt}\left\{\frac{\partial}{\partial \dot{q}_j}\left[\sum_i \frac{1}{2} m_i \left(\frac{dx_i}{dt}\right)^2\right]\right\} \\ & - \frac{\partial}{\partial q_j}\left[\sum_i \frac{1}{2} m_i \left(\frac{dx_i}{dt}\right)^2\right] \end{aligned} \tag{A.13.9}$$

Let me now return to the normal x-, y-, z-Cartesian coordinates.

The kinetic energy T in Cartesian coordinates for an ensemble of N particles is

$$\begin{aligned}T &= \frac{1}{2}m_1\left(\dot{x}_1^2+\dot{y}_1^2+\dot{z}_1^2\right)\cdots+\frac{1}{2}m_N\left(\dot{x}_N^2+\dot{y}_N^2+\dot{z}_N^2\right)\\ &= \frac{1}{2}\sum_{i=1}^{M} m_i\left(\dot{x}_i^2+\dot{y}_i^2+\dot{z}_i^2\right)\end{aligned} \tag{A.13.10}$$

The potential energy V of the particles in a conservative system is a function of the coordinates of all N particles in the ensemble and the force acting on each particle may be decomposed along the Cartesian axes according to

$$\left.\begin{aligned}f_{i,x} &= -\frac{\partial V}{\partial x_i}\\ f_{i,y} &= -\frac{\partial V}{\partial y_i}\\ f_{i,z} &= -\frac{\partial V}{\partial z_i}\end{aligned}\right\}\quad i = 1, 2, \ldots, N \tag{A.13.11}$$

The *Newton's equations of motion* expressed in Cartesian coordinates are then

$$\begin{aligned}\frac{d}{dt}\frac{\partial T}{\partial \dot{x}_i} + \frac{\partial V}{\partial x_i} &= 0\\ \frac{d}{dt}\frac{\partial T}{\partial \dot{y}_i} + \frac{\partial V}{\partial y_i} &= 0\\ \frac{d}{dt}\frac{\partial T}{\partial \dot{z}_i} + \frac{\partial V}{\partial z_i} &= 0\end{aligned} \tag{A.13.12}$$

The *Lagrangian function* L for Newtonian systems is the difference of the kinetic and potential energies: L = T − V. Because in Cartesian coordinates T is a function of the velocities only and (for the systems in our interest) V is a function of the coordinates only, Eq. A.13.12 assumes the form

$$\begin{aligned}\frac{d}{dt}\frac{\partial L}{\partial \dot{x}_i} - \frac{\partial L}{\partial x_i} &= 0\\ \frac{d}{dt}\frac{\partial L}{\partial \dot{y}_i} - \frac{\partial L}{\partial y_i} &= 0\\ \frac{d}{dt}\frac{\partial L}{\partial \dot{z}_i} - \frac{\partial L}{\partial z_i} &= 0\end{aligned} \tag{A.13.13}$$

Equation A.13.13 is the *Lagrangian equation of motion* expressed in Cartesian coordinates.

In generalized coordinates, taking into account Eqs. A.13.7 and A.13.10, we have

$$\frac{d}{dt}\frac{\partial T}{\partial \dot{q}_j} - \frac{\partial T}{\partial q_j} = Q_j \tag{A.13.14}$$

then applying Eq. A.13.11:

$$Q_j = -\sum_i \frac{\partial V}{\partial x_i}\frac{\partial x_i}{\partial q_j} = -\frac{\partial V}{\partial q_j} \tag{A.13.15}$$

$$\frac{d}{dt}\frac{\partial T}{\partial \dot{q}_j} - \frac{\partial}{\partial q_j}(T - V) = 0 \tag{A.13.16}$$

$$\frac{d}{dt}\frac{\partial L}{\partial \dot{q}_j} - \frac{\partial L}{\partial q_j} = 0 \tag{A.13.17}$$

Equation A.13.17 is the *Lagrangian equation of motion* in generalized coordinates.

A.14 Hermite Polynomials

The Probabilistic Hermite polynomial of degree n on the variable ξ is defined by the relation

$$K_n(\xi) = (-1)^n \exp\left(\frac{\xi^2}{2}\right)\frac{d^n}{d\xi^n}\exp\left(-\frac{\xi^2}{2}\right) \tag{A.14.1}$$

A.14.1 Recurrence of Consecutive Polynomials

Consider the polynomial of degree n + 1. From (A.14.1) we obtain

$$K_{n+1}(\xi) = (-1)^{n+1}\exp\left(\frac{\xi^2}{2}\right)\frac{d^n}{d\xi^n}\left[\xi\exp\left(-\frac{\xi^2}{2}\right)\right] \tag{A.14.2}$$

Applying to Eq. A.14.2 the Leibnitz formula

$$\frac{d^n}{dx^n}xf(x) = x\frac{d^n}{dx^n}f(x) + \frac{d^{n-1}}{dx^{n-1}}f(x) \tag{A.14.3}$$

we have

$$(-1)^{n+1}K_{n+1}(\xi) = \exp\left(\frac{\xi^2}{2}\right)\left[-\xi\frac{d^n}{d\xi^n}\exp\left(-\frac{\xi^2}{2}\right) - n\frac{d^{n-1}}{d\xi^{n-1}}\exp\left(-\frac{\xi^2}{2}\right)\right] \tag{A.14.4}$$

It follows that

$$(-1)^{n+1}K_{n+1}(\xi) = -(-1)^{n}\xi K_{n}(\xi) - (-1)^{n-1}\xi K_{n-1}(\xi) \tag{A.14.5}$$

and

$$K_{n+1}(\xi) = \xi K_{n}(\xi) - nK_{n-1}(\xi) \tag{A.14.6}$$

Thanks to the recurrence formula (A.14.6), it is sufficient to know the first three Hermite polynomials to determine all the others. The Probabilistic Hermite polynomials up to the 10th power are as follows:

$$\begin{aligned}
K_0(\xi) &= 1\\
K_1(\xi) &= \xi\\
K_2(\xi) &= 2\xi^2-1\\
K_3(\xi) &= \xi^3-3\xi\\
K_4(\xi) &= \xi^4-6\xi^2+3\\
K_5(\xi) &= \xi^5-10\xi^3+15\xi\\
K_6(\xi) &= \xi^6-15\xi^4+45\xi^2-15\\
K_7(\xi) &= \xi^7-21\xi^5+105\xi^3-105\xi\\
K_8(\xi) &= \xi^8-28\xi^6+210\xi^4-420\xi^2+105\\
K_9(\xi) &= \xi^9-36\xi^7+378\xi^5-1260\xi^3+945\xi\\
K_{10}(\xi) &= \xi^{10}-45\xi^8+630\xi^6-3150\xi^4+4725\xi^2-945
\end{aligned}$$

A.14.2 Recurrence Between Hermite Polynomial and Their Derivatives

By making the derivative of Eq. A.14.1, we obtain

$$\dot{K}_n(\xi) = (-1)^n \exp\left(\frac{\xi^2}{2}\right)\frac{d^n}{d\xi^n}\exp\left(-\frac{\xi^2}{2}\right) + (-1)^n \exp\left(\frac{\xi^2}{2}\right)\frac{d^n}{d\xi^n}\left[-\xi\exp\left(-\frac{\xi^2}{2}\right)\right] \tag{A.14.7}$$

which, based on Leibnitz formula, gives

$$\dot{K}_n(\xi) = nK_{n-1}(\xi) \tag{A.14.8}$$

A.14.3 Orthogonality of Hermite Polynomials

The Hermite Polynomials are orthogonal in the $-\infty \leq \xi \leq \infty$ real space, respectively, to the weight function $\exp\left(-\frac{\xi^2}{2}\right)$.

Following Guilpin (1979), we may compute the integral I_{nk} in the assumption that $k > n$ without loss of generality

$$I_{nk} = \int_{-\infty}^{\infty} K_n(\xi)K_k(\xi)\exp\left(-\frac{\xi^2}{2}\right)d\xi \tag{A.14.9}$$

and

$$I_{nk} = (-1)^{n+k}\int_{-\infty}^{\infty} \exp\left(\frac{\xi^2}{2}\right)\frac{d^n}{d\xi^n}\exp\left(-\frac{\xi^2}{2}\right)\frac{d^k}{d\xi^k}\exp\left(-\frac{\xi^2}{2}\right)d\xi \tag{A.14.10}$$

Performing an integration by part, and posing

$$dv = \frac{d^k}{d\xi^k}\exp\left(-\frac{\xi^2}{2}\right)d\xi \tag{A.14.11}$$

we have

$$v = \frac{d^{k-1}}{d\xi^{k-1}}\exp\left(-\frac{\xi^2}{2}\right) \tag{A.14.12}$$

$$u = \exp\left(\frac{\xi^2}{2}\right)\frac{d^n}{d\xi^n}\exp\left(-\frac{\xi^2}{2}\right) \tag{A.14.13}$$

$$du = -n\exp\left(\frac{\xi^2}{2}\right)\frac{d^{n-1}}{d\xi^{n-1}}\exp\left(-\frac{\xi^2}{2}\right) \tag{A.14.14}$$

When $n \neq k$, the $u \cdot v$ product is null in the $-\infty \leq \xi \leq \infty$ real space because it contains $\exp\left(-\frac{\xi^2}{2}\right)$ as factor. We may thus write (Guilpin 1999):

$$I_{nk} = I_{n-1,k-1} \tag{A.14.15}$$

$$I_{nk} = n!\int_{-\infty}^{\infty} \frac{d^{k-n}}{d\xi^{k-n}}\exp\left(-\frac{\xi^2}{2}\right)d\xi \tag{A.14.16}$$

$$\int_{-\infty}^{\infty} \frac{\mathrm{d}^{k-n}}{\mathrm{d}\xi^{k-n}} \exp\left(-\frac{\xi^2}{2}\right) \mathrm{d}\xi = 0 \quad \text{for} \quad k \neq n \tag{A.14.17}$$

Hence the Hermite polynomials are orthogonal with respect to the normal probability distribution.

A.14.4 Physical Hermite Polynomials

The *Physical* Hermite Polynomials adopted in Quantum Mechanics differ from the *Probabilistic* ones being

$$H_n(\xi) = (-1)^n \exp\left(\frac{\xi^2}{1}\right) \frac{\mathrm{d}^n}{\mathrm{d}\xi^n} \exp\left(-\frac{\xi^2}{1}\right) \tag{A.14.18}$$

The two forms are related by

$$H_n(\xi) = 2^{n/2} K_n\left(\sqrt{2}\xi\right) \tag{A.14.19}$$

The Physical Hermite polynomials up to the 10th power are as follows:

$$\begin{aligned}
H_0(\xi) &= 1 \\
H_1(\xi) &= 2\xi \\
H_2(\xi) &= 4\xi^2 - 2 \\
H_3(\xi) &= 8\xi^3 - 12\xi \\
H_4(\xi) &= 16\xi^4 - 48\xi^2 + 12 \\
H_5(\xi) &= 32\xi^5 - 160\xi^3 + 120\xi \\
H_6(\xi) &= 64\xi^6 - 480\xi^4 + 720\xi^2 - 120 \\
H_7(\xi) &= 128\xi^7 - 1344\xi^5 + 3360\xi^3 - 1680\xi \\
H_8(\xi) &= 256\xi^8 - 3584\xi^6 + 13440\xi^4 - 13440\xi^2 + 1680 \\
H_9(\xi) &= 512\xi^9 - 9216\xi^7 + 48384\xi^5 - 80640\xi^3 + 30240\xi \\
H_{10}(\xi) &= 1024\xi^{10} - 23040\xi^8 + 161280\xi^6 - 403200\xi^4 + 302400\xi^2 - 30204
\end{aligned}$$

Some of them are drawn in Fig. A.6 together with their probabilistic counterparts. The orthogonality property holds also for the Physical Hermite Polynomials.

As we may see in Fig. A.6, the Hermite polynomials are either *even* or *odd* functions being reflected or changing sign across the y-axis. That is

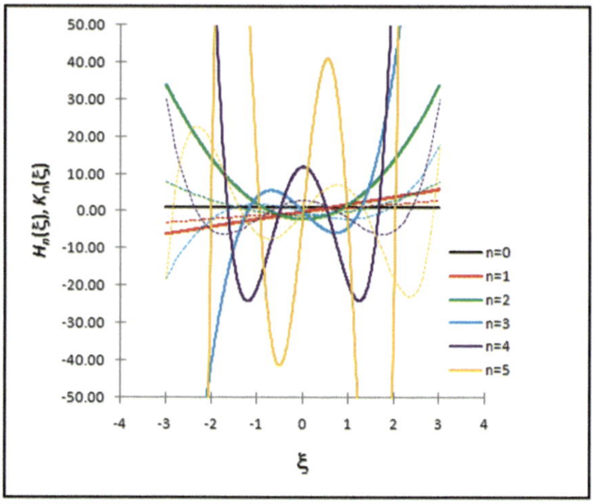

Fig. A.6 Hermite polynomials of various order. The physical ones (continuous lines) are compared with their probabilistic counterparts (dotted lines)

$$f(\xi) = f(-\xi) \quad \text{even} \quad (n = 0, 2, 4)) \tag{A.14.20}$$

$$f(\xi) = -f(-\xi) \quad \text{odd} \quad (n = 1, 3, 5)) \tag{A.14.21}$$

A.15 Legendre Polynomials and Associated Legendre Functions

We may generally state that the Legendre polynomials of all degree are orthogonal polynomials in the fundamental interval (−1, +1) returning the value +1 for the +1 value of the variable. Legendre conceived these functions in the aim of exploring the consequences of the Newton attraction law in $1/r^2$. In 1783, he extended thus the Mc Laurin–Clairault treatment of a fluid with a central charge to points external to the rotational ellipsoid. It should be stressed that the application to the Legendre polynomials to the gravitational field preceded the application to the electrical field (Coulomb's law appeared roughly 2 years later; Guilpin 1999). In the following, I will use the mathematical notation in the identification of the degree of polynomials (n, m) and I will associate to them a precise significance (*l, m*) only later on.

Denoting as $P_n(x)$ the Legendre polynomial of degree n, the definition imposes that

$$\int_{-1}^{+1} P_n(x) P_m(x) dx = 0 \tag{A.15.1}$$

with $P_m(x)$ as Legendre polynomial of degree m ≠ n. Proceeding with an integration by parts, one obtains a general expression of the Legendre polynomial known as Rodriguez formula (Rodriguez 1816):

$$P_n(x) = \frac{1}{2^n n!} \frac{d^n}{dx^n} \left(x^2 - 1\right)^n \tag{A.15.2}$$

The first Legendre polynomials up to the 6th degree are as follows:

$$P_0(x) = 1 \tag{A.15.3}$$

$$P_1(x) = x \tag{A.15.4}$$

$$P_2(x) = \left(3x^2 - 1\right)/2 \tag{A.15.5}$$

$$P_3(x) = \left(5x^3 - 3x\right)/2 \tag{A.15.6}$$

$$P_4(x) = \left(35x^4 - 30x^2 + 3\right)/8 \tag{A.15.7}$$

$$P_5(x) = \left(63x^5 - 70x^3 + 15x\right)/8 \tag{A.15.8}$$

$$P_6(x) = \left(231x^6 - 315x^4 + 105x^2 - 5\right)/16 \tag{A.15.9}$$

The C program of the procedure necessary to compute the coefficients may be found at http://www.edpsciences.com/guilpin/.

A.15.1 Recursion Formula

Applying the Rodriguez formula (A.15.2) it may be shown as well that a recurrence relation exists among the consecutive polynomials. Writing in fact Eq. A.15.2 for the highest degree (n + 1):

$$P_{n+1}(x) = \frac{1}{2^{n+1}} \frac{1}{(n+1)!} \frac{d^{n+1}}{dx^{n+1}} \left(x^2 - 1\right)^{n+1} \tag{A.15.10}$$

One arrives through several passages (Guilpin 1999) to

$$(n+1)P_{n+1}(x) - x(2n+1)P_n(x) + nP_{n-1}(x) = 0 \tag{A.15.11}$$

A.15.2 Generating Formulas

Without any demonstration we admit now the following equalities (Guilpin 1979):

$$\frac{1}{\sqrt{1-2hx+h^2}} = P_0(x) + hP_1(x) + h^2P_2(x) + \cdots h^nP_n(x) \qquad (A.15.12)$$

$$P_n(x) = \frac{1}{\pi}\int_0^{\pi}\left(x \pm \sqrt{(x^2-1)}\cos\theta\right)^n d\theta \qquad (A.15.13)$$

Denoting as T(h, x) the generating formula, i.e.

$$T(h,x) = \frac{1}{\sqrt{1-2hx+h^2}} = \sum_{n=0}^{\infty} P_n(x)h^n \qquad (A.15.14)$$

and differentiating it with respect to h one has (Guilpin, 1999)

$$\frac{\partial T(h,x)}{\partial h} = \sum_{n=0}^{\infty} nP_n(x)h^{n-1} = \frac{x-h}{\left(1-2xh+h^2\right)^{3/2}} \qquad (A.15.15)$$

$$(x-h) = \sum_{n=0}^{\infty} P_nh^n = \left(1-2xh+h^2\right)\sum_{n=0}^{\infty} nP_nh^{n-1} \qquad (A.15.16)$$

It may be easily seen that, by equating (for a given power n) the coefficients of h on the two sides of Eq. A.15.16 one obtains again Eq. A.15.11.

By differentiation in dx one gets

$$\frac{\partial T}{\partial x} = \sum_{n=0}^{\infty} \dot{P}_nh^n = \frac{1}{\left(1-2xh+h^2\right)^{3/2}} \qquad (A.15.17)$$

$$h\sum_{n=0}^{\infty} P_nh^n \equiv \left(1-2xh+h^2\right)\sum_{n=0}^{\infty} \dot{P}_nh^n \qquad (A.15.18)$$

$$\dot{P}_{n+1}(x) - 2x\dot{P}_n(x) + \dot{P}_{n-1}(x) = P_n(x) \qquad (A.15.19)$$

A.15.3 Quantum–Mechanical Application: First Solution and Legendre's Associated Equations

In the treatment of the harmonic oscillator, we met an Eq. (3.1.7) obtained by a change of variable ($x = \cos\theta$) operated on the Sommerfeld's Eq. (3.1.6). The second-order differential equation was in the form:

$$\left(1-x^2\right)\frac{d^2y}{dx^2}-2x\frac{dy}{dx}+\lambda y=0 \tag{A.15.20}$$

with λ as "*separation constant*". This is Legendre's equation. Because it generates from a change of variable (i.e. $x = \cos\theta$) we are concerned mainly in solutions defined in the interval $(-1, +1)$ and at $x = 0$ which is an ordinary point (Mc Weeny 1979). It should be stressed, however, that the variable of physical interest in the Legendre polynomials and the Associated Legendre functions is θ and not x (McQuarrie 2008).

Let me assign now to the degree n of the polynomials a precise significance, namely, the azimuthal (l) and the magnetic (m) quantum numbers. In fact, in finding the solution of the θ equation we made a substitution $z = \cos\theta$ and introduced the Azimuthal quantum number l. We remind now that with n as total quantum number ($n = 1, 2, 3, \ldots$), the azimuthal quantum number may attain the values $l = 0, 1, 2, \ldots. n - 1$ and the magnetic quantum number m_l may attain $m_l = -l, -l + 1, \ldots -1, 0, +1, +l - 1, +l$. There are thus $2l + 1$ independent wave functions for given values of n and l (*completed subgroup*) and n^2 independent functions for a given value of n (*completed group*). Proceeding as done for the Hermite polynomials we may produce a descending series, valid for l odd or even:

$$P_l(x)=\frac{(2l)!}{2^l(l!)^2}\left[x^l-\frac{l(l-1)}{2(2-1)}x^{l-2}+\frac{l(l-1)(l-2)(l-3)}{2.4(2l-1)(2l-3)}x^{l-4}-\ldots\right] \tag{A.15.21}$$

Posing the separation constant λ equal to $l(l + 1)$ the *Legendre's Associated equation* has the form

$$\left(1-x^2\right)\frac{d^2y}{dx^2}-2x\frac{dy}{dx}+\left[l(l+1)-\frac{m^2}{1-x^2}\right]y=0 \tag{A.15.22}$$

$$\begin{aligned} y &= a_0\left[1+\frac{\overline{\lambda}}{2!}x+\frac{\left(2.3+\overline{\lambda}\right)}{4!}x^2+\cdots\right] \\ &+a_1\left[\frac{x+\left(1.2+\overline{\lambda}\right)x^3}{}3!+\frac{\overline{\lambda}\left(3.4+\overline{\lambda}\right)}{5!}x^5+\cdots\right] \end{aligned} \tag{A.15.23}$$

with a_0 and a_1 indeterminate and $\overline{\lambda} = -\lambda$. It is rather obvious that, when $m = 0$ the above equation reduces to Legendre's equation (A.15.20).

Consider a possible solution near the singularities at $x = \pm 1$. Introducing the substitution

$$y = \left(1 - x^2\right)^{m_l/2} u \tag{A.15.24}$$

we get

$$\left(1 - x^2\right)\frac{d^2u}{dx^2} - 2(m_l + 1)\frac{du}{dx} + (l - m_l)(l + m_l + 1)u = 0 \tag{A.15.25}$$

Comparing the above equation with the equation satisfied by $P_n(x)$:

$$\left(1 - x^2\right)\frac{d^2P_n}{dx^2} - 2x\frac{dP_n}{dx} n(n + 1)P_n = 0 \tag{A.15.26}$$

We deduce that

$$u = \frac{d^m P_n}{dx^m} \tag{A.15.27}$$

and that the Associated Legendre's function has a solution

$$P_l^{m_l}(x) = \left(1 - x^2\right)^{m_l/2} \frac{d^m P_l(x)}{dx^m} \tag{A.15.28}$$

The first Associated Legendre's $P_l^{m_l}(x)$ functions up to $l = 3$ and with the change of variable $x = \cos\theta$ are the following (Mc Quarrie 2008):

$$P_0^0(x) = 1 \tag{A.15.29}$$

$$P_1^0(x) = x \equiv \cos\theta \tag{A.15.30}$$

$$P_1^1(x) = \left(1 - x^2\right)^{1/2} \equiv \sin\theta \tag{A.15.31}$$

$$P_2^0(x) = \frac{1}{2}\left(3x^2 - 1\right) \equiv \frac{1}{2}(3\cos\theta - 1) \tag{A.15.32}$$

$$P_2^1(x) = 3x\left(1 - x^2\right)^{1/2} \equiv 3\cos\theta\sin\theta \tag{A.15.33}$$

$$P_2^2(x) = 3\left(1 - x^2\right) \equiv 3\sin^2\theta \tag{A.15.34}$$

$$P_3^0(x) = \frac{1}{2}\left(5x^3 - 3x\right) \equiv \frac{1}{2}\left(5\cos^3\theta - 3\cos\theta\right) \tag{A.15.35}$$

$$P_3^1(x) = \frac{3}{2}\left(5x^2 - 1\right)\left(1 - x^2\right)^{1/2} \equiv \frac{3}{2}\left(5\cos^2\theta - 1\right)\sin\theta \tag{A.15.36}$$

$$P_3^2(x) = 15x\left(1 - x^2\right) \equiv 15\cos\theta\sin^2\theta \tag{A.15.37}$$

$$P_3^3(x) = 15\left(1 - x^2\right)^{3/2} \equiv 15\sin^3\theta \tag{A.15.38}$$

A.16 Laguerre Polynomials and Associated Laguerre Functions

The Laguerre polynomials are adopted in numerical analysis to find the roots of a continuous function. We may generally state that the Laguerre polynomials are orthogonal with respect to the weight function exp(−x) in the interval (0, ∞) (Guilpin 1999).

Denoting as $L_n(x)$ the Laguerre polynomial of degree n, the polynomials are

$$L_n(x) = \exp(x)\frac{d^n}{dx^n}\left[\exp(-x)x^n\right] \tag{A.16.1}$$

The first Laguerre polynomials up to the 5th degree are the following (Guilpin 1979):

$$L_0(x) = 1 \tag{A.16.2}$$

$$L_1(x) = -x + 1 \tag{A.16.3}$$

$$L_2(x) = x^2 - 4x + 2 \tag{A.16.4}$$

$$L_3(x) = -x^3 + 9x^2 - 18x + 6 \tag{A.16.5}$$

$$L_4(x) = x^4 - 16x^3 + 72x^2 - 96x + 24 \tag{A.16.6}$$

$$L_5(x) = -x^5 + 25x^4 - 200x^3 + 600x^2 - 600x + 120 \tag{A.16.7}$$

As anticipated for the Legendre polynomials, also the Laguerre polynomials may be generated with the aid of a Fortran-C program (http://www.edpsciences.com/guilpin/).

A.16.1 Recursion Formula

The definition formula (A.16.1) is written in a more general form (Guilpin 1999):

$$L_n^\alpha(x) = \exp(x)\frac{x^{-\alpha}}{n!}\frac{d^n}{dx^n}\left[\exp(-x)x^{n+\alpha}\right] \tag{A.16.8}$$

where the real part of α is intended to be higher than −1.

Equation A.16.8, when written for the polynomial $L_{n+1}^\alpha(x)$, gives

$$L_{n+1}^\alpha(x) = \exp(x)\frac{x^{-\alpha}}{(n+1)!}\frac{d^{n+1}}{dx^{n+1}}\left[\exp(-x)x^{n+\alpha+1}\right] \tag{A.16.9}$$

After various passages, involving the Leibnitz relation, the following recursion formula may be obtained (Guilpin 1999):

$$(n+1)L_{n+1}^\alpha(x) - (2n+1+\alpha-x)nL_n^\alpha(x) + (n+\alpha)L_{n-1}^\alpha(x) = 0 \tag{A.16.10}$$

A.16.2 Associated Laguerre Polynomials and Functions

The Associated Laguerre Polynomial of degree m and order n is the mth derivative of the Laguerre polynomial of degree n:

$$L_n^m(x) = \frac{d^m}{dx^n}L_n(x) \tag{A.16.11}$$

The differential equation satisfied by $L_n^m(x)$ is (Pauling and Bright Wilson 1935)

$$x\ddot{L}_n^m(x) + (m+1-x)\dot{L}_n^m(x) + (n-m)L_n^m(x) = 0 \tag{A.16.12}$$

Replacing n with $n + l$ and m with $2l + 1$, we have

$$x\ddot{L}_{n+l}^{2l+1}(x) + [2(l+1)-x]\dot{L}_{n+l}^{2l+1}(x) + (n-l+1)L_{n+l}^{2l+1}(x) = 0 \tag{A.16.13}$$

The above equation is adopted in Quantum Mechanics when solving the *r* equation (see Sect. 5.2.4). The polynomials obtained for the hydrogen atom are in this case the associated Laguerre polynomials of degree $n-l-1$ with order $2l + 1$. The wave functions in ρ are (when disregarding the normalizing factors) the *Associated Laguerre Functions*:

$$\exp\left(-\frac{\rho}{2}\right)\rho^l L_{n+l}^{2l+1}(\rho) \tag{A.16.14}$$

A.17 Laplace's Spherical Harmonics

From the Laplace equation in spherical coordinates:

$$\begin{aligned}\nabla^2 = &\frac{1}{r^2}\frac{\partial}{\partial r}\left(r^2\frac{\partial f}{\partial r}\right)_{\theta,\phi} + \frac{1}{r^2\sin\theta}\frac{\partial}{\partial\theta}\left(\sin\theta\frac{\partial f}{\partial\theta}\right)_{r,\phi} \\ &+ \frac{1}{r^2\sin^2\theta}\left(\frac{\partial^2 f}{\partial\phi^2}\right)_{r,\theta} = 0\end{aligned} \tag{A.17.1}$$

two differential equations are obtained by separation of variables:

$$\frac{1}{R}\frac{d}{dr}\left(r^2\frac{dR}{dr}\right) = \lambda \tag{A.17.2}$$

$$\frac{1}{Y}\frac{1}{\sin\theta}\frac{\partial}{\partial\theta}\left(\sin\theta\frac{\partial Y}{\partial\theta}\right) + \frac{1}{Y}\frac{1}{\sin^2\theta}\left(\frac{\partial^2 Y}{\partial\phi^2}\right) = -\lambda \tag{A.17.3}$$

Assuming that Y has the form:

$$Y(\theta,\phi) = \Theta(\theta)\Phi(\phi) \tag{A.17.4}$$

separating again the variables, other two differential equations are obtained:

$$\lambda\sin^2\theta + \frac{\sin\theta}{\Theta}\frac{d}{d\theta}\left(\sin\theta\frac{d\Theta}{d\theta}\right) = m^2 \tag{A.17.5}$$

$$\frac{1}{\Phi}\frac{d^2\Phi}{d\phi^2} = -m^2 \tag{A.17.6}$$

The solution function $Y(\theta, \varphi)$ is regular at the poles, where $\theta = 0, \pi$; m is a real integer; Φ is a linear combination of the complex exponentials $\exp(\pm im\phi)$; and, at the light of the Sturm–Liouville problem, the parameter λ is in the form $\lambda = \ell\,($ $+ 1)$ with $\ell \geq |m|$. For a given value of ℓ, there are $2\ell + 1$ independent solutions of the form (A.17.5), one for each integer m with $-\ell \leq m \leq \ell$. These angular solutions are a product of trigonometric functions, which may be represented as the product of a normalization function *N*, a complex exponential, and an associated Legendre polynomial:

$$Y_\ell^m(\theta,\phi) = N\exp(im\phi).P_\ell^m\cos(\theta) \tag{A.17.7}$$

In the quantum mechanical application, Eq. (A.17.4) has solutions:

$$\Phi(\phi) = \frac{1}{\sqrt{2\pi}} \exp(\mathrm{i} m_l \phi). \quad \text{(A.17.8)}$$

$$\Theta_l^{m_l}(\theta) = \left[\frac{2l+1}{2} \frac{(l-|m_l|)!}{(l-|m_l|)} \right]^{1/2} . P_l^{|m_l|} \cos(\theta) \quad \text{(A.17.9)}$$

where $m_l = 0, \pm1, \pm2, \ldots, l$ is the azimuthal quantum number and $P_l^{|m_l|}$ is the Associated Legendre Function. The spherical harmonics are

$$\mathrm{Y}_l^{m_l}(\theta, \phi) = \mathrm{i}^{m_l + |m_l|} \left[\frac{2l+1}{2} \frac{(l-|m_l|)!}{(l-|m_l|)} \right]^{1/2} . P_l^{|m_l|} \cos(\theta) \quad \text{(A.17.10)}$$

A.18 Bessel Functions

A first introduction of Bessel functions in wave mechanics is due to Sommerfeld (1937) in describing the wave equation of a point mass under no forces. Slater (1937) adopted spherical Bessel functions for wave functions inside the spheres in the APW procedure (cf. Sect. 12.6.1 and Eq. 12.6.13). A more recent implementation of spherical Bessel functions is due to Weinert (1981) who adopted it in the PW expansion necessary to get the multi-pole moments of the total charge in the sphere (Eq. 13; see also Jackson 1962; to this purpose).

A.18.1 Bessel Functions of Integer Order

A.18.1.1 Notation

Following the British Association and Royal Society Mathematical Tables, the following notation is adopted:

$$\mathrm{z} = \mathrm{x} + \mathrm{iy} \quad \text{(A.18.1)}$$

with x, y real;

$$\mathrm{n} = \text{positive integer or zero} \quad \text{(A.18.2)}$$

$$\nu = \text{unrestricted} \quad \text{(A.18.3)}$$

A.18.1.2 Bessel Functions of the First (*J*), Second (*Y*), and Third (*H*) Kind

Consider the differential equation (Abramovitx and Stegun 1972):

$$z^2\frac{d^2w}{dz^2}+z\frac{dw}{dz^2}+(z-\nu)w=0 \tag{A.18.4}$$

The Bessel functions of the first kind $J_{\pm\nu}(z)$, of the second kind $Y_\nu(z)$ (Weber's functions), and of the third kind $H^1_\nu(z), H^2_\nu(z)$ (Hankel functions) are solutions of the above equation. Each is a regular (holomorphic) function of z throughout the z-plane cut along the negative real axis. For fixed $z \neq 0$, each is an entire (integral) function of ν.

For each kind, the following relations exist:

$$J_{-n}(z)=(-)^n J_n(z) \tag{A.18.5}$$

$$Y_{-n}(z)=(-)^n Y_n(z) \tag{A.18.6}$$

$$H^{(1)}_{-\nu}(z)=\exp(\nu\pi i)H^{(1)}_\nu(z) \tag{A.18.7}$$

$$H^{(2)}_{-\nu}(z)=\exp(\nu\pi i)H^{(2)}_\nu(z) \tag{A.18.8}$$

Between Y and J, the following relation exists:

$$Y_\nu(z)=\frac{J_\nu(z)\cos(\nu\pi)-J_{-\nu}(z)}{\sin(\nu\pi)} \tag{A.18.9}$$

If ν is an integer or zero the right of Eq. (A.18.9) is replaced by its limiting value. Between Y, J, and H the following relation exists:

$$\begin{aligned}H^{(1)}_\nu(z)&=J_\nu(z)+iY_\nu(z)\\&=i\csc(\nu\pi)\{\exp(-\nu\pi i)J_\nu(z)-J_{-\nu}(z)\}\end{aligned} \tag{A.18.10}$$

$$\begin{aligned}H^{(2)}_\nu(z)&=J_\nu(z)-iY_\nu(z)\\&=i\csc(\nu\pi)\{J_{-\nu}(z)-\exp(-\nu\pi i)J_\nu(z)\}\end{aligned} \tag{A.18.11}$$

A.18.2 Bessel Functions of Fractional Order

A.18.2.1 Differential Equation

$$z^2\ddot{w} + 2z\dot{w} + \left[z^2 - n(n-1)\right]w = 0 \quad (n = 0, \pm 1, \pm 2, \ldots) \tag{A.18.12}$$

The Spherical Bessel function of the first kind is the particular solution:

$$j_n(z) = \sqrt{\frac{1}{2}\pi \Big/ z}\, J_{n+1/2}(z) \tag{A.18.13}$$

The Spherical Bessel function of the second kind is the particular solution:

$$y_n(z) = \sqrt{\frac{1}{2}\pi \Big/ z}\, Y_{n+1/2}(z) \tag{A.18.14}$$

The Spherical Bessel functions of the third kind are

$$h_n^{(1)}(z) = j_n(z) + iy_n(z) = \sqrt{\frac{1}{2}\pi \Big/ z}\, H_{n+1/2}^{(1)}(z) \tag{A.18.15}$$

$$h_n^{(2)}(z) = j_n(z) - iy_n(x) = \sqrt{\frac{1}{2}\pi \Big/ z}\, H_{n+1/2}^{(2)}(z) \tag{A.18.16}$$

A.19 Gaussian Functions

The Gaussian (or normal) probability function along the x-axis [Z(x)] is defined by Abramoviz and Stegun (1972)

$$Z(x) = \frac{1}{\sqrt{2\pi}} \exp\left(-\frac{z^2}{2}\right) \tag{A.19.1}$$

Posing

$$P(x) = \frac{1}{\sqrt{2\pi}} \int_{-\infty}^{z} \exp\left(-\frac{t^2}{2}\right) dt = \int_{-\infty}^{z} Z(t) dt \tag{A.19.2}$$

$$Q(x) = \frac{1}{\sqrt{2\pi}} \int_{x}^{\infty} \exp\left(-\frac{t^2}{2}\right) dt = \int_{x}^{\infty} Z(t) dt \tag{A.19.3}$$

$$A(x) = \frac{1}{\sqrt{2\pi}} \int_{-x}^{x} \exp\left(-\frac{t^2}{2}\right) dt = \int_{-x}^{x} Z(t) dt \tag{A.19.4}$$

with t positive constant, we have:

$$P(x) + Q(x) = 1 \tag{A.19.5.1}$$

$$P(-x) = Q(x) \tag{A.19.5.2}$$

$$A(x) = 2P(x) - 1 \tag{A.19.5.3}$$

If the probability (Pr) that $X \leq x$ is given by

$$\Pr\{X \leq x\} = \frac{1}{\sigma\sqrt{2\pi}} \int_{-\infty}^{z} \exp\left[\frac{(t-m)^2}{2\sigma^2}\right] dt \tag{A.19.6}$$

then the random variable X is said to be "*normally distributed*" with mean m and variance σ^2.

The corresponding probability density function is

$$\frac{\partial}{\partial x} P\left(\frac{x-m}{\sigma}\right) = \frac{1}{\sigma} Z\left(\frac{x-m}{\sigma}\right) = \frac{1}{\sigma\sqrt{2\pi}} \exp\left[-\frac{(x-m)^2}{2\sigma^2}\right] \tag{A.19.7}$$

The inflection points of the probability density function are located at $m \pm \sigma$ and the function is symmetric around m, i.e.

$$Z\left(\frac{m+x}{\sigma}\right) = Z\left(\frac{m-x}{\sigma}\right) \tag{A.19.8}$$

Moreover, the maximum is observed at m, the variance is σ^2, and the Full Width at Half Maximum is $\text{FWHM} = 2\sigma\sqrt{2\ln 2} \cong 2.35482\sigma$.

Applying the Gaussian integral $\int_{-\infty}^{\infty} \exp(-x^2)dx$ to the probability density function (A.19.7), and posing $a = \frac{1}{c\sqrt{2\pi}}$; $b = m$; $c = \sigma$, we obtain

$$\int_{-\infty}^{\infty} a \exp\left[-\frac{(x-b)^2}{2c^2}\right] dx = ac\sqrt{2\pi} = \frac{\sigma\sqrt{2\pi}}{\sigma\sqrt{2\pi}} = 1 \tag{A.19.9}$$

In other terms, the integral below a Gaussian probability function is always 1, independent of the values assigned to m and σ (see also Eq. A.19.5.1).

It should be stressed that, while the product of two Gaussian functions is a Gaussian, the product of two Gaussian probability functions is not a Gaussian probability function.

Posing $I_{-1}(x) = Z(x)$ it may be shown moreover that (Abramovitz and Stegun 1972)

$$I_n(x) = \int_x^\infty I_{n-1}(x)dx \quad (n \geq 0) \tag{A.19.10}$$

$$I_{-n}(x) = \left(-\frac{d}{dx}\right)^{n-1} Z(x) = (-1)^{n-1} Z^{(n-1)}(x) \quad (n \geq -1) \tag{A.19.11}$$

A.20 Green's Function

Consider a rope of length *l* stretched tight between two limiting boundaries by two identical forces *H* applied horizontally in opposite direction (Fig. A.7).

Apply then a constant load *F* at some point (x) the rope is deflected to x′. The deflection of the rope u(x) obeys the differential equation:

$$-H\frac{\partial^2 u(x)}{\partial x^2} = F(x) \tag{A.20.1}$$

Posing $u(0) = u(l) = 0$ for $0 < x < l$ both u(x) and its Green's function G(x, y) may be written explicitly as

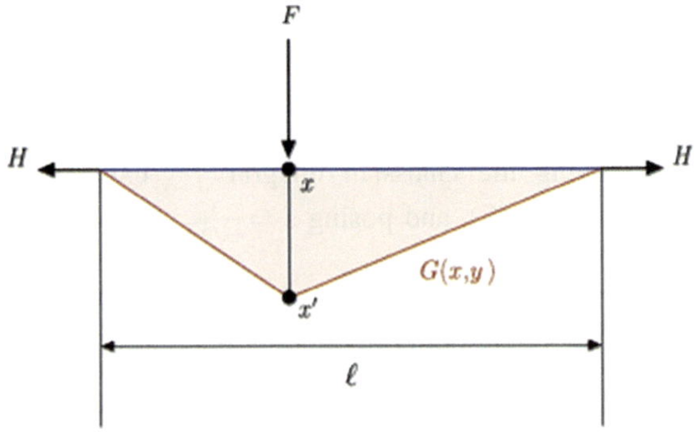

Fig. A.7 Physical interpretation of Green's function (from Stover 2018)

$$u(x) = \frac{F}{2H}\left(lx - x^2\right) \tag{A.20.2}$$

and

$$G(x, y) = \frac{1}{Hl}\begin{cases} y(l - x) & y \le x \\ y(l - y) & x \le y \end{cases} \tag{A.20.3}$$

The G = G(x, y) format in the above function indicates that it is piecewise linear (i.e. a two segment in the above example) and Green's function represents thus fully the effect of the action of the horizontal rope subjected to stretching forces. Given a linear differential operator $\mathsf{L} = \mathsf{L}(x)$ acting on the collection of charges, Green's function $G = G(x, s)$ at the point $s \in \Omega$ corresponding to L is any solution of

$$\mathsf{L}G(x, s) = \delta(x - s) \tag{A.20.4}$$

where δ denotes Dirac's delta function. Multiplying the above identity by a function f(s) and integrating in s, one has

$$\int \mathsf{L}G(x, s)f(s)ds = \int \delta(x - s)f(s)ds \tag{A.20.5}$$

Due to properties of the delta function, the right-hand side reduces merely to f(x). Since the linear operator L acts on x but not on s, the left-hand side can be rewritten as

$$\mathsf{L}\left[\int G(x, s)f(s)ds\right] \tag{A.20.6}$$

Consider now the differential of u = u(x). We may write the differential equations as

$$\mathsf{L}u(x) = f(x) \tag{A.20.7}$$

It follows that

$$\mathsf{L}u(x) = \mathsf{L}\left[\int G(x, s)f(s)ds\right] \tag{A.20.8}$$

Hence *u*, once integrated obeys

$$u(x) = \int G(x, s)f(s)ds \tag{A.20.9}$$

We have seen the utilization of Green's functions to visualize the effect of a charge distribution with charge density $\rho(\mathbf{r})$, with $\mathbf{r}$ belonging to a subset Ω of the Cartesian space corresponding to the limiting surface of a solvated molecule (Sect. 7.7.3). The potential at r_1 [$V(r_1)$] generated by all elements of charge $\rho(r_2)$ d^3r_2 is, by application of Poisson's equation,

$$V(r_1) = \frac{1}{4\pi\varepsilon}\int d^3r_2 \frac{\rho(r_2)}{|r_1 - r_2|} \quad (A.20.10)$$

The right-hand side of Eq. (A.20.10) can be seen as an integral operator that converts the charge density $\rho(r_2)$ into the potential $V(r_1)$ and the corresponding Green's function $G = G(r_1, r_2)$ has form:

$$G(r_1, r_2) = \frac{1}{4\pi\varepsilon}\frac{1}{|r_1 - r_2|} \quad (A.20.11)$$

One may then write (Arfken 1985b; Stover 2018)

$$V(r_1) = \int d^3r_2 G(r_1, r_2)\rho(r_2) \quad (A.20.12)$$

Another important application of Green's function in quantum chemistry is the study of surfaces and point defects in crystals (Sects. 8.12.7 and 8.12.8). Denoting in this case with V a "*perturbation matrix*" corresponding to the difference in the Hamiltonians of the perfect and imperfect crystals (H^0, H; resolved with the same basis sets and theory level):

$$V = \mathsf{H} - \mathsf{H}^0 \quad (A.20.13)$$

and with [G(z)] the "*Green matrix*" defined in terms of localized basis functions (ϕ_μ):

$$G(z) = (zS - \mathsf{H})^{-1} \quad (A.20.14)$$

where z is a complex energy value:

$$z = \varepsilon + is \quad (A.20.15)$$

one has

$$[G(z)]^{-1} = \left[G^0(z)\right]^{-1} - V \quad (A.20.16)$$

which through elementary matrix algebra is transformed into the identity:

$$G(z) = \left[I - G^0(z)V\right]^{-1} G^0(z) \tag{A.20.17}$$

Because the secular equation is satisfied when det(zS-H) = 0, and for real values of z ($z = \varepsilon_j$):

$$\varepsilon_j = \sum\nolimits_{\lambda} S_{\mu,\lambda} c_{\lambda,j} \Big/ \sum\nolimits_{\mu} H_{\mu,\nu} c_{\lambda,j} \tag{A.20.18}$$

with eigenfunctions

$$\psi_j(\mathbf{r}) = \sum_{\mu} c_{\mu,j} \phi_{\mu}(\mathbf{r}) \tag{A.20.19}$$

$V_{\mu,} \neq 0$ only when both μ and belong to a finite subset C of the representative basis set (i.e. the perturbation is "local") the "*embedding equation*" (Pisani 1996b) is

$$G_C(z) = \left\{ \left[G_C^0(z)\right]^{-1} - V_C \right\} \tag{A.20.20}$$

Equation A.20.20 allows to limit the computation of the electronic structure in the defect region, while taking into account "exactly" of the boundary condition.

I remand to Appendix A in Pisani (1976b) for an extensive account of the utilization of Green's functions in imperfect crystals.

A.21 Wannier Functions

Wannier functions are the localized molecular orbitals of crystalline systems. Wannier (1937) conceived these functions as a method apt "*to study the energy spectrum for an excited electron configuration in an ideal crystal*". The motivation was basically that, although "*It would no doubt be more satisfactory for insulating crystals, to discuss the Hamiltonian using atomic functions rather than Bloch functions*" this cannot be done because atomic functions are not orthogonal. One can however "*build up orthogonal functions having all the advantages of atomic ones by starting out from a Bloch approximation*". The orthogonality of Wannier functions for different lattice sites allows the investigation of excited states in certain regimes. They find application in the investigation of electronic transport, polarization processes, spectral properties, thermal electricity, and many other cases. We found them in the Quantum Mechanical treatment of surface energy and defect energy (see Sects. 8.12.7 and 8.12.8).

For a wave function obeying the Block condition

$$\psi_{\mathbf{k}}(\mathbf{r}) = \mathrm{u}_{\mathbf{k}}(\mathbf{r}) \exp(\mathrm{i}\mathbf{k} \cdot \mathbf{r}) \tag{A.21.1}$$

where $\mathrm{u}_{\mathrm{k}}(\mathbf{r})$ has the same periodicity of the crystal, the Wannier function is

$$\phi_{\mathbf{R}}(\mathbf{r}) = \frac{1}{\sqrt{\mathrm{N}}} \sum_{\mathbf{k}} \psi_{\mathbf{k}}(\mathbf{r}) \exp(-\mathrm{i}\mathbf{k} \cdot \mathbf{R}) \tag{A.21.2}$$

where **R** is any lattice vector, N is the number of primitive cells in the crystal, and the **k** summation includes all the values of **k** spanning the Brillouin zone and consistent with the periodic boundary conditions. From (A.21.1) and (A.21.2), it follows that any Block state is related to Wannier functions by

$$\psi_{\mathbf{R}}(\mathbf{r}) = \frac{1}{\sqrt{\mathrm{N}}} \sum_{\mathbf{R}} \phi_{\mathbf{R}}(\mathbf{r}) \exp(\mathrm{i}\mathbf{k} \cdot \mathbf{r}) \tag{A.21.3}$$

with the summation extended to all lattice vectors **R**.

For any lattice vector **R′**, one has moreover

$$\phi_{\mathbf{R}+\mathbf{R}'}(\mathbf{r} + \mathbf{R}') = \phi_{\mathbf{R}}(\mathbf{r}) \tag{A.21.4}$$

The most convenient set of Wannier functions is usually the maximally localized set, in which the Wannier function $\phi_{\mathbf{R}}$ is localized around the point **R** and rapidly goes to zero away from **R**. Kohn (1959) has shown that for a 1D lattice there is always a unique choice that gives these properties. A localization algorithm for 3D crystals that allows one to iteratively transform the extended Bloch orbitals of a first-principles calculation into a unique set of maximally localized Wannier functions (MLWFs) has been developed by Marzari et al. (2003), to whom I remand for an extensive review of the problem.

References

Abragam A (1961) The principles of nuclear magnetism. Clarendon Press, Oxford

Abraham RJ, Bretschneider E (1974) In: Orville-Thomas WJ (ed) Internal Rotation in Molecules, Wiley, London-New York-Sidney-Toronto, pp 481–584

Abramowitz M, Stegun I A (1972) Handbook of Mathematical Functions with Formulas, Graphs, and Mathematical Tables Dover Publications Inc, New York

Accad Y, Pekeris CL, and Schiff B (1975) Two-electron S and P term values with smooth Z dependence. Phys Rev A 11:1479–1481

Ackermann RJ, Sorrell CA (1974) Thermal expansion and the high-low transformation in quartz I High-temperature X-ray studies. J Appl Cryst 7:461–467

Adamkoviková K, Kosa L, Proks I (1980) The heat of fusion of $CaSiO_3$. Silikaty 24:193–201

Adamo C, Barone V (1998) Exchange functionals with improved long-range behavior and adiabatic connection methods without adjustable parameters: The mPW and mPW1PW models. J Chem Phys 108: 664–675

Adamo C, Barone V (1999) Toward reliable density functional methods without adjustable parameters: The PBE0 model. J Chem Phys 110:6158–6170

Adams DJ, Oganov AR (2006) Ab initio molecular dynamics study of $CaSiO_3$ perovskite at P-T conditions of Earth's lower mantle. Phys Rev B 73:184106-1-8

Aggarwal KG (1967) Lattice dynamics of diamond. Proc Phys Soc 91:381–389

Ahlrichs R, Penco R, Scoles G (1976) Intermolecular forces in simple systems. Chem Phys 19: 119–130

Ahuja R, Belonoshko AB, Johansson B (1998) Melting and liquid structure of aluminum oxide using a molecular-dynamics simulation. Phys Rev E 57:1673–1676

Akaogi M, Ito E, Navrotsky A (1989) Olivine-modified spinel–spinel transitions in the system Mg_2SiO_4–Fe_2SiO_4: calorimetric measurements, thermochemical calculation geophysical application. J Geophys Res 94:15671–15686

Akaogi M, Yusa H, Shiraishi K, Suzuki T (1995) hermodynamic properties of α-quartz, coesite and stishovite and equilibrium phase relations at high pressures and high temperatures. J Geophys Res-Sol Ea 100:22337–22347

Akaogi M, Takayama H, Kojitani H, Kawaji H, Atake T (2007) Low-temperature heat capacities, entropies and enthalpies of Mg_2SiO_4 polymorphs, and α–β–γ and post-spinel phase relations at high pressure. Phys Chem Minerals 34:169–183

Akber-Knutson S, Bukowinski M S T, Matas J (2002) On the structure and compressibility of $CaSiO_3$ perovskite. Geophys Res Lett 29:4-1-4

Al'tshuler L V, Brusnikin S E, Kuzmenkov E A (1987) Isotherms and Grüneisen functions for 25 metals. J Appl Mech Tech Phys 28:129–141

G. A. Ottonello, *Quantum Geochemistry*, Springer Geochemistry,
https://doi.org/10.1007/978-3-031-21837-8

Aleonard, R (1960) Etude paramagnetique des ferrites d'yttrium. J Phys Chem Solids 15:167–182
Alfè D (2009) PHON: A program to calculate phonons using the small displacement method. Comput Phys Commun 180:2622–2633
Alfè D, Price G, Gillan M (2001) Thermodynamics of hexagonal-close-packed iron under Earth's core conditions. Phys Rev B 64:45123-1-16
Alia, Matysik J, Soede-Huijbregts C, Baldus M, Raap J, Lugtenburg J, Gast P, van Gorkom HJ, Hoff AJ, and de Groot HJM (2001) Ultrahigh Field MAS NMR Dipolar Correlation Spectroscopy of the Histidine Residues in Light-Harvesting Complex II from Photosynthetic Bacteria Reveals Partial Internal Charge Transfer in the B850/His Complex. J Am Chem Soc 123:4803–4809
Allan G, Lenglart P (1968) Electron surface states and surface entropy for a simple cubic lattice. Surf Sci 15:101–108
Allard F, Hauschildt PH, Alexander DR, Starrfield S (1997) Model atmosphere of very low mass stars and brown dwarfs. Ann Rev Astr Astrophys 35:137–177
Amovilli C, Mennucci B (1997) Self-Consistent-Field Calculation of Pauli Repulsion and Dispersion Contributions to the Solvation Free Energy in the Polarizable Continuum Model. J Phys Chem B 101:1051–1057
Anderson OL (1959) The Debye temperature of vitreous silica. J Phys Chem Solids 12:41–52
Andersen OK (1975) Linear methods in band theory. Phys Rev B 12:3060–3083
Andersen HC (1980) Molecular dynamics simulations at constant pressure and/or temperature. J Chem Phys 72:2384–2393
Anderson OL (1995) Equations of State of Solids for Geophysics and Ceramic Science, Oxford Monographs on Geology and Geophysics, Vol 21, Oxford University Press, New York, Oxford
Anderson OL, Sumino Y (1980) The thermodynamic properties of the Earth's lower mantle. Phys Earth Planet In 23:314–331
Anderson OL, Zou K (1990) Thermodynamic functions and properties of MgO at high compression and high temperature. J Phys Chem Ref Data 19:69–81
Anderson OL, Isaak DG, Oda H (1992) High temperature elastic constant data on minerals relevant to geophysics. Rev Geophys 30:57–90
Anderson SM, Hülsebusch D, Mauersberger K (1997) Surprising rate coefficients for four isotopic variants of O + O_2 + M. J Chem Phys 107:5385–5392
Angel RJ, Hugh-Jones DA (1994) Equations of state and thermodynamic properties of enstatite pyroxenes. J Geophys Res 99:19777–19783
Angel RJ, Finger LW, Hazen RM, Kanzaki M, Weidner DJ, Liebermann RC, Veblen DR (1989) Structure and twinning of single-crystal $MgSiO_3$ garnet synthesized at 17 GPa and 1800 °C. Am Mineral 74:509–512
Angel RJ, Chopelas A, Ross NL (1992) Stability of high-density clinoenstatite at upper-mantle pressures. Nature 358:322–324
Angell CA (1982) Supercooled water. Amer Rev Phys Chem 34:593–630
Angyan JG, Loos M, Mayer I (1994) Covalent Bond Orders and Atomic Valence Indices in the Topological Theory of Atoms in Molecules. J Phys Chem 98:5244–5248
Angyan JG, Rosta E, Surjan PR (1999) Covalent bond orders and atomic valences from correlated wavefunctions. Chem Phys Lett 299:1–8
Anisimov VI, Zaanen J, Andersen OK (1991) Band theory and Mott insulators: Hubbard *U* instead of stoner *I*. Physical Review B 44:943–954
Aprà E (1996) Hartree-Fock Treatment of Spin-Polarized Crystals In: Pisani C (ed) Quantum-Mechanical Ab-initio calculations of the Properties of Crystalline Materials, Lecture Notes in Chemistry, 67, Springer, Berlin-Heidelberg-New York pp 101–112
Arfken G (1985a) Circular Cylindrical Coordinates. In: Mathematical Methods for Physicists, 3rd ed, Academic Press, Orlando, FL, p 95–101
Arfken G (1985b) Nonhomogeneous Equation—Green's Function Green's Functions- One Dimension, and Green's Functions- Two Dimensions. In: Mathematical Methods for Physicists, 3rd ed: Academic Press, Orlando, FL, p 480–491

Arfken, G (1985c) Nonhomogeneous Equation—Green's Function Green's Functions- Three Dimensions. In: Mathematical Methods for Physicists, 3rd ed, Academic Press, Orlando, FL, p 897–924
Aroyo MI, Müller U, Wondratschek H (2006) Historical introduction. In: International Tables for Crystallography A1:2–5
Arrhenius SV (1896) On the influence of carbonic acid in the air upon the temperature of the ground. Phil Mag 41:237–276
Ashida T, Kume S, Ito E, Navrotsky A (1988) $MgSiO_3$ ilmenite heat capacity, thermal expansivity, and enthalpy of transformation. Phys Chem Minerals 16:239–245
Austin JB (1954) The coefficient of linear thermal expansion of tridymite. J Am Chem Soc 76: 6019–6020
Authier A, Zarembovitch A (2006) Elastic Properties, In: International Tables for Crystallography, Vol D: Physical Properties of Crystals, IUCr-Springer p 72–98
Baber WG, Hasse HR (1935) The Two Center Problem in Wave Mechanics. Proc Camb Phil Soc 31:564–566
Baboul AG, Curtiss LA, Redfern PC, Raghavachari K (1999) Gaussian-3 theory using density functional geometries and zero-point energies. J Chem Phys 110:7650–7657
Bachelet GB, Hamann DR, Schlüter M (1982) Pseudopotentials that work: From H to Pu. Phys Rev B 26:4199–4228
Bader, RFW (1994) Atoms in Molecules A Quantum Theory, Clarendon Press, Oxford, UK
Bader RFW, Jones GA (1963) The electron density distribution in hydride molecules, I: The water molecule. Can J Chem 41:586–606
Bader HP, Schilling R (1979) Conditions for a ferromagnetic ground state of Heisenberg Hamiltonians. Phys Rev B 19:3556–3560
Bader RFW, Stephens ME (1974) Fluctuation and Correlation of Electrons in Molecular Systems. Chem Phys Lett 26:445–449
Bader RFW, Stephens ME (1975) Spatial localization of the electronic pair and number distributions in molecules. J Amer Chem Soc 97:7391–7399
Badyal YS, Saboungi M-L, Price DL, Shastri SD, Haeffner DR, Soper AK (2000) Electron distribution in water. J Chem Phys 112:9206–9208
Baes CF, Mesmer RE (1976) The Hydrolysis of Cations. Wiley, New York
Bai Q, Kohlstedt DL (1993) Effects of Chemical Environment on the Solubility and Incorporation Mechanism for Hydrogen in Olivine. Phys Chem Minerals 19:460–471
Baima J, Erba A, Rérat M, Orlando R, Dovesi R (2013) Beryllium oxide nanotubes and their connection to the flat monolayer. J Phys Chem C 117:12864–12872
Bajgain SK, Ghosh DB, Karki BB (2015) First-principles simulations of CaO and $CaSiO_3$ liquids: structure, thermodynamics and diffusion. Phys Chem Minerals 42:393–404
Balian R, De Dominicis D (1964) Quasi-particle formulation of quantum statistics. Physica 30: 1927–1932
Ball JW, Nordstrom DK (1998) Critical evaluation and selection of standard state thermodynamic properties for chromium metal and its aqueous ions, hydrolysis species, oxides and hydroxides. J Chem Eng Data 43:895–918
Barbe A, Secroun C, Jouve P (1974) Infrared Spectra of $^{16}O_3$ and $^{18}O_3$: Darling and Dennison Resonance and Anharmonic Potential Function of Ozone. J Mol Spec 49:171–182
Barin, I (1995) Thermochemical Data of Pure Substances. VCH, Weinheim, NY
Barin I, Knacke O (1973) Thermochemical Properties of Inorganic Substances. Springer, Berlin-Heidelberg-New York
Barin I, Knacke O, Kubaschewsky O (1977) Thermochemical Properties of Inorganic Substances, Supplement. Springer, Berlin-Heidelberg-New York
Barker Jr AS (1963) Infrared Lattice Vibrations and Dielectric Dispersion in Corundum. Phys Rev 132:1474–1481
Barone V, Cossi M, Tomasi J (1997) A new definition of cavities for the computation of solvation free energies by the polarizable continuum model. J Chem Phys 107:3210–3221

Baroni S, Giannozzi P, Testa A (1987) Green's-function approach to linear response in solids. Phys Rev Lett 58:1861–1864

Baroni S, de Gironcoli S, Dal Corso A, Giannozzi P (2001) Phonons and related crystal properties from density functional perturbation theory. Rev Modern Phys 73:515–562

Barr LW, Liliard AB (1971) Defects in ionic crystals In: Jost W (series ed): Physical Chemistry, an Advanced Treatise, Vol 10, Academic Press Inc, New York, pp 151–228

Barron THK, Berg WT, Morrison JA (1959) On the heat capacity of crystalline magnesium oxide. Proc R Soc 250:70–83

Barrow GM (1962) Introduction to Molecular Spectroscopy. McGraw-Hill Book Company, Inc, New York – San Francisco – Toronto – London

Bartlett R J, Silver DM (1975) Many-body perturbation theory applied to electron pair correlation energies I Closed-shell first-row diatomic hydrides. J Chem Phys 62:3258–3268

Bartmess JE (1994) Thermodynamics of the electron and proton. J Phys Chem 98:6420–6424

Basak D, Ghose J (1994) Infrared studies on some substituted copper chromite spinels. Spectrochim Acta 50A:713–718

Bass JD (1995) Elasticity of Minerals, Glasses and Melts In: TJ Arhens (Ed) A Handbook of Physical Constants: Mineral Physics and Crystallography, AGU Reference Shelf Series, Vol 2, American Geophysical Union, pp 9789–9803

Bates DR (1988) Suggested explanation of heavy ozone. Geophys Res Lett 15:13–16

Bates FS, Fredrickson GH (1999) Block copolymers – designer soft materials. Physics Today 52: 32–38

Bates DR, Ledsham K, Stewart L (1953) Wave functions of the hydrogen molecular ion. Phil Trans Roy Soc London 246:215–240

Battino R, Rettich TR, Tominaga T (1983) The Solubility of Oxygen and Ozone in Liquids. J Phys Chem Ref Data 12:163–177

Battino R, Rettich TR, Tominaga T (1984) The Solubility of Nitrogen and Air in Liquids. J Phys Chem Ref Data 13:563–600

Baty SR, Burakovsky L, Errandonea D (2021) Ab Initio Phase Diagram of Copper. Crystals 11: 537-1-14

Baur WH (1972) Computer simulated crystal structures of observed and hypothetical Mg_2SiO_4 polymorphs of low and high density. Am Mineral 57:709–731

Baur WH, Kahn AA (1971) Rutile-type compounds IV SiO_2, GeO_2 and a comparison with other rutile-type compounds. Acta Cryst B 27:2133–2139

Becke AD (1986) Density functional calculations of molecular bond energies. J Chem Phys 84: 4524–4529

Becke AD (1988) Density-functional exchange-energy approximation with correct asymptotic behavior. Phys Rev A 38:3098–3100

Becke AD (1993a) A new mixing of Hartree–Fock and local density-functional theories. J Chem Phys 98:1372–1377

Becke AD (1993b) Density-functional thermochemistry III The role of exact exchange J Chem Phys 98:5648–5652

Becke AD (1997) Density-functional thermochemistry V Systematic optimization of exchange-correlation functional: J Chem Phys 107:8554–8560

Becke AD, Johnson ER (2006) A simple effective potential for exchange J Chem Phys 124: 221101-1-4

Beckmann E, Waentig P (1910) Cryoscopic Measurements at Low Temperatures. Z Anorg Chem 67:17–61

Bekker A, Holland HD, Wang PL, Rumble D 3rd, Stein HJ, Hannah JL, Beukes NJ (2004) Dating the rise of atmospheric oxygen. Nature 427:117–120

Bell RJ, Dean P (1970) Atomic vibrations in vitreous silica. Discuss Faraday Soc 50 :55–71

Bell DR, Rossman G (1992) Water in earth's mantle: The role of nominally anhydrous minerals. Science 255:391–1397

Bell RJ, Bird NF, Dean P (1968) The vibrational spectra of vitreous silica, germania and beryllium fluoride. J Phys C 1:299–303

Belmonte D (2013) Ab initio thermodynamics of deep mantle minerals: the system MgO-SiO_2. Dissertation, University of Genova

Belmonte D (2017) First Principles Thermodynamics of Minerals at HP–HT Conditions: MgO as a Prototypical Material. Minerals 7:183–217

Belmonte D, Ottonello G, Vetuschi Zuccolini M (2013) Melting of α-Al_2O_3 and vitrification of the undercooled alumina liquid: Ab initio vibrational calculations and their thermodynamic implications. J Chem Phys 138:064507-1-7

Belmonte D, Ottonello G, Vetuschi Zuccolini M (2014) Ab-initio thermodynamic and thermophysical properties of sapphirine end-members in the join $Mg_4Al_8Si_2O_{20}$ – $Mg_3Al_{10}SiO_{20}$. Am Mineral 99:1449–1461

Belmonte D, Gatti C, Ottonello G, Richet P, Vetuschi Zuccolini M, (2016) Ab Initio Thermodynamic and Thermophysical Properties of Sodium Metasilicate, Na_2SiO_3, and Their Electron-Density and Electron-Pair-Density Counterparts. J Phys Chem A 120:8881–8895

Belmonte D, Ottonello G, Vetuschi Zuccolini M (2017a) Ab initio -assisted assessment of the CaO-SiO_2 system under pressure. CALPHAD 59:12–30

Belmonte D, Ottonello G, Vetuschi Zuccolini M, Attene M (2017b) The system MgO-Al_2O_3-SiO_2 under pressure: A computational study of melting relations and phase diagrams. Chem Geol 461:54–64

Belonoshko AB, Ahuja R, Eriksson O, Johansson B (2000) Quasi ab initio molecular dynamic study of Cu melting Phys Rev B 61:3838–3844

Belonoshko AB, Skorodumova NV, Rosengren A, Johansson B (2006) Melting and critical superheating. Phys Rev B 73:012201-1-3

Belousov R (2011) Ab initio Quantum-Mechanical simulation of Earth minerals at estreme conditions of temperature and pressure. Dissertation, University of Torino

Belov NV, Belova EN, Andrianova NH, Smirnova PF (1951) Determination of the parameters in the olivine (forsterite) structure with the harmonic three-dimensional synthesis. Dokladi Akademii Nauk SSSR 81 :399–402

Ben Naim, A (1987) Solvation Thermodynamics. Plenum Press, New York

Benassi P, Krisch M, Masciovecchio C, Mazzacurati V, Monaco G, Ruocco G, Sette F, Signorelli G, Verbeni R (1996) Evidence of High Frequency Propagating Modes in Vitreous Silica. Phys Rev Letters 77: 3835–3838

Benkova Z, Sadlej AJ (2004) Electric Moments of carbon dichalcogenides. Mol Phys, 102: 687–699

Ben-Naim A, Stillinger FH (1972) In: RA Horne (ed) Structure and transport processes in water and aqueous solutions. Wiley Interscience, New York

Benson GC, Wyllie G (1951) A Quantum Mechanical Treatment of the Lithium Fluoride Crystal. Proc Phys Soc London A64:276–282

Benson BB, Krause D, Peterson MA (1979) The solubility and isotopic fractionation of gases in dilute aqueous solution I Oxygen. J Soln Chem 8:655–690

Berger AL (1978) Long-term variations of daily insolation and Quaternary climatic change. J Atmos Sci 35:2362–2367

Berman, RG (1988) Internally-consistent thermodynamic data for minerals in the system Na_2O-K_2O-CaO-MgO-FeO-Fe_2O_3-Al_2O_3-SiO_2-TiO_2-H_2O-CO_2. J Petrol 29:445–522

Berman, RG, and Brown TH (1985) Heat capacity of minerals in the system Na_2O-K_2O-CaO-MgO-FeO-Fe_2O_3-Al_2O_3-SiO_2-TiO_2-H_2O-CO_2: Representation, estimation, and high temperature extrapolation. Contrib Mineral Petr 89:168–183

Berman, RG, and Brown TH (1986) Erratum Heat capacity of minerals in the system Na_2O-K_2O-CaO-MgO-FeO-Fe_2O_3-Al_2O_3-SiO_2-TiO_2-H_2O-CO_2: Representation, estimation, and high temperature extrapolation. Contrib Mineral Petr 94:262

Berman R, Hudson P R W, and Martinez M (1975) Nitrogen in diamond: evidence from thermal conductivity, J Phys C: Solid State 8:L430–L434

Berman R G, Brown T H, and Greenwood H J (1985) An internally consistent thermodynamic data base for minerals in the system Na_2O-K_2O-CaO-MgO-FeO-Fe_2O_3-Al_2O_3-SiO_2-TiO_2-H_2O-CO_2 Report N° TR-377, Atomic Energy of Canada Ltd, Ottawa
Bernal JD (1924) The structure of graphite P R Soc London A106:749–773
Bernal JD (1964) The structure of liquids P R Soc London A280:299–322
Bernal JD, Fowler RH (1933) A Theory of Water and Ionic Solution, with Particular Reference to Hydrogen and Hydroxyl Ions. J Chem Phys 1:515-548
Bernas A, Ferradini C, Jay-Gerin JP (1997) On the electronic structure of liquid water: Facts and reflections. Chem Phys 222:151–160
Bernholc J, Lipari NO, Pantelides ST (1978) Self-Consistent Method for Point Defects in Semiconductors: Application to the Vacancy in Silicon. Phys Rev Letters 41:895–899
Bertaut F (1952) L'énergie électrostatique de réseaux ioniques. J Phys Radium 13:499–505
Bethe H (1929) Splitting of terms in crystals. Amer Phys 3:133–206
Bethe HA (1997) Selected Works of Hans A Bethe. Word Scientific Pub Co, Singapore, River Edge, NY
Beyer, W H (1987) Standard Mathematical Tables, CRC Press, Boca Raton, FL
Bhattarai D, Karki BB, Stixrude L (2006) Space-time multiresolution atomistic visualization of MgO and $MgSiO_3$ liquid data. Vis Geosci 11:1–11
Bichowsky FR, Urey HC (1926) A Possible Explanation of the Relativity Doublets and Anomalous Zeeman Effect by Means of a Magnetic Electron. Proc Nat, Acad Sci 12:80–85
Bidwell WE (1901) Vector Analysis: A text-book for the use of students of mathematics and physics, founded upon the lectures of J Willard Gibbs. Yale University Press
Biegler-König W, Bader RFW, Tang TH (1982) Calculation of the average properties of atoms in molecules II. J Comput Chem 3:317–328
Bigeleisen J, Mayer MG (1947) Calculation of equilibrium constants for isotopic exchange reactions. J Chem Phys 13:261–267
Bina CR (1998) Lower mantle mineralogy and the geophysical perspective In: Hemley RJ (ed) Ultrahigh-Pressure Mineralogy: Physics and Chemistry of the Earth's Deep Interior Reviews in Mineralogy; Mineralogical Society of America Chantilly, VA, USA, Volume 37, p 205–240
Binkley JS, Pople JA, Hehre WJ (1980) Self-consistent molecular orbital methods 21 Small split-valence basis sets for first-row elements. J Am Chem Soc 102:939–947
Binning RC Jr, Curtiss LA (1990) Compact contracted basis sets for 3rd-row atoms: Ga-Kr. J Comput Chem 11:1206–1216
Bjerrum N (1929) Neuere Anschauungen über Elektrolyte. Deuts Chem Gesell Ber 62:1091–1103
Blaha P, Schwarz K, Tran F, Laskowski R, Madsen GKH, Marks LD (2020) WIEN2k: An APW +lo program for calculating the properties of solids. J Chem Phys 152:074101-1-30
Blaudeau J, McGrath MP, Curtiss LA, Radom L (1997) Extension of Gaussian-2 (G2) theory to molecules containing third-row atoms K and Ca. J Chem Phys 107:5016–5021
Bloch F (1928) Über die Quantenmechanik der Elektronen in Kristallgittern. Zeits Phys 52:555–600
Blöchl P E (1994) Projector augmented-wave method. Phys Rev B 50:17953–17979
Blöchl P E, Jepsen O, Andersen OK (1994) Improved tetrahedron method for Brillouin-zone integrations. Phys Rev B 49:16223–16233
Bochicchio C, Lain L, Torre A (2003) On the definition of bond orders at correlated level. Chem Phys Lett 374:567–571
Bockris JO'M, Reddy AKN (1970) Modern Electrochemistry. Plenum Press, New York
Boehler R (1982) Adiabats of quartz, coesite, olivine, and magnesium oxide to 50 kbar and 1000 K, and the adiabatic gradient in the Earth's mantle. J Geophys Res 87:5501–5506
Boehler R, Chopelas A (1991) A new approach to laser heating in high pressure mineral physics. Geophys Res Lett 18:1147–1150
Boehler R, Kennedy GC (1977) Pressure dependence of the thermodynamical Grüneisen parameter of fluids. J Appl Phys 48:4183–4186

Boehler R, Ramakrishnan J (1980) Experimental results on the pressure dependence of the Grüneisen parameter: A review. J Geophys Res 85:6996–7002
Bolesta AV, Fomin VM (2014) Molecular dynamics simulations of polycrystalline copper. J Appl Mech Tech Phys 55:800–811
Bondi A (1964) van der Waals Volumes and Radii. J Phys Chem 68:441–451
Borges PD, Cott J, Pinto FG, Tronto J, Scolfaro L (2016) Native defects as sources of optical transitions in $MgAl_2O_4$ spinel. Mater Res Express 3:076202-1-7
Born M (1920) Volumen und Hydratationswärme der Ionen. Zeits Phys 1:45–48
Born M, Huang K (1954) Dynamical Theory of Crystal Lattices. Oxford at the Claredon Press
Born M, von Karman T (1912) Uber Schwingungen im Raumgittern. Phys Zeits, 13:297–309
Born M, von Karman T (1913) Hüber die Verteilung der Eigenschwingungen von Punktgittern. Phys Zeits 14:65–71
Boswara IM, and Franklyn AD (1968) Theory of the energetics of simple defects in oxides In: Wachtman JB and Franklyn AD (eds) Mass Transport in Oxides, NatlBurStand (US) Spec Publ.
Bottinga Y (1969a) Calculated fractionation factors for carbon and hydrogen isotope exchange in the system calcite-carbon dioxide-graphite-methane-hydrogen-water vapor. Geochim Cosmochim Acta 33:49–64
Bottinga Y (1969b) Carbon isotope fractionation between graphite, diamond and carbon dioxide. Earth Planet Sc Lett 5:301–307
Bourdonnay H, Bousquet A, Cotton JP, Cribier D, Farnoux B, Feuillatre R, Hennion B, Jacquier J, Jannink G, Kahn R, Mons D, Mom J, Parette G, Perpy G, Rouleau L, Trotin J P, Moussa Mrs (1971) Experimental determination of exchange integrals in magnetite. Journal de Physique Colloque C1:1182–1183
Bourova E, Richet P (1998) Quartz and cristobalite: high-temperature cell parameters and volumes of fusion. Geophys Res Lett 25:2333–2336
Bourova E, Richet P, Petitet J-P (2006) Coesite (SiO_2) as an extreme case of superheated crystal: an X-ray diffraction study up to 1776 K. Chem Geol 229:57–63
Bovey YFA (1969) Nuclear Magnetic Resonance Spectroscopy, Academic Press, New York
Bowen NL, Andersen O (1914) The binary system MgO - SiO_2. Amer Jour Sci 37:487–500
Boys SF (1950) Electronic Wave Functions I A General Method of Calculation for the Stationary States of Any Molecular System. P R Soc London A200:542–544
Boys SF, Bernardi F (1970) The calculation of small molecular interactions by the differences of separate total energies Some procedures with reduced errors. Mol Phys 19:553–566
Boys SF, Handy NC (1969) A First Solution, for LiH, of a Molecular Transcorrelated Wave Equation by Means of Restricted Numerical Integration. P R Soc London A311:309–329
Bracewell, R (1999) The Sifting Property In: The Fourier Transform and Its Applications, (3rd Ed) McGraw-Hill, New York: pp 74–77
Brack A (2007) From interstellar amino acids to prebiotic catalytic peptides: A review. Chem Biodiversity 4:665–679
Bragg WH, Bragg WL (1913) The Structure of the Diamond. Nature 91:557–557
Bragg WL, Brown GB (1926) Die Structur des Olivine. Z Kristallogr 63:538–556
Brand JCD, Cross KJ, Hoy AR (1978) The Huggins bands of ozone. Can J Phys 56:327–333
Brandenburg JG, Alessio M, Civalleri B, Peintinger MF, Bredow T, Grimme S (2013) Geometrical Correction for the Inter- and Intramolecular Basis Set Superposition Error in Periodic Density Functional Theory Calculations. J Phys Chem A117:9282–9292
Brazhkin VV, Lyapin SG, Trojan IA, Voloshin RN, Lyapin AG, Mel'nik NN (2000) Anharmonicity of short-wavelength acoustic phonons in silicon at high temperatures. JETP Letters 72:195–198
Brazhkin VV, McNeil LE, Grimsditch M, Bendeliani NA, Dyuzheva TI, and Lityagina LM (2005) Elastic constants of stishovite up to its amorphization temperature. J Phys-Condens Matt 17:1869–1875

Brennan BJ, Stacey FD (1979) A thermodynamically based equation of state for the lower mantle. J Geophys Res 84:5535–5539

Bright Wilson E (1977) Impact of the Heitler-London Hydrogen Molecule. Paper on Chemistry International Journal of Quantum Chemistry Quantum Chemistry Symposium #11 pp 17–28

Bright Wilson E, Decius JC, Cross PC (1955) Molecular Vibrations, Dover Publications Inc, New York

Brillouin L (1946) Wave propagation in Periodic Structures Dover, New York

Brito F, Ascanio J, Mateo S, Hernàndez C, Araujo L, Gili P, Martin-Zarza P, Dominguez S, Mederos A (1997) Equilibria of chromate (VI) species in acid medium and ab initio studies of these species. Polyhedron 16:3835–3846

Brodholt J, Patel A, Refson K (1996) An ab initio study of the compressional behavior of forsterite. Am Miner 76:1100–1109

Broecker WS (1970) Man's oxygen reserves. Science 168:1537–1538

Broecker WS (1975) Climate change: Are we on the brink of a pronounced global warming? Science 401:460–463

Broecker WS (1985) Wow to Build a Habitable Planet. Eldigio Press, Lamont-Doherty Earth Observatory, Palisades, NY

Brookins DG (1988) Eh-pH Diagrams for Geochemistry. Springer, Berlin-Heidelberg-New York

Brown JM, Shankland TJ (1981) Thermodynamic parameters in the Earth as determined from seismic profiles. Geophys J Int 66:579–596

Broyden CG (1970) The converge of a class of double-rank minimization algorithms 1 General considerations. IMA J Appl Math 6:76–90

Bucaro A, Dardy HD (1976) Equilibrium compressibility of glassy SiO_2 between the transformation and melting temperature. J Non-Cryst Solids 20:149–151

Buchenau U, Prager M, Nücker N, Dianoux AJ, Ahmad N, Phillips WA (1986) Low-frequency modes in vitreous silica. Phys Rev B34:5665–5673

Bueso-Sanllehí F (1941) Rotational Analysis of the 2900A Band of C. Phys Rev 60:556–570

Bultinck, P (2007) Critical analysis of the local aromaticity concept in polyaromatic hydrocarbons. Faraday Discuss 135:347–365

Burakovsky L, Burakovsky N, Preston DL (2015) Ab initio melting curve of osmium. Phys Rev B 92:174105-1-7

Burkel E (1991) Inelastic Scattering of X Rays with Very High Energy Resolution. Springer Tracts in Modern Physics, Vol 125, Springer, Berlin

Burnham CW (1975) Water and magmas: A mixing model. Geochim Cosmochim Acta 39:1077–1084

Burns RG (1970) Mineralogical Application of Crystal Field Theory. Cambridge University Press, Cambridge

Burns RG (1993) Mineralogical Applications of Crystal Field Theory Cambridge University Press, Second Edition

Burrau Ø (1927) Berechnung des Energiewertes des Wasserstoffmolekel-Ions (H_2^+) im Normalzustand. Det Kgl Videnskabernes Selskab Mat Meddel 14:1–18

Caillet J, Claverie P, Pullman B (1978) On the conformational varieties of acetylcholine in the crystals of its halides Acta Cryst. B 34:3266–3272

Callen HB, Harrison SE, Kriessman CJ (1956) Cation distributions in ferrospinels: theoretical. Phys Rev 103:851–856

Cammi R, Tomasi J (1994) Analytical derivatives for molecular solutes I Hartree–Fock energy first derivatives with respect to external parameters in the polarizable continuum model. J Chem Phys 100:7495–7502

Cancès E, Mennucci B (1998a) Analytical derivatives for geometry optimization in solvation continuum models I Theory. J Chem Phys 109:249–259

Cancès E, Mennucci B (1998b) New applications of integral equations methods for solvation continuum models: ionic solutions and liquid crystals. J Math Chem 23:309–326

Cancès MT, Mennucci B, Tomasi J (1997) A new integral equation formalism for the polarizable continuum model: Theoretical background and applications to isotropic and anisotropic dielectrics. J Chem Phys 107:3032–3041

Car R, Parrinello M (1985) Unified Approach for Molecular Dynamics and Density-Functional Theory. Phys Rev Letters 55:2471–2474

Car R, Kelly PJ, Oshiyama A, Pantelides ST (1984) Microscopic Theory of Atomic Diffusion Mechanisms in Silicon. Phys Rev Letters 52:1814–1817

Car R, Kelly PJ, Oshiyama A, Pantelides ST (1985) Microscopic Theory of Impurity-Defect Reactions and Impurity Diffusion in Silicon. Phys Rev Letters 54:360-363

Car R, Parrinello M, Payne M (1991) Comment on 'error cancellation in the molecular dynamics method for total energy calculations'. J Phys Cond Matt 3:9539–9543

Caracas R, Hirose K, Nomura R, Ballmer MD (2019) Melt–crystal density crossover in a deep magma ocean. Earth Planet Sc Lett 516: 202–211

Carlsson JM (2007) Pseudopotentials: From Ultrahard to Ultrasoft CASTEP talks-07 http://www.tcm.phy.cam.ac.uk/

Caro-Lopera F, Leiva V, Balakrishnan N (2012) Connection between the Hadamard and matrix products with an application to matrix-variate Birnbaum–Saunders distributions. Journal of Multivariate Analysis 104:126–139

Carr WJ, Maradudin AA (1964) Ground-State Energy of a High-Density Electron Gas. Phys Rev 133:371–374

Carroll MR, Rutherford MJ (1988) Sulfur speciation in hydrous experimental glasses of varying oxidation state: Results from measured wavelength shifts of sulfur X-rays. Am Mineral 73: 845–849

Carroll MR, Webster JD (1994) Solubilities of sulfur, noble gases, nitrogen, chlorine and fluorine in magmas In: Carroll MR and Holloway JR (eds), Volatiles in Magmas. Reviews in Mineralogy Vol 30, Mineralogical Society of America pp 231–279

Carson TR (2011) Steps to the Planck Function : A Centenary Reflection School of Physics and Astronomy, University of St Andrews Cornell University Library, arXiv:astro-ph/0011219

Carter PT, MacFarlane TG (1957) Thermodynamics of slags systems: Part II The thermodynamic properties of $CaO–SiO_2$ slags. JISI 185:62–66

Cartz L, Jorgensen JD (1981) The high-pressure behavior of α-quartz, oxynitride, and nitride structures. J Appl Phys 52:236–244

Castanet R (1984) Selected data on the thermodynamic properties of α-alumina. High Temp High Press 16:449–457

Catti M (1978) Electrostatic lattice energy in ionic crystals: Optimization of the convergence of Ewald series. Acta Cryst A34:974–979

Catti M (1981) A generalized Born-Mayer parametrization of the lattice energy in orthorhombic ionic crystals. Acta Cryst A37:72–76

Catti M, Valerio G, Dovesi R, Causà M (1994) Quantum-mechanical calculation of the solid-state equilibrium $MgO + \alpha\text{-}Al_2O_3 \Leftrightarrow MgAl_2O_4$ (spinel) versus pressure. Phys Rev B 49:14179–14187

Catti M, Freyria Fava F, Zicovich C, Dovesi R (1999) High-pressure decomposition of MCr_2O_4 spinels (M = Mg, Mn, Zn) by ab initio methods. Phys Chem Minerals 26:389–395

Causà M, Dovesi R, Pisani C, Colle R, Fortunelli A (1987) Correlation correction to the Hartree-Fock total energy of solids. Phys Rev B 36:891–897

Causà M, Colle R, Fortunelli A, Dovesi R, Pisani C (1988) Correlation correction to the Hartree-Fock total energy of solids II. Phys Scr 38:194–198

Causà M, Dovesi R, and Roetti C (1991) Pseudopotential Hartree-Fock study of seventeen III-V and IV-IV semiconductors. Phys Rev B 43:11937–11943

Ceperley DM (1978) Ground state of the fermion one-component plasma: A Monte Carlo study in two and three dimensions. Phys Rev B 18:3126–3138

Ceperley DM, Alder BJ (1980) Ground State of the Electron Gas by a Stochastic Method. Phys Rev Letters 45:566–569

Cernik RJ, Murray PK, Pattison P, Fitch AN (1990) A two-circle powder diffractometer for synchrotron radiation with a closed loop encoder feedback system. J Appl Cryst 23 :292–296
Chamberlain JW (1961) Physics of the Aurora and Airglow. Academic Press, London
Chandrasekhar J, Andrade JG, von Ragué Schleyer P (1981) Efficient and accurate calculation of anion proton affinities. J Amer Chem Soc 18:5609–5612
Chang ZP, Barsch GR (1973) Pressure dependence of single-crystal elastic constants and anharmonic properties of spinel. J Geophys Res 78:2418–2433
Chapman S (1930) A Theory of Upper-Atmosphere Ozone. Mem Roy Meteorol Soc 3:103–125
Chappellaz J, Barnola JM, Raynaud D, Korotkevich YS, Lorius C (1990) Ice-core record of atmospheric methane over the past 160,000 years. Nature 345:127–131
Charles RJ (1969) The origin of immiscibility in silicate solutions. Phys Chem Glasses 10: 169–178
Charlier J-C, Michenaud J-P, Gonze X, Vigneron J-P (1991) Tight-binding model for the electronic properties of simple hexagonal graphite. Phys Rev B 44:13237–13249
Charlier J-C, Gonze X, Michenaud J-P (1994) First-principles study of the stacking effect on the electronic properties of graphite(s). Carbon 32:289–299
Chase MW Jr (1998) NIST-JANAF Thermochemical Tables Fourth Edition. J Phys Chem Ref Data, Monograph 9:1–1951
Chase MW Jr, Davies CA, Downey JR Jr, Frurip DJ, McDonald RA, Syverud AN (1985) JANAF Thermochemical Tables, Third Edition. J Phys Chem Ref Data 14, Suppl 1, 1–927
Chen YY, Lodge TP, Bates FS (2000) Entropically driven phase separation of highly branched/linear polyolefin blends. J Polym Sci B 38:2965–2975
Chen YY, Lodge TP, Bates FS (2002) Influence of long-chain branching on the miscibility of poly (ethylene-r-ethylethylene) blends with different microstructures. J Polym Sci B 40:466–477
Chen M, Ko H-Y, Remsing RC, Calegari Andrade MF, Santra B, Sun Z, Selloni A, Car R, Klein ML, Perdew JP, Xua X (2017) Ab initio theory and modeling of water. PNAS 114:10846–10851
Cherezov V, Clogston J, Papiz MZ, Caffrey M (2006) Room to Move: Crystallizing Membrane Proteins in Swollen Lipidic Mesophases. J Mol Biol 357:1605–1618
Chong DP, Van Lenthe E, Van Gisbergen S, Baerends EJ (2004) Even-tempered slater-type orbitals revisited: From hydrogen to krypton. J Comput Chem 25:1030–1036
Chopelas A (1991) Single crystal Raman spectra of forsterite, fayalite and monticellite. Am Mineral 76:1100–1109
Chopelas A (1999) Estimates of mantle relevant Clapeyron slopes in the $MgSiO_3$ system from high-pressure spectroscopic data. Am Mineral 84: 233–244
Chopelas A (2000) Thermal expansivity of mantle relevant magnesium silicates derived from vibrational spectroscopy at high pressure. Am Mineral 85: 270–278
Chopelas A, Boehler R, Ko T (1994) Thermodynamics and behavior of γ-Mg_2SiO_4 at high pressure: implications for Mg_2SiO_4 phase equilibrium. Phys Chem Minerals 21:351–359
Christiansen PA, Lee YS, Pitzer KS (1979) Improved ab initio effective core potentials for molecular calculations. J Chem Phys 71:4445–4450
Cicerone RJ, McCrumb JL (1980) Photodissociation of isotopically heavy O_2 as a source of atmospheric O_3. Geophys Res Lett 7:251–254
Cioslowski J, Mixon ST (1991) Covalent bond orders in the topological theory of atoms in molecules. J Am Chem Soc 113:4142–4145
Civalleri B (2017) Pairwise dispersion and BSSE corrections to HF and DFT Methods https://www.crystal.unito.it/
Civalleri B, Garrone E, Ugliengo P (1998) Vibrational modes of isolated hydroxyls of silica computed ab initio in a cluster approach. Chem Phys Letters 294:103–108
Civalleri B, D'Arco P, Orlando R, Saunders VR, Dovesi R (2001) Hartree-Fock geometry optimization of periodic system with the CRYSTAL code. Chem Phys Lett 348:131–138

Clark T, Chandrasekhar J, Günther W Spitznagel GW, Von Ragué Schleyer P (1983) Efficient diffuse function-augmented basis sets for anion calculations III The 3–21+G basis set for first-row elements, Li–F. J Comput Chem 3:294–301

Clarke LJ, Stich I, Payne MC (1992) Large-scale ab initio total energy calculations on parallel computers. Comput Phys Commun 72:14–28

Clementi E (1965) Ab Initio Computations in Atoms and Molecules. IBM Res Develop 9:2–19

Clementi E (1974) From quantum mechanics to statistical thermodynamics (V Coloquio Internacional de Quimicos Teoricos de Expresion Latina IBM Research Laboratory, San Jose, California, USA

Clementi E, Davis DR (1966a) Barrier to Internal Rotation in Ethane. J Chem Phys 45:2593–2599

Clementi E, Davis DR (1966b) Electronic structure of large molecular systems. J Comput Phys 1:223–244

Clementi E and Roetti C (1974) Roothaan-Hartree-Fock atomic wavefunctions: Basis functions and their coefficients for ground and certain excited states of neutral and ionized atoms, $Z \leq 54$. At Data Nucl Data Tables 14:177–478

Clever HL, Battino R, Saylor JH, Gross PM (1957) The Solubility of Helium, Neon, Argon and Krypton in Some Hydrocarbon Solvents. J Phys Chem 61:1078–1082

Clusius K, Weigand K (1940) Melting Curves of the Gases A, Kr, Xe, CH_4, CH_3D, CD_4, C_2H_4, C_2H_6, COS, and PH_3 to 200 Atmospheres Pressure The Change of Volume on Melting. Z Phys Chem Abt B 46:1–37

Cochran W, Cowley RA (1967) Phonons in Perfect Crystals In: Genzel L (ed) Light and Matter Ia. Encyclopedia of Physics, Vol 5/25/2/2a Springer, Berlin, Heidelberg

Cococcioni M (2012) The LDA+U approach: A simple Hubbard correction for correlated ground states. Correlated Electrons: From Models to Materials Modeling and Simulation. Forschungszentrum Julich, Germany

Cococcioni M, Dal Corso A, de Gironcoli S (2003) Fayalite: Comparison of LDA and GGA results. Phys Rev B 67:094106-1-7

Cohen LH (1979) High-low cristobalite transitions in SiO_2, $AlPO_4$ and $GaPO_4$ Investigations by differential thermal analysis under hydrostatic pressure < 6 kbar. Phil Mag 39A:399–404

Cohen RE (1991) Bonding and elasticity of stishovite SiO_2 at high pressure: Linearized augmented plane wave calculations. Am Mineral 76:733–742

Cohen RE (2000) MgO-The simplest oxide In Aoki, H, Syono, Y, Hemley, RJ (eds). Physics Meets Mineralogy: Condensed-Matter Physics in Geosciences. Cambridge University Press, Cambridge, UK, pp 95–123

Cohen, LH, Klement W Jr (1967) High-low quartz inversion: Determination to 35 kilobars. J Geophys Res 72:4245–4251

Cohen LH, Klement W Jr (1980) Tridymite: effect of hydrostatic pressure to 6 kbars on temperatures of two rapidly reversible transitions. Contrib Mineral Petr 71:401–405

Cohen-Tannoudji C, Diu B, Laloë F (1977) Quantum Mechanics, Vol I. Wiley, New York – London – Sidney – Toronto

Colle R, Salvetti D (1975) Approximate Calculation of the Correlation Energy for the Closed Shells. Theor Chim Acta 37: 329–334

Colle R, Salvetti D (1979) Approximate Calculation of the Correlation Energy for the Closed and Open Shells. Theor Chim Acta 53: 55–63

Colle R, Moscardo F, Riani P, Salvetti O (1977) V-N Vertical Transition of Planar Ethylene. Theor Chim Acta 44:1–7

Colle R, Montagnani R, Riani P, Salvetti O (1978) Calculation of some electronic excited states of formaldehyde. Theor Chim Acta 49:37–44

Collins GAD, Cruikshank DWJ, Breeze A (1972) Ab initio calculations on the silicate ion, orthosilicic acid and their L2,3 X-ray spectra. J Chem Soc Faraday Trans II 68:1189–1195

Coltorti M, Girardi V, Schoerster J (1987) Liquid immiscibility in the Archean Greenstone Belt of Piumhi (Minas Gerais, Brazil). Lithos 20:77–91

Condon EU, Shortley GH (1935) The Theory of Atomic Spectra. Cambridge University Press, Cambridge, England

Cook RL, De Lucia FC, Helminger P (1975) Molecular Force Field and Structure of Hydrogen Sulfide: Recent Microwave Results. J Mol Str 28:237–246

Coolidge AS (1932) A Quantum Mechanical Treatment of the Water Molecule. Phys Rev 42:189–209

Cossi M, Barone V (2000) Solvent effect on vertical electronic transitions by the polarizable continuum model. J Chem Phys 112:2427–2435

Cossi M, Barone V, Mennucci B, Tomasi J (1998) Ab initio study of ionic solutions by a polarizable continuum dielectric model. Chem Phys Lett 286:253–260

Cossi M, Mennucci B, Cammi R (1996) Analytical first derivatives of molecular surfaces with respect to nuclear coordinates. J Comput Chem 17:57–73

Cossi M, Scalmani G, Rega N, Barone V (2002) New developments in the polarizable continuum model for quantum mechanical and classical calculations on molecules in solution. J Chem Phys 117:43–54

Cotterill RMJ (1980) The Physics of Melting. J Cryst Growth 48:582–588

Cotton FA, Wilkinson G (1972) Advanced Inorganic Chemistry. Third Edition, Interscience Publishers, New York

Courant R, Hilbert D (1953) Methods of Mathematical Physics Volume I. Wiley Classics Library; Wiley

Cowley RA (1963) The lattice dynamics of an anharmonic crystal. Adv Phys 12:421–480

Cox, JD, Wagman DD, Medvedev VA (1984) CODATA Key Values for Thermodynamics. Hemisphere Publishing Corp, New York

Cripps-Clark CJ, Sridhar R, Jeffes JHE, Richardson FD (1974) Chain distribution and transition temperatures for phosphate glasses In: Jeffes JHE and Tait RJ (eds). Physical Chemistry of Process Metallurgy. Inst Mining Met, London

Curtiss LA, Frurip DJ, Blander M (1979) Studies of molecular association in H_2O and D_2O vapors by measurement of thermal conductivity. J Chem Phys 71:2703–2711

Curtiss LA, Jones C, Trucks GW, Raghavachari K, Pople JA (1990) Gaussian-1 theory of molecular energies for second-row compounds. J Chem Phys 93:2537–2545

Curtiss LA, Raghavachari, K, Trucks, GW, Pople, JA (1991) Gaussian-2 theory for molecular energies of first- and second-row compounds. J Chem Phys 94:7221–7230

Curtiss LA, McGrath M P, Blaudeau J, Davis NE, Binning RC Jr, Radom L (1995) Extension of Gaussian-2 theory to molecules containing third-row atoms Ga–Kr. J Chem Phys 103:6104–6113

Curtiss LA, Raghavachari K, Redfern P C, Pople JA (1997) Assessment of Gaussian-2 and density functional theories for the computation of enthalpies of formation. J Chem Phys 106:1063–1079

d'Amour H, Schiferl D, Denner W, Schulz H, Holzapfel WB (1978) High-pressure single-crystal structure determinations for ruby up to 90 kbar using an automatic diffractometer. J Appl Phys 49:4411–4416

d'Amour H, Denner W, Schultz H (1979) Structure determination of α-quartz up to 68×10^8 Pa. Acta Crist B35:550–555

Da Rocha S, Thibaudeau P (2003) Ab initio high-pressure thermodynamics of cationic disordered $MgAl_2O_4$ spinel. J Phys-Condens Matt 15:7103–7115

Da Silva C, Stixrude L, Wentzcovitch RM (1997) Elastic anisotropy of forsterite at high pressure. Geophys Res Lett 24:1963–1966

Da Silva CO, Da Silva EC, Nascimento MAC (1999) Ab initio calculations of absolute pK_a values in aqueous solution I Carboxylic acids. J Phys Chem A103:11194–11199

Dal Corso A (1996) A pseudopotential Plane Waves Program (PWSCF) and some case studies In: Pisani C (ed). Quantum-Mechanical Ab-initio calculations of the Properties of Crystalline Materials. Lecture Notes in Chemistry, 67, Springer, Berlin-Heidelberg-New York pp 155–178

Dal Corso A (2011) Modern pseudopotentials: an introduction. SISSA and Democritos, Trieste, Italy https://people.sissa.it/

Dal Corso A, Baroni S, Resta R, de Gironcoli S (1993) Ab initio calculation of phonon dispersions in II–VI semiconductors. Phys Rev B 47: 3588–3592

Dalton JA, Presnall DC (1997) No liquid immiscibility in the system $MgSiO_3$-SiO_2 at 50 GPa. Geochim Cosmochim Acta 61:2367–2373

D'Arco P, Freyria Fava F, Dovesi R, Saunders VR (1996) Structural and electronic properties of pyrope garnet ($Mg_3Al_2Si_3O_{12}$): an ab initio study. J Phys Cond Matter 8:8815–8828

Darling BT, Dennison BM (1940) The Water Vapor Molecule. Phys Rev 57:128–139

Davidson ER, Hagstrom SA, Chakravorty SJ, Umar VM, Froese C (1991) Ground-state correlation energies for two- to ten-electron atomic ions. Phys Rev A 44:7071–7083

de Boer JH, Verwey EJW (1937) Semi-conductors with partially and with completely filled 3d-lattice bands. Proc Phys Soc 49:59–71

de Groot RA, Buschow KHJ (1986) Recent developments in half-metallic magnetism. J Magn Magn Mater 54-57:1377–1380

de Koker N, Stixrude L (2009) Self-consistent thermodynamic description of silicate liquids, with application to shock melting of MgO periclase and $MgSiO_3$ perovskite. Geophys J Int 178: 162–179

de Koker N, Stixrude L, Karki BB (2008) Thermodynamics, structure, dynamics, and freezing of Mg_2SiO_4 liquid at high pressure. Geochim Cosmochim Acta 72:1427–1441

de Koker N, Karki BB, Stixrude L (2013) Thermodynamics of the MgO–SiO_2 liquid system in Earth's lowermost mantle from first principles. Earth Planet Sc Lett 361: 58–63

de Launay J (1956) The theory of specific heats and lattice vibrations. Solid State Phys 2:219–303

De La Pierre M, Orlando R, Maschio L, Doll K, Ugliengo P, Dovesi R (2011) Performance of six functionals (LDA, PBE, PBESOL, B3LYP, PBE0 and WC1LYP) in the simulation of vibrational and dielectric properties of crystalline compounds The case of forsterite Mg_2SiO_4. J Comput Chem 32:1775–1784

De Ligny D, Westrum EF Jr (1996) Entropy of calcium and magnesium aluminosilicate glasses. Chem Geol 128:113–128

De Vita A, Manassidis I, Lin JS, Gillan MJ (1992) The Energetics of Frenkel Defects in Li_2O from First Principles. EPL (Europhys Letters) 19:605–610

Debye P (1912) Zur Theorie der spezifischen warmer. Ann Physik 344:789–839

Debye P, Scherrer P (1916) Interferenzen an regellos orientierten Teilchen im Röntgenlicht. Physik Z 17:277–283

Decleva P (2013) Algebretta per la Meccanica Quantistica. https://moodle2unitsit/mod/resource/viewphp?id=12604

Deisenhofer J, Epp O, Miki K, Huber R, Michel H (1985) Structure of the protein subunits in the photosynthetic reaction centre of Rhodopseudomonas viridis at 3 Å resolution. Nature 318:618–624

Delano JW (2001) Redox History of the Earth's Interior since ~3900 Ma: Implications for Prebiotic Molecules Origin. Life Evol Biosph 31:311–341

Della Giusta A, Ottonello G (1993) Energy and long-range disorder in simple spinels. Phys Chem Minerals 20:228–241

Della Giusta A, Ottonello G, Secco L (1990) Precision estimates of interatomic distances using site occupancies, ionization potentials and polarizability in Pbnm silicate olivines. Acta Cryst B46: 160–165

Della Giusta A, Carbonin S, Ottonello G (1996) Temperature-dependent disorder in a natural Mg-Al-Fe^{2+}-Fe^{3+} spinel. Min Mag 60:603–616

Dellien I, Hall FM, Hepler GL (1976) Chromium, Molybdenum, and Tungsten: Thermodynamic properties, chemical equilibria, and standard potentials. Chem Rev 76:283–310

Demichelis R, Civalleri B, Ferrabone M, Dovesi R (2010) On the performance of eleven DFT functionals in the description of the vibrational properties of aluminosilicates. Int J Quantum Chem 110:406–415

Demuth Th, Jeanvoine Y, Hafner J, Angyan J G (1999) Polymorphism in silica studied in the local density and generalized-gradient approximations. J Phys-Condens Matt 11:3833–3874

Dent, B M (1929) LV The effect of boundary distortion on the surface energy of a crystal. Phil Mag Series 5 8:530–539

Derzsi M, Piekarz P, Jochym PT, Ła˙zewski J, Sternik M, Ole´s AM, Parlinski K (2009) Effects of Coulomb interaction on the electronic structure and lattice dynamics of the Mott insulator Fe_2SiO_4 spinel. Phys Rev B 79:205105-1-10

Derzsi M, Piekarz P, Tok´arl K, Jochym PT, Ła˙zewski J, Sternik M, Parlinski K (2011) Comparative ab initio study of lattice dynamics and thermodynamics of Fe_2SiO_4- and Mg_2SiO_4-spinels. J Phys-Condens Matt 23:105401-1-6

Dewaele A, Loubeyre P, Mezouar M (2004) Equations of state of six metals above 94 GPa. Phys Rev B 70:094112-1-8

Dill D (2006) In: Chemical aspects of the World Chapter 35, Many-electron atoms: Fermi holes and Fermi heaps http://www.bu.edu/quantum/notes/GeneralChemistry/

Dirac PAM (1926) On the Theory of Quantum Mechanics. P R Soc London 112:661–677

Dirac PAM (1930) Note on exchange phenomena in the Thomas atom. Proc Camb Phil Soc 26: 376–385

Dirac PAM (1957) The Principles of Quantum Mechanics, IV Edition (Revised). Oxford University Press, London

DiStasio RA Jr, Santra B, Li Z, Wu X, Car R (2014) The individual and collective effects of exact exchange and dispersion interactions on the ab initio structure of liquid water. J Chem Phys 141: 084502-1-16

Distin PA, Whiteway SG, Masson CR (1971) Solubility of oxygen in liquid iron from 1785° to 1960 °C New technique for the study of slag-metal equilibrium. Can Meteorol Q 10:73–78

Ditchfield R (1974) Self-consistent perturbation theory of diamagnetism I A gauge-invariant LCAO method for NMR chemical shifts. Mol Phys 27:789–807

Ditchfield R, Ellis PD (1972) ^{13}C NMR chemical shifts for fluoroethylenes and fluoroacetylene. Chem Phys Lett 17:342–344

Ditchfield R, Hehre WJ, Pople JA (1970) Self-Consistent Molecular Orbital Methods VI Energy Optimized Gaussian Atomic Orbitals. J Chem Phys 52:5001–5007

Ditchfield R, Hehre WJ, Pople JA (1971) Self-Consistent Molecular-Orbital Methods IX An Extended Gaussian-Type Basis for Molecular-Orbital Studies of Organic Molecules. J Chem Phys 54:724–728

Ditmars DA, Douglas TB (1971) Measurement of the relative enthalpy of pure α-Al_2O_3 (NBS heat capacity and enthalpy Standard Reference Material No 720) from 273 to 1173 K. J Res Natl Bur Stand 75A:401–420

Ditmars DA, Ishihara S, Chang SS, Bernstein G, West ED (1982) Enthalpy and Heat-Capacity Standard Reference Material: Synthetic Sapphire (Alpha-Al_2O_3) From 10 to 2250 K. J Res Natl Bur St 87:159–163

Dmitriev V, Torgashev V, Tolédano P, Salje EKH (1997) Theory of SiO_2 polymorphs. Europhys Lett 37:553–558

Dole M (1935) The relative atomic weight of oxygen in water and in air. J Am Chem Soc 57:2731

Dolg M, Wedig U, Stoll H, Preuss H (1987) Energy-adjusted ab initio pseudopotentials for the first row transition elements. J Chem Phys 86:866–872

Doll K (2001) Implementation of analytical Hartree_Fock gradients for periodic systems. Comput Phys Comm 137:74–88

Doll K, Harrison NM, Saunders VR (2001) Analytical Hartree-Fock gradients for periodic systems. Int J Quantum Chem 82:1–13

Doll K, Dovesi R, Orlando R (2004) Analytical Hartree-Fock gradients with respect to the cell parameter for systems periodic in three dimensions. Theor Chem Accounts 112:394–402

Dolling G (1963) In: Ekland S (ed) Inelastic Scattering of Neutrons in Solids and Liquids, IAEA, Vienna, Vol II

Doltsinis NL, Marx D (2002) First principles molecular dynamics involving excited states and nonadiabatic transitions. J Theor Comp Chem 1:319–349

Dorogokupets PI, Oganov AR (2007) Ruby, metals, and MgO as alternative pressure scales: A semiempirical description of shock-wave, ultrasonic, X-ray, and thermochemical data at high temperatures and pressures. Phys Rev B 75:024115

Dove M T (1993) Introduction to Lattice Dynamics, Cambridge University Press, Cambridge, England

Dovesi R (1996) The language of Band Theory In: Pisani C (ed) Quantum-Mechanical Ab-initio calculations of the Properties of Crystalline Materials. Lecture Notes in Chemistry, 67, Springer, Berlin-Heidelberg-New York pp 31–46

Dovesi R, Ermondi E, Ferrero E, Pisani C, Roetti, C (1983) Hartree-Fock study of lithium hydride with the use of a polarizable basis set. Phys Rev B 29:3591–3600

Dovesi R, Roetti C, Freyria-Fava C, Prencipe M (1991) On the elastic properties of lithium, sodium and potassium oxide An ab initio study. Chem Phys 156:11–19

Dovesi R, Saunders V R, Roetti C, Orlando R, Zicovich-Wilson CM, Pascale F, Civalleri B, Doll K, Harrison NM, Bush IJ, D'Arco P, and Llunell M (2009) CRYSTAL09 user's manual. Universita` di Torino: Torino, Italy

Downs JW (1989) Possible sites for protonation in β-Mg_2SiO_4 from an experimentally derived electrostatic potential. Amer Min 74: 1124–1129

Downs RT, Hall-Wallace M (2003) The American Mineralogist Crystal Structure Database. Amer Min 88:247–250

Downs RT, Palmer DC (1994) The pressure behavior of α cristobalite. Amer Min. 79:9–14

Downs RT, Zha CS, Duffy TS, and Finger LT (1996) The equation of state of forsterite to 172 GPa and effects of pressure media. Amer Min 81:51–55

Driessens FCM (1968a) Thermodynamics and defect chemistry of some oxide solid solutions Part I: Nearest neighbour interactions and the effects of substitutional disorder. Ber Buns 72:764–772

Driessens FCM (1968b) Thermodynamics and defect chemistry of some oxide solid solutions Part II: Pair interactions. Ber Buns 72:764–772

Driessens FCM (1968c) Thermodynamics and defect chemistry of some oxide solid solutions Part III: Defect equilibria and the formalism of pair interactions. Ber Buns 72:1123–1168

Duan W, Karki B, Wentzcovitch RM, Gu BL (1991) Ab initio study of $MgSiO_3$ low-clinoenstatite at high pressure. Amer Min 86:762–766

Dubrovinsky LS, Saxena SK (1997) Thermal expansion of periclase (MgO) and tungsten (W) to melting temperatures. Phys Chem Miner 24:547–550

Duffin RJ, Schaeffer AC (1952) A class of nonharmonic Fourier series. Trans Amer Math Soc 72:341–366

Duffy JA (1989) Use of refractivity data for obtaining optical basicity of transition metal oxides. Ironmaking Steelmaking 16:426–432

Duffy JA (1990) Bonding, Energy Levels and Bands. Longmans, London

Duffy JA (1993) A review of optical basicity and its applications to oxidic systems. Geochim Cosmochim Acta 57:3961–3970

Duffy JA, Ingram MD (1970) A new correlation between s-p spectra and the nephelauxetic ratio: Applications in molten salt and glass chemistry. J Chem Phys 64:443–444

Duffy JA Ingram MD (1971) Establishment of an optical scale for Lewis basicity in inorganic oxyacids, molten salts, and glasses. J Amer Chem Soc 93:6448–6454

Duffy JA, Ingram D (1973) Nephelauxetic Effect and Pauling Electronegativity. J Chem Soc Chem Comm 17:635–636

Duffy JA, Ingram MD (1974) Optical basicity-IV: Influence of electronegativity on the Lewis basicity and solvent properties of molten oxianion salts and glasses. J Inorg Nucl Chem 17, 1203–1206

Duffy JA Ingram MD (1976) An interpretation of glass chemistry in terms of the optical basicity concept. J Non Cryst Solids 21:373–410

Duncan TM (1997) Principal Components of Chemical Tensors: A Compilation, Farragut Press

Duncan ABF, Pople JA (1953) The structure of some simple molecules with lone pair electrons. Trans Faraday Soc 49:217–227

Dunitz JD, Orgel LE (1957) Electronic properties of transition metal oxides–II Cation distribution amongst octahedral and tetrahedral sites. J Phys Chem Solids 3:318–323

Dunning TH Jr (1989) Gaussian basis sets for use in correlated molecular calculations I The atoms boron through neon and hydrogen. J Chem Phys 90:1007–1023

Dunning TH Jr, Hay PJ (1977) Gaussian Basis Sets for Molecular Calculations In: Schaefer HF (ed) Methods of Electronic Structure Theory Modern Theoretical Chemistry, Vol 3, Springer, Boston, Massachussets

Dunning, TH Jr, Pitzer RM, Aung S (1973) Near Hartree-Fock Calculations on the Ground State of the Water Molecule: Energies, Ionization Potentials, Geometry, Force Constants, and One-Electron Properties. J Chem Phys 57:5044–5051

Dyke TR, Mack KM, Muentner JS (1977) The structure of water dimer from molecular beam electric resonance spectroscopy. J Chem Phys 66:498–510

Eckert F, Klamt A (2002) Fast Solvent Screening via Quantum Chemistry: COSMO-RS Approach. AIChE Journal 48:369–385

Edwards HGM, Lewis IR (1993) Vibrational spectroscopic studies of iron (II) acetate J Mol Struct 296:15–20

Edwards S, Roncin JY, Launay F, Rostas F (1993) The electronic ground state of molecular nitrogen J Mol Spec 162:257–267

Ehrenberg H, Fuess H (1993) Analytical interpretation and simulation of the static magnetic properties of synthetic α-Fe_2SiO_4. J Phys-Condens Matt 5:3663–3672

Ehrenberg H, Knapp M, Baehtz C, Klemme S (2002) Tetragonal low-temperature phase of $MgCr_2O_4$. Powder Diffraction 17:230–233

Ehrenfest P (1927) Bemerkung über die angenäherte Gültigkeit der klassischen Meckanik innerhalb der Quantenmechanik. Zeits Phys 45:455–457

Ehrenreich H, Cohen MA (1951) The SCF approach to the manyelectron problem. Phys Rev B 115:789–790

Einstein A (1907) Die Plancksche Theorie der Strahlung und die Theorie der spezifischen Wärme. Ann Physik 22:180–190

Eisemberg D, Kauzmann W (1969) The Structures and Properties of Water. Oxford University Press, London

Elliott JF(1955) Activities in the Iron Oxide-Silica-Lime System. Trans AIME 203:485–488

Ellis AJ (1967) The chemistry of some explored hydrothermal systems In: Barnes HL (ed) Geochemistry of Hydrothermal Ore Deposits Holt, Rinehart and Winston, pp 465–514

Ellis A J (1970) Quantitative interpretation of chemical characteristics of hydrothermal systems. Geothermics 2:516–528

Emsley YJW, Feeney J, Sutcliffe LH (1966) High Resolution Nuclear Magnetic Resonance Spectroscopy, Vol 2. Pergamon Press, Oxford

Erba A, Mahmoud A, Belmonte D, Doves, R (2014) High pressure elastic properties of minerals from ab-initio simulations: The case of pyrope, grossular and andradite silicate garnets. J Chem Phys, 140:124703

Essene E (1974) High-pressure transformations in $CaSiO_3$. Contrib Mineral Petr 45:247–250

Eugster HP, Baumgartner L (1987) Mineral solubilities and speciation in supercritical metamorphic fluids In: Ribbe PH (series ed), Reviews in Mineralogy, Vol 17, Mineralogical Society of America, pp 367–404

Evarestov RA, Platonenko A, Zhukovskii YF (2018) Site symmetry approach applied to the supercell model of $MgAl_2O_4$ spinel with oxygen interstitials: Ab initio calculations. Comput Mat Sci 150: 517–523

Ewald PP (1921) Die Berechmmg optischer und electrostaticher Gitterpotentiale. Ann Phys 64: 253–287

Eyring H, Walter J, Kimball GE (1944) Quantum Chemistry. Wiley New York and London

Fahey PM, Griffin PB, Plummer JD (1989) Point defects and dopant diffusion in silicon. Rev Mod Phys 61:289–384
Fegley B Jr, Prinn RG, Hartman H, Watkins GH (1986) Chemical effects of large impacts on the Earth's primitive atmosphere. Nature 319:305–308
Fei Y (1995) Thermal Expansion In: Arhens TJ (ed), A Handbook of Physical Constants: Mineral Physics and Crystallography, AGU Reference Shelf Series, Vol 2, American Geophysical Union
Fei Y, Bertka CM (1999) Phase transitions in the Earth's mantle and mantle mineralogy, In Fei Y, Bertka CM, Mysen BO (eds) Mantle Petrology: Field Observations and High Pressure Experimentation, Special Publication No 6, Geochemical Society, Houston, pp 189–207,
Fei Y, Saxena SK (1986) A thermochemical data base for phase equilibria in the system Fe-Mg-Si-O at high pressure and temperature. Phys Chem Minerals 13:311–324
Fei Y, Saxena SK, Navrotsky A (1990) Internally consistent thermodynamic data and equilibrium phase relations for compounds in the system MgO-SiO_2 at high pressure and high temperature. J Geophys Res 95:6915–6928
Fei Y, Mao HK, Shu J, Parthasarathy G, Bassett WA (1992) Simultaneous high-P, high-T X ray diffraction study of β-(Mg, $Fe)_2SiO_4$ to 26 GPa and 900 K. J Geophys Res 97:4489–4495
Fei Y, Frost DJ, Mao HK, Prewitt CT, Häusermann D (1999) In situ determination of the high-pressure phase of Fe_3O_4 . Amer Min 84:203–206
Fermi E (1927) Un Metodo Statistico per la Determinazione di alcune Proprietà dell'Atomo. Rend Accad Naz Lincei 6:602–607
Fermi E (1928) Statistische Berechnung der Rydbergkorrektionen der s-Terme. Zeits Phys 49:550–554
Feyereisen MW, Feller D, Dixon DA (1996) Hydrogen bond energy of the water dimer. J Phys Chem 100:2993–2997
Feynman RP (1939) Forces in Molecules. Phys Rev 56:340–343
Feynman RP, Leighton RB, Sands M (1965) The Feynman Lectures on Physics – Volume III: The new Millenium Edition Basic Books, California Institute of Technology
Filippi C, Singh DJ, Umrigar CJ (1994) All-electron local-density and generalized-gradient calculations of the structural properties of semiconductors. Phys Rev B 50:14947–14951
Fincham CJB, Richardson FD (1954) The behaviour of sulphur in silicate and aluminate melts. Proc R Soc London, Series A 223:40–62
Finger LW,, and Hazen RM (1978) Crystal structure and compression of ruby to 46 kbar. J Appl Phys 49:5823–5826
Fitzpatrick R (2010) Representation of waves via complex functions http:// farsidephutexasedu/teaching/qmech/Quantum/node17html
Fletcher R (1987) Practical methods of optimization, (2nd ed). Wiley, New York
Flood H, Förland T (1947) The acidic and basic properties of oxides. Acta Chem Scand 1:952–1005
Flood H, Grjotheim T (1952) Thermodynamic calculation of slag equilibria. J Iron Steel Inst 171:64–80
Floris FM, Tomasi J (1986) Evaluation of the dispersion contribution to the solvation energy A simple computational model in the continuum approximation. J Comput Chem 10:616–627
Floris FM, Tomasi J, Pascual-Ahuir JL (1991) Dispersion and repulsion contributions to the solvation energy: Refinements to a simple computational model in the continuum approximation. J Comput Chem 12:784–791
Flory PJ (1953) Principles of Polymer Chemistry, Cornell University Press, Ithaca, NY
Flubacher P, Leadbetter AJ, Morrison JA, Stoicheff BP (1959) The low-temperature heat capacity and the Raman and Brillouin spectra of vitreous silica. J Phys Chem Solids 12:53–65
Fock V (1930) Näherungsmethode zur Lösung des quantenmechanischen Mehrkörperproblems Zeits Phys 61:126–148
Foresman JB, Frisch Æ (1993) Exploring Chemistry with Electronic Structure Methods, 2nd Edition, Gaussian Inc, Pittsburg PA

Foresman JB, Frisch Æ (2015) Exploring Chemistry with Electronic Structure Methods, 3rd Edition, Gaussian Inc Pittsburg PA

Foresman JB, Head-Gordon M, Pople JA, Frisch MJ (1992) Toward a systematic molecular orbital theory for excited states. J Phys Chem 96:135–149

Foresman B, Keith TA, Wiberg KB, Snoonian J, and Frisch M J (1996) Solvent Effects 5 Influence of Cavity Shape, Truncation of Electrostatics, and Electron Correlation on ab Initio Reaction Field Calculations. J Phys Chem 100:16098–16104

Foster JM, Boys FS (1960) A Quantum Variational Calculation for HCHO. Rev Mod Phys 32:303–304

Fourier JB (1822) Théorie analytique de la chaleur. Firmin Didot père et fils, Paris

Fourier JB (1827) Mémoire sur les températures du globe terrestre et des espaces planétaires. Mém Acad Sci 7:569–604

Fradera X, Austen MA, Bader RFW (1999) The Lewis Model and Beyond. J Phys Chem A 103:304–314

Francl MM, Pietro WJ, Hehre WJ, Binkley JS, DeFrees DJ, Pople JA, Gordon MS (1982) Self-consistent molecular orbital methods XXIII A polarization-type basis set for second-row elements. J Chem Phys 77:3654–3665

Frank HS, Evans MW (1945) Free Volume and Entropy in Condensed Systems III Entropy in Binary Liquid Mixtures; Partial Molal Entropy in Dilute Solutions; Structure and Thermodynamics in Aqueous Electrolytes. J Chem Phys 13:507–532

Frank W, Elsässer C, Fähnle M (1995) Ab initio Force-Constant Method for Phonon Dispersions in Alkali Metals. Phys Rev Lett 74:1791–1794

Frankiss SG (1963) Nuclear Magnetic Resonance Spectra of Some Substituted Methanes. J Phys Chem 67:752–755

Fraser DG (1977) Thermodynamic properties of silicate melts In: Fraser DG (ed), Thermodynamics in Geology. D Reidel Pub Co, Dortrecht-Holland, pp 301–325

Fraser GT (1991) $(H_2O)_2$: spectroscopy, structure and dynamics. Int Rev Phys Chem 10:189–206

Fratanduono DE, Smith RF, Ali SJ, Braun DG, Fernandez-Pañella A, Zhang S, Kraus RG, Coppari F, McNaney JM, Marshall MC, Kirch LE, Swift DC, Millot M, Wicks JK, Eggert JH (2020) Probing the solid phase of noble metal copper at terapascal conditions. Phys Rev Lett 124: 015701-1-6

Fredrickson GH, Liu AJ (1995) Design of miscible polyolefin copolymer blends. J Polym Sci B 33:1203–1212

Fredrickson GH, Liu AJ, Bates FS (1994) Entropic corrections to the Flory-Huggins theory of polymer blends: architectural and conformational effects. Macromolecules 27:2503–2511

French, RH (1990) Electronic Band Structure of Al_2O_3, with Comparison to Alon and AIN. J Am Ceram Soc 73:477–489

Freyria-Fava C, Dovesi F, Saunders VR, Leslie M, Roetti C (1993) Ca and Be substitution in bulk MgO: 'ab initio' Hartree-Fock and ionic model supercell calculations. J Phys Cond Matter 5:4793–4804

Friedel J (1964) Dislocations, Addison-Wesley, Reading, Massachusetts

Frisch MJ, Trucks GW, Schlegel HB, Scuseria GE, Robb,MA, Cheeseman JR, Montgomery JJA, Vreven T, Kudin KN, Burant JC, Millam JM, Iyengar SS, Tomasi J, Barone V, Mennucci B, Cossi M, Scalmani G, Rega N, Petersson GA, Nakatsuji H, Hada M, Ehara M, Toyota K, Fukuda R, Hasegawa J, Ishida M, Nakajima T, Honda Y, Kitao O, Nakai H, Klene M, Li X, Knox JE, Hratchian HP, Cross JB, Bakken V, Adamo C, Jaramillo J, Gomperts R,; Stratmann RE, Yazyev O, Austin AJ, Cammi R, Pomelli C, Ochterski JW, Ayala PY, Morokuma K,Voth GA, Salvador P, Dannenberg JJ, Zakrzewski VG, Dapprich S, Daniels AD, Strain MC, Farkas O, Malick DK,; Rabuck AD, Raghavachari K, Foresman JB, Ortiz JV, Cui Q, Baboul AG, Clifford S, Cioslowski J, Stefanov BB, Liu G, Liashenko A, Piskorz P, Komaromi I, Martin RL, Fox DJ, Keith T, Al-Laham MA, Peng CY, Nanayakkara A, Challacombe M, Gill PMW, Johnson B, Chen W, Wong MW, Gonzalez C, Pople JA (2004) Gaussian03; Revision B05 ed; Gaussian, Inc: Wallingford, CT

Fuchs M, Scheffler M (1999) Ab initio pseudopotentials for electronic structure calculations of poly-atomic systems using density-functional theory. Comp Phys Comm 119:67–98

Fujino K, Sasaki S, Takeuchi Y, Sadanaga R (1981) X-ray determination of electron distribution in forsterite, faialite and tephroite. Acta Cryst B37 :513–518

Fuller CS, Ditzenberger JA (1956) Diffusion of Donor and Acceptor Elements in Silicon. J Appl Phys 27:544–553

Fulton RL (1993) Sharing of electrons in molecules. J Phys Chem 97:7516–7529

Fulton RL, Mixon ST (1993) Comparison of covalent bond indexes and sharing indexes. J Phys Chem 97:7530–7534

Fumi FG, Tosi MP (1957) Naor relations between Madelung constants for cubic ionic lattices. Phil Mag 2:284–285

Galeener F L (1982) Planar rings in glasses. Solid State Comm 44:1037–1040

Galkin VM, Dorochev AM, Babitch Yu V (1987) Termitcheskoe raschirenie koesita. Izvestia AN SSSR, Geochimia 1645–1646

Ganguly J (2008) Thermodynamics in Earth and Planetary Sciences. Springer, Berlin-Heidelberg

Ganguly J, Saxena SK (1987) Mixtures and Mineral Reactions. Springer, Berlin-Heidelberg-New York

Garrels RM, Thompson ME (1962) A chemical model for seawater at 25°C and one atmosphere total pressure. Amer Jour Sci 260:57–66

Gaskell PH (1970) Vibrational spectra of simple silicate glasses. Discuss Faraday Soc 50:82–93

Gaskell DR (1982) On the correlation between the distribution of phosphorous between slag metal and the theoretical optical basicity of the slag. Trans Iron Steel Inst Jpn 22:997–1000

Gaskell (2000) Early model of the thermodynamic behavior of slags and salts In: Seetharaman S and Sichen DU (eds), Proceedings of the Sixth International Conference on Molten Slags, Fluxes and Salts, Stockholm, Sweden—Helsinki, Finland, ISBN 91-7170-606-2

Gasparik T (2003) Phase Diagrams for Geoscientists An Atlas of the Earth's Interior. Springer, Berlin-Heidelberg-New York

Gasparik T, Wolf K and Smith CM (1994) Experimental determination of phase relations in the $CaSiO_3$ system from 8 to 15 GPa. Amer Min 79:1219–1222

Gatti C, Lasi D (2007) Source function description of metal–metal bonding in d-block organometallic compounds. Faraday Discuss 135:55–78

Gatti C, Macchi P (2012) A guided tour through modern charge density analysis. In: Modern Charge Density Analysis. Gatti C and Macchi P (eds) Springer, Dordrecht Heidelberg London New York, pp 1.78

Gatti C, Ottonello G, Richet P, (2012) Energetics and Bonding in Aluminosilicate Ring with Alkali Metal and Alkaline-Earth Metal Charge-Compensating Cations. J Phys Chem A 116:8584–8598

Gauss J (1992) Calculation of NMR chemical shifts at second-order many-body perturbation theory using gauge-including atomic orbitals. Chem Phys Lett 191:614–620

Geisinger KL, Spackman MA, Gibbs GV (1987) Exploration of structure, electron density distribution, and bonding in coesite with Fourier and pseudo atom refinement methods using single-crystal X-ray diffraction data. J Phys Chem 91:3237–3244

Gell-Mann M, Brueckner K (1957) Correlation Energy of an Electron Gas at High Density. Phys Rev 106:364–368

George AM, Richet P, Stebbins JF (1998) Cation dynamics and premelting in lithium metasilicate (Li_2SiO_3) and sodium metasilicate (Na_2SiO_3); a high-temperature NMR study. Amer Min 83: 1277–1284

Gerlach TM, Nordlie BE (1975a) The C-O-H-S- gaseous systems, Part I: composition limits and trends in basaltic cases. Amer Jour Sci 275:353–376

Gerlach TM, Nordlie BE (1975b) The C-O-H-S- gaseous systems, Part II: temperature, atomic composition, and molecular equilibria in volcanic gases. Amer Jour Sci 275:377–394

Gerlach TM, Nordlie BE (1975c) The C-O-H-S- gaseous systems, Part III: magmatic gases compatible with oxides and sulfides in basaltic magmas. Amer Jour Sci 275:395–410

Gerlach W, Stern O (1922a) Der experimentelle Nachweis des magnetischen Moments des Silber atoms. Zeits Phys 8:110–111

Gerlach W, Stern O (1922b) Der experimentelle Nachweis der Richtungsquantelung im Magnetfeld. Zeits Phys 9:349–352

Ghiorso MS, Carmichael ISE, Rivers ML, Sack RO (1983) The Gibbs free energy of natural silicate liquids; an expanded regular solution approximation for the calculation of magmatic intensive variables. Contrib Mineral Petr 84:107–145

Ghita I, Bell HB (1982) Sulphide capacity of slags containing FeO, CaO, TiO_2 and SiO_2. Ironmaking Steelmaking 9:239–243

Giannozzi P, de Gironcoli S, Pavone P, Baroni S (1991) Ab initio calculation of phonon dispersions in semiconductors. Phys Rev B 43:7231–7242

Giannozzi P, Baroni S, Bonini N, Calandra M, Car R, Cavazzoni C, Ceresoli D, Chiarotti G L, Cococcioni M, Dabo I, Dal Corso A, de Gironcoli S, Fabris S, Fratesi G, Gebauer R, Gerstmann U, Gougoussis C, Kokalj A, Lazzeri M, Martin-Samos L, Marzari N, Mauri F, Mazzarello R, Paolini S, Pasquarello A, Paulatto L, Sbraccia C, Scandolo S, Sclauzero G, Seitsonen AP, Smogunov A, Umari P, Wentzcovitch RM (2009) QUANTUM ESPRESSO: a modular and open-source software project for quantum simulations of materials. J Phys-Condens Matt 21: 395502-1-19

Giauque WF (1949) An Example of the Difficulty in Obtaining Equilibrium Corresponding to a Macrocrystalline Non-volatile Phase The Reaction $Mg(OH)_2 \rightleftarrows MgO + H_2O(g)$. J Am Chem Soc 71:3192–3194

Gierke TD, Flygare WH (1972) Empirical evaluation of the individual elements in the nuclear diamagnetic shielding tensor by the atom dipole method. J Am Chem Soc 94:7277–7283

Giggenbach WF (1980) Geothermal gas equilibria. Geochim Cosmochim Acta 44, 2021–2032

Giggenbach WF (1996) Chemical Composition of Volcanic Gases In: Monitoring and Mitigation of Volcano Hazards Springer, Berlin, Heidelberg https://doiorg/101007/978-3-642-80087-0_7

Gill PMW (1996) A new gradient-corrected exchange functional. Mol Phys 89:433–445

Gill PMW, Pople JA (1993) Exact exchange functional for the hydrogen atom. Phys Rev A 47:2383–2385

Gill PMW, Johnson BG, Pople JA, Frisch MJ (1992) An investigation of the performance of a hybrid of Hartree-Fock and density functional theory. Int J Quantum Chem Supplement 26:319–331

Gill PMW, Johnson BG, Pople JA (1993) A standard grid for density function calculations. Chem Phys Lett 209:506–512

Glasser D (2014) Thermodynamic consistencies and anomalies among end-member lanthanoid garnets. J Chem Thermodynamics 78:93–98

Glasser ML, Milford FJ (1963) Spin Wave Spectra of Magnetite. Phys Rev 130:1783–1789

Glew DN, Moelwyn-Hughes EA (1953) Chemical statics of the methyl halides in water. Discuss Faraday Soc 15:150–161

Glorieux B, Millot F, Rifflet J-C, Coutures J-P (1999) Density of Superheated and Undercooled Liquid Alumina by a Contactless Method. Int J Thermophys 20:1085–1094

Goedecker S, Teter M, Hutter J (1996) Separable dual-space Gaussian pseudopotentials. Phys Rev B 54:1703–1710

Goeppert-Mayer M, Sklar AL (1938) Calculations of the Lower Excited Levels of Benzene. J Chem Phys 6:645–652

Gogonea V, Băleanu-Gogonea C, Osawa E (1998), Solvent hard sphere diameter from van der Waals volume A statistical analysis of computed and solubility determined solvent diameters. J Mol Struct (Theochem) 432:177–189

Gole, JL, Zare RN (1972) Determination of D_0^0(AlO) from Crossed-Beam Chemiluminescence of Al + O_3. J Chem Phys 57:5331–5335

Golub GH, Van Loan CF (1983) Matrix Computation (third edition). The Johns Hopkins University Press, Baltimore & London

Gombàs P (1943) Erweiterung der statistischen Theorie des Atoms. Zeits Phys 121:13232–13542

Goodman NR (1960) Measuring amplitude and phase. J Franklyn Inst 270:437–450
Goodwin RD (1983) Hydrogen sulfide provisional thermophysical properties from 188 to 700K at pressures to 75 MPa. Report NBSIR-83-1694; NTIS No PB84-122704, 177
Gordon MS, Binkley JS, Pople JA, Pietro WJ, Hehre WJ (1982) Self-consistent molecular-orbital methods 22 Small split-valence basis sets for second-row elements. J Am Chem Soc 104:2797–2803
Gottardi G (1972) I Minerali. Serie Biologia e Geologia, Boringhieri, Bologna
Goudsmit S (1928) Multiplet Separations for Equivalent Electrons and the Roentgen Doublet Law. Phys Rev 31:946–959
Gray DL, Robiette AG (1979) The anharmonic force field and equilibrium structure of methane. Mol Phys 37:1901–1920
Grayson-Smith H, Stanley JP (1950) Note on the Derivation of the Frequency Spectrum of a Crystal from Specific Heat Measurements. J Chem Phys 18:236
Green AES, Peterson LR, and Prasad SS (1969) Semi-Empirical Cross Sections and the Airglow and Aurora. In: McCormac BM and Omholt A (eds), Atmospheric Emissions, Van Nostrand Reinhold, New York
Greenberg E, Dubrovinsky LS, McCammon C, Rouquette J, Kantor I, Prakapenka V, Rozenberg GK, Pasternak M (2011) Pressure-induced structural phase transition of the iron end-member of ringwoodite (γ-Fe_2SiO_4) investigated by X-ray diffraction and Mössbauer spectroscopy. Amer Min 96:833–840
Greenwood NN (1970) Ionic Crystals, Lattice Defects and Non-stoichiometry Chemical Pub Co, New York
Grigera T S, Martin-Mayor V, Parisi G, and Verrocchio P (2003) Phonon interpretation of the 'boson peak' in supercooled liquids. Nature 422:289–292
Grimme S (2004) Accurate description of van der Waals complexes by density functional theory including empirical corrections. J Comput Chem 25:1463–1473
Grimme S (2006a) Semiempirical GGA-Type Density Functional Constructed with a Long-Range Dispersion Correction. J Comp Chem 27:1787–1799
Grimme S (2006b) Semiempirical hybrid density functional with perturbative second-order correlation. J Chem Phys 124:34108-1-16
Grønvold F, Sveen A (1974) Heat capacity and thermodynamic properties of synthetic magnetite (Fe_3O_4) from 300 to 1050 K Ferromagnetic transition and zero-point entropy. J Chem Thermodyn 6:859–872
Guillot B J (2002) A reappraisal of what we have learnt during three decades of computer simulation of water. J Mol Liq 101:219–260
Guilpin C (1999) Manuel de Calcul Numérique Appliqué. EDP Sciences, Les Ulis Cedex A, France
Gunnarsson O, Lundqvist B (1976) Exchange and correlation in atoms, molecules, and solids by the spin-density-functional. Phys Rev B 13:4274–4298
Gunnarsson O, Lundqvist B (1977) Erratum: Exchange and correlation in atoms, molecules, and solids by the spin-density-functional. Phys Rev B 15:6006
Gupta PK, Mauro JC (2007) The laboratory glass transition. J Chem Phys 126:224504-1-9
Gupta HC, Parashar A, Gupta VB, Tripathi BB (1990) Lattice vibrations in spinel-type compounds $CdCr_2S_4$ and $CdCr_2Se_4$. Physica Status Solidi B 160:K19–K24
Gupta HC, Sinha MM, Chand KB, Balram (1992) Zone-centre phonon frequency calculation of the spinel structure compounds $MnCr_2S_4$ and $FeCr_2S_4$. Physica Status Solidi B 160:K65–K68
Gupta HC, Sinha MM, Balram, Tripathi BB (1993) Lattice dynamics of oxide spinel $ZnCr_2O_4$. J Solid State Chem 102:315–317
Gussoni M, Rui R, Zerbi G (1998) Electronic and relaxation contribution to linear molecular polarizability An analysis of the experimental values. J Mol Struct 447:163–215
Haar L, Gallager JG, Kell GS (1979) Thermodynamic properties of fluid water In: Contributions to the 9th Int Conf on Properties of Steam, Munich, Western Germany

Haar L, Gallager JG, Kell GS (1984) NBS/NRC Steam Tables Thermodynamic and transport properties and computer programs for vapor and liquid states of water in SI units. Hemisphere Pub Co, McGraw Hill, New York

Haas JL (1971) Effect of salinity on the maximum thermal gradient of a hydrothermal system at hydrostatic pressure. Econ Geol 66:940–946

Haavik C, Stolen S, Fjellvag H, Hanfland M, Hausermann D (2000) Equation of state of magnetite and its high-pressure modification: Thermodynamics of the Fe-O system at high pressure. Amer Min 85:514–523

Hackbusch W (1995) Integral Equations: Theory and Numerical Treatment. Birkhauser Verlag, Basel, Switzerland

Haering RR (1958) Band structure of rhombohedral graphite. Can J Phys 36:352–362

Hafner St (1961) Ordnung/Unordnung und Ultrarotabsorption IV Die Absorption einiger Metalloxyde mit Spinellstruktur. Z Kristallogr 115:331–358

Hageman VBM, Oonk, HAJ (1986) Liquid immiscibility in the SiO_2 + MgO, SiO_2 + SrO, SiO_2 + LaO_3, and SiO_2 + Y_2O_3 systems. Phys Chem Glasses 27:194–198

Haiber M, Ballone P, and Parrinello M (1997) Structure and dynamics of protonated Mg_2SiO_4: ab-initio molecular dynamics study. Am Mineral 82: 913–922

Hall GG (1951) The Molecular Orbital Theory of Chemical Valency VIII A Method of Calculating Ionization Potentials. Proc Roy Soc A 205:541–552

Hamann DR (1989) Generalized norm-conserving pseudopotentials. Phys Rev B 40:2980–2987

Hamann DR (1996) Generalized gradient theory for silica phase transitions. Phys Rev Lett 76:660–663

Hamann DR, Schlüter M, and Chiang C (1979) Norm-Conserving Pseudopotentials. Phys Rev Lett 43:1494–1497

Hameka HF (1958) On the nuclear magnetic shielding in the hydrogen molecule. Mol Phys 1:203–215

Hameka HF (1962) Theory of Magnetic Properties of Molecules with Particular Emphasis on Hydrogen Molecule. Rev Mod Phys 34:87–101

Hammes-Schiffer S (2009) Theory of Proton-Coupled Electron Transfer in Energy Conversion Processes. Acc Chem Res 42:1881–1889

Handy NC and Cohen AJ (2001) Left-right correlation Energy. Mol Phys 99:403–412

Hanel RA, Conrath BJ, Kunde VG, Prabhakara C, Revah I, Salomonson VV, Wolford G (1972) The Nimbus 4 infrared spectroscopy experiment: 1 Calibrated thermal emission spectra. J Geophys Res 15:2629–2641

Hariharan PC, Pople JA (1973) The influence of polarization functions on molecular orbital hydrogenation energies. Theor Chim Acta 28:213–222

Harris, AB, Meyer H (1962) Calorimetric determination of energy levels in rare-earth and yttrium-iron garnets. Phys Rev 127:101–118

Harrison NM, Saunders VR, Aprà E, Causà M, Dovesi R (1992) Correlation functional estimates of the dispersion interaction in semi-ionic compounds. J Phys Cond Mat 4:L261–L264

Hartree DR (1928) The Wave Mechanics of an Atom with a Non-Coulomb Central Field Part I Theory and Methods. Proc Camb Phil Soc 24:89–110

Haser M, Alrichs R, Baron HP, Weis P, Hom H (1992) Direct computation of second-order SCF properties of large molecules on workstation computers with an application to large carbon clusters. Theor Chim Acta 83:455–470

Hashimoto GL, Abe Y, Sugita S (2007) The chemical composition of the early terrestrial atmosphere: Formation of a reducing atmosphere from CI-like material. J Geophys Res 112: E05010-1-12

Hassel O, Mark H (1924) Über die Kristallstruktur des Graphites. Z Physik 25:317–337

Hay PJ (1973) On the calculation of natural orbitals by perturbation theory. J Chem Phys 59:2468–2476

Hay PJ, Wadt WR (1985a) Ab initio effective core potentials for molecular calculations Potentials for the transition metal atoms Sc to Hg. J Chem Phys 82:270–283

Hay PJ, Wadt WR (1985b) Ab initio effective core potentials for molecular calculations Potentials for main group elements Na to Bi. J Chem Phys 82:284–298

Hay PJ, Wadt WR, Kahn LR (1978) Ab initio effective core potentials for molecular calculations II All-electron comparisons and modifications of the procedure. J Chem Phys 68:3059–3066

Haynes WM (2005) CRC Handbook of Chemistry and Physics 86th Ed CRC, Boca Raton, FL

Hays JD, Imbrie I, Shackelton NJ (1976) Variations in Earth's orbit: pacemaker of ice ages. Science 194:1121–1132

Hazen R (1976a) Effects of temperature and pressure on the cell dimension and X-ray temperature factors of periclase. Amer Min 61:266–271

Hazen R (1976b) Effect of temperature and pressure on the crystal structure of forsterite. Amer Min 61:1280–1293

Hazen RM, Zhang J, Ko J (1990) Effects of Fe/Mg on the compressibility of synthetic wadsleyite: β-(Mg1–xFex)SiO_4 (x<025). Phys Chem Minerals 17:416–419

Hazen RM, Downs RT, Finger LW, Ko J (1993) Crystal chemistry of ferromagensian silicate spinels Evidence of Mg-Si disorder. Amer Min 78:1320–1323

Hazen RM, Weinberger MB, Yang H, Prewitt CT (2000) Comparative high-pressure crystal chemistry of wadsleyite, β-(Mg1-xFex)SiO_4, with x=0 and 025. Amer Min 85:770–777

Head-Gordon M, Rico RJ, Oumi M, Lee TJ (1994) A doubles correction to electronic excited states from configuration interaction in the space of single substitutions. Chem Phys Lett 219:29–29

Head-Gordon M, Maurice D, Oumi M (1995) A perturbative correction to restricted open shell configuration interaction with single substitutions for excited states of radicals. Chem Phys Lett 246:114–121

Hedin L, Lundqvist BI (1971) Explicit local exchange-correlation potentials. J Phys C 4:2064–2083

Heid R, Strauch D, Bohnen, K-P (2000) Ab initio lattice dynamics of sapphire. Phys Rev B 61: 8625–8627

Heidenreich JE III, Thiemens MH (1986) A non-mass-dependent oxygen isotope effect in the production of ozone from molecular oxygen: The role of molecular symmetry in isotope chemistry. J Chem Phys 84:2129–2136

Heisenberg W (1926) Mehrkörperproblem und Resonanz in der Quantenmechanik. Zeits Phys 38: 411–426

Heisenberg W (1927a) Mehrkörperproblem und Resonanz in der Quantenmechanik II. Zeits Phys 41: 239–267

Heisenberg W (1927b) Über den anschaulichen Inhalt der quantentheoretischen Kinematik und Mechanik. Zeits Phys 43:172:198

Heisenberg W (1928) Zur Theorie des Ferromagnetismus. Zeits Phys 49:619–636

Helffrich G (1999) Practical use of Suzuki's thermal expansivity formulation. Phys Earth Planet In 116:133–136

Helgaker T, Gauss J, Jorgensen P, Olsen J (1997) The prediction of molecular equilibrium structures by the standard electronic wave functions. J Chem Phys 106:6430–6440

Helgeson HC (1969) Thermodynamics of hydrothermal system at elevated temperatures and pressures. Amer Jour Sci 267:729–804

Helgeson HC, Kirkham DH (1974a) Theoretical prediction of thermodynamic behavior of aqueous electrolytes at high pressures and temperatures: I Summary of the thermodynamic/electrostatic properties of the solvent. Amer Jour Sci 274:1089–1198

Helgeson HC, Kirkham DH (1974b) Theoretical prediction of thermodynamic behavior of aqueous electrolytes at high pressures and temperatures: II Debye-Hückel parameters for activity coefficients and relative partial molal properties. Amer Jour Sci 274:1199–1261

Helgeson HC, Kirkham DH (1976) Theoretical prediction of thermodynamic behavior of aqueous electrolytes at high pressures and temperatures: III Equation of state for aqueous species at infinite dilution. Amer Jour Sci 276:97–240

Helgeson HC, Delany J, Bird DK (1978) Summary and critique of the thermodynamic properties of rock-forming minerals. Amer Jour Sci 278A:1–229

Helgeson HC, Kirkham DH, Flowers GC (1981) Theoretical prediction of the thermodynamic behavior of aqueous electrolytes at high pressures and temperatures: IV Calculation of activity coefficients, osmotic coefficients and apparent molal and standard and relative partial molal properties to 600 °C and 5 kbar. Amer Jour Sci 281:1249–1516

Hellman O, Abrikosov IA, Simak SI (2011) Lattice dynamics of anharmonic solids from first principles. Phys Rev B 84:180301-1-4

Hellmann H (1937) Einführung in die Quantenchemie. Franz Deuticke, Leipzig und Wien

Hellmann R, Bich E, Vogel E (2008) Ab initio intermolecular potential energy surface and second pressure virial coefficients of methane. J Chem Phys 128:214303-1,9

Hemingway BS, Bohlen SR, Hankins WB, Westrum Jr EF, Kuskov OL (1998) Heat capacity and thermodynamic properties for coesite and jadeite, reexamination of the quartz-coesite equilibrium boundary. Amer Min 83:409–418

Henley RW (1984) Chemical structure of geothermal systems In: Robertson JM (series ed), Reviews in Economic Geology, Vol I, Society of Economic Geologists, Chelsea MI, pp 9–28

Herman F (1959) Lattice vibrational spectrum of germanium. J Phys Chem Solids 8:405–418

Herman F, Van Dyke JP, Ortenburger IB (1969) Improved statistical exchange approximation for inhomogeneous many-electron systems. Phys Rev Letters 22:807–811

Hernandez JA, Caracas R (2016) Superionic-superionic phase transitions in body-centered cubic H_2O ice. Phys Rev Lett 117:135503-1-5

Hernández ER, Brodholt J, Alfè D (2015) Structural, vibrational and thermodynamic properties of Mg_2SiO_4 and $MgSiO_3$ minerals from first-principles simulations. Phys Chem Minerals 240:1–24

Herzberg G (1950) Spectra of Diatomic Molecules. Van Nostrand Company, New York

Herzberg G (1966) Electronic spectra and electronic structure of polyatomic molecules. Van Nostrand, New York

Hess PC (1971) Polymer models of silicate melts. Geochim Cosmochim Acta 35:289–306

Heyd J, Scuseria GE, Ernzerhof M (2003) Hybrid functionals based on a screened Coulomb potential. J Chem Phys 118:8207–8215

Hill R (1952) The Elastic Behaviour of a Crystalline Aggregate. Proc Phys Soc London A65:359–354

Hill RJ, Craig JR, Gibbs GV (1979) Systematics of the spinel structure type. Phys Chem Minerals 4:317–339

Hilleard RJ (1975) Concerning the structure of $MgCr_2O_4$. J Phys C Solid State 8:193–194

Himmetoglu B, Floris A, De Gironcoli S, Cococcioni M (2013). Hubbard-corrected DFT Energy functionals: The LDA+U description of correlated systems. Int. Jour. Quantum Chem. 114:14–49

Himmrich J, Lutz HD (1991) Normal coordinate analyses and lattice dynamical calculations of spinel-type $ZnCr_2O_4$. Solid State Comm 79:447–452

Hiyagon H, Ozima M (1986) Partition of noble gases between olivine and basalt melt. Geochim Cosmochim Acta 50:2045–2057

Hoffmann R, Kabanov A, Golov A, Proserpio D (2016) Homo citans and carbon allotropes: For an ethics of citation. Angewandte Chemie 55:10962–10976

Hofmeister A (1987) Single-crystal absorption and reflection infrared spectroscopy of forsterite and fayalite. Phys Chem Minerals 14:499–513

Hofmeister AM, Campbell KR (1992) Infrared spectroscopy of yttrium aluminum, yttrium gallium, and yttrium iron garnets. J Appl Phys 72:638–646

Hofmeister AM, Ito E (1992) Thermodynamic properties of $MgSiO_3$ ilmenite from vibrational spectra. Phys Chem Minerals 18:423–432

Hohenberg P, Kohn W (1964) Inhomogeneous Electron Gas. Phys Rev B 136:864–871

Hohm U (2013) Experimental static dipole–dipole polarizabilities of molecules. J Mol Str 1054–55: 282–292

Holder J, Granato AV (1969) Thermodynamic properties of solid containing defects. Phys Rev 182: 729–741

Holland HD (1978) The Chemistry of Atmosphere and Oceans Wiley Interscience, New York

Holland HD (2006) The oxygenation of the atmosphere and oceans. Phil Trans R Soc B 361:903–915

Holland TJB, Powell R (1990) An enlarged and updated internally consistent thermodynamic dataset with uncertainties and correlations: the system K_2O–Na_2O–CaO–MgO–MnO–FeO–Fe_2O_3–Al_2O_3–TiO_2–SiO_2–C–H_2–O_2. J Metamorph Geol 8:89–124

Holland TJB, Powell R (1998) An internally consistent thermodynamic data set for phases of petrological interest. J Metamorph Geol 16:309–343

Hollister C, Sinanoğlu O (1966) Molecular Binding Energies. J Amer Chem Soc 88:13–21

Holloway JR (1977) Fugacity and activity of molecular species in supercritical fluids In: Fraser DG (ed), Thermodynamics in Geology, Reidel Pub Co, Dordrecht, pp 161–181

Holloway-Phillips M (2018) Photosynthetic Oxygen Production: New Method Brings to Light Forgotten Flux. Plant Physiology 177:7–9

Holm B, Ahuja R (1999) Ab initio calculation of elastic constants of SiO_2 stishovite and α-quartz. J Chem Phys 111 :2071–2074

Holm JL, Kleppa OL, Westrum EF (1967) Thermodynamics of polymorphic transitions in silica Thermal properties from 5 to 1070 °K and pressure-temperature stability fields for coesite and stishovite. Geochim Cosmochim Acta 31:2289–2307

Hong KM, Noolandi J (1981) Theory of inhomogeneous multicomponent polymer systems Macromolecules 14:727–736

Hooke R, Jeeves TA (1961) "Direct search" solution of numerical and statistical problems. J ACM 8:212–229

Hoover W (1985) Canonical dynamics: Equilibrium phase-space distributions. Phys Rev A 31:1695–169

Horadam KJ (2006) Hadamard Matrices and Their Applications. Princeton University Press, Princeton, New Jersey

Horiuchi H, Sawamoto H (1981) β-$(Mg,Fe)_2SiO_4$: Single crystal X-ray diffraction study. Am Mineral 66:568–575

Horiuchi H, Hirano M, Ito E, Matsui Y (1982) $MgSiO_3$ (ilmenite-type): Single crystal X-ray diffraction study. Am Mineral 67:788–793

Horiuchi H, Ito E, Weidner DJ (1987) Perovskite-type $MgSiO_3$: Single-crystal X-ray diffraction study. Am Mineral 72:357–360

Hornung EW, Fisher RA, Brodale GE, Giauque WF (1969) Isotope Sieve for ^4He vs ^3He The Low-Temperature Heat Capacity of SiO_2 Glass before and after Contact with ^4He and ^3He. J Chem Phys 50:4878–4886

Hosieni KR, Howald RA, Scanlon MW (1985) Thermodynamics of the lambda transition and the equation of state of quartz. Am Mineral 70:782–793

Hotop H, Lineberger W C (1985) Binding Energies in Atomic Negative Ions: II. J Phys Chem Ref Data 14:732–750

Hsu CS, Rahman A (1979a) Crystal nucleation and growth in liquid rubidium. J Chem Phys 70: 5234–5240

Hsu CS, Rahman A (1979b) Interaction potentials and their effect on crystal nucleation and symmetry. J Chem Phys 71:4974–4986

Hubbard J (1964) Electron correlations in narrow energy bands II. The degenerate band case, Proc. R. Soc. London A 277:237–259

Hubbard J (1963) Electron Correlations in Narrow Energy Bands. Proc R Soc Lond A, 276, 238–257

Huber KP, and Herzberg G (1979) Molecular Spectra and Molecular Structure IV Constants of Diatomic Molecules, Van Nostrand Reinhold, New York, 689 pp

Hudon P, Jung I-H, Baker DR (2005) Experimental investigation and optimization of thermodynamic properties and phase diagrams in the systems $CaO-SiO_2$, $MgO-SiO_2$, $CaMgSi_2O_6-SiO_2$ $CaMgSi_2O_6-Mg_2SiO4$ to 10 GPa. J Petrol 46:1859–1880

Hughes, RIG (2007) Theoretical Practice: the Bohm-Pines Quartet. Perspectives on Science, The Massachusetts Institute of Technology, 14:457–524

Hugh-Jones DA, and Angel RJ (1994) A compressional study of $MgSiO_3$ orthoenstatite up to 85 GPa Am Mineral, 79, 405–410

Hui JCK, and Allen PB (1975) Thermodynamics of anharmonic crystals with application to Nb. J Phys C Solid State 8:2923–2935

Hull AW (1917) A new method of X-ray crystal analysis. Physic Rev 10:661–696

Hund F (1927) Linienspektren und periodisches System der Elemente Springer, Berlin Heidelberg GmbH

Hunter EP, Lias SG (1998) Evaluated Gas Phase Basicities and Proton Affinities of Molecules: An Update. J Phys Chem Ref Data 27:413–656

Hutter J, Alavi A, Deutsch T, Bernasconi M, Goedecker S, Marx D, Tuckerman M, Parrinello M, (1995) CPMD Version 33 MPI für Festkörperforschung and IBM Zürich Research Laboratory, Zürich, Swiss

Huzinaga S (1965) Gaussian-Type Functions for Polyatomic Systems I. J Chem Phys 42:1293–1302

Hylleraas EA (1931) Über die Elektronenterme des Wasserstoffmoleküls. Zeits Phys A 71:739–763

Ihm J, Zunger A, Cohen ML (1979) Momentum-space formalism for the total energy of solids. J Phys C 12:4409–4422

Iizumi M, Koetzle TF,Shirane G, Chikazumi S, Matsui M, Todo S (1982) Structure of magnetite (Fe_3O_4) below the Verwey transition temperature. Acta Cryst B 38:2121–2133

Iishi K (1978) Lattice dynamics of forsterite. Am Mineral 63:1198–1208

Imada M., Fujimori A, Tokura Y. (1998) Metal-insulator transitions. Rev. Mod. Phys. 70:1039–1263

Imbrie J, Imbrie JZ (1980) Modeling the Climatic Response to Orbital Variations Science 207:943–953

Imbrie J, Palmer Imbrie K (1979) Ice Ages Solving the Mystery. McMillan, London

Inoue T (1994) Effect of water on melting phase-relations and melt composition in the system $Mg_2SiO_4-Mg_2SiO_3-H_2O$ up to 15GPa. Phys Earth Planet In 85:237–263

Inoue T, Irifune T, Higo Y, Sanchira T, Sueda Y, Yamada A, Shinmei T, Yamazaki D, Ando J, Funakoshi K, Utsumi W (2006) The phase boundary between wadsleyite and ringwoodite in Mg_2SiO_4 determined by in situ X-ray diffraction. Phys Chem Minerals 33:106–114

Iona M (1941) On the Distribution of Lattice Vibrations of the KCl Crystal. Pys Rev 60:822–826

Isaak DG (1991) Elasticity of single crystal forsterite measured to 1700 K. J Geophys Res 95:5895–5906

Isaak DG, Anderson OL, Goto T (1989) Elasticity of single-crystal forsterite measured to 1700 K. J Geophys Res 94:5895–5906

Ito T, Sadanaga R, Takeuchi Y, Tokonami M (1969) The existence of partial mirrors in wollastonite. Proc Jpn Acad 45:913–918

Jacobs MHG, De Jong BWHS (2003) The high-temperature and high-pressure behavior of MgO derived from lattice vibration calculations Kieffer's model revisited. Phys Chem Chem Phys 5: 2056–2065

Jacobs MHG, De Jong BWHS (2005) An investigation into thermodynamic consistency of data for the olivine, wadsleyite and ringwoodite form of $(Mg,Fe)_2SiO_4$. Geochim Cosmochim Acta 69: 4361–4375

Jacobs MHG, van der Berg AP, De Jong BWHS (2006) The derivation of thermo-physical properties and phase equilibria of silicate materials from lattice vibrations: application to convection in the Earth's mantle. CALPHAD 30:131–146

Jaffé G (1934) Zur Theorie des Wasserstoffmolekülions. Zeits Phys A 87:535–544

Jahn S (2008), High-pressure phase transitions in $MgSiO_3$ orthoenstatite studied by atomistic computer simulation. Am Mineral 93:528–532
Jahn S, Martoňak R (2009) Phase behavior of protoenstatite at high pressure studied by atomistic simulations. Am Mineral 94:950–956
Jahnke E, Emde F (1945) Tables of Functions , 4th Edition, Dover Publications
Jakobsen H, Skibsted J, Bildsoe H, and Nielsen H (1988) Magic-angle spinning NMR spectra of satellite transitions for quadrupolar nuclei in solids. J Magn Res 85:173–180
James HM, Coolidge AA (1933) The Ground State of the Hydrogen Molecule. J Chem Phys 1:825–835
James F, Roos M (1977) MINUIT: A system for function minimization and analysis of the parameter errors and correlations CERN Computer Center, Geneva, Switzerland
Janbon A, Weber H, Braun O (1986) Solubility of He, Ne, Ar, Kr and Xe in a basalt melt in the range 1250–1600°C Geochemical implications. Geochim Cosmochim Acta 50:401–408
Jankowski K, Malinowski P (1980) Application of symmetry-adapted pair functions in atomic structure calculations: A variational-perturbation treatment of the Ne atom. Phys Rev A 21:20–65
Japel S, Schwager B, Boehler R, Ross M (2005) Melting of Copper and Nickel at High Pressure: The Role of d Electrons. Phys Rev Lett 95:167801-1-4
Jeanloz R, Roufosse M (1982) Anharmonic properties: Ionic model of the effects of compression and coordination change. J Geophys Res 87:10763–10772
Jensen H (1932) Die Ladungsverteilung in Ionen und die Gitterkonstante des Rubidiumbromids nach der statistischen Methode. Zeits Phys A 77:722–745
Jensen H (1936a) Über die Existenz negativer Ionen im Rahmen des statistischen Modells. Zeits Phys A 101:141–163
Jensen H (1936b) Quantentheoretische Berechnung der Alkalihalogenidgitter. Zeits Phys A 101: 164–185
Joback KG, Reid RC (1987) Estimation of Pure-Component Properties from Group-Contributions. Chem Eng Commun 57:233–243
Jochym PT, Parlinski K, Krzywiec P (2004) Elastic tensor of the forsterite (Mg_2SiO_4) under pressure. Comput Mat Sci 29:414–418
Johnson KH (1966) Multiple-scattering model for polyatomic molecules. J Chem Phys 45:3085–3098
Johnson JW, Norton D (1991) Critical phenomena in hydrothermal systems: state, thermodynamic, electrostatic, and transport properties of H_2O in the critical region. Amer Jour Sci 291:541–648
Johnson BG, Gill PMW, Pople JA (1993) The performance of a family of density functional methods. J Chem Phys 98:5612–5626
Jones P, Haggett ML, Longridge JL (1964) The hydration of carbon dioxide: A double clock experiment. J Chem Edu 41:610–612
Jørgensen, CK, 1962 Absorption spectra and chemical bonding in complexes Pergamon, Oxford
Jørgensen CK (1969) Oxidation Numbers and Oxidation States, Springer, New York
Jorgensen JD (1978) Compression mechanisms in α-quartz structures – SiO_2 and GeO_2. J Appl Phys 49:5473–5478
Jorgensen WL, Chandrasekhar J, Jeffry D, Madura JD (1983) Comparison of simple potential functions for simulating liquid water. J Chem Phys 79:926–935
Jost W (1960) Diffusion in Solid State Physics Academic Press, New York
Joswig W, Stachel Th, Harris JW, Baur WH, Brey G (1999) New Ca-silicate inclusions in diamonds – tracer from the lower mantle. Earth Planet Sc Lett 173:1–6
Joswig W, Paulus EF, Winkler B, Milman V (2003) The crystal structure of $CaSiO_3$-walstromite, a special isomorph of wollastonite-II. Z Kristallogr 218:811–818
Jura G, Garland CW (1952) The Experimental Determination of the Surface Tension of Magnesium Oxide. J Am Chem Soc 74:6033–6034
Jura G, Pitzer KS (1952) The Specific Heat of Small Particles at Low Temperatures. J Am Chem Soc 74:6030–6032

Kahn L R, Baybutt P, Truhlar DG (1976) Ab initio effective core potentials: Reduction of all-electron molecular structure calculations to calculations involving only valence electrons. J Chem Phys 65:3826–3853

Kajiyoshi K (1986) High – temperature equations of state for mantle mineralsand their anharmonic properties, Dissertation, Okayama University, Japan

Kalampounias AG, Yannopoulos SN, Papatheodorou GN (2006) A high-temperature Raman spectroscopic investigation of the potassium tetrasilicate in glassy, supercooled, and liquid states. J Chem Phys 124:014504-1-8

Kalkstein D, Soven P (1971) A Green's function theory of surface states. Surf Sci 26:85–99

Kalyanram MR, MacFarlane TG, Bell HB (1960) The activity of calcium oxide in slags in the systems CaO-MgO-Al_2O_3-SiO_2 at 1500°C. J Iron Steel Inst 195:58–64

Kamensky IL, Tokarev IV, Tolstikhin IN (1991) ^3H-^3He dating: a case for mixing of young and old groundwaters. Geochim Cosmochim Acta 55:2895–2899

Kanke Y, Nawrotsky A (1998) A Calorimetric Study of the Lanthanide Aluminum Oxides and the Lanthanide Gallium Oxides: Stability of the Perovskites and the Garnets. J Solid State Chem 141:424–436

Kanzaki M (1990) Melting of silica up to 7 GPa. J Am Ceram Soc 73:3706–3707

Kanzaki M, Stebbins JF, Xue X (1991) Characterization of quenched high pressure phases in $CaSiO_3$ system by XRD and ^{29}Si NMR. Geophys Res Letters 18:463–466

Karki BB, Stixrude L (2010) Viscosity of $MgSiO_3$ Liquid at Earth's Mantle Conditions: Implications for an Early Magma Ocean. Science 328:740–742

Karki BB, Wentzcovitch R (2002) First-principles lattice dynamics and thermoelasticity of $MgSiO_3$ ilmenite at high pressure. J Geophys Res 107:2267-1-6

Karki BB, Stixrude L, Crain J (1997a) Ab initio elasticity of three high-pressure polymorphs of silica. Geophys Res Lett 24:3269–3272

Karki, BB, Stixrude L, Clark SJ, Warren MC, Ackland GJ, Crain J (1997b) Structure and elasticity of MgO at high pressure. Am Mineral 82:51–60

Karki BB, Wentzcovitch RM, de Gironcoli S, Baroni S (2000) High-pressure lattice dynamics and thermoelasticity of MgO. Phys Rev B 61:8793–8800

Karki BB, Bhattarai D, Stixrude L (2006) First principles calculations of the structural, dynamical and electronic properties of liquid MgO. Phys Rev B 73:174208-1-7

Karki BB, Bhattarai D, Stixrude L (2007) First-principles simulations of liquid silica: structural and dynamical behavior at high pressure. Phys Rev B 76:104205-1-12

Kasting JF, Ono S (2006) Paleoclimates: the first two billion years. Phil Trans R Soc B 361:917–929

Kasting JF, Tazewell Howard M (2006) Atmospheric composition and climate on the early Earth. Phil Trans R Soc B 361:1733–1742

Kato T, Kumazawa M (1985) Garnet phase of $MgSiO_3$ filling the pyroxene-ilmenite gap at very high temperature. Nature 316:803–805

Kato K, Nukui A (1976) Die Kristallstructur des monoclinen Tief-trydymits. Acta Cryst B 32:2486–2491

Katsura T, Ito E (1989) The system Mg_2SiO_4-Fe_2SiO_4 at high pressures and temperatures: precise determination of stabilities of olivine, modified spinel, and spinel. J Geophys Res 94:15663–15670

Katsura T, Nagashima S (1974) Solubility of sulfur in some magmas at 1 atmosphere. Geochim Cosmochim Acta 38:517–531

Katsura T, Yamada H, Nishikawa O, Song MS, Kubo A, Shinmei T, Yokoshi S, Aizawa Y, Yoshino T, Walter MJ, Ito E, Funakoshi K (2004) Olivine-wadsleyite transition in the system $(Mg,Fe)_2SiO_4$. J Geophys Res-Sol Ea 109:B02209

Kay DAR, Taylor J (1960) Activities of silica in the lime + alumina + silica system. Trans Faraday Soc 56:1372–1386

Kaye JA, Strobel DF (1983) Enhancement of Heavy Ozone in the Earth's Atmosphere? J Geophys Res 88:8447–8452

Kell GS (1967) Precise Representation of Volume Properties of Water at One Atmosphere. J Chem Eng Data 12:66–69
Kern DM (1960) The hydration of carbon dioxide. J Chem Educ 37:14–23
Kerrick DM, Jacobs GK (1982) A modified Redlich-Kwong equation for H_2O, CO_2 and H_2O-CO_2 mixtures at elevated pressures and temperatures. Amer Jour Sci 281:735–767
Keskar NR, Chelikowsky JR (1992) Structural-properties of 9 silica polymorphs. Phys Rev B 46: 1–13
Kiefer B, Stixrude L, Wentzcovitch R (1999) Normal and inverse ringwoodite at high pressures. Am Mineral 84:288–293
Kiefer B, Stixrude L, Hafner J, Kresse G (2001) Structure and elasticity of wadsleyite at high pressures. Am Miner 86:1387–1395
Kieffer SW (1979a) Thermodynamics and lattice vibrations of minerals: 1 Mineral heat capacities and their relationships to simple lattice vibrational models. Rev Geophys Space Phys 17:1–19
Kieffer SW (1979b) Thermodynamics and lattice vibrations of minerals: 2 Vibrational characteristics of silicates. Rev Geophys Space Phys 17:20–34
Kieffer SW (1979c) Thermodynamics and lattice vibrations of minerals: 3 Lattice dynamics and an approximation for minerals with application to simple substances and framework silicates. Rev Geophys Space Phys 17:35–59
Kieffer SW (1980) Thermodynamics and lattice vibrations of minerals: 4 Application to chain and sheet silicates and orthosilicates. Rev Geophys Space Phys 18:862–886
Kieffer SW (1982) Thermodynamics and lattice vibrations of minerals: 5 Applications to phase equilibria, isotopic fractionation, and high-pressure thermodynamic properties. Rev Geophys Space Phys 20:827–849
Kieffer SW (1985) Heat capacity and entropy: systematic relations to lattice vibrations In: Ribbe PH (series ed), Reviews in Mineralogy, Vol 14, Mineralogical Society of America, pp 65–126
Kikuchi R (1954) Gaussian Functions in Molecular Integrals. J Chem Phys 22:148
Kimizuka N, Yamamoto A, Ohashi H, Sugihara T, and Sekine T (1983) The stability of the phases in the Ln_2O_3-FeO-Fe_2O_3 systems which are stable at elevated temperatures (Ln: Lanthanide elements and Y). J Solid State Chem 49:65–76
Kirfe A, Will G, and Arndt J (1979) A new phase of coesite SiO_2. Z Krist 149:315–326
Kirkpatrick JD, Reid IN, Liebert J, Cutri RM, Brant N, Beichman CA, Dahn CC, Monet DG, Gizis JE, Skrutskie MF (1999) Dwarfs Cooler than "M": The Definition of Spectral Type "L" Using Discoveries from the 2 Micron All-Sky Survey (2MASS) Astrophys J 519:802–833
Kirsten T (1968) Incorporation of rare gases in solidifying enstatite melts. J Geophys Res 73:2807–2810
Kistenmacher H, Lie GC, Popkie H, Clementi E (1974) Study of the structure of molecular complexes VI Dimers and small clusters of water molecules in the Hartree Fock approximation. J Chem Phys 61:546–561
Kittel C (1976) Introduction to Solid State Physics. 5th edition, Wiley, New York, London, Sydney, Toronto
Klamt A (1995) Conductor-like Screening Model for Real Solvents: A New Approach to the Quantitative Calculation of Solvation Phenomena. J Phys Chem 99:2224–2235
Klamt A, and Schüürmann G (1993) COSMO: A New Approach to Dielectric Screening in Solvents with Explicit Expressions for the Screening Energy and its Gradient. J Chem Soc Perkin Trans 2:799–805
Klamt A, Eckert F, Diedenhofen M, Beck ME (2003) First principles calculations of aqueous pK_a values for organic and inorganic acids using COSMO-RS reveal an inconsistency in the slope of the pK_a scale. J Phys Chem A 107:9380–9386
Klein C (2005) Some Precambrian banded iron-formations (BIFs) from around the world: Their age, geologic setting, mineralogy, metamorphism, geochemistry, and origin. Am Miner 90: 1473–1499

Kleinman L, Bylander DM (1982) Efficacious Form for Model Pseudopotentials. Phys Rev Lett 48:1425–1428

Klopper W, van Duijneveldt-van de Rijdt JGC M, van Duijneveldt FB (2000) Computational determination of equilibrium geometry and dissociation energy of the water dimer: Phys Chem Chem Phys 2:2227–2234

Koelling DD, Arbman GO (1975) Use of energy derivative of the radial solution in an augmented plane wave method: application to copper. J Phys F Met Phys 5:2041–2054

Koenigsberger JG (1938a) Natural residual magnetism of eruptive rocks: Part I. J Geophy Res 43: 119–130

Koenigsberger JG (1938b) Natural residual magnetism of eruptive rocks: Part II. J Geophy Res 43: 299–320

Kohn W (1959) Analytic Properties of Bloch Waves and Wannier Functions. Phys Rev 115:809–821

Kohn W, Sham LJ (1965) Self-Consistent Equations Including Exchange and Correlation Effects. Phys Rev A 140:1133–1138

Kolesov B, Geiger C (2004) A Raman spectroscopic study of Fe-Mg olivines. Phys Chem Minerals 31:142–154

Kolos, W, Wolniewicz, L (1965) Potential-Energy Curves for the X 1Σg+, b3Σu+, and C 1Πu States of the Hydrogen Molecule. J Chem Phys 43:2429–2441

Kondrat'ev SN, Strizhov BV (1985) Pression de vapeur saturée et enthalpies des transitions de phases du trioxyde de soufre. Žurnal Fizičeskoj Himii 59:2933–2938

Koster GF, Slater JC (1954) Wave Functions for Impurity Levels. Phys Rev 95:1167–1176

Kotani M, Mizuno Y, Kayama K, Ishiguro E (1957) Electronic Structure of Simple Homonuclear Diatomic Molecules I Oxygen Molecule. J Chem Phys Soc Japan 12:707–736

Krane KS (1987) Introductory Nuclear Physics Wiley, New York

Krasovska OV, Winkler B, Krasovskii EE, Yaresko AN, Antonov VN, Langer N (1997) Ab initio calculation of the pleochroism of fayalit. Amer Min 82:672–676

Kress VC, Carmichael ISE (1991) The compressibility of silicate liquids containing Fe_2O_3 and the effect of composition, temperature, oxygen fugacity and pressure on their redox states. Contrib Mineral Petr 108:82–92

Kresse G, Furthmüller J (1996) Efficiency of ab-initio total energy calculations for metals and semiconductors using a plane-wave basis set. Comput Mater Sci 6:15–50

Kresse G, Hafner J (1993a) Ab initio molecular dynamics for liquid metals. Phys Rev B 47:558–561

Kresse G, Hafner J (1993b) Ab initio molecular dynamics for open-shell transition metals. Phys Rev B 47:558–561

Kresse G, Hafner J (1994) Ab initio molecular dynamics for open-shell transition metals. Phys Rev B 49:14251–14269

Kresse G, Hafner J, Needs RJ (1992) Optimized norm-conserving pseudopotentials. J Phys-Condens Matt 4:7451–7468

Kresse G, Furthmüller J, Hafner J (1995) Ab initio Force Constant Approach to Phonon Dispersion Relations of Diamond and Graphite. Europhys Lett 32:729–734

Krishnan R, Binkley JS, Seeger R, Pople JA (1980a) Self-consistent molecular orbital methods XX. A basis set for correlated wave functions. J Chem Phys 72:650–654

Krishnan R, Schlegel HB, Pople JA (1980b) Derivative studies in configuration–interaction theory. J Chem Phys 72:4654–4655

Kröger FA (1964) The Chemistry of Imperfect Crystals North Holland Pub Co, Amsterdam

Krol DM, Lyons KB, Brawer SA, Kurkjian CR (1986) High-temperature light scattering and the glass transition in vitreous silica. Phys Rev B 33:4196–4202

Kroopnick P, Craig H (1972) Atmospheric oxygen: Isotopic composition and solubility fractionation. Science 175:54–55

Kruger MB, Nguyen JH, Caldwell W, Jeanloz R (1997) Equation of State of $MgAl_2O_4$ spinel to 65 GPa. Phys Rev B 56:1–4

Kruse H, Grimme S (2012) A geometrical correction for the inter- and intra-molecular basis set superposition error in Hartree-Fock and density functional theory calculations for large systems. J Chem Phys 136:154101-1-16

Kübler J, Sandratskii LM, Uhl M (1992) Noncollinear spiral magnetic order in itinerant-electron magnets. J Magn Magn Mater 104–107:695–696

Kumazawa M (1969) The elastic constants of polycrystalline rocks and nonelastic behavior inherent to them. J Geophys Res 74:5311–5320

Kunc K, Martin R (1982) Ab Initio Force Constants of GaAs: A New Approach to Calculation of Phonons anti Dielectric Properties. Phys Rev Lett 48:406–409

Kuno H, Hess HH (1953) Unit cell dimensions of clinoenstatite and pigeonite in relation to other common clinopyroxenes. Amer Jour Sci 251:741–752

Kuo, I-F W, Mundy CJ, McGrath MJ, Siepmann JI, Vande Vondele J, Sprik M, Hutter J, Chen B, Klein ML, Mohamed F, Krack M, Parrinello M (2004) Liquid Water from First Principles: Investigation of Different Sampling Approaches. J Phys Chem B 108:12990–12998

Kurashina T, Hirose K, Ono S, Sata N, Ohishi Y (2004) Phase transition in Al-bearing $CaSiO_3$ perovskite: implications for seismic discontinuities in the lower mantle. Phys Earth Planet In 145:67–74

Kuroda K, Irifune T, Inoue T, Nishiyama N, Miyashita M, Funakoshi K, Utsumi W (2000) Determination of phase boundary between ilmenite and perovskite in $MgSiO_3$ by in situ X-ray diffraction and quench experiments Phys Chem Miner, 27, 523–532

Kushiro I (1964) Wollastonite-pseudowollastonite inversion. Carnegie Inst Wash Yb 63:83–84

Laasonen K, Sprik M, Parrinello M, Car R (1993) "Ab initio" liquid water. J Chem Phys 99: 9080–9089

Lager GA, Ross FK, Rotella FJ, Jorgensen JD (1981) Neutron powder diffraction of forsterite Mg_2SiO_4 : a comparison with single crystal investigations. J Appl Cryst 14 :137–139

Landau LD, Lifshitz EM (1980) Statistical Physics Part 1, 3rd ed; Butterworth-Heinemann: Oxford, UK

Landron C, Hennet L, Jenkins TE, Greaves GN, Coutures JP, Soper AK (2001) Liquid Alumina: Detailed Atomic Coordination Determined from Neutron Diffraction Data Using Empirical Potential Structure Refinement. Phys Rev Letters 86:4839–4842

Lange RA, Carmichael ISE (1987) Densities of Na_2O-K_2O-MgO-MgO-FeO-Fe_2O_3-Al_2O_3-TiO_2-SiO_2 liquids: New measurements and derived partial molar properties. Geochim Cosmochim Acta 51:2931–2946

Lange MA Lambert P, Ahrens TJ (1985) Shock effects on hydrous minerals and implications for carbonaceous meteorites. Geochim Cosmochim Acta 49:1715–1726

Langhoff SR, Davidson ER (1974) Configuration interaction calculations on the nitrogen molecule. Int J Quant Chem 8:61–72

Larson JW, Cerutti P, Garber HK, Hepler LG (1968) Electrode potentials and thermodynamic data for aqueous ions Copper, Zinc, Cadmium, Iron, Cobalt and Nickel. J Phys Chem 78:2902–2907

Lasaga AC (1981) The atomistic basis of kinetics: defects in minerals In: Reviews in Mineralogy, Vol 8, PH Ribbe (series ed), Mineralogical Society of America, pp 261–319

Lasaga AC (1998) Kinetic Theory in Earth Sciences.. Princeton University Press, New Jersey

Lasaga AC, Cygan RT (1982) Electronic and ionic polarizabilities of silicate minerals. Am Mineral 67:328–334

Latimer WM (1952) The oxidation states of the elements and their potentials in aqueous solutions Englewood Cliffs, NJ, Prentice Hall

Laves P, Baskin Y (1956) On the Formation of the Rhombohedral Graphite Modification. Z Kristallogr 107:337–356

Le Page Y, Calvert LD, Gabe EJ (1980) Parameter variation in low-quartz between 94 and 298 K. Phys Chem Solids 41:721–725

Leadbetter AJ (1969) Inelastic Cold Neutron Scattering from Different Forms of Silica. J Chem Phys 51:779–786

Lebowitz JL, Helfand E, Praestgaard E (1965) Scaled particle theory of fluid mixtures. J Chem Phys 43:774–779

Ledbetter HM, Naimon ER (1974) Elastic Properties of Metals and Alloys II Copper. J Phys Chem Ref Data 3:897–935

Lee TJ (1990) On the energy separation between the open and cyclic forms of ozone. Chem Phys Lett 169:529–533

Lee C, Gonze X (1994) Lattice dynamics and dielectric properties of SiO_2 stishovite. Phys Rev Lett 72:1686–1689

Lee C, Gonze X (1995) Ab initio calculation of the thermodynamic properties and atomic temperature factors of SiO_2 α-quartz and stishovite. Phys Rev B 51:8610–8613

Lee TJ, Scuseria GE (1990) The vibrational frequencies of ozone. J Chem Phys 93:489–494

Lee C, Yang W, Parr RG (1988) Development of the Colle-Salvetti correlation-energy formula into a functional of the electron density. Phys Rev B 37:785–789

Lee TJ, Martin JML, and Taylor PR (1995) An accurate ab initio quartic force field and vibrational frequencies for CH_4 and isotopomers. J Chem Phys 102:254–261

Lehman G, Taut M (1972) On the numerical calculation of the density of states and related properties. Phys Status Solidi B 54:469–476

Lennard-Jones FRS (1949) The Molecular Orbital Theory of Chemical Valency II Equivalent Orbitals in Molecules of Known Symmetry. P R Soc London A 198:14–26

Lennard-Jones JE, Dent B M (1928) The Change in Lattice Spacing at a Crystal Boundary. P R Soc London A 121:247–259

Lennard-Jones FRS, Pople JA (1950) The molecular orbital theory of chemical valency IV The significance of equivalent orbitals. P R Soc London A 202:166–180

Lennard-Jones JE, Taylor PA (1925) Some Theoretical Calculations of the Physical Properties of Certain Crystals. P R Soc London A 109:476–508

Lenz W (1932) Über die Anwendbarkeit der statistischen Methode auf Ionengitter. Zeits Phys A 77: 713–721

Leopold DG, Pendley RH, Roebber JL, Hemley RJ, Vaida V (1984) Direct absorption spectroscopy of jet-cooled polyenes II The $1^1B_u^+ \leftarrow 1^1A_g^-$ transitions of butadienes and hexatrienes. J Chem Phys 81:4128–4229

Leslie M, Gillan NJ (1985) The energy and elastic dipole tensor of defects in ionic crystals calculated by the supercell method. J Phys C Solid State 18:973–982

Levelt Sengers JMH, Kamgar-Parsi B, Balfour FW, Sengers JV (1983) Thermodynamic properties of steam in the critical region. J Phys Chem Ref Data 12:1–28

Levien L, and Prewitt CT (1981) High-pressure crystal structure and compressibility of coesite. Am Mineral 66:324–333

Levien L, Prewitt CT, Weidner DL (1980) Structure and elastic properties of quartz at pressure. Am Mineral 65:920–930

Levin, I, Gemming T, Brandon DG (1993) Some Metastable Polymorphs and Transient Stages of Transformation in Alumina. Phys Stat Sol A 166:197–218

Levy M, Perdew JP (1993) Tight bound and convexity constraint on the exchange-correlation-energy functional in the low-density limit, and other formal tests of generalized-gradient approximations. Phys Rev B 48:11638–11645

Lewis GN (1916) The Atom and the Molecule. J Am Chem Soc 38:762–785

Lewis GN, Randall M (1921) The activity coefficient of strong electrolytes. J Amer Chem Soc 43: 1112–1154

Li B, Gwanmesia GD, Liebermann RC (1996) Sound velocities of olivine and beta polymorphs of Mg_2SiO_4 at Earth's transition zone pressures. Geophys Res Letters 23:2259–2262

Li B, Liebermann RC, Weidner D (1998) Elastic moduli of wasdleyite (β-Mg_2SiO_4) to 7 gigapascals and 873 kelvin. Science 281:675–677

Li L, Weidner DJ, Wentzcovitch RM, Da Silva CRS (2007) Vibrational and thermodynamic properties of forsterite at mantle conditions. J Geophys Res 112:B05206-1-8

Lide DR (1958) Internal Barrier in Ethane. J Chem Phys 29:1426–1427

Lide DR (2002) Handbook of Chemistry and Physics, 83rd edition, CRC Press, Boca Raton

Lie GC, Clementi E, Yoshimine M (1976) Study of the structure of molecular complexes XIII Monte Carlo simulation of liquid water with a configuration interaction pair potential. J Chem Phys 64:2314–2323

Lieb EH, Oxford S (1981) Improved lower bound on the indirect Coulomb energy. Int J Quantum Chem 19:427–439

Liechtenstein AI, Anisimov VI, Zaanen J (1995) Density-functional theory and strong interactions: Orbital ordering in Mott-Hubbard insulators. Phys Rev B 52:R5467-R5470

Lin I-C, Coutinho-Neto MD, Felsenheimer C, von Lilienfeld OA, Tavernelli I, Rothlisberger U (2007) Library of dispersion-corrected atom-centered potentials for generalized gradient approximation functionals: Elements H, C, N, O, He, Ne, Ar, and Kr. Phys Rev B 75:205131-1-5

Lin IC, Seitsonen AP, Tavernelli I, Coutinho-Neto MD, Rothlisberger U (2008) Importance of van der Waals Interactions in Liquid Water. J Phys Chem B 113:1127–1131

Lin IC, Seitsonen AP, Tavernelli I, Rothlisberger U (2012) Structure and Dynamics of Liquid Water from ab Initio Molecular Dynamics - Comparison of BLYP, PBE, and revPBE Density Functionals with and without van der Waals Corrections. J Chem Theory Comp 8:3902–3910

Lindemann FA (1910) The calculation of molecular vibration frequencies. Phys Zeits 11:609–612

Lindsley DH (1976) The crystal chemistry and structure of oxide minerals as exemplified by the Fe-Ti oxides In: Rumble D III (ed), Oxide Minerals. Mineralogical Society of America Short Course Notes, Southern Printing Company, Blacksburg, Virginia, pp 1–60

Lindsley DH (1982) Phase equilibria of pyroxenes at pressure > 1 atmosphere In: Ribbe PH (series ed), Reviews in Mineralogy, Vol 7 (2nd edition), Mineralogical Society of America, pp 209–308

Linnett JW, Poë AJ (1951) Directed Valency in Elements of the First Short Period. Trans Faraday Soc 47:1033–1041

Lipparini F; Scalmani G; Mennucci B; Cances E; Caricato M; Frisch MJ (2010) A variational formulation of the polarizable continuum model. J Chem Phys 133:014106-1-11

Lipsett SG, Johnson FMG, Maas O (1927) The surface energy and the heat of solution of solid sodium chloride. J Am Chem Soc 49:925–943

Lipsett SG, Johnson FMG, Maas O (1928) The surface energy of solid sodium chloride III The heat of solution of finely ground sodium chloride. J Am Chem Soc 50:2701–2703

Liu L-G, Ringwood AE (1975) Synthesis of a perovskite-type polymorph of $CaSiO_3$. Earth Planet Sc Lett 28:209–211

Liu F, Garofalini SH, King-Smith D, Vanderbilt D (1994) First-principles study of crystalline silica. Phys Rev B 49:12528–12534

Liu J, Zhang J, Flesh L, Li B, Weidner DJ, Liebermann D (1999) Thermal equation of state of stishovite. Phys Earth Planet In 112:257–266

Łodziana Z, Parliński K (2003) Dynamical stability of the α and θ phases of alumina. Phys Rev B 67:174106-1-7

Lofthus A, Krupenie PH (1977) The Spectrum of Molecular Nitrogen. J Phys Chem Ref Data 6:113–307

Lohr LL, Helman AJ (1987) Ab initio synthesis of the ozone ultraviolet continuum. J Chem Phys 86:5329–5336

London F (1930) Zur Theorie und Systematik der Molekularkräfte. Zeits Phys 63:245–279

London F (1937) Théorie quantique des courants interatomiques dans les combinaisons aromatiques. J Phys Radium 8:397–409 DANS LES COMBINAISONS AROMATIQUES

Lorius C, Jouzel J, Raynaud D, Hansen J, Le Treut H (1990) The ice-core record: climate sensitivity and future greenhouse warming. Nature 347:139–145

Louie SG, Ho K-M, Cohen ML (1979) Self-consistent mixed-basis approach to the electronic structure of solids. Phys Rev B 19:1774–1782

Lui CH, Li Z, Chen Z, Klimov PV, Brus LE, Heinz TF (2011) Imaging Stacking Order in Few-Layer Graphene. Nano Lett 11:164–169

Lukesh J S (1950) On the symmetry of graphite. Phys Rev 80:226–229
Lukesh JS, Pauling L (1950) The problem of the graphite structure. Amer Min 35: 125
Lummer O, Pringsheim E (1901) Kritisches zur schwarsen Strahlung. Ann Phys 311:192–210
Lüthi D, Le Floch M, Bereiter B, Blunier T, Barnola JM, Siegenthaler U, Raynaud D, Jouzel J, Fische H, Kawamura K, Stocker TF (2008) High-resolution carbon dioxide concentration record 650,000–800,000 years before present. Nature 453:379–382
Lutz, HD, Muller, B, Steiner, HJ (1991) Lattice vibration spectra LIX Single-crystal infrared and Raman studies of spinal type oxides. J Solid State Chem 90:54–60
Lux HZ (1939) Acid and bases in fused salt bath: the determination of oxygen-ion concentration. Z Elektrochem 45:303–309
Lux G (1987) The behavior of noble gases in silicate liquids: Solution, diffusion, bubbles and surface effects, with application to natural samples. Geochim Cosmochim Acta 51:1549–1560
Luz B, Barkan E (2011) The isotopic composition of atmospheric oxygen. Global Biogeochem Cycles 25:1–14
Lyons TW, Reinhard CT, Planavsky NJ (2014) The rise of oxygen in Earth's early ocean and atmosphere. Nature 506:307–315
Ma C, Tschauner O, Beckett JR, Liu Y, Rossman GR, Sinogeikin SV, Smith JS, Taylor LA (2016) Ahrensite, γ-Fe_2SiO_4, a new shock-metamorphic mineral from the Tissint meteorite: Implications for the Tissint shock event on Mars. Geochim Cosmochim Acta 184:240–256
Macchi P, Sironi A (2003) Chemical bonding in transition metal carbonyl clusters: complementary analysis of theoretical and experimental electron densities. Coord Chem Rev 238–239:383–412
Mack KM, Muenter JS (1977) Stark and Zeeman properties of ozone from molecular beam spectroscopy. J Chem Phys 66:5278–5283
Mackenzie JK (1950) The Stresses and Energies associated with Inter-Crystalline Boundaries. Proc Phys Soc London A 63:1370
Mackwell SJ, Kohlstedt DL, Paterson MS (1985) The role of water in the deformation of olivine single crystals. J Geophys Res 90:11319–11333
MacRury TB, Steele WA (1974a) Quantum effects on the dielectric virial coefficients of polar gases. J Chem Phys 61:3352–3365
MacRury TB, Steele WA (1974b) Calculations of the second virial coefficients for water and ammonia. J Chem Phys 61:3366–3375
Magyari-Köpe B, Vitos L, Grimvall G, Johansson B, Kollar J (2002) perovskite: An ab initio total energy study. Phys Rev B 65:193107-1-4
Majer V, Svoboda V (1985) Enthalpies of Vaporization of Organic Compounds: A Critical Review and Data Compilation. Blackwell Scientific Publications, Oxford
Makri N, Sim E, Makarov DE, Topaler M (1996) Long-time quantum simulation of the primary charge separation in bacterial photosynthesis. PNAS 93:3926–3931
Makris JS, Masters BJ (1973) Phosphorus Isoconcentration Diffusion Studies in Silicon. J Electrochem Soc 120:1252–1255
Manara D, Böhler R, Capriotti L, Quaini A, Bao Z, Boboridis K, Luzzi L, Janssen A, Pöml P, Eloirdi R, Konings RJM (2014) On the melting behaviour of calcium monoxide under different atmospheres: A laser heating study. J Eur Ceram Soc 34:1623–1636
Manghnani MH, Matsui T (1981) Temperature dependence of pressure derivatives of single-crystal elastic constants of pure forsterite (abstract) ISPEI Symposium of Properties of Materials at High Pressures and High Temperatures, Toronto, Canada
Manghnani MH, Vijayakumar V, Bass JD (1998) High-pressure Raman scattering study of majorite-garnet solid solutions in the system $Mg_4Si_4O_{12}$-$Mg_3Al_2Si_3O_{12}$. In: Manghnani MH and Yagi T (eds), Properties of Earth and Planetary Materials at High Pressure and Temperature, Geophysical Monograph No 101, American Geophysical Union, Washington DC, pp 129–138
Mao HK, Chen LC, Hemley RJ, Jephcoat AP, Wu Y (1989) Stability and equation of state of $CaSiO_3$-Perovskite to 134 GPa. J Geophys Res 94:17889–17894

Mao H, Sundman B, Wang Z, Saxena SK (2001) Volumetric properties and phase relations of silica – thermodynamic assessment. J Alloys Comp 327:253–262
Maradudin AA, Vosko SH (1968) Symmetry Properties of the Normal Vibrations of a Crystal. Rev Mod Phys 40:1–37
Marenich AV, Cramer CJ, Truhlar DG (2009) Universal solvation model based on solute electron density and a continuum model of the solvent defined by the bulk dielectric constant and atomic surface tensions. J Phys Chem B 113:6378–96
Martinez JM, Hernàndez-Cobos J, Saint-Martin H, Pappalardo RR, Ortega-Blake I, Sanchez Marcos E (2000) Coupling a polarizable water model to the hydrated ion-water interaction potential: a test on the Cr^{3+} hydration. J Chem Phys 112:2339–2347
Marx D, Hutter J (2010) Ab Initio Molecular Dynamics Basic Theory and Advanced Methods Cambridge University Press, Cambridge, UK
Marzari N, Souza I, Vanderbilt D (2003) An Introduction to Maximally-Localized Wannier Functions. Psi-K Newsletter 57:129–168
Marzari N, Mostofi AA, Yates JR, Souza I, Vanderbilt D (2012) Maximally localized Wannier functions: Theory and applications. Rev Mod Phys 84:1419–1475
Massiot D (1996) Sensitivity and Lineshape Improvements of MQ-MAS by Rotor-Synchronized Data Acquisition. J Magn Res 122:240–244
Masson CR (1965) An approach to the problem of ionic distribution in liquid silicates. P R Soc London A 287:201–221
Masson CR (1968) Ionic equilibrium in liquid silicates. J Am Ceram Soc 51:134–143
Masson CR, Smith IB, Whiteway SG (1970) Activities and ionic distributions in liquid silicates: application of polymer theory. Can J Chem 48:1456–1464
Matito E, Solà M, Salvador P, Duran M (2007) Electron sharing indexes at the correlated level Application to aromaticity calculations. Faraday Discuss 135:325–345
Matsen MW (1998) Gyroid versus double-diamond in ABC triblock copolymer melts. J Chem Phys 108:785–796
Matsen MW, Bates FS (1996) Unifying weak- and strong-segregation block copolymer theories. Macromolecules 29:1091–1098
Matsen MW, Schick M (1994a) Microphase separation in starblock copolymer melts Macromolecules 27:6761–6767
Matsen MW, Schick M (1994b) Stable and unstable phases of a diblock copolymer melt. Phys Rev Letters 72:2660–2663
Matsui Y, Manghnani MH (1985) Thermal expansion of single-crystal forsterite to 1023 K by Fizeau interferometry. Phys Chem Minerals 12:201–210
Matsui M, Price GD (1992) Computer Simulation of the MgSiO Polymorphs. Phys Chem Minerals 18:365–372
Matsui M, Parker SC, Leslie M (2000) The MD simulation of the equation of state of MgO: Application as a pressure calibration standard at high temperature and high pressure. Am Mineral 85:312–316
Matsuoka O, Clementi E, Yoshimine M (1976) CI study of the water dimer potential surface. J Chem Phys 64:1351–1361
Matthews GP, Smith EB (1976) An intermolecular pair potential energy function for methane. Mol Phys 32:1719–1729
Mauersberger K (1981) Measurements of heavy ozone in the stratosphere. Geophys Res Lett 8: 935–937
Mauersberger K (1987) Ozone isotope measurements in the stratosphere. Geophys Res Lett 14:80–83
Mauersberger K, Krankovsky D, Janssen C (2003) Oxygen isotope processes and transfer reactions. Space Sci Rev 106:265–269
Mauro JC, Loucks RJ, Gupta PK (2009) Fictive Temperature and the Glassy State. J Am Ceram Soc 92:75–86

Mayer JG (1933) Dispersion and polarizability and the Van Der Waals potential in the alkali halides. J Chem Phys 1:270–279
Mayer I (1983) Charge, bond order and valence in the AB initio SCF theory. Chem Phys Lett 97: 270–274
McCarthy MI, Harrison NM (1994) Ab initio determination of the bulk properties of MgO. Phys Rev B 49:8574–8582
McCarty MI, Babu SVK (1970) First quantum corrections to the second virial coefficient of a Stockmayer gas. J Phys Chem 74:1113–1115
McClure DS (1957) The distribution of transition metal cations in spinels. J Phys Chem Solids 3: 311–317
McClure DS (1965) The effects of inner-orbitals on thermodynamic properties In: Dunn TM, McClure DS and Pearson RG (eds), Crystal field theory, Harper and Row, New York, pp 77–95
McClure JW (1969) Electron energy band structure and electronic properties of rhombohedral graphite. Carbon 7:125–132
McCormac BM (1971) The Radiating Atmosphere Springer, New York, and Reidel, Dordrecht, Holland
McDevitt N (1969) Infrared lattice spectra of Rare Earth aluminum, gallium and iron garnets. J Opt Soc Am 59:1240–1244
McGrath MP, Radom L (1991) Extension of Gaussian-1 (G1) theory to bromine-containing molecules. J Chem Phys 94:511–516
McLean AD, Chandler GS (1980) Contracted Gaussian basis sets for molecular calculations I 2nd row atoms, Z = 11 – 18. J Chem Phys 72:5639–5648
McLean AD, McLean RS (1981) Roothaan-Hartree-Fock atomic wave functions Slater basis-set expansions for Z = 55–92. At Data Nucl Data Tables 26:197–381
McMillan P, Akaogi M, Ohtani E, Williams Q, Nieman R, Sato R (1989) Cation disorder in garnets along the $Mg_3Al_2Si_3O_{12}$–$Mg_4Si_4O_{12}$ join: an infrared, Raman and NMR study. Phys Chem Minerals 16:428–435
McMillan PF, Akaogi M, Sato RK, Poe B, Foley J (1991) Hydroxyl groups in β-Mg_2SiO_4. Amer Min 76:354–360
McNaught AD, Wilkinson A (1997) Compendium of Chemical Terminology, 2nd Edition (the "Gold Book") Blackwell Scientific Publications, Oxford
McQuarrie DA (1984) Statistical Mechanics, University Science Books, Sausalito, CA
McQuarrie DA (2008) Quantum Chemistry, 2nd Edition, University Science Books, Mill Valley, California
McQuarrie DA, Simon JD (1997) Physical Chemistry: A Molecular Approach. University Science Books
McSkimin HJ (1953) Measurement of Elastic Constants at Low Temperatures by Means of Ultrasonic Waves–Data for Silicon and Germanium Single Crystals, and for Fused Silica. J Appl Phys 24 : 988–997
McWeeny R (1953) X-Ray Scattering by Aggregates of Bonded Atoms III The Bond Scattering Factor: Simple Methods of Approximation in the General Case. Acta Cryst 53:631–637 g
McWeeny R. (1954) On the Basis of Orbital Theories. Proc. Roy. Soc. London A 232:114–135
McWeeny R (1960) Some Recent Advances in Density Matrix Theory. Rev Mod Phys 32:335–369
McWeeny R (1972) Quantum Mechanics: Principles and Formalism Pergamon Press, Oxford - New York – Toronto – Sydney – Braunschweig
Meade C, Jeanloz R (1987) Frequency-dependent equation of state of fused silica to 10 GPa. Phys Rev B 35:236–244
Meadowcroft TR, Richardson FD (1965) Structural and thermodynamic aspects of phosphate glasses. Trans Faraday Soc 61:54–70
Meckler A (1953) Electronic energy levels of molecular oxygen. J Chem Phys 21:1750–1762

Meng Y, Weidner DJ, Gwanmesia DG, Liebermann RC, Vaughan MT, Wang Y, Leinenweber K, Pacalo RE, Yaganeh-Haeri A, Zhao Y (1993) In situ high P-T X Ray diffraction studies on three polymorphs (α, β, γ) of Mg_2SiO_4. J Geophys Res 98:22199–22207

Meng Y, Fei Y, Weidner D, Gwanmesia GD, Hu J (1994) Hydrostatic compression of γ-Mg_2SiO_4 to mantle pressures and 700 K: thermal equation of state and related thermodynamic properties. Phys Chem Minerals 21:407–412

Mennucci B, Tomasi J (1997) Continuum solvation models: A new approach to the problem of solute's charge distribution and cavity boundaries. J Chem Phys 106:5151–5158

Mennucci B, Cancès E, Tomasi J (1997) Evaluation of Solvent Effects in Isotropic and Anisotropic Dielectrics and in Ionic Solutions with a Unified Integral Equation Method: Theoretical Bases, Computational Implementation, and Numerical Applications. J Phys Chem B 101:10506–10517

Mermin N (1965) Thermal Properties of the Inhomogeneous Electron Gas. Phys Rev 137:A1441–1443

Methfessel M, Paxton AT (1989) High-precision sampling for Brillouin-zone integration in metals. Phys Rev B 40:3616–3621

Meyer V, Sutter D, Dreizler H (1991) The Centrifugally Induced Pure Rotational Spectrum and the Structure of Sulfur Trioxide A Microwave Fourier Transform Study of a Nonpolar Molecule. Z Naturforsch 46a:710–714

Miehlich B, Savin A, Stoll H, Preuss H (1989) Results obtained with the correlation energy density functionals of Becke and Lee, Yang and Parr. Chem Phys Lett 157:200–206

Miertuš S, Tomasi J (1982) Approximate evaluations of the electrostatic free energy and internal J energy changes in solution processes. Chem Phys 65:239–245

Milankovitch M (1913) Über ein Problem der Warmeleitung und dessen Anwendung auf die Theorie des solaren Klimas. Zeits Math Phys 62:63–77

Miller SL, Urey HC (1959) Organic compound synthesis on the primitive Earth. Science 130:245–251

Millero FJ (1972) The partial molal volumes of electrolytes in aqueous solutions In: Horne RA (series ed), Water and Aqueous Solutions, Wiley Interscience, New York, pp 519–597

Mills R (1973) Self-diffusion in normal and heavy water in the range 1–45 deg. J Phys Chem 77: 685–688

Ming LC, Manghnani MH, Kim YH, Usha-Devi S, Xu JA, Ito E (1992) Thermal expansion studies of (Mg, Fe)2SiO4-spinels using synchrotron radiation In: Saxena SK (Ed) Thermodynamic data, systematics and estimation Advances in Physical Geochemistry, vol 10 Springer, New York, pp 315–334

Mirgorodsky AP, Smirnov MB, Quintard PE, Merle-Méjean T (1995) Strain-induced destabilization of crystals: Lattice dynamics of the cubic-tetragonal phase transition in ZrO_2 . Phys Rev B 52:9111–9114

Moffitt W (1949) The electronic structures of carbon monoxide and carbon dioxide P R Soc London A 196:524–539

Möglestue KT (1968) Exchange integrals in magnetite Fourth IAEA Symposium on Neutron Inelastic Scattering Vol II, International Atomic Energy Agency, Vienna. pp 117–122

Møller, Chr, Plesset, MS (1934) Note on an Approximation Treatment for Many-Electron Systems. Phys Rev 46:618–622

Monkhorst HJ, and Pack JD (1976) Special points for Brillouin-zone integrations. Phys Rev B 13: 5188–5192

Montanari B, Civalleri B, Zicovich-Wilson CM, Dovesi R (2006) Influence of the exchange-correlation functional in all-electron calculations of the vibrational frequencies of corundum (α-Al_2O_3). Int J Quantum Chem 106:1703–1714

Montwill A, Breslin A (2012) Does God Play Dice?, Imperial College Press, London

Mookherjee M, Stixrude L, Karki B (2008) Hydrous silicate melt at high pressure. Nature 458:983–986

Moore PB, Smith JV (1970) Crystal structure of β-Mg_2SiO_4: crystal-chemical and geophysical implications. Phys Earth Planet In 3:166–177

Moretti R, Ottonello G (1998) An appraisal of thermodynamic properties of rare earth garnets. Geochim Cosmochim Acta 62:1147–1173

Moretti R, Ottonello G (2003a) A polymeric approach to the sulfide capacity of silicate slags and melts. Metall Mater Trans B 34:399–410

Moretti R, Ottonello G (2003b) Polymerization and disproportionation of iron and sulfur in silicate melts: Insights from an optical basicity-based approach. J Non-Cryst Sol 323:111–119

Moretti R, Ottonello G (2005) Solubility and speciation of sulfur in silicate melts: the Conjugated Toop-Samis-Flood-Grjotheim (CTSFG) model. Geochim Cosmochim Acta 69:801–823

Moretti R, Ottonello G (2022) Silicate melt thermochemistry and the redox state of magmas Rev Mineral Geochem 87:339–403

Morimoto N, Appleman DE, Evans HT Jr (1959) The crystal st6ructures of clinoenstatite and pigeonite. Z Kristallogr 114:120–147

Morley LW, and Larochelle A (1964) Paleomagnetism as a means of dating geological events. Geochronology in Canada 8:39–51

Morlotti R, Ottonello G (1982) Solution of rare earth elements in silicate solid phases: Henry's law revisited in light of defect chemistry Phys Chem Minerals 8:87–97

Morlotti R, Ottonello G (1984) The solution of trace amounts of Sm in forsterites olivine: an experimental study by EMF galvanic cell measurements. Geochim Cosmochim Acta 48:1173–1181

Morrison TJ, Billett FB (1952) The salting out of non-electrolytes II The effect of variation of non-electrolyte. J Chem Soc (Resumed) 0:3819–3822

Morse PM (1929) Diatomic Molecules according to the Wave Mechanics II Vibrational Levels. Phys Rev 34:57–64

Morton J, Barnes J, Schueler B, Mauersberger K (1990) Laboratory Studies of Heavy Ozone. J Geophys Res 95: 901–907

Morton D C, Wu Q, Drake G W F (2006) Energy Levels for the Stable Isotopes of atomic Helium (^{4}He I and ^{3}He I). Can J Phys 84:83–105

Mosesman MA, Pitzer KS (1941) Thermodynamic properties of the crystalline forms of silica. J Am Chem Soc 63:2348–2356

Mott NF (1949) The Basis of the Electron Theory of Metals, with Special Reference to the Transition Metals. Proc Phys Soc A 62:416–422

Mott NF (1967) Electrons in disordered structures. Adv Phys 16:49–144

Mott NF, Littleton MS (1938) Conduction in polar crystals I Electrolytic conduction in solid salts. Trans Faraday Soc 34:485–499

Mott NF, Peierls R (1937) Discussion of the paper by de Boer and Verwey. Proc Phys Soc 49:72–73

Mott NF, Twose WD (1961) The theory of impurity conduction. Adv Phys 10:107–163

Moynihan CT, Easteal AJ, Wilder J, Tucker J (1974) Dependence of the glass transition temperature on heating and cooling rate. J Phys Chem 78:2673–2677

Muller P (1994) Glossary of Terms used in Physical Organic Chemistry (IUPAC Recommendations 1994). Pure & Appl Chem 66:1077–1184

Müller T, Sotiris SX, Dachsel H, Harrison RJ, Nieplocha J, Shepard R, Kedziora GS, Lischka H (1998) A systematic ab initio investigation of the open and ring structures of ozone. Chem Phys Lett 293:72–80

Müller T, Dallos M, Lischka H, Dubrovay Z, Szalay PG (2001) A systematic theoretical investigation of the valence excited states of the diatomic molecules B_2, C_2, N_2 and O_2. Theor Chim Acta 105:227–243

Mulliken RS (1934) Electronic Structures of Polyatomic Molecules and Valence. Phys Rev 40:55–62

Mulliken RS (1935) Electronic Structures of Molecules XIV Linear Triatomic Molecules, Especially Carbon Dioxide. J Chem Phys 3:720–739

Mulliken RS (1951) Overlap and Bonding Power of 2s,2p-Hybrid Orbitals. J Chem Phys 19:900–912

Munowitz M, Bachovchin WW, Herzfeld J, Dobson CM, Griffin RG (1982) Acid-base and tautomeric equilibriums in the solid state: nitrogen-15 NMR spectroscopy of histidine and imidazole. J Am Chem Soc 104:1192–1196

Murray CN, Riley JP (1971) The solubility of gases in distilled water and sea water - IV Carbon dioxide. Deep Sea Res Ocean Abs 18:533–541

Myers WD, Swiatecki WJ (1966) Nuclear masses and deformations. Nucl Phys 81:1–16

Myhre GD, Shindell FM, Bréon W, Collins J, Fuglestvedt J, Huang D, Koch JF, Lamarque D, Lee B, Mendoza T, Nakajima A, Robock G, Stephens T, Takemura, Zhang H (2013) Anthropogenic and Natural Radiative Forcing In: Stocker TF, Qin D, Plattner G-K, Tignor M, Allen SK, Boschung J, Nauels A, Xia Y, Bex V and Midgley PM (eds). Climate Change 2013: The Physical Science Basis Contribution of Working Group I to the Fifth Assessment Report of the Intergovernmental Panel on Climate Change. Cambridge University Press, Cambridge, United Kingdom and New York, NY, USA. pp 659–740

Nada R, Saunders VR, Pisani C (1993) Super-cell versus embedded cluster simulation of a lithium vacancy in a lithium fluoride monolayer. Chem Phys 169:297–303

Nada R, Nicholas JB, McCarthy MI, Hess AC (1996) Basis sets for ab initio periodic Hartree_Fock studies of zeolite/adsorbate interactions: He, Ne, and Ar in silica sodalite. Int J Quantum Chem 60:809–820

Nakazawa H, Sekine T, Kakegawa T, Nakazawa S (2005) High yield shock synthesis of ammonia from iron, water and nitrogen available on the early Earth. Earth Planet Sc Lett 235:356–360

Namazian M, Halvani S (2006) Calculations of pKa values of carboxylic acids in aqueous solution using density functional theory. J Chem Thermodyn 38:1495–1502

Navrotsky A (1994) Physics and Chemistry of Earth Materials Cambridge Univ Press, Cambridge, UK

Néel ML (1948) Proprietés magnétiques des ferrites; ferrimagnetisme et antiferromagnetisme. Ann Phys 3:137–198

Néel ML (1950) Magnetische Eigenschaften der Ferrite und der Ferrimagnetismus. Zeits Anorg Chemie 262:175–184

Negita K (1989) Lattice vibration and cubic to tetragonal phase transition in ZrO_2. Acta Metall 37:313–317

Nemeth, P, Garvie, LAJ, Aoki, T, Dubrovinskaia N, Dubrovinsky L, Buseck PR (2014) Lonsdaleite is faulted and twinned cubic diamond and does not exist as a discrete material. Nature Comm 5: 5447-1-5

Nery J-P, Calandra M, Mauri F (2021) Ab-initio energetics of graphite and multilayer graphene: stability of Bernal versus rhombohedral stacking. Cond Mat Mtrl Sci, arXiv: 201109614v3

Nesbet RK(1960) Ground State Electronic Wave Function of Methane. J Chem Phys 32:1114–1122

Nestola F, Periotto B, Andreozzi GB, Bruschini E, Bosi F (2014) Pressure-volume equation of state for chromite and magnesiochromite: A single-crystal X-ray diffraction investigation. Amer Min 99:1248–1253

Newmann DB, Moskowitz JW (1968) One-Electron Properties of Near-Hartree–Fock Wavefunctions II HCHO, CO. J Chem Phys 50:2216–2236

Nichols CS, Van de Walle CG, Pantelides ST (1989) Mechanisms of dopant impurity diffusion in silicon. Phys Rev B 8:5484–5496

Nielsen OH, Martin RM (1985) Quantum-mechanical theory of stress and force. Phys Rev B 32: 3780–3791

Nielsen IMB, Seidl ET, Janssen CL (1999) Accurate structures and binding energies for small water clusters: The water trimer. J Chem Phys 110:9435–9442

Nilsson G, Nelin G (1972) Study of the Homology between Silicon and Germanium by Thermal-Neutron Spectrometry. Phys Rev B 6:3777–3786

Nilsson G, Rolandson S (1973) Lattice Dynamics of Copper at 80 K. Phys Rev B 7:2393–2400

NIST (2020) Chemistry Webbook (http://webbooknistgov/chemistry)
Noel Y, Catti M, D'Arco Ph, Dovesi R (2006) The vibrational frequencies of forsterite Mg_2SiO_4: an all-electron ab-initio study with the CRYSTAL code. Phys Chem Miner 33:383–393
Nordlie BE (1971) The composition of the magmatic gas of Kilauea and its behavior in the near surface environment. Amer Jour Sci 271:417–473
Nordstrom DK, Plummer LN, Wigley TML, Wolery TJ, Ball JW, Jenne EA, & Thrailkill J (1979) A comparison of computerized chemical models for equilibrium calculations in aqueous systems. In: Jenne EA (ed), Chemical Modeling in Aqueous Systems, ACS Symposium, Series, v 93, American Chemical Society, Washington, DC, pp 857–892
Nosé S (1984) A unified formulation of the constant temperature molecular dynamics methods. J Chem Phys 81:511–519
Nyden MR, Petersson GA (1981) Complete basis set correlation energies I The asymptotic convergence of pair natural orbital expansions. J Chem Phys 75:1843–1862
Nye JF (1969) Physical properties of crystals: Their representatives by tensors and matrices Clarendon Press, Oxford, UK
Ochterski JW (2000) Thermochemistry in Gaussian, © 2000 Gaussian Inc
Ochterski JW, Petersson GA, Montgomery JA Jr (1996) A complete basis set model chemistry V Extensions to six or more heavy atoms. J Chem Phys 104:2598–2619
Odutola JA, Dyke TR (1980) Partially deuterated water dimers: Microwave spectra and structure. J Chem Phys 72:5062–5070
Oganov AR, Dorogokupets PI (2003) All-electron and pseudopotential study of MgO: Equation of state, anharmonicity, and stability. Phys Rev B 67:224110
Oganov AR, Brodholt JP, Price G D (2001) The elastic constants of$MgSiO_3$ perovskite at pressures and temperatures of the Earth's mantle. Nature 411:934–936
Oh M, Nuckolls KP, Wong D, Lee RL, Liu X, Watanabe K, Taniguchi T, Yazdani A (2021) Evidence for unconventional superconductivity in twisted bilayer graphene. Nature 600:240–245
Ohanian HC (1986) What is spin? Amer Jour Phys 54:500–504
Ohashi Y (1984) Polysynthetically-twinned structures of enstatite and wollastonite Phys Chem Minerals 10:217–229
Ojovan J (2008) Configurons: Thermodynamic Parameters and Symmetry Changes at Glass Transition. Entropy 10:334–364
Ol'shanskii Ya I (1951) Equilibrium of two immiscible liquids in silicate systems of alkali earth metals. Dokl Akad Nauk SSSR 76:93–96
Oleś A M, Khaliullin G, Horsch P, Feiner L F (2005) Fingerprints of spin-orbital physics in cubic Mott insulators: Magnetic exchange interactions and optical spectral weights. Phys Rev B 72: 214431-1-32
Olinger B (1976) The compression of stishovite. J Geophys Res 81:5241–5248
Olinger B, Halleck PM (1976) The compression of α-quartz. J Geophys Res 81:5711–5714
Olney TN, Cann NM, Cooper G, Brion CE (1997) Absolute scale determination for photoabsorption spectra and the calculation of molecular properties using dipole sum-rules. Chem Phys 223:59–98
Olson JR, Pohl RO, Vandersande JW, Zoltan A, Anthony TR, Banholzer WF (1993) Thermal conductivity of diamond between 170 and 1200 K and the isotope effect. Phys Rev B 47:14850–14856
O'Neill HStC, Navrotsky A (1983) Simple spinels: crystallographic parameters, cation radii, lattice energies and cation distribution. Amer Min 68:181–194
Onn D G, Witek A, Qiu Y Z, Anthony T R, Banholzer W F (1992) Some aspects of the thermal conductivity of isotopically enriched diamond single crystals. Phys Rev Lett 68:2806–2809
Ono S, Ohishi Y, Mibe K (2004) Phase transition of Ca-perovskite and stability of Al-bearing Mg-perovskite in the lower mantle. Am Mineral 89:1480–1485
Onsager L (1936) Electric Moments of Molecules in Liquids. J Am Chem Soc 58:1486–1493

O-ohata K, Taketa H, Huzinaga S (1966) Gaussian Expansions of Atomic Orbitals. J Phys Soc Japan 21:2306–2313
Osborn EF (1942) The system $CaSiO_3$-diopside-anorthite. Amer Jour Sci 240:751–788
Osborn EF, Schairer JF (1941) The ternary system pseudowollastonite-akermanite-gehlenite. Amer Jour Sci 239:715–763
Ottonello G (1986) Energetics of multiple oxides with spinel structure. Phys Chem Minerals 13:79–90
Ottonello G (1987) Energies and interactions in binary (Pbnm) orthosilicates: a Born parametrization. Geochim Cosmochim Acta 51:3119–3135
Ottonello G, (1991) Principi di Geochimica. Zanichelli, Bologna
Ottonello G, (1997) Principles of Geochemistry. Columbia University Press, New York-Chichester-West Sussex
Ottonello G (2001) Thermodynamic constraints arising from the polymeric approach to silicate slags: the system CaO-FeO-SiO_2 as an example. J Non-Cryst Solids 282:72–85
Ottonello G (2005) Chemical interactions and configurational disorder in silicate melts. Ann Geophy-Italy 48:561–581
Ottonello G, (2018) Ab initio reactivity of Earth's materials Riv Nuovo Cimento 41:225–288
Ottonello G, Moretti R (1998) On the significance of static interactions in garnets. J Phys Chem Solids 59:893–901
Ottonello G, Moretti R (2004) Lux-Flood basicity of binary silicate melts. J Phys Chem Solids 65: 1609–1614
Ottonello G, Richet P (2014) The solvation radius of silicate melts based on the solubility of noble gases and scaled particle theory. J Chem Phys 140:044506-1-9
Ottonello G, Vetuschi Zuccolini M (2005a) Ab-initio structure, energy and stable Cr isotopes equilibrium fractionation of some geochemically relevant H-O-Cr-Cl complexes. Geochim Cosmochim Acta 69:851–874
Ottonello G, Vetuschi Zuccolini M (2005b) Ab initio thermochemistry of some geochemically relevant molecules in the system Cr-O-H-Cl. Geochim Cosmochim Acta 69:3505–3518
Ottonello G, Vetuschi Zuccolini M (2008) The iron-isotope fractionation dictated by the carboxylic functional: An ab-initio investigation. Geochim Cosmochim Acta 72:5920–5934
Ottonello, G, Vetuschi Zuccolini M (2009) Ab-initio structure, energy and stable Fe isotope equilibrium fractionation of some geochemically relevant H–O–Fe complexes. Geochim Cosmochim Acta 73:6447–6469
Ottonello G, Princivalle F, Della Giusta A (1990) Temperature, composition and fO_2 effects on intersite distribution of Mg and Fe^{2+} in olivines. Phys Chem Minerals 17:301–312
Ottonello G, Bokreta M, Sciuto PF, (1996) Parameterization of energy and interactions in garnets: end-member properties. Am Mineral 81:429–447
Ottonello G, Civalleri B, Vetuschi Zuccolini M, Zicovich-Wilson CM (2007) Ab-initio thermal physics and Cr-isotopic fractionation of $MgCr_2O_4$. Amer Min 92:98–108
Ottonello G, Civalleri B, Ganguly J, Vetuschi Zuccolini M, Noel Y (2009a) Thermophysical properties of α–β–γ polymorphs of Mg_2SiO_4: a computational study. Phys Chem Minerals 36: 87–106
Ottonello G, Vetuschi Zuccolini M, Civalleri B (2009b) Thermo-chemical and thermo-physical properties of stishovite: An ab-initio all-electron investigation. CALPHAD 33:457–468
Ottonello G, Civalleri B, Ganguly J, Perger WF, Belmonte D, Vetuschi Zuccolini M (2010a) Thermo-chemical and thermo-physical properties of the high-pressure phase anhydrous B ($Mg_{14}Si_5O_{24}$): An ab-initio all-electron investigation. Amer Min 95:563–573
Ottonello G, Vetuschi Zuccolini M, Belmonte D (2010b) The vibrational behavior of silica clusters at the glass transition: Ab initio calculations and thermodynamic implications. J Chem Phys 133: 104508-1-11
Ottonello G, Richet P, Vetuschi Zuccolini M (2015) The wet solidus of silica: Predictions from the scaled particle theory and polarized continuum model. J Chem Phys 142:54503-1-9

Ozhabes Y., Gunceler D., and Arias T.A. (2015) Stability and surface diffusion at lithium-electrolyte interphases with connections to dendrite suppression. Cond-mat Mtrl Sci arXiv:1504.05799

Ozoliņš V, Körling M (1993) Full-potential calculations using the generalized gradient approximation: Structural properties of transition metals. Phys Rev B 48:12304–12307

Pacchioni G, Mayer M, Krüger S, Rösch N (1999) Bonding and adsorbate core level shifts of transition metal atoms on the Al(100) surface from density functional calculations. Chem Phys Letters 299:137–144

Palmer RG (1982) Broken ergodicity. Adv Phys 31:669–675

Palmer DC, Downs RT (1991) High pressure behaviour of cristobalite revealed by single crystal X-ray diffraction Eos Trans AGU 72 (44) Fall Meeting Suppl, pp 486–638

Pandey KC (1986) Diffusion without Vacancies or Interstitials: A New Concerted Exchange Mechanism. Phys Rev Letters 57:2287–2290

Papale P (1997) Modeling of the solubility of one component H_2O or CO_2 fluid in silicate liquid. Contrib Mineral Petr 126:237–251

Papale P, Moretti R, Barbato D (2006) The compositional dependence of the saturation surface of $H_2O + CO_2$ fluids in silicate melts. Chem Geol 229:78–95

Parida SC, Rakshit SK, Singh Z (2008) Heat capacities, order–disorder transitions, and thermodynamic properties of rare-earth orthoferrites and rare-earth iron garnets. J Solid State Chem 181:101–121

Park J, Rhee PC (2001) Ionic properties of oxygen in slags. J Non Cryst Solids 282:7–14

Parlinski K, Li ZQ, Kawazoe Y (1997) First-principles determination of the soft mode in cubic ZrO_2. Phys Rev Lett 78:4063–4066

Parsafar G, Mason EA (1994) Universal equation of state for compressed solids. Phys Rev B 49:3049–3060

Pascale F, Zicovich-Wilson CM, López Gejo F, Civalleri B, Orlando R, Dovesi R (2004) The calculation of the vibrational frequencies of crystalline compounds and its implementation in the CRYSTAL code. J Comput Chem 25:888–897

Pascale, F, Zicovich-Wilson CM, Orlando R, Roetti C, Ugliengo P, Dovesi R, Pascual-Ahuir JL, Silla E (2005) Vibration Frequencies of $Mg_3Al_2Si_3O_{12}$ Pyrope. An ab Initio Study with the CRYSTAL Code. J Phys Chem 109:6146–6152

Pasquarello A, Car R (1998) Identification of Raman Defect Lines as Signatures of Ring Structures in Vitreous Silica. Phys Rev B 80:5145–5147

Pasternak MP, Nasu S, Wada K, Endo S (1994) High-pressure phase of magnetite. Phys Rev B 50: 6446–6449

Pastore G, Smargiassi E, Buda F (1991) Theory of ab initio molecular-dynamics calculations. Phys Rev A 44:6334–6347

Paterson MS (1982) The determination of hydroxyl by infrared absorption in quartz, silicate glasses and similar materials. Bull Mineral 105:20–29

Pauling L (1932) Nature of the chemical bond IV The energy of single bonds and the relative electronegativity of atoms. Jour Amer Chem Soc 54:3570–3582

Pauling L (1933) The Calculation of Matrix Elements for Lewis Electronic Structures of Molecules. J Chem Phys 1:280–283

Pauling L (1949) The Nature of the Chemical Bond Cornell University Press, Ithaca, NY

Pauling L (1980) The nature of silicon-oxygen bonds. Amer Min 65:321–323

Pauling L, Bright Wilson E Jr (1935) Introduction to Quantum Mechanics. McGraw-Hill Books, New York

Pauling L, Goudsmit S (1930) The structure of Line Spectra. McGraw–Hill Book Company Inc, New York Toronto London

Pauthenet R (1959) Les proprétés magnetiques des ferrites d'yttrium et des terres rares de formule $5Fe_2O_3{\cdot}3M_2O_3$. Ann Phys 3:424–462

Pavone P, Karch K, Schiitt O, Windl W, Strauch D, Giannozzi P, Baroni S (1993) Ab initio lattice dynamics of diamond. Phys Rev B 48:3156–3163

Payne MC (1989) Error cancellation in the molecular dynamics method for total energy calculations. J Phys Cond Matt 1:2199–2210

Peacor DR (1973) High temperature single-crystal study of the cristobalite inversion. Zeit Krist 138: 274–298

Pearson RG (1986) Ionization potentials and electron affinities in aqueous solution. J Amer Chem Soc 108:6109–6114

Peintinger MF, Vilela Oliveira D, Bredow T (2013) Consistent Gaussian basis sets of triple-zeta valence with polarization quality for solid-state calculations. J Comput Chem 34:45–49

Pekeris CL (1934) The Rotation-Vibration Coupling in Diatomic Molecules. Phys Rev 45:98–103

Pelton A D, Eriksson G, Romero-Serrano A (1993) Calculation of sulfide capacities of multicomponent slags. Metall Trans B 24:817–825

Pénicaud M, Siberchicot B, Sommers CB, Kübler J (1992) Calculated electronic band structure and magnetic moments of ferrites. J Magn Magnetic Mat 103:212–220

Perdew JP, Wang Y (1986) Accurate and simple density functional for the electronic exchange energy: Generalized gradient approximation. Phys Rev B 33:8800–8802

Perdew JP, Wang Y (1992) Accurate and simple analytic representation of the electron-gas correlation energy. Phys Rev B 45:13244–13249

Perdew JP, Zunger A (1981) Self-interaction correction to density-functional approximations for many-electron systems. Phys Rev B 23:5048–5079

Perdew JP, Burke K, Ernzerhof M (1996) Generalized Gradient Approximation Made Simple. Phys Rev Letters 77:3865–3868

Pereira Gomes AS, Custodio R (2002) Exact Gaussian expansions of Slater-type atomic orbitals. J Comput Chem 23:1007–1012

Peterson RC, Lager GA, and Hitterman RL (1991) A time-of-flight neutron powder diffraction study of MgAl2O4 at temperatures up to 1273 K. Amer Min 76:1455–1458

Peterson KA, Figgen D, Goll E, Stoll H, Dolg M (2003) Systematically convergent basis sets with relativistic pseudopotentials II Small-core pseudopotentials and correlation consistent basis sets for the post-d group 16–18 elements. J Chem Phys 119:11113–11123

Petersson GA (1977) The pairwise correlated generalized valence bond model of electronic structure II A simple physical model for electron correlation. Int J Quantum Chem 11:775–812

Petersson GA, Al-Laham MA (1991) A complete basis set model chemistry II Open-shell systems and the total energies of the first-row atoms. J Chem Phys 94:6081–6090

Petersson GA, Frisch MI (2000) A journey from generalized valence bond theory to the full CI complete basis set limit. J Phys Chem A 104:2183–2190

Petersson GA, Licht SL (1981) Complete basis set correlation energies II The beryllium isoelectronic series. J Chem Phys 75:4556–4566

Petersson GA, Nyden MR (1981) Interference effects in pair correlation energies: Helium L-limit energies. J Chem Phys 75:3423–3425

Petersson GA, Yee AK, Bennett A (1985) Complete basis set correlation energies III The total correlation energy of the neon atom. J Chem Phys 83:5105–5128

Petersson GA, Bennett A, Tensfeldt TG, Al-Laham MA, Shirley WA, Mantzaris J (1988) A complete basis set model chemistry I The total energies of dosed-shell atoms and hydrides of the first-row elements. J Chem Phys 89:2193–2218

Petit JR, Jouzel J, Raynaud D, Barkov NI, Barnola JM, Basile I, Bender M, Chappellaz J, Davisk M, Delaygue G, Delmotte M, Kotlyakov VM, Legrand M, Lipenkov VY, Lorius C, Pépin L, Ritz C, Saltzmank E, Stievenard M (1999) Climate and atmospheric history of the past 420,000 years from the Vostok ice core, Antarctica. Nature 399:429–436

Phillips JC (1970) Ionicity in the chemical bond in crystals. Rev Mod Phys 42:317–319

Phillips J, Kleinman L (1959) New Method for Calculating Wave Functions in Crystals and Molecules. Phys Rev 116:287–294

Pick R, Cohen MH, Martin RM (1970) Microscopic Theory of Force Constants in the Adiabatic Approximation. Phys Rev B 1:910–920

Piekarz P, Jochym PT, Parlinski K, Lazewski J (2002) High-pressure and thermal properties of γ-Mg_2SiO_4 from first-principles calculations. J Chem Phys 117:3340–3344
Piekarz P, Parlinski K, Ole´s AM (2006) Mechanism of the Verwey transition in magnetite. Phys Rev Lett 97:156402-1-4
Pierotti RA (1963) The solubility of gases in liquids. J Phys Chem 67:1840–1845
Pierotti RA (1965) Aqueous solutions of nonpolar gases. J Phys Chem 69:281–288
Pierotti RA (1976) A scaled particle theory of aqueous and nonaqueous solutions. Chem Rev 76: 717–726
Pies W, Weiss A, Will G (1979) Crystal structure data of inorganic compounds – Key element: Si Part I In: Hellwege K-H, Hellwege AM (eds), Landolt-Börnstein New Series – Group III Condensed Matter, Vol 7D1α, Springer, Berlin
Pisani C (1996a) Ab-initio treatment of periodic systems In: Pisani C (ed) Quantum-Mechanical Ab-initio calculations of the Properties of Crystalline Materials Lecture Notes in Chemistry, 67, Springer, Berlin-Heidelberg-New York, pp 47–76
Pisani C (1996b) Loss of symmetry in crystals: Surfaces and local defects In: Pisani C (ed) Quantum-Mechanical Ab-initio calculations of the Properties of Crystalline Materials Lecture Notes in Chemistry, 67, Springer, Berlin-Heidelberg-New York, pp 227–244
Pisani C, Dovesi R, Nada R, Kantorovich L N (1990) Ab initio Hartree–Fock perturbed-cluster treatment of local defects in crystals. J Chem Phys 92:7448–7460
Pisani C, Corà F, Nada R, Orlando R (1994a) Hartree—Fock perturbed-cluster treatment of local defects in crystals I The EMBED program: general features. Comp Phys Comm 82:139–156
Pisani C, Corà F, Dovesi R, Orlando R (1994b) Embedded-cluster and super-cell treatment of charged defects in crystals. J Elec Spect 69:1–12
Pisani C, Biurkenheuer F, Corà F, Nada R, Casassa S (1996) EMBED96 User Documentation University of Torino, Torino
Pitzer KS (1973) Thermodynamics of electrolytes I Theoretical basis and general equations. J Phys Chem 77:268–277
Pitzer KS (1983) Dieletric constants of water at very high temperature and pressure. Proc Natl Acad Sci USA 80:4575–4576
Planck M (1900) Zur Theorie des Gesetzes der Energieverteilung im Normalenspektrum. Verh Dtsch Phys Ges 2:237–245
Planck M (1901) Ueber das Gesetz der Energieverteilung im Normalspectrum. Ann Phys 309:553–563
Pliego JR Jr (2003) Thermodynamic cycles and the calculation of pK_a. Chem Phys Lett 367:145–149
Pliego JR Jr, Riveros JM (2002) Gibbs energy of solvation of organic ions in aqueous and dimethyl sulfoxide solutions. Phys Chem Chem Phys 4:1622–1627
Pluschkell W, Engell HJ (1968) Ionen und Elektronenleitung in Magnesiumorthosilikat. Ber Dtsch Keram Ges 45:388–394
Poater J, Duran M, Solà M, Silvi B (2005) Theoretical Evaluation of Electron Delocalization in Aromatic Molecules by Means of Atoms in Molecules (AIM) and Electron Localization Function (ELF) Topological Approaches. Chem Rev 105:3911–3947
Pöhlmann M, Benoit M, Kob1 W (2004) First-principles molecular-dynamics simulations of a hydrous silica melt: Structural properties and hydrogen diffusion mechanism. Phys Rev B 70: 184209-1-14
Poirier JP, Tarantola A (1998) A logarithmic equation of state. Phys Earth Planet In 109:1–8
Polyakov VB, Kharlashina NN (1995) The use of heat capacity data to calculate carbon isotope fractionation between graphite, diamond, and carbon dioxide: A new approach. Geochim Cosmochim Acta 59:2561–2572
Popkie H, Kistenmacher H, Clementi E (1973) Study of the structure of molecular complexes IV The Hartree Fock potential for the water dimer and its application to the liquid state. J Chem Phys 59:1325–1336

Pople JA (1950) The Molecular Orbital Theory of Chemical Valency V The Structure of Water and Similar Molecules. P R Soc London A 202:323–336

Pople JA (1962) I Nuclear Magnetic Resonance in Diamagnetic Materials The Theory of Chemical Shifts. Discuss Faraday Soc 34:7–14

Pople JA, Nesbet RK (1954) Self-Consistent Orbitals for Radicals. J Chem Phys 22:571–572

Pople JA, Binkley J S, Seeger R (1976) Theoretical models incorporating electron correlation. J Quantum Chem S10:1–19

Pople JA, Seeger R, Krishnan R (1977) Variational configuration interaction methods and comparison with perturbation theory. Int J Quantum Chem S11:149–163

Pople JA, Schlegel HB, Krishnan R, Defrees DJ, Binkley JS, Frisch MJ, Whiteside RA, Hout RF, Hehre WJ (1981) Molecular orbital studies of vibrational frequencies. Int J Quantum Chem20, Supplement 15:269–278

Pople JA, Head-Gordon M, Raghavachari K (1987) Quadratic configuration interaction A general technique for determining electron correlation energies. J Chem Phys 87:5968–5975

Pople JA, Head-Gordon M, Fox DJ, Raghavachari K, Curtiss LA (1989) Gaussian-1 theory: A general procedure for prediction of molecular energies. J Chem Phys 90:5622–5629

Porto SPS, Krishnan RS (1967) Raman Effect of Corundum. J Chem Phys 47:1009–1012

Potter RW, Clynne MA (1978) The solubility of noble gases He, Ne, Ar, Kr and Xe in water up to the critical point. J Soln Chem 7:837–844

Presnall DC, Dixon JR, O'Donnell TH, Dixon SA (1979) Generation of Mid-ocean Ridge Tholeiites. J Petrol 20:1–35

Press WH, Flannery BP, Teukolsky SA, Vetterling WT (1986) Numerical Recipes in Fortran 77, 1st Edition, Numerical Recipes: The Art of Scientific Computing Vol 1, Cambridge University Press ,Cambridge, England

Preudhomme J, Tarte P (1971a) Infrared studies of spinels-I A critical discussion of the actual interpretations. Spectrochim Acta 27A:961–968

Preudhomme J, Tarte P (1971b) Infrared studies of spinels-II The experimental bases for solving the assignment problem. Spectrochim Acta 27A:845–851

Preudhomme J, Tarte P (1971c) Infrared studies of spinels-III The normal II-III spinels. Spectrochim Acta 27A:1817–1835

Pugliano N, Saykally RJ (1992) Measurement of the ν_8 intermolecular vibration of $(D_2O)_2$ by tunable far infrared laser spectroscopy. J Chem Phys 96:1832–1839

Putnis A, McConnell JDC (1980) Principles of Mineral Behavior. Blackwell Scientific Publications, Oxford

Qin Q (2014) Green's Functions of Magneto-Electro-Elastic Plate Under Thermal Loading. In: Hetnarski RB (ed), Encyclopedia of Thermal Stresses, Springer, Netherlands, pp 2096–2103

Raffenetti RC (1973) Even-tempered atomic orbitals II Atomic SCF wavefunctions in terms of even-tempered exponential bases. J Chem Phys 59:5936–5949

Raghavachari K, Pople JA (1981) Calculation of one-electron properties using limited configuration interaction techniques. Int J Quant Chem 20:1067–1071

Rappe, AM, Rabe, KM, Kaxiras, E, Joannopoulos, JD (1990) Optimized pseudopotentials. Phys Rev B 41:1227–1230

Rappé AK, Casewit CJ, Colwell KS, Goddard III WA, Skiff WM (1992) UFF, a Full Periodic Table Force Field for Molecular Mechanics and Molecular Dynamics Simulations. J Am Chem Soc 114:10024–10035

Rappoport D, Furche F (2010) Property-optimized Gaussian basis sets for molecular response calculations. J Chem Phys 133:134105-1-11

Rasmussen B, Fletcher JR, Brocks JJ, Kilburn MR (2008) Reassessing the first appearance of eukaryotes and cyanobacteria. Nature 455:1101–1109

Rassolov VA, Pople JA, Ratner MA, and Windus TL (1998) 6-31G∗ basis set for atoms K through Zn. J Chem Phys 109:1223–1229

Rauch M, Keppler H, Häfner W, Poe B, Wokaun A (1996) A pressure-induced phase transition in $MgSiO_3$-rich garnet revealed by Raman spectroscopy. Am Mineral 81:1289–129

Raynes WT, Lazzeretti P, Zanasi R, Sadlej AJ, Fowler PW (1987) Calculations of the force field of the methane molecule. Mol Phys 60:509–525

Raz U (1983) Thermal and Volumetric Measurements on Quartz and Other Substances at Pressures up to 6 kbars and Temperatures up to 700°C. Dissertation, ETH Research Collection, https://doiorg/103929/ethz-a-000306771

Redfern SAT, Harrison RJ, O'Neill HStC, Wood DRR (1999) Thermodynamics and kinetics of cation ordering in $MgAl_2O_4$ spinel from 1600 °C from in situ neutron diffraction. Am Mineral, 84:299–310

Reed EJ, Fried LE, Joannopoulos JD (2003) A method for tractable dynamical studies of single and double shock compression. Phys Rev Lett 90:235503-1-4

Reeves CM, Fletcher R (1965) Use of Gaussian Functions in the Calculation of Wavefunctions for Small Molecules III The Orbital Basis and Its Effect on Valence. J Chem Phys 42:4073–4081

Reimers J, Watts R, Klein M (1982) Intermolecular potential functions and the properties of water. Chem Phys 64:95–114

Rein RH, Chipman J (1965) Activities in the liquid solution SiO_2–CaO-MgO-Al_2O_3 at 1600°. Trans AIME 233:415–425

Reiss, H (1965) Scaled Particle Methods in the Statistical Thermodynamics of Fluids. Adv Chem Phys 9:1–84

Reiss H, Frisch HL, Lebowitz JL (1959) Statistical mechanics of rigid spheres. J Chem Phys 31: 369–380

Reiss H, Frisch HL, Helfand E, Lebowitz JL (1960) Aspects of the statistical mechanics of real fluids. J Chem Phys 32:119–124

Resta R (1985) Microscopic electronic screening in semiconductors. In: Grosse P (ed) Festkörperprobleme Advances in Solid State Physics, Vol 25, Viewer, Braunschweig. pp 183–194

Rettich TR, Battino R, Wilhelm E (1984) Solubility of gases in liquids XVI Henry's law coefficients for nitrogen in water at 5 to 50°C. J Soln Chem 13:335–347

Reyes RA, Gaskell DR (1983) The thermodynamic activity of ZnO in silicate melts. Metall Trans B 17B:805–815

Reynard B (1991) Single crystal infrared reflectivity of pure Mg_2SiO_4 forsterite and $(Mg_{086}Fe_{014})SiO_4$ olivine. Phys Chem Minerals 718:19–25

Reynard B, Rubie D (1996) High-pressure, high-temperature Raman spectroscopic study of ilmenite-type $MgSiO_3$. Am Mineral 81:1092–1096

Rezáč JR, Kevin E, Hobza P (2011) S66: A Well-balanced Database of Benchmark Interaction Energies Relevant to Biomolecular Structures. J Chem Theory Comp 8:2427–2438

Řezníček R, Chlan V, Štěpánková H, Novák P, Żukrowski J, Kozłowski A, Kakol Z, Tarnawski Z, Honig JM (2017) Understanding the Mössbauer spectrum of magnetite below the Verwey transition: Ab initio calculations, simulation, and experiment. Phys Rev B 96:195124-1-10

Rhodes C (2017) Radiation Physics http://www.xylenepower.com/

Richardson FD (1956) Activities in ternary silicate melts. Trans Faraday Soc 52:1312–1324

Richardson OW, Davidson PM (1929) The spectrum of H_2 The bands analogous to the Parhelium line spectrum Part II. P R Soc London A123: 466–488

Richet P (1982) Proprietes Thermodynamiques des silicates fondus. Dissertation, University of Paris VII, Paris, France

Richet NF (2009) Heat capacity and low-frequency vibrational density of states Inferences for the boson peak of silica and alkali silicate glasses. Physica B 404:3799–3806

Richet P (2021) The temperature - CO_2 climate connection: an epistemological reappraisal of ice-core messages. Hist Geo Space Sci 12:97–110

Richet P, Bottinga Y (1984) Glass transitions and thermodynamic properties of amorphous SiO_2, $NaAlSi_nO_{2n+2}$ and $KAlSi_3O_8$. Geochim Cosmochim Acta 48:453–470

Richet P, Bottinga Y (1985) Heat capacity of aluminum-free liquid silicates. Geochim Cosmochim Acta 49:471–486

Richet P, Ottonello G (2014) The Earth as a multiscale quantum-mechanical system. CR Geosci 346:317–325

Richet P, Bottinga Y, Deniélou L, Petitet J P, Téqui C (1982) Thermodynamic properties of quartz, cristobalite and amorphous SiO_2: drop calorimetry measurements between 1000 and 1800 K and a review from 0 to 2000 K. Geochim Cosmochim Acta 46:2639–2658

Richet P, Robie RA, Hemingway BS (1991) Thermodynamic properties of wollastonite, pseudowollastonite and $CaSiO_3$ glass and liquid. Eur J Mineral 3 :475–484

Richet P, Nidaira A, Neuville DR, Atake T, (2009) Aluminum speciation, vibrational entropy and short-range order in calcium aluminosilicate glasses. Geochim Cosmochim Acta 73:3894–3904

Ridgen SM, Jackson I (1991) Elasticity of germanate and silicate spinels at high pressure. J Geophys Res 96:9999–10006

Rienstra-Kiracofe JC, Tschumper GS, Schaefer III HF, Nandi S, Ellison GB (2002) Atomic and Molecular Electron Affinities: Photoelectron Experiments and Theoretical Computations. Chem Rev 102:231–282

Rigby GR, Lovell GHB, Green AT (1946) Some properties of the spinels associated with chrome ores. Trans Br Ceram Soc 45:137–148

Ringwood AE (1958) The constitution of the mantle-II Further data on the olivine-spinel transition. Geochim Cosmochim Acta 15:18–29

Ringwood AE (1962) A Model for the Upper Mantle. J Geophys Res 67:857–867

Ringwood AE, Major A (1967) Some high-pressure transformations of geophysical significance. Earth Planet Sc Lett 2:106–110

Rittner ES, Hutner RA, Du Pré FK (1949) Concerning the work of polarization in ionic crystals of the NaCl type II Polarization around two adjacent charges in the rigid lattice. J Chem Phys 17: 198–203

Robie RA, Bethke PM (1962) Molar Volumes and Densities of Minerals US Geological Survey, Report TEI-882

Robie RA, Hemingway BS, Fisher JR (1978) Thermodynamic properties of minerals and related substances at 29815 K and 1 bar (105 pascal) pressure and at higher temperatures. US Geol Surv Bull, 1452, United States Government Printing office, Washington

Robie RA, Hemingway SB (1995) Thermodynamics properties of minerals and related substances at 29815 K and 1 bar (10^5 Pascal) pressure and at the high temperatures US Geol Surv Bull, 2131, United States Government Printing office, Washington

Robinson HA (1965) On the structure of vitreous SiO_2— I A new pentagonal dodecahedral model. J Phys Chem Solids 26:209–222

Robinson P (1982) The composition space of terrestrial pyroxenes Internal and external limits In: Ribbe PH (series ed), Reviews in Mineralogy, Vol 7, Mineralogical Society of America, pp 419–494

Robinson DB, Senturk NH (1979) The vapor pressure and critical properties of carbonyl sulfide. J Chem Thermodyn 11:461–467

Rocha J (1998) Direct observation of highly distorted hexa-coordinated aluminium in andalusite by very fast ^{27}Al MAS NMR. Chem Comm 22:2489–2490

Rodrigues O (1816) De l'attraction des spheroids. Dissertation for the Faculty of Science of the University of Paris. Correspondence sur l'École Impériale Polytechnique 3:361–385

Roedder E, Wieblen PW (1971) Lunar petrology of silicate melt inclusions, Apollo 11 rocks. Proc Apollo 11 Lunar Sci Conf 1:801–837

Roetti C (1996) The CRYSTAL code In: Pisani C (ed) Quantum-Mechanical Ab-initio calculations of the Properties of Crystalline Materials. Lecture Notes in Chemistry, 67, Springer, Berlin-Heidelberg-New York, pp 125–137

Roothaan CCJ (1951) New developments in Molecular Orbital Theory. Rev Modern Phys 23:69–89

Ropp J, Lawrence C, Farrar T, Skinner J (2001) Rotational motion in liquid water is anisotropic: A nuclear magnetic resonance and molecular dynamics simulation study. J Am Chem Soc 123: 8047–8052

Rosen N (1931) The normal state of the hydrogen molecule. Phys Rev 38:2099–2113

Rosinger W, Grade M, Hirschwald W (1983) *Detection of ion states of S_2 to S_8 by electron impact.* Int J Mass Spectrom 47:239–242

Rosman KJR, Taylor PDP (1998) Isotopic composition of the elements 1997 (Technical Report). Pure & Appl Chem 70:217–235

Ross NL, Shu JF, Hazen RM, Gasparik T (1990) High-pressure crystal chemistry of stishovite. Am Mineral 75:739–747

Ross M, Yang LH, Boehler R (2004) Melting of aluminum, molybdenum, and the light actinides Phys Rev B 70:184112-1-8

Rothman LS, Gamache RR, Tipping RH, Rinsland CP, Smith MAH, Benner DC, Devi VM, Flaud J-M, Camy-Peyret C, Perrin A, Goldman A, Massie ST, Brown LR, Toth RA (1992) The HITRAN molecular database: Editions of 1991 and 1992. J Quant Spectrosc Radiat Transfer 48: 469–507

Rothman LS, Rinsland CP, Goldman A, Massie ST, Edwards DP, Flaud-Perrin A, Camy-Peyret C, Dana V, Mandin J-Y, Schroeder J, McCann A, Gamache RR, Wattson RB, Yoshino K, Chance KV, Jucks KW, Brown LR, Nemtchinov V, Varanasi P (1998) The HITRAN Molecular Spectroscopy Database and HAWKS (HITRAN Atmospheric Workstation): 1996 Edition. J Quant Spectrosc Radiat Transfer 60:665–710

Rowlinson JS (1949) The second virial coefficients of polar gases. Trans Faraday Soc 45:974–984

Rowlinson JS (1951) The lattice energy of ice and the second virial coefficient of water vapour. Trans Faraday Soc 47:120–129

Rüdiger Siebert R, Fleurat-Lessard P, Schinke R, Bittererová M, Farantos SC (2002) The vibrational energies of ozone up to the dissociation threshold: Dynamics calculations on an accurate potential energy surface. J Chem Phys 116:9749–9767

Ruud K, Helgaker T, Bak KL, Poul Joergensen P, Jensen HJA (1993) Hartree–Fock limit magnetizabilities from London orbitals. J Chem Phys 99:3847–3859

Rysselberghe PV (1932) Remarks concerning the Clausius–Mossotti Law. J Phys Chem 36:1152–1155

Salje E, Wruck B, Marais S (1991) Order parameter saturation at low temperatures – Numerical results for displacive and O/D systems. Ferroelectrics 124:185–188

Samuelsen EJ, Steinsvoll O (1974) Low-Energy Phonons in Magnetite. Phys Status Solidi B 61: 615–620

Sánchez-Baracaldo P, Cardona T (2020) On the origin of oxygenic photosynthesis and Cyanobacteria. New Phytologist 225:1440–1446

Santoro RP, Newnham RE, and Nomura S (1966) Magnetic properties of Mn_2SiO_4 and Fe_2SiO_4. J Phys Chem Solids 27:655–666

Santra B, DiStasio RAJr, Martelli F, Car R (2015) Local structure analysis in ab initio liquid water. Mol Phys 113:2829–2841

Saracino GAA, Improta R, Barone V (2003) Absolute pK_a determination for carboxylic acids using density functional theory and the polarizable continuum model. Chem Phys Lett 373:411–415

Saunders VR, Dovesi R, Roetti C, Orlando R, Zicovich-Wilson CM, Harrison NM, Doll K, Civalleri B, Bush IJ, D'Arco P, Llunell M (2003) CRYSTAL03 User's Manual, Università di Torino, Torino

Savrasov SY (1996) Linear-response theory and lattice dynamics: A muffin-tin-orbital approach. Phys Rev B 54:16470–16486

Sawamoto H, Weidner DJ, Sasaki S, Kumazawa M (1984) Single-crystal elastic properties of the modified spinel (beta) phase of Mg_2SiO_4. Science 224:749–751

Saxena SK, Chatterjee N, Fey Y, Shen G (1993) Thermodynamic data on oxides and silicates Springer, Berlin-Heidelberg-New York

Scalmani G, Frisch MJ (2010), Continuous surface charge polarizable continuum models of solvation I General formalism. J Chem Phys 132:114110-1-15

Schaefer L, Fegley B Jr (2007) Outgassing of ordinary chondritic material and some of its implications for the chemistry of asteroids, planets, and satellites. Icarus 186:462–483
Schaefer L, Fegley B Jr (2010) Chemistry of atmospheres formed during accretion of the Earth and other terrestrial planets. Icarus 208:438–448
Schäfer A, Horn H, Ahlrichs R (1992) Fully optimized contracted Gaussian basis sets for atoms Li to Kr. J Chem Phys 97:2571–2577
Schirmeister BE, Sánchez-Baracaldo P, Wacey D (2016) Cyanobacterial evolution during the Precambrian. Int J Astrobiol 15:187–204
Schlegel HB (1982) Optimization of equilibrium geometries and transition structures. J Comput Chem 3:214–218
Schmider HL, Becke AD (1998) Optimized density functionals from the extended G2 test set. J Chem Phys 108:9624–9631
Schneider WG, Bernstein HJ, Pople JA (1958) Proton Magnetic Resonance Chemical Shift of Free (Gaseous) and Associated (Liquid) Hydride Molecules. J Chem Phys 28:601–607
Schober H, Strauch D, Dorner B (1993) Lattice dynamics of sapphire (Al_2O_3). Eur Phys J B Condens Matter Complex Syst 92:273–283
Scholz E (2006) Introducing groups into quantum theory (1926–1930). Hist Math 33:440–490
Schreinemakers FAH (1915a) In-, mono-, and di-variant equilibria I. Proc Sect Sci K Ned Akad Wet 18:116–126
Schreinemakers FAH (1915b) In-, mono-, and di-variant equilibria II Proc Sect Sci K Ned Akad Wet. 18:532–542
Schreinemakers FAH (1915a) In-, mono-, and di-variant equilibria III Proc Sect Sci K Ned Akad Wet. 18:820–828
Schrödinger E (1926) Quantisierung als Eigenwertproblem. Ann Phys 79:489–527
Schubert M, Tiwald TE, Herzinger CM (2000) Infrared dielectric anisotropy and phonon modes of sapphire. Phys Rev B 61:8187–8201
Schumann R Jr, Ensio PJ (1951) Thermodynamics of iron-silicate slags: slags saturated with gamma iron. Trans AIME 191:401–411
Schwartz C (1962) Importance of Angular Correlations between Atomic Electrons. Phys Rev 126: 1015–1019
Schwenke DW, Partridge H (2001) Vibrational energy levels for CH_4 from an ab initio potential. Spectrochim Acta A 57:887–895
Senn MS, Wright JP, Attfield JP (2012) Charge order and three-site distortions in the Verwey structure of magnetite. Nature 481:173–176
Sergeant EP, Dempsey B (1979) Ionization Constants of Organic Acids in Aqueous Solution. Pergamon Press, Oxford, UK
Servoin JL, Piriou B (1973) Infrared reflectivity and Raman scattering of magnesium silicate single crystal. Phys Status Solidi B 55:677–686
Sette F, Krisch MH, Masciovecchio C, Ruocco G, Monaco G (1998) Dynamics of Glasses and Glass-Forming Liquids Studied by Inelastic X-ray Scattering. Science 280:1550–1555
Seward TM (1981) Metal complex formation in aqueous solutions at elevated temperatures and pressures. Phys Chem Earth 13:113–132
Sharma RA, Richardson FD (1962) The solubility of calcium sulphide and activities in lime-silica melts. JISI 200:373–379
Sharma SM, Turneaure SJ, Winey JM, Gupta YM (2020) Transformation of shock-compressed copper to the body-centered-cubic structure at 180 GPa. Phys Rev B 102:020103-1-6
Shaw RW, Harrison WA (1967) Reformulation of the Screened Heine-Abarenkov Model Potential. Phys Rev 163:604–611
Shen G, Lazor P (1995) Measurement of melting temperatures of some minerals under lower mantle pressures. J Geophys Res-Sol Ea 100:17699–17713
Shibata T, Takahashi E, Matsuda JI (1998) Solubility of neon, argon, krypton and xenon in binary and ternary silicate systems: A new view on noble gas solubility. Geochim Cosmochim Acta 62: 1241–1253

Shimamura Ki, Shimojo F, Nakano A, Tanaka S (2016) Meteorite Impact-Induced Rapid NH_3 Production on Early Earth: Ab Initio Molecular Dynamics Simulation. Scientific Reports 6: 38953-1-9

Shimanouchi T (1967) Tables of Molecular Vibration Frequencies, Part I. Publication NSRDS-NBS 39, National Bureau of Standards, Washington, DC

Shinmei, T, Tomioka, N, Fujino, K, Kuroda, K, Irifune, T (1999) In situ X-ray diffraction study of enstatite up to 12 GPa and 1473 K and equations of state. Am Mineral 84:1588–1594

Shiratani E, Sasai M (1998) Molecular scale precursor of the liquid–liquid phase transition of water. J Chem Phys 108:3264–3276

Shock EL, Helgeson HC (1988) Calculations of the thermodynamic and transport properties of aqueous species at hight pressures and temperatures: correlation algorithms for ionic species and equations of state predictions to 5Kb and 1000°C. Geochim Cosmochim Acta 52:2009–2036

Shpil'rain EE, Kagan DN, Barkhatov LS (1972) Thermodynamic properties of the condensed phase of alumina near the melting point. High Temp-High Press 4:605–609

Sievers AJ, Tinkam M (1961) Far-Infrared exchange resonance in Ytterbium Iron Garnet. Phys Rev 124:321–325

Simons JW, Paur RJ, Webster HA III, Bair EJ (1973) Ozone ultraviolet photolysis VI The ultraviolet spectrum. J Chem Phys 59:1203–1208

Singh BP, Srivastava SK, Dinesh K (2004) Relation between thermal expansivity and bulk modulus for ionic solids at high temperatures. Phys B 349:401–407

Sinha MM (1999) Vibrational analysis of optical phonons in mixed cromite spinels. Nucl Inst Meth Phys Res B 153:183–185

Sinn H, Glorieux B, Hennet L, Alatas A, Hu M, Alp EE, Bermejo FJ, Price DL, Saboungi M-L (2003) Microscopic Dynamics of Liquid Aluminum Oxide. Science 299:2047–2049

Skinner BJ (1962) Thermal expansion of ten minerals. U S Geol Surv Res Prof Pap 450-D:D109–D112

Skinner, BJ (1966) Thermal expansions In: Clark SP (ed), Handbook of physical constants, Geological Society of America 97:75–96

Slack GA, Bartram SF (1975) Thermal expansion of some diamondlike crystals. J Appl Phys 46: 89–98

Slater JC (1929) The Theory of Complex Spectra. Phys Rev 34:1293–1322

Slater JC (1930) Atomic Shielding Constants. Phys Rev 36:57–64

Slater JC (1931) Directed Valence in Polyatomic Molecules. Phys Rev 37:481–497

Slater, JC (1937) Wave Functions in a Periodic Potential. Phys Rev 51:846–851

Slater JC, Johnson KH (1972) Self-Consistent-Field Xα Cluster Method for Polyatomic Molecules and Solids. Phys Rev B 5:844–853

Slater JC (1963) Quantum Theories of Molecules and Solids Vol I Electronic Structure of Molecules. McGraw-Hill Book Company Inc, New York - San Francisco – Toronto – London

Slater JC (1974) Quantum Theories of Molecules and Solids Vol 4 The Self-Consistent Field for Molecules and Solids. McGraw-Hill Book Company Inc, New York - San Francisco – Toronto – London

Sleep NH (2010) The Hadean-Achaean environment. Cold Spring Harb Perspect Biol, 2, a002527 https://doi.org/10.1101/cshperspect.a002527

Sleep NH, Zahnle KJ, Neuhoff PS (2001) Initiation of clement surface conditions on the earliest Earth Proc Natl Acad Sci 98:3666–3672

Slonczewski J, Weiss P (1958) Band Structure of Graphite. Phys Rev 109:272–279

Smart JS (1955) The Néel Theory of Ferrimagnetism. Am J Phys 23:356–370

Smirnov NA (2021) Relative stability of Cu, Ag, and Pt at high pressures and temperatures from ab initio calculations. Phys Rev B 103:064107-1-6

Smith DM, Stocker RL (1975) Point defects and non-stoichiometry in fosterite. Phys Earth Planet In 10:183–192

Smyth JR (1987) β-Mg_2SiO_4—A potential host for water in the mantle? Amer Min 72:1051–1055

Smyth JR (1994) A crystallographic model for hydrous wadsleyite (β-Mg_2SiO_4): An ocean in the earth's interior? Amer Min 79:1021–1024
Smyth JR, Bish DL (1988) Crystal Structures and Cation Sites of the Rock-Forming Minerals Allen and Unwin, Boston
Smyth JR, Hazen RM (1973) The crystal structures of forsterite and hortonolite at several temperatures up to 900 C. Amer Min 58:588–593
Smyth JR, Smith JV, Artioli G, Kvick A (1987) Crystal structures of coesite at 15 K and 298 K from neutron and X-ray single crystal data: test of bonding models. J Phys Chem 91:988–992
Solomatova NV, and Caracas R (2021) Buoyancy and Structure of Volatile-Rich Silicate Melts J Geophys Res-Sol Ea 126, e2020JB021045 https://doi.org/10.1029/2020JB021045
Sommerfeld A (1930) Wave – Mechanics, English Translation by Henry Brose, Melhuen & Co LTD, London
Sommerfeld A, Runge J (1911) Anwendung der Vektorrechnung auf die Grundlagen der geometrischen Optik. Ann. Physik, 340, 277–298
Song TRA, Helmberger DV, Grand SP (2004) Low-velocity zone atop the 410-km seismic discontinuity in the northwestern United States. Nature 427:530–533
Soper A, Benmore C (2008) Quantum differences between heavy and light water. Phys Rev Lett 101:065502-1-4
Sosinsky DJ, Sommerville ID (1986) The composition and temperature dependence of the sulfide capacity of metallurgical slags. Metall Trans 17B:331–337
Speziale S, Zha C-S, Duffy TS, Hemley RJ, and Mao H (2001) Quasi-hydrostatic compression of magnesium oxide to 52 GPa: Implications for the pressure-volume-temperature equation of state. J Geophys Res-Sol Ea 106:515–528
Spiesecke H, Schneider WG (1961) Effect of Electronegativity and Magnetic Anisotropy of Substituents on C^{13} and H^1 Chemical Shifts in CH_3X and CH_3CH_2X Compounds. J Chem Phys 35:722–730
Sponer H, Teller E (1941) Electronic Spectra of Polyatomic Molecules Rev Modern Phys 13:75–170
Srivastava SK (2006) Analysis of thermoelastic constants of solids and minerals at high temperatures. J Phys Chem Solids 67:2275–2281
Stacey FD (1977) A thermal model of the Earth. Phys Earth Planet In 15:341–348
Stalder R, Kronz A, Schmidt BC (2009) Raman spectroscopy of synthetic (Mg,Fe)SiO_3 single crystals An analytical tool for natural orthopyroxenes. Eur J Min 21:27–32
Stangarone C (2016) Ab initio calculations of Raman and IR spectra of orthoenstatite and forsterite: lattice dynamics and modelling for planetary remote sensing. Dissertation, University of Parma. wwwrepositoryuniprit › bitstream
Stangarone C, Tribaudino M, Prencipe M, Lottici PP (2016) Raman modes in Pbca enstatite ($Mg_2Si_2O_6$): an assignment by quantum mechanical calculation to interpret experimental results J Raman Spectrosc, 47, 1247–1258
Stearn AE, Eyring H (1937) The deduction of reaction mechanisms from the theory of absolute rates. J Chem Phys 5:113–124
Stebbins JF, Carmichael ISE, Moret LK (1984) Heat capacities and entropies of silicate liquids and glasses. Contrib Mineral Petr 86:131–148
Steinbach O (2008) Numerical Approximation Methods for Elliptic Boundary Value Problems Springer Science + Business Media LLC
Stephens PJ, Devlin FJ, Chabalowski CF, Frisch MJ (1994) Ab initio calculation of vibrational absorption and circular dichroism spectra using density functional force fields. J Phys Chem, 98: 11623−11627
Sternik M, Parlinski K (2005) Lattice vibrations in cubic, tetragonal, and monoclinic phases of ZrO_2 J Chem Phys 122:064707-1,6
Stevens RM, Pitzer RM, Lipscomb WNJ (1963) Perturbed Hartree-Fock Calculations I Magnetic Susceptibility and Shielding in the LiH Molecule. J Chem Phys 38:550–560
Stewart RF (1970a) Small Gaussian Expansions of Slater-Type Orbitals. J Chem Phys 52:431–438

Stewart RF (1970b) Small Gaussian Expansions of Atomic Orbitals: Second-Row Atoms. J Chem Phys 52:5243–5247

Stewart RF, Whitehead MA, Donnay D (1980) The ionicity of the Si–O bond in low-quartz. Amer Min 65:324–326

Stixrude L, Karki B (2005) Structure and Freezing of $MgSiO_3$ Liquid in Earth's Lower Mantle. Science 310:297–299

Stixrude L, Cohen R, Yu R, Krakauer H (1996) Prediction of phase transition in CaSiO3 perovskite and implications for lower mantle structure. Am Mineral 81:1293–1296

Stixrude L, de Koker N, Sun N, Mookherjee M, Karki B (2009) Thermodynamics of silicate liquids in the deep Earth. Earth Planet Sc Lett 278:262–232

Stockmayer WH (1941) Second virial coefficients of polar gases. J Chem Phys 9:398–402

Stover C (2018) Green's Function From MathWorld—A Wolfram Web Resource, created by Eric W Weisstein http://mathworld.wolfram.com/GreensFunction.html

Stover C, Weisstein EW (2018) Polar Coordinates, From MathWorld—A Wolfram Web Resource http://mathworld.wolfram.com/PolarCoordinates.html

Strip KF, Kirkwood JG (1951) Asymptotic expansion of the rotational partition function of the asymmetric top. J Chem Phys 19:1131–1133

Struik, DJ (1969) A Source Book in Mathematics 1200–1800. Harvard University Press Cambridge, Massachusetts

Stull DR (1937) A Semi-micro Calorimeter for Measuring Heat Capacities at Low Temp. J Am Chem Soc 59:2726–2733

Stumm W, Morgan JJ (1981) Acquatic Chemistry. Wiley, New York

Stumpf R, Gonze X, Scheffler M (1990) A list of separable, normconserving ab-initio pseudopotentials. Fritz-Haber-Institute Research Report, unpublished

Sugiura Y (1927) Uber die Eigenschaften des Wasserstoffmoleküls in Grundzustande. Z Phys 45:484–492

Sun T, Shen X, Allen PB (2010) Phonon quasiparticles and anharmonic perturbation theory tested by molecular dynamics on a model system. Phys Rev B 82:224304-1-10

Sun XW, Song T, Chu YD, Liu ZJ, Zhang ZR, Chen QF (2010) The high-pressure melting curve of CaO. Solid State Comm 150:1785–1788

Sun T, Zhang D-B, Wentzcovitch RM (2014) Dynamic stabilization of cubic $CaSiO_3$ perovskite at high temperatures and pressures from ab initio molecular dynamics. Phys Rev B 89:094109-1-9

Suto H, Koike C, Sogawa H, Tsuchiyama A, Chihara H, Mizutani K (2002) Infrared spectra of fayalite crystal. Astron Astrophys 389:568–571

Suzuki I (1975) Thermal expansion of periclase and olivine, and their anharmonic properties. J Phys Earth 23:145–159

Suzuki I, Ohtani E, Kumazawa M (1979) Thermal expansion of γ-Mg_2SiO_4. J Phys Earth 27:53–61

Suzuki I, Ohtani E, Kumazawa M (1980) Thermal expansion of modified spinel β-Mg_2SiO_4. J Phys Earth 28:273–280

Swainson IP, Dove MT (1995) On the thermal expansion of b-cristobalite. Phys Chem Miner 22: 61–65

Swamy V, Dubrovinsky LS (1997) Thermodynamic data for the phases in the $CaSiO_3$ system, Geochim Cosmochim Acta 61:1181–1191

Tan H, Dai CD, Zhang LY, Xu CH (2005) Method to determine the melting temperatures of metals under megabar shock pressures. Appl Phys Lett 87:221905-1-3

Tanaka T, Morino Y (1970) Coriolis interaction and anharmonic potential function of ozone from the microwave spectra in the excited vibrational states. J Mol Spect 33:538–551

Tanaka S, Sawamoto H, Fujimura A, Akamatsu T, Hashizume H, Shimomura O (1987) Precise measurement of compressibility of β–Mg_2SiO_4 using synchrotron radiation. Paper presented at the 28th High Pressure Conference Japan, Kobe

Tange Y, Nishihara Y, and Tsuchiya T (2009) Unified analyses for P–V–T equation of state of MgO: A solution for pressure-scale problems in high P–T experiments. J Geophys Res-Sol Ea 114:B03208-1-16

Tanger JC, Helgeson HC (1988) Calculation of the thermodynamic and transport properties of aqueous species at high pressures and temperatures: revised equation of state for the standard partial molal properties of ions and electrolytes. Amer Jour Sci 288:19–98

Taylor B (1715) Methodus Incrementorum Directa et Inversa (Prop VII, Thm 3, Cor 2) London

Taylor D (1984) Thermal expansion IV Binary oxides with the silica structures. Br Ceram Trans J 83:129–134

Temkin M (1945) Mixtures of fused salts as ionic solutions. Acta Phys Chim URSS 20:411–420

Teter DM, Gibbs GV, Boisen MB Jr, Allan DC, Teter MP (1995) First-principles study of several hypothetical silica framework structures. Phys Rev B 52:8064–8073

Thakkar AJ, Wu T (2015) How well do static electronic dipole polarizabilities from gas-phase experiments compare with density functional and MP2 computations. J Chem Phys 143:144302-1-8

Thibaudeau P, Gervais F (2002) Ab initio investigation of phonon modes in the $MgAl_2O_4$ spinel. J Phys-Condens Matt 14:3543–3452

Thomas LH (1927) The calculation of atomic fields. Proc Camb Phil Soc 23:542–548

Thompson P, Grimes NW (1977) Madelung calculations for the spinel structure. Phil Mag 36:501–505

Thompson AB, Wennemer M (1979) Heat capacities and inversions in tridymite, cristobalite and tridymite-cristobalite mixed phases. Am Mineral 64:1018–1026

Tinkam, M (1961) Low-lying spectrum of Rare-Earth iron garnets. Phys Rev 124:311–320

Tissandier MD, Cowen KA, Feng WY, Gundlach E, Cohen MH, Earhart AD, Coe JV, Tuttle TR (1998) The Proton's Absolute Aqueous Enthalpy and Gibbs Free Energy of Solvation from Cluster-Ion Solvation Data. J Phys Chem 102:7787–7794

Tolba SA, Gameel KM, Ali BA, Almossalami HA, Allam NK (2018) The DFT+U: Approaches, Accuracy, and Applications. In: Density Functional Calculations - Recent Progresses of Theory and Application. Gang Yang G (ed) 1–30, IntechOpen

Tomasi J, Bonaccorsi R (1987) Electrostatic Interaction of a Solute with a Continuum Improved Description of the Cavity and of the Surface Cavity Bound Charge Distribution. J Comput Chem 8:778–787

Tomasi J, Persico M (1994) Molecular Interactions in Solution: An Overview of Methods Based on Continuous Distributions of the Solvent. Chem Rev 94:2027–2094

Tomasi J, Mennucci B, Cancès E, (1999) The IEF version of the PCM solvation method: an overview of a new method addressed to study molecular solutes at the QM ab initio level. J Mol Struct (Theochem) 464:211–226

Toop GW, Samis CS (1962a) Some new ionic concepts of silicate slags. Can Met Quart 1:129–152

Toop GW, Samis CS (1962b) Activities of ions in silicate melts. Trans Met Soc AIME 224:878–887

Topol IA, Tawa GJ, Burt SK, Rashin AA (1999) On the structure and thermodynamics of solvated monoatomic ions using a hybrid solvation model. J Chem Phys 111:10998–11014

Tosi MP (1964) Cohesion of ionic solids in the Born model. Solid State Phys 16:1–120

Tossell JA (1975) The electronic structures of Mg, Al and Si in octahedral coordination with oxygen from SCF Xα calculations. J Phys Chem Solids 36:1373–1280

Tossell JA (1976) SCF-Xα studies of the electron structures of C, Si and Ge oxides. J Phys Chem Solids 37:1043–1050

Tossell JA (1977) SCF-Xα scattered wave MO studies of the electronic structure of ferrous iron in octahedral coordination with sulfur. J Chem Phys 66:5712–5719

Tossell JA (1978) Self-consistent field Xα study of one electron energy levels in Fe_3O_4. Phys Rev B 17:484–487

Tossell JA, Vaughan DJ (1992) Theoretical Geochemistry: Applications of Quantum Mechanics in the Earth and Mineral Sciences Oxford University Press, New York & Oxford

Tossell JA, Vaughan DJ, Johnson KH (1973) X-ray photoelectron, X-ray emission and UV spectra of SiO_2 calculated by the SCF Xα scattered wave method. Chem Phys Letters 20:329–334
Trojer FJ (1969) The crystal structure of a high-pressure polymorph of $CaSiO_3$. Z Kristallogr 130: 185–206
Troullier N, Martins JL (1991) Efficient pseudopotentials for plane-wave calculations. Phys Rev B 43:1993–2006
Tsuchiya T, Kawamura K (2001) Systematics of elasticity: Ab initio study in B1-type alkaline earth oxides. J Chem Phys 114:10086–10093
Turekian KK (1969) The oceans, streams and atmosphere In: Wedepohl KH (ed), Handbook of Geochemistry, Vol 1, Springer, Berlin-Heidelberg-New York, pp 297–323
Tuzson B (2005) Symmetry Specific Study of Ozone Isotopomer Formation, Dissertation, Rupertus Carola University, Heidelberg, Germany, http://archive.ub.uni-heidelberg.de/
Tyuterev VG, Tashkun SA, Jensen P, Barbe A, Cours T (1999) Determination of the Effective Ground State Potential Energy Function of Ozone from High-Resolution Infrared Spectra. J Mol Spec 198:57–76
Tyuterev VG, Tashkun SA, Schwenke DW, Jensen P, Cours T, Barbe A, Jacon M (2000) Variational EKE-calculations of rovibrational energies of the ozone molecule from an empirical potential function. Chem Phys Lett 316:271–279
Ugliengo P, Viterbo D, Chiari G (1993) MOLDRAW: Molecular graphics on a personal computer. Z Cristallogr 207:9–23
Uhl M, Siberchicot B (1995) A first-principles study of exchange integrals in magnetite. J Phys-Condens Matt 7:4227–4237
Uhlembeck GE, Goudsmit S (1925) Ersetzung der Hypothese vom unmechanischen Zwang durch eine Forderung bezfiglich des inneren Verhaltens jedes einzelnen Elektrons. Naturwissenshaften 13:953–954
Uhlembeck GE and Goudsmit S (1926) Spinning Electrons and the Structure of Spectra. Nature 117:264–265
Umari P, Gonze X, and Pasquarello A (2003) Concentration of Small Ring Structures in Vitreous Silica from a First-Principles Analysis of the Raman Spectrum. Phys Rev Lett 90:027401-1-4
Urbain G, Bottinga Y, Richet P (1982) Viscosity of liquid silica, silicates and alumino-silicates. Geochim Cosmochim Acta 46:1061–1072
Urey HC (1951) The origin and development of the earth and other terrestrial planets. Geochim Cosmochim Acta 1:209–277
Urey HC, Brickwedde FG, Murphy GM (1932) An isotope of hydrogen of mass 2 and its concentration (abstract). Phys Rev 39:864
Vaidya SN, Bailey S, Pasternack T, Kennedy GC (1973) Compressibility of fifteen minerals to 45 kilobars. J Geophys Res 78:6893–6898
Valentini JJ (1987) Mass-independent isotopic fractionation in nonadiabatic molecular collisions. J Chem Phys 12:6757–6765
Valiev M, Bylaska EJ, Govind N, Kowalski K, Straatsma TP, van Dam HJJ, Wang D, Nieplocha J, Apra E, Windus TL, and de Jong WA (2010) NWChem: A comprehensive and scalable open-source solution for large scale molecular simulations. Comp Phys Comm 181:1477–1489
Vallance Jones A (1971) Auroral Spectroscopy. Sci Rev 11:776–826
Van der Hoff BME, Benson GC (1953) A method for the evaluation of some lattice sums occurring in calculations of physical properties of crystals. Can J Phys 31:1087–1094
Van der Hoff BME, Benson GC (1954) A Quantum-Mechanical Calculation of the Surface Energy of Crystalline Lithium Fluoride. J Chem Phys 22:475–480
Van der Wal RJ, Vos A, Kirfel A (1987) Conflicting results for the deformation properties of Forsterite, Mg_2SiO_4. Acta Cryst 43:132–143
Van Vleck JH (1929) The new quantum mechanics. Chem Rev 4:467–507
Van Vleck, JH (1932) Theory of the Variations in Paramagnetic Anisotropy Among Different Salts of the Iron Group. Phys Rev 41:208–215

Van Vleck JH (1978) Quantum Mechanics: The Key to Understanding Magnetism. Science 210: 113–120

Van Gunsteren WF, Berendsen HJC (1990) Computer Simulation of Molecular Dynamics: Methodology, Applications, and Perspectives in Chemistry. Angew Chem Int Ed Engl 29:992–1023

Van Vleck JH, Cross PC (1934) A Calculation of the Vibration Frequencies and Other Constants of the H_2O Molecule. J Chem Phys 1:357–361

Vanderbilt D (1990) Soft self-consistent pseudopotentials in a generalized eigenvalue formalism. Phys Rev B 41:7892–7895

Vayssilov GN, Rösch N (1999) Density Functional Studies of Alkali-Exchanged Zeolites: Basicity and Core-Level Shifts of Framework Oxygen Atoms. J Catal 186:423–432

Vayssilov GN, Staufer M, Belling T, Neyman KM, Knozinger H, Rösch N (1999) Density Functional Studies of Alkali-Exchanged Zeolites Cation Location at Six-Rings of Different Aluminum Content. J Phys Chem B 103:7920–7928

Vayssilov GN, Lercher JA, Rösch N (2000a) Interaction of Methanol with Alkali Metal Exchanged Molecular Sieves 2 Density Functional Study. J Phys Chem B 104:8614–8623

Vayssilov GN, Hu A, Birkenheuer U, Rösch N (2000b) Dinitrogen as probe molecule of alkali-exchanged zeolites: A density functional study. J Mol Cat A Chem 162:135–145

Velde G, Bickelhaupt FM, Baerends EJ, Fonseca Guerra C, van Gisbergen SJA, Snijders JG, Ziegler T (2001) Chemistry with ADF. J Comput Chem 22:931–967

Venuti E, Halonen L, Della Valle RG(1999) High dimensional anharmonic potential energy surfaces: The case of methane. J Chem Phys 110:7339–7347

Verwey, EJW (1939) Electronic conduction of magnetite (Fe_3O_4) and its transition point at low temperatures. Nature 144:327–328

Verwey EJW (1941) The charge distribution in the water molecule and calculation of the intermolecular forces. Recl Trav Chim Pays Bas Belg 60:887–896

Verwey EJ, Haayman PW, Romeijn FC (1947) Physical Properties and Cation Arrangement of Oxides with Spinel Structures II Electronic Conductivity. J Chem Phys 15:181–187

Vetuschi Zuccolini M, Ottonello G, Belmonte D (2011) Ab-initio assessment of conventional standard-state thermodynamic properties of geochemically relevant gaseous and aqueous species. Comp Geosci 37:646–661

Viellard A, Clementi E (1968) Correlation Energy in Atomic Systems V Degeneracy Effects for the Second-Row Atoms. J Chem Phys 49:2415–2412

Vine FJ, Matthews DH (1963) Magnetic Anomalies Over Oceanic Ridges. Nature 199:947–949

Viterbo D (1996) Crystal Lattice and Crystal Symmetry In: Pisani C (ed) Quantum-Mechanical Ab-Initio Calculation of the Properties of Crystalline Materials, Lecture Notes in Chemistry, 67, Springer, Berlin-Heidelberg-New-York pp 1–29

Vočadlo L, Brodholt J, Alfè D, Gillan MJ, and Price GD (2000) Ab initio free energy calculations on the polymorphs of iron at core conditions. Phys Earth Planet In 117:123–137

Vočadlo L, Alfé D, Gillan MJ, Price GD (2003) The properties of iron under core conditions from first principles calculations. Phys Earth Planet In 140:101–125

Vočadlo L, Alfè D, Price GD, Gillan M J (2004) Ab initio melting curve of copper by the phase coexistence approach. J Chem Phys 120:2872–2878

von Barth U, Hedin L (1972) A local exchange-correlation potential for the spin polarized case: I. J Phys C Solid State 5:1629–1652

Vosko S H, Wilk L, and Nusair M (1980) Accurate spin-dependent electron liquid correlation energies for local spin density calculations: a critical analysis. Can J Phys 58:1200–1211

Vo-Thanh D, Bottinga Y, Polian A, Richet P (2005) Sound velocity in alumino-silicate liquids determined up to 2550K from Brillouin spectroscopy: glass transition and crossover temperatures. J Non-Cryst Solids 351:61–68

Vuilleumier R, Sator N, Guillot B (2011) Electronic redistribution around oxygen atoms in silicate melts by ab initio molecular dynamics simulation. J Non-Cryst Solids 375:2555–2561

Wachters A J H (1970) Gaussian Basis Set for Molecular Wavefunctions Containing Third-Row Atoms. J Chem Phys 52:1033–1036

Wadt WR, Hay PJ (1985) Ab initio effective core potentials for molecular calculations Potentials for K to Au including the outermost core orbitals. J Chem Phys 82:299–310

Wagman, DD, Evans, WH, Parker, VB, Schumm, RH, Halow, I, Bailey, SM, Churney, KL and Nuttall, RL (1982) The NBS tables of chemical thermodynamic properties Selected values for inorganic and C1 and C2 organic substances in SI units J Phys Chem Ref Data, 11, Supplement N2

Wagner SC (2011) Biological nitrogen fixation Nat Edu Knowledge, 3, 15 http://wwwnaturecom/scitable/knowledge/library/biological-nitrogen-fixation-23570419

Waldron RD (1955) Infrared spectra of ferrites. Physical Rev 99:1727–1735

Wallace PR (1947) The Band Theory of Graphite. Phys Rev 71:622–634

Wallace DC (1972) Thermodynamics of Crystals Wiley, New York

Wallace DC (2002) Statistical Physics of Crystals and Liquids World Scientific, Singapore

Wang SC (1928) The problem of the normal hydrogen molecule in the new quantum mechanics. Phys Rev 31:579–586

Wang Y, Perdew JP (1991) Spin scaling of the electron-gas correlation energy in the high-density limit. Phys Rev B 43:8911–8916

Wang Y, Weidner DJ, Guyot F (1996) Thermal equation of state of $CaSiO_3$ perovskite. J Geophys Res – Sol Ea 101:661–672

Wang Y, Uchida T, Zhang J, Rivers ML, Sutton SR (2004) Thermal equation of state of akimotoite $MgSiO_3$ and effects of the akimotoite–garnet transformation on seismic structure near the 660 km discontinuity. Phys Earth Planet In 143–144:57–80

Wannier GH (1937) The structure of electronic excitation levels in insulating crystals. Phys Rev 52: 191–197

Ward A, Broido D, Stewart D, Deinzer G (2009) Ab initio theory of the lattice thermal conductivity in diamond. Phys Rev B 80:125203-1-8

Warren J L, Yarnell J L, Dolling G, Cowley R A (1967) Lattice Dynamics of Diamond. Phys Rev 158:805–808

Wasserman E, Yuen DA, and Rustad JR (1993) Compositional effects on the transport and thermodynamic properties of MgO-SiO2 mixtures using molecular dynamics. Phys Earth Planet In 77:189–193

Watanabe H (1982) Thermochemical properties of synthetic high-pressure compounds relevant to the Earth's mantle In: Akimoto H, Manghnani MH (eds) High Pressure Research in Geophysics. Center for Academic Publications, Tokio, pp 441–464

Watanabe H, Brockhouse BN (1962) Observation of optical and acoustical magnons in magnetite. Phys Lett 1:189–190

Watson RE (1958) Analytic Hartree-Fock Solutions for $O^{=}$. Phys Rev 111:108–110

Watson JN, Craven IE, Ritchie GLD (1997) Temperature dependence of electric field-gradient induced birefringence in carbon dioxide and carbon disulfide. Chem Phys Lett 274:1–6

Watts RO (1977) An accurate potential for deformable water molecules. Chem Phys 26: 367–377

Wawrzyniak P K (2011) Ab initio modeling of primary processes in photosynthesis Protein induced activation of bacteriochlorophylls for efficient light harvesting and charge separation, Dissertation, Leiden University

Wawrzyniak PK, Alia A, Schaap RG, Heemskerk MM, de Groot HJM, Buda F (2008) Protein-induced geometric constraints and charge transfer in bacteriochlorophyll-histidine complexes in LH2. Phys Chem Chem Phys 10:6971–6978

Wei L, Kuo PK, Thomas RL, Anthony TR, Banholzer WF (1993) Thermal conductivity of isotopically modified single crystal diamond. Phys Rev Lett 70:3764–3767

Weidner DJ, Bass JS, Ringwood AE, Sinclair W (1982) The single-crystal elastic moduli of stishovite. J Geophys Res 87 :4740–4746

Weidner DJ, Sawamoto H, Sasaki S, Kumazawa M (1984) Single-crystal elastic properties of the spinel phase of Mg_2SiO_4. J Geophys Res 89:7852–7860

Weigend F, Ahlrichs R (2005) Balanced basis sets of split valence, triple zeta valence and quadruple zeta valence quality for H to Rn: Design and assessment of accuracy. Phys Chem Chem Phys 7:3297–3305

Weinert M (1981) Solution of Poisson's equation: Beyond Ewald-type methods. J Math Phys 22: 2433–2439

Weinert M, Wimmer E, Freeman AJ (1982) Total-energy all-electron density functional method for bulk solids and surfaces. Phys Rev B 26:4571–4578

Weiss P (1907) L'hypothèse du champ moléculaire et la propriété ferromagnétique. Journal de Physique Théorique et Appliquée 6:661–690

Weiss RF (1974) Carbon dioxide in water and seawater: the solubility of a non-ideal gas. Mar Chem 2:203–215

Weisstein EW (2018) Cylindrical Coordinates From MathWorld—A Wolfram Web Resource, hhttp://mathworld.wolfram.com/CylindricalCoordinates.html

Weisstein EW (2020) Delta Function From MathWorld—A Wolfram Web Resource, http://mathworld.wolfram.com/DeltaFunction.htm

Wendler K, Thar J, Zahn S, Kirchner B (2010) Estimating the Hydrogen Bond Energy. J Phys Chem A 114:9529–9536

Wenthold PG, Jonas KL, Lineberger WC (1997) Ultraviolet photoelectron spectroscopy of the chromium dioxide negative ion. J Chem Phys 106:9961–9962

Wentzcovitch R M (1991) Invariant molecular-dynamics approach to structural phase transitions. Phys Rev B 44:2358–2361

Wentzcovitch RM, Martins JL (1991) First principles molecular dynamics of Li: Test of a new algorithm. Solid State Commun 78:831–834

Wentzcovitch RM, Karki BB, Cococcioni M, de Gironcoli S (2004) Perovskite: Insights on the Nature of the Earth's Lower Mantle. Phys Rev Lett 92:018501-1-4

Wentzcovitch RM, Wu Z, and Carrier P (2010) First principles quasiharmonic thermoelasticity of mantle minerals In: Wentzcovitch RM and Stixrude L (eds) Theoretical and Computational Methods in Mineral Physics: Geophysical Applications. Reviews in Mineralogy and Geochemistry Vol 71, Mineralogical Society of America and Geochemical Society Chantilly, VA, USA, pp 99–128

Werthamer NR (1969) Light Scattering from Solid Helium. Phys Rev B 1:572–581

Westrum EF Jr (1956) The Low Temperature Heat Capacity of Neutron Irradiated Quartz Proc IVe Congrés International du Verre, Paris, pp 396–399

White WB, De Angelis BA (1967) Interpretation of the vibrational spectra of spinels. Spectrochim Acta 23A:985–995

Whitfield M (1975a) An improved specific interaction model for sea water at 25°C and 1 atmosphere total pressure. Mar Chem 3:197–213

Whitfield M (1975b) The extension of chemical models for sea water to include trace components at 25°C and 1 atm Pressure. Geochim Cosmochim Acta 39:1545–1557

Wickersheim, KA (1961) Spectroscopic study of the ytterbium-iron exchange interaction in ytterbium iron garnet. Phys Rev 122:1376–1381

Wigner E (1927) Einige Folgerungen aus der Schrödingerschen Theorie für die Termstrukturen. Zeits Phys A 43:624–652

Wigner E, Seitz F (1933) On the Constitution of Metallic Sodium. Phys Rev 43:804–810

Wigner E Seitz F (1934) On the Constitution of Metallic Sodium II. Phys Rev 46:509–524

Wilhelm E, Battino R, Wilcock RJ (1976) Low-pressure solubility of gases in liquid water. Chem Rev 77:219–262

Williams RW, Malhotra D (2006) van der Waals corrections to density functional theory calculations: Methane, ethane, ethylene, benzene, formaldehyde, ammonia, water, PBE, and CPMD. Chem Phys 327:54–62

Wilson CW, Hopper DG (1981) Theoretical studies of the ozone molecule I Ab initio MCSCF/CI potential energy surfaces for the X 1A1 and a 2B2 states. J Chem Phys 74:595–607

Wimmer E, Krakauer H, Weinert M, and Freeman AJ (1981) Full-potential self-consistent linearized-augmented-plane-wave method for calculating the electronic structure of molecules and surfaces: O_2 molecule. Phys Rev B 24:864–875

Wolery TJ (1983) EQ3NR A Computer Program for Geochemical Aqueous Speciation-Solubility Calculations: User's Guide and Documentation Lawrence Livermore Laboratory, Livermore, CA, UCRL-53414

Wolinski K, Hinton JF, Pulay P (1990) Efficient implementation of the gauge-independent atomic orbital method for NMR chemical shift calculations. J Am Chem Soc 112:8251–8260

Wong MW, Frisch MJ, Wiberg KB (1991) Solvent effects 1 The mediation of electrostatic effects by solvents. J Am Chem Soc 113:4776–4782

Woods RD, and Saxon DS (1954) Diffuse Surface Optical Model for Nucleon-Nuclei Scattering. Phys Rev 95:577–578

Woolley HW (1980) Thermodynamic properties for H_2O in the ideal gas state. In: Straub J, Scheffler K (eds) Proceedings of the Ninth International Conference on the Properties of Steam, Munich, 1979. Pergamon Press, Oxford, New York, pp. 166–175

Woon DE, Dunning TH Jr (1993) Gaussian basis sets for use in correlated molecular calculations III The atoms aluminum through argon. J Chem Phys 98:1358–1371

Woon DE, Dunning TH Jr (1995) Gaussian basis sets for use in correlated molecular calculations IV Calculation of static electrical response properties. J Chem Phys 100:2975–2988

Wright JP, Attfield JP, Radaelli PG (2001) Long Range Charge Ordering in Magnetite Below the Verwey Transition. Phys Rev Lett 87:266401-1-4

Wu Z, Wentzcovitch RM (2007) Vibrational and thermodynamic properties of wadsleyite: a density functional study. J Geophys Res 112:B12202-1-11

Wu Z, Cohen RE (2006) Generalized Gradient Approximation Made More Accurate for Solids. Phys Rev B 73:235116-1-6

Wu, CC, Mason TO (1981) Thermopower measurement of cation distribution in magnetite. J Amer Ceram Soc 64:520–522

Wu Z, Wentzcovitch RM, Umemoto K, Li B, Hirose K, Zheng J-C (2008) Pressure-volume-temperature relations in MgO: An ultrahigh pressure-temperature scale for planetary sciences applications. J Geophys Res-Sol Ea 113:B06204-1-12

Wyckoff R (1963) Crystal Structures I, Second Edition Interscience Publishers New York, NY

Xiao L, Li X, Yang X (2018) Electronic and optical properties of Fe_2SiO_4 under pressure effect: ab initio study. Eur Phys J B 91:85-1-7

Xie D, Guo H, Peterson KA (2000) Accurate ab initio near-equilibrium potential energy and dipole moment functions of the ground electronic state of ozone. J Chem Phys 112:8378–8386

Xie Z, Tomioka N, Sharp T G (2002) Natural occurrence of Fe_2SiO_4-spinel in the shocked Umbarger L6 chondrite. Am Mineral 87:1257–1260

Yagi T, Akaogi M, Shimomura O, Suzuki T, and Akimoto S (1987) In Situ Observation of the Olivine-Spinel Phase Transformation in Fe_2SiO_4 Using Synchrotron Radiation. J Geophys Res B92:6207–6213

Yamanaka T, Kyono A, Nakamoto Y, Kharlamova S, Struzhkin VV, Gramsch SA, Mao H, Hemley RJ (2015) New structure of high-pressure body-centered orthorhombic Fe2SiO4. Amer Min 100: 1736–1743

Yamashita K, Morokuma K, Le Quéré , Leforestier C (1992) New ab initio potential surfaces and three-dimensional quantum dynamics for transition state spectroscopy in ozone photodissociation. Chem Phys Lett 191:515–520

Yamazaki D, Karato SI (2001) Some mineral physics constraints on the rheology and geothermal structure of Earth's lower mantle. Am Mineral 86:385–391

Yanase A, Siratori K (1984) Band Structure in the High Temperature Phase of Fe_3O_4. J Phys Soc Jpn 53:312–317

Yang J, Epstein S (1987a) The effect of the isotopic composition of oxygen on the non-mass dependent isotopic fractionation in the formation of ozone by discharge of O_2. Geochim Cosmochim Acta 51:2011–2017

Yang J, Epstein S (1987b) The effect of pressure and excitation energy on the isotopic fractionation in the formation of ozone by discharge of O_2. Geochim Cosmochim Acta 51:2019–2024

Yang H, Prewitt CT (1999a) On the crystal structure of pseudowollastonite ($CaSiO_3$). Amer Mineral 84:929–932

Yang H, Prewitt CT (1999b) Crystal structure and compressibility of a two-layer polytype of pseudowollastonite (CaSiO3). Amer Mineral 84:1902–1905

Yoneda A (1990) Pressure derivatives of elastic constants of single crystal MgO and $MgAl_2O_4$. J Phys Earth 38:19–55

Yong W, Dachs E, Withers AC (2007) Heat capacity of γ-Fe_2SiO_4 between 5 and 303 K and derived thermodynamic properties. Phys Chem Minerals 34:121–127

Yong WJ, Botis S, Shieh SR, Shi WG, Withers AC (2012) Pressure-induced phase transition study of magnesiochromite ($MgCr_2O4$) by Raman spectroscopy and X-ray diffraction. Phys Earth Planet In 196–197:75–82

Yoshimori A, Kitano Y (1956) Theory of the lattice vibration of graphite. J Phys Soc Japan 11:352–361

Young CL (1983) IUPAC Solubility Data Series Volume 12 Pergamon Press Ltd, Headington Hill Hall, Oxford OX3 OBW, England, ISBN 0-08·026218·X

Young JA, Koppel JU (1965) Phonon spectrum of graphite. J Chem Phys 42:357–364

Young RW, Duffy JA, Hassall GJ, Xu Z (1992) Use of optical basicity concept for determining phosphorous and sulphur slag–metal partitions. Ironmaking Steelmaking 19:201–219

Young, TE, Green II HW, Hofmeister AM, Walker D (1993) Infrared spectroscopic investigation of hydroxyl in β-$(Mg,Fe)_2SiO_4$ and coexisting olivine: Implications for mantle evolution and dynamics. Phys Chem Minerals 19:409–422

Yu X, Dongbin D (2014) Ab initio Calculations of Electron-Impact Excitation Cross Sections for N_2 Molecule. Plasma Sci Tec 16:104–109

Yu YG, Wentzcovitch RM (2006) Density functional study of vibrational and thermodynamic properties of ringwoodite. J Geophys Res 111:B12202-1-8

Yu YG, Wentzcovitch RM (2009) Low-pressure clino- to high-pressure clinoenstatite phase transition: A phonon-related mechanism. Am Mineral 94:461–466

Yu YG, Wentzcovitch RM, Angel RJ (2010) First principles study of thermodynamics and phase transition in low-pressure (P21/c) and high-pressure (C2/c) clinoenstatite $MgSiO_3$. J Geophys Res 115:B02201-1-10

Yu YG, Wentzcovitch RM, Vinograd VL, Angel RJ (2011) Thermodynamic properties of $MgSiO_3$ majorite and phase transitions near 660 km depth in $MgSiO_3$ and Mg_2SiO_4: a first principles study. J Geophys Res 116:B02208-1-19

Zaanen J, Sawatzky GA (1990) Systematics in band gaps and optical spectra of 3D transition metal compounds. J Solid State Chem 88:8–27

Zahnle KJN, Arndt N, Cockell C, Halliday A, Nisbet E, Selsis F, Sleep NH (2007) Emergence of a habitable planet. Space Sci Rev 129:35–78

Zahnle K, Schaefer L, Fegley B (2010) Earth's Earliest Atmospheres. Cold Spring Harb Perspect Biol 2:4895-1-17

Zener C (1930) Analytic Atomic Wave Functions. Phys Rev 36:51–56

Zha CS, Duffy TS, Downs RT, Mao HK, Hemley RJ (1996) Sound velocity and elasticity of single-crystal forsterite to 16 GPa. J Geophys Res 101:17535–17545

Zhang LY, Friesner, RA (1998) Ab initio calculation of electronic coupling in the photosynthetic reaction center. Proc Natl Acad Sci USA 95:13603–13605

Zhang Y, Yang W (1998) Comment on "Generalized Gradient Approximation Made Simple". Phys Rev Lett 80:890

Zhang J, Liebermann RC, Gasparik T, Herzberg CT (1993) Melting and subsolidus relations of SiO_2 at 9–14 GPa. J Geophys Res-Sol Ea 98:19785–19793

Zhang LY, Friesner, RA, Murphy RB (1997) Ab initio quantum chemical calculation of electron transfer matrix elements for large molecules. J Chem Phys 107:450–459

Zhang JS, Reynard B, Montagnac G, Wang RC, Bass JD (2013) Pressure-induced Pbca-P21/c phase transition of natural orthoenstatite: Compositional effect and its geophysical implications. Am Mineral 98:986–992

Zhang YY, Liu X, Xiong ZH, Zhang ZG (2016) Compressional behavior of $MgCr_2O_4$ spinel from first-principles simulation. Science China Earth Sci 59:989–996

Zhang JS, Hu Y, Shelton H, Kung J, Dera P (2017) Single-crystal X-ray diffraction study of Fe_2SiO_4 fayalite up to 31 GPa. Phys Chem Minerals 44:171–179

Zicovich-Wilson C M, Pascale F, Roetti C, Saunders V R, Orlando R, Dovesi R (2004) Calculation of the vibration frequencies of α-quartz: The effect of Hamiltonian and basis set. J Comput Chem 25:1873–1881

Ziemniak SE, Jones ME, Combs KES (1998) Solubility and phase behavior of Cr(III) oxides in lakline media at elevated temperatures. J Soln Chem 27:33–66

Zill DG, Cullen MR (2006) Advanced Engineering Mathematics (3rd ed) Jones & Bartlett Learning ISBN 0-7637-4591-X

Zotov N, Ebbsjö I, Timpel D, Keppler H (1999) Calculation of Raman spectra and vibrational properties of silicate glasses: Comparison between $Na_2Si_4O_9$ and SiO_2 glasses. Phys Rev B 60: 6383–6397

Zou Y, Zhou J, Xu Y, Zao P, Zhang Z (1982) Some aspects of the thermodynamics of melts. Acta Metall Sinica 18:127–140

Zunger A (1978) Self-consistent LCAO calculation of the electronic properties of graphite I The regular graphite lattice. Phys Rev B 17:626–641

Zupan A, Blaha P, Schwartz K, Perdew JP (1998) Pressure-induced phase transitions in solid Si, SiO_2 and Fe Performance of local-spin-density and generalized-gradient-approximation density functional. Phys Rev B 58:11266–12272

Subject Index

G. A. Ottonello, *Quantum Geochemistry*, Springer Geochemistry,
https://doi.org/10.1007/978-3-031-21837-8

F

T

GPSR Compliance

The European Union's (EU) General Product Safety Regulation (GPSR) is a set of rules that requires consumer products to be safe and our obligations to ensure this.

If you have any concerns about our products, you can contact us on ProductSafety@springernature.com

In case Publisher is established outside the EU, the EU authorized representative is:

Springer Nature Customer Service Center GmbH
Europaplatz 3
69115 Heidelberg, Germany

Batch number: 10732592

Printed by Printforce, the Netherlands